U.R. Christensen • A. Balogh • D. Breuer •
K.-H. Glaßmeier
Editors

Planetary Magnetism

Previously published in *Space Science Reviews* Volume 152, Issues 1–4, 2010

Springer

U.R. Christensen
Max Planck Institute for Solar
System Research
Katlenburg-Lindau, Germany

A. Balogh
International Space Science Institute
Bern, Switzerland

D. Breuer
DLR–German Aerospace Center
Berlin, Germany

K.-H. Glaßmeier
Institut für Geophysik und
extraterrestriche Physik
Braunschweig, Germany

ISBN 978-1-4419-5900-3
Springer New York Dordrecht Heidelberg London

Library of Congress Control Number: 2010929762

Cover illustration: Images taken an hour apart show variations in the auroras at Saturn's poles (blue), caused by electrons impacting the upper atmosphere of Saturn. Aurorae (well documented at Earth, Jupiter and Saturn) are spectacular examples of processes caused by the dynamic interaction of a planet's magnetic field with the solar wind. (Photos taken by the UVIS instrument of the Cassini Saturn Orbiter spacecraft, used courtesy of Prof. Larry Esposito, Laboratory for Atmospheric and Space Physics, University of Colorado at Boulder.)

Printed on acid-free paper

Springer is part of Springer Science+Business Media (www.springer.com)

Contents

Space Sci Rev (2010) 152: 1–3
DOI 10.1007/s11214-010-9651-1

Planetary Magnetism—Foreword

André Balogh · Doris Breuer · Ulrich Christensen · Karl-Heinz Glassmeier

Published online: 30 April 2010

The solar system consists of a wide diversity of objects: other than our central star, the Sun, there are eight planets, six of these with satellites of their own, as well as the innumerable smaller objects we call asteroids and comets. Understanding their structure, composition, environment and internal processes is important, primarily, for understanding the origin and evolution of our own solar system. Increasingly, however, interest in planetary systems beyond ours is becoming a powerful motivation for comparative studies.

The magnetic properties of solar system objects provide important clues to their origin, evolution, parentage as well as to their interior structure and its dynamic aspects. In planetary sciences, the measurement of magnetic fields and their description in terms of models that relate the observations to the objects' material properties and thus explain their origin are important research tools. These are now routinely applied to all solar system objects. Almost all classes of objects have been visited by spacecraft that could measure the magnetic field in their environment. These measurements have been interpreted in the various cases in terms of internally generated magnetic fields or in terms of the interaction of the body with the solar wind, or both.

There has never been a comprehensive volume on planetary magnetism, covering the whole range of solar system objects, although of course there are numerous sources for specific objects, particularly of course the planets. The Earth's magnetic field has been well

A. Balogh (✉)
International Space Science Institute, Bern, Switzerland
e-mail: balogh@issibern.ch

D. Breuer
DLR—German Aerospace Center, Berlin, Germany

U. Christensen
Max Planck Institute for Solar System Research, Katlenburg-Lindau, Germany

K.-H. Glassmeier
Institut für Geophysik und extraterrestrische Physik, Braunschweig, Germany

documented over the decades, but for other bodies in the solar system, the discovery of a planetary scale magnetic field awaited, in general, visits by spacecraft equipped with magnetometers. The exception was Jupiter: its kilometric radiation, detected from the Earth, was correctly interpreted as evidence for an intense magnetic field of planetary origin. The first two planets visited, Venus and Mars, have no internal sources. The interpretation of the early measurements was hotly debated, but, eventually, the absence of significant magnetic fields was generally accepted. In the case of Mars, a long-extinct dynamo had left some locally strong crustal magnetization that was only identified in the mid-1990s. Crustal magnetism on the Moon was discovered by the magnetometers carried to the surface by the Apollo landers. Venus is still known as an unmagnetised planet.

The other planets have internally-generated intrinsic magnetic fields, surprisingly so in the case of Mercury, more understandably so in the case of Jupiter, Saturn, Uranus and Neptune (and of course the Earth). Also surprising are the magnetic properties of Jupiter's Galilean satellites: Ganymede with its internal dynamo and Europa with its large magnetic field, interpreted as caused by the currents induced in the electrically conductive medium of a (sub-surface) ocean in Jupiter's fast rotating magnetic field. Some of the asteroids visited so far appear to have had an intrinsic magnetic field shortly after their formation, while others have not. In all cases, the measured magnetic fields are a complex superposition of any internally generated component and of the fields generated by currents that arise from the interaction of the planet or other object with the solar wind.

A Workshop organised by the International Space Science Institute in Bern, Switzerland, held on 1 to 5 September 2008, brought together over forty scientists with an interest in measuring, analysing, modelling and explaining magnetic fields in and around solid bodies in the solar system. The coverage was wide: all objects were reviewed; comparing what we know about the different planets and smaller bodies led to a very rich harvest in insight into the origin and nature of planetary magnetism. The present volume contains a collection of papers that summarise the insight gained and the progress made in understanding the many aspects of planetary magnetism and their inter-relations. While planetary dynamo theory remains a dynamic and evolving topic, the mutual understanding among observers, modellers and theorist that was achieved during the Workshop and that is embodied in several papers in this volume is expected to lead to the better understanding that is needed for future progress in this field.

This Workshop was one of a series of three held by ISSI on magnetism in the solar system. The other two Workshops covered the topics of solar and terrestrial magnetism, respectively. While there is a strong overlap between the scientific communities involved in planetary and terrestrial magnetism, there is generally less contact with the solar magnetism community. To help cross-fertilisation, joint meetings were held by the Convenors of the three Workshops and talks about solar magnetism and the solar dynamo were presented at both the planetary and terrestrial magnetism Workshops, while a talk about planetary dynamos was also presented to the solar magnetism community. In all three cases, these talks were not only well received and discussed, but also led to papers that have been included in the respective volumes. A later Workshop, held in March 2010, addressed large-scale magnetic fields in the Universe, thus extending the coverage of magnetism to a much broader astrophysical context.

The Convenors of this Workshop wish to thank ISSI, in particular its Executive Director Roger-Maurice Bonnet, for the initiative and support for the Workshop and its participants. Essential support was provided by Brigitte Shutte, Saliba Saliba, Silvia Wenger, Andrea Fischer and Katja Schüpbach. The Editors also wish to thank the authors of the papers in this volume who collaborated so splendidly to present the synthesis of current research, based on the presentations and discussions at the Workshop. Thanks are also due to the reviewers of the papers who helped to make the completed volume to be of the highest standard and one that the Editors hope will be of lasting use to the community.

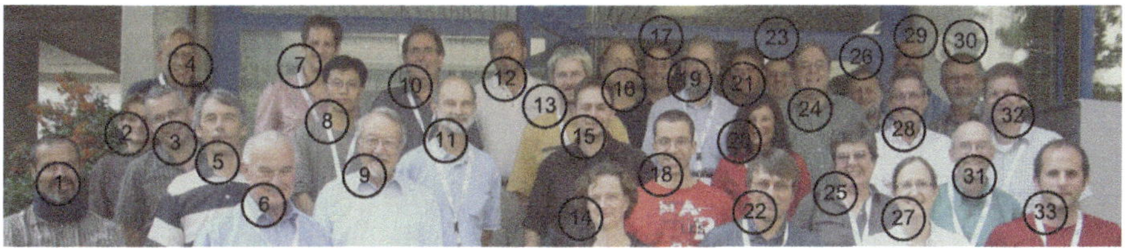

Participants in the Workshop on Planetary Magnetism, held at the International Space Science Institute, Bern, Switzerland, 1 to 5 September 2008. (1) Pierre Rochette, (2) Nadine Nettelmann, (3) Gauthier Hulot, (4) Wolfgang Baumjohann, (5) Michel Blanc, (6) André Balogh, (7) Benjamin Weiss, (8) Xianzhe Jia, (9) Norman Ness, (10) Tilman Spohn, (11) Gary Glatzmaier, (12) Jack Connerney, (13) Benoit Langlais, (14) Doris Breuer, (15) Joachim Saur, (16) Johannes Wicht, (17) Chris Finley, (18) Gautier Verhille, (19) Ulrich Christensen, (20) Mioara Mandea, (21) Vincent Lesur, (22) Stephane Labrosse, (23) Jonathan Fortney, (24) David Stevenson, (25) Catherine Constable (26) Chris Jones, (27) Sabine Stanley, (28) Jean-Francois Pinton, (29) Nils Olsen (30) Karl-Heinz Glassmeier, (31) Brian Anderson, (32) Andreas Tilgner, (33) Daniel Heyner.

Space Sci Rev (2010) 152: 5–22
DOI 10.1007/s11214-009-9567-9

Space Exploration of Planetary Magnetism

Norman F. Ness

Received: 13 April 2009 / Accepted: 26 June 2009 / Published online: 5 August 2009

Abstract This chapter gives a brief overview of the major observational advances in our quantitative knowledge of the intrinsic magnetic fields of the 8 planets, except Earth, from Mercury to Neptune, since "The Space Age" began on 4 October 1957 with the USSR launching of the world's first artificial satellite SPUTNIK I.

Keywords Intrinsic magnetic fields · Planetary magnetism

1 Introduction

Spacecraft of the USA, USSR/Russia and ESA have explored and identified the seven magnetized planets, either with flyby trajectories or from orbits about the target planet. A number of reviews of studies of planetary (Connerney 1993; Ness 1994, 1970; Stevenson 1983; Schulz and Paulikas 1990) and lunar magnetism (Dyal et al. 1974; Ness 2001) exist at various epochs in the more than 50 years of spacecraft studies, which should be consulted for more detailed discussion of the actual observations. Computer numerical simulation of the geodynamo has also begun (see Glatzmaier and Roberts 1996) and several chapters in this book which discuss similar developments for other planets. This review is only meant to highlight the exciting quantitative results which these experimental studies have discovered about the intrinsic magnetic fields of the 7 magnetized planets.

These space missions have been conducted with large multi-instrumented complex spacecraft, most of which use solar cell arrays for power sources. Some have used "atomic batteries" also known more correctly as RTG's (Radio-isotope Thermo-electric Generators) for orbits when beyond 5 AU. Table 1 gives a summary of the spacecraft names and years of discovery for those spacecraft which have added to our quantitative knowledge of planetary magnetism.

The origin of these intrinsic planetary fields is due mainly to a dynamo process (Roberts 1995; Procter and Gilbert 1996; Busse 1978) which is active in the planets'

N.F. Ness (✉)
University of Delaware, Newark, DE 19716, USA
e-mail: nfness@udel.edu

Table 1 Summary of spacecraft launched 1973–2006 which have discovered and investigated, in situ, the magnetic fields of the planets. Number following year indicates closest approach distance as measured in units of target planet's radii. At Jupiter and Saturn, the plus sign (+) refers to fact that those spacecraft had multiple close approaches because they were in orbit about the parent planet and therefore had multiple periapses at variable distances

Year of encounter or discovery (•) and/or orbit/periapis in R_p (+ and thereafter)						
	Mercury	Mars	Jupiter	Saturn	Uranus	Neptune
Mariner 10	•1974/1.30					
	1975/1.13					
Pioneer 10			1973/2.84			
Pioneer 11			1974/1.31	•1979/1.35		
Voyager 1			1979/4.88	1980/3.07		
Voyager 2			1979/10.1	1981/2.69	•1986/4.18	•1989/1.18
Ulysses			1992/6.3			
Galileo (orbiter)			1995 + /6–15			
Cassini (orbiter)				[2004/1+]		
Mars Global Surveyor		1997/1.03				
Messenger	2008a/1.08					
(flybys)	2008b					

interior, driven by various energy sources which later chapters on individual planets' fields will discuss. It should be noted that there is ample and persuasive evidence for contributions from magnetized solid crustal rock materials in two of the four terrestrial planets: Mars and Earth. The origin of Mercury's field (Ness et al. 1974; Ness 1978; Connerney and Ness 1988) is identified as due primarily to an active dynamo. Mars' field (Acuña et al. 1998, 1999, 2001) is most assuredly associated with intensely magnetized regions of its sub-Curie point crust.

In the case of the four giant or outer planets, Jupiter, Saturn, Uranus and Neptune, there is no evidence in the spacecraft observations or with our general understanding of planetary dynamos to suggest that there is any contribution from magnetized material due to traditional models of these planets internal structures and composition.

The general goal in studies of planetary magnetism is to develop a mathematical representation of the 3-dimensional vector field surrounding the planet. A major complicating factor in this goal is due to the existence of the solar wind, the magnetized collisionless plasma flowing supersonically away from the Sun. Various electrical current systems develop well outside the planets in their ionospheres and magnetospheres which generate magnetic fields and add to the internal field sources. Some of these electrical current systems are confined within the ionospheres of the planets driven by solar radiations which also generate magnetic fields close to the planets' surfaces just above the neutral atmospheres. See Chap. 4 for a discussion of the details and physics of these current systems.

The mathematical representation of the vector magnetic field of a planet is traditionally derived from the gradient of a scalar potential function which contains both internal and external sources. The mathematical formalism is described as a spherical harmonic expansion in terms of planetocentric multipoles of order n and degree m from which it is possible to determine the vector field B at any point of space surrounding the planet which is devoid or empty of any electrical currents. This development goes back to the era of Gauss and very early observational studies of the geomagnetic field both on land and at sea. The use of the

Table 2 Summary of magnetic fields, solar wind sunward stagnation point, and rotation periods of the planets (Note: 1 G cm^3 = 10^{-3} A m^2; 1 nT = 10^{-5} G = 10^{-9} T)

Planet	Dipole moment (G cm^3)	Tilt & Sense	Dipole equatorial field (nT)	Average stagnation point distance (R_p)	Rotation period
Mercury	5×10^{22}	+14°	330	1.4	58.7^d
Venus	$< 4 \times 10^{21}$	–	< 2	1.0+	-243^d
Earth	8.0×10^{25}	+11.7°	31 000	10.4	23.9^h
Moon	$< 1 \times 10^{19}$	–	< 0.2	None	
Mars	$< 2 \times 10^{20}$	–	< 0.5	1.29	24.6^h
Jupiter	1.6×10^{30}	−9.6°	428 000	65 15±	9.92^h
Saturn	4.7×10^{28}	−0.0°	21 200	20 ± 3	10.66^h
Uranus	3.8×10^{27}	−58.6°	23 000	20	17.24^h
Neptune	2.0×10^{27}	−46.9°	14 000	26	16.1^h

concept of magnetic multipoles of higher order than a simple dipole is an essential feature of Gaussian harmonic expansion. See Chap. 5 regarding separation of internal and external contributions.

A summary of the salient observed features of 8 planetary fields is given in Table 2. As it turns out, the strongest multipole moment observed on most planets is the dipole. The orientation of the magnetic axis of that multipole assumes a directional convention in which at the north geographic pole of Earth, the magnetic field points away from the surface. There are wide ranges of dipole tilt angles seen in this table as well as equatorial field strengths. The table includes the stagnation point distances to the sub-solar point of the magnetopause, which is the boundary between the planets' field and the deflected solar wind flow. There is also a wide variation in this stand-off distance as measured in units of planetary radii. Of all the 8 planets, only Venus appears to be devoid of any intrinsic field of any source, dynamo or remanence, which is sometimes ascribed as being due to its slow rotation.

It should be noted here that spacecraft observations of non terrestrial planetary magnetic fields were preceeded by radio astronomers in 1955 when non-thermal radio emissions from the planet Jupiter were discovered. Following the discovery of the Earth's radiation belts in 1958 by the US Explorers 1 and 2, it was then realized that Jupiter must also possess trapped radiation belts in its own magnetic field. Estimates of the field intensity and the dipole orientation were made which were eventually confirmed and considerably expanded upon by the Pioneer 10 and 11 and then the Voyager 1 and 2 flyby spacecraft.

Figure 1 presents a summary of the spectra of non-thermal radiation which has been observed from the 4 giant planets by spacecraft. There are periodic time variations of these signals which are identified as being controlled by and assumed to be equal to the rotation of the planetary interior's dynamo. The solid surface of these planets are not readily identified and only estimates of the planetary rotation rates have been possible to be made from the study of the motions of cloud structures. However, recent time variations of Saturn's periodic radio emissions have introduced questions about the structure of Saturn's magnetosphere and the variable planetary rotation rate.

2 Terrestrial Planets

Only Venus appears to lack any intrinsic magnetic field. All observations by the several spacecraft with magnetometers and plasma and energetic particles instrumentation which

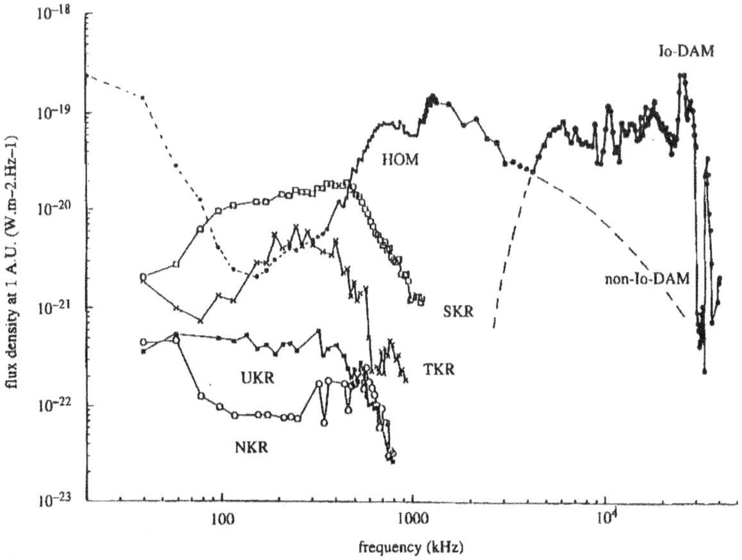

Fig. 1 Non-thermal auroral radio emission spectra of the four giant plants Jupiter, Saturn, Uranus and Neptune, and Earth as determined by the Voyager 1 and 2 spacecraft. Abbreviations SKR, UKR, NKR represent individual giant planets' Kilometric Radiation and TKR for Earth's. HOM is Jovian Hectometric and Io-DAM and non-Io-DAM are Jovian Decametric Radiation associated or not with the foot of the magnetic field line threading the Jovian moon Io

have flown close by the planet and the Pioneer Venus Orbiter have shown no evidence for any fields other than those induced magnetic fields associated with solar wind interaction processes. This planet is sometimes referred to as the Earth's twin because its size and average density are close to those of Earth. The very slow rotation of the planet and very high surface temperatures are reasons given for lack of any fields due to either an internal dynamo or remanence of crustal materials.

The intrinsic magnetic field of Earth has been well studied for several hundred years (Cain 1995; Glatzmaier and Roberts 1996; Jacobs 1994; Campbell 1997; Merrill et al. 1996). Its field sources are an example of a combination of both an active internal dynamo, which has reversed its polarity multiple times in the ancient past, and a complex pattern of remanence magnetization in the sub-Curie point crustal materials. The complexity of the pattern is due, in part, to the geological structure of the crustal materials and their composition as well as the state of the dynamo at the time residual permanent magnetization was acquired.

In the exploration of planets, both Mercury and Mars were surprisingly found to possess intrinsic fields due to either an active dynamo or remanence associated with an inactive or extinct dynamo. The global magnetic field of Mercury was discovered in 1974 by the Mariner 10 (Ness et al. 1974) in the 1st of 3 successive and very successful close flybys.

2.1 Mercury

Mariner 10 was the first spacecraft to use the gravitational attraction of another planet other than the Earth, Venus, to shape its heliocentric trajectory after launch so as to enable it to accomplish its primary goal of reaching the target planet, Mercury. It was purely a surprise, serendipity on a fortunate celestial conjunction of circumstances, that following its

Table 3 Summary of MARINER 10's trajectory characteristics (CA = closest approach) during the 2 close approaches to Mercury which yielded insitu observations of the Herman magnetic field and those fields resulting from interactions with the solar wind. Temporal variations of the interplanetary magnetic field (IMF) led to different types of bow shock crossings, perpendicular or parallel. Standoff distance of sub-solar magnetopause, Rmp, was estimated to be the same, within uncertainties. Maximum measured magnetic fields differed by factor of 4, 98 nT versus 400 nT

	#1	#2	#3
Encounters	3/29/1974	9/21/1974	3/16/1975
IMF	18 nT, variable		20 nT, steady
Bow Shock:		Not close enough	
Inbound		to planet	//
Outbound	// waves upstream	to detect	
Rmp (est.)	1.6 0.2 R_m		1.4 0.2 R_m
BMAX (Obs.)	98 nT		400 nT
CA	.705 km = 0.29 R_m		327 km = 0.13 R_m
Latitude	2° S		68° N

close flyby of Venus, the heliocentric orbital period of Mariner 10 was nearly exactly twice that of Mercury's heliocentric orbital period. Fine tuning and continued adjustments of the spacecraft's trajectory permitted three successive close flybys to occur with different specific science emphasis for the instrument complement onboard.

Summarized in Table 3 are the salient features of the first and third flybys of Mercury. The interplanetary magnetic field was somewhat different during each of these but the standoff distance of the sun-ward stagnation point of solar wind flow was estimated to be almost the same; 1.5+ or -0.2 R_m. But the maximum Hermean field measured at closest approach was different by a factor of 4. This was a direct result of the significant change made in the targeting of the latitude and altitude of the periapsis on this flyby following the analysis and interpretation of the first flyby magnetic field data.

Quantitative studies showed that the observed field, close to the planet, was well represented by an internal field of a simple centered dipole term with an axis tilted at about 14 degrees off from the planet's axis of rotation. Unlike Earth, however, the magnetopause was much closer to the planet so that, to a large degree, the planet could be described as almost filling its own magnetosphere. That is the primary reason why there were no trapped radiation belts observed at Mercury since charged energetic particles could not circulate in their drift motions around the planet's field lines without impacting and being absorbed on its surface.

The third flyby vector magnetic field data is shown in Fig. 2 in three projections in a Cartesian coordinate system in which the $+X_{\mathrm{me}}$ axis points generally toward the Sun in the ecliptic and the $-Y_{\mathrm{me}}$ axis points generally along the direction of the planet's heliocentric motion parallel to the ecliptic. Superimposed on the plots is the planet and the estimated positions of the average magnetopause surface and detached bow shock wave scaled from the case at Earth and projected on the X_{me}–Y_{me} plane. The observed positions of the magnetopause boundaries are shown as filled triangles along the trajectory. The individual vectors illustrate the smooth variation in orientation and magnitude to be expected from a slightly tilted but centered source dipole.

The spherical harmonic analysis of the 3rd flyby data including axi-symmetric internal terms up to 2nd order, g_1^0 and g_2^0, along with external terms of only 1st order, G_1^0 and G_1^1, is shown in Fig. 3 (Connerney and Ness 1988). It illustrates the close correlations of the

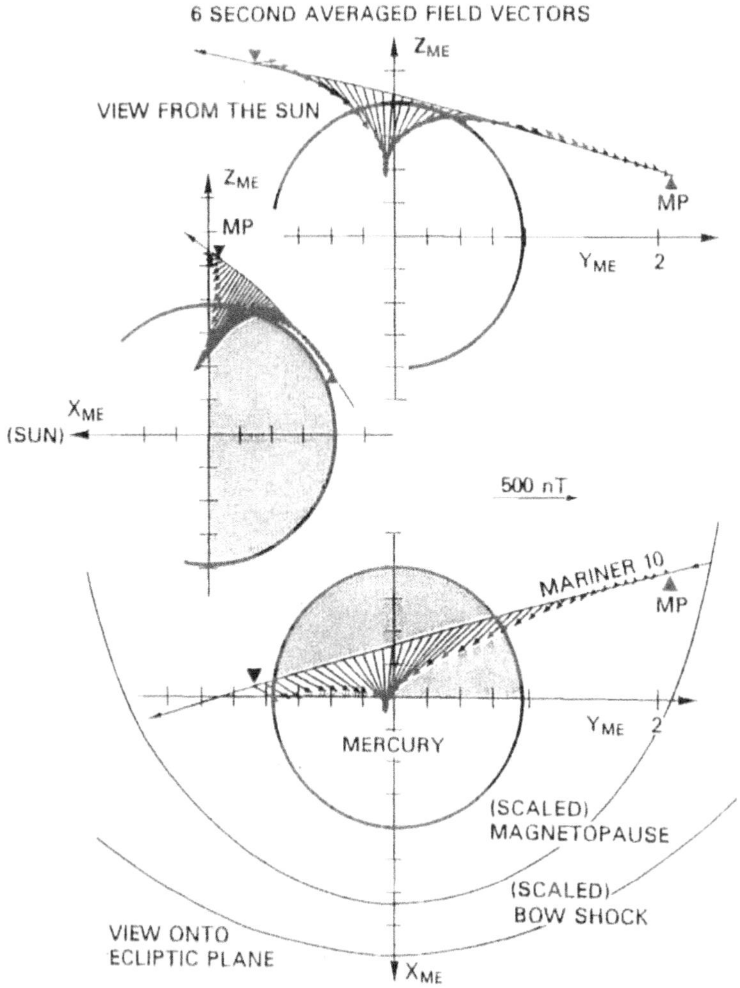

Fig. 2 Summary 3-D vector representation of the MARINER 10 magnetic field measured during the 3rd and closest encounter with Mercury in 1975 (Ness 1978)

estimates of internal and external terms as the size of the data set is changed and the quality of the fit varies.

The source origin of the intrinsic global field of Mercury is still uncertain. If it were due to remanence, it would require a very special evolutionary history such that materials throughout the interior all became intensely magnetized in such a configuration so as to be externally equivalent to an internal dipole. This seems to be highly unlikely. If the field is due to an active internal dynamo, the internal structure and dynamics must be somewhat special.

Chapter 9 in this book deals with possible Hermean dynamos using numerical modeling techniques and should be consulted directly to understand the advanced state of such theoretical studies. Note should also be made that Mercury was visited twice in 2008 by the MESSENGER spacecraft with flyby trajectories prior to eventual insertion into a captured

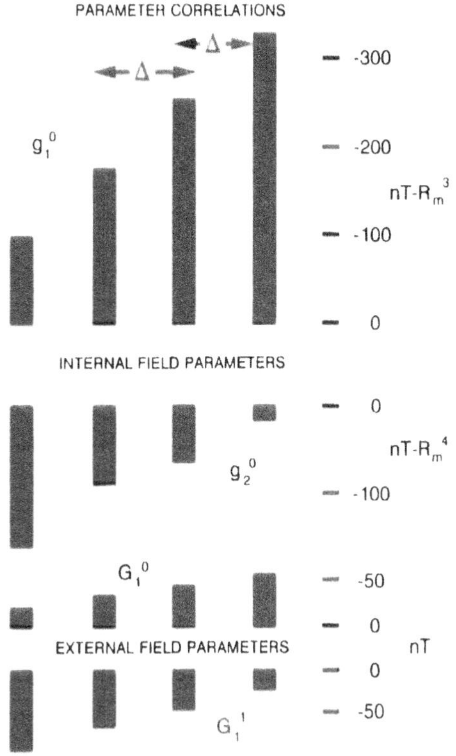

Fig. 3 Illustration of coherent variation of internal and external coefficients for Mercury derived from different sets of flyby data with different criteria for best fit representations. (← Δ →) refers to the variations which result therefrom (Connerney and Ness 1988)

Hermean orbit in 2011. See Anderson et al. (2008) and Slavin et al. (2008) for initial results of the first flyby. Chapter 10 in this book deals with the results of those two flybys and should be read for the analyses and interpretations of the data obtained from them. Finally, the ESA mission Bepi Colombo is dedicated to placing a spacecraft into a captured Hermean orbit in 2019.

2.2 Mars

Mars is a planet which has held a special high priority position in the era of space exploration for a long time. It has been the target for more missions than any other planet than Earth. Table 4 summarizes all those missions which the world's space-faring nations or consortia have launched with the goal of achieving a close flyby or orbiting trajectory or even landing on the surface. The table also has a compilation of the successes or failures of these 24 missions. The large fraction of failures, 42%, is due to several causes, most being errors in tracking and/or guidance, under performance or even simple but stupid confusion in communications or lack thereof about which system of units are being utilized, English or metric!

There are also failures which are not fully understood and explained. Since telemetric communication was lost prior to attempts for final trajectory corrections or injection activities.

The reason for so many missions to Mars has been the goal to search for any evidence that there exists or has existed on Mars any liquid water or other possible life-as-we-know-it forming or derived materials or structures.

Table 4 Launch dates (MOI) of spacecraft to Mars (** = failure after launch)

USA	LAUNCH YEAR	USSR
Mariner 3—5 November	** 1962	Mars 1—November
Mariner 4—18 November	** 1964	
Mariner 6—24 February	1964	
Mariner 7—27 March	1969	
	1969	
	* 1971 L	Mars 2
	1971 L	Mars 3
Mariner 8—8 May	** 1971 O	
Mariner 9—30 May	1971 O	
	** 1973 O	Mars 4
	1973 O	Mars 5
	1973 L	Mars 6
	1973 L	Mars 7
Viking 1—20 August	1975 O, L	
Viking 2—9 September	1975 O, L	
	** 1988 O	Phobos 1
	1988 O	Probos 2—29 January 89
Observer	** 1992 O	
Global Surveyor—12 Sept 97	1996 O	
	** 1996 O, L	Mars 8
Path Finder—4 July 97	1996 L	
	** 1998 O	JAPAN: Planet B/NOZOMI
Climate Orbiter	** 1999 O	
Polar Lander & Deep Space 2	** 1999 L + Probes	
Mars Odyssey	2001 O	
Mars Twin Rovers	2003 L	ESA: Mars Express
Mars Reconnaissance Orbiter	2005 O	

Score Card

	Successes	Failures	
US	9	5	14
USSR	5	5	10
TOTALS	14	10	24

Only a few spacecraft have carried magnetometers to explicitly study any magnetic fields which might exist. Some landers have carried special small circular relatively strong magnets on their "landing" feet which have been able to be viewed by cameras and the images transmitted back to earth. They reveal the collection of ensembles of small particles or grains of materials on the Martian surface but reveal little, if anything, about Martian magnetic fields.

A few of the early USA Mariner and USSR Mars flyby missions did carry magnetometers. Their data revealed the presence of the expected detached bow shock wave associated with the solar wind interaction with the atmosphere and ionosphere of Mars. Their data were often contaminated by recognized but imprecisely estimated known magnetic fields of spacecraft origin. They also did not get close enough to Mars to definitively sample the intrinsic Martian crustal magnetic fields that were discovered in 1997 by the Mars Global Surveyor (MGS) orbiting spacecraft, forty years after the SPACE AGE began!

Serendipity was responsible for the tremendous success of the MGS mission in elaborating on details of the complex Martian field of remanence origin. MGS was a replacement/repeat mission for the earlier failed Mars Observer (MO) mission of 1992. MO was a mission which carried enough on-board propellant to inject the spacecraft directly into its targeted nearly circular, polar mapping orbit at ~400 km above the surface. The MGS mission was less robust with insufficient on-board propulsion capability for direct injection. Instead, it was designed to use the technique of aero-braking via many repeated periapsis passes at altitudes as low as 200 km in the northern hemisphere following injection into an initially elliptical orbit before reaching the final mapping orbit at ~400 km.

The magnetic fields discovered by MGS (Acuña et al. 2001; Ness et al. 1999) revealed that there was a complex array of intensely magnetized and very localized regions of the Martian crust in the Southern hemisphere with fields as large as many hundreds of nT at an altitude of ~200 km. These fields were obviously associated with an ancient dynamo which was no longer active (Curtis and Ness 1988). Various authors have derived spherical harmonic representations of these observed fields from both the mapping orbit and also the many aero-braking orbits with differing assumptions about the nature of the sources (Voorhies et al. 2002).

Figure 4 presents a comparison of the magnitude of the power R_n in each of the multipoles of degree n up to 65 for earth and 90 for Mars. Earth's magnetic spectrum is distinguished by having a steeply declining spectrum up to order ~18 which is assumed to be

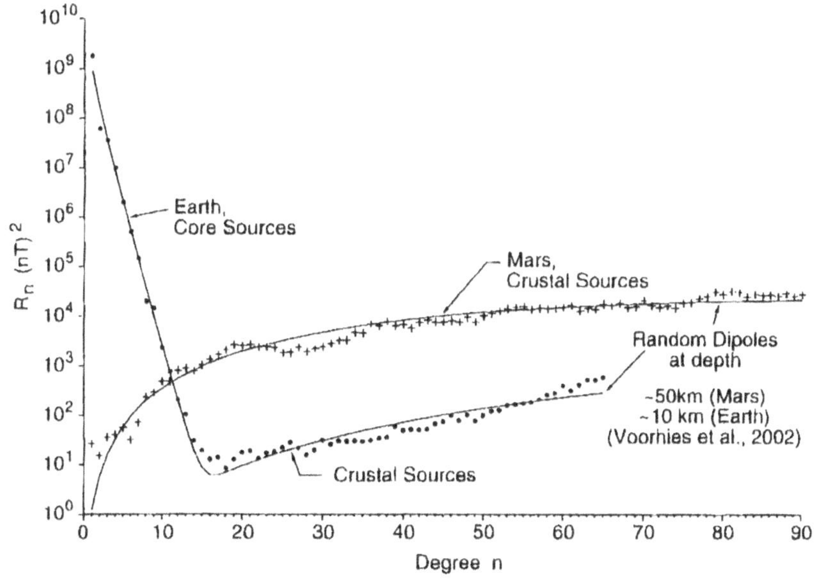

Fig. 4 Mean square amplitude of the magnetic field on the surface of a sphere (radius a) from spherical harmonics of order n for Earth ($a = 6371$ km) and Mars ($a = 3394$ km). The fitted curve for the Earth is the sum of spectra from core sources (dynamo) and that of a crustal source shell of random dipoles at shallow depth. The fitted curve for Mars is that of a source shell of random dipoles in the crust only (after Voorhies et al. 2002)

representative of the core dynamo. Then the spectrum increases slowly at higher values of n and the spectrum there is identified as due to the crustal remanence sources. Mars' spectrum rapidly increases after the dipole term and then more slowly increases out to order 90 although it is somewhat questionable whether or not higher order terms greater than $n = 60$ are physically meaningful.

There is no clear association of the locations of the intense magnetic fields with the topography or surface features such as valleys or craters, which may be due to impacts or volcanoes. It is important to note that all the strong fields regions are observed south of the dichotomy line which generally separates the large smooth plains of the northern hemisphere from the heavily crated southern hemisphere regions.

There are many paired features in the Southern hemisphere with opposite signs of vertical field components indicating near surface regions of substantially but oppositely magnetized materials. Some are elongated and are suggestive of those terrestrial magnetic crustal anomalies which are identified as due to ocean floor spreading. Figure 5 illustrates an attempt by Connerney et al. (2005) to match the B_x and B_z magnetic field components orthogonal to the linear axis of the anomaly detected in one of many aero-braking orbital passes. The model is based upon a set of linear crustal strips of variable magnetization (J_x and J_z). Those modeling efforts yield a remarkably good fit with the observations. The magnetizations so derived are consistent with the physical process of ocean floor spreading. These efforts suggest the possibility that a process similar to plate tectonics was active on Mars in forming its upper crustal features.

The inversion of these Martian magnetic field data for interpretations of the upper Martian crustal structures are limited by the unknown details of the subsurface structures and the associated morphological processes active in their formation. Most unfortunately, the 2005 USA's Mars Reconnaissance Orbiter (MRO), which used southern hemisphere periapsis passes for aero-braking, did not carry any magnetometer so there will be no new orbital magnetic field data in the near-term future to assist in improving our understanding of the history of Martian mantle and core dynamo dynamics. There was a very close flyby of Mars with the ROSETTA spacecraft whose results have just appeared (Boesswetter et al. 2009).

3 Giant Planets

The four most distant planets, Jupiter, Saturn, Uranus and Neptune, represent a special class of planets not only because of their large size and relatively low average density but also because they all have robust global intrinsic magnetic fields. It has been mentioned previously that the Jovian field was inferred to exist as a result of the discovery of its non-thermal radiation in 1955 which was studied extensively by ground based radio astronomers. It was to be many years before in-situ studies by spacecraft occurred and revealed the several unique characteristics of some of these global fields, quite unanticipated by any prior theoretical studies.

This section will only survey the early exploratory missions which provided all of the general knowledge of each of their fields. The details of their fields and especially the intrinsic and induced fields of their moons are discussed in more detail in Chaps. 8 and 9 in this book. See also Connerney (1993) for a comprehensive summary of studies of these 4 planets' fields.

3.1 Jupiter

The first spacecraft to study insitu the field of Jupiter was the flyby mission of Pioneer 10 in 1973 with a closest approach of 2.84 R_j. This was followed by Pioneer 11 in 1974 which

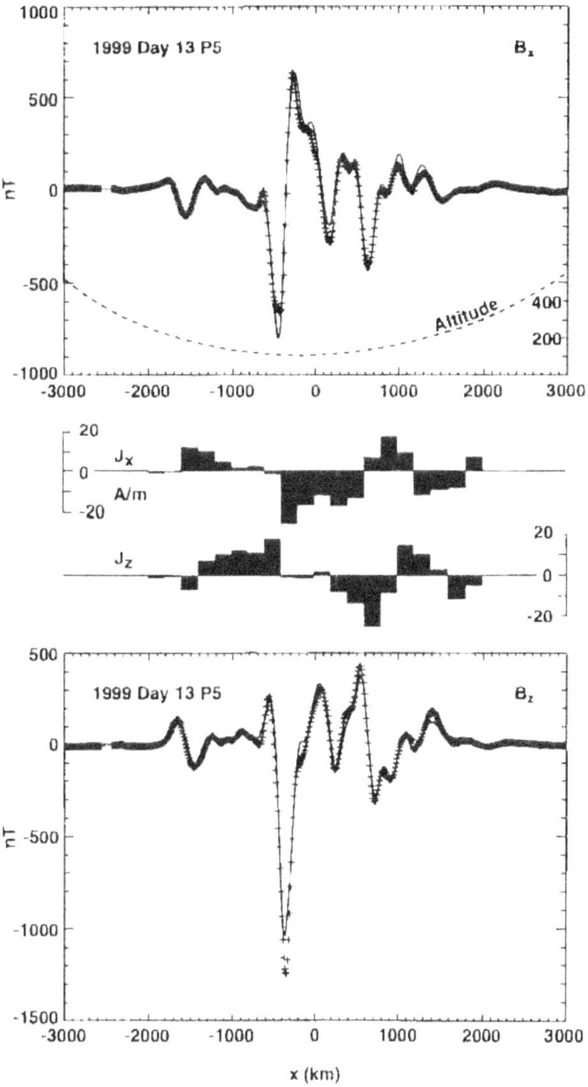

Fig. 5 Representative MGS low altitude (periapsis aerobraking phase) horizontal, B_X, and vertical, B_Z, observed component magnetic field data (+) compared to theoretical model (*continuous lines*) using a parallel set of thin elongated, shallow areas of intensely magnetized crustal material (J_X and J_Z in amps/meter) (Connerney et al. 2005)

had a much closer flyby at 1.31 R_j. Data from both these missions confirmed the general characteristics of the intrinsic field of the planet which radio astronomers had estimated; a magnetic dipole with polar fields approximately 10 Gauss with a tilt angle relative to the rotation axis of approximately 10 degrees. It also showed that the polarity of the north-south poles was opposite to that at Earth and quantified the equatorial dipole field intensity as 4.3 Gauss on the surface of the planet with a small offset of the dipole center from the planetary center.

It was not until 1977 that the twin Voyager spacecraft, V1 and V2, were launched and later conducted their flyby missions of Jupiter in 1979 while enroute to subsequent additional giant planet encounters. It should be noted that because of the relative celestial positions of the four giant planets in 1977 and thereafter, the possibility of conducting a GRAND TOUR of all four giant planets was achieved in the Voyager 2 mission. This was most fortunate and again serendipitous for space-exploration in general.

Due to the plans to have V1 and V2 continue on to Saturn, the flyby trajectories were constrained at Jupiter so as to deflect their pathways onward in a technique referred to as GRAVITATIONAL ASSIST. This can change appreciably both the flyby periapsis and the post flyby trajectory direction as well as the spacecraft velocity. What is actually happening is that the gravity field of the planet does work on the spacecraft and thereby transfers energy to it.

The net result of the demands of trajectory modifications required that the Jovian flybys be at somewhat greater distances than the previous 2 Pioneers. The V1 closest approach was at 4.88 R_j and that of V2 was at 10.1 R_j. It should also noted that because of the relatively rapid rate of rotation of Jupiter, the trajectories of both spacecraft flybys actually appear to be spirals inward and outward from closest approach and concentric about the planet in fixed planetary coordinates. This is especially important for mapping studies of the internal field and helps considerably in the determination of the higher order magnetic multipole moments. The large magnetosphere of Jupiter also contributes to this favorable situation.

Presented in Fig. 6 is one of several maps of the global field of Jupiter on its surface that has been developed from the four flyby trajectories of the Pioneer 10 and 11 and Voyager 1 and 2 missions. The maximum field estimated is in the north polar region is 14.0 Gauss while the minimum field is 3.3 Gauss at two locations on the surface. The dashed lines represent the footprints of the field line threading through the moon Io. Jovian decametric radio emissions are known, from historical observations, to be enhanced at frequencies corresponding to the field magnitude of the surface Jovian field at the foot of the field line threading the Galilean moon, Io. See Chaps. 8 and 9 in this book for more detailed presentations and discussions of the derivation of the main field of Jupiter from the observations and the intrinsic or induced fields of the Galilean moons in their interactions with the co-rotating magnetosphere of the parent planet Jupiter.

3.2 Saturn

Saturn has a most unique global intrinsic magnetic field of all planets. The Pioneer 11 discovery flyby at 1.35 R_s in 1979 was among the closest of early flybys of any of the giant planets. The uniqueness property is that after analysis of the observed magnetic field data, the planet appeared to possess an axi-symmetric intrinsic field. An enigmatic aspect of Saturn, in this regard, is that it also possesses a northern polar region which is the source of non-thermal Saturnian Kilometric Radiation(SKR) which is time varying. As with Jupiter, the SKR are assumed to be due to the intermittent precipitation of energetic particles from the naturally occurring radiation belts into the polar field region.

The SKR are observed to be periodically modulated at what has been assumed to be the intrinsic rotation rate of the internal dynamo region of the planet, which generates its global field. Latest results from the CASSINI orbiter mission, launched in 1997 and injected into a captured orbit in 2004, have created some doubt on this simple assumption and much study and work needs to be done to explain this conundrum. The periodicity of the SKR, previously assumed relatively constant at 10.66 hours during and after the Voyager encounters, now seems to have a slight variable or another periodicity within 1% of the Voyager established period. See the recent summary of all the SKR studies by V1, V2, Cassini and the ULYSSES missions (Kurth et al. 2008 and the many references therein) and the changing estimates of Saturn's rotation rate.

Closely following in time the Pioneer 11 flyby, both V1 and V2 had flybys which provided additional observational data with closest approaches in 1980 at 3.07 R_s and in 1981

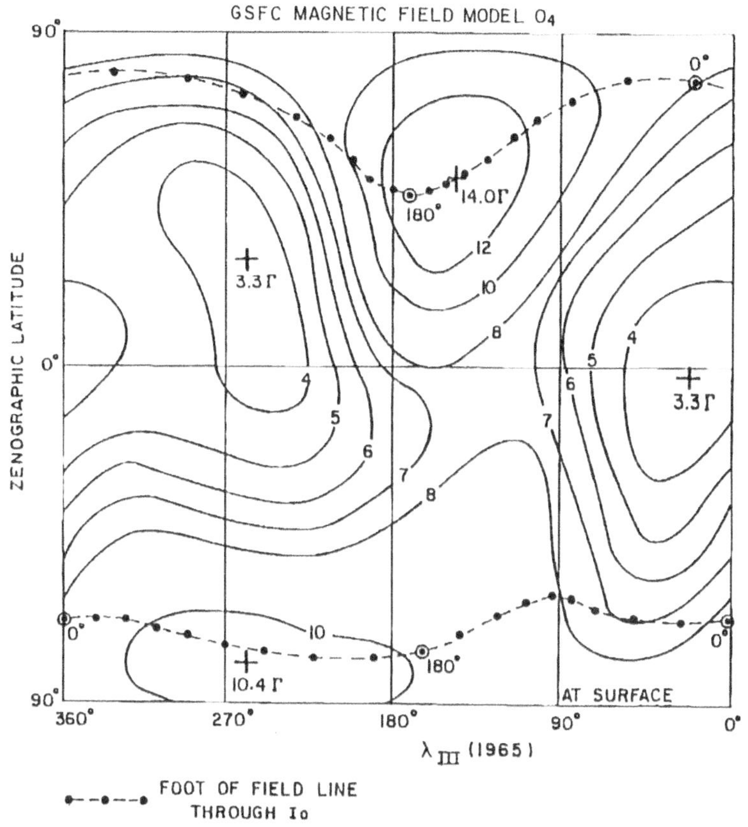

Fig. 6 Isointensity plot of magnitude of surface magnetic field, in gauss, of Jupiter from theoretical representation (GSFC-O4) derived from several spacecraft data sets. Trace of magnetic field line threading through the moon Io, at 5.95 R_j is illustrated by dashed line at 10 degree intervals in Io's orbit about parent planet (Connerney 1993)

V2 at 2.69 R_s. At the present moment, the quantitative multipole model of the Saturnian field which best fits all three encounters is referred to as Z3. This means only 3 zonal coefficients g_1^0, g_2^0 and g_3^0 are required to optimumly match the observations of the three flyby spacecraft which have studied the field. It should be noted that such an axisymmetric field is a problem for any internal dynamo since it is ruled out by Cowling's Theorem. Figure 7 portrays the Saturnian magnetic field at various latitudes for different models of the field. The lack of any non axi-symmetric terms implies that there is no local deviation of the field from its axisymmetric characteristics which would be a primary driving reason for the periodic modulation of any radio source region in the northern polar region.

3.3 Uranus and Neptune

The intrinsic global magnetic fields discovered by V2 in its GRAND TOUR trajectory encounters with each of these planets presented surprises yet again. As was the case with Saturn, there were no ground based observations of non-thermal radio emissions which would have indicated the presence of trapped energetic particles in radiation belts contained by a

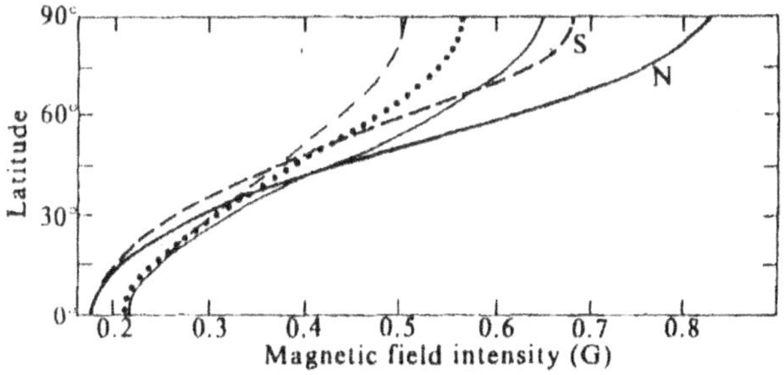

Fig. 7 Magnetic field intensity as a function of latitude on Saturn's oblate (flattening = 10.6) surface. *Bold solid (dashed) line* is the field intensity at north (south) latitudes corresponding to the Z_3 model. *Light lines* indicate the field intensity of the offset (~0.04 R_s) dipole model and the *dotted line* corresponds to a simple centered dipole with 0.21 G – R_s^3 moment. N represents the Northern hemisphere, S the Southern (Connerney et al. 1982)

global field. V2 discovered the Uranian field in its 1986 flyby at 4.18 R_u and the Neptunian field in its 1989 flyby at 1.18 R_n.

Both planets possess global fields and radiation belts which have only been studied once. There are no firm plans by any space faring nation or consortium for any future missions to these two planets which share special and most interesting global fields. From the analysis of the V2 data, it has been determined that each planet has a global field which can be approximated in first order by a magnetic dipole which has a very large spatial offset from the center of each planet and a very large tilt angle with respect to the rotation axis of each planet. There are referred to as Offset Tilted Dipoles. Such OTD models are simple but useful approximations when considering solar wind interaction effects on the 3-D configurations of the real planetary fields.

This is portrayed in Fig. 8 where the Uranian large offset of 0.31 R_u and the large tilt of the dipole axis of 60 degrees is shown. The larger offset of 0.55 R_n but smaller tilt angle of 47 degrees is illustrated for Neptune in the discovery epochs. There is an important aspect of these large offsets and tilts with respect to the structure and dynamics of their magnetospheres as the planets orbit the Sun. During certain phases of their heliocentric motion, these two planets will present nearly pole-on configurations of their main magnetic fields to solar wind flow. In those periods, as the planet rotates, the magnetic tails will contain magnetic neutral zones separating tail regions of opposite magnetic field polarity which are cylindrical in shape as opposed to the usual sheet-like cross tail structure at Earth, Jupiter or Saturn, a theta configuration in cross-section. See Voigt et al. (1987) for a discussion of this configuration. Another serendipitous Voyager happening was that the Neptune encounter occurred exactly when the polar cusp region of its magnetosphere was pointed sunward. Thus, V2 observed the only set of such pole-on observations in situ ever recorded (Lepping et al. 1992).

A special aspect of the unusual magnetospheric structure for these 2 planets resulting from the large spatial and tilt angle offsets is that the auroral regions for each of them are far removed from the rotation axis poles. The end result of that fact is why there were few observations of auroral phenomena seen during either encounter by V2. The scan platform with the UV instrument had been programmed to search for and spend most of its observational time looking for auroral signatures hear the rotation axis poles of the planets.

Figure 9 illustrates the location of the auroral zones on Uranus for 2 different models of its main field, a quadrupole configuration up to multipole order 2 and an octupole configuration up to order 3. In both models the auroral zones are in mid-latitudes with one even

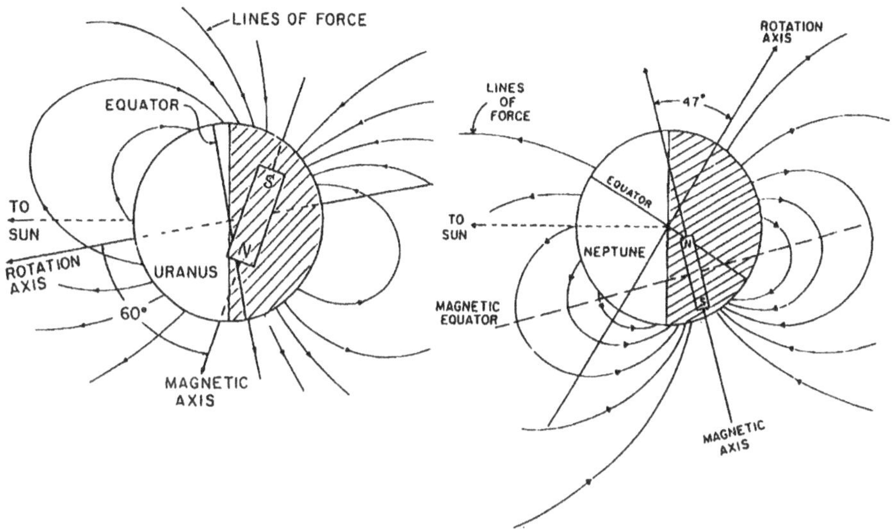

Fig. 8 Schematic representation of the offset tilted dipole model magnetic fields of Uranus and Neptune. Note the large spatial offsets of the magnetic centers from the center of each planet as well as the large angular offsets of the magnetic axes from the rotational axes (Ness 1994)

Fig. 9 Isointensity Mercator plots on surface of planet of Q3 and O3 models of Uranus' magnetic field. *Cross hatched areas* represent theoretical location of auroral and polar cap zones (Ness 1994)

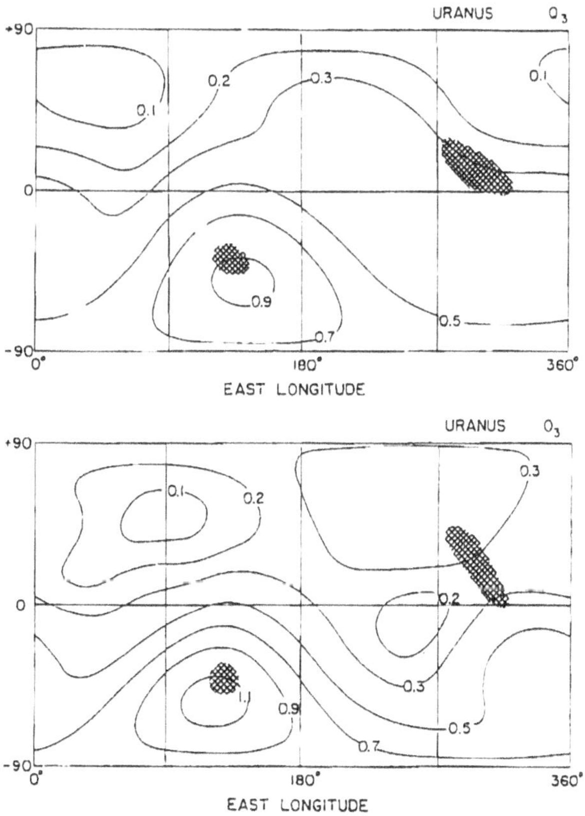

including the equatorial region. Even the magnetic dip equators are far removed from the normal region to be expected because of the large offsets and tilts. Figure 10 illustrates the location of the auroral zones and the magnetic equator for an octupole model of Neptune's field.

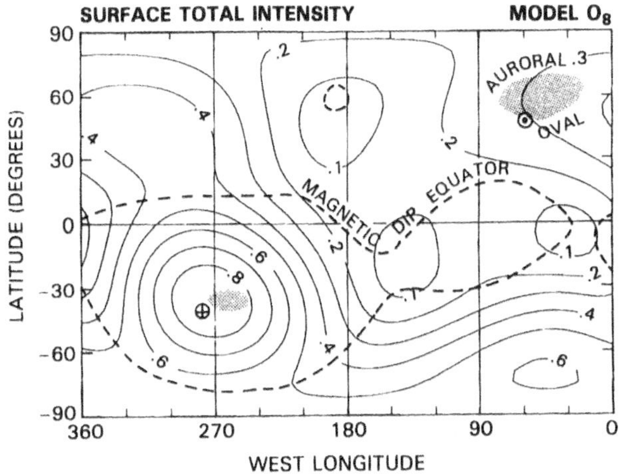

Fig. 10 Isointensity Mercator plot of magnitude of surface magnetic field, in gauss, of Neptune from theoretical representation derived from V2 data using model with up to octupole magnetic harmonic terms. *Shaded areas* represent estimated auroral ovals and *dashed line* magnetic dip equator locations on surface (Connerney et al. 1992)

Table 5 Spherical harmonic coefficients (in Gauss) or planetocentric multipole parameters for Earth and the four giant planets: Jupiter, Saturn, Uranus and Neptune. Models are identified by conventional notation common in the literature, some of which are defined in this article (OTD = Offset Tilted Dipole Best Fit)

Planet	Earth	Jupiter	Saturn	Uranus	Neptune
(Radius in km)	(6378)	(71 372)	(60 330)	(25 600)	(24 765)
Model	IGRF 85	O6	Z3	O3	O8
$g(1,0)$	−0.29877	4.24202	+0.21535	+0.11893	+0.09732
$g(1,1)$	−0.01903	−0.65929	0	+0.11579	+0.03220
$h(1,1)$	+0.05497	0.24116	0	−0.15685	−0.09889
$g(2,0)$	−0.02073	−0.02181	+0.01642	−0.06030	+0.07448
$g(2,1)$	+0.03045	−0.71106	0	−0.12587	+0.00664
$h(2,1)$	−0.02191	−0.40304	0	+0.06116	+0.11230
$g(2,2)$	+0.01691	0.48714	0	+0.00196	+0.04499
$h(2,2)$	−0.00309	0.07179	0	+0.04759	−0.00070
$h(3,0)$	+0.01300	0.07565	+0.02743	+0.02705	−0.06592
$g(3,1)$	−0.02208	−0.15493	0	+0.01188	+0.04098
$h(3,1)$	−0.00312	−0.38824	0	−0.07095	−0.03669
$g(3,2)$	+0.01244	0.19775	0	−0.04808	−0.03581
$h(3,2)$	+0.00284	0.34243	0	−0.01616	+0.01791
$g(3,3)$	+0.00835	−0.17958	0	−0.02412	+0.00484
$h(3,3)$	−0.00296	−0.22439	0	−0.02608	−0.00770
Dipole moment	$0.304\ R_E^3$ G	$4.28\ R_J^3$ G	$0.215\ R_S^3$ G	$0.288\ R_U^3$ G	$0.142\ R_N^3$ G
Dipole tilt	+11.4°	−9.6°	−0.0°	−58.6°	−46.9°
OTD offset	$0.08\ R_E$	$0.07\ R_J$	$0.04\ R_S$	$0.31\ R_U$	$0.55\ R_N$

4 Summary

Since the SPACE AGE began in 1957, spacecraft have sampled and explored every planet for its intrinsic magnetic properties. The discovery and mapping of global magnetic fields at Mercury, Saturn, Uranus and Neptune along with Earth and Jupiter now provides the community of scientists studying the origin of magnetic field dynamo generation in condensed rapidly rotating celestial bodies a wealth of real naturally occurring examples.

Similarly, the presence of remnant paleomagnetic fields on Mars demonstrates the previously unknown existence of a Martian field due to an internal dynamo which is now extinct. Studies of the dynamo processes discussed in other chapters in this book have been stimulated and benefited greatly from these experimental observations and interpretations. There remain a number of unanswered questions about planetary magnetism but the many successes of missions to the planets in the SPACE AGE has contributed much to our basic knowledge of facts upon which to build a better understanding of planetary origins and dynamics.

Table 5 provides a summary listing of the spherical harmonic coefficients or the magnetic multipoles for each of the 5 planets for which such a representation has been reliably and accurately derived. They should prove valuable in related studies of the origins of magnetic fields from internally operating dynamo regions.

References

M.H. Acuña, J.E.P. Connerney, P. Wasilewski, R.P. Lin, K.A. Anderson, C.W. Carlson, J. McFadden, D.W. Curtis, D. Mitchell, H. Rème, C. Mazelle, J.A. Sauvaud, C. d'Uston, A. Cros, J.L. Médale, S.J. Bauer, P. Cloutier, M. Mayhew, D. Winterhalter, N.F. Ness, Magnetic field and plasma observations at Mars: Preliminary results of the Mars Global Surveyor mission. Science **279**, 1676–1680 (1998)

M.H. Acuña, J.E.P. Connerney, N.F. Ness, R.P. Lin, D. Mitchell, C.W. Carlson, J. McFadden, K.A. Anderson, H. Rème, C. Mazelle, D. Vignes, P. Wasilewski, P. Cloutier, Global distribution of crustal magnetization discovered by the Mars Global Surveyor MAG/ER experiment. Science **284**, 794–798 (1999)

M.H. Acuña, J.E.P. Connerney, P. Wasilewski, R.P. Lin, D. Mitchell, K.A. Anderson, C.W. Carlson, J. McFadden, H. Rème, C. Mazelle, D. Vignes, S.J. Bauer, P. Cloutier, N.F. Ness, Magnetic field of Mars: Summary of results from the aerobraking and mapping orbits. J. Geophys. Res. **106**, 23403–23417 (2001)

B.J. Anderson, M.H. Acuña, H. Korth, M.E. Purucker, C.L. Johnson, J.A. Slavin, S.C. Solomon, R.L. McNutt, The structure of Mercury's magnetic field from MESSENGER's first flyby. Science **321**(5885), 82–85 (2008)

A. Boesswetter, U. Auster, I. Richter, M. Fränz, B. Langlais, S. McKenna-Lawlor, S. Simon, U. Motschmann, K.H. Glassmeier, N.J.T. Edberg, R. Lundin, Rosetta swing-by at Mars—an analysis of the ROMAP measurements in comparison with results of 3d multi-ion hybrid simulations and MEX/ASPERA-3 data. Ann. Geophys. (2009, in press)

F. Busse, Theory of planetary dynamos, in *Solar System Plasma Physics—A 20th Anniversary Review*, ed. by C.F. Kennel, L.J. Lanzerotti, E.N. Parker (North-Holland, Amsterdam, 1978), pp. 293–318

J.C. Cain, Main field and secular variation. Rev. Geophys. Suppl. **33**, 145–152 (1995)

W.H. Campbell, *Introduction to Geomagnetic Fields* (Cambridge University Press, Cambridge, 1997)

J.E.P. Connerney, Magnetic fields of the outer planets. J. Geophys. Res. **98**(E10), 18659–18679 (1993)

J.E.P. Connerney, N.F. Ness, Mercury's magnetic field and interior, in *Mercury*, ed. by M. Matthews, C. Chapman, F. Vilas (University of Arizona Press, Tucson, 1988), pp. 494–513

J.E.P. Connerney, N.F. Ness, M.H. Acuña, Zonal harmonic model of Saturn's magnetic field from Voyager 1 and 2 observations. Nature **298**(5869), 44–46 (1982)

J.E.P. Connerney, M.H. Acuña, N.F. Ness, The magnetic field of Neptune. Adv. Space Res. **12**(8), 239–248 (1992)

J.E.P. Connerney, M.H. Acuña, N.F. Ness, G. Kletetschka, D.L. Mitchell, R.P. Lin, H. Rème, Tectonic implications of Mars crustal magnetism. Proc. Natl. Acad. Sci. **102**(42), 14970–14975 (2005)

S.A. Curtis, N.F. Ness, Remanent magnetism at Mars. Geophys. Res. Lett. **15**, 737–739 (1988)

P. Dyal, C.W. Parkin, W.D. Daily, Magnetism and the interior of the Moon. Rev. Geophys. Space Phys. **12**, 568–591 (1974)

G.A. Glatzmaier, P.H. Roberts, Rotations and magnetism of Earth's inner core. Science **274**, 1887–1891 (1996)

J.A. Jacobs, *Reversals of Earth's Magnetic Field*, 2nd edn. (Cambridge University Press, Cambridge, 1994)

W.S. Kurth, T.F. Averkamp, D.A. Gurnett, J.B. Groene, A. Lecacheux, An update to a Saturnian longitude system based on kilometric radio emissions. J. Geophys. Res. **113**, AO5232 (2008)

R.P. Lepping, L.F. Burlaga, A.J. Lazarus, V.M. Vasyliunas, A. Szabo, J. Steinberg, N.F. Ness, S.M. Krimigis, Neptune's polar cusp region: Observations and magnetic field analysis. J. Geophys. Res. **97**, 8135–8144 (1992)

R.T. Merrill, M.W. McElhinny, P.L. McFadden, *The Magnetic Field of the Earth*, 2nd edn. (Academic Press, San Diego, 1996)

N.F. Ness, Magnetometers for space research. Space Sci. Rev. **11**, 111–222 (1970)

N.F. Ness, Mercury: Magnetic field and interior. Space Sci. Rev. **21**, 527–554 (1978)

N.F. Ness, Intrinsic magnetic fields of the planets: Mercury to Neptune. Philos. Trans. R. Soc. A **349**, 249–260 (1994)

N.F. Ness, Planetary and lunar magnetism, in *The Century of Space Science* (Springer, Berlin, 2001), pp. 1479–1492

N.F. Ness, K.W. Behannon, R.P. Lepping, K.H. Schatten, Y.C. Whang, Magnetic field observations near Mercury: Preliminary results from Mariner 10. Science **185**, 151–160 (1974)

N.F. Ness, M.H. Acuña, J. Connerney, P. Wasilewski, C. Mazelle, J. Sauvaud, D. Vignes, C. d'Uston, H. Rème, R. Lin, D.L. Mitchell, J. McFadden, D. Curtis, P. Cloutier, S.J. Bauer, MGS magnetic fields and electron reflectometer investigation: Discovery of paleomagnetic fields due to crustal remanance. Adv. Space Res. **23**(11), 1879–1886 (1999)

M.R.E. Procter, A.D. Gilbert, *Lectures on Solar and Planetary Dynamos* (Cambridge University Press, Cambridge, 1996)

P.H. Roberts, Dynamics of the core, geodynamo. Rev. Geophys. Suppl. **33**, 443–450 (1995)

M. Schulz, G.A. Paulikas, Planetary magnetic fields: A comparative view. Adv. Space Res. **10**(1), 55–64 (1990)

J.A. Slavin, M.H. Acuña, B.J. Anderson, D.N. Baker, M. Benna, G. Gloeckler, R.E. Gold, G.C. Ho, R.M. Killen, H. Korth, S.M. Krimigis, R.L. McNutt, L.R. Nittler, J.M. Raines, D. Schriver, S.C. Solomon, R.D. Starr, P. Trávníček, T.H. Zurbuchen, Mercury's magnetosphere after MESSENGER's first flyby. Science **321**, 85–89 (2008)

D.J. Stevenson, Planetary magnetic fields. Physics **46**, 555–620 (1983)

G.-H. Voigt, K.W. Behannon, N.F. Ness, Magnetic fields and current structures in the magnetosphere of Uranus. J. Geophys. Res. **92**, 15337–15346 (1987)

C.V. Voorhies et al., On magnetic spectra of Earth and Mars. J. Geophys. Res. **107**, 148–227 (2002). doi:10.1029/2001JE001534

Space Sci Rev (2010) 152: 23–97
DOI 10.1007/s11214-010-9643-1

Planetary Magnetic Field Measurements: Missions and Instrumentation

André Balogh

Received: 7 January 2010 / Accepted: 26 February 2010 / Published online: 30 April 2010

Abstract The nature and diversity of the magnetic properties of the planets have been investigated by a large number of space missions over the past 50 years. It is clear that without the magnetic field measurements that have been carried out in the vicinity of all the planets, the state of their interior and their evolution since their formation would not be understood even though questions remain about how the different planetary dynamos (in six of the eight planets) work. This paper describes the motivation for making magnetic field measurements, the instrumentation that has been used and many of the missions that carried out the pioneering observations. Emphasis is given to the historically important early missions even if the results from these have been in some cases bettered by later missions.

Keywords Planetary magnetism · Planetary space missions · Space magnetometers

1 Motivation for Measuring the Magnetic Field of Planets and Satellites

The presence or absence of a planetary scale magnetic field places a strong constraint of the state of a planet's interior. The nature of the field (dipolar, multipolar, crustal, induced or, in general, the combination of all such terms in different proportions) provides further indication of the details of the thermal evolution and current state and has been used extensively in constraining models of planetary structure. If a planet's magnetic field is dominated by dipolar and low-order multipolar terms, the presumed existence of a magnetohydrodynamic dynamo in the planet's interior is an additional, important factor in its internal dynamics.

For planets in the solar system, the interaction of the planetary obstacle with the all-pervasive solar wind critically influences the magnetic environment of planets. The interaction depends strongly on the nature of the magnetism of the planetary body, as well as on

A. Balogh (✉)
International Space Science Institute, Bern, Switzerland
e-mail: balogh@issibern.ch

A. Balogh
Imperial College, London, UK

its neutral and ionized atmosphere. The formation of magnetospheres with their often large and usually highly variable current systems (see Baumjohann et al. 2010a, this issue) leads to the magnetic field on and above the surface being a complex resultant from internal and external terms. The interpretation of magnetic field measurements needs to take into account the complex interplay of the different constituents of the measured field vector. Separation of terms of different origins is essential. The inversion problem is always a formidable task and remains impossible or ambiguous if there are not enough measurements. Because of the temporal variability of external terms, measurements not only need to cover a fine spatial grid, but have to provide data as a function of time with an appropriate temporal resolution. The problem of separating internal and external terms is discussed in more detail in this volume (Olsen et al. 2009, this issue).

The Earth's magnetic field is of course the best known in the solar system, thanks to the extensive ground-based measurements carried out with an increasing coverage and accuracy over the past centuries. For planets, on the other hand, measurements of their magnetic field is (almost) exclusively reserved for magnetometers onboard space probes that fly by or orbit the planet at a sufficiently small distance so that its magnetic environment can be detected, measured and modelled. (The exception is Jupiter; the existence of its magnetic field had been deduced from Earth-based and space-based radio astronomy measurements. These could provide an indirect estimate of some parameters of the planet's magnetic field, see, e.g. Smith and Gulkis 1979.)

In the solar system, it is now known that there is a wide variety of magnetized and unmagnetised bodies of very different sizes and different internal and external properties. All bodies for which magnetic field measurements are available present drastically different internal and external characteristics. At the start of the space age, however, thoughts about the magnetic fields of planets were based on comparisons with the Earth. This meant that Venus was considered similar to Earth in terms of its internal state, but the existence of a magnetic field had remained open, in fact quite doubtful due to the very slow rotation rate of the planet. On the other hand, Mars and Mercury, because of their smaller sizes, would have frozen out early in their history and would now have a solid core, hence unable to support a magnetohydrodynamic dynamo. The Moon would also be solid, without a large-scale magnetic field. The outer planets, starting with Jupiter, are clearly different from the terrestrial planets and the origin of their magnetic field cannot be deduced from what we know about the Earth. The status of the early results and theoretical considerations was reviewed by Stevenson (1975), after Mercury's magnetic field was discovered, and Jupiter's was first measured in situ, but before further observations allowed a much more detailed assessment of the theoretical explanations in terms of the planetary interiors and the operation of the planetary scale dynamos. The present understanding and future prospects of planetary magnetic fields are discussed in a review by Stevenson (2010, this issue).

It is remarkable that most planetary spacecraft, from the earliest missions in the 1960s, carried magnetometers, a confirmation that the importance of the magnetic field for characterizing planetary bodies was widely recognised. An example is provided by the Pioneer 10 and 11 spacecraft, the first missions to Jupiter and Saturn, which were equipped with magnetometers as part of their core payload (see Fig. 1). In this paper, the history of magnetic field instrumentation on planetary missions is reviewed, with an outline of the major milestones in both planetary exploration and magnetometer development. The first decade of magnetometry in the space age was reviewed by Smith (1969), Ness (1964, 1970) and Smith and Sonett (1976). The following decades were reviewed by Snare (1998) and Acuña (2002). Key moments in planetary magnetism are sketched out in this volume from the perspective of a leading figure in planetary missions over the past decades (Ness 2010, this issue).

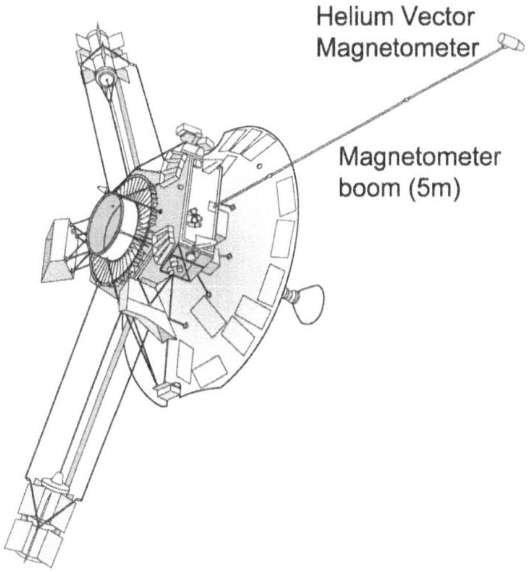

Fig. 1 Drawing of the Pioneer 10 and 11 spacecraft. The drawing shows the 5 m long magnetometer boom and the Helium Vector Magnetometer that was the first to measure in situ the magnetic fields of Jupiter and Saturn

A brief technical review of the key instrument types used on planetary missions is also presented. Only patchy instrumental information is available for some of the early missions, in particular for instruments on the Soviet programs to Mars and Venus. We also review the requirements on the instruments and on spacecraft, as well as on mission type (flyby or orbiter) to meet the scientific objectives of the missions. In order to best structure the presentation of the material, first the more general considerations about instrument type and instrument/mission requirements are described (Sect. 2). This introductory material is followed by the description of instruments and missions to specific planets in Sect. 3 for the terrestrial planets and in Sect. 4 for the outer planets.

2 Magnetometers: Sensors, Electronics, Data Processing and Calibration

The general requirements on magnetometers on different kinds of missions are listed in Table 1. While there has been improvement in the performance of magnetometers, even earlier magnetometers met, in general, these requirements (see e.g. Acuña 2002). The requirements are however not simply on the instruments, but on the measurements; stated like this, the overall performance is assessed end-to-end, combining the instruments with the spacecraft on which they are flown. Generally speaking, the limiting factor on the measurement accuracies tends to be the magnetic environment of the instrument generated by the spacecraft at the location of the magnetometer sensor. In the following, after a few general points concerning the overall functionality of magnetometers, the magnetic cleanliness issues that affect the measurements and the measures to ensure that the magnetic environment of the sensors should be compatible with the mission scientific requirements will be described briefly.

There are two important classes of magnetometers that have been flown on planetary missions. These are the fluxgate type, the most widely used space magnetometer, first at Mercury, the Moon, Venus, Uranus and Neptune, and the vector helium magnetometer that has flown on fewer missions, but in fact was the first to provide usable measurements near Mars, and the first to describe the fields of Jupiter and Saturn. Fluxgate magnetometers subsequently flew also on missions to Mars (determining on the Mars Global Surveyor the origin of Mars' magnetic field) and to Jupiter and Saturn, increasing the observational data base for their magnetic fields.

Table 1 Requirements on the accuracy of magnetic field measurements

Environment	Range	Required accuracy
Earth's field and low Earth orbit	0–45,000 nT	~0.1 to 1 nT
Magnetospheres	0–10,000 nT	~0.05 to 0.2 nT
Planetary orbit/flyby		
Mercury	0–1,000 nT	~0.05 to 1 nT
Venus	0–200 nT	
Moon	0–200 nT	
Mars	0–4,000 nT	
Jupiter	0–100,000 nT	
Saturn	0–20,000 nT	
Inner heliosphere	0–100 nT	~0.05 to 0.1 nT
Outer heliosphere	0–30 nT	~0.01 nT

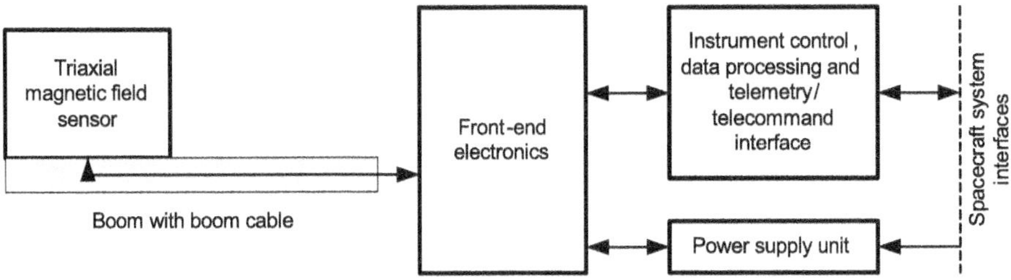

Fig. 2 Basic elements of a magnetometer. The sensor is usually mounted on a rigid, deployable boom to remove it from the close vicinity of the spacecraft and the magnetic background that it generates. The front-end electronics is specific to the sensor and its design has changed little over the past decades. The digital part of the instrument for control, onboard data processing and interfacing with the spacecraft telemetry and telecommand system is usually highly mission specific and has evolved significantly with the development of digital components and systems

These two types of magnetometer are known as DC instruments; their frequency response covers a range from (practically) 0 Hz to an upper limit which is usually in the few tens of Hz, determined by the response of the electronics. Their sensitivity decreases as a function of frequency; for the magnetic component of electromagnetic waves from ~1 Hz to several kHz, search coil magnetometers are used. Although these instruments have been traditionally included in most recent planetary missions (see, e.g. Gurnett et al. 2004 for the Cassini mission and references therein), their scientific objectives are related to processes and phenomena in the planetary environment and are only indirectly related to the internal fields of the planets. In usual operation, these AC magnetic sensors are considered as antennas and their output is combined with electric field measurements for the characterization of electromagnetic waves in the neighbourhood of planets. These magnetometers will not be discussed in this paper.

As the magnetic field is a vector quantity, instruments have to measure three components, normally in an orthogonal coordinate frame. Both the fluxgate and the helium type magnetometers have a special arrangement to enable the three-axial operation, although by radically different designs. All magnetometers share a common basic functional block diagram, although there are considerable variations in the implementation. The common block diagram is shown in Fig. 2. The sensor and the front-end electronics are intimately linked and although often separated by a significant physical distance (the length of the boom), they are closely tuned to operate specifically together. The sensor electronics is different for the two types of magnetometer, the outline of their operation is given below. In both cases, however, the analogue voltage or current signal that is proportional to the measured magnetic field is transformed into a digital signal for further processing and transmission through the telemetry to the ground.

The digital processing unit has evolved most drastically during the space age and is the most mission-specific part of the magnetometer. Its role is to execute onboard instrument control functions, such as setting the vector sampling rate and selecting the measurement range. It is also this unit that performs the interface functions with the spacecraft, such as sending the data to the spacecraft telemetry system and receiving and interpreting telecommands that provide the means remotely to control the experiment. The range of onboard digital functions that can be performed on the data has increased considerably with the increase in the performance of, and decrease in the resources needed by digital electronics. Among these functions are filtering, averaging, event-recognition, burst data storage and several more that can be included according to the mission requirement and available resources.

Both types of magnetometer sensors work at the null point, i.e. the ambient magnetic field measured by the sensor is used to generate a feedback of an equal and oppositely directed magnetic field around the sensor and it is the amount of feedback that is used as a measure of the ambient field. This ensures the greatest linearity in the measurements.

2.1 The Fluxgate Magnetometer

Fluxgate magnetometers are the most widely used sensors for space applications in general and for planetary missions in particular (see, e.g. Primdahl 1982; Acuña 2002; Musmann and Afanassiev 2010 and references therein). Several fluxgate magnetometers used on planetary missions are described in the following sections, from the early Mariner series to the most recent planetary magnetometers on missions to Mercury, Venus, Mars, Jupiter and Saturn.

The most frequently used fluxgate magnetometer in space applications uses ringcore sensors. Early on, a single ring core with two sets of sense winding (see below) was used for a two-axis sensor by Acuña and Pellerin (1969). This type of construction (two axes per core) was not generally taken up in subsequent designs; three separate single-axis sensors are normally used for the three components of the magnetic field. (This design has been used more recently by Carr et al. (2005) for a magnetospheric mission and is being used for the BepiColombo Mercury Orbiter mission by Glassmeier et al. (2010). Most currently used fluxgate sensors can trace back their origins to the early work on magnetic material for the cores of the sensors and their general construction to the sensors developed for the Apollo Lunar Magnetometer (Gordon and Brown 1972, also see Snare 1998 for a review of fluxgate magnetometers). As shown in Fig. 3, the fluxgate sensor has a ring-shaped high-permeability ferromagnetic core which has a toroidal drive winding around it. The sensor core and the drive winding are enclosed in a square coil former around which another, sense winding is placed. The two coils (drive and sense) are effectively orthogonal, so that the magnetic coupling between them is minimised, and is ideally zero. Bipolar, symmetric current pulses in the drive winding are used to drive the core material deep into saturation around the hysteresis loop, at a frequency usually about $f_o = 7$ to 15 kHz. A triaxial vector fluxgate magnetometer is constructed from an orthogonal arrangement of three single-axis sensors (Fig. 3, right).

In the absence of an external field, as shown in Fig. 4 (A), the sum of the magnetic flux in the two halves of the toroidal core sums to zero, due the symmetry of the hysteresis loop.

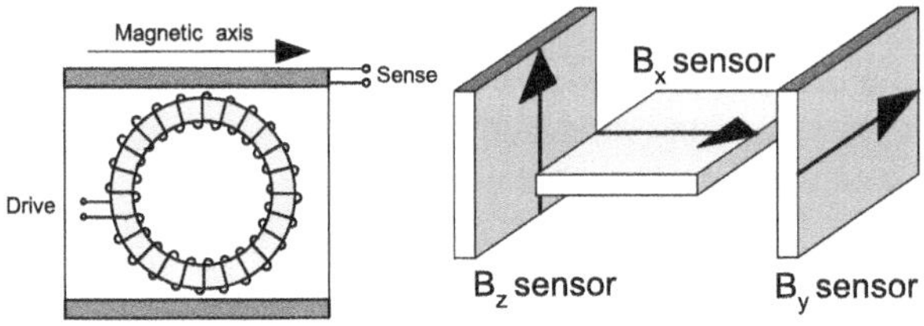

Fig. 3 (*Left*) Sketch of a single axis ring-core fluxgate sensor. The toroidal drive winding is around a ring consisting several turns of a very thin tape of high-permeability magnetic material (such as, for instance, molybdenum-pemalloy). The sense winding is around a rectangular coil former that defines the magnetic axis of the sensor. (*Right*) Arrangement of three single axis sensors to form an orthogonal triad for the measurement of the magnetic field vector. It is possible, by installing a second sense winding perpendicular to the first around one of the sensors, to use a single core for two orthogonal measurements

The signal in the sense winding, due to the magnetic induction law, is the time derivative of the magnetic flux and therefore in this case it is zero. However, in the presence of a non-zero component of an external magnetic field along the axis of the sense winding (case B), the hysteresis loop is slightly displaced, and the flux in the two halves of the core is no longer equal, so the sum of the flux generated in the two halves does no longer add to zero, and the variation of the flux induces a voltage signal as shown in Fig. 4. The voltage signal is the derivative of the periodically variable flux. This induced voltage signal in the sense winding has a frequency of twice the drive signal. This induced signal, of order <1 μV, is proportional to the component of the magnetic field along the axis of the sense winding. It is first amplified, then detected, using a synchronous detector. After some further amplification, the resultant voltage signal is fed back, through a transconductance (voltage-to-current) amplifier and the sense winding, as a feedback current counteracting the effect of the external field in the core. The schematic drawing of the electronics associated with a single-axis ringcore sensor is show in Fig. 5. As shown in the figure, the output signal is an analogue voltage proportional to the magnetic field component along the sensor axis. This signal is suitable for conversion to a digital number and further digital processing and transmission to the ground via the telemetry. The noise performance of fluxgate magnetometers depends on both the magnetic sensor itself (material and construction) and the front end electronics. A set of noise spectral density measurements is shown in Fig. 6.

In the fluxgate instruments described above, the feedback signal to null the field at the sensor was used at the level of the single-axis sensors: the feedback current was added to the pick-up coil to null the field along the axis of the sensor. It is generally recognised that, in high fields, there can be cross-talk between coils on the sensors and also uncompensated transverse fields that can adversely affect the vector measurements. A solution to improve the quality and accuracy of measurements in this case is to enclose the sensor triad inside a three-axis coil system so that the feedback is applied vectorially to the sensors. An elegant solution for minimising the size of the external, vector-feedback coil system was proposed by Primdahl and Jensen (1982) and used on Earth-orbiting missions as described below in Sect. 3.4 (Nielsen et al. 1995). A vector feedback arrangement, using a three-axis Helmholtz-type coil system is also used in the BepiColombo magnetometer intended for Mercury described below in Sect. 3.1 (Glassmeier et al. 2010).

In more recent fluxgate magnetometers, the analogue electronics has been replaced by a digital version. The digital magnetometer presents many advantages; for the measurements of magnetic fields, it provides a significantly greater flexibility in the sensor control and its feedback loop (Auster et al. 1995; Magnes et al. 2003, 2008; O'Brien et al. 2007).

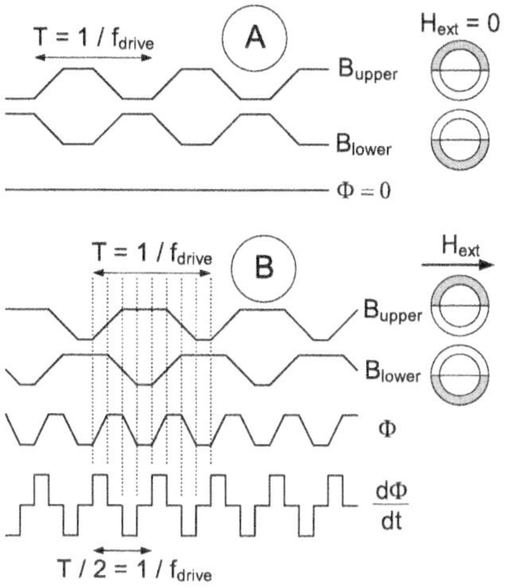

Fig. 4 Simplified principle of the operation of a fluxgate magnetic sensor. *Panel (A)* shows the magnetic flux generated in the two halves of the ringcore magnetic material by the drive current waveform in the absence of an external field. The total flux is zero (the flux generated in the two halves cancel each other). In the presence of an external field, *panel (B)*, the magnetising field due to the drive current waveform is added to the external field in one half of the ringcore and is subtracted from it in the other half. This creates a flux imbalance (there is no cancellation between the two half rings) that has a period half that of the drive waveform. The derivative of the flux signal induces a voltage in the sense coil that surrounds the ringcore

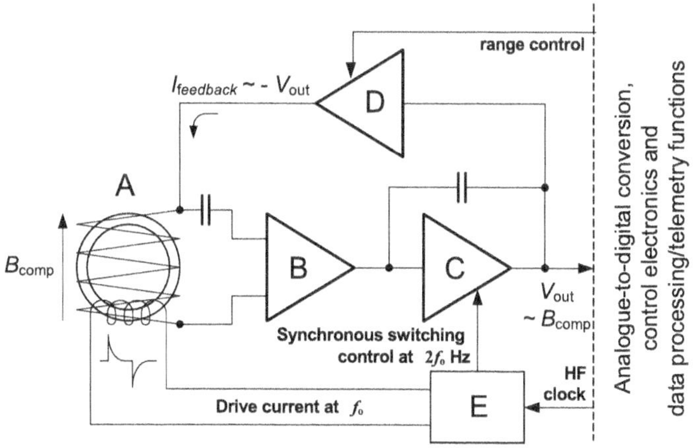

Fig. 5 A single axis fluxgate magnetometer sensor with its associated front end analogue electronics. (*A*) ringcore sensor with drive and sense/feedback windings, (*B*) pre-amplifier, (*C*) integrator with phase synchronous detector at twice the drive waveform frequency, (*D*) feedback transconductance (voltage-to-current converter) amplifier, (*E*) drive current waveform generator. In this configuration, the measurement range of the sensor is controlled through the gain of the transconductance amplifier. Normally three such units make up a complete magnetometer, with the sensors arranged in an orthogonal triad configuration as illustrated in Fig. 3

In the earlier versions of the digital magnetometer, an analogue-to-digital converter was used to transform the sensor output directly into a digitally sampled form so that filtering and phase-sensitive detection can be performed in a dedicated digital device in which the

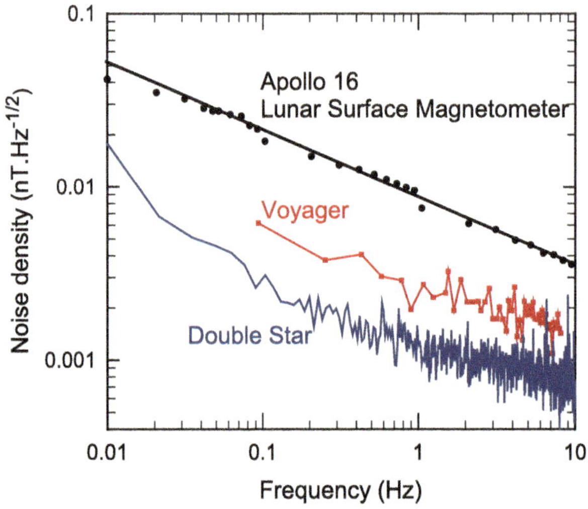

Fig. 6 The noise density of fluxgate magnetometers as a function of frequency. The three instruments illustrated span about 40 years of development. All three use ringcore sensors of very similar dimensions and the same magnetic material, although they are of different construction details. The front-end electronics designs are based on the same schematic but use increasingly better, lower noise amplifiers and other components. Much of the order-of-magnitude improvement in noise performance is attributed to the evolution of the electronics associated with the sensors (after Dyal and Gordon 1973; Behannon et al. 1977; Carr et al. 2005)

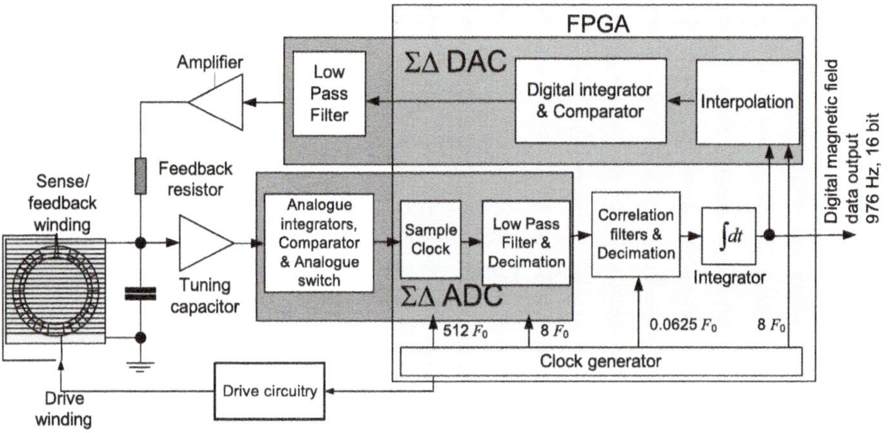

Fig. 7 A schematic diagram of the digital magnetometer, using a second-order sigma-delta modulator instead of the more classical Analogue-to-Digital Converter design (after O'Brien et al. 2007)

field-proportional signal is already in digital form. The same device then also controls the feedback to the sensor through a digital-to-analogue converter. The gain of this design is primarily in the resources needed for the instrument; the noise level of the magnetometer remains limited by the sensor performance.

The more recent version of the digital magnetometer is illustrated in Fig. 7. This design uses a sigma-delta single bit modulator both for digitising the sensor signal and to provide the feedback to the sensor as illustrated in Fig. 7. One of the main motivations for this design is the ability to implement the electronics of the sensor with components that can tolerate the very high doses of radiation that will be experienced in planned Jupiter missions.

2.2 The Vector and Scalar Helium Magnetometers

A class of magnetometers used on space missions since the 1960s has applied the effect of magnetic fields on the magnetic state of atoms for measuring with high precision and accuracy either the magnitude of magnetic fields (scalar magnetometers) or their three components (vector magnetometers). In general, magnetometers using Nuclear Magnetic Resonance (NMR) or Electron Spin Resonance (ESR) are used to measure the scalar value of magnetic fields, in particular the Earth's field in ground-based geomagnetic observatories. The operation of such resonance magnetometers were generally described by Hartmann (1972). A few such magnetometers have been used in space in conjunction with vector magnetometers to provide high accuracy measurements of the Earth's magnetic field (Primdahl 1998). The helium magnetometer described below can in fact also be configured to operate in a scalar mode that provides a high accuracy measurement of the magnitude of the magnetic field (Slocum and Reilly 1963; Slocum et al. 1971). A helium magnetometer operating in the scalar mode was included in the Cassini Saturn Orbiter magnetometer (Smith et al. 2001).

However, it is also possible to build a quantum-effect based magnetometer in a tri-axial vector-measuring configuration. A particularly successful three-axis (vector) magnetometer based on the quantum-mechanical properties of helium has been used on several key planetary and space missions, such as Mariner 4 and 5 (launched in 1964, Connor 1968; Smith 1969), Pioneer 10 and 11 (launched in 1972 and 1973, Smith et al. 1975a, 1975b), ISEE-3 (launched in 1978, Frandsen et al. 1978), Ulysses (launched in 1990, Balogh et al. 1992a, 1992b) and Cassini-Huygens (launched in 1997, Smith et al. 2001; Dougherty et al. 2004). The basic principle of the helium magnetometer is that in the presence of an external magnetic field the efficiency of optical pumping in a cell containing He is reduced and this reduction can be related quite accurately to determine the components and/or the magnitude of the external field.

In optical pumping, a non-thermodynamic equilibrium distribution of atoms among the various energy substates is produced by incident polarised radiation. The relevant energy levels in ^{4}He are shown in Fig. 8. The circularly polarised light at 1.08 μm passing through the He cell induces optical pumping of the metastable He population in the triplet 2^3S_1 "ground" state (Colegrove and Franken 1960). The change in optical pumping efficiency depends on the intensity of the magnetic field and its angle with respect to the optical axis of the sensor. The polarization of the incident light is an important factor in the way the magnetometer operates and linear polarization has also been used in another implementation of the magnetometer. The general arrangement of a vector helium magnetometer that was flown on several planetary missions is shown in Fig. 9, illustrating the components of the sensor. When a rotating sweep field of ~300 Hz is applied using the Helmholtz coil that surrounds the optical cell, the pumping efficiency, and therefore the IR throughput detected by the IR sensor vary as the vector sum of the sweep field and the component of the steady ambient field in the sweep plane. Under these conditions, the signal in the IR detector contains a sinusoidal component at the sweep frequency whose magnitude is proportional to the ambient field strength and whose phase relative to the sweep waveform depends on the direction of the ambient field in the sweep plane (see, e.g. Slocum and Reilly 1963 and Smith et al. 1975a). The pT-level performance of an early prototype helium magnetometer is illustrated in Fig. 10.

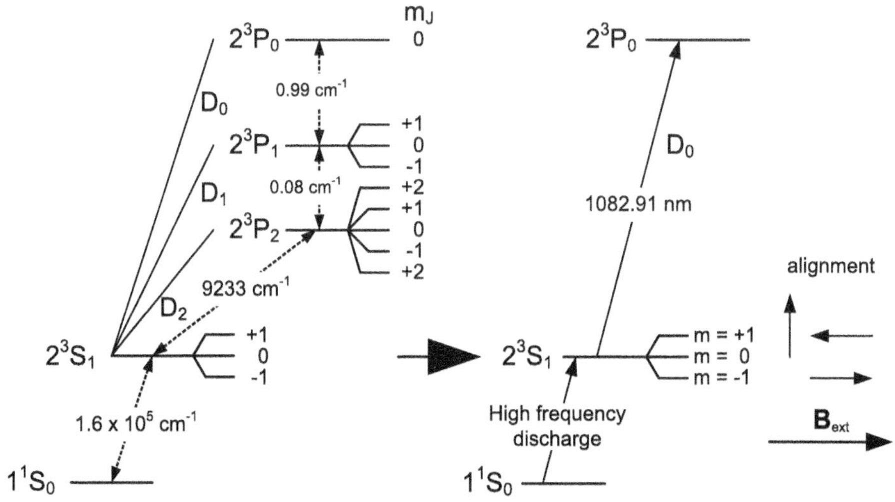

Fig. 8 The energy level diagram for helium used in the optically pumped helium magnetometer. Of particular interest are the 2^3S_1 terms that form a "ground" state for higher levels in optical pumping. *On the left of the figure*, all terms from the ground state to the first excited (^3P) are shown (Colegrove and Franken 1960). *On the right*, the key transitions used in the optically pumped magnetometer are shown, together with the alignment of the spit levels if the 2^3S_1 state in the presence of an external magnetic field

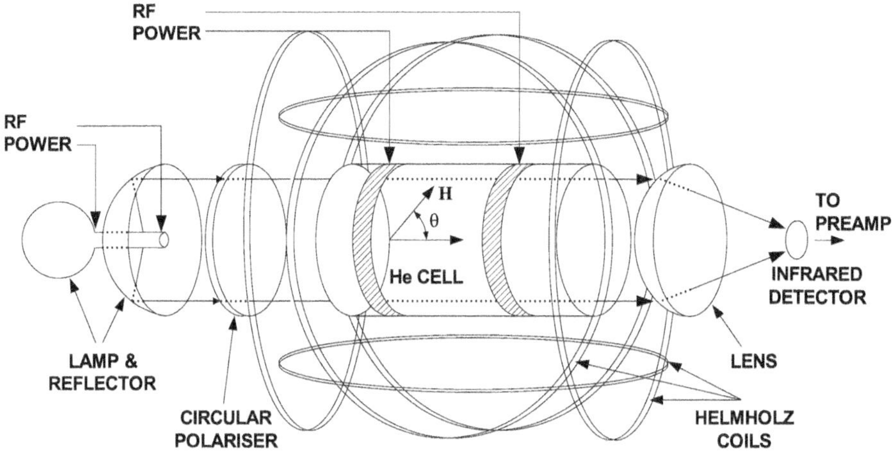

Fig. 9 Schematic diagram of the Vector Helium Magnetometer used on Pioneer 10, 11, ISEE-3, Ulysses and Cassini missions. For the Cassini Saturn Orbiter mission, the magnetometer has been configured to be used either as a scalar sensor (measuring the magnitude of the magnetic field) or as a vector magnetometer (figure courtesy of E.J. Smith)

Phase coherent detection of the output of the IR sensor in the magnetometer electronics produces voltages representing the ambient field components along the optical and transverse axes. These voltages are used to generate feedback currents in the sensor coils that null the ambient field on both axes. The feedback currents are highly linear measures of the two field components. Triaxial (vector) measurements are obtained by alternating the sweep field between two orthogonal planes that intersect along the optical axis. The performance of

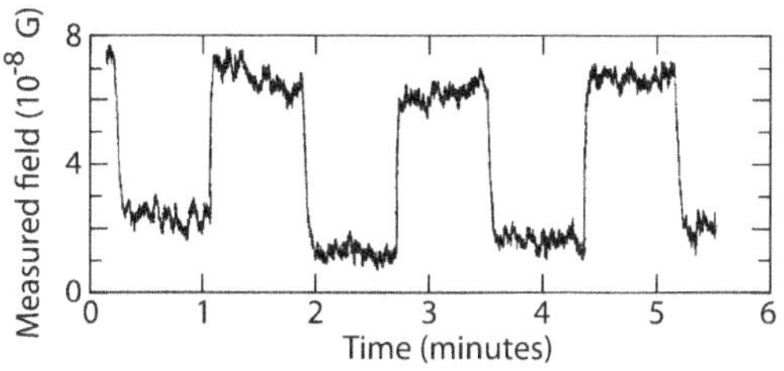

Fig. 10 The output of an early development model of the helium magnetometer demonstrating its excellent low-field sensitivity. A square calibration pulse of amplitude 5×10^{-8} G (or 5 pT) was applied along the sensor optical axis to the instrument contained in a magnetic shield (from Slocum 1972)

the VHM is characterised by its very stable offsets, with variations of no more than ~20 pT over several years, and its very low noise, ~3 pT rms/Hz$^{1/2}$; the best current fluxgates have an offset stability of only about 0.2 nT, although their noise performance has been improving (see Fig. 6) and is close to that of the VHM as used on recent missions.

More recent developments include the use of a laser instead of the He lamp used in the original design for providing optical pumping of the helium gas in the sensor cell (Slocum et al. 1988; Slocum 1991). As reported by Slocum (2002), the new design includes vector and scalar operation in the same instrument that can achieve accuracies of better than 1 part in 10,000 (better than ~1 nT in the Earth's magnetic field) and a sensitivity of ~0.2 pT/Hz$^{1/2}$. A parallel development of a new generation of helium magnetometers has been announced (Guttin et al. 1994; Gravrand et al. 2001) but no planetary mission has been found as yet for these instruments. (Such a magnetometer had been envisaged as part of the payload for NASA's Juno mission to Jupiter, but its implementation was later abandoned.) A particular advantage of this instrument is that it avoids, through the use of a controlled linear polarizer, the directional sensitivity of the circularly polarized design (Chaillout et al. 1993; Leger 1995). A scalar version of this magnetometer has been incorporated into the payload of the 3-spacecraft Swarm mission (see Sect. 3.4) that is planned for launch in 2012 and will study the Earth's magnetic field (Merayo et al. 2008).

2.3 Magnetic Background, Magnetic Cleanliness

A key characteristic of the performance of magnetometers in space is the magnetic environment at the location of the sensors. For space missions, the accuracy with which the ambient natural magnetic field due to the planet and its environment needs to be measured is between 0.01 to 1 nT, depending on the mission target. Even though spacecraft that have carried magnetometers in the past had been the subject of magnetic cleanliness programmes that enforced upper values of magnetic fields generated by subsystems and payload instruments, the outcome was a spacecraft-generated magnetic field that, on or close to the spacecraft, was significantly higher than the requirement by the magnetometer. This was partly or wholly resolved by placing the magnetometer sensor(s) on a boom that removed them from the vicinity of magnetic sources on the spacecraft (Mehlem 1978), thanks to the inverse cube law of the fall-off of magnetic field intensity as a function of distance from a dipole (and higher exponents for higher order poles). Boom lengths have varied from ~1 m (Rosetta, Venus Express) to ~5 m (Pioneer 10 and 11, Pioneer Venus Orbiter, Ulysses), to 12 m (Selene, JAXA's mission to the Moon) and to 13 m (Voyager 1 and 2). Although the use of booms is not by itself a complete solution to magnetic cleanliness, their length (determined from the technical and financial resources that are available) is generally a trade-off between

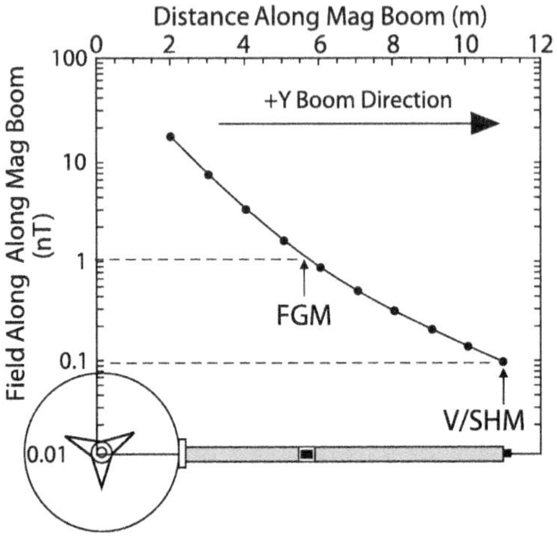

Fig. 11 Background magnetic fields measured (*full circles*) and modelled (*line*) along the magnetometer boom on the Cassini Saturn Orbiter spacecraft after implementing a strict magnetic cleanliness programme (from Narvaez 2004)

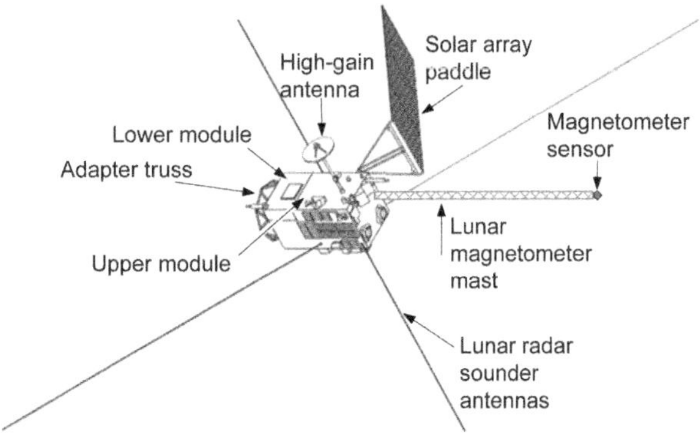

Fig. 12 The Kaguya (Selene) spacecraft of the Japanese JAXA/ISAS was launched in September 2007 to orbit the Moon and make very low altitude (~50 km) observations, including the detailed mapping of the Moon's crustal magnetic field (drawing courtesy of JAXA/ISAS)

the cost of a strict magnetic cleanliness programme and a tolerable level of background at the location of the magnetometer sensor. Figure 1 in the Introduction illustrates the Pioneer 10/11 spacecraft with its 5 m boom and the vector helium sensor located at the boom tip; the fall-off of the spacecraft background field along the 11 m magnetometer boom, as the outcome of the magnetic cleanliness programme for the Cassini Saturn Orbiter spacecraft, is shown in Fig. 11. Such lengths are adequate when combined with a spacecraft-level test programme.

The much shorter booms used on Rosetta and Venus Express, for instance, and the absence in these cases of a significant magnetic cleanliness programme make the data processing and in-flight calibration challenging. This was already the case for Mars Global Surveyor (see the description of the magnetic background removal program in Sect. 3.3) and represents a general trend by space agencies not to accept the requirements of magnetic field investigations for an adequate cleanliness level at the location of the sensors. A remarkable recent exception to the general tendency of leaving the spacecraft background removal to the magnetic field experimenters is the Kaguya (Selene) spacecraft of JAXA/ISAS, with its 12 m magnetometer boom as illustrated in Fig. 12. The spacecraft, launched on 14 September 2007 was placed into an orbit around the Moon on 4 October 2007 and after a highly successful mission, it was crashed into the Moon on 10 June 2009. When the Kaguya boom was deployed in space, the magnetometer monitored the decrease in the background at its

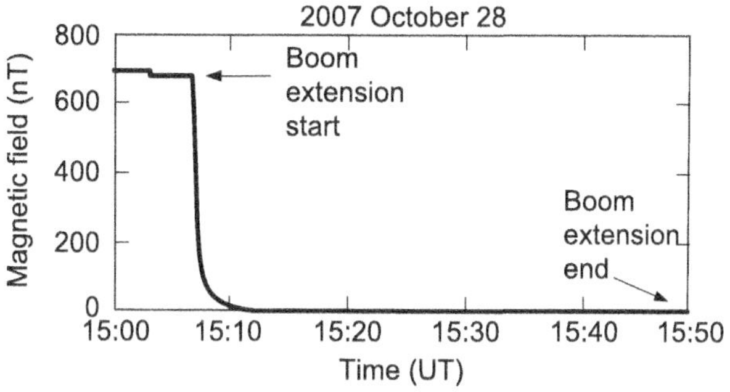

Fig. 13 Magnetic field measured by the boom-end magnetometer on JAXA's Kaguya lunar mission spacecraft as the 12-m boom was extended. The spacecraft background is effectively zero at the location of the magnetometer sensor (from Nazakawa et al. 2009)

location; the resulting data, as the spacecraft background turned to close to zero, are shown in Fig. 13.

The required magnetic cleanliness at the sensors on space missions to planets and their satellites is usually specified as an absolute value, for instance less than 1 nT in the ideal case, and an upper limit on the variability, such as 0.5 nT in 100 s. Such values are in practice difficult to achieve and compromises need to be made, in particular concerning the absolute value of the upper limit of the background. Meeting the stability requirement is in fact more important; a constant but known DC offset at the location of the sensor would be acceptable, but it is difficult to isolate from the varying component of the spacecraft background, and even from the sensor offset itself. The challenges of magnetic field measurements on spacecraft and the techniques used to deal with them have been described by, among others, Iufer (1970), Sanders et al. (1972), Ness et al. (1971), Davis et al. (1973), Neubauer and Schatten (1974), Musmann (1988), Auster et al. (1990) and Narvaez (2004). The topic has remained very active (see further description of the technique in Sect. 2.4). A recent example when a highly sophisticated use of the dual magnetometer (or gradiometer) technique that was applied to the magnetic field investigation on a magnetospheric spacecraft (TC-1 of the Double Star mission) was presented by Georgescu et al. (2008).

The preliminary steps to ensure a magnetically clean spacecraft consist in specifying and verifying by tests the magnetic state of individual components of the spacecraft: subsystem and payload boxes, moving parts and other mechanisms. A magnetic budget is usually drawn up and by using a digital model of the locations of the units and their magnetic characteristics, a magnetic model of the spacecraft can be derived. An early study of modelling and predicting spacecraft magnetic fields was published for NASA by Halacsi (1969). This model can be regarded, however, only as an estimate, as mathematical models can only be approximate, due principally to the complex self-compensation and mutual induction effects among the different units on the spacecraft. In some cases, where full testing was not possible, such as for the very important and successful Mars Global Surveyor (Acuña et al. 1996, see below), there was a special effort to implement self-compensation, in particular of the currents in the solar panels, followed by modelling with special care and in greater detail than usual, so that despite the unusual mounting of the sensors at the outboard ends of the solar panels (see Fig. 33 in Sect. 3.3), the background field could be, after added in-flight calibration, removed from the measurements. For several spacecraft with demanding mission objectives, a long magnetometer boom, a strict magnetic cleanliness program, and a pre-launch testing and modelling program were implemented to ensure the spacecraft's compatibility with the magnetic cleanliness requirements. An example is the Cassini Saturn Orbiter spacecraft, for which the challenge was to ensure the magnetic cleanliness of the Radioisotope Thermoelectric Generators (RTGs), potentially significant sources of background contamination (Mehlem and Narvaez 1999). As shown in Fig. 11, the final pre-launch magnetic cleanliness of the Cassini spacecraft met the requirements of the mission, thanks in large part to the 11-m magnetometer boom (Narvaez 2004).

A difficult, but critically important aspect of magnetic cleanliness is the pre-flight testing of the magnetic properties of the spacecraft as a whole. For high-performance planetary missions, this testing is an important requirement. There are only very few facilities worldwide that are able to perform whole spacecraft magnetic testing. The Magnetfeldsimulationsanlage (MFSA, or "magnetic field simulation facility", see Kügler 2004), operated by the German industrial testing consortium IABG and situated near Ottobrunn, Germany and the Spacecraft Magnetic Test Facility (SMTF) of NASA's Goddard Space Flight Center in Maryland, USA are the two most used and most important such facilities.

The test facilities contain large three-axial coil systems with the purpose of compensating the Earth's magnetic field, its variations and fluctuations within a central volume so that a spacecraft placed in that volume is in close to zero magnetic field, as it would be in space. The three orthogonal axes each have four co-axial coils, circular Braunbek (1934) coils of 13 m diameter in the case of the NASA facility and square Helmholtz coils of 15 m for the European facility. The central, controlled volume has linear dimensions of about 2 m in the case of the SMTF and about 4 m in the case of the MFSA.

The recently refurbished capabilities of the MFSA are described in Kügler (2004) and a typical application to a spacecraft test is described by Kügler (2001). The controlled volume can simulate either a zero field environment or some other value that can be controlled by the currents in the coils. The stability is very good at $\sim$0.5 nT/hour in both facilities, even when the Earth's field varies—although testing becomes significantly more difficult during geomagnetic storms when the performance of the servo control of the coil currents performs less well. The reference to compensate the Earth's magnetic field is given by a high quality, high stability magnetometer at some distance from the facility, to avoid the reference signal to be affected by the coil currents.

A facility of more recent construction is the Japanese magnetic test facility of JAXA/ISAS in which the lunar spacecraft Kaguya (Selene) was tested (Shimizu et al. 2008). This facility also uses a Braunbek circular coil system of 15 m diameter. The volume of good uniformity, where the magnetic field can be maintained stably to an accuracy of $\pm$2.5 nT, is a 2.3 m diameter sphere centrally located in the coil system.

Two earlier, no longer active but historically important spacecraft-scale magnetic test facilities were located in California, USA. The first of these, at NASA Ames Research Center, Moffett Federal Airfield, was used in the early 1960s to test the early Mariner spacecraft (Mariners 2, 4 and 5), also the OGO (Orbiting Geophysical Observatory) series, and Pioneer 6 (Iufer 1970). It is interesting to note the typical values for the spacecraft fields obtained at the location of the magnetometers for these early missions, as quoted by Iufer (1970): Mariner 2 (130 nT), Mariner 4 (35 nT), Mariner 5 (12 nT), OGO (2 nT), Explorer 18 (0.5 nT), Pioneer 6 (0.25 nT). The second of these facilities, at TRW Inc., Redondo Beach, was used for the Pioneer 10 and 11 spacecraft (Sanders et al. 1972).

Magnetic cleanliness programmes and pre-launch testing are necessary for ensuring accurate measurements in space. However, the resources needed for pre-launch tests are not always available and, even for missions that have completed extensive pre-light magnetic tests, the evaluation of the spacecraft-induced background remains a routine part of in-flight calibration. In the following, specific missions and instruments are highlighted that have proved a challenge to remove the spacecraft background from the measurements. For many missions, two magnetometers, at different distances from the spacecraft body, have been used so that a gradiometrically calculated estimate of the background could be subtracted from the observations (e.g. Ness et al. 1971); this is discussed further in the next section on magnetometer calibration. Specific cases are also described below. Neither an intensive magnetic cleanliness programme, nor extensive pre-launch testing has been carried out on some of the planetary missions described below: examples that are discussed in some detail are Mars Global Surveyor, the Rosetta cometary mission and Venus Express for which a reduced spacecraft-level magnetic cleanliness programme was carried out, in addition to very comprehensive instrument-level programmes. In the last two of these, only a short boom was used, with two magnetometers, while in the case of MGS, the magnetometers were mounted on the solar panels. Remarkably, however, significant discoveries have been made with both Venus Express and with MGS and, similarly, in-flight testing of the Rosetta magnetometer promises good results once the spacecraft reaches its target comet in 2014.

2.4 Magnetometer Calibration

The calibration of the magnetic field data measured by the instruments is one of the major challenges facing instrument designers and operators. Magnetometers in space or in planetary environments measure field component values over very wide dynamic ranges, with required accuracies (as shown in Table 1), up to five or six orders of magnitude smaller than the ambient magnetic field of the Earth in the laboratory. This necessitates that the calibration process be a conceptually two-step procedure, the first pre-launch, the second in-flight. In the following, a brief description is given of the main elements in the calibration process. The notation selected is not unique, as there are many different descriptions of the calibration techniques, but the principles listed are very general. (This description has been based on that found in Balogh et al. 2001; see also Gloag et al. 2010. An alternative, well documented description of a calibration programme for a terrestrial magnetism mission can be found in Risbo et al. 2003 and Olsen et al. 2003.) Specific references to the calibration of the data are also made in the description of instruments and missions throughout the paper.

The magnetic field to be measured is denoted $\mathbf{B}_{\mathrm{actual}}$; this is a vector quantity and is defined in a physically meaningful coordinate system associated, for instance, with the target planet. The telemetered output of the magnetometer as received on the ground is denoted $\mathbf{V}$; this is also a vector quantity, i.e. an ordered set of three numbers. The relationship between $\mathbf{V}$ and $\mathbf{B}_{\mathrm{actual}}$ can be described as a set of consecutive linear transformations in which the input is $\mathbf{B}_{\mathrm{actual}}$ and the output is $\mathbf{V}$

$$\mathbf{V} = \underline{\underline{c}}^{(\mathrm{sensor})}\underline{\underline{c}}^{(\mathrm{transfer})}\underline{\underline{c}}^{(\mathrm{att})}\mathbf{B}_{\mathrm{actual}} + \mathbf{c}_0 \qquad (1)$$

The transformation matrix $\underline{\underline{c}}^{(\mathrm{att})}$ is simply the rotation of the vector from the physical (planetary) coordinate system into a coordinate system rigidly tied to the spacecraft. The transformation $\underline{\underline{c}}^{(\mathrm{transfer})}$ represents the rotation from the rigid spacecraft coordinates into the coordinates tied to the sensor. This transformation can be included in $\underline{\underline{c}}^{(\mathrm{att})}$ if the sensor is mounted rigidly so that its reference coordinate system is strictly tied to the spacecraft and its attitude determination system, within the limits of the required error budget ($\sim$6 arcmin for many space physics missions). In many early planetary missions this was assumed, as the rigidity of the sensor mounts, often on a rigid boom, was ensured and fully tested before flight. In that case, $\underline{\underline{c}}^{(\mathrm{att})}$ is a set of two consecutive rotations, one from the physical frame into the frame in which the spacecraft attitude is defined and measured and one that transforms from that frame into the frame of the magnetometer sensor mounting. However, when the booms are not sufficiently rigid and the reproducibility of the deployment (as is normally the case) does not by itself meet the requirements on the knowledge of the angular alignment of the sensor mounting, other measures need to be implemented to determine $\underline{\underline{c}}^{\text{(transfer)}}$.

Three different methods have been used for a precise determination of $\underline{\underline{c}}^{\text{(transfer)}}$ on planetary missions. The Voyager and Galileo spacecraft, described in Sect. 4.1, used a single calibrated coil at the boom root, referenced rigidly to the spacecraft, that when energised by a calibrated DC current, produced a well-defined and precise magnetic field vector at the location of the magnetometer sensors. The Kaguya lunar spacecraft (described in Sect. 3.5) and the Cassini Saturn orbiter spacecraft (described in Sect. 4.2) used two orthogonal coils mounted on the spacecraft which generated known magnetic field vectors that have been used to determine the precise orientation of the sensors with respect to the spacecraft. Another method, using an optical Attitude Transfer System between the spacecraft body and the magnetometer platform was used on the Magsat mission, as described in Sect. 3.4 (Mobley et al. 1980). In addition, star cameras were used on the spacecraft for determining the attitude of the spacecraft accurately. For terrestrial magnetic field measurements, the accuracy requirement is normally greater than for planetary and space physics missions; it is of order 1 part in 10^5, to reach the required sub-nT accuracy in the earth's magnetic field. For Ørsted and subsequent missions (see also in Sect. 3.4) with the objective to measure accurately the Earth's magnetic field, an Advanced Stellar Compass (ASC) was developed that enabled the absolute attitude of the magnetometer sensor itself to be determined (Jørgensen et al. 1997). By mounting the magnetometer sensor and the ASC on a common optical bench, the absolute determination of the magnetometer attitude no longer relied on the spacecraft attitude measurement, but could be taken directly from the attitude determined by the star camera.

The matrix $\underline{\underline{c}}^{\text{(sensor)}}$ represents the properties of the sensor and magnetometer electronics, as determined by pre-launch calibration procedures. It includes the linear scale factors in units of voltage or counts at the output of the instrument per units of the magnetic field component (in nT) measured for each axis. It also includes the rotation from the assumed orthogonal axes defined for the sensor and the (generally nearly, but not strictly orthogonal) magnetic axes of the vector sensor.

The vector $\mathbf{c}_0$ added in (1) results from the offset of the sensor itself and of the sum of the background contributions due to the spacecraft.

Calibration of the received data usually proceeds in consecutive steps, to determine the measured field vector $\mathbf{B}_S$ in the orthogonal coordinate system tied to the sensor, by performing first a transformation (using a diagonal matrix $\underline{\underline{M}}$ with nominal scale factors) from telemetered values to values of the components in physical units (nT), followed by the application of the calibration matrix $\underline{\underline{c}}^{\text{cal}}$ which incorporates all the scale factor corrections as well as the orthogonalisation of the axes from the magnetic measurement coordinate system to the sensor coordinate system;

$$\mathbf{B}_S = \underline{\underline{c}}^{\text{cal}} \underline{\underline{M}} \mathbf{V} - \mathbf{o}^{\text{cal}} \tag{2}$$

The vector $\mathbf{o}^{\text{cal}}$ represents the total offset determined usually in flight; it includes contributions by the sensor and from all external sources. The determination of $\underline{\underline{c}}^{\text{cal}}$ and $\mathbf{o}^{\text{cal}}$ is always a major component of the scientific operation of a magnetometer on a planetary (or other) spacecraft. Pre-launch calibration tests at sensor level (see, e.g. Risbo et al. 2003) cover measurements of the scale factors and their linearity; the frequency response (including the electronics and any digital filtering or other manipulation); the noise of the sensor and electronics as a function of temperature; the sensor offset vector and its stability vs. time and temperature; the orthogonality of the triaxial sensor; and finally any crosstalk between axes of the sensor (very important for sensors in large fields close to planets and in Earth orbit, cf. Brauer et al. 1997). The parameters thus determined are used as a baseline for the determination of the calibration parameters in flight. Generally speaking, all the parameters in the calibration matrix $\underline{\underline{c}}^{\text{cal}}$ should be considered as time dependent, and the calibration effort needs to be continued throughout the flight operations phase of the instruments.

The determination of the compound offset vector $\mathbf{o}^{\text{cal}}$ remains similarly a continuous task. The reason for this is that the sensor offset, particularly for fluxgate sensors, is variable in time as well as in temperature. In the descriptions of specific instruments in this paper, factors affecting the sensor offset and that due to the spacecraft background are outlined in several instances. For the determination of the offsets, a distinction needs to be made between spinning and three-axis stabilised spacecraft. While most planetary missions use three-axis stabilisation to satisfy the requirements of imaging investigations, a few (e.g. Pioneer 10 and 11) were spin stabilised. In the case of spinning spacecraft, the offsets in the spin plane can be determined relatively easily, as two components of the magnetic field (if one of the axes of the sensor is aligned with the spin-axis) are modulated by the spin and should, for complete spins, average to zero. A departure from the zero average corresponds to the offset for the given axis. As the alignment of the sensor magnetic axes and even its coordinate system in the spacecraft coordinate system aligned with the spin-axis is never perfect, this provides only a partial determination of the zero levels even of the two axes in the spin plane.

Procedures have been developed for the determination of magnetometer offsets relying on the statistical properties of the magnetic field in the solar wind (Davis and Smith 1968; Belcher 1973; Hedgecock 1975). Most changes in the interplanetary magnetic field are changes in direction rather than in magnitude. The method proposed by Davis and Smith (1968) uses this property and minimizes the variance of the squared magnitude of the magnetic field over a suitabl interval to determine the offsets. Belcher (1973) proposed a variant of the Davis and Smith method. As it became known that there are many intervals in the solar wind when the fluctuations are primarily transverse (Alfvénic), Belcher (1973) proposed that the component offsets could be determined by maximising the maximum variance vector to be perpendicular to the average magnetic field. The method proposed by Hedgecock (1975) implemented the fluctuating property of the interplanetary magnetic field to determine the offsets by imposing that the correlation between changes in the inclination of the magnetic field to the coordinate axes and the magnitude of the vector field should be a minimum over a suitably chosen data interval. These methods have been widely used over the past 40 years. As an example, Fig. 14 shows the long term behaviour of the zero level offsets of the Vector Helium Magnetometer on the Ulysses mission (Balogh et al. 1992a, 1992b). The offsets were determined by taking advantage of the spacecraft spin as well as the application of the Hedgecock (1975) method. The extremely high long term stability of the VHM sensor is well illustrated; the systematic variations were due to the spacecraft warming up as it approached perihelion in 1995 and 2001. More recently, Leinweber et al. (2008) have revisited these three methods and have concluded that a suitably modified variant of the Davis and Smith (1968) method is the best for determining magnetometer zero levels.

Terrestrial missions for measuring the magnetic field of the Earth (as described in detail below, in Sect. 3.4) have used the vector measurement of the field in combination with a scalar measurement of its magnitude (Merayo et al. 2000). Given that the scalar magnetometer provides an absolute measurement of the magnetic field magnitude, the methodology proposed by Merayo et al. (2000) allows the determination of nine parameters for calibrating the vector measurements of a tri-axial fluxgate sensor. These are the three zero

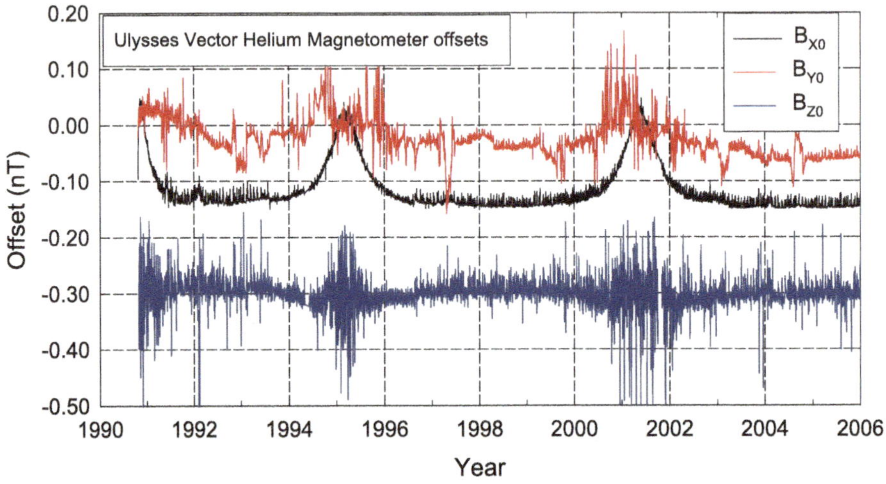

Fig. 14 The zero level offsets of the Vector Helium Magnetometer on the Ulysses spacecraft determined over the first 15 years of its operation in flight using the technique first proposed by Hedgecock (1975). The very high long tern stability of the offset is well illustrated, with a small sensitivity to temperature (notable in 1995 and 2001) as the spacecraft went through perihelion in its nearly polar orbit around the Sun

level offsets on the three axes, three scale factors and three angles that define the misalignment of the sensor axes from an orthogonal reference system. As the resources needed by scalar magnetometers are generally significant, the only planetary mission so far that has used a similar configuration is the Cassini Saturn Orbiter (see Sect. 4.2) that used a triaxial vector fluxgate, together with a vector/scalar helium sensor (Smith et al. 2001; Dougherty et al. 2004). A potential alternative to the use of a scalar magnetometer is an independent measurement of the magnetic field magnitude. This was done on ESA's Cluster spacecraft (not a planetary mission as such) by an electron gun instrument that measured the gyroradius of electrons of calibrated energy and thus the magnetic field strength. The intercalibration with the fluxgate vector magnetometer on the same spacecraft has been investigated by Georgescu et al. (2006).

The determination of the background field at the location of the magnetometer sensor due to the spacecraft is an important part of the in-flight calibration of the magnetic field data. The topic of magnetic cleanliness has been described above, in Sect. 2.3. In the following sections, several further mentions are made to the dual magnetometer technique, whereby two sensors are used along a boom at different distances to use gradiometry for deducing the spacecraft field. Early application of the technique proposed by Ness et al. (1971) was made on the Mariner 10 spacecraft as discussed in Sect. 3.1. However, the technique needs to be used with some care, as was pointed out by Davis et al. (1973), Neubauer and Schatten (1974), Neubauer (1975) also examined the fall-off of the spacecraft signature with distance that's highly relevant to the application of the method. Further use of the technique was made on the Voyager mission (see Sect. 4.1). The gradiometer technique has had to be brought up to date in recent years as the precautions for minimising the spacecraft fields (the magnetic cleanliness programmes described in Sect. 2.3) have not been fully carried out on some of the missions like Rosetta and Venus Express. For these, the gradiometer technique was the only tool that allowed the recovery of a clean magnetic field signal from the measurements (Delva et al. 2003; Pope et al. 2009). Important, if unintended experience was gained in the application of the technique to the Double Star TC-1 mission (Georgescu et al. 2006).

Use of the gradiometer technique for terrestrial missions in which use is made of a scalar magnetometer in combination with a vector sensor was examined by Primdahl et al. (2006). Using the example of Ørsted (see Sect. 3.4), they developed a procedure for assessing the spacecraft field at the location of the triaxial fluxgate sensor, using the measurements of the magnitude of the field at the end of the boom by a scalar sensor. However, their results showed one particular potential limitation of the technique. This is due, as concluded by Primdahl et al. (2006) to an unknown shift in the offset values of the magnetometer itself that is caused by the launch stresses. Because of this, the pre-fight determination of the sensor offset cannot be used in the gradiometer equations, at least not without some considerable uncertainty.

3 The Terrestrial Planets

The four terrestrial planets, Mercury, Venus, the Earth and Mars, together with the Moon are solid bodies, composed of a mostly iron core and a silicate mantle (Stevenson et al. 1983). Their magnetic properties, despite the similarities of their constitution, are very different. Apart from the Earth which has a very well documented magnetic field and magnetic environment, the other three (plus the Moon) are less well known, but their main magnetic field (if any) has been observed and modelled. Historically, the earliest planetary missions explored Venus and Mars; this was somewhat unfortunate, since, as we now know, Venus appears not to possess a global magnetic field at all and Mars' is a complex remnant crustal field confined to a part of the planet. Mercury provided the greatest surprise by having a small but very interesting planetary scale magnetic field. The moon has no large scale magnetic field but its local concentrations of magnetized material have interesting implications for its history.

3.1 Mercury

The small planet Mercury, the terrestrial planet closest to the Sun, was not considered to have a planetary dynamo and was assumed to be Moon-like. The reason for this assumption was that, given the size of Mercury, its interior would have cooled and frozen rapidly after its formation, not allowing a dynamo to start and operate. Even Mercury's known, very high density that implied a large iron core was considered an inadequate argument for any part of Mercury's core to remain liquid to allow a dynamo to operate. Given this consensus about Mercury's likely internal state, it is very surprising that a magnetometer was included in the payload of Mariner 10, the mission known as Mariner Venus/Mercury 73 prior to its launch in 1973. However, the magnetic field measurements, demonstrating the existence of a global planetary field, led to a fundamental re-assessment of the planet's thermal evolution and the current state of its interior (Ness 1978).

The magnetometer on Mariner 10 consisted of two triaxial sensors, one at the end of a 5.8 m boom, the other at 2.3 m inboard from the boom tip (Ness et al. 1974, 1975). The sensors were built by Schonstedt Instrument Company, manufacturers of several of the early space-borne magnetometers. The performance of the sensors enabled the measurement of the unexpected magnetic field and magnetosphere of Mercury during the two flybys close to the planet, on 29 March 1974 and 16 March 1975. A sketch of one of the two triaxial fluxgate sensor units is shown in Fig. 15.

This was the first time that a dual magnetometer system was used to measure the magnetic field of a planet (not counting the Mariner 10 flyby of Venus), and it is of interest to note the method used to subtract the disturbance magnetic field generated by the spacecraft

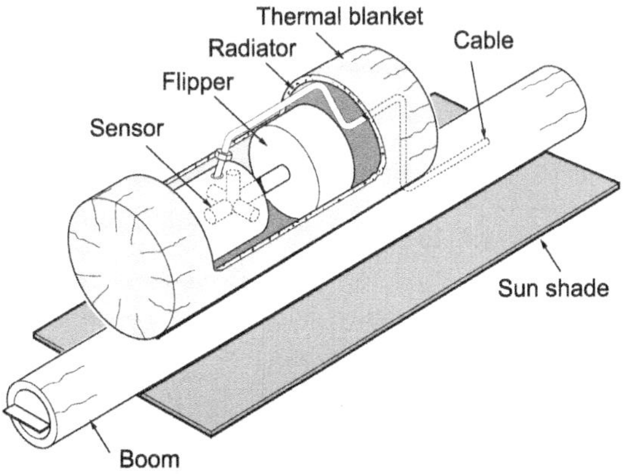

Fig. 15 A cutaway sketch of one of the two Mariner 10 triaxial sensor units, showing the flipper assembly next to the three orthogonal single axis Schonstedt fluxgate sensors (redrawn by N. Powell after http://nssdc.gsfc.nasa.gov/nmc/experimentDisplay.do?id=1973-085A-04). The mass of the two magnetometer instruments including the analogue electronics was 4.3 kg and its average power consumption was 13.6 W. The instrument had a data rate of 1052 bps. The complete instrument was designed at the Goddard Space Flight Center

from the measurements. Following the description of the post-launch calibration of the data by Ness et al. (1974), the first step was to calculate the vector contribution of the spacecraft $\mathbf{B}_{SC}$ at the location of the outboard sensor by combining the measurements at the inboard and outboard sensors through $\mathbf{B}_{SC} = \alpha/(1-\alpha)[\mathbf{B}_{IB} - \mathbf{B}_{OB}]$ where $\alpha = 0.3$ is a coupling coefficient between the sensors assuming a spacecraft-centred dipole (calculated from the distances along the boom and adding the distance of the boom root from the spacecraft centre); $\mathbf{B}_{IB}$ and $\mathbf{B}_{OB}$ are the raw measurements at the inboard and outboard sensors, respectively. It was found that at the location of the sensors the dipole assumption for the spacecraft background was adequate, although the strength and orientation of the dipole was variable. The spacecraft background field $\mathbf{B}_{SC}$ is then averaged over 3 seconds and then a corrected vector $\mathbf{B}_{OBC}$ is calculated by subtracting from each measurement sample the averaged value of the spacecraft-generated background $\mathbf{B}_{OBC} = \mathbf{B}_{OB} - \langle \mathbf{B}_{SC} \rangle$. The sampling rate for the outboard sensor was 25 vectors/s; for the inboard sensors three such vectors sampled at 40 ms intervals were averaged to give a time resolution of 120 ms. The magnetometer offset was determined from the pre-flight calibration tests and from in-flight assessments using the periods when the spacecraft was rolled and also when the flipper that exchanged mechanically the sensor axes was activated. The accuracy of the measurements following the correction for the spacecraft background and its variable component (due, among other causes, to the operation of the scan platform on which the dual television cameras were mounted) was assessed to be ± 1 nT.

This accuracy was more than sufficient for the results from two flybys of Mercury to provide the intriguing but remarkable result that there was a strong likelihood for the existence of a global, planetary scale dynamo in Mercury's interior. As detailed by Connerney and Ness (1988) in their definitive analysis of the Mariner 10 data, the limitation on describing the planetary field arose from the limited set of observations from the two flyby trajectories, rather than the magnetometer, its performance and any spacecraft generated backgrounds. The time between the first and last bow shock crossings was 33 minutes on the first flyby on 29 March 1974 and 27 minutes on the third flyby on 16 March 1975. The intervals spent between magnetopause crossings in and out of Mercury's magnetosphere was even

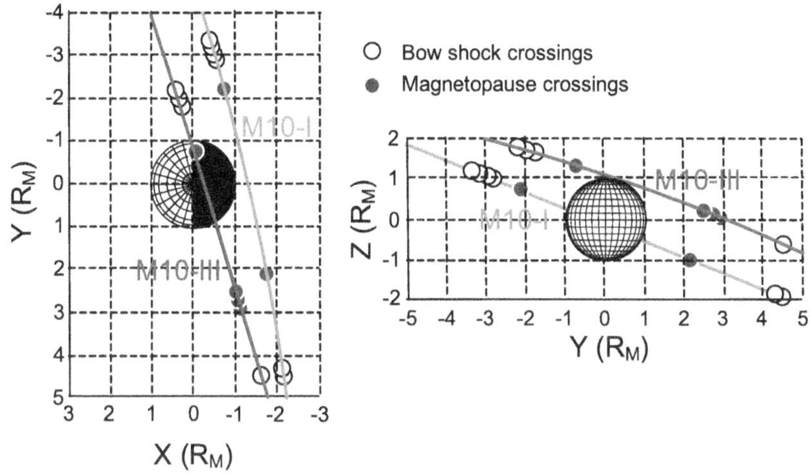

Fig. 16 Flyby trajectories of the Mariner 10 spacecraft at Mercury in 1974 (M10-I) and 1975 (M10-III), with indications of the locations of the bow shock and magnetopause crossings. *The diagram on the left* is the view from above the north pole of Mercury, with the Sun on the left, *the diagram on the right* is the view of the planet and the paths of the spacecraft seen from the direction of the Sun

shorter, 18 minutes on the first flyby and 13 minutes on the third flyby, giving a total of 31 minutes of data that were associated directly with the planetary field. The flyby trajectories are shown in Fig. 16, indicating the limited spatial coverage. The magnitude of the magnetic field during the two close flybys is shown in Fig. 17. The analysis of the data, despite the limited data sets, showed (Connerney and Ness 1988) that the equivalent equatorial dipole field was between about 200 and 350 nT. This range of figures was confirmed by modelling the magnetosphere of Mercury, based on the boundary crossings observed by Mariner 10. The problem of identifying and quantifying the planetary field against the background of the fields of external origin, due to the interaction with the solar wind was widely recognised and addressed using different models and assumptions concerning the details of the interaction. This problem, however, could not be resolved using the Mariner 10 data; the resolution will come from the next generation of missions to Mercury, MESSENGER, soon to be in orbit around the planet and the two-spacecraft Bepi-Colombo mission that will explore Mercury towards the end of the next decade.

The MESSENGER spacecraft has already provided, during its two flybys of Mercury in January and October 2008, ample confirmation of the global magnetic field of the planet and the very likely existence of a dynamo operating in the planet's interior (Anderson et al. 2008, 2009; Slavin et al. 2009). MESSENGER, only the second space mission to reach Mercury, will be the first to orbit around the planet. Launched on 3 August 2004, the spacecraft needed an Earth flyby a year after launch, followed by two Venus flybys on 24 October 2006 and 5 June 2007, and then three flybys of Mercury. All three flybys have already been successfully completed on 14 January, 6 October 2008 and 29 September 2009. The spacecraft will be finally injected into an eccentric orbit on 18 March 2011. Mercury is deep in the gravitational well of the Sun and a large amount of (breaking) energy is required to place a spacecraft in orbit around it; the use of planetary flybys is a cost-effective solution to the problem, at the cost of a very long transfer time from Earth to the planet (McAdams 2003).

The trajectories of the two first flybys were similar, as shown in Fig. 18. (The third flyby had also a very similar geometry, but the results have not been published before this paper

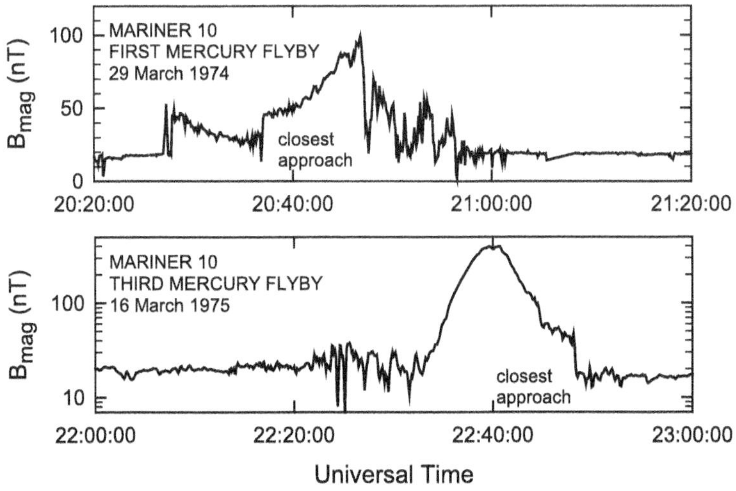

Fig. 17 The magnitude of the magnetic field measured by Mariner 10 on the two close flybys of Mercury in 1974 and 1975. Note the different profiles of the field on the two flybys. The first flyby was marked by a sudden collapse of the field followed by a highly disturbed interval that has been attributed to a major reconfiguration of the magnetosphere possibly caused by a Hermean version of a substorm. *In the lower panel*, the dominant global dipole term is evident; note that the field magnitude for this flyby has been plotted on a logarithmic scale to follow the representation originally published by Ness et al. (1976a, 1976b)

was completed.) Both were in a plane close to the equator and, although the closest approach was, in both cases, at a height of 200 km over the planet's surface, the maximum magnetic field measured was about 160 nT, less than the maximum field (about 400 nT) measured by Mariner 10 on the third flyby at a closest approach distance of 327 km, but at high latitude as shown in Fig. 16. The magnetic field magnitude measured during the two flybys by MESSENGER as shown in Fig. 19; the differences are due mainly to the different dynamic state of the magnetosphere at the two epochs (after Anderson et al. 2008 and Slavin et al. 2009). The two flybys differed by their relative phasing in Mercury longitude that allowed different regions of the surface to be imaged further to complement the earlier Mariner 10 coverage. The successful third flyby that took place on 29 September 2009 was also in Mercury's equatorial plane, at an altitude, at closest approach, of 231 km, although the data coverage was interrupted because the spacecraft was automatically switched into its safe operating mode. Mercury orbit insertion will take place between 18 and 21 March 2011 into a high inclination (~80°) eccentric orbit with a period of 12 hours, a perihem height of 199 km at latitude ~60°N and an apoherm at a maximum distance of ~15,200 km from the planet's surface. The dynamo hypothesis for the origin of Mercury's magnetic field, already favoured after Mariner 10, has been considerably strengthened by the consistent observations during the first two MESSENGER flybys (Anderson et al. 2009) as the analysis of the data confirms that the value of a pure dipole would be ~250 nT R_M^3, but with an apparently weaker value near the equator (matching in magnitude the equally equatorial measurement by Mariner 10 on its first flyby) which implies the influence of higher order moments and/or magnetospheric (tail) currents that are stronger or closer than expected. The equatorial surface field has been confirmed to be in the range 250 to 290 nT, roughly about mid-range of the earlier Mariner 10 estimates. Further refinement will now come from the one year long orbiting phase of MESSENGER in 2011–2012.

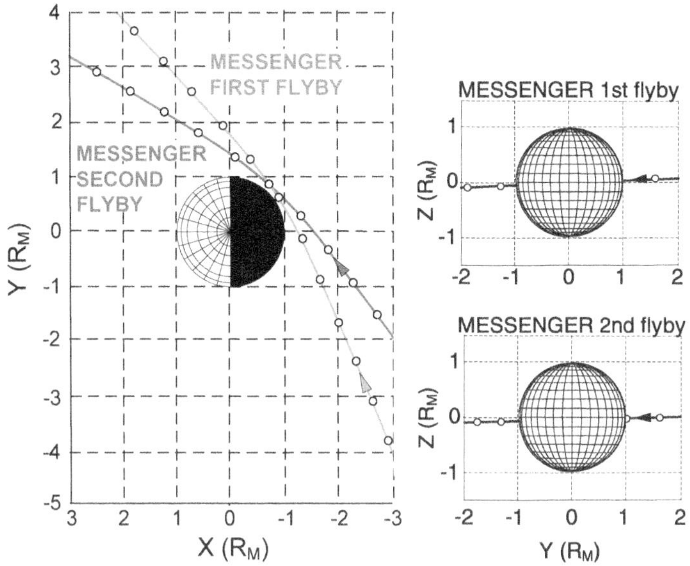

Fig. 18 The geometries of the first two flybys of Mercury by MESSENGER on 14 January and 6 October 2008. Both flyby trajectories were close to Mercury's equatorial plane at a closes approach altitude of 200 km s. *On the left*, the trajectories are shown in Mercury's equatorial plane with the X axis pointing toward the Sun. *On the right*, the view is from the Sun in the plane perpendicular to the equatorial plane. The markers on the flyby trajectory paths indicate 5 minute intervals. The third flyby on 29 September 2009 is to have a similar geometry, at an altitude of 228 km

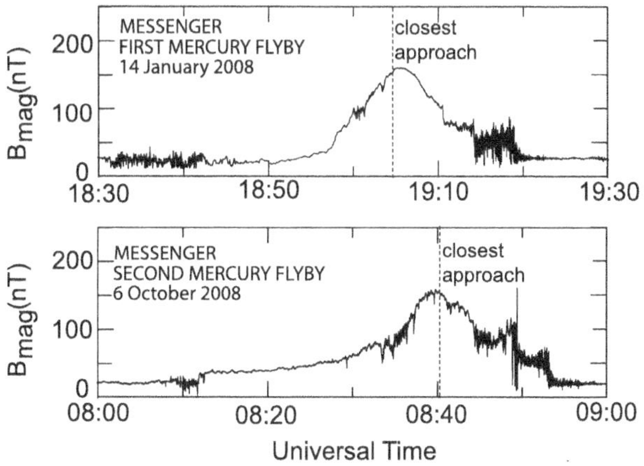

Fig. 19 The magnitude of the magnetic field measured by MESSENGER on its first two flybys of Mercury. The dominant dipole component of the internally generated field is evident, as is the strong variability introduced by the externally generated currents in Mercury's magnetosphere. After (*top panel*) Anderson et al. (2008), and (*lower panel*) Slavin et al. (2009)

The resolution of the details of Mercury's magnetic field between internal and external effects will remain a difficult task with a single-spacecraft mission, because of the very dynamic nature of the magnetosphere (see reviews and assessments by Russell et al. 1988; Slavin 2004; Scuffham and Balogh 2006; Alexeev et al. 2008). The separation of external and internal terms in the measurements has been discussed in general terms by Olsen et al. (2009), and specifically for Mercury by Connerney and Ness (1988), Korth et al. (2004), Scuffham et al. (2006), Anderson et al. (2008, 2009) and Johnson et al. (2009). An additional factor especially important for Mercury is the role of induced magnetic fields (Grosser et al.

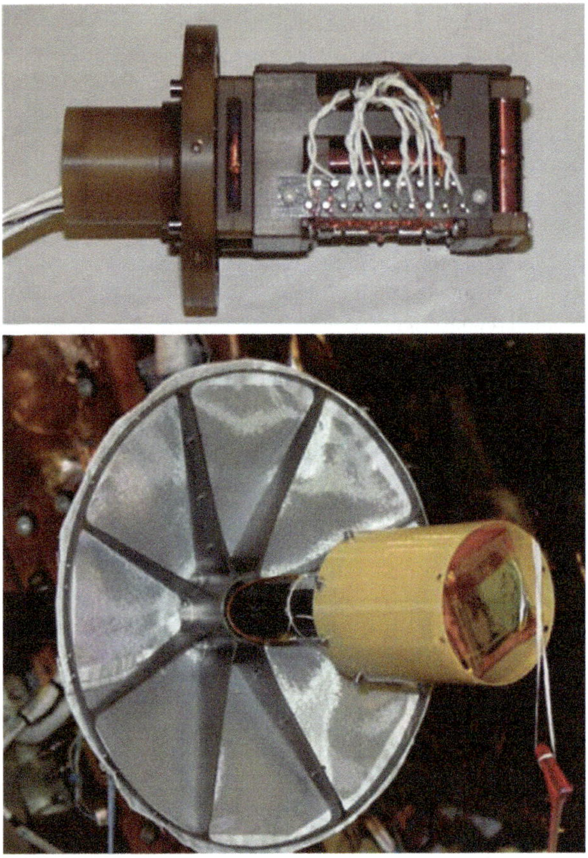

Fig. 20 *Upper panel*: The MESSENGER tri-axial fluxgate magnetometer sensor. *The boom* mounting is on the left. The three single axis sensors can be seen in profile on the right. *Lower panel*: The MAG sensor mounted on the boom tip behind its special, additional heat shield. The sensor is inside a protective cover that was covered with a thermal blanket prior to launch (photos courtesy of M.H. Acuña and B.J. Anderson)

2004; Glassmeier et al. 2007a; Saur et al. 2009), due to the large iron core and the high variability of the external magnetic environment.

The magnetometer on board the MESSENGER spacecraft has been described in detail by Anderson et al. (2007). The magnetometer sensor (see Fig. 20, upper panel) is placed at the end of a deployable boom of 3.6 m. The scientific requirements impose an absolute accuracy of 3 nT in measuring the magnetic field. This implies that the spacecraft field at the location of the sensor should not vary by more than 1 nT. A mission to Mercury, particularly an orbiter, is exposed to a very harsh thermal environment both from the factor 10 increase in solar input and from the infrared radiation of the planet's sunlit side. MESSENGER carries a sun-shield protecting the spacecraft; the magnetometer boom is extended in the anti-sunward position. An additional, more local shield was also needed to further protect the magnetometer when during orbital manoeuvres there will be intervals when the magnetometer is not in the shade of the sun-shield (Fig. 20, lower panel). Particular care had to be applied to the design and construction to this local shield, in particular to the materials used, to avoid the generation of thermo-electric currents that would affect the performance of the magnetometer. As the sensor is operated in the shadow of the spacecraft and its protective shield, its temperature in orbit will be less than 20°C even when illuminated by the planet, in fact the sensor carries a non-magnetic heater (already used on previous missions an thus fully flight-proven) which operate in a proportional mode from −15°C and −40°C.

The MESSENGER magnetometer is a three-axis ringcore fluxgate, originally developed in the 1970s by Acuña (1974) and successfully flown in continually improved versions and with different analogue and digital electronics on a large number of missions. Other planetary magnetometers by M.H. Acuña will be cited extensively in this paper. The sensor is operated during the mission in a single range of ±1530 nT per sensor axis; using a 20-bit analogue-to-digital conversion but transmitting only 16 bits gives a digital resolution of 0.047 nT. (There was an additional, high-field range of ±51,300 nT built into the instrument for ground-testing before launch.)

The analogue outputs of the three sensors are first filtered by 10 Hz low-pass filters to avoid aliasing and then digitized at the rate of 50 ms per sample in three parallel analogue-to-digital converters (ADCs). The 20-bit ADCs are synchronized and transmit the digitized values of the three magnetic field components to the instrument's Event Processing Unit (EPU) that takes care of instrument-specific control functions such as triggering the synchronization of the samples. The EPU interfaces with the Digital Processing Unit of the spacecraft to receive commands and to transmit the data to the telemetry. The optimized functionality of the instrument in matching the transmitted data (in terms of using matched analogue and digital filters) and data rate (in units of vectors/s) is very complex but results in 11 possible data rates, from 0.01 to 20 samples/s, covering all the possible data rates required by the scientific requirements. The performance of the MAG instrument on MESSENGER has been excellent through to date and it is expected to complete the survey of Mercury's magnetic field during the orbiting phase of the mission.

The highly variable contribution, already emphasised, that the external currents make to measurements in Mercury orbit will limit the scope of single-point measurements in determining the fine structure of the planet's internal magnetic field. Models of the origin of Mercury's internal field and the relationship of the presumed dynamo to the internal structure, state and material properties of the planet may only be distinguished by more accurate values that can be attributed to the internal field. This accuracy is limited by the need to remove the variable external component from the observed data. The joint BepiColombo mission of the European Space Agency and Japan's Aerospace Exploration Agency (JAXA)/Institute of Space and Astronautical Science (ISAS), due to be launched in 2014, will consist of two orbiters around Mercury, operating at the same time (Benkhoff et al. 2010). The Mercury Planetary Orbiter (MPO) supplied by ESA will have a low altitude, nearly circular polar orbit, while the Mercury Magnetospheric Orbiter (MMO) of JAXA/ISAS will have an eccentric orbit synchronized to the MPO orbital period. The currently planed orbits are shown in Fig. 21. The periods of the orbits are, respectively, 2.3 hours for the MPO and 9.2 hours for the MMO, allowing coordinated phasing for simultaneous measurements.

Both spacecraft carry magnetometers. The magnetometer on the MPO has been described by Glassmeier et al. (2010) and the one on the MMO by Baumjohann et al. (2010b). On the

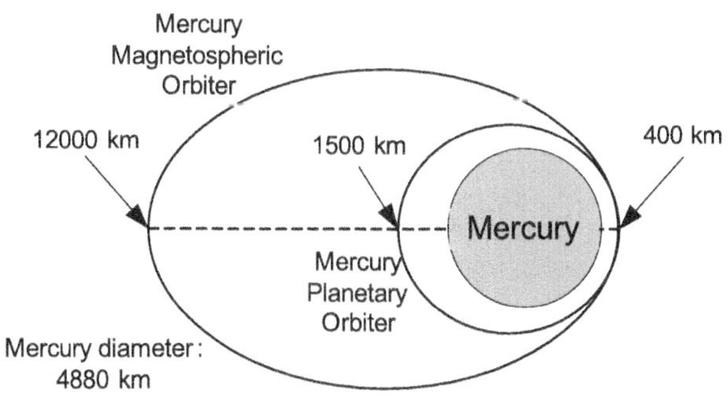

Fig. 21 Orbits of the two spacecraft, Mercury Planetary Orbiter and Mercury Magnetospheric Orbiter of the BepiColombo mission

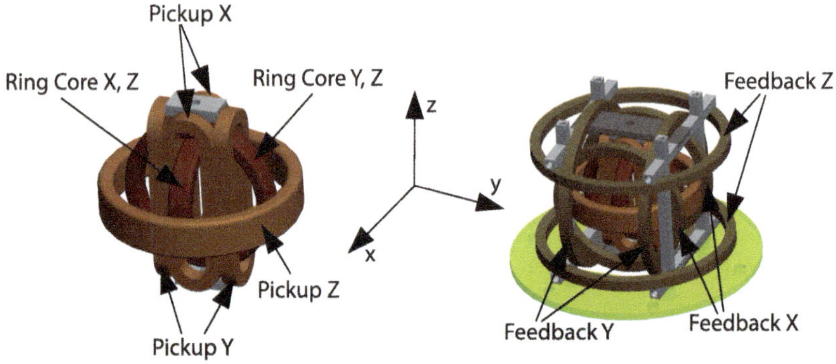

Fig. 22 Schematic drawing of one the two triaxial fluxgate sensors used on the BepiColombo Mercury Planetary Orbiter. *On the left*, the inner part of the sensor is shown, consisting of the two orthogonal ringcores surrounded by the sense or pick up coils for each axis. *On the right*, the same sensor is shown enclosed in the triaxial coil system that generates the feedback field to null the field at the sensor (from Glassmeier et al. 2010)

MPO, two identical triaxial fluxgate sensors are used on a 2.8 m boom extending from the spacecraft, to enable the use of the dual magnetometer/gradiometer technique to determine and remove the spacecraft background field (Ness et al. 1971; Neubauer and Schatten 1974), even though the technique needs to be applied with considerable care (Davis et al. 1973). The geometrical arrangement of the sensors and pick up and feedback coils is different from the arrangement used, for instance, on MESSENGER, and has the advantage of being compact and very light. The sensors use two ringcores in orthogonal planes; by placing orthogonal pickup or sense coils around each ringcore as shown in Fig. 22 the components of the magnetic field in all three orthogonal directions can be measured. In this design, the feedback coils (two circular coils per axis) are separated from the sense coils and use a quasi-Helmholtz geometry.

The feedback coils are also used for a vector compensation of the sensors whereby a magnetic field of programmable magnitude and direction can be applied, up to $\pm 5{,}000$ nT per axis, to null the effect of static background fields generated by the spacecraft.

A further feature of the MPO magnetometer is the use of digital front-end electronics in lieu of the more classical analogue design. As described in more detail above in Sect. 2.1, in the digital magnetometer electronics, the sensor signal is digitized at a high rate ($\sim$40 kHz) and processed to generate the feedback current signal through a precision digital-to-analogue converter. The controller associated with this function handles at the same time the sensor and the interface to further digital processing of the instrument data for transmission to the telemetry. Digital magnetometers of similar design are already used in the instruments on the Rosetta Comet Lander (Auster et al. 2007), the Venus Express mission (Zhang et al. 2006), and planned missions to Jupiter (O'Brien et al. 2007). There is a single measurement range of $\pm 2{,}000$ nT per axis, with a resolution of 2 pT, and an instrument noise of $\sim$10 pT/Hz$^{1/2}$. The telemetry depends on the different operating modes of the spacecraft and the instrument; magnetic field samples will be taken at rates of 0.5 to 64 vectors/s, with a short-duration burst mode when 128 vectors/s can be recorded.

As for the MESSENGER magnetometer described above, a key consideration of the accommodation of the magnetometer sensors on the boom of BepiColombo has been the thermal protection against both direct sunlight (which is up to $\sim$10 times more intense than in Earth orbit) and illumination in the infrared from the sunlit surface of the planet. The sensor has been designed to operate in the unusually wide temperature range of −100°C to +200°C and has a low (∼50 pT/°C) thermal offset drift coefficient. A protective cover envelops the sensor, consisting of a box shaped structure supporting Optical Solar Reflectors (OSRs), a frequently used device for thermal protection which has a low absorptance ($\alpha = 0.1$ to 0.2 at the end-of-life) and high emittance ($\varepsilon > 0.8$). Also, because of the high temperatures of surfaces exposed to the Sun, large temperature gradients are expected and, unless special care is taken, the possibility of thermoelectric currents. The potential variability of these currents is a serious hazard to the objectives of the investigation, and both careful design and extensive testing are required to avoid their harmful effects.

The second component of the BepiColombo mission is the Mercury Magnetospheric Orbiter (MMO), a spinning spacecraft in a more eccentric orbit than the MPO, concentrating on a suite of magnetospheric instruments, although with some planet-targeted telescopes as well. The magnetometer on the MMO (Baumjohann et al. 2010b) is also a dual tri-axial fluxgate, the outboard one mounted at the tip of a 4.4 m boom, while the inboard one is located at 1.6 m from the boom tip. The two sensors are different, the outboard one closely resembles the sensors on the MPO, including the front-end digital version of the electronics, while the inboard one is a classical sensor consisting of three ringcores arranged in an orthogonal triad, with an analogue version of the front-end electronics, similar to the schematic shown in Fig. 5 in Sect. 2.1.

In order to use the two sensors for implementing the background measurement and elimination using the dual-magnetometer method (Ness et al. 1971; Neubauer and Schatten 1974), it is important that the two magnetometers be fully synchronized for simultaneous measurement of the magnetic field. This is ensured by their common use of a 128 Hz synchronization signal. Otherwise, the magnetometers are fully independent of each other; each is operated on a different electronics and data chain on board to provide full redundancy.

The operating range of the magnetometers is ±2000 nT with a 3.8 pT digital resolution. Each samples the magnetic field at the rate of 128 vectors/s. This high sampling rate provides the capability to observe that high variability of Mercury's plasma environment; from an instrumental point of view, it also allows a high level of correlation of the spacecraft-generated background so that it can be removed with high accuracy from the measurements.

The principal advantage of the two-spacecraft approach to determine Mercury's magnetic field is that the external conditions, the very highly variable magnetosphere, can be better characterised and its effects better identified in the measurements. A more accurate and more detailed determination of both the magnetospheric and the planetary magnetic field is expected from the mission.

3.2 Venus

There are close similarities between Venus and the Earth in terms of size and density, although there are also significant differences, not only because of Venus' atmosphere and surface temperature, but also because of its very slow and retrograde rotation rate. It is this slow rotation rate that already implies that if there is a dynamo in the interior of Venus, it is much weaker than that of the Earth, so that no Earth-sized dipole field was expected. Scaling laws led Busse (1976) to predict a surface field for Venus of about a fraction 1/430 (or less) of the Earth's surface field. Consequently, the scale (and nature) of the interaction of the solar wind with Venus was certain to be different from that of the Earth. However, given the uncertainties in both the knowledge of Venus's interior state and its thermal evolution, as well as in dynamo theory, measurements were essential to confirm the existence or absence of an intrinsic magnetic field for Venus.

In the 1960s, several early space probes attempted to approach Venus carrying a magnetometer in their payload. Mariner 2, launched in 1962, was the first to report an absence of magnetic signature as it flew by on 14 December 1962 on a trajectory with a closest approach of ~41,000 km or 6.77 Venus radii (Smith et al. 1963). The absence of any signature that could be associated with the flyby of the planet excluded an intrinsic magnetic field at all comparable to that of the Earth. During the flyby, the components of the measured magnetic field digitized with relatively large steps (using an 8-bit analogue-to-digital converter) so that the total change in the magnetic field amplitude was less than 10 nT (Smith et al. 1965a, 1965b). Changes of this magnitude were seen by Mariner 2 away from the encounter, so even this small value could not be attributed to the planet. An upper limit of the planetary field estimated by Smith et al. (1965a, 1965b) was at most about 5% of that of the Earth; it was estimated, that an identical trajectory to that of Mariner 2 around the Earth would have yielded a signal of amplitude ~125 nT. The magnetometer used on the Mariner 2 mission was a three axis fluxgate (FGM) instrument (fabricated by Institut Dr Förster, Germany) with the electronics built under the supervision of the Jet Propulsion Laboratory.

Mariner 5 had been the backup spacecraft for the Mariner 4 mission to Mars. However, as Mariner 4 was successful (see Sect. 3.2), Mariner 5 was sent to Venus instead. Launched on 14 June 1967, the spacecraft flew by Venus on 19 October 1967 at a closest approach distance of 4,000 km. This allowed a good set of measurements of Venus' plasma and magnetic field environment (Bridge et al. 1967) that provided the basis for a valid first model of the characteristics of the planet-solar wind interaction features, dominated by the Venus ionosphere. However, It was not possible to exclude the possibility of a weak intrinsic magnetic field, but the close approach allowed to establish a much lowered upper limit, less than a fraction 1/400 of that of the Earth, than was possible with Mariner 2. The magnetometer on Mariner 5 was identical to the one flown on Mariner 4; it was the first time a quantum effect instrument, a Vector Helium Magnetometer was embarked on a planetary mission (Slocum and Reilly 1963; Connor 1968). This instrument is described in more detail in Sect. 2.2, see also Fig. 28.

A series of Soviet spacecraft were aimed at Venus in the 1960s, 1970s and 1980s, with the primary aim of delivering entry probes and landers; however, some magnetic field measurements were also made. A value of $\sim 3 \times 10^{12}$ T m^3 for the intrinsic magnetic field was proposed by Dolginov et al. (1978), so about a fraction 1/3000 or less of the Earth's dipole. Also, on its way to Mercury in early 1974, NASA's Mariner 10 spacecraft flew by Venus at a closest approach of ~5800 km and confirmed the previous upper limit for the magnetic field through the observation and positioning of a bow shock (Ness et al. 1974).

Pioneer Venus Orbiter, launched on 20 May 1978 and inserted into orbit around Venus on 4 December 1978 provided the best comprehensive set of observations in the complex plasma environment to determine Venus' intrinsic magnetic field. The instrument has been described by Snare and Means (1977) and Russell et al. (1980a). The magnetometer was of the fluxgate type, with three single axis sensors, but located somewhat unusually along a 4.7 m long boom. Two of the single axis sensors were located at the boom tip, to minimise the spacecraft-induced background, while the third sensor of the tri-axial set was located at about 1.7 m inboard on the boom, inclined at 45° from the spacecraft spin axis. The arrangement is shown schematically in Fig. 23.

Initial observations suggested that it was much less than 10^{12} T m^3 (Donahue 1979). A more detailed analysis yielded 4.3×10^{11} T m^3 (Russell et al. 1980b, 1980c) as an upper limit. A significant improvement on this value could only be reached by a more extended spatial coverage around the planet and an in-depth examination of all the possible sources

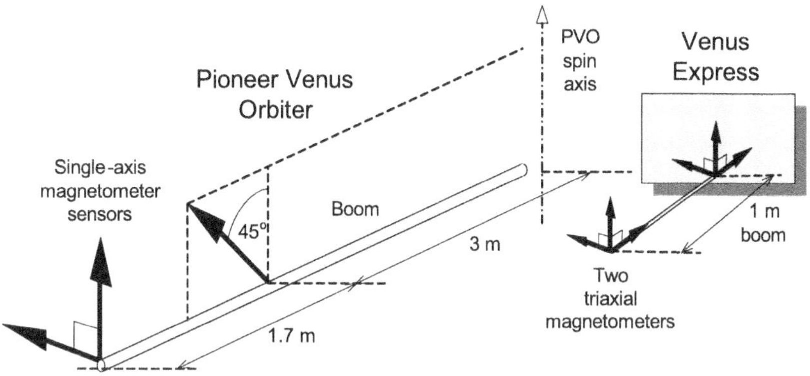

Fig. 23 The geometry of the fluxgate magnetometers used on Pioneer Venus Orbiter (Russell et al. 1980a, 1980b) and Venus Express (Zhang et al. 2008). The three-axis magnetometer on PVO was notable for the unique geometry of removing one of the sensors to a third of the way from the boom tip and installing it at an angle with respect to the spin axis to act also as a gradiometer monitoring the background field of the spacecraft. The dual tri-axial senor arrangement on VeX is also unique in that one set is on the tip of the 1 m long boom, while the other is installed on the spacecraft at the boom root. A highly complex and novel data processing method uses the output of the two tri-axial sensors to eliminate the background caused by the spacecraft

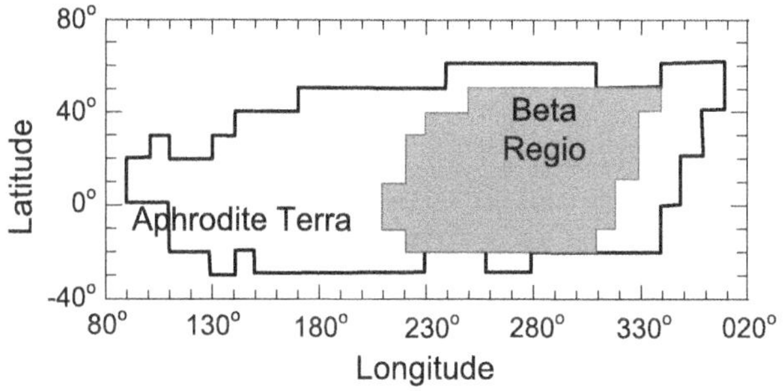

Fig. 24 Pioneer Venus Orbiter coverage used for the definitive data analysis to establish the likely absence of any intrinsic planetary field and to lower the limits of uncertainty. The original analysis by Russell et al. (1980b) used the darker shaded area, while the later analysis by Phillips and Russell (1987) used the larger, lighter shaded area that led to a reduction of by a factor 3 in the upper limit for the intrinsic magnetic field. The limitations that defined the spatial extent of the usable data were the altitude (North and South) and the flow zenith angle (East and West), defined to exclude the regions affected by the solar wind interaction with the ionosphere of the planet (after Phillips and Russell 1987)

of error, including instrumental errors (Phillips and Russell 1987). In particular, given the longer data set in the latter work, the coverage over the planetary surface was significantly increased, as shown in Fig. 24. As a result of this work, the deduced value for the intrinsic magnetic field of the planet is consistent with zero, with an upper limit (based on the uncertainties in the determination) 8.4×10^{10} T m^3, so at least a factor 100,000 weaker than that of the Earth.

The selection of data that can be used from the orbiter data that was collected is a critical step in the determination of the planetary field (Russell et al. 1980c). First, the data set used needed to be restricted to a volume near the planet in which magnetic field measurements could be considered to include the intrinsic field of the planet without being overwhelmed by other sources of magnetic fields, such as currents resulting from the interaction of the ionosphere with the solar wind. The strongest currents are on the sunward side of the planet; the analysis excluded the dayside region by limiting the data set to solar zenith angles greater than 120°. Additionally, the altitude above the surface was limited to a maximum of 600 km from the periapsis height of 150 km. The combination of the height and solar zenith angle constraints restricts the usable seasons when the usable data were collected. As the orbit was fixed in inertial space, and the rotation period of Venus is very long (243 days in a retrograde direction) there was an overlap between the planetary longitudes covered from one night-time season to the next. The later re-examination of the usable volume by Phillips and Russell (1987) led to increasing the volume used by raising the maximum altitude of the volume to 750 km and relaxing the condition on the solar zenith angle from 120° to 105°.

The usable volume during the night-time portions of the orbit was divided into 10° × 10° bins in planetary latitude and longitude. These bins contained data from a variable number of orbits and there fore varying quantities of data; in the first study by Russell et al. (1980c) two seasons were included in which the periapsis fell within the restricted volume (one of these in fact was used only partially because of limited ground coverage of the telemetry), while in the later study, Phillips and Russell (1987) could use three additional seasons of data from 378 orbits, giving a total of 18,000 vectors, each a 12-second averaged value that were used in the analysis.

The data in each bin was averaged and used to calculate the equivalent value of the planetary dipole that would yield the observed magnetic field value in the centre of the bin. Thus a large number of "equivalent" dipole moments, a total of 181 were calculated that were used to derive a mean value for the components of a equivalent planet-centred dipole. Error calculations, taking into account that not all moment determinations were independent, yielded error values that were commensurate with the values of the components of the planetary moment determined from the measurements (that would yield a surface field of ~0.35 nT). The Pioneer Venus Orbiter results are therefore consistent with a zero intrinsic planetary magnetic field. The reasons for the apparently complete absence of a global magnetic field source within Venus is not yet understood, although modelling its thermal evolution it is possible to derive scenarios and parameters that would lead to such a non-magnetic state.

The next mission to Venus equipped with a magnetometer is the currently operating Venus Express of the European Space Agency, launched on 9 November 2005 and inserted into an eccentric 24-hour orbit around Venus on 11 April 2006. The spacecraft, a copy of ESA's Mars Express spacecraft, had not been designed to be magnetically clean (Mars Express did not carry a magnetometer), nor to have a magnetometer boom, so the instrument selected to be included in the payload was a dual tri-axial fluxgate magnetometer that comprised also a boom to carry one of the vector sensors. The boom could only be 1 m long and the second vector magnetometer had to be mounted on the spacecraft platform immediately next to the mounting of the boom. The arrangement is shown schematically in Fig. 23, next to the PVO sensor arrangement.

Although PVO had shown that Venus does not possess a measurable global internal magnetic field, its limited coverage (see Fig. 24), particularly at high latitudes, leaves the opportunity to Venus Express to search for local magnetization, given its periapsis at high latitude (80°). Even though surface temperatures are above the Curie temperature for basalt rock, pure iron or material under high pressure may preserve historic magnetization (Tielong Zhang, personal communication). The orbits of PVO and Venus Express are illustrated in Fig. 25.

For the Venus Express magnetometer, producing usable data in the presence of the very high spacecraft magnetic background at the magnetometers sensors has been a unique challenge (Zhang et al. 2008; Pope et al. 2009). Full use is made of the dual magnetometer

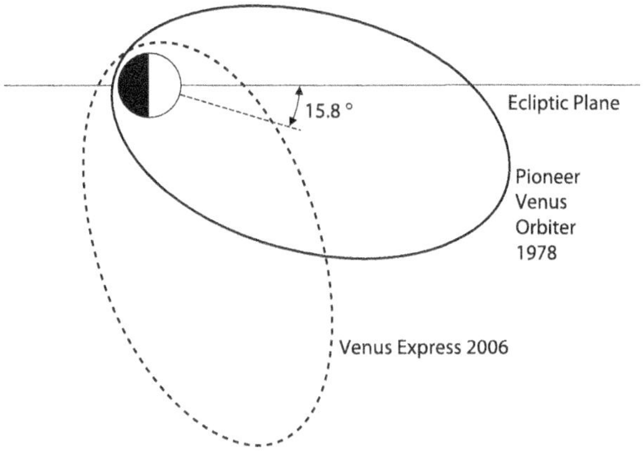

Fig. 25 The orbits of Pioneer Venus Orbiter and Venus Express. Both orbits have a period of 24 hours, with a periapsis of 150 km for PVO and 250 km for Venus Express. The periapsis latitude for Venus Express is 80°. The high latitude of the periapsis of Venus Express provides opportunities for exploring potential local sources of magnetisation

technique first proposed by Ness et al. (1971) but rarely used to its full capability; even in this case, the technique is used in a highly sophisticated mathematical environment to avoid the potential pitfalls pointed out by Davis et al. (1973). For the Venus Express instrument, one of the key features is the simultaneous high speed sampling of the ambient field vector, at 128 Hz, at the location of the two sensors. Both sensors are affected by large stray fields, typically ~200 nT at the outboard location and several thousand nT at the boom root. The basic measurement ranges are, respectively, ±262 nT and ±524 nT, but both sensors can be automatically compensated whenever necessary by an artificially imposed field of up to ±10,000 nT. The relatively low value ranges are justified by the high resolution that can be achieved when the magnetic field at the outboard sensor is within the measurement range. There are several operational features of the spacecraft related to the solar array driving mechanism, the antenna and the reaction wheels, among others, that generate large transient background effects. The data processing and calibration software operates on the differences between the values measured by the outboard and inboard sensors. The software incorporates self-adapting features, such as neural networks and fuzzy logic to characterise the disturbances that need to be removed. The methodology used does not aim to identify the sources of the disturbances but only their effects at the location of the outboard sensor. It has been estimated (Zhang et al. 2008) that after the data cleaning operations, the absolute value of the magnetic field can be determined to an accuracy ~1 nT, and that the differences can be determined to ~0.1 nT.

3.3 Mars

For a surprisingly long time, for more than thirty years after the first missions to Mars, it remained unclear whether Mars had an intrinsic magnetic field. The measurements by the American probe Mariner 4 and by several Soviet probes (Mars 2, 3 and 5 and Phobos) showed a magnetic signature that could be associated with the interaction of the solar wind with Mars. However, it was not possible to conclude if the obstacle was only Mars's ionosphere or the ionosphere combined with a weak planetary field. The upper limit (or perhaps the value) of Mars's magnetic moment was stated to be about 10^{12} T m^3, or about a fraction 10^{-4} of the Earth's field. The failure to determine the magnetic field of Mars was at first the lack of a sufficiently close-by orbiter data, and also the somewhat limited resolution of the magnetometers on the Soviet probes. But possibly the most important factor was the difficulty to identify and isolate in the observations a non-dipolar, yet important

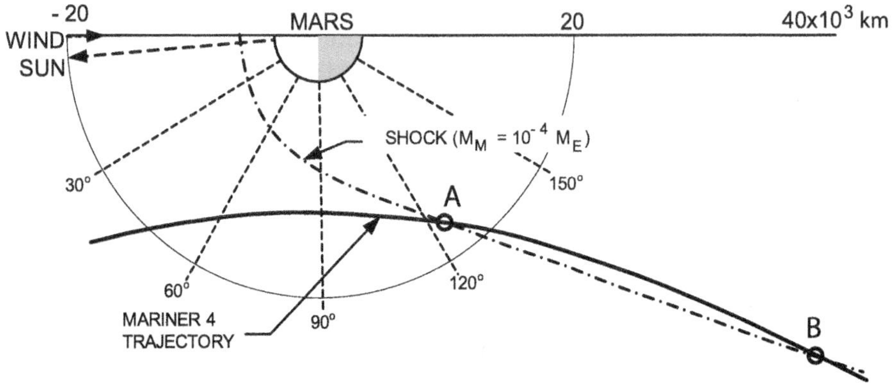

Fig. 26 The flyby trajectory of Mariner 4 at Mars. The contour of a bow shock scaled from that of the Earth, assuming that the magnetic moment of Mars is a fraction of 1/10,000 of that of the Earth is also shown, together with the two points marked A and B that correspond to the sudden increase and sudden decrease in the magnetic signal shown in Fig. 27 (after Smith 1969)

magnetic field against a background dominated by the interaction of the solar wind with the ionosphere in somewhat similar manner to what happens at Venus.

The first mission with an interesting, but far from definitive result was Mariner 4 which was launched on 28 November 1964 and flew by Mars on 15 July 1965. The flyby orbit is shown in Fig. 26; the closest approach was at 9,846 km from the Martian surface at 01:00:57 UT. The mission was successful, particularly in providing the first close-up photographs of Mars. The magnetometer operated throughout the flyby, the data it provided are shown in Fig. 27. Just after closest approach (at 13,200 km or $3.9R_M$), there was first an abrupt increase in the magnitude of the magnetic field at 01:23 on 15 July 1965, followed by an equally abrupt decrease three hours later, as described by Smith et al. (1965a, 1965b) and by Smith (1969). (There was a 54 minute gap in the telemetry during this interval, due to Mars eclipsing the spacecraft from Earth visibility.) These two events, marked A and B in Fig. 27, corresponded to the points also marked A and B along the flyby trajectory. A bow shock, scaled from that of the Earth assuming a magnetic moment $\sim 1/10^4$ smaller for Mars is shown in Fig. 26, matched to the two points along the trajectory with the abrupt changes in the magnetic field magnitude. Smith et al. (1965a, 1965b) and Smith (1969) pointed out that although the interaction of Mars with the solar wind could be similar to that of Venus, i.e. mostly/wholly with the ionosphere of the planet, the observations were also consistent with a small Martian intrinsic field. This point was emphasised by the modelling comparisons between Earth and Mars by Dryer and Heckman (1967) who concluded that the magnetic dipole moment of Mars was about 3×10^{12} T m^3, close to the possible value suggested by Smith et al. (1965a, 1965b). The magnetometer used on this mission was the first vector helium instrument (Connor 1968) that was to lead to a series of instruments on key planetary missions for the next 40 years. The schematic of the instrument is shown in Fig. 28.

The interpretations of the observations of the Soviet probes Mars 2, 3 and 5 in the 1970s remained controversial, given that it was possible to argue that the ionospheric interaction model, as for Venus, could explain qualitatively and quantitatively the observed size and character of the obstacle to the solar wind flow (Russell 1978). That interpretation itself was challenged on the basis of the comparison between respective abilities of the ionospheres of Venus and Mars to hold off the solar wind and act as the primary obstacle to the flow (Slavin and Holzer 1981) who concluded that unless the solar wind pressure is exceptionally high,

Fig. 27 The sudden increase followed by a sudden decrease in the magnetic field (marked A and B) appears to correspond to the inbound and outbound crossings of the Martian bow shock as illustrated by the open circles at points A and B on the flyby trajectory in Fig. 26. This is likely to be the first evidence, long disregarded, that Mars has a small magnetic field, although its nature as a remnant crustal field could not be appreciated (after Smith 1969)

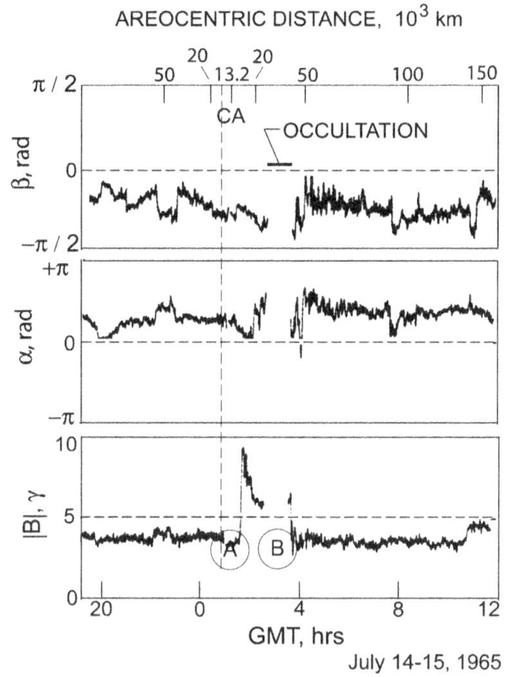

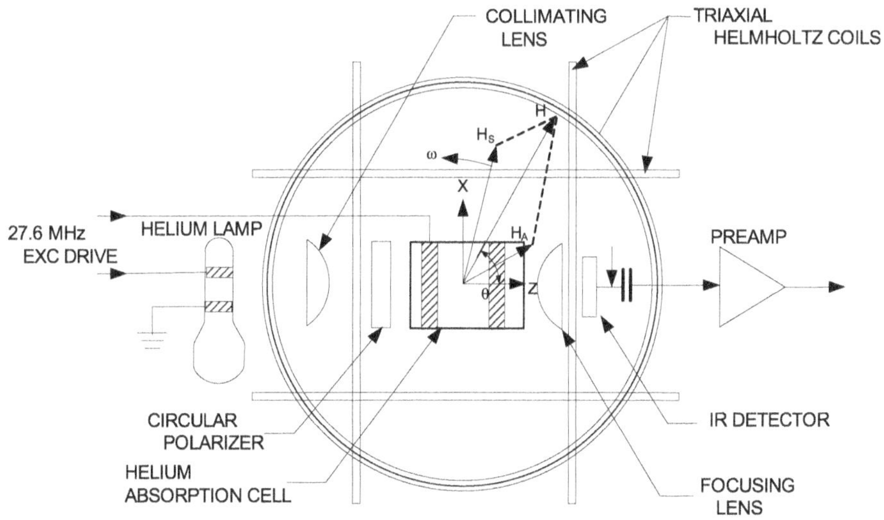

Fig. 28 The vector helium magnetometer used on the Mariner 4 spacecraft that was the first to measure a significant magnetic signal in the vicinity of Mars. An identical instrument was flown on the Mariner 5 mission to Venus (after Connor 1968)

the size of the obstacle in the case of Mars did imply an intrinsic field somewhat in excess of 10^{12} T m^3.

However, Dolginov et al. (1973) insisted on a weak intrinsic field of about 2.4×10^{12} T m^3 that implied an equatorial intensity of about 60 nT, based on the observations of the bow

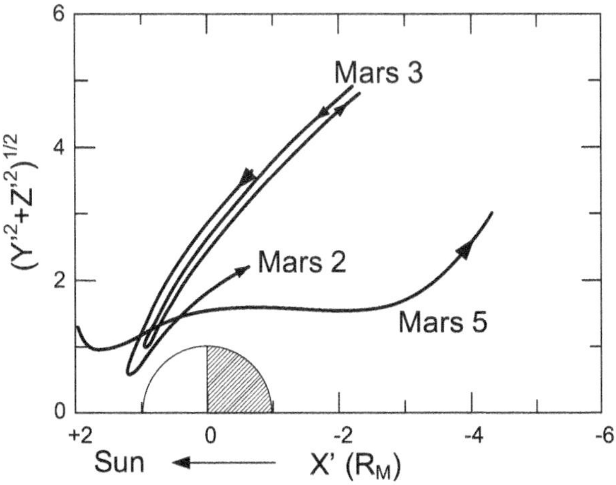

Fig. 29 Trajectories of the Soviet Mars probes that reached the planet in the 1970s and were able to detect the bow shock but could not decisively determine the nature of the interaction with the solar wind, in particular whether a small planetary magnetic field contributed to the interaction in addition to the Venus-like ionospheric interaction type (after Slavin et al. 1991)

shock by the Mars 2 and 3 probes. These authors came to an interestingly insightful conclusion: "Such a small dipole moment can be understood if it is assumed that the observed field is an ancient field, being a trace of a magnetic dynamo existing in the past. In any case, it can be expected that, on the surface of Mars, regions with fields of high intensity can exist." Following the next successful space probe, Mars 5 that was placed in orbit around Mars on 12 February 1974, the arguments in favour of a small intrinsic field were reinforced by the observations (Dolginov et al. 1976). The trajectories of the Mars 2, 3 and 5 spacecraft are shown in Fig. 29; these were sufficiently close to the planet to observe the effects of the solar wind interaction with the ionosphere, but (as has been made clear by the Mars Global Surveyor measurements described below) were not able to detect the highly localised crustal magnetic sources.

The last exchange of arguments on whether Mars had an intrinsic field or not occurred following the Phobos-2 mission (Möhlmann et al. 1991; Dolginov 1992; Russell et al. 1992). The trajectory of Phobos 2 around Mars is illustrated in Fig. 30 (after Slavin et al. 1991); Möhlmann et al. (1991) concluded that the solar wind interaction is primarily with the ionosphere/upper atmosphere of Mars, but they also found some periodic signatures in the magnetic field that corresponded to the rotation of the planet. It was this signature that was interpreted by them as evidence of an intrinsic magnetic field. Slavin et al. (1991) examined data from all the previous Mars missions, from Mariner 4, to establish the position and variability of the Martian bow shock. Both the mean distance of the bow shock at about 1.5 planetary radii corresponding to an "obstacle" to the solar wind at a height about 500 km and its lack of dependence on the phase of the solar cycle were arguments for the interaction with the solar wind to be different from that of Venus. This therefore led to the conclusion of a small intrinsic field of $\sim 10^{12}$ T m^3, as concluded by the majority of the previous studies. Slavin et al. (1991) also noted the unusually large bow shock distances observed occasionally by Mars 2, 3 and Phobos 2. These distances were compatible with a height of the Martian obstacle of 2000 to 4000 km, incompatible with the ionospheric-only interaction model. Furthermore, an explanation based on low solar wind pressure was also unlikely, as the solar wind parameters implied were quite unrealistic but not impossible. There was only a limited sample of such extreme bow shock crossings that could not be statistically related to the planetary rotation and remained at that stage without a satisfactory explanation. However, the later observations of Mars Global Surveyor confirmed the likelihood that periodic asymmetries in the bow shock location corresponded to the phasing of the most

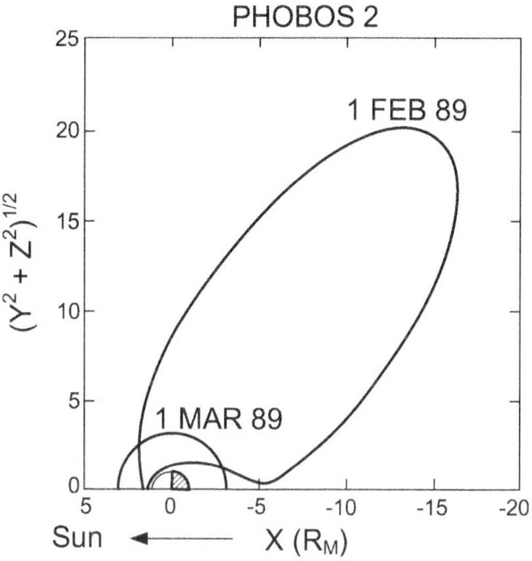

Fig. 30 Two typical orbits of the Phobos 2 spacecraft that allowed the determination of the solar wind interaction with the ionosphere of Mars, but was not sufficiently close to the planet to detect the high intensity, highly localised crustal sources (after Slavin et al. 1991)

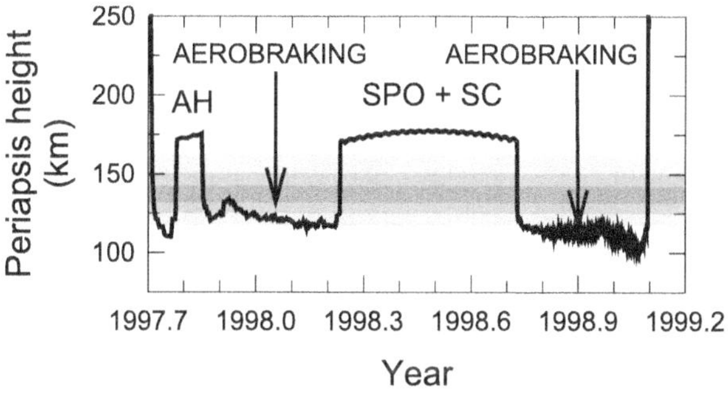

Fig. 31 The height of the periapsis of the Mars Global Surveyor Orbit during the first, aerobraking phase of the mission, below the ionosphere, shown in *the grey shaded area* (according to a model derived from Mars Express observations). The periapsis was raised twice, once for troubleshooting (AH: Aerobraking Hiatus) and once for awaiting for the phasing of the orbit (SPO: Science Phasing Orbits) and because of solar conjunction (SC). The lowest periapsis phases were the best suited to map Mars's crustal magnetic field. The periapsis was raised to ~380 km (off the scale of the figure) in March 1999 for the start of the mapping phase of the mission

intense crustal magnetic sources that extended the size of the Martian obstacle to the solar wind (Acuña et al. 1999).

The resolution of the question of Mars' planetary magnetic field came from the magnetic field observations by the Mars Global Surveyor (MGS) spacecraft (Dallas 1997). MGS was launched from Earth on 7 November 1996, placed into orbit around Mars on 11 September 1997 and operated until 2 November 2006. The eventual operational orbit, the circular mapping orbit, at a height of ~380 km above the Martian surface) was achieved in March 1999. In the early part of the mission, the original high eccentricity orbit with a period of ~44 hours was gradually modified using aerobraking, exploiting the drag of the spacecraft at periapsis to lower the apoapsis and circularise the orbit. The history of the periapsis height during the aerobraking phase of the mission is shown in Fig. 31, together with a grey-shaded representation of the ionospheric electron density, based on Mars Express measurements of the electron density. The peak density is at ~140 km, above the periapsis during aerobraking, as can be seen in Fig. 31.

The observations during this early phase of the mission, particularly those during the lowest periapsis, aerobraking phases have led to the definitive result that Mars has no measurable global magnetic dipole and to the discovery that part of the Martian crust is magnetised (Acuña et al. 1998, 1999, 2001; Ness et al. 1999). The latitude range of the periapsis around which the measurements were made was first in the northern hemisphere but, during the second aerobraking phase, moved to the southern hemisphere close to the polar region. A reasonably comprehensive coverage in latitude could therefore be performed. As concluded by Acuña et al. (1998), the magnetic field observations show an interaction of the atmosphere/ionosphere with the solar wind, similar to that on Venus. However, MGS also detected highly localised, intense sources of magnetisation. One of these measurements, during the first aerobraking phase is shown in Fig. 32 (from Acuña et al. 1998). The measurements, made at ~100 km, show the source with magnetic field strengths of 400 nT, compared to the upper limit of a few nT expected from a possible global field.

The comprehensive survey through the whole aerobraking phase of the mission was reported by Acuña et al. (1999) who provided a map and a description, in terms of Martian geology, of the sources of crustal magnetisation. During the second aerobraking phase, magnetic fields up to ~1600 nT were measured, representing intense sources that are likely to be the causes of the increased asymmetric Martian obstacle size first noted in the Phobos 2 data. Having established the extent and topography of the crustal magnetic field sources, the second phase of the MGS mission, the mapping phase at a close to circular orbit of ~400 km height provided the confirmation of the earlier results and enabled a full description of the nature of the magnetism of Mars and the planet's interaction with the solar wind that shapes its environment.

As reported in detail by Acuña et al. (2001) who included observations from the mapping phase of MGS as well, it is now certain that Mars had a planetary scale dynamo in its very early history, but that the dynamo stopped operating at about 3.9 Gy ago. The crustal magnetisation is effectively restricted to the very old, heavily cratered terrain in the highlands in the southern hemisphere. There are numerous implications for the evolution and geology of Mars in the magnetic field maps that have been extensively discussed in the literature (see Connerney et al. 1999; Nimmo and Stevenson 2000).

There were two factors in the great achievement of Mars Global Surveyor in providing answers to the long-standing questions concerning the magnetic field of Mars. One was undoubtedly the mission orbit design which included a low periapsis phase, essentially below the ionosphere for a significant interval of time early in the mission (see Fig. 31). (Note that this phase was primarily introduced as part of the strategy to lower the apoapsis through repeated aerobraking passages in the upper atmosphere near periapsis, but it proved to be crucial in resolving the question of the Martian magnetic field.) An example of the very highly localised intense crustal filed signatures is shown in Fig. 32 (after Acuña et al. 1998); these fields could only be measured at the altitudes the MGS altitudes during the mission phase when it was very close to the surface of the planet. The second was the remarkable implementation of the magnetometer experiment on the MGS spacecraft. The instrument itself was identical to that flown on the Mars Observer spacecraft launched in 1991; however, that spacecraft failed just before Mars orbit insertion. MGS was conceived as a replacement for Mars Observer, but one that had a radically different configuration to match a less capable launcher that had been used for Mars Observer. Due to this reduction in resources, the experiment package with the purpose of discovering the true nature and extent of Mars's magnetic field, the magnetometer and the electron reflectometer could not be mounted on a

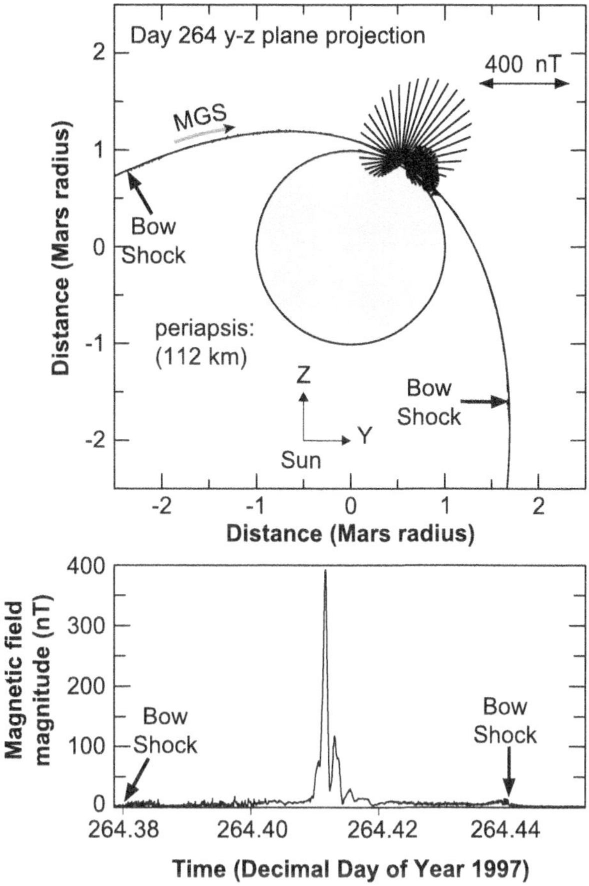

Fig. 32 The magnetometer on the Mars Global Mission discovered isolated magnetic anomalies that gave sudden, very large increases in the measured magnetic field, showing that they corresponded to highly magnetised, but locally narrow crustal features. *Upper panel*: projection of the near-periapsis orbit of MGS with a projection of the measured magnetic field vector along the orbit. The view is from the Sun, in a plane perpendicular to Mars' equatorial plane. *Lower panel*: the measured magnitude of the magnetic field through the periapsis pass. Two distinct anomalies can be seen, the first, larger one near 32.9°N latitude, 22.4°W longitude, the other near 22.8°N latitude and 23.6°W longitude (after Acuña et al. 1998)

boom as they had been on Mars Observer (which had a deployable 6 m boom), so another accommodation was needed that would enable the magnetometer to measure the expected nT range magnetic fields at Mars in the presence of a magnetically uncontrolled spacecraft. The configuration that was devised (Acuña et al. 1996) was to place the two triaxial fluxgate magnetometer sensors at the ends of the two solar panels (see Fig. 33). While this mounting removed the sensors from the spacecraft to a distance of 4.5 m, the length of the extended solar panels, it was important to minimise the magnetic field generated by the currents flowing in the solar panels. An ingenious wiring scheme was devised and extensively modelled that led to the nominal self-cancellation of the background fields at the magnetometer sensors (Acuña et al. 1996).

The in-flight performance of the magnetometer sensors in the spacecraft environment needed careful analysis to determine the sources of noise/background at the location of the sensors (Acuña et al. 2002). While the solar array performed close to the levels predicted by the pre-flight modelling (with a background field at the sensors <0.2 nT), other subsystems of the spacecraft did introduce significant levels of background that needed to be monitored and corrected. The Travelling Wave Tube Amplifier (TWTA) that was mounted with the high gain antenna of the spacecraft, in particular, introduced a background that proved to be a significant source of variable offset and needed to be characterised. During the first,

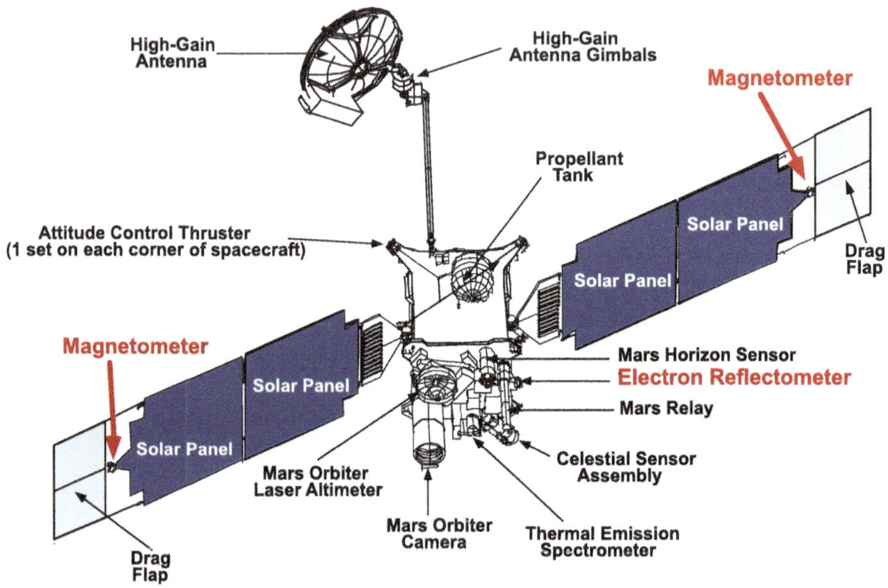

Fig. 33 General view of the Mars Global Surveyor spacecraft, indicating its main components, including the two tri-axial magnetometers at the ends of the solar panels and the electron reflectometer (after MGS—NASA Facts)

aerobraking phase of the mission, through to March 1999, the High Gain Antenna remained in a fixed position, so that its magnetic field seen at the location of the sensors was constant. However, during the mapping phase of the mission, the High Gain Antenna was constantly articulated and thus introduced a variable background at the location of the sensors that was synchronous with the orbit.

Modelling and subtracting the background due to the TWTA and the High Gain Antenna were essential. Measuring the magnetic fields simultaneously with the two sensors, at the ends of the two solar panels allowed to identify the "common mode" signal that was the ambient field due to Mars and the difference signal that was due to the spacecraft. As described by Acuña et al. (2002), the modelling of the differences led to an rms deviation between measurements and model of $\sim$0.5 nT with the solar panel in the shadow of Mars, and double that value when the solar panels were illuminated.

These outside contributions to the magnetometer zero level and its measurement noise have been significantly greater than the magnetometer's own zero-level offsets and noise. The stability of the zero offset for the Mars Observer magnetometer (identical to the one mounted on MGS) was better than 0.15 nT over a temperature range of −40°C to +60°C. Based on previous magnetometer sensors of closely similar materials and construction used on missions such as Voyager, the estimated zero-level drift was determined to be about 0.2 nT/year. The noise of the magnetometer was 0.006 nT in the frequency band 0 to 10 Hz. These values (Acuña et al. 1992, 2002) show that the performance of magnetometers far surpasses the constraints on the measurements generated by external sources. This is the case for most missions, to the planets and elsewhere; the platform noise dominates the background in the data.

3.4 The Earth

The best known, longest studied planetary magnetic field is that of the Earth (see, for an up-to-date review, Hulot et al. 2010, this issue). Complementary to the very long series of ground-based observations, extending now for several centuries, the space age has brought a significant new tool in dedicated space missions that have provided extensive maps of the Earth's magnetic field from low-Earth orbit. Given the emphasis in the current review to planets other than the Earth, the following is intended to provide a brief summary of what has been achieved by spacecraft measurements and the missions and instruments that have been used to date. A review of space data and their contribution to geomagnetic field modelling from the early Sputnik 3 (1957) up to Magsat (1979–1980) was given by Langel and Baldwin (1992). A more up-to-date summary survey of the relevant missions, in particular Magsat, Ørsted and CHAMP, can be found in Kramer (2002) and Olsen et al. (2002).

An early series of low-orbiting satellites with a scalar magnetometer to measure the Earth's magnetic field were the three Polar Orbiting Geophysical Observatory (POGO) satellites launched and operated between 1965 and 1971. The POGO satellites (the low orbit members of the six OGO satellite series) were in polar orbits, with a perigee close to 400 km and an apogee between 900 and 1500 km. The satellites were large and extensively instrumented (each carrying in excess of 20 instruments). The payload included a rubidium-vapour scalar magnetometer (Farthing and Folz 1967). Measurements were made over a range of 15,000 to 64,000 nT and with a precision of 0.5 to 1.5 nT. The objective was to make refined measurements to improve the description of the magnetic field and to monitor secular variations. Although the precision of the magnetometer was not matched by the positional uncertainties (contributing maybe as much as a 7 nT error to the data), the three POGO satellites made a first significant and systematic contribution to the acquisition and use of space data for the main geomagnetic field (Cain 1971; Langel 1974). In addition, the measurements contributed to a better understanding of the structure and behaviour of the magnetic field in the inner magnetosphere (Cain and Sweeney 1970).

Magsat was the first spacecraft dedicated by NASA to the study of the Earth's magnetic field (Mobley et al. 1980; Langel et al. 1982). The spacecraft carried both a scalar and a vector magnetometer. It was launched on 30 October 1979 into a sun-synchronous orbit with an inclination of 97°, a perigee of 352 km and an apogee of 578 km. The orbital period was 94 minutes. The mission lasted seven and a half months before orbit instabilities led to the re-entry of the spacecraft into the atmosphere.

The Magsat instrumentation consisted of a caesium vapour scalar magnetometer that was due to deliver the magnitude of the magnetic field to an accuracy of 1.5 nT or better (Farthing 1980). Unfortunately, a partial failure of the scalar magnetometer limited its scientific usefulness, although it was able to contribute to the in flight calibration of the vector magnetometer. This latter was a triaxial fluxgate magnetometer (Acuña et al. 1978; Acuña 1980), built to have an especially high level of mechanical, thermal and magnetic stability. The triaxial fluxgate assembly is shown schematically in Fig. 34. The single axis sensors were of the classical ring core type, with the magnetic material being a specially selected molybdenum permalloy developed to have low noise and high stability, originally for the Voyager missions. The vector feedback was applied to each single-axis sensor separately, around a high-stability coil-former. The alignment and temperature stability of the feedback coil are critical for meeting the measurement requirements, as potentially the largest error sources. The reasons why the solution of a feedback coil system around the triaxial sensor assembly was not retained (that was to be used for Ørsted, see below) were partly the bulk of the

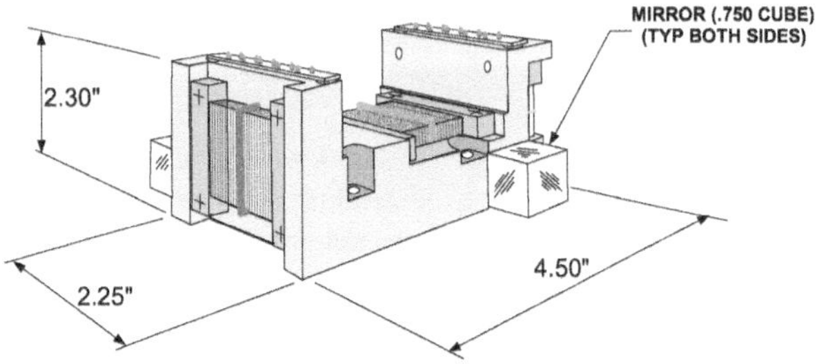

Fig. 34 Schematic diagram of the triaxial fluxgate magnetometer for Magsat (after Acuña et al., 1978)

external coil system and partly the excessive stray fields that would have been generated by the coils that would have disrupted the measurements of the scalar magnetometer that was physically relatively close to the vector fluxgate (see Fig. 34).

The triaxial assembly, as a whole, was manufactured out of a single block of machinable ceramic material characterised by its very low thermal expansion coefficient ($\sim 10^{-6}$), matched to that of the sensor core, its supporting structure and the feedback coil and its support. Dimensional stability and accuracy were of major importance to meet the demanding requirements: mounting surfaces were polished to an accuracy better than 10 μm and all structural deformations had to be less than 50 μm to ensure the alignment stability of 5 arcsecs. The magnetometer electronics, described in detail by Acuña et al. (1978), had to match the accuracy of the sensor. A fully redundant system was designed, with high precision components.

It was recognised from the outset that two of the key requirements for the spacecraft were a very precise attitude determination for the vector measurements, and a magnetically clean environment for both scalar and vector magnetometers. The attitude determination at the location of the three-axis fluxgate sensor could not be carried out simply, as the star sensors used on the spacecraft for the attitude determination were highly magnetic and could not be placed in the proximity of the sensors. The sensors were located at the end of a deployable boom that could not be constructed to ensure the necessary rigidity to ensure the required ±5 arc sec accuracy, as illustrated in Fig. 35. Hence a complex Attitude Transfer System (ATS) was designed to implement the requirements (Mobley et al. 1980). A further requirement on the boom was to ensure that the ATS remained in its linear range for the determination of the orientation of the vector measurements. The total error budget for the vector magnetometer was ~6 nT (rms) along each axis. On the magnetometer itself, the accuracy requirement was 3.75 nT (including 0.5 nT for the spacecraft stray field). The error budget related to the attitude determination translated to ~4 nT (rms), taking into account the angular uncertainties from the star cameras (~10 arc sec) and from the ATS (7 arc sec). There was also an error budget of 1.5 nT associated with the position determination of the spacecraft. The overall performance measured in orbit depended on the use of the star cameras and the sun sensor for attitude determination; small inconsistencies resulted in occasional small discontinuities in the data. The in-flight calibration of the measurements effectively confirmed the pre-flight expectations (Lancaster et al. 1980). The near-polar, sun-synchronous orbit allowed a complete latitudinal coverage, but within a restricted range of magnetic local times that needed to be taken into consideration for the analysis of the data. Magsat, as a dedicated mission to survey the geomagnetic field was successful in providing high precision data to extend the harmonic terms of the Earth's magnetic field and the calibrated data collected the mission remained the basic data set until the Ørsted and CHAMP missions, 20 years later, came to provide new and more extensive data.

The Ørsted satellite was launched on 23 February 1999 into a near-polar orbit of inclination 96.5°, with a perigee of 638 km and an apogee of 849 km. Contrary to the fixed local

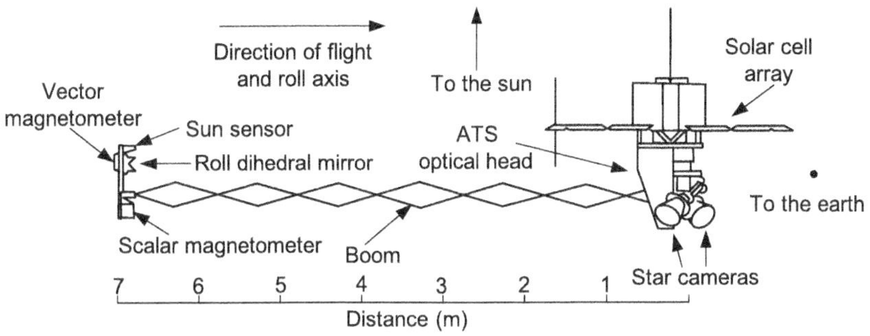

Fig. 35 Sketch of the Magsat spacecraft in flight configuration, with the boom that carried the scalar and vector magnetometers extended

time sun synchronous orbit of Magsat, the orbit of Ørsted drifted in local time by ~0.46 hours per standard month. This allowed a survey of local times over the mission and several studies of the local-time dependent external contributions (such as ionospheric currents). It was equipped with a scalar and a vector magnetometer. The magnetometers were mounted on an 8 m boom, with the scalar at the boom tip and the vector fluxgate sensor at ~2 m inboard. The Overhauser scalar magnetometer (for its operating details, see Kernevez and Glénat 1991) as used on Ørsted was described by Duret et al. (1995, 1996). The operation of the Overhauser magnetometer is based on proton Larmor precession, the Larmor frequency is proportional to the magnetic field strength. The measurement range of the Overhauser scalar magnetometer was 16,000 to 64,000 nT with a resolution of 0.1 nT and an absolute accuracy ~0.5 nT. The absolute magnitude measurements were used to calibrate the measurements by the vector magnetometer.

The vector magnetometer on Ørsted was a new development (Nielsen et al. 1995). It was a triaxial fluxgate sensor, using for its magnetic core material, for the first time, a metallic glass, instead of the more conventional crystalline permalloy type magnetic alloy (Nielsen et al. 1991). This material was developed to make the sensors less sensitive to acoustic vibrations and mechanical deformations. Additional heat treatment reduced the hysteresis of the material to close to zero. The single sensor assembly, consisting of the toroidal core with the tape of metallic glass material, the drive winding around the core and a sense or pickup winding on a rectangular former was of classical design. The three orthogonally mounted single axis sensors, however, were enclosed in a spherically mounted three-axis coil system that provided the feedback to null the field at the sensor assembly (Primdahl and Jensen 1982). The vector magnetometer sensitivity was limited by the sensor noise (~15 pT rms in the bandwidth 0.06–10 Hz). With a typically $1/f$ noise spectrum, this corresponded to ~6 pT/Hz$^{1/2}$ at 1 Hz. The operating range of the instrument was ±65,536 nT; using an 18-bit analogue-to-digital converter this corresponded to a resolution of 0.5 nT. The bandwidth of the instrument was set at 250 Hz. Its temperature stability was better than 1 nT in the range −20° to +60°C.

A key component of the magnetic field instrumentation is its association with a star camera, the Advanced Stellar Compass (ASC) that enables the absolute attitude to be determined to the required accuracy (Jørgensen et al. 1997). The ASC and the vector magnetometer were mounted on a common optical bench on the spacecraft boom. This solution, different from that implemented on Magsat, had a requirement to enable the vector magnetometer to achieve an attitude determination accuracy of ~20 arc sec per axis. The ASC differed from more standard star trackers by the way it transferred the image on its CCD camera to a Data Processing Unit which used the Hipparcos star catalogue as a stored data base for establishing the exact pointing attitude of the ASC and hence the optical bench on which the vector magnetometer was mounted. An important feature of the ASC was its close to zero stray magnetic field, so that it was possible to co-mount it with the vector magnetometer.

An Argentinean satellite, SAC-C, in cooperation with NASA, was launched on 21 November 2000 into a circular sun-synchronous polar orbit of height ~700 km, with equator crossing (local) times of 10:15 and 22:30 (Colomb et al. 2004). The satellite carried a magnetic field investigation package similar to that of Ørsted, with a vector-compensated vector fluxgate magnetometer, co-mounted on an optical bench with an ASC, both identical to the instrument on Ørsted. The scalar magnetometer was, however, a scalar helium magnetometer, a model identical to that flown on Cassini (Smith et al. 2001). However, the ASC failed to function in orbit and therefore the attitude information that can be used with the vector data are not sufficiently accurate for precise measurements of the Earth's magnetic field. The scalar magnetometer was shown to be accurate to better than 2.5 nT; however, there were concerns that the stray field of the complex spacecraft may affect the measurements and these cannot meet the target of 1 nT accuracy. However, by cross-calibrating between CHAMP (see below) and Ørsted on the one hand and SAC-C on the other when the spacecraft came sufficiently close in orbit, the SAC-C data could be sufficiently calibrated to enable their use in conjunction with the other two spacecraft.

The Ørsted satellite was a significant success and contributed a large amount of data for the very detailed study of the earth's magnetic field (Olsen 2002; Mandea 2006). Only a year and a half after the launch of Ørsted, the German CHAMP satellite was launched on 15 July 2000 into near-circular orbit of initial altitude 454 km, with a near-polar inclination of 87° (Reigber et al. 2002). (By December 2009, the altitude of the orbit decayed to ~300 km.) Similarly to Orsted, CHAMP's orbit was not sun-synchronous (at constant local time) but it advanced one hour in local time within eleven days, thus was able to cover all local times. In order to ensure a longer duration for the mission, while retaining an altitude less than that of Ørsted, the spacecraft was made deliberately heavy (522 kg) and a small along-track cross section to minimize atmospheric drag. The mission was dedicated to measuring the Earth's gravitational field as well as its magnetic field. The satellite carried a very similar magnetic instrument package to Ørsted on a 4 m boom: an Overhauser scalar magnetometer of identical design and construction to that flown on Ørsted, two vector fluxgate magnetometers (for redundancy and gradiometry) identical to the one on Ørsted, and an Advanced Stellar Compass with two optical heads, one each for the two vector magnetometers, but with a common Data Processing Unit. Because of the possible magnetic interference from the vector feedback system with the operation of the Overhauser scalar sensor, the two sensor systems were mounted at a distance of about 2 m from each other along the boom. The fluxgate sensors were rigidly mounted on an optical bench with the ASC optical heads; this provided a mechanical stability better than 10 arc sec. (Another dual-head ASC was used in the spacecraft for attitude control.)

The vector magnetometer, although using identical sensors to Ørsted, had a higher resolution (24-bit ADC, 10 pT/bit). The noise level was stated to be <100 pT (rms), with a linearity of the same order and an offset drift about 0.5 nT. The sampling rate was 50 Hz, but several different data schemes could be used, employing both data compression and reduced sampling rates. The high quality of the measurements, combined with an orbit lower than that of Ørsted, enabled CHAMP to contribute to investigate both new current-related phenomena in the ionosphere (Lühr et al. 2002) and new details of the Earth's magnetic field, in particular its lithosphere (Maus et al. 2002). CHAMP, combined with Ørsted, have made the current greatly refined model of the Earth's magnetic field possible (Olsen et al. 2009), thanks to the unexpected longevity of these missions.

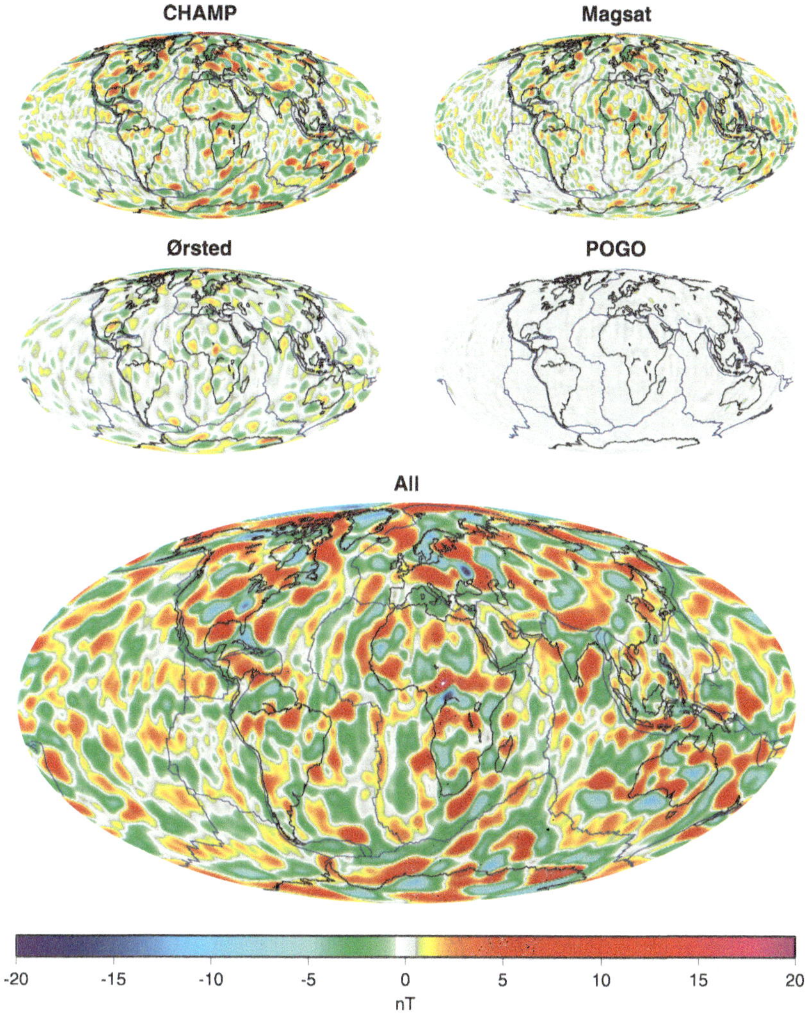

Fig. 36 Using a comprehensive model of the Earth's magnetic field, the graphs show the portion of the radial component of the lithospheric magnetic field at 400 km altitude as resolved by the satellites CHAMP, Magsat, Ørsted and POGO. Using all data, the resolution and coverage of the lithospheric field is greatly increased. The major tectonic boundaries are also shown for reference (from Sabaka et al. 2004)

Remarkable progress has been made in collecting large amounts of data to model not just the Earth's main magnetic field, but its lithospheric field as well as several aspects of the (variable) external contributions. The improvement that has been possible from combining data from all dedicated terrestrial magnetic field missions is illustrated in Fig. 36. However, further progress requires a simultaneous and coordinated set of measurements at several locations. In particular, simultaneous coverage at different longitudes is necessary for separating the terms that vary as a function of local time. The next major mission to study the Earth's magnetic field is ESA's three-spacecraft SWARM, to be launched in 2012 (Friis-Christensen et al. 2006, 2008). Two of the spacecraft, SWARM-A and -B will be in identical orbits, at $\sim$450 km height, but separated in longitude by $\sim 1^{\circ}$. The third, SWARM-C will be in a higher orbit (~530 km initially) with the right ascension of the ascending node close to the other two satellites.

The three spacecraft will have a shape and layout similar to CHAMP. The magnetic field instrumentation will also be similar (Merayo et al. 2008). There will be a single vector fluxgate magnetometer with vector feedback in a spherical configuration as for Ørsted and CHAMP, with redundant electronics and performance figures comparable or better than previous comparable missions. The magnetometer will be mounted together with the Advanced Stellar Compass star camera on an optical bench. Three orthogonally mounted optical heads will be used to minimise attitude errors affecting the vector measurements. In addition, the sensor itself will be mounted at a distance ~40 cm from the ASC heads to minimise further the stray fields at the location of the vector magnetometer. The absolute magnitude of the magnetic field is measured by two optically pumped helium 4 scalar instruments. Compared to the Overhauser magnetometer used on Ørsted and CHAMP, the helium magnetometer offers reduced noise (higher sensitivity, ~1 pT/Hz$^{1/2}$), an absolute accuracy ~0.1 nT and a much higher rate of measurements (in principle several hundred Hz).

3.5 The Moon

The Moon belongs to the family of terrestrial planets by association and its magnetism can be considered in the same terms as those of the terrestrial planets. While it was discovered early that the Moon has no global scale, internally generated magnetic field (none was expected), the weak but still significant magnetic fields that have been observed have many important implications for models of its early history, particularly its cratering history, and its material properties.

The first mission to reach the Moon, the Soviet Luna 2, was launched on 12 September 1959 and crashed on the surface two days later. The data returned showed that the Moon was non-magnetic, with a relatively crude upper limit for its surface field of ~100 nT. The instrument sensor itself, of the fluxgate type, was placed on a boom protruding from the spacecraft (see Fig. 37). The previous mission attempt, Luna 1 that flew by at a distance of 6000 km from the Moon also carried an identical magnetometer, except that the measurement range for Luna 1 was ±3000 nT, while it was reduced to ±750 nT for Luna 2 that also had a digitization uncertainty of 12 nT and an offset/background between 50 and 100 nT. The null result from the instrument was thus compatible with an upper limit of 100 nT as the Moon's magnetic field.

There were several other Luna missions in the early to mid-1960s, but no definitive value was found for the lunar magnetic field, in fact not even any evidence of its existence but only upper limits, due to a combination of relatively high orbits and the sensitivity of the magnetometers. This was reported by Dolginov et al. (1966) for the Luna missions up to Luna 10 and by Sonett et al. (1967) and Ness et al. (1967) based on the findings of the first NASA lunar orbiter Explorer 35 that carried two magnetometers. The Explorer 35 magnetometer operated into the Apollo era when, as described below, it was used in conjunction with the lunar Surface Magnetometers to investigate induction effects in the planetary body and thereby to deduce the electrical properties of the Moon's interior. A revisit of the observations following the first Moon landings also allowed an interpretation of some of the variability observed on the far side of the Moon to be ascribed to patchy but strong crustal magnetization (Mihalov et al. 1971).

The first evidence for remnant magnetisation was found in the lunar samples returned by the Apollo 11 astronauts in July 1969 (see, e.g., Runcorn et al. 1970 and the retrospective review by Fuller 1998). More such evidence was collected from the samples brought

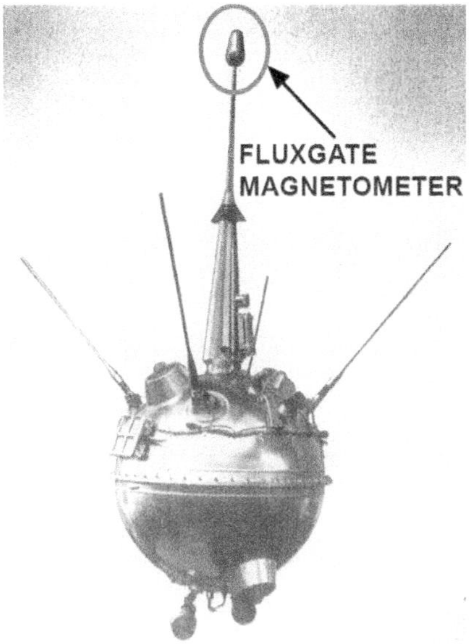

Fig. 37 A photo of the Soviet Luna 2 spacecraft that was the first to reach the Moon in 1959; this mission showed that, as expected, the Moon did not have a significant global magnetic field. The instrument and impact orbit did not allow for the detection of the crustal magnetic field signatures identified on the Moon during Apollo programme and later missions

back by Apollo 12 (Runcorn et al. 1971) and subsequent Apollo missions. (Some of the interpretation of the early evidence for the presence of magnetised material in the samples was controversial, see Fuller 1974.) However, direct evidence of the magnitude of the surface field was obtained through the use of the first lunar surface magnetometer delivered by Apollo 12 (Dyal et al. 1970a). Using similar surface magnetometers on subsequent Apollo missions described by Dyal et al. (1974) around the landing sites provided data that confirmed the existence of crustal magnetism. The surface magnetometers found a variable level of natural remnant magnetisation. There is uncertainty even today whether the magnetisation is compatible with models of the Moon's core that could operate an early dynamo in its history. Additionally, particles and fields sub-satellites for Apollo 15 and 16 measured magnetic fields in orbit which supported the overall picture of non-uniform crustal magnetisation with a hint of trends related to cratering combined with an induced magnetic field in the lunar interior that is by definition highly variable.

The Lunar Surface Magnetometers and the way they were used have been described by Dyal et al. (1970b) and Dyal and Gordon (1973). The basic design of the static magnetometers used on the Apollo 12, 14, 15 and 16 missions was the same, although detailed improvements were made from mission to mission. Portable magnetometers were also carried on the Apollo 14 and 16 missions. The Apollo 16 static magnetometer fluxgate sensors used a new design based on highly stable ringcores. This development by the US Naval Ordnance Laboratory of the special high-permeability material for the fluxgate sensors with low intrinsic noise when used in a magnetometer, the construction of the sensor itself with its coil formers and drive and sense windings was to bring a general improvement in magnetometer performance for all space applications. Indeed this sensor, as described by Gordon and Brown (1972) was to be a template for fluxgate magnetometers for the next three and a half decades; most of these used the same high permeability magnetic tape from the material developed by the Naval Ordnance Laboratory on the ringcores.

In the light of later developments of operating fluxgate sensors, it is interesting to note that the sensor was driven by a 6 kHz constant voltage sine wave. (In the lunar portable magnetometers used on Apollo flights 14 and 16, the drive waveform was a constant voltage

Fig. 38 The Lunar Surface Magnetometer deployed on the Moon by the astronauts of the Apollo 12, 14, 15 and 16 missions. These magnetometers first measured the steady crustal magnetization of the Moon but that was found to be highly variable from site to site (photo: NASA)

square wave.) The signal from the pick-up or sense coil did not contain the fundamental of the drive frequency because of the orthogonal geometry of the winding and it was the second harmonic and its phase with respect to the drive waveform that was used for measuring the ambient field component magnitude and sign. In later fluxgate sensors, the drive waveform used routinely is approximately a shaped bipolar exponential current waveform.

The three single axis sensors were located on separate booms of about 1 m each that formed an orthogonal triad. The sensor booms were oriented at an angle approximately 35° from the lunar surface. In order to reference the measurements in lunar coordinates, the magnetometer assembly was aligned with its Z axis to the East by an astronaut using a shadowgraph, the X sensor arm was aligned to the Northwest, and the Y sensor made up the right-handed orthogonal triad. In addition, level sensors were used to determine the orientation of the assembly with respect to the horizontal plane. Each sensor had thermally operated flippers that rotated the sensors by 90° to calibrate the zero levels. The magnetometers were built for operation for one year and were remotely operated from the Manned Spacecraft Center, Houston. The overall accuracy was $\sim$1° in lunar coordinates. The field values measured by the static lunar magnetometers were 38 nT, 103 nT, 3 nT and 327 nT, respectively from the four missions. A photograph of the setup of the Lunar Surface Magnetometer is shown in Fig. 38.

Results from the observations by the magnetometers on the Apollo sub-satellites 15 and 16 provided a detailed and convincing view of the crustal magnetisation of the Moon and induction effects in its interior (Coleman et al. 1972, 1972b). The sub-satellites were released from the Apollo Service Module in Moon orbit intended for a nominal mission duration of one year. The first was launched into a low eccentricity first orbit 102 km × 139 km, but with rapidly changing orbital parameters because of gravitational perturbations. The second was placed in a similarly low orbit; the decay of the orbit to low altitudes above the lunar surface allowed some detailed mapping of anomalies in crustal magnetisation and associating these with surface/geological features (see, e.g. Hood et al. 1979). Combining the in-orbit data from Explorer 35 and the data from the surface, it was possible to estimate parameters of the induction caused in the lunar crust by sudden changes in the interplanetary magnetic field. One such simultaneous set of observations is shown in Fig. 39.

There were identical magnetometers on the two sub-satellites. The design was a two-axis fluxgate. One of the sensors was aligned with the spin axis of the spacecraft, the other was aligned in the spin plane. The spin rate was $\sim$12 rpm; demodulation of the spin-plane sensor provided the direction and magnitude of the spin-plane component of the magnetic field (Coleman et al. 1972). The observations made by the magnetometers of the nature and geological distribution of crustal magnetism formed the historical data base for lunar magnetic fields for 25 years.

A powerful technique for the study of small-scale planetary crustal magnetic fields was developed using the energetic electron instrument on the two Apollo sub-satellites (Anderson et al. 1972). The technique, named Electron Reflectometry, is based on the effect of localized magnetic fields on the pitch angle distribution of energetic electrons that are reflected from close to the planetary surface by the crustal field concentrations (Anderson et al. 1974; Howe et al. 1974). This technique was used for matching the fine-scale magnetic crustal anomalies detected by the sub-satellites' magnetometers to the geological formations of different structures, origins and ages (McCoy et al. 1975; Lin et al. 1976). The applicability of the technique to both the lunar and Martian crustal magnetisation was exploited on the Lunar Prospector and Mars Global Surveyor mission which both flew magnetometer—electron reflectometer combinations.

After the first decade of space missions to the Moon in the Apollo era, there was a hiatus of more that 25 years before a magnetometer-carrying spacecraft again visited the Moon. One of NASA's Discovery missions, Lunar Prospector was launched on 6 January 1998 and reached a circular lunar polar orbit at a height of 100 km above the surface on 11 January 1998. The orbit had to be restored periodically, because of its degradation due to gravitational perturbations. After one year, the orbit was lowered to a much lower, 15 km (minimum) and 45 km (maximum) height orbit. On 31 July, 1999, the spacecraft impacted the lunar surface, ending the mission.

Magnetic fields were measured by a magnetometer—electron reflectometer combination similar to the one used on Mars Global Surveyor and also following the serendipitous discovery of the technique based on the Apollo 15 and 17 sub-satellites observations (Howe et al. 1974). The two instruments were mounted on a 2.5 m boom, with the magnetometer sensor (a tri-axial fluxgate) on an additional 0.8 m extension arm outboard of the reflectometer. Given the polar orbit of the spacecraft and the relatively long duration of the mission, very comprehensive mapping of the lunar crustal magnetic fields could be achieved (Halekas et al. 2001; Hood et al. 2001; Mitchell et al. 2008). The most recent work (Richmond and Hood 2008) has used a considerable amount of data from the low-altitude, 30 to 40 km, phase of the Lunar Prospector mission, in particular 329 passes at the nightside and the terminator that are best suited for analysis without solar wind and other background effects, to generate detailed maps of the crustal anomalies. The low altitude Lunar Prospector observations have also made possible an evaluation of the very weak internal magnetic field of the Moon (Purucker 2009) that can be applied to refine the model of the lunar interaction with the solar wind.

The latest mission to the Moon was the Japanese Kaguya (Selene), launched on 14 September 2007. It was placed initially into a circular polar orbit on 18 October 2007 at an

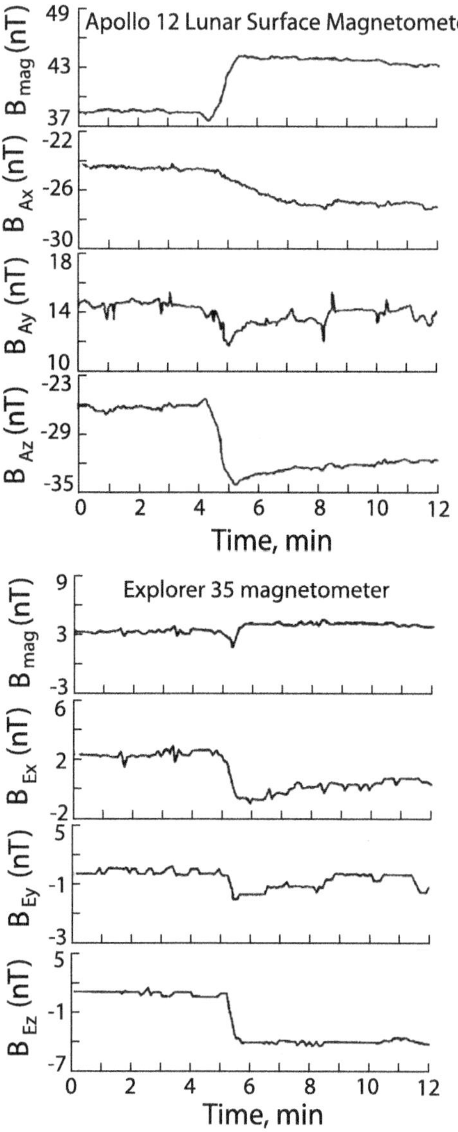

Fig. 39 Simultaneously measured magnetic field variations at the Apollo 12 landing site by the Lunar Surface Magnetometer (*upper panels*) and at Explorer 35 in orbit around the Moon (*lower panels*). The measurements start on 8 December 1969 at 13:38 UT. The large amplitude angular discontinuity at Explorer 35 at about 5 minutes after the start of the measurements caused a variation in the field measured by the LSM that is superposed on the steady ∼38 nT crustal field background (from Dyal et al. 1972a)

altitude of 100 km for a 10-months long operational phase, followed by a mission extension that came to an end on 11 June 2009 when the spacecraft was deorbited to impact the Moon. Kaguya carried a magnetometer that had been very carefully calibrated (Shimizu et al. 2008), and a long, 12-m magnetometer boom that was successfully deployed after launch. The Lunar Magnetometer (LMAG) is a triaxial fluxgate sensor of classical construction. The frequency range of the instrument is limited to 10 Hz and a 32 Hz sampling rate. The accuracy of the measurements is better than 0.1 nT. It has four operating ranges, ±64 nT, ±256 nT, ±1024 nT, and a range that had been used for ground testing before launch, in the Earth's field. A special feature of the in-flight calibration of the instrument is the use of two orthogonal coils near the boom root that generate calibrated 2 nT fields at the location of the sensor at the boom tip, similar to the coils used on the Cassini spacecraft described in Sect. 4.2. (The Voyager spacecraft had a single calibration coil at its magnetometer boom root.)

Preliminary results of the magnetic field mapping and the modelling of the data were reported by Shibuya et al. (2008). It is expected that following this development of the methodology for assessing the crustal fields, the observations of Kaguya will add further data to resolve the problems raised by the complexities of lunar magnetism. (The Lunar Reconnaissance Orbiter of NASA, launched in 2009, does not carry a magnetometer.)

3.6 Comets and Asteroids

Comets are not expected to have magnetic fields, but the outgassing that occurs in a complex manner as comets approach the sun during their perihelion passage creates a complex plasma-magnetic field environment as the cometary material interacts with the ambient solar wind flow. Studying this complex and dynamic environment is primarily of interest for learning more about comets. So far four comets have been visited by spacecraft equipped with magnetometers: 21P/Giacobini-Zinner in 1985 (Smith et al. 1986), 1P/Halley's comet in 1986 (Neubauer et al. 1986; Riedler et al. 1986a; Saito et al. 1986), 26P/Grigg-Skjellerup in 1992 (Neubauer et al. 1993) and 19P/Borelly (Glassmeier et al. 2007b). The magnetometer on the ISEE-3/ICE mission that flew by Giacobini-Zinner in 1985 was a Vector-Helium magnetometer that has already been described (Frandsen et al. 1978). The two spacecraft that flew close to Halley's comet in 1986 were ESA's Giotto and the Soviet Vega spacecraft, both equipped with fluxgate magnetometers (Neubauer et al. 1987; Riedler et al. 1986b). The Giotto magnetometer sensors were mounted very close to the spacecraft body and in particular to the despin motor of the antenna that was made to point at Earth on the spinning spacecraft. The data were contaminated by the spacecraft's stray magnetic fields; however, extensive and careful assessment of the noise background has led to valuable science data and results concerning the comet's environment. The Vega magnetometer consisted of a triaxial fluxgate sensor mounted on a boom at the end of the extended solar panels, and another, single axis fluxgate about 1 m inboard, used as a gradiometer to resolve the spacecraft-generated background field at the outboard sensor. The Vega magnetometer was closely similar to the instruments flown on the Venera 13 and 14 missions to Venus. After the Halley encounter, the Giotto spacecraft was redirected to make a close flyby (at only 200 km) of comet 26P/Grigg-Skjellerup in 1992. The magnetometer and the plasma instrumentation had survived the encounter with Halley and provided valuable data about the plasma environment of this comet (Glassmeier and Neubauer 1993). A further cometary flyby, within a distance of 2200 km, was carried out by NASA's Deep Space 1 mission on 22 September 2001. The target comet was 19 P/Borelly.

The dedicated comet rendezvous mission of ESA, Rosetta, was launched in 2004 and will reach Comet 67 P/Churyumov-Gerasimenko, after a 10-year complex trajectory, in 2014 (Glassmeier et al. 2007c; Schulz 2009). The spacecraft will then be captured into an orbit at about 25 km height around the comet and will accompany it on its journey towards perihelion. A five-instrument plasma science package includes a magnetometer (Glassmeier et al. 2007b) for the study of the plasma environment and its dynamical processes through the developing activity phase of the comet. Rosetta also carries a lander that will be released from the main spacecraft bus once a suitable landing site has been identified. The lander carries a miniature magnetometer that will measure the magnetic field during the descent and also on the surface to try and identify the structure of the comet's remanent magnetisation, if any (Auster et al. 2007). In addition, coordinated measurements between the magnetometers on the orbiter and lander under changing interplanetary magnetic field conditions can be used, as was done for the Moon (Dyal et al. 1972b, 1974) to study the conductivity of the comet nucleus.

Asteroids represent a vast population of objects presenting a very great variety in terms of size, origin, and material composition. Learning more about the material properties of asteroids, in particular though their magnetic properties is important for better understanding the formation of the solar system. The magnetism of asteroids can only be measured in situ, on a sufficiently close flyby or by an orbiter and/or lander. There have been a number of serendipitous asteroid flybys over the past two decades that have resulted in an interesting set of properties. Of the three asteroid-focussed missions flown so-far, only NASA's NEAR to 433 Eros, launched on 17 February 1996, carried a magnetometer (Lohr et al. 1997; Acuña et al. 1997). Japan's Hayabusa launched on 9 May 2003 to asteroid 25143 Itokawa and NASA's Dawn mission to the largest asteroids, 1 Ceres and 4 Vesta, launched on 27 September 2007 do not carry a magnetometer. This is particularly disappointing for the Vesta flyby, as it appears to be an evolved body with, potentially, an important magnetic signature.

The Galileo spacecraft, on its way to Jupiter, flew by two asteroids. The first was 951 Gaspra, on 29 October 1991, at a closest approach distance of 1600 km; the second was 243 Ida, on 28 August 1993, at a distance of 2400 km. The Galileo magnetic field investigation already described in Sect. 4.1 (Kivelson et al. 1992) carried out measurements during both flybys, to study the interaction of the asteroids with the interplanetary medium and to detect any intrinsic magnetic field. The flyby of Gaspra yielded a surprising result: the surface magnetic field value deduced from the signature of the magnetosphere-like interaction was commensurate with that of the earth (Kivelson et al. 1993). The asteroid is classed as an S-type, potentially within a family of the parent bodies of stony-iron meteorites (covered, probably, by regolith). The magnetic field measurements were consistent with this interpretation, yielding a magnetic moment per unit mass characteristic of iron meteorites and highly magnetized chondrites. This also supports the hypothesis that Gaspra is a fragment of a previously differentiated and magnetised parent body. Measurements during the flyby of Ida did not imply an intrinsic magnetic field, but signatures of the interaction of the asteroid with the solar wind were still noted in the data. A comparison of the different signatures at Gaspra and Ida have been investigated and interpreted as whistler waves generated by the interaction of the asteroids with the solar wind (Kivelson et al. 1995).

NASA's Deep Space-1 mission, a technological demonstrator for an ion drive orbital booster, flew by asteroid 9969 Braille on 29 July 1999, at a closest approach distance of only 28 km, on the night side of the asteroid. Although the magnetic signature needed to be recovered from a very noisy background (the magnetometer was primarily included as a diagnostic sensor for the ion engine), there remained a significant residual signal that could be associated with the asteroid. The two miniature tri-axial fluxgate sensors included in the spacecraft were in fact the prototypes of the sensors used for the Rosetta cometary mission, described above (Glassmeier et al. 2007b). The magnetic signal has been interpreted as evidence for an intrinsic magnetic field, with an upper limit value for an equivalent dipole moment of 2.1×10^{11} Am^2 (Richter et al. 2001).

The Cassini approach to asteroid 2685 Masursky on 23 January 2000 was not suited to observing any aspect of its interaction with the solar wind, as it occurred at 1.5 million km from the asteroid. On June 27, 1997 the NEAR spacecraft passed within 1200 km of main belt asteroid 253 Mathilde, a C-type asteroid. No magnetic signature was observed during the flyby that could have been attributed to the asteroid.

The prime objective of the Near-Earth Asteroid Rendezvous (NEAR) mission of NASA was the asteroid 433 Eros. NEAR was launched on 17 February 1996. There were several close orbiting phases, at 200 km, 50 km and 35 km heights. Eventually, the spacecraft touched down on Eros 12 February 2001. Contrary to the other S-type asteroids (Gaspra, Braille) that had been found to have a magnetic field, Eros was found to be completely unmagnetised (Acuña et al. 2002). The result is robust as the NEAR magnetometer was extensively calibrated in flight (Anderson et al. 2001). The interaction of Eros with the ambient solar wind was also found to be devoid of any signature that may have indicated the presence of a magnetic field (Anderson and Acuña 2004).

The Rosetta mission includes two asteroid flybys (Barucci et al. 2007); the first with the rare E-type 2867 Steins has already taken place on 5 September 2008 (Auster et al. 2010). The flyby with the much larger (~100 km) C-type 21 Lutetia is expected on 10 July 2010. C-type asteroids are considered to be the most primitive type that had not undergone heating or differentiation. Consequently, no magnetic signature is expected during the flyby, although the magnetometer will make measurements in the asteroid environment.

4 The Outer Planets

4.1 Jupiter and Its Moons

Analysis of radio emissions from Jupiter that were first detected from Earth (Burke and Franklin 1955) indicated that the source of the radio waves was emission by energetic charged particles in a strong magnetic field. This implied a large planetary scale magnetic field as exists for the Earth, except much stronger. The radio measurements and their implications in terms of the magnetic field of Jupiter have been reviewed by Berge and Gulkis (1976) and Smith and Gulkis (1979). However, the details of the field (magnitude, dipolar or higher order polarity, direction of axes) needed confirming to establish its origin and, in general, the internal structure of the planet.

The first spacecraft to fly by Jupiter and measure its magnetic field was Pioneer 10 (see the spacecraft in Fig. 1 in the Introduction). It was launched on 3 March 1972 and flew by Jupiter on 3 December 1973. The identical Pioneer 11 spacecraft was launched on 6 April 1973 and reached Jupiter on 4 December 1974. Pioneer 11 was targeted to Jupiter in such a way that the flyby would redirect it to a close flyby of Saturn that took place on 1 September 1979. After the planetary flybys, the trajectories of both Pioneer 10 and 11 were directed to the outer heliosphere; both remained operational into the 1990s.

The magnetometer sensors were carried on a boom of 6.6 m from the main body of the spacecraft. The primary instruments on both missions were high sensitivity Vector Helium Magnetometers (Smith et al. 1975a) that have been described in Sect. 2.3. Pioneer 11 also carried a vector fluxgate magnetometer designed for measuring the high fields in the vicinity of Jupiter and Saturn (Acuña and Ness 1975a, 1975b). The results from the magnetometer of the first flyby were reported by Smith et al. (1974a, 1974b) and the second flyby by Smith et al. (1975b) and Acuña and Ness (1976). The Pioneer results were reviewed by Smith et al. (1976). Plots of the magnetic field magnitude from the two Pioneer flybys measured by the VHM, together with the jovicentric distance during the flybys, are shown in Fig. 40. Models of Jupiter's very large internal magnetic field were published by the same authors, although differences were reported between the model based on the VHM and High Field FGM data.

The radio measurements that implied much higher magnetic field values at Jupiter than at the Earth set a requirement for a broader measurement range than for near-Earth missions. For the magnetometers on missions to Jupiter the expected maximum field was of order of several Gauss, or of order of mT. At the same time, the interplanetary magnetic field at the orbit of Jupiter that needed to be measured was of order of 1 nT. So a measurement range

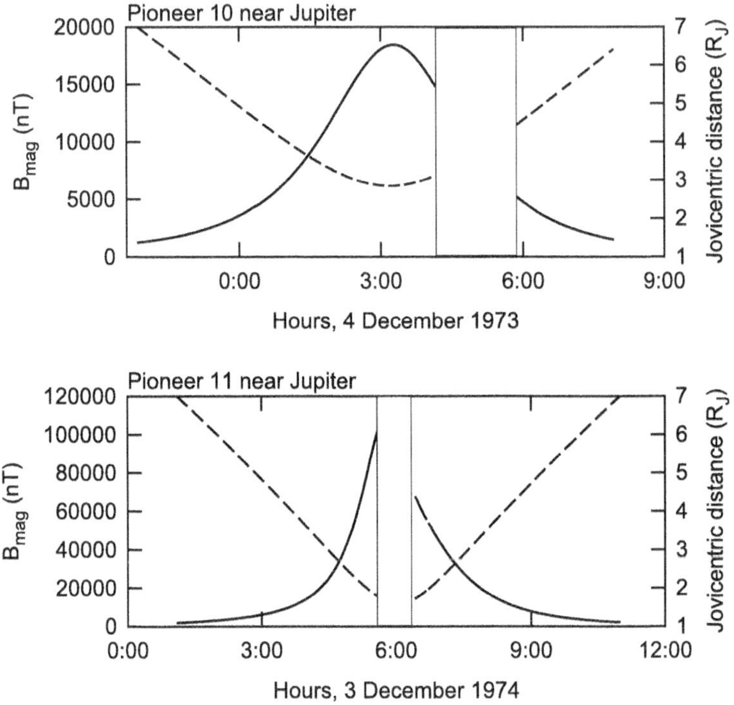

Fig. 40 The magnetic field intensities measured by the Vector Helium Magnetometer during the hours of closest approach to Jupiter during the flybys of Pioneer 10 and 11 (*solid lines*) and the jovicentric distance of the spacecraft (*dashed lines*). *The blanked out* intervals correspond to the two spacecraft being occulted from the Earth by the planet on the two flybys. The closest approach distances were $2.84R_J$ for Pioneer 10 and $1.62R_J$ for Pioneer 11. For Pioneer 11, the closest approach occurred during the occultation interval

covering about 6 orders of magnitude was needed. The VHM designed for Pioneers 10 and 11 had eight ranges, from ±4 nT per vector component to ±140,000 nT (Smith et al. 1975a). Switching between the ranges was either automatic or manual, the mode being selected by ground command. The digital resolution was determined by the 9-bit Analogue-to-Digital Converter (ADC) that included 1 bit for the sign of the signal; thus for the ±4 nT range the resolution was 0.016 nT, while in the highest range it was 547 nT. As shown in Fig. 40, the maximum field measured and telemetered from Jupiter by Pioneer 11 was about 110,000 nT, so that the relative accuracy of the measurements close to Jupiter was about 0.5%. It was this instrument that, developed from the early Mariner 4 and 5 magnetometers using improved electronic components and adapted to the different mission requirements, was to be used on the ISEE-3/International Cometary Explorer mission, on Ulysses and on the Cassini Saturn Orbiter.

The fluxgate magnetometer that was carried on Pioneer 11 was a unique design for that mission that was severely limited in resources as a late addition to the payload (Acuña 1974; Acuña and Ness 1975a, 1975b). It shared its resources with an experiment that measured cosmic rays and was mounted on the spacecraft experiment platform, rather than at the boom tip as the VHM. Because of the expected high fields at Jupiter, the instrument was operated, uniquely for a spacecraft FGM, in open rather than in closed loop, so that there was no attempt to null the field at the location of the sensor. The instrument, in addition, used only two ringcore sensors, but each of these was designed to measure two orthogonal components

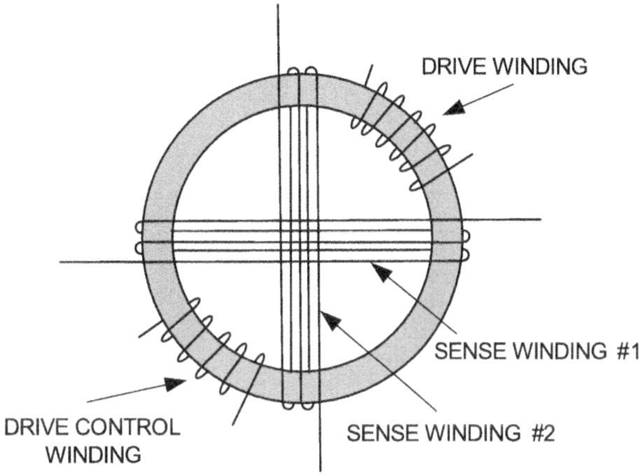

Fig. 41 Schematic arrangement of the ringcore sensor used on the High Field FGM on Pioneer 11. The sensor used two orthogonal sense windings so that the same ringcore could be used to measure two components of the magnetic field vector, thus making it possible to measure the three components of the magnetic field with two such sensors mounted orthogonally to each other (after Acuña 1974)

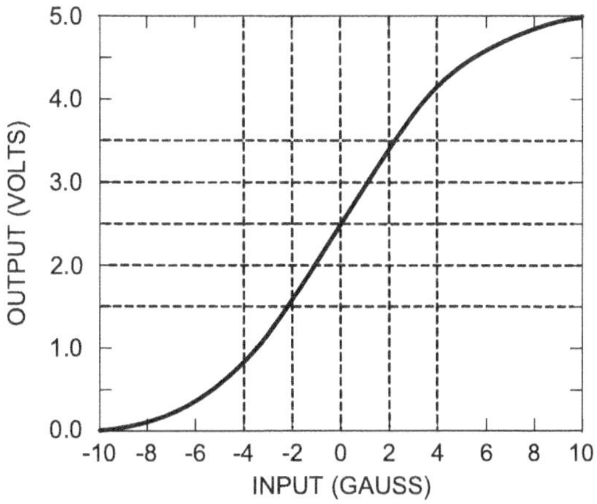

Fig. 42 Response function of the open-loop High Field FGM used on Pioneer 11. The output has adequate linearity below about 200,000 nT (after Acuña 1974)

of the ambient magnetic field by the use of two orthogonally placed sense windings as illustrated in Fig. 41 (Acuña 1974). The measurement range for each component was 1 mT (10 Gauss), with a resolution of 1 μT. It was thought that the spacecraft background at the body-mounted location of the sensor was less than that value; the two sensors needed for the vector measurements were in fact mounted as an integral part of the electronics box. As the sensors are operated without feedback, the response function (in terms of the voltage output for a given magnetic field) is in principle inherently non-linear; however, as illustrated in Fig. 42, for input fields less than about 200,000 nT (2 Gauss), the instrument was considered to be adequate given its resolution.

The two-component measurement geometry of the two ringcore sensors was arranged such that one of the sensors had sense windings along the X and Z axes, the other along the Y and Z axes, with the second Z axis signal (called Z_R, for redundant) not being processed by the instrument's own ADC, but by that of the spacecraft. The first analysis of the Jupiter flyby using the High Field FGM (Acuña and Ness 1975b) indicated that, contrary to the Pioneer 10 flyby analysis by Smith et al. (1974b), there were very large quadrupole and octupole moments in Jupiter's main magnetic field. However, a later analysis of the measurements by the redundant Z_R sensor through the spacecraft ADC indicated that a 10%

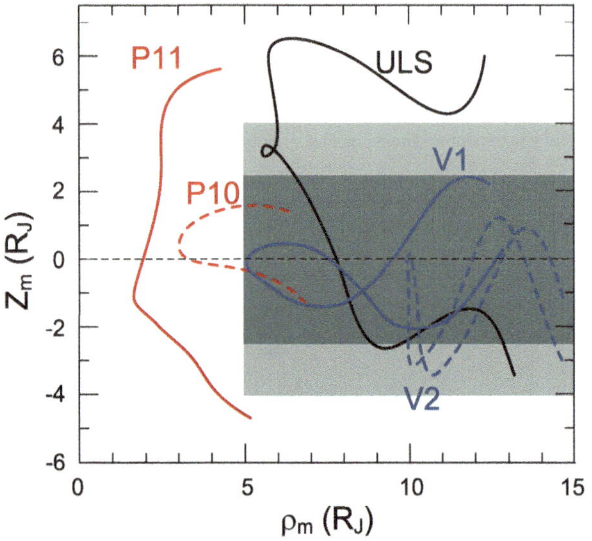

Fig. 43 The trajectories of the five spacecraft that made magnetic field measurements near their closest approach to Jupiter (P10 and P11 = Pioneer 10 and 11, V1 and V2 = Voyagers 1 and 2, ULS = Ulysses). *The vertical axis* is the distance along the dipole axis and *the horizontal axis* is the distance normal to the dipole axis, hence the oscillations in the trajectory plots, due to the 10-hour rotation period of the planet. *The darker shaded area* represents the Jovian current sheet in the O6 model of the magnetic field in Jupiter's vicinity (Connerney et al. 1981a, 1981b; Connerney 1992), the lighter shaded area is the current sheet determined from the Ulysses measurements. The figure has been adapted from Dougherty et al. (1996)

sensitivity correction had to be carried out on the data from the instrument's other, X–Z sensor, considerably reducing the values of the higher order terms (Acuña and Ness 1976) and better matching the data obtained by the other instruments on the spacecraft.

The results of the Pioneer 10 and 11 flybys remain important for the internal magnetic field of Jupiter, because to date no other spacecraft approached Jupiter closer. The following four flybys were more distant from the planet: Voyagers 1 and 2 in March and July 1979 (at periapsis distances of 4.9 and 10.1 R_J), Ulysses in February 1992 (at a periapsis distance of 6.3 R_J) and Cassini-Huygens in December 2000 at a distance of 138 R_J. The Galileo mission targeted primarily the moons of Jupiter and will be referred to in more detail below.

The two Voyager spacecraft were launched on 20 August 1977 (Voyager 1) and on 5 September 1977 (Voyager 2). Both were targeted to flyby first Jupiter and then Saturn. However, after the Saturn flybys, Voyager 1 was sent on to encounter Uranus and Neptune. The trajectories of both spacecraft were to take them subsequently to the outermost regions of the heliosphere. The spacecraft crossed the termination shock of the heliosphere in late 2004 and in 2007, respectively.

The Jovian flybys took place on 5 March 1979 at a closest approach distance of 4.89 R_J (Voyager 1) and 9 July 1979 at 10.1 R_J (Voyager 2). The close-in flyby trajectories of the first five spacecraft that flew past Jupiter are illustrated in Fig. 43, in a Jupiter centred coordinate system that has its vertical axis aligned with Jupiter's dominant dipole axis. The measurements of the magnetic field in the planetary environment were reported by Ness et al. (1979a, 1979b). The planetary field of Jupiter, based on the Voyager and Pioneer results was presented by Connerney et al. (1982a), the Jovian current sheet and the inner magnetosphere were described by Connerney et al. (1981a).

The magnetic field investigation on the two Voyager spacecraft was described by Behannon et al. (1977). It consists of two sets of triaxial (vector) fluxgate sensors that are boom-mounted and the corresponding electronics in the spacecraft instrument bay. The deployable Voyager boom is 13 m in length from the spacecraft body to the boom tip. The two Low-Field Magnetometer (LFM) triaxial sensors are mounted on the boom tip and at a distance of 7.4 m from the spacecraft, respectively. The LFM measures the magnetic field vector in eight ranges, from ±8.8 nT to ±50,000 nT on each orthogonal axis; the corresponding digitisation errors are ±2.2 pT to ±12.2 nT. The sensor noise was ~6 pT RMS; it was shown as a function of frequency in Fig. 6 in Sect. 2.2. The vector sampling rate was 16.67 Hz and the bandwidth of the instrument was 8.3 Hz to satisfy the Nyquist criterion. The two High-Field Magnetometer (HFM) triaxial sensors are mounted about 1 m apart, close to the spacecraft along the canister that contained the expandable boom. The HFM used specially developed small cores in the sensors, to minimise the power consumption that scales with the intensity of the field being measured. Its two measurement ranges were conservatively designed for the potentially large fields encountered near Jupiter and Saturn, covering ±50 μT and ±2 mT in each orthogonal axis, at a sampling rate of 1.67 vectors/s. During the Jupiter flyby, the Voyager 1 LFM operated in seven of its eight possible ranges, up to ±6, 400 nT per axis, so that there was no need to use the HFM data which had coarser resolution.

There were two specific aspects related to the Voyager magnetometer investigation that can be mentioned in addition. One is the use of the dual magnetometer technique applied to the two LFM sensor data. For Voyager, the dual magnetometer technique originally proposed by Ness et al. (1971) and refined by Neubauer and Schatten (1974) was written by Behannon et al. (1977) as

$$\mathbf{B}_{sc}^{est}(r_2) = \alpha \frac{[\mathbf{B}_{obs}(r_1) - \mathbf{B}_{obs}(r_2)]}{1-\alpha}$$

where $\mathbf{B}_{sc}^{est}(r_2)$ is the estimated value of the value of the magnetic field due to the spacecraft at the location (r_2) of the outboard sensor, $\mathbf{B}_{obs}(r_1)$ and $\mathbf{B}_{obs}(r_2)$ are the measured magnetic field values at the location of the inboard and outboard sensors, respectively, and $\alpha = (r_1/r_2)^3$ is the correction factor. Subtracting the estimated value of the spacecraft-caused background field $\mathbf{B}_{sc}^{est}(r_2) \sim 0.2$ nT from the measured value, $\mathbf{B}_{obs}(r_2)$, at the outboard sensor is taken to provide the true value of the ambient field.

The second aspect is related to the knowledge of the orientation of the magnetometer axes with respect to the spacecraft reference frame on the 13 m expandable boom. The boom construction is a form of coiled spring which, when expanded, is quite stable against bending (±0.5° around axes perpendicular to the boom axis), but prelaunch tests have shown that the twist angle (rotation around the boom axis) is only reproducible to about 7°. To determine the orientation of the sensor axes after deployment of the boom in space, a calibration coil was installed around the high gain antenna dish of Voyager which, when energised with a 0.5 A current, generated a known magnetic field of amplitudes 6.1 nT and 33.4 nT respectively at the outboard and inboard LFM sensors. While this is not sufficient to provide an unambiguous calibration of the orientation of the sensor axes, appropriate assumptions and approximations made the calibration process of sufficient accuracy for the measurement of the planetary fields encountered by the two Voyager spacecraft (Behannon et al. 1977). A dual coil system was used for the same purpose of determining the magnetometer sensor alignment on the Cassini spacecraft (Dougherty et al. 2004) which has a boom of similar construction to that of the Voyager boom and is described below.

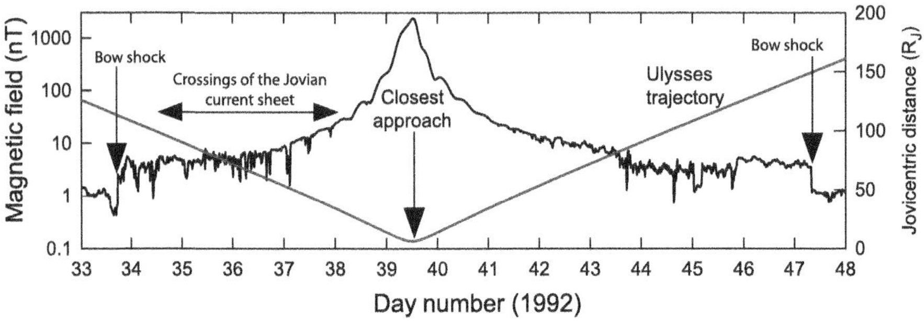

Fig. 44 The Jovian magnetic field during the Ulysses flyby in February 1992. Note the repeated crossings of the Jovian current sheet during the inbound part of the trajectory and the disturbed, dynamic nature of the magnetospheric magnetic field away from the closest approach to the planet. The jovicentric distance is shown in *red*

The Ulysses spacecraft flew by Jupiter in February 1992 for a gravity assist manoeuvre to deflect the trajectory of the spacecraft out of the ecliptic plane. The magnetic field observations during the flyby have been described by Balogh et al. (1992a), see in Fig. 44 the magnitude of the magnetic field through the 15 days interval while the spacecraft was inside the magnetosphere of Jupiter. Closest approach was at $6.3R_J$ on 8 February 1992. Given the closest part of the flyby trajectory (Fig. 43) that covered higher Jovian latitudes than either Voyager or Pioneer 10, but crossed the current sheet near the equator, Ulysses could assess the extent of the current sheet near its inner edge. It was found that for the Ulysses epoch, the current sheet was somewhat thicker (about $4R_J$ half width) than during the Voyager epoch (Connerney 1981) but with a reduced current density. Variations in the current sheet parameters are not surprising even near Jupiter, in response to the highly variable conditions that have been observed in the Jovian magnetosphere (Smith et al. 1975b; Connerney 1981; Dougherty et al. 1996).

In the Ulysses case, the sunward bow shock was encountered at $113R_J$ followed at $110R_J$ by an encounter with the magnetopause. The much larger distances seen by Ulysses than by any of the previous flybys by Pioneer and Voyager spacecraft indicate a much expanded magnetosphere during the Ulysses epoch, probably due to a sudden drop in solar wind pressure also noted by Ulysses (Balogh et al. 1992b). The bow shock and magnetopause encounters were also very close, given the size of the magnetosphere; this indicated a still expanding magnetosphere. It is generally recognised, that the outer part of the Jovian magnetosphere is subject to very large-scale dynamic effects dependent mostly on variations in the dynamic pressure of the solar wind.

The prominent feature of the middle and inner magnetosphere is the current sheet that was first discovered by Pioneer 10 (Smith et al. 1974a, 1974b) and observed by all subsequent spacecraft that reached the inner part of the Jovian magnetosphere. Observations from the last spacecraft to sample it (Russell et al. 1999a, 1999b) reported on its structure and summarized its variable features in the context of previous observations. The current sheet, extending from $5R_J$ to about $50R_J$ from the planet but with its outer edge ill-defined and variable, dominates the structure of the magnetosphere inside its outer edge. In addition, the current strength is strong enough to make a significant contribution to the measured magnetic field close to the planet (see Fig. 43) so that it makes the task of separating the internal planetary field considerably harder. Jupiter is, in this way, quite different from the other planets and it is difficult to apply concepts of external current systems derived from similarity with earth-like magnetospheres (see review by Baumjohann et al. 2010a, 2010b, this issue).

The Ulysses magnetic field instrument was described by Balogh et al. (1992a). The instrument was optimized for the primary mission of Ulysses, exploring the heliospheric medium over the poles of the Sun. It consisted in two tri-axial sensors mounted on a 5 m long boom attached in the spin plane of the spacecraft. The outboard sensor is a Vector Helium Magnetometer (VHM) following the Pioneer 10 and 11 and ISEE-3 heritage, the inboard sensor is a ring-core vector Fluxgate Magnetometer (FGM). Because of the expected very weak magnetic fields in the heliosphere at 5 AU, an extremely careful magnetic cleanliness programme was carried out for the Ulysses spacecraft. It involved the magnetic compensation and test of the Radioisotope Thermoelectric Generator, as well as a full spacecraft level test programme in the MFSA facility (see Sect. 2.1) that resulted in a spacecraft background field of $\sim$0.03 nT at the location of the VHM and $\sim$0.1 nT at the location of the FGM. The VHM and FGM both had operating ranges of $\pm$8 nT and $\pm$64 nT per axis intended for the heliospheric medium and also for the outer regions of the Jovian magnetosphere; in addition, the FGM also had operating ranges for the Jovian flyby of $\pm$2048 nT (resolution 1 nT) and $\pm$44,000 nT (resolution $\sim$11 nT). The time resolution of the measurements was 1 vector/s for each of the magnetometers. As the spacecraft was spinning, the spin plane axes of the two magnetometers were easily calibrated for zero offset; for the spin axis offset, the method developed by Hedgecock (1975) was used. (The sensor offsets have been discussed in Sect. 2.4 above, see also Fig. 14.) By matching the two Jupiter ranges of the FGM during the flyby, the overall accuracy of the measurements (once the highest range data were averaged over 1 minute) was about 1 nT even for the data taken at closest approach ($\sim$2,500 nT).

Although the published models of Jupiter's planetary magnetic field differ in some detail, the value of the dipole is well established within $\sim$5% from 1.46×10^{20} T m^3 corresponding to the original value reported by Smith et al. (1974a, 1974b) as $4.0GR_{\mathrm{J}}^3$ to about 1.55×10^{20} T m^3 ($4.23 \times GR_{\mathrm{J}}^3$, quoted by Connerney 1993 for Pioneer 11). The Ulysses value was 1.48×10^{20} T m^3 (Dougherty et al. 1996). The range of values for the dipole moment deduced from radio observations and from the direct spacecraft observations were compared by Smith and Gulkis (1979); the two sets generally agree within the variability of the values obtained either way. The angle between the dipole axis and the rotation axis is $\sim$10° and there are significant higher order multipoles in the spherical expansion of the internal field (see, e.g. Connerney 1993; Russell and Dougherty 2010). In fact, the ratios of the dipole, quadrupole and octupole terms for Jupiter are, about 1:0.25:0.2 compared to Earth for which these ratios are 1:0.14:0.1, indicating that the region of field generation in the interior of Jupiter extends out to about $0.8R_{\mathrm{J}}$ (see, e.g. Stevenson 1983).

The Galileo spacecraft, the only orbiter so far around Jupiter, followed a complex mission plan with orbits targeted for a comprehensive exploration of the Jovian magnetosphere and the Galilean satellites. It was launched in 1989 and was inserted into Jupiter orbit in 1995 then repeatedly flew to $\sim 10R_{\mathrm{J}}$ perijove during its orbiting mission (see Fig. 45 and discussion below). The orbits were modified by targeted flybys of the satellites, to set up successive phases of the exploration. The orbits during the first two years of operations in Jupiter orbit are shown in Fig. 45. The jovicentric distance of perijove passes is also shown in Fig. 45. However, the perijove distances of the orbits remained too high (Fig. 45) for higher order terms of the planetary magnetic field to be determined. Analysis of the magnetic field observations close to Jupiter have shown that there may have been some secular change in the dipole term (Russell et al. 2001, 2002), consistent with the type of change observed for the Earth's magnetic field.

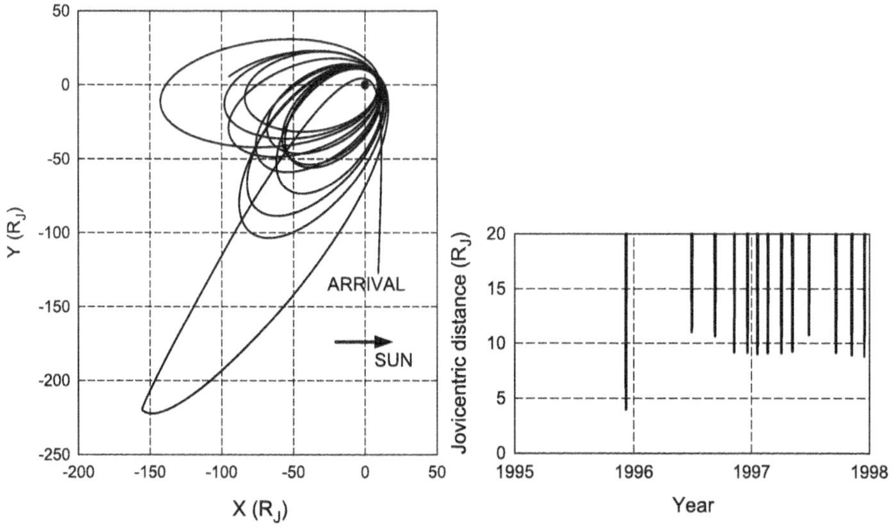

Fig. 45 *Left panel*: The operational orbits of the Galileo spacecraft in 1995–1997, following the arrival at Jupiter. The orbits were designed and optimised for close encounters with the Galilean moons. *Right panel*: Perijove passages of Galileo. The closest approach to Jupiter occurred on arrival to the planet

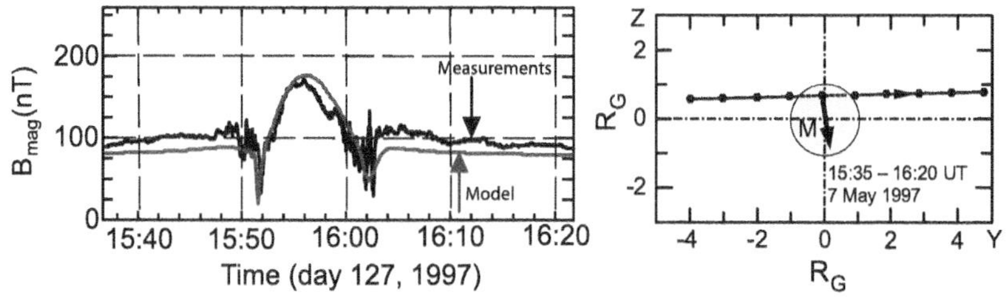

Fig. 46 (*Left panel*) The magnetic field magnitude measured by the Galileo spacecraft on one of its close flybys (G8) of the Galilean moon Ganymede (after Jia et al. 2010). (*Right panel*) The flyby geometry close to Ganymede. The trajectory is shown projected in the Jovian meridian plane passing through the centre of Ganymede; units along the axes are Ganymede radii. Markers along the trajectory are at intervals of 5 minutes. The trajectory passes in front of the moon, so that it samples the upstream region of Ganymede's magnetosphere. The modelled magnetic dipole of Ganymede is illustrated (after Kivelson et al. 1997)

The magnetic properties of the four Galilean moons were extensively investigated during the Galileo mission from December 1995 to September 2003. For a review of the magnetic fields of the satellites of Jupiter and Saturn, see Jia et al. (2010). The Galileo magnetometer discovered the internal, dynamo-generated magnetic field of Ganymede (Kivelson et al. 1996, 1997). Data from the eight close flybys of Ganymede were presented and analysed in terms of an internal dynamo component and a component due to induction by Kivelson et al. (2002). Magnetic field observations during one of the close passes by Ganymede are shown in Fig. 46, together with a model fit that estimates the dipole moment of Ganymede to be 1.4×10^{13} T m^3.

After six close flybys of the volcanic moon Io, the Galileo magnetic field observations did not show any evidence of an internal magnetic field. The magnetic signatures, however, showed the high level of complexity and variability of the interaction of the Jovian magnetic field with Io's ionosphere that originates from the volcanic emissions (Khurana et al. 2002). Galileo also discovered the induced magnetic field in the interiors of Europa and Callisto (Khurana et al. 1998), both assumed to originate in subsurface oceans.

The induced magnetic field signatures observed by Galileo during the close flybys of Callisto and Europa (Khurana et al. 1998) and their interpretation has been extensively discussed (Schilling et al. 2004, see also for up-to-date assessments Jia et al. 2010; Saur et al. 2009). In particular, Europa has received close attention as there are indications that under its icy surface there may be a liquid ocean that provides the conductivity for its inductive signatures (Neubauer 1998). Europa is differentiated, with a metallic core, a silicate mantle and water-ice with the likelihood of liquid water covered by the ice sheets seen in images by the Galileo spacecraft. The magnetic field data acquired by Galileo during close flybys of Europa show that the induction signatures are clear when the moon is outside the Jovian current sheet, but that the magnetic signatures are more complex within the sheet due to the interaction of the induced field inside the moon with the current sheet.

The Galileo magnetic field investigation has been described by Kivelson et al. (1992). Two triaxial vector fluxgate magnetometer sensors are mounted on an extendable 11 m boom, one at the boom tip, the other at 6.87 m from the spacecraft spin axis. The Galileo spacecraft consisted of two sections, one section (with the in situ instruments including the magnetometer sensors) spinning at 3 rpm, the other despun for fixed direction pointing of the remote sensing instruments such as the camera system. One of the axes of the magnetometer sensors was aligned with the spacecraft spin axis, the other two were orthogonal in the spin plane. There were several sampling schemes foreseen, including a despun data stream with a resolution of $\sim$30 s/vector and a small data store capable of storing 200 averaged vectors for later transmission. Care was taken in the electronics and data sampling modes to include data filtering matching the Nyquist criterion.

As an in-flight calibration feature of the sensors, a thermally activated, bimetallic strip operated flipper mechanism was built into the sensors which allowed rotating the spin-axis oriented ring-core sensor into the spin plane, thus replacing the spin-axis sensor with one of the spin-plane sensors. The objective of such a flipper mechanism (first used on the lunar-orbiting Explorer 35 spacecraft, launched in 1967) is to determine the zero level of all three single-axis sensors. Offsets of the two sensors in the spin plane can be determined with great accuracy, as the signal from these two sensors are two sine waves in quadrature, synchronous with the spin (when the magnetic field is nearly steady over a spin period); the offset of the sine waves from a zero average is the value of the zero offset of the sensor. This, however, is not the case for the spin axis sensor for which the zero level determination is more complex (see, e.g. Hedgecock 1975). Additional in-flight calibration was foreseen using a MAG Cal Coil, a 0.5 m $\times$ 0.45 m square coil near the root of the boom which was to generate a calibration field of amplitude $\sim$4.5 nT at the outboard sensor with a sequence of on-off cycles at several frequencies. Synchronous detection of the calibration signal allowed the very precise determination of the orientation of the sensors (thus safeguarding against possible temperature-induced deformations in the boom) and their scale factors. The magnetometer analogue electronics was calibrated using internal calibration signals.

The measurement ranges on the Galileo magnetometer were ± 32 nT and ± 512 nT on each axis on the boom-tip magnetometer and ± 512 nT and $\pm 16,384$ nT on the inboard magnetometer. Some of the operations modes planned pre-launch (including planned high temporal resolution coverage) could not be carried out because of the non-deployment of the Galileo high-gain antenna after launch. However, a closely planned and highly targeted operations schedule for the magnetometer yielded extensive data during the mission, from both the magnetosphere and from the close flybys of the Galilean moons.

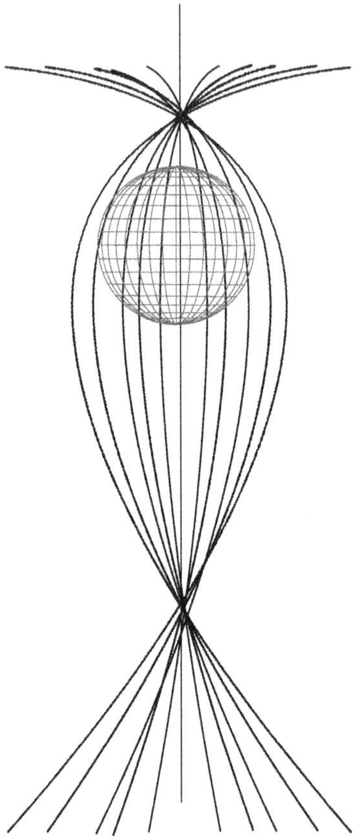

Fig. 47 Typical orbits of the Juno spacecraft close to perijove. A total of 30, 11-day period high inclination orbits will be spaced at 12° longitude intervals during the planned operational mission phase

The next mission to Jupiter will be NASA's Juno spacecraft (Bolton et al. 2006; Dodge et al. 2007; Matousek 2007). The spacecraft will be launched in 2011 and will, after a gravity assist flyby of Earth late in 2013, arrive at Jupiter in late 2016. The nominal orbit around Jupiter is very highly eccentric, with a polar inclination and has a period of 11 days. The apoapsis is at $\sim 39 R_J$, while periapsis distance is nominally $1.06 R_J$, and the orbit will vary between heights of 4200 and 5200 km above the planet at closest approach at about 30° north of the Jovian equator. The primary science operations will be concentrated in ± 3 hours around perijove. In all 32 such orbits are foreseen, with 30 of these orbits devoted to science data acquisition. The first 15 science orbits are phased with a separation of 24°, then, following an adjustment of 12°, the next 15 orbits, again at 24° phase separation, will complete the complete coverage of the planet at a longitudinal phasing of 12°. Typical orbits near perijove are sketched in Fig. 47.

The spacecraft is a spinner and has three solar panels. At the outboard edge of one of the solar panels, there is a special mounting platform for the magnetic field investigation. This investigation consists of two tri-axial fluxgate magnetometers, each co-mounted on an optical bench with two star cameras (Advanced Stellar Compasses) to determine with high accuracy the orientation of the magnetic field vector measurements (Connerney and Acuña 2008). The mounting platform and the locations of the two magnetometers is illustrated in Fig. 48. (The figure also shows the location of a planned Scalar helium Magnetometer that was originally foreseen for the mission but was descoped due to limitation of resources.)

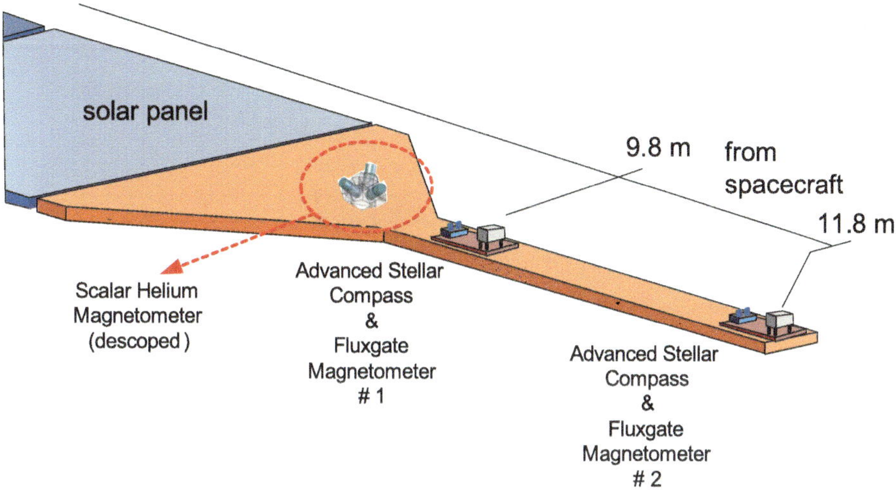

Fig. 48 The accommodation of the two triaxial Fluxgate Magnetometers on an extension to one of the solar panels on NASA's forthcoming Juno mission to Jupiter. The sensors are co-mounted on optical benches with the Advanced Stellar Compasses that provide very high accuracy pointing information. This is to minimise the attitude error in the magnetic field measurements. The originally planned Scalar Helium Magnetometer was descoped late in the development programme due to resources limitations

Given the very close passage of the spacecraft over the northern polar regions, the prime scientific objectives of the magnetic field investigation include detailed measurements of the jovian auroral system, field aligned currents and magnetosphere-ionosphere coupling.

The magnetometer is expected to measure field strengths of up to 12 Gauss (1.2 mT) near perijove, within a very rapidly changing magnetic field environment. The highest sampling frequency of the magnetometer is 64 Hz, used near the planet. The magnetometers have to be able to cover the variation field strength expected around the orbit, down to the level of a few nT, or at least six orders of magnitude.

4.2 Saturn

Unlike in the case of Jupiter, no kilometric radiation had been detected from Earth to conclude about the existence and size of its magnetic field (for an extensive discussion see, e.g. Kaiser et al. 1984). The discovery of modulation in the radio emissions observed by Voyager 1 before encountering Saturn (Kaiser et al. 1980) revealed features in the planet's magnetic field that needed closer investigation by the magnetometers on the flyby missions. Scaling laws based on planetary models comparing Jupiter and Saturn had predicted a larger internal dipole moment than was found by in situ measurements. The differences with the predictions and their consequences for the structure and composition of Saturn's interior were first discussed by Stevenson (1980) and, more recently, by Christensen and Wicht (2008). The first measurements of Saturn's magnetic field were taken on 1 September 1979 when Pioneer 11 flew by the planet, in a close to equatorial orbit, making a closest approach at a Kronocentric distance of $1.35 R_S$, or about 21,000 km above the planet's surface.

The measurements made by the Pioneer 11 VHM instrument (Smith et al. 1980a, 1980b), illustrated in Fig. 49, were used to construct a model of Saturn's magnetic field. Thanks to the very close flyby distance, the data provided sufficient resolution to the model so

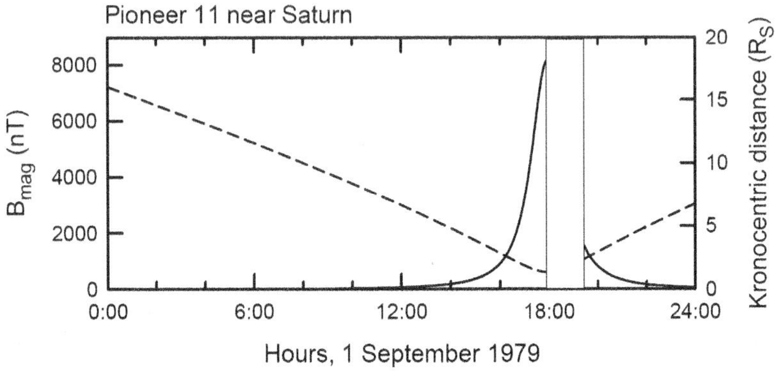

Fig. 49 The magnitude of the magnetic field measured by the VHM instrument on the Pioneer 11 spacecraft on the day of closest approach to Saturn (1 September 1979). The Kronospheric distance is shown in *dashed lines*

that differences between the measurements and the model field were found to be consistently less than 0.5%. In particular, the Pioneer 11 measurements showed the remarkable and unexpected axisymmetry of the internal magnetic field. A similar conclusion was drawn from the High Field Magnetometer measurements (Acuña and Ness 1980a; Acuña et al. 1980b). However, when comparing the models derived from Pioneer 11 to the models based on the subsequent Voyager flybys (see below), it was found that due to the very small angle between the spacecraft-Sun direction and the spacecraft spin axis during the encounter, the modulation signal provided by the spacecraft sun-sensor from which the spin phase angle of the spacecraft was derived was probably in error by about 1.4° (Connerney et al. 1984). As the spin phase angle is used for the demodulation of the magnetometer signal, this error was found to be a source of a small discrepancy between the models derived from Pioneer 11 and Voyager 1 and 2 (Acuña et al. 1983; Connerney et al. 1984; Davis and Smith 1986, 1990).

The axisymmetry of the magnetic field found by Pioneer 11 and subsequently by Voyagers 1 and 2 (Connerney et al. 1982b) was difficult to reconcile not only with the dynamo theorem that excluded the existence of such dynamos Cowling 1933, but also with the modulation detected in the Saturn kilometric radiation (SKR, e.g. Kaiser et al. 1984) that showed a strong periodicity at what was interpreted as the internal rotation rate of the planet. Following the analysis of Ulysses observations of the SKR, Galopeau and Lecacheux (2000) found that there was a variability in the modulation of the radio signal, contrary to what had been found by Voyager. The results of the Cassini orbiter concerning the rotation rate of Saturn, using the magnetic field data (Giampieri et al. 2006) or the radio data or both (Gurnett et al. 2007) have not attributed a firm period but have shown instead a variability that remains fully to be explained (Kurth et al. 2008; Burton et al. 2009). However, the modulation of the Saturn kilometric radiation is sufficiently strong and has been used to establish a kronographic longitude system that is now in general use (Kurth et al. 2008).

Voyagers 1 and 2 reached their closest approach to Saturn on 12 November 1980 at $3.07 R_S$ and on 26 August 1981, at a distance of $2.69 R_S$, respectively. Contrary to Pioneer 11 which had a closer approach than the Voyager spacecraft but remained close to Saturn's equatorial plane, the Voyager flyby trajectories covered a wider range of kronospheric latitudes and their longitude coverage was largely complementary around the planet (Fig. 50).

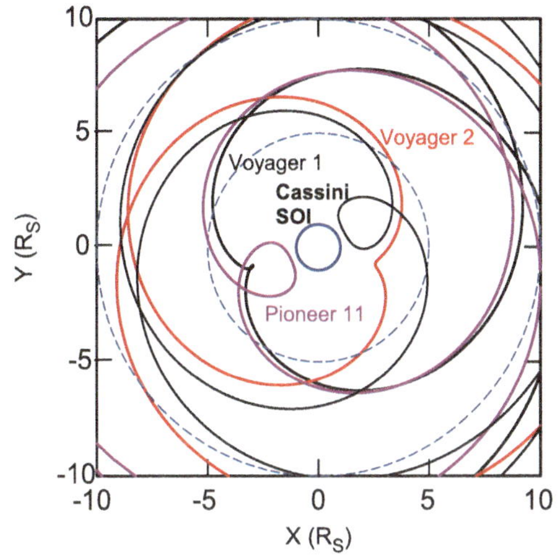

Fig. 50 Flyby orbits of Pioneer 11, Voyager 1 and 2 spacecraft close to Saturn in the kronographic equatorial plane and the Saturn Orbit Injection trajectory of Cassini. (Saturn is the blue circle at the centre of the coordinate system, the dashed blue circles show kronocentric distances of 5 and $10R_S$, respectively.) The closest approaches were, respectively, 1.35 (P11), 3.07 (V1), 2.69 (V2) and $1.33R_S$ (Cassini). While Pioneer 11 covered a limited range of latitudes close to the equatorial plane, the two Voyager spacecraft covered complementary kronographic longitude ranges and latitude ranges of more than $\pm 30°$. Cassini's close approaches during its orbiting phase have also covered a broad range of latitudes (see Fig. 51)

The magnetic field observations, made with the same instrumentation that had already visited Jupiter, were reported by Ness et al. (1981, 1982).

The Voyager measurements confirmed the near-axial orientation of the dominant dipole moment of Saturn's internal magnetic field as well as the magnitude of the dipole moment as 4.3×10^{18} T m^3 (or, as quoted by Ness et al. 1982, ~0.21 Gauss R_S^3). For detailed confirmation and analysis of the Voyager 2 data, an initial error in the spacecraft roll orientation needed correcting (Ness et al. 1982; Connerney et al. 1982a). Such a correction, even if small, affects some of the detailed interpretation of the data; this is particularly relevant near Saturn, because of the complex deciphering needed to assess the respective contributions of the internal field and the spatially and temporally varying external current systems (Connerney et al. 1983, and for the latest assessment of the problem, see Arridge et al. 2008, Blanc et al., 2009). The axisymmetry of Saturn's magnetic field remains to be fully understood. Models of its interior structure that explain the axisymmetry have been proposed (Stevenson 1982); in such models, the dynamo generation region is masked by an overlaying stratified layer still deep in the planet's interior.

Following the analyses carried out first on the data from Pioneer 11 and then of the two Voyagers, a comprehensive review of the combined data set was carried out by Davis and Smith (1990). This review confirmed the previous findings concerning the axisymmetry of the magnetic field. It also considered the higher order, axial quadrupole and octupole terms and examined the sensitivity of these higher order terms, and in particular any non-axisymmetric terms and found that the flyby data were not sufficient to constrain them uniquely. As in all cases, orbiter spacecraft are, in principle required for a comprehensive determination of planetary magnetic fields; the arrival of the Cassini orbiter around Saturn on 1 July 2004 (Dougherty et al. 2005) was the start of a new phase in the exploration of the magnetic fields and the environment of Saturn. The still-unresolved question of Saturn's

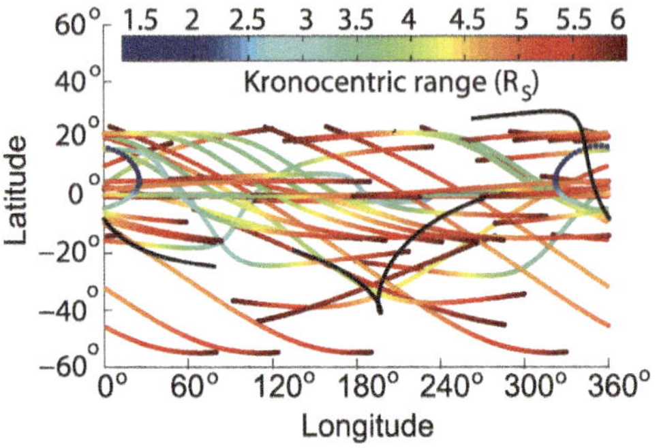

Fig. 51 Orbital coverage of the Cassini spacecraft in kronographic coordinates from Saturn Orbit Insertion on 1 July 2004 up to 12 June 2007. The kronocentric radial distance is colour coded on the tracks and covers the range up to $6R_S$. Data acquired during these orbits have been used to determine Saturn's internal magnetic field by Burton et al. (2009). *The heavy black lines* show the coverage of the Voyager 1 and 2 spacecraft also up to $6R_S$ (after Burton et al. 2009)

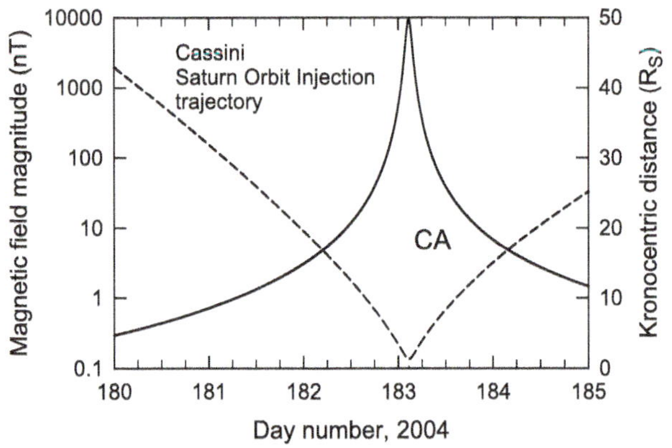

Fig. 52 The magnetic field magnitude measured by the Cassini magnetometer prior and following the closest approach passage to Saturn during orbit injection. The Kronocentric distance of the spacecraft is shown as a *dashed line*. In contrast to the close jovian passages when the current sheet noticeably affected the magnetic field on similar timescales (Fig. 43), note the absence of a significant contribution from Saturn's external environment to the field magnitude

rotation period and the processes that it generates to yield complex periodic signals in radio waves and the magnetic field has already been mentioned. The internal field has recently been the subject of a comprehensive reappraisal by Burton et al. (2009), using those parts of the Cassini orbits which were within $6R_S$ of the planet over a close to three year interval from Saturn Orbit Injection; the projections of these orbit segments in cronographic latitude and longitude are shown in Fig. 51, together with the two orbit segments from the Voyager flybys that were also within $6R_S$.

The Cassini magnetometer investigation has been described in detail by Kellock et al. (1996), Smith et al. (2001) and Dougherty et al. (2004). It consists of two sensors mounted on an 11 m boom that was extended during the cruise phase of the mission, prior to the spacecraft's swingby of the Earth on 18 August 1999. The results of the Earth's flyby, primarily intended to boost the spacecraft's speed on its way to Jupiter and Saturn, was also useful in providing an opportunity to calibrate the magnetometers; the results of the flyby have been described by Smith et al. (2001) and Southwood et al. (2001). The sensors are a Vector/Scalar Helium Magnetometer (V/SHM) mounted at the tip of the 11-m long magnetometer boom and a Fluxgate Magnetometer at 5.5 m inboard from the boom tip (see above, Fig. 11 in Sect. 2.3).

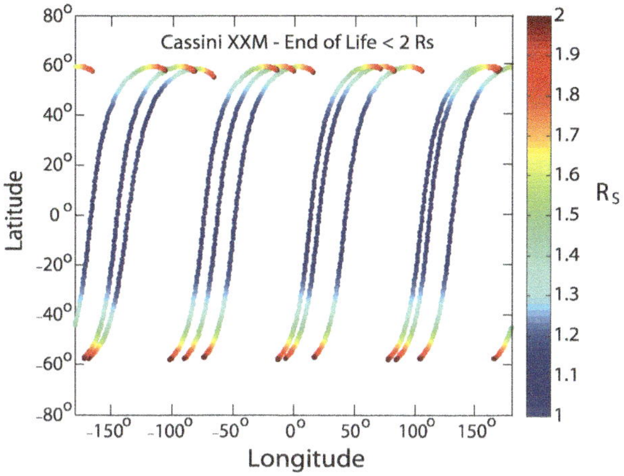

Fig. 53 Planned end-of-life orbits for the Cassini mission at the conclusion of its Extended-Extended Mission (XXM) in 2017. The wide latitude coverage enabled by the high latitude inclination orbit with a very low periapsis near the equatorial plane will lead to a significant addition to the magnetic field measurements from which the internal field of Saturn can be better characterised figure courtesy of M. Burton)

The Fluxgate magnetometer (FGM) is of classical construction using three orthogonally arranged ringcore sensors mounted on a ceramic holder chosen for its low thermal expansion coefficient and machined for dimensional stability. The noise performance of the FGM sensors is better than 5 $pT/Hz^{1/2}$ at 1 Hz. The FGM measures the magnetic field vector at a rate of 32 vectors/s. The operating ranges in normal use are ± 40 nT, ± 400 nT and $\pm 10{,}000$ nT per axis with corresponding resolutions of ~ 5 pT, 49 pT and 1.2 nT. The instrument also has a range of $\pm 44{,}000$ nT that was used for ground testing. It is noted that the maximum value of the magnetic field amplitude at closest approach (at $1.33 R^{S}$ planetocentric distance) during Saturn orbit injection was 9,420 nT (see Fig. 52), thus maximising the relative resolution through the operating range.

The V/SHM is an optically pumped helium magnetometer, built using the inheritance of previous missions, in particular Ulysses, described in Sect. 2.3. It normally operated in vector mode, measuring the ambient field at the rate of 2 vectors/s through the cruise phases of the mission.

A particularly interesting possibility for extending the magnetic field measurements is being planned at the conclusion of the so-called Extended-Extended Mission (XXM) of Cassini. This phase of the mission will cover the years 2011 to 2017, with, at its conclusion, 28 orbits reaching inclinations $\sim 63°$, with a height above the 1 bar pressure level ("the cloud tops") of only 2000 to 5000 km. These orbits are sketched in Fig. 53. This set of orbits would allow a significant increase in the accuracy of the magnetic field measurements that could be used to refine the internal field model that is offset to the northern hemisphere and also the rotation rate of the planet.

4.3 Uranus and Neptune

The two outermost planets in the solar system, the icy giants, have only been visited once, by Voyager 2 when the spacecraft trajectory took advantage of the special alignment of the planets that only occurs every 187 years. The Uranus flyby took place on 24 January 1986, at a closest approach of $4.19 R_U$ where $R_U = 25{,}559$ km. Voyager 2 flew by Neptune on 25 August 1989 at a closest approach distance of $1.18 R_N$ where $R_N = 24{,}760$ km.

The magnetic field measurements around the closet approach and their interpretations were described by Ness et al. (1986) for the Uranus flyby and by Ness et al. (1989) for the Neptune flyby. The results are recalled and summarised by Ness (2010). The Voyager

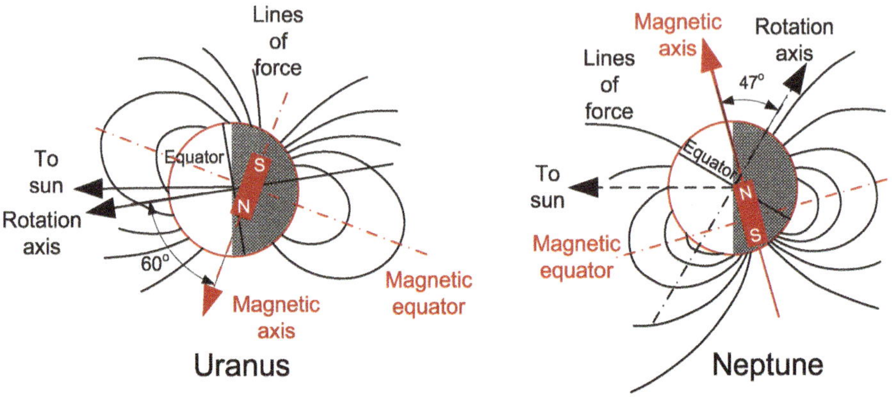

Fig. 54 Equivalent dipoles representing the internal magnetic fields of Uranus and Neptune (after Ness et al. 1986, 1989, see also Ness 2010)

magnetometers have been described above in Sect. 4.1. The following brief descriptions follow the two discovery papers (Ness et al. 1986, 1989).

In the case of both these planets, significant magnetic fields of internal origin were observed. The maximum field measured at Uranus was 413 nT, a few minutes away from closest approach. Uranus' rotation axis is fact close to the planet's orbital plane, so that the spacecraft's flyby trajectory, not far off Uranus' orbital plane, covered a wide range of latitudes. Both the magnetosphere of Uranus and its internal magnetic field have orientations which are far from that of the other planets (except Neptune). The analysis of the flyby data was best fitted (Connerney et al. 1987) with an offset, tilted dipole as illustrated in Fig. 54. A very special feature of Uranus' internal magnetic field, when represented by the equivalent offset, tilted dipole is the large angle (~60°) between the rotation axis of the planet and the dipole axis.

During the much closer approach (when compared to that near Uranus) during the flyby of Neptune the maximum magnetic field observed by the Voyager magnetometer was 9,700 nT. Similarly to Uranus, the observations could be fitted with an offset, tilted dipole with an axis at ~47° from the rotation axis, shown schematically in Fig. 54.

In addition to the very different orientation and location of the internal dipoles of Uranus and Neptune, when compared to the other two giant planets, Jupiter and Saturn, large quadrupolar terms were found as well (Connerney et al. 1991; Connerney 1993). Clearly, the very different internal structure and composition in the case of the icy giants leads to dynamo mechanisms and geometries which are unique to these two planets. Detailed modelling and numerical simulations have led to a better understanding of the dynamo processes (Stanley and Bloxham 2004, 2006; Stanley and Glatzmaier 2010) as well as the internal structure and material characteristics of these planets. Given the challenges to reach both the icy giants, it is unlikely that follow-up missions will be envisaged for a considerable time, so that the Voyager 2 observations will remain the benchmark against which continued understanding of these planets will be tested.

5 Conclusion

In the past five decades, about five dozen space missions have made useful magnetic field measurements in the vicinity of the eight planets of the solar system as well as close to some of the smaller bodies. As reviewed by Stevenson (2010), there is a good basic understanding and detailed knowledge of the magnetic properties of all the planets, despite their considerable diversity. The degree of similarity is small among the usual grouping of the planets (terrestrial planets, gas giants, ice giants).

It is recognised that magnetic field measurements contribute information about the interiors of planets that is not accessible by other measurements. Furthermore, the magnetic state of planets is also an important factor in understanding the evolution of planets from the time of their formation. As this importance is recognised by planetary scientists, there is a requirement for future missions that will provide even more detailed magnetic field measurements to constrain planetary interiors and their evolution. While it is unlikely that the planets will be as well characterised as the Earth is now, but enough uncertain parameters remain concerning the operation of planetary dynamos to make more precise observational constraints necessary.

Planetary spacecraft that measure magnetic fields precisely need to have low-altitude and high-inclination orbits to ensure appropriate coverage. In several cases, this is difficult: at Mercury, the thermal constraints are very demanding and expensive, at Jupiter and Saturn there are challenges in terms of both orbital dynamics and tolerable radiation dose. Simply reaching the ice giants within the next generation will be unlikely, due to the combination of relatively small interest and the high cost. A possible future motivation for better understanding the planets in our solar system (which involves better understanding their magnetic properties) will be the need to make comparisons with extrasolar planetary systems as these will be discovered and characterised in increasing detail.

In terms of instrumentation, the magnetometers available today have the required performance to make the high resolution measurements that will be required. This applies equally to the fluxgate magnetometers and to the quantum (Helium 4) magnetometers. There will be an effort to further optimise resources required by these instruments, but it is likely that appropriate performance will be achieved by resources (mass, power) that are similar to contemporary instruments. Another resource, telemetry bandwidth, needs to be increased (higher vector measurement rates). There has been some erosion of telemetry rates: it is symptomatic that the vector rate allocated to the magnetometer on BepiColombo is less than that used on Mariner 10 almost 40 years ago. The need for higher telemetry rates arises partly from the requirement for denser spatial coverage and partly for the identification of the time variable component of the magnetic field at the location of the measurements due to external sources. Understanding the nature of the external sources that in all cases influence the in situ measurements is a necessary requirement for identifying the internal sources of planetary magnetic fields. These external terms are always significant and, because of their origin in dynamic plasma interactions, are always variable.

The greatest challenge to useful magnetic field measurements comes from the spacecraft-generated background at the location of the magnetometer sensor. Many ground-breaking early missions had a long boom for mounting the magnetometer away from the spacecraft. This was combined with a magnetic cleanliness programme that ensured that both the static and dynamic components of the disturbing background were minimised. Due to the pressure on costs, booms and the magnetic cleanliness programme are increasingly considered to be luxuries on planetary missions and often inadequate alternatives are pressed on experimenters who are then faced with an almost impossible challenge to clean the measurements from the spacecraft interference. There have been notable successes: the Giotto mission to comet Halley, the Mars Global Surveyor that identified the source of Mars' magnetism and Venus Express that is able to provide detailed characterisation of the Venus magnetic environment are examples where the experimenters were able to clean the raw data sufficiently for major discoveries to be made. However, the need for future measurements is for accuracies that can only achieved by space missions on which the magnetometer can measure the ambient magnetic field without any significant background from the spacecraft.

Acknowledgements The author thanks his many colleagues who over the years have helped him understand magnetometers and their measurements in planetary environments. Particular thanks are due to Dr Ed Smith (JPL) who has been a great source of knowledge and wisdom on magnetic field measurements and missions that carried magnetometers. The author is also privileged to have known and collaborated with the late Dr Mario Acuña who has had the opportunity to visit with his magnetometers all the planets as well as comets and asteroids. Thanks are also due to the International Space Science Institute, its Founding Director, Professor Johannes Geiss and its current Executive Director Professor Roger Bonnet for creating a highly successful environment in which scientists can make significant progress in understanding a wide range of space science topics.

References

M.H. Acuña, IEEE Trans. Magn. **MAG-10**, 519–523 (1974)
M.H. Acuña, John Hopkins APL Tech. Dig. **1**, 210–213 (1980)
M.H. Acuña, Rev. Sci. Instrum. **73**, 3717–3735 (2002)
M.H. Acuña, C.J. Pellerin, IEEE Trans. Geosci. Electron. **GE-7**, 252–260 (1969)
M.H. Acuña, N.F. Ness, Space Sci. Instrum. **1**, 177–188 (1975a)
M.H. Acuña, N.F. Ness, Nature **253**, 327 (1975b)
M.H. Acuña, N.F. Ness, J. Geophys. Res. **81**, 2917–2922 (1976)
M.H. Acuña, N.F. Ness, Science **207**, 444 (1980a)
M.H. Acuña, C.S. Scearce, J. Seek, J. Scheifele, NASA Technical Memorandum 79656, NASA/GSFC, 1978
M.H. Acuña, N.F. Ness, J.E.P. Connerney, J. Geophys. Res. **85**, 5675–5678 (1980b)
M.H. Acuña, J.E.P. Connerney, N.F. Ness, J. Geophys. Res. **88**, 8771–8778 (1983)
M.H. Acuña, J.E.P. Connerney, P. Wasilewski, R.P. Lin, K.A. Anderson, C.W. Carlson, J. McFadden, D.W. Curtis, H. Réme, A. Cros, J.L. Médale, J.-A. Sauvaud, C. d'Uston, S. Bauer, P. Cloutier, M. Mayhew, N.F. Ness, J. Geophys. Res. **97**, 7799–7814 (1992)
M.H. Acuña, J.L. Scheifele, P. Stella, C. Kloss, B. Smith, G. Heinshohn, K. Sharmit, in *Photovoltaic Specialists Conference, Conference Record of the Twenty Fifth IEEE* (1996), pp. 325–328
M.H. Acuña, C.T. Russell, L.J. Zanetti, B.J. Anderson, J. Geophys. Res. **102**, 23751–23760 (1997)
M.H. Acuña, J.E.P. Connerney, P. Wasilewski, R.P. Lin, K.A. Anderson, C.W. Carlson, J. McFadden, D.W. Curtis, D. Mitchell, H. Reme, C. Mazelle, J.A. Sauvaud, C. d'Uston, A. Cros, J.L. Medale, S.J. Bauer, P. Cloutier, M. Mayhew, D. Winterhalter, N.F. Ness, Science **279**, 1676–1680 (1998)
M.H. Acuña, J.E.P. Connerney, N.F. Ness, R.P. Lin, D. Mitchell, C.W. Carlson, J. McFadden, K.A. Anderson, H. Rème, C. Mazelle, D. Vignes, P. Wasilewski, P. Cloutier, Science **284**, 790–793 (1999)
M.H. Acuña, J.E.P. Connerney, P. Wasilewski, R.P. Lin, D. Mitchell, K.A. Anderson, C.W. Carlson, J. McFadden, H. Réme, C. Mazelle, D. Vignes, S.J. Bauer, P. Cloutier, N.F. Ness, J. Geophys. Res. **106**, 23,403–23,417 (2001)
M.H. Acuña, B.J. Anderson, C.T. Russell, P. Wasilewski, G. Kletetshka, L. Zanetti, N. Omidi, Icarus **155**, 220–228 (2002)
I.I. Alexeev, E.S. Belenkaya, S.Yu. Bobrovnikov, J.A. Slavin, M. Sarantos, J. Geophys. Res. **113**, CiteID A12210 (2008). doi:10.1029/2008JA013368
B.J. Anderson, L.J. Zanetti, D.A. Lohr, J.R. Hayes, M.H. Acuña, C.T. Russell, T. Mulligan, IEEE Trans. Geosci. Remote Sens. **39**, 907–917 (2001)
B.J. Anderson, M.H. Acuña, D.A. Lohr, J. Scheifele, A. Raval, H. Korth, J.A. Slavin, Space Sci. Rev. **131**, 417–450 (2007)
B.J. Anderson, M.H. Acuña, Adv. Space Res. **33**, 1989–1995 (2004)
B.J. Anderson, M.H. Acuña, H. Korth, M.E. Purucker, C.L. Johnson, J.A. Slavin, S.C. Solomon, R.L. McNutt Jr., Science **321**, 82–85 (2008)
B.J. Anderson, M.H. Acuña, H. Korth, J.A. Slavin, H. Uno, C.L. Johnson, M.E. Purucker, S.C. Solomon, J.M. Raines, T.H. Zurbuchen, G. Gloeckler, R.L. McNutt Jr., Space Sci. Rev. (2009). doi:10.1007/s11214-009-9544-3, this issue
K.A. Anderson, L.M. Chase, R.P. Lin, J.E. McCoy, R.E. McGuire, J. Geophys. Res. **77**, 4611–4626 (1972)
K.A. Anderson, H.C. Howe, R.P. Lin, R.E. McGuire, L.M. Chase, J.E. McCoy, Lunar Planet. Sci. Conf. **5**, 18 (1974)

C.S. Arridge, C.T. Russell, K.K. Khurana, N. Achilleos, S.W.H. Cowley, M.K. Dougherty, D.J. Southwood, E.J. Bunce, J. Geophys. Res. **113**, 4214 (2008)
H.-U. Auster, K.-H. Fornacon, J. Rustenbach, Th. Roatsch, R. Schoredter, K. Schwingenschuh, V. Auster, Geophys. Res. Lett. **17**, 881–884 (1990)
H.-U. Auster, A. Lichopoj, J. Rustenbach, H. Bitterlich, K.-H. Fornacon, O. Hillenmaier, R. Krause, S. Schenk, V. Auster, Meas. Sci. Technol. **6**, 477–481 (1995)
H.-U. Auster, I. Apathy, G. Berghofer, A. Remizov, R. Roll, K.H. Fornacon, K.H. Glassmeier, G. Haerendel, I. Hejja, E. Kührt, W. Magnes, D. Moehlmann, U. Motschmann, I. Richter, H. Rosenbauer, C.T. Russell, J. Rustenbach, K. Sauer, K. Schwingenschuh, I. Szemerey, R. Waesch, Space Sci. Rev. **128**, 221–240 (2007)
H.-U. Auster, I. Richter, K.-H. Glassmeier, G. Berghofer, C.M. Carr, U. Motschmann, Magnetic field investigations during Rosetta's 2867 Steins flyby. Planet. Space Sci. (2010, in press)
A. Balogh, T.J. Beek, R.J. Forsyth, P.C. Hedgecock, R.J. Marquedant, E.J. Smith, D.J. Southwood, B.T. Tsurutani, Astron. Astrophys. Suppl. Ser. **92**, 221–236 (1992b)
A. Balogh, M.K. Dougherty, R.J. Forsyth, D.J. Southwood, E.J. Smith, B.T. Tsurutani, N. Murphy, M.E. Burton, Science **257**, 1515–1518 (1992a)
A. Balogh, C.M. Carr, M.H. Acuña, M.W. Dunlop, T.J. Beek, P. Brown, K.-H. Fornacon, E. Georgescu, K.-H. Glassmeier, J. Harris, G. Musmann, T. Oddy, K. Schwingenschuh, Ann. Geophys. **19**, 1207–1217 (2001)
M.A. Barucci, M. Fuchignoni, A. Rossi, Space Sci. Rev. **128**, 67–78 (2007)
W. Baumjohann, M. Blanc, A. Fedorov, K.-H. Glassmeier, Space Sci. Rev. (2010a). doi:10.1007/s11214-010-9629-z
W. Baumjohann, A. Matsuoka, W. Magnes, K.-H. Glassmeier, R. Nakamura, H. Biernat, M. Delva, K. Schwingenschuh, T. Zhang, H.U. Auster, K.-H. Fornacon, U. Motschmann, I. Richter, A. Balogh, C. Carr et al., Planet. Space Sci. **58**, 279–286 (2010b)
K.W. Behannon, M.H. Acuña, L.F. Burlaga, R.P. Lepping, N.F. Ness, F.M. Neugebauer, Space Sci. Rev. **21**, 235–257 (1977)
J.W. Belcher, J. Geophys. Res. **78**, 6480–6490 (1973)
J. Benkhoff, J. van Casteren, H. Hayakawa, M. Fujimoto, H. Laakso, M. Novara, P. Ferri, Planet. Space Sci. **58**, 2–20 (2010)
G.L. Berge, S. Gulkis, in *Jupiter*, ed. by T. Gehrels (University of Arizona Press, Tucson, 1976), pp. 621–692
S. Bolton (The Juno Science Team), in *1st European Planetary Science Conference*, Berlin (2006)
P. Brauer, J.M.G. Merayo, O.V. Nielsen, F. Primdahl, J.R. Petersen, Sens. Actuators A **59**, 70–74 (1997)
W. Braunbek, Z. Phys. **88**, 399–402 (1934)
H.S. Bridge, A.J. Lazarus, C.W. Snyder, E.J. Smith, L. Davis Jr., P.J. Coleman Jr., D.E. Jones Mariner V, Science **158**, 1669–1673 (1967)
B.F. Burke, K.L. Franklin, J. Geophys. Res. **60**, 213–217 (1955)
M.E. Burton, M.K. Dougherty, C.T. Russell, Planet. Space Sci. **57**, 1706–1713 (2009)
F.H. Busse, Phys. Earth Planet. Inter. **12**, 350–358 (1976)
J.C. Cain, R.E. Sweeney, J. Geophys. Res. **75**, 4360–4362 (1970)
J.C. Cain, Rev. Geophys. Space Phys. **9**, 259–272 (1971)
C.M. Carr, P. Brown, T.L. Zhang et al., Ann. Geophys. **23**, 2713–2732 (2005)
J.J. Chaillout, N. Kernevez, J.-M. Leger, Optical pumping, resonance magnetometer using a light beam with controlled polarization, US Patent, No. 5,272,436 (1993)
U.R. Christensen, J. Wicht, Icarus **196**, 16–34 (2008)
F.D. Colegrove, P.A. Franken, Phys. Rev. **119**, 680–690 (1960)
P.J. Coleman Jr., G. Schubert, C.T. Russell, L.R. Sharp, in *Apollo 15 Preliminary Science Report*, NASA SP-289, pp. (22)1–(22)9 (1972a)
P.J. Coleman Jr., B.R. Lichtenstein, G. Schubert, C.T. Russell, L.R. Sharp, in *Apollo 16 Preliminary Science Report*, NASA SP-315, pp. (23)1–(23)13 (1972b)
F.R. Colomb, C. Alonso, C. Hofmann, I. Nollmann, Adv. Space Res. **34**, 2194–2199 (2004)
J.E.P. Connerney, J. Geophys. Res. **86**, 7679–7693 (1981)
J.E.P. Connerney, in *Planetary Radio Emissions III*, ed. by H.O. Rucker, S.J. Bauer, M.L. Kaiser (Austrian Acad. of Sci. Press, Vienna, 1992), pp. 13–33
J.E.P. Connerney, J. Geophys. Res. **98**, 18,659–18,679 (1993)
J.E.P. Connerney, M.H. Acuña, The Juno magnetic field investigation (MAG): exploration of the polar magnetosphere, American Geophysical Union, Fall Meeting, abstract #SM41B-1679 (2008)
J.E.P. Connerney, M.H. Acuña, N.F. Ness, J. Geophys. Res. **86**, 8370–8384 (1981a)
J.E.P. Connerney, M.H. Acuña, N.F. Ness, Nature **292**, 724–726 (1981b)
J.E.P. Connerney, M.H. Acuña, N.F. Ness, J. Geophys. Res. **87**, 3623–3627 (1982a)
J.E.P. Connerney, N.F. Ness, M.H. Acuña, Nature **298**, 44 (1982b)

J.E.P. Connerney, M.H. Acuña, N.F. Ness, J. Geophys. Res. **88**, 8779–8789 (1983)
J.E.P. Connerney, M.H. Acuña, N.F. Ness, J. Geophys. Res. **89**, 7541–7544 (1984)
J.E.P. Connerney, M.H. Acuña, N.F. Ness, J. Geophys. Res. **92**, 15329–15336 (1987)
J.E.P. Connerney, M.H. Acuña, N.F. Ness, J. Geophys. Res. **96**, 19023–19042 (1991)
J.E.P. Connerney, N.F. Ness, in *Mercury*, ed. by F. Vilas, C.R. Chapman, M.S. Matthews (University of Arizona Press, Tucson, 1988), pp. 494–513
J.E.P. Connerney, M.H. Acuña, P. Wasilewski, N.F. Ness, H. Rème, C. Mazelle, D. Vignes, R.P. Lin, D. Mitchell, P. Cloutier, Science **284**, 794–798 (1999)
B.V. Connor, IEEE Trans. Magn. **MAG-4**, 391–397 (1968)
T.G. Cowling, Mon. Not. R. Astron. Soc. **94**, 39–48 (1933)
S.S. Dallas, IEEE Aerospace Conf. Proc. **4**, 173–189 (1997)
L. Davis Jr., E.J. Smith, Trans. AGU **49**, 257 (1968)
L. Davis Jr., E.J. Smith, D.E. Jones, J. Geophys. Res. **78**, 4803–4808 (1973)
L. Davis Jr., E.J. Smith, J. Geophys. Res. **91**, 1373–1380 (1986)
L. Davis Jr., E.J. Smith, J. Geophys. Res. **95**, 15257–15261 (1990)
M. Delva, H. Feldhofer, K. Schwingenschuh, K. Mehlem, Multiple magnetic sensor technique for field measurements in space, EGS–AGU–EUG Joint Assembly, *Abstracts from the meeting held in Nice*, France, 6–11 April 2003, abstract #5053 (2003)
R. Dodge, M.A. Boyles, C.E. Rasbach, Key and driving requirements for the Juno Payload of instruments. Paper AIAA 2007-6111, *AIAA Space Conference & Exposition*, Long Beach, California, September 18–20 (2007)
Sh.Sh. Dolginov, E.G. Eroshenko, L.N. Zhuzgov, N.V. Pushkov, Dokl. Akad. Nauk SSSR **170**, 574–577 (1966)
Sh.Sh. Dolginov, Ye.G. Yeroshenko, L.N. Zhuzgov, J. Geophys. Res. **78**, 4779–4886 (1973)
Sh.Sh. Dolginov, Ye.G. Yeroshenko, L.N. Zhuzgov, J. Geophys. Res. **81**, 3353–3362 (1976)
Sh.Sh. Dolginov, L.N. Zhuzgov, V.A. Sharova, V.B. Buzin, Kosm. Issled. **16**, 827–863 (1978)
Sh.Sh. Dolginov, Adv. Space Res. **12**(8), (8)187–(8)211 (1992)
T.M. Donahue, Science **205**, 41–44 (1979)
M.K. Dougherty, A. Balogh, D.J. Southwood, E.J. Smith, J. Geophys. Res. **101**, 24929–24942 (1996)
M.K. Dougherty, S. Kellock, D.J. Southwood, A. Balogh, E.J. Smith, B.T. Tsurutani, B. Gerlach, K.-H. Glassmeier, F. Gliem, C.T. Russell, G. Erdos, F.M. Neubauer, S.W.H. Cowley, Space Sci. Rev. **114**, 331–383 (2004)
M.K. Dougherty, N. Achilleos, N. Andre, C.S. Arridge, A. Balogh, C. Bertucci, M.E. Burton, S.W.H. Cowley, G. Erdos, G. Giampieri, K.-H. Glassmeier, K.K. Khurana, J. Leisner, F.M. Neubauer, C.T. Russell, E.J. Smith, D.J. Southwood, B.T. Tsurutani, Science **307**, 1266 (2005)
M. Dryer, G.R. Heckman, Solar Phys. **2**, 112–124 (1967)
D. Duret, J. Bonzom, M. Brochier, M. Francès, J.M. Léger, R. Odru, C. Salvi, T. Thomas, A. Perret, Overhauser magnetometer for the Danish Oersted satellite. IEEE Trans. Magn. 3197–3199 (1995)
D. Duret, J.M. Léger, M. Francès, J. Bonzom, F. Alcouffe, A. Perret, J.C. Llorens, C. Baby, Performances of the OVH magnetometer for the Danish Oersted satellite. IEEE Trans. Magn. 4935–4937 (1996)
P. Dyal, C.W. Parkin, C.P. Sonett, Science **169**, 762–764 (1970a)
P. Dyal, C.W. Parkin, C.P. Sonett, IEEE Trans. Geosci. Electron. **GE-8**, 203–215 (1970b)
P. Dyal, C.W. Parkin, C.P. Sonett, in *Apollo 15 Preliminary Science Report*, NASA SP-289, pp. (9)1–(9)16 (1972a)
P. Dyal, C.W. Parkin, P. Cassen, in *Proc. Third Lunar Sci. Conf.*, 3 (Suppl. 3, Geochim. Cosmochem. Acta), pp. 2287–2307 (1972b)
P. Dyal, D.I. Gordon, IEEE Trans. Magn. **Mag-9**, 226–231 (1973)
P. Dyal, C.W. Parkin, W.D. Daily, Rev. Geophys. **12**(4), 568–591 (1974)
W.H. Farthing, W.C. Folz, Rev. Sci. Instrum. **38**, 1023–1030 (1967)
W.H. Farthing, John Hopkins APL Tech. Dig. **1**, 205–209 (1980)
A.M.A. Frandsen, B.V. Connor, J. van Amersfoort, E.J. Smith, IEEE Trans. Geosci. Electron. **GE-16**, 195–198 (1978)
E. Friis-Christensen, H. Lühr, G. Hulot, Earth Planets Space **58**, 351–358 (2006)
E. Friis-Christensen, H. Lühr, D. Knudsen, R. Haagmans, Adv. Space Res. **41**, 210–216 (2008)
M. Fuller, Rev. Geophys. **12**, 23–70 (1974)
M. Fuller, Phys. Chem. Earth **23**, 725–735 (1998)
P.H.M. Galopeau, A. Lecacheux, J. Geophys. Res. **105**, 13089 (2000)
E. Georgescu, H. Vaith, K.-H. Fornaçon, U. Auster, A. Balogh, C.M. Carr, M. Chutter, M.W. Dunlop, M. Foerster, K.-H. Glassmeier, J. Gloag, G. Paschmann, J. Quinn, R. Torbert, in *Cluster and Double Star Symposium*, ed. by K. Fletcher. ESA SP-598, p. 63.1 (2006)

E. Georgescu, H.U. Auster, T. Takada, J. Gloag, H. Eichelberger, K.-H. Fornaçon, P. Brown, C.M. Carr, T.L. Zhang, Adv. Space Res. **41**, 1579–1584 (2008)
J.M. Gloag, E.A. Lucek, L.-N. Alconcel, A. Balogh, P. Brown, C.M. Carr, C.N. Dunford, T. Oddy, J. Soucek, in *The Cluster Active Archive, Studying the Earth's Space Plasma Environment*, ed. by H. Laakso, M.G.T.T. Taylor, C.P. Escoubet. Astrophysics and Space Science Proceedings (Springer, Berlin, 2010), pp.109–128
G. Giampieri, M.K. Dougherty, E.J. Smith, C.T. Russell, Nature **441**, 62 (2006)
K.-H. Glassmeier, F.M. Neubauer, J. Geophys. Res. **98**, 20921–20935 (1993)
K.-H. Glassmeier, J. Grosser, U. Auster, D. Constantinescu, Y. Narita, S. Stellmach, Space Sci. Rev. **132**, 511–527 (2007a)
K.-H. Glassmeier, I. Richter, A. Diedrich, G. Musmann, U. Auster, U. Motschmann, A. Balogh, C. Carr, E. Cupido, A. Coates, M. Rother, K. Schwingenschuh, K. Szego, B. Tsurutani, Space Sci. Rev. **128**, 649–670 (2007b)
K.-H. Glassmeier, H. Boehnhardt, D. Koschny, E. Kürth, I. Richter, Space Sci. Rev. **128**, 1–21 (2007c)
K.-H. Glassmeier, H.-U. Auster, D. Heyner, K. Okrafka, C. Carr, G. Berghofer, B.J. Anderson, A. Balogh, W. Baumjohann et al., Planet. Space Sci. **58**, 287–299 (2010)
D. Gordon, R. Brown, IEEE Trans. Magn. **8**, 76–82 (1972)
O. Gravrand, A. Khokhlov, J.L. Le Mouël, J.M. Léger, Earth Planets Space **53**, 949–958 (2001)
J. Grosser, K.-H. Glassmeier, A. Stadelmann, Planet. Space Sci. **52**, 1251–1260 (2004)
D.A. Gurnett, W.S. Kurth, D.L. Kirchner et al., Space Sci. Rev. **114**, 395–463 (2004)
D.A. Gurnett, A.M. Persoon, W.S. Kurth, J.B. Groene, T.F. Averkamp, M.K. Dougherty, D.J. Southwood, Science **316**, 442 (2007)
C. Guttin, J.M. Léger, E. Stoeckel, J. Phys. III, Suppl. **4**, C4-655–C4-659 (1994)
A.A. Halacsi, Study to develop methods of predicting spacecraft magnetic fields, *NASA* CR-73256 (1969)
J.S. Halekas, D.L. Mitchell, R.P. Lin, S. Frey, L.L. Hood, M.H. Acuña, A.B. Binder, J. Geophys. Res. **106**, 27841–27852 (2001)
F. Hartmann, IEEE Trans. Magn. **MAG-8**, 66–75 (1972)
P.C. Hedgecock, Space Sci. Instrum. **1**, 83–90 (1975)
L.L. Hood, P.J. Coleman, D.E. Wilhelms, Science **204**, 53–57 (1979)
L.L. Hood, A. Zakharian, J. Halekas, D.L. Mitchell, R.P. Lin, M.H. Acuña, A.B. Binder, J. Geophys. Res. **106**, 27,825–27,840 (2001)
H.C. Howe, R.P. Lin, R.E. McGuire, K.A. Anderson, Geophys. Res. Lett. **1**, 101–104 (1974)
G. Hulot, C.C. Finlay, C.G. Constable, N. Olsen, M. Mandea, Space Sci. Rev. (2010). doi:10.1007/s11214-010-9644-0, this issue
E.J. Iufer, Rev. Phys. Appl. **5**, 169–174 (1970)
X. Jia, M.G. Kivelson, K.K. Khurana, R.J. Walker, Space Sci. Rev. (2010). doi:10.1007/s11214-009-9507-8, this issue
C.L. Johnson, H. Uno, M.E. Purucker, B.J. Anderson, H. Korth, J.A. Slavin, S.C. Solomon, in *40th Lunar Planet. Sci. Conf.* Id. 1385 (2009)
J.L. Jørgensen, C.C. Liebe, A.R. Eisenman, G.B. Jensen, in *Spacecraft Guidance, Navigation and Control Systems, ESA SP-381*, ed. by B. Kaldeich-Schuermann (European Space Agency, Paris, 1997), pp. 303–309
M.L. Kaiser, M.D. Desch, J.W. Warwick, J.B. Pearce, Science **209**, 1238 (1980)
M.L. Kaiser, M.D. Desch, W.S. Kurth, A. Lecacheux, F. Genova, B.M. Pedersen, D.R. Evans, in *Saturn*, ed. by T. Gehrels, M.S. Matthews (University of Arizona Press, Tucson, 1984), pp. 378–415
S. Kellock, P. Austin, A. Balogh, B. Gerlach, R. Marquedant, G. Musmann, E. Smith, D. Southwood, S. Szalai, in *Cassini/Huygens: A Mission to the Saturnian Systems*, ed. by L. Horn. Proc. SPIE, vol. 2803 (SPIE, Bellingham, 1996), pp. 141–152
N. Kernevez, H. Glénat, IEEE Trans. Magn. **27**, 5402–5404 (1991)
K.K. Khurana, M.G. Kivelson, D.J. Stevenson, G. Schubert, C.T. Russell, R.J. Walker, C. Polanskey, Nature **395**, 777–780 (1998)
K.K. Khurana, M.G. Kivelson, C.T. Russell, R.J. Walker, S. Joy, in *EGS XXVII General Assembly*, abstract #5119, Nice, 21–26 April 2002
M.G. Kivelson, K.K. Khurana, J.D. Means, C.T. Russell, R.C. Snare, Space Sci. Rev. **60**, 357–383 (1992)
M.G. Kivelson, L.F. Bargatze, K.K. Khurana, D.J. Southwood, R.J. Walker, P.J. Coleman Jr., Science **261**, 331–334 (1993)
M.G. Kivelson, K.K. Khurana, C.T. Russell, R.J. Walker, J. Warnecke, F.V. Coroniti, C. Polanskey, D.J. Southwood, G. Schubert, Nature **384**, 537–541 (1996)
M.G. Kivelson, Z. Wang, S. Joy, K.K. Khurana, C. Polanskey, D.J. Southwood, R.J. Walker, Adv. Space Res. **16**, (4)59–(4)68 (1995)

M.G. Kivelson, K.K. Khurana, F.V. Coroniti, S. Joy, C.T. Russell, R.J. Walker, J. Warnecke, L. Bennett, C. Polanskey, Geophys. Res. Lett. **24**, 2155–2158 (1997)
M.G. Kivelson, K.K. Khurana, M. Volwerk, Icarus **157**, 507–522 (2002)
H. Korth, B. Anderson, M.H. Acuña, J.A. Slavin, N.A. Tsyganenko, S.C. Solomon, R.L. McNutt, Planet. Space Sci. **52**, 733–746 (2004)
H.J. Kramer, *Observation of the Earth and Its Environment: Survey of Missions and Sensors*, 4th edn. (Springer, Berlin, 2002), pp. 595–600
W.S. Kurth, T.F. Averkamp, D.A. Gurnett, J.B. Groene, A. Lecacheux, J. Geophys. Res. **113** (2008). doi:10.1029/2007JA012861
H. Kügler, in *Proceedings 4th International Symposium on Environmental Testing for Space Programmes*, ESA SP-467, pp. 69–72 (2001)
H. Kügler, in *Proc. 5th Int. Symp. on Env. Testing for Space Programmes*, ESA SP-558, pp. 407–411 (2004)
E.R. Lancaster, T. Jennings, M. Morrissey, R. Langel, *NASA Tech. Mem.* 82046, NASA/GSFC (1980)
R.A. Langel, J. Geophys. Res. **79**, 2363–2371 (1974)
R.A. Langel, R.T. Baldwin, in *Types and Characteristics of Data for Geomagnetic Field Modelling, NASA CP-3153*, ed. by R.A. Langel, R.T. Baldwin (NASA, Washington, 1992), pp. 75–136
R. Langel, G. Ousley, J. Berbert, J. Murphy, M. Settle, Geophys. Res. Lett. **9**, 243–245 (1982)
J.-M. Leger, Resonance magnetometer with optical pumping using a monolithic laser. US Patent 5,436,561 (1995)
H.K. Leinweber, C.T. Russell, K. Torkar, T.L. Zhang, V. Angelopoulos, Meas. Sci. Technol. **19**, 055104 (2008). doi:10.1088/0957-0233/19/5/055104
R.P. Lin, K.A. Anderson, R. Bush, R.E. McGuire, J.E. McCoy, in *Proc. Lunar Sci. Conf. 7th*, pp. 2691–2703 (1976)
D.A. Lohr, L.J. Zanetti, B.J. Anderson, T.A. Potemra, J.R. Hayes, R.E. Gold, R.M. Henshaw, F.F. Mobley, D.B. Holland, M.H. Acuña, J.L. Scheifele, Space Sci. Rev. **82**, 255–281 (1997)
H. Lühr, S. Maus, M. Rother, Res. Lett. **29**, 1489 (2002). doi:10.1029/2001GL013845
W. Magnes, D. Pierce, A. Valavanoglou, J. Means, W. Baumjohann, C.T. Russell, K. Schwingenschuh, G. Graber, Meas. Sci. Technol. **14**, 1003–1012 (2003)
W. Magnes, M. Oberst, A. Valavanoglou, H. Hauer, C. Hagen, I. Jernej, H. Neubauer, W. Baumjohann, D. Pierce, J. Means, P. Falkner, Meas. Sci. Technol. **19**, 115801–13 (2008). doi:10.1088/0957-0233/19/11/115801
M. Mandea, C. R. Geosci. **338**, 1002–1011 (2006)
S. Maus, M. Rother, R. Holme, H. Lühr, N. Olsen, V. Haak, Geophys. Res. Lett. **29**, 1702 (2002). doi:10.1029/2001GL013685
S. Matousek, Acta Astron. **61**, 932–939 (2007)
J.W. McAdams, Adv. Astronaut. Sci. **114**, 1549–1567 (2003). Part III: Spaceflight Mechanics
J.E. McCoy, K.A. Anderson, R.P. Lin, H.C. Howe, R.E. McGuire, The Moon **14**, 35–47 (1975)
K. Mehlem, IEEE Trans. Magn. **MAG-14**, 1064–1071 (1978)
K. Mehlem, P. Narvaez, in *Proc. IEEE EMC Symposium*, Seattle, WA, pp. 899–904 (1999)
J.M.G. Merayo, P. Brauer, F. Primdahl, J.R. Petersen, O.V. Nielsen, Meas. Sci. Technol. **11**, 120–132 (2000)
J.M.G. Merayo, J.L. Jørgensen, E. Friis-Christensen, P. Brauer, F. Primdahl, P.S. Jørgensen, T.H. Allin, T. Denver, in *Small Satellites for Earth Observation*, ed. by R. Sandau, H.-P. Röser, A. Valenzuela (Springer, Dordrecht, 2008), pp. 143–151
J.D. Mihalov, C.P. Sonett, J.H. Binsack, M.D. Moutsoulas, Science **171**, 892–895 (1971)
D.L. Mitchell, J.S. Halekas, R.P. Lin, S. Frey, L.L. Hood, M.H. Acuña, A. Binder, Icarus **194**, 401–409 (2008)
F.F. Mobley, L.D. Eckard, G.H. Fountain, G.W. Ousley, IEEE Trans. Magn. **MAG-16**, 758–760 (1980)
D. Möhlmann, W. Riedler, J. Rustenbach, K. Schwingenschuh, J. Kurths, U. Motschmann, T. Roatsch, K. Sauer, H.T.M. Lichtenegger, Planet. Space Sci. **39**, 83–88 (1991)
G. Musmann, Ocean Dyn. **41**, 265–276 (1988)
G. Musmann, Y. Afanassiev, Fluxgate Magnetometers for Space Research. Books on Demand GmbH (2010). ISBN-13: 978-3-8391-3702-4
S. Nazakawa, Y. Iijima, H. Tsunakawa, M. Matsushima, T. Ono, A. Kumamoto, Y. Kasahara, S. Ikegamis, T. Ishikawa, Electromagnetic compatibility of SELENE (KAGUYA). Preprint (2009)
P. Narvaez, Space Sci. Rev. **114**, 385–394 (2004)
N.F. Ness, *NASA/GSFC Report* X-612-64-389 (1964)
N.F. Ness, Space Sci. Rev. **11**, 459–554 (1970)
N.F. Ness, Space Sci. Rev. **21**, 527–553 (1978)
N.F. Ness, Space Sci. Rev. (2010). doi:10.1007/s11214-009-9567-9, this issue
N.F. Ness, B.W. Behannon, C.S. Scearce, S.C. Cantarano, J. Geophys. Res. **72**, 5769–5778 (1967)
N.F. Ness, K.W. Behannon, R.P. Lepping, K.H. Schatten, J. Geophys. Res. **76**, 3564–3573 (1971)

N.F. Ness, K.W. Behannon, R.P. Lepping, Y.C. Whang, K.H. Schatten, Science **183**, 1301–1306 (1974)
N.F. Ness, K.W. Behannon, R.P. Lepping, Y.C. Whang, K.H. Schatten, Science **185**, 151–160 (1975)
N.F. Ness, K.W. Behannon, R.P. Lepping, Y.C. Whang, J. Geophys. Res. **80**, 2708–2716 (1976a)
N.F. Ness, K.W. Behannon, R.P. Lepping, Y.C. Whang, Icarus **28**, 479–488 (1976b)
N.F. Ness, M.H. Acuña, R.P. Lepping, L.F. Burlaga, K.W. Behannon, F.M. Neubauer, Science **204**, 982–987 (1979a)
N.F. Ness, M.H. Acuña, R.P. Lepping, L.F. Burlaga, K.W. Behannon, F.M. Neubauer, Science **206**, 966–972 (1979b)
N.F. Ness, M.H. Acuña, R.P. Lepping, J.E.P. Connerney, K.W. Behannon, L.F. Burlaga, F.M. Neubauer, Science **212**, 211–217 (1981)
N.F. Ness, M.H. Acuña, K.W. Behannon, L.F. Burlaga, J.E.P. Connerney, R.P. Lepping, F.M. Neubauer, Science **215**, 588 (1982)
N.F. Ness, M.H. Acuña, K.W. Behannon, L.F. Burlaga, J.E.P. Connerney, R.P. Lepping, Science **233**, 85–89 (1986)
N.F. Ness, M.H. Acuña, L.F. Burlaga, J.E.P. Connerney, R.P. Lepping, Science **246**, 1473–1478 (1989)
N.F. Ness, M.H. Acuña, J. Connerney, P. Wasilewski, C. Mazelle, J. Sauvaud, D. Vignes, C. d'Uston, H. Reme, R. Lin, D.L. Mitchell, J. McFadden, D. Curtis, P. Cloutier, S.J. Bauer, Adv. Space Res. **23**, 1879–1886 (1999)
F.M. Neubauer, J. Geophys. Res. **80**, 3235–3240 (1975)
F.M. Neubauer, Nature **395**, 749–751 (1998)
F.M. Neubauer, K.H. Schatten, J. Geophys. Res. **79**, 1550–1554 (1974)
F.M. Neubauer, K.H. Glassmeier, M. Pohl, J. Raeder, M.H. Acuña, L.F. Burlaga, N.F. Ness, G. Musmann, F. Mariani, M.K. Wallis, E. Ungstrup, H.U. Schmidt, Nature **321**, 352–355 (1986)
F.M. Neubauer, M.H. Acuña, L.F. Burlaga, B. Franke, B. Gramkow, F. Mariani, G. Musmann, N.F. Ness, H.U. Schmidt, R. Terenzi, E. Ungstrup, M. Wallis, J. Phys. E, Sci. Instrum. **20**, 714–720 (1987)
F.M. Neubauer, H. Marschall, M. Pohl, K.-H. Glassmeier, G. Musmann, F. Mariani, M.H. Acuña, L.F. Burlaga, N.F. Ness, M.K. Wallis, H.U. Schmidt, E. Ungstrup, Astron. Astrophys. **268**, L5–L8 (1993)
O.V. Nielsen, J.R. Petersen, A. Femindez, B. Hernando, P. Spisak, F. Primdahl, N. Moser, Meas. Sci. Technol. **2**, 435–440 (1991)
O.V. Nielsen, J.R. Petersen, F. Primdahl, P. Brauer, B. Hernando, A. Fernandez, J.M.G. Merayo, P. Ripka, Meas. Sci. Technol. **6**, 1099–1115 (1995)
F. Nimmo, D.J. Stevenson, J. Geophys. Res. **105**, 11969–11979 (2000)
H. O'Brien, P. Brown, T. Beek, C. Carr, E. Cupido, T. Oddy, Meas. Sci. Technol. **18**, 3645–3650 (2007)
N. Olsen, Geophys. J. Int. **149**, 454–462 (2002)
N. Olsen, T. Moretto, E. Friis-Christensen, J. Geodyn. **33**, 29–41 (2002)
N. Olsen, L. Tøffner-Clausen, T.J. Sabaka, P. Brauer, J.M.G. Merayo, J.L. Jørgensen, J.-M. Leger, O.V. Nielsen, F. Primdahl, T. Risbo, Earth Planets Space **55**, 11–18 (2003)
N. Olsen, K.-H. Glassmeier, X. Jia, Space Sci. Rev. (2009). doi:10.1007/s11214-009-9563-0, this issue
J.L. Phillips, C.T. Russell, J. Geophys. Res. **92**, 2253–2263 (1987)
S.A. Pope, T.L. Zhang, M. Delva, M. Balikhin, L. Hvizdoš, K. Kudela, Z. Vörös, in *ESA Workshop on Aerospace EMC*, Florence, Italy, 2009
F. Primdahl, J. Phys. E, Sci. Instrum. **12**, 241–253 (1982)
F. Primdahl, in *Measurement Techniques in Space Plasmas: Fields*, ed. by R.F. Pfaff, J.E. Borovsky, D.T. Young. Geophys. Monograph, vol. 103 (American Geophysical Union, Washington, 1998), pp. 85–99
F. Primdahl, P.A. Jensen, J. Phys. E, Sci. Instrum. **15**, 221–226 (1982)
F. Primdahl, T. Risbo, J.M.G. Merayo, P. Brauer, L. Tøffner-Clausen, Meas. Sci. Technol. **17**, 1563–1569 (2006)
M.E. Purucker, Icarus **197**, 19–23 (2009)
Ch. Reigber, H. Luhr, P. Schwintzer, Adv. Space Res. **30**, 129–134 (2002)
N.C. Richmond, L.L. Hood, A preliminary global map of the vector lunar crustal magnetic field based on Lunar Prospector magnetometer data. J. Geophys. Res. **113**, CiteID E02010 (2008). doi:10.1029/2007JE002933
I. Richter, D.E. Brinza, M. Cassel, K.-H. Glassmeier, F. Kuhnke, G. Musmann, C. Othmer, K. Schwingenschuh, B.T. Tsurutani, Geophys. Res. Lett. **28**, 1913–1916 (2001)
W. Riedler, K. Schwingenschuh, Y.G. Yeroshenko, V.A. Styashkin, C.T. Russell, Nature **321**, 288–289 (1986a)
W. Riedler, K. Schwingenschuh, Y.G. Yeroshenko, V.A. Styashkin, C.T. Russell, in *Field, Particle and Wave Experiments on Cometary Missions*, ed. by K. Schwingenschuh, W. Riedler (Verlag der Osterreichischen Akademie der Wissenschaften, Wien, 1986b), pp. 155–168.

T. Risbo, P. Brauer, J.M.G. Merayo, O.V. Nielsen, J.R. Petersen, F. Primdahl, I. Richter, Meas. Sci. Technol. **14**, 674–688 (2003)
S.K. Runcorn, D.W. Collinson, W. O'Reilly, A. Stephenson, N.N. Greenwood, M.H. Battey, Science **167**, 697–699 (1970)
S.K. Runcorn, D.W. Collinson, W. O'Reilly, A. Stephenson, M.H. Battey, A.J. Manson, P.W. Readman, Proc. R. Soc. Lond. Ser. A **325**, 157–174 (1971)
C.T. Russell, Geophys. Res. Lett. **5**, 81–84 (1978)
C.T. Russell, J.G. Luhmann, K. Schwingenschuh, Planet. Space Sci. **40**, 707–710 (1992)
C.T. Russell, R.C. Snare, J.D. Means, R.C. Elphic, IEEE Trans. Geosci. Remote Sens. **GE-18**, 32–35 (1980a)
C.T. Russell, R.C. Elphic, J.G. Luhmann, J.A. Slavin, J. Geophys. Res. **85**, 8319–8332 (1980b)
C.T. Russell, R.C. Elphic, J.G. Luhmann, J.A. Slavin, in *Proc. 11th Lunar Planet. Sci. Conf.*, pp. 1897–1906 (1980c)
C.T. Russell, D.N. Baker, J.A. Slavin, in *Mercury*, ed. by F. Vilas, C.R. Chapman, M.S. Matthews (University of Arizona Press, Tucscon, 1988), pp. 514–561
C.T. Russell, D.E. Huddleston, K.K. Khurana, M.G. Kivelson, Planet. Space Sci. **47**, 521–527 (1999a)
C.T. Russell, D.E. Huddleston, K.K. Khurana, M.G. Kivelson, Planet. Space Sci. **47**, 1101–1109 (1999b)
C.T. Russell, Z.J. Yu, K.K. Khurana, M.G. Kivelson, Adv. Space Res. **28**, 897–902 (2001)
C.T. Russell, Z.J. Yu, K.K. Khurana, S.P. Joy, M.G. Kivelson, in *EGS XXVII General Assembly*, Nice, 21–26 April 2002, abstract #609 (2002)
C.T. Russell, M.K. Dougherty, Space Sci. Rev. (2010). doi:10.1007/s11214-009-9621-7, this issue
T.J. Sabaka, N. Olsen, M.E. Purucker, Geophys. J. Int. **159**, 521–547 (2004)
T. Saito, K. Yumoto, K. Hirao, T. Nagakawa, K. Saito, Nature **321**, 303–307 (1986)
N. Sanders, R. Brocc, G. Inouyc, IEEE Trans. Magn. **MAG-8**, 591–593 (1972)
J. Saur, F.M. Neubauer, K.-H. Glassmeier, Space Sci. Rev. (2009). doi:10.1007/s11214-009-9581-y, this issue
N. Schilling, K.K. Khurana, M.G. Kivelson, J. Geophys. Res. **109**, CiteID E05006 (2004). doi:10.1029/2003JE002166
R. Schulz, Solar Syst. Res. **43**, 343–352 (2009)
J. Scuffham, A. Balogh, Adv. Space Res. **38**, 616–626 (2006)
J. Scuffham, G. Giampieri, A. Balogh, Adv. Space Res. **38**, 610–615 (2006)
H. Shibuya, M. Toyoshima, M. Matsushima, H. Shimizu, F. Takahashi, H. Tsunakawa, *American Geophysical Union*, abstract #P31B-1399, Fall Meeting 2008
H. Shimizu, F. Takahashi, N. Horii, A. Matsuoka, M. Matsushima, H. Shibuya, H. Tsunakawa, Earth Planets Space **60**, 353–363 (2008)
J.A. Slavin, Adv. Space Res. **33**, 1859–1874 (2004)
J.A. Slavin, R.E. Holzer, in *Proc. Third International Colloquium on Mars, published by the Lunar and Planetary Institute*, pp. 239–241 (1981)
J.A. Slavin, K. Schwingenschuh, W. Riedler, Y. Yeroshenko, J. Geophys. Res. **96**, 11,235–11,241 (1991)
J.A. Slavin, M.H. Acuña, B.J. Andreson et al., Science **324**, 606–610 (2009)
R.E. Slocum, Phys. Rev. Lett. **29**, 1642–1645 (1972)
R.E. Slocum, Radiation source for helium magnetometers. US Patent, No. 5,036,278 (1991)
R.E. Slocum, in *American Geophysical Union, Spring Meeting 2002*, abstract #GP51A-07 (2002)
R.E. Slocum, F.N. Reilly, IEEE Trans. Nucl. Sci. **10**, 165–171 (1963)
R.E. Slocum, P.C. Cabiness, S.L. Blevins, Rev. Sci. Instrum. **42**, 763–766 (1971)
R.E. Slocum, L.D. Schearer, P. Tin, R. Marquedant, J. Appl. Phys. **64**, 6615–6617 (1988)
E.J. Smith, Adv. Space Experiments **25**, pp. 103–130 (1969), AAS 68-186
E.J. Smith, L. Davis Jr., P.J. Coleman Jr., C.P. Sonett, Science **139**, 909 (1963)
E.J. Smith, L. Davis Jr., P.J. Coleman Jr., C.P. Sonett, J. Geophys. Res. **70**, 1571–1586 (1965a)
E.J. Smith, L. Davis Jr., P.J. Coleman Jr., D.E. Jones, Science **149**, 1241–1242 (1965b)
E.J. Smith, L. Davis Jr., D.E. Jones, D.S. Colburn, P.J. Coleman, P. Dyal, C.P. Sonett, Science **183**, 305–306 (1974a)
E.J. Smith, L. Davis Jr., D.E. Jones, P.J. Coleman, D.S. Colburn, P. Dyal, C.P. Sonett, A.M.A. Frandsen, J. Geophys. Res. **79**, 3501–3513 (1974b)
E.J. Smith, B.V. Connor, G.T. Foster Jr., IEEE Trans. Magn. **MAG-11**, 962–980 (1975a)
E.J. Smith, L. Davis Jr., D.E. Jones, P.J. Coleman, D.S. Colburn, P. Dyal, C.P. Sonett, Science **188**, 451–455 (1975b)
E.J. Smith, C.P. Sonett, IEEE Trans. Geosci. Electron. **GE-14**, 154–171 (1976)
E.J. Smith, S. Gulkis, Annu. Rev. Earth Planet. Sci. **7**, 385–415 (1979)
E.J. Smith, L. Davis Jr., D.E. Jones, in *Jupiter*, ed. by T. Gehrels (University of Arizona Press, Tuscon, 1976), pp. 788–829

E.J. Smith, L. Davis Jr., D.E. Jones, P.J. Coleman Jr., D.S. Colburn, P. Dyal, C.P. Sonett, Science **207**, 407–410 (1980a)
E.J. Smith, L. Davis Jr., D.E. Jones, P.J. Colman Jr., D.S. Colburn, P. Dyal, C.P. Sonett, J. Geophys. Res. **85**, 5655–5674 (1980b)
E.J. Smith, B.T. Tsurutani, J.A. Slavin, D.E. Jones, G.L. Siscoe, D.A. Mendis, Science **232**, 382–385 (1986)
E.J. Smith, M.K. Dougherty, C.T. Russell, D.J. Southwood, J. Geophys. Res. **106**, 30,129–30,139 (2001)
R.C. Snare, in *Measurement Techniques in Space Plasmas—Fields*, ed. by R.F. Pfaff, J.E. Borovsky, D.T. Young. Geophys. Monograph, vol. 103 (American Geophysical Union, Washington, 1998), p. 101
R.C. Snare, J.D. Means, IEEE Trans. Magn. **MAG-13**, 1107–1109 (1977)
C.P. Sonett, D.S. Colburn, R.G. Currie, J. Geophys. Res. **72**, 5503–5507 (1967)
D.J. Southwood, M.K. Dougherty, A. Balogh, S.W.H. Cowley, E.J. Smith, B.T. Tsurutani, C.T. Russell, G.L. Siscoe, G. Erdos, K.-H. Glassmeier, F. Gliem, F.M. Neubauer, J. Geophys. Res. **106**, 30,109–30,128 (2001)
S. Stanley, J. Bloxham, Nature **428**, 151–153 (2004)
S. Stanley, J. Bloxham, Icarus **184**, 556–572 (2006)
S. Stanley, G.A. Glatzmaier, Space Sci. Rev. (2010). doi:10.1007/s11214-009-9573-y, this issue
D.J. Stevenson, AIP Conf. Proc. **24**, 781–784 (1975)
D.J. Stevenson, Science **208**, 746–748 (1980)
D.J. Stevenson, Geophys. Astrophys. Fluid Dyn. **21**, 113–127 (1982)
D.J. Stevenson, Rep. Progr. Phys. **46**, 555–620 (1983)
D.J. Stevenson, Space Sci. Rev. (2010). doi:10.1007/s11214-009-9572-z, this issue
D.J. Stevenson, T. Spohn, G. Schubert, Icarus **54**, 466–489 (1983)
T.L. Zhang, W. Baumjohann, M. Delva, H.-U. Auster, A. Balogh, C.T. Russell et al., Planet. Space Sci. **54**, 1336–1343 (2006)
T.L. Zhang, M. Delva, W. Baumjohann, M. Volwerk et al., Planet. Space Sci. **56**, 785–789 (2008)

Space Sci Rev (2010) 152: 99–134
DOI 10.1007/s11214-010-9629-z

Current Systems in Planetary Magnetospheres and Ionospheres

Wolfgang Baumjohann · Michel Blanc · Andrei Fedorov · Karl-Heinz Glassmeier

Received: 23 November 2009 / Accepted: 7 January 2010 / Published online: 20 February 2010

Abstract The interaction of planets with the solar wind produces a diversity of current systems, yet these can be classified into only a few different types, which include ionospheric currents, currents carried by magnetospheric boundaries like the magnetopause or ionopause, magnetotail currents, and currents flowing inside the magnetospheres, like ring currents, plasma sheet currents and currents aligned to the magnetic field lines (or field-aligned currents).

Keywords Planetary magnetospheres · Planetary ionospheres · External planetary current systems

W. Baumjohann (✉)
Space Research Institute, Austrian Academy of Sciences, 8042 Graz, Austria
e-mail: baumjohann@oeaw.ac.at

W. Baumjohann
Technische Universität Graz, 8010 Graz, Austria

M. Blanc
Ecole Polytechnique, Route de Saclay, 91128 Palaiseau Cedex, France

A. Fedorov
Centre d'Etude Spatiale des Rayonnements, 31028 Toulouse, France

K.-H. Glassmeier
Technische Universität Braunschweig, 38106 Braunschweig, Germany

K.-H. Glassmeier
Max Planck Institut für Sonnensystemforschung, 37191 Katlenburg-Lindau, Germany

1 Introduction

1.1 Sources of Planetary Magnetic Fields

Significant magnetic fields are permanent features of some planets' interiors, and of mostly all of their ionized environments, their ionospheres and magnetospheres. They originate from dynamo action and electromotive forces generating and/or maintaining magnetic fields either inside the Sun and some classes of stars, in planetary interiors, or finally in the Galaxy where a large scale dynamo is also active (e.g., Ferrière 2001). The magnetic configuration of planetary environments generally results from the interplay of fields generated or maintained in the planet's interior, and fields transported from the solar corona by the solar wind. In planetary ionospheres and magnetospheres, this interplay is complex, because they are filled with plasmas, e.g. macroscopically neutral admixtures of positive and negative charges which react to the applied field by generating additional electrical currents, hence also additional locally generated magnetic fields. These additional external magnetic fields can be of the same order of magnitude as the originally applied field of internal or solar origin.

1.2 Sources of Planetary Plasmas

There are mainly three sources of plasmas in planetary environments (e.g., Blanc et al. 2005):

- the *solar wind plasma*, convected from the solar corona to the planetary environment, which is mostly made of atomic hydrogen H^+ with a few percent He^{++};
- the *ionospheric plasma*, which is the plasma generated in the upper atmosphere of the planet, when there is one, by the ionizing agents acting on the atmospheric gas (mainly solar photons and energetic particle impact). The composition of this ionospheric plasma usually reflects the composition of the atmospheric gas via the action of ion-neutral chemical reactions. This plasma is gravitationally bound to the planet, but it may diffuse outward and upward along magnetic field lines due to different physical processes, forming an extended region of plasma called a plasmasphere.
- finally, in giant planet systems, *orbital plasma sources* also exist and contribute significantly to feeding the magnetosphere. They may be generated as ionospheric plasma in the upper atmosphere of a moon (e.g., Titan), or from the ionization of neutral gas tori that are generated and maintained by the activity of some of the planet's moons (e.g., Io at Jupiter and Enceladus at Saturn). Just as in the latter two cases, these plasmas of satellite origin are often the dominant source of the magnetospheric plasma.

1.3 Diversity of Magnetic Field Configurations

The final magnetic configuration in a given planetary or moon environment results from the interplay between the different configurations and intensities of the magnetic field sources on the one hand (essentially, of solar, planetary or moon origin) and the different intensities and configurations of the plasma sources on the other.

One key element which distinguishes the two major types of magnetospheric configurations is the nature of the obstacle opposed to the solar wind flow impinging on the planet (its surface, its atmosphere, or its magnetic field): when the solar wind flow is opposed by a planetary magnetic field, this field digs a long cavity in the solar wind flow, which is filled by the planetary magnetic field and for this reason is called the "magnetosphere". In this case, which applies to Mercury, Earth and all giant planets, one speaks of an "intrinsic magnetosphere". In the case when the solar wind interacts directly with the ionized upper atmosphere of an obstacle (like Venus, Mars, and comets), a similar cavity is formed by the effects of the currents induced in the planetary or cometary ionosphere. In the latter case one speaks of an "induced magnetosphere", whose size is comparable to the obstacle's size. When the solar wind directly strikes a body that lacks both atmosphere and magnetic field (like some moons and asteroids), just a small void is created in the solar wind on the downstream side.

How can we separate the parametric domains of intrinsic and induced magnetospheres? The solar-wind stand-off distance, upstream of the planet's obstacle, is the distance at which the solar wind total pressure (mainly flow dynamic pressure) is matched by the opposing pressure of the planetary obstacle. This obstacle can be one of two objects: the object's atmosphere/ionosphere, or the planetary magnetic field itself.

When the planetary magnetic field is dominated by its dipole component, the calculation of the distance at which this field can stand off the solar wind pressure is well known, and leads to the so-called Chapman-Ferraro distance R_{CF}:

$$R_{\mathrm{CF}} = R_{\mathrm{P}}(B^2{}_{\mathrm{surf}}/\mu_0 \rho V_{\mathrm{sw}})^{1/6} \tag{1}$$

which is the distance at which the magnetic pressure of the planetary field on the sun-planet line (taking into account the additional effect of currents flowing on the magnetopause) balances the total solar wind pressure. In this formula R_{P} represents the planetary radius. The different types of obstacles met by the solar wind depend on the relative magnitudes of R_{P} and R_{CF}:

- if $R_{\mathrm{CF}} \gg R_{\mathrm{P}}$, the solar wind interacts with the planetary field, we have an "intrinsic magnetosphere";
- if $R_{\mathrm{CF}} \ll R_{\mathrm{P}}$, the solar wind is not deviated by a planetary magnetic field and it interacts directly with the planet's atmosphere/ionosphere. We then speak of an "induced magnetosphere", as it is the draping of the solar wind magnetic field around the planetary obstacle which creates a cavity and a wake;

The case $R_{\mathrm{CF}} \sim R_{\mathrm{P}}$ is an interesting one: the planet and its magnetic field can both contribute to the planetary obstacle. Mars, where magnetic anomalies extending into the ionosphere have been detected, is of this type. We will elaborate on this specific case later.

Table 1 summarizes the different cases of planetary obstacles to the solar wind, and how they are determined by the relative magnitudes of R_{P} and R_{CF} and by the nature of the planet's envelope. The solar system nicely offers to us all different cases to explore.

1.4 Global Magnetic Configuration of Intrinsic Magnetospheres

When the obstacle is the planetary magnetic field, the solar wind plasma is mostly deflected around the region of space dominated by the planetary magnetic field (after having been decelerated to from super-magnetosonic to sub-magnetosonic speed at the planetary bow shock). This is a consequence of the frozen-in characteristics of highly conducting plasmas. The boundary separating the two different regions is called the magnetopause. The cavity within which the planetary field remains confined is called magnetosphere (cf. Fig. 1), and the magnetosphere is characterized by the fact that all field lines have at least one foot point in the planetary ionosphere and the planetary body. At low and middle latitudes, these field lines keep a basically dipolar topology: they are "closed" field lines connected to the two magnetic hemispheres. But the kinetic pressure of the solar wind plasma, acting through

Table 1 Classification scheme of planetary magnetospheres

Planet's or satellite's envelope:	$R_{CF} \gg R_P$	$R_{CF} \ll R_P$
Solid surface	Obstacle = planetary field	Obstacle = solid surface
	Mercury	Moon
	(Ganymede)	(Europa, Callisto, Saturnian icy satellites)
Dense atmosphere	Obstacle = planetary field	Obstacle = planetary atmosphere/ionosphere
	Earth	Venus
	Giant planets	(Io, Titan)

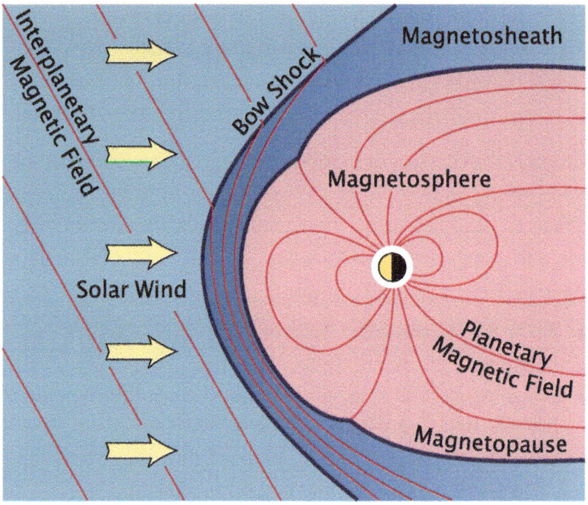

Fig. 1 Solar wind interaction with a planetary magnetic field (adapted from Baumjohann and Treumann 1996)

the ionosphere, distorts the outer part of this planetary dipolar field. On the dayside it compresses the field, while on the night side the dipolar magnetic field lines are stretched out into a long magnetotail.

1.5 Global Magnetic Configuration of Induced Magnetospheres

In the case of un-magnetized bodies like Venus and Mars, the obstacle to the external flow is the object's atmosphere itself. In that case, the solar wind and its frozen-in interplanetary magnetic field interact directly with the dayside ionosphere. The outer ionosphere is an almost perfect conductor thus excluding any external magnetic field by a current system generated on the surface (called the "ionopause") separating the external magnetic field and the field-free ionosphere interior. The resulting magnetic field created by this current system is called an "induced magnetosphere". The properties of the magnetospheres of Mars and Venus are very similar, but since Mars has a crustal magnetic field distributed over the planet's surface, only Venus demonstrates a pure example of the induced magnetosphere.

The interplanetary magnetic field piles up to form a magnetic barrier in the inner dayside magnetosheath. This region of interaction is confined on the outside by the bow shock. The ionosphere that bounds this region from the inside manifests itself as an abrupt drop

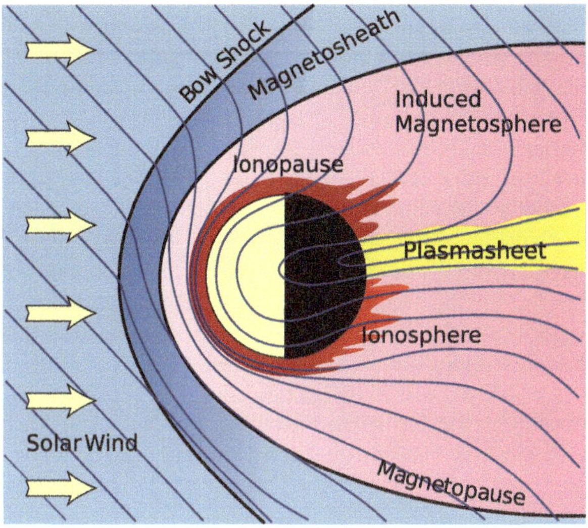

Fig. 2 Schematic of the Venusian magnetosphere. (Adopted from Zhang et al. 2008)

in the magnetic field strength. Between these two boundaries the magnetic field gradually increases, reaching a magnitude ten times higher than the incident solar wind magnetic field just outside the ionospause. In the plane perpendicular to the solar wind convection electric field, the magnetic field lines are bent and diverge as shown in Fig. 2, squeezing out the solar wind plasma from the dayside equatorial region and creating a void of charged particles in the magnetic barrier. Inside the magnetic barrier the ram and thermal pressure ($\rho V^2/2 + nkT$) of the incident solar wind are balanced by magnetic pressure ($B^2/2\mu_0$). There is only thermal pressure below the ionopause, and in the space between the bow shock and the obstacle the combination of ram, thermal and magnetic pressure corresponds to the normal gas-dynamics flow of a magnetized plasma. However, in the vicinity of the ionopause the magnetic field pressure balances the entire incident kinetic pressure of the solar wind. The boundary between the magnetic barrier proper and the magnetosheath flow is called "induced magnetospheric boundary" or "magnetic pileup boundary". This boundary is the real obstacle for the solar wind plasma. It extends into the planetary wake and, finally, confines a vast solar wind void downstream of the planet.

The magnetic topology of the wake is tail-like, as shown in Fig. 2. Such a magnetic tail, where the magnetic field vector is approximately co-aligned with the solar wind velocity vector, is divided into two "lobes" with opposite magnetic field directions. A thin current sheet separates the lobes, and the plane of the current sheet contains the solar wind velocity vector and the solar wind convection electric field $-\mathbf{V} \times \mathbf{B}$. Since the interplanetary magnetic field is, on average, at a low inclination to the ecliptic plane, the current sheet mid plane tends to be orthogonal or at a high inclination to the planet's orbital plane. The induced magnetotail is very long and extends downtail dozens of planetary radii.

1.6 Magnetic Configurations in the Environments of Planetary Moons

For the magnetic configuration of planetary moons around giant planets, a similar division exists between intrinsic and induced magnetospheres. In that case one just has to replace the solar wind by the planetary magnetosphere's plasma flow. Around most giant planets, this plasma essentially flows along closed flow lines to large radial distances. It interacts with each moon in a similar way as the solar wind interacts with planets, creating a specific magnetic configuration in the vicinity of the moon (but not a bow shock). If the moon has no intrinsic magnetic field, the configuration is that of "induced magnetosphere". Titan is the best documented example in this category, and for this reason it bears some strong similarities with the cases of Venus and Mars. If the moon possesses an intrinsic magnetic field, we have an "intrinsic magnetosphere". The only known case of a moon's intrinsic magnetosphere is that of Ganymede, one of the four Galilean moons of Jupiter, which adds a lot to the interest and uniqueness of this object, which is also the largest moon in the solar system.

In the following, we will describe the different magnetic configurations generated by the interplay of plasma and magnetic field sources around solar system objects. First, we will briefly describe the different types of planetary plasmas. Then we will describe the different types of magnetospheric current systems generated in these plasma populations, with specific descriptions of each planet's case. Finally, once each type of current system has been described, we will illustrate how global current systems connecting these different components are established and maintained in each planetary environment, and we will explain their intrinsic relation to the global dynamics and flow systems of planetary magnetospheres.

2 Planetary Plasmas and Associated Currents

A plasma is a gas composed of a mixture of positively and negatively charged particles, which consists of equal numbers of positive and negative elementary charges. On average, a plasma looks electrically neutral to the outside. However this fourth state of matter behaves quite differently from a neutral gas. In the presence of electric and magnetic fields, like there always are in planetary environments, and of additional forces acting differentially on positive and negative charges, planetary plasmas generate and carry electric currents, which in turn generate additional magnetic fields.

In general, the dynamics of a plasma can be described by solving the equation of motion for each individual particle. Since the electric and magnetic fields appearing in each equation include the internal fields generated by every other moving particle, all equations are coupled and have to be solved simultaneously. Such a full solution is not only too difficult to obtain, but also is not really needed to describe the basic behaviour and effects of magnetospheric and ionospheric currents. A much simpler approach can be used, namely the single-particle motion or guiding-center description. It describes the motion of a particle under the influence of external electric and magnetic fields. This approach neglects the collective behaviour of a plasma, but is useful when studying a low-density plasma threaded by strong magnetic fields, like in inner planetary magnetospheres (actually, in case of nearly steady plasma-generated mean fields this approach can be used, too).

Let us first consider in this simplified approach the case of a plasma in the presence of electromagnetic forces only. The equation of motion for a particle of charge q under the action of the Coulomb and Lorentz forces can be written as

$$m\frac{d\mathbf{v}}{dt} = q(\mathbf{E} + \mathbf{v} \times \mathbf{B}) \tag{2}$$

where m represents the particle mass and $\mathbf{v}$ the particle velocity. Under the absence of an electric field and in the presence of a homogeneous magnetic field, (2) describes a circular orbit of the particle around the magnetic field, with the sense of rotation depending on the sign of the charge. The center of this orbit is called the guiding center. A possible constant velocity of the particle parallel to the magnetic field will make the actual trajectory of the particle three-dimensional and look like a helix. Taking the electric field into consideration will result in a drift of the particle superimposed onto its gyratory motion. This $\mathbf{E} \times \mathbf{B}$ drift is independent of the sign of the charge: in steady state electrons and ions move together with the same speed in the same direction and thus no current is generated.

Now, if we add a force $\mathbf{F}$ on the right-hand side of (2), such that

$$m\frac{\mathrm{d}\mathbf{v}}{\mathrm{d}t} = q(\mathbf{E} + \mathbf{v} \times \mathbf{B}) + \mathbf{F} \tag{3}$$

an additional charge-dependent drift ($\mathbf{F} \times \mathbf{B}/eB^2$) is imposed on particles of opposite signs, and currents flow in the plasma. In practice, this force can be:

- *gravity* (though it is mostly negligible),
- *inertial forces* acting on the particles, such as the centrifugal force in the fast rotating magnetospheres of giant planets,
- *pressure gradients*, which in the fluid analog of (3) will appear in any situation where the plasma or the ambient magnetic field are inhomogeneous: this is the origin of ring currents and plasma sheet currents in intrinsic magnetospheres,
- *collisions with neutrals*, which are translated in the fluid equation as an additional "collision term" $-m\upsilon_{\mathrm{n}}\mathbf{v}$ on the right-hand side of (3), where $\mathbf{v}$ is the drift velocity of the charged species relative to the neutral gas and υ_{n} is the collision frequency. This is the origin of current flows in ionospheric plasmas.

Finally, when *newly created electrons and ions* are added to the plasma flow, for instance by the effect of ionization of neutrals or of the exchange of charges between ions and neutral particles, an additional $\mathbf{F}$ term also appears in the particle-averaged equation of motion for the plasma, which correspond to the rate of momentum added to the plasma flow by the newborn particles. These newborn charged particles therefore carry a specific current, the so-called *pick-up current*, which plays an important role in all media where the plasma flow is superimposed on and coupled to a background neutral gas. This includes ionospheres, neutral gas tori in giant planets magnetospheres, and extended exospheres in the environments of un-magnetized bodies with an atmosphere like Venus, Mars and cometary environments. The pick-up current can usually be expressed as an Ohmic term:

$$\mathbf{J} = \sigma_{\mathrm{pickup}}\mathbf{E} = \frac{m_i\dot{n}}{B^2}\mathbf{E} \tag{4}$$

where $\mathbf{E}$ is the ambient electric field, and the so-called "pick-up conductivity" σ_{pickup} is proportional to the time rate of addition of new-born ions to the flow, $\dot{n}$.

2.1 Planetary Ionospheres and Associated Plasma Domains

The solar ultraviolet and X-ray light impinging on a planet's or moon's atmosphere ionizes a fraction of the atmosphere's neutral gas. The balance between this ionization source and the recombination of oppositely charged particles into neutral atoms or molecules produced by random collisions results in the maintenance of free positive and negative charges at a certain density level. Since collisions decrease with increasing altitude due to the related decrease of atmospheric gas density, the plasma density is generally negligible up to a certain altitude (about 80 or 100 kilometres in the Earth case), and then increases upwards, forming a plasma layer called the ionosphere. Ionospheres then generally extend to rather high altitudes, about a thousand kilometers in the Earth's case. Above a certain altitude, the atmospheric gas is so tenuous that its contribution to the total altitude-integrated source of ions becomes completely negligible, the vertical distribution of the different constituents of the ionospheric gas is governed by hydrostatic balance, and plasma densities essentially exponentially decrease with increasing altitude. In the case of a planet with an intrinsic magnetic field, ionospheric plasma diffuses and extends into closed magnetic flux tubes under this law of hydrostatic equilibrium, forming a torus-shaped volume called the plasmasphere. The plasmasphere is usually identified as a separate plasma domain, containing relatively cool but dense plasma of ionospheric origin, which merely corotates with the planet. In the Earth's case the plasmasphere exists at middle and low latitudes. At high and polar latitudes planetary field lines become very extended or open, and ionospheric plasma is lost into distant space before reaching a steady hydrostatic equilibrium. At these latitudes, or, more generally, at locations where magnetic field lines are aligned near-vertically, magnetospheric electrons can precipitate along magnetic field lines down to ionospheric altitudes, where they collide with and ionize neutral atmospheric particles, thus providing an additional source for the generation of the ionospheric plasma. For this reason, the high-latitude and polar ionospheres of the Earth and giant planets may be more dense than their low-latitude counterparts, despite a lower solar illumination. A by-product of this magnetospheric particle source is the collisional excitation of atmospheric neutrals into unstable excited electronic and vibrational states, which ultimately emit photons and create the aurora.

2.2 Solar Wind and Magnetosheath Plasma

The Sun emits highly conducting plasma into interplanetary space as a result of the supersonic expansion of the solar corona. This plasma is called the solar wind. It flows with a supersonic speed of about 500 km s^{-1} and consists mainly of electrons and protons, with an admixture of 5% helium ions (alpha particles). Because of the high conductivity, the solar magnetic field is 'frozen' in the plasma (as in a perfect conductor) and drawn outward by the expanding solar wind. When the supermagnetosonic solar wind impinges on a planetary obstacle, a bow shock wave is generated (see Figs. 1 and 2). The region of thermalized subsonic plasma behind it is the magnetosheath. Its plasma is denser and hotter, and its magnetic field strength higher, than in the solar wind.

2.3 Orbital Plasma Populations and Sources

Giant planets are sort of small planetary systems embedded in the solar system. Their moons orbit to a large fraction inside their own magnetospheres, so that each moon interacts directly with the planetary magnetospheric flow and with the planet's radiation belt. Irradiation of the moon's atmosphere (in the case of Titan, of the tenuous exospheres of the Galilean moons, and of Neptune's moon Triton), or directly of its surface (for all icy moons) produces free molecules, radicals, or ion–electron pairs by irradiation, sputtering and desorption.

Furthermore, a few of the giant planets' moons are themselves intense sources of gas and plasma and play a major role in the generation of the magnetospheric plasma population. Jupiter and Saturn both harbour one of these unique and spectacular satellite sources. Jupiter's volcanic satellite Io (Fig. 3) releases huge amounts of gas via its numerous volcanic plumes. In addition to generating a sporadic atmosphere, this gas emission expands all around Io's orbit to generate a permanent gas torus which is observed from Earth. Ionization of this neutral torus produces on the order of one ton per second of fresh plasma, which happens to be by far the dominant source of plasma for the whole Jovian magnetosphere. At the level of Io, its tenuous atmosphere and ionosphere creates an electric conductor which, by moving through Jupiter's rotating magnet, generates a 200 kV e.m.f. and drives an electric circuit which flows along the Io flux tube, generates an aurora at its foot, and closes via the Jovian ionosphere.

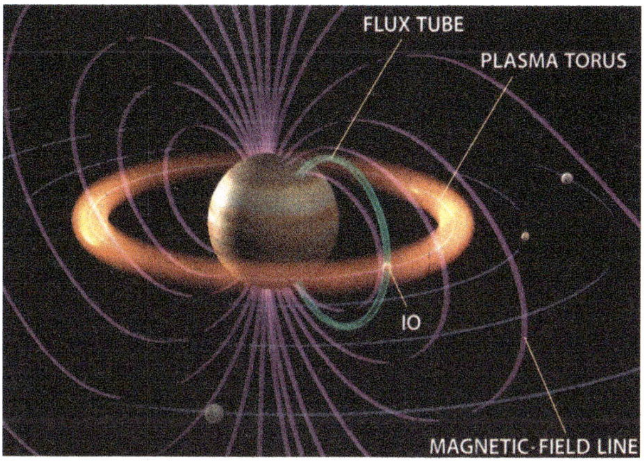

Fig. 3 Schematic of Jupiter's Io torus, Io flux tube and its electrical connection to Jupiter's upper atmosphere (From Johnson, Sci. American, 2000)

In a somewhat similar manner, Saturn's satellite Enceladus releases intense plumes of ice water and other minor components through its systems of gigantic southern hemisphere geysers. The origin of this moon's intense activity is not yet properly understood, but it is now established that the Enceladus plume is the major internal source of plasma at Saturn, in just the same way as Io is at Jupiter. The plume first feeds a permanent and extended water torus along the Enceladus orbit. Under the effect of irradiation by Saturn's magnetospheric energetic electrons, it generates on the order of 100 kg of water ion plasma per second, which later diffuses radially and populates an extended plasma sheet in the vicinity of Saturn's equatorial plane.

The Io and Enceladus neutral and plasma tori are examples of plasma sources and reservoirs which orbit inside their host magnetosphere, move at different speeds (the Keplerian velocity for the neutral gas, the corotation velocity of the planet for the plasma which is trapped in the planet's rotating magnetic field) and interact with the giant planet's magnetosphere.

3 Different Components of Magnetospheric Currents

3.1 Bow Shock and Magnetosheath Currents

Approaching a planet and its magnetosphere from interplanetary space, the first signature of its existence is the bow shock, a shock wave standing in the supersonic solar wind flow in front of the magnetosphere. Parameters like flow velocity, plasma density, and magnetic field all change significantly across the bow shock. According to Ampère's law the jump in the magnetic field across the bow shock is associated with an electric current flowing in the bow shock region. A typical bow shock thickness of 1000 km and a jump of the B_Z component across it of 5 nT gives an electric current density of 4 $\mathrm{nA/m^2}$. Along the bow shock the current density varies, which implies the need for closure of these currents via the magnetosheath.

Detailed observational analysis of bow shock and magnetosheath current systems is not yet available. However, numerical simulations by, e.g., Janhunen and Koskinen (1997), Siscoe and Siebert (2006), and Guo et al. (2008) indicate that electric currents generated at the bow shock significantly contribute to the overall magnetospheric current system, in particular the region 1 field-aligned current system. Depending on solar wind conditions the total current ranges between 0.03 MA and 4.62 MA (Guo et al. 2008).

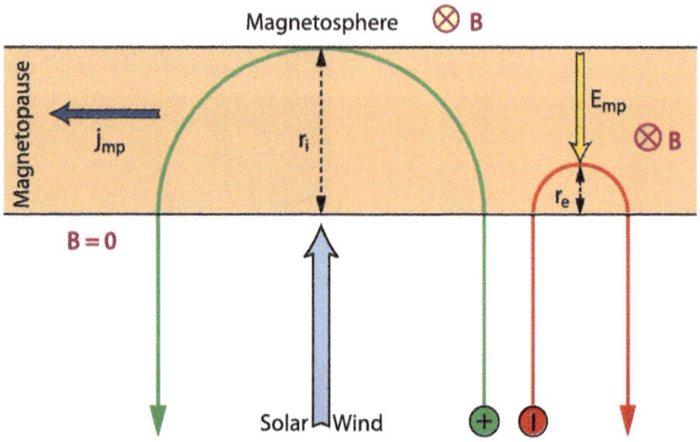

Fig. 4 Specular reflection off a magnetopause (adapted from Baumjohann and Treumann 1996)

3.2 Magnetopause and Ionopause Currents

The distortion of the internal dipole field into the typical shape of a magnetosphere produced by the interaction with the solar wind is accompanied by electrical currents. The compression of the internal magnetic field on the dayside is associated with current flow across the magnetopause surface, the magnetopause current. The tail-like field of the nightside magnetosphere is accompanied by a current flowing on the tail magnetopause surface and the cross-tail neutral sheet current in the central plasma sheet, both of which are connected and form a Θ-like current system, if seen from along the Earth–Sun line.

Separating the shocked solar wind, i.e., the magnetosheath plasma, from the magnetospheric magnetic field and being a surface across which the magnetic field strength jumps from its low interplanetary value to the high magnetospheric field strength, the magnetopause represents a surface current layer. The origin of this current can be understood from Fig. 4.

Ions and electrons hitting the magnetospheric field inside the magnetopause boundary will perform half a gyro-orbit inside the magnetic field before escaping with reversed normal velocity from the magnetopause back into the magnetosheath. The thickness of the solar wind-magnetosphere transition layer under such idealized conditions becomes of the order of the ion gyro radius. Electrons also perform half gyro orbits, but with much smaller gyro radii. The sense of gyration inside the boundary is opposite for both kinds of particles leading to the generation of a narrow surface current layer. This current provides an additional magnetic field, which compresses the magnetospheric field in the magnetosphere and at the same time annihilates its external part. It is a diamagnetic current caused by the perpendicular density gradient at the magnetopause.

3.2.1 Earth

The terrestrial magnetopause is typically located at a height of 10 (subsolar point) to 15 R_E and carries a current with a density of about 10^{-6} A m^{-2}. The total current flowing in the magnetopause is of the order of 10 MA. In the equatorial plane the magnetopause current flows from dawn to dusk. It closes on the tail magnetopause, where it splits into northern and southern parts flowing across the lobe magnetopause from dusk to dawn. The tail magnetopause current is additionally fed by the cross-tail neutral sheet current which flows from dawn to dusk.

Magnetograms often also show a positive excursion of the horizontal field magnitude at the beginning of a magnetic storm. This excursion is the magnetic signature of the solar wind impinging faster than usual onto the magnetopause. The position of the dayside magnetopause is essentially determined as the surface of equilibrium between the magnetic pressure of the terrestrial magnetic field and the kinetic pressure of the solar wind. Whenever the speed of the solar wind increases, the terrestrial field is compressed and the magnetopause recedes to a new equilibrium position.

3.2.2 Mercury

The Mariner 10 and Messenger flybys at Mercury demonstrate the existence of a Hermean magnetopause (Ness et al. 1974; Slavin et al. 2008). Over this magnetopause the magnetic field jump is of the order of a few tens of nT. The magnetopause thickness is estimated at a few hundred km (e.g. Russell and Walker 1985). Using the Ampère-Maxwell law this implies a magnetopause sheet current density of the order of $5 \cdot 10^{-7}$ A m^{-2}, a value comparable to the current strength at the terrestrial magnetopause. The total current is estimated at $2.5 \cdot 10^5$ A. Though almost two orders of magnitude smaller than the terrestrial Chapman-Ferraro current, its magnetic effect at the surface and in the interior of Mercury is much more pronounced than at Earth, where Chapman-Ferraro currents cause a surface magnetic field perturbation of the order of 10 nT or 0.3% of the dynamo generated magnetic field. At Mercury this ratio is of the order of 10%. The magnetopause current itself may have an impact on the dynamo action in Mercury's interior (e.g. Glassmeier et al. 2007).

3.2.3 Venus and Mars

The abrupt vanishing of the external magnetic field on entering the ionosphere proves that the ionosphere of Venus may be approximated as a perfect conductor. The ionospheric current system associated with such a magnetic field drop is somewhat similar to the Chapman-Ferraro current system bounding the Earth's magnetosphere, but in an inverse sense. This current system occupies the thin layer at the ionopause and the $\mathbf{J} \times \mathbf{B}$ force transfers the pressure of the external magnetic field to the ionosphere. Ionopause currents diverge on the ionopause surface and close upstream of it, in the induced magnetosphere and on the magnetic pileup boundary, where they contribute to breaking and diverting the flow of the impinging shocked solar wind.

3.2.4 Jupiter and Saturn

Thanks to the Voyager, Galileo and Cassini missions, empirical models of the magnetopause currents and fields have been developed for both Jupiter and Saturn. For Jupiter (e.g., Khurana et al. 2004), a good model of magnetopause currents is the one by Engle (1992). Because it uses a spherical harmonic expansion, it is limited to the dayside and does not cover the tail. It gives a good representation of the diurnal variation of the field due to the tilt of Jupiter's dipole, but using a description of the Jovian current sheet which will need to be improved in future works. For Saturn, a similar, pre-Cassini model of magnetopause currents was published by Maurice et al. (1996). It uses the Z3 internal field model and the Saturnian plasma sheet model of Connerney et al. (1981). More recently, an improved model using Tsyganenko's analytical technique has been developed using Voyager data and the first 25 Cassini orbits (e.g., Gombosi et al. 2009); this model successfully describes the magnetotail and Saturn's bowl-shaped current sheet.

3.3 Magnetotail Currents

3.3.1 Magnetotails in Intrinsic Magnetospheres

Besides the magnetopause current sheet, another typical example of a diamagnetic current is the neutral sheet current in the geomagnetic tail which divides the tail into northern and southern lobes with their stretched magnetic field lines. In the southern lobe the field lines extend from the southern polar cap and, in the Earth's case, point antisunward, while in the northern lobe they come from the distant tail pointing sunward ion the terrestrial case and ending in the northern polar cap. This stretching of the otherwise approximately dipolar terrestrial magnetic field can be accounted for by a diamagnetic current flowing across the magnetospheric tail. Such a current transports positive charges from one flank to the other (from dawn to dusk in the terrestrial case) and negative charges in the opposite direction across the tail and, because of its stationarity and its macroscopic magnetic effect, cannot be anything else but a diamagnetic current. Its cause is a gradient in the plasma pressure perpendicular to the current layer pointing from north to south in the upper (northern) half and from south to north in the lower (southern) half of the current layer. Hence, the current layer is a concentration of dense and hot plasma which is called the neutral sheet because of the weak magnetic field it contains.

3.3.1.1 Earth Spacecraft measurements have revealed that the neutral sheet in the geomagnetic tail contains a quasi-neutral ion–electron plasma of roughly 1–10 keV temperature and a density of about 1 cm^{-3}. The transverse magnetic field in the neutral sheet is not zero but rather weak, of the order of 1–5 nT. The main current sheet has a typical thickness of 1–2 R_E and the maximum current density is of the order of a few nA/m^2. However, especially during disturbed times and before substorm onset, the current sheet can be much thinner and the current density much higher (e.g., Nakamura et al. 2006; Baumjohann et al. 2007). Moreover, recent four-point measurements with the Cluster spacecraft have revealed that the magnetotail current sheet is often not a planar sheet, but is often wavy and twisted, exhibiting large-scale kink-like oscillations and flapping motions, that propagate from the midnight sector towards the flanks and toward Earth (e.g., Zhang et al. 2002, 2005; Volwerk et al. 2003; Sergeev et al. 2004).

3.3.1.2 Mercury Mercury exhibits a magnetosphere similar to that of the Earth, but of a much smaller size. Therefore, magnetotail currents are also present in the magnetotail of Mercury. This tail current sheet serves to establish the typical magnetic topology of a magnetospheric tail, that is an elongated structure with reversed magnetic fields above and below a sheet current system. The first detection of the Hermean neutral sheet current was made by the Mariner 10 spacecraft. The tail current closes either within the magnetotail or via currents flowing on the tail magnetopause.

3.3.1.3 Giant Planets All giant planets magnetospheres display elongated tails which have been explored by the Pioneer, Voyager and (for Jupiter and Saturn) Galileo and Cassini spacecraft, respectively. Their geometry is very similar to the Earth's tail, with varying sizes according to their dipole field intensity. As for Earth, their extended lobes are threaded by magnetic field lines rooted into their northern and southern polar caps, and are separated by a plasma sheet extending in the antisolar direction, first along the planet's magnetic equator, then along the solar wind direction beyond a "hinge point". As for Earth, their plasma sheets seem to experience sporadic reconnection events during which plasma is accelerated along the tail axis, partly in the planetward direction, partly in the antisunward direction. These events participate in the so-called "Vasyliunas cycle" of giant planets (e.g., Sect. 4), but also represent giant planet analogues to the terrestrial geomagnetic substorms. The differences in size between these various tail systems result in different characteristic time constants for the development and repeat rate of these events, while the fast rotation rates of these planets induce specific features in the magnetospheric flows associated to these reconnection events. One striking feature of the Jovian tail (e.g., Khurana et al. 2004) appears to be a strong dawn–dusk asymmetry: the mid-tail current sheet is significantly thicker on the dusk side, and the tail lobes more developed on the dawn side, a feature related by the authors to the interplay between convection and corotation flows producing a prevalent dawn-side Dungey–type convection cell (see Sect. 4.1 below). Cowley et al. (2005), using Cassini data and HST observations, established a consistent picture of reconnection field and flow geometry in Saturn's rotation-dominated magnetosphere.

3.3.2 Tail Formation in Induced Magnetospheres

As previously indicated, the Venusian ionosphere tends to behave as a superconducting sphere. Such a perfectly conductive sphere would have a current system closed entirely on its surface if it were immersed in a stationary magnetized plasma, as illustrated in Fig. 5A. However in the fast-flowing plasma of the solar wind the magnetic configuration would be more like that shown in Fig. 5B. In this case flowing magnetized plasma encounters the obstacle boundary where the magnetic field and plasma flow normal components are forced to vanish. The magnetic field then piles up on the windward side of the sphere and weakens on the lee side. So, the flow of magnetized solar wind plasma around a perfect conductive planetary ionosphere explains the generation of a magnetic barrier, but cannot entirely explain the formation of the tail-like wake (see the difference between Figs. 2 and 5B). The only possibility to generate such a tail is to assume an additional momentum exchange mechanism between the solar wind and the planetary plasma. Since the induced magnetotail is composed of inner magnetosheath flux tubes that slip over the obstacle to fill the flow wake, we need to brake and almost stop the solar wind flow with its frozen-in magnetic field in the vicinity of the ionosphere. If the central part of the flux tube (that is passing near the planet) is "hung up" and the periphery parts of the tube continue moving with solar wind speed, the intermediate parts of the field line are stretched antisunward to the sides of the planetary

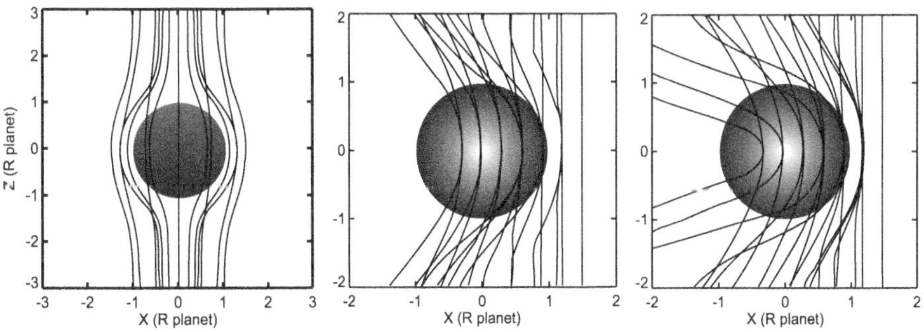

Fig. 5 **A**: Draping of magnetic field lines around perfectly conductive sphere in case of stationary plasma; **B**: same but when magnetized plasma flow from the right to the left; **C**: same but with mass-loading by continuously ionizing exospheric atoms. (Adopted from Luhmann et al. 2004)

magnetotail. The kink point between the solar wind part of the field line and its magnetotail part corresponds to the extension of the induced magnetospheric boundary to the night side of the planet. The final magnetic field topology then corresponds to Figs. 2 and 5C.

There are two mechanisms of momentum transfer between the solar wind and the planetary ions. The first one was proposed by Alfvén in 1957 to explain the formation of cometary tails. Above the ionopause there is still a "corona" of neutral gas of planetary origin. These atoms are being continuously ionized by solar ultraviolet radiation and direct impact of the solar wind electrons. When ionized, the new ion is "picked-up", i.e. accelerated by the convection electric field of the solar wind $-\mathbf{V}_{SW} \times \mathbf{B}_{SW}$ and dragged into the bulk motion of the solar wind following a cycloidal trajectory. The additional momentum transferred to these pick-up ions is extracted from the ambient bulk flow, and tends to slow it down. This process is called "mass loading".

Another similar process (though of different origin) occurs if there is a slow diffusion of the field of the magnetic barrier into the ionosphere. Such a feature is expected at both Venus and Mars especially at the time of solar minimum when the planetary ionospheres are weak. A magnetized ionosphere was observed when the solar wind pressure exceeded the ionospheric pressure near the exobase. Under these conditions, the ionopause shifts below the exobase down to the collision regime of the plasma. The planet obstacle apparently looses the perfect conductivity and the external magnetic fields diffuse into the ionosphere. The viscous forces produce a similar result as mass loading and the central practically steady part of the flux tube is filled by pickup and ionospheric ions. The stretched tail parts of the field line create the tail lobes separated by a thin current layer. The slowing of the central part of the field line is so strong that a thin current sheet is already formed at the terminator (see Fig. 2). Two points where the current sheet crosses the dawn–dusk meridian are called "the magnetic poles" to distinguish them from magnetic equator (the plane of Fig. 2). The intensive erosion of ionospheric material by this process leads to the elimination of the ionosphere in the nightside polar regions and to the generation of so-called "ionospheric holes".

The total magnetic flux in the Venusian tail is 10 times greater than the flux of the magnetic barrier. This means that most of the tail flux tubes are not connected to the ionosphere or the magnetic barrier (see Fig. 7). A part of this flux is created by magnetic loops that have already passed the planet and are freely convecting tailward, another part of the magnetic flux could be related to reconnection in the tail.

3.4 Ring Current and Magnetodisk Currents

In a magnetosphere the magnetic field has gradients and field lines are curved. This inhomogeneity of the magnetic field leads to a 'magnetic' drift of charged particles. In a magnetic field configuration with a gradient in field strength, ions and electrons drift into opposite directions, perpendicular to both $\mathbf{B}$ and ∇B. The opposite drift directions of electrons and ions lead to a transverse current. When the field lines are curved, a 'curvature' drift appears. The curvature drift is perpendicular to the magnetic field and its curvature. It again creates a transverse current since ion and electron drifts have opposite signs. In a cylindrically symmetric field, like in a dipole field, gradient and curvature drifts can be combined, resulting in circular drift orbits around the planet, but with ions and electrons drifting in opposite directions. The resulting net motion of current charges plus a so-called magnetization current due to particle gyration about the field lines produces a circumplanetary current, the "ring current", whose radial profile, intensity and thickness are functions of the plasma populations of ions and electrons which carry the current. While at Earth the intensity of the ring

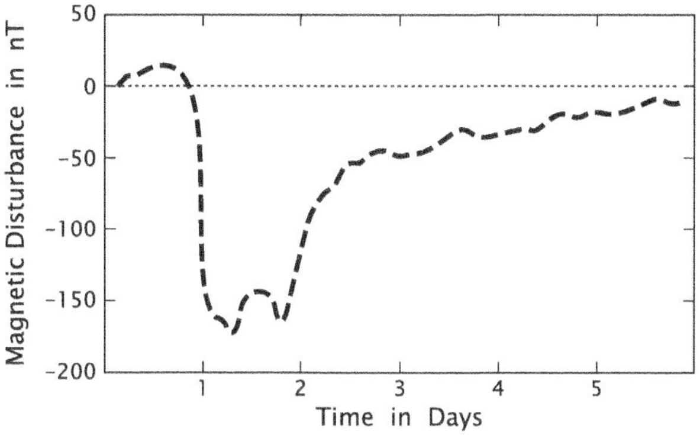

Fig. 6 Magnetic field variation during a magnetic storm (adapted from Baumjohann and Treumann 1996)

current remains modest as are its plasma sources, at Jupiter and Saturn the equivalent of the terrestrial ring current is fed by the dominant orbital plasma sources of the planet produced by its satellites. A very intense and extended ring current is produced, which stretches magnetic field lines in the vicinity of the equator and, in the case of Jupiter at least, produces a spectacular magnetodisk. Furthermore, in the Jovian and Kronian case, the inertia current due to partial corotation makes a significant contribution to the total magnetodisk current.

3.4.1 Earth

In the terrestrial magnetosphere, the ions drift westward around the Earth while electrons drift eastward, constituting a giant current loop of 1–10 MA, that can significantly alter the terrestrial field even at the Earth's surface (cf., for example, Kamide et al. 1998; Daglis et al. 1999). Even so, the ring current and its associated disturbance field are not stationary features. At times, during enhanced solar wind-magnetosphere coupling, more particles than usual are injected from the magnetotail into the ring current. This way the total energy of the ring current is increased and the additional depression of the surface magnetic field can clearly be seen in near-equatorial magnetograms, as shown in Fig. 6. Strong depressions of the terrestrial field, up to 2–3% of the total surface field in extreme cases, have been noticed in magnetograms long before the ring current was known; these intervals have been termed magnetic storms. During a storm like the one shown in Fig. 6, the total ring current reaches more than 10^7 A.

A magnetic storm has two distinct phases. For some hours or days, more and more particles are injected into the inner magnetosphere, building up the strong storm-time ring current and the associated magnetic disturbance field. After a day or two, the rate of injection returns to the normal level. The disturbance field starts to recover, since the ring current loses more and more storm-time particles by charge exchange processes with exospheric neutrals. This recovery phase typically lasts several days.

3.4.2 Mercury

The planetary magnetic field of Mercury is rather weak, only about 290 nT at the equatorial surface of the planet (e.g. Anderson et al. 2008). Due to this the magnetosphere of Mercury is rather small, the sub-solar magnetopause distance is only at about 1.7 R_M (Mercury's equatorial radius: 1 $R_M = 2440$ km). Most of the magnetosphere is thus filled by the planet itself. In this small magnetosphere trapped particles cannot exist to the extent observed in the

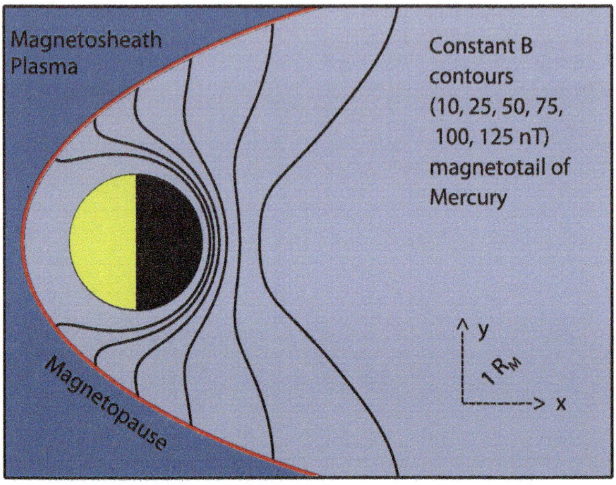

Fig. 7 Contours of constant magnetic field strength in the equatorial plane of the Hermean magnetosphere. *Closed contours* represent closed drift paths of charged particles with pitch angle 90°

terrestrial magnetosphere. Therefore, no classical radiation belts exist and no ring current is expected, as we are just about to explain.

First, the gyromotion of most charged particle is associated with a longitudinal motion along its guiding magnetic field line, extending from one mirror point to the other one. The particle mirror point can be located within a planet's atmosphere or below the planetary surface. If this is the case the particle is not trapped, but is lost from the magnetospheric particle population (e.g., Roederer 1970). As the solid body of Mercury occupies most of the Hermean magnetosphere, the vast majority of the particles have their mirror point formally at or below the planetary surface, except for those with equatorial pitch angles close to 90°. Thus they cannot be stably trapped in the Hermean magnetic field.

Even the remaining trapped particles do not constitute a major ring current. The gradient-curvature drift of the trapped particle population is perpendicular to both the magnetic field $\mathbf{B}$ and $\nabla_\perp B$. Therefore, particles always drift along a $B =$ const line in a plane perpendicular to $\mathbf{B}$. Figure 7 displays contours of constant magnetic field strength in the equatorial plane of the Hermean magnetosphere. To determine these contours the magnetic field model of Korth et al. (2004) has been used. The contours represent drift paths of particles with a pitch angle of 90°. As one can see, there are not many closed contours or drifts paths. Most of the contours run into the magnetopause boundary. Thus the particles drifting along these paths do not perform complete motions around Mercury, and instead they are lost through the magnetopause. They do not form a ring current as known from the terrestrial magnetosphere.

3.4.3 Jupiter and Saturn

At Jupiter and Saturn, the main magnetospheric plasma source is of satellite origin. At Jupiter, for instance, Io's volcanic activity results in the pick-up and injection of about one ton of fresh ions (mainly sulphur and oxygen) per second into the equatorial magnetosphere around 6 R_J from Jupiter's center. The effect of this intense mass loading of the magnetospheric flow, which near-corotates with Jupiter at this distance, is to create a population of ions with the local corotation velocity. This population experiences gradient and curvature drifts, as previously explained, and carries an intense ring current which slightly increases the magnetic field inside Io's orbit, but also tends to decrease the field intensity outside of Io's orbit. This results in a cylindrically symmetric magnetic field configuration in which the field lines are increasingly stretched outside with increasing radial distances.

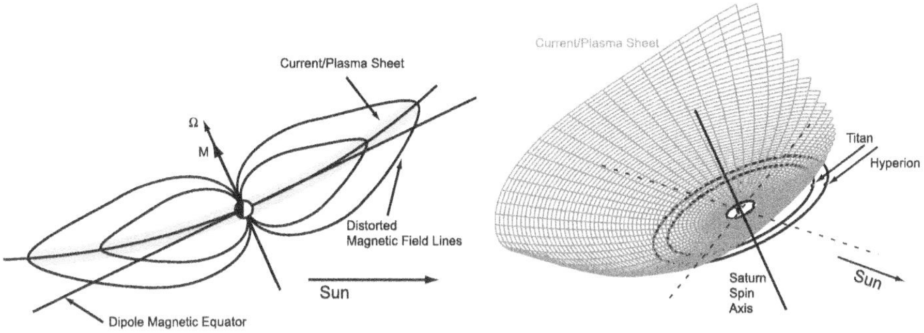

Fig. 8 Configuration of the magnetodisk at Saturn as revealed by the Cassini magnetometer (Arridge et al. 2008): a magnetodisk similar to the Jovian one can be seen in the noon–midnight meridian for low solar wind pressures only (*left-hand panel*). For solstice conditions, under the effect of the solar wind, the median surface of this magnetodisk is distorted into a "bowl shape". This bowl shape is expected to become planar at equinox, before reversing around the next solstice

Since the Iogenic plasma diffuses radially outwards to form an extended plasma population, the stretching of magnetic field lines actually extends from about 20 R_J to 100 R_J or beyond (depending on local time and solar wind velocity, and forms an extended magnetodisk which is the dominant feature of Jupiter's magnetosphere.

The formation of this magnetodisk can also be understood in simple MHD terms. The mass loading of Jovian magnetic flux tubes in the vicinity of Io's orbit, and beyond, combined with the fast rotation of these flux tubes which are dragged into Jupiter's atmosphere fast rotation by the motion of their ionospheric "roots", generates a strong centrifugal force acting on the trapped plasma. This centrifugal force acts on the flux tube to stretch it outside, while at the same time the trapped plasma is confined in the vicinity of the equator by the effect of this same centrifugal force, thus increasing the distorting effect of the centrifugal force even more near the equator.

At Saturn, the plasma generated by the icy satellites, and most noticeably by the Enceladus water source, generates in a similar way an inner torus of water ions which extends from about 4 to 12 Saturn radii, and is continued to larger distances into an extended plasma sheet. Because of the more moderate size of its magnetosphere, however, Saturn's ring current and plasma sheet does not develop such a spectacular magnetodisk all the time as we observe at Jupiter. A magnetodisk is indeed permanently present in the dawn sector (Arridge et al. 2007). But because of the confining effect of the solar wind pressure on the magnetopause, it extends into the dayside only for sub-solar magnetopause distances larger than about 23 R_S, i.e. for low solar wind pressures when the magnetosphere is expanded. The left panel of Fig. 8 shows the geometry of this Saturnian magnetodisk for such conditions. In addition, Cassini magnetometer observations also revealed a strong seasonal effect in the magnetodisk configuration. Around solstice, as at the time of Cassini arrival at Saturn, the effect of the solar wind, which flows at an angle to the magnetic equator, is to distort the magnetodisk into a "bowl shape", as shown on the right-hand side of Fig. 8 (Arridge et al. 2008).

3.5 Ionospheric Currents

The partially ionized plasma present in a planet's ionosphere can exhibit a differential motion of ions and electrons and thus a current, under the effect of the existence of a large scale electric field in the rest frame of the neutral gas. This is due to the resistivity existing in the gas in the presence of collisions. Indeed, at certain ionospheric altitudes, the ions and, to a lesser degree, also the electrons are coupled by collisions to the neutral components of the upper atmosphere and follow their dynamics. When the atmosphere is magnetized, atmospheric winds and tidal oscillations of the atmosphere force the ion component to move across the magnetic field lines, while the electrons move much more slowly at right angles to both the field and the neutral wind. This relative movement constitutes an additional electric current driven by the neutral wind, and such a region bears the name "dynamo layer", the generator of which is the atmospheric wind motion.

3.5.1 Collisions, Conductivities and Ionospheric Currents

In the presence of collisions between charged and neutral particles, a term has to be added to (2)

$$m\frac{\mathrm{d}\mathbf{v}}{\mathrm{d}t} = q(\mathbf{E}' + \mathbf{v} \times \mathbf{B}) - m\nu_{\mathrm{n}}\mathbf{v} \tag{5}$$

The collisional term on the right-hand side describes the momentum lost through collisions with neutrals occurring at a frequency ν_{n}. It is often called "frictional term" since it impedes motion. An important point is that the electric field $\mathbf{E}'$ is the electric field measured in the centre-of-mass frame of the system, in other words more or less exactly *in the rest frame of the neutral gas* (for a weakly ionized gas as we have in the upper atmosphere-ionosphere).

The friction term introduces a differential motion between electrons and ions and thus a current, even in homogeneous magnetic fields. In fact, when abundant collisions between the ionized and the neutral part of an upper atmosphere interrupt the cyclotron motion of electrons and/or ions the above equation reduces to an anisotropic Ohm's law

$$\mathbf{j} = \sigma_{\|}\mathbf{E}_{\|} + \sigma_{\mathrm{P}}\mathbf{E}'_{\perp} - \sigma_{\mathrm{H}}(\mathbf{E}'_{\perp} \times \mathbf{B})/B \tag{6}$$

The Hall conductivity, σ_{H}, determines the Hall current in the direction perpendicular to both the electric and magnetic field. The Hall conductivity maximizes at a height where the ions collide so frequently with the neutrals that they are essentially at rest, while the electrons already undergo a somewhat impeded $\mathbf{E} \times \mathbf{B}$ drift. The Pedersen conductivity, σ_{P}, governs the Pedersen current in the direction of that part of the electric field, $\mathbf{E}'_{\perp}$, which is transverse to the magnetic field. The Pedersen conductivity maximizes typically at a somewhat higher altitude than the Hall conductivity, namely where the ions are scattered in the direction of the electric field before they can start to gyrate about the magnetic field. The quantity $\sigma_{\|}$ is called the parallel conductivity since it governs the magnetic field-aligned current driven by the parallel electric field component, $E_{\|}$.

The fully developed relation between current, conductivity, electric field, and neutral winds can now be seen by replacing $\mathbf{E}'_{\perp}$ with its neutral-wind-dependent expression

$$\mathbf{E}'_{\perp} = \mathbf{E}_{\perp} + \mathbf{v}_{\mathrm{n}} \times \mathbf{B} \tag{7}$$

in (6). For dynamo currents, the dominant driving force, or electromotive force in the classical terminology of electrodynamics, is actually the $\mathbf{v}_{\mathrm{n}} \times \mathbf{B}$ term induced by the motion of ions, which are coupled to the neutral atmosphere via collisions and thus move with the neutral wind, across the magnetic field. At low magnetic latitudes, the most important dynamo effect is the daily variation of the atmospheric motion caused by

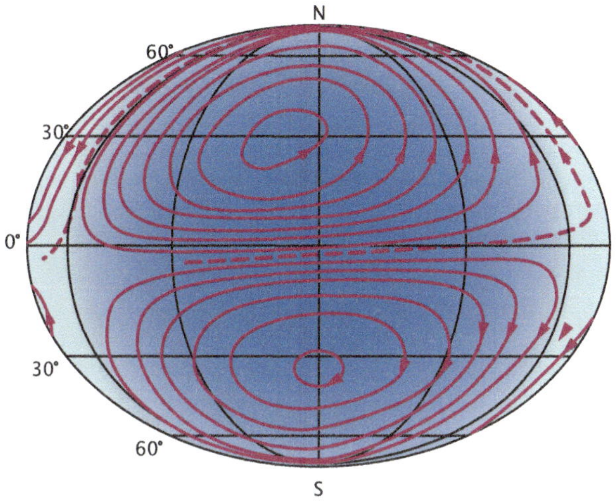

Fig. 9 Dayside view of the Sq current system (adapted from Baumjohann and Treumann 1996)

the tides of the atmosphere, that is, diurnal and semi-diurnal oscillation, which are excited by the heating of the atmosphere due to solar radiation. At auroral and polar latitudes, neutral winds generated by auroral heating may actually be much stronger, introducing even more complex coupling and feed-back processes between magnetospheric, ionospheric and upper atmosphere motions (see, e.g., Kamide and Baumjohann 1993; Untiedt and Baumjohann 1993).

3.5.2 Earth

The current system created by tidal motion of the Earth's atmosphere at typical altitudes of 100–130 km is called the "solar quiet" or Sq current. This current system creates daily magnetic variations. Figure 9 presents a global view of the average Sq current system from above the terrestrial ionosphere: the lines give the direction of the current while the distance between the lines is inversely proportional to the height-integrated current density. The Sq currents form two vortices, one in the Northern and the other in the Southern Hemisphere, which touch each other at the geomagnetic equator. In accordance with the day–night contrast in the ionospheric conductivities, the Sq currents are concentrated on the dayside.

At the geomagnetic equator, the Sq current vortices of the southern and northern hemispheres touch each other and form an extended nearly jet-like current in the ionosphere, the equatorial electrojet. However, this electrojet would not be so strong if it were formed only by the concentration of the Sq current. The special geometry of the magnetic field at the equator together with the nearly perpendicular incidence of solar radiation causes an equatorial enhancement in the effective conductivity which leads to an amplification of the jet current.

3.5.3 Mercury

Mercury does not have a classical atmosphere. The gravitational field of the planet is too weak and the surface temperature too high to allow a gas envelope to stably exist. Mercury is surrounded by an active exosphere, whose particles originate from the interaction between the solar wind and magnetospheric particles directly impinging onto the surface where significant sputtering occurs. Sodium, calcium, and magnesium have been detected in this exosphere by ground observations and in the recent Messenger measurements (e.g. McClintok et al. 2009). The role these species play in magnetospheric dynamics is not yet clear. In the terrestrial case magnetosphere–ionosphere coupling plays a major role in regulating magnetospheric dynamics by, e.g. electric current-closure accomplished via ionospheric currents. In the Hermean magnetosphere, if such a coupling exists, it must be either directly with the upper planetary interior, with a photo-ionization layer, or current closure must occur entirely in the magnetosphere proper (e.g. Glassmeier 2000; Milillo et al. 2005).

3.5.4 Venus and Mars

As was stressed previously, the Venusian ionosphere is essentially un-magnetized under low solar-wind pressure conditions (as shown in the left-hand panel of Fig. 10). In the absence of a magnetic field, the ionospheric conductivity is isotropic, with a scalar conductivity σ_0 equal to $\sigma_{\parallel}$. The Pedersen conductivity is also equal to σ_0, and the Hall conductivity goes to zero. Since the Venusian ionosphere near its ionopause is collisionless, σ_0 is very large, and the ionosphere is a very efficient conductor, and in practice behaves as a perfect conductor which prevents the diffusion of any external magnetic electric field into the ionospheric layer. Hence the pile-up of solar-wind magnetic, fields in the magnetic barrier above the ionopause described in Sect. 1.5.

This, however, does not prevent the Venusian ionosphere from carrying electrical currents, if only the ionopause currents located at the top of the layer which cancel the effect of external magnetic fields inside the ionosphere. In the case of low solar wind pressure, the upper layer of the ionosphere must therefore carry "image currents" induced by the sum of all the currents flowing outside of it.

When the solar wind pressure increases (from left to right in Fig. 10), the ionopause position moves down below the exobase and ion–neutral collisions become important. Under these conditions, the solar wind magnetic field diffuses more and more into the ionospheric layer, which becomes magnetized and recovers electrical properties closer to the Earth case,

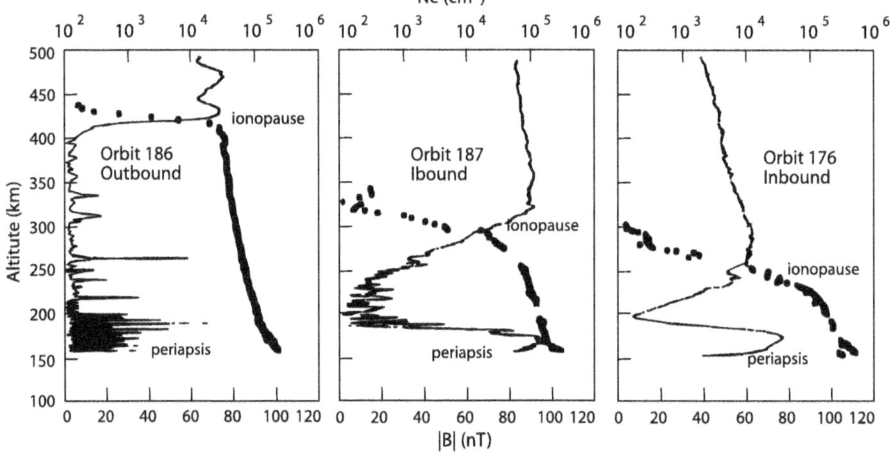

Fig. 10 Vertical profile of magnetization state and electron density of Venusian ionosphere measured by Pioneer Venus Orbiter for different levels of solar wind pressure. From left to right: low (orbit 186), moderate (orbit 177) and high (orbit 176) dynamic pressure (from Elphic et al. 1981)

with an anisotropic conductivity tensor and probably electric currents induced either by the solar wind interaction, or by the effect of thermospheric winds. There is, however, no observational signature of these currents available.

3.5.5 Jupiter and Saturn

Our knowledge of giant planets ionospheres is strongly limited by the sparsity of data. The main data source is provided by radio occultation measurements from interplanetary (Pioneer, Voyager) or orbiting planetary probes (Galileo, Cassini), with the addition of some data on infrared emissions of H_3^+ ions, one of the main ion species in giant planets' ionospheres. No direct measurement of ionospheric currents is available to our knowledge, so our understanding of ionospheric currents there essentially relies on separate data concerning the ionospheric layers, neutral and plasma winds, and electric fields, and on the use of models to retrieve the currents from these elements. Since very little is known about the Uranus and Neptune ionospheres, we will focus on the cases of Jupiter and Saturn.

Both planets display ionospheric layers with peak electron densities in the range of 10^3 to 10^4 cm^{-3}, and their striking common feature (in contrast to Earth) is the dominance of highly structured, multi-layer vertical ionospheric profiles with considerable variability from one occultation profile to another, even at similar latitudes. At Saturn, using the set of occultation data provided by the Cassini Radio Science experiment, some general trends have been established by, e.g., Kliore et al. (2009). There appears to be a tendency for the peak in ionospheric density to increase with increasing latitude. The large day–night variation speaks in favour of an ionosphere dominated by relatively short-lived ions such as H_3^+ or H_3O^+. Another important feature of the Saturnian ionosphere is the importance of its coupling to the rings system. Both direct shadowing by the rings and influx of water from the rings to the conjugate ionosphere are suggested by the data. Figure 11 (from Kliore et al. 2009) shows average density profiles for three different latitude ranges. An increase of the peak electron density with increasing latitude is clearly evidenced.

Also at Jupiter there is an extreme variability from one profile to another, which suggests that solar UV plays a limited role in the balance of ionospheric layers, and that atmospheric gas dynamics and the effects of auroral precipitation and Joule heating at high latitudes must be very important. In such a context, models are key to providing a global picture of giant planets ionospheres and their current systems. Hence the development of thermosphere/ionosphere general circulation models, in which the coupling to the magnetosphere, which maximises at auroral latitudes, is taken into account via the imposition of energy deposition to the thermosphere by particle precipitation and Joule heating, and of collisional coupling of the thermosphere to the sub-corotational auroral ionospheric plasma. All models show that energy and momentum deposited at auroral latitudes drives a strong thermospheric wind system and associated current systems which in turn redistribute energy far beyond the auroral region. Model results show that this redistribution is global at Jupiter, extending down to the equator, whereas the latitude extent of this global aurora-induced wind system and heating is still controversial at Saturn.

3.6 Field-Aligned Currents

As was shown previously, magnetospheric current systems can mainly be described as a sum of elementary current systems flowing in specific regions, such as the magnetopause and tail, the ring current and magnetodisk, and the ionospheric current system. Though a large fraction of these currents flows in closed loops (i.e. is divergence-free), some fraction of it may

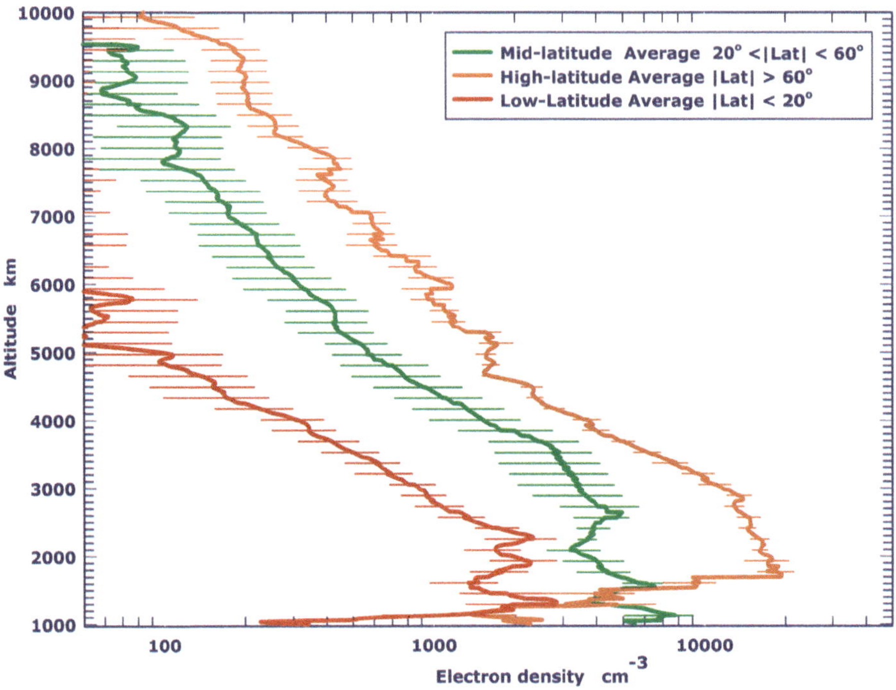

Fig. 11 Average ionospheric electron density profiles determined from Cassini radio occultation measurements near dawn and dusk within three latitude ranges. From Kliore et al. (2009)

accumulate charges in specific regions, thus generating electric potential drops between different regions, or be connected to permanent sources of electric potential difference, like the solar wind when the planetary field is reconnected with interplanetary field lines (e.g., Sect. 4.1 below). In such a situation, charge accumulation or existing potential drops generate electric current flows along conducting paths connecting regions of different potential. Such paths exist in planetary intrinsic magnetospheres: if magnetic field lines, which are near perfect conductors for cold ionospheric electrons, connect a region of (typically weak) charge accumulation to the planetary ionosphere, these field lines can carry so-called "field-aligned currents", which flow along magnetic field lines and close horizontally through the ionospheric conductor. These field-aligned currents are of the utmost importance at high latitudes at Earth and the giant planets, where near-vertical ionospheric field lines provide a direct electrical connection between the auroral ionosphere and distant magnetospheric regions.

One frequent (if not systematic) visible manifestation of these field-aligned current flows is the generation of intense auroral emissions. Indeed, in magnetospheric regions where the density is low, the reservoir of free current-carrying electrons is limited, and the mirror force along converging field lines limits the access of electrons to the ionosphere, upward current flow along field lines requires the generation of limited voltage drops along field lines. The current-carrying electrons are thus accelerated along their guiding field line and precipitated into the ionosphere, where they produce an aurora. For this reason, as we will show in Sect. 4, auroral displays at the various planets are a good first-order tracer of the ionospheric roots of upward field-aligned currents.

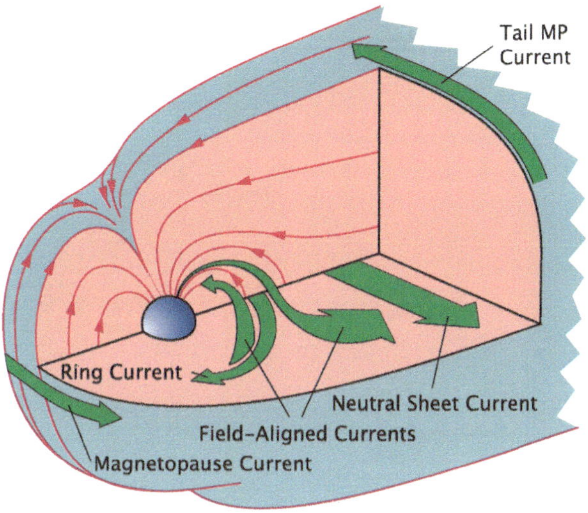

Fig. 12 Magnetospheric current systems (adapted from Baumjohann and Treumann 1996)

4 Global Current Systems and Magnetospheric Dynamics

4.1 Intrinsic Magnetospheres

The global configuration of current systems in intrinsic magnetospheres, as described in the previous section, is shown in Fig. 12 for the Earth's case. However this is a static view. Let us describe now how these connected current systems, magnetospheric, field-aligned and ionospheric, are related to the overall dynamics and flow of the plasma in the magnetospheric cavity. For this, we will consider the different momentum sources acting on plasmas and magnetic field lines.

4.1.1 Solar Wind–Magnetosphere Interaction: The Dungey Cycle

The concurrent drift of plasma and field lines as one entity is called convection. Due to the near-infinite conductivity in a collisionless plasma, the electric field is zero in the frame of reference moving with the plasma at a velocity $\mathbf{v}_c$. However, an observer in a fixed frame of reference will measure a convection electric field

$$\mathbf{E}_c = -\mathbf{v}_c \times \mathbf{B} \tag{8}$$

Hence, the flow of the magnetized solar wind around a magnetosphere represents an electric field in the planet's frame of reference. Since the solar wind cannot penetrate the magnetopause, this electric field cannot directly penetrate into the magnetosphere. However, when the interplanetary magnetic field has a component that is antiparallel to the planetary field lines at the dayside magnetopause, planetary and interplanetary field lines can merge.

As shown in Fig. 13 for the terrestrial case, when a southward directed interplanetary field line encounters the magnetopause, it can merge with a closed terrestrial field line, which has both foot points on the Earth. The merged field lines will split into two open field lines, each with one end connected to the Earth and the other stretching out into the solar wind. Subsequently, the solar wind will transport this field line across the polar cap down the tail and due to magnetic tension, the magnetospheric part of the field line, will

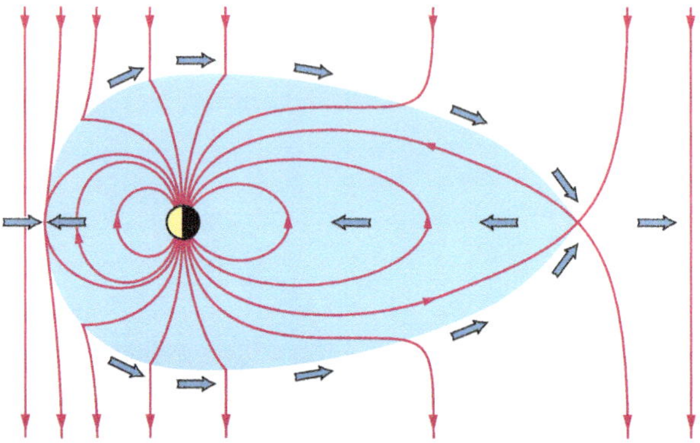

Fig. 13 Reconnection and convection in a magnetosphere (adapted from Baumjohann and Treumann 1996)

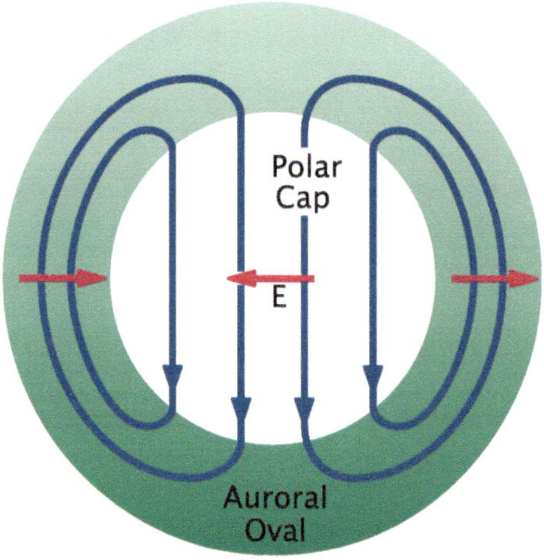

Fig. 14 Convection and electric field in the polar ionosphere (adapted from Baumjohann and Treumann 1996)

also be transported tailward. At the night side end of the magnetosphere the two open field line halves will meet again and reconnect, leaving a closed but stretched terrestrial field line in the magnetotail and an open solar wind field line down tail of the magnetosphere. The stretched tail field line will relax and shorten in the earthward direction. During this relaxation it transports the plasma, to which it is frozen, toward the Earth.

The motion of flux tubes across the polar cap due to magnetic merging also moves the ionospheric foot point of the flux tube and the plasma tied to it across the polar cap to the nightside. Similarly, the sunward convection of magnetospheric flux tubes leads to a sunward convection of the foot points of these flux tubes on the dawn- and dusk-side high-latitude ionospheres. This leads to a two-cell convection pattern in the polar ionosphere (Fig. 14).

The convection pattern is equivalent to an electric potential pattern. Hence, we can take the two-cell convection pattern as a two-cell pattern of equipotential contours, which is equivalent to an ionospheric electric field that is directed toward dusk in the northern polar cap. Inside the Northern Hemisphere auroral oval the electric field is directed toward the pole on the dusk side, while it has a southward direction in the morning hours.

Since the ionospheric conductivity has three different components, three types of currents will be generated by the convection electric field. The first type is the field-aligned currents flowing parallel to the magnetic field into and out of the ionosphere. Second, there are the Pedersen currents which flow perpendicular to the magnetic field lines and parallel to the ionospheric convection field. Finally, Hall currents will flow perpendicular to both the magnetic and the electric field.

This type of solar-wind induced convection pattern, called the "Dungey cycle", is dominant at Earth, and may partly exist at Mercury and Saturn.

Earth In the terrestrial magnetosphere, the total potential difference between the dawn and dusk magnetopause, or equivalently across the polar cap, corresponds to about 50–100 kV. For a cross section of the magnetosphere of about 30 R_E, this amounts to a dawn-to-dusk directed electric field of some 0.2–0.5 mV m^{-1}.

Since energetic particles precipitating from the magnetotail into the auroral oval (the green shaded ring in Fig. 8) cause significant ionization, its conductivity is much higher than that of the polar cap, which is threaded by open field lines. As a result, the high latitude current flow is concentrated inside the auroral oval, where it forms the auroral electrojets.

The auroral electrojets carry a total current of some million amperes. They are primarily Hall currents which originate around noon where they are fed by downward field-aligned currents. Typical sheet current densities range between 0.5 and 1 A m^{-1}. The eastward electrojet flows in the afternoon sector and terminates in the pre-midnight region where it partially flows up magnetic field lines and partially rotates northward, joining the westward electrojet. The westward electrojet flows through the morning and midnight sector and typically extends into the evening sector along the poleward border of the auroral oval where it also diverges as upward field-aligned currents.

Convection is not a stationary process: magnetic merging between interplanetary and terrestrial field lines at the dayside magnetopause does not occur all the time, but mainly for southward-oriented interplanetary field lines, and is typically not in equilibrium with reconnection in the magnetotail. Only part of the flux transported into the tail is reconnected instantaneously in the deep tail and convected back to the dayside. The remaining field lines are added to the tail lobes, where they increase the magnetic flux density and, hence, enhance the cross-tail current in the neutral sheet. After some tens of minutes these intermediately stored field lines are suddenly reconnected at tail distances of 20–25 Earth radii and their magnetic energy is explosively released. The sudden reconnection of previously stored flux tubes has rather dramatic effects on the magnetospheric plasma and associated phenomena like aurora and magnetospheric and ionospheric currents. These effects are summarized as a "magnetospheric substorm".

A substorm starts when the dayside merging rate is distinctively enhanced, typically due to a southward turning of the interplanetary magnetic field. The flux eroded on the dayside magnetopause is transported into the tail. Part of the flux is reconnected and convected back to the dayside magnetosphere. The enhanced convection causes enhanced current flow in the convection electrojets. The other part of the flux is added to the tail lobes. After 30–60 min, too much magnetic flux and thus magnetic energy has been accumulated in the tail. The tail becomes unstable and must release the surplus energy. This is the time of substorm onset and the beginning of the substorm expansion phase. At substorm onset, the aurora suddenly brightens and fills the whole sky. During the following 30–60 min, rather dramatic changes are seen in the auroral zone currents.

The unloading of magnetic flux previously stored in the magnetotail leads to the formation of a substorm electrojet with strongly enhanced westward current flow in the midnight sector. The substorm electrojet is concentrated in the region of active aurora and expands westward during the course of the expansion phase. While in the case of the convection electrojets, the field-aligned currents are distributed over a wide local time range, for the substorm electrojet, the jet itself and its field-aligned currents are much more concentrated

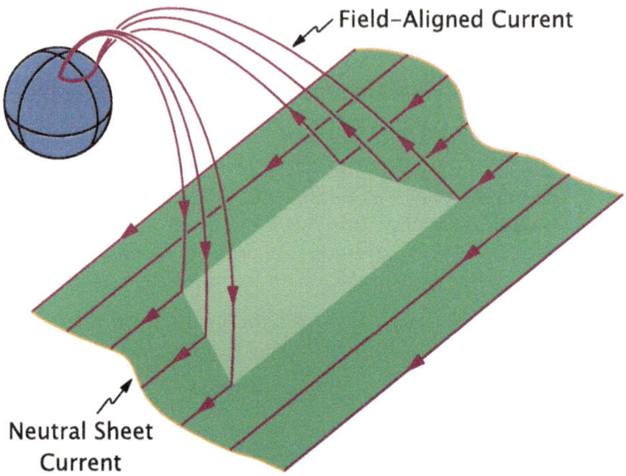

Fig. 15 Substorm current wedge (adapted from Baumjohann and Treumann 1996)

in the midnight sector, forming a current wedge as depicted in Fig. 15. The brightest aurora, at the western edge of the current wedge is associated and caused by upward field-aligned currents, i.e., by the energetic electrons precipitating here into the ionosphere and carrying the upward field-aligned current.

Mercury Field-aligned currents possibly associated with a "Dungey cycle" have also been observed in the Hermean magnetosphere (Slavin et al. 1997). Current densities are of the order of 10^{-7} A m^{-2}, a value again comparable with terrestrial values. How these field-aligned currents are closed within the magnetospheric system and what role they play is yet unexplored.

Jupiter and Saturn At Jupiter, the solar-wind driven Dungey cycle plays a minor role. The planet is dominated by planetary rotation to large radial distances. Saturn seems to be in intermediate case, as shown by the Saturn aurora which is clearly modulated by the solar wind (e.g., Cowley et al. 2005). A Dungey cycle is likely present there, superimposed onto the so-called Vasyliunas cycle, which we will now describe. Extensive description of the magnetospheric configuration and dynamics at giant planets can be found in Khurana et al. (2004) for Jupiter, and in Gombosi et al. (2009) for Saturn.

4.1.2 Magnetosphere–Planetary Rotation Interaction: The Vasyliunas Cycle

Let us first look at the Jovian aurora, which traces the basic features of Jovian magnetospheric dynamics as observed in UV light by the Hubble Space Telescope (Fig. 16). This aurora displays three distinct components. At the highest latitudes, a series of faint and time-variable auroras are believed to be related to the solar-wind interaction and possibly to reconnection processes in the magnetotail, indicating the possible presence of a Dungey cycle confined to the outermost regions of the magnetosphere. At the lowest latitudes, a series of three bright localized spots are connected among magnetic field lines to the three Galilean moons Io, Europa and Ganymede. They are generated by the electrons accelerated in the moon-magnetosphere interaction, and can also be seen as tracers of the ionospheric closure of the Jupiter/moon current systems. But the most prominent feature of the Jovian aurora is the stable, circumpolar and bright auroral feature called the main oval, which runs approximately along a magnetic shell at a colatitude of 16° and is about 1000 km wide.

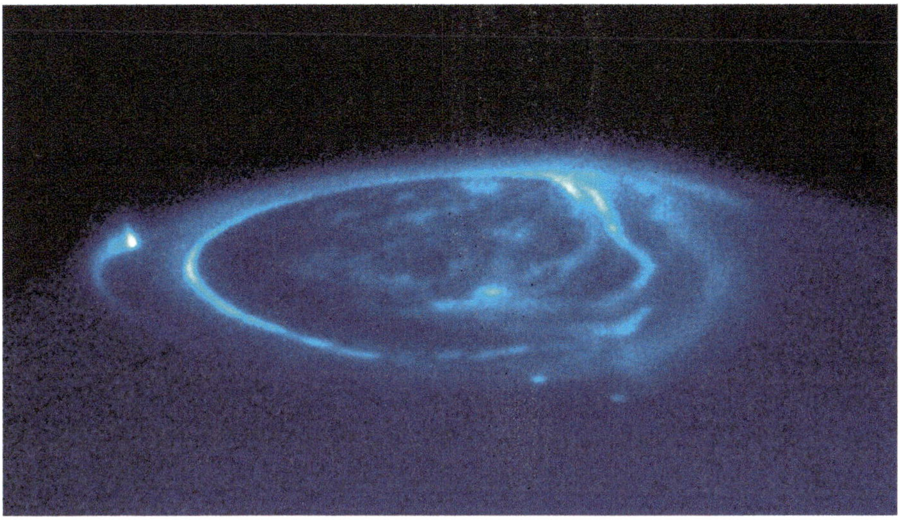

Fig. 16 Jovian aurora seen in UV light by the Hubble Space Telescope, revealing three distinct components

Voyager and Galileo plasma flow observations near Jupiter's equatorial plane make it possible to connect this oval to the specific region in the middle magnetosphere where rigid corotation of Jovian plasma and magnetic field lines with the planet starts to break down, at radial distances of about 20 Jovian radii and beyond. Inside of that distance, the magnetospheric plasma near-rigidly corotates with the planet, while outside of it the efficiency of rotational coupling between the outer magnetosphere and the planet's upper atmosphere decreases with increasing distance, and magnetospheric plasma flow increasingly lags behind corotation. The model of ionosphere–magnetosphere electrical coupling developed by Cowley and Bunce (2001) has been able to explain this feature as a consequence of centrifugally-driven outward transport of Iogenic plasma (Fig. 17). The plasma generated in the Io neutral torus by ionization and pick-up experiences a strong outward centrifugal force due to the large corotational flow (on the order of 100 km/s at Io's orbit), under which it diffuses outward by some still only partly understood mechanism. As it is transported outward, the Iogenic plasma has to gain angular momentum if rigid corotation with the planet is to be maintained, and that is a natural result of electrodynamic coupling between the ionosphere and magnetosphere. With magnetic field lines playing the role of electric connectors, a current circuit is established in which an equatorward horizontal current flows in the ionosphere, and an outward radial current flows near the equatorial plane in the magnetospheric plasma. The $\mathbf{J} \times \mathbf{B}$ forces associated with these two current segments work in such a way as to brake the ionosphere and thermosphere, thus extracting angular momentum from them, and to accelerate the rotation of the magnetospheric plasma to maintain corotation, thus transferring angular momentum to it. As one sees, this large-scale azimuthally symmetric current system has to be closed by magnetic-field aligned currents flowing out of the ionosphere in the inner part of the region where corotation starts to break down and angular momentum must be supplied from the Jovian atmosphere, and into the ionosphere at the outer edge of the region throughout which this current system extends. Using empirical models of the plasma azimuthal flow and of the magnetic field, Cowley and Bunce showed that the main oval is produced by the precipitating electrons which carry the upward field-aligned currents. These

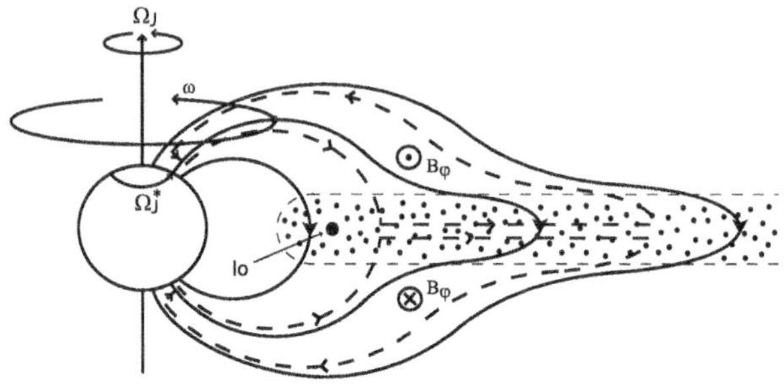

Fig. 17 Schematic of the current system and magnetic field deviations which maintain corotation of Iogenic plasma and flux tubes in the Jovian magnetosphere. The flux tube rotation speed ω lags behind the planetary rotation speed Ω_J, while the ionospheric rotation speed Ω_J^* is intermediate. From Cowley and Bunce (2001)

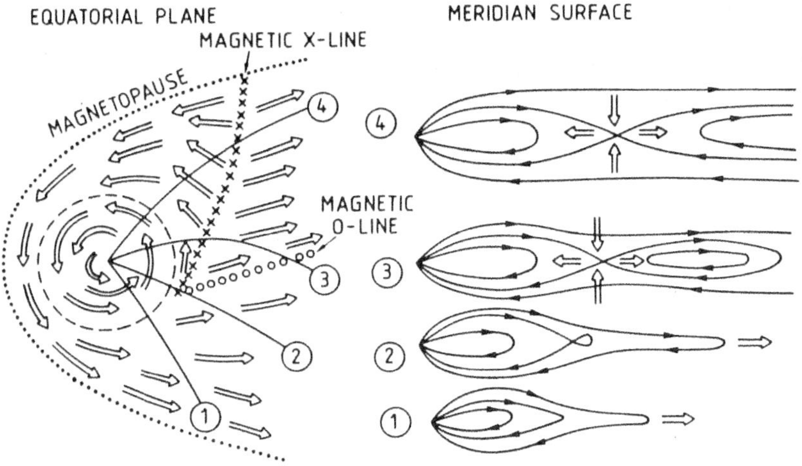

Fig. 18 Cartoon of the "Vasyliunas cycle", believed to be the dominant plasma and flux tube circulation mode in the Jovian magnetosphere (from Vasyliunas 1983)

electrons have to be accelerated by ~100 kV potential drops along auroral field lines, and they carry a 1 to 2 μA m^{-2} current density which maximizes around 16° colatitudes.

While the Iogenic plasma can flow radially outward in an azimuthally average sense, that is not the case for the advection of magnetospheric flux tubes which carry that plasma, the net radial transport of magnetic flux must be zero over time, even if the net transport of plasma is not. So, in reality, the transport scheme shown in Fig. 17 must be expanded in three dimensions in such a way that the Iogenic plasma flows from its source region (the Io torus) to some sink region, while magnetic flux can be recirculated inward. Vasyliunas (1983) provided a solution to this difficult problem by describing a large-scale convection system specific to Jupiter which is illustrated in Fig. 18. In the equatorial plane, the combination of corotation and the centrifugal force drives an outward plasma flow which develops preferentially in the afternoon and evening sectors and expands into the magnetic tail. The strongly elongated flux tubes which are convected in the magnetotail region, according to Vasyliunas, encounter several discontinuities and finally and X-line where they experience magnetic reconnection as shown by the meridional cross-sections in the figure. Tailward of the X line a plasma island is formed and detached from the Jovian field, and it is accelerated tailward, carrying with it a fraction of the Iogenic plasma. This constitutes the necessary plasma sink. Closer to Jupiter, nearly empty flux tubes are accelerated back towards the

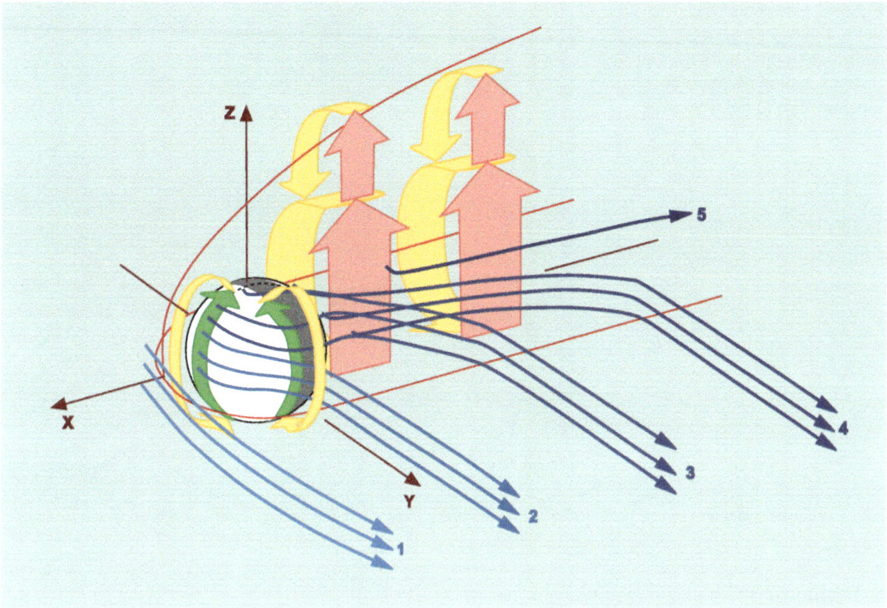

Fig. 19 Currents and magnetic field in a non-magnetic planet environment. *Digits* and different *shades of blue* show the successive draping of the initial (*1*) solar wind magnetic field lines. *Green arrows* show the ionospheric induced currents, *pink arrows* corresponds to the current in the current sheet, and *yellow arrows* show the currents in the magnetic barrier and on the induced magnetospheric boundary

planet and drift sunward via the morning sector. This so-called "Vasyliunas cycle" has essentially been validated both by Galileo observations and, more recently, by global MHD simulations.

4.2 Induced Magnetospheres

A global picture of magnetospheric and ionospheric currents in induced magnetospheres is presented in Fig. 19. The induced ionospheric currents (green arrows) are closed by the currents flowing on the induced magnetospheric boundary (yellow arrows) which is the ramp of the magnetic barrier on the dayside. The cross-tail current carried by the pressure gradient in the plasma sheet (pink arrows) can be considered as an extension of the ionospheric currents. This current is closed by currents (yellow arrows) on the nightside induced magnetospheric boundary. The last one is a natural tailward extension of the magnetic barrier. The cross-tail current is very asymmetric. On the south side of the midnight meridional plane the current is abruptly confined by induced magnetospheric boundary, but on the north side there is a more gradual divergence of the cross-tail current. The average cross-tail current density is $3\ \mathrm{nA\,m^{-2}}$.

Recent observations suggest the existence of reconnection in the Venusian magnetotail (Volwerk et al. 2009), with formation of an X-like magnetic field configuration at $X = -1.5\ R_{\mathrm{V}}$. If this is the case, the post-reconnection magnetic and current configuration would look as shown in Fig. 20. Just as in the Earth magnetotail, the cross tail current diverges from the reconnection site, creating a "wedge" containing the field aligned currents and ionospheric electrojet. This scenario, which lacks direct experimental evidence, will have to be checked by future space missions to our sister planet.

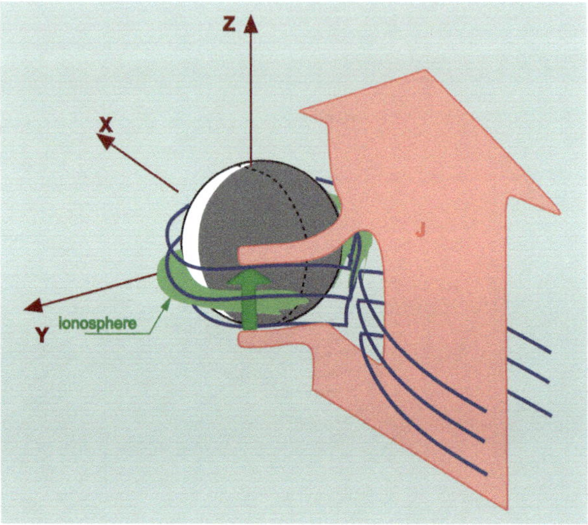

Fig. 20 Cartoon of the field-aligned current and additional ionospheric current (*green*) created by possible reconnection in the Venusian magnetotail

4.3 Moon-Associated Current Systems

While the small moons around Mars and even the big terrestrial Moon have no or hardly any effect on the Martian and terrestrial magnetospheres, respectively, the situation is different for the larger moons of the giant gas planets. As already mentioned (e.g., Sect. 2.3), many of them orbit inside the planetary magnetosphere and are geologically active bodies delivering material to their space environment. Io's active volcanism, cryovolcanism at Enceladus, and geyser activity at Triton are the key examples for the Jupiter, Saturn, and Neptune systems. Jupiter's moon Ganymede remains unique as possessing an intrinsic magnetic field and magnetosphere, and Saturn's largest moon Titan is unique for its dense atmosphere of nitrogen and methane.

The configurations and objects cover a variety of possible cases for the interaction between orbiting moons and the ambient magnetospheric plasma flow (e.g., Sect. 1.3 and Table 1), and for the establishment of the resulting current and magnetic field systems. We will describe this variety only briefly here, and refer the reader to another ISSI Space Science Series book ("Moons of the outer solar system: exchange processes involving the interior", to be published 2009), as well as to previous review papers on the subject (e.g., Kivelson et al. 2004), for extensive descriptions.

A unique case of an *intrinsic satellite magnetosphere*, approximately the size of Mercury's magnetosphere can be found at Ganymede, which interacts with the Jovian plasma flow. It was discovered by the Galileo magnetometer (Kivelson et al. 1996), and has been the subject of several modelling studies. Figure 21 shows the model Ganymede magnetospheric field and flow computed by Jia et al. (2009). One sees how well this magnetosphere corresponds to the Dungey model, in which the Jovian magnetospheric flow replaces the solar wind flow. Further efforts will be necessary to prepare and optimize observation strategies of the next generation of missions to the Jovian system and its satellites (e.g., Blanc et al. 2009), which are likely to include a dedicated Ganymede orbiter and will provide a comprehensive description of this mini-magnetosphere.

Saturn's moon Titan represents thanks to Voyager and now to more than four years of Cassini observations, a very well documented case of an *induced satellite magnetosphere*

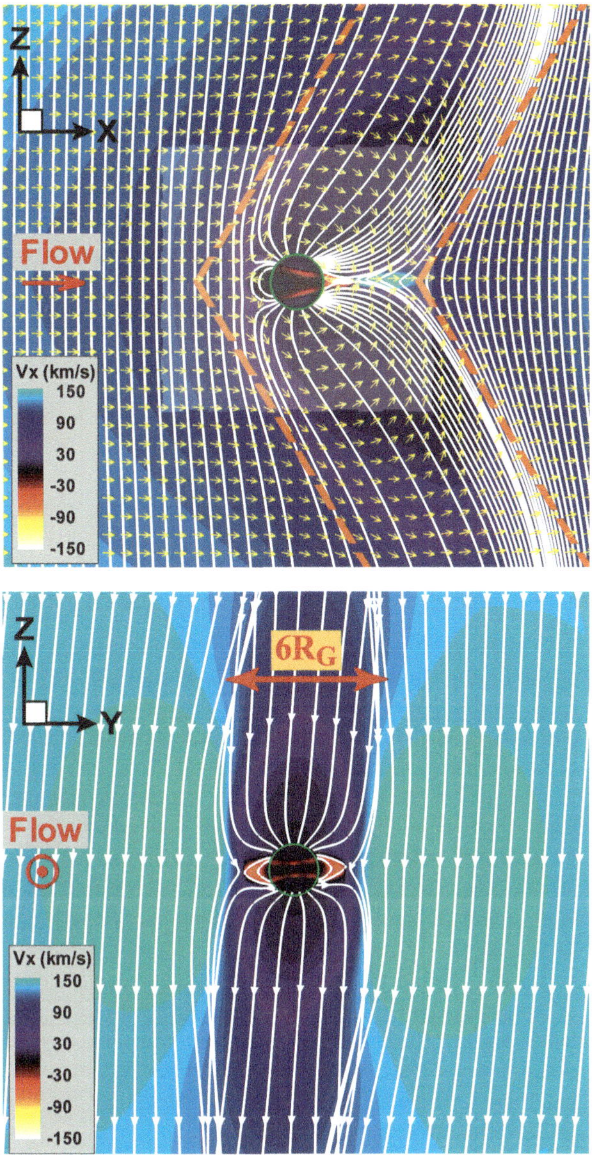

Fig. 21 Model of Ganymede's magnetosphere, its magnetic field lines (*white continuous curves*) and its plasma flow (*yellow arrows*) seen in the XZ and YZ planes (Z is roughly along the undisturbed Jovian main field, X is aligned to the incident flow direction; from Jia et al. 2009)

around a planetary moon, which bears many similarities with Venus though in the context of a sub-Alfvénic interaction (see Sittler et al. 2009a for a detailed description). In addition to the fact that it represent the best documented satellite induced magnetosphere, Titan's magnetospheric interaction has another outstanding interest: Due to irradiation of the N_2–CH_4 upper atmosphere, Titan produces via photolysis and recombination a rich variety of heavy hydrocarbon species which might be the seed particles for Titan's organic haze, and in that sense could be one of the main sources of the satellite's intriguing prebiotic chemistry (Sittler et al. 2009b).

A common feature of moon-magnetosphere interactions is the generation of Alfvén wings, a characteristic feature of low Alfvénic Mach number interactions between a mov-

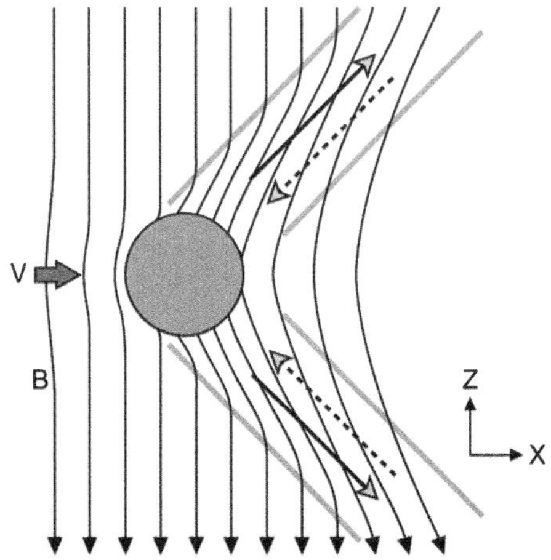

Fig. 22 (*Left*) Geometry of Alfvén wings in the plane containing the flow direction (X-axis) and the background magnetic field (Z-axis). (*Right*) Alfvén wings in the *YZ* plane where Y axis points in the direction of Jupiter and Z axis is parallel to the background magnetic field. From Khurana et al. (2009)

ing conductor and a magnetized plasma flow, as illustrated in Fig. 22. Such Alfvén wings develop around the Galilean satellites, and at least three of them extend down to the Jovian upper atmosphere and ionosphere, where they close their currents and generate the auroral spots mentioned previously (Fig. 16). As such, Alfvén wings and the associated field-aligned current systems represent important mechanisms for the exchange of momentum and energy between the magnetized environments of giant planets and those of their moons.

One final, very important aspect of magnetism at giant planet moons must be mentioned here. It is the presence of induced currents and induced fields in the sub-surface oceans of these moons. Indeed, for Europa and Callisto the likely presence of sub-surface oceans has been inferred or detected by the Galileo spacecraft using magnetic measurements (Kivelson et al. 1996; Khurana et al. 1998). The interaction of Jupiter's tilted magnetic dipole, which follows the planet's 10-hour rotation, with these conducting ocean layers indeed produces within the satellite bodies and in their immediate environment an additional component of induced magnetic fields which modifies both the total field and the configuration of the moon-magnetosphere interaction, and provides a powerful tool to characterize these internal ocean layers. Future missions to the Jupiter system and its moons (e.g., Blanc et al. 2009) will for this reason make extensive use of magnetic field measurements as a key contribution to a full characterization of these oceans.

4.4 Mars: An Intermediate and Unique Case

As a final example, let us emphasize the unique case of the Martian magnetic environment. Mars Global Surveyor measurements have revealed a system of magnetic anomalies which are most likely the result of remanent rock magnetism at the Martian surface, which extend into the Martian ionosphere, upper atmosphere and exosphere. These magnetic anomalies originate essentially from the older Martian terrains. The consequence is that, unlike Venus which has an essentially unmagnetized ionosphere (at least for low solar wind pressure), the Martian ionosphere is magnetized, and may establish magnetic connections between the Martian remanent planetary field below the ionosphere and the field from the induced magnetosphere above it.

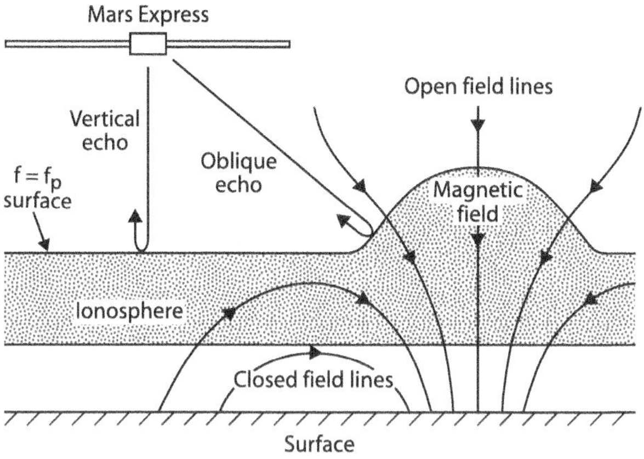

Fig. 23 Cartoon illustrating the magnetic geometry above Martian magnetic anomalies in a region of reversal of horizontal magnetic polarity. The field becomes vertical at the center of the reversal structure, thus opening a cusp structure to the entry and precipitation of electrons from the induced magnetosphere. The resulting ionospheric density enhancement is detected by the MARSIS radar on board Mars Express as a set of oblique echoes and then as an increase of the cut-off frequency above the cusp. From Duru et al. (2006)

Using the Mars Express MARSIS radar, Duru et al. (2006) and Gurnett et al. (2008) conducted a systematic study of ionospheric echoes and the corresponding density structures. They found a set of localized ionospheric density enhancements which coincide rather well with regions of reversals of the horizontal component of the Martian remanent field. Figure 23 illustrates their interpretation of these radar observations. The regions of horizontal polarity reversal correspond to regions where the field is locally vertical, and displays a cusp structure open to the regions above the ionosphere. Electron precipitation from the region of the induced magnetosphere into the ionosphere via this cusp structure can locally enhance the ionospheric plasma density, and result in the system of oblique radar echoes and increased cut-off frequency which is detected by the MARSIS instrument. As illustrated in the figure, the Martian ionosphere thus displays, at a local and regional scale, some of the features characterizing the auroral ionospheres of Earth and giant planets, making the Martian plasma environment an intermediate case between induced and intrinsic magnetospheres.

5 Conclusions

The interaction of planetary environments with the solar wind, as well as the interplay of the different components of giant planet systems, produce a diversity of current systems which we have reviewed in this chapter. The broad diversity of these currents can be explained by:

- the diversity of the types of interactions between planets and the solar wind, leading to "intrinsic magnetospheres" for magnetized planets and moons, and to "induced" magnetospheres for unmagnetized objects;
- the existence and importance of an atmosphere;
- the intensity and variety of solar-wind and internal plasma sources;
- the intensity and geometry of the internal planetary magnetic field.

Despite this diversity, all current components can be classified into only a few different types, which include ionospheric currents, currents carried by the magnetospheric boundaries like the magnetopauses or ionopauses, magnetotail currents, and currents flowing inside the magnetospheres, like ring currents, plasma sheet currents and currents aligned to the magnetic field lines (or field-aligned currents). The flow of these external currents has to be divergence-free, so that they are actually organized into global current systems which connect different regions via field-aligned currents, and play an important role in the transfer of energy and momentum between the solar wind and the different regions of planetary environments. In current research, and thanks in particular to planetary orbiters, we are starting to elucidate how these transfers of energy and momentum in planetary environments are related to the local and global dynamical regimes of their planetospheres.

At Earth, ionospheric currents have been detected and studied since the 19th century, using their magnetic signatures at the surface of our planet. Magnetospheric currents were characterized only during the space age, and even today they are not so well known as ionospheric current systems. At other planets, the situation as of today is just the opposite, magnetospheric magnetic fields and by extension magnetospheric currents start to be described and partly understood by means of interplanetary probes and of planetary orbiters, but there is no or little "ground truth" for ionospheric currents, so that one of the challenges for the coming decades is to describe these currents. This will take orbiters with specific orbits coming very close to the planets' surface, like the Juno mission does for the investigation of polar current systems at Jupiter, and moon orbiters dedicated to the exploration of giant planet systems. Much progress will have to be accomplished, given this limiting element, before our knowledge of global current systems at planets comes close to the one we have achieved for Earth.

Still, progress on these open questions of planetary external magnetism will be a key ingredient towards achieving a better understanding of how planetary environments work. Not only do we need to better understand external current systems for their own interest and for our own physical understanding, but in addition they play an important role for progress in other areas of planetary sciences and exploration. At Saturn's satellite Titan, the induced magnetosphere seems to play a triggering role in the initial formation of complex organic molecules, and therefore of subsequent prebiotic chemistry. And the exploration and characterization of the putative subsurface oceans of Jupiter's moons Europa, Ganymede and Callisto has already and will even more in the future make use of in-orbit magnetic field measurements as a powerful means of sounding the mysterious interiors of these fascinating bodies.

References

B.J. Anderson, M.H. Acuña, H. Korth, M.E. Purucker, C.J. Johnson, J.A. Slavin, S.C. Solomon, R.L. McNutt, Science **321**, 82–86 (2008)

C.S. Arridge, C.T. Russell, K.K. Khurana, N. Achilleos, N. Andre, A.M. Rymer, M.K. Dougherty, A.J. Coates, Geophys. Res. Lett. **34**, L09108 (2007)

C.S. Arridge, K.K. Khurana, C.T. Russell, D.J. Southwood, N. Achilleos, M.K. Dougherty, A.J. Coates, H.K. Leinweber, J. Geophys. Res. **113**, A08217 (2008)

W. Baumjohann, R.A. Treumann, *Basic Space Plasma Physics* (Imperial College Press, London, 1996)

W. Baumjohann, A. Roux, O. LeContel, R. Nakamura, J. Birn, M. Hoshino, A.T.Y. Lui, C.J. Owen, J.-A. Sauvaud, A. Vaivads, D. Fontaine, A. Runov, Ann. Geophys. **25**, 1365–1389 (2007)

M. Blanc, R. Kallenbach, N.V. Erkaev, Space Sci. Rev. **116**, 227–298 (2005). doi:10.1007/s11214-005-1958-y

M. Blanc, Y. Alibert, N. André, S. Atreya, R. Beebe, W. Benz, S.J. Bolton, A. Coradini, A. Coustenis, V. Dehant, M. Dougherty, P. Drossart, M. Fujimoto, O. Grasset, L. Gurvits, P. Hartogh, H. Hussmann, Y. Kasaba, M.G. Kivelson, K.K. Khurana, N. Krupp, P. Louarn, J. Lunine, M. McGrath, D. Mimoun, O. Mousis, J. Oberst, T. Okada, R. Pappalardo, O. Prieto-Ballesteros, D. Prieur, P. Regnier, M. Roos-Serote, S. Sasaki, G. Schubert, C. Sotin, T. Spilker, Y. Takahashi, T. Takashima, F. Tosi, D. Turrini, T. van Hoolst, L. Zelenyi, Exp. Astron. **23**, 849–892 (2009). doi:10.1007/s10686-008-9127-4

J.E.P. Connerney, M.H. Acuña, N.F. Ness, Nature **292**, 724–726 (1981)

S.W.H. Cowley, E.J. Bunce, Planet. Space Sci. **49**, 1067–1088 (2001)

S.W.H. Cowley, S.V. Badman, E.J. Bunce, J.T. Clarke, J.-C. Gérard, D. Grodent, C.M. Jackman, S.E. Milan, T.K. Yeoman, J. Geophys. Res. **110**, A02201 (2005). doi:10.1029/2004JA010796

I.A. Daglis, R.M. Thorne, W. Baumjohann, S. Orsini, Rev. Geophys. **37**, 407–438 (1999)

F. Duru, D.A. Gurnett, T.F. Averkamp, D.L. Kirchner, R.L. Huff, A.M. Persoon, J.J. Plaut, G. Picardi, J. Geophys. Res. **111**, A12204 (2006). doi:10.1029/2006JA011975

I.M. Engle, J. Geophys. Res. **97**, 17169 (1992)

R.C. Elphic, C.T. Russell, J.G. Luhmann, F.L. Scarf, L.H. Brace, J. Geophys. Res. **86**, 11430–11438 (1981)

K. Ferrière, Rev. Mod. Phys. **73**, 1031–1066 (2001)

K.H. Glassmeier, Currents in Mercury's magnetosphere, in *Magnetospheric Current Systems*. Geophysical Monograph, vol. 118 (American Geophysical Union, Washington, 2000), pp. 371–380

K.H. Glassmeier, U. Auster, U. Motschmann, Geophys. Res. Lett. **34**, L22201 (2007). doi:10.1029/2007GL031662

T.I. Gombosi, T.P. Armstrong, C.S. Arridge, K.K. Khurana, S.M. Krimigis, N. Krupp, A.M. Persoon, M.F. Thomsen, Saturn's magnetospheric configuration, in *Saturn* (2009, in press)

X.C. Guo, C. Wang, Y.Q. Hu, J.R. Kan, Geophys. Res. Lett. **35**, L03108 (2008). doi:10.1029/2007GL032713

D.A. Gurnett, R.L. Huff, D.D. Morgan, A.M. Persoon, T.F. Averkamp, D.L. Kirchner, F. Duru, F. Akalin, A.J. Kopf, E. Nielsen, A. Safaeinili, J.J. Plaut, G. Picardi, Adv. Space Res. **41**, 1335–1346 (2008)

P. Janhunen, H.E.J. Koskinen, Geophys. Res. Lett. **24**, 1419–1422 (1997)

X.Z. Jia, R.J. Walker, M.G. Kivelson, K.K. Khurana, J.A. Linker, J. Geophys. Res. **114**, A09209 (2009). doi:10.1029/2009JA014375

Y. Kamide, W. Baumjohann, *Magnetosphere–Ionosphere Coupling* (Springer, Heidelberg, 1993)

Y. Kamide, W. Baumjohann, I.A. Daglis, W.D. Gonzalez, M. Grande, J.A. Joselyn, R.L. McPherron, J.L. Phillips, E.G.D. Reeves, G. Rostoker, A.S. Sharma, H.J. Singer, B.T. Tsurutani, V.M. Vasyliunas, J. Geophys. Res. **103**, 17705–17728 (1998)

K.K. Khurana, M.G. Kivelson, D.J. Stevenson, G. Schubert, C.T. Russell, R.J. Walker, C. Polanskey, Nature **395**, 777–780 (1998)

K. Khurana, M.G. Kivelson, V.M. Vasyliunas, N. Krupp, J. Woch, A. Lagg, B.H. Mauk, W.S. Kurth, The configuration of Jupiter's magnetosphere, in *Jupiter, The Planet, Satellites and Magnetosphere*, ed. by F. Bagenal, T. Dowling, W. McKinnon (Cambridge University Press, Cambridge, 2004)

Khurana, et al., *Moons of The Outer Solar System: Exchange Processes Involving the Interior*. ISSI Space Science Series (2009, to be published)

M.G. Kivelson, K.K. Khurana, C.T. Russell, R.J. Walker, J. Warnecke, F.V. Coroniti, C. Polanskey, D.J. Southwood, G. Schubert, Nature **384**, 537 (1996)

M.G. Kivelson, F. Bagenal, W.S. Kurth, F.M. Neubauer, C. Paranicas, J. Saur, Magnetospheric Interactions with Satellites, in *Jupiter, The Planet, Satellites and Magnetosphere*, ed. by F. Bagenal, T. Dowling, W. McKinnon (Cambridge University Press, Cambridge, 2004)

A.J. Kliore, A.F. Nagy, E.A. Marouf, A. Anabtawi, E. Barbinis, D.U. Fleischman, D.S. Kahan, J. Geophys. Res. **114**, A04315 (2009). doi:10.1029/2008JA013900

H. Korth, B. Anderson, M.H. Acuña, J.A. Slavin, N.A. Tsyganenko, S.C. Solomon, R.L. McNutt, Planet. Space Sci. **52**, 733–746 (2004)

J.G. Luhmann, S.A. Ledvina, C.T. Russell, Adv. Space Res. **33**, 1905–1912 (2004)

S. Maurice, I. Engle, M. Blanc, M. Skubis, J. Geophys. Res. **101**, 27053–27059 (1996)

W.E. McClintok, R.J. Vervack Jr., E.T. Bradley, R.M. Killen, N. Mouawad, A.L. Sprague, M.H. Burger, S.C. Solomon, N.R. Izenberg, Science (2009, in press)

A. Milillo, P. Wurz, S. Orsini, D. Delcourt, E. Kallio, R.M. Killen, H. Lammer, S. Massetti, A. Mura, S. Barabash, G. Cremonese, I.A. Daglis, E. Angelis, A.M. Lellis, S. Livi, V. Mangano, K. Torkar, Space Sci. Rev. **117**, 397–443 (2005)

R. Nakamura, W. Baumjohann, A. Runov, Y. Asano, Space Sci. Rev. **122**, 29–38 (2006)

N.F. Ness, K.W. Behannon, R.P. Lepping, Y.C. Whang, K.H. Schatten, Science **185**, 131–135 (1974)

J.G. Roederer, *Dynamics of Geomagnetically Trapped Radiation* (Springer, Berlin, 1970)

C.T. Russell, R.J. Walker, J. Geophys. Res. **90**, 11067–11071 (1985)

V.A. Sergeev, A. Runov, W. Baumjohann, R. Nakamura, T.L. Zhang, A. Balogh, P. Louarn, J.A. Sauvaud, H. Rème, Geophys. Res. Lett. **31**, L05807 (2004). doi:10.1029/2003GL019346

G.L. Siscoe, K.D. Siebert, J. Atmos. Sol. Terr. Phys. **68**, 911–920 (2006)
E.C. Sittler Jr., A. Ali, J.F. Cooper, R.E. Hartle, R.E. Johnson, A.J. Coates, D.T. Young, Planet. Space Sci. **57**, 1547–1557 (2009a)
E.C. Sittler Jr., R.E. Hartle, C. Bertucci, A.J. Coates, T. Cravens, I.A. Dandouras, D. Shemansky, Energy deposition processes in Titan's upper atmosphere and its induced magnetosphere, in *Titan from Cassini-Huygens*, ed. by R.H. Brown et al. (Springer, Berlin, 2009b)
J.A. Slavin, J.C.J. Owen, J.E.P. Connerney, S.P. Christon, Planet. Space Sci. **45**, 133–141 (1997)
J.A. Slavin, M.H. Acuña, B.J. Anderson, D.N. Baker, M. Benna, G. Gloeckler, R.E. Gold, G.C. Ho, R.M. Killen, H. Korth, S.M. Krimigis, R.L. McNutt, L.R. Nittler, J.M. Raines, D. Schriver, S.C. Solomon, R.D. Starr, P. Travnicek, T.H. Zurbuchen, Science **321**, 85 (2008)
J. Untiedt, W. Baumjohann, Space Sci. Rev. **63**, 245 (1993)
V.M. Vasyliunas, Plasma distribution and flow, in *Physics of the Jovian Magnetosphere*, ed. by A.J. Dessler (Cambridge University Press, New York, 1983)
M. Volwerk, K.-H. Glassmeier, A. Runov, W. Baumjohann, R. Nakamura, T.L. Zhang, B. Klecker, A. Balogh, H. Rème, Geophys. Res. Lett. **30**, 1320 (2003). doi:10.1029/2002GL016467
M. Volwerk, M. Delva, Y. Futaana, A. Retino, Z. Vörös, T.L. Zhang, W. Baumjohann, S. Barabash, Ann. Geophys. **27**, 2321–2330 (2009)
T.L. Zhang, W. Baumjohann, R. Nakamura, A. Balogh, K.-H. Glassmeier, Geophys. Res. Lett. **29**, 1899 (2002). doi:10.1029/2002GL-015544
T.L. Zhang, R. Nakamura, M. Volwerk, A. Runov, W. Baumjohann, H.U. Eichelberger, C. Carr, A. Balogh, V. Sergeev, J.K. Shi, K.-H. Fornacon, Ann. Geophys. **23**, 2909–2914 (2005)
T.L. Zhang, M. Delva, W. Baumjohann, M. Volwerk, C.T. Russell, H.Y. Wei, C. Wang, M. Balikhin, S. Barabash, H.-U. Auster, K. Kudela, J. Geophys. Res. **113**, E00B20 (2008). doi:10.1029/2008JE003215

Space Sci Rev (2010) 152: 135–157
DOI 10.1007/s11214-009-9563-0

Separation of the Magnetic Field into External and Internal Parts

N. Olsen · K.-H. Glassmeier · X. Jia

Received: 27 March 2009 / Accepted: 2 June 2009 / Published online: 25 July 2009

Abstract The magnetic field of a planet or a planetary moon contains contributions from a variety of sources in the environment of the body (external sources) and its interior (internal sources). This chapter describes different methods that have been developed for the separation of external and internal source contributions, and their application to selected planets and one of Jupiter's moons, Ganymede.

Keywords Spherical harmonic analysis · External current systems · Magnetospheric currents · Earth · Mars · Mercury · Ganymede

1 Introduction

Magnetic fields play an important role in physical processes throughout the Universe. In our own solar system the planetary magnetic fields are surprisingly different, even for planets of similar composition. In addition to being evidence of the evolution of the planet the magnetic field exerts a very direct control of the electrodynamic environment.

N. Olsen (✉)
DTU Space and Niels Bohr Institute of Copenhagen University, Juliane Maries Vej 30, 2100 Copenhagen Ø, Denmark
e-mail: nio@space.dtu.dk

K.-H. Glassmeier
Institut für Geophysik und extraterrestrische Physik, TU Braunschweig, Mendelssohnstraße 2-3, 38106 Braunschweig, Germany

K.-H. Glassmeier
Max-Planck-Institut für Sonnensystemforschung, Max-Planck-Strasse 2, 37191 Katlenburg-Lindau, Germany

X. Jia
Department of Earth and Space Sciences and Institute of Geophysics and Planetary Physics, University of California, Los Angeles, 405 Hilgard Ave., Los Angeles, CA, 90095-1567, USA

However, the magnetic field measured at or near the surface of a planet is the superposition of contributions from a variety of sources in the planetary environment (e.g. electric currents flowing in the ionosphere and magnetosphere) and its interior (e.g. fluid core, magnetization of rocks in the crust and induced currents in the planetary interior by the time-variations of the external fields). The scientific challenge is the sophisticated separation of these various sources and the accurate determination of the spatial and temporal structure of them.

The first attempt to separate internal and external magnetic field contributions has been performed about 180 years ago by Carl-Friedrich Gauss (1839). He developed the concept of spherical harmonic analysis, applied it to observations of the Earth's magnetic field, and came to the conclusion that by far the largest part of the field originates inside the Earth. The theory of external-internal field separation based on a spherical harmonic expansion will be treated in Sect. 2.1, and application to Earth and Mars (two planets for which globally distributed magnetic field observations obtained from satellites are available) will be discussed in Sects. 3.1 and 3.2.

As an alternative to such a separation by means of spherical harmonic expansion (which does not require a priori knowledge on the spatial structure of the magnetic field since this is estimated from the data), a separation based on physically motivated models of external sources (i.e. knowledge of the time-space structure of the magnetic field and/or the currents generating it) is often more efficient in terms of the required numbers of model parameters to describe the observations. Especially in the case of sparse observations, for instance from a satellite flyby rather than an orbiting satellite, this approach helps to reduce the interpretation ambiguity. An external-internal field separation based on this approach is discussed in Sect. 2.3, and application to Earth and Mercury is treated in Sects. 3.1 and 3.3. Field separation in the presence of plasma interactions, with application to one of Jupiter's moons, Ganymede, is discussed in Sect. 4.

2 Mathematical Separation of Internal and External Contributions

2.1 Separation by Means of Spherical Harmonics and the Theory of Gauss

In regions without electric currents, and for sufficiently slowly changing fields, the magnetic field $\mathbf{B} = -\nabla V$ can be expressed as the negative gradient of a scalar potential V. Because of $\nabla \cdot \mathbf{B} = 0$, the potential V has to be a solution of Laplace's equation: $\nabla^2 V = 0$. Under these assumptions the magnetic field is a Laplacian potential field, which puts considerable constraints on the spatial dependency of $\mathbf{B}$.

It is advantageous to use planetocentric spherical coordinates r, θ, ϕ for describing the magnetic field of a planet. Here r is the radius of the point in consideration, θ is its colatitude, and ϕ is its longitude. Assuming that a region of radius $a < r < c$ is free of electric currents (i.e. $\nabla \times \mathbf{B} = 0$ everywhere in that shell, cf. Fig. 1a), $\mathbf{B} = -\nabla V$ in that region, and the potential V can be expanded into series of spherical harmonics.

Solutions of Laplace's equation in spherical harmonics consist of terms with radial dependency $1/r^{n+1}$ and r^n, respectively. For $r > a$, only terms $\propto 1/r^{n+1}$ are physical meaningful solutions and lead to that part, V^{int}, of the potential which describes sources of internal origin (due to electric currents $\mathbf{J}^{\text{int}}$ in the region $r < a$). Contrary, in the region $r < c$ only terms $\propto r^n$ are physically meaningful; they correspond to the potential part V^{ext} that describes sources of external origin (due to electric currents $\mathbf{J}^{\text{ext}}$ in the region $r > c$). Adding

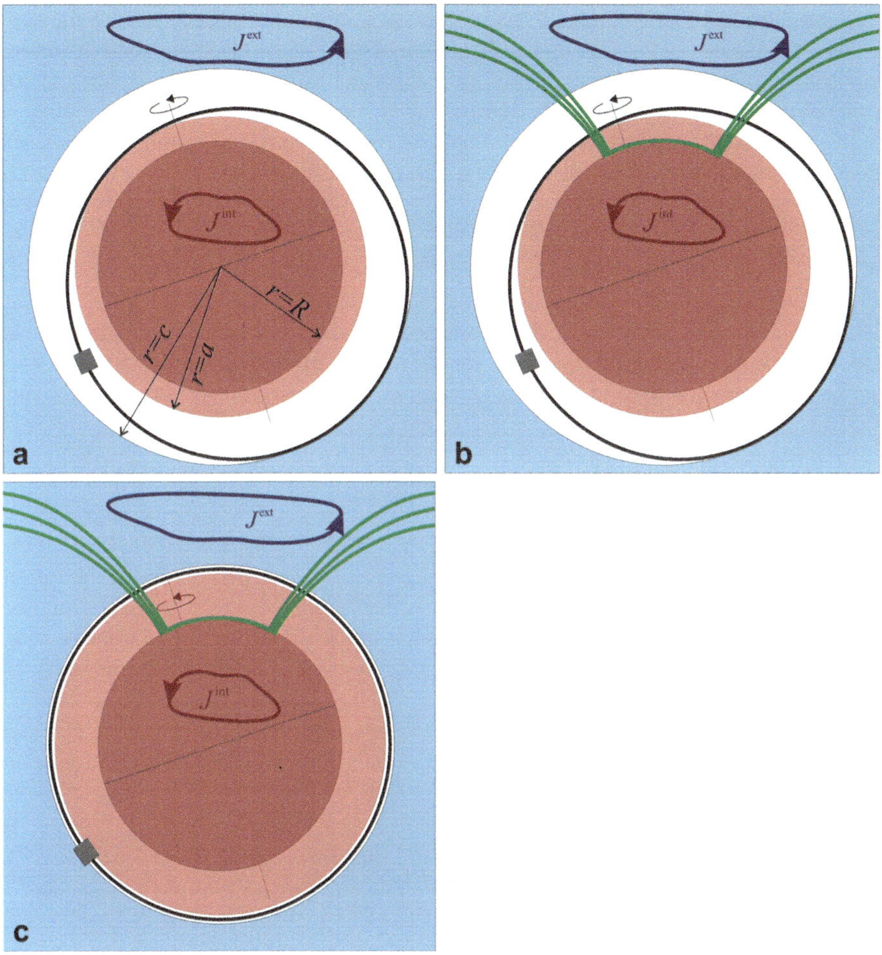

Fig. 1 Scheme of the various source regions. *Blue* indicates regions that cover the external currents $\mathbf{J}^{\text{ext}}$, *red* indicates regions that include the internal currents $\mathbf{J}^{\text{int}}$. *White* regions are free of currents. *The black thick circle* illustrates the orbit of the satellite (*grey box*). The planet is indicated by the sphere of radius R. Currents that cross satellite altitude are shown in *green* in panels **b** and **c**. Their existence violates the assumption of a shell $a < r < c$ free of currents shown in panel **a**

both parts together results in an expansion of V which is valid in the whole current-free region $a < r < c$:

$$\begin{aligned} V &= V^{\text{int}} + V^{\text{ext}} \\ &= R \sum_{n=1}^{N_{\text{int}}} \sum_{m=0}^{n} \left(g_n^m \cos m\phi + h_n^m \sin m\phi \right) \left(\frac{R}{r} \right)^{n+1} P_n^m(\cos\theta) \quad \text{(1a)} \\ &\quad + R \sum_{n=1}^{N_{\text{ext}}} \sum_{m=0}^{n} \left(q_n^m \cos m\phi + s_n^m \sin m\phi \right) \left(\frac{r}{R} \right)^{n} P_n^m(\cos\theta) \quad \text{(1b)} \end{aligned}$$

(Chapman and Bartels 1940; Langel 1987), where R is a reference radius (typically the mean radius of the planet), P_n^m are the associated Schmidt semi-normalized Legendre functions, N_{int} is the maximum degree and order of the internal potential coefficients g_n^m, h_n^m, and N_{ext} is that of the external potential coefficients q_n^m, s_n^m.

In complex notation (and setting for simplicity $N = N_{\text{int}} = N_{\text{ext}}$ as maximum degree and order for both internal and external sources), this equation reads

$$V = \text{Re}\left\{ R \sum_{n=1}^{N} \sum_{m=0}^{n} \left[\iota_n^m \left(\frac{R}{r}\right)^{n+1} + \epsilon_n^m \left(\frac{r}{R}\right)^n \right] P_n^m e^{im\phi} \right\} \tag{2}$$

where $\iota_n^m = g_n^m - ih_n^m$ and $\epsilon_n^m = q_n^m - is_n^m$ are the (complex) coefficients describing internal, resp. external, sources, and Re$\{\cdot\}$ stands for the real part. In following we will drop Re$\{\cdot\}$ and assume implicitly that only the real part of measured quantities like the magnetic field is taken.

The magnetic field components $\mathbf{B} = (B_r, B_\theta, B_\phi)^T$ (where B_r is pointing radially outward, B_θ is pointing Southward, and B_ϕ is pointing Eastward) follow as

$$\begin{aligned} B_r &= -\frac{\partial V}{\partial r} = \sum_{n,m} \left((n+1)\iota_n^m \left(\frac{R}{r}\right)^{n+2} - n\epsilon_n^m \left(\frac{r}{R}\right)^{n-1} \right) P_n^m e^{im\phi} \\ B_\theta &= -\frac{1}{r}\frac{\partial V}{\partial \theta} = \sum_{n,m} \left(\iota_n^m \left(\frac{R}{r}\right)^{n+2} + \epsilon_n^m \left(\frac{r}{R}\right)^{n-1} \right) \frac{dP_n^m}{d\theta} e^{im\phi} \\ B_\phi &= -\frac{1}{r\sin\theta}\frac{\partial V}{\partial \phi} = \sum_{n,m} \left(\iota_n^m \left(\frac{R}{r}\right)^{n+2} + \epsilon_n^m \left(\frac{r}{R}\right)^{n-1} \right) \frac{im}{\sin\theta} P_n^m e^{im\phi}. \end{aligned} \tag{3}$$

A separation of internal and external sources, i.e. a determination of the expansion coefficients ι_n^m and ϵ_n^m, is possible from a combined analysis of radial and horizontal magnetic field components. This becomes obvious when looking at the magnetic field components at the planetary surface ($r = R$):

$$B_r = \sum_{n,m} \left((n+1)\iota_n^m - n\epsilon_n^m \right) P_n^m e^{im\phi} \tag{4a}$$

$$B_\theta = \sum_{n,m} \left(\iota_n^m + \epsilon_n^m \right) \frac{dP_n^m}{d\theta} e^{im\phi} \tag{4b}$$

$$B_\phi = \sum_{n,m} \left(\iota_n^m + \epsilon_n^m \right) \frac{im}{\sin\theta} P_n^m e^{im\phi} \tag{4c}$$

Thus the magnetic horizontal components B_θ, B_ϕ provide information on the *sum* of the internal and external field, $\iota_n^m + \epsilon_n^m$, while the radial component B_r provides information on their *difference*, $(n+1)\iota_n^m - n\epsilon_n^m$. An analysis of all three vector components therefore allows to separate sources of internal and external origin, as first done by Gauss (1839).

Of special interest are the terms with $n = 1$; they have a simple geometric meaning. As can be seen from Eq. 3, there is no radial dependency of the magnetic field of external origin for $n = 1$, and therefore this part corresponds to a uniform field. If x, y and z define a Cartesian coordinate system with origin in the planetary center, z pointing towards the North pole, x pointing towards the zero meridian ($\phi = 0°$) and y pointing towards the meridian

$\phi = 90°$, then the magnetic field component in z direction is given by $q_1^0 = \mathrm{Re}\{\epsilon_1^0\}$, while those in the directions of x and y are given by $q_1^1 = \mathrm{Re}\{\epsilon_1^1\}$ and $s_1^1 = -\mathrm{Im}\{\epsilon_1^1\}$, respectively. For the internal part, $g_1^0 = \mathrm{Re}\{\iota_1^0\}$ represents a dipole at the planets center that is aligned with z, while $g_1^1 = \mathrm{Re}\{\iota_1^1\}$ and $h_1^1 = -\mathrm{Im}\{\iota_1^1\}$ are dipoles in the equatorial plane pointing towards x, resp. y.

It is important to notice that the above described separation into internal and external sources is done with respect to the altitude of the observations, not with respect to the planetary surface. As a consequence, sources between surface and satellite altitude (e.g. ionospheric currents) are seen as internal fields by satellites, although they are external to the planet. Contributions internal to the planet may originate from dynamo action in planetary cores ("core field"), from magnetized material, typically near the surface ("crustal field"), and from electromagnetically induced currents in the conducting planetary interior ("induced fields", cf. Saur et al. 2009).

A representation of the magnetic field by means of a spherical harmonic expansion of the scalar potential, as given by Eqs. 1–4, is only possible if the whole region in which the data are acquired is free of electric currents. In practice this is often not the case when satellite data are analyzed, since electric currents in regions sampled by the satellite violate this condition (cf. the high-latitude field-aligned currents shown in Fig. 1b). Even if *most* of the region (for instance the non-polar region) is free of electric currents (i.e. $\nabla \times \mathbf{B} = 0$), the existence of currents at some latitudes violates the assumption that the *whole* region has to be source free, and therefore a spherical harmonic representation of the magnetic field is not possible. (Note that a local representation of the magnetic field $\mathbf{B} = -\nabla V$ in the current free regions is still achievable although the global expansion of V into series of spherical harmonics is not.)

2.2 Separation in Current-Carrying Regions

In regions with electric currents the magnetic field $\mathbf{B}$ cannot be expressed as the negative gradient of a scalar potential only, but is defined by the expression $\mathbf{B} = -\nabla V + \nabla \times \mathbf{A}$, where $\mathbf{A}$ is a vector potential. In this case the Gauss algorithm outlined above cannot be used and a separation of the measured field into its various contributions is more difficult. Assuming that a region of radius $a < r < c$ is filled with electric currents a representation of the magnetic field in this shell by means of poloidal and toroidal vector field $\mathbf{B}^{\mathrm{pol}}$ and $\mathbf{B}^{\mathrm{tor}}$ is useful:

$$\mathbf{B} = \mathbf{B}^{\mathrm{pol}} + \mathbf{B}^{\mathrm{tor}} \tag{5}$$

where

$$\mathbf{B}^{\mathrm{pol}} = \nabla \times (\mathbf{r} \times \nabla P) \tag{6a}$$

$$\mathbf{B}^{\mathrm{tor}} = \mathbf{r} \times \nabla T. \tag{6b}$$

The two scalar fields P and T are called the poloidal and toroidal scalars of $\mathbf{B}$. Representing a magnetic vector field by its poloidal and toroidal vector fields is called the Mie representation of $\mathbf{B}$. It has extensively been discussed by Backus (1986) and is formally not restricted to spherical shells and volume regions between spherical shells, but is also applicable on more general surfaces and volumes generated by two such surfaces.

Consider magnetic field measurements taken on a spherical shell, either a planetary surface or the shell described by a polar orbiting spacecraft. Following Backus (1986), the observed magnetic field $\mathbf{B}$ can be decomposed into a poloidal part of internal origin (sources internal to the observation altitude), $\mathbf{B}^{\text{pol,int}}$, a poloidal part of external origin, $\mathbf{B}^{\text{pol,ext}}$, and a third (toroidal) part, $\mathbf{B}^{\text{tor}}$. This toroidal part is due to radial currents at the shell. If $\mathbf{B}$ is known everywhere on the shell, then also the three contributions can be found everywhere on the shell. The internal and external poloidal vector fields $\mathbf{B}^{\text{pol,int}}$ and $\mathbf{B}^{\text{pol,ext}}$ can be determined, much as the Gauss algorithm can be used to separate internal and external contributions in case of a current-free situation. However, it should be noted that the internal contribution is due to all the electric currents flowing inside the sphere. Nothing can be said about the currents in the dynamo region, where the planetary field is generated. All the currents flowing inside the sphere add up in a poloidal contribution at the surface of the sphere.

If there are currents in a thick shell formed by e.g. spheres with the radius of the apoapsis and the periapsis of a spacecraft orbiting a planet (cf. Fig. 1b) a Mie representation can still be used. In this case the scalar fields P and T are functions in the shell uniquely determined by the magnetic field $\mathbf{B}$ in the shell. An expansion of P and T into spherical harmonics is possible, but requires expansion coefficients which depend on the radial distance r within the shell. Without further constrains determination of the expansion coefficients requires knowledge of $\mathbf{B}$ everywhere in the shell. An *a priori* physical model, based on e.g. simulation results of the interaction of the solar wind with the planet and/or its internal magnetic field, could provide such constrains.

However, if the data are acquired in a thin shell (of thickness $c - a \ll b = (a+c)/2$, cf. Fig. 1c), an extension of the classical spherical harmonic representation as developed by Gauss is possible. As pointed out by Backus (1986), the observed magnetic field $\mathbf{B}$ can be decomposed into a part of internal origin (sources internal to the observation altitude), $\mathbf{B}^{\text{int}} = -\nabla V^{\text{int}}$, a part of external origin, $\mathbf{B}^{\text{ext}} = -\nabla V^{\text{ext}}$, and the third (toroidal) part, $\mathbf{B}^{\text{tor}}$, that is due to in-situ (radial) currents at satellite altitude:

$$\begin{aligned}
\mathbf{B} &= \mathbf{B}^{\text{int}} + \mathbf{B}^{\text{ext}} + \mathbf{B}^{\text{tor}} \\
&= -\nabla\left[R\sum_{n,m}\left(\frac{R}{r}\right)^{n+1} \iota_n^m P_n^m e^{im\phi}\right] && \text{(7a)} \\
&\quad -\nabla\left[R\sum_{n,m}\left(\frac{r}{R}\right)^{n} \varepsilon_n^m P_n^m e^{im\phi}\right] && \text{(7b)} \\
&\quad -\nabla\times\left[\mathbf{r}\sum_{n,m}\left(\frac{R}{r}\right) \psi_n^m P_n^m e^{im\phi}\right]. && \text{(7c)}
\end{aligned}$$

Once the expansion coefficients ϕ_n^m are known, the radial current density through the shell $r = b$ can be found from

$$J_r(b,\theta,\phi) = \frac{1}{\mu_0 b}\sum_{n,m} n(n+1)\left(\frac{R}{b}\right)\psi_n^m P_n^m e^{im\phi}. \tag{8}$$

The mean-square value (over the sphere with radius r) of the magnetic field intensity produced by spherical harmonics of degree n has been introduced independently by Mauersberger (1956) and Lowes (1966) and is denoted by W_n. For a magnetic field containing internal, external and toroidal parts, W_n is given by Backus (1986), Eq. 81

$$\begin{aligned}
W_n(r) &= W_n^{\text{int}}(r) + W_n^{\text{ext}}(r) + W_n^{\text{tor}}(r) \\
&= (n+1)\left(\frac{r}{R}\right)^{2n+4} \sum_{m=0}^{n} |\iota_n^m|^2 && \text{(9a)}
\end{aligned}$$

$$+n\left(\frac{R}{r}\right)^{2n-2}\sum_{m=0}^{n}|\epsilon_n^m|^2 \tag{9b}$$

$$+\frac{n(n+1)}{2n+1}\left(\frac{r}{R}\right)^{2}\sum_{m=0}^{n}|\psi_n^m|^2. \tag{9c}$$

At the planetary surface, $r = R$, this equation reduces to

$$W_n = (n+1)\sum_{m=0}^{n}|\iota_n^m|^2 \tag{10a}$$

$$+n\sum_{m=0}^{n}|\epsilon_n^m|^2 \tag{10b}$$

$$+\frac{n(n+1)}{2n+1}\sum_{m=0}^{n}|\psi_n^m|^2. \tag{10c}$$

Olsen (1997) made extensive use of the above described approach when studying F region currents in the terrestrial ionosphere. Analysing vector magnetic field observations taken by the Magsat satellite in 1979 and 1980, he determined the toroidal expansion coefficients ψ_n^m separately for morning and evening in local time, and for two months of data centered at December 21, 1979, and March 21, 1980, respectively. Global maps of the radial current density J_r were then synthesized using Eq. 8. These maps show the well-known field-aligned currents at polar latitudes (average current density $J_r \lesssim 200$ nA/m^2) and the meridional current system of the terrestrial equatorial electrojet ($J_r \approx 10 - 20$ nA/m^2) during dusk. In addition, these maps revealed for the first time a midlatitude interhemispheric current system in the terrestrial ionosphere, with current direction from the winter to the summer hemisphere during dusk and in opposite direction during dawn. This is in agreement with predictions by three-dimensional models of the ionospheric dynamo.

An application of the poloidal-toroidal decomposition approach to Mars is discussed in Sect. 3.2.

2.3 Internal-External Field Separation Based on Models of External Source Fields

If magnetic field measurements are made in current-carrying regions a detailed knowledge of the electric current distribution is necessary for a final separation of the measured field into its internal and external parts. Such a knowledge requires, for example, a larger number of spacecraft onboard which magnetic field measurements are made. Currently only the four Cluster spacecraft, orbiting round Earth, provide such a possibility to a limited extend (Dunlop et al. 2002). (As a near-Earth equivalent, the two side-by-side flying satellites of the *Swarm* constellation mission (Friis-Christensen et al. 2006) to be launched in 2011 will allow for an instantaneous determination of the radial current density in the terrestrial F region ionosphere.)

However, in the terrestrial magnetosphere the distribution of electric currents is not uniform, but largely localized in the magnetopause and ring current regions as well as in the ionosphere. Therefore, most of the measurements are done in almost current-free regions, where the Gauss algorithm can be used and the description of the magnetic field by a scalar magnetic potential is justified. However, it is practise not to determine the external field directly from the actual data, but to make extensive use of parameterized external magnetic field models.

There are three categories of knowledge on external field geometry that can be used in this situation to achieve an external-internal field separation:

- choice of appropriate coordinate systems;
- parameterized models of ionospheric sources (Lühr et al. 2009);
- parameterized models of magnetospheric sources, for instance the Tsyganenko model series (e.g., Tsyganenko 1989, 1995, 2002a, 2002b; Tsyganenko and Sitnov 2007).

Concerning the first category, it is useful to remember that internal sources are often fixed with respect to the planet (magnetic fields due to induced currents in the planetary interior are an exception) and thus follow its rotation, while many external fields are often fixed with respect to the sun. Describing external fields in a coordinate frame that follows the (apparent) movement of the sun is therefore advantageous. On planets with Earth-like magnetospheres the use of Solar Magnetospheric (SM) coordinates for describing near magnetospheric currents like the ring-current, and of Geocentric Solar Magnetospheric (GSM) coordinates for describing far magnetospheric current systems like the tail currents has been turned out to be useful (Maus and Lühr 2005; Olsen et al. 2006). A definition of the SM and GSM coordinate frames can be found in Kivelson and Russell (1995).

Concerning the second category, extensive modelling of the ionospheric current systems and their magnetic fields have been done. For a recent study and review reference is made to Lühr et al. (2009) and references therein. Such modelling is indispensable for a determination of the spherical harmonic expansion coefficients of the geomagnetic field up to very large degree.

The third category, magnetospheric magnetic field modelling, uses parameterized field models based on theoretical models of the magnetosphere. The model parameter are determined from actually measured field values. The models suggested by Tsyganenko and co-workers (Tsyganenko 1989, 1995, 2002a, 2002b; Tsyganenko and Sitnov 2007) are the most widely used models in this respect. In these models the external magnetic field of magnetospheric origin, $\mathbf{B}^{\text{ext}}$, is described as the sum

$$\mathbf{B}^{\text{ext}} = \mathbf{B}^{\text{CF}} + \mathbf{B}^{\text{RC}} + \mathbf{B}^{\text{T}} + \mathbf{B}^{\text{FAC}} + \mathbf{B}^{\text{OF}}, \tag{11}$$

where $\mathbf{B}^{\text{CF}}$, $\mathbf{B}^{\text{RC}}$, $\mathbf{B}^{\text{T}}$, and $\mathbf{B}^{\text{FAC}}$ denote field contributions from the Chapman-Ferraro currents flowing in the magnetopause, the magnetospheric ring current, magnetotail currents, and field-aligned currents, respectively. The contribution $\mathbf{B}^{\text{OF}}$ denotes any other field contribution one would like to incorporate into the model. For the various contributions analytical descriptions are derived with parameters depending on solar wind dynamic pressure, the interplanetary magnetic field, or magnetospheric activity as described by the D_{st} or Kp indices.

The magnetic field $\mathbf{B}^{\text{CF}}$ of the Chapman-Ferraro currents is described as the gradient of a scalar potential

$$\mathbf{B}^{\text{CF}} = -\nabla V^{\text{CF}} \tag{12}$$

where the scalar potential V^{CF} can be derived by solving the Neumann problem

$$\nabla^2 V^{\text{CF}} = 0 \tag{13}$$

$$\left.\frac{\partial V^{\text{CF}}}{\partial n}\right|_S - B_n^{\text{int}} = 0. \tag{14}$$

Here the subscript n denotes the component normal to the magnetopause surface S. For this surface S a suitable description as an axially symmetric ellipsoid or paraboloid needs to be found. The Neumann condition also implies that the normal component of the magnetic field at the boundary vanishes. This corresponds to a so-called closed magnetosphere. Other conditions are possible, depending on the physical system to be modelled.

For the scalar potential V^{CF} a suitable analytical representation is necessary. A flexible choice is a representation (Tsyganenko 1995)

$$V^{\mathrm{CF}} = \sum_{i,k} a_i \chi(b_k, \mathbf{r}), \tag{15}$$

where a_i are the expansion coefficients and $\chi(b_k, \mathbf{r})$ can be any harmonic function of the position vector $\mathbf{r}$ and some parameter b_k.

The magnetic field contribution of the cross-tail current, for example, can be described by a vector potential $\mathbf{A} = A_\phi \hat{\mathbf{e}}_\phi$ with a suitable parameterized description of the azimuthal component A_ϕ of the vector potential; usually cylindrical coordinates are used. In similar ways the further contributions can be represented.

These parameterized models are to a certain extend comparable to the spherical harmonic expansion introduced by Carl-Friedrich Gauss as in both descriptions the entire field is decomposed into its various contributions. However, the Gauss description uses a set of complete orthogonal functions, spherical harmonics, to describe the field. This description is a purely mathematical, not requiring any physical argument for its usefulness. A description as used by Tsyganenko and others relies on physical insight into the problem, as knowledge about the form of the magnetopause, its current distribution, or the spatial structure of the ring current distribution is required. This *a priori*, physical input to the model allows to reduce the number of independent parameters. But, in a mathematical sense the description is not complete. Nevertheless, parameterized models are very successful and allow already a very good description of the external field contribution.

3 Application to Selected Planets

3.1 Earth

After the pioneering work of Gauss (1839), attempts to separate external and internal contributions have been performed by Schmidt (1895, 1898), Fritsche (1900) and Bauer (1923). These early determinations found external field contributions of the order of few hundred nT, which is far too high for average geomagnetic activity conditions. The large values are due to spatial aliasing caused by the very uneven distribution of the ground data that have been used in these investigations. Only satellites are able to provide the uniform data coverage that is necessary for a reliable external-internal field separation. Although magnetic intensity data taken by the POGO satellites improved the situation considerably, it was not before the availability of vector satellite observations that a reliable determination of the large-scale magnetospheric field of about 20 nT (Langel and Estes 1985a, 1985b) during geomagnetic quiet times was possible. More recent satellite data confirmed this value (Maus et al. 2005; Olsen et al. 2006), and furthermore demonstrated that about half of it is due to near magnetospheric sources (best described in SM coordinates) while the other half is due to more distant sources (and follow the GSM frame).

The most complete external-internal separation of the Earth's magnetic field is given by the Comprehensive Model by Sabaka and co-workers. In its present version called CM4

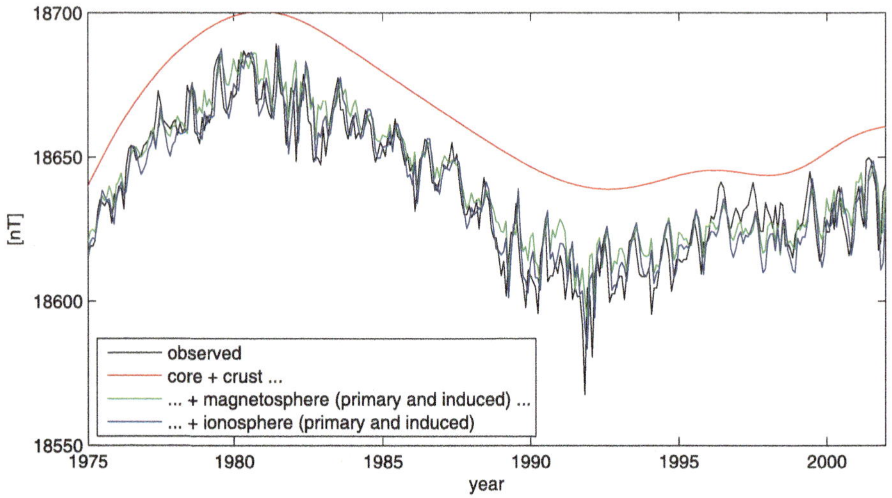

Fig. 2 Observed and predicted (from CM4) monthly mean values of the *X* component at Niemegk/Germany

(Sabaka et al. 2004), the model attempts to describe the various components of the near-Earth magnetic field by co-estimating the internal, magnetospheric and ionospheric fields (with their secondary, Earth-induced counterparts), in one huge inversion process.

By a joint analysis of ground-based and satellite data it was possible to separate Earth-internal sources, ionospheric sources (which are external to ground, but internal to satellites) and magnetospheric sources (external to ground and low-orbiting satellites).

As an example for the ability of CM4 to separate the various sources, Fig. 2 shows observed monthly mean values of the *X* (North) component at the Niemegk observatory in Germany (black line) together with CM4 predictions. The red curve shows the internal (core and crustal) part; adding the magnetospheric field (including its induced counterpart) results in the green curve, and adding the ionospheric (plus induced) contribution leads finally to the blue curve. Internal and external field contributions together (blue curve) nicely fits the observed field (black curve).

The comprehensive approach has recently been applied to synthetic satellite data (Sabaka and Olsen 2006), which confirmed that a proper separation of the various source contributions is possible.

3.2 Mars

Models of the Martian magnetic field usually describe only internal sources, given by the first part of Eq. 7. However, the simultaneous estimation of internal, external and toroidal magnetic field contributions has turned out to be an advantage for studying the terrestrial magnetic field (Sabaka et al. 2004), and an application to Mars, a planet for which magnetic data from the mapping-quality satellite Mars Global Surveyor (MGS) are available, is straightforward.

To demonstrate this, we used MGS mapping data (taken at about 400 km altitude) and performed a separate analysis of dayside and nightside data, respectively. Data were sampled between March and November 1999, which yields about 50000 vector triplets in each data set. Following Eq. 7, we solved for internal (up to degree/order $n = 40$), external (up to $n = 20$), and toroidal (up to $n = 20$) field contributions. The external current systems are expected to be highly time variable, but here we only consider the time-independent part of the field.

The radial magnetic field component at 400 km altitude due to external, respectively internal, sources as estimated from night-side data only are shown in Figs. 3a and 3b. The internal field is concentrated in the southern highlands where it is about 10 times stronger than the external field, which, however, is of approximately similar strength everywhere but especially dominant at polar latitudes. Figure 3c shows the external field obtained from the day-side data; the lower right panel presents the difference between the external field determined from day-side data minus that derived from night-side data. Note that "internal" refers to sources below satellite altitude (400 km) and includes contributions from ionospheric currents. This is confirmed by the fact that the additional "internal" field on the day-side is about twice as strong as the corresponding "external" field, which is not possible from purely electromagnetic induction in the Martian interior.

Figure 4 shows magnetic power-spectra of the various contributions (internal, external and toroidal) to the Martian magnetic field at surface (left) and 400 km altitude (right), determined from day-side and night-side observations, respectively. Also shown is the spectrum of the model of Cain et al. (2003). At satellite altitude, contributions from in-situ currents (toroidal magnetic field) exceed those from external sources, even during night-time.

3.3 Mercury

Currently the planetary magnetic field of Mercury is not well known. The planet was first visited by the Mariner 10 spacecraft in 1974. The discovery of a magnetic field was certainly a major surprise (Ness et al. 1974). Recently, Mercury was visited again by the MESSENGER spacecraft and additional magnetic field measurements have been possible (Anderson et al. 2008). Due to the limited number of flybys determination of the planetary magnetic field is still in its infancy. This will only change when the MESSENGER spacecraft will start to orbit planet Mercury in 2011 and once the European-Japanese dual spacecraft mission BepiColombo with its two spacecraft Mercury Planetary Orbiter (MPO) and Mercury Magnetospheric Orbiter (MMO), both equipped with magnetometer systems (Glassmeier et al. 2009; Baumjohann et al. 2009), are inserted into polar orbits around Mercury in 2019.

The dipole moments estimated up to now using different approaches significantly differ (the minus sign indicates southward moment): -170 nT R_{M}^3 (Jackson and Beard 1977; Whang 1977), -192 nT R_{M}^3 (Alexeev et al. 2008), -287 nT R_{M}^3 (Korth et al. 2004), $-(230-290)$ nT R_{M}^3 (Anderson et al. 2008). The most recent joint analysis of Mariner 10 and Messenger observations by Uno et al. (2009) suggest a dipole moment -227 nT R_{M}^3. Higher order moments are to a large extend unconstrained.

The differences mainly result from insufficient data coverage, but also indicate incomplete separation of internal and external contributions by the different magnetospheric magnetic field models used. Knowledge of the external fields is certainly a primary limiting factor in extracting reliable information from the data on Mercury's internal magnetic field.

First models of the Hermean magnetospheric field were provided by e.g. Jackson and Beard (1977) and Whang (1977) with significant recent improvements reached by Giampieri and Balogh (2001), Korth et al. (2004), Grosser et al. (2004), Scuffham and Balogh (2006), and Alexeev et al. (2008). The model used by Jackson and Beard (1977) is a scaled version of the terrestrial magnetic field model of Choe and Bearda (1974a, 1974b). The subsolar magnetic field magnitude ratio is used as a scaling parameter.

The field model of Whang (1977) uses an image dipole to describe the magnetic effect of magnetopause currents and a finite-thickness tail current sheet to model the tail magnetic

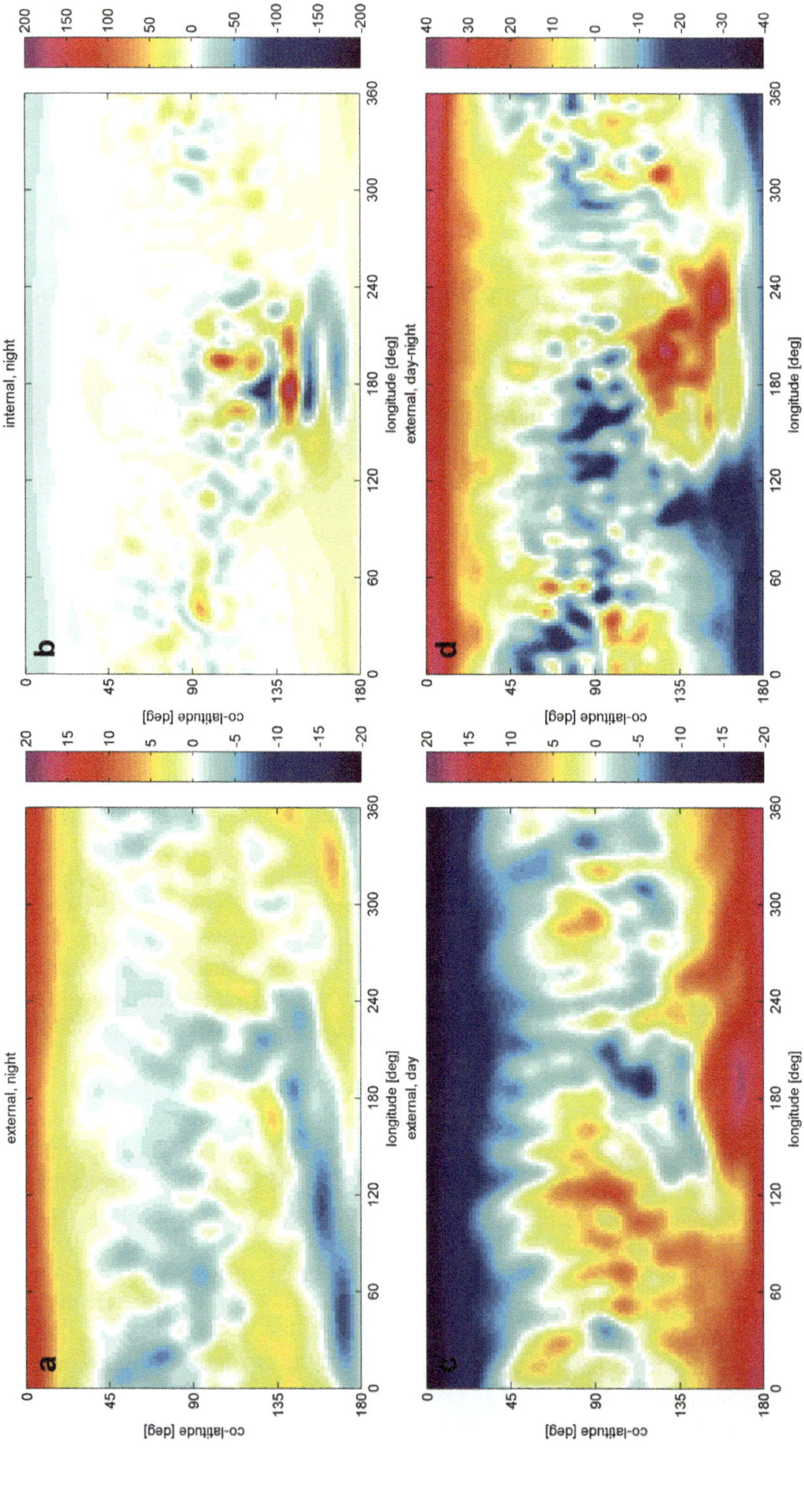

Fig. 3 Global maps of the magnetic radial component at 400 km altitude (in nT) estimated from MGS data

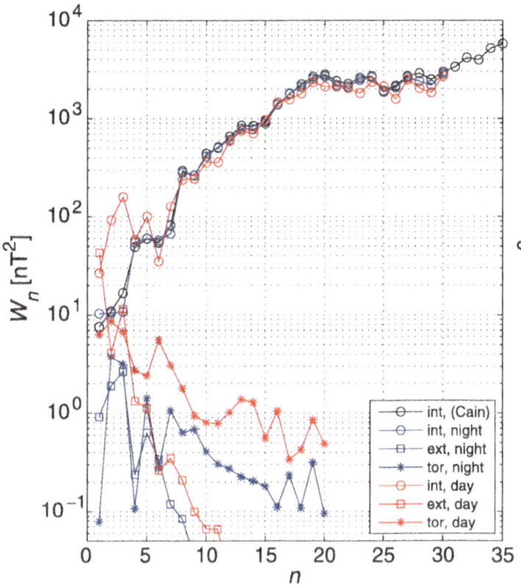

Fig. 4 Power spectra of internal, external and toroidal contributions to the Martian magnetic field at surface (*left*) and 400 km altitude (*right*)

field. Korth et al. (2004) use a scaled version of the Tsyganenko-96 model (Tsyganenko 1995) with any ring current contribution neglected. A somewhat different approach have Alexeev et al. (2008), who use a magnetopause which is a paraboloid of rotation around the Sun-planet line. Magnetic fields due to electric currents on this magnetopause and the tail current system are obtained by appropriate shielding fields to confine the various fields of magnetospheric origin inside the region defined by the magnetopause.

All these models assume sheet current distributions generating the magnetospheric magnetic field. Therefore a scalar potential representation as in Eq. 1 is possible in most regions of the magnetosphere. However, this is the limiting factor of the field models used so far as the weak planetary magnetic field causes a rather small magnetosphere with the planetocentric stand-off distance of the magnetopause at $R_{\mathrm{MP}} \approx 1.7 R_{\mathrm{M}}$ (where $R_{\mathrm{M}} = 2440$ km is the Hermean planetary radius) (Siscoe and Christopher 1975). The magnetic field of the Chapman-Ferraro currents flowing in this close-to-the-planet magnetopause causes a surface magnetic field of the order of 100 nT (e.g. Grosser et al. 2004). As the thickness of the Hermean magnetopause is of the order of a few hundred kilometers (e.g. Russell and Walker 1985) this implies that the magnetopause current layer at Mercury is not just a current sheet, but a broad current-carrying layer.

The complexity and yet unexplored nature of the Hermean magnetopause is also demonstrated by the recent Messenger observations (Slavin et al. 2008), where indications for a double-magnetopause type structure were identified. This poses another major problem for the separation of internal and external field contributions as the regions where magnetic field measurements are taken is not necessarily current-free. Thus, the Gauss algorithm is strictly not applicable in most of the magnetospheric regions at Mercury. Also the current parameterized models fail in describing this new situation. The more general Mie representation (see Eq. 5) and related separation algorithms are a more appropriate tools to describe and analyze the available observations.

Incorrectly assuming a current-free shell in which data are taken and application of the Gauss algorithm has major effects on the estimated spherical harmonic expansion coefficients. Using a spherically shaped magnetosphere and azimuthally flowing magnetopause

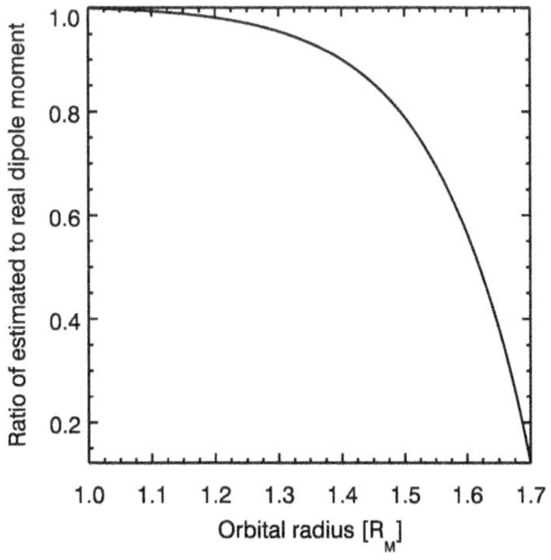

Fig. 5 Comparison of the estimated dipole field coefficient with respect to the true internal value. A spherical magnetosphere with a finite magnetopause and azimuthally flowing electric currents is assumed. The estimate is based on erroneously assuming a scalar potential field description of the magnetic field

current density with a Rayleigh distribution-type variation in radial direction Heyner (2007) and Glassmeier et al. (2009) modelled external magnetic fields to which an internal dipole field was added. The current distribution decays towards the planet with a scale of 500 km and towards the magnetosheath with 50 km; maximum current density reaches 2.8×10^{-7} Am^2. The magnetopause is located at $1.7 R_{\mathrm{M}}$.

Figure 5 displays the results of an inversion of magnetic field observations taken along circular polar orbits around the model planet at various radii. The closer the spacecraft is flying to the current-carrying magnetopause the less reliable is the estimated dipole moment. Magnetic fields of the inner current layers actually shield the internal dipole. In the terrestrial magnetosphere one can avoid this by not taking into account measurements made close to the magnetopause. However, at Mercury this possibility does not exists. For example, the MPO spacecraft of the BepiColombo mission will take magnetic field measurements in a polar orbit around Mercury with a periherm of $1.16 R_{\mathrm{M}}$ and an apoherm of $1.6 R_{\mathrm{M}}$. MPO will fly at a relative distance to the magnetopause between 0.68–$0.94 R_{\mathrm{MP}}$. In the terrestrial magnetosphere this would correspond to an orbit with a perigee of about 43 000 km and an apogee of about 59 000 km.

Due to the sparsity of available data the problem of a proper separation of the fields of internal and external origin has not yet been solved. New tools need to be developed for an efficient exploration of the future magnetic field data from the Messenger and BepiColombo magnetometer experiments (Anderson et al. 2007; Baumjohann et al. 2009; Glassmeier et al. 2009). In a first step, one needs to estimate the strength of any local electric current flows. The BepiColombo spacecraft will provide dual-point magnetic field observations. Around periherm passage MPO and MMO will be separated by a distance of 50–100 km in radial direction. This allows for a determination of the radial gradient. Neglecting displacement currents we have $\nabla \times \mathbf{B} = \mu_0 \mathbf{J}$, which reads in spherical coordinates close to the equator ($\theta \approx 90°$):

$$
\begin{aligned}
\mu_0 J_r &= \frac{1}{r}\left(\frac{\partial B_\phi}{\partial \theta} - \frac{\partial B_\theta}{\partial \phi}\right), \\
\mu_0 J_\theta &= \frac{1}{r}\left(\frac{\partial B_r}{\partial \phi} - \frac{\partial r B_\phi}{\partial r}\right),
\end{aligned}
\qquad (16)
$$

$$\mu_0 J_\phi = \frac{1}{r}\left(\frac{\partial r B_\theta}{\partial r} - \frac{\partial B_r}{\partial \theta}\right).$$

To first order it is justified to neglect the azimuthal and field-aligned variations $\partial/\partial\phi$ and $\partial/\partial\theta$ of the magnetic field. Thus, MPO/MMO two-point magnetic field measurements will provide a first-order estimate of azimuthal and field-aligned current flows in the magnetosphere of Mercury. If these currents are large, the Gauss algorithm and any similar approach can not be used.

4 External-Internal Field Separation in the Presence of Plasma Interactions

The interaction between a flowing plasma and a magnetized obstacle generates currents that cause the magnetic field in the vicinity of the obstacle to deviate significantly from the superposition of the internal field and the background field. Description of the external field perturbations caused by plasma interactions can be provided by physically motivated models of the global interaction. For instance, in an attempt to improve the constraints on the inductive response of Europa, one of the Jovian moons that generates an induced magnetic field through electromagnetic induction, Schilling et al. (2007) applied a global MHD model to account for the effects arising from plasma interactions. Here we use Ganymede, another Galilean satellite of Jupiter, as an example to discuss some issues related to the separation of an internal field from the field arising from external sources including the perturbations introduced by plasma interactions.

Ganymede, the largest moon in the solar system, has an intrinsic magnetic field that is sufficiently strong to stand off Jupiter's flowing magnetospheric plasma above its surface and form a magnetosphere (Kivelson et al. 1996). Similar to that of the planet Mercury, Ganymede's magnetosphere is rather small with the typical stand-off distance of the magnetopause $R_{\mathrm{MP}} \approx 1.8 R_{\mathrm{G}}$ (where $R_{\mathrm{G}} = 2634$ km is the radius of Ganymede) measured from the moon's center into the direction of the incident flow. Our knowledge of Ganymede's magnetosphere and its internal magnetic field comes from the observations obtained by the Galileo spacecraft during its six close encounters with the moon (Kivelson et al. 1996, 1998). However, it was suggested by Kivelson et al. (2002) that only three of the six passes contain information relevant for determination of the internal field through quadrupole order. This was because two passes did not approach closely enough to detect higher order moments and one pass remained close to the boundary of Ganymede's magnetosphere where perturbations from large localized currents dominated the signal.

The magnetic field measured in the vicinity of Ganymede consists of several parts: Jupiter's ambient magnetospheric field, Ganymede's internal field and Ganymede's magnetospheric field arising from plasma interactions. The separation of external and internal field contributions was performed by using the data acquired in regions inside of Ganymede's magnetosphere that are relatively far from the magnetopause boundary where strong magnetospheric currents are present (Kivelson et al. 2002). The approach taken by Kivelson et al. (2002) was based on the classical spherical harmonic expansion described in Sect. 2 and based on the assumption that the data used for the analysis were collected in regions free of electric currents. Evident problems with this approach are that the spacecraft trajectory was not current-free and that it remained above the ionosphere, so ionospheric sources were combined with internal sources. The external contributions were solved only for the first-order ($n = 1$) terms of a spherical harmonic expansion, which corresponds to a uniform external field, but were allowed to vary from pass to pass. The internal contributions were solved for a set of fixed internal moments through quadrupole ($n = 2$) order and also for a combination of a fixed dipole and an induced dipole, whose orientation depends upon the instantaneous external driving field. The analysis imposed the obvious requirement that internal moments be identical on all passes. Both models provided equally satisfactory fits to the magnetic field data on multiple passes. The two different characterizations of the internal field could not be unambiguously distinguished based on the accuracy with which they reproduced the available flyby data. However, the model of permanent and induced dipole moments was favored because it requires fewer free parameters than does the model of a permanent dipole moment and quadrupole moments (Kivelson et al. 2002).

Uncertainties in the separation of external and internal contributions arise in part from the assumption that the external fields resulting from Ganymede's magnetospheric currents are spatially uniform. A more realistic representation of the magnetospheric fields can be provided by a physically motivated model of the global magnetosphere. Three-dimensional global MHD simulations, including single-fluid resistive MHD simulations (Kopp and Ip 2002; Ip and Kopp 2002; Jia et al. 2008) and multi-fluid MHD simulations (Paty and Winglee 2004, 2006; Paty et al. 2008), have been carried out to understand Ganymede's plasma and field environment. Among these, Jia et al. (2008) developed an MHD model that includes the moon's interior in the simulation domain and, therefore, has the advantage of including the magnetic diffusion effects in the moon's mantle. Moreover, their model used boundary conditions that are consistent with the available observations and a high resolution grid that enabled them to resolve the magnetospheric boundaries. The simulation results, in general, agree well with the magnetic field observations on multiple passes. However, the simulated magnetosphere is slightly smaller than that inferred from the measurements. Subsequently, Jia et al. (2009b) improved their model by applying an ionospheric boundary condition more appropriate for Ganymede and found that the simulated magnetosphere in the improved model is in extremely good agreement with the observations on multiple passes. The work therefore demonstrated that an MHD model can consistently describe the external plasma currents that contribute to the magnetic field measurements in the near-Ganymede environment. The new work also illustrated the sensitivity of global simulations to internal boundary conditions.

The external current system in Ganymede's magnetosphere identified from the MHD simulation is illustrated in Fig. 6. At low latitudes near Ganymede, there are two regions of intense currents flowing perpendicular to the magnetic field, one at the upstream magnetopause and the other in the downstream magnetotail (Fig. 6a). These are analogous to the classic *Chapman-Ferraro* magnetopause currents and the tail currents found in the terrestrial magnetosphere. Present at high latitudes are field-aligned currents (Fig. 6b), also called the Alfvén wing currents, which are generated mainly by the strong flow shear between the connected and external field lines. This current system is similar to the Region-1 current system in the Earth's magnetosphere. The currents flow towards Ganymede on the side closer to Jupiter and away from the moon on the side away from the planet. As shown by the current streamlines in Fig. 6c, the Alfvén wing currents close not only through the magnetopause and tail current sheets, but also through the moon and its ionosphere.

Although Ganymede's magnetosphere shares many properties with the planetary magnetospheres, there are noticeable differences in the global configuration resulting from different properties of the ambient plasma flow. In the super-Alfvénic solar wind, planetary magnetospheres form bullet-shape cavities in which the two tail lobes, normally containing open field lines, are roughly aligned with the direction of the incident solar wind. In contrast, in the sub-Alfvénic plasma flow typical at Ganymede's orbit, the magnetosphere forms into a roughly cylindrical shape with two nearly vertically extended Alfvén wings containing open field lines (see field lines in Fig. 6a). As a result, the Alfvén wing currents which

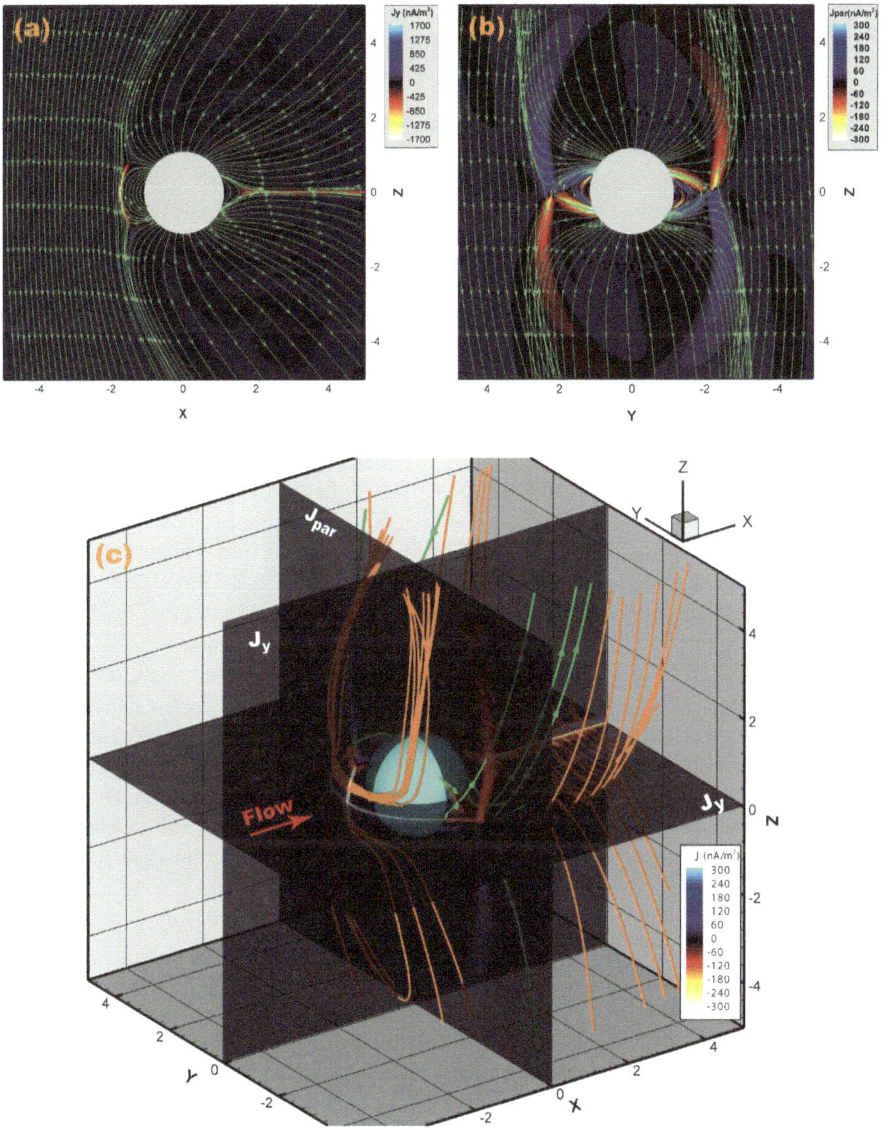

Fig. 6 (**a**) *Color contours* of the current density (J_Y) in the $\hat{Y}$ direction plotted in the XZ plane at $Y = 0$. The projection of magnetic field lines in this plane is superposed as *green lines* with arrowheads representing the field orientation. Note that density of field lines is arbitrary. (**b**) As for (**a**) but for the field-aligned current density (J_{par}) in the YZ plane. Flow is into the plane. (**c**) A global view of the current flowlines in the magnetosphere from the upstream flank side. Results are shown in a Ganymede-centered Cartesian coordinates ("GPhiO"), where $\hat{X}$ is along the incident flow direction, $\hat{Y}$ is along the Ganymede-Jupiter vector, positive towards Jupiter, and $\hat{Z}$ is parallel to Jupiter's spin axis. The axes are labelled in units of Ganymede's radius R_G. The figure is reproduced from Jia et al. (2008)

go through Ganymede and its magnetosphere close far from Ganymede. Therefore, directly scaling empirical models of the magnetospheric currents developed in the context of a planetary magnetosphere to Ganymede's magnetosphere is not appropriate. In particular, a model such as that of Stone and Armstrong (2001) in which the external currents are specified by modifying empirically a model of Earth's magnetosphere, with the magnetopause and tail currents included, produces a magnetosphere whose global configuration differs greatly from that obtained in the MHD simulations (Jia et al. 2008, 2009b). In the Stone and Armstrong (2001) model, the magnetospheric currents, which separate the unperturbed external field and the magnetospheric field lines, close near Ganymede (within several Ganymede radii) whereas open flux tubes at Ganymede appear to link to Jupiter's ionosphere. The latter assertion is based on the auroral footprints (Clarke et al. 2002) that couple directly to Ganymede.

The perturbation field arising from external currents can be extracted directly from the MHD model. Figure 7 shows the magnetic field comparisons between the magnetometer measurements and the MHD simulations for two Galileo flybys, G2 and G28, that were included in the Kivelson et al. (2002) analysis. The G2 flyby is a polar pass with a closest approach altitude of ~264 km ($0.10R_G$) and the G28 flyby is an upstream low-latitude pass with a closest approach altitude of ~900 km ($0.34R_G$). Good agreement between the model results and the measurements allows us to identify the perturbation field quite accurately and thus to establish the degree of validity of the approximation in which the perturbation field is represented as a uniform field. The initial input in the MHD model consists of the background field (Jupiter's magnetospheric field) and Ganymede's internal field, including both the permanent and the induced dipole moments given by Kivelson et al. (2002). The difference between the simulated fields and the initial input observed along the spacecraft trajectory can then be considered as the perturbation field imposed by the external current system. For both flybys, the perturbation fields extracted from the MHD model deviate from uniform fields. In particular, for the G28 flyby, whose trajectory lies close to the upstream magnetopause, the perturbation field deviates considerably from a uniform field because the magnetopause shape and current density change significantly over small spatial scales. It should also be noted that the perturbation fields extracted from the comparison between the actual and the simulated data includes the contribution arising from the ionospheric currents, thereby avoiding a problem inherent in the spherical harmonic expansion of combining the contributions of ionospheric sources with sources internal to the moon.

The fact that the MHD simulation, which used the internal field model of Kivelson et al. (2002) as initial input, yields a rather good agreement with the observations implies that the dominant component of the internal field basically is reasonably well captured by the present internal field model. However, use of a good MHD model to extract plasma-driven effects should improve the determination of internal moments and the inductive response efficiency. Nonetheless, several factors need to be considered in such an approach. Firstly, results of MHD simulations depend largely on the boundary conditions at both the upstream and the inner boundary near the obstacle. In the planetary case, such as for Mercury, the upstream solar wind condition is highly variable and an accurate description of the magnetospheric currents requires knowledge of the instantaneous solar wind conditions. Fortunately in Ganymede's case, the upstream Jupiter's field and plasma conditions are relatively steady and predictable. However, properties of Ganymede's ionosphere, such as the conductivity, are poorly constrained observationally. Therefore, model results must be validated by comparing with spacecraft observations before being used to interpret the observations. Secondly, during the course of spacecraft measurements, the magnetosphere and the magnetospheric currents may exhibit temporal variations on time scales shorter than the time

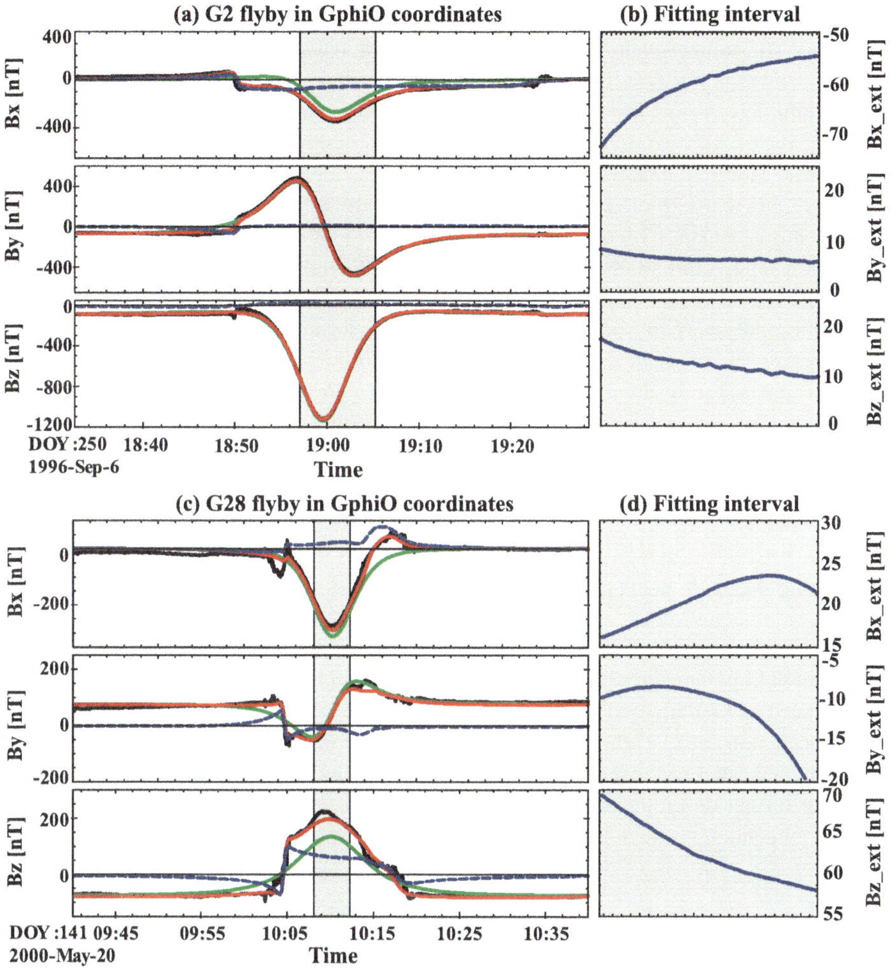

Fig. 7 Magnetic field comparisons between the Galileo measurements and MHD simulations for two Galileo flybys, G2 (**a** and **b**) and G28 (**c** and **d**). In panel (**a**) and (**c**), *black solid lines* are the Galileo data, *green lines* are the initial input in the MHD model and red traces are the simulation results. In order to account for the rapid temporal variations of Ganymede's magnetosphere resulting from reconnection on the magnetopause, simulation results shown in the comparisons are averaged over time steps that span a time interval of ~2 minutes after the system stabilizes. *Blue dashed lines* are the differences between *red and green traces* and they represent the perturbation fields due to the external current system produced in the MHD model. *Grey shaded regions* are the time intervals used by Kivelson et al. (2002) for the separation of internal-external fields. An expanded view of the external fields during the marked intervals is also shown in panel (**b**) and (**d**). Magnetic field vectors are shown in a Ganymede-centered Cartesian coordinates (GPhiO) defined in Fig. 6

interval of data collection owing to changes in the external plasma and field conditions or to dynamical processes in the magnetosphere (such as reconnection) that may be unsteady even under relatively steady external conditions. Under such circumstances, a steady state solution does not fully represent the variable external perturbations. Space-time aliasing of observations must be considered when the perturbation fields are inferred from simulations.

Another major limiting factor in fully characterizing Ganymede's internal field is insufficient data coverage. The magnetic field measurements at Ganymede were acquired during a limited number of flybys, which provided only a sparse coverage of the whole magnetosphere. The southern hemisphere, especially on the downstream side, has not been fully explored yet (see Fig. 9 in Jia et al. 2009a) and any magnetic anomaly localized in this region would have been "invisible". Even if the external plasma currents were fully known, the sparsity of data would remain a source of uncertainty. A complete determination of Ganymede's internal field requires more magnetic field measurements from future missions, especially in the unexplored high latitude regions in the southern hemisphere.

An optimal situation for accurately determining the internal field can be achieved by a polar orbiting spacecraft at altitudes below the current flowing region in the ionosphere. However, because the nature of Ganymede's ionosphere and atmosphere is not well constrained, it remains unclear where such a current-free region is located or even if it is present at all in Ganymede's environment. Until data from such an orbiter can be obtained, global models of the magnetospheric environment undoubtedly will play an essential role in characterizing the internal field and the inductive response of the moon.

5 Conclusions

Separating magnetic fields of internal and external origin is essential to infer details on the internally generated fields. The classical Gauss algorithm turns out to be very successful for studying the geomagnetic field, mainly because the field is dominated by a Laplacian potential field in the region where the measurements are taken. However, at other planetary bodies such as Mercury and Ganymede more elaborate techniques must be used before any firm conclusions can be drawn on the magnetic properties of these bodies. Measuring the magnetic field in a planetary environment is essential but not sufficient, and there is urgent need for further development of advanced analysis schemes. Field separation as discussed in this publication is an integral part of the interpretation process.

Acknowledgements NO highly appreciates valuable discussions with Michael E. Purucker on Mars data processing and model interpretation. XJ would like to thank Margaret Kivelson, Raymond Walker and Krishan Khurana for useful discussions. His work has been supported by NASA grant NNG05GB82G. The work by KHG was financially supported by the German Ministerium für Wirtschaft und Technologie and the German Zentrum für Luft- und Raumfahrt under contract 50 QW 0602.

References

I.I. Alexeev, E.S. Belenkaya, S.Yu. Bobrovnikov, J.A. Slavin, M. Sarantos, Paraboloid model of Mercury's magnetosphere. J. Geophys. Res. **113**(A12), 12210 (2008)

B.J. Anderson, M.H. Acuña, D.A. Lohr, J. Scheifele, A. Raval, H. Korth, J.A. Slavin, The magnetometer instrument on MESSENGER. Space Sci. Rev. **131**, 417–450 (2007). doi:10.1007/s11214-007-9246-7

B.J. Anderson, M.H. Acuña, H. Korth, M.E. Purucker, C.L. Johnson, J.A. Slavin, S.C. Solomon, R.L. McNutt, The structure of Mercury's magnetic field from MESSENGER's first flyby. Science **321**, 82 (2008). doi:10.1126/science.1159081

G. Backus, Poloidal and toroidal fields in geomagnetic field modeling. Rev. Geophys. **24**, 75–109 (1986)

L. Bauer, Chief results of a preliminary analysis of the Earth's magnetic field for 1922. Terr. Magn. Atmos. Electr. **28**(1), 1–28 (1923)

W. Baumjohann, A. Matsuoka, W. Magnes, K.H. Glassmeier, R. Nakamura, H. Biernat, M. Delva, K. Schwingenschuh, T. Zhang, H.H.U. Auster, K. Fornaçon, U. Motschmann, I. Richter, A. Balogh, C. Carr, P.J. Cargill, M. Dougherty, T.S. Horbury, E.A. Lucek, T. Takahashi, M. Tanaka, T. Nagai, H. Tsunakawa, M. Matsushima, M. Shinohara, H. Kawano, A. Yoshikawa, H. Shibuya, T. Nakagawa, M. Hoshino, Y. Tanaka, R. Kataoka, B. Anderson, C. Russell, Magnetic field investigation of Mercury's magnetosphere and inner heliosphere environment by MMO/MGF. Planet. Space Sci. **57**, (2009)

J.C. Cain, B.B. Ferguson, D. Mozzoni, An $n = 90$ internal potential function of the Martian crustal magnetic field. J. Geophys. Res. (Planets) **108**, 2–1 (2003). doi:10.1029/2000JE001487

S. Chapman, J. Bartels, *Geomagnetism*, vols. I+II (Clarendon Press, Oxford, 1940)

J.Y. Choe, D.B. Beard, The compressed geomagnetic field as a function of dipole tilt. Planet. Space Sci. **22**, 595–608 (1974a)

J.Y. Choe, D.B. Beard, The near Earth magnetic field of the magnetotail current. Planet. Space Sci. **22**, 609–615 (1974b). doi:10.1016/0032-0633(74)90094-4

J.T. Clarke, J. Ajello, G. Ballester, L.B. Jaffel, J. Connerney, J.C. Gerard, G.R. Gladstone, D. Grodent, W. Pryor, J. Trauger, J.H. Waite Jr., Ultraviolet emissions from the magnetic footprints of Io, Ganymede and Europa on Jupiter. Nature **415**, 997–1000 (2002)

M.W. Dunlop, A. Balogh, K.H. Glassmeier, P. Robert, Four-point cluster application of magnetic field analysis tools: The curlometer. J. Geophys. Res. **107**, 23–12314 (2002). doi:10.1029/2001JA005088

E. Friis-Christensen, H. Lühr, G. Hulot, *Swarm*: A constellation to study the Earth's magnetic field. Earth Planets Space **58**, 351–358 (2006)

H. Fritsche, *Die Elemente des Erdmagnetismus und ihre säkularen Änderungen während des Zeitraumes 1550 bis 1900* (St. Petersburg, 1900)

C.F. Gauss, *Allgemeine Theorie des Erdmagnetismus. Resultate aus den Beobachtungen des Magnetischen Vereins im Jahre 1838* (Göttinger Magnetischer Verein, Leipzig, 1839), pp. 1–52

G. Giampieri, A. Balogh, Modelling of magnetic field measurements at Mercury. Planet. Space Sci. **49**, 1637–1642 (2001)

K.H. Glassmeier, H. Auster, D. Heyner, K. Okrafka, C. Carr, G. Berghofer, B.J. Anderson, A. Balogh, W. Baumjohann, P.J. Cargill, U. Christensen, M. Delva, M. Dougherty, K. Fornaçon, T.S. Horbury, E.A. Lucek, W. Magnes, M. Mandea, A. Matsuoka, M. Matsushima, U. Motschmann, R. Nakamura, Y. Narita, I. Richter, K. Schwingenschuh, H. Shibuya, J.A. Slavin, C. Sotin, B. Stoll, H. Tsunakawa, S. Vennerstrom, J. Vogt, T. Zhang, The fluxgate magnetometer of the BepiColombo planetary orbiter. Planet. Space Sci. **57**, (2009)

J. Grosser, K.H. Glassmeier, A. Stadelmann, Induced magnetic field effects at planet Mercury. Planet. Space Sci. **52**, 1251–1260 (2004). doi:10.1016/j.pss.2004.08.005

D. Heyner, Magnetfeldmessungen am Merkur: Einfluss externer Stromsysteme. Diploma thesis, Techn. Univ. Braunschweig (2007)

W.H. Ip, A. Kopp, Resistive MHD simulations of Ganymede's magnetosphere: 2. Birkeland currents and particle energetics. J. Geophys. Res. **107**(A12), 1491 (2002). doi:10.1029/2001JA005072

D.J. Jackson, D.B. Beard, The magnetic field of Mercury. J. Geophys. Res. **82**, 2828–2836 (1977). doi:10.1029/JA082i019p02828

X. Jia, R.J. Walker, M.G. Kivelson, K.K. Khurana, J.A. Linker, Three-dimensional MHD simulations of Ganymede's magnetosphere. J. Geophys. Res. **113**, 6212 (2008). doi:10.1029/2007JA012748

X. Jia, M.G. Kivelson, K.K. Khurana, R.J. Walker, Magnetic fields of the satellites of Jupiter and Saturn. Space Sci. Rev. (2009a, this issue). doi:10.1007/s11214-009-9507-8

X. Jia, R.J. Walker, M.G. Kivelson, K.K. Khurana, J.A. Linker, Properties of Ganymede's magnetosphere inferred from improved three-dimensional MHD simulations. J. Geophys. Res. (2009b) doi:10.1029/2009JA014375

M.G. Kivelson, C.T. Russell, *Introduction to Space Physics* (Cambridge University Press, Cambridge, 1995)

M.G. Kivelson, K.K. Khurana, M. Volwerk, The permanent and inductive magnetic moments of Ganymede. Icarus **157**, 507–522 (2002)

M.G. Kivelson, K.K. Khurana, C.T. Russell, R.J. Walker, J. Warnecke, F.V. Coroniti, C. Polanskey, D.J. Southwood, G. Schubert, Discovery of Ganymede's magnetic field by the Galileo spacecraft. Nature **384**, 537–541 (1996)

M.G. Kivelson, J. Warnecke, L. Bennett, S. Joy, K.K. Khurana, J.A. Linker, C.T. Russell, R.J. Walker, C. Polanskey, Ganymede's magnetosphere: Magnetometer overview. J. Geophys. Res. **103**, 19963–19972 (1998)

A. Kopp, W.H. Ip, Resistive MHD simulations of Ganymede's magnetosphere: 1. Time variabilities of the magnetic field topology. J. Geophys. Res. **107**(A12), 1490 (2002). doi:10.1029/2001JA005071

H. Korth, B.J. Anderson, M.H. Acuña, J.A. Slavin, N.A. Tsyganenko, S.C. Solomon, R.L. McNutt, Determination of the properties of Mercury's magnetic field by the MESSENGER mission. Planet. Space Sci. **52**, 733–746 (2004). doi:10.1016/j.pss.2003.12.008

R.A. Langel, The main field, in *Geomagnetism*, vol. 1, ed. by J.A. Jacobs (Academic Press, London, 1987), pp. 249–512

R.A. Langel, R.H. Estes, Large-scale, near-Earth magnetic fields from external sources and the corresponding induced internal field. J. Geophys. Res. **90**, 2487–2494 (1985a)

R.A. Langel, R.H. Estes, The near-Earth magnetic field at 1980 determined from MAGSAT data. J. Geophys. Res. **90**, 2495–2509 (1985b)

F.J. Lowes, Mean-square values on sphere of spherical harmonic vector fields. J. Geophys. Res. **71**, 2179 (1966)

H. Lühr, M. Korte, M. Mandea, The recent magnetic field and its variations, in *Geomagnetic Variations*, ed. by K. Glassmeier, H. Soffel, J.W. Negendank (Springer, Berlin, 2009), pp. 25–63

P. Mauersberger, Das Mittel der Energiedichte des geomagnetischen Hauptfeldes an der Erdoberfläche und seine säkulare Änderung. Gerl. Beitr. Geophys. **65**, 207–215 (1956)

S. Maus, H. Lühr, Signature of the quiet-time magnetospheric magnetic field and its electromagnetic induction in the rotating Earth. Geophys. J. Int. **162**, 755–763 (2005)

S. Maus, H. Lühr, G. Balasis, M. Rother, M. Mandea, Introducing POMME, the Potsdam magnetic model of the Earth, in *Earth Observation with CHAMP, Results from Three Years in Orbit*, ed. by C. Reigber, H. Lühr, P. Schwintzer, J. Wickert (Springer, Berlin, 2005), pp. 293–298

N.F. Ness, K.W. Behannon, R.P. Lepping, Y.C. Whang, K.H. Schatten, Magnetic field observations near Mercury: Preliminary results from Mariner 10. Science **185**, 131–135 (1974)

N. Olsen, H. Lühr, T.J. Sabaka, M. Mandea, M. Rother, L. Tøffner-Clausen, S. Choi, CHAOS—a model of Earth's magnetic field derived from CHAMP, Ørsted, and SAC-C magnetic satellite data. Geophys. J. Int. **166**, 67–75 (2006). doi:10.1111/j.1365-246X.2006.02959.x

N. Olsen, Ionospheric F region currents at middle and low latitudes estimated from Magsat data. J. Geophys. Res. **102**(A3), 4563–4576 (1997)

C. Paty, R. Winglee, Multi-fluid MHD simulations of Ganymede's magnetosphere. Geophys. Res. Lett. **31** (2004). doi:10.1029/2004GL021220

C. Paty, R. Winglee, The role of ion cyclotron motion at Ganymede: magnetic morphology and magnetospheric dynamics. Geophys. Res. Lett. **33** (2006). doi:10.1029/2005GL025273

C. Paty, W. Paterson, R. Winglee, Ion energization in Ganymede's magnetosphere: Using multifluid simulations to interpret ion energy spectrograms. J. Geophys. Res. **113** (2008). doi:10.1029/2007JA012848

C.T. Russell, R.J. Walker, Flux transfer events at Mercury. J. Geophys. Res. **90**, 11067–11072 (1985)

T.J. Sabaka, N. Olsen, M. Purucker, Extending comprehensive models of the Earth's magnetic field with Ørsted and CHAMP data. Geophys. J. Int. **159**, 521–547 (2004). doi:10.1111/j.1365-246X.2004.02421.x

T.J. Sabaka, N. Olsen, Enhancing comprehensive inversions using the *Swarm* constellation. Earth Planets Space **58**, 371–395 (2006)

J. Saur, F.M. Neubauer, K.H. Glassmeier, Induced magnetic field in solar system bodies. Space Sci. Rev. (2009, this issue)

N. Schilling, F.M. Neubauer, J. Saur, Time-varying interaction of Europa with the Jovian magnetosphere: Constraints on the conductivity of Europa's subsurface ocean. Icarus **192**(1), 41–55 (2007). doi:10.1016/j.icarus.2007.06.024

A. Schmidt, Der magnetische Zustand der Erde zur Epoche 1885.0. Archiv der deutschen Seewarte, 21–75 (1898)

A. Schmidt, Mitteilungen äber eine neue Berechnung des erdmagnetischen Potentials. Abh. K. Bayrisch. Akad. Wiss. **19**(1), 1–66 (1895)

J. Scuffham, A. Balogh, A new model of Mercury's magnetospheric magnetic field. Adv. Space Res. **38**, 616–626 (2006). doi:10.1016/j.asr.2005.08.052

G. Siscoe, L. Christopher, Variations in the solar wind stand-off distance at Mercury. Geophys. Res. Lett. **2**, 158–160 (1975)

J.A. Slavin, M.H. Acuña, B.J. Anderson, D.N. Baker, M. Benna, G. Gloeckler, R.E. Gold, G.C. Ho, R.M. Killen, H. Korth, S.M. Krimigis, R.L. McNutt, L.R. Nittler, J.M. Raines, D. Schriver, S.C. Solomon, R.D. Starr, P. Trávníček, T.H. Zurbuchen, Mercury's magnetosphere after MESSENGER's first flyby. Science **321**, 85 (2008). doi:10.1126/science.1159040

S.M. Stone, T.P. Armstrong, Three-dimensional magnetopause and tail current model of the magnetosphere of Ganymede. J. Geophys. Res. **106**(A10), 21263–21275 (2001)

N.A. Tsyganenko, A magnetospheric magnetic field model with a warped tail current sheet. Planet. Space Sci. **37**, 5–20 (1989). doi:10.1016/0032-0633(89)90066-4

N.A. Tsyganenko, Modeling the Earth's magnetospheric magnetic field confined within a realistic magnetopause. J. Geophys. Res. **100**, 5599–5612 (1995). doi:10.1029/94JA03193

N.A. Tsyganenko, A model of the near magnetosphere with a dawn-dusk asymmetry 1. Mathematical structure. J. Geophys. Res. **107**(A8), 12-1 (2002a)

N.A. Tsyganenko, A model of the near magnetosphere with a dawn-dusk asymmetry 2. Parameterization and fitting to observations. J. Geophys. Res. **107**(A8), 10-1 (2002b)

N.A. Tsyganenko, M.I. Sitnov, Magnetospheric configurations from a high-resolution data-based magnetic field model. J. Geophys. Res. **112**(A11), 6225 (2007). doi:10.1029/2007JA012260

H. Uno, C.L. Johnson, B.J. Anderson, H. Korth, S.C. Solomon, Modeling Mercury's internal magnetic field with smooth inversions. Earth Planet. Sci. Lett. (2009) doi:10.1016/j.epsl.2009.02.032

Y.C. Whang, Magnetospheric magnetic field of Mercury. J. Geophys. Res. **82**, 1024–1030 (1977). doi:10.1029/JA082i007p01024

Space Sci Rev (2010) 152: 159–222
DOI 10.1007/s11214-010-9644-0

The Magnetic Field of Planet Earth

G. Hulot · C.C. Finlay · C.G. Constable · N. Olsen · M. Mandea

Received: 22 October 2009 / Accepted: 26 February 2010 / Published online: 9 April 2010

Abstract The magnetic field of the Earth is by far the best documented magnetic field of all known planets. Considerable progress has been made in our understanding of its characteristics and properties, thanks to the convergence of many different approaches and to the remarkable fact that surface rocks have quietly recorded much of its history. The usefulness of magnetic field charts for navigation and the dedication of a few individuals have also led to the patient construction of some of the longest series of quantitative observations in the history of science. More recently even more systematic observations have been made possible from space, leading to the possibility of observing the Earth's magnetic field in much more details than was previously possible. The progressive increase in computer power was also crucial, leading to advanced ways of handling and analyzing this considerable corpus

G. Hulot (✉)
Equipe de Géomagnétisme, Institut de Physique du Globe de Paris (Institut de recherche associé au CNRS et à l'Université Paris 7), 4, Place Jussieu, 75252, Paris, cedex 05, France
e-mail: gh@ipgp.jussieu.fr

C.C. Finlay
ETH Zürich, Institut für Geophysik, Sonneggstrasse 5, 8092 Zürich, Switzerland

C.G. Constable
Cecil H. and Ida M. Green Institute of Geophysics and Planetary Physics, Scripps Institution of Oceanography, University of California at San Diego, 9500 Gilman Drive, La Jolla, CA 92093-0225, USA

N. Olsen
DTU Space and Niels Bohr Institute of Copenhagen University, Juliane Maries Vej 30, 2100 Copenhagen Ø, Denmark

M. Mandea
Helmholtz-Zentrum Potsdam Deutsches GeoForschungsZentrum, 14473 Potsdam, Germany

Present address:
M. Mandea
Equipe de Géophysique Spatiale et Planétaire, Institut de Physique du Globe de Paris (Institut de recherche associé au CNRS et à l'Université Paris 7), Bâtiment Lamarck, 5 rue Thomas Mann, 75013 Paris, France

of data. This possibility, together with the recent development of numerical simulations, has led to the development of a very active field in Earth science. In this paper, we make an attempt to provide an overview of where the scientific community currently stands in terms of observing, interpreting and understanding the past and present behavior of the so-called main magnetic field produced within the Earth's core. The various types of data are introduced and their specific properties explained. The way those data can be used to derive the time evolution of the core field, when this is possible, or statistical information, when no other option is available, is next described. Special care is taken to explain how information derived from each type of data can be patched together into a consistent description of how the core field has been behaving in the past. Interpretations of this behavior, from the shortest (1 yr) to the longest (virtually the age of the Earth) time scales are finally reviewed, underlining the respective roles of the magnetohydodynamics at work in the core, and of the slow dynamic evolution of the planet as a whole.

Keywords Earth · Geomagnetism · Archeomagnetism · Paleomagnetism · Magnetic observations · Archeomagnetic records · Paleomagnetic records · Spherical harmonic magnetic field models · Statistical magnetic field models · Geomagnetic secular variation · Geomagnetic reversals · Core magnetohydrodynamics · Numerical dynamo simulation · Geodynamo · Earth's planetary evolution

1 Introduction

Earth's magnetism has been known to man for a very long time. The successive discoveries of the needle's rough orientation towards the geographic North, of the concepts of declination, inclination, and intensity, and of the fact that the Earth's magnetic field changed through time (in a process known as secular variation) progressively led to a growing body of magnetic observations. The usefulness of a precise knowledge of the declination for navigation purposes and the need to monitor the secular variation of the field to regularly update maps, also quickly led to systematic observations all over the globe, both at sea and on land, and to the establishment of permanent magnetic observatories. Nowadays, these observatories and satellite observations make it possible to closely monitor and investigate the various fields that add up to produce the observed field, with sources in the core (which produces by far the largest component of the field), the crust, the ionosphere, the magnetosphere and, to a lesser extent, the mantle and oceans.

Such direct observations only extend back approximately four centuries. This is long enough for significant changes to be observed in "movies" of the reconstructed core field evolution, but far too short to infer anything about the long-term field behaviour. Fortunately considerable additional information is also available from indirect observations provided by the magnetization of both human artefacts (such as bricks, tiles, potteries), and various types of rocks (mainly basalts and sediments) that can be sampled and measured. Each such magnetized sample carries some quantitative information about the field it experienced at the time it acquired its magnetization. These indirect observations can be used to extend our knowledge of Earth's ancient field very far back in time.

The accuracy with which those samples can be measured is however limited and only information about the dominant core field can reasonably be recovered (other contributions being at best of comparable magnitude to measurement errors). Dating accuracy is also an important limiting factor for spatio-temporal analysis of the ancient field behavior, which requires temporal synchronization of the information provided by distant samples. This important constraint is the main reason our knowledge of the historical patterns of field evolution cannot yet be expanded very far back in time. Nonetheless, the time evolution of the largest scales of the core field can, and has been, reconstructed over the past few millennia, using techniques akin to those used for historical data analysis.

Information concerning the more ancient geomagnetic field can also be recovered, but this requires other types of analysis. In fact, and as we shall see, quite a few different types of analysis can be used, depending on the sample studied, the time span considered and the geomagnetic information to be recovered. Perhaps the single most important information revealed by the paleomagnetic record in general, is that the core field has always remained predominantly an axial dipole of comparable magnitude to the present field (which varies within the range 25.000–65.000 nT at the Earth's surface), except during relatively short periods of time (on order 10 ky) when the field dropped significantly and became dominated by non-dipole components. Those short periods of time always led the field to grow back to its usual axial dipole dominated structure either with the same polarity (in which case one usually refers to these events as "excursions"), or with the opposite polarity (events known as "reversals"). Between two such events, the field is then said to have been of "stable polarity".

Because of their continuous nature, sediment records are particularly well suited for the investigation of the long-term temporal behaviour of the field at a particular location. But only relatively recent sediments (up to a few million years old) with fast sedimentation rates can provide high-resolution temporal information. This is enough to provide very useful information with respect to recent excursions and reversals. Unfortunately, acquiring high accumulation rate sediment records with good global coverage on much longer time scales (several hundreds of millions of years) is much more difficult. The available records heavily smooth the magnetic signal, and mainly provide information about the rate at which the dipole component of the field evolved and reversed in the past.

Very useful complementary information is also provided by rocks that acquired so-called thermo-remanent magnetization (TRM), some of which testify for a very ancient field (over 3 Gy old). The bulk of this data comes from lava flows. Their main advantage is that they provide instantaneous spot readings of the ancient core field. Their main disadvantage is that they do not sample time in a regular way. Such data can nevertheless catch the field at times of excursions or reversals. More often however, they provide information about the field at times of stable polarity. All those data can be used to characterize long-term statistical properties of the core field, including the rate of reversals in the past.

Another efficient way of recovering information about past reversal rates is the analysis of the extensive record provided by the magnetized ocean crust, at least over the time period covered by the seafloor age range (back to a couple hundred million years). As new sea floor is created at ridge crests because of sea floor spreading, it cools, acquires a TRM and therefore captures a record of past field variation. The beauty of this specific record is that it is directly available in the form of the worldwide distribution of ocean crust magnetization with alternating polarities, the signal of which produces characteristic linear magnetic anomalies (parallel to the ridges) in marine magnetic surveys. Although the detailed analysis of such signals is far from being trivial, appropriate procedures can be used to recover the polarity, and to some extent the intensity, of the field that produced the magnetization.

All those different types of data complement each other. They have led to a fairly comprehensive view of the various sources that contribute to the Earth's magnetic field and of the way this field evolved in time. Here, however, we will only focus on the field produced within the core. We first provide an overview of the various types of data routinely used to investigate this field (Sect. 2), next describe the way these data can be used to recover the behavior of the present and past core field (Sect. 3), and finally review the way this behavior is currently understood in terms of planetary core dynamics (Sect. 4). External sources will be briefly mentioned only to the extent those need to be taken into account in the analysis of modern data. Likewise, crustal sources will be mentioned only to the extent they provide a record of the ancient core field. For more details about those and other non-core sources in a planetary context, the reader is referred to the companion papers of Baumjohann et al. (2010), Olsen et al. (2009b), Langlais et al. (2009) and Saur et al. (2009).

2 Observations

2.1 Satellite Observations

The biggest advantage of measuring Earth's magnetic field from space with low Earth-orbiting satellites is that it yields an excellent spatial data coverage, an important prerequisite for obtaining good global models of the geomagnetic field. It also ensures that different regions of Earth are sampled with the same instrumentation. However, because satellites are moving fast (at typically 8 km/s for low-Earth orbiting satellites), the field changes they sense are a combination of both changes due to the movement of the satellite within the field, and actual temporal changes of the field. As we will later see, this makes the identification of the contributions of the various sources of the magnetic field quite challenging.

The first global satellite observations of the Earth's magnetic field were taken by the POGO satellites that operated between 1965 and 1971. POGO measured only the scalar field (magnetic intensity) but not the vector components (Fig. 1). This, we now know, unfortunately only provides partial information about the field, and leads to a fundamental ambiguity in its determination (Backus 1970; Lowes 1975). Although such ambiguity can be overcome with the help of additional information (Khokhlov et al., 1997, 1999; Ultré-Guérard et al. 1998; Holme et al. 2005), the need for measurements of the full vector magnetic field quickly became obvious.

The first such vector satellite mission was Magsat, which flew for 8 months in 1979–80 at an altitude of 300 to 550 km. After this very successful (see e.g. Langel and Hinze 1998) but short-lived mission, quite a few satellite missions were proposed. But it was not until 20 years later that these efforts payed off, with the successful launch of the Ørsted satellite in February 1999, which marked the beginning of a new era of continuous space magnetometry. Being the first satellite of the *International Decade of Geopotential Research*, Ørsted and its instrumentation (in particular, its combined set of an absolute scalar magnetometer, vector magnetometer and star tracker to achieve high precision oriented vector magnetic field measurements at 1 Hz and 50 Hz sampling rates, see e.g. Neubert et al. 2001) has since become a model for other missions such as CHAMP (launched in July 2000, Reigber et al. 2002) and SAC-C (November 2000–December 2004). More than 10 years of continuous measurements of the Earth's magnetic field from space are now available, with a typical accuracy of 0.5 nT for intensity measurements, and somewhat less good (2nT for the best CHAMP data) for individual field component measurements.

The low altitude (350–450 km) of CHAMP has proved extremely useful for the investigation of the ionospheric and crustal fields, while the combination of simultaneous observations taken by Ørsted (650–850 km altitude), CHAMP and SAC-C ($\approx$ 700 km altitude) led to considerable progress in the investigation of the temporal behavior of the core field. Building on this past experience, ESA's *Swarm* satellite constellation to be launched

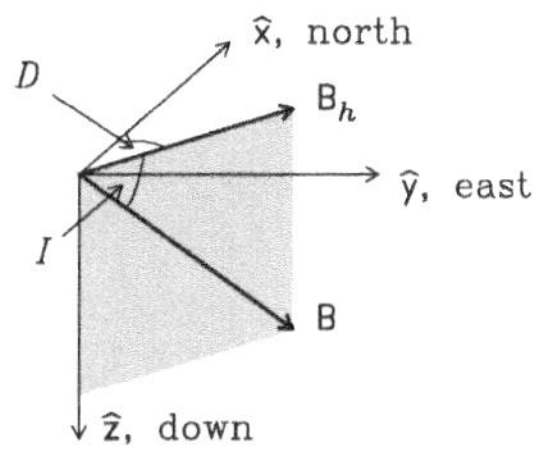

Fig. 1 Geomagnetic elements in a local coordinate system: *D* declination, *I* inclination; *B* magnetic field strength; B_h horizontal component of magnetic field

in 2011, will consist of a pair of side-by-side satellites at an initial altitude of 450 km, and a third satellite orbiting at higher altitude (530 km) with a different orbital drift rate. This configuration will allow for an even better separation of internal and external fields, and an enhanced sensitivity to small-scale structures of the crustal field (Olsen et al. 2006; Sabaka and Olsen 2006). Such an improved continuation of magnetic field observation from space is thus expected to lead to even more progress in our understanding of all sources of the Earth's magnetic field (Friis-Christensen et al. 2006, 2009).

2.2 Magnetic Observatories

Before the advent of satellites, all magnetic measurements were carried out on ground and at sea. Because it was recognized that the magnetic field displayed significant changes on historical time scales, regular measurements of field elements (at first only Declination and Inclination, see Fig. 1) at fixed known locations were soon carried out for instance in London (Malin and Bullard 1981), Paris (Alexandrescu et al. 1996a), Rome (Cafarella et al. 1992) and Edinburgh (Barraclough 1995). But it was not until the 1840's, under the impulse of Gauss and Weber, that a global network of fully dedicated Magnetic Observatories (by then measuring all magnetic field elements, including its intensity) started to develop to monitor temporal changes of the Earth's magnetic field (see Fig. 2, which shows the current distribution observatories, and Fig. 3, which shows records of the magnetic field as measured in the Niemegk observatory since 1890). Measurements carried out in such magnetic observatories have generally involved regular *absolute measurements* to monitor instrumental drift of variometers, which otherwise provided continuous *variation measurements* of the three components of the magnetic field.

Nowadays, variations are continuously measured and digitally recorded, either by three-component fluxgate magnetometers or by magnetometers based on sensors which measure field components by a scalar sensor equipped with coil systems. The elements of the geomagnetic field vector are then recorded in instrument-related coordinate systems. Such variometers are unfortunately subject to drifts arising both within the instrument (e.g., temperature effects) and because of the limited stability of the instrument mounting. To monitor and correct for those drifts, and also to convert such measurements into absolute units in the geographical reference frame, additional absolute measurements are carried out. For the field direction, this is usually done with the help of a flux-gate theodolite, searching for the plane perpendicular to the field (which is detected when the highly sensitive single-component flux-gate sensor sees no more field). Absolute measurements of the field intensity are otherwise directly measured with the help of an absolute scalar magnetometer. Several measurements are usually carried out using appropriate procedures to remove all systematic instrument errors (for more details see e.g. Jankowski and Sucksdorff 1996; Turner et al. 2007). This requires well-trained personnel and one complete measurement takes about 30 min. Such absolute measurements are typically performed on a weekly basis.

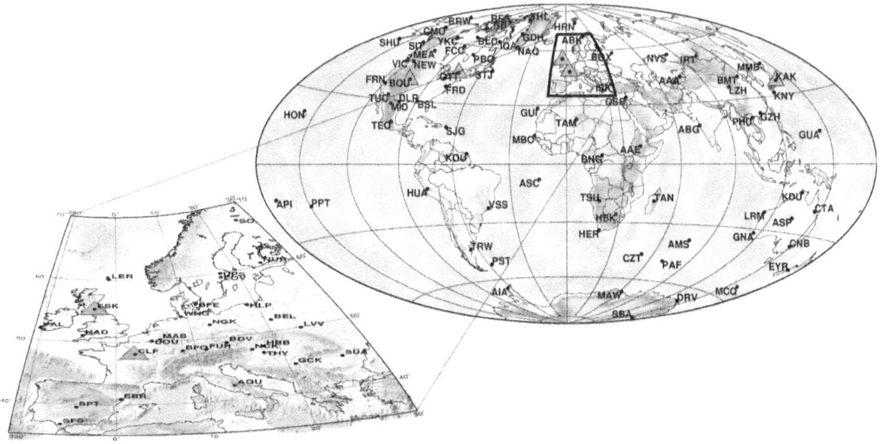

Fig. 2 Distribution of currently operating geomagnetic observatories from the INTERMAGNET worldwide network

Current quality standards for geomagnetic observatory data ask for an accuracy better than ±5 nT, including a long-term stability of variation recordings better than 5 nT/yr. An accuracy of 1 nT can be achieved for absolute measurements by well-trained observatory staff. Recently, however, the wish to record one-second data has been expressed by the community, especially in view of the upcoming *Swarm* mission, and indeed some observatories are already able to provide such high-resolution data. Many of the better magnetic observatories, which maintain a higher-level standard for data measurement and provide near real-time distribution, collectively form the INTERMAGNET worldwide network of observatories.[1]

As is clear from Fig. 2, one important drawback of the current network of magnetic observatories is that it is unevenly distributed, with high concentration in Europe and North America, and a dearth in the Southern Hemisphere and over the oceans. Efforts to correct for this drawback is currently oriented towards developing fully automated observatories that could be installed in remote areas (e.g., Gravrand et al. 2001; Van Loo and Rasson 2006; Auster et al. 2006, 2007).

2.3 Historical Records

Going further back in time the importance of Earth's magnetic field as a navigational tool, together with the intrigue it generated amongst prominent early scientists, result in large numbers of well documented direct field observations spanning the past four centuries. This period is commonly referred to as the 'historical era' in the geomagnetic literature. Here only a brief summary of the most important historical sources are given; for further details readers should consult the landmark paper of Bloxham et al. (1989) and the review article of Jonkers et al. (2003) where a comprehensive database comprising 151,560 declination, 19,525 inclination and 16,219 intensity observations made between 1510 and 1930 (available from the World Data Centre for Geomagnetism at the British Geological Survey,

[1] www.intermagnet.org.

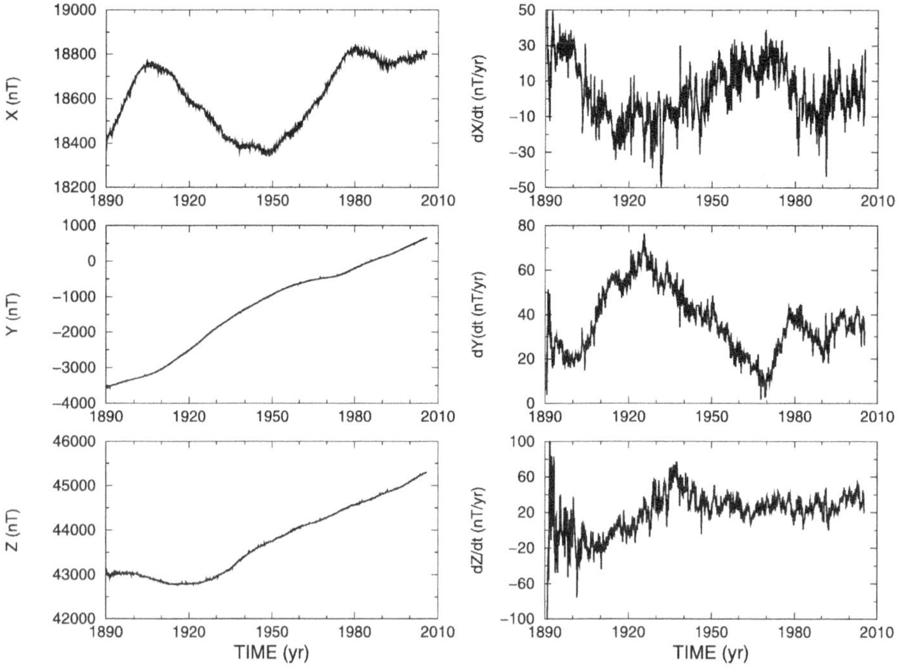

Fig. 3 Monthly means for the magnetic field and its corresponding first time derivative, recorded at the Niemegk Observatory (Early observations were actually carried out in Potsdam between 1890 and 1907, and Seddin between 1907 and 1930, and are here reported after correcting for the current location of the Niemegk observatory). Note the sudden changes of trends in the secular variation, best seen in the East component dY/dt, for instance in 1970. Those events are known as geomagnetic jerks (Courtillot et al. 1978, see Sect. 4.1.2)

Edinburgh) is described. Interesting accounts of the history of geomagnetism are otherwise given by Schröder (2000), Stern (2002) and Courtillot and Le Mouël (2007).

Historical observations of the geomagnetic field are dominated by directional measurements. By the second half of the 16th century compasses were widely employed to measure declination. Inclination, which required measuring the dip of the magnetic vector below the horizontal plane (Fig. 1), was determined on a number of vessels in the late 16th and early 17th centuries. However since it never attained a place in standard navigational practice, inclination measurements are much more scarce. Useful relative intensity measurements were made only after 1790 while absolute intensity measurements were first carried out by Gauss in 1832.

The majority of the useful historical geomagnetic observations were made by mariners involved in merchant and naval shipping during their travels across the globe (Records from more than 2000 such voyages are included in Jonkers et al. 2003). There are only a small number of observations available prior to AD1590. Between AD1590 and AD1700 many more of observations exist, thanks particularly to records made by mariners working for the Dutch and English East Indian companies. From AD1700 to AD1800 the number of observations again increased due to a dramatic expansion of naval traffic especially along Atlantic and Arctic trade routes. 18th century declination observations are plotted geographically in Fig. 4. Between 1800 and 1930, in addition to observations made on the oceans by mariners, extensive land surveys were also carried out in continental interiors. All those observations

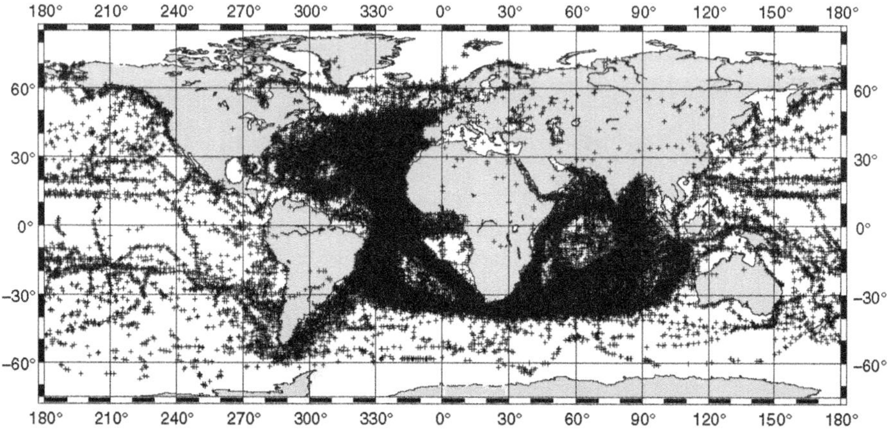

Fig. 4 Geographical distribution of the 68,076 declination observations made from AD1700–1799; some points may overlap; cylindrical equidistant projection (after Jonkers et al. 2003)

nicely complement the long time series provided by magnetic observatories at fixed locations.

It remains possible that major new archives of historical records could be unearthed. However the majority of recently discovered historical data are comparatively modest, for example in newly discovered records made by explorers crossing continental interiors (Vaquero and Trigo 2006). Another possibility is that accurate indirect archeomagnetic data (see next section) could be used to supplement the historical observations especially during the 16th and 17th centuries when direct observations are scarce.

The heterogeneous origin of historical observations dictates that there are significant variations in the number of observations as a function of time. In Fig. 5 the number of historical data per 5 years is plotted together with a selection of modern data used by Jackson et al. (2000) to construct the *gufm1* field model (see Sect. 3.2). Note there are rather few data available pre-AD1650. In the mid-19th century there is a sharp increase in the number of available data, thanks in part to the magnetic endeavours of Gauss and Sabine. Clearly the number of observations available in the 20th century dwarfs the number of direct observations available at earlier times. Note that it is not just the number of available data but also the type of measurement (from declination and inclination to three component vector measurements) that changes with time.

Also noteworthy are the major variations in the density of measurements with geographical location (recall Fig. 4). A bias towards commercially and militarily important shipping routes is obvious, with trans-Atlantic paths very well covered. In contrast Pacific and Polar regions are sparsely covered. There also are few observations in continental interiors, especially outside Europe. In addition, the vast majority of the maritime observations are of declination with inclination and intensity measurements much rarer due to the greater difficulties involved in their measurement.

Such a heterogeneous distribution of historical data must clearly be borne in mind when carrying out historical field modeling. Somewhat fortunately however, potential field theory shows this not to be so severe an issue, if the goal is to recover the large scales of the field produced at the core surface (Gubbins and Roberts 1983).

In order to extract the maximum amount of information from historical observations an understanding of their inherent errors is also of great importance. Jackson et al. (2000)

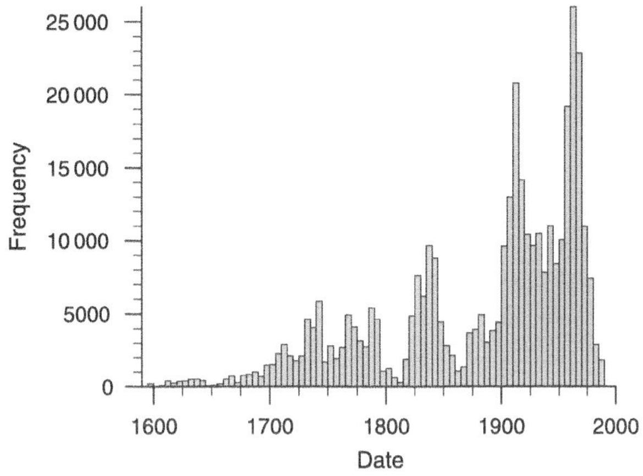

Fig. 5 Overall number historical data (as described by Jonkers et al. 2003), together with observatory annual means, twentieth century survey data, repeat station data and satellite data used in the construction of *gufm1* (Jackson et al. 2000). Note that this depicts a subset of data available, as some data selection has taken place, based on criteria designed to avoid the effect of the correlation in errors due to the crust

studied this issue in depth, developing error budgets accounting for observational errors, errors due to the influence of unknown crustal magnetic fields, and errors in positions due to positional uncertainty. Jackson et al. (2000) also showed that historical declination measurements were surprisingly accurate, with a typical error of only 0.5 degrees; the total error budget for such data is consequently dominated by the unknown crustal field in a manner similar to modern survey measurements. This surprising accuracy of the historical measurements, together with their worldwide extent, are the crucial factors allowing detailed reconstruction of the evolution of Earth's magnetic field over the past 400 years.

2.4 Archeomagnetic Records

The indirect observations of the magnetic field characterized as archeomagnetic records comprise the recovery of at least one of the magnetic elements, namely declination, inclination, or field strength (D, I, or B, recall Fig. 1) with accompanying age information either from a man-made structure or archeological artifact or from a relatively young volcanic flow. The term archeomagnetic usually carries with it an implicit restriction on the age of the record under consideration. Most of the archeological artifact and accurately dated lava flows are younger than 10 ka (see e.g. Korte et al. 2005; Genevey et al. 2008). But it is reasonable to suggest a temporal range of 0–50 ka, corresponding to the upper age limit on all records included in GEOMAGIA50, currently the most comprehensive database of such records (http://geomagia.ucsd.edu/, Donadini et al. 2006, 2009; Korhonen et al. 2008). In archeological samples or lava flows the recording mechanism for the magnetic field is typically a thermal remanent magnetization (TRM), acquired as the material cools through its magnetic blocking temperature spectrum in an ambient magnetic field (e.g. Dunlop and Özdemir 2007). Standard sampling techniques that preserve the orientation in geographic coordinates, followed by laboratory cleaning procedures (described in e.g. Constable 2007; Turner et al. 2007) allow the recovery of the ancient field directions and often the field strength too (e.g. Tauxe and Yamazaki 2007). The usual assumption is that the resulting magnetization will be aligned with the ambient field and its intensity will be linearly dependent on its strength. In some cases one or both of these assumptions will be violated and some care is required to detect this: correction for an anisotropic response to the ancient field can be accomplished for both directions and field strength; non-linearity is harder to detect, and complications in recovering the ancient field strength using the most widely used Thellier-Thellier (and related) methods (see e.g. Tauxe and Yamazaki 2007 and Genevey et al. 2008 for details) are common since the technique requires reheating the sample in the laboratory and may produce undesired alterations to the magnetic mineralogy (not to mention that the material under study may have also suffered chemical alteration affecting the magnetic mineralogy prior to sample collection). The strength of the thermal remanence also depends on the cooling rate when it was acquired. Although corrections are possible to account for the more rapid cooling rate in the laboratory (see e.g. Genevey et al. 2003), the original rate is often difficult to estimate accurately. Considerable effort has been invested in developing new and improved laboratory procedures over the past decade, so that well-documented experimental data are easier to evaluate than they used to be. It is generally expected that declinations and inclinations can in principle be recovered to within a few degrees, and intensities to within 10%.

As with the historical geomagnetic data set there are large variations in the number of archeomagnetic data available as a function of time and place. These reflect the development of human settlements and associated artifacts and geophysical constraints on temporal and spatial distributions of lava flows. Europe for instance has a very rich archeological record that makes it possible to reconstruct the field directional behavior fairly continuously since 1000 BC (Fig. 6, Gallet et al. 2002). Considerable efforts are also put into reconstructing similar continuous regional records of the field intensity, which are generally more difficult to recover (see Fig. 7, and e.g. Genevey et al. 2009). It is worth also noting that the past decade has led to new projects targeting the construction of regional records outside the European region. In particular the use of more novel materials such as lime plasters (e.g., in Mexico, Hueda-Tanabe et al. 2004), non-welded pyroclastic deposits (e.g., in West Indies, Genevey et al. 2002) and slag deposits from copper mining (e.g., in the middle east, Ben-Yosef et al. 2008a, 2008b, 2009), offer promise of extending the archeomagnetic record in both time and space. Figure 7 demonstrates that large changes in field strength (10–15 μT) commonly occur on time-scales of just a few hundred years. Initial paleointensity results from the slag deposits even suggest that on occasion the local field strength may have been twice its current strength, and subject to rapid change (Ben-Yosef et al. 2009). Unfortunately, the number of archeomagnetic data falls off sharply prior to about 3 ka and even since that time the spatial coverage is very inhomogeneous, with almost no southern hemisphere data (e.g. Korte et al. 2005; Genevey et al. 2008). Figure 8(b) and (c) illustrate the spatial distribution for the CALS7K.2 data set of Korte et al. (2005) covering the past 7 kyr.

These archeomagnetic data can be supplemented with sedimentary records from more homogeneously distributed locations (Fig. 8(a)). Such records are fortunately available thanks to the fact that the magnetic mineral grains contained in the sediments settle under the influence of the geomagnetic field, thus producing a weak but measurable continuous magnetization (and therefore geomagnetic record) in the sedimentary section (see e.g. Dunlop and Özdemir 2007).

Archeological artifacts are the main contribution to the rise in total number of data since 1000 BC seen in Fig. 9 while the time series of variations acquired from sediments are generally much more uniform in temporal coverage. Ongoing efforts with data gathering and compilations are generating significantly larger global data sets (Donadini et al. 2009), but with similar intrinsic limitations.

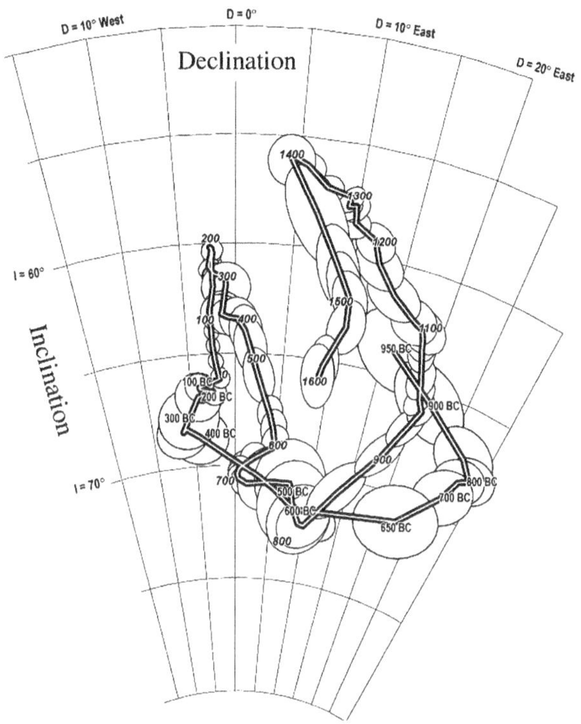

Fig. 6 Directional variations of the Earth's magnetic field in France since 1000 BC, as recovered from French archeological artifacts (all data were reduced to Paris), adapted from Gallet et al. (2002). Note the occurrence of sharp changes, or cusps, at roughly 800 BC, AD 200, AD 800 and AD 1400, known as "archeomagnetic jerks" (Gallet et al. 2003) (see Sect. 4.1.3)

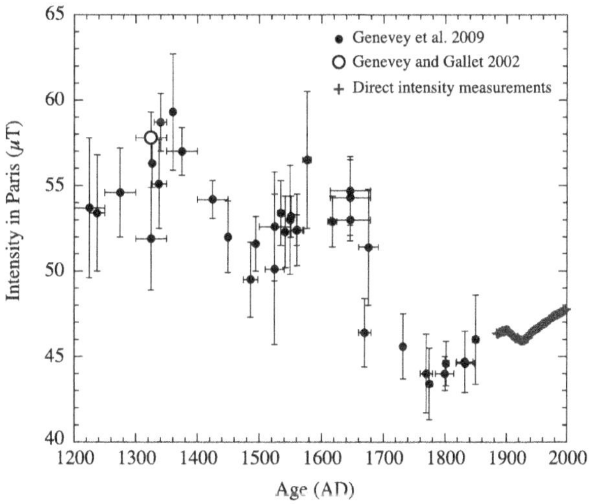

Fig. 7 Intensity of the Earth's magnetic field in France since AD 1200, as recovered from French archeological artifacts (all data were reduced to Paris), after Genevey et al. (2009). The *single open circle* is from a previous study by Genevey and Gallet (2002). Direct observatory measurements for recent epochs are also shown for reference

2.5 Paleomagnetic and Seafloor Records

For longer term variations of the geomagnetic field we distinguish three major sources of data, igneous rocks, sediment records, and marine magnetic anomalies, each with their own advantages and limitations.

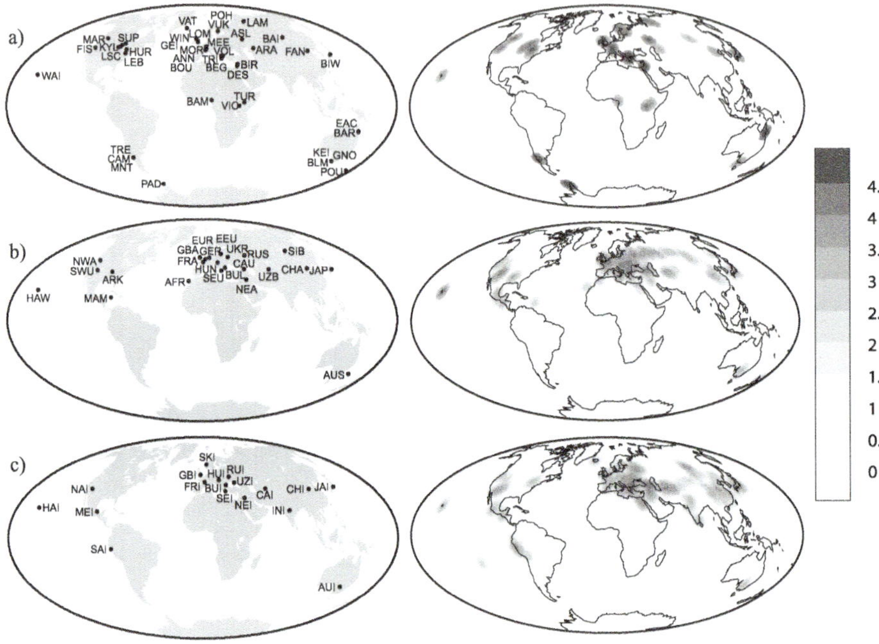

Fig. 8 Locations represented in the Korte et al. (2005) global data compilation for the 0–7 ka time interval. Sites of (**a**) lakes, (**b**) archeomagnetic directional data, and (**c**) archeomagnetic intensity data. *Left side* gives locations for individual sediment records, and average locations for archeomagnetic regions, *right side* contours of data concentration

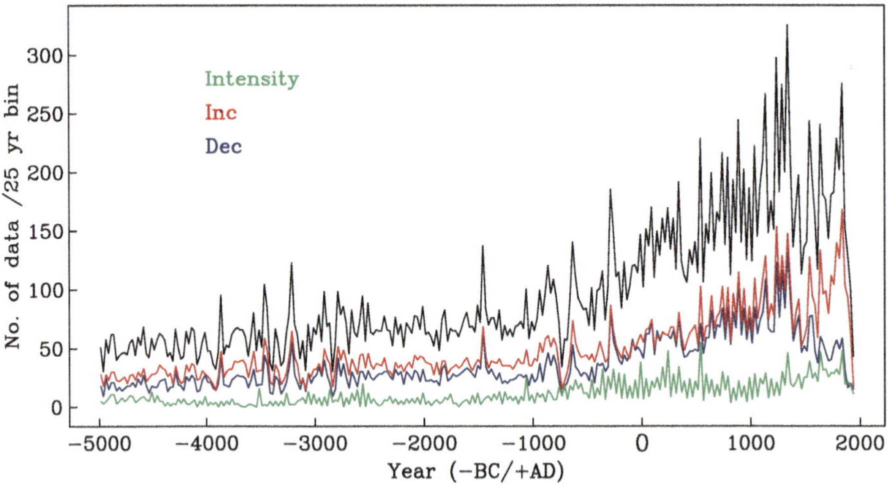

Fig. 9 Numbers of each element type available in the Korte et al. (2005) global data compilation for the 0–7 ka time interval (*blue*, declination; *red*, inclination; *green*, intensity; *black*, total number of data)

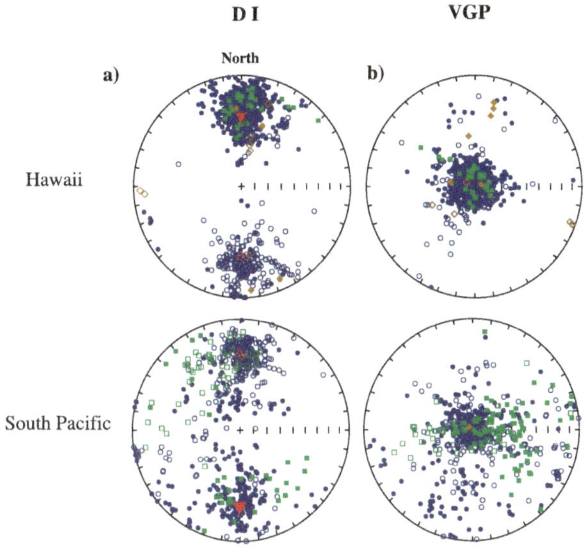

Fig. 10 Equal area projections of directional data compilations from 0–5 Ma from Hawaii and South Pacific region near 20° latitude to show (D, I) projection (*left*). Also shown (*red triangles*), the direction a pure geocentric axial dipole field would predict. *Green symbols* reflect sites affected by post-emplacement tectonic rotation, and *brown symbols* are directions derived from less than 3 samples per flow. *Solid* (*open*) *circles* represent normal (reverse) directions. The *right panels* show the same data plotted in terms of VGP positions (see Sect. 3.3). After Lawrence et al. (2006)

Igneous rock data, and in particular lava flow data which make the bulk of such data, comprise spot records (from a geological perspective) in time and space of the direction and/or intensity of the geomagnetic field. Figure 10 illustrates the nature of typical directional lava flow data and several concerns using regional compilations for two general locations, the Hawaiian islands and volcanic islands in the South Pacific region near 20° latitude. Those data sets cover the 0–5 Ma time period and are discussed in some detail by Lawrence et al. (2006), where it is noted that some flows may have undergone post-emplacement rotations that render the recovered directions unsuitable for geomagnetic field studies. Much care must be taken to avoid such lava flows. Other limitations worth noting are that the accuracy of the directions recovered depends on acquiring average results from multiple independently oriented samples distributed across each flow ($<$ 5 samples is now generally considered marginal), and that, contrary to the much younger lava flows that qualify as archeomagnetic records, the age control usually provided by radioisotopic dating is most often not adequate to construct a time series of variations. As we shall later see (Sect. 3.3), this major limitation is one that will force a different (i.e. statistical) analysis of the paleomagnetic data compared to the one (deterministic) used when considering historical and archeomagnetic data (Sect. 3.2).

Also of some concern is the fact that as a result of plate tectonics, sites of old lava flows will usually have moved and rotated since the magnetization was acquired. Obviously, the initial location and orientation of such flows must also be recovered for their optimal use in paleomagnetic field modeling. Recent plate tectonic motions are fortunately well enough known (e.g. DeMets et al. 1994) that the bulk of the data (most of which is less than 5 Ma) can be assigned to their correct locations for when the magnetization was acquired. But such corrections grow more problematic as one goes further back in time, not least because as we shall later see (Sect. 3.3) directional information provided by such lava flows are used in ancient plate tectonic reconstructions. This will again force a different approach to the oldest of the available lava flow data. Note that this issue will also affect any other type of very ancient paleomagnetic data.

As already pointed out in the previous section, lava flows can also be used to recover intensity of the ancient field. But alteration issues (even more critical for old samples) make paleointensities difficult to recover and lava flows that provide paleodirections often fail to provide reliable paleointensities. The same issue affects other igneous rocks, such as plutonic rocks, which are also subject to additional uncertainties with respect to evaluating their cooling rates. This unfortunate state of affairs has led to intense search for improved measurement strategies less prone to alteration issues and better suited to the recovery of paleointensities. One interesting strategy has been proposed by Cottrell and Tarduno (1999) (see also Tarduno et al. 2006 for a recent review), which consists in using plagioclase crystals, extracted from igneous rocks that otherwise provide poor paleointensity, but good paleodirection records. When the plagioclase contains single domain magnetic inclusions, these are less susceptible to *in situ* chemical alteration than magnetic minerals that form part of the rock ground mass. Single crystal intensity estimates of this kind have been benchmarked using historical lava flows from Hawaii. They are especially useful for very old materials, including studies of the field during Proterozoic and Archean times. Submarine basaltic glasses have also been confirmed as good candidates for paleointensity work (Pick and Tauxe 1993; Carlut and Kent 2000; Tauxe and Staudigel 2004) despite some criticism (e.g. Heller et al. 2002). But such deep-sea submarine samples are difficult to orient with respect to geographical coordinates (Cogne et al. 1995) and therefore often fail to provide associated paleodirections. Discussions of these and other recent methods can be found in Valet (2003), Tauxe and Yamazaki (2007).

Figure 11 shows locations with igneous data ranging from 0–2 Ma in age that have been used in various recent studies of geomagnetic field structure and variability (see Johnson and McFadden 2007). As in the case of archeomagnetic data, one can see that such data are once again affected by uneven geographic sampling, with large areas of the globe lacking information. This problematic issue has recently prompted a major multi-institutional effort, the Time-Averaged Field Investigations (TAFI) project, to improve on this situation, at least as far as data younger than 5 Ma are concerned (Johnson et al. 2008). Efforts to improve the collection of igneous data of older ages (up to 3.2 Ga so far; Tarduno et al. 2007) are also ongoing (e.g. Smirnov and Tarduno 2004; Biggin et al. 2008a, 2008b), particularly in view of improving the existing paleointensity IAGA (International Association of Geomagnetism and Aeronomy) data basis (Perrin and Schnepp 2004; Biggin et al. 2009). Both the newer direction and intensity data and legacy collections are now the subject of a systematic archival project under the Magnetics Information Consortium (MagIC) at http://earthref.org/MAGIC/.

Marine sediment records extending to million year time scales can also be used with the advantage that they provide nominally continuous records, with well-defined stratigraphy and (barring anomalous variations in sedimentation rate) a reasonably uniform temporal sampling. However, individual sediment cores are generally less accurately oriented than the average of multiple samples from a single lava flow, there is no independence in the orientation errors among successive directions in the stratigraphic column, and in many cases only relative declination is acquired. Since there is no adequate general theory or laboratory mechanism for replicating the acquisition of remanence in sediments, only relative variations in geomagnetic field strength can be recovered, and even these rely on assumptions of uniformity in magnetic mineralogy and appropriate normalization for concentration variations (see e.g. Levi and Banerjee 1976; Valet 2003; Tauxe and Yamazaki 2007). It is likely that these assumption are violated at some level, leading to systematic bias in individual relative paleointensity estimates. An assessment of regional and global consistency among records thus plays an important role in evaluating the validity of sedimentary paleomagnetic records. The calibration of relative paleointensity variations is usually accomplished by a scaling inferred from comparison with globally

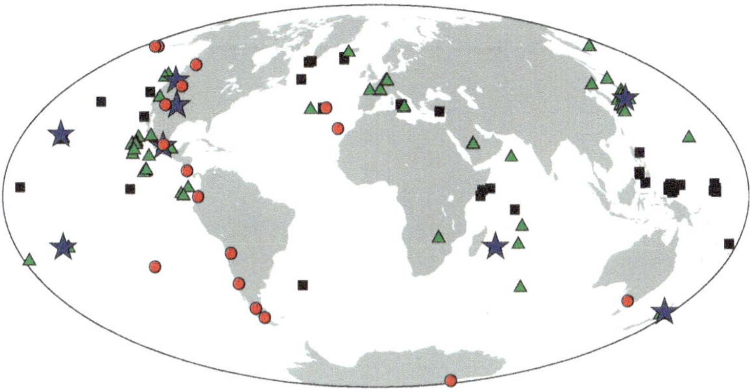

Fig. 11 Current status of global paleomagnetic data sets for 0–2 Ma paleomagnetic field modeling. The figure includes studies of lava flows (mainly directional data), that were part of the recent TAFI project (*red circles*), but lacks data from some published individual flows. *Blue stars* indicate published regional compilations of directional information from lava flows. The *green triangles* are absolute paleointensity data sites, where Thellier-Thellier measurements with specific alteration checks have been performed. Also shown as *black squares* are sediment cores included in either the Sint800 stack (Guyodo and Valet 1999) or the Sint2000 stack (Valet et al. 2005) (see Johnson and McFadden 2007 for details)

distributed absolute intensities derived from igneous rocks. The absolute values are usually converted to virtual axial dipole moments (VADM, see Sect. 3.3.1) an instantaneous measure of global axial dipole moment variation, albeit contaminated with non-axial-dipole field contributions. The resultant scaling for sedimentary records can produce intensity values that are uncertain by 10–25%. Stacking and averaging of globally distributed records (see Fig. 11) over time intervals ranging from some tens of thousands of years up to 2 My improves the reliability of the intensity record. But it also smoothes the signal which is then expected to mainly reflect variations in the axial dipole moment (Guyodo and Valet 1996, 1999; Laj et al. 2000, 2004; Valet et al. 2005). The temporal resolution in these stacks is determined by the sedimentation rate, the quality of the age control, and ability to match coeval events in different cores. Mismatches in age contribute to smoothing and bias in the results. The sint800 stacked sediment record of paleointensity variations (Guyodo and Valet 1999) for the time interval 10–800 ka is shown as the lowermost trace in Fig. 12.

Figure 12 also draws on work by Gee et al. (2000) that shows sea-surface and near-bottom marine magnetic anomaly profiles from the East Pacific Rise, along with an estimate of the sea-floor magnetization that accounts for those. Such profiles are obtained by towing a scalar magnetometer behind a ship (or a submarine), and processing the measurements to remove contributions from the external and core fields (by relying on nearby magnetic observatory or temporary fixed based station synchronous measurements, and using a contemporary IGRF field model, see Sect. 3.2). The resulting magnetic anomaly profiles are thus indeed expected to reflect contributions from the magnetized ocean crust below the ship (see e.g. Tivey 2007a). This magnetization is known to result from the oceanic crust acquiring an essentially TRM type of magnetization when it forms from rising magma at ridge axis, before moving away from those ridges (with its frozen-in magnetization) in the general context of sea floor spreading, as had originally been proposed by Vine and Matthews (1963), Morley and Larochelle (1964) (see e.g. Tivey 2007b). Provided the local ocean spreading rate can be recovered, and the process of oceanic crust production has been regular enough over the time period of interest, magnetic anomaly profiles can thus directly be interpreted in terms of records of the ancient magnetic field variations as a function of time (see e.g. Gee

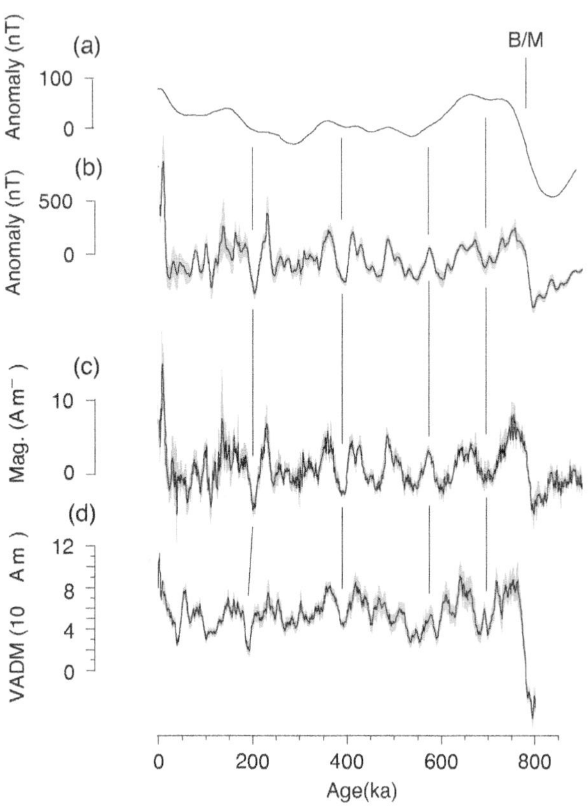

Fig. 12 Comparison of geomagnetic intensity variations over the past 800 ky from sedimentary records and in seasurface and near-bottom magnetic anomalies from the East Pacific Rise at 19° S. (**a**) Stack of sea-surface anomaly profiles coincident with the near-bottom magnetic anomaly in (**b**) and inversion solution stacks in (**c**) (see Gee et al. 2000 for details of inversion). Ages calculated assuming constant spreading rate and an age of 780 ka for Brunhes/Matuyama (B/M) boundary. Lower panel (**d**), shows Sint800 sedimentary relative paleointensity stack for 10–800 ka (Guyodo and Valet 1999) combined with global archeomagnetic data for past 10 ky (Merrill et al. 1996), all scaled as virtual axial dipole moment (VADM). Modified from Gee et al. (2000)

and Kent 2007). In particular, it is clear from Fig. 12 that detailed magnetic anomaly profiles have the ability to provide a measure of relative geomagnetic paleointensity variations with broad similarities to the SINT800 record. Note that just like SINT800, such measures however mainly capture low frequency intensity variations, essentially dominated by variations in the axial dipole field. Substantial efforts are currently under way to try and recover similar detailed information about earlier field intensity variations from both sea-surface and near-bottom marine magnetic anomaly profiles (e.g. Pouliquen et al. 2001; Bowers et al. 2001; Bouligand et al. 2006; Tivey et al. 2006; Tominaga et al. 2008).

The strongest of the marine magnetic anomalies however, and those that are therefore best known and understood, are those reflecting the occurrence of magnetic field reversals, the times at which the geomagnetic field has changed polarity in the past (see Sect. 4.2). The signature of the most recent reversal (the Bruhnes/Matuyama reversal, which occurred some 780 kyr ago) can also be seen in Fig. 12. Such reversals produce very strong marine magnetic anomaly signatures because of the opposite signs of the magnetization recorded in the ocean crust before and after the reversal. Extensive marine magnetic anomaly records extending back as far as the oldest sea-floor (roughly 180 Ma) have been used in successively more refined constructions of the geomagnetic polarity times scale (GPTS). Considerable work has also been devoted to providing independent checks of the GPTS with the help of magnetostratigraphy, i.e. piecewise continuous sedimentary records, and radiometrically dated igneous rocks, both of which obviously also have the ability to provide direct evidence of geomagnetic reversals. Combining all this information, the calibration of sea-floor spreading to absolute age makes it possible to provide a reliable record of reversals for the past 160 My (see Fig. 13 and Gee and Kent 2007 for a recent detailed review). Most sea-floor magnetic anomalies from earlier times have unfortunately been subducted along with the oceanic crust so that longer term data only comes to us in a piecemeal fashion from the available geological record. Nevertheless, there are records of the reversal history quite far back, thanks in particular to the availability of very ancient sediment records (up to almost 2 Ga, see e.g. Gallet et al. 2000; Elston et al. 2002; Dunlop and Yu 2004; Pavlov and Gallet 2005, 2010).

3 Core Field Models

3.1 Satellite Era

The challenge of geomagnetic field modeling is that of converting a (sometimes very large) database of magnetic observations into a set of mathematical descriptions of the various magnetic fields that add up to produce the observed geomagnetic field. In the case of very precise satellite observations, many different sources contribute significantly. Those sources can be above the satellite (in the magnetosphere), well below the satellite (in the core, the magnetized crust or the slightly conducting mantle), but also in the immediate environment of the satellite which orbits in the upper layers of the ionosphere. The most serious issues in producing geomagnetic field models from satellite data are related to local small-scale irregularly fast-changing sources, mainly currents the satellite is bound to cross at high latitudes (where magnetospheric currents connect to the ionosphere). Several strategies can be used to avoid the signal produced by such sources, ranging from data-selection to avoid contaminated data (relying on e.g. night-side quiet-time data, see e.g. Thomson and Lesur 2007), to only using the least-affected intensity data at high-latitude. Simplified mathematical modeling of the local sources encountered by the satellite can also be used. Details of the way this can be achieved can be found in Hulot et al. (2007) and Olsen et al. (2009b). For the sake of simplicity, and since most published models actually rely on selection procedures that avoid local ionospheric sources, we will now briefly describe how models of the core field can be recovered from satellite data, assuming the data are acquired in a shell devoid of local sources.

In that case the magnetic field $\mathbf{B} = -\nabla V$ can be expressed as the negative gradient of a scalar potential V. Expanding V into series of spherical harmonics yields

$$V = V^{\text{int}} + V^{\text{ext}}$$

$$= a \sum_{l=1}^{L_{\text{int}}} \sum_{m=0}^{l} \left(g_l^m \cos m\phi + h_l^m \sin m\phi\right) \left(\frac{a}{r}\right)^{l+1} P_l^m(\cos\theta) \tag{1a}$$

$$+ a \sum_{l=1}^{L_{\text{ext}}} \sum_{m=0}^{l} \left(q_l^m \cos m\phi + s_l^m \sin m\phi\right) \left(\frac{r}{a}\right)^{l} P_l^m(\cos\theta) \tag{1b}$$

(Chapman and Bartels 1940; Langel 1987), where $a = 6371.2$ km is a reference radius, (r, θ, ϕ) are geographic coordinates, P_l^m are the associated Schmidt semi-normalized Legendre functions, L_{int} is the maximum degree and order of the internal potential coefficients g_l^m, h_l^m, and L_{ext} is that of the external potential coefficients q_l^m, s_l^m.

The corresponding internal potential recovered from satellite data may also include some signal from the ionospheric sources below the satellite (which the satellite indeed sees as internal sources). This issue is well-recognized. But most of this signal can be avoided through

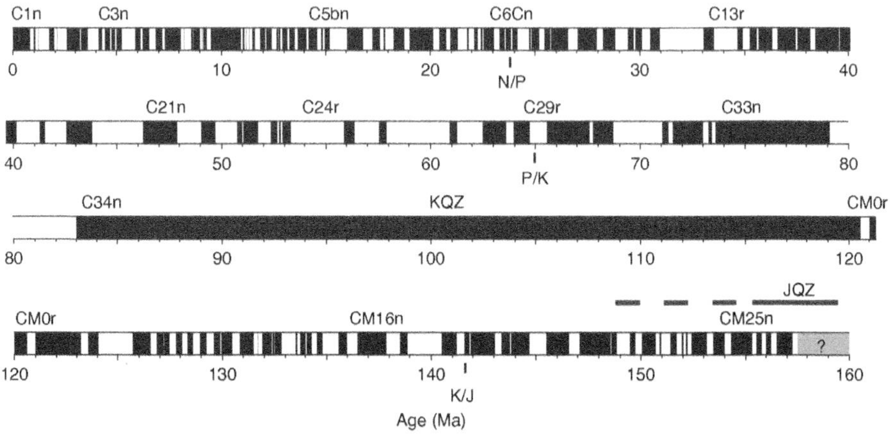

Fig. 13 Geomagnetic polarity timescale from marine magnetic anomalies for 0–160 Ma, after Lowrie and Kent (2004); largely based on Cande and Kent (1995) and Channell et al. (1995). Filled and open blocks represent intervals of normal and reverse geomagnetic field polarity. Those intervals, known as "chrons", are labelled in an elaborate way to account for the fact that shorter chrons (subchrons) and possible but not firmly identified even shorter chrons (cryptochrons) have progressively been included (for details see e.g. Gee and Kent 2007). Only key chrons that were used as calibration tiepoints are identified by their names above the bar graph (C1n, C3n, etc.). Correlated positions of geologic period boundaries are otherwise indicated by ticks below the bar graph (N/P, Neogene/Paleogene; P/K, Paleogene/Cretaceous, K/J, Cretaceous/Jurassic). KQZ is the Cretaceous Quiet Zone, an unusually long chron also known as the Cretaceous Normal Superchron. JQZ is the Jurassic Quiet Zone, corresponding to times with low field strength (see Sect. 4.3.2)

data selection (by selecting night-time data when ionospheric sources are weakest, see however Gillet et al. 2009b), or by using so-called comprehensive modeling approaches (see e.g. Sabaka et al. 2004).

The time change of the field of truly internal origin is then modeled either by a Taylor expansion of each Gauss coefficient g_l^m, h_l^m around a given epoch

$$g_l^m(t) = g_l^m\,|_{t_0} + \dot{g}_l^m\,|_{t_0} \cdot (t - t_0) + \frac{1}{2}\ddot{g}_l^m\,|_{t_0} \cdot (t - t_0)^2 + \cdots \tag{2}$$

(and similar for h_l^m) or by means of a spline representation (see next section where this representation is further discussed). The time variation of the external field (i.e. of the expansion coefficient q_l^m, s_l^m) is typically parameterized by proxies of the large-scale magnetospheric field variations like the *Dst* index (which is a measure of the strength of the dynamic magnetospheric ring-current) derived from observatory data. Those proxies are also used to correct the field of internal origin for signals produced by externally induced currents within the slightly conducting mantle. The model parameters (i.e. the expansion coefficients $g_l^m, h_l^m, q_l^m, s_l^m$ including their temporal representation) are finally estimated from the magnetic field observations using standard inverse methods (e.g. Parker 1994; Tarantola 2005).

Such procedures then lead to Gauss coefficients g_l^m, h_l^m describing the field of internal origin, with sources in the core and the crust. Potential theory does not provide any further formal way of separating the signal of each of those two sources. However, as demonstrated from models derived from MAGSAT data in particular (Langel and Estes 1982), plotting the so-called spatial Lowes-Mauersberger power spectrum of the field of internal origin (Fig. 14, which shows the contribution of each degree l to the surface average value of

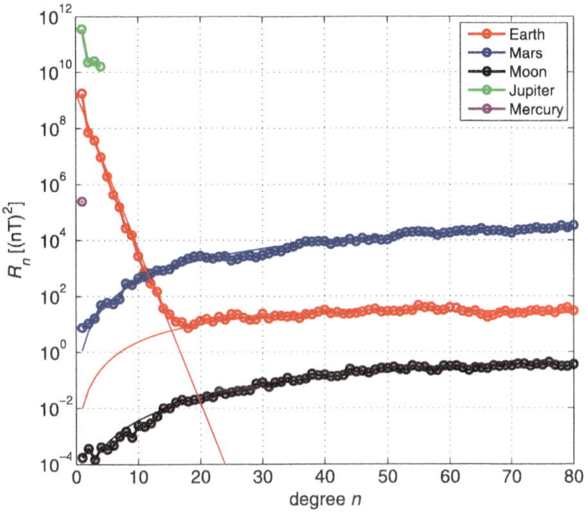

Fig. 14 Lowes-Mauersberger power spectra of the field of internal origin for the Earth (after Olsen et al. 2009a and Maus et al. 2008), Mars (after Cain et al. 2003), Jupiter, Mercury (after Connerney 2008) and the Moon (after Purucker 2008) at their respective surface reference radius. Also shown are theoretical crustal spectra (thin curves, Voorhies et al. 2002) for the Earth, Mars and the Moon. Note the lack of any significant core field in the case of Mars and the Moon, which display pure crustal types of spectra

$\mathbf{B}^2$ at a given reference radius, Lowes 1974) clearly suggests that its large-scale decreasing segment up to spherical harmonic degree $l = 13$ is dominated by the field from the remote core, while its fairly flat segment beyond degree $l = 16$ is dominated by the field of the nearby crust (which indeed is expected to produce such a spectrum, see e.g. Jackson 1994; Voorhies et al. 2002 and Fig. 14). Likewise, it can be argued that detectable time changes in the large scale field of internal origin most certainly reflect core field changes, while yet undetected crustal field changes likely dominate the signal beyond degree 22 (Hulot et al. 2009a; Thébault et al. 2009).

This natural separation of the field of internal origin into a large scale component mainly produced by the core, and a small scale component mainly produced by the crust, is an essential property. It implies that only the largest scales of the field of internal origin can be associated with the core field and down-continued to the core-mantle boundary (CMB) with the help of (1a), which only holds where no sources lie. Thus models of the field of internal origin inferred from satellite data can be used to infer the core field at the CMB where it originates, provided however that one restricts those models to degree $l = 13$ or less for the field, to degree 22 or less for its first-time derivative.

Several core field models derived from Ørsted, CHAMP and SAC-C satellites data have recently been published, for which the first time derivative is now determined up to perhaps degree $l = 14$–16 (e.g. Maus et al. 2006; Lesur et al. 2008; Olsen et al. 2009a). Current efforts are directed towards also better constraining the higher derivatives of the field, to better detect possible fast core field changes (see e.g. Olsen and Mandea 2008). Figure 15 shows maps of the radial component of the present core field and of its first time-derivative at the Earth's surface and at the CMB.

3.2 Time-Dependent Models Over Historical and Archeological Times

Building geomagnetic field models that span longer time intervals presents additional technical challenges. The simplest procedure, which has for example been used in the construction of the International Geomagnetic Reference Field (IRGF) model series (see Barton 1997 and Macmillan and Maus 2005 for the most recent revision), consists of a series of snapshots of

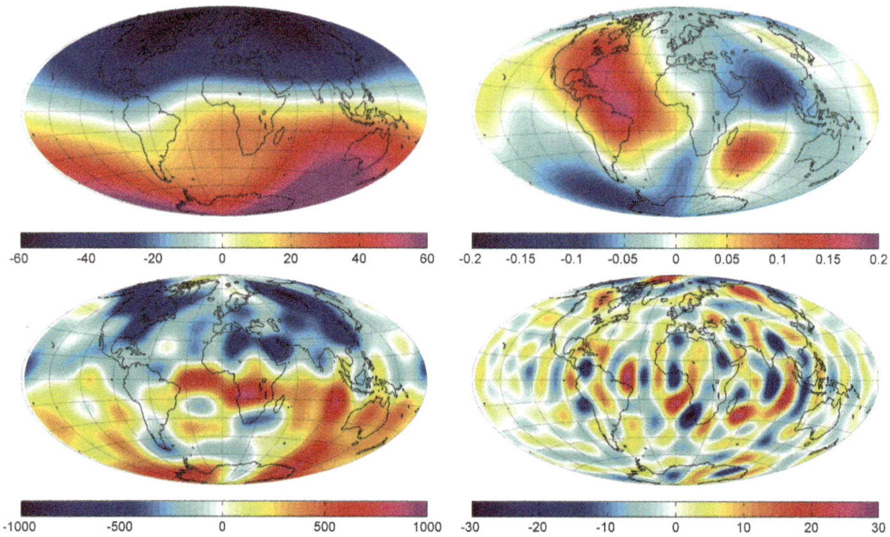

Fig. 15 Maps of B_r (*left column*, in units of μT), resp. dB_r/dt (*right column*, in units of μT/yr) at the Earth's surface (*top row*), and at the CMB (*bottom row*), according to model CHAOS-2s Olsen et al. 2009a for epoch 2004. B_r is tapered at degree $n = 13$ ((8) of Olsen et al. 2009a with $\mu = 1.4 \cdot 10^{-8}$), while dB_r/dt is tapered at degree $n = 16$ ($\mu = 3.5 \cdot 10^{-10}$)

the internal field. These snapshots are available at five year intervals since 1900 and are updated every five years. The IGRF model is designed to estimate robustly the internal (core) field and is used for a wide variety of industrial and societal applications. However, due to the limitations of its linear interpolation temporal representation, it is not suitable for detailed scientific study of secular variation.

A more sophisticated approach is to invert for core field models that are by construction continuously time-dependent. Early such models relied on polynomial representations of time (e.g. Bloxham 1987; Bloxham and Jackson 1989). However the most widely used historical core field model *gufm1* (Jackson et al. 2000) is based on a cubic (4th order) spline temporal representation of the Gauss coefficients first introduced by Bloxham and Jackson (1992). Under this framework each spherical harmonic coefficient g_l^m is considered to be time-dependent and is expanded as

$$g_l^m(t) = \sum_n g_l^{mn} M_n(t), \tag{3}$$

where the $M_n(t)$ are B-spline basis functions (e.g. Lancaster and Salkauskas 1986) and g_l^{mn} are the coefficients defining the time-dependency of the Gauss coefficients that must be determined from the observations.

Adopting a B-spline temporal representation has several advantages. In particular, the B-splines provide a natural basis for a smoothly varying description of noisy data. It can be shown that of all the interpolators passing through a time-series of points (say $f(t_i), i = 1, N$), an expansion in B-splines of order 4 ($\hat{f}(t)$ say) is the unique interpolator which minimizes the following measure of roughness (see for example De Boor 2001)

$$\int_{t_s}^{t_e} \left[\frac{\partial^2 \hat{f}(t)}{\partial t^2} \right]^2 dt. \tag{4}$$

This form of optimally smooth representation is designed to avoid that too extra detail be present in the solution other than that truly demanded by the data.

The inverse problem of determining the model parameters g_l^{mn} can then be addressed in several ways. In the case of *gufm1* Jackson et al. (2000) rely on a regularized least-squares approach that involves minimizing an objective function of the form,

$$\Theta(\mathbf{m}) = [\mathbf{d} - \mathbf{f}(\mathbf{m})]^T \mathbf{C}_e^{-1} [\mathbf{d} - \mathbf{f}(\mathbf{m})] + \mathbf{m}^T \mathbf{C}_m^{-1} \mathbf{m}, \tag{5}$$

where $\mathbf{d}$ is a vector of the magnetic observations, $\mathbf{f}(\mathbf{m})$ is a vector of the observations predicted by the field model, $\mathbf{C}_e$ is the data covariance matrix and $\mathbf{C}_m$ is a model covariance matrix that includes norms measuring the spatial and temporal complexity of the field

$$\mathbf{C}_m^{-1} = \left(\lambda_S \mathbf{S}^{-1} + \lambda_T \mathbf{T}^{-1}\right), \tag{6}$$

where λ_S and λ_T are spatial and temporal damping parameters. But other objective functions than (5) can be used, assuming e.g. Laplace rather than Gaussian error distributions (e.g. Walker and Jackson 2000), or maximum entropy rather than quadratic regularisation in both space (Jackson et al. 2007a) and time (Gillet et al. 2007a). In the case of *gufm1*, the temporal norm is further chosen to be the square of the temporal curvature of the radial magnetic field integrated over the CMB and over time,

$$\mathbf{m}^T \mathbf{T}^{-1} \mathbf{m} = \frac{1}{t_e - t_s} \int_{t_s}^{t_e} \oint_{CMB} (\partial_t^2 B_r)^2 \, d\Omega dt. \tag{7}$$

Here t_s and t_e are the start and end of the time interval being modeled. Note the close correspondence of this norm with the roughness measure mentioned above; this illustrates that choice of a cubic B-spline temporal basis is optimal when the temporal curvature norm is employed.

In addition to temporal regularization, spatial regularization is also an essential ingredient in historical core field modeling where data spatial coverage can be very sparse. Bloxham and Jackson (1992) and later Jackson et al. (2000) employed a measure of model spatial complexity based on minimizing the Ohmic heating due to poloidal magnetic field inferred at the CMB (since we are dealing with core sources),

$$\mathbf{m}^T \mathbf{S}^{-1} \mathbf{m} = \frac{4\pi}{t_e - t_s} \int_{t_s}^{t_e} \sum_{l=1}^{L} f(l) \sum_{m=0}^{l} \left[(g_l^m)^2 + (m_l^m)^2\right] dt, \tag{8}$$

$$\text{with} \quad f(l) = \frac{(l+1)(2l+1)(2l+3)}{l} \left(\frac{a}{c}\right)^{2l+4}. \tag{9}$$

The optimization problem of minimizing Θ can then be solved numerically via an iterative quasi-Newton scheme (LSQN); an iterative approach is necessary when using inclination, declination and intensity data because these depend non-linearly on the model parameters g_l^{mn}. This methodology has been successfully applied by Bloxham and Jackson (1992) and Jackson et al. (2000) to compute core field models from the historical data sources described in Sect. 2.3, together with more recent observatory, survey and satellite data.

Most historical field models including the IGRF series and *gufm1* (Jackson et al. 2000) rely heavily on survey and observatory data collected at Earth's surface. Since such observations are made below the ionosphere in an approximately source free environment the

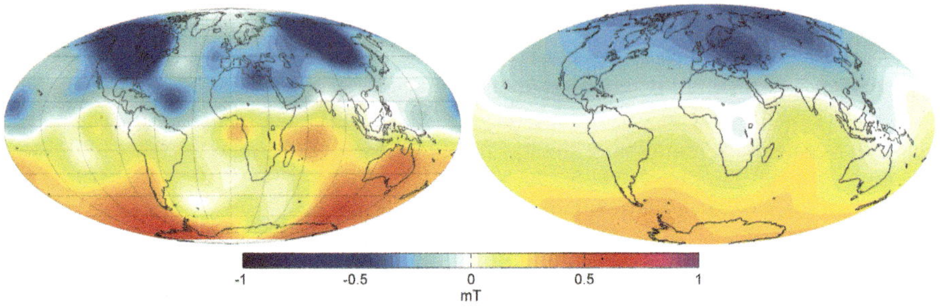

Fig. 16 Maps of time-averaged B_r at the CMB, averaged over the historical 1590–1990 period (*left*, as inferred from the *gufm1* model of Jackson et al. 2000), and over the past 7000 years (*right*, as inferred from model CALS7K.2 of Korte and Constable 2005). Units in mT

modeling strategies invert only for the Gauss coefficients of the internal field. To avoid signal from externally induced currents within the slightly conducting mantle, data selection procedures are used (such as choosing magnetically quiet days), together with temporal filtering (possible because observatories are fixed points in space) where monthly or annual means are employed. In addition to careful data selection, error budget assessment (particularly to account for the non-modelled crustal contributions) and spatial and temporal regularizations are found to be essential in the production of structurally simple, smoothly evolving, field models.

One weakness of historical core field models is that they only rely on directional observations prior to AD1840; before this no intensity observations were carried out. Directional observations alone provide enough information to constrain the core field morphology, but not its absolute magnitude (Hulot et al. 1997). Historical models such as *gufm1* have thus far assumed a simple linear trend for the axial dipole (which effectively defines the magnitude of the field model) from AD1590 to AD1840. This assumption is however rather arbitrary and attempts have recently been made to directly constrain the trend from archeomagnetic intensity data covering AD1590-1840 (see Sect. 4.1.3). Figure 16 (left), shows a map of the radial component of the average core field between AD1590 and AD1990, as inferred from the *gufm1* historical model.

Models describing the core field evolution before AD1590 can also be built using analogous modeling strategies and using the declination, inclination and intensity provided by archeomagnetic records. The magnetic field recording process in archeological artifacts and young volcanic flows is indeed fast (a few days at most). It provides a good record of past core field values. Again, however, contributions from external and crustal fields must be considered as part of the error budget. An important additional specificity of such records is the fairly large uncertainty (50 years, if not more) with which the age of each sample is known, from historical accounts, or isotopic methods. Those uncertainties are usually converted into additional contributions to the error budget. But they also imply that observations from different locations cannot be used to constrain phenomena occurring on time scales of less than say, a century. Similar dating errors affect sediment data which further suffer the effect of temporal smoothing associated with their magnetic recording process. This sets an important intrinsic limit to our ability to recover information about medium to small scale core field variations, which mainly occur on such time scales (see Fig. 19 in Sect. 4.1). This issue only mildly affects the recovery of the largest scales of the core field, which are also those best resolved by the still limited geographical distribution of data.

Early archeomagnetic field models were built as a sequence of snapshots and restricted to only the first few degrees of the field (Hongre et al. 1998) and next to somewhat higher degrees (by introducing some spatial regularization, Constable et al. 2000). More recently, the B-spline technique described above has also been introduced. But the problems in producing trustworthy field models are even more challenging than in the case of historical data. Iterative data rejection together with strong spatial and temporal regularization were key ingredients used to combat these difficulties in producing the most widely used CALS7K.2 field model spanning the interval BC5000 to AD1950 (Korte and Constable 2005). CALS7K.2 is believed to be a good representation of Earth's magnetic field up to perhaps spherical harmonic degree $l = 4$, with temporal resolution of approximately 300 years and agrees satisfactorily with a spatially truncated and temporally smoothed version of *gufm1* during the time when the models overlap. Figure 16 (right), shows a map of the radial component of the average core field over the 5000 BC to AD 1950 time period as inferred from CALS7K.2. Unfortunately the scarcity of data from low latitudes and the southern hemisphere, and a general bias of data towards Europe and the near-East makes the study of global patterns of field evolution difficult at present. In recent years the available of suitable data sources has expanded considerably and associated new field models have recently been published (Donadini et al. 2009; Korte et al. 2009).

3.3 Paleomagnetic Field Models

When moving back in time deterministic, time-dependent, spherical harmonic field modeling is usually no longer possible, because dating errors become larger than the dominant time-scales of the secular variation (see Fig. 19). Each data must then be seen as a sample of the core field value at a given known location and at a roughly known time. Fortunately this time is often known with enough accuracy that the data belonging to a common chron in the geomagnetic polarity times scale (GPTS, recall Fig. 13) can still be used for some statistical analysis of the field at times of stable polarity. Of course, not all data correspond to such stable polarity periods and some will correspond to times of transitional (reversals) or unstable polarity (excursions). Fortunately these times can be identified and the behavior of the field during such events investigated separately (see Sect. 4.2).

3.3.1 The Geocentric Axial Dipole Hypothesis and Related Concepts

A very useful concept since the early days of paleomagnetism has been the "Virtual Geomagnetic Pole" (VGP). Its usefulness is related to the fact that the core field happens to always have been essentially consistent with the so-called "Geocentric Axial Dipole" (GAD) hypothesis which states that the field has always been dominated by its axial dipole component (g_1^0), with either the present ("normal") or opposite ("reverse") polarity. Starting from any paleodirectional data which provides a record of I and D at a given location, the corresponding VGP is defined as being the pole of the pure dipole field that would have produced the observed I and D at this location (see e.g. Merrill et al. 1996; McElhinny 2007 for details and formulae). Figure 10 shows an example of such a conversion of directional data into VGPs for data covering the 0–5 Ma period. As can be seen, VGPs do cluster about the location of either the North or the South geographical pole as expected from the GAD hypothesis. Scatter about those poles can then be understood in terms of additional contributions from equatorial dipole and non-dipole components, each datum being affected by different amounts of such fields as a result of secular variation acting during the time elapsed between the various data samples. This scatter can be measured and is usually referred to as the "VGP scatter".

The fact that VGPs do cluster about the geographical poles is confirmed by all igneous directional data that are young enough to not have been affected by any significant plate tectonic motions. This provides support for the GAD hypothesis up to at least 5 Ma. For earlier epochs, and not surprisingly, VGPs from different sites will usually cluster about different poles. But extensive work has shown that those different paleopoles can be reconciled, and brought back to the geographical poles, if appropriate motions and rotations are applied to the tectonic plates to which the various sites belong. It is important to stress that such paleopoles are part of the input data used to carry out such plate motion reconstructions, and that the validity of the GAD hypothesis stems from the internal consistency of those reconstructions with all observed paleopoles, and with independent information recovered from ocean magnetic anomalies. These directly provide an image of the history of ocean floor spreading associated with plate motions (for more details see e.g. McElhinny and McFadden 2000). For even earlier epochs, testing the GAD hypothesis becomes much more difficult. But it is fair to state along with McElhinny (2007), that all tests done so far suggest that the GAD hypothesis is a reasonable first-order approximation for the time-averaged field at least for the past 400 My (see also Perrin and Shcherbakov 1997) and probably for the whole of the geological time. It will thus not come as a surprise to the reader that when investigating the times of stable polarity, much of the paleomagnetic field modeling strategy is being geared towards first, quantifying the relative amount of additional non GAD field component needed to properly account for the time-averaged field (TAF), and second, quantifying the amplitude of field fluctuations about this TAF, the so-called PaleoSecular Variation (PSV).

Before getting into the details of these TAF and PSV modeling strategies, it is useful to introduce a number of other GAD related concepts. Paleointensity data are for instance often converted into so-called Virtual Dipole Moment (VDM) values, a concept closely related to that of VGP in that it also converts local observations into information about the virtual dipole field that would have produced those observations. Whereas the VGP is the pole of this virtual dipole field, the VDM is its dipole moment. Its computation requires the knowledge of both the paleointensity B and the inclination I. Then indeed the angular distance λ_{VGP} from the sampling site to the VGP can be inferred from (see e.g. Merrill et al. 1996):

$$\tan \lambda_{VGP} = \frac{1}{2} \tan I \tag{10}$$

and the VDM from

$$VDM = \frac{4\pi a^3}{\mu_0} B \left(1 - 3\sin^2 \lambda_{VGP}\right)^{-1/2} \tag{11}$$

where μ_0 is the magnetic permeability and a the Earth's mean radius. Note that both λ_{VGP} and the VDM can be computed without any knowledge of the declination D. This is an important property that makes it possible to compute VDMs even when considering old samples from sites that may have experienced considerable (possibly unknown) plate tectonic displacement and be affected by systematic declination (but hopefully no inclination) biases. VDMs provide estimates of the dipole moment $M_D = (4\pi a^3/\mu_0)((g_1^0)^2 + (g_1^1)^2 + (h_1^1)^2)^{1/2}$ of the paleomagnetic field to within the (quite large) uncertainty introduced by the non-dipole field contributions to the data.

Equation (10) can also be used to recover an estimate of the paleolatitude of a sampling site from directional data, if enough such data are available, so that an average direction can be computed, hopefully reflecting the local direction of the TAF at the site under consideration (possibly to within some systematic declination bias due to plate tectonic motion, which again is not an issue). Using the inclination of this average direction in (10) then leads to an estimate of the paleogeographic latitude of the site, under the assumption that the averaging properly removed the effect of the secular variation, and that the TAF of the time was very close to satisfy the GAD hypothesis. This estimate is known as the paleomagnetic latitude of the site.

When paleointensities are available alone, neither VDMs nor paleomagnetic latitudes can be computed. But if the geographic latitude of the site happens to be known directly (when the data is young enough) or can be recovered by independent means (thanks to plate tectonic reconstruction, for instance), then a so-called Virtual Axial Dipole Moment (VADM) can still be computed with the help of (11) by just changing λ_{VGP} into the geographic latitude. Such VADMs are quite similar to VDMs, except for the fact that they now provide local estimates of the axial (and not the full) dipole moment $M_{AD} = (4\pi a^3/\mu_0)|g_1^0|$ of the paleomagnetic field to within the (larger) uncertainties introduced by both the non-dipole field and the equatorial dipole field contribution to the data.

Both VDMs and VADMs are commonly used to compare paleointensity data from distant sites, to calibrate sedimentary relative paleointensity records such as the one shown in Fig. 12, and to reconstruct the past variations of the dipole moments of the geomagnetic field.

3.3.2 Time-Averaged Field and Paleosecular Variation Models

Current time-average field (TAF) and paleosecular variation (PSV) modeling strategies are best understood in terms of the so-called Giant Gaussian Process (GGP) statistical description of the field, first introduced by Constable and Parker (1988) and next generalized by Hulot and Le Mouël (1994). The GGP description consists of considering that at times of stable polarity, the core field can be described in terms of a multidimensional stationary random Gaussian process governing a point of coordinates $\mathbf{x}(t)$ defined by the time-varying Gauss coefficients $g_l^m(t)$ and $h_l^m(t)$ in a multidimensional space. At any given instant t, this point completely characterizes the core field (by virtue of (1a)). It evolves through time about a mean point $\mu = E\{\mathbf{x}(t)\}$ with coordinates (i.e. Gauss coefficients) fluctuating about their mean values $G_l^m = E\{g_l^m(t)\}$ (resp. $H_l^m = E\{h_l^m(t)\}$). These fluctuations are statistically described by a covariance matrix $\gamma(t'-t) = E\{[\mathbf{x}(t)-\mu][\mathbf{x}(t')-\mu]^T\}$ defining the correlation times $\tau(g_l^m)$ (resp. $\tau(h_l^m)$) and variances $\sigma^2(g_l^m)$ (resp. $\sigma^2(h_l^m)$) of those fluctuations, as well as the possible cross-correlations two different Gauss coefficients may experience (see Bouligand et al. 2005 for details).

Such a GGP formalism provides a very decent statistical description of the field produced by geodynamo numerical simulations (McMillan et al. 2001; Kono et al. 2000a; Bouligand et al. 2005) and analysis of such simulations have even shown that interesting symmetry breaking properties can be detected (Hulot and Bouligand 2005). But paleomagnetic data are not as numerous as synthetic data provided by simulations and in practice a number of simplifying assumptions must be introduced. Even so, simplified GGP analysis of the historical, archeomagnetic and paleomagnetic fields have proven very useful. Hulot and Le Mouël (1994) and Hongre et al. (1998) have for instance shown that the dominant correlation time scales in the historical and archeomagnetic fields is on the order of a few centuries and decreases fast as a function of the degree l of the Gauss coefficients (as illustrated in e.g. Fig. 19). Since paleomagnetic data, and particularly those from igneous rocks, are frequently separated in time by more than a millenium, temporal correlations and time issues are often ignored altogether. Each such data (say D_i, I_i) can then be seen as a local measure of a single independent field realization $\mathbf{x}_i$ from a Gaussian distribution with means G_l^m (resp. H_l^m) and covariance matrix $\gamma = E\{[\mathbf{x} - \mu][\mathbf{x} - \mu]^T\}$. Additional simplifications are usually introduced by further assuming a lack of cross-correlations among different Gauss coefficients. Then γ becomes diagonal and is entirely defined by just the variances $\sigma^2(g_l^m) = E\{(g_l^m - G_l^m)^2\}$ and $\sigma^2(h_l^m) = E\{(h_l^m - H_l^m)^2\}$. Although reasonable to first order, this simplification carries a number of hidden assumptions, and one should be aware of these (see Bouligand et al. 2005; Hulot and Bouligand 2005).

Within the GGP framework, and assuming the above simplifications, all TAF and PSV modeling carried out so far can then be understood in terms of attempts to recover the mean Gauss coefficients (G_l^m, H_l^m) which define the TAF over the period considered, and the variances $(\sigma^2(g_l^m), \sigma^2(h_l^m))$ which characterize the PSV.

TAF and PSV Models for the Past 5 My Lava flows less than 5 My old, which have not significantly been affected by plate tectonic motions, are particularly well suited for this type of modeling since they directly comply with the underlying assumptions of the above simplified GGP approach. This is the time period most extensively investigated so far and the one we will now focus on. Provided enough such data are available at a given site, local statistical distributions of field parameters (usually D, I and occasionally B) can be computed. Those reflect the underlying parameters of the TAF and PSV.

In particular, the distribution of vector field values $\mathbf{B}_i$ observed at a given location during a given chron is expected to be that of a 3D Gaussian distribution centered on the vector field value $\overline{\mathbf{B}}$ the TAF would produce (see e.g. Khokhlov et al. 2001). In this ideal situation, exactly the same modeling methods can be used as in the case of historical data, to recover estimates of G_l^m and H_l^m. In addition, moments of the local 3D Gaussian distributions of the $\mathbf{B}_i$ values at each site can also be inverted for the variances $(\sigma^2(g_l^m), \sigma^2(h_l^m))$ of the PSV, at least in principle. In practice however, this turns out to be a difficult endeavor and only one study so far has looked into this (Kono et al. 2000b). In fact, even just inverting for the TAF turns out to be problematic, both because of the limited amount of such data, and because of the still large uncertainties affecting paleointensity data (again, see Kono et al. 2000b).

Most investigations of the TAF over the past 5 My have therefore focused on the much more numerous and accurately recovered paleodirectional data. Those studies again consist in computing averages $\overline{D}$ and $\overline{I}$ of the declination and inclination values available at each sampling site, and assuming that those averages reflect the declination and inclination the TAF would predict at those sites. Then the G_l^m and H_l^m can again be recovered, as if dealing with historical data (see e.g. Gubbins and Kelly 1993). Because, as already noted in Sect. 3.2, such directional data can only define the morphology of the field, the absolute value of the field is usually defined by assuming $G_1^0 = -30.000$ nT (roughly the modern value of g_1^0) when considering TAF models for normal polarity chrons, and the opposite value when considering TAF models for reverse polarity chrons. This choice is arbitrary and has in fact been challenged (Tauxe and Kent 2004) (see Sect. 4.3.2). Such models have been build for the present normal chron (Bruhnes, up to roughly 780 kyr ago), the previous reverse chron (Matuyama, between 780 ky and 990 ky ago), and for all combined normal or reverse chrons over the past 5 Ma, under the assumption that the Earth's dynamo is likely to have produced the same normal (resp. reverse) TAF during this period of time. Some models have also been built by combining all chrons (reversing the orientation of data for reverse chrons), under the additional assumption that the reverse TAF must exactly be the opposite of the normal TAF.

Not all models have been built in the same way and based on the same data basis (usually one of the two data basis set by Quidelleur et al. 1994 and Johnson and Constable 1996,

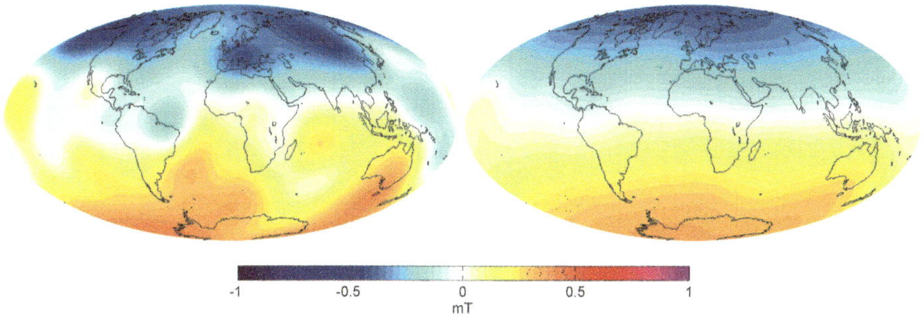

Fig. 17 Maps of the Time-Averaged Field B_r at the CMB for the past 5 My, as inferred from normal lava flow data (*left*, model LN1 of Johnson and Constable 1995), and as inferred from both normal lava flow data and additional normal marine sediment data (*right*, model LSN1 of Johnson and Constable 1997). Units in mT

both of which have been reanalyzed and updated to form the most recent TAFI database of Johnson et al. 2008). Several TAF models have been developed since the work of Gubbins and Kelly (1993) (see Johnson and Constable 1995, 1997, 1998; Kelly and Gubbins 1997; Carlut and Courtillot 1998 and the recent review of Johnson and McFadden 2007). Some of those models have also used additional marine sediment data (such as those compiled by Schneider and Kent 1988, 1990) to improve the geographical coverage of the sampling sites, in which case only average inclinations are considered (because of coring related orientation issues with respect to the declination). Figure 17 shows maps of the radial component B_r for two such TAF models plotted at the core surface: model LN1 of Johnson and Constable (1995) built from normal lava flow data for the past 5 My, and model LSN1 of Johnson and Constable (1997) built with additional normal marine sediment data. The comparison of these two maps perfectly illustrates the difficulty of recovering TAF models. Whereas LN1 would suggest fairly strong non-zonal structures in the TAF (as originally suggested by Gubbins and Kelly 1993), LSN1 clearly suggests far less structure. In fact a number of authors have argued that no significant non-zonal features can yet be recovered from normal lava flow data (McElhinny et al. 1996b; Carlut and Courtillot 1998). Clearly, the exact amount of non-zonal structure present in the TAF over the past 5 My is still a matter of debate.

Most of the problem lies in the relatively poor geographical coverage of sites and the non-uniform temporal sampling associated with volcanic processes (see e.g. Johnson and McFadden 2007). But one must also be aware that computing $\overline{D}$ and $\overline{I}$ independently from one another (as is done by most authors) can introduce some biases, even if the original directional data distribution can be assumed isotropic (see e.g. Love and Constable 2003). In addition, GGP models usually predict local directional distributions that are not isotropic, and this may introduce further biases (Khokhlov et al. 2001). Relying on just $\overline{D}$ and $\overline{I}$ for TAF modeling is thus a questionable choice. Recent methodological progress have however been made by Khokhlov et al. (2006) who showed how a given joint TAF and PSV model could be tested against any directional data without having to resort to questionable averaging procedures. They too concluded that no non-zonal structures are needed in the TAF to account for the Quidelleur et al. (1994) data they tested.

It thus seems safe to conclude that for the time being, and as far as the past 5 My are concerned, only the zonal (axisymmetric) structure of the TAF can be recovered with some certainty, suggesting a TAF with at least a G_2^0 component of 2–4% of G_1^0 and perhaps some G_3^0 of similar or less relative magnitude. Still, there are some good reasons to believe that

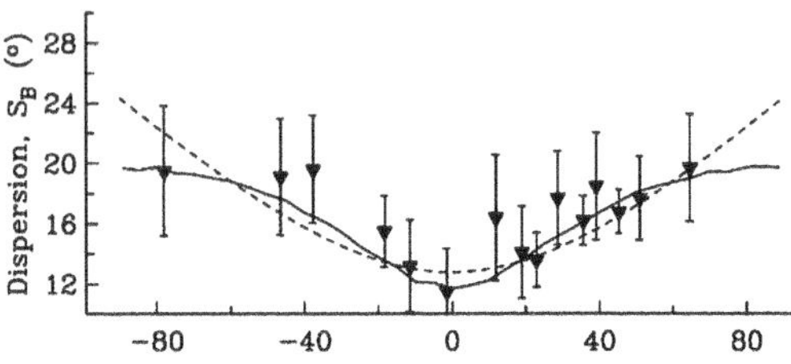

Fig. 18 VGP scatter over the past 5 My as a function of site latitude, as inferred from normal lava flow data selected within the PSVRL database of McElhinny and McFadden (1997). Also shown the prediction from the TK03 GGP model of Tauxe and Kent (2004) (*solid line*) and model G of McFadden et al. (1988) (*dashed line*). After Johnson et al. (2008)

some amount of non-zonal structure must be present in the TAF (see Sect. 4.3.1) and it may be that the one seen in Fig. 17, though not proven robust yet, provides some hints of it. Also of fundamental interest is the fact that the most recent investigation of Johnson et al. (2008) suggests that significantly different relative magnitudes might hold for G_2^0 and G_3^0 when considering the Bruhnes normal TAF and the Matuyama reverse TAF (a challenging result also to be discussed in Sect. 4.3).

What about the PSV over the past 5 My? Studies of this PSV have usually been carried out by investigating the way lava flow paleodirectional data scatter about their mean direction at each site. Because, as we have described, the TAF is found to mainly be axisymmetric, the PSV is also most often assumed to only be a function of the site latitude. It is traditionally measured in either of two ways: by measuring the dispersion of the direction of the field at each site, the "directional scatter", or by measuring the dispersion of the corresponding VGPs, the "VGP scatter" already introduced in Sect. 3.3.1. Both are measured by assuming that the corresponding dispersions follow a Fisherian distribution (for details see e.g. Merrill et al. 1996). Unfortunately the transformation of field directions into VGPs does not transform a Fisherian distribution into another Fisherian distribution (Cox 1970) and this has led to quite some discussion about which measure of PSV is most appropriate (again, see e.g. Merrill et al. 1996). More recent work based on the GGP formalism has however brought very useful clarification. In particular, and as already mentioned, this formalism predicts that the distribution of field directions will usually not be Fisherian. In contrast VGP scatter, though not strictly Fisherian, can formally (albeit approximately) be related to the variances $(\sigma^2(g_l^m), \sigma^2(h_l^m))$ defining the PSV (Kono and Tanaka 1995; Hulot and Gallet 1996). Figure 18 shows a typical VGP scatter curve for normal polarity lava flow data over the past 5 My. This figure clearly suggests an increase of the VGP scatter with latitude. Although perhaps exaggerated by some possible inclusion of low-quality data (as the recent investigation of Johnson et al. 2008 suggests might have been the case, at least to some extent), such a trend brings important information: it shows that the PSV produced by the geodynamo somewhat "senses" the Earth's rotation axis, and breaks the spherical symmetry (Hulot and Gallet 1996). Although this does not come as a surprise, it nevertheless shows that VGP scatter curves can be used to investigate how and how strongly this symmetry is broken. This is not a trivial exercise, as the geodynamo, and the many variances $(\sigma^2(g_l^m), \sigma^2(h_l^m))$ of a PSV model, have plenty of options for producing such a curve.

Two alternative PSV models have been proposed to account for this curve. Both start from the a priori assumption that the present field has little reason to be significantly different from the field over the past 5 My. This field must then be seen as one realization **x** of the GGP process that has been governing the field over the recent past. Since the observed TAF contributes little beyond the axial dipole, it is reasonable to assume that the core field Lowes-Mauersberger power spectrum shown in Fig. 14 reflects the PSV. This spectrum can then be used to construct a simple baseline PSV model by setting $\sigma(g_l^m) = \sigma(h_l^m) = \sigma_l$ and appropriately scaling σ_l, as originally proposed by Constable and Parker (1988). By construction, such a PSV satisfies the spherical symmetry (see Hulot and Bouligand 2005). It predicts the right order of magnitude for the VGP scatter curve, but not surprisingly fails to account for its increasing trend with latitude. However, just increasing the relative contribution of order one ($m = 1$) variances (Hulot and Gallet 1996), possibly mainly in the degree two ($l = 2$) (Kono and Tanaka 1995; Quidelleur and Courtillot 1996) can account for this trend. Alternatively, as originally proposed by McFadden et al. (1988) (though via some different more empirical means) and recently discussed by Tauxe and Kent (2004), one may also increase the relative contributions from the so-called "Dipole family" (or "antisymmetric family", with $l - m$ odd) and decrease those from the "Quadrupole family" (or "symmetric family", with $l - m$ even) variances. This too leads to a satisfactory fit to the VGP Scatter curve (Fig. 18). More detailed joint TAF and PSV tests by Khokhlov et al. (2006) however suggest that the increased degree two order one assumption is more compatible with the normal lava flow data set (at least as provided by the Quidelleur et al. 1994 database). These alternative suggestions will be discussed further in Sect. 4.3.1.

VGP scatter curves assume the PSV to be axisymmetric. But just as in the case of the TAF, some amount of non-axisymmetric structure might also affect the PSV. Can this be assessed? This is clearly an even more challenging task (see discussion in e.g. Constable and Johnson 1999; Hulot and Bouligand 2005; Bouligand et al. 2005). Indeed, even though some claims are regularly made that regionally low PSV might have affected the Pacific (e.g. Lawrence et al. 2006), those claims are just as often refuted on the basis that the data analyzed might not have properly sampled the PSV (McElhinny et al. 1996a; Johnson and McFadden 2007). This issue is also still rather open.

TAF and PSV Models Prior to 5 Ma Inferring departures of the TAF from the GAD geometry and recovering the PSV for even earlier epochs is severely limited by both the fewer data available within a given reasonably narrow time period, and the fact that plate tectonic motions must be taken into account. But the internal consistency of the TAF geometry required for plate tectonic reconstruction does provide some constraints. We already pointed out that these constraints provide the main proof that the TAF has always been dominated by a GAD in the geological past. In fact these constraints further suggest that on average over the past 200 My, some G_2^0 is also present, on order of 3% of G_1^0 (Besse and Courtillot 2002; Courtillot and Besse 2004). A number of claims have also been made that at least on some occasions in the past, the TAF could have also included additional terms (e.g. Thomas et al. 1993; Chauvin et al. 1996), and in particular some axial octupole (G_3^0) component (Kent and Smethurst 1998; Van der Voo and Torsvik 2001; Si and Van der Voo 2001; Torsvik and Van der Voo 2002). Quite a few of those claims are however based on data coming from Central Asia that can also be interpreted in terms of extreme internal deformation of the Eurasian plate (Cogne et al. 1999; Hankard et al. 2007). They also very often rely on data recovered from sediments (in particular redbeds) which may suffer from so-called inclination flattening. This flattening tends to bias inclinations towards shallower values (as a result of compaction in the sedimentation process, see e.g. Dunlop and Özdemir 2007) and map into a G_3^0 signature in the TAF (e.g. Gilder et al. 2003). Elegant methods developed by Kodama and Sun (1992) and Tauxe and Kent (2004) to detect and correct for such inclination flattening tend to confirm this interpretation (Tauxe 2005; Tauxe et al. 2008). But it should be emphasized that no such explanation can account for shallow inclinations observed in igneous rocks (e.g. Kent and Smethurst 1998). It thus is still unclear whether the ancient TAF truly involves more than a G_2^0 of relative amount similar to that required for the recent TAF. (Note, as a final comment, that as all the above results are derived from directional data, they only constrain the relative value of the TAF with respect to G_1^0, the long term variations of which will be discussed in Sect. 4.3.2).

Recovering information about the ancient PSV is also of prime interest. This is achieved by building VGP scatter curves similar to the one seen in Fig. 18. But just as in the case of the TAF, one has to deal with the fact that site latitudes must first be recovered. For most of the past 20 years, the reference work for such PSV studies has been the one by McFadden et al. (1991), who investigated the PSV over the 0–195 Ma time period with the help of the now rather out of date Lee (1983) lava flow data set. Site latitudes were reconstructed by directly inferring continental drift from the same data set. Recent studies either rely on more recent and precise plate motion reconstructions (when e.g. simultaneously investigating the TAF and the PSV), or directly recover the paleolatitude from the mean inclination (as described in Sect. 3.3.1), the latter option being the only one available when investigating the very ancient (e.g. Archean) PSV. This then leads to VGP scatter curves similar in shape to the one shown in Fig. 18, with a minimum scatter at the equator, and a maximum at the poles, though those extrema, and the resulting trend in between, may differ. Most investigations have followed the lead of McFadden et al. (1991) and characterized those curves via a best fit to the so-called model G of McFadden et al. (1988), which assumes a VGP scatter of the form:

$$S^2 = (\alpha\lambda)^2 + \beta^2 \tag{12}$$

Although the rationale behind this empirical model is now known to be questionable (Hulot and Gallet 1996), it does provide a very useful means (via the two parameters α and β) of quantifying the changes of the PSV through geological times (see Sect. 4.3.2).

4 Geophysical Interpretation

4.1 Core Field Changes on the Annual to Millennial Time Scales

Interpreting the manner in which the core field has changed over annual to millennial time scales requires consideration of how motional induction occurs in the outer core, where liquid iron alloy is undergoing vigorous convection driven by the cooling of the planet. The theoretical framework for describing these processes, magnetohydrodynamics, is described in some detail by Gubbins and Roberts (1987).

The fluid is assumed to satisfy the Navier-Stokes equation (in the Boussinesq approximation)

$$\rho_0\left(\frac{\partial \mathbf{u}}{\partial t} + \mathbf{u}.\nabla\mathbf{u} + 2\boldsymbol{\Omega}\wedge\mathbf{u}\right) = -\nabla p + \rho'\mathbf{g} + \mathbf{J}\wedge\mathbf{B} + \rho_0\nu\nabla^2\mathbf{u} \tag{13}$$

where ρ_0 and ρ' are the hydrostatic density and departure from hydrostatic density respectively, $\mathbf{u}$ is the fluid velocity, $\mathbf{B}$ is the magnetic field, $\boldsymbol{\Omega}$ is the Earth's rotation vector, p is the non-hydrostatic part of the pressure, $\mathbf{g}$ the acceleration due to gravity, ν the kinematic viscosity, and $\mathbf{J}$ the current density.

The evolution of the magnetic field is assumed to follow the magnetic induction equation

$$\frac{\partial \mathbf{B}}{\partial t} = \nabla\wedge(\mathbf{u}\wedge\mathbf{B}) + \eta\nabla^2\mathbf{B} \tag{14}$$

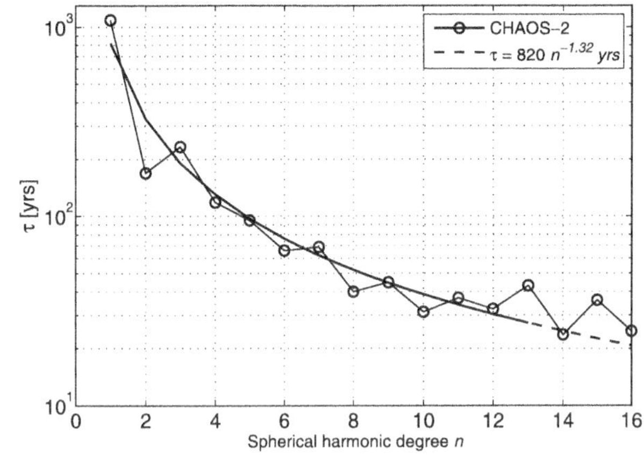

Fig. 19 Time-scales of the secular variation, as defined by Hulot and Le Mouël (1994), using $\tau_n = (W_n / W_n')^{1/2}$ where W_n and W_n' are the degree n contributions to the Lowes-Mauersberger spectra of the core field and its first time derivative. Estimates derived from the CHAOS-2 model of Olsen et al. (2009a) for epoch 2004. *Solid line* shows a two-parameter exponential fit to the data

where $\eta = 1/(\mu_0 \sigma)$ the magnetic diffusivity, with μ_0 the magnetic permeability and σ the electrical conductivity of the core fluid. This equation follows from Maxwell's equations of electrodynamics and Ohm's law applied to moving conductors, under the magnetohydrodynamic approximation that we are considering fluid motions much slower than the speed of light.

Interpretation of the changes in Earth's magnetic field is often simplified by neglecting the contribution of magnetic diffusion (last term on the RHS of (14))—this is known as the Frozen Flux Hypothesis (Roberts and Scott 1965; Backus 1968). This hypothesis turns out to be a good approximation if the length scale of the magnetic field feature is sufficiently large and the time scale of their variations sufficiently short. This is reasonably the case when considering changes in the observed core field that occur on the dominant secular variation time scales (as characterized by $\tau_n = (W_n / W_n')^{1/2}$ where W_n and W_n' are the degree n contributions to the Lowes-Mauersberger spectra of the core field and its first time derivative, see Hulot and Le Mouël 1994 and Fig. 19). Then, the time evolution of the down-continued radial component B_r of the field at the CMB can be understood as the consequence of the equation

$$\frac{\partial B_r}{\partial t} = -\nabla_H \cdot (\mathbf{u} B_r) \tag{15}$$

where ∇_H is the horizontal component of the gradient operator ∇, and $\mathbf{u}$ is the flow at the top of the core. This equation is derived from the radial component of (14), assuming $\eta = 0$ and that the flow $\mathbf{u}$ is tangent to the CMB. It also assumes that B_r is continuous across the CMB. This indeed is the only component of the field that can be assumed continuous across the CMB (Jault and Le Mouël 1991b). Note more generally that because of (1a), all the core field behavior we may witness at the Earth's surface is the direct consequence of the way B_r behaves at the core surface, which is where the core field must therefore be investigated.

4.1.1 Large Scale Core Flows

In Earth's core because the viscosity of liquid iron alloys at high pressures and temperatures is small, and because the rotation time scale is much faster than the time scale of observed field changes, the influences of viscosity and inertia are often neglected in (13). This leads to the so-called magnetostrophic approximation of core dynamics (Taylor 1963; Moffatt 1978)

$$\rho_0 (2\boldsymbol{\Omega} \wedge \mathbf{u}) = -\nabla p + \rho' \mathbf{g} + \mathbf{J} \wedge \mathbf{B} \tag{16}$$

The term $\mathbf{J} \wedge \mathbf{B}$ on the right hand side represents the Lorentz force through which the magnetic field influences the fluid motions. It is also often assumed to be reasonable to further neglect the Lorentz force in the horizontal force balance at the core surface, because the horizontal component of the magnetic field and its radial gradient will likely be small at this location. This is known as the tangentially geostrophic assumption (Le Mouël 1984; Bloxham and Jackson 1991).

Both the frozen-flux and the tangentially geostrophic assumptions imply some constraints on the way the radial component of the field behaves at the core surface (Backus 1968; Gubbins 1991; Jackson and Hide 1996; Jackson 1996; Chulliat and Hulot 2001; Chulliat 2004), and in principle some estimates of the amount of diffusion and of the strength of the Lorentz force can thus be computed to check the validity of those assumptions (see e.g., Hulot and Chulliat 2003). Indeed, various attempts have been made to quantify the amount of diffusion that may have occurred through historical times (see e.g. Bloxham et al. 1989). But the issue is complicated by our lack of knowledge of the small scales of the core field, and it has been argued that diffusion detected in this way remains within the error bounds implied by the quality of the available core field models (Backus 1988, see also Gillet et al. 2009a). As a matter of fact, several recent studies have shown that models describing the core field evolution over the past century can be built that comply with the frozen-flux constraints (Constable et al. 1993; O'Brien et al. 1997; Jackson et al. 2007b, but see also Chulliat and Olsen 2010). Over the past four centuries however, the growth of the large reversed field patch currently seen below the South Atlantic (recall Fig. 15) and which was much smaller in the early historical period (Bloxham et al. 1989; Jackson et al. 2000), is not compatible with the frozen-flux constraints (which imply the field flux to be conserved within such a patch). As first noted by Gubbins (1987) (see also Gubbins 1996), it must involve some mechanism of flux expulsion from within the core. This suggests that although negligible when considering decade to century time scales, diffusion must play a significant role when longer time scales are considered (see also Jackson and Finlay 2007 where this issue is discussed in some detail).

Adopting the frozen flux hypothesis and a core dynamic assumption (most commonly the tangentially geostrophic assumption), it is then possible to use (15) to estimate the large scale motions at the core surface that can account for observed geomagnetic secular variation (see e.g., Bloxham and Jackson 1991; Chulliat and Hulot 2000; Holme 2007). This is a highly non-unique inverse problem—regularisation is required to produce large scale flows and several choices of dynamic constraint are possible. Furthermore there are difficulties concerning the large scale secular variation produced by the small scale field and flow (Hulot et al. 1992; Eymin and Hulot 2005).

Nonetheless, the dominant features that emerge from such inversion seem fairly robust (Amit and Olson 2006; Holme 2007); a typical example of a large scale core surface flow is presented in Fig. 20. Note the intense westward gyre that runs from under west of Australia, under southern Africa through to under southern America. Another prominent feature is the anticylonic vortex below Asia. There is also some evidence for the existence of vortices in the polar regions (Olson and Aurnou 1999; Pais and Hulot 2000; Hulot et al. 2002).

It is important to stress that such computations only give access to (estimates of) the large scale core surface flows. An important question is the extent to which the inferred core surface flows reflect deeper core flows. Early attempts of down-continuing this flow within the rest of the core were based on the assumption that the Lorenz force could also be neglected (or balanced by pressure forces) within the deep core and essentially organized along so-called Busse (1970) rolls parallel to the Earth's rotation axis (Hulot et al. 1990). Recently Jault (2008), building on the ideas of Hide (1966), suggested that for flows evolving

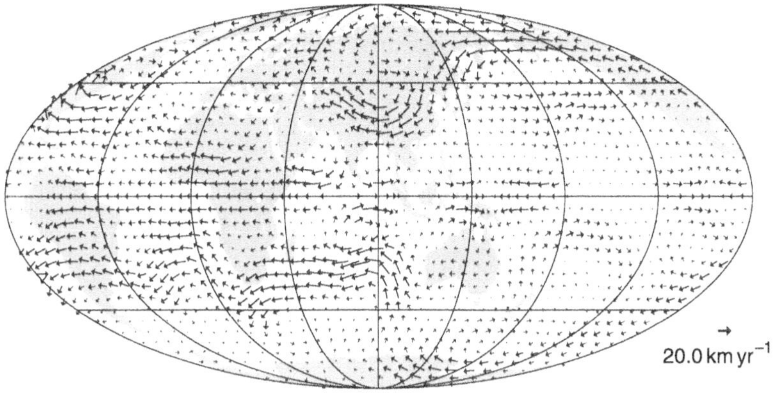

Fig. 20 Large-scale core surface flow constructed by Holme and Olsen (2006) under the frozen-flux and tangential geostrophy assumptions from a satellite observation-derived secular variation model

on time scales much shorter than the diffusive time $\tau_D = r_c^2/\eta$, where r_c is the radius of Earth's core, similar down-continuation in the form of quasi-geostrophic flows could also be performed in the presence of a relatively strong magnetic field, provided the so-called Lehnert number (Lehnert 1954) $\lambda = B/(\Omega(\mu_0\rho_0)^{1/2}r_c)$ is small enough, which is indeed likely the case in the Earth's core. Note that such assumptions also imply that the flow should then be symmetric with respect to the equator, an assumption that has the advantage of further reducing the non-uniqueness of the core surface flow determination, and which indeed seems to comply with the observations when considering fast changing flows (Pais and Jault 2008; Gillet et al. 2009a). On longer time scales, however, the dynamics may very well be different, and it has indeed been pointed out that core surface flows averaged over centuries might rather reflect some influence of the asymmetric thermal boundary conditions imposed by the (very slowly) convecting mantle (Aubert et al. 2007; Amit et al. 2008, see also Sect. 4.3.1).

One important aspect of the magnetostrophic approximation (see (16)) is that it requires the average torque exerted by the Lorentz force on each axisymmetric cylinder (about the Earth's rotation axis, see Fig. 21) to be exactly zero. This is known as the Taylor constraint (Taylor 1963). But the dynamo may very well produce magnetic fields that do not exactly comply with that constraint. Furthermore, slight departures of the CMB shape and of the gravitational field **g** from axisymmetry may also lead to additional torques. If such torques arise, the cylinders will start accelerating. This a mechanism by which the core and the mantle can exchange axial angular momentum on short time scales (Braginsky 1970, 1984; Jault and Le Mouël 1991a), with the solid inner core possibly also playing an important role (Mound and Buffett 2005). Because the surface expression of such cylindrical accelerations must show up as part of the estimated core surface flows, those can be used to infer the amount of axial angular momentum the core is exchanging with the mantle (Jault et al. 1988). Indeed, numerous studies have shown that observed length of day variations on decade time scales can be accounted for by this mechanism (Jault et al. 1988; Jackson et al. 1993; Jackson 1997; Holme 1998; Pais and Hulot 2000). Note that core-mantle exchange of equatorial angular momentum must of course also occur. But because equatorial torques do not break the Taylor constraint, their consequence on the core surface flow cannot be easily identified. Nonetheless their magnitude can be roughly estimated and shown to also be compatible with the weak and poorly resolved decade

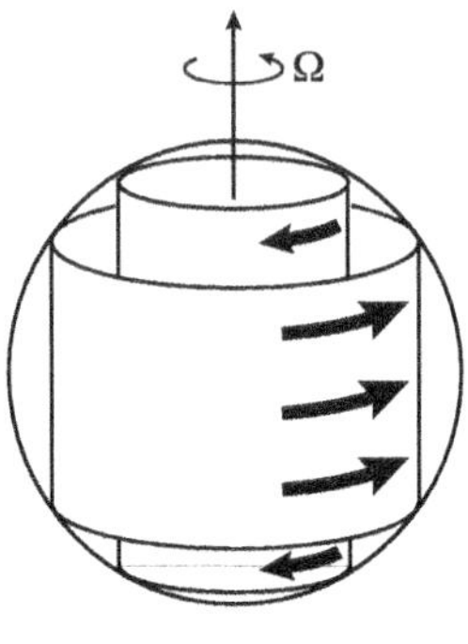

Fig. 21 Torsional oscillations after Dumberry (2008b): coaxial cylinders on which torsional oscillations occur

time-scale motion of the Earth's pole of rotation (Hide et al. 1996; Hulot et al. 1996; Dumberry 2008a).

4.1.2 Torsional Oscillations and Magnetostrophic Waves

The axisymmetric cylindrical differential rotation produced by the breaking of the Taylor condition within the fluid core cannot act for too long, as this stretches the magnetic field lines (by virtue of (14)) and leads to electromagnetic restoring torques. This interplay between (13) and (14) then leads to so-called torsional oscillations (Taylor 1963; Braginsky 1970). These are a special type of Alfvén wave that propagate along the cylindrically radial component of the magnetic field perpendicular to Earth's rotation axis (see e.g. Dumberry 2008b where an accessible introductory account is given). Figure 21 shows schematically the geometrical form of torsional oscillations.

The fundamental period T_M of Alfvén waves propagating as torsional oscillation in Earth's core scales as $T_M \sim r_c\sqrt{\rho_0\mu_0/\{B_s^2\}}$ where $\{B_s^2\}$ is the average value of the square of the cylindrically radial component of the magnetic field in Earth's core. This quantity is poorly known; if $\sqrt{\{B_s^2\}}$ is 0.2 mT (roughly the order of magnitude of the observed rms amplitude of the large scale radial field at the core surface) then $T_A \sim 60$ years which is compatible with the observed timescale of decadal changes in zonal core flows. On the other hand if $\sqrt{\{B_s^2\}}$ is 5 mT, as might be the case if there is a strong toroidal field in the core, then $T_A \sim 2.5$ years, and torsional oscillations would be involved in more rapid core dynamics. Most tentative observations of torsional oscillations in Earth's core have to date been based on an interpretation of decadal variations of the equatorially symmetric, axisymmetric component of the core surface flow inversions, precisely to investigate the unknown $\sqrt{\{B_s^2\}}$ quantity (see e.g. Zatman and Bloxham 1997, 1999; Buffett et al. 2009). More detailed, dynamically-consistent, models of torsional oscillations are now being developed and promise exciting insights in the next few years.

Torsional oscillations have also been proposed as the origin of a peculiar phenomena known as geomagnetic jerks (Bloxham et al. 2002). These jerks are traditionally defined as sudden changes of trends in the secular variation recorded in observatories (recall Fig. 3, Courtillot et al. 1978). They often occur worldwide, though slightly earlier (by one or two years) in the northern than in the southern hemisphere (Alexandrescu et al. 1996b). This delay might be caused by the slightly conducting mantle, though this has recently been shown to be unlikely (Pinheiro and Jackson 2008). It is still unclear what may cause those jerks, and a variety of instability mechanisms have been proposed (Desjardins et al. 2001; Bellanger et al. 2001). However it seems unavoidable that geomagnetic jerks must be related to core surface flows (Hulot et al. 1993) and involve sudden changes in the acceleration of part of those flows (Le Huy et al. 1998), including flows associated with torsional oscillations. This could explain why the occurrence of geomagnetic jerks appear to be associated with changes of trends in the length of day variations (Holme and de Viron 2005).

Another variety of hydromagnetic wave can arise from the coupling of (13) and (14) when the Magnetic (Lorentz) forces act to oppose the restoring action of the Coriolis forces. In this scenario rather slow oscillations of fluid parcels can occur and the resulting waves are referred to as magnetostrophic (Lehnert 1954; Acheson and Hide 1973). Magnetostrophic waves are also sometimes referred to in the literature as MC or magneto-inertial waves. In the simplest (plane layer) models, magnetostrophic waves have a period T_{MC} that scales as $T_{MC} \sim \Omega\rho\mu_0 r_c / B_0^2 m^2$ where Ω is the rotation rate of the fluid, B_0 is the background magnetic field strength and m is the azimuthal wavenumber of the disturbance. Taking $m =$ 8 and $B_0 = 5$ mT yields a period of 300 years, while taking a weaker field strength of $B_0 = 0.5$ mT yields a period of 30,000 years. For comparable strengths of magnetic fields, magnetostrophic waves are thus typically much slower than either torsional oscillations or simple inertial waves; in geomagnetism they are therefore sometimes called 'slow waves'. As first noted by Hide (1966) and Braginsky (1967) magnetostrophic waves possess the correct timescale to contribute to the observed secular variation. For a review of the diverse forms of wave motion possible in rapidly-rotating MHD fluids see Finlay (2008b).

The advent of modern core field models has led to observational evidence consistent with the presence of magnetostrophic waves in Earth's core. Using high quality satellite data Jackson (2003) pinpointed the existence of wave-like features at low latitudes under the Atlantic hemisphere. Finlay and Jackson (2003) investigated the historical evolution of these wave-like features in more detail (building on the earlier work of Bloxham et al. 1989 and Jackson et al. 2000) and found a particularly distinct westward moving wave pattern centered on the equator with period of approximately 270 years and azimuthal wavenumber of $m = 5$ (Fig. 22). This spatially and temporally coherent feature possesses a significant equatorially symmetric component that is not usually reproduced in numerical simulations of the geodynamo.

Despite this recent progress, precise models of hydromagnetic waves in Earth's core remain elusive. The primary difficulty is our ignorance of the structure and magnitude of the magnetic field within Earth's core. Knowledge of this is a prerequisite for calculating hydromagnetic wave properties that can be compared to observations; perhaps methods of data assimilation (such as proposed by e.g., Fournier et al. 2007; Kuang et al. 2008; Canet et al. 2009, see also Fournier et al. 2010 for a review) may in the future allow a resolution of such issues. Another major unknown is the precise mechanism generating the waves in the Earth's core. These could be driven by convection (as originally proposed by Braginsky 1964, 1967—he called such motions MAC waves), but also by shear instability, magnetic instability, hydromagnetic boundary layer instability, or forced by core-mantle topography.

4.1.3 Interpretation of Archeomagnetic Field Behaviour

Interpretations of the archeomagnetic field behaviour in terms of large scale core flows have also been proposed (Dumberry and Bloxham 2006; Wardinski and Korte 2008). Such interpretations must acknowledge the fact that archeomagnetic field models only give information about the very large scale core field and cannot account for temporal variations with time scales less than a century (recall Sect. 3.2). As a result most geophysical assumptions used in the context of the investigation of the historical field must also be reconsidered with care. Dumberry and Bloxham (2006) for instance used the frozen-flux approximation to account for the archeomagnetic field behavior in terms of the superposition of a stationary

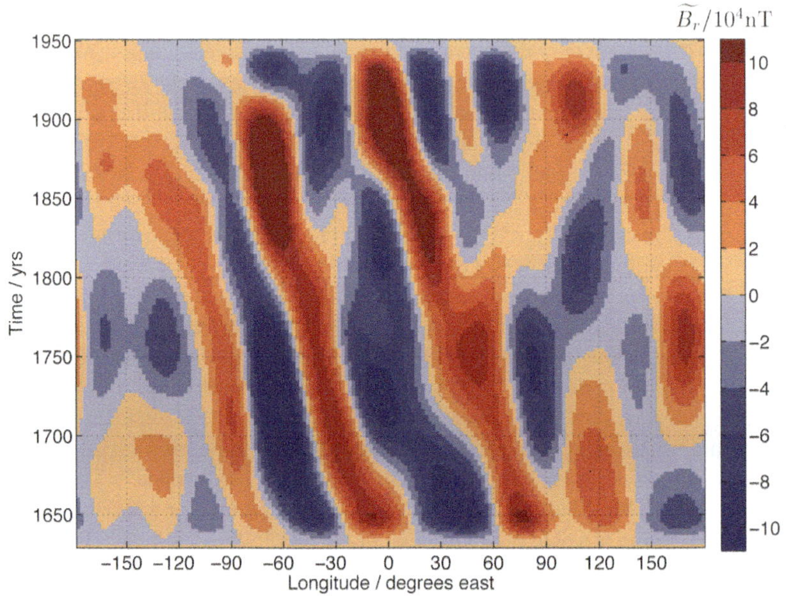

Fig. 22 Time-longitude plot of historical wave-like variations in B_r. This plot presents B_r at the equator after removal of both the time-averaged axisymmetric field and signals with time-scales larger than 400 years as a function of longitude and time. The analysis reveals oblique bands indicating the westward motion of an equatorial wave-like structure with azimuthal wavenumber $m = 5$. After Finlay and Jackson (2003)

flow and of time-dependent axisymmetric zonal flows. Those were then interpreted in terms of axial cylindrical motions to predict the possible contribution of the core to the length of day variations on millennial time scales (in very much the same way core flows derived from historical field models had successfully been used to predict length of day variations on decade time scales, recall Sect. 4.1.1). This led to a prediction with the correct amplitude, suggesting that the core could indeed be responsible for some of the length of day variations also on those time scales. However the prediction also appeared not to be correlated with the observed length of day variations. As noted by Dumberry and Bloxham (2006), this could precisely be because assuming axial cylindrical motions when investigating such long time scales is questionable, though we further note that the frozen-flux assumption could also be an issue (recall Sect. 4.1.1).

Interpreting the archeomagnetic field behaviour without invoking any of those two assumptions is also possible. Dumberry and Finlay (2007) for instance used the same type of analysis as Finlay and Jackson (2003) to directly investigate the drift of prominent magnetic field features at the core surface in the CALS7K.2 model of Korte and Constable (2005) over the past 3000 yr. They noted some wave-like westward motions close to the equator, akin to those found by Finlay and Jackson (2003) in the historical field (recall Fig. 22), mainly in the Atlantic hemisphere, but somewhat weaker and slower. They also noted a much stronger signal at mid- to high latitudes in the Northern hemisphere, where eastward and westward motions occur. These motions appear to correspond to the slow displacements and distortions of the two main high-latitude Northern normal magnetic flux patches best seen in the radial component of the time-average historical field at the core surface (Fig. 16). Those flux patches are thought to be the consequence of core downwelling flows associated with prograde vortices that concentrate the field in those regions, as suggested by both high reso-

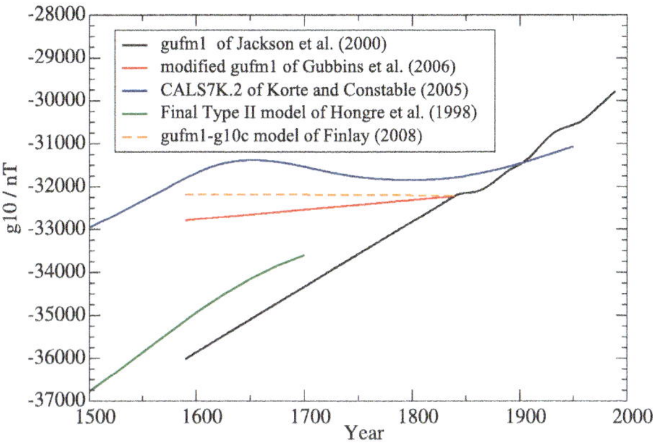

Fig. 23 Estimates of the g_1^0 Gauss coefficient over the past 500 years, from the gufm1 historical model of Jackson et al. (2000) which linearly extrapolates between 1590 and 1840 the well-constrained 1840 to 1990 historical trend (*black*); from the modified gufm1 models of Gubbins et al. (2006) (*red*) and Finlay (2008a) (*dashed yellow*), which use the Korte et al. (2005) archeomagnetic data to rescale the 1590–1840 gufm1 g_1^0 linear trend; from the CALS7K.2 archeomagnetic field model of Korte and Constable (2005) (*blue*), which only relies on the Korte et al. (2005) archeomagnetic data; and by the earlier Hongre et al. (1998) archeomagnetic field model (*green*). Note that the most recent models of Korte et al. (2009) based on the latest archeomagnetic data of Donadini et al. (2009) again predict a trend close to that of Hongre et al. (1998) (not shown). After Finlay (2008a)

lution core surface flow estimates (e.g. Hulot et al. 2002) and dynamo numerical simulations (e.g. Olson et al. 1999). Interestingly, Dumberry and Finlay (2007) also noted that changes in the motions they detected could be associated with so-called "archeomagnetic jerks" (Gallet et al. 2003), i.e. sharp changes, or cusps, in directional archeomagnetic plots such as the one shown in Fig. 6. In a more recent analysis of the same CALS7K.2 model, Gallet et al. (2009) further showed that such archeomagnetic jerks tend to occur at times when a simple description of the core field in terms of an eccentric (i.e. off-centre) dipole reveals the center of this eccentric dipole to be at a maximum distance away from the Earth's center. They noted this could be a consequence of the high latitude normal flux patches drifting closer to each other, and producing a field stronger in one hemisphere (say the Pacific hemisphere, as is presently the case), and weaker in the opposite hemisphere. More generally, and despite the limited resolution of the CALS7K.2 model, it is clear that high latitude normal flux patches must have undergone significant motions over the past millennia, as is apparent from the very weak signature they leave when the archeomagnetic field is averaged over thousands of years (recall Fig. 16). On the million year time scale however, and as will later be discussed in more detail (see Sect. 4.3.1), high latitude flux patches seem to display a statistical preference for being located where they presently stand.

A number of studies have also looked into the interpretation of the recent variations in the axial dipole moment of the core field. We know for sure that this moment (proportional to the absolute value $|g_1^0|$ of the g_1^0 Gauss coefficient, recall Sect. 3.3.1) has been decreasing fast since 1840, when systematic direct observations of the magnetic field intensity were first introduced. Its earlier evolution however can only be estimated from archeomagnetic data and is much harder to resolve accurately (Fig. 23). Some studies (e.g. Hongre et al. 1998; Genevey et al. 2008; Valet et al. 2008; Korte et al. 2009) would suggest a general decreasing trend over the past millenium (amounting to an increasing trend in the presently negative g_1^0 coefficient), while others (Korte and Constable 2005; Gubbins et al. 2006; Finlay 2008a) would favor a more stable behavior over the AD1500–1800 time period, if not an oscillatory behavior (Genevey et al. 2009). Whatever the exact long-term trend, there is little doubt that the axial dipole moment decrease observed since 1840 is related to the evolution of the reverse patch currently seen below the South Atlantic (recall Fig. 15, Gubbins 1987; Bloxham and Jackson 1992). How such a large reverse patch can develop is still an open issue. Hulot et al. (2002) noted that it is located where retrograde vortices are to be seen in their detailed core surface flow calculations and pointed out that such vortices are found to indeed be associated with flux expulsions in many numerical simulations. More recently, however, Olson and Amit (2006) pointed out that more complex mechanisms must be at work, involving both flux expulsion and meridional advection of magnetic flux by the core flow.

4.2 Reversals and Excursions

One of the motivations for investigating the current fast decrease of the axial dipole is that, were it to go on for another two millenia, it would result in a polarity reversal. Whether this has any chance of happening has been the subject of much speculation (see e.g. Constable and Korte 2006 for a review). Although a very recent investigation of a numerical dynamo simulation suggests that some reversal precursors could possibly be found by inspection of the field morphology at the core surface, no such candidate precursor was found when inspecting the present field (Olson et al. 2009). This might not be a surprise since results from a more systematic scaling study of the limit of predictability of dynamos suggest that no reversal can possibly be predicted so far ahead of time (Hulot et al. 2009b). In fact, as can be seen in Fig. 12, it often happened in the past that the Earth's dipole field experienced similar fast drops without undergoing a reversal.

Paleomagnetic data more generally bring important information about the way the field behaves during a reversal. For the most recent reversals (over the past few My) sediment data can provide relatively high-resolution time series of the inclination, declination and relative intensity at given locations, with some reasonable time-control. Lava flow data are quite complementary. They can provide ordered sequences (as defined by the flow stratification) of spot readings of the absolute intensity (though not always), inclination and declination. But the time elapsed between two successive flows is impossible to measure with enough accuracy. Synchronizing information from various types of data and sites is thus extremely difficult. Interesting attempts to produce time-varying spherical harmonic field models of reversals analogous to those produced from archeomagnetic data have recently been published (Leonhardt and Fabian 2007; Ingham and Turner 2008; Leonhardt et al. 2009). Such models can account for the data used to build them, and provide interesting insight (see below). But they rely on many free parameters and involve arbitrary adjustments for precise synchronization of data from different sites and to ensure uniqueness of the model. Less detailed, but perhaps more robust information about reversals can be recovered by relying on simpler tools, such as the VADM value and the VGP position that can be computed from any time series or sequence available at a given site (recall Sect. 3.3.1).

Considerable work has been devoted to this type of investigation, and the interested reader is referred to reviews such as those by Merrill and McFadden (1999), Coe and Glen (2004) or Glatzmaier and Coe (2007). Here we will focus on what appears to be the most important characteristics of reversals inferred so far. The first of these is that reversals only occur once the field intensity has already dropped to a value on order 10–20% of its present value, comparable to the average intensity we would witness if the axial dipole field was to vanish, keeping the rest of the non-dipole field to its current magnitude. The intensity might drop to an even lower value, but this is difficult to assess because of the intrinsic

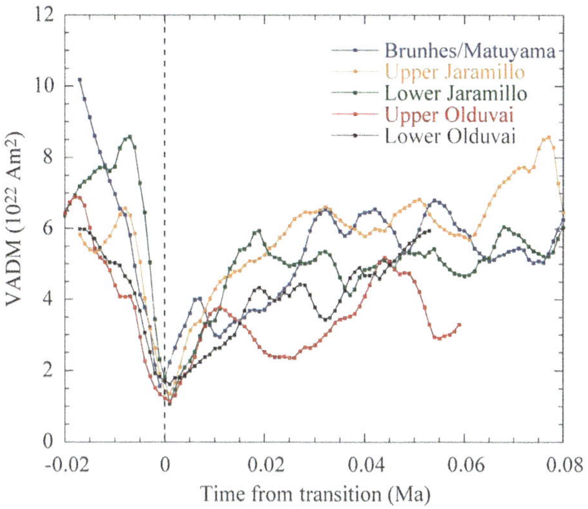

Fig. 24 Field intensity variations at times of reversals, across the five last reversals, as estimated from sediment data, after Valet et al. (2005)

limitations of paleointensity records. The way the field decreases to such a low value has been found to be complicated. Relative sedimentary paleointensity records built in very much same way as the one shown in Fig. 12d, but for the 0–4 Ma time period, initially suggested that just after recovering from a previous reversal, the VADM (a measure of the dipole field strength), would start to progressively decrease in a staggered manner until the next reversal, resulting in a so-called saw-tooth pattern (Valet and Meynadier 1993; Meynadier et al. 1994). However, this has been much disputed ever since (see e.g. Valet 2003 and Tauxe and Yamazaki 2007 for recent discussions). Although there is still some evidence of a weak correlation between the average intensity of the field during a given chron, and the length of the chron (Tauxe and Hartl 1997; Constable et al. 1998), more recent data suggest that conditions for a reversal to occur are usually met only after some rather erratic intensity decline ending with a final drop over several millennia. This can be seen in Fig. 24, which would also suggest that the recovery of the field strength after the reversal is perhaps somewhat quicker (though this also happens to be disputed see e.g. Tauxe and Yamazaki 2007). Note that this figure also shows that the time-scales involved in the reversal process are short compared to the time (on order 40 ky, see e.g. Gubbins and Roberts 1987) it would take for the dipole field to freely decay because of core convection becoming quiescent for some time (with just the diffusive term on the RHS of (14) governing the field evolution). This already shows that reversals are the result of some active dynamo process.

The second important characteristic of reversals is that the field appears to go on being dominated by its axial dipole component until quite close to the reversal itself. Recent investigations of the four most recent reversals, again from sedimentary records, clearly show that the VGP latitudes remain close to the geographical pole until they suddenly switch to the opposite geographical pole, within a matter of 2000 years for sites close to the equator, but substantially more slowly, within 10,000 years, for sites closer to the geographic poles (Fig. 25, Clement 2004). This dependence of the length of the reversal event on the site latitude is a manifestation of the third important characteristic, namely that, the field is usually not dominated by its dipole component during the reversal itself.

This finding is also evident from the observation that VGP paths (i.e. the sequence of successive VGP locations as recovered from both continuous sedimentary records and lava flow irregular sequences) inferred from different sites will usually be very different. Obvi-

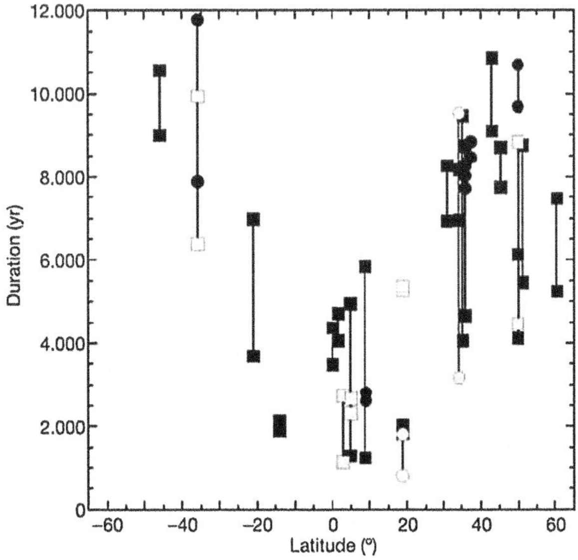

Fig. 25 Estimates of the local duration of reversals as a function of site latitude for the four last reversals: Bruhnes-Matuyama (*solid squares*), Upper Jaramillo (*open circles*), Lower Jaramillo (*open squares*), Upper Oluvai (*solid circles*), as inferred from sediment data. After Clement (2004)

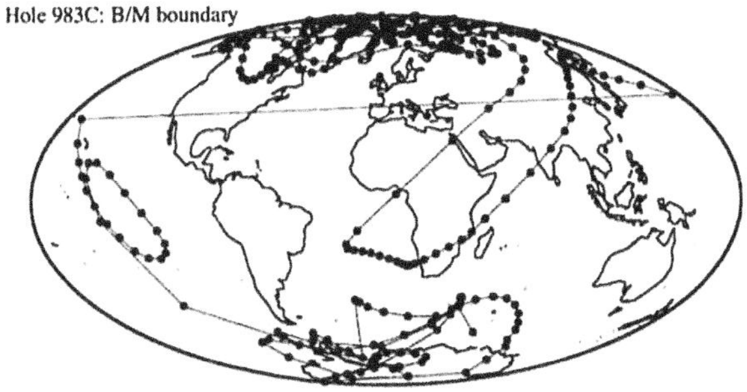

Fig. 26 Example of a VGP path for the last (Bruhnes-Matuyama) reversal, as inferred from sediment data (ODP hole 983C). After Channell and Lehman (1997)

ously, had the field remained mainly dipolar, all VGP paths would follow roughly (to within accuracy, and because of the non-negligible contribution of the non-dipole field) the same paths. This is not the case. Furthermore, even when analyzing data from a single site, VGP paths can be very complicated (Fig. 26). There is thus no doubt that the field becomes much less dipolar and has a very dynamical behavior during a reversal, presumably because of the fast changing non-dipole components of the field (as the present dominant short time scales of the non-dipole field would suggest, recall Fig. 19). This is also what tentative time-varying spherical harmonic models of the last reversal (Leonhardt and Fabian 2007; Ingham and Turner 2008), and simple forward analysis (Brown et al. 2007; Valet and Plenier 2008) would suggest.

Complex VGP paths reaching low latitudes are not observed only at times of reversals, but also each time the field intensity gets low enough. From a pure observational point of view it is usual to define such events as "excursions" as soon as latitudes reached by VGPs are less than 45° (Jacobs 2007), to distinguish them from the regular VGP scatter associated with PSV at times of stable polarity (recall Sect. 3.3.2). But this is obviously an arbitrary choice which can lead to the classification of one such event as being an excursion when the data analyzed come from one site, and not when they come from another site. This issue is well recognized and more sophisticated ways of distinguishing excursions from regular

PSV have been proposed (e.g. Vandamme 1994). It also raises the interesting question of the pertinence of considering excursions differently from regular PSV. When excursions are unambiguously identified in many sites, this is usually because all VGP paths go way below the 45° latitude limit, often reaching the opposite hemisphere and getting close to the opposite geographical pole, suggesting that such excursions could then amount to failed reversals.

To better understand the origin of reversals and excursions, and the possible link between excursions and PSV, the fast growing body of 3D numerical simulations initiated by Glatzmaier and Roberts (1995) turns out to be particularly useful. Those simulations model thermo-chemical dynamos driven by the slow cooling of the Earth. Heat extracted at the core surface, and crystallization of the inner-core (which releases both latent heat and light elements at the base of the outer core) provide the conditions required for maintaining the liquid outer core in convective motion. These motions then interact in a constructive way with the magnetic field (via (13) and (14)) to permanently regenerate it, despite magnetic diffusion (i.e., ohmic dissipation). The detailed way such self-consistent simulations are carried out can be found in e.g. Christensen and Wicht (2007). Although these simulations are run in parameter regimes that are still very remote from that of the geodynamo (see also Dormy et al. 2000), they show in particular that by varying the control parameters, but keeping the same fundamental equations, a wide range of different dynamo behaviors can be found. Some regimes display no reversal at all. Others display so many reversals that the distinction between PSV, excursions and reversals is virtually irrelevant (see e.g. Christensen and Aubert 2006). This then suggests that the geodynamo regime could currently be intermediate, with excursions and reversals simply rare enough that periods of stable polarity can be defined from an observational point of view, without necessarily implying that PSV, excursions and reversals result from distinctly different processes. Detailed analysis of numerical simulations indeed suggest that reversals occur when in the course of their spontaneous evolution dynamos reach a state that meets specific conditions. Those conditions are found to be subtle (e.g. Wicht and Olson 2004; Aubert et al. 2008a), so subtle that even marginally perturbing a dynamo that is apparently bound to reverse can change the course of the field evolution, produce an excursion, or even prevent any such event (Hulot et al. 2009b). This then suggests that reversals could result from a late option sometimes taken by the field when a particularly extreme excursion is already under way. What then makes a reversal a reversal and not just an excursion is that the subsequent evolution of the geodynamo enables the new polarity to establish itself in the entire core, and in particular within the inner core, as suggested by e.g. Gubbins (1999) following Hollerbach and Jones (1993, 1995) (though for an alternative view of the importance of magnetic diffusion in the inner core, see Wicht 2002). This would explain why substantially more excursions are observed than reversals, as is also found in numerical simulations (e.g. Wicht 2005). Finally, simulations also show that other events that would qualify as excursions from an observational point of view are in fact manifestations of extreme PSV (Wicht 2005). Numerical simulations thus suggest that excursions could indeed form an intermediate class of phemonena ranging from an extreme expression of PSV to failed reversals.

Reversals produced by numerical dynamo simulations can also be used to produce plots such as those shown in Figs. 24–26 (see e.g. Coe et al. 2000; Coe and Glen 2004; Wicht 2005). One particularly important lesson learned from such investigations is that reversals do not always occur in the same way, even within a single dynamo run with fixed parameters and boundary conditions (Coe et al. 2000). VGP paths are site dependent, and generally as complicated as the one shown in Fig. 26. In addition two successive reversals can lead to very different paths even when considering a given fixed site, as is observed in the data (see e.g. Channell and Lehman 1997). In contrast however, when heterogeneous thermal boundary conditions are imposed on such dynamos, some statistical preference for specific (but rather broad) bands of longitudes can be found in the VGP paths (e.g. Coe et al. 2000; Kutzner and Christensen 2004). When the imposed thermal boundary conditions are inferred from seismic tomography of the present lower-most mantle (such as that of Masters et al. 1996) assuming more heat than average flows out the core into seismically faster-than-average and therefore cooler regions of the mantle (but note that this is not a trivial assumption, as some of the seismic structure may well be of compositional, rather than thermal, origin), those bands roughly lie at American and East Asian longitudes. This result is interesting because it shows that some signature of the mantle influence on the core might be obtained by investigating the statistics of VGP paths. Such preferential bands have been reported by several authors when analyzing sediment data (Laj et al. 1991; Clement 1991), and some lava flow data have even provided evidence of temporary clustering of VPGs within more specific regions inside those bands (Hoffman 1992). But the robustness of those early results has since been questioned (e.g. Langereis et al. 1992; Valet et al. 1992; McFadden et al. 1993; Prévot and Camps 1993; Barton and McFadden 1996) and even though particularly careful more recent analysis of reversals and excursions recorded in lava flows over the past 20 My do lend support to some of those earlier results (Love 1998), it is perhaps best to view the question of the two preferred longitudinal bands as still open from an observational perspective (Mazaud 2007).

4.3 Core Field Long-Term Behavior

4.3.1 The Past 5 My

Guidance from numerical simulation has also proven very useful for the interpretation of the long-term behavior of the Earth's dynamo. Many simulations have for instance looked into the possible influence of heterogeneous thermal boundary conditions on the structure of the TAF, for comparisons with the 0–5 Ma TAF inferred from paleomagnetic data (recall Sect. 3.3.2 and Fig. 17). This has led to very stimulating results (e.g. Bloxham 2000; Olson and Christensen 2002; Christensen and Olson 2003; Gubbins et al. 2007; Willis et al. 2007; Aubert et al. 2008b; Davies et al. 2008). Such simulations usually impose heterogeneous thermal (heat flux) boundary conditions based on the Masters et al. (1996) shear-wave mantle tomography model (as in the VGP preferred longitudinal bands investigations mentioned in the previous section). They do however differ in both their choice of the magnitude of the thermal contrast to be applied with this morphology, and in their choice of dynamo control parameters (which we stress are anyway very remote from those of the geodynamo). Such differences can lead to a variety of geomagnetic field behaviors. Interestingly, all such "tomographic" simulations lead to a TAF with some amount of non-zonal structure, reminiscent of what can be seen in Fig. 17. To our knowledge however, none have yet succeeded in recovering the G_2^0 Gauss coefficient considered most robust in the observations (recall Sect. 3.3.2). In fact, the only simulation so far that did produce a TAF with G_2^0 and G_3^0 coefficients of the right sign and magnitude is one which simply assumes a Y_1^0 thermal boundary condition at the core surface, forcing a higher heat flow in the northern hemisphere, than in the southern hemisphere (Olson and Christensen 2002). Perhaps this is a suggestion that a more antisymmetric heat flow component is imposed by the mantle than suggested by the Masters et al. (1996) tomography model.

Another intriguing result to come from dynamo simulations is model *g* of Glatzmaier et al. (1999), run with perfectly homogeneous thermal boundary conditions and analyzed in detail by Bouligand et al. (2005). This simulation also produced a G_2^0 of a relative magnitude comparable to that of the observed TAF, but with the wrong sign. This wrong sign might lead one to reject such a simulation as being unrealistic. But it turns out that for such homogeneous dynamos, if a solution is found with a TAF displaying a mix of G_1^0 and G_2^0 coefficients, then a "mirror symmetry" solution can always be found with a TAF displaying the same G_1^0, but the opposite G_2^0 (see Hulot and Bouligand 2005 for details). Thus simulation g of Glatzmaier et al. (1999) actually shows that even homogeneous thermal boundary conditions can produce a TAF with the right G_2^0 coefficient. It is worth further pointing out that the existence of "mirror symmetries" in such dynamos more generally implies that four different fundamental states can be sustained (Hulot and Bouligand 2005). To see this, first recall that changing **B** into −**B** in a solution always leads to a second solution (since such a sign change is compatible with (13) and (14), and all other dynamo equations, see e.g. Christensen and Wicht 2007). This is formally why dynamos may experience magnetic field reversals. But as noted by Hulot and Bouligand (2005), "mirror symmetries", when applicable, imply in much the same way that a third and a fourth state can also be found by either changing the sign of all coefficients of the "Dipole family" (as defined in Sect. 3.3.2) or changing the sign of all coefficients of the "Quadrupole family". Of course, it may be argued that mirror symmetries cannot strictly apply to the geodynamo, which does not see exactly homogeneous boundary conditions. But it could be that two closely related pairs of equivalent states exist, leading to the possibility of observing "Dipole family" dominated and "Quadrupole family" dominated reversals, in addition to the better known full field reversals. As explained by Hulot and Bouligand (2005), this could provide a natural explanation for the fact that observed normal and reverse TAF field models, computed by ignoring the possibility of such partial reversals, seem to differ.

Several important recent studies indicate that the non-zonal structure in the Masters et al. (1996) tomography model perhaps captures some of the true heat flow pattern imposed by the mantle on the geodynamo. Aubert et al. (2008b) noted that such boundary conditions would lead to a TAF displaying some of the non-zonal features found in TAF models (particularly that of Kelly and Gubbins 1997), as well as to a statistical preference for core flow patterns bearing some resemblance with the time-averaged core surface flow inferred from historical data (Amit and Olson 2006). In their simulations, the preferred convection patterns are further shown to be associated with some asymmetric growth of the inner core, compatible with the asymmetric pattern seismically observed in the upper layers of the inner core (e.g. Tanaka and Hamaguchi 1997; Cao and Romanowicz 2004).

Strong locking of the geomagnetic field morphology itself can also be achieved if appropriate thermal contrasts and control parameters are imposed. Such simulations produce high-latitude patches strikingly similar to those found in the time-averaged historical field shown in Fig. 16 (see e.g. Gubbins et al. 2007; Willis et al. 2007). Whether such strong locking of the field actually occurs is however debatable as it would also imply that the archeomagnetic field and the TAF be very similar to the time-averaged historical field. This does not seem to be the case (compare maps in Figs. 16 and 17). Although one could argue that this is simply because the archeomagnetic field and TAF models are not yet sufficiently resolved, it seems more likely that no strong locking has permanently been at work. One possible scenario is that the field can often be temporarily locked in the configuration it has a had over the past centuries (with characteristic high-latitude patches below the Americas, Asia and South of Australia), while still regularly moving away from that preferred configuration on millennial timescales, as CALS7K.2 suggests has been the case over the previous six millenia. This would explain why some weaker high-latitude flux patches can be found in some of the TAF models (most strikingly in the Gubbins and Kelly (1993) model, and to a lesser extent in model LN1 shown in Fig. 17) at locations close to those seen in the time-averaged historical field. Several numerical dynamo simulations also display such behavior (e.g. Bloxham 2002; Amit et al. 2010).

Paleosecular variation produced by numerical simulations have also been investigated for comparison with the PSV recovered from paleomagnetic data over the past 5 My. A few publications have also looked into the possible influence of heterogeneous thermal boundary conditions, particularly with the aim of assessing how much non-zonal PSV those conditions could lead the geodynamo to produce (e.g. Bloxham 2000; Christensen and Olson 2003; Bouligand et al. 2005; Davies et al. 2008). Those studies show that quite different types of signature can be expected. For the time being however, comparisons with observations are much less conclusive than in the case of the TAF, not least because observational evidence of a non-zonal PSV pattern is still much debated (recall Sect. 3.3.2). More generally, all numerical dynamo simulations seem to fail to reproduce the observed magnitude of the VGP scatter. They do however often predict an increase of the scatter with the latitude, as observed (recall Fig. 18). Understanding the cause of this trend in those simulations does provide valuable insight into the origin of such trends found in the data.

As noted in Sect. 3.3.2, the behavior of VGP scatter curves is entirely dictated by the way each Gauss coefficient of the field contributes to the Lowes-Mauersberger power spectrum (Fig. 14). We also noted that the VGP scatter curve for the past 5 My could be accounted for by either assuming a stronger than average contribution from the order 1 Gauss coefficients, possibly mainly in the degree 2 (as proposed by e.g. Kono and Tanaka 1995; Hulot and Gallet 1996; Quidelleur and Courtillot 1996), or a significant imbalance between contributions from the "Dipole" and "Quadrupole" family Gauss coefficients (as proposed by e.g. McFadden et al. 1988; Tauxe and Kent 2004). Most numerical dynamos in a dipole-dominated regime have a Lowes-Mauersberger power spectrum roughly similar to that of the Earth shown in Fig. 14 (though the dipole field itself is then often too dominant, see e.g., Christensen and Wicht 2007). But the detailed way each Gauss coefficient contributes changes significantly from one simulation to the next. Interestingly, in a number of simulations, those differences arise in the form of a stronger contribution from the "Dipole" family than from the "Quadrupole" family (Bouligand et al. 2005; Coe and Glatzmaier 2006), confirming that interpreting VGP scatter curves in terms of "Dipole" versus "Quadrupole" family contributions is a possibility.

Interpreting the recent PSV (0–5 Ma) along these lines is however not without controversy. As noted by Hulot and Gallet (1996) and Tauxe and Kent (2004), this would imply a non-dipole contribution of the "Dipole" family to the field spectrum of about an order of magnitude larger than that of the "Quadrupole" family. Yet, no such imbalance is to be found in the present geomagnetic field (Hulot and Gallet 1996) which would thus have to be considered as being in a very unusual state (as also pointed out by Tauxe and Kent 2004). In contrast, the alternative suggestion that over the past 5 My the field has experienced a stronger than average contribution from the order 1 Gauss coefficients is compatible with the historical field, which displays such an imbalance (Hulot and Gallet 1996). In addition, as noted by Gallet et al. (2009), interpreting archeomagnetic field behavior in terms of high-latitude flux patches dynamics (recall Sect. 4.1.3) can provide an explanation for what may cause this imbalance, particularly in the degree 2 coefficients. It will be of great interest in the future to investigate whether this scenario can be realized in numerical dynamo simulations.

4.3.2 Geological Time Scales

Evidence for very long time scale changes in geomagnetic field behavior is primarily provided by three different types of paleomagnetic observations: the sequence of reversals as provided by the GPTS over the past 160 My (recall Fig. 13) and by additional more ancient, but more sparse magnetostratigraphic sequences (see e.g. Pavlov and Gallet 2005, 2010); a collection of paleointensity measurements (see e.g. Biggin et al. 2009), a few of which date back to 3.2 Ga (e.g. Tarduno et al. 2007); and PSV estimates (up to 2.8 Gy ago, see e.g. Smirnov and Tarduno 2004; Biggin et al. 2008a). As noted in Sect. 3.3, very little is otherwise known with respect to the geometry of the very ancient TAF, except for the fact that it must have been dominated by its GAD component since at least 400 My ago. There is no compelling evidence that this was not the case for earlier epochs. Of course, this should not be taken as evidence that the TAF has always had the same geometry, particularly as the TAF is apparently sensitive to the boundary conditions imposed by the mantle on the core (see previous section). Rather, it illustrates that current paleomagnetic data are still too few to reveal the subtle changes that are likely to have occurred in the TAF.

There are many reasons to expect the geomagnetic field to have experienced long-term changes in its behavior over geological times. The Earth is known to be $\sim$4.5 Gy old (Allègre et al. 1995) and its core is thought to have formed very early on (to within roughly 100 My, see e.g. Allègre et al. 2008). But thermal models of the Earth show that the heat extracted from the core must have significantly evolved since that time (e.g. Gubbins et al. 1979; Buffett et al. 1996; Labrosse et al. 2007; Nimmo 2007). They also suggest that the inner core only started its growth (by crystallization of the central part of the core (Jacobs 1953)) quite late, probably not earlier than 1 Gy ago (Buffett et al. 1992; Labrosse et al. 2001, but see Buffett 2003). In addition, mantle convection must have always imposed time-varying heterogeneous thermal boundary conditions at the top of the core. Finally the Earth's rotation rate is known to have slowly decreased over geological times (see e.g. Varga et al. 1998). Since all of those changes affect parameters that are important in defining planetary dynamo behavior (see e.g. Christensen and Wicht 2007), it is inevitable that they must have produced some signature in the paleomagnetic data. Both the GPTS (recall Fig. 13) and the temporal power spectrum shown in Fig. 27, reveal long-term changes in the field behavior over periods of time commensurate with mantle convection timescales (Schubert et al. 2001).

The possibility that mantle convection could control the geomagnetic reversal rate has been proposed more than thirty years ago (Jones 1977) and much investigated since (for a review of the early work, see e.g. Merrill et al. 1996). Most recent investigations rely on the statistical tools introduced by McFadden (1984) (but see also Marzocchi 1997; Constable 2000; McFadden and Merrill 2000). Those have led McFadden and Merrill (1984, 1993) to conclude that reversals within the GPTS behave as if produced by a Gamma process (as originally proposed by Naidu 1971) characterized by a time-varying statistical reversal rate, defining the probability for a reversal to occur at any time, and an inhibition time, defining the short period of time after a reversal during which no other reversal can occur. The inhibition time was initially estimated to be of order 40 ky by McFadden and Merrill (1993). But this might well be an artifact linked to the limited resolution of the GPTS. Marzocchi (1997), and most recently Lowrie and Kent (2004), indeed concluded that no significant inhibition time seems to be required by the data, if additional relevant short polarity chrons (cryptochrons) are taken into account. In this case the GPTS could simply be described in terms of a time-varying Poisson process, as originally proposed by Cox (1968) and in agreement with the fact that reversals happen to be unpredictable until very shortly before they occur (recall Sect. 4.2). Such a description has the advantage

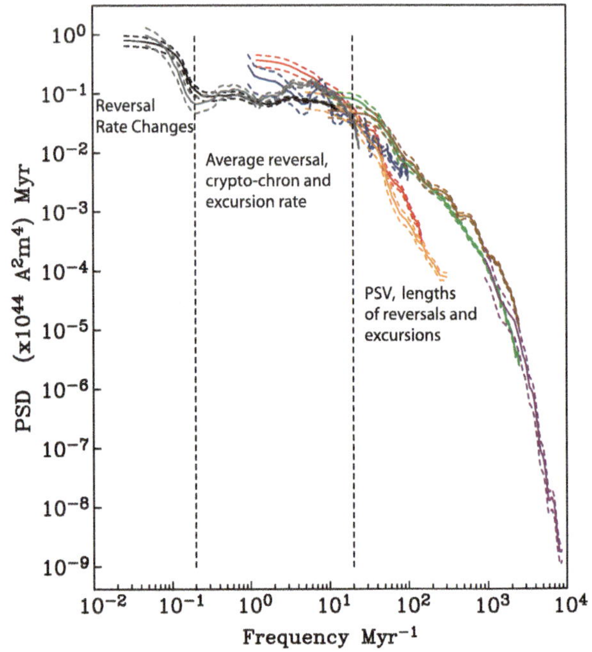

Fig. 27 Estimate of the temporal power spectrum of the Earth's dipole moment for the time interval 0–160 Ma, as inferred from a composite analysis of the GPTS (including cryptochrons, *black and grey*), of various slowly depositing ocean sediment cores (*blue, red*), of faster depositing sediment cores (*blue, brown, orange*), and of the archeomagnetic field (*pink*). After Constable and Johnson (2005) where details can be found. Note that this spectrum focusses on the dipole moment and does not include historical data, which is dominated by the much shorter time-scales of the secular variation (recall Fig. 19)

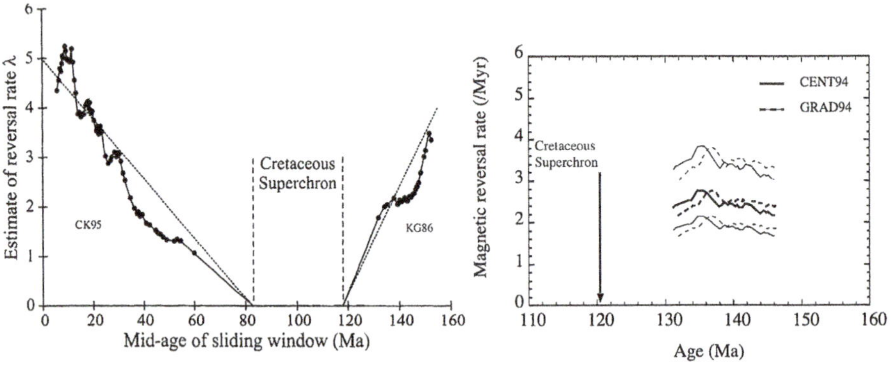

Fig. 28 Estimates of the reversal rate (in My^{-1}) for the time interval 0–160 Ma, following the method of McFadden (1984) and using a sliding window over 50 successive intervals. *Left*: estimate based on the Cande and Kent (1995) GPTS (CK95) for the Upper Cretaceous to Cenozoic and on the Kent and Gradstein (1986) GPTS (KG86) for the Upper Jurassic to Lower Cretaceous, with the simple linear model for the evolution of this rate proposed by McFadden and Merrill (2000) (after which the plot is adapted); *Right*: Revised estimates for the Upper Jurassic to Lower Cretaceous reversal rate based on the GPTS of Gradstein et al. (1994) (GRAD94) and Channell et al. (1995) (CENT94), with 2σ curves between which the true reversal rate lies, after Hulot and Gallet (2003)

that estimates of the reversal rate can then be readily computed (see e.g. McFadden 1984; Merrill et al. 1996). Note that such calculations usually involve some averaging of the raw information provided by the GPTS which may smooth out important sudden changes in the GPTS behavior.

Figure 28 (left) shows such a smooth estimate of the reversal rate based on an early version of the GPTS, using a sliding window over 50 successive intervals and following the method of McFadden (1984). This estimate progressively decreases between 160 Ma and the onset of the Cretaceous Normal Superchron, when the reversal process ceased, and then progressively increases again after the superchron. As noted by most authors, following McFadden and Merrill (1984, 2000), this behavior suggests that slow changes in the boundary conditions imposed by the mantle on the core could have progressively led the geodynamo to reach a non-reversing regime, and next led the geodynamo to progressively return to a frequently reversing regime. That this could have been the case is supported by some numerical simulations which show that the tendency of dynamos to produce reversals is indeed sensitive to the pattern of inhomogeneous thermal boundary conditions imposed at the core surface (Glatzmaier et al. 1999). However, an alternative interpretation of the observed transition to the superchron can also be proposed, based on the more recent reversal rate estimate proposed by Hulot and Gallet (2003) for the Upper Jurassic to Lower Cretaceous time period (Fig. 28 (right), consistent with the GPTS shown in Fig. 13). As can be seen, this revised estimate no longer shows any unambiguous sign of decrease before the superchron. A careful look at the GPTS itself (Fig. 13) shows that the transition to the superchron may well have been very sudden. This suggests an alternative interpretation that the geodynamo could have entered the superchron as a result of a sudden change in its boundary conditions (for example, following the rapid arrival of a cold subducted slab at the core surface, the most effective way of quickly and significantly changing thermal boundary conditions at the core surface (Gallet and Hulot 1997)), or perhaps as a result of a spontaneous transition from a reversing to a non-reversing state (as proposed by Hulot and Gallet 2003 and supported by the analysis of Lowrie and Kent 2004). Although no such transition has yet been observed in numerical simulations (possibly because no very long runs displaying as many reversals as observed in the GPTS have yet been simulated), it is worth noting that numerical dynamos displaying occasional transitions between two distinctly different states have recently been found (Simitev and Busse 2009), and that simplified dynamo models can indeed spontaneously produce superchrons (Ryan and Sarson 2007, 2008). This interpretation does not deny that mantle convection could modulate the reversal rate of the geodynamo while in its reversing state (though some have argued that all of the temporal power spectrum of the field, shown in Fig. 27, from the shortest to the superchron time scales, could result from pure core dynamics, see e.g., Jonkers 2007).

Further interesting insight into the possible cause of superchrons can be gained from inspection of more ancient magnetostratigraphic data. These data reveal that at least four other superchrons have occurred over the past 1.4 Ga: the $\sim$50 My long Kiaman reverse superchron, from $\sim$310 to $\sim$260 Ma (Opdyke and Channell 1996); the $\sim$30 My long Moyero reversed superchron, from $\sim$490 to $\sim$460 Ma (Pavlov and Gallet 2005); and two more recently recognized (and not yet baptized) even older superchrons, one dated $\sim$1 Ga and at least 16 My long (if not 30 My or more) (Pavlov and Gallet 2010), and another one dated $\sim$1.4 Ga and $\sim$30 My long (Elston et al. 2002). Some information is also often available with respect to the sequence of reversals that preceded or followed these superchrons. For the past 550 My, this information is summarized in Fig. 29, analogous to Fig. 28, except for the fact that reversal rates have been estimated by geological stage, rather than by relying on a moving window average, to harmonize post- and pre- 150 Ma data for which much less data is available. This figure shows that superchrons tend to occur with a rough periodicity of about 200 My, again suggesting a possible global link between the occurrence of superchrons, mantle convection and related phenomena, such as true polar wander, mantle plumes, exceptional volcanism, and even mass species extinctions

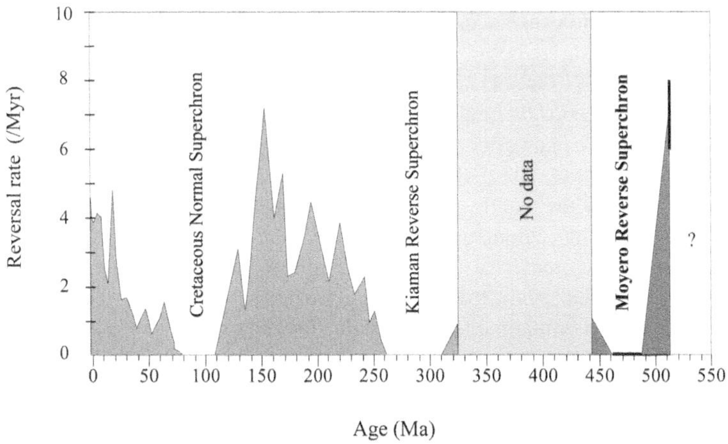

Fig. 29 Estimate of the reversal rate (in My^{-1}) for the time interval 0–500 Ma, where reversal rates are estimated by geological stage, rather than by using a moving window average (as in Fig. 28), to harmonize post- and pre- 150 Ma data for which much less data is available. After Pavlov and Gallet (2005)

(see e.g. Loper and McCartney 1986; Courtillot and Besse 1987; Larson and Olson 1991; Courtillot and Olson 2007).

When considering the possible cause of such a succession of superchrons over a period of time of 1.4 Gy, one should also consider the possible consequences of changes in the size of the inner core, in the global amount of heat extracted from the core, and in Earth's rotation rate. Although rather few simulation studies have looked into this so far (Driscoll and Olson 2009b, 2009a; Aubert et al. 2009), they all conclude that to produce superchrons as a result of mantle control, the geodynamo must have operated in a regime close to a transition between reversing and non-reversing regimes, as had been proposed by Courtillot and Olson (2007). Then, the time-varying heterogeneous thermal boundary conditions produced by mantle convection, combined with the slow thermal evolution of the Earth, growth of the inner core, and increase of the Earth's rotation rate, could have conspired to occasionally shift the geodynamo from a reversing regime to a non-reversing regime. However, this scenario is not without difficulties. As noted by Aubert et al. (2009), models of the thermal evolution of the Earth also likely imply that the transition from the reversing to the non-reversing regime was very close to the transition from the non-reversing regime to the non-dynamo regime, particularly over the past 500 My, making it quite remarkable that superchrons could have been produced in this way without shutting down the dynamo altogether. Furthermore this scenario predicts a progressive decrease in the reversal rate before reaching each superchron, followed by a progressive increase after the superchron. In contrast, the transition to the Cretaceous Normal Superchron seems to have been sudden and, although little is known about the reversal rate before the Kiaman reverse superchron, there also are clear indications that the transition towards the Moyero reverse superchron could have been as sudden (Pavlov and Gallet 2001). Finally, and as noted by Pavlov and Gallet (2010), both of the two previous superchrons strongly suggest even more sudden transitions between periods of frequent reversals and superchrons. Such sudden transitions seem to favor a spontaneous origin of superchrons, as proposed by Hulot and Gallet (2003).

As noted earlier, the sequence of reversals is not the only information available to investigate the very long term behavior of the geomagnetic field. Additional important information can be recovered from the investigation of the ancient PSV. As explained in Sect. 3.3.2, this information is usually recovered in the form of VGP scatter curves formally converted into best fits to the model G (see (12)). This was first done by McFadden et al. (1991) who produced model G estimates for six broad epochs covering the past 195 My. Figure 30 shows the evolution of the α and β best-fit parameters defining the evolution of those VGP scatter curves, as originally proposed by McFadden et al. (1991). As noted by these authors, comparing this Figure to Fig. 28 suggests a connection between the changes in the VGP scatter and in the reversal rate over the past 160 My. Quite a few investigations have since been conducted (e.g. Cronin et al. 2001; Tarduno et al. 2002; Mankinen 2008; Biggin et al. 2008b) and it now appears that this connection is not as clear-cut as initially envisioned by McFadden et al. (1991) (see e.g. Biggin et al. 2008b). All these investigations nevertheless confirm the most striking fact that the PSV has been behaving very differently during the Cretaceous Normal Superchron compared to other times when the field was reversing, particularly compared to the best documented past 5 My. As can be seen in Fig. 30, the parameter β (a direct measure of the VGP scatter at the equator, recall (12)) was low. This result is interesting because VGP scatter at the equator can only be produced by Gauss coefficients belonging to the "Quadrupole" family (e.g. Kono and Tanaka 1995; Hulot and Gallet 1996). At this time, the contribution of the "Quadrupole" family to the Lowes-Mauersberger spectrum (relative to the axial dipole) must thus have been significantly less than over the past 5 My. Parameter α, a measure of the increasing trend of the scatter with latitude, was high. This corresponds to the fact that the VGP scatter at high latitudes was then about the same as over the past 5 My (as confirmed by the more recent studies of Tarduno et al. 2002; Biggin et al. 2008b). This now implies that contributions from the "Dipole" family must have remained large enough. Whereas the most plausible explanation for the VGP scatter curve over the past 5 My seems related to the order one Gauss coefficients behavior (recall Sect. 4.3.1), it now seems much more natural to invoke an imbalance between the "Dipole" and "Quadrupole" families to account for the VGP scatter curve during the Cretaceous Normal Superchron, as envisioned by McFadden et al. (1991).

Two additional lines of evidence support such an interpretation. First, the few numerical dynamos displaying such a strong imbalance (with a weak "Quadrupole" family field) are particularly stable with respect to reversals (Coe and Glatzmaier 2006). Second, the VGP scatter during the previous Kiaman Reverse Superchron was quite similar to the one observed during the Cretaceous Normal Superchron (Haldan et al. 2009). Taken together the above results suggest that during superchrons the geodynamo operated in a very different way than at other times, with no reversals and a significantly depleted "Quadrupole" family field.

Just as in the case of the reversal rate, some PSV observations are also available for even earlier epochs. But the data are sparse and those most appropriate for investigating the sequence of reversals rarely provide suitable information for PSV studies. In addition, rather contradictory results have been reported so far. Smirnov and Tarduno (2004) have for instance found that Late Archean-Early Proterozoic data from dikes dated ~ 2.5 Ga and ~ 2.7 Ga would favor a VGP scatter comparable to that observed over the past 5 My while Biggin et al. (2008a) concluded that the VGP scatter was similar to that observed during the Cretaceous Normal Superchron. Biggin et al. (2008a) argued that the reasons for the discrepancy were the stricter selection criteria they applied and the larger data set they investigated, sampling the 2.82 to 2.45 Ga time period. As noted by Hulot (2008) this time period is indeed so long (longer than 190 My) that some time evolution should be expected (recall Fig. 30) which could result in such disagreement between the two studies. The results of Biggin et al. (2008a) nevertheless indicate that during the Late Archean-Early Proterozoic the geodynamo produced a PSV much more often comparable to that of the Cretaceous

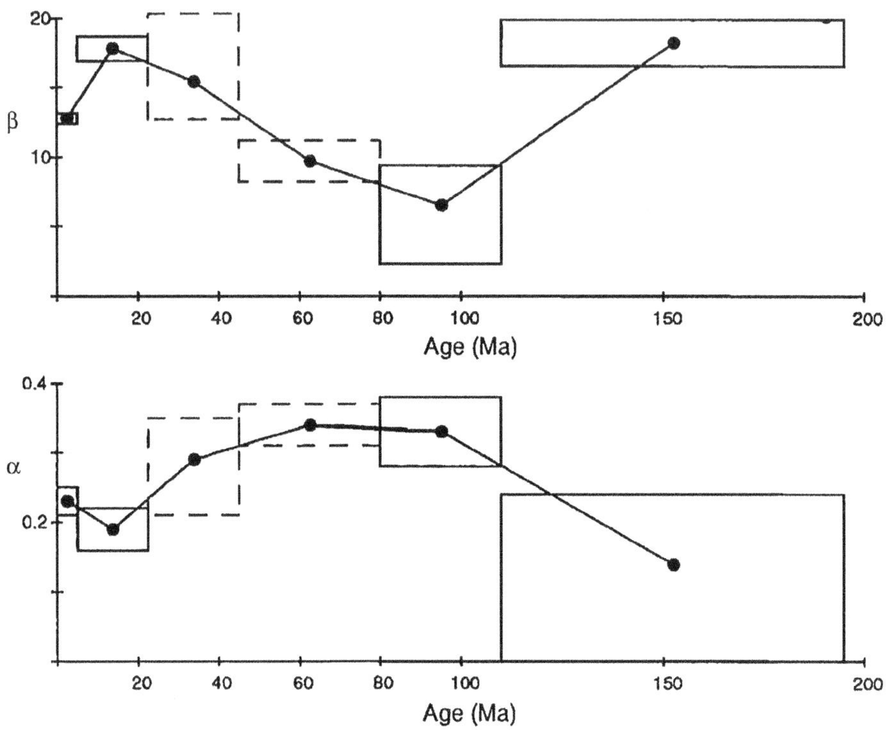

Fig. 30 VGP scatter Model G parameters as a function of time over the past 190 My. Parameters α (*lower panel*, dimensionless) and β (*top panel*, in degrees), as defined by (12). Boxes define range of age contributing to the estimated parameter and 95% confidence limits, dotted boxes indicates that the fit to (12) was poor. After McFadden et al. (1991)

Normal Superchron, than to that of other recent epochs when the field was reversing. This observation led Biggin and co-workers to speculate that the geodynamo could have been less prone to reversals during the Late Archean-Early Proterozoic. Their hypothesis is consistent with the data available over the 2.775 to 1.05 Ga time period (Halls 1991; Gallet et al. 2000; Elston et al. 2002; Strik et al. 2003) during which the few time intervals investigated so far, totaling 225–250 My, document only 45 reversals and suggest an average reversal rate of only 0.2 My^{-1} (Coe and Glatzmaier 2006).

Both Coe and Glatzmaier (2006) and Biggin et al. (2008a) interpret their observations as a consequence of the inner core being smaller (or even absent) at the time, based on results from a very stable small inner-core simulation by Roberts and Glatzmaier (2001), which also displayed the weak "Quadrupole" family field characteristic that we saw could account for the PSV at times of superchrons. But it should be noted that reversing dynamos with no inner core can also be found (e.g. Sakuraba and Kono 1999), and that other factors may have influenced the dynamo on such long time scales. Indeed Aubert et al. (2009) recently confirmed that small inner-core dynamos can reverse frequently. They also noted that the thermal evolution of the Earth (and changes in the rotation rate) could have led the geodynamo to lie closer to the transition between the reversing and non-reversing regimes at the time. If superchrons are caused by the geodynamo occasionally crossing this transition, they could then have occurred more often in the past. If superchrons are spontaneous, it could alternatively be that the regime was more prone to superchrons at the time. Whatever the

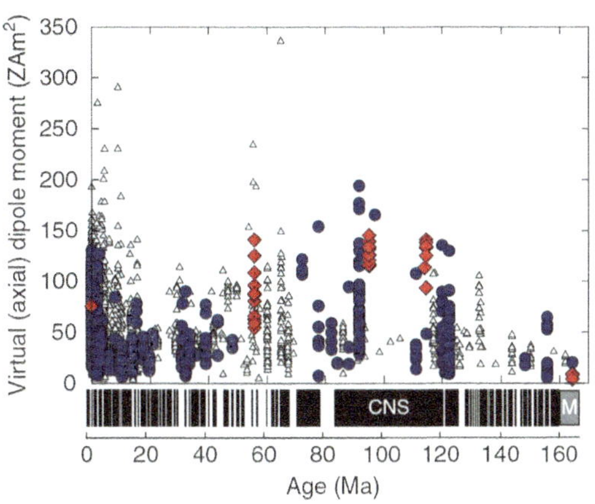

Fig. 31 Paleointensities over the past 170 My, expressed in terms of Virtual Axial Dipole Moments (VADM, see Sect. 3.3.1). Data extracted from the PINT06 database of Tauxe and Yamazaki (2007) after some minimum quality criteria selection. Blue dots are submarine basaltic glass data. Red diamonds are single-crystal results. Triangles are all other data (whole rocks). Units ZAm^2 stand for 10^{21} Am^2. After Tauxe and Yamazaki (2007) where more details can be found

cause of superchrons, it is finally worth recalling that both the data used by Coe and Glatzmaier (2006) to estimate a low 0.2 My^{-1} reversal rate, and the more recent data published by Pavlov and Gallet (2010) primarily reveal transitions between periods with frequent reversals and superchrons. The main cause of the apparent low reversal rate observed by Coe and Glatzmaier (2006) and of the superchron type of VGP scatter found by Biggin et al. (2008a) could thus be the more frequent occurrence of superchrons before 1 Ga than during the past 160 My (or even 300 My, recall Figs. 28–29).

Finally we turn to the paleointensity observations. As already mentioned in Sect. 2.5, these data are particularly difficult to recover and most prone to artifacts, especially when very ancient. Essentially four different types of data can be used: relative paleointensity records from sediments, absolute intensity estimates from whole igneous rocks, submarine basaltic glass samples, or single silicate crystals extracted from igneous rocks. Relative paleointensity records have already been much discussed in Sect. 4.2 where we noted that they suggest some weak correlation between the average field intensity during a chron and the length of the chron (Tauxe and Hartl 1997; Constable et al. 1998, Sect. 4.2). Do the other types of paleointensity data confirm this correlation or reveal other correlations with e.g., the reversal rate or the paleosecular variation?

Answers to these questions unfortunately remain rather unclear. Figure 31 perfectly illustrates the complexity of the information provided by these various types of data over the past 170 My. This figure was plotted from the PINT06 database of Tauxe and Yamazaki (2007) after some minimum quality criteria selection and conversion of the data in terms of VADM (as defined in Sect. 3.3.1). It first shows that much more data are available for recent epochs than for ancient epochs (with roughly 40% of the data younger than 1 My). It also suggests that the field intensity experienced considerable variability at all times and shows that defining long-term trends in the geomagnetic field intensity is no simple matter. Considerable effort has been put into deciphering the messages possibly embedded in these data.

Selecting data from six polarity intervals of known durations (with ages ranging from 3.3 Ma to 121 Ma), Tauxe and Yamazaki (2007) for instance argued that a weak correlation between the average field intensity during a chron and the length of the chron could again be found. But the correlation is very weak (as is in fact obvious from the fact that the Cretaceous Normal Superchron, the longest of all chrons by far, does not display such an outstanding VADM, see Fig. 31). It also is not clear why such a correlation should apply between 3.3 Ma and 121 Ma, and not to more recent data, when both short intervals and large intensities are to be found.

Identifying epochs when the field was high or low on average has also been the subject of many studies. Some authors (e.g. Juarez et al. 1998; Selkin and Tauxe 2000; Juarez and Tauxe 2000), relying on submarine basaltic glasses, have argued that significantly higher values are found only when considering data younger than 300 ky. [This, incidentally, is the reason why Tauxe and Kent (2004) chose a weaker value for the G_1^0 TAF Gauss coefficient than usually assumed in their GGP model for the past 5 My, recall Sect. 3.3.2]. But this view is not supported by the latest basaltic glass data plotted in Fig. 31, which now suggest that the average VADM during the Normal Cretaceous Superchron was quite similar to today's dipole moment (Tauxe 2006). These findings are more consistent with previous results based on whole igneous rock data (e.g. Prévot et al. 1990; Tanaka et al. 1995; Perrin and Shcherbakov 1997) which suggested that the only time the field was truly low was during the Mesozoic (before 120 Ma, as is suggested by Fig. 31). Interestingly, independent indications that this may well have been the case can also be found in marine magnetic anomaly profiles which happen to be particularly weak during the 167–155 Ma time period (e.g., Tivey et al. 2006; Tominaga et al. 2008). This portion of the magnetic anomaly record, found in the Pacific, displays many small amplitude fluctuations and is known as the Pacific Jurassic Quiet Zone (JQZ in Fig. 13). However, this is also an epoch for which magnetostratigraphic evidence for polarity reversals is ambiguous (as reflected by the grey area in the GTPS of Fig. 13, also reproduced below the intensity plot in Fig. 31), and it is still unclear whether the observed anomalies truly reflect fluctuations of a weak field with few reversals, or a rapid succession of reversals (Tivey et al. 2006; Tominaga et al. 2008).

More recently Heller et al. (2002, 2003), following earlier work by Perrin and Shcherbakov (1997), pointed out an intriguing property of whole igneous rock data selected using their own criteria. Investigating the longer 320–0 Ma time period, which they split in time intervals of up to a few tens of millions of years, they noted that the distribution of intensities could be fit by a bimodal distribution with the same two (broad) peaks (at roughly 9×10^{22} Am^2 and 4–5 $\times 10^{22}$ Am^2). The only temporal variation they could find over the corresponding 320 My was in the relative distribution of the data within this distribution, which reduced to a unimodal distribution on only two occasions: over the past 5 My, when it only displayed the high 9×10^{22} Am^2 peak; between $\sim$250 Ma and $\sim$20 Ma, when it only displayed the low 4–5 $\times 10^{22}$ Am^2 peak. The reason for such a behavior is unclear. Heller et al. (2003) suggests that perhaps the geodynamo has the choice to operate in two different states, each defined by a different (either high, or low) average dipole moment, and constantly switched (at an unknown rate) between those two states over the past 320 My, except during the 250–120 Ma and 5–0 Ma time periods. But it is puzzling that such a change of behavior could have remained unaffected by the occurrence of the Cretaceous Normal Superchron. More likely, and as also noted by Heller et al. (2003), this observation may result from some artifacts related to the selection criteria, undersampling of the data for the earliest epochs, and possibly unidentified rock magnetic issues. That the most recent and much more numerous data only display a unimodal distribution is perhaps an indication that this is the more likely explanation.

Particularly suspicious of whole rock data, Tarduno (2009) (see also Tarduno et al. 2006) argues that perhaps one should rather concentrate on single-silicate crystal data (red diamonds in Fig. 31). These data show the same kind of general trends as the other data, but in a clearer way, with a maximum during the Cretaceous Normal Superchron, and a somewhat weaker field at times of frequent reversals. That the field could usually be strong at times of superchrons is also supported by the single-silicate crystal results recently published by Cottrell et al. (2008), who found VADM estimates on order 9×10^{22} Am^2 during the Kiaman Reverse Superchron. It should however be emphasized that so far only two (mutually consistent) data points have been obtained in this way and that the inferred VADM is similar to that of the recent frequently reversing field. Perhaps what controls the average field paleointensity is simply unrelated to what controls the occurrence of superchrons.

Some information concerning the very ancient field intensity is also available. Most of the data are from whole rock studies, though quite a few, among which the oldest data, are from single crystal silicates (up to 3.2 Ga, see Tarduno et al. 2007). These data have also been scrutinized for long-term evolution, particularly in view of identifying some possible signature associated with the onset of the inner-core crystallization (see e.g. Hale 1987; Labrosse and Macouin 2003; Macouin et al. 2004 and most recently Biggin et al. 2009). Unfortunately, this has again led to rather contradicting results and interpretations. This is because of the still very limited database currently available. It may also be because, as illustrated by the recent numerical simulations of Aubert et al. (2009), only weak long-term changes in the paleointensity should be expected as a result of the Earth's thermal history and inner core growth. Such a scenario is consistent with the undisputed fact that the field has always been of the same order of magnitude and displaying the same kind of fluctuations, since at least 3.2 Gy ago (the earliest record Tarduno et al. 2007).

5 Concluding Remarks

As must now be obvious to the reader, the past decade has been a remarkable period of progress in our knowledge and understanding of the magnetic field of the Earth. This we largely owe to the considerable success of recent magnetic satellite missions, to the emergence of new archeomagnetic and paleomagnetic measurement techniques, to the development of advanced geomagnetic, archeomagnetic and paleomagnetic modeling methods, and to the coming of age of fully consistent numerical dynamo simulations. Of course, much remains to be done, and many issues are still open.

From an observational point of view, there is little doubt that a lot could be learned from more, ideally uninterrupted, magnetic observations from space, combined with synchronous observations from an improved network of ground based observatories. Such observations are crucial for better characterizing and understanding the sometimes surprisingly fast changing phenomena that occur within the core. Fortunately, more such missions are now scheduled (such as the ESA *Swarm* mission, Friis-Christensen et al. 2006, 2009) or in preparation, and more observatories are joining the INTERMAGNET worldwide network standards.

Technical progress is also under way in archeomagnetism and paleomagnetism, and databases are continually building up, thanks to patient collective endeavors. Such improvements are still badly needed to improve geographical coverage, to enable better documentation of field variations at all time scales, to enable more reliable reconstructions of the detailed behavior of the field during excursions and reversals, and to better assess the influence of the heterogeneous mantle. A major challenge is also to further improve our understanding of paleointensity data and of the temporal filtering involved in sedimentary data, not to mention the permanent need to improve the time control on the age of those samples.

From a field modeling point of view the next step, already initiated (recall Sect. 4.1.3), is to introduce dynamical considerations in the currently purely descriptive time varying spherical harmonic models of the geomagnetic and archeomagnetic fields, using e.g. data assimilation techniques. This is crucial to better taking advantage of the recent increase in data quality, and for improved understanding of the short to medium-term dynamics inside the core. Also of importance in relation to better understanding the links between decadal field fluctuations and variations in Earth's rotation, is the still unclear role of gravitational coupling between the inner core and the mantle. Gravitational coupling may both prevent the inner core from freely rotating within the outer core on the long term and be an important ingredient in the short-term dynamics of the core (see Buffett 2007 for a recent discussion).

Possibly just as important is the fact, that we did not discuss, that the Earth is not an isolated body within the solar system. Its mechanical interactions with the Moon and the Sun, but also with other planets, are responsible for phenomena such as tides and precession which affect the coupled (because of the slight ellipticity of the structure of the Earth) inner-core—core-mantle system. It has indeed been suggested that the energy provided by such precession phenomena could play a significant role in the geodynamo process (see e.g. Tilgner 2007 for a recent review), and it may even be that some resonance arose in the geological past, which could have very significantly affected the geodynamo (Greff-Lefftz and Legros 1999). Evidence for possible orbital influence in the magnetic field data (in sediment data) has been searched for, but only with rather contradicting results so far (see e.g. Roberts et al. 2003; Xuan and Channell 2008; Thouveny et al. 2008). The manner in which precession and related phenomena (such as so-called parametric instabilities, see e.g. Aldridge and Baker 2003) could interfere with the main geodynamo process (driven by thermo-chemical convection) needs to be assessed more precisely.

More progress is also to be expected in the numerical simulation of the geodynamo. Running simulations with more Earth-like control parameters remains a permanent challenge, particularly because of the extremely high resolution required by the very small viscosity of the outer core liquid iron. Several avenues are currently being explored, with some taking advantage of the latest advances in computer power (e.g. Kageyama et al. 2008), while others look for advanced ways of dealing with unresolved length-scales (e.g. Matsui and Buffett 2005, 2007; Matsushima 2006), or investigate whether using more appropriate boundary conditions could be important with improving numerical resolution (Sakuraba and Roberts 2009). Further exploration and development of dynamo scaling laws derived from simulations with more moderate numerical resolution is also an interesting avenue (e.g. Christensen 2009).

But perhaps the next great advances will come instead from laboratory experiments. Although running such experiments is very challenging, a number of teams now have operational set-ups that have already brought very interesting results (see the recent reviews of Cardin and Olson 2007 and Verhille et al. 2009). Some of these have succeeded at producing magnetic fields from energy drawn from conducting fluids. But the way this has been achieved so far is by mechanically forcing the conducting fluid (liquid sodium) through well-designed pipes (e.g. Gailitis et al. 2001; Stieglitz and Müller 2001), or by using counter-rotating propellers in a cylinder (Monchaux et al. 2007). No experimental dynamo has yet been found that directly compares to the geodynamo (though some experiments already brought interesting insight into the way reversals can occur, see Berhanu et al. 2007; Petrelis et al. 2009). Whether fast-rotating, convecting, experimental dynamos can be achieved some day remains an open question. Nonetheless other set-ups are available that already make it possible to investigate the specific dynamics of fast rotating flows (in particular in connection with precession, see Tilgner 2007 for a review), sometimes in the presence of thermal convection (see Cardin and Olson 2007 for a review), and with imposed magnetic fields (e.g. Brito et al. 1995; Aurnou and Olson 2001; Nataf et al. 2006, 2008; Gillet et al. 2007b;

Schmitt et al. 2008). Such experiments bring very interesting insight into the dynamics relevant for the investigation of the historical field behavior (recall Sect. 4.1.2).

No doubt the combination of results from such improved experiments, more advanced numerical simulations, and better modeling of more high-quality observations promises to make the next decade just as productive as the one covered by the present review.

Acknowledgements The authors are very grateful to Nicolas Gillet for his very thorough review of the present paper and to Andre Balogh, Roger-Maurice Bonnet, and the International Space Science Institute for inviting them to take part in the Planetary Magnetism Workshop held in Bern in September 2008. CC acknowledges support from US National Science Foundation Geophysics Program Grants EAR-0537986 and EAR-0809709. This is IPGP contribution 2602.

References

D.J. Acheson, R. Hide, Rep. Prog. Phys. **36**, 159–221 (1973)

K. Aldridge, R. Baker, Phys. Earth Planet. Int. **140**(1–3), 91–100 (2003). doi:10.1016/j.pepi.2003.07.005

M. Alexandrescu, V. Courtillot, J.L. Le Mouël, Phys. Earth Planet. Int. **98**, 321–360 (1996a)

M. Alexandrescu, D. Gibert, G. Hulot, J.L. Le Mouël, G. Saracco, J. Geophys. Res. **101**(B10), 21975–21994 (1996b)

C.J. Allègre, G. Manhès, C. Göpel, Geochim. Cosmochim. Acta **59**(8), 1445–1456 (1995)

C.J. Allègre, G. Manhès, C. Göpel, Earth Planet. Sci. Lett. **267**(1–2), 386–398 (2008). doi:10.1016/j.epsl.2007.11.056

H. Amit, P. Olson, Phys. Earth Planet. Int. **155**, 120–139 (2006)

H. Amit, J. Aubert, G. Hulot, P. Olson, Earth Planets Space **60**(8), 845–854 (2008)

H. Amit, J. Aubert, G. Hulot, J. Geophys. Res. (2010). doi:1029/2009JB006542

J. Aubert, H. Amit, G. Hulot, Phys. Earth Planet. Int. **160**, 143–156 (2007)

J. Aubert, J. Aurnou, J. Wicht, Geophys. J. Int. **172**(3), 945–956 (2008a). doi:10.1111/j.1365-246X.2007.03693.x

J. Aubert, H. Amit, G. Hulot, P. Olson, Nature **454**(7205), 758–80 (2008b). doi:10.1038/nature07109

J. Aubert, S. Labrosse, C. Poitou, Geophys. J. Int. **179**, 1414–1428 (2009). doi:10.1111/j.1365-246X.2009.04361.x

J.M. Aurnou, P.L. Olson, J. Fluid Mech. **430**, 283–307 (2001)

H. Auster, M. Mandea, A. Hemshorn, E. Pulz, M. Korte, Earth Planets Space **59**(9), 1007–1014 (2007)

V. Auster, O. Hillenmaier, R. Kroth, M. Weideman, in *XIIth IAGA Workshop on Geomagnetic Observatory Instruments, Data Acquisition and Processing*, Abstract volume 21, 2006

G. Backus, Philos. Trans. R. Soc. Lond. A **263**, 239–266 (1968)

G.E. Backus, J. Geophys. Res. **75**(31), 6339–6341 (1970)

G.E. Backus, Geophys. J. **92**, 125–142 (1988)

D. Barraclough, Trans. R. Soc. Edinb. Earth Sci. **85**, 239–252 (1995)

C.E. Barton, J. Geomagn. Geoelectr. **49**, 123–148 (1997)

C.E. Barton, P.L. McFadden, Earth Planet. Sci. Lett. **140**(1–4), 147–157 (1996)

W. Baumjohann, M. Blanc, A. Fedorov, K.H. Glassmeier, Space Sci. Rev. (2010). doi:10.1007/s11214-010-9629-z

E. Bellanger, J.L. Le Mouël, M. Mandea, S. Labrosse, Phys. Earth Planet. Int. **124**(1–2), 95–103 (2001)

E. Ben-Yosef, L. Tauxe, H. Ron, A. Agnon, U. Avner, M. Najjar, T.E. Levy, J. Archeol. Sci. **35**(11), 2863–2879 (2008a). doi:10.1016/j.jas.2008.05.016

E. Ben-Yosef, H. Ron, L. Tauxe, A. Agnon, A. Genevey, T.E. Levy, U. Avner, M. Najjar, J. Geophys. Res. **113**(B8), B08101 (2008b). doi:10.1029/2007JB005235

E. Ben-Yosef, L. Tauxe, T.E. Levy, R. Shaar, H. Ron, M. Najjar, Earth Planet. Sci. Lett. **287**(3–4), 529–539 (2009). doi:10.1016/j.epsl.2009.09.001

M. Berhanu, R. Monchaux, S. Fauve, N. Mordant, F. Petrelis, A. Chiffaudel, F. Daviaud, B. Dubrulle, L. Marie, F. Ravelet, M. Bourgoin, P. Odier, J.F. Pinton, R. Volk, Europhys. Lett. **77**(5), 59001 (2007). doi:10.1209/0295-5075/77/59001

J. Besse, V. Courtillot, J. Geophys. Res. **107**(B11), 2300 (2002). doi:10.1029/2000JB000050
A.J. Biggin, G.H.M.A. Strik, C.G. Langereis, Nat. Geosci. **1**(6), 395–398 (2008a). doi:10.1038/ngeo181
A.J. Biggin, D.J.J. van Hinsbergen, C.G. Langereis, G.B. Straathof, M.H.L. Deenen, Phys. Earth Planet. Int. **169**(1–4, Sp. Iss.), 3–19 (2008b). doi:10.1016/j.pepi.2008.07.004
A.J. Biggin, G.H.M.A. Strik, C.G. Langereis, Earth Planets Space **61**(1), 9–22 (2009)
J. Bloxham, J. Geophys. Res. **92**(B11), 11597–11608 (1987)
J. Bloxham, Philos. Trans. R. Soc. Lond. A **358**(1768), 1171–1179 (2000)
J. Bloxham, Geophys. Res. Lett. **29**(18), 1854 (2002). doi:10.1029/2001GL014543
J. Bloxham, A. Jackson, J. Geophys. Res. **94**(B11), 15753–15769 (1989)
J. Bloxham, A. Jackson, Rev. Geophys. **29**, 97–120 (1991)
J. Bloxham, A. Jackson, J. Geophys. Res. **97**, 19537–19563 (1992)
J. Bloxham, D. Gubbins, A. Jackson, Philos. Trans. R. Soc. Lond. A **329**(1606), 415–502 (1989)
J. Bloxham, S. Zatman, M. Dumberry, Nature **420**, 685–687 (2002)
C. Bouligand, G. Hulot, A. Khokhlov, G.A. Glatzmaier, Geophys. J. Int. **161**(3), 603–626 (2005). doi:10.1111/j.1365-246X.2005.02613.x
C. Bouligand, J. Dyment, Y. Gallet, G. Hulot, Earth Planet. Sci. Lett. **250**(3–4), 541–560 (2006). doi:10.1016/j.epsl.2006.06.051
N.E. Bowers, S.C. Cande, J.S. Gee, J.A. Hildebrand, R.L. Parker, J. Geophys. Res. **106**(B11), 26379–26396 (2001)
S.I. Braginsky, Geomagn. Aeron. **4**, 698–712 (1964)
S.I. Braginsky, Geomagn. Aeron. **7**, 851–859 (1967)
S.I. Braginsky, Geomagn. Aeron. **10**, 1–8 (1970)
S.I. Braginsky, Geophys. Astrophys. Fluid Dyn. **30**, 1–78 (1984)
D. Brito, P. Cardin, H.C. Nataf, G. Marolleau, Phys. Earth Planet. Int. **91**(1–3), 77–98 (1995)
M.C. Brown, R. Holme, A. Bargery, Geophys. J. Int. **168**(2), 541–550 (2007). doi:10.1111/j.1365-246X.2006.03234.x
B.A. Buffett, Science **299**(5613), 1675–1677 (2003)
B.A. Buffett, in *Treatise on Geophysics*, vol. 8, ed. by P. Olson (Elsevier, Amsterdam, 2007)
B.A. Buffett, H.E. Huppert, J.R. Lister, A.W. Woods, Nature **356**(6367), 329–331 (1992)
B.A. Buffett, H.E. Huppert, J.R. Lister, A.W. Woods, J. Geophys. Res. **101**(B4), 7989–8006 (1996)
B. Buffett, J. Mound, A. Jackson, Geophys. J. Int. **177**, 878–890 (2009)
F.H. Busse, J. Fluid Mech. **44**, 441 (1970)
L. Cafarella, A. De Santis, A. Meloni, Phys. Earth Planet. Int. **73**, 206–221 (1992)
J.C. Cain, B.B. Ferguson, D. Mozzoni, J. Geophys. Res. **108**, 2 (2003). doi:10.1029/2000JE001487
S.C. Cande, D.V. Kent, J. Geophys. Res. **100**(B4), 6093–6095 (1995)
E. Canet, A. Fournier, D. Jault, J. Geophys. Res. **114**(B11), 11101 (2009). doi:10.1029/2008JB006189
A. Cao, B. Romanowicz, Earth Planet. Sci. Lett. **228**(3–4), 243–253 (2004). doi:10.1016/j.epsl.2004.09.032
P. Cardin, P. Olson, in *Treatise on Geophysics*, vol. 8, ed. by P. Olson (Elsevier, Amsterdam, 2007)
J. Carlut, V. Courtillot, Geophys. J. Int. **134**(2), 527–544 (1998)
J. Carlut, D.V. Kent, Earth Planet. Sci. Lett. **183**(3–4), 389–401 (2000)
J.E.T. Channell, B. Lehman, Nature **389**(6652), 712–715 (1997)
J.E.T. Channell, E. Erba, M. Nakanishi, T. Tamaki, in *Geochronology, Timescales, and Stratigraphic Correlation*, ed. by W.A. Berggren, D.V. Kent, M. Aubry, J. Hardenbol (SEPM, Tulsa, 1995), pp. 51–64
S. Chapman, J. Bartels, *Geomagnetism*, vol. I, II (Clarendon Press, Oxford, 1940)
A. Chauvin, H. Perroud, M.L. Bazhenov, Geophys. J. Int. **126**(2), 303–313 (1996)
U.R. Christensen, Space Sci. Rev. (2009). doi:10.1007/s11214-009-9553-2
U.R. Christensen, J. Aubert, Geophys. J. Int. **166**(1), 97–114 (2006). doi:10.1111/j.1365-246X.2006.03009.x
U.R. Christensen, P. Olson, Phys. Earth Planet. Int. **138**(1), 39–54 (2003). doi:10.1016/S0031-9201(03)00064-5
U.R. Christensen, J. Wicht, in *Treatise on Geophysics*, vol. 8, ed. by P. Olson (Elsevier, Amsterdam, 2007)
A. Chulliat, Geophys. J. Int. **157**(2), 537–552 (2004). doi:10.1111/j.1365-246X.2004.02216.x
A. Chulliat, G. Hulot, Phys. Earth Planet. Int. **117**(1–4), 309–328 (2000)
A. Chulliat, G. Hulot, Geophys. J. Int. **147**(2), 237–246 (2001)
A. Chulliat, N. Olsen, J. Geophys. Res. (2009). doi:10.1029/2009JB006994
B.M. Clement, Earth Planet. Sci. Lett. **104**(1), 48–58 (1991)
B.M. Clement, Nature **428**(6983), 637–640 (2004). doi:10.1038/nature02459
R.S. Coe, G.A. Glatzmaier, Geophys. Res. Lett. **33**(21), L21311 (2006). doi:10.1029/2006GL027903
R.S. Coe, J.M.G. Glen, in *Timescales of the Paleomagnetic Field*, ed. by J.E.T. Channell, D.V. Kent, W. Lowrie, J.G. Meert, Geophysical Monograph Series, vol. 145 (2004), pp. 221–232. 0-87590-410-6. doi:10.1029/145GM16
R.S. Coe, L. Hongre, G.A. Glatzmaier, Philos. Trans. R. Soc. Lond. A **358**(1768), 1141–1170 (2000)

J.P. Cogne, J. Francheteau, V. Courtillot, R. Armijo, M. Constantin, J. Girardeau, R. Hekinian, R. Hey, D.F. Naar, R. Searle, Earth Planet. Sci. Lett. **136**(3–4), 213–222 (1995)
J.P. Cogne, N. Halim, Y. Chen, V. Courtillot, J. Geophys. Res. **104**(B8), 17715–17734 (1999)
J. Connerney, in *Treatise on Geophysics*, vol. 10, ed. by T. Spohn (Elsevier, Amsterdam, 2008). Chap. 7
C. Constable, Phys. Earth Planet. Int. **118**(3–4), 181–193 (2000)
C. Constable, C. Johnson, Phys. Earth Planet. Int. **153**(1–3, Sp. Iss.), 61–73 (2005). doi:10.1016/j.pepi.2005.03.015
C.G. Constable, in *Treatise on Geophysics*, vol. 5, ed. by M. Kono (Elsevier, Amsterdam, 2007)
C.G. Constable, C.L. Johnson, Phys. Earth Planet. Int. **115**(1), 35–51 (1999)
C. Constable, M. Korte, Earth Planet. Sci. Lett. **246**(1–2), 1–16 (2006). doi:10.1016/j.epsl.2006.03.038
C.G. Constable, R.L. Parker, J. Geophys. Res. **93**(B10), 11569–11581 (1988)
C.G. Constable, R.L. Parker, P.B. Stark, Geophys. J. Int. **113**(2), 419–433 (1993)
C.G. Constable, L. Tauxe, R.L. Parker, J. Geophys. Res. **103**(B8), 17735–17748 (1998)
C.G. Constable, C.L. Johnson, S.P. Lund, Philos. Trans. R. Soc. Lond. A **358**, 991–1008 (2000)
R.D. Cottrell, J.A. Tarduno, Earth Planet. Sci. Lett. **169**(1–2), 1–5 (1999)
R.D. Cottrell, J.A. Tarduno, J. Roberts, Phys. Earth Planet. Int. **169**(1–4, Sp. Iss.), 49–58 (2008). doi:10.1016/j.pepi.2008.07.041
V. Courtillot, J. Besse, Science **237**(4819), 1140–1147 (1987)
V. Courtillot, J. Besse, in *Timescales of the Paleomagnetic Field*, ed. by J. E.T. Channell, D.V. Kent, W. Lowrie, J.G. Meert, Geophysical Monograph Series, vol. 145 (2004), pp. 59–74. 0-87590-410-6. doi:10.1029/145GM05
V. Courtillot, J.L. Le Mouël, Rev. Geophys. **45**(3), RG3008 (2007). doi:10.1029/2008GC002295
V. Courtillot, P. Olson, Earth Planet. Sci. Lett. **260**(3–4), 495–504 (2007). doi:10.1016/j.epsl.2007.06.003
V. Courtillot, J. Ducruix, J.L. Le Mouël, C. R. Acad. Sci. D **287**(12), 1095–1098 (1978)
A. Cox, J. Geophys. Res. **73**(10), 3247 (1968)
A. Cox, Geophys. J. R. Astron. Soc. **20**(3), 253 (1970)
M. Cronin, L. Tauxe, C. Constable, P. Selkin, T. Pick, Earth Planet. Sci. Lett. **190**(1–2), 13–30 (2001)
C.J. Davies, D. Gubbins, A.P. Willis, P.K. Jimack, Phys. Earth Planet. Int. **169**(1–4, Sp. Iss.), 194–203 (2008). doi:10.1016/j.pepi.2008.07.021
C. De Boor, *A Practical Guide to Splines* (Springer, New York, 2001)
C. DeMets, R.G. Gordon, D.F. Argus, S. Stein, Geophys. Res. Lett. **21**(20), 2191–2194 (1994)
B. Desjardins, E. Dormy, E. Grenier, Phys. Earth Planet. Int. **123**(1), 15–26 (2001)
F. Donadini, K. Korthonen, P. Riisager, L. Pesonen, EOS Trans. AGU **87**(14), 137 (2006)
F. Donadini, M. Korte, C. Constable, Geochem. Geophys. Geosyst. **10**(6), Q06007 (2009). doi:10.1029/2008GC002295
E. Dormy, J.P. Valet, V. Courtillot, Geochem. Geophys. Geosyst. **1**(10), 2000GC000062 (2000)
P. Driscoll, P. Olson, Geophys. J. Int. **178**(3), 1337–1350 (2009a). doi:10.1111/j.1365-246X.2009.04234.x
P. Driscoll, P. Olson, Earth Planet. Sci. Lett. **282**(1–4), 24–33 (2009b). doi:10.1016/j.epsl.2009.02.017
M. Dumberry, Geophys. J. Int. **172**(3), 903–920 (2008a). doi:10.1111/j.1365-246X.2007.03653.x
M. Dumberry, in *Les Houches Preceedings: Session LXXXVIII—Dynamos*, ed. by P. Cardin, L. Cugliandolo (Elsevier, Amsterdam, 2008b), pp. 383–401
M. Dumberry, J. Bloxham, Geophys. J. Int. **165**(1), 32–46 (2006). doi:10.1111/j.1365-246X.2006.02903.x
M. Dumberry, C.C. Finlay, Earth Planet. Sci. Lett. **254**(1–2), 146–157 (2007). doi:10.1016/j.epsl.2006.11.026
D.J. Dunlop, Y. Yu, in *Timescales of the Paleomagnetic Field*, ed. by J.E.T. Channell, D.V. Kent, W. Lowrie, J.G. Meert, Geophysical Monograph Series, vol. 145 (2004), pp. 85–100. 0-87590-410-6. doi:10.1029/145GM07
D. Dunlop, Ö. Özdemir, in *Treatise on Geophysics*, vol. 5, ed. by M. Kono (Elsevier, Amsterdam, 2007), pp. 277–336
D.P. Elston, R.J. Enkin, J. Baker, D.K. Kisilevsky, Bull. Geol. Soc. Am. **114**(5), 619–638 (2002)
C. Eymin, G. Hulot, Phys. Earth Planet. Int. **152**(3), 200–220 (2005)
C.C. Finlay, Phys. Earth Planet. Int. **170**, 1–14 (2008a). doi:10.1016/j.pepi.2008.06.029
C.C. Finlay, in *Les Houches Proceedings: Session LXXXVIII—Dynamos*, ed. by P. Cardin, L. Cugliandolo (Elsevier, Amsterdam, 2008b), pp. 407–447
C.C. Finlay, A. Jackson, Science **300**, 2084–2086 (2003)
A. Fournier, C. Eymin, T. Alboussiere, Nonlinear Process. Geophys. **14**(2), 163–180 (2007)
A. Fournier, G. Hulot, D. Jault, W. Kuang, A. Tangborn, N. Gillet, E. Canet, J. Aubert, F. Lhuillier, Space Sci. Rev. (2010)
E. Friis-Christensen, H. Lühr, G. Hulot, Earth Planets Space **58**, 351–358 (2006)

E. Friis-Christensen, H. Lühr, G. Hulot, R. Haagmans, M. Purucker, EOS Trans. AGU **90**(25), 213–215 (2009)
A. Gailitis, O. Lielausis, E. Platacis, S. Dement'ev, A. Cifersons, G. Gerbeth, T. Gundrum, F. Stefani, M. Christen, G. Will, Phys. Rev. Lett. **86**(14), 3024–3027 (2001)
Y. Gallet, G. Hulot, Geophys. Res. Lett. **24**(15), 1875–1878 (1997)
Y. Gallet, V.E. Pavlov, M.A. Semikhatov, P.Y. Petrov, J. Geophys. Res. **105**(B7), 16481–16499 (2000)
Y. Gallet, A. Genevey, M. Le Goff, Phys. Earth Planet. Int. **131**(1), 81–89 (2002)
Y. Gallet, A. Genevey, V. Courtillot, Earth Planet. Sci. Lett. **214**(1–2), 237–242 (2003). doi:10.1016/S0012-821X(03)00362-5
Y. Gallet, G. Hulot, A. Chulliat, A. Genevey, Earth Planet. Sci. Lett. **284**(1–2), 179–186 (2009). doi:10.1016/j.epsl.2009.04.028
J.S. Gee, D.V. Kent, in *Treatise on Geophysics*, vol. 5, ed. by M. Kono (Elsevier, Amsterdam, 2007)
J.S. Gee, S.C. Cande, J.A. Hildebrand, K. Donnelly, R.L. Parker, Nature **408**(6814), 827–832 (2000)
A. Genevey, Y. Gallet, J. Geophys. Res. **107**(B11), 2285 (2002)
A. Genevey, Y. Gallet, G. Boudon, Earth Planet. Sci. Lett. **201**(2), 369–382 (2002)
A. Genevey, Y. Gallet, J.C. Margueron, J. Geophys. Res. **108**(B5), 2228 (2003). doi:10.1029/2001JB001612
A. Genevey, Y. Gallet, C.G. Constable, M. Korte, G. Hulot, Geochem. Geophys. Geosyst. **9**(4), Q04038 (2008). doi:10.1029/2007GC001881
A. Genevey, Y. Gallet, J. Rosen, M. Le Goff, Earth Planet. Sci. Lett. **284**(1–2), 132–143 (2009). doi:10.1016/j.epsl.2009.04.024
S. Gilder, Y. Chen, J.P. Cogne, X.D. Tan, V. Courtillot, D.J. Sun, Y.G. Li, Earth Planet. Sci. Lett. **206**(3–4), 587–600 (2003)
N. Gillet, A. Jackson, C.C. Finlay, Geophys. J. Int. **171**(3), 1005–1016 (2007a)
N. Gillet, D. Brito, D. Jault, H.C. Nataf, J. Fluid Mech. **580**, 123–143 (2007b). doi:10.1017/S0022112007005289
N. Gillet, M.A. Pais, D. Jault, Geochem. Geophys. Geosyst. **10**(6), Q06004 (2009a). doi:10.1029/2008GC002290
N. Gillet, V. Lesur, N. Olsen, Space Sci. Rev. (2009b). doi:10.1007/s11214-009-9586-6
G.A. Glatzmaier, R.S. Coe, in *Treatise on Geophysics*, vol. 8, ed. by P. Olson (Elsevier, Amsterdam, 2007)
G.A. Glatzmaier, P.H. Roberts, Phys. Earth Planet. Int. **91**(1–3), 63–75 (1995)
G.A. Glatzmaier, R.S. Coe, L. Hongre, P.H. Roberts, Nature **401**(6756), 885–890 (1999)
F.M. Gradstein, F.P. Agterberg, J.G. Ogg, S. Hardenbol, P. Vanveen, J. Thierry, Z.H. Huang, J. Geophys. Res. **99**(B12), 24051–24074 (1994)
O. Gravrand, A. Khokhlov, J.L. Le Mouël, J.M. Léger, Earth Planets Space **53**, 949–958 (2001)
M. Greff-Lefftz, H. Legros, Science **286**(5445), 1707–1709 (1999)
D. Gubbins, Nature **326**(6109), 167–169 (1987)
D. Gubbins, Phys. Earth Planet. Int. **68**(1–2), 170–182 (1991)
D. Gubbins, Phys. Earth Planet. Int. **98**, 193–206 (1996)
D. Gubbins, Geophys. J. Int. **137**(1), 1–3 (1999)
D. Gubbins, P. Kelly, Nature **365**(6449), 829–832 (1993)
D. Gubbins, N. Roberts, Geophys. J. R. Astron. Soc. **73**, 675–687 (1983)
D. Gubbins, P.H. Roberts, Geomagnetism **2**, 1–183 (1987)
D. Gubbins, T.G. Masters, J.A. Jacobs, Geophys. J. R. Astron. Soc. **59**(1), 57–99 (1979)
D. Gubbins, A. Jones, C.C. Finlay, Science **312**, 900–902 (2006)
D. Gubbins, A.P. Willis, B. Sreenivasan, Phys. Earth Planet. Int. **162**(3–4), 256–260 (2007). doi:10.1016/j.pepi.2007.04.014
Y. Guyodo, J.P. Valet, Earth Planet. Sci. Lett. **143**(1–4), 23–36 (1996)
Y. Guyodo, J.P. Valet, Nature **399**(6733), 249–252 (1999)
M.M. Haldan, C.G. Langereis, A.J. Biggin, M.J. Dekkers, M.E. Evans, Geophys. J. Int. **177**(3), 834–848 (2009). doi:10.1111/j.1365-246X.2009.04124.x
C.J. Hale, Nature **329**(6136), 233–237 (1987)
H.C. Halls, Earth Planet. Sci. Lett. **105**(1–3), 279–292 (1991)
F. Hankard, J.P. Cogne, V.A. Kravchinsky, L. Carporzen, A. Bayasgalan, P. Lkhagvadorj, J. Geophys. Res. **112**(B2), 02101 (2007). doi:10.1029/2006JB004488
R. Heller, R.T. Merrill, P.L. McFadden, Phys. Earth Planet. Int. **131**(3–4), 237–249 (2002)
R. Heller, R.T. Merrill, P.L. McFadden, Phys. Earth Planet. Int. **135**(2–3), 211–223 (2003). doi:10.1016/S0031-9201(03)00002-5
R. Hide, Philos. Trans. R. Soc. Lond. A **259**, 615–647 (1966)
R. Hide, D.H. Boggs, J.O. Dickey, D. Dong, R.S. Gross, A. Jackson, Geophys. J. Int. **125**(2), 599–607 (1996)
K.A. Hoffman, Nature **359**(6398), 789–794 (1992)
R. Hollerbach, C.A. Jones, Nature **365**(6446), 541–543 (1993)

R. Hollerbach, C.A. Jones, Phys. Earth Planet. Int. **87**(3–4), 171–181 (1995)
R. Holme, Geophys. J. Int. **132**(1), 167–180 (1998)
R. Holme, in *Treatise on Geophysics*, vol. 8, ed. by P. Olson (Elsevier, Amsterdam, 2007), pp. 107–130
R. Holme, O. de Viron, Geophys. J. Int. **160**(2), 435–439 (2005). doi:10.1111/j.1365-246X.2004.02510.x
R. Holme, N. Olsen, Geophys. J. Int. **166**(2), 518–528 (2006). doi:10.1111/j.1365-246X.2006.03033.x
R. Holme, M.A. James, H. Lühr, Earth Planets Space **57**, 1203–1209 (2005)
L. Hongre, G. Hulot, A. Khokhlov, Phys. Earth Planet. Int. **106**, 311–335 (1998)
Y. Hueda-Tanabe, A.M. Soler-Arechalde, J. Urrutia-Fucugauchi, L. Barba, L. Manzanilla, M. Rebolledo-Vieyra, A. Goguitchaichvili, Phys. Earth Planet. Int. **147**(2–3), 269–283 (2004). doi:10.1016/j.pepi.2004.06.006
G. Hulot, Nat. Geosci. **1**(6), 345–346 (2008). doi:10.1038/ngeo213
G. Hulot, C. Bouligand, Geophys. J. Int. **161**(3), 591–602 (2005). doi:10.1111/j.1365-246X.2005.02612.x
G. Hulot, A. Chulliat, Phys. Earth Planet. Int. **135**(1), 47–54 (2003)
G. Hulot, Y. Gallet, Phys. Earth Planet. Int. **95**(1–2), 37–53 (1996)
G. Hulot, Y. Gallet, Earth Planet. Sci. Lett. **210**(1–2), 191–201 (2003). doi:10.1016/S0012-821X(03)00130-4
G. Hulot, J.L. Le Mouël, Phys. Earth Planet. Int. **82**, 167–183 (1994). doi:10.1016/0031-9201(94)90070-1
G. Hulot, J.L. Le Mouël, D. Jault, J. Geomagn. Geoelectr. **42**(7), 857–874 (1990)
G. Hulot, J.L. Le Mouël, J. Wahr, Geophys. J. Int. **108**, 224–246 (1992)
G. Hulot, M. Le Huy, J.L. Le Mouël, C. R. Acad. Sci. II **317**(3), 333–341 (1993)
G. Hulot, M. Le Huy, J.L. Le Mouël, Geophys. Astrophys. Fluid Dyn. **82**(1–2), 35–67 (1996)
G. Hulot, A. Khokhlov, J.L. Le Mouël, Geophys. J. Int. **129**, 347–354 (1997)
G. Hulot, C. Eymin, B. Langlais, M. Mandea, N. Olsen, Nature **416**, 620–623 (2002)
G. Hulot, T.J. Sabaka, N. Olsen, in *Treatise on Geophysics*, vol. 5, ed. by M. Kono (Elsevier, Amsterdam, 2007)
G. Hulot, N. Olsen, E. Thébault, K. Hemant, Geophys. J. Int. **177**, 361–366 (2009a). doi:10.1111/j.1365-246X.2009.04119.x
G. Hulot, F. Lhuillier, J. Aubert, Geophys. Res. Lett. (2009b). doi:10.129/2009GL041869
M. Ingham, G. Turner, Phys. Earth Planet. Int. **168**(3–4), 163–178 (2008). doi:10.1016/j.pepi.2008.06.008
A. Jackson, Geophys. J. Int. **119**, 991–998 (1994)
A. Jackson, Philos. Trans. R. Soc. Lond. A **452**(1953), 2195–2201 (1996)
A. Jackson, Phys. Earth Planet. Int. **103**, 293–311 (1997)
A. Jackson, Nature **424**, 760–763 (2003)
A. Jackson, C.C. Finlay, in *Treatise on Geophysics*, vol. 5, ed. by M. Kono (Elsevier, Amsterdam, 2007)
A. Jackson, R. Hide, Geophys. J. Int. **125**(3), 925–927 (1996)
A. Jackson, J. Bloxham, D. Gubbins, in *Dynamics of Earth's Deep Interior and Earth Rotation*, Geophysical Monograph, vol. 72 (1993), pp. 97–107
A. Jackson, A.R.T. Jonkers, M.R. Walker, Philos. Trans. R. Soc. Lond. A **358**, 957–990 (2000)
A. Jackson, C. Constable, N. Gillet, Geophys. J. Int. **171**(3), 995–1004 (2007a). doi:10.1111/j.1365-246X.2007.03530.x
A. Jackson, C.G. Constable, M.R. Walker, R.L. Parker, Geophys. J. Int. **171**(1), 133–144 (2007b). doi:10.1111/j.1365-246X.2007.03526.x
J.A. Jacobs, Nature **172**, 297–300 (1953)
J.A. Jacobs, in *Encyclopedia of Geomagnetism and Paleomagnetism*, ed. by D. Gubbins, E. Herrero-Bervera (Springer, Heidelberg, 2007)
J. Jankowski, C. Sucksdorff, *IAGA Guide for Magnetic Measurements and oBservatory Practice* (IAGA, Warszawa, 1996)
D. Jault, Phys. Earth Planet. Int. **166**(1–2), 67–76 (2008)
D. Jault, J.L. Le Mouël, J. Geomagn. Geoelectr. **43**(2), 111–129 (1991a)
D. Jault, J.L. Le Mouël, Phys. Earth Planet. Int. **68**(1–2), 76–84 (1991b)
D. Jault, C. Gire, J.L. Le Mouël, Nature **333**, 353–356 (1988)
C.L. Johnson, C.G. Constable, Geophys. J. Int. **122**(2), 489–519 (1995)
C.L. Johnson, C.G. Constable, Philos. Trans. R. Soc. Lond. A **354**(1704), 89–141 (1996)
C.L. Johnson, C.G. Constable, Geophys. J. Int. **131**(3), 643 (1997)
C.L. Johnson, C.G. Constable, Geophys. Res. Lett. **25**(7), 1011–1014 (1998)
C.L. Johnson, P. McFadden, in *Treatise on Geophysics*, vol. 5, ed. by M. Kono (Elsevier, Amsterdam, 2007)
C.L. Johnson, C.G. Constable, L. Tauxe, R. Barendregt, L.L. Brown, R.S. Coe, P. Layer, V. Mejia, N.D. Opdyke, B.S. Singer, H. Staudigel, D.B. Stone, Geochem. Geophys. Geosyst. **9**(4), Q04032 (2008). doi:10.1029/2007GC001696
G.M. Jones, J. Geophys. Res. **82**(11), 1703–1709 (1977)
A.R.T. Jonkers, Geophys. J. Int. **171**(2), 581–593 (2007). doi:10.1111/j.1365-246X.2007.03551.x
A.R.T. Jonkers, A. Jackson, A. Murray, Rev. Geophys. **41**(2), 1006 (2003). doi:10.1029/2002RG000115

M.T. Juarez, L. Tauxe, Earth Planet. Sci. Lett. **175**(3–4), 169–180 (2000)
M.T. Juarez, L. Tauxe, J.S. Gee, T. Pick, Nature **394**(6696), 878–881 (1998)
A. Kageyama, T. Miyagoshi, T. Sato, Nature **454**(7208), 1106–1109 (2008). doi:10.1038/nature07227
P. Kelly, D. Gubbins, Geophys. J. Int. **128**(2), 315–330 (1997)
D.V. Kent, F.M. Gradstein, in *The Geology of North America*, vol. M, ed. by P.R. Vogt, B.E. Tucholke (Geological Society of America, Boulder, 1986), pp. 45–50
D.V. Kent, M.A. Smethurst, Earth Planet. Sci. Lett. **160**(3–4), 391–402 (1998)
A. Khokhlov, G. Hulot, J.L. Le Mouël, Geophys. J. Int. **130**, 701–703 (1997)
A. Khokhlov, G. Hulot, J.L. Le Mouël, Geophys. J. Int. **137**, 816–820 (1999)
A. Khokhlov, G. Hulot, J. Carlut, Geophys. J. Int. **145**(1), 157–171 (2001)
A. Khokhlov, G. Hulot, C. Bouligand, Geophys. J. Int. **167**(2), 635–648 (2006). doi:10.1111/j.1365-246X.2006.03133.x
K.P. Kodama, W.W. Sun, Geophys. J. Int. **111**(3), 465–469 (1992)
M. Kono, H. Tanaka, J. Geomagn. Geoelectr. **47**(1), 115–130 (1995)
M. Kono, A. Sakuraba, M. Ishida, Philos. Trans. R. Soc. Lond. A **358**(1768), 1123–1139 (2000a)
M. Kono, H. Tanaka, H. Tsunakawa, J. Geophys. Res. **105**(B3), 5817–5833 (2000b)
K. Korhonen, F. Donadini, P. Riisager, L. Pesonen, Geochem. Geophys. Geosyst. **9**(4), Q04029 (2008). doi:10.1029/2007GC001893
M. Korte, C. Constable, Geochem. Geophys. Geosyst. **6**(1), Q02H16 (2005). doi:10.1029/2004GC000801
M. Korte, A. Genevey, C.G. Constable, U. Frank, E. Schnepp, Geochem. Geophys. Geosyst. **6**(2), Q02H15 (2005). doi:10.1029/2004GC000800
M. Korte, F. Donadini, C. Constable, Geochem. Geophys. Geosyst. **10**(6), Q06008 (2009). doi:10.1029/2008GC002297
W. Kuang, A. Tangborn, W. Jiang, D. Liu, Z. Sun, J. Bloxham, Z. Wei, Commun. Comput. Phys. **3**(1), 85–108 (2008)
C. Kutzner, U.R. Christensen, Geophys. J. Int. **157**(3), 1105–1118 (2004). doi:10.1111/j.1365-246X.2004.02309.x
S. Labrosse, M. Macouin, C. R. Geosci. **335**(1), 37–50 (2003). doi:10.1016/S1631-0713(03)00013-0
S. Labrosse, J.P. Poirier, J.L. Le Mouël, Earth Planet. Sci. Lett. **190**(3–4), 111–123 (2001)
S. Labrosse, J.W. Hernlund, N. Coltice, Nature **450**(7171), 866–869 (2007). doi:10.1038/nature06355
C. Laj, A. Mazaud, R. Weeks, M. Fuller, E. Herrero-Bervera, Nature **351**(6326), 447 (1991)
C. Laj, C. Kissel, A. Mazaud, J.E.T. Channell, J. Beer, Philos. Trans. R. Soc. Lond. A **358**(1768), 1009–1025 (2000)
C. Laj, C. Kissel, J. Beer, in *Timescales of the Paleomagnetic Field*, ed. by J.E.T. Channell, D.V. Kent, W. Lowrie, J.G. Meert, Geophysical Monograph Series, vol. 145 (2004), pp. 255–265. 0-87590-410-6. doi:10.1029/145GM19
P. Lancaster, K. Salkauskas, *Curve and Surface Fitting* (Academic Press, San Diego, 1986)
R.A. Langel, in *Geomagnetism*, vol. 1, ed. by J.A. Jacobs (Academic Press, London, 1987), pp. 249–512
R.A. Langel, R.H. Estes, Geophys. Res. Lett. **9**, 250–253 (1982)
R.A. Langel, W.J. Hinze, *The Magnetic Field of the Earth's Lithosphere: The Satellite Perspective* (Cambridge University Press, Cambridge, 1998)
C.G. Langereis, A.A.M. van Hoof, P. Rochette, Nature **358**(6383), 226–230 (1992)
B. Langlais, V. Lesur, M.E. Purucker, J.E.P. Connerney, M. Mandea, Space Sci. Rev. (2009). doi:10.1007/s11214-009-9557-y
R.L. Larson, P. Olson, Earth Planet. Sci. Lett. **107**(3–4), 437–447 (1991)
K.P. Lawrence, C.G. Constable, J.L. Johnson, Geochem. Geophys. Geosyst. **7**(7), Q07007 (2006). doi:10.1029/2005GC001181
M. Le Huy, M. Alexandrescu, G. Hulot, J.L. Mouël, Earth Planets Space **50**(9), 723–732 (1998)
S. Lee, PhD thesis, Austr. Nat. Univ., Canberra, 1983
B. Lehnert, Astrophys. J. **119**, 647–654 (1954)
J.L. Le Mouël, Nature **311**, 734–735 (1984)
R. Leonhardt, K. Fabian, Earth Planet. Sci. Lett. **253**(1–2), 172–195 (2007). doi:10.1016/j.epsl.2006.10.025
R. Leonhardt, K. Fabian, M. Winklhofer, A. Ferk, C. Laj, C. Kissel, Earth Planet. Sci. Lett. **278**(1–2), 87–95 (2009). doi:10.1016/j.epsl.2008.11.028
V. Lesur, I. Wardinski, M. Rother, M. Mandea, Geophys. J. Int. **173**, 382–294 (2008)
S. Levi, S.K. Banerjee, Earth Planet. Sci. Lett. **29**(1), 219–226 (1976)
D.E. Loper, K. McCartney, Geophys. Res. Lett. **13**(13), 1525–1528 (1986)
J.J. Love, J. Geophys. Res. **103**(B6), 12435–12452 (1998)
J.J. Love, C.G. Constable, Geophys. J. Int. **152**(3), 515–565 (2003)
F.J. Lowes, Geophys. J. R. Astron. Soc. **36**, 717–730 (1974)
F.J. Lowes, Geophys. J. R. Astron. Soc. **42**, 637–651 (1975)

W. Lowrie, D.V. Kent, in *Timescales of the Paleomagnetic Field*, ed. by J.E.T. Channell, D.V. Kent, W. Lowrie, J.G. Meert, Geophysical Monograph Series, vol. 145 (2004), pp. 117–129. 0-87590-410-6. doi:10.1029/145GM09
S. Macmillan, S. Maus, Earth Planets Space **57**, 1135–1140 (2005)
M. Macouin, J.P. Valet, J. Besse, Phys. Earth Planet. Int. **147**(2–3), 239–246 (2004). doi:10.1016/j.pepi.2004.07.003
S.R.C. Malin, E.C. Bullard, Philos. Trans. R. Soc. Lond. A **299**, 357–423 (1981)
E.A. Mankinen, Geochem. Geophys. Geosyst. **9**(5), Q05017 (2008). doi:10.1029/2008GC001957
W. Marzocchi, J. Geophys. Res. **102**(B3), 5157–5171 (1997)
G. Masters, S. Johnson, G. Laske, H. Bolton, Philos. Trans. R. Soc. Lond. A **354**(1711), 1385–1410 (1996)
H. Matsui, B.A. Buffett, Phys. Earth Planet. Int. **153**(1–3, Sp. Iss.), 108–123 (2005). doi:10.1016/i.pepi.2005.03.019
H. Matsui, B.A. Buffett, Geophys. Astrophys. Fluid Dyn. **101**(5–6), 451–468 (2007). doi:10.1080/03091920701562145
M. Matsushima, Geophys. Astrophys. Fluid Dyn. **100**(4–5), 363–377 (2006). doi:10.1080/03091920600768397
S. Maus, M. Rother, C. Stolle, W. Mai, S. Choi, H. Lühr, D. Cooke, C. Roth, Geochem. Geophys. Geosyst. **7**(7), Q07008 (2006). doi:10.1029/2006GC001269
S. Maus, F. Yin, H. Lühr, C. Manoj, M. Rother, J. Rauberg, I. Michaelis, C. Stolle, R. Müller, Geochem. Geophys. Geosyst. **9**(7), Q07021 (2008). doi:10.1029/2008GC001949
A. Mazaud, in *Encyclopedia of Geomagnetism and Paleomagnetism*, ed. by D. Gubbins, E. Herrero-Bervera (Springer, Heidelberg, 2007)
M.W. McElhinny, in *Encyclopedia of Geomagnetism and Paleomagnetism*, ed. by D. Gubbins, E. Herrero-Bervera (Springer, Heidelberg, 2007)
M.W. McElhinny, P.L. McFadden, Geophys. J. Int. **131**(2), 240–252 (1997)
M.W. McElhinny, P.L. McFadden, *Paleomagnetism: Continents and Oceans* (Academic Press, London, 2000)
M.W. McElhinny, P.L. McFadden, R.T. Merrill, Earth Planet. Sci. Lett. **143**(1–4), 13–22 (1996a)
M.W. McElhinny, P.L. McFadden, R.T. Merrill, J. Geophys. Res. **101**(B11), 25007–25027 (1996b)
P.L. McFadden, J. Geophys. Res. **89**(NB5), 3363–3372 (1984)
P.L. McFadden, R.T. Merrill, J. Geophys. Res. **89**(NB5), 3354–3362 (1984)
P.L. McFadden, R.T. Merrill, J. Geophys. Res. **98**(B4), 6189–6199 (1993)
P.L. McFadden, R.T. Merrill, J. Geophys. Res. **105**(B12), 28455–28460 (2000)
P.L. McFadden, R.T. Merrill, M.W. McElhinny, J. Geophys. Res. **93**(B10), 11583–11588 (1988)
P.L. McFadden, R.T. Merrill, M.W. McElhinny, S.H. Lee, J. Geophys. Res. **96**(B3), 3923–3933 (1991)
P.L. McFadden, C.E. Barton, R.T. Merrill, Nature **361**(6410), 342–344 (1993)
D.G. McMillan, C.G. Constable, R.L. Parker, G.A. Glatzmaier, Geochem. Geophys. Geosyst. **2**(11), 2000GC000130 (2001)
R. Merrill, P. McFadden, M. McElhinny, *The Magnetic Field of the Earth: Paleomagnetism, the Core, and the Deep Mantle* (Academic Press, San Diego, 1996)
R.T. Merrill, P.L. McFadden, Rev. Geophys. **37**(2), 201–226 (1999)
L. Meynadier, J.P. Valet, F.C. Bassinot, N.J. Shackleton, Y. Guyodo, Earth Planet. Sci. Lett. **126**(1–3), 109–127 (1994)
H.K. Moffatt, *Magnetic Field Generation in Electrically Conducting Fluids* (Cambridge University Press, Cambridge, 1978)
R. Monchaux, M. Berhanu, M. Bourgoin, M. Moulin, P. Odier, J.F. Pinton, R. Volk, S. Fauve, N. Mordant, F. Petrelis, A. Chiffaudel, F. Daviaud, B. Dubrulle, C. Gasquet, L. Marie, F. Ravelet, Phys. Rev. Lett. **98**(4), 44502 (2007). doi:10.1103/PhysRevLett.98.044502
L.W. Morley, A. Larochelle, R. Soc. Can. Spec. Publ. **8**, 39–50 (1964)
J.E. Mound, B.A. Buffett, J. Geophys. Res. **110**(B8), 08103 (2005). doi:10.1029/2004JB003555
P.S. Naidu, J. Geophys. Res. **76**(11), 2649 (1971)
H.C. Nataf, T. Alboussiere, D. Brito, P. Cardin, N. Gagniere, D. Jault, J.P. Masson, D. Schmitt, Geophys. Astrophys. Fluid Dyn. **100**(4–5), 281–298 (2006). doi:10.1080/03091920600718426
H.C. Nataf, T. Alboussiere, D. Brito, P. Cardin, N. Gagniere, D. Jault, D. Schmitt, Phys. Earth Planet. Int. **170**(1–2), 60–72 (2008). doi:10.1016/j.pepi.2008.07.034
T. Neubert, M. Mandea, G. Hulot, R. von Frese, F. Primdahl, J.L. Joergensen, E. Friis-Christensen, P. Stauning, N. Olsen, T. Risbo, EOS Trans. AGU **82**(7), 81–88 (2001)
F. Nimmo, in *Treatise on Geophysics*, vol. 8 (Elsevier, Amsterdam, 2007), pp. 31–65
M.S. O'Brien, C.G. Constable, R.L. Parker, Geophys. J. Int. **128**(2), 434–450 (1997)
N. Olsen, M. Mandea, Nat. Geosci. **1**(6), 390 (2008)

N. Olsen, R. Haagmans, T.J. Sabaka, A.V. Kuvshinov, S. Maus, M.E. Purucker, V. Lesur, M. Rother, M. Mandea, Earth Planets Space **58**, 359–370 (2006)
N. Olsen, M. Mandea, T.J. Sabaka, L. Tøffner-Clausen, Geophys. J. Int. **179**(3), 1477–1487 (2009a). doi:10.1111/j.1365-246X.2009.04386.x
N. Olsen, K. Glassmeier, X. Jia, Space Sci. Rev. (2009b). doi:10/1007/s11214-009-9563-0
P. Olson, H. Amit, Naturwissenschaften **93**(11), 519–542 (2006). doi:10.1007/s00114-006-0138-6
P. Olson, J. Aurnou, Nature **402**(6758), 170–173 (1999)
P. Olson, U.R. Christensen, Geophys. J. Int. **151**(3), 809–823 (2002)
P. Olson, U. Christensen, G.A. Glatzmaier, J. Geophys. Res. **104**(B5), 10383–10404 (1999)
P. Olson, P. Driscoll, H. Amit, Phys. Earth Planet. Int. **173**(1–2), 121–140 (2009). doi:10.1016/j.pepi.2008.11.010
N. Opdyke, J. Channell, *Magnetic Stratigraphy* (Academic Press, London, 1996)
A. Pais, G. Hulot, Phys. Earth Planet. Int. **118**, 291–316 (2000)
A. Pais, D. Jault, Geophys. J. Int. **173**, 421–443 (2008)
R. Parker, *Geophysical Inverse Theory* (Princeton University Press, Princeton, 1994)
V. Pavlov, Y. Gallet, Earth Planet. Sci. Lett. **185**(1–2), 173–183 (2001)
V. Pavlov, Y. Gallet, Episodes **28**(2), 78–84 (2005)
V. Pavlov, Y. Gallet, Geochem. Geophys. Geosyst. **11**(1), Q01Z10 (2010). doi:10.1029/2009GC002583
M. Perrin, E. Schnepp, Phys. Earth Planet. Int. **147**(2–3), 255–267 (2004). doi:10.1016/j.pepi.2004.06.005
M. Perrin, V. Shcherbakov, J. Geomagn. Geoelectr. **49**(4), 601–614 (1997)
F. Petrelis, S. Fauve, E. Dormy, J.P. Valet, Phys. Rev. Lett. **102**(14), 144503 (2009). doi:10.1103/PhysRevLett.102.144503
T. Pick, L. Tauxe, Nature **366**(6452), 238–242 (1993)
K. Pinheiro, A. Jackson, Geophys. J. Int. **173**(3), 781–792 (2008). doi:10.1111/j.1365-246X.2008.03762.x
G. Pouliquen, Y. Gallet, J. Dyment, P. Patriat, C. Tamura, J. Geophys. Res. **106**(B12), 30549 (2001)
M. Prévot, P. Camps, Nature **366**(6450), 53–57 (1993)
M. Prévot, M.E. Derder, M. McWilliams, J. Thompson, Earth Planet. Sci. Lett. **97**(1–2), 129–139 (1990)
M.E. Purucker, Icarus **197**(1), 19–23 (2008)
X. Quidelleur, V. Courtillot, Phys. Earth Planet. Int. **95**(1–2), 55–77 (1996)
X. Quidelleur, J.P. Valet, V. Courtillot, G. Hulot, Geophys. Res. Lett. **21**(15), 1639–1642 (1994)
C. Reigber, H. Lühr, P. Schwintzer, Adv. Space Res. **30**, 129–134 (2002)
A.P. Roberts, M. Winklhofer, W.T. Liang, C.S. Horng, Earth Planet. Sci. Lett. **216**(1–2), 187–192 (2003). doi:10.1016/S0012-821X(03)00480-1
P.H. Roberts, G.A. Glatzmaier, Geophys. Astrophys. Fluid Dyn. **94**(1–2), 47–84 (2001)
P.H. Roberts, S. Scott, J. Geomagn. Geoelectr. **17**(2), 137–151 (1965)
D.A. Ryan, G.R. Sarson, Geophys. Res. Lett. **34**(2), L02307 (2007). doi:10.1029/2006GL028291
D.A. Ryan, G.R. Sarson, Europhys. Lett. **83**(4), 49001 (2008). doi:10.1209/0295-5075/83/49001
T.J. Sabaka, N. Olsen, M.E. Purucker, Geophys. J. Int. **159**, 521–547 (2004). doi:10.1111/j.1365-246X.2004.02421.x
T.J. Sabaka, N. Olsen, Earth Planets Space **58**, 371–395 (2006)
A. Sakuraba, M. Kono, Phys. Earth Planet. Int. **111**(1–2), 105–121 (1999)
A. Sakuraba, P.H. Roberts, Nat. Geosci. **2**(11), 802–805 (2009). doi:10.1038/ngeo643
J. Saur, V. Neubauer, K.H. Glassmeier, Space Sci. Rev. (2009). doi:10.1007/s11214-009-9581-y
D. Schmitt, T. Alboussiere, D. Brito, P. Cardin, N. Gagniere, D. Jault, H.C. Nataf, J. Fluid Mech. **604**, 175–197 (2008). doi:10.1017/S0022112008001298
D.A. Schneider, D.V. Kent, J. Geophys. Res. **93**(B10), 11621–11630 (1988)
D.A. Schneider, D.V. Kent, Rev. Geophys. **28**(1), 71–96 (1990)
W.E. Schröder, *Geomagnetism Research; Past and Present* (2000)
G. Schubert, D.L. Turcotte, P. Olson, *Mantle Convection in the Earth and Planets* (Cambridge University Press, Cambridge, 2001)
P.A. Selkin, L. Tauxe, Philos. Trans. R. Soc. Lond. A **358**(1768), 1065–1088 (2000)
J.W. Si, R. Van der Voo, Terra Nova **13**(6), 471–478 (2001)
R.D. Simitev, F.H. Busse, Europhys. Lett. **85**(1), 19001 (2009). doi:10.1209/0295-5075/85/19001
A.V. Smirnov, J.A. Tarduno, Geophys. Res. Lett. **31**(16), L16607 (2004). doi:10.1029/2004GL020333
D.P. Stern, Rev. Geophys. **40** (2002). doi:10.1029/2000RG000097
R. Stieglitz, U. Müller, Phys. Fluids **13**(3), 561–564 (2001)
G. Strik, T.S. Blake, T.E. Zegers, S.H. White, C.G. Langereis, J. Geophys. Res. **108**(B12), 2551 (2003). doi:10.1029/2003JB002475
H. Tanaka, M. Kono, H. Uchimura, Geophys. J. Int. **120**(1), 97–102 (1995)
S. Tanaka, H. Hamaguchi, J. Geophys. Res. **102**(B2), 2925–2938 (1997)

A. Tarantola, *Inverse Problem Theory and Methods for Model Parameter Estimation* (SIAM, Philadelphia, 2005)
J.A. Tarduno, Elements **5**(4), 217–222 (2009). doi:10.2113/gselements.5.4.217
J.A. Tarduno, R.D. Cottrell, A.V. Smirnov, Proc. Natl. Acad. Sci. **99**(22), 14020–14025 (2002). doi:10.1073/pnas. 222373499
J.A. Tarduno, R.D. Cottrell, A.V. Smirnov, Rev. Geophys. **44**(1), RG1002 (2006). doi:10.1029/2005RG000189
J.A. Tarduno, R.D. Cottrell, M.K. Watkeys, D. Bauch, Nature **446**(7136), 657–660 (2007). doi:10.1038/nature05667
L. Tauxe, Earth Planet. Sci. Lett. **233**(3-4), 247–261 (2005). doi:10.1016/j.epsl.2005.01.027
L. Tauxe, Phys. Earth Planet. Int. **156**(3–4), 223–241 (2006). doi:10.1016/j.pepi.2005.03.022
L. Tauxe, P. Hartl, Geophys. J. Int. **128**(1), 217–229 (1997)
L. Tauxe, D.V. Kent, in *Timescales of the Paleomagnetic Field*, ed. by J.E.T. Channell, D.V. Kent, W. Lowrie, J.G. Meert, Geophysical Monograph Series, vol. 145 (2004), pp. 101–115. 0-87590-410-6. doi:10.1029/145GM08
L. Tauxe, H. Staudigel, Geochem. Geophys. Geosyst. **5**(2), Q02H06 (2004). doi:10.1029/2003GC000635
L. Tauxe, T. Yamazaki, in *Treatise on Geophysics*, vol. 5, ed. by M. Kono (Elsevier, Amsterdam, 2007)
L. Tauxe, K.P. Kodama, D.V. Kent, Phys. Earth Planet. Int. **169**(1–4, Sp. Iss.), 152–165 (2008). doi:10.1016/j.pepi.2008.05.006
J. Taylor, Proc. R. Soc. Lond. A **9**, 274–283 (1963)
E. Thébault, K. Hemant, G. Hulot, N. Olsen, Geophys. Res. Lett. **36**, L01307 (2009). doi:10.1029/2008GL036416
J.C. Thomas, H. Perroud, P.R. Cobbold, M.L. Bazhenov, V.S. Burtman, A. Chauvin, E. Sadybakasov, J. Geophys. Res. **98**(B6), 9571–9589 (1993)
A.W.P. Thomson, V. Lesur, Geophys. J. Int. **169**(3), 951–963 (2007)
N. Thouveny, D.L. Bourles, G. Saracco, J.T. Carcaillet, F. Bassinot, Earth Planet. Sci. Lett. **275**(3–4), 269–284 (2008). doi:10.1016/j.epsl.2008.08.020
A. Tilgner, in *Treatise on Geophysics*, vol. 8, ed. by P. Olson (Elsevier, Amsterdam, 2007)
M.A. Tivey, in *Encyclopedia of Geomagnetism and Paleomagnetism*, ed. by D. Gubbins, E. Herrero-Bervera (Springer, Heidelberg, 2007a)
M.A. Tivey, in *Encyclopedia of Geomagnetism and Paleomagnetism*, ed. by D. Gubbins, E. Herrero-Bervera (Springer, Heidelberg, 2007b)
M.A. Tivey, W.W. Sager, S.M. Lee, M. Tominaga, Geology **34**(9), 789–792 (2006). doi:10.1130/G22894.1
M. Tominaga, W.W. Sager, M.A. Tivey, S.M. Lee, J. Geophys. Res. **113**(B7), B07110 (2008). doi:10.1029/2007JB005527
T.H. Torsvik, R. Van der Voo, Geophys. J. Int. **151**(3), 771–794 (2002)
G.M. Turner, J.L. Rasson, C.V. Reeves, in *Treatise on Geophysics*, vol. 5, ed. by M. Kono (Elsevier, Amsterdam, 2007)
P. Ultré-Guérard, M. Hamoudi, G. Hulot, Geophys. Res. Lett. **22**(16), 3201–3204 (1998)
J.P. Valet, Rev. Geophys. **41**(1), 1004 (2003). doi:10.1029/2001RG000104
J.P. Valet, L. Meynadier, Nature **366**(6452), 234–238 (1993)
J.P. Valet, G. Plenier, Phys. Earth Planet. Int. **169**(1–4, Sp. Iss.), 178–193 (2008). doi:10.1016/j.pepi.2008.07.031
J.P. Valet, P. Tucholka, V. Courtillot, L. Meynadier, Nature **356**(6368), 400–407 (1992)
J.P. Valet, L. Meynadier, Y. Guyodo, Nature **435**(7043), 802–805 (2005). doi:10.1038/nature03674
J.P. Valet, E. Herrero-Bervera, J.L. Le Mouël, G. Plenier, Geochem. Geophys. Geosyst. **9**(1), Q01008 (2008). doi:10.1029/2007GC001728
R. Van der Voo, T.H. Torsvik, Earth Planet. Sci. Lett. **187**(1–2), 71–81 (2001)
S.A. Van Loo, J.L. Rasson, in *XIIth IAGA Workshop on Geomagnetic Observatory Instruments, Data Acquisition and Processing*, Abstract volume 21, 2006
D. Vandamme, Phys. Earth Planet. Int. **85**(1–2), 131–142 (1994)
J.M. Vaquero, R.M. Trigo, Phys. Earth Planet. Int. **157**, 8–15 (2006)
P. Varga, C. Denis, T. Varga, J. Geophys. **25**(1–2), 61–84 (1998)
G. Verhille, N. Plihon, M. Bourgoin, P. Odier, J.F. Pinton, Space Sci. Rev. (2009). doi:10.1007/s11214-009-9546-1
F.J. Vine, D.H. Matthews, Nature **199**(489), 947–949 (1963)
C.V. Voorhies, T.J. Sabaka, M. Purucker, J. Geophys. Res. **107**(E6), 5034 (2002). doi:10.1029/2001JE001534
M.R. Walker, A. Jackson, Geophys. J. Int. **143**, 799–808 (2000)
I. Wardinski, M. Korte, J. Geophys. Res. **113**(B5), B05101 (2008). doi:10.1029/2007JB005024
J. Wicht, Phys. Earth Planet. Int. **132**(4), 281–302 (2002)

J. Wicht, Geophys. J. Int. **162**(2), 371–380 (2005). doi:10.1111/j.1365-246X.2005.02665.x
J. Wicht, P. Olson, Geochem. Geophys. Geosyst. **5**(3), Q03H10 (2004). doi:10.1029/2003GC000602
A.P. Willis, B. Sreenivasan, D. Gubbins, Phys. Earth Planet. Int. **165**(1–2), 83–92 (2007). doi:10.1016/j.pepi.2007.08.002
C. Xuan, J.E.T. Channell, Phys. Earth Planet. Int. **169**(1–4, Sp. Iss.), 140–151 (2008). doi:10.1016/j.pepi.2008.07.017
S. Zatman, J. Bloxham, Nature **388**, 760–763 (1997)
S. Zatman, J. Bloxham, Geophys. J. Int. **138**(3), 679–686 (1999)

Space Sci Rev (2010) 152: 223–249
DOI 10.1007/s11214-009-9557-y

Crustal Magnetic Fields of Terrestrial Planets

Benoit Langlais · Vincent Lesur · Michael E. Purucker · Jack E. P. Connerney · Mioara Mandea

Received: 2 March 2009 / Accepted: 15 June 2009 / Published online: 3 July 2009

Abstract Magnetic field measurements are very valuable, as they provide constraints on the interior of the telluric planets and Moon. The Earth possesses a planetary scale magnetic field, generated in the conductive and convective outer core. This global magnetic field is superimposed on the magnetic field generated by the rocks of the crust, of induced (i.e. aligned on the current main field) or remanent (i.e. aligned on the past magnetic field). The crustal magnetic field on the Earth is very small scale, reflecting the processes (internal or external) that shaped the Earth. At spacecraft altitude, it reaches an amplitude of about 20 nT. Mars, on the contrary, lacks today a magnetic field of core origin. Instead, there is only a remanent magnetic field, which is one to two orders of magnitude larger than the terrestrial one at spacecraft altitude. The heterogeneous distribution of the Martian magnetic anomalies reflects the processes that built the Martian crust, dominated by igneous and cratering processes. These latter processes seem to be the driving ones in building the lunar magnetic field. As Mars, the Moon has no core-generated magnetic field. Crustal magnetic features are very weak, reaching only 30 nT at 30-km altitude. Their distribution is heterogeneous too, but the most intense anomalies are located at the antipodes of the largest impact basins.

B. Langlais
Laboratoire de Planétologie et Géodynamique, CNRS UMR 6112, 2 rue de la Houssinière, 44000 Nantes, France

B. Langlais (✉)
Laboratoire de Planétologie et Géodynamique, Université de Nantes, 2 rue de la Houssinière, 44000 Nantes, France
e-mail: benoit.langlais@univ-nantes.fr

V. Lesur · M. Mandea
Helmholtz Centre Potsdam, German Research Centre for Geosciences, Potsdam, Germany

M.E. Purucker
Raytheon at Planetary Geodynamics Laboratory, Code 698, NASA Goddard Space Flight Center, Greenbelt, MD 20771, USA

J.E.P. Connerney
Planetary Magnetospheres Laboratory, Code 695, NASA Goddard Space Flight Center, Greenbelt, MD 20771, USA

The picture is completed with Mercury, which seems to possess an Earth-like, global magnetic field, which however is weaker than expected. Magnetic exploration of Mercury is underway, and will possibly allow the Hermean crustal field to be characterized. This paper presents recent advances in our understanding and interpretation of the crustal magnetic field of the telluric planets and Moon.

Keywords Magnetic Field · Telluric Planets · Crust · Measurements · Modeling Techniques · Interpretation

1 Introduction

The description and the understanding of the magnetism associated with rocks has always been a very important topic, for both fundamental and applied sciences. The Chinese were probably the first to use compasses for orientation purposes (Needham 1962), with a south-pointing iron needle, or "Si Nan". The use of the compass for marine navigation was introduced in Europe during the 12th century, even if the reason why it was indicating the North pole (or the South one) was not well understood. For some, compasses were showing the polar star, and for others it was indicating the location of a close-to-the-pole island, where magnetite was mined. The first scientific explanation came in 1600, when William Gilbert published the results of his experiments with a spherical magnets called "terrelae" (Gilbert 1600). Among other conclusions, he proposed that the magnetic field of the Earth was not stationary, but that instead it was rotating together with the Earth. He also correctly assumed that the center of the Earth was made of iron. However, as shown in Fig. 1, he erroneously attributed the deviation of the compass to the presence of positive or negative imperfections on the sphere, such as oceanic areas or land masses.

William Gilbert was not completely wrong. Land masses may indeed deviate the compass needle from the pure intern geomagnetic pole. Rocks of the upper lithosphere (or crust) of

a)

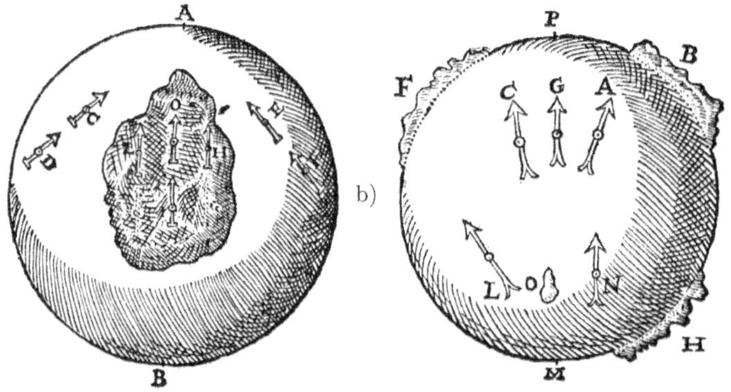

b)

Fig. 1 Behavior of the magnetic compass or vector as a function of land masses, as imagined by Gilbert (1600). (**a**) A and B denote North and South poles, respectively. Compasses C, D, E and F show the true magnetic pole, because they are distant from the "imperfect and weaker" part (the imperfect part is meant to resemble an oceanic area). Compass O also shows the pole, as it is located in the middle of the imperfection. On the contrary, H and L compasses deviate towards the edges of the imperfect area. (**b**) P and M denote North and South poles, respectively. Imperfections in B, F, H, and O are positive anomalies, or land masses. Compass G shows the pole, as it is equidistant from F and B. On the contrary, A and C compasses deviate toward B and F, respectively. Compass L shows pole M, as O is too small to affect it. But compass N is deviated by H the Earth and other planets such as the Moon and Mars may produce a magnetic field, whether it is induced by an internal or external source, or remanent. Actually, everything is magnetic, from a piece of wood to a plate of steel. This effect, known as diamagnetism, is a result of the interaction of orbiting electrons within an external magnetic field. This produces an opposite direction magnetic field of relatively small magnitude. More important is the paramagnetism effect. It affects only atoms which possess an atomic magnetic moment. The induced magnetization is aligned and parallel to the applied magnetic field, but it will return to zero if the applied magnetic field is removed. This effect is inversely proportional to temperature. The most important effect is ferromagnetism: this also affects atoms with magnetic moment, but in this case, adjacent atomic magnetic moments strongly interact, allowing the material to retain a magnetic field even in the absence of an external magnetic field.

Ferromagnetic materials are mainly iron, cobalt and nickel. Of these, iron is by far the most common in nature. Combined with oxygen, iron is found as hematite and magnetite in common rock types. Titanium often replaces some of the iron in both the more oxidized hematite, and the less oxidized magnetite. These Fe–Ti oxides are also common magnetic phases in many rock types. Some iron sulfides (e.g. pyrrhotite) and oxyhydroxides (e.g. goethite) are also common magnetic phases in rocks (Weiss et al. 2009). Such minerals are capable of bearing both remanent and induced magnetization. The Koenigsberger ratio measures the relative importance of the induced M_i to the remanent M_r contribution. To first order, induced magnetization is parallel and proportional to the direction and intensity of the external field **B**. The proportionality is expressed through χ, the magnetic susceptibility, which mainly depends on the magnetic mineralogy. The relation between the remanent and the original (i.e. the one that gave the minerals their remanence) is less simple. It depends on several factors, among which is the Curie temperature. This is the temperature above which magnetic minerals loose their remanent magnetization. The Curie temperature of magnetite and hematite, the most commonly found terrestrial magnetic minerals, is 580°C and 670°C, respectively. For titanomagnetites and titanohematites, the Curie temperature decreases as the Ti content increases. Other important parameters are the magnetic mineralogy, the grain size and shape. The M_r versus **B** behavior can be described by an hysteresis loop (see Dunlop and Özdemir 1997 for a review).

The Earth and Mars are very alike in the sense that they are basically made of similar components. Together with Venus and Mercury, they constitute the telluric planets, with are primarily composed of silicate rocks. These planets are differentiated, with a core, a mantle and a crust. For simplicity, we include the Moon in that category too, although the existence of a Lunar core is debated (Lognonné et al. 2003).

On the Earth, the main present-day magnetic field has a deep origin. It originates from the outer liquid core, where convection of a conducting fluid create a (mostly) dipolar, axial and centered magnetic field. At the Earth's surface, this main field ranges from about 20,000 nT at the magnetic equator to 70,000 nT at the magnetic poles. It is time variable, on different time scales. The most dramatic variations correspond to the inversion of the polarity of the field. Such inversions are recorded by magnetic minerals when they cool down below their Curie temperature. Intensity variation (or secular variation) exists too and may also be recorded by minerals. Shorter terms variations are most of the time related to external sources. The magnetic field of the crust is often referred as the anomaly field. These anomalies are actually fields in excess or in deficit of the main field. Terrestrial magnetic anomalies are therefore positive or negative, while they are only positive for Mars and the Moon, where there is no core field. Fields of crustal origin can be very intense, up to 200,000 nT near Kursk (Jankowski and Sucksdorff 1996), but they decrease very fast as the distance to the source increase. At 400 km altitude, crustal fields are estimated to range between ±20 nT (Mandea and Purucker 2005). In the following, the words crust and lithosphere are often used interchangeably. 'Crust' is differentiated from 'mantle' on the basis of its chemical composition, while 'lithosphere' is a rheological term and usually comprises the strong crust and uppermost mantle. Magnetic rocks are confined to regions of the crust and upper mantle cooler than the Curie temperature, defining a magnetic crust or lithosphere. Most upper mantle rocks are non-magnetic, even if they are cooler than the Curie temperature, because iron is overwhelmingly in silicates, not in oxides. Serpentinite with magnetite is an exception to this rule, and can be found in the mantle overlying subduction zones.

In this review paper, we first introduce the modeling and representing techniques, on both global and local scales, with associated limits and caution. We then summarize magnetic measurements and maps that were acquired around the Earth, Mars, Mercury, Venus and the Moon. We will focus only on the magnetic field of crustal and lithospheric origin; those interested in the core magnetic field and in the external magnetic field are referred to Hulot et al. (2009) and to Baumjohann et al. (2009). We finally review recent developments and interpretations of magnetic measurements in terms of planetary dynamics or surface processes, and conclude with open issues and perspectives.

2 Modeling Techniques and Issues

The acquisition of magnetic measurements above the surface of a planet is challenging, but the interpretation of these measurements in terms of magnetic field model or magnetization distribution is also difficult. In the following, we briefly review the modeling techniques used. We make the distinction between techniques used to describe the field and those used to interpret the field in terms of source properties.

2.1 Modeling the Field

The magnetic field is measured on a discrete basis, along the spacecraft trajectory, in the air, or on the ground. It is of course possible to average measurements onto regular grids, but a more powerful technique consists in building a continuous description of the field on the surface of a sphere. This is commonly done with spherical harmonics, but other methods exist.

Spherical Harmonics (SH) form the natural basis to describe a potential field on a spherical surface because they are the eigen functions of the Laplacian in spherical geometry. As such, they form the smallest set of functions that can describe all possible potential fields up to a given wavelength. Furthermore, if a magnetic field is described in terms of spherical harmonics, the separation of the field due to sources internal to the observation (internal dynamo or magnetized rocks), from those due to external sources (magnetospheric field) can in theory be performed. In geo- and planetary magnetism, the Schmidt semi-normalized spherical harmonics are commonly used.

Away from the magnetic sources, the magnetic field is therefore described as the gradient of a potential, following $\mathbf{B} = -\nabla V(\theta, \phi, r, t)$. The potential is a function of colatitude θ, longitude ϕ, radius r and time t. An internal and an external potential are associated with this poloidal field, V_i and V_e respectively. They are described on a spherical surface by:

$$V_i(\theta, \phi, r, t) = a \sum_{n=1}^{N_i} \sum_{m=0}^{n} \left(\frac{a}{r}\right)^{n+1} \left(g_n^m(t) \cos(m\phi) + h_n^m(t) \sin(m\phi)\right) P_n^m(\cos\theta) \quad (1)$$

$$V_e(\theta,\phi,r,t) = a\sum_{n=1}^{N_e}\sum_{m=0}^{n}\left(\frac{r}{a}\right)^n \left(q_n^m(t)\cos(m\phi) + s_n^m(t)\sin(m\phi)\right) P_n^m(\cos\theta), \qquad (2)$$

where a is the reference radius, $P_n^m(\cos\theta)$ are the Schmidt quasi-normalized associated Legendre functions of degree and order n and m, and $(g_n^m(t), h_n^m(t))$, $(q_n^m(t), s_n^m(t))$ are the Gauss internal and external coefficients, respectively. As the upper bound N_i and N_e of these summations increase, the magnetic field is described, globally, in more details, i.e. with shorter wavelengths. For a given value of N_i the number of Gauss coefficients describing the internal field is $N_i(N_i+2)$.

The drawbacks in describing the field in terms of spherical harmonics comes from the global nature of these functions. A data set rarely covers the whole planet, e.g. for satellite data a hole generally remains close to the poles, and therefore the spherical harmonic model is prone to oscillations. Furthermore, over a planet the spectral content of the observed magnetic field may vary, and therefore the required resolution of the model may not be the same everywhere. Typically for the Earth, at satellite altitudes, the oceanic crust is nearly free of sharp and strong anomalies and does not need high degree spherical harmonics, whereas over continental areas very small details are present and require a sharper description. Constraining locally the spherical harmonic model is possible (Lesur 2006) but this has been infrequently used.

Wavelets or localized basis are an interesting alternative to SH. This representation technique replaces the usual global spherical harmonic expansion of the potential by a summation over functions of the type:

$$\mathcal{F}_i(\theta,\phi) = \sum_{n,m}^{N} f_n\, Y_n^m(\theta_i,\phi_i) Y_n^m(\theta,\phi). \qquad (3)$$

In this expression, (θ_i,ϕ_i) is the center of the function. The upper bound to the summation, N, can tend to infinity if required, and the f_n are adjusted to "concentrate" the function in both spatial and spectral domains. These techniques have been infrequently used for modeling the crustal field even if they allow for a varying resolution over the sphere. They can also be used for building localized models. Localized functions have been applied by Lesur and Maus (2006). One can refer to the work of Chambodut et al. (2005) for wavelets. Also, vector scaling wavelets have been used for modeling the Earth's crust (Mayer and Maier 2006).

Spherical Cap Harmonics Analysis (SCHA) is possibly the most frequently used method on the sphere that uses local basis functions. It was developed by Haines (1985) to model the main magnetic field over small areas and has, since then, been applied in a large variety of circumstances (see Torta et al. 2006 for a review). As for SH, the internal potential is expressed as a sum:

$$V_i(\theta,\phi,r,t) = a\sum_{k\geq m}\sum_{m\geq 0}\left(\frac{a}{r}\right)^{n_k+1} \left(G_{n_k}^m(t)\cos(m\phi) + H_{n_k}^m(t)\sin(m\phi)\right) P_{n_k}^m(\cos\theta). \qquad (4)$$

In this expression, the order m is integer, and the degree n_k is generally real, with $n_k \geq m$. SCHA has some fundamental limitations, in particular with respect to the separation of internal from external fields. Also of concern is the assimilation of multi-level data (Thébault and Gaya-Piqué 2008). A revised SCHA was proposed recently (Thébault et al. 2006a), addressing the problem of altitude and multi-level data compatibility. This was used to merge ground and satellite data (Thébault et al. 2006b). A specific formalism was recently developed to process ground data only (Thébault 2008).

Equivalent Source Dipole (ESD) is the alternate and often used technique for global description of magnetic field around planets. This method was introduced by Mayhew (1979). Considering the magnetic moment **M** of a given dipole located at (r_d, θ_d, ϕ_d), the magnetic potential observed at (r, θ, ϕ) is expressed as

$$V_i = -\mathbf{M} \cdot \nabla \frac{1}{l}. \tag{5}$$

This relation is valid provided that there are no sources between the dipole and the observation location. The distance l between the dipole and the observation location is written:

$$l = \left(r_d^2 + r^2 - 2r_d r \left[\cos(\theta)\cos(\theta_d) + \sin(\theta)\sin(\theta_d)\cos(\phi - \phi_d)\right]\right)^{\frac{1}{2}}. \tag{6}$$

The observed magnetic field **B** is the gradient of the potential V_i due to a series of dipole sources located inside the planet. This method was primarily designed to reduce scattered measurements to a common altitude.

This method has several advantages. First, it associates with a magnetic data set, information about the magnetization distribution. If carefully used, this information can be useful but a simplistic direct interpretation may lead to erroneous conclusions as described below. Second, as for the localized methods, the resolution in both spatial and spectral domain can be changed by modifying the density of the dipoles or their depths. This freedom carries also difficulties because, for a given source depth, the density has to be high enough in order to avoid spurious model magnetic field behavior. As a result, the number of required dipoles is usually relatively high. For a given resolution (i.e. a given maximum spherical harmonic degree), the number of parameters is as least twice as large as the number of requested Gauss' coefficients. When used to model terrestrial magnetic field anomalies, one can use *a priori* information to constrain the model (and to reduce the number of parameters). For instance a purely induced magnetization aligned onto the core field is often assumed over the continents (Purucker et al. 1998).

2.2 Modeling the Magnetization

We now turn to the methods for extracting the magnetization information from the data. The magnetic field **B** is linked to the rock magnetization by the relation:

$$\mathbf{B}(\mathbf{r}) = \int_v \mathbf{G}(\mathbf{r}, \mathbf{s}) \cdot \mathbf{M}(\mathbf{s}) dv. \tag{7}$$

where **r** and **s** are two points in space, outside and inside the magnetized volume v respectively. $\mathbf{G}(\mathbf{r}, \mathbf{s})$ is the usual Green tensor given by:

$$\mathbf{G}(\mathbf{r}, \mathbf{s}) = -\frac{\mu_0}{4\pi} \nabla_\mathbf{r} \nabla_\mathbf{s} \frac{1}{|\mathbf{r} - \mathbf{s}|} \tag{8}$$

The most often used techniques consist in describing the magnetization in term of spherical uniformly magnetized bodies (i.e. dipoles), sometimes uniformly magnetized cylinders for

2-D geometry, or as a discrete sum of the Green tensors themselves (Parker et al. 1987). Other techniques are described in Blakely (1995), Purucker and Whaler (2007).

It is clear that the inverse problem of finding a distribution of magnetization that exactly explains the measured magnetic field suffers from fundamental non-uniqueness. There indeed exist magnetization distributions, also known as magnetic annihilators, which do not produce significant magnetic field outside of the magnetized area (Runcorn 1975; Lesur and Jackson 2000). As a simple example given by Parker et al. (1987), if f is an arbitrary continuously differentiable function, the magnetization $\mathbf{M} = \nabla f$ defined in a volume v does not produced any external field when the function f vanishes on the boundary of the magnetized volume. This is a simple implication of the Gauss' theorem. Even if very strong constraints are applied on the magnetization distribution, as in the dipole representation, the magnetization distribution is not unique: a sufficient number of radially magnetized dipoles can always explain a finite set of vector magnetic measurements (Purucker et al. 2000).

A possible way for deriving information about the magnetization is to search for minimum norm solutions. Instead of looking for the best fit to the data without imposing any constraints on the solution, the continuously varying magnetization is bounded so that its rms amplitude is minimal (Whaler and Langel 1996). For the Earth's case the hypothesis of purely induced magnetization is often used. This is generally a sufficient hypothesis to resolve the non-uniqueness of the magnetization, although on global scale the inverse problem still present difficulties (Maus and Haak 2003). On more local scale, it is possible to resolve the non uniqueness by imposing strict source geometries, such as disks, cylinders or spheres (Quesnel et al. 2008). In this case, a unique magnetization solution can be found to explain magnetic field measurements. However, it must be kept in mind that the magnetization estimates actually depends on the *a priori* imposed source geometry and location.

3 Magnetic Measurements and Global Maps

Magnetic measurements made around the Earth, Mars, Venus, Mercury and the Moon are described in the following. For the Earth, only the lithospheric part is discussed.

3.1 The Earth

Relative to other planets, the amount of data available for studying the Earth's lithospheric field is very large. For these studies three types of data are available, namely: aeromagnetic data, marine data and survey satellite data. The quality of the measurements depends of course on the experiment itself, but also on the capacity to remove transient variations of external origin (i.e. contributions of the ionospheric and magnetospheric current systems). This is typically done by monitoring these variations at a nearby magnetic observatory.

Airborne magnetic surveys have been commonly used for mapping the crustal magnetic field in view of mineral exploration or for studying regional crustal structures. However, because of the cost involved and other technical difficulties, these surveys are of limited extent and the information for wavelength larger than 200 km are seldom reliable. They nonetheless offer accurate mapping of the magnetic field and, since the 1950's, the accumulated set of local surveys have allowed a significant part of the northern hemisphere continental area to be covered. Australia is also fully covered and long-range surveys have been organized there to preserve the long wavelength information. However, over most of Africa and South-America, very little data is freely available. Similarly, Antarctica is only partially covered.

Over the oceanic area, away from the coast, there is little aeromagnetic survey data but marine data can be used. The processing of such survey is rather difficult because there is usually no nearby monitoring of the external field variations. Furthermore, outside small areas where specific surveys have been undertaken, the ship tracks do not form a dense and regular set of data. The interpolation process is then difficult (Dyment et al. 1995) and the accuracy of the obtained maps may be sometimes debatable. As for aeromagnetic data, the Northern Hemisphere is much better covered by marine data than the Southern Pacific, Atlantic and Indian Oceans.

Because of these uncovered areas, geomagnetic measurements on board artificial satellites were soon envisaged. The first magnetometer flew on Sputnik 3, between May 1958 and April 1960 (Dolginov et al. 1961). Satellites of the Polar Orbiting Geophysical Observatories (POGO) series were launched during the late 1960's. They carried out only scalar measurements (Langel 1990), but they provided the first maps of the crustal anomalies at mid latitudes (Regan et al. 1975). These results motivated the MAGSAT satellite mission (November 1979 to April 1980) which measured the vector component of the magnetic field. The first global SH description of the magnetic field of crustal origin was that of Cain et al. (1984). Their model is based on MAGSAT vector measurements, and described the field up to degree 29. In the following twenty years, the only satellite data available were obtained for the POGs satellite (1990–1993) launched by the US Navy, although only scalar data were returned. In February 1999 the Ørsted satellite was launched carrying both vector and scalar magnetometers (Olsen et al. 2000). Its high altitude (between 643 km and 881 km after launch) however precludes an accurate mapping of the crustal field beyond degree 30. During 2000 two satellites carrying magnetometers were launched, the CHAMP German satellite at an altitude of 450 km (Reigber et al. 2002) and the SAC-C satellite at a higher altitude (Colomb et al. 2004). The advantage of the CHAMP satellite is twofold: the satellite has acquired high quality vector data, and its low-orbit is optimized for crustal studies. In 2009, CHAMP is flying at a lower altitude of about 330 km, which, combined with the solar minimum activity, will make it possible to increase the resolution of the crustal field description.

If the quality of these modern satellites is such that the core and crustal fields can be accurately mapped, the crustal anomaly field with wavelength longer than 2500 km (i.e. SH degree < 16) remains unknown because the main core field is much stronger than the crustal field and overlaps it at these length scales: their respective contributions cannot be separated from magnetic data alone. Several models of the crustal field have been produced. The latest and best models of the crustal field have been derived from the CHAMP satellite data set.

The main difficulty when dealing with satellite magnetic data arises from the contribution of the magnetic fields generated in the magnetosphere, ionosphere and by Field Aligned Currents. Of particular concern is the part generated in the ionosphere that is seen as an internal source by satellites. To circumvent these problems one approach is to process (sometimes referred to as filtering) the survey data and to remove, as much as possible, these undesirable contributions. This is the approach used for the `MF` series of crustal magnetic field models. The first of these very successful models was the `MF1`, released on 2002 (Maus et al. 2002). It goes up to SH degree 80, and is based on CHAMP scalar magnetic data. The `MF2` came one year later and included vector satellite data. It was soon followed by the `MF3` and `MF4` versions, but none of these models have an acceptable behavior everywhere at the Earth's surface. The `MF4x` (Lesur and Maus 2006) model was built from exactly the same data set as the `MF4`, but the system of representation used localized functions which allow a varying resolution of the model depending on location. This new way of regularizing the model led to a model with an acceptable behavior at ground level. It goes up to SH degree 60 at high latitudes and degree 90 at mid and low latitudes. In 2007 and 2008 were released the latest versions of the `MF` series models, the versions `MF5` (Maus et al. 2007) and `MF6` (Maus et al. 2008) respectively. This last model provides a SH representation of the crustal field up to a maximum degree 120 but is regularized from degree 80 onwards. The data used are scalar and vector CHAMP measurement from 2003 up to mid 2007.

An alternate method to model the crustal field can be described as the "comprehensive" approach. It consists in modeling as accurately as possible all the main sources of the magnetic field (Sabaka et al. 2004). Even though this approach has been very efficient in providing accurate models of the core magnetic field, it has been less successful regarding the crustal field. Nonetheless, the two latest of these models, `GRIMM` (Lesur et al. 2008) and `xCHAOS` (Olsen and Mandea 2008), present a remarkable agreement up to SH degree 45 (Lesur et al. 2008).

At spacecraft altitude, the magnetic field anomalies are relatively smoothed (see Fig. 2). The model of Lesur et al. (2008) predicts a field ranging between $\pm$ 20 nT. The largest fields are mostly found above the continental cratons. The most noticeable anomalies are those of Bangui (Central Africa) and Kursk (Ukraine).

One of the most important achievements in describing the Earth's crustal field was the publication of the World Digital Magnetic Anomaly Map (WDMAM) (Korhonen et al. 2007). The project was the realization of an international scientific joint effort—supported by the International Association of Geomagnetism and Aeronomy (IAGA), as well as by the Commission for the Geological Map of the World (CGMW)—to compile and publish a reliable world map of magnetic anomalies that are attributable to the Earth's uppermost lithosphere. This map is derived from a multitude of aeromagnetic surveys acquired over continents or from ship cruises, during the past five decades. For the first time, these data are referenced to satellite magnetic measurements and geomagnetic observatories in a comprehensive way. The resulting product is a printed magnetic anomaly map of the World at scale 1:50,000,000 (uniform with Geological Map of the World by CGMW) and a digital database that includes anomaly values on a grid of resolution 3 arc minutes (about 5 km at the equator). The nominal observation altitude is defined as 5 km above the geoid.

3.2 Mars

The exploration of planet Mars began in 1960 with the launch of two Soviet probes (Marsnik 1 and Marsnik 2) that failed shortly after launch (Perminov 1999). The first successful mission to Mars was the USA's Mariner 4 spacecraft, in 1965. It approached Mars within 4 radii of the planet, but did not detect anything but a bow shock (Smith et al. 1965).

Despite numerous opportunities over the next 32 years, no space probe instrumented to measure magnetic fields flew close enough to the planet's surface to establish the presence of an intrinsic magnetic field. Estimates of a Mars' magnetic dipole ranged from 0.8 to 2.55 10^{22} $G \cdot cm^3$, equivalent to an equatorial surface field of 20 to 65 nT, but its nature and origin was highly debated (Ness 1979). The USSR's Phobos 2 mission provided observations as close as 800 km altitude. No conclusive evidence for a magnetic field of internal origin emerged (Riedler et al. 1989), but some argued in favor of a small planetary field (Dolginov and Zhuzgov 1991; Slavin et al. 1991; Moehlmann et al. 1991) or localized magnetic anomalies (Moehlmann 1992). The controversy did not end until Mars Global Surveyor (MGS) entered Mars orbit in September 1997.

MGS was instrumented with a magnetometer and electron reflectometer experiment. Two triaxial fluxgate magnetometers were mounted at the outer extremity of the solar array panels, and an electron reflectometer was mounted on the spacecraft body. The vector magne-

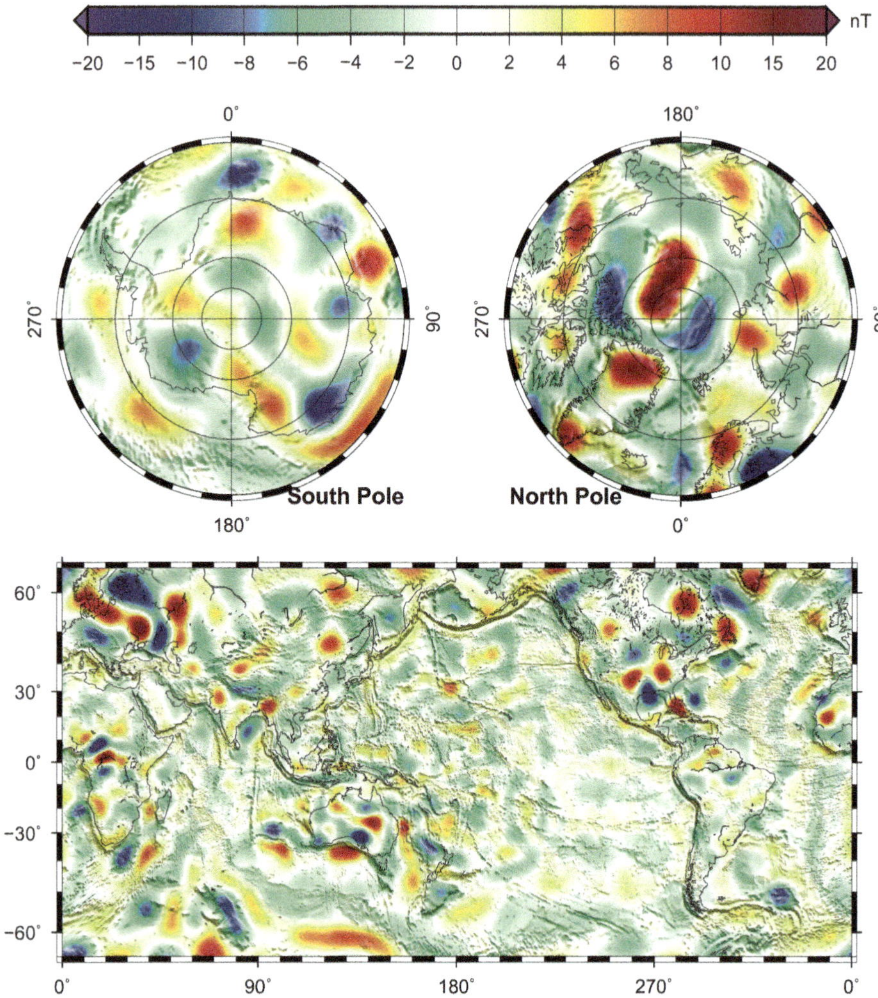

Fig. 2 Predicted radial component of the terrestrial crustal magnetic field at an altitude of 350 km above the reference radius, based on the SH model of Lesur et al. (2008) up to degree 45, superposed onto a terrestrial shaded relief (ETOPO5 1988). *Top*: South and North pole (orthographic projection), down to 60° S latitude. *Bottom*: Mercator projection between ±65° S latitude. Coastlines are plotted for clarity

tometers provided in situ measurement of the ambient magnetic field with 12-bits resolution over 8 automatically-selected dynamic ranges from ±4 nT to ±65,536 nT full scale. The electron reflectometer measured the local electron distribution function in the range of ∼10 eV to 20 keV. These measurements can be used to remotely sense the magnetic field magnitude down to altitudes of approximately 170 km, where atmospheric absorption of electrons limits the application of the electron reflection method (Mitchell et al. 2001). A detailed description of the instrumentation is found in (Acuña et al. 1992), and a discussion of the MGS spacecraft, spacecraft magnetic field mitigation, and early results are given in (Acuña et al. 2001; Connerney et al. 2004).

The MGS mission was designed to recover a subset of the science objectives of the Mars Observer mission, which ended prematurely in August, 1993, with an unsuccessful orbit insertion maneuver. To save fuel, MGS was supposed to reach its high inclination, circular mapping orbit at 400 km altitude by aerobraking in the Mars atmosphere. This approach uses atmospheric drag to reduce apoapsis; over time, and with each periapsis pass, the elliptical orbit becomes more and more circular. As a result of a spacecraft anomaly experienced early in the mission, MGS used many more drag passes than originally intended to slow the spacecraft (Albee et al. 2001). This unanticipated development allowed a far greater than expected sampling of the magnetic field well below the nominal 400 km mapping phase altitude.

MGS observations are best described by reference to three distinct mission modes. These are the aerobraking orbits (AB), science phasing orbits (SPO), and mapping orbits (MO). The two aerobraking phases (AB1 and AB2) occurred with the spacecraft in elliptical orbit about Mars. These two phases were separated by SPO1 and SPO2, during which the spacecraft was "parked" at a fixed periapsis altitude. These four mission phases, extending through March, 1999, provided in situ observations extending down to altitudes of as little as 100 km at periapsis. During almost all AB and SPO passes, MGS acquired measurements of the vector magnetic field and electron distributions along the orbit at varying altitude above the surface. The latitude of periapsis at the beginning of AB1 was about 30°N. Subsequent orbits evolved such that the latitude at periapsis progressed slowly northward toward and over the pole through SPO2. During AB2, the latitude of periapsis progressed southward, reaching a maximum of 87°S at the end of aerobraking. At that time, 1023 aerobraking passes (subset for which magnetic field data were obtained) were distributed more or less randomly in longitude, completed by 60 (out of 130) passes for SPO1 and 211 (out of 244) passes for SPO2. By the end of pre-mapping about half of Mars' surface had been sparsely sampled at low altitude (Acuña et al. 1999).

MO observations began in March, 1999 and continued until November 2, 2006, at which time the spacecraft was lost after a series of errors that left its battery depleted, and the spacecraft unable to maintain attitude. The primary (one full Mars year, until the end of January, 2001) and the extended phases over 3 Mars years provided abundantly oversampled coverage of the magnetic field at a nominal altitude of 400 km (370–438 km) and fixed (2 am–2 pm) local time. During the mission, MGS completed 100 mapping cycles before it was lost. Each cycle lasted 28 days, providing coverage of the entire planet by using a 7-day, 88-orbit repeat geometry.

At the 400 km nominal mapping altitude, magnetic fields generated by the interaction of Mars' atmosphere with the solar wind can at times be appreciable. To first approximation, external fields draped over a conducting obstacle will align with the conducting surface (ionosphere), appearing largely in the horizontal component of the field. External fields are time variable, reflecting variations in the solar wind, and greater in magnitude near the sub-solar point. The external field is both quieter and weaker in magnitude over the darkened hemisphere (Vennerstrom et al. 2003).

The MGS vector magnetic field observations from all phases were compiled into different maps and models of the three components of the martian magnetic field. At an altitude of 400 km (see Fig. 3), the martian magnetic field varies between $\pm$ 250 nT. The largest magnetic fields are located in the Southern Hemisphere. A map at nominal mapping altitude around 400 km was produced (Connerney et al. 2001), based on night side observations only. Measurements were sorted onto 1° latitude-longitude bins, keeping only the median value to minimize transient variations. An improved map was published later (Connerney et al. 2005), using more data and a more elaborate technique for removal of external fields. However, these maps did not incorporate the early measurements of the AB and SPO low altitude phases. The first model based on those measurements was produced by Purucker

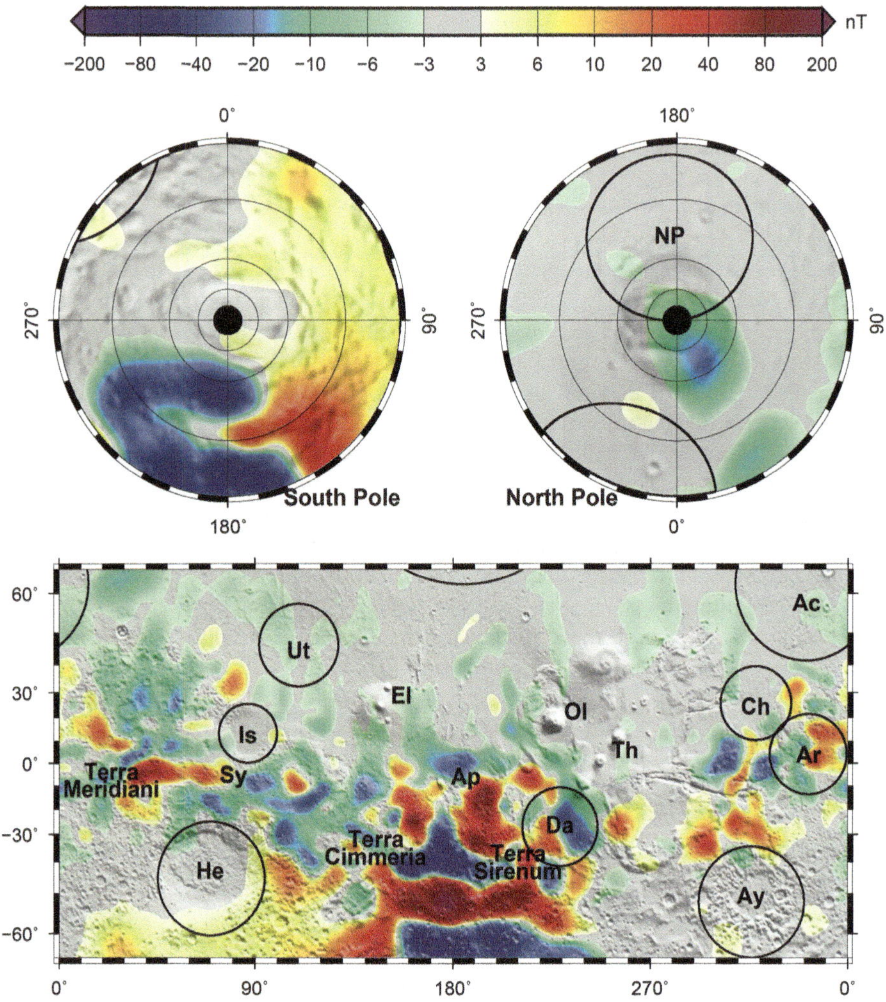

Fig. 3 Predicted radial component of the martian magnetic field at an altitude of 400 km above the martian mean radius, based on the equivalent source dipole model of Langlais et al. (2004), overlapped onto a martian shaded relief (Smith et al. 2003). *Top*: South and North pole (orthographic projection), down to 60°S latitude. *Bottom*: Mercator projection between ±65°S latitude. The main rims of the largest impact basins are plotted, with labels: He Hellas, Ar Ares, Ac Acidalia, Da Daedalia, Ay Argyre, Ch Chryse, NP North Pole, Is Isidis, and Ut Utopia. Also shown is the location of major volcanoes Sy Syrtis Major, Th Tharsis Montes, Ol Olympus Mons, El Elysium Mons, and Ap Apollinaris Patera

et al. (2000), based on binned measurements of the radial field only and an ESD approach. The first SH model (up to degree 50) of the martian magnetic field was based on the three measured components of the MO phases (Arkani-Hamed 2001a). Later models were more elaborate, and used measurements of all three mission phases, to produce SH models up to degree 90 (Cain et al. 2003; Arkani-Hamed 2004), and constant altitude maps through ESD (Langlais et al. 2004) or continuous magnetization solutions (Whaler and Purucker 2005). The MGS electron reflectometer observations have also been compiled to produce maps of the magnetic field magnitude at a reference altitude of about 170 km (Mitchell et al. 2007;

Lillis et al. 2008b) although with a somewhat lower spatial resolution. Data were also used to built local models over localized magnetic anomalies, using various forward and inverse techniques (Hood and Zakharian 2001; Arkani-Hamed 2001b; Frawley and Taylor 2004; Quesnel et al. 2007; Milburya et al. 2007; Langlais and Quesnel 2008).

3.3 The Moon, Venus and Mercury

Magnetic fields of near-lunar space, and the lunar surface, were first measured in the 1970's during the initial US-Soviet exploration of the Moon. Early measurements showed that the Moon has no global, core-generated magnetic field. Magnetometers, and early electron reflectometers, were place in low-inclination orbits and recognized the internal magnetic signature originating in the South Polar-Aitken basin region on the far side of the Moon. Magnetometers were carried on both US and Soviet landed vehicles, and measured fields of up to several hundred nT. The first global magnetic field measurement of the Moon was carried out by Lunar Prospector in 1998 and 1999, using both a magnetometer and an electron reflectometer (Binder 1998).

Lunar Prospector's magnetic field experiment was flown at low altitudes (30 km average altitude) between 19 December 1998 and 29 July 1999. The magnetometer was on a 2.5 m boom, and was a low-noise (6 pT RMS) triaxial vector fluxgate magnetometer that sampled at rates up to 18 Hz. Spin-averaged, calibrated, data was produced at 5 s intervals. All data are on NASA's Planetary Data System node. The fluxgate magnetometer senses three orthogonal components of the field at the spacecraft, while the electron reflectometer measures the magnitude of the magnetic field near the surface. Recent global maps of the internal magnetic field of the Moon include those of Purucker (2008) and Richmond and Hood (2008) using the vector fluxgate magnetometer, and the map of Mitchell et al. (2008) using the electron reflectometer.

The most detailed map (see Fig. 4) of the Lunar magnetic field is that of Purucker (2008) which describes the magnetic field to spherical degree 150, corresponding to wavelengths of 73 km. The map was made from radial and north–south horizontal components of the field measured in the lunar wake and tail regimes, after first removing a simple model of the external magnetic field. The technique involves isolating the correlative parts of three adjacent passes using a space domain formalism employing ESD (Purucker et al. 1996). This method allows for efficient field calculation and altitude normalization from relatively narrow N–S swaths, which are then assembled into a global mosaic. The density of the coverage was such that 99.2% of the one degree by one-half degree bins are populated, the remaining bins being in the polar regions.

Although Venus is often referred as the twin sister to the Earth, it does not possess an intrinsic magnetic field. The lack of a magnetic field is often interpreted as the result of a non convecting core inside Venus (Stevenson et al. 1983; Nimmo 2002). The surface of Venus is relatively young ($\sim$ .5 Gyr), and may be the result of a planet-scale resurfacing event (Stevenson 2003). The surface temperature of Venus (470°C) is below the Curie temperature of some magnetic minerals, but the titanium content deduced from X-ray fluorescence measurements made by Venera 13, 14 and Vega 2 (Fegley et al. 1997) suggests the presence of titanomagnetite, with associated lower Curie temperature. The lithospheric magnetic field of Venus, if any, is still yet to be detected.

The magnetic field of Mercury remains mysterious. Mariner 10 in 1974/5 reported evidences that Mercury possesses an internal magnetic field. It was measured to be as large as 400 nT at 330 km altitude (Ness et al. 1975). However, only two flybys were performed, opening a 30-year long debate on the origin of the Hermean magnetic field (Wicht

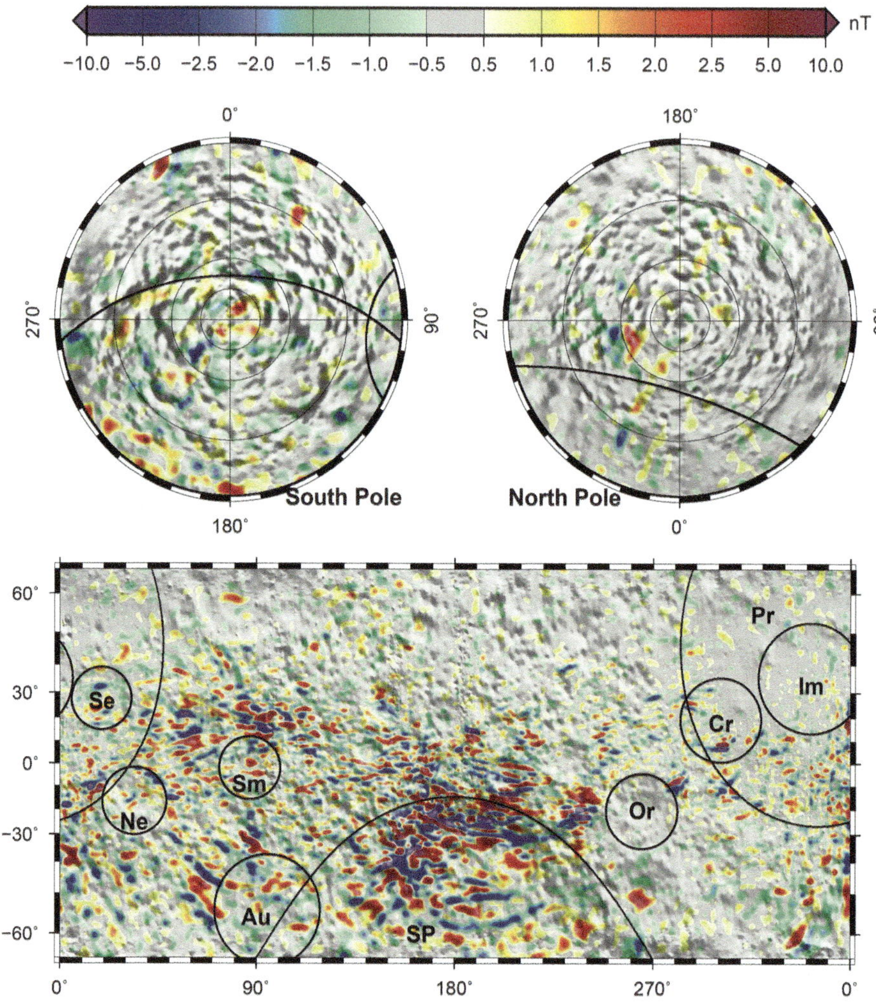

Fig. 4 Predicted radial component of the lunar magnetic field at an altitude of 30 km above the lunar mean radius, based on the spherical harmonic model of Purucker (2008), overlapped onto the ULCN lunar shaded relief. *Top*: South and North pole (orthographic projection), down to 60° S latitude. *Bottom*: Mercator projection between ±65° S latitude. The main rims of the largest impact basins are shown, with labels: Pr Procellarum, SP South Pole, Im Imbrium, Cr Crisium, Or Orientale, Au Australe, Ne Nectaris, Sm Smythii, and Se Serenitatis

et al. 2007). Many hypotheses were formulated, including thermoelectric dynamo (Stevenson 1987; Giampieri and Balogh 2002), or earth-like core dynamo acting under different conditions, with for instance a small inner core (Heimpel et al. 2005) or to the opposite a large inner core (Stanley et al. 2005; Takahashi and Matsushima 2006). Other explanations invoke a convection in a layered core, affecting only the innermost layer of the outer core (Christensen 2006). There may exist a lithospheric field on Mercury (Srnka 1976; Stephenson 1976; Aharonson et al. 2004), but flyby observations from the current MESSENGER mission (Anderson et al. 2008; Purucker et al. 2009) have yet to conclusively observe a lithospheric magnetic field. The interested reader is referred to Anderson et al. (2009) for a detailed review on the Hermean magnetic field.

4 Origin(s) of the Crustal Field(s)

The Earth, the Moon and Mars present very different crustal magnetic fields, both in amplitude and in length scale. Their magnetic fields reflect the cumulative results of both magnetization and demagnetization processes which took place on these planets. In the following, we first discuss the remanent and the induced magnetization and then review the geodynamic properties and processes associated with large-scale magnetic signatures that possibly shaped the magnetic figure of these planets.

4.1 Remanent vs. Induced Origin?

On Mars and the Moon, the crustal magnetic field is primarily of remanent origin. On the Earth, to first order, most of the remanent magnetic contributions cancel out at large wavelength (i.e. those measurable at the spacecraft altitude), and the induced part dominates above continental domains.

The exact contribution of the remanent magnetization is actually difficult to estimate, unless the physical properties of the magnetic sources are known through rock sampling and analysis (Dunlop and Özdemir 2007). It is possible to estimate what is their relative contribution by using forward approaches. Global forward approaches mostly rely on the distribution of the vertically integrated magnetization. A simple estimate is based on the lateral magnetization contrast associated with the boundary between oceanic and continental domains. Counil et al. (1991) showed that abrupt boundaries between the thin oceanic crust (on average 7 km) and the thicker continental crust (on average 40 km) translates into a large scale magnetic signal which is present above the continental margins, and also above continental and oceanic basins. This does not suffice to explain all the observed magnetic anomalies. Taking into account the topography of the oceanic crust (Cohen and Achache 1994), as well as the oceanic remanent signature associated with the Cretaceous Quiet Zone (Dyment and Arkani-Hamed 1998), increases the correlation between forward predictions and actual observations. More complexity can be added to the forward modeling approach. Lateral thickness variations of the continental crust can be introduced as an initial parameter. This is what was done by Purucker et al. (2002). They computed the magnetic field associated with an *a priori* model of the magnetic crust, using a global seismic tomography model (Nataf and Ricard 1996) for the thickness of the crust completed by a sediment thickness model on top of it (Mooney et al. 1998). Two additional assumptions were made: (i) remanent magnetization is solely associated with oceanic crust, and (ii) relatively close susceptibility values for the continents and oceans (0.04 and 0.035 SI, respectively) were assumed (Purucker et al. 1998). The comparison of that forward predicted magnetic field (including terms between degrees 15 and 26) to those of MAGSAT and Ørsted-based models showed some differences in the location of some anomalies (see Fig. 5). These differences could of course be explained by unmodeled remanent contributions, but Purucker et al. (2002) instead suggested to shift the boundary between thick and thin magnetic layer inland, closer to the inboard Coastal plain boundary. The slightly corrected forward model then matched more closely actual observations.

The induced contribution is also the largest time-variable contribution of internal, non-core origin. In theory, it can be separated from the static remanent contribution, as pointed

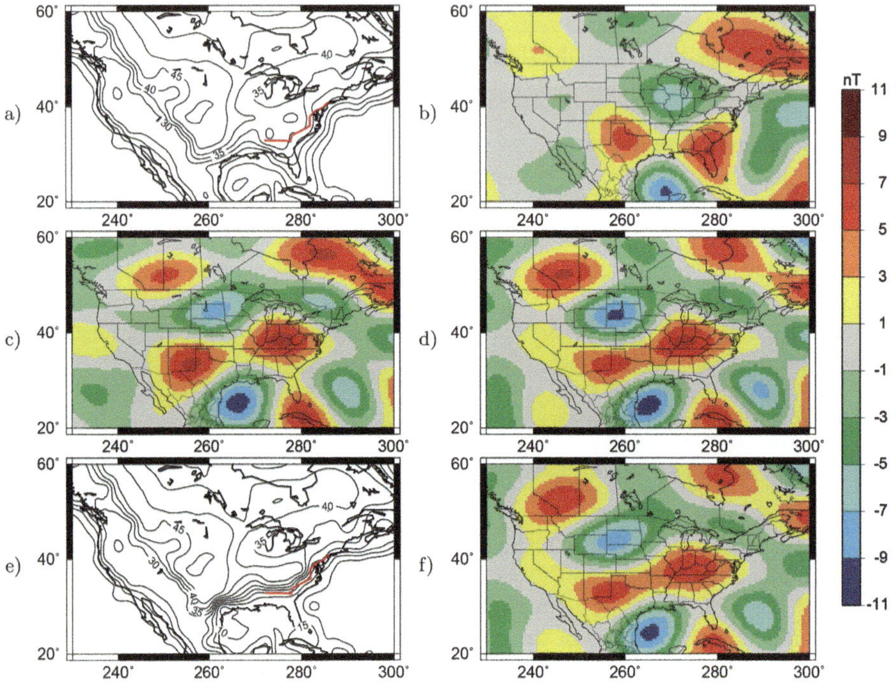

Fig. 5 (**a**) *A priori* thickness of the crust above the north American craton. The *solid red line* corresponds to the boundary between non-magnetic sediments of the Coastal Plain and more magnetic igneous and metamorphic continental rocks. (**b**) Predicted magnetic field at 400 km altitude from (**a**), between degrees 15 and 26. (**c**) and (**d**) 400-km altitude magnetic anomalies from MAGSAT-based and Ørsted-based models, respectively. Note the locations of the two positive lobes, centered above Kentucky and Texas, which differs from locations found in (**b**). (**e**) *A posteriori* thickness of the crust: main difference with (**a**) is the thinner thickness below Florida, Georgia and the rest of the Atlantic and Gulf Coastal Plains. (**f**) Predicted magnetic field at 400 km altitude from (**e**). After (Purucker et al. 2002)

by McLeod (1996): "crustal-source secular variation should dominate core-source secular variation for degrees greater than 22". This was recently challenged by Hulot et al. (2009), who showed that the time-varying crustal field is likely to hide the core field time variations beyond degree 18 or so. Lesur and Gubbins (2000) found that a time-dependent induced magnetic field would better explain the observed crustal biases than a static remanent field. Time variation of the crustal fields is predicted to be on average 0.06 to 0.12 $\mathrm{nT{\cdot}yr^{-1}}$ at the Earth's surface between degrees 15 and 90 (Thébault et al. 2009). In some places, it could reach up to 1.3 $\mathrm{nT{\cdot}yr^{-1}}$. Such a variation at spacecraft altitude would be lower, of the order of 0.02 $\mathrm{nT{\cdot}yr^{-1}}$, and therefore can not be directly detected without long time series of magnetic measurements. This crustal secular variation may for instance explain the difference observed between the observatory crustal biases (i.e. the non-core, lithospheric magnetic field at observatory locations) computed in 1980 and in 2000, using MAGSAT and Ørsted measurements (Mandea and Langlais 2002). The magnetic field of induced origin may indeed dominate the remanent magnetic field. These results, and others, make the remanent magnetic field of Mars even more intriguing, as the intensity of the remanent field on Mars exceeds by three orders of magnitude that of the Earth.

4.2 Impact Structures and Signatures

Impacts have been recognized as a major building and modifying process of planet's shapes and interiors. Impacts are usually associated with destructive effects, which depend on the size, velocity, trajectory angle and composition of the impactor, as well as on the nature of the impacted material. These parameters not only constrain the final shape of the crater (diameter, depth, single vs. complex crater), but also the peak pressure and the released energy associated with the impact. Their effects, as well as those associated with volcanoes and other deeper processes, can actually both (re-)magnetize and demagnetize rocks, through thermal, pressure or fracturing processes. The magnetic signature of an impact will therefore be different depending on whether it is emplaced while a core dynamo is active or not.

The pre-impact magnetization (if any) can be modified through mainly three processes. First, the impact is always associated with an excavation area, material being removed and spread away around the crater (Croft 1980). Second, the shock wave associated with the impact generates high pressures, which may demagnetize material, depending on their magnetic coercitivity (Cisowski and Fuller 1978). Such high pressures are also associated with local (re)magnetization processes (Gattacceca et al. 2008). Thermal effects are also possible, which will demagnetize (and possibly re-magnetize) those minerals which have a Curie temperature lower than the high temperatures associated with impacts. New minerals can also form, and acquire a magnetic remanence. Finally, impacts can cause plasmas, which may generate transient, very intense, magnetic fields (Crawford and Schultz 1999).

On the Earth, 176 impact structures have been identified so far (PASSC, Planetary and Space Science Centre 2009), ranging from 150 m to 300 km in diameter. This may be seen as a small number, especially when compared to other planets such as Mars or the Moon. But the identification of impact structures on the Earth is made difficult because of alteration processes. The largest conclusively identified impact structure, Vredefort, is located in South Africa. It is about 2.0 Ga old and is emplaced in a granitic basement. At the surface, magnetic fields are found to range between 10,000 and 75,000 nT, with horizontal gradients reaching a maximum of 9,000 $nT{\cdot}m^{-1}$ (Muundjua et al. 2007). Aeromagnetic surveys over this structure showed that magnetic anomalies range between $\pm$ 1500 nT at 300-m altitude (Corner and Wilsher 1989). These are mostly organized on a circular annulus, inside the crater rim as shown in Fig. 6. The very intense, small scale, magnetic fields found above Vredefort are very enigmatic, as this is highlighted by the extensive literature on the subject (see for instance (Reimold et al. 2008; Muundjua et al. 2008) for a recent debate). Most of the terrestrial craters exhibit magnetic anomalies weaker than their surroundings, with circular structures dominating (Grieve and Pilkington 1996). This is often interpreted as a local decrease in the magnetic susceptibility. However, this is not an absolute rule, and some craters are associated with magnetic highs. It must also be noted that none of the largest impact craters on the Earth (Vredefort, Chicxulub or Sudbury) produce magnetic signatures that are visible from orbit (see Fig. 2). It may also very well be that some of the observed large-scale magnetic anomalies are associated with subdued impact structures, as it has been suggested for Bangui (Central Africa), one of the largest (and yet unexplained) magnetic anomaly on the Earth (Girdler et al. 1992).

Magnetic effects of impact structures would theoretically be easier to study on other planets, because erosion effects have been less important, keeping a structure that is closer to the original one. But surface surveys are not possible, and one must rely on high altitude measurements. Comparison with terrestrial craters is therefore not directly possible.

Early studies of the lunar magnetic field suggested the importance of plasma-generated magnetic fields as the origin of observed magnetic anomalies (Gold and Sorer 1976). This

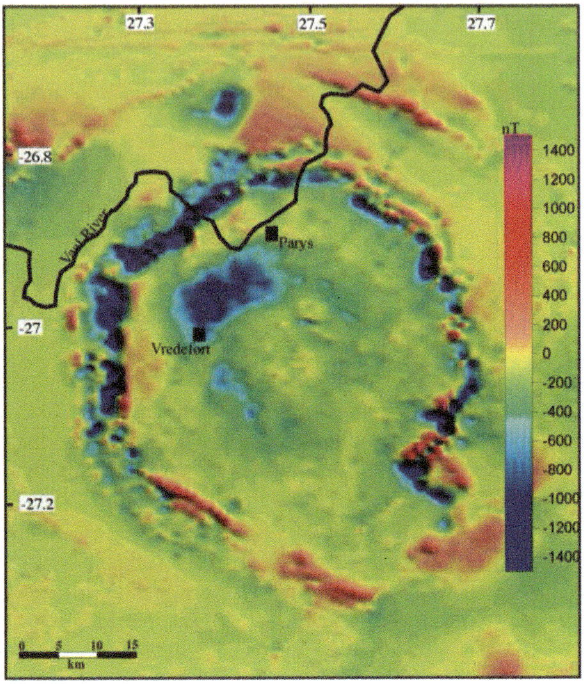

Fig. 6 Aeromagnetic anomaly map above Vredefort crater, at a nominal altitude of 150 m. Two partial circular features are visible, the outer one being located above iron-rich shales and the inner one above a facies transition. After (Muundjua et al. 2007)

is because lunar magnetic field are preferentially found on the antipodes to Crisium, Serenitatis, Imbrium, and Orientale basins (see Fig. 4). The strongest anomalies are actually found on the northwest edge of South Pole—Aitken crater, which is antipodal to Imbrium. Theoretical simulations (Hood and Huang 1991) showed that convergence of impact-related vapor clouds to the antipode of an impact can result into enhanced magnetic fields, if an ambient magnetic field existed at the time of the impact, of internal or external origin. Impacts are also associated with ejectas, which too converge to the antipodes. Although their thickness has been predicted to be small, they could contribute to a local increase of a pre-existing magnetization (Hood and Artemieva 2008). This later would for instance be due to an ancient lunar paleodynamo, as this was recently suggested (Garrick-Bethell et al. 2009).

On Mars, the absence of magnetic field above the largest recognized impact structures (Hellas, Argyre, Isidis, see Fig. 3) has been interpreted as the proof that the core dynamo had already stopped at the time when this impacts took place (Acuña et al. 1999). This simplistic view is however incomplete, as craters are not isolated structures. More complexity comes from the secondary craters, associated with the largest impacts, which were suggested to be too associated with demagnetization processes (Artemieva et al. 2005). Studying the correlation between impact structures and magnetic fields requires some caution. Demagnetization effects often extend very far away from the crater rims, as this was suggested by Hood et al. (2003). For instance, pyrrhotite would loose about 50% of its magnetization under a 1 GPa pressure, which horizontally translates into 4 radii distance away from Hellas and Argyre (Rochette et al. 2003). Using different assumptions, Mohit and Arkani-Hamed (2004) concluded that only the area located inside the crater rims would be demagnetized, with partial demagnetization up to 1.4 radii away from the crater. The apparent demagnetization observed above large craters is used to estimate the time at which the dynamo stopped. Based on the magnetic signature of visible and buried impact structures (Frey 2008), Lillis et al. (2008a) estimated that the dynamo ceased most likely 4.115–4.13 Ga ago, and that this cessation was rather rapid. Such a termination could be attributed to the Late Heavy Bombardment on Mars (Roberts et al. 2009). These different results highlight the importance of studying impact craters on other planets.

4.3 The Magnetic Signature of Other Processes

Many other processes can affect the creation, destruction, and mobilization of magnetic materials within the lithosphere of a planet. In the near surface, volcanism and related igneous processes such as dike emplacement, rifting, and faulting act to mobilize and create magnetic materials.

Generally speaking, magnetization is associated with minerals cooling down below their Curie temperature in the presence of a magnetic field. On the Earth, the magnetic signature associated with seafloor spreading has been recognized for a long time (Vine and Matthews 1963). This signature is used to constrain and reconstruct plate tectonic patterns. On Mars, a sea-floor spreading like scenario was proposed to explain the southern hemisphere intense anomalies (Connerney et al. 1999). Transformed faults have also been suggested in Terra Meridiani (Connerney et al. 2005), but other elements associated with plate tectonics such as subduction zones have yet not been identified. Remote magnetic measurements made above volcanic constructs can be used to estimate the magnetic properties of a volcano (Parker et al. 1987). Taking into account the precise shape of the structure, as well as information obtained from other techniques such as drilling, can help to understand the structure of the underlying volcano (Blanco-Montenegro et al. 2007). Such studies are also useful to constrain the behavior of the paleodynamo. Volcanic edifices and associated magnetic signatures are also studied on Mars, whether they are magnetized, like Apollinaris Patera (Langlais and Purucker 2007), or not, like Hadriaca Patera (Lillis et al. 2008b). Recently, the magnetic field around Arsia Mons volcano was analyzed. Results suggested an important intrusive regime, required to explain the partial demagnetization (Lillis et al. 2009).

Deeper processes can also be characterized through their magnetic signature. This is the case for granulite-grade metamorphism (Clark 1997), and for serpentinization. This later is recognized as an important source for secondary magnetization. It is a metamorphic reaction, occurring at low-to-medium temperature and low pressure. It corresponds to the hydration of mafic or ultramafic rocks of the crust and mantle, and produces serpentine, and other minerals such as talc, quartz or magnetite. The newly formed magnetite acquires a stable remanent magnetization in the presence of an intense magnetic field. On the Earth, serpentinization is commonly found at mid-ocean ridges (Mével 2003). Fresh basalts are rapidly altered, which may reduce the initial thermoremanent magnetization, but the chemical remanent magnetization associated with new minerals could contribute to as much as 80% of the observed magnetic anomalies (Raymond and Labrecque 1987). Serpentinization is one of the processes that is suggested as the origin of the intense Martian magnetic field anomalies (Hood et al. 2005), it is also invoked to explain the crustal dichotomy (Quesnel et al. 2009). Another setting for serpentinization reaction is the mantle wedge of subduction where water is released from the subducted cooler crust. On the Earth the lithospheric magnetic field anomalies observed above the Cascadia subduction zone (Fig. 7) were interpreted by a hydrated mantle, this hypothesis being also supported by gravity field interpretations and seismic velocities (Blakely et al. 2005). Unaltered subdued material can also increase the thickness of the magnetized layer, as this was suggested above the Sumatra subduction area (Purucker and Ishihara 2005; Mandea and Thébault 2007).

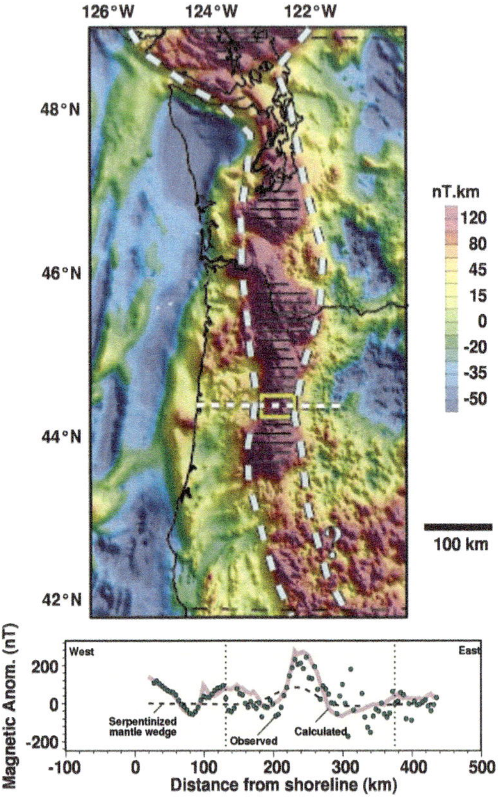

Fig. 7 *Top*: aeromagnetic anomalies (transformed to magnetic potential in nT·km) above the Cascadia subduction zone. *Black horizontal line* pattern shows location of the most intense magnetic anomalies. The *white East–West dashed line* is the location of a teleseismic transect, showing evidence of serpentinized forearc mantle (*yellow rectangle*). *Light blue dashed line* bounds magnetic anomalies interpreted as partially caused by hydrated mantle. *Bottom*: stacked magnetic profiles and predicted magnetic field associated to the Oregon forearc model. *Vertical dotted lines* demotes location of the teleseismic transect. Also shown the magnetic field associated with the mantle wedge alone. Adapted from (Blakely et al. 2005)

5 Conclusions and Open Issues

The magnetic signature of the lithosphere of the telluric planets and of the Moon is related to all the processes that affected and shaped the figure of these planets. The Earth, the Moon, Mars, Venus and Mercury present five different magnetic faces, with overlapping induced and remanent contributions, a weak remanent field of uncertain origin, a strong remanent field very likely related to a past dynamo, a lack of (so far) detectable magnetic field, and a possibly weak lithospheric component overlapped by a weak dynamo field, respectively.

It is recognized that magnetization within these telluric planets and the Moon is controlled in part by the amount of available iron within the crust. Iron is partitioned among oxide, sulfide, and silicate phases in the crust (Clark 1997), and only the first and second of these phases can retain significant remanent or induced magnetization. The relationship between iron abundance and minimum magnetization can be estimated, using techniques developed by Parker (2003). Results are shown on Fig. 8, assuming a 40-km thick magnetized layer. There is a broad correlation between crustal iron content and magnetization for Mars, the Earth and Moon. Venus, and to some extent, Mercury, may appear as anomalous. But the assumed 40-km thick magnetized layer for these two planets is probably wrong by a factor of magnitude; more realistic values would increase the magnetization contrast by a similar factor. Additional influences on the magnetization include the strength of the dynamo field in which the magnetization was acquired, and the mineralogy of the magnetic phases.

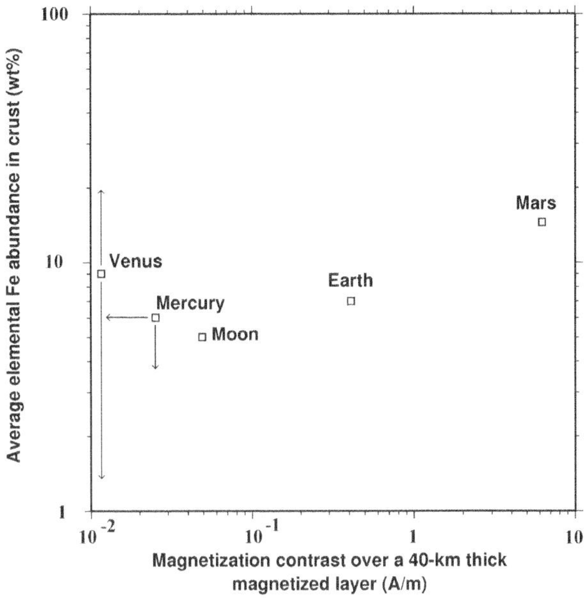

Fig. 8 Minimum magnetization contrast (A/m) versus Fe content of crust (wt%) for the Earth, Mars, Moon, Mercury and Venus, assuming a 40-km-thick magnetized layer. Fe content taken from Hahn et al. (2007), Lodders and Fegley (1998), Grimm and Hess (1997), Solomon et al. (2008). *Arrows on Mercury* indicate that upper bounds for Fe content and magnetization contrast are considered. *Arrows on Venus* indicate the poorly constrained Fe content. Adapted from (Purucker et al. 2009)

The magnetic field of lithospheric origin has a different behavior close to the surface or at spacecraft altitude. Simply imagine the Earth (or its twin) without a dynamo: a MGS-like spacecraft would have measured very little magnetic fields, if any! It is mandatory to pursue our efforts in measuring, modeling and interpreting the magnetic field at various wavelength on the Earth and on other planets.

On the Earth, the WDMAM 2.0 is currently in progress; improvements will be made with the addition of both surface and spacecraft measurements. This will be possible thanks to the forthcoming ESA's Swarm mission, a novel constellation comprising three identical satellites carrying magnetometers. Two of the Swarm satellites will fly close to each other at lower altitudes, measuring the East–West gradient of the magnetic field, while the third one will fly at a higher altitude (Friis-Christensen et al. 2006). These further improvements in describing and understanding the magnetic field anomaly source properties will also benefit from joint analysis of geophysical and geological datasets.

The flying NASA MESSENGER mission, and the planned ESA's Bepi Colombo mission will characterize the Hermean magnetic field, and possibly put limits on the lithospheric field of Mercury. On Mars, following the surprising discovery of Mars crustal magnetism by MGS in 1996, ten more spacecraft and landers have been launched with no experiment capable of measuring magnetic fields, with the exception of Rosetta which made one flyby close to Mars in 2007 (Boesswetter et al. 2009). Hopefully, the next NASA's SCOUT mission, MAVEN, will provide new measurements of the Martian magnetic field. Such measurements, combined with monitoring of the current atmospheric escape on Mars (Langlais et al. 2009; Leblanc et al. 2009) could be used to estimate what has been the fate of water on Mars along its past history.

More than 400 year ago, Gilbert concluded from his observations that the compass was deviated by "imperfections" (i.e. oceanic areas or continental masses) on the sphere. Now we know that the deviation of a compass is related to changes in the magnetic content, whether it is the magnetization components or the amount of magnetized material. The next step will be to evaluate which one of these two properties governs the magnetic signature of the lithosphere of terrestrial planets.

Acknowledgements All geographical figures were produced with GMT (Wessel and Smith 1991). Support was provided by Helmholtz-Zentrum Potsdam Deutsches GeoForschungsZentrum, Germany; Institut National des Sciences de la Terre/CNRS, France. We would like to thank E. Thébault for discussions on Spherical Cap Harmonic Analysis, R. Hart for providing figure on Vredefort crater, R. Balogh and the staff at ISSI for organizing a fruitful workshop. This work is dedicated to the memory of Mario H. Acuña, who inspired so many of us.

References

M.H. Acuña, J.E.P. Connerney, P. Wasilewski, R.P. Lin, K.A. Anderson, C.W. Carlson, J. McFadden, D.W. Curtis, H. Réme, A. Cros, J.L. Médale, J.A. Sauvaud, C. d'Uston, S.J. Bauer, P. Cloutier, M. Mayhew, N.F. Ness, Mars Observer magnetic fields investigation. J. Geophys. Res. **97**, 7799–7814 (1992)

M.H. Acuña, J.E.P. Connerney, N.F. Ness, R.P. Lin, D. Mitchell, C.W. Carlson, J. McFadden, K.A. Anderson, H. Reme, C. Mazelle, D. Vignes, P. Wasilewski, P. Cloutier, Global distribution of crustal magnetization discovered by the Mars global surveyor MAG/ER experiment. Science **284** (1999). doi:10.1126/science.284.5415.790

M.H. Acuña, J.E.P. Connerney, P. Wasilewski, R.P. Lin, D. Mitchell, K.A. Anderson, C.W. Carlson, J. McFadden, H. Rème, C. Mazelle, D. Vignes, S.J. Bauer, P. Cloutier, N.F. Ness, Magnetic field of Mars: Summary of results from the aerobraking and mapping orbits. J. Geophys. Res. **106** (2001). doi:10.1029/2000JE001404

O. Aharonson, M.T. Zuber, S.C. Solomon, Crustal remanence in an internally magnetized non-uniform shell: a possible source for Mercury's magnetic field? Earth Planet. Sci. Lett. **218**, 261–268 (2004)

A.L. Albee, R.E. Arvidson, F. Palluconi, T. Thorpe, Overview of the Mars global surveyor mission. J. Geophys. Res. **106** (2001). doi:10.1029/2000JE001306

B.J. Anderson, M.H. Acuña, H. Korth, M.E. Purucker, C.L. Johnson, J.A. Slavin, S.C. Solomon, R.L. McNutt, The structure of Mercury's magnetic field from MESSENGER's first flyby. Science **321**, 82–85 (2008). doi:10.1126/science.1159081

Anderson et al., Space Sci. Rev. (2009, this issue)

J. Arkani-Hamed, A 50-degree spherical harmonic model of the magnetic field of Mars. J. Geophys. Res. **106** (2001a). doi:10.1029/2000JE001365

J. Arkani-Hamed, Paleomagnetic pole positions and pole reversals of Mars. Geophys. Res. Lett. **28** (2001b). doi:10.1029/2001GL012928

J. Arkani-Hamed, A coherent model of the crustal magnetic field of Mars. J. Geophys. Res. **109** (2004). doi:10.1029/2004JE002265

N. Artemieva, L.L. Hood, B.A. Ivanov, Impact demagnetization of the Martian crust: Primaries versus secondaries. Geophys. Res. Lett. **32** (2005). doi:10.1029/2005GL024385

Baumjohann et al., Space Sci. Rev. (2009, this issue)

A.B. Binder, Lunar Prospector: Overview. Science **281**, 1475–1476 (1998)

R.J. Blakely, *Potential Theory in Gravity and Magnetic Applications* (Cambridge University Press, Cambridge, 1995), p. 441

R.J. Blakely, T.M. Brocher, R.E. Wells, Subduction zone magnetic anomalies and implications for hydrated forearc mantle. Geology **33**, 445–448 (2005)

I. Blanco-Montenegro, R. De Ritis, M. Chiappini, Imaging and modelling the subsurface structure of volcanic calderas with high-resolution aeromagnetic data at vulcano (Aeolian islands, Italy). Bull. Volcanol. **69** (2007). doi:10.1007/s00445-006-0100-7

A. Boesswetter, U. Auster, I. Richter, M. Fränz, B. Langlais, S. McKenna-Lawlor, S. Simon, U. Motschmann, K.H. Glassmeier, N.J.T. Edberg, R. Lundin, Rosetta swing-by at Mars - An analysis of the ROMAP measurements in comparison with results of 3d multi-ion hybrid simulations and MEX/ASPERA-3 data. Ann. Geophys. **27**, 2383–2398 (2009)

J.C. Cain, D.R. Schmitz, L. Muth, Small-scale features in the Earth's magnetic field observed by Magsat. J. Geophys. Res. **89**, 1070–1076 (1984)

J.C. Cain, B.B. Ferguson, D. Mozzoni, An $n = 90$ internal potential function of the Martian crustal magnetic field. J. Geophys. Res. **108** (2003). doi:10.1029/2000JE001487

A. Chambodut, I. Panet, M. Mandea, M. Diament, M. Holschneider, O. Jamet, Wavelet frames: an alternative to spherical harmonic representation of potential fields. Geophys. J. Int. **163** (2005). doi:10.1111/j.1365-246X.2005.02754.x

U.R. Christensen, A deep dynamo generating Mercury's magnetic field. Nature **444** (2006). doi:10.1038/nature05342
S.M. Cisowski, M. Fuller, The effect os shock on the magnetism of terrestrial rocks. J. Geophys. Res. **83**, 3441–3458 (1978)
D. Clark, Magnetic petrophysics and magnetic petrology: aids to geological interpretation of magnetic surveys. AGSO J. Aust. Geol. Geophys. **2**, 83–103 (1997)
Y. Cohen, J. Achache, Contribution of induced and remanent magnetization to long-wavelength oceanic magnetic anomalies. J. Geophys. Res. **99**, 2943–2954 (1994)
F.R. Colomb, C. Alonso, C. Hofmann, I. Nollmann, SAC-C mission, an example of international cooperation. Adv. Space Res. **34** (2004). doi:10.1016/j.asr.2003.10.039
J.E.P. Connerney, M.H. Acuna, P.J. Wasilewski, N.F. Ness, H. Reme, C. Mazelle, D. Vignes, R.P. Lin, D.L. Mitchell, P.A. Cloutier, Magnetic lineations in the ancient crust of Mars. Science **284**, 794 (1999)
J.E.P. Connerney, M.H. Acuña, P.J. Wasilewski, G. Kletetschka, N.F. Ness, H. Rème, R.P. Lin, D.L. Mitchell, The global magnetic field of Mars and implications for crustal evolution. Geophys. Res. Lett. **28** (2001). doi:10.1029/2001GL013619
J.E.P. Connerney, M.H. Acuña, N.F. Ness, T. Spohn, G. Schubert, Mars crustal magnetism. Space Sci. Rev. **111** (2004). doi:10.1023/B:SPAC.0000032719.40094.1d
J.E.P. Connerney, M.H. Acuña, N.F. Ness, G. Kletetschka, D.L. Mitchell, R.P. Lin, H. Reme, Tectonic implications of Mars crustal magnetism. Proc. Nat. Acad. Sci. **102** (2005). doi:10.1073/pnas.0507469102
B. Corner, W.A. Wilsher, Structure of the Witwatersrand basin derived from interpretation of aeromagnetic and gravity data. Exploration '87 Geol. Surv. Can. Spec. **3**, 523–546 (1989)
J. Counil, Y. Cohen, J. Achache, The global continent-ocean magnetization contrast. Earth Planet. Sci. Lett. **103**, 354–364 (1991)
D.A. Crawford, P.H. Schultz, Electromagnetic properties of impact-generated plasma, vapor and debris. Int. J. Impact Eng. **23**, 169–180 (1999)
S.K. Croft, Cratering flow fields – Implications for the excavation and transient expansion stages of crater formation, in *Proc. Lunar Planet. Sci. Conf.*, vol. 11, 1980, pp. 2347–2378
S.S. Dolginov, L.N. Zhuzgov, The magnetic field and magnetosphere of the planet Mars. Planet. Space Sci. **39**, 1493–1510 (1991)
S.S. Dolginov, L.N. Zhuzgov, N.V. Pushkov, Preliminary report on geomagnetic measurements on the third Soviet artificial Earth satellite. Planet. Space Sci. **5**, 244–247 (1961)
D.J. Dunlop, O. Özdemir, *Rock Magnetism: Fundamentals and Frontiers* (Cambridge University Press, Cambridge, 1997), p. 573
D.J. Dunlop, O. Özdemir, Magnetizations of rocks and minerals, in *Treatise on Geophysics*, vol. 5, ed. by M. Kono (Elsevier, Amsterdam, 2007), pp. 277–336
J. Dyment, J. Arkani-Hamed, Equivalent source magnetic dipoles revisited. Geophys. Res. Lett. **25**, 2003–2006 (1998)
J. Dyment, S.C. Cande, J. Arkani-Hamed, Skewness of marine magnetic anomalies created between 85 and 40 Ma in the Indian ocean. J. Geophys. Res. **99**, 24121–24134 (1995)
ETOPO5, Data Announcement 88-MGG-02, Digital relief of the surface of the Earth. NOAA. National Geophysical Data Center, Boulder, Colorado, USA (1988)
B.J. Fegley, G. Klingelhfer, K. Lodders, T. Widemann, Geochemistry of surface-atmosphere interactions on Venus, in *Venus II*, ed. by W. Boucher, D. Hunten, R. Phillips (Univ. of Arizona Press, Tucson, 1997), pp. 591–636
J.J. Frawley, P.T. Taylor, Paleo-pole positions from martian magnetic anomaly data. Icarus **172**, 316–327 (2004)
H.V. Frey, Ages of very large impact basins on Mars: Implications for the late heavy bombardment in the inner solar system. Geophys. Res. Lett. (2008). doi:10.1029/2008GL033515
E. Friis-Christensen, H. Lühr, G. Hulot, A constellation to study the Earth's magnetic field. Earth Planets Space **58**, 351–358 (2006)
I. Garrick-Bethell, B.P. Weiss, D.L. Shuster, J. Buz, Early lunar magnetism. Science **323** (2009). doi:10.1126/science.1166804
J. Gattacceca, L. Berthe, M. Boustie, F. Vadeboin, P. Rochette, T. de Resseguier, On the efficiency of shock magnetization processes. Phys. Earth Planet. Int. **166** (2008). doi:10.1016/j.pepi.2007.09.005
G. Giampieri, A. Balogh, Mercury's thermoelectric dynamo model revisisted. Planet. Space Sci. **50**, 757–762 (2002)
W. Gilbert, *De magnete* (Translation by Silvanus Phillips Thompson and the Gilbert Club, Chiswick Press, 1900, 337 pp., London, UK, 1600)
R.W. Girdler, P. Taylor, J.J. Frawley, A possible impact origin for the Bangui magnetic anomaly (Central Africa). Tectonophysics **212**, 45–58 (1992)
T. Gold, S. Sorer, Cometary impact and the magnetization of the Moon. Planet. Space Sci. **24**, 45–54 (1976)

R.A.F. Grieve, M. Pilkington, The signature of terrestrial impacts. AGSO J. Geol. Geophys. **16**, 399–420 (1996)
R.E. Grimm, P.C. Hess, The crust of Venus, in *Venus II*, ed. by W. Boucher, D. Hunten, R. Phillips (Univ. of Arizona Press, Tucson, 1997), pp. 1205–1244
B.C. Hahn, S.M. McLennan, G.J. Taylor, W.V. Boynton, J.M. Dohn, M.J. Finch, D.K. Hamara, D.M. Janes, S. Karunatillake, J.M. Keller, K.E. Kerry, A.E. Metzger, R.M.. Williams, Mars Odyssey Gamma Ray Spectrometer elemental abundances and apparent relative surface age: implications for Martian crustal evolution. J. Geophys. Res. **112** (2007). doi:10.1029/2006JE002821
G.V. Haines, Spherical cap harmonic analysis. J. Geophys. Res. **90**, 2583–2591 (1985)
M.H. Heimpel, J.M. Aurnou, F.M. Al-Shamali, N. Gomez Perez, A numerical study of dynamo action as a function of spherical shell geometry. Earth Planet. Sci. Lett. **236** (2005). doi:10.1016/j.epsl.2005.04.032
L.L. Hood, N.A. Artemieva, Antipodal effects of lunar basin-forming impacts: initial 3D simulations and comparisons with observations. Icarus **193** (2008). doi:10.1016/j.icarus.2007.08.023
L.L. Hood, Z. Huang, Formation of magnetic anomalies antipodal to lunar impact basins – Two-dimensional model calculations. J. Geophys. Res. **96**, 9837–9846 (1991)
L.L. Hood, A. Zakharian, Mapping and modeling of magnetic anomalies in the northern polar region of Mars. J. Geophys. Res. **106** (2001). doi:10.1029/2000JE001304
L.L. Hood, N.C. Richmond, E. Pierazzo, P. Rochette, Distribution of crustal magnetic fields on Mars: Shock effects of basin-forming impacts. Geophys. Res. Lett. **30** (2003). doi:10.1029/2002GL016657
L.L. Hood, C.N. Young, N.C. Richmond, K.P. Harrison, Modeling of major martian magnetic anomalies: Further evidence for polar reorientations during the Noachian. Icarus **177** (2005). doi:10.1016/j.icarus.2005.02.008
Hulot et al., Space Sci. Rev. (2009, this issue)
G. Hulot, N. Olsen, E. Thébault, K. Hemant, Crustal concealing of small scale core field secular variation. Geophys. J. Int. **177** (2009). doi:10.1111/j.1365-246X.2009.04119.x
J. Jankowski, C. Sucksdorff, *Guide for Magnetic Measurements and Observatory Practice* (Int. Assoc. Geomag. Aeronomy, Warsaw, 1996), p. 235
J.V. Korhonen, J.D. Fairhead, M. Hamoudi, K. Hemant, V. Lesur, M. Mandea, S. Maus, M.E. Purucker, D. Ravat, T. Sazonova, E. Thébault, Magnetic anomaly map of the world—Carte des anomalies magnétiques du monde, Scale: 1:50,000,000, 1st edition. Commission for the Geological Map of the World (2007)
R.A. Langel, Global magnetic anomaly maps derived from POGO spacecraft data. Phys. Earth Planet. Int. **62**, 208–230 (1990)
B. Langlais, M.E. Purucker, A polar magnetic paleopole associated with Apollinaris Patera. Planet. Space Sci. **55** (2007). doi:10.1016/j.pss.2006.03.008
B. Langlais, Y. Quesnel, New perspectives on Mars' crustal magnetic field. Comptes Rendus Geosci. **340** (2008). doi:10.1016/j.crte.2008.08.006e
B. Langlais, M.E. Purucker, M. Mandea, Crustal magnetic field of Mars. J. Geophys. Res. **109** (2004). doi:10.1029/2003JE002048
B. Langlais, F. Leblanc, T. Fouchet, S. Barabash, D. Breuer, E. Chassefière, A. Coates, V. Dehant, F. Forget, H. Lammer, S. Lewis, M. Lopez-Valverde, M. Mandea, M. Menvielle, A. Pais, M. Paetzold, P. Read, C. Sotin, P. Tarits, S. Vennerstrom, G. Branduardi-Raymont, G. Cremonese, J.G.M. Merayo, T. Ott, H. Rème, J.G. Trotignon, J.E. Walhund, Mars Environment and Magnetic Orbiter: model payload. Exp. Astron. **22** (2009). doi:10.1007/s10686-008-9101-1
F. Leblanc, B. Langlais, T. Fouchet, S. Barabash, D. Breuer, E. Chassefière, A. Coates, V. Dehant, F. Forget, H. Lammer, S. Lewis, M. Lopez-Valverde, M. Mandea, M. Menvielle, A. Pais, M. Paetzold, P. Read, C. Sotin, P. Tarits, S. Vennerstrom, Mars environment and magnetic orbiter: Science and measurement objectives. Astrobiology (2009). doi:10.1089/AST.2007.022
V. Lesur, Introducing localized constraints in global geomagnetic field modelling. Earth Planets Space **58**, 477–483 (2006)
V. Lesur, D. Gubbins, Using geomagnetic secular variation to separate remanent and induced sources of the crustal magnetic field. Geophys. J. Int. **142**, 889–897 (2000)
V. Lesur, A. Jackson, Exact solutions for internally induced magnetization in a shell. Geophys. J. Int. **140**, 453–459 (2000)
V. Lesur, S. Maus, A global lithospheric magnetic field model with reduced noise level in the polar regions. Geophys. Res. Lett. **33** (2006). doi:10.1029/2006GL025826
V. Lesur, I. Wardinski, M. Rother, M. Mandea, GRIMM: the GFZ Reference Internal Magnetic Model based on vector satellite and observatory data. Geophys. J. Int. **173** (2008). doi:10.1111/j.1365-246X.2008.03724.x
R.J. Lillis, H.V. Frey, M. Manga, Rapid decrease in Martian crustal magnetization in the Noachian era: Implications for the dynamo and climate of early Mars. Geophys. Res. Lett. **35** (2008a). doi:10.1029/2008GL034338

R.J. Lillis, H.V. Frey, M. Manga, D.L. Mitchell, R.P. Lin, M.H. Acuña, S.W. Bougher, An improved crustal magnetic field map of Mars from electron reflectometry: Highland volcano magmatic history and the end of the martian dynamo. Icarus **194** (2008b). doi:10.1016/j.icarus.2007.09.032

R.J. Lillis, J. Dufek, J.E. Bleacher, M. Manga, Demagnetization of crust by magmatic intrusion near the arsia mons volcano: Magnetic and thermal implications for the development of the tharsis province, mars. J. Volcanol. Geotherm. Res. (2009, in press). doi:10.1016/j.jvolgeores.2008.12.007

K. Lodders, J.B. Fegley, *The Planetary Scientist's Companion* (Oxford University Press, New York, 1998), p. 371

P. Lognonné, J. Gagnepain-Beyneix, H. Chenet, A new seismic model of the Moon: implications for structure, thermal evolution and formation of the Moon. Earth Planet. Sci. Lett. **211** (2003). doi:10.1016/S0012-821X(03)00172-9

M. Mandea, B. Langlais, Observatory crustal magnetic biases during MAGSAT and Ørsted satellite missions. Geophys. Res. Lett. **29** (2002). doi:10.1029/2001GL013693

M. Mandea, M.E. Purucker, Observing, modeling, and interpreting magnetic fields of the solid Earth. Surv. Geophys. **26** (2005). doi:10.1007/s10712-005-3857-x

M. Mandea, E. Thébault, *The Changing Faces of the Earth's Magnetic Field* (Commission for The Geological Map of the World, France, 2007), p. 49

S. Maus, V. Haak, Magnetic field annihilators: invisible magnetization at the magnetic equator. Geophys. J. Int. **155**, 509–513 (2003)

S. Maus, M. Rother, R. Holme, H. Lühr, N. Olsen, V. Haak, First scalar magnetic anomaly map from champ satellite data indicates weak lithospheric field. Geophys. Res. Lett. **29** (2002). doi:10.1029/2001GL013685

S. Maus, H. Lühr, M. Rother, K. Hemant, G. Balasis, P. Ritter, C. Stolle, Fifth-generation lithospheric magnetic field model from CHAMP satellite measurements. Geochem., Geophys., Geosyst. **8** (2007). doi:10.1029/2006GC001521

S. Maus, F. Yin, H. Lühr, C. Manoj, M. Rother, J. Rauberg, I. Michaelis, C. Stolle, R.D. Müller, Resolution of direction of oceanic magnetic lineations by the sixth-generation lithospheric magnetic field model from CHAMP satellite magnetic measurements. Geochem., Geophys., Geosyst. **9** (2008). doi:10.1029/2008GC001949

C. Mayer, T. Maier, Separating inner and outer Earth's magnetic field from CHAMP satellite measurements by means of vector scaling functions and wavelet. Geophys. J. Int. **167** (2006). doi:10.1111/j.1365-246X.2006.03199.x

M.A. Mayhew, Inversion of satellite magnetic anomaly data. J. Geophys. **45**, 119–128 (1979)

M.G. McLeod, Spatial and temporal power spectra of the geomagnetic field. J. Geophys. Res. **101**, 2745–2763 (1996)

C. Mével, Serpentinization of abyssal peridotites at mid-ocean ridges. Comptes Rendus Geosci. **335**, 825–852 (2003). doi:10.1016/j.crte.2003.08.006

C.A.E. Milburya, S.E. Smrekar, C.A. Raymond, G. Schubert, Lithospheric structure in the eastern region of Mars' dichotomy boundary. Planet. Space Sci. **55** (2007). doi:10.1016/j.pss.2006.03.009s

D.L. Mitchell, R.P. Lin, C. Mazelle, H. Rème, P.A. Cloutier, J.E.P. Connerney, M.H. Acuña, N.F. Ness, Probing Mars' crustal magnetic field and ionosphere with the MGS Electron Reflectometer. J. Geophys. Res. **106** (2001). doi:10.1029/2000JE001435

D.L. Mitchell, R.J. Lillis, R.P. Lin, J.E.P. Connerney, M.H. Acuña, A global map of Mars' crustal magnetic field based on electron reflectometry. J. Geophys. Res. **112** (2007). doi:10.1029/2005JE002564

D.L. Mitchell, J.S. Halekas, R.P. Lin, S. Frey, L.L. Hood, M.H. Acuña, A. Binder, Global mapping of lunar crustal magnetic fields by Lunar Prospector. Icarus **194** (2008). doi:10.1016/j.icarus.2007.10.027

D. Moehlmann, The question of a Martian planetary magnetic field. Adv. Space Res. **12**, 213–217 (1992)

D. Moehlmann, W. Riedler, J. Rustenbach, K. Schwingenschuh, J. Kurths, U. Motschmann, T. Roatsch, K. Sauer, H.T.M. Lichtenegger, The question of an internal Martian magnetic field. Planet. Space Sci. **39**, 83–88 (1991)

P.S. Mohit, J. Arkani-Hamed, Impact demagnetization of the martian crust. Icarus **168** (2004). doi:10.1016/j.icarus.2003.12.005

W.D. Mooney, G. Laske, T.G. Masters, CRUST 5.1: A global crustal model at 5 degrees x 5 degrees. J. Geophys. Res. **103**, 727–747 (1998)

M. Muundjua, R.J. Hart, S.A. Gilder, L. Carporzen, A. Galdeano, Magnetic imaging of the vredefort impact crater, south Africa. Earth Planet. Sci. Lett. **261** (2007). doi:10.1016/j.epsl.2007.07.044

M. Muundjua, A. Galdeano, L. Carporzen, S.A. Gilder, R.J. Hart, M.A.G. Andreoli, M. Tredoux, Reply to scientific comment by W.U. Reimold, R.L. Gibson and H. Henkel on Muundjua et al. (2007), magnetic imaging of the vredefort impact crater, south Africa, epsl 261, pp. 456–468. Earth Planet. Sci. Lett. **273** (2008). doi:10.1016/j.epsl.2008.06.044

H.C. Nataf, Y. Ricard, 3SMAC: An a priori tomographic model of the upper mantle based on geophysical modeling. Phys. Earth Planet. Int. **95**, 101–122 (1996)

J. Needham, Science and civilisation in China, vol. 4 *Physics and Physical Technology*, Part 1 *Physics* (Cambridge University Press, Cambridge, 1962), pp. 314–334

N.F. Ness, The magnetic fields of Mercury, Mars, and moon. Ann. Rev. Earth Planet. Sci. **7**, 249–288 (1979)

N.F. Ness, K.W. Behannon, R.P. Lepping, Y.C. Whang, Magnetic field of Mercury confirmed. Nature **255**, 204 (1975)

F. Nimmo, Why does Venus lack a magnetic field? Geology **30**, 989–990 (2002)

N. Olsen, R. Holme, G. Hulot, T. Sabaka, T. Neubert, L. Toffner-Clausen, F. Primdahl, J. Jorgensen, J.-M. Leger, D. Barraclough, J. Bloxham, J. Cain, C. Constable, V. Golovkov, A. Jackson, P. Kotze, B. Langlais, S. Macmillan, M. Mandea, J. Merayo, L. Newitt, M. Purucker, T. Risbo, M. Stampe, A. Thomson, C. Voorhies, Orsted initial field model. Geophys. Res. Lett. **27**(22), 3607–3610 (2000)

N. Olsen, M. Mandea, Rapidly changing flows in the Earth's core. Nature Geosci. **1** (2008). doi:10.1038/ngeo203

R.L. Parker, Ideal bodies for Mars magnetics. J. Geophys. Res. **108** (2003). doi:10.1029/2001JE001760

R.L. Parker, L. Shure, J.A. Hildebrand, The application of inverse theory to seamount magnetism. Rev. Geophys. **25**, 17–40 (1987)

PASSC, Planetary and Space Science Centre, *Earth impact database*, 2009. http://www.unb.ca/passc/impactdatabase/

V.G. Perminov, The Difficult Road to Mars: A Brief History of Mars Exploration in the Soviet Union (Monographs in aerospace history; no. 15, NASA Headquarters, Washington, D.C., 1999)

M.E. Purucker, A global model of the internal magnetic field of the Moon based on Lunar Prospector magnetometer observations. Icarus **197** (2008). doi:10.1016/j.icarus.2008.03.016

M.E. Purucker, T. Ishihara, Magnetic images of the Sumatra region crust. EOS, Trans. Am. Geophys. Union **86** (2005). doi:10.1029/2005EO100002

M.E. Purucker, K. Whaler, Crustal magnetism, in *Treatise on Geophysics*, vol. 5, Geomagnetism, ed. by M. Kono (Elsevier, Amsterdam, 2007), pp. 195–237

M.E. Purucker, T.J. Sabaka, R.A. Langel, Conjugate gradient analysis: A new tool for studying satellite magnetic data sets. Geophys. Res. Lett. **23**, 507–510 (1996)

M.E. Purucker, R.A. Langel, M. Rajaram, C. Raymond, Global magnetization models with a priori information. J. Geophys. Res. **103**, 2563–2584 (1998)

M.E. Purucker, D. Ravat, H. Frey, C. Voorhies, T. Sabaka, M.H. Acuña, An altitude-normalized magnetic map of Mars and its interpretation. Geophys. Res. Lett. **27** (2000). doi:10.1029/2000GL000072

M.E. Purucker, B. Langlais, N. Olsen, G. Hulot, M. Mandea, The southern edge of cratonic North America: Evidence from new satellite magnetometer observations. Geophys. Res. Lett. **29** (2002). doi:10.1029/2001GL0136450

M.E. Purucker, T.J. Sabaka, S.C. Solomon, B.J. Anderson, H. Korth, M.T. Zuber, G.A. Neumann, Mercury's internal magnetic field: Constraints on large- and small-scale fields of crustal origin. Earth Planet. Sci. Lett. (2009). doi:10.1016/j.epsl.2008.12.017

Y. Quesnel, B. Langlais, C. Sotin, Local inversion of magnetic anomalies: Implication for Mars' crustal evolution. Planet. Space Sci. **55** (2007). doi:10.1016/j.pss.2006.02.004

Y. Quesnel, B. Langlais, C. Sotin, A. Galdéano, Modeling and inversion of local magnetic anomalies. J. Geophys. Eng. **5** (2008). doi:10.1088/1742-2132/5/4/003

Y. Quesnel, C. Sotin, B. Langlais, S. Costin, M. Mandea, M. Gottschalk, J. Dyment, Serpentinization of the martian crust during Noachian. Earth Planet. Sci. Lett. **277** (2009). doi:10.1016/j.epsl.2008.10.012

C.A. Raymond, J.L. Labrecque, Magnetization of the oceanic crust: thermoremanent magnetization or chemical remanent magnetization. J. Geophys. Res. **92**, 8077–8088 (1987)

R.D. Regan, J.C. Cain, W.M. Davis, Global magnetic anomaly map. J. Geophys. Res. **80**, 794–802 (1975)

C. Reigber, H. Lühr, S. P., CHAMP Mission Status. Adv. Space Res. **30** (2002). doi:10.1016/S0273-1177(02)00276-4

W.U. Reimold, R.L. Gibson, H. Henkel, Scientific comment on Muundjua et al., 2007: Magnetic imaging of the Vredefort impact crater, South Africa, EPSL 261, 456–468. Earth Planet. Sci. Lett. **273** (2008). doi:10.1016/j.epsl.2008.06.046

N.C. Richmond, L.L. Hood, A preliminary global map of the vector lunar crustal magnetic field based on Lunar Prospector magnetometer data. J. Geophys. Res. **113** (2008). doi:10.1029/2007JE002933

W. Riedler, K. Schwingenschuh, D. Moehlmann, V.N. Oraevskii, E. Eroshenko, J. Slavin, Magnetic fields near Mars – First results. Nature **341**, 604–607 (1989)

J.H. Roberts, R.J. Lillis, M. Manga, Giant impact on early Mars and the cessation of the Martian dynamo. J. Geophys. Res. **114** (2009). doi:10.1029/2008JE003287

P. Rochette, G. Fillion, R. Ballou, F. Brunet, B. Ouladdiaf, L.L. Hood, High pressure magnetic transition in pyrrhotite and impact demagnetization on Mars. Geophys. Res. Lett. **30** (2003). doi:10.1029/2003GL017359

S.K. Runcorn, On the interpretation of lunar magnetism. Phys. Earth Planet. Int. **10**, 327–335 (1975)
T.J. Sabaka, N. Olsen, M.E. Purucker, Extending comprehensive models of the Earth's magnetic field with Ørsted and CHAMP data. Geophys. J. Int. **159** (2004). doi:10.1111/j.1365-246X.2004.02421.x
J.A. Slavin, K. Schwingenschuh, W. Riedler, E. Eroshenko, The solar wind interaction with Mars – Mariner 4, Mars 2, Mars 3, Mars 5, and PHOBOS 2 observations of bow shock position and shape. J. Geophys. Res. **96**, 11235 (1991)
E.J. Smith, L.J. Davis, P.J. Coleman Jr., D.E. Jones, Magnetic field measurements near Mars. Science **149**, 1241–1242 (1965)
D. Smith, G. Neumann, R.E. Arvidson, E.A. Guinness, S.Slavney, Mars global surveyor laser altimeter mission experiment gridded data record. NASA Planetary Data System, MGS-M-MOLA-5-MEGDR-L3-V1.0 (2003)
S.C. Solomon, R.L. McNutt Jr., T.R. Watters, D.J. Lawrence, W.C. Feldman, J.W. Head, S.M. Krimigis, S.L. Murchie, R.J. Phillips, J.A. Slavin, M.T. Zuber, Return to Mercury: A global perspective on MESSENGER's first Mercury flyby. Science **321**, 59–62 (2008)
L.J. Srnka, Magnetic dipole moment of a spherical shell with TRM acquired in a field of internal origin. Phys. Earth Planet. Int. **11**, 184–190 (1976)
S. Stanley, J. Bloxham, W.E. Hutchison, M.T. Zuber, Thin shell dynamo models consistent with Mercury's weak observed magnetic field [rapid communication]. Earth Planet. Sci. Lett. **234** (2005). doi:10.1016/j.epsl.2005.02.040
A. Stephenson, Crustal remanence and the magnetic moment of Mercury. Earth Planet. Sci. Lett. **28**, 454–458 (1976)
D.J. Stevenson, Mercury's magnetic field – A thermoelectric dynamo? Earth Planet. Sci. Lett. **82**, 114–120 (1987)
D.J. Stevenson, Planetary magnetic fields. Earth Planet. Sci. Lett. **208** (2003). doi:10.1016/S0012-821X(02)01126-3
D.J. Stevenson, T. Spohn, G. Schubert, Magnetism and thermal evolution of terrestrial planets. Icarus **54**, 466–489 (1983)
F. Takahashi, M. Matsushima, Dipolar and non-dipolar dynamos in a thin shell geometry with implications for the magnetic field of Mercury. Geophys. Res. Lett. **33** (2006). doi:10.1029/2006GL025792
E. Thébault, A proposal for regional modelling at the Earth's surface, R-SCHA2D. Geophys. J. Int. **174** (2008). doi:10.1111/j.1365-246X.2008.03823.x
E. Thébault, L. Gaya-Piqué, Applied comparisons between SCHA and R-SCHA regional modeling techniques. Geochem., Geophys., Geosyst. **9** (2008). doi:10.1029/2008GC001953
E. Thébault, J.J. Schott, M. Mandea, Revised Spherical Cap Harmonic Analysis (R-SCHA): Validation and Properties. J. Geophys. Res. **111** (2006a). doi:10.1029/2005JB003836
E. Thébault, J.J. Schott, M. Mandea, Modeling the lithospheric magnetic field over France by means of revised spherical cap harmonic analysis (R-SCHA). J. Geophys. Res. **111** (2006b). doi:10.1029/2005JB004110
E. Thébault, K. Hemant, G. Hulot, N. Olsen, On the geographical distribution of induced time-varying crustal magnetic fields. Geophys. Res. Lett. **36** (2009). doi:10.1029/2008GL036416
J.M. Torta, L.R. Gaya-Piqué, A. De Santis, Spherical cap harmonic analysis of the geomagnetic field with application for aeronautical mapping, in *Geomagnetics for Aeronautical Safety: A Case Study in and Around the Balkans*, ed. by J.L. Rasson, T. Delipetrov (Springer, Dordrecht, 2006), pp. 291–307
S. Vennerstrom, N. Olsen, M.E. Purucker, M.H. Acuña, J.C. Cain, The magnetic field in the pile-up region at Mars, and its variation with the solar wind. Geophys. Res. Lett. **30** (2003). doi:10.1029:2003GL016883
F.J. Vine, D.H. Matthews, Magnetic anomalies over oceanic ridges. Nature **199**, 947–949 (1963)
Weiss et al., Space Sci. Rev. (2009, this issue)
P. Wessel, W.H.F. Smith, Free software helps map and display data. EOS, Trans. Am. Geophys. Union **72**, 441 (1991)
K.A. Whaler, R.A. Langel, Minimal crustal magnetization from satellite data. Phys. Earth Planet. Int. **48**, 303–319 (1996)
K.A. Whaler, M.E. Purucker, A spatially continuous magnetization model for Mars. J. Geophys. Res. **110** (2005). doi:10.1029/2004JE002393
J. Wicht, M. Mandea, F. Takahashi, U.R. Christensen, M. Matsushima, B. Langlais, The origin of Mercury's internal magnetic field. Space Sci. Rev. **132** (2007). doi:10.1007/s11214-007-9280-5

Space Sci Rev (2010) 152: 251–269
DOI 10.1007/s11214-009-9621-7

Magnetic Fields of the Outer Planets

C.T. Russell · M.K. Dougherty

Received: 20 August 2009 / Accepted: 3 December 2009 / Published online: 20 January 2010

Abstract The rapidly rotating giant planets of the outer solar system all possess strong dynamo-driven magnetic fields that carve a large cavity in the flowing magnetized solar wind. Each planet brings a unique facet to the study of planetary magnetism. Jupiter possesses the largest planetary magnetic moment, 1.55×10^{20} Tm^3, 2×10^4 times larger than the terrestrial magnetic moment whose axis of symmetry is offset about 10° from the rotation axis, a tilt angle very similar to that of the Earth. Saturn has a dipole magnetic moment of 4.6×10^{18} Tm^3 or 600 times that of the Earth, but unlike the Earth and Jupiter, the tilt of this magnetic moment is less than 1° to the rotation axis. The other two gas giants, Uranus and Neptune, have unusual magnetic fields as well, not only because of their tilts but also because of the harmonic content of their internal fields. Uranus has two anomalous tilts, of its rotation axis and of its dipole axis. Unlike the other planets, the rotation axis of Uranus is tilted 97.5° to the normal to its orbital plane. Its magnetic dipole moment of 3.9×10^{17} Tm^3 is about 50 times the terrestrial moment with a tilt angle of close to 60° to the rotation axis of the planet. In contrast, Neptune with a more normal obliquity has a magnetic moment of 2.2×10^{17} Tm^3 or slightly over 25 times the terrestrial moment. The tilt angle of this moment is 47°, smaller than that of Uranus but much larger than those of the Earth, Jupiter and Saturn. These two planets have such high harmonic content in their fields that the single flyby of Voyager was unable to resolve the higher degree coefficients accurately.

The four gas giants have no apparent surface features that reflect the motion of the deep interior, so the magnetic field has been used to attempt to provide this information. This approach works very well at Jupiter where there is a significant tilt of the dipole and a long baseline of magnetic field measurements (Pioneer 10 to Galileo). The rotation rate is 870.536° per day corresponding to a (System III) period of 9 h 55 min 26.704 s. At Saturn, it has been much more difficult to determine the equivalent rotation period. The most probable

C.T. Russell (✉)
Institute of Geophysics and Planetary Physics, Department of Earth and Space Sciences, University of California, Los Angeles, CA 90095, USA
e-mail: ctrussel@igpp.ucla.edu

M.K. Dougherty
Department of Physics, Blackett Laboratory, Imperial College, London SW7 2AZ, UK

rotation period of the interior is close to 10 h 33 min, but at this writing, the number is still uncertain. For Uranus and Neptune, the magnetic field is better suited for the determination of the planetary rotation period but the baseline is too short. While it is possible that the smaller planetary bodies of the outer solar system, too, have magnetic fields or once had, but the current missions to Vesta, Ceres and Pluto do not include magnetic measurements.

Keywords Planetary magnetism · Planetary rotation · Jupiter · Saturn · Uranus · Neptune

1 Introduction

The existence of the Earth's magnetic field has been known and used in navigation for at least a millennium, but the realization that other planets similarly possess a magnetic dynamo was not achieved until the middle of the 20th century when radio telescopes discovered electromagnetic emissions from Jupiter (Burke and Franklin 1955). After this early success, terrestrial radio astronomy has played, at most, a minor role in the exploration of planetary magnetic fields because the electromagnetic emissions from the other magnetized planets cannot be detected at Earth. With the advent of in situ observations of planetary magnetic fields on flybys, orbiters and landers in some cases, the near-ubiquity of current or ancient dynamo action in bodies of almost any size from that of the Earth's Moon to that of Jupiter became evident. Only Venus and the smaller moons and asteroids seemed to be devoid of current or remanent magnetic fields.

A particularly powerful tool for increasing our understanding of planetary processes is comparative planetology, where one takes a common process and examines the behavior of this process across a number of planets, under different boundary conditions. The four gas giants, Jupiter, Saturn, Uranus, and Neptune, form a good set of planets for studying how planetary magnetic fields arise because of their varying interior properties. Both Jupiter and Saturn have interiors consisting principally of hydrogen and helium under high pressure, but Saturn has only 30% of Jupiter's mass. Uranus and Neptune have a core, a shell of salty water, and a thick, largely helium and hydrogen atmosphere. All four planets generate a magnetic field, but we believe that the fields of Jupiter and Saturn may be generated quite differently than those of Uranus and Neptune. Certainly their outward manifestations are different. Jupiter has the largest magnetic dipole moment and has a series of higher moments of decreasing size as does the Earth. Its dipole field is tilted by close to 10°, also like the Earth's. Saturn, too, possesses a multipole magnetic field similar to that of the Earth and Jupiter, but different in one important aspect. There is very little tilt to the dipole. In contrast, the magnetic fields of Uranus and Neptune have dipole axes that are tilted far from the rotation axes of the planet.

Tilted dipole moments can be especially useful for determining the rotation period of the interior of a planet when there is no visible surface feature that rotates with the period of the interior of the planet. In this review, we will show how this technique returns a very accurate rotation rate for the interior of Jupiter. We also review the attempts to do the same for Saturn where the tilt of the dipole is much smaller. For a survey of the early measurements of planetary magnetic fields, the interested reader is referred to the chapter entitled "Space Exploration of Planetary Magnetism" (Ness 2009). We begin our review with the planet for which we have the longest set of measurements, Jupiter.

2 Jovian Magnetic Field

2.1 Early Observations

The solar-system record holder in almost every category is Jupiter. Its radius of 71,400 km makes it the largest of the planets. Its IAU-defined (System III) period, 09 h 55 min 29.7 s, makes it the fastest rotating plasma. It has the most mass, the strongest radio emissions, and not surprisingly, the strongest magnetic dipole moment. The existence of its magnetic field was inferred from its polarized radio emissions. These were first detected by Burke and Franklin (1955) with radio telescopes measuring megahertz frequencies. The changing location of the source in the sky clearly identified this planetary source. Later, synchrotron waves were identified at decimetric wavelengths due to energetic electrons gyrating in Jupiter's magnetic field (Sloanaker 1959; McClain and Sloanaker 1959).

The radio data provided many constraints on the nature of Jupiter's magnetic field. The dipole tilt was expected to be close to 9.5° (Roberts and Komesaroff 1965; Komesaroff and McCullough 1967; Morris et al. 1968; Whiteoak et al. 1969; Gardner and Whiteoak 1977). The field was estimated to be $0.04\ \text{mT} < B < 0.1\ \text{mT}$ in the radiating region (Komesaroff et al. 1970) with the magnetic moment directed northward, opposite to the direction of the terrestrial dipole moment (Dowden 1963; Berge 1965).

This magnetic field arises in magnetic dynamo acting in a core of liquid hydrogen and helium (cf. Hide and Stannard 1976). The pressure of the interior of Jupiter is so great that it is expected that in the deep interior the hydrogen forms a metallic liquid. The transition between the outer molecular state and the inner metallic state is thought now to be gradual with sufficient conductivity for dynamo operation at a radius of $\sim 0.8\ R_J$, but substantial uncertainty in this value (Guillot et al. 2004).

2.2 Flyby Measurements

Based on the number of high-quality radio observations that had been obtained prior to the advent of in situ observations with spacecraft, one would have expected few surprises upon the arrival of the Pioneer and Voyager probes. Nevertheless, there were surprises and much excitement when these spacecraft first arrived. Pioneer 10 flew by Jupiter on 4 December 1973, passing within 2.9 R_J (Jovian radii) of the center of the planet (Smith et al. 1974). The magnetosphere itself was huge, extending to about 100 Jovian radii at the subsolar point, about 100 times more distant than the subsolar point of the Earth's magnetosphere, making the volume of the magnetosphere a million times greater than that of the Earth. The structure of the magnetosphere was also quite different from the terrestrial magnetosphere. In both magnetospheres, the rotation of the planet is important. In the terrestrial magnetosphere, the ionosphere fills the innermost magnetosphere with cold plasma that co-rotates with the planet, but the centrifugal force is small and the cold plasma is mainly gravitationally bound and not magnetically bound. In the Jovian magnetosphere, the volcanic moon, Io, adds plasma to the equatorial magnetosphere directly, but here, the addition is beyond synchronous orbit. Like the plasma in the Earth's magnetosphere, this newly added plasma is accelerated to the co-rotation speed from the Keplerian orbital speed of Io. At Io's location, the centrifugal force exceeds the gravitational force, and it is the magnetic stress that binds the plasma to Jupiter. Gravity is not sufficient. As a result, the density builds up in the magnetosphere from Io's orbit outward stretching the magnetic field into a disk-shaped configuration. The electrically-conducting ionosphere both attempts to enforce co-rotation and to anchor the field lines to particular latitude, but it partially fails at both. The plasma rotates at a speed less than that of the planetary rotation and the plasma moves slowly outward,

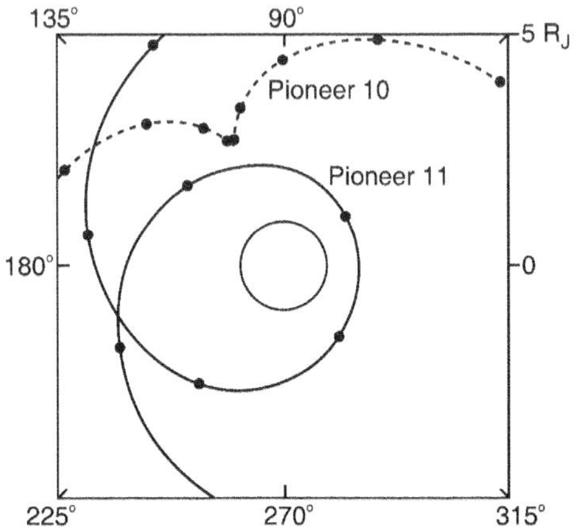

Fig. 1 Pioneer 10 and 11 trajectories near Jupiter. Angles shown in System III longitude. Distance is measured from center of planet and is not a projection. Dots show the position every hour. Pioneer 10 traveled around Jupiter in a prograde sense and Pioneer 11 in a retrograde sense (after Smith et al. 1976)

spiraling eventually into the tail where the plasma content of the field lines is unloaded, and emptied magnetic flux tubes can return (floating buoyantly through the slow outward radial flow) into the inner magnetosphere to be eventually refilled near Io and repeat their journey. This very dynamic transport process and strong distortion of the magnetic field make the distant magnetosphere (outside of the orbit of Io at 5.0 R_J) unsuitable for precision determinations of the interior magnetic field of Jupiter. Thus, measurements have concentrated on the interior region. Figure 1 shows the trajectories of Pioneer 10 and 11 inside 5 R_J. While Pioneer 10 did not cover all planetary longitudes inside 5 R_J, Pioneer 11 arriving on 3 December 1974, approaching within 1.6 R_J of the center of Jupiter and flying by in a retrograde direction, did cover all longitudes (Smith et al. 1975). Pioneer 10 and 11 carried an accurate vector helium magnetometer. A fluxgate magnetometer was included in the Pioneer 11 payload for redundancy. Initially, the fluxgate magnetometer measurements differed significantly from those of the helium magnetometer, but eventually the fluxgate magnetometer measurements were recalibrated to agree with the helium data. The consensus dipole to quadrupole to octopole moment ratio was 1.00:0.24:0.21, compared to that of the Earth of 1.00:0.14:0.10. The dipole moment was set at 1.55×10^{20} Tm^3, 20,000 times larger than that of the Earth.

Voyager 1 arrived at Jupiter on 5 March 1979, passing within 4.9 R_J of Jupiter's center, with Voyager 2 arriving only five months later on 20 August, passing within 10.1 R_J of the planet (Ness et al. 1979). Neither measurement could contribute much to the determination of the internal magnetic field, and the baseline from the Pioneer 10 and 11 data was too short to detect any secular variation of even the dipole magnetic field. In February 1992, the Ulysses spacecraft also flew within 6 R_J of the planet, carrying both a vector helium and a fluxgate magnetometer (Balogh et al. 1992), but it too did not constrain the magnetic dipole moment accurately enough to study the secular variation of the magnetic field from the time of the Pioneer and Voyager measurements (Dougherty et al. 1996). Finally, on December 7, 1995, the Galileo spacecraft was inserted into Jovian orbit and remained in orbit until September 2003.

2.3 Galileo Orbiter Measurements

The Galileo orbiter carried two fluxgate magnetometers with flippers mounted on booms on the spinning portion of the spacecraft. This configuration allowed determination of the zero levels of the magnetometers continuously in the spin plane. The sensor along the spin axis could be interchanged with one in the spin plane to allow its zero level to be monitored (Kivelson et al. 1992). The outbound sensor had dynamic ranges of ±32 nT and ±512 nT and ±16,384 nT. The early orbits, G1 to C9, where the letter designates the moon visited (Ganymede, Europa and Callisto) and the number indicates the orbit number, did not venture inside 9 R_J except on the insertion pass I0, where the spacecraft traveled far inside Io's orbit to 5.0 R_J, but orientation data were unavailable. On orbits G10 to C20, the spacecraft again stayed at or beyond 9 R_J. Finally, beginning on C21, Galileo's perijove was lowered to Io (I27) and kept near 6 R_J until the final plunge on September 21, 2003.

2.4 Inversion Results

Most studies of the Jovian magnetic field have followed a very similar rubric, solving an overdetermined linear system

$$y = Ax \tag{1}$$

where y is a column matrix of the $3N$ magnetic field observations (3 components observed N times) and A is a $3N$ by M matrix relating the observation to the model parameters x. The vector x is arranged as a column vector of length m.

The magnetic field at any point is a sum of coefficients dependent on locations and Schmidt-normalized Legendre polynomials

$$B_r = \sum_{n=1}^{\infty}\sum_{m=0}^{n}\left\{(n+1)\left(\frac{a}{r}\right)^{n+2}[g_n^m\cos(m\phi)+h_n^m\sin(m\phi)]P_n^m(\cos\theta)\right\}$$

$$B_\theta = \sum_{n=1}^{\infty}\sum_{m=0}^{n}\left\{\left(\frac{a}{r}\right)^{n+2}[g_n^m\cos(m\phi)+h_n^m\sin(m\phi)]\frac{dP_n^m(\cos\theta)}{d\theta}\right\}$$

$$B_\theta = \frac{1}{\sin(\theta)}\sum_{n=1}^{\infty}\sum_{m=0}^{n}\left\{m\left(\frac{a}{r}\right)^{n+2}[g_n^m\sin(m\phi)-h_n^m\cos(m\phi)]P_n^m(\cos(\theta))\right\}$$

The ith observation, y_i, is related to the model parameters by the function $P_i(x_i)$. The functions $P_i(x_j)$ can be Taylor expanded around some initial parameter set, x_j:

$$y_i = P_i(x_j^0) + \left.\frac{\partial P_i}{\partial x_j}\right|_{x_j^0}\Delta x_j + \cdots$$

We can proceed by calculating the residual from an existing model, e.g. the O6 model of Connerney (1992):

$$\begin{aligned}\Delta y_i &= y_i - P_i(x_j^0)\\ &= A'\Delta x_j\\ \text{where } A' &= \left.\frac{\partial P_i}{\partial x_j}\right|_{x_j^0}\end{aligned}$$

The matrix A' is the same as A and is determined by the spacecraft trajectory independent of which model being used.

To solve this matrix, we use singular value decomposition to write

$$y = USV^T x$$

where U is a $3N$ by M matrix consisting of M orthonormalized eigenvectors associated with the M largest eigenvalues of the AA^T as columns. S is an M by M diagonal matrix consisting of the eigenvalues, xv_i, of $A^T A$. The S matrix is assembled with the largest sv_i in the upper left, all elements positive in the order $sv_i > sv_2 > sv_3 > \cdots > sv_n$.

Multiplying both sides of the equation by U_T, we obtain

$$(U^T y) = S(V^T x)$$

Since S is an $M \times M$ diagonal matrix, we regard the above equation as M independent equations relating eigendata (on the left), through the eigenvalues sv_i, to eigenvectors of parameter space, the linear combination of the original parameters $V^T x$. The solution to $y = Ax$, that is, the parameter vector x minimizing in a least-squares sense (Lanczos 1971) the difference between the model and the observations, is given by

$$x = VS'U^T y$$

Writing $\beta = U^T y$, the solution can be constructed by a summation over the orthonormalized V_i of parameter space:

$$x = \sum_{i=1}^{M} (\beta_i / sv_i) V_i$$

Magnetic field observations along a certain trajectory are insensitive to certain linear combination of parameters. One advantage of the SVD is that the parameter vectors which are poorly constrained by the available observations, are explicitly identified. They are the eigenvectors associated with the small eigenvalues of $A^T A$.

One way to examine the condition, or stability, of a linear system is to calculate the "condition number," *CN*, defined as the ratio of the largest and smallest singular values (the square root of the eigenvalue) (Lanczos 1971):

$$CN = sv_1 / sv_m$$

Errors in the mth generalized parameter can be expected to be about *CN* times larger in magnitude than errors in the first generalized parameter. So, the condition number gives a very good estimate of the "quality for inversion" for a system.

Figure 2 shows the condition number for different spacecraft trajectories with different eigenvectors included. The 15 ev solution includes the octupole terms while the 8 ev solutions contain up to the quadrupole (Yu 2004; Yu et al. 2009).

The condition numbers vary for the different trajectories depending on whether the 15 ev or 8 ev representation is used. For 15 ev solutions, the condition numbers vary from below 100 to above 1000; while for the 8 ev solutions, the variation of condition numbers are smaller and remain of the order of 10. That presents a very good reason to select only the 8 ev solutions to compare measurements: the 8 ev solution has similar inversion qualities for the different trajectories while higher order solutions do not. Used by itself, the P11 pass has a better condition number for the 8 parameter inversion than any other pass including the I27 Galileo pass.

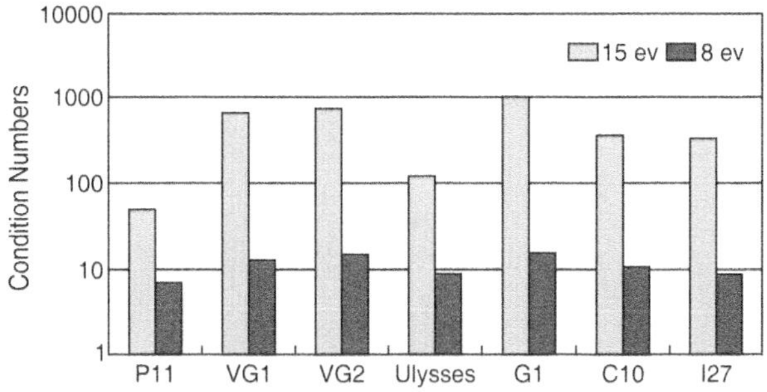

Fig. 2 Condition numbers for the spacecraft trajectories of Pioneer 11, Voyager 1, Voyager 2, Ulysses, Galileo's Ganymede 1 orbit, Callisto 10 orbit and Io 27 orbit. Condition numbers for 15 eigenvalues (dipole, quadrupole and octupole) and 8 eigenvalues (dipole plus quadrupole) are shown

Table 1 Octupole model from Galileo observations. Full octupole model with 15 coefficients and partial octupole model with 13 coefficients are compared with the GSFC O6 model which is mainly inverted from Pioneer 11 observations. In the Galileo 13 model, the g31 and h31 coefficients that have minimum field contributions near the orbital plane of Galileo are held fixed at their O6 value (Yu et al. 2009)

	1975	1995–2003	
	GSFC O6	Galileo 15	Galileo 13
g10	4.242	4.258	4.273
g11	−0.659	−0.725	−0.716
h11	0.241	0.237	0.235
g20	−0.022	0.212	0.270
g21	−0.711	−0.592	−0.593
g22	0.487	0.517	0.523
h21	−0.403	−0.448	−0.442
h22	0.072	0.152	0.157
g30	0.076	−0.013	−0.092
g31	−0.155	−0.764	−0.155
g32	0.198	0.292	0.274
g33	−0.180	−0.095	−0.096
h31	−0.388	0.950	−0.388
h32	0.342	0.521	0.506
h33	−0.224	−0.309	−0.299
tilt	9.40	10.16	10.00
long	159.9	161.9	161.8

Condition number is not the only parameter useful in judging the accuracy of an inversion. If the noise level of the data set is known, one can develop a parameter resolution matrix (Jackson 1972) and calculate the accuracy of each parameter. Using the parameter, one can choose to maximize the accuracy of parameters by truncating an inversion. For example, truncating a quadrupole inversion at 7 coefficients as opposed to 8 could significantly increase the accuracy of the 7 solved coefficients over their values obtained in the field dipole plus quadrupole inversion.

The Galileo measurements have been inverted using those techniques together with the use of robust estimators (Yu 2004). More recently, Yu et al. (2009) have re-examined observations during the two Galileo Earth flybys to verify the calibrations used in the Galileo processing in the PDS and used by Yu (2004). They find that the three sensor gains were originally miscalibrated. To correct the Galileo gains, one must multiply them by 0.9907 ± 0.0006, which has been done in presenting the Galileo data that follows. Table 1 shows a comparison of the best inversions of the data from the Galileo mission with the O6 model obtained mainly from Pioneer 11. The 15 terms of the full octupole inversion agree

Table 2 Seven-coefficient fit for various measurement epochs compared with O6 model (Yu et al. 2009)

	O6	P11-FGM	Voyager 1	Voyager 2	Ulysses	Galileo
g10	4.242	4.279	4.236	4.406	4.116	4.272
g11	−0.659	−0.627	−0.658	−0.659	−0.636	−0.694
h11	0.241	0.217	0.247	0.225	0.223	0.228
g21	−0.711	−0.655	−0.630	−0.355	−0.836	−0.572
g22	0.487	0.435	0.493	0.410	0.463	0.541
h21	−0.403	−0.295	−0.420	−0.524	−0.291	−0.439
h22	0.072	0.004	0.070	0.235	0.183	0.159
tilt	9.39	8.81	9.42	8.97	9.30	9.71

moderately well with O6 except for the g31 and h31 terms. Since Galileo measurements are obtained in the rotational equator, they cannot resolve well the g20, g31, or h31 terms. Thus, in the solution in the right-hand column of Table 1, we have fixed the g31 and h31 coefficients at their O6 values. We note that the longitude of the dipole axis has changed 2 degrees between 1975 and the Galileo epoch. This is due to a slight inaccuracy of the IAU-defined System III period, as discussed in the following section. This small period shift will also mask small changes in the other coefficients as well as cause changes where values should be constant.

Table 2 shows a comparison of the O6 model with models based on the Pioneer 11, Voyager 1, Voyager 2, Ulysses and Galileo measurements. We use the 7-eigenvalue solution in which g20 is held fixed at the O6 value because Galileo in the orbital plane cannot well resolve this value. There is no statistically significant change in the coefficients between the Pioneer 11 epoch and the Galileo epoch. The tilt angles are all within two standard deviations of the mean value with four of the six within one standard deviation. Since all of the Galileo data used in this study were obtained outside the orbit of the moon Io that controls the dynamics of the magnetosphere and produces a mass loaded plasma disk, the tilt angle for Galileo might be less accurate than, say, the Pioneer 11 model which is constructed from data largely obtained at lower altitudes.

2.5 Rotation Period of Jupiter

The rotation rate of the Earth is unambiguous. The rotation rate of the surface is the same everywhere and the interior and crust are for all practical purposes locked together. Different parts of the system may have slightly different speeds with respect to the surface, such as the winds in the atmosphere or the fluid motions in the core, but we know clearly what to define as the speed of rotation or the length of the day on Earth. It may vary over time due to tides and changes in the distribution of angular momentum between the fluid core, the solid planet and the atmosphere (Roberts et al. 2007), but at any moment in time, it has a fixed planet-wide value and changes with time are very small. On the gas giants, there is no solid surface. There are no observatories that rotate with the interior of the planet nor any surface features that are locked to the interior. Presently there are four rotation systems defined for Jupiter: Systems I to IV, whose specifications are listed in Table 3 (Dessler 1983). System I applies to cloud features within about 10° of the equator. System II applies to high-latitude clouds. System III is a measure of the periodicity of a certain class of radio emissions controlled by the magnetic field. This system is believed to rotate with the interior of Jupiter where the magnetic field is generated. This system has been revised using later radio data and could

Table 3 The four rotational systems of Jupiter

System	Epoch	Rotation Rate [d^{-1}]	Period [h, min, s]
I	Noon GMT July 14, 1897	877.90°*	9 h 50 min 30.0034 s
II	Noon GMT July 14, 1897	870.27°*	9 h 55 min 40.6322 s
III (1957.0)	0000 UT Jan. 1, 1957	870.544°	9 h 55 min 29.37 s*
III (1965)	0000 UT Jan. 1, 1957	870.536°	9 h 55 min 29.71 s*
IV	0000 UT Jan. 1, 1979+	845.057°×	10 h 13 min 27 s

*Value in IAU definition. Note that the IAU defines these longitudes in a left-handed sense, i.e. increasing westward. Many researchers prefer to use a right-handed system with longitude increasing eastward in the direction of rotation

+At this time, the central meridian longitude is set to be $\lambda_{IV} = 126°$ (Sandal and Dessler 1988)

×Defined period but expected to be temporally varying

be revised again using the Galileo measurements of dipole longitude given above. Finally, there is a proposed System IV that better organizes some magnetospheric phenomena. In this section, we will first use the data discussed in the previous section to update the System III period and then say a few words about the reality of System IV and the possible physical processes behind this periodicity.

If we reanalyze the Pioneer 11, Voyager 1, Voyager 2, and Ulysses magnetometer data from the publicly available databases using the same software and data preparation techniques as we used for the Galileo data, we get the longitudes shown in Fig. 3 (Yu and Russell 2009). Here we have grouped the Galileo data into six groups of four orbits. The slope of the line is non-zero with a probability of 93% using the standard F-test. The slope corresponds to a decrease of the IAU System III period by 6 ± 3 ms, for a period of 9 h 55 min 27.704 ± 0.003 s. This change is within the accuracy expected for the IAU defined period. This difference will be important in comparing future Jovian magnetic measurements with those of past missions. Users of planetary ephemerides produced by JPL should be cautious of the current Jupiter longitudes because the IAU changed their defined rotation period in 2000 and this erroneous period found its way to the SPICE system in 2003. Work is currently under way to get a new IAU-defined period and to correct the SPICE kernels but at this writing both are still incorrect.

The case for the existence of yet a fourth rotation period has been made by Sandal and Dessler (1988). Their proposed System IV period is 10 h 13 min 27 s almost 18 min longer than the System III period. It is reasonable that magnetospheric phenomena would require longer to "co-rotate" than the interior of the planet because slippage would be expected in the coupling process between the ionosphere and the magnetosphere as momentum is exchanged to speed up the mass added by Io to the magnetosphere and to maintain that "co-rotational" speed as the material convects or diffuses outward from its source region. The only surprise is that a single rotation value is a unifying rate for many magnetospheric phenomena. The only other periodic rate associated with Io is its Keplerian orbital period of 42.456 h. The fourth harmonic of the corresponding frequency has a period of 10.614 h which is longer than the observed System IV period of 10.224 h. If there were a resonance with orbiting material, the material would have to be at a distance of 5.75 R_J not Io's 5.9 R_J. Surprisingly, the Io ribbon (Trauger 1984) lies very close to this distance. It is possible that in the interaction of the co-rotating plasma with Io, the slowed flow moves inward because the magnetic stresses in place for co-rotating speeds now apply an overpressure to the more

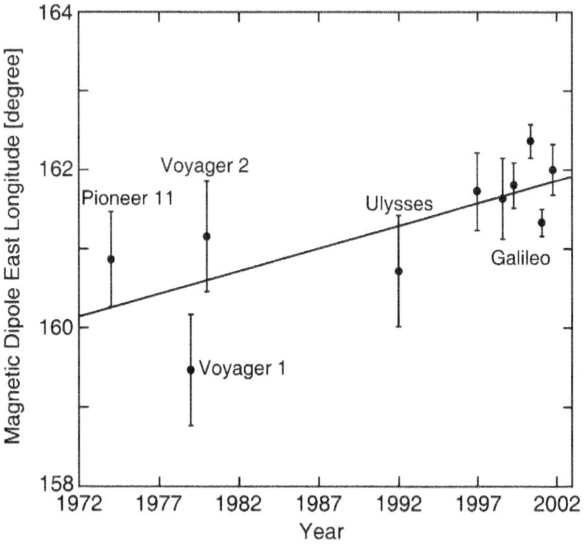

Fig. 3 System III longitude of the Jovian magnetic dipole axis on Pioneer 11, Voyager 1, Voyager 2, and Ulysses flybys, together with 6 groups of four orbits of Galileo. Error bars are given for each measurement and the least-square, best-fit straight line to the data points is shown (Yu and Russell 2009)

slowly rotating post-interaction plasma. If this is confirmed by simulations, it could explain both the location of the Io ribbon and the System IV period. A circulating flow would be built up driven by Io. The quasi-harmonic resonance would lead to density asymmetries in the flow as the interaction periodically reinforced previously formed density enhancements. Such a quasi-resonance also appears to occur at Saturn.

3 Saturnian Magnetic Field

3.1 Early Observations

Saturn is Jupiter's smaller sibling with less than one third the mass and 60% of its volume, rotating about 10% slower and with a magnetic moment of only 3% of that of Jupiter, but these slights of nature are more than compensated by Saturn's spectacular ring system and two of the most exotic moons of the solar system, Enceladus and Titan. The Saturn radius is 60,268 km and the rotational period of its interior roughly 10 h 33 min, but as we will discuss below, is not decisively determined at present. Saturn radio emissions cannot be detected from Earth, and properties of the Saturnian magnetic field remained hidden until Pioneer 11 arrived on September 1, 1979, passing within 1.4 R_S of the center of the planet. Voyager 1 soon followed, passing within 3.1 R_S on August 22, 1980, and then Voyager 2 on 26 August 1981, passing within 2.7 R_S. The observed field was surprising. First, it was much weaker than expected with a dipole moment of only 4.6×10^{18} Tm^3 and an equatorial surface field of about 20,000 nT. The quadrupole field relative to the dipole field on the surface is only half that in the Earth. A simple explanation of this is that the source is relatively deeper inside Saturn than the dynamos of the Earth and of Jupiter (cf. Elphic and Russell 1978). The next and perhaps the major surprise was the tilt of the dipole axis which is less than 1° compared to the near 10° tilts of the dipole axes for Earth and Jupiter (Smith et al. 1980; Ness et al. 1981, 1982). To resolve this conundrum, Stevenson (1982) proposed that the internal field is tilted, but the tilted component of the field is shielded from the external observer by spin-axis-symmetric differential rotation of a conducting layer between the helium-rich

Table 4 Zonal spherical harmonic coefficients

Multipole Term	Cassini	SPV	Z3	GD
g_{10} [nT]	21,162	21,225	21,248	21,232
g_{20} [nT]	1514	1566	1613	1563
g_{30} [nT]	2283	2332	2683	2821

core and the helium-depleted molecular mantle. This hypothesis has not been confirmed or refuted. Finally, the various components of the spin-axis-symmetric magnetic field combine in such a way that a northward displacement of the dipole moment is a good approximation to the magnetic field. This displacement is seen throughout the magnetosphere, causing a discernible offset of the magnetic and rotational equators.

On June 30, 2004, Cassini was inserted into orbit, carrying in its payload a fluxgate magnetometer and a scalar/vector helium magnetometer (Dougherty et al. 2004).This instrument has the capability of defining the magnetic field much more precisely than on the earlier flybys. Table 4 shows the zonal dipole, quadrupole and octupole coefficients for the SPV (Acuna et al. 1983), Z3 (Connerney et al. 1983), GD (Giampieri and Dougherty 2004) models compared with the Cassini measurements (Burton et al. 2009a). The differences between models are not large, but since the Cassini analyses are based on close to three years of orbital data, they are to be preferred.

Only the zonal harmonics are given in Table 4 because it soon became realized that the rotation rate of the interior is not manifested by the period of the radio emissions as they are on Jupiter. Any error in the rotation rate will smear out the non-zonal harmonics and over time average them to zero, and the rotation period is poorly known. As at Jupiter, initially, the rotation rate was chosen based on periodicities in the radio emissions that are detectable from nearby satellites, initially Voyager and later Ulysses and the Cassini. These waves indicate a period of about 10 h 47 min, but the period is not fixed (Kurth et al. 2004, 2007). These periodic signals are accompanied by periodic modulation of the magnetospheric magnetic field at the same period. It was only when a large shift in period between Voyager/Pioneer days and the Cassini epoch was noticed that the community became suspicious of the SKR radio period. In fact, it was only after continued variation in the period after several years of Cassini's data that the diehards gave up. This saga continues at this writing and deserves some discussion as there are important lessons for Saturn from the Jovian situation discussed above.

3.2 Rotation Period of Saturn

Since the dipole magnetic moment is nearly aligned with the rotation axis of Saturn, it does not produce a significant pulse for the timing of the rotation of the interior. While this observation was obvious to all observers, the hope existed that some asymmetry was strong enough to control magnetospheric processes such as the generation of radio emissions. Thus, many sought to study radio emissions as a measure of the rotation rate. While such emissions are not visible from Earth, they are detectable by nearby interplanetary spacecraft at a period around 10 h 47 min. Galopeau and Lecacheux (2000) noted however that the periodicity of the Saturn kilometric radiation had changed between the time of Voyager 1 and 2, Saturn flybys in the early 1980s and Ulysses observations in the period 1994 to 1997. The period continued to vary when Cassini arrived and a longitude system was developed based on this measurement (Kurth et al. 2004). Most recently, Gurnett et al. (2009) have reported two simultaneously varying different periods in the north and the south. Thus this signal cannot be due to the rotation of the interior of Saturn.

These periodic radio signals are accompanied by periodic modulation of the magnetospheric magnetic field. This was first noted by Espinosa et al. (2003a, 2003b) who reanalyzed the magnetic field data from the Pioneer and Voyager flybys. Giampieri and Dougherty (2004) modeled these data and proposed that there was a tilted dipole at an angle of 0.17° rotating within a second of the radio signals. Originally, this was linked to the interior with a camshaft model which was purported to launch a periodic signature into the magnetosphere. Initial analysis of Cassini's magnetometer data from the first two years of orbital tour seemed to confirm that the magnetic period measured was very similar to that observed by the radio measurements (Giampieri et al. 2006), but these periods were observed to change also and do not pertain to the deep interior. Hence it is clear that the link between the internal rotation of Saturn and its external magnetic field is much more complex than had previously been recognized (Dougherty et al. 2005). Gurnett et al. (2007) correctly deduced that the radio emission had its origin in the inner magnetosphere, and that this system slips slowly in phase to Saturn's internal rotation. Thus this scenario clearly resembles the relationship between Jupiter's System IV (plasma phenomena) and System III (internal field), but contrary to the initial interpretation, the Saturn kilometric radiation period is the analogue of System IV and there are no radio signals marking the period of Saturn's System III, its internal rotation.

For the same reason, as noted above for Jupiter, it is likely that the System III interior rotation period is significantly shorter than that of the System IV period of the magnetosphere which must be "spun up" by the ionosphere. A completely different approach using Pioneer and Voyager radio occultation and wind data has estimated the interior period to be 10 h 32 min 35 $\pm$ 13 s (Anderson and Schubert 2007). More recently, Read et al. (2009) have estimated a period of 10 h 34 min based on its planetary-wave configuration, and Burton et al. (2009b) have presented evidence for a possible spin rate of 10 h 34 min by examining the non-axial power in their model inversions as a function of rotation period as well as examining the root-mean-square misfit field as a function of period. The coincidence of a broad maximum in the non-axial power and a minimum in the misfit at 10 h 34 min points to this signal as due to the interior field. We note, however, that the properties of the inverted magnetic fields depend very sensitively on the accuracy of the rotation rate used. An error of one minute in rotation period can totally erase the dipole component in the rotational equator in 180 days. All current models are affected to some extent by smearing of the coefficients due to the choice of an inaccurate rotation rate. The convergence of these independent approaches to determining the rotation period of the interior is heartening and should soon allow accurate field models to be obtained as well as a consensus spin period.

Returning to the correspondence between Jupiter's System IV and the SKR period at Saturn, we note that, like the Jovian magnetosphere, Saturn has a mass-loading body Enceladus well beyond synchronous orbit so that plasma at near the co-rotational speed will exact an outward force stretching the magnetic field (Dougherty et al. 2005). When this near corotating plasma encounters Enceladus, the plasma slows down as it loses momentum to fast neutrals and picks up slow new ions from the Enceladus plume. The plasma will then be pulled toward Saturn by the magnetic stress where it will be sped up by the field lines connected to the ionosphere. Thus, the Enceladus interaction sets up a global circulation pattern. The flows reported by Tokar et al. (2007) were probably generated in this way. Like in the Io interaction, there is a quasi-resonance (but this time 3 to 1) between the Keplerian rotation rate at Enceladus and the material circulating at these distances. Thus, we would expect that density asymmetries would develop in the flow as density enhancements reinforced themselves when they re-encountered Enceladus. In short, the Jupiter–Io coupling is probably

Table 5 Uranus Q_3 magnetic field model Schmidt-normalized spherical harmonic coefficients (Uranus radius of 25,600 km)

n	m	g_n^m [nT]	h_n^m [nT]
1	0	11,900	
1	1	11,580	−15,680
2	0	−6030	
2	1	−12,590	6120
2	2	200	4760

After Connerney et al. (1987)

very similar to the Saturn–Enceladus interaction. The difference in the Jovian System III and System IV periods is 18 min and the difference between the Burton–Anderson–Read period and the SKR period is a similar 14 min. It seems there is much to learn about each system by their intercomparison.

4 The Magnetic Fields of Uranus and Neptune

Uranus and Neptune have been explored by only one spacecraft, Voyager 2, passing within 4.2 R_u of Uranus in January 1986, and within 1.18 R_n of Neptune in August 1989. The different flyby distances resulted in quite different sensitivities of the observations to higher degrees and order components of the magnetic field. Thus, as shown in Table 5, the modeled Uranian field consists of only dipole and quadrupole terms while in Table 6, the modeled Neptunian field includes dipole, quadrupole and octupole components. As noted in the tables, some of the components are poorly resolved, limited by the flyby geometry. In these two tables, the coefficients are listed to the nearest ten nT in the absence of information on their individual accuracies, which are probably less than 10 nT.

Uranus has a radius of 25,600 km and an obliquity of 98°. Thus, it is a retrograde rotator with its spin axis tilted below its orbit plane. Its rotation axis can be almost aligned with the planet–Sun line and Voyager 2 encountered it at such a time. The magnetic inversion process revealed the best-fit rotation period to be 17.29 ± 0.01 h (Ness et al. 1986). A more recent inversion (Herbert 2009) reveals a 17.21 ± 0.02 h period, more in accord with the radio period of 17.24 ± 0.01 h (Desch et al. 1986). Figure 4 shows an offset tilted dipole model of the Uranian magnetosphere, illustrating this configuration of the rotation axis and the very large 60° tilt of the dipole axis to the rotation axis (Ness et al. 1986). The surface magnetic field from the model given in Table 5 is shown in Fig. 5 (Connerney et al. 1987). A more recent model derived using both magnetic observations and auroral data is given in Table 7. Comparison with Table 5 reveals only qualitative agreement. Orbiter measurements will be needed before we have a definitive model of the Uranian magnetic field.

Neptune has a radius of 24,765 km and is a prograde rotator with an obliquity of 30°, comparable to that of Saturn. Figure 6 shows the offset tilted dipole model of Neptune's magnetic field, illustrating its large tilt, 47°, with respect to the rotation axis (Ness et al. 1989). The rotation rate used to determine the moments was 16 h 63 min, and was not derived from the magnetic field data. Figure 7 shows the surface magnetic field contours (Connerney et al. 1991). It is clear that the magnetic fields of Uranus and Neptune are quite comparable. Finally, in Fig. 8 we show the relative contributions of the dipole, quadrupole, and octupole terms to the magnetic field along the trajectory past Neptune. Near periapsis, the quadrupole and octupole contributions are each greater than that of the dipole.

The magnetic fields of Uranus and Neptune are thus mutually similar and qualitatively different from those of Jupiter and Saturn. The simplest explanation of the high harmonic

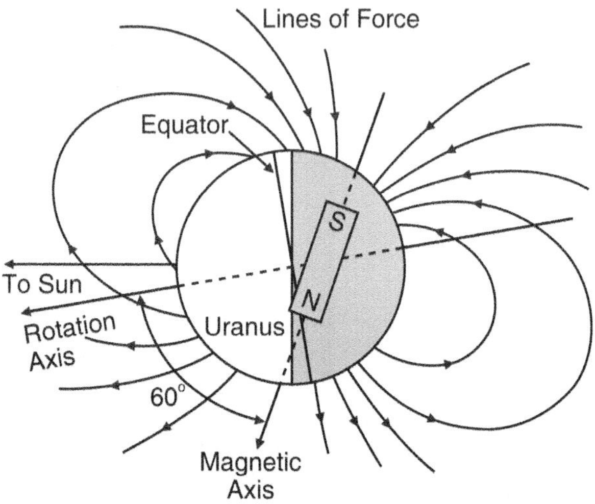

Fig. 4 The offset tilted dipole model for Uranus (after Ness et al. 1986). The rotation axis of Uranus is tilted south of its orbital plane. Its magnetic dipole axis is at 60° to its spin axis

Table 6 Neptune O_8 magnetic field model Schmidt-normalized spherical harmonic coefficients (Neptune radius of 24,765 km)

n	m	g_n^m [nT]	h_n^m [nT]
1	0	9730*	
1	1	3220*	−9890*
2	0	7450†	
2	1	660†	11,230*
2	2	4500*	−70*
3	0	−6590	
3	1	4100	−3670†
3	2	−3580	1790
3	3	480†	770†

Coefficients are poorly resolved or unresolved unless noted otherwise. After Connerney et al. (1991)

*Coefficient well resolved ($Rxx > 0.95$)

†Coefficient marginally resolved ($0.75 < Rxx < 0.95$)

Table 7 Uranus AH_5 magnetic field model up to quadrupole terms. After Herbert (2009)

n	m	g_n^m [nT]	h_n^m [nT]
1	0	11,278	
1	1	10,928	−16,049
2	0	−9648	
2	1	−12,284	6405
2	2	1453	4220
3	0	−1265	
3	1	2778	−1548
3	2	−4535	−2165
3	3	−6297	−3036

content of their magnetic fields is that the source of the fields is a dynamo much nearer the surface than at Jupiter and Saturn, i.e. one in the global ocean of both of these gas giants. We stress that while these two unusual magnetic fields have given us much to think about,

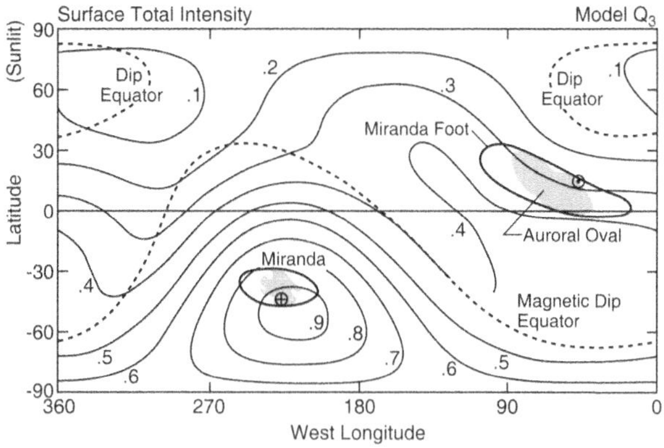

Fig. 5 Total surface intensity (Gauss) of Uranus' magnetic field on a dynamically flattened surface (Connerney et al. 1987)

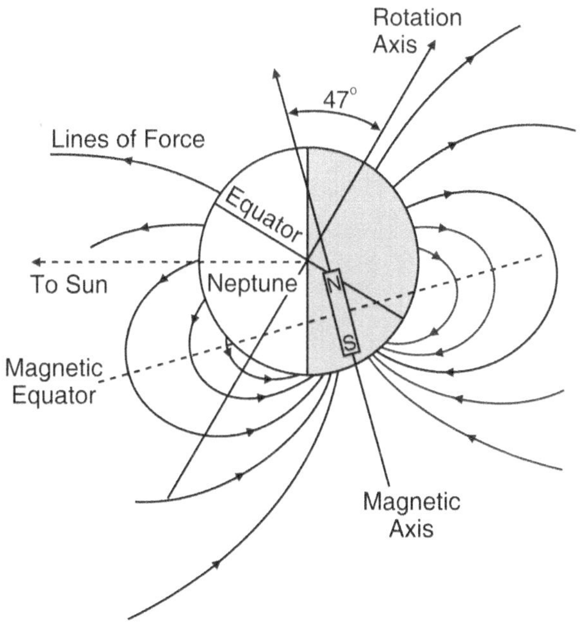

Fig. 6 Offset tilted dipole field lines of Neptune in the plane containing the rotation axis and the dipole center. The axis of the offset tilted dipole is inclined at 22° to the plane of the page (Ness et al. 1989)

their current best models are severely underdetermined. We look forward to the day when these magnetic fields can be explored with orbiting spacecraft.

5 Summary and Conclusions

The four gas giants have strong magnetic dynamos despite their varying internal structures. Because of their different settings, each behaves in a different manner. Jupiter has a tilted magnetic field like the Earth, and the rotation of this dipole field controls much of the dynamic behavior that we see at Jupiter. Over the two decades between Pioneer/Voyager and Galileo, there has been no unambiguous secular variation of the Jovian magnetic field. Similarly, in the same interval, there has been no unambiguous change in Saturn's internal mag-

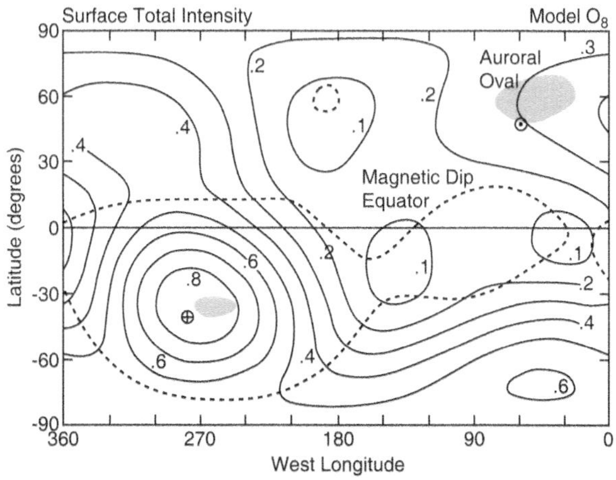

Fig. 7 Magnetic field intensity (Gauss) on dynamically flattened surface of Neptune using the O_8 model (Connerney et al. 1991)

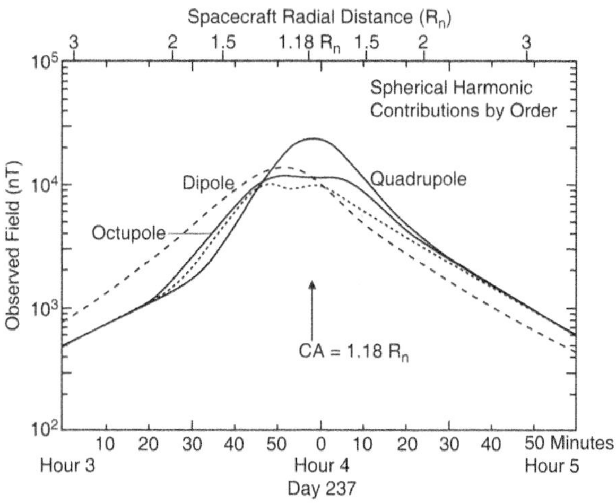

Fig. 8 Magnitude of the field (*short dashed line*) as a function of time and spacecraft radial distance compared that due to partial solutions: the dipole coefficients, the quadruple coefficients and the octupole coefficients calculated separately (after Connerney et al. 1983)

netic field, but here we emphasize that analysis with the incorrect spin period has distorted the field models reported to date.

The big surprise about Saturn's magnetic field is its azimuthal symmetry. It is possible that when the period of its interior rotation is better determined, this may change a little but it is not likely to change qualitatively. There must be little field in the non-axial coefficients and a small dipole tilt. Since this is so unlike the other planets and appears to violate Cowlings theorem, one must wonder if Saturn has a dynamo. While Stevenson (1983) argues that the diffusive time for the decay of Saturn's field is too short for it to be primordial, and while he also proposes a mechanism to hide a tilted interior field (Stevenson 1982), the small tilt angle and the weakness of the field are consistent with the field being primordial. If these properties persist in the improved studies of the Saturn magnetic field now underway, then perhaps it is slowly decaying and once was much stronger like Jupiter's main field. However, if this is occurring, the conductivity of Saturn's interior must be much greater than currently estimated.

In contrast to the role of Jupiter's tilted dipole in controlling the dynamics of the magnetosphere, Saturn's internal magnetic field with its small tilt plays no significant role in the dynamics of its magnetosphere which, in turn, is dominated by the Enceladus mass loading. The resultant coupling of the magnetospheric plasma to the ionosphere via the planetary magnetic field controls the plasma circulation. This coupling allows slippage and, hence, other (time-varying) periods of rotation dominate the interaction. These periods seem to involve a complex interplay between mass loading by the moon Enceladus and the generation of a global circulation pattern in the inner magnetosphere. The quasi-resonance of the third harmonic of the orbital frequency of Enceladus with the frequency of this magnetospheric circulation may allow the build-up of density enhancements that drive dynamic processes throughout the Saturnian magnetosphere. At Jupiter, the dynamics associated with the System IV circulation pattern, probably generated in an analogous fashion near the fourth harmonic of Io's orbital period, is much weaker than that associated with the tilted dipole, but still quite measurable. At Saturn, we finally have an estimate of the true spin period of the interior with a period about 14 min shorter than the SKR period and not unlike the 18 min difference between System III and System IV at Jupiter.

As exciting as the magnetospheres of Jupiter and Saturn might be, and as mysterious may be the process that generates the fields at these two planets, the field configurations of Uranus and Saturn baffle us even more. They both have very strong contributions at high order and degree and they have large tilts of the dipole moment. The high contribution of the higher order and degree components to the surface field indicates immediately the likelihood of a dynamo source close to the surface. Thus, we must look to sources in the water layer and not deep in the core.

Our experience at Saturn in being stymied so long by our ignorance of the rotation rate of the interior provides lessons for Uranus and Neptune as well. In order to make progress here, we need orbiters that provide both complete coverage of the body and a long temporal baseline of measurements.

Finally, our long sequence of surprises in planetary magnetism should be a lesson not to make assumptions about what will be seen at a planetary body in the absence of any a priori information. Thus, the recent tendency to not make magnetic measurements on certain missions to small bodies including, notably, (early) Mars, Vesta, Ceres and Pluto, while it ensures that surprises will cease, is not to be encouraged. We need to understand the magnetism of all bodies in the outer solar system that may have once had or even possibly today have, convecting interiors.

Acknowledgements We gratefully acknowledge the assistance of Z.J. Yu in the analysis of the Galileo data. The preparation of this review was supported by the National Aeronautics and Space Administration under contract 1236948 from the Jet Propulsion Laboratory.

References

M.H. Acuna, J.E.P. Connerney, N.F. Ness, The Z3 zonal harmonic model of Saturn's magnetic field – Analyses and implications. J. Geophys. Res. **88**, 8771 (1983)

J.D. Anderson, G. Schubert, Saturn's gravitational field, internal rotation and interior structure. Science **317**(5843), 1384–1387 (2007). doi:10/1126/Science.1144835

A. Balogh, M.K. Dougherty, R.J. Forsyth, D.J. Southwood, B.T. Tsurutani, N. Murphy, M.E. Burton, Magnetic field observations in the vicinity of Jupiter during the Ulysses flyby. Science **257**, 1515–1518 (1992)

G.L. Berge, Circular polarization of Jupiter's decimetric radiation. Astrophys. J. **142**, 1688 (1965)

B.F. Burke, K.L. Franklin, Observations of variable radio source associated with the planet Jupiter. J. Geophys. Res. **60**, 213–217 (1955)

M.E. Burton, M.K. Dougherty, C.T. Russell, Model of Saturn's internal planetary magnetic field based on Cassini observations. Planet. Space Sci. **57**, 1706–1713 (2009a)

M.E. Burton, M.K. Dougherty, C.T. Russell, Saturn's rotation rate as determined from its non-axisymmetric magnetic field, Eos Trans. AGU **90**(52), Fall Meeting Suppl., Abstract P53C-03 (2009b)
J.E.P. Connerney, Doing more with Jupiter's magnetic field, in *Planetary Radio Emissions*, vol. III, ed. by H.O. Rucher, S.J. Bauer, M.L. Kaiser (Austrian Academy of Science, Vienna, 1992), pp. 13–33
J.E.P. Connerney, M.H. Acuna, N.F. Ness, J. Geophys. Res. **88**, 8779 (1983)
J.E.P. Connerney, M.H. Acuna, N.F. Ness, The magnetic field of Uranus. J. Geophys. Res. **92**, 15329–15336 (1987)
J.E.P. Connerney, M.H. Acuna, N.F. Ness, The magnetic field of Neptune. J. Geophys. Res. **96**, 19023–19042 (1991)
M.D. Desch, J.E.P. Connerney, M.K. Kaiser, The rotation period of Uranus. Nature **322**, 42–43 (1986)
A.J. Dessler, Appendix B: Coordinate systems, in *Physics of the Jovian Magnetosphere*, ed. by A.J. Dessler (Cambridge Univ. Press, Cambridge, 1983), pp. 498–504
M.K. Dougherty, D.J. Southwood, A. Balogh, E.J. Smith, The Ulysses assessment of the Jovian planetary field. J. Geophys. Res. **101**, 24929–24941 (1996)
M.K. Dougherty, S. Kellock, D.J. Southwood, A. Balogh, E.J. Smith, B.T. Tsurutani, B. Gerlach, K.-H. Glassmeier, F. Gleim, C.T. Russell, G. Erdos, F.M. Neubauer, S.W.H. Cowley, The Cassini magnetic field investigation. Space Sci. Rev. **114**, 331–383 (2004)
M.K. Dougherty et al., Cassini magnetometer observations during Saturn orbit insertion. Science **307**, 1266–1270 (2005)
R.K. Dowden, Polarization measurements of Jupiter radio bursts. Aust. J. Phys. **16**, 398–410 (1963)
R.C. Elphic, C.T. Russell, On the apparent source depth of planetary magnetic fields. Geophys. Res. Lett. **5**, 211–214 (1978)
S.A. Espinosa, D.J. Southwood, M.K. Dougherty, Reanalysis of Saturn's magnetospheric field data view of spin-periodic perturbations. J. Geophys. Res. **108**, 1085 (2003a)
S.A. Espinosa, D.J. Southwood, M.K. Dougherty, How can Saturn impose its rotation period in a noncorotating magnetosphere. J. Geophys. Res. **108**, 1086 (2003b)
P.H.M. Galopeau, A. Lecacheux, Variations of Saturn's radio rotation period measured at kilometer wavelengths. J. Geophys. Res. **105**, 13089–13101 (2000)
F.F. Gardner, J.B. Whiteoak, Linear polarization observations of Jupiter at 6, 11, and 21 cm wavelengths. Astron. Astrophys. **60**, 369 (1977)
G. Giampieri, M.K. Dougherty, Rotation rate of Saturn's interior from magnetic field observations. Geophys. Res. Lett. **31**, L16701 (2004). doi:10.1029/2004GL020194
G. Giampieri, M.K. Dougherty, E.J. Smith, C.T. Russell, A regular period for Saturn's magnetic period that may track its internal rotation. Nature **441**(7089) (2006)
T. Guillot, D.J. Stevenson, W.B. Hubbard, D. Saumon, The interior of Jupiter, in *Jupiter*, ed. by F. Bagenal et al. (Cambridge University Press, Cambridge, 2004), pp. 33–57
D.A. Gurnett, A.M. Persoon, W.S. Kurth, J.B. Groene, T.F. Averkamp, M.K. Dougherty, D.J. Southwood, The variable rotation and period in the inner region of Saturn's plasma disk. Science **315**(52823), 442–445 (2007). doi:10/1126/science.1138562
D.A. Gurnett, A.M. Persoon, J.B. Greene, A.J. Kopf, G.B. Hospodarsky, W.S. Kurth, A north–south difference in the rotation rate of auroral hiss at Saturn: Comparison to Saturn's kilometric radiation. Geophys. Res. Lett. **36**, L21108 (2009). doi:10/1029/2009GL040774
F. Herbert, Auroral and magnetic field of Uranus. J. Geophys. Res. **114**, A11206 (2009). doi:10.1029/2009JA014394
R. Hide, D. Stannard, Jupiter's magnetism: Observation and theory, in *Jupiter*, ed. by T. Gehrels (The University of Arizona Press, Tucson, 1976), pp. 767–787
D.D. Jackson, Interpretation of inaccurate, insufficient, and inconsistent data. Geophys. J. R. Astron. Soc. **28**, 97–109 (1972)
M.G. Kivelson, K.K. Khurana, J.D. Means, C.T. Russell, R.C. Snare, The Galileo magnetic field investigation. Space Sci. Rev. **60**, 357–383 (1992)
M.M. Komesaroff, P.M. McCullough, The radio rotation period of Jupiter. Astrophys. Lett. **1**, 39 (1967)
M.M. Komesaroff, D. Morris, J.A. Roberts, Circular polarization of Jupiter's decimetric emission and the Jovian magnetic field strength. Astrophys. Lett. **7**, 31 (1970)
W.S. Kurth, A. Lecacheux, T.F. Averkamp, J.B. Groene, D.A. Gurnett, A Saturnian longitude system based on a variable kilometric radiation period. Geophys. Res. Lett. **34**, L02201 (2004). doi:10.1029/2004GL020194
W.S. Kurth, A. Lecacheux, T.F. Averkamp, J.B. Groene, D.A. Gurnett, A Saturn longitude system based on a variable radiation period. Geophys. Res. Lett. **34**, L02201 (2007). doi:10.1029/2006GL028336
C. Lanczos, *Linear Differential Operators* (Van Nostrand, Princeton, 1971), 564 pp
E.F. McClain, R.M. Sloanaker, Preliminary observations at 10 km wavelength using the NRL 84-foot radio telescope, in *Proc. IAU Symp. No. 9 URSA Symp. No. 1*, ed. by R. Bracewell (Stanford University Press, Stanford, 1959), pp. 61–68

D. Morris, J.B. Whiteoak, F. Tonking, The linear polarization of radiation from Jupiter at 6 cm wavelength. Aust. J. Phys. **21**, 337 (1968)

N.F. Ness, Space exploration of planetary magnetism. Space Sci. Rev. (2009). doi:10.1007/s11214-009-9567-9

N.F. Ness, M.H. Acuna, R.P. Lepping, L.F. Burlaga, K.W. Behannon, F.M. Neubauer, Magnetic field studies at Jupiter by Voyager 1: Preliminary results. Science **204**, 982–987 (1979)

N.F. Ness et al., Magnetic field studies by Voyager 1—Preliminary results at Saturn. Science **212**, 211–217 (1981)

N.F. Ness, M.H. Acuña, K.W. Behannon, L.F. Burlaga, J.E.P. Connerney, R.P. Lepping, F.M. Neubauer, Magnetic field studies by Voyager 2: Preliminary results at Saturn. Science **215**, 558–563 (1982)

N.F. Ness, M.H. Acuna, K.W. Behannon, L.F. Burlaga, J.E.P. Connerney, R.P. Lepping, F.M. Neubauer, Magnetic fields at Uranus. Science **233**(4759), 85–89 (1986)

N.F. Ness, M.H. Acuna, L.F. Burlaga, J.E.P. Connerney, R.P. Lepping, F.M. Neubauer, Magnetic fields at Neptune. Science **246**(4936), 1473–1478 (1989)

P.L. Read, T.E. Dowling, G. Schubert, Saturn's rotation period from its atmospheric planetary-wave configuration. Nature **460**, 606–610 (2009). doi:10.1038/nature08194

J.A. Roberts, M.M. Komesaroff, Observations of Jupiter's radio spectrum and polarization in the range from 6 to 100 cm. Icarus **4**, 127–156 (1965)

P.H. Roberts, Z.J. Yu, C.T. Russell, On the 60-year signal from the core. Geophys. Astrophys. Fluid Dyn. **101**, 11–35 (2007). doi:10.1080/03091920601083820

B.R. Sandal, A.J. Dessler, Dual periodicity of the Jovian magnetosphere. J. Geophys. Res. **93**, 5487–5504 (1988)

R.M. Sloanaker, Apparent temperature of Jupiter at a wavelength of 10 cm. Astron. J. **64**, 346 (1959)

E.J. Smith, L. Davis, D.E. Jones, P.J. Coleman, D.S. Colburn, P. Pyal, C.P. Sonett, A.M.A. Frandsen, The planetary magnetic field and magnetosphere of Jupiter: Pioneer 10. J. Geophys. Res. **79**, 3501–3513 (1974)

E.J. Smith, L. Davis, D.E. Jones, P.J. Coleman et al., Jupiter's magnetic field, magnetosphere, and interaction with the solar wind: Pioneer II. Science **188**, 451–454 (1975)

E.J. Smith, L. Davis Jr., D.E. Jones, Jupiter's magnetic field and magnetopause, in *Jupiter*, ed. by T. Gehrels (The University of Arizona Press, Tucson, 1976), pp. 788–829

E.J. Smith, L. Davis, D.E. Jones, P.J. Coleman, D.S. Colburn, C.P. Sonett, Saturn's magnetic field and magnetosphere. Science **207**, 407–410 (1980)

D.J. Stevenson, Reducing the non-axis symmetry of a planetary dynamo and an application to Saturn. Geophys. Astrophys. Fluid Dyn. **21**, 113–127 (1982)

D.J. Stevenson, Rep. Prog. Phys. **46**, 555–620 (1983). doi:10.1088/0034-4885/46/5/001

R. Tokar, R.E. Johnson, T.W. Hill, D.H. Pontius, W.S. Kurth, F.J. Crary, D.T. Young, M.F. Thomsen, D.B. Reisenfeld, A.J. Coates, G.R. Lewis, E.C. Sittler, D.A. Gurnett, The interaction of the atmosphere of Enceladus with Saturn's plasma. Science **311**, 1409–1412 (2007)

J. Trauger, The Jovian nebula: A post-Voyager perspective. Science **226**, 337–341 (1984)

J.B. Whiteoak, F.F. Gardner, D. Morris, Jovian linear polarization at 6 cm wavelength. Astrophys. Lett. **3**, 81 (1969)

Z.J. Yu, Spacecraft magnetic field observations as a probe of planetary interiors—Methodology and application to Jupiter and Saturn. Thesis submitted in partial fulfillment of requirements for the Ph.D. Degree, UCLA, 2004

Z.J. Yu, C.T. Russell, Rotation period of Jupiter from observations of its magnetic field. Geophys. Res. Lett. **36**, L20202 (2009). doi:10.1029/2009GL040094

Z.J. Yu, H. Leinweber, C.T. Russell, Galileo constraints on the secular variation of the Jovian magnetic field. J. Geophys. Res. (2009, in press)

Space Sci Rev (2010) 152: 271–305
DOI 10.1007/s11214-009-9507-8

Magnetic Fields of the Satellites of Jupiter and Saturn

Xianzhe Jia · Margaret G. Kivelson · Krishan K. Khurana · Raymond J. Walker

Received: 18 March 2009 / Accepted: 3 April 2009 / Published online: 24 April 2009

Abstract This paper reviews the present state of knowledge about the magnetic fields and the plasma interactions associated with the major satellites of Jupiter and Saturn. As revealed by the data from a number of spacecraft in the two planetary systems, the magnetic properties of the Jovian and Saturnian satellites are extremely diverse. As the only case of a strongly magnetized moon, Ganymede possesses an intrinsic magnetic field that forms a mini-magnetosphere surrounding the moon. Moons that contain interior regions of high electrical conductivity, such as Europa and Callisto, generate induced magnetic fields through electromagnetic induction in response to time-varying external fields. Moons that are non-magnetized also can generate magnetic field perturbations through plasma interactions if they possess substantial neutral sources. Unmagnetized moons that lack significant sources of neutrals act as absorbing obstacles to the ambient plasma flow and appear to generate field perturbations mainly in their wake regions. Because the magnetic field in the vicinity of the moons contains contributions from the inevitable electromagnetic interactions between these satellites and the ubiquitous plasma that flows onto them, our knowledge of the magnetic fields intrinsic to these satellites relies heavily on our understanding of the plasma interactions with them.

Keywords Magnetic field · Plasma · Moon · Jupiter · Saturn · Magnetosphere

1 Introduction

Satellites of the two giant planets, Jupiter and Saturn, exhibit great diversity in their magnetic properties. Magnetic fields at moons have multiple sources, both internal and external. An internal magnetic field can arise from a core dynamo, remanent magnetization or

X. Jia (✉) · M.G. Kivelson · R.J. Walker
Department of Earth and Space Sciences, Univ. of California, Los Angeles, CA 90095-1567, USA
e-mail: xzjia@igpp.ucla.edu

X. Jia · M.G. Kivelson · K.K. Khurana · R.J. Walker
Institute of Geophysics and Planetary Physics, Univ. of California, Los Angeles, CA 90095-1567, USA

the inductive response of a conducting layer within the moon (such as a shell of conducting liquid) to any time-varying external field. External magnetic sources associated with a moon arise through interactions between the ambient flowing plasma and the moon's intrinsic magnetic field, atmosphere and ionosphere. Knowledge of the magnetic field associated with a moon when combined with data on the gravity field can provide useful clues to and place significant constraints on models of the moon's composition and interior structure. In the following, we outline some general properties of the internal and external sources that contribute to the magnetic fields associated with moons.

1.1 Internal Sources: Permanent vs. Inductive

Remanent magnetization, magneto-convection and dynamo action can produce quasi-stable internal magnetic fields. Internal fields arising from these sources are expected to be steady over a relatively long time period and therefore, are sometimes referred to as "permanent" internal fields. Remanent magnetization develops if the magnetic materials of the moon capture and retain the magnetic field imposed at some past time by either an internal dynamo or an ambient field in which the moon is embedded. In magneto-convection, the convective motions of a conducting fluid in the presence of the imposed external magnetic field generate electric currents that modify the internal magnetic field. The magnetic field produced by magneto-convection is not self-sustaining. If the external field vanishes, convection itself does not generate a new magnetic field. In contrast, a dynamo, which also requires an internal region that contains electrically conducting fluid, can self-generate magnetic fields through the convective motion of the conducting fluid and maintain such magnetic fields as long as the convection continues. A dynamo does not require the presence of an external field, but may be affected by such a field. Time-varying internal fields arise if there is a conducting layer within the moon, such as a subsurface ocean, that can carry currents driven by a time-varying external field. The property of the induced field, such as its field strength and orientation, depends on both the external driving field and the properties of the conducting layer. An induced magnetic field tends to exhibit variations on relatively short time scales because its strength and orientation vary with changes of the external driving field.

Time-varying external fields are present at the orbits of the Galilean satellites of Jupiter. Because Jupiter's dipole is tilted with respect to its rotational axis by roughly 10°, the Jovian dipole equator sweeps over the Galilean satellites at their synodic periods (the rotation period of Jupiter seen in the rest frame of the moon). Effectively, these moons wobble up and down through Jupiter's plasma sheet sampling regions of different plasma and field conditions. The periodically changing background field at the orbits of these moons provides a strong driving signal for magnetic induction but a global response is expected only if the moon contains a global shell of high conductivity.

Unlike Jupiter's magnetospheric field at the orbits of the Galilean satellites, Saturn's magnetospheric field at the orbits of its inner icy moons does not contain a substantial time-varying component that can drive strong magnetic induction because Saturn's internal field is axisymmetric. Temporal variations in the magnetic field may arise because of local plasma interactions at the moons or dynamic changes in Saturn's global magnetosphere. Nonetheless, the induction effects at Saturn's moons are expected to be much less prominent than those at Jupiter's moons and correspondingly difficult to detect.

1.2 External Sources: Plasma Interaction with Moons

Magnetic perturbations can arise through the interaction between the moons and the ambient flowing plasma within the magnetospheres of their parent planets or the solar wind. The magnetic field perturbations add to the field generated within the moon. Magnetic field perturbations resulting from plasma interactions are referred to as fields from external sources.

Plasma interactions at the moons can have various forms depending on both the moons' magnetic character and the nature of the ambient plasma. A magnetized moon with a sufficiently strong internal magnetic field can stand off the incident plasma above its surface and form its own magnetosphere analogous to that formed in the interaction between its parent planet and the flowing solar wind. Ganymede, one of Jupiter's moons, is the only known example of such a case; it will be discussed in detail later in this Chapter. An unmagnetized moon can also interact with the incident plasma through the mass-loading process and, if it serves as a substantial source of neutrals, can produce observable signatures in the magnetic field. A few moons in the two planetary systems have been found to belong to this category and their interactions with the magnetized plasma generally can be described by an Alfvén wing model, which will be introduced below. Moons with neither strong internal magnetic field nor significant neutral sources (so-called "inert" moons) perturb the ambient plasma and field principally by absorbing incident plasma in a manner analogous to that of Earth's Moon. Magnetic perturbations associated with inert moons are observed mainly in their wake regions where the shadowing of the incident plasma modifies the field.

It is useful to characterize the plasma interaction with an obstacle by introducing some dimensionless parameters, such as the Alfvénic Mach number, $M_A = V_{flow}/V_A$, the sonic Mach number, $M_s = V_{flow}/C_s$, the magnetosonic Mach number, $M_{ms} = V_{flow}/\sqrt{C_s^2 + V_A^2}$ and the plasma beta, $\beta = P_{th}/(B^2/2\mu_0)$, where V_{flow} is the unperturbed bulk flow velocity, $V_A = B/\sqrt{\mu_0 n_i m_i}$ is the Alfvén speed given in terms of the ambient unperturbed magnetic field, B, the number density, n_i, and the mass, m_i, of the dominant ions, $C_s = \sqrt{\gamma P_{th}/n_i m_i}$ is the sound speed, a function of the plasma thermal pressure, P_{th}, the mass density and the ratio of specific heats, γ. The Mach numbers indicate whether or not perturbations carried by different wave modes can propagate upstream in the flow. In addition, the Alfvén Mach number M_A and the sonic Mach number M_s are closely related to the ratio of the dynamic pressure of the flow to the magnetic pressure and the plasma thermal pressure, respectively. The quantity β directly measures whether thermal plasma effects or magnetic field effects dominate the interaction.

Except for some of Saturn's outer moons, which spend some part of their orbital motion outside of Saturn's magnetosphere, the other moons that will be discussed here always remain embedded within a planetary magnetosphere. At the orbits of these moons, the ambient low energy magnetospheric plasma approximately corotates with the planet at bulk flow speeds much larger than the Keplerian speeds of the moons (Kivelson et al. 2004; Khurana et al. 2008). In the rest frame of these moons, therefore, the corotating plasma continually overtakes them from their trailing sides (in the sense of moon's rotation). Figure 1 shows a schematic of the Alfvén wing structure that results from the interaction between a sub-Alfvénic flow and a conducting obstacle. As a magnetized plasma flows towards a conducting obstacle, such as a moon with an ionosphere, it is slowed down and diverted around the moon because the incoming plasma and field are prevented from penetrating into the conductor by the induced currents. The slowing of the flow generates perturbations carried by different wave modes, among which compressional perturbations can propagate effectively across the field. These waves produce small amplitude disturbances in front of the moon if the flow is sub-magnetosonic ($M_{ms} < 1$), as is the case for most of the moons discussed here. Conversely, if the ambient flow is super-magnetosonic ($M_{ms} > 1$), the wave fronts of these perturbations steepen to form a standing shock upstream of the moon. The perturbations caused by the slowing of the flow near the obstacle also generate Alfvén waves

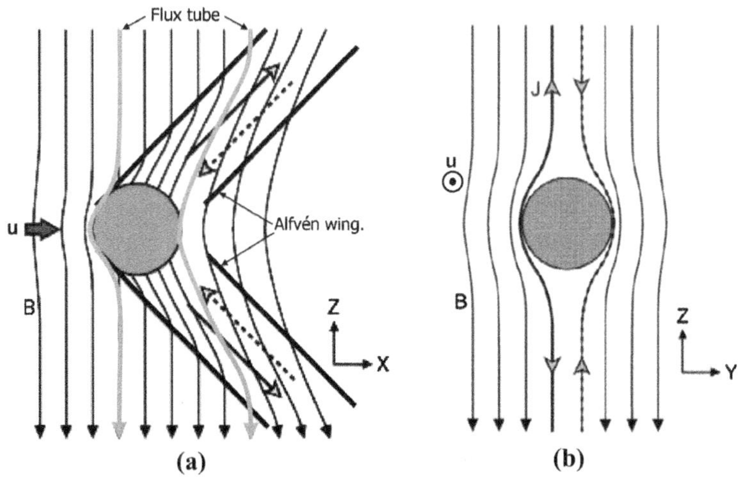

Fig. 1 A schematic illustration of the Alfvén wing structure arising from the interaction between a sub-Alfvénic ($M_A < 1$) flow and a conducting obstacle. Panel (**a**) represents a plane (XZ plane) that contains the directions of the unperturbed flow ($+\hat{X}$) and the background magnetic field ($-\hat{Z}$). Panel (**b**) shows the cross-section of the interaction region in a plane (YZ plane) normal to the plasma flow direction. The Alfvén wing currents flowing along the field lines are represented by arrows. *Solid* and *dashed lines* represent the currents flowing on the $-\hat{Y}$ and $+\hat{Y}$ side, respectively. The figure is adapted from Kivelson et al. (2004)

propagating away from the obstacle along magnetic field lines and slowing the flow at large distances from the moon. Field-aligned currents associated with these Alfvén waves close through perpendicular currents in the conducting region near the obstacle, such as the Pedersen currents in the ionosphere or through the moon itself. In the rest frame of the moon, the perturbations associated with the field-aligned currents cause the field lines to bend back within regions called the "Alfvén wing" whose boundaries (shown as heavy solid lines in Fig. 1(a)) are tilted with respect to the background field at an angle $\theta = \tan^{-1}(u/V_A)$, where u and V_A are the flow speed and the Alfvén speed of the unperturbed flow, respectively.

If there is a source of neutrals present at the moon, pickup ions may be produced by the ionization of neutrals through various processes, such as photo-ionization, electron impact ionization and ion-neutral interaction (charge exchange). When the newly generated ions are added to the ambient flow, they acquire kinetic energy of flow and of thermal motion. The energy is extracted from the background plasma. In order to conserve momentum, the resulting bulk flow slows down. The addition of newly ionized ions and the associated slowing of the flow are described by the term "mass-loading". Associated with this mass-loading process, there is a so-called "pickup current" (Goertz 1980), which arises from the effective charge separation between positively and negatively charged particles as they are accelerated by the motional electric field in the ambient flow. The pickup current, along with any associated Pedersen currents in the moon and its ionosphere, flows across the field and is closed by field-aligned currents (or the so-called "Alfvén wing" currents) owing to the requirement that current be divergence-less. It has been shown that the contribution of mass-loading to the plasma currents in the interaction can be equivalently described as a Pedersen conductance (Neubauer 1998a; Hill and Pontius 1998).

The Alfvén wing model mentioned above originally was applied to the sub-Alfvénic ($M_A < 1$) interaction between Io and its plasma torus (Goertz 1980; Neubauer 1980; Southwood et al. 1980), but it also can be applied to the interactions of the flowing plasma with other moons that have ionospheres and/or a source of neutrals. Even if the flow is super-Alfvénic ($M_A > 1$) but sub-magnetosonic ($M_{ms} < 1$), as for ambient conditions at Titan's orbit, the Alfvén wing model remains a valid description of the interaction. However, there are some unique aspects of the interaction, which will be described in the section on Titan.

Numerical simulations are valuable tools for studying plasma interactions with an obstacle. Several types of simulation models are widely used in the study of plasma-moon interactions, i.e., magnetohydrodynamic (MHD) simulation, hybrid simulation and test particle simulation. Each of the approaches has advantages and limitations and may be appropriate only in certain situations. Most of the Galilean satellites of Jupiter (except Callisto) and the inner icy satellites of Saturn inside the orbit of Dione are significantly larger than the characteristic scale lengths of the thermal plasma, such as the gyroradii of heavy ions (Kivelson et al. 2004; Khurana et al. 2008). MHD simulation, in which both ions and electrons are treated as fluid, is suitable for studying the global plasma interactions at these moons, where the effect of particle gyromotion is not important. MHD simulation usually can provide a description of the global interaction over a reasonably large region around the obstacle and with relatively high resolution at a feasible computational cost. However, at some moons, such as Saturn's icy moon Rhea, the gyroradii of heavy ions can become a significant fraction of the moon's radius (Khurana et al. 2008). Under such circumstances, kinetic effects due to the finite gyroradii of heavy ions need to be taken into account when considering plasma interactions with the moon. Hybrid simulation, in which ions are treated as kinetic and electrons are considered as fluid, is more appropriate for studying plasma interactions with such objects than is MHD simulation. However, hybrid simulation normally needs relatively expensive computational resources to achieve reasonably good resolution and to reduce system noise (Ledvina et al. 2008 and references therein). As an alternative, at relatively low computational expense, test particle simulation often is used in studying plasma transport when finite gyroradius effects are important. In a test particle simulation, the motion of non-interacting test particles is driven by the electric and magnetic fields output from other global models, such as MHD simulations. However, it should be kept in mind that modifications of the electromagnetic fields due to test particles are not fed back to the global electric and magnetic fields and wave-particle interactions usually are not self-consistently included in test particle simulations.

Planetary magnetic fields are reviewed in other chapters (Connerney 2009; Hulot et al. 2009; Anderson 2009) in this issue/book; in this chapter we focus on the magnetic fields of the Jovian and Saturnian moons and their surroundings. Jupiter and Saturn have many satellites (more than 60 have been identified for each planet), most of which are small bodies with radii less than 200 km. These small moons are either collisional products of larger moons or captured asteroids and comet nuclei and they are in general irregular in shape. Little is known about their physical properties because there have been few close encounters with any of these small moons. Consequently, in this paper we shall concentrate on the major satellites (whose radius exceeds about 200 km) for which planetary spacecraft have made one or more close flybys. Moons of Jupiter and Saturn that will be discussed in this paper are listed in Table 1, which gives an overview of properties that are relevant to the topic of this chapter.

The following sections discuss the moons of interest in groups based on the origin of their magnetic field. In each section, the available magnetic field observations and some basic features of the plasma interactions are reviewed. Inert moons (Tethys and Rhea) are discussed in Sect. 2 and moons that have internally induced magnetic fields (Europa and Callisto) are considered in Sect. 3. Moons with strong plasma interactions (Io, Enceladus

Table 1 General properties of the principal satellites of Jupiter and Saturn

Satellite	Radius (km)	Orbital distance (planetary radii)	Orbital eccentricity	Density (g/cm^3)	Differentiation	Neutral source[1]	Source(s) of magnetic field[2]
				Jupiter			
Io	1821.6	5.91	0.0041	3.53	Yes[3]	Yes(SO_2)[4]	Plasma(Ind?)[5]
Europa	1560.8	9.40	0.0094	3.01	Yes[6]	Yes(O_2)[4]	Ind, Plasma[7]
Ganymede	2631.2	14.97	0.0011	1.94	Yes[8]	Yes(O_2)[4]	Intrinsic, Ind, Plasma[9]
Callisto	2410.3	26.33	0.0074	1.83	Incomplete[10]	Yes(O_2/CO_2?)[4]	Ind, Plasma[11]
				Saturn			
Enceladus	252.1	3.95	0.0045	1.61	Probably[12]	Yes(H_2O)[13]	Plasma (Ind?)[14]
Tethys	533.0	4.89	0.0000	0.97	No[15]	No[16]	Inert[16]
Dione	561.7	6.26	0.0022	1.48	?	?	?
Rhea	764.3	8.74	0.0010	1.23	No[17]	No[16]	Inert[16]
Titan	2575.5	20.27	0.0292	1.88	Yes(?)[18]	Yes(N_2)[19]	Plasma(Ind?)[20]
Iapetus	735.6	59.09	0.0283	1.08	?	No	?

[1] This column indicates whether or not a moon possesses a substantial source of neutrals and only major neutral species is listed in the parentheses

[2] This column describes the origin of magnetic fields in the vicinity of the moons. "Intrinsic" refers to intrinsic magnetic field, "Ind" refers to internally induced field through electromagnetic induction, "Plasma" refers to plasma-induced magnetic field and "Inert" refers to moons that are unmagnetized and do not possess substantial sources of neutrals. "?" means that situation is unclear

[3] Anderson et al. (2001a) [4] McGrath et al. (2004) and references therein [5] Kivelson et al. (2004); Saur et al. (2004) and references therein [6] Anderson et al. (1998b) [7] Kivelson et al. (2000) [8] Anderson et al. (1996); Schubert et al. (1996) [9] Kivelson et al. (1996); Kivelson et al. (2002) [10] Anderson et al. (2001b) [11] Khurana et al. (1998); Kivelson et al. (1999) [12] Schubert et al. (2007) [13] Hansen et al. (2006); Waite et al. (2006) [14] Dougherty et al. (2006); Khurana et al. (2007) [15] Thomas et al. (2007) [16] Khurana et al. (2008) [17] Anderson and Schubert (2007) [18] Grasset et al. (2000); Sohl et al. (2003) [19] Tyler et al. (1981); Hunten et al. (1984) [20] Ness et al. (1982); Backes et al. (2005); Neubauer et al. (2006)

and Titan) are reviewed in Sect. 4 and the only case of an intrinsically magnetized moon, Ganymede, is presented in Sect. 5. In Sect. 6 we describe two of Saturn's moons (Dione and Iapetus) for which the situation is unclear. Section 7 gives a brief summary.

2 Inert Moons: *Tethys and Rhea*

Tethys, with a mean radius of 533 km, is the fifth largest moon of Saturn. The interior of Tethys is believed to be homogeneous and undifferentiated, composed mostly of water ice as suggested by its low density ($\sim$0.97 g/cm^3) and its ellipsoidal shape (Thomas et al. 2007). Rhea, which has a mean radius of 764 km and a mean density of 1.23 g/cm^3, is the second largest moon of Saturn. An interior model of Rhea developed by Anderson and Schubert (2007) using measurements from Cassini suggest that, like Tethys, this moon is probably also undifferentiated and is a homogeneous mixture of ice (75%) and rock (25%). No observational evidence suggests that either of these moons is geologically active at the present time. Neither of the moons appears to retain an appreciable atmosphere or to be a significant source of neutrals. Direct in-situ magnetic field measurements (Dougherty et al. 2004) in the close vicinity of these two icy moons were acquired by the Cassini spacecraft, which made a single close encounter with each of them in 2005. The Cassini observations show no evidence of any appreciable internal magnetic field for either moon (Khurana et al. 2008). Because the background field of Saturn is nearly constant around the orbits of these moons, no inductive response is possible regardless of the interior structure.

Some important features in the Cassini magnetic field observations that help us identify the moons' magnetic properties are worth emphasizing. During each close flyby, the spacecraft flew through regions downstream of the moon with closest approach distances of 3.83 for Tethys and 1.67 for Rhea (in moon radii), respectively. The Rhea flyby occurred very close to the moon's central wake region while the Tethys' flyby took place south of the central wake (shown in Fig. 2(a)). A portion of the magnetic field data near Rhea is shown in Fig. 2(b), in which the background magnetic field of Saturn is removed in order to reveal the weak perturbations associated with the moon. During the Rhea flyby, the field strength is slightly enhanced near closest approach with depressions on both sides. The field enhancement in the central wake is consistent with absorption of the incident flow by the moon on the upstream side and plasma depletion in the wake. Thus the magnetic field strength increase in the wake region develops in order to maintain the balance of total pressure (sum of plasma and magnetic pressure). In addition, the bi-polar signature (from negative to positive) in the B_y component near closest approach arises because field lines are sucked into the wake (shown in Fig. 2(a)). In passes by other moons, we will identify a field minimum in the central wake and a change in the sign of B_y from positive to negative traced along an analogous trajectory (see Fig. 1(b)). We will interpret such a signature as evidence that the moon provides a substantial neutral source that slows the plasma flow either by mass-loading (as in the case of Io, discussed below) or because the moon and its ionosphere are conducting. In either of such cases, field lines in the wake bulge out. The Tethys flyby data basically show features similar to those described in Rhea's case even though the Cassini trajectory was south of Tethys. The available Cassini observations suggest that neither Rhea nor Tethys produces a substantial neutral source that can mass-load Saturn's plasma flow and generate measurable effects. However, it is possible that neutral material is sputtered primarily from the upstream side of the moons' surfaces and that weak mass-loading would be evident only on upstream flybys (Khurana et al. 2008). Such flybys of these moons are needed to complete our knowledge of their interactions with Saturn's magnetosphere.

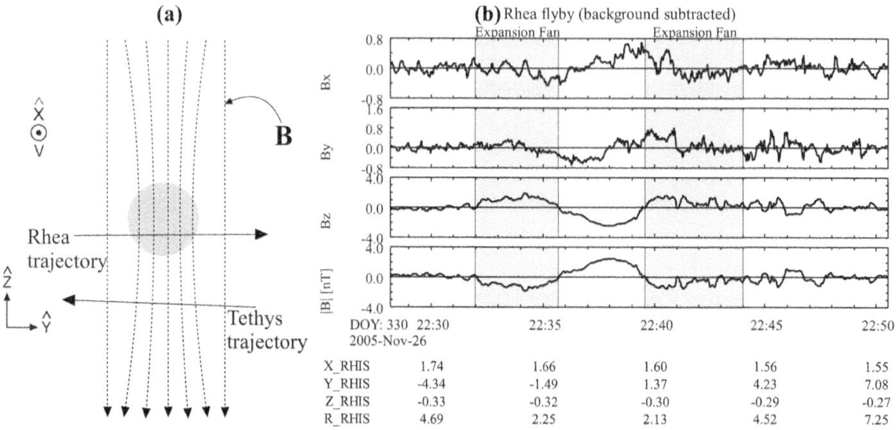

Fig. 2 (**a**) Projections of the Cassini trajectories during close flybys of Tethys and Rhea, shown in a plane in the wake region and normal to the incident plasma flow direction. Also drawn is a schematic of magnetic field lines (in *dotted lines*) based on the Cassini observations shown in (**b**). In this plot, the $\hat{x}$-axis points along the corotation direction of the plasma, the $\hat{y}$-axis points towards Saturn and the $\hat{z}$-axis lies along the spin axis of Saturn. (**b**) Detrended magnetic field data (shown in a moon interaction coordinate system illustrated in (**a**)) from Cassini during a close flyby to Rhea. Figures are adapted from Khurana et al. (2008)

A close analogy to the interaction of Tethys and Rhea with Saturn's magnetospheric plasma is found in the interaction of Earth's moon (no appreciable global magnetic field, no mass-loading and non-conducting) with the plasma that engulfs it. The Moon's interaction has been extensively studied in the past several decades through observations and modeling (Schubert and Lichtenstein (1974) and references therein; Bosqued et al. (1996); Ogilvie et al. (1996); Owen et al. (1996); Halekas et al. (2005)). Although Tethys and Rhea share many similarities with the Earth's Moon, their inert interactions with flowing plasma differ from the lunar case because of the properties of the ambient plasma. One feature of plasma interaction that is especially relevant to mass-loading at an inert moon is the net flux of sputtering particles that reaches the surface during the time required for a flux tube to flow across the moon's diameter. The flux depends not only on the plasma density but also on the ratio of the sound speed (which determines the flow speed along the flux tube) to the flow speed (which establishes the duration of the transit across the moon). The Earth's Moon spends much of its orbital time in the solar wind, in which that ratio normally is $\ll 1$ (or $M_s \gg 1$). For Tethys and Rhea, which orbit around Saturn in the inner part of the Kronian magnetosphere with mean orbital distances of 4.9 and 8.7 Saturn radii (1 $R_S = 60{,}268$ km), respectively, the ambient plasma flow typically is transonic ($M_s \sim 1$) (Acuna et al. 1983; Sittler et al. 2006; Khurana et al. 2008; Wilson et al. 2008). As a consequence, the absorption of particles moving along a flux tube that is in contact with the moon leads to a more extended plasma depletion region (in which magnetic field strength increases) around the moon in a transonic interaction than in a super-sonic interaction. Details of the comparison between the inert moon interactions in the two distinct plasma regimes are given by Khurana et al. (2008).

Roussos et al. (2008) studied the plasma interaction at Rhea by using a 3D hybrid simulation, which takes into account ion gyromotion. Their simulation, in which Rhea is treated as a pure plasma-absorbing body, yields a satisfactory agreement with the Cassini magnetic field observations. Because the gyroradii of the dominant water-group ions at Rhea's orbit with typical thermal energy ($\sim$100 eV) become a significant fraction of the moon's radius, it would be interesting to investigate how the plasma interaction with an inert moon is modified by the finite gyroradii effects of ions by comparing results from both hybrid and MHD simulations.

In summary, Tethys and Rhea are examples of inert moons, which possess neither an appreciable internal magnetic field nor a significant neutral halo that can generate substantial magnetic field perturbations through plasma interactions. To a good approximation, they act as absorbing obstacles. Their interactions with Saturn's sub-magnetosonic plasma under relatively steady magnetic field and plasma conditions are analogous to the inert interaction of the Earth's moon with the super-magnetosonic and highly variable solar wind, albeit that interaction is greatly affected by the variability and the supersonic flow of the solar wind.

3 Satellites with Internally Induced Magnetic Fields: *Europa and Callisto*

When a conductor is placed in an environment with a time-varying external magnetic field, it responds to the driving (or primary) field by generating eddy currents that flow near the surface of the conductor. An induced magnetic field perturbation that acts to prevent the driving field from penetrating into the conductor then arises from the eddy currents. Such a process can be described by the diffusion equation for the magnetic field:

$$\nabla^2 \mathbf{B} = \mu\sigma \frac{\partial \mathbf{B}}{\partial t} \tag{1}$$

a relationship that applies when plasma convection is neglected. Here μ is the permeability (normally taken to be the permeability of vacuum, μ_0) and σ is the electrical conductivity. Based on (1), an important scale length, the skin depth $\delta = \sqrt{\mu\sigma\omega/2}$, which varies with the frequency (ω) of the driving signal, can be defined to describe the depth to which the power of a time-varying external signal can penetrate into a conductor. A steady external field ($\omega = 0$) does not induce currents in a conductor and consequently diffuses into the conductor over the diffusion time scale given by $\tau = R^2\mu\sigma$, where R is the scale size of the conductor. For a perfect conductor ($\sigma \sim \infty$), the skin depth $\delta \sim 0$ meaning that an incoming wave signal is fully excluded from the conductor.

Most of planetary satellites of interest are embedded in plasmas dominated by the planetary magnetic field (except Iapetus, whose orbital distance is larger than the spatial scale of Saturn's magnetosphere and Titan, which can spend part of its orbital period beyond Saturn's magnetopause). Typically, then, the scale lengths of the field variations in the background plasma are much larger than the spatial scales of the satellites. Therefore, the external large-scale fields that these satellites experience can, to a good approximation, be considered as spatially uniform fields. If a moon with a spherically symmetric region of high conductivity, such as an ionosphere or a subsurface ocean, is embedded in a region of uniform time-varying external magnetic field (as shown in Fig. 3), an induced field will be produced. The induced field has the structure of an internal dipole outside of the conductor and the sum of the induced and the primary fields is approximately zero inside the conductor. As a result, the external field will be excluded by the conductor, with magnetic field lines draped tangentially over its surface. In reality, however, the conductivity of the conductive layer is probably finite. In this case, the driving field can penetrate into the conductor with amplitude falling exponentially with depth at a rate described by the skin depth δ. The instantaneous induced field will have a phase lag from the driving signal and an amplitude that is a fraction of the maximum response for a perfect conductor (Lahiri and Price 1939; Rikitake 1966; Zimmer et al. 2000).

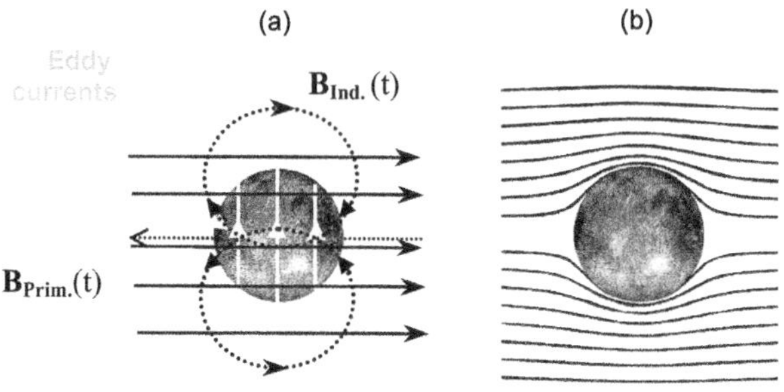

Fig. 3 Schematics adapted from Khurana et al. (2002) showing (**a**) magnetic field lines of the dipolar response ($\mathbf{B}_{ind}(t)$) of a spherical perfect conductor (in which induced eddy currents are flowing) to a uniform time-varying external field ($\mathbf{B}_{prim}(t)$), (**b**) magnetic field lines of the resulting total field ($\mathbf{B}_{prim}(t) + \mathbf{B}_{ind}(t)$), which are tangent to the surface of the sphere and do not penetrate into the conductor

Detailed discussions of the induced magnetic fields associated with various objects in the solar system, including planetary satellites, are given by Saur et al. (2009) in another chapter in this issue. In this section, we briefly review the magnetic field observations at the two Galilean satellites of Jupiter, Europa and Callisto, where strong evidence of induced magnetic fields have been found.

3.1 Europa

Europa, with a radius (R_E) of ~1560 km, is the smallest of the four Galilean satellites of Jupiter, orbiting at a distance of ~9.4 R_J (R_J, radius of Jupiter = 71,492 km). Gravitational measurements from four Galileo flybys suggest that Europa's interior is most likely to be a differentiated structure consisting of a metallic core with a rocky mantle covered by a water-ice outer shell of thickness between 80 and 170 km (Anderson et al. 1998b). Europa's surface is geologically young compared to the surfaces of Ganymede and Callisto. Voyager observations reveal global fracture-like patterns and few impact craters on its icy surface (Smith et al. 1979a, 1979b), features confirmed by the high-resolution images from Galileo (Carr et al. 1998).

Europa does not appear to possess any appreciable permanent internal magnetic field (Khurana et al. 1998; Kivelson et al. 2000). Rather, observations obtained from multiple passes have clearly established that Europa generates an induced magnetic field in response to the external time-varying field of Jupiter. The induced magnetic fields basically can account for the large-scale (relative to the size of Europa) field perturbations observed near Europa. Europa possesses a tenuous atmosphere (Hall et al. 1995) and an ionosphere (Kliore et al. 1997), both of which can interact with the ambient magnetospheric plasma producing observable magnetic field perturbations.

Our knowledge of Europa's magnetic fields and its associated plasma interactions is based on the measurements obtained by Galileo on eight Europa flybys from which magnetic field data were acquired. During its first several flybys of Europa (between 1996 and 2000), the Galileo magnetometer observed strong field perturbations near the moon. The localized perturbations were attributed to Europa's interaction with the corotating Jovian plasma since similar large fluctuations were not observed outside of the close encounter intervals (Kivelson et al. 1997). Several different possibilities regarding the source of the observed field perturbations were initially proposed. The perturbations could have been interpreted as evidence of the plasma currents arising from the electromagnetic interactions of the moon with the Jovian plasma or of an internal magnetic field from either remanent magnetization or self-generated dynamo action.

Plasma interactions at Europa are important and magnetic field perturbations localized near the moon arise in part because of the existence of an oxygen atmosphere and an ionosphere. Magnetic field enhancement occurs upstream of Europa, a signature of the slowing of the incident flow by mass-loading. The sub-Alfvénic ($M_A < 1$) plasma interaction at Europa generates an Alfvén wing structure similar to that formed when Io interacts with its plasma torus (Neubauer 1980; Southwood et al. 1980). Field lines are bent back above and below the moon by the field-aligned Alfvén wing currents that link to Jupiter's ionosphere. However, analysis of the magnetic field data from multiple flybys indicates that plasma effects alone cannot account for the large-scale (relative to the size of Europa) features of the observed magnetic field (Kivelson et al. 1999). A substantial source of a dipolar field whose axis lies roughly in the moon's equatorial plane is needed to interpret the observed magnetic field variations on multiple passes. A core-dynamo is not a favored source of an internal field whose dipole axis is nearly perpendicular to the moon's rotation axis, but that possibility initially could not be ruled out (Kivelson et al. 1997).

Another possible source of an internal field is magnetic induction. Because Jupiter's dipole is tilted with respect to its rotational axis by roughly 10°, the Jovian dipole equator sweeps Europa at the synodic period of ~11.1 hours. Consequently, Europa experiences a strongly time-varying external field, whose major temporal variations are in the component that lies roughly in Europa's equatorial plane. As first proposed by Colburn and Reynolds (1985), the periodically changing background field at Europa's orbit could drive currents inside of Europa if the moon contains a global region of high conductivity. Indeed, when an induced dipole moment, whose strength and orientation are determined by the instantaneous value of the periodically varying component of Jupiter's field, is used to model the flyby data assuming that the highly conducting region is a shell lying close to the moon's surface, it yields a reasonably good fit to the observations for passes during which plasma effects are relatively weak (Kivelson et al. 1999).

If the observed perturbations arise from a permanent internal source, the inferred dipole moment should not vary in orientation and strength over a time scale as short as the intervals between Galileo flybys of Europa. Conversely, if an induced dipole moment is the dominant source, the inferred dipole moment would change in orientation and strength as the external driving field varies. A crucial test of the induction model would be provided by comparing perturbations present on close flybys north and south of the Jovian central plasma sheet, because the orientation of the time-varying component of the driving field has opposite polarities on the two sides. However, all the passes that occurred before 2000 and had magnetometer data available took place when Europa was located either near or above Jupiter's central plasma sheet. The time-varying component of the external field during these passes was either too weak to drive induction with measurable effects or pointed roughly in the same direction (radially outward from Jupiter). Moreover, on passes near the center of the plasma sheet, the field perturbations arising from plasma effects are large and obscure a weak internal source. Thus, either a dynamo source or an induced source was possible.

The last close encounter ("E26" flyby, Galileo flybys of the Galilean moons are labeled by the moon's name and the orbit number) of Europa occurred in early 2000 when Europa was far south of Jupiter's central current sheet and Jupiter's field pointed inward toward the planet. This pass provided an optimal flyby geometry for removing the ambiguity. As shown in Fig. 4, the predictions computed from the dipole moment fit to a previous pass (E4) are in antiphase to the general trend of the observed field perturbations. The E26 observations demonstrate that an internal dipole with the opposite orientation as that inferred from previous passes is required to interpret the field measurements and therefore provides convincing evidence for the existence of an induced internal field at Europa (Kivelson et al.

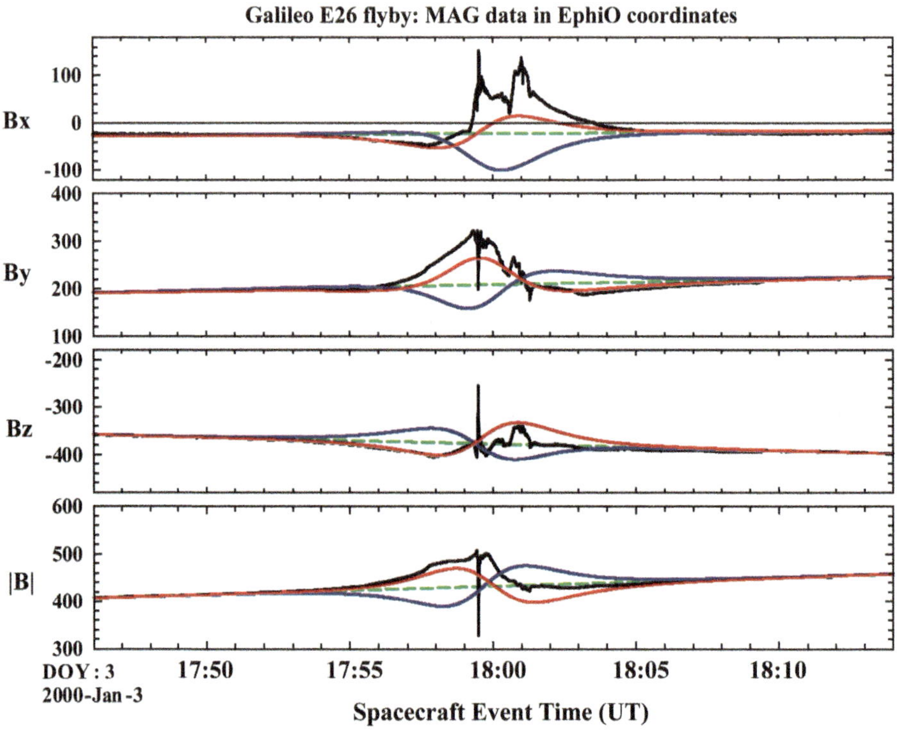

Fig. 4 Galileo magnetometer data (three components and the field magnitude in nT) from the E26 flyby of Europa for a range of $\sim 5\ R_E$ from Europa's center. Data are shown in the EphiO coordinate system, in which $\hat{x}$ is along the corotating plasma flow, $\hat{y}$ is radially in toward Jupiter and $\hat{z}$ is parallel to Jupiter's spin axis. *Black traces* are the Galileo measurements. *Green dashed curves*, obtained from a polynomial fit to the field components measured before and after the encounter, represent the background field without Europa perturbations. *Blue traces* are computed from the dipole moment fit to the E4 pass. *Red traces* are predictions from an induced dipole model without plasma effects. The figure is adapted from Kivelson et al. (2000)

2000). Data from all passes reveal that the internal field of Europa is dominated by a magnetically induced field and that the contribution from a permanent field is at most a minor component (Schilling et al. 2004).

The induced dipole model consistent with the Galileo observations requires a global shell of high conductivity near the moon's surface capable of carrying substantial currents (Khurana et al. 1998; Kivelson et al. 1999, 2000; Zimmer et al. 2000). Detailed constraints on the properties of the conducting layer, such as the total conductance depending on the thickness and the conductivity of the layer, can be derived from the inferred induced dipole moment (Zimmer et al. 2000). If the induced field were generated in a conducting core with an assumed size $\leq 0.5\ R_E$, then the induced field strength would be only $\leq 1/8$ of the signal from a near-surface conductor. Such a weak near-surface field is inconsistent with the observations. Another possible candidate for the required conductor is Europa's ionosphere, which has been observed by the radio occultation measurements (Kliore et al. 1997). However, the ionosphere is so tenuous that the estimated conductance, including both the Pedersen conductance and the conductance due to ion pickup, integrated over the whole ionosphere is too small (by several orders of magnitude) to carry the current needed to produce the observed magnetic field perturbations. On the other hand, a layer of melted

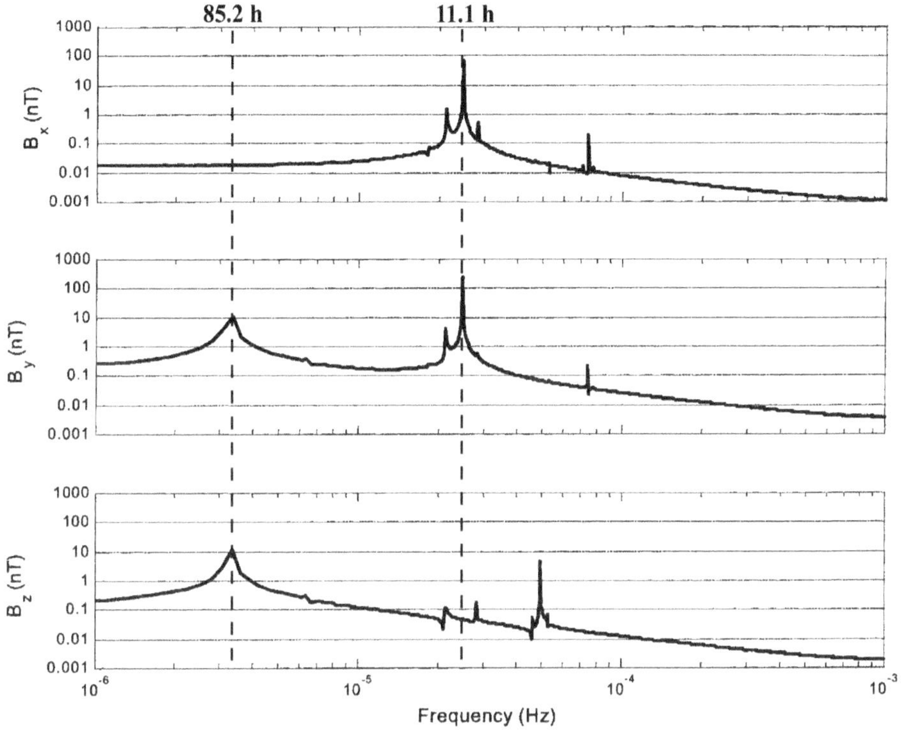

Fig. 5 The amplitude spectrum of the primary field observed by Europa calculated from an empirical model of Jupiter's magnetic field (Khurana 1997). The figure is reproduced from Khurana et al. (2002)

ice beneath the moon's icy surface can account for the inferred induction response. Features on Europa's icy surface strongly support the presence of such a subsurface ocean (Carr et al. 1998). Analysis from Zimmer et al. (2000) shows that a subsurface ocean (at depth <200 km) composed of terrestrial seawater with thickness of ~ 10 km can produce the magnetic perturbations observed on several passes. In an attempt to improve the constraints on the properties of the conducting region, Schilling et al. (2007) studied the induction effects at Europa by using a time-dependent MHD model to account for effects arising from plasma interactions. They found that when plasma effects are included, a higher conductivity of the conducting shell is needed to interpret the magnetic field observations, implying that the existence of a subsurface ocean is even more likely. However, the complete knowledge of the properties of Europa's interior structure must rely on more in-situ observations and accurate description of the plasma interactions at Europa.

Ideally one would hope to constrain both the thickness and the conductivity of the conducting layer. It would be possible to do so if one could measure the magnetic induction at multiple frequencies (Khurana et al. 2002). However, the Galileo data reveal only the response at a single frequency (the 11.1 hour synodic period), and in this case either the thickness or the conductivity must be assumed. In the future, it should be possible to measure the inductive response over times long compared with the synodic period. Then the data will contain lower frequency signals as well. A low frequency signal arises because of Europa's orbital eccentricity (see Table 1). As the moon moves around Jupiter, its radial motion through Jupiter's field imposes a signal at its orbital period (85.2 hours). As shown in Fig. 5,

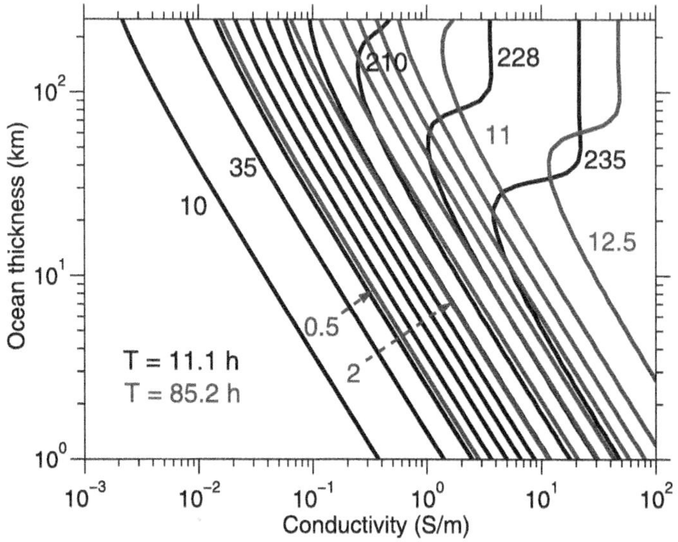

Fig. 6 Figure adapted from Khurana et al. (2002) showing contours of the maximum surface strength (in nT) of the induced field generated in response to the 11.1 h wave (*black lines*) at the synodic rotation period of Jupiter and the 85.2 h wave (*grey lines*) at the orbital period of Europa as a function of the ocean thickness and the ocean conductivity. It is assumed that the wave amplitude at the synodic rotation period of Europa is 250 nT and the wave amplitude at the orbital period is 14 nT

the amplitude of this variation is relatively small compared to that of the aforementioned 11.1-hour periodicity wave but should drive a response that could be extracted from long-term (a month or more) continuous magnetic field observations at Europa. Figure 6 shows the contours of the maximum induction field strength generated at Europa's surface in response to Jupiter's varying field at the two main frequencies. It is shown that when the ocean thickness is >100 km and the conductivity is > 0.2 S/m, the amplitude curves of the two frequencies begin to intercept each other. From such multi-frequency driving signals it will be possible to identify independently the thickness and the conductivity of the subsurface layer.

3.2 Callisto

Callisto, the outermost Galilean satellite of Jupiter, orbiting around the planet at a distance of 26.33 R_J is the second largest Jovian moon (radius of ~2410 km). Similar to Europa, Callisto also experiences a periodically changing external field as it wobbles through the Jovian plasma sheet. Galileo observations suggest that Callisto also generates an induced internal field through a near-surface conducting layer, possibly a subsurface ocean.

The source of Callisto's magnetic field was identified mainly through the data from two Galileo flybys (C3 and C9), one of which (C3) occurred above and the other (C9) below Jupiter's plasma sheet. During the C3 flyby, Galileo flew by the moon in the wake region at a distance of ~ 1139 km above the surface. Magnetic field variations on the scale of the moon's radius along with a small field enhancement (only a few nT above an averaged ambient field of ~35 nT) near the closest approach were detected. If the field enhancement and the large-scale variations were produced by an internal dipole, the inferred dipole moment should have a surface field strength of ~ 15 nT and an orientation that is approximately opposite to the dominant component (radially outward from Jupiter) of the ambient Jupiter's field and nearly orthogonal to the moon's spin axis (Khurana 1997). As for Europa, a highly tilted and rather weak dipole moment is more likely to be generated by induction than by internal dynamo action (Khurana 1997; Neubauer 1998a). The distinction between a permanent and an induced dipole moment is supported by the observations from a subsequent flyby (the C9 flyby) during which Callisto was located well below the Jovian plasma sheet, a situation opposite to that of the C3 flyby (Kivelson et al. 1999). The inferred dipole moments, which well represent the large-scale field variation in the two individual flyby data, flip sign from the C3 to the C9 flyby. This result rules out the possibility that a permanent internal field is the dominant source that generates the observed large-scale perturbations centered around the moon; it leads to the conclusion that Callisto's internal field originates predominantly from magnetic induction produced by a highly conducting layer associated with the moon.

Callisto has an ionosphere whose electron density exceeds that of Europa by almost two orders of magnitude (Kliore et al. 2002). The estimated ionospheric conductance based on the radio occultation measurements can carry the current needed to account for the inferred induced field (Zimmer et al. 2000). However, Callisto's ionosphere is detected only when the trailing (or ram) side of the moon is illuminated. The transient nature of the ionosphere makes it difficult to attribute the inductive response to the ionosphere. Therefore, as for Europa, a subsurface ocean of high electrical conductivity seems to be a more plausible source (Kivelson et al. 1999; Zimmer et al. 2000). For Callisto, the estimated depth to the top of the ocean is $\sim$300 km. The finding of an induced field arising from an internal ocean provides useful insights into Callisto's interior structure and imposes valuable constraints on models of the interior structure. Gravitational experiments from several Galileo flybys suggest that Callisto's interior probably is in a state of partial differentiation (Anderson et al. 1998a, 2001b). The appearance of Callisto's surface, full of impact craters, indicates that this moon has been geologically inactive over eons. A dilemma arises in understanding heat transport. A subsurface ocean that can inductively generate the observed field perturbations requires a substantial internal heat source to melt a portion of the outer ice layer. However, if the heat transport in the deep interior is associated with convection, the heat loss rate is too high to continue over geological time. Recent thermal models have shown that the cooling of the interior through convection can be reduced if the viscosity of the convecting fluid is considered. This means that it is possible for a liquid-water layer surrounded by ice both above and below to survive to the present time (Ruiz 2001; Spohn and Schubert 2003).

In summary, magnetic field observations from Galileo show that both Europa and Callisto lack appreciable intrinsic magnetic fields. However, both of them appear to generate induced magnetic fields in response to the periodically changing field imposed by the motion of Jupiter's magnetic equator relative to their orbits. The observed large-scale magnetic field perturbations near both of the moons can be understood if there are highly conducting layers buried just beneath their surfaces. The most plausible source of such a conducting layer is a subsurface ocean.

4 Satellites with Significant Neutral Sources and Strong Plasma Interactions: *Io, Enceladus and Titan*

4.1 Io

Io is the innermost Galilean satellite, orbiting Jupiter at a distance of $\sim$5.9 R_J. Io appears to have the youngest surface in our solar system because of its extraordinarily active volcanoes and other strong geological activity (Morabito et al. 1979; Smith et al. 1979a). Materials erupted from the volcanoes and sublimed frost form a thin and patchy neutral atmosphere (mainly composed of SO_2) surrounding Io (Spencer and Schneider 1996). The ionization of the neutrals by photoionization and electron impact ionization supports an ionosphere on Io. Neutrals are sputtered off of the atmosphere and escape into Jupiter's magnetosphere, where they are ionized by multiple processes including charge-exchange at a rate of about 1 ton/second, forming a plasma torus around Io's orbit. The complex interaction of Io with the plasma of the torus has been extensively studied in the past several decades and recent reviews can be found in Kivelson et al. (2004) and Saur et al. (2004). Here we focus on the magnetic fields in the vicinity of Io, which can be affected by Jupiter's field, plasma interactions including mass-loading, the ionosphere and internal sources. Galileo data from several flybys reveal no appreciable internal field and indicate that the observed magnetic field perturbations at Io can be largely attributed to the plasma interaction with Io's ionosphere and its neutral cloud.

Difficulties in separating the signature of an internal field from the signature of plasma interactions arise because the complex nature of the interacting system. Consequently, even after the first few flybys, the critical question of whether or not Io has an intrinsic field remained unanswered. Voyager 1 made the first relatively close flyby of Io at a distance of about 11 R_{Io} ($R_{\mathrm{Io}} = 1821.6$ km, the radius of Io). The spacecraft trajectory passed south of Io and upstream of the southern Alfvén wing. The perturbation magnetic fields can be attributed to the anticipated Alfvén wing current system. The trajectory did not approach close enough to Io to obtain data constraining an internal field.

One of the high priorities of the Galileo mission was to probe Io's environment with additional close flybys. Most of our knowledge of Io's magnetic field and its plasma interaction come from the Galileo observations. Magnetic field and particle data were obtained on five flybys. Determination of Io's magnetic field based on the three low-latitude passes (I0, I24 and I27) was ambiguous because near Io's equator, plasma interactions obscure signatures of an internal field. For example, during the I0 flyby, Galileo went through Io's geometric wake region (downstream of Io) near the equator at an altitude of ~900 km and it detected a field strength depression of ~40% in the background field of ~1800 nT. Several mechanisms can produce large field depressions. As suggested by Kivelson et al. (1996a, 1996b), an internal dipole whose axis is approximately anti-aligned with Io's spin axis could depress the field in the downstream equatorial region. On the other hand, the observations of cool and dense plasma (Frank et al. 1996; Gurnett et al. 1996a) in the field depression region led to an alternate interpretation that the Alfvén wing currents resulting from the interaction between a sub-Alfvénic flow and Io (Neubauer 1980; Southwood et al. 1980), which close through Io's ionosphere and/or an extended region of ion pickup, also could produce field perturbations qualitatively consistent with the observations (Neubauer 1998b; Linker et al. 1998; Combi et al. 1998; Hill and Pontius 1998; Saur et al. 1999).

During the other two low-latitude flybys (I24 and I27), Galileo passed from upstream to downstream along the flanks of the interaction region at altitudes of ~600 km and ~200 km, respectively. Field and particle measurements along these two passes show that plasma flows were strongly diverted around Io and accelerated above the background flow speed near the flank region (Frank and Paterson 2000; Kivelson et al. 2001a). The diversion of the flow could be produced if Io has an intrinsic magnetic field or by a highly conducting region associated with Io such as its ionosphere and mass-loading region. During these two low-latitude upstream passes, Galileo spent most of its time on flux tubes external to Io and its ionosphere and thus the flyby geometry was not optimal for resolving the ambiguity of Io's internal field. However, the observations did impose useful constraints on any existing

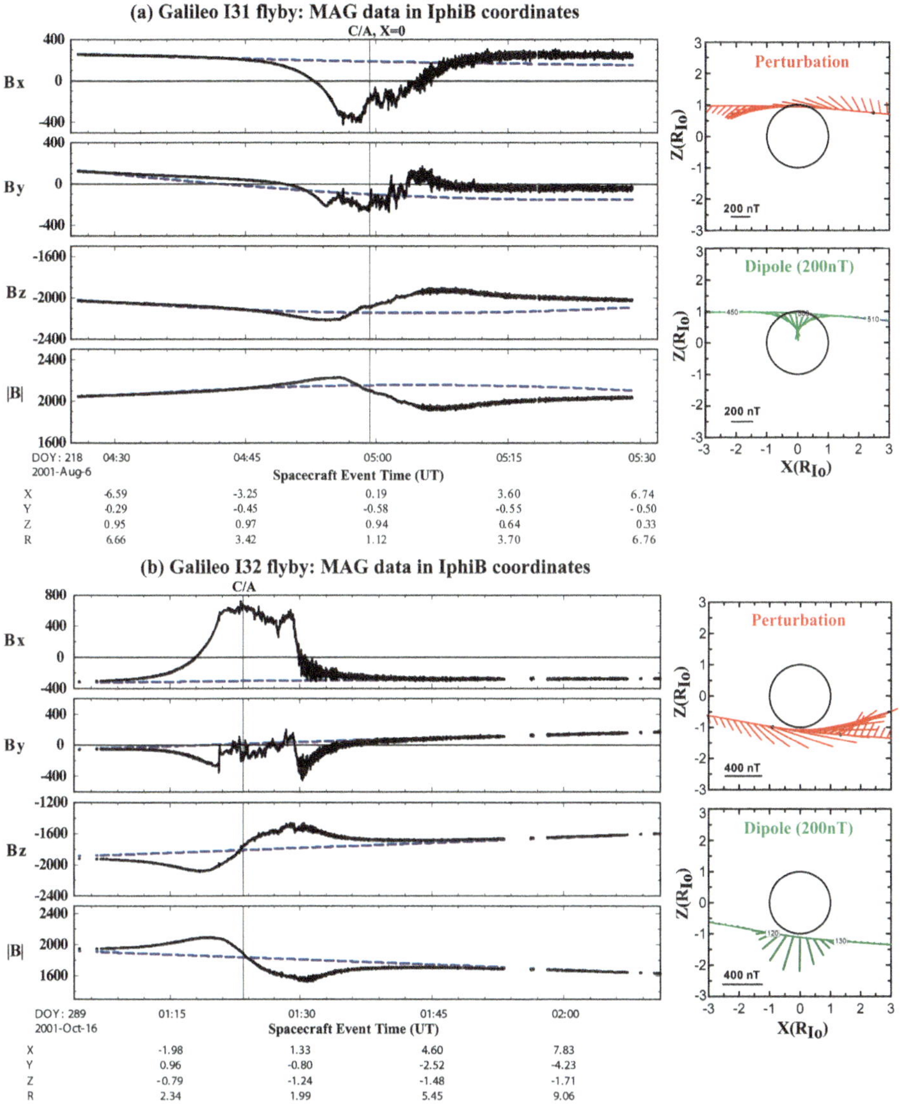

Fig. 7 Galileo magnetometer data (three components and the field magnitude in nT) for the closest approach (C/A) intervals of the (**a**) I31 and (**b**) I32 flybys. *Blue dashed traces* represent the background field expected in the region when Io is not nearby. Data are shown in the Io-centric IphiB coordinate system, in which $\hat{x}$ is along the flow direction of the corotating plasma, $\hat{y}$ is orthogonal to both the background field near C/A and the incident flow direction ($\hat{x}$), and $\hat{z} = \hat{x} \times \hat{y}$. On the *right side of each panel*, the projected field vectors (2 min averages on 1 min centers) of the perturbation field (the residual between the measurement and the background field) are shown in the XZ plane along the spacecraft trajectories for the two flybys. Also shown are the field vectors (*in green*) of an internal dipole that is anti-aligned with Io's rotational axis and has an equatorial surface field strength of 200 nT

internal magnetic field. In an attempt to model the observed field perturbations from the I27 flyby by using a global MHD simulation of Io's plasma interaction, it was found that a strongly magnetized Io (with surface equatorial strength comparable to the ambient field) could be ruled out but a relatively weak internal dipole at Io with surface equatorial strength comparable to that of Ganymede ~750 nT (see Sect. 5) still remained possible (Kivelson et al. 2001a).

More definitive information capable of distinguishing between a magnetized and a non-magnetized Io finally were obtained on the two polar flybys (I31 and I32), during which Galileo went through the northern and the southern polar regions, respectively (Kivelson et al. 2001b; Frank and Paterson 2002; Chust et al. 2005). As shown in Fig. 7, magnetic field observations exhibited negative perturbations in the component (B_x) along the direction of the background flow over the northern pole (I31 flyby) and positive perturbations over the southern pole (I32 flyby). The internal field model based on the previous flybys could not reproduce the general features of the observed perturbations. In contrast, the features of the field perturbations are consistent with those expected from an Alfvén wing model. It was, therefore, concluded that Io does not possess an appreciable internal field and the observed magnetic field perturbations at Io can be produced predominantly by the plasma interaction with Io's ionosphere and its neutral cloud.

4.2 Enceladus

Despite its small size (radius (R_{En}) $\sim$ 252.1 km), Enceladus is one of the most geologically dynamic bodies in the solar system, similar in many aspects to Io. Features on its extraordinarily bright icy surface suggest that this tiny moon experiences strong endogenic activity (Smith et al. 1982). Furthermore, the correlation between the peak density of Saturn's E-ring and Enceladus' orbit ($\sim 3.95 R_S$) has led to the speculation that Enceladus is the source of the E-ring particles (Horanyi et al. 1992). The first three flybys of Enceladus made by Cassini in 2005 have confirmed this speculation and revealed surprising new aspects of the moon's enigmatic properties (Spencer et al. 2006). It has been found that a water plume consisting of multiple jets is being vented from several warm troughs (named "tiger-stripes") located near the south pole of Enceladus (Waite et al. 2006; Tokar et al. 2006; Porco et al. 2006; Spahn et al. 2006; Hansen et al. 2006). The Cassini magnetometer observations suggest that Enceladus lacks an appreciable internal field but perturbs the ambient magnetic field through an interaction between its neutral cloud arising from the water plume and Saturn's magnetospheric plasma (Dougherty et al. 2006).

As will be shown below, magnetic field measurements played a key role in discovering the existence of a dynamic atmosphere at Enceladus. On the first flyby, Cassini's closest approach was at a relatively large distance (1265 km $\approx$ 5 R_{En}) from the surface on the upstream side and well above the moon's equatorial plane. Magnetic field perturbations measured by Cassini show signatures consistent with a slowing and diversion of the plasma flow near Enceladus, as expected for a moon that acts as an obstacle to the ambient flowing plasma. The slowing and deflection can arise for various reasons. If Enceladus is intrinsically magnetized as a result of either dynamo action or remanent magnetization, the observed features in the magnetic field can be accounted for. A rough estimate of the internal dipole strength required to generate the perturbation field at Cassini's location requires a surface field strength on the order of 1000 nT. However, a small moon is unlikely to be capable of generating such a field. An alternate interpretation based on plasma-neutral interactions seems more reasonable. If Enceladus provides a substantial neutral source, currents can be driven through the cloud of ionized neutrals by the background motional electric field due to the electric conductivity arising from an ionosphere or ion pickup (via charge-exchange and electron impact ionization (Sittler et al. 2004)). The plasma interaction with a conducting obstacle, which can be described by the Alfvén wing model mentioned above, also can generate the features in the perturbation field observed by Cassini.

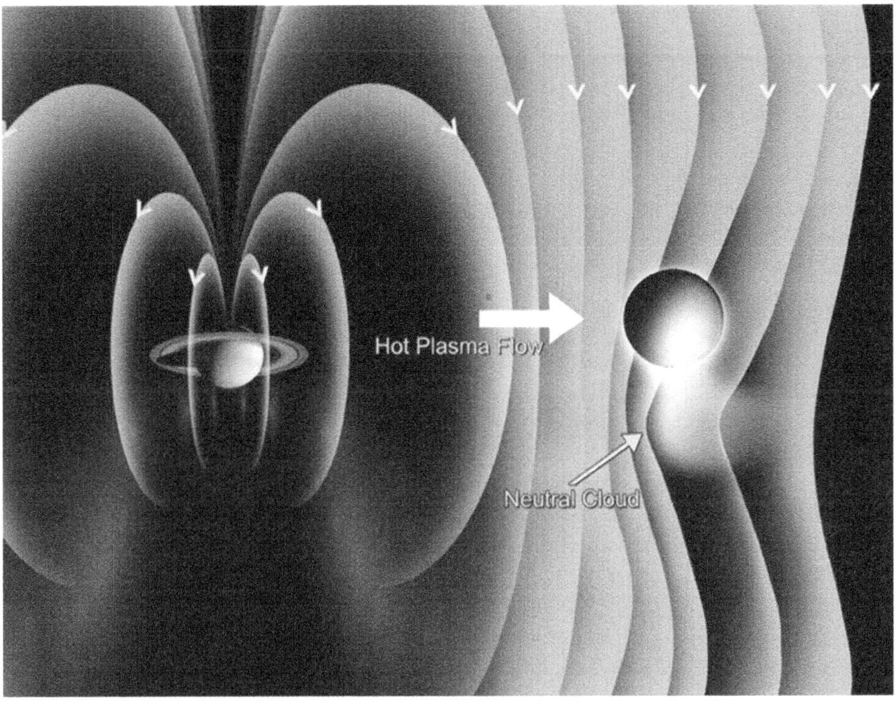

Fig. 8 A cartoon reproduced from Dougherty et al. (2006) showing Enceladus' water plume located near the south pole and its interaction with Saturn's magnetosphere

Data from the second flyby confirmed that Enceladus acts as an obstacle to the ambient flow but also presented some surprising features about the nature of the mass-loading region. This pass was also an upstream pass but the spacecraft passed south of Enceladus at an altitude of $\sim$500 km ($\approx$2 R_{En}). The magnetometer observed features in the perturbation field similar to those observed during the first flyby. This was quite surprising because the neutral source was expected to peak near the moon's equatorial plane and the magnetic perturbations of the components transverse to the background field were expected to be antisymmetric about the equator as they are in Fig. 1. Instead, the perturbations are consistent with the picture (Fig. 8) in which the dominant mass-loading region is located somewhere below Enceladus.

In order to confirm the interpretation based on the magnetometer measurements and to further identify the nature of the possible neutral source, the third flyby was designed to approach Enceladus' south pole at the much lower altitude of $\sim$173 km. Multiple in-situ measurements and remote sensing observations from this flyby established the presence of a water plume, whose dimensions are on the order of the moon's size, located near Enceladus' south pole. This established that Enceladus is the dominant source of the material forming Saturn's E-ring. Based on the Cassini observations (Hansen et al. 2006) and subsequent modeling work (Burger et al. 2007), the H_2O escape rate from Enceladus is estimated as 150–350 kg/s.

Of particular interest here is the total production rate of pickup ions, which gives insight into the strength of the plasma-neutral interaction and its consequences for Saturn's magnetosphere (Kivelson 2006). The total mass loading rate has been inferred by several means, which result in significantly different values. Pontius and Hill (2006) used an analytic model of the electrodynamic coupling between Enceladus and Saturn's ionosphere to model the perturbations in plasma velocity at relatively large distances (beyond 10 R_{En}) measured by the Cassini Plasma Spectrometer during the third flyby (Tokar et al. 2006). They inferred that a total mass loading rate >100 kg/s is required to explain the flow perturbations over an extended region of ~30 R_{En}. Khurana et al. (2007) applied a current wire model to interpret the magnetic field perturbations for all three flybys and found that the plume exhibited considerable variability on time scales of months in terms of its location and the total production rate, which is related to the total current required to interpret the field observations. They obtained a total mass-loading rate between 1 and 3 kg/s. The dramatic difference (almost two orders of magnitude) in the total mass-loading rate estimated using different approaches probably results from the assumptions regarding the spatial distribution of the neutral cloud and the plasma properties in the vicinity of Enceladus. More observations on the nature of the neutral cloud together with global numerical simulations of plasma interactions, which can self-consistently take into account the non-linear process of ion pickup, are needed to fully understand the plasma and field environment at Enceladus.

Interior models based on the recent Cassini observations suggest that Enceladus is most likely to be differentiated, with a rock-metal core with radius greater than half of the moon's radius surrounded by a water-ice shell (Schubert et al. 2007). If there is liquid water beneath the surface, a time-varying external field could drive magnetic induction as at Europa and Callisto (see Sect. 3). However, the symmetry of Saturn's field, which dominates the field in Saturn's inner magnetosphere, makes the approach of using magnetic induction to diagnose the moon's interior structure challenging.

4.3 Titan

As the largest (1 $R_{\mathrm{T}} = 2575.5$ km) satellite of Saturn and the second largest satellite in the solar system, Titan is a unique object because of its extensive atmosphere mainly composed of molecular nitrogen with a substantial contribution of methane (Tyler et al. 1981; Hunten et al. 1984). Before spacecraft measurements became available, Titan's large size suggested that it might have an intrinsic magnetic field. However, in-situ observations of Titan from both Voyager (Ness et al. 1982) and Cassini (Backes et al. 2005; Neubauer et al. 2006) found no evidence of any appreciable internal magnetic field. Voyager 1 observations gave an upper bound of ~4 nT for the equatorial surface strength of an intrinsic magnetic field (Neubauer et al. 1984).

Because of the absence of an appreciable internal field, the magnetic field and plasma environment at Titan are governed predominantly by the plasma interaction with the dense atmosphere that creates what has been called an induced magnetosphere (Ness et al. 1982). Such a magnetosphere is similar to that produced by the solar wind interaction with Venus and comets except that no bow shock forms in front of Titan (Ness et al. 1981) because the ambient flow typically is sub-magnetosonic ($M_{ms} < 1$) (Hartle et al. 1982). Both Titan's ionosphere and its exosphere appear as obstacles to the flowing plasma. As the flux tubes carrying the incident plasma approach Titan, they first slow down upstream owing to the extraction of momentum by pickup ions created mainly by photoionization and electron impact ionization of Titan's exosphere (hydrogen and nitrogen corona). As the flux tubes continue to move towards Titan, they are further slowed down by the forces exerted by the pressure gradient and the induced currents in the ionosphere. Magnetic flux can, to a good approximation, be considered frozen into the plasma except at low-altitudes (<1000 km, below the typical altitude of the peak ionospheric density ~1200 km) where the transport of magnetic flux is governed mainly by magnetic diffusion. The perturbations caused by the slowing of the flow near Titan generate Alfvén waves propagating away from Titan along the magnetic field lines and cause the flow at large distances from Titan to slow. As a result, those field lines are bent and draped around Titan. This is the same picture as the Alfvén wing model discussed in the previous sections. However, the super-Alfvénic ($M_A > 1$) condition of the typical ambient flow (Hartle et al. 1982) suggests that the Alfvén wing must have a large bendback angle compared to the case of the Galilean satellites (except perhaps Callisto) where the ambient flows are typically sub-Alfvénic ($M_A < 1$). This interaction eventually leads to the formation of an induced magnetotail with a central neutral sheet on the downstream side separating oppositely oriented magnetic fields (Ness et al. 1982). It is generally believed that those draped field lines divert around Titan as they move downstream and eventually are detached from the tail and accelerated back to the ambient flow velocity by the tension force arising from the highly kinked field geometry (Neubauer et al. 1984). Another mechanism that involves magnetic reconnection may contribute to the evolution of the tail structure. During its passage through the tail, the Voyager magnetometer detected a region in which the projection of the magnetic fields onto the ambient field was negative. This was considered by Kivelson and Russell (1983) as an indication of the plasma flowing towards the tail current sheet, which could possibly lead to tail reconnection and subsequent release of plasma and magnetic flux down tail.

The global configuration of Titan's induced magnetosphere depends on the properties of both its ionosphere and the incoming magnetospheric plasma. The day-side ionosphere is produced predominantly by solar radiation with some contribution from the impact of the magnetospheric electrons, whereas the night-side ionosphere is mainly created by the magnetospheric electrons. The imposed day-night asymmetry in the ionosphere has important consequences for the configuration of the interaction region observed during the Voyager 1 encounter (Ness et al. 1982; Hartle et al. 1982). On the other hand, the ram direction plays a significant role in controlling the global configuration of the induced magnetosphere. Titan is orbiting around Saturn in the outer region of the Kronian magnetosphere at $\sim$20 R_S, which is near the nominal standoff distance of Saturn's magnetosphere. Therefore, Titan spends most of its orbital time inside Saturn's magnetosphere but may occasionally (especially near Saturn's local noon sector) find itself outside of Saturn's magnetosphere embedded either in the magnetosheath or even in the solar wind depending on the incident solar wind conditions. Even inside of Saturn's magnetosphere, Saturn's magnetospheric plasma and field conditions at Titan's orbit may contain variations depending on Titan's local time. Therefore, unlike the relatively steady ambient conditions at the Galilean satellites, which are always inside the Jovian magnetosphere, the background plasma and field conditions at Titan may vary dramatically with the location of the moon within (or outside of) Saturn's magnetosphere. The ambient flow typically is transonic and trans-Alfvénic; however, its direction is somewhat variable and may sometimes deviate from the nominal corotation direction even when Titan is inside the magnetosphere because Saturn's magnetosphere expands and contracts in response to solar wind disturbances. Deviations in the tail configuration from the symmetry expected for the nominal corotation direction were observed during the Voyager encounter (Ness et al. 1982). It was found that the ambient flow at Voyager's encounter deviated from the corotation direction with an aberration of $\sim 20°$ radially inward towards Saturn as measured by the plasma instrument (Hartle et al. 1982) and inferred from the magnetic field observations by invoking the symmetry of the tail around the incident flow direction (Kivelson and Russell 1983).

All of the factors mentioned above lead to a complex interaction that depends strongly on Titan's orbital phase. The ongoing Cassini mission has been designed to include numerous close encounters optimized to collect data on Titan and its near environment. Through

2008, Cassini completed more than 40 close flybys of Titan that covered different regions of the Titan system under a wide range of the ambient field and plasma conditions. Although many features of the Cassini field and plasma observations can be interpreted reasonably well in the framework of an induced magnetosphere based on the Voyager 1 observations (Backes et al. 2005; Neubauer et al. 2006; Ma et al. 2006), many new aspects of the Titan system have been revealed by Cassini. In particular, Titan's atmosphere appears to be both a source of and a sink for the ambient flowing plasma. A recent review by Ma et al. (2008a) discusses consequences for plasma transport, a key to our understanding of Titan's plasma environment and hence the evolution of the atmosphere. In addition, radio occultation measurements (Kliore et al. 2008) showed that well below the nominal altitude (near 1200 km) of the main density peak of the ionosphere there is a second peak of electron density at the altitude of about 500 km, suggesting that Titan's ionosphere may be more complicated than anticipated on the basis of the previous observations and modeling.

One new aspect of the Cassini magnetic field observations is the detection of "fossil" fields in Titan's ionosphere (Bertucci et al. 2008). Cassini happened to catch Titan outside Saturn's magnetosphere in the magnetosheath during the T32 flyby. Earlier Voyager 1 and all 31 Cassini close flybys had occurred when Titan was inside Saturn's magnetosphere. On T32, at high altitudes, the Cassini magnetometer observed layers of draped magnetic fields consistent with the orientation of the ambient magnetosheath field surrounding Titan. However, deep in the ionosphere (between 1200 and 1600 km), the magnetometer observed fields of opposite orientation. These have been interpreted as fossil Saturnian magnetospheric fields deposited in the ionosphere before Titan entered the magnetosheath. Nearly opposite orientations between the fossil fields and the ambient field strongly imply that reconnection may be responsible for replacing the fossil fields in the ionosphere (Bertucci et al. 2008; Ma et al. 2008b). The memory of previous environments captured by Titan's ionosphere adds another ingredient that further complicates the Titan interaction.

Knowledge of Titan's inductive response to time-varying external field can provide important clues to our understanding of the moon's interior structure. It is likely that Titan, which is intermediate between Ganymede and Callisto in bulk properties (e.g., radius and mean density), is differentiated into a rocky and metallic core with an outer water-ice layer as suggested by interior models (Grasset et al. 2000; Sohl et al. 2003). The presence of even a small amount of volatile material, such as ammonia, in the ice makes the existence of a water layer more likely. In principle, induced currents can be driven in a hypothetical interior ocean by any time-varying external fields sensed along Titan's orbit. The amplitude of possible induced fields at Titan, however, is likely to be at most comparable to the perturbation field arising from plasma interactions and, therefore, hard to detect. The presence of Titan's thick atmosphere precludes low altitude ($<$950 km) orbits for Cassini. It seems unlikely that Cassini measurements will be able to provide clear evidence of either a possible inductive response or a weak intrinsic field from the measurements dominated by contributions from Titan's ionosphere and plasma interactions.

5 A Satellite with an Intrinsic Magnetic Field: *Ganymede*

Ganymede, with a radius (R_G) of $\sim$2631 km, is the largest satellite in the solar system, even larger than the planet Mercury. It earns its unique place among the planetary satellites not only because of its great size but also because it is the only satellite in our solar system known to possess an intrinsic magnetic field. In addition, like its neighboring moons Europa and Callisto, Ganymede appears to possess a subsurface ocean that generates an inductive response to the external Jovian field.

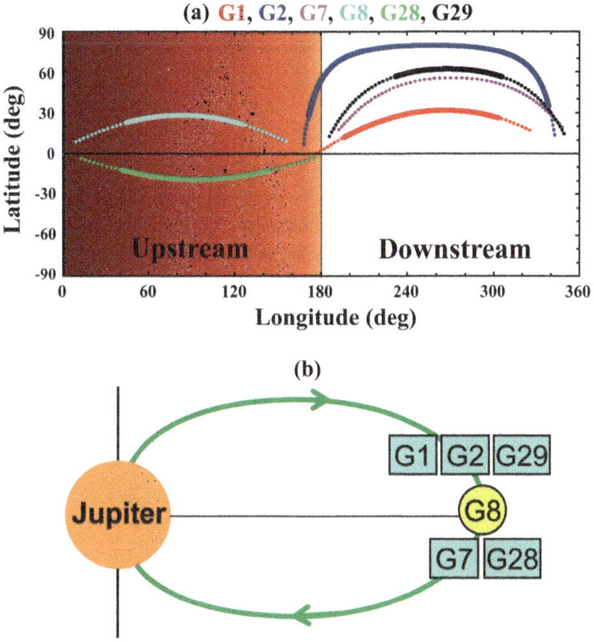

Fig. 9 (**a**) Plot of latitude versus longitude of the six Galileo trajectories (color-coded) in Ganymede-centered spherical coordinates. Longitude is measured from the Jupiter-facing meridian in a right-handed sense and latitude is measured from Ganymede's equator. The *solid traces* represent the intervals when the radial distance of the spacecraft was within 2 R_G, beyond which the internal field strength becomes less than the external field strength. (**b**) A schematic showing Ganymede's location relative to Jupiter's magnetospheric current sheet during Galileo's six flybys. The G1, G2 and G29 flybys occurred well above the current sheet while the G7 and G28 flybys occurred well below the current sheet. The G8 pass is the only one that occurred near the central current sheet

It was not anticipated that Ganymede would have a strong intrinsic field before the arrival of the Galileo spacecraft in Jupiter's system, although Neubauer (1978) speculated that the moon might have a relatively weak dynamo field (with an equatorial surface strength of ~200 nT). Kivelson et al. (1979) recognized that the proposed field could carve out a magnetosphere within Jupiter's plasmas. However, during the first two Galileo flybys (G1 and G2), the measured magnetic field increased dramatically near Ganymede (Kivelson et al. 1996). In particular, during the pass (G2) over the north pole at a closest approach altitude of ~260 km, the field increased above the ambient field (~110 nT) by about an order of magnitude, varying with distance as required for an intrinsic magnetic field. The intrinsic field is sufficiently strong to shield the moon from direct impact of the incident flow of Jupiter's magnetospheric plasma. A mini-magnetosphere embedded in Jupiter's giant magnetosphere forms around the moon (Kivelson et al. 1996). Evidence for the existence of a magnetosphere associated with Ganymede was also clearly present in observations from other instruments, such as the plasma wave system (Gurnett et al. 1996b; Kurth et al. 1997) and the particle measurements (Frank et al. 1997a, 1997b; Williams et al. 1997b). In addition to the in-situ observations, the Hubble Space Telescope observed oxygen (OI) airglow features at Ganymede (Hall et al. 1998; Feldman et al. 2000) associated with a Ganymede aurora.

In order to understand the observed magnetic fields obtained from the first two passes, Kivelson et al. (1996) first developed a vacuum-superposition model in which the internal field of Ganymede was represented by a fixed dipole field. The modeled internal dipole, which reproduces the field observations near Ganymede reasonably well, has an equatorial field strength of ~750 nT and a tilt angle of about 170° from the rotational axis (Kivelson et al. 1996). Four additional passes were completed during the Galileo mission. The distribution of all the six flyby trajectories in Ganymede-centered spherical coordinates is shown in Fig. 9(a). The six passes took place at significantly different locations relative to the moon, with four passes in the downstream region covering a wide latitude range and two low-latitude passes on the upstream side, providing us with a comprehensive sampling of Ganymede's magnetospheric system. Moreover, the six close encounters occurred at different locations relative to Jupiter's dipole equator (shown in Fig. 9(b)), with three passes above, two passes below and one pass near Jupiter's central current sheet. Therefore, the Galileo data were acquired in Ganymede's magnetosphere under various external particle and field conditions, including distinctly different polarities in the radial component of the external field. Given the discoveries of induced fields at Europa and Callisto, it is of special interest to find out if an induced field is driven in Ganymede in response to Jupiter's time-varying external field. Even if present, an induced field could not be the dominant source of Ganymede's internal field because of the substantial difference in field strength between the inferred internal field and the ambient field.

In their continuing studies, Kivelson et al. (2002) used magnetic field data acquired during multiple passes to refine the evaluation of the internal moments by looking for a solution consistent with measurements on multiple passes and also allowing for a time variable inductive component. Three of the six flybys (G1, G2 and G28) were considered to be relevant to the determination of the internal field while the other three (G7, G8 and G29) flybys were excluded from the analysis because the flyby data were obtained either too far from the surface (G7) or too close to the magnetospheric boundary where strong localized currents are flowing (G8), or did not add meaningful information for the analysis of Ganymede's internal magnetic field (G29). The fitting procedure used to invert the internal moments was based on the basic principle of separating internal and external magnetic field contributions, which is discussed at length by Olsen et al. (2009) in another chapter in this issue. Different combinations of a permanent dipole, a permanent quadrupole and an induced dipole were considered and carefully examined in comparison with the data. Kivelson et al. (2002) concluded that the available observations of Ganymede's internal field are most likely to be the signatures of a permanent dipole plus an induced dipole. A combination of an intrinsic dipole and a quadrupole also reproduced the observations reasonably well. Unfortunately, the available flyby data does not contain adequate information to distinguish between an inductive response and fixed quadrupole moments. However, the model with magnetic induction was favored because it provides equally satisfactory fits to the data but requires fewer free parameters (five for the quadrupole model and only one for the induction model). The inferred permanent dipole has an equatorial surface strength of 719 nT, slightly smaller than the initially proposed value, and a tilt angle of $\sim 176^\circ$ with respect to the rotational axis (Kivelson et al. 2002). The inductive response efficiency was inferred to be slightly smaller than that from a perfect conductor of the moon's size. This result can be well accounted for if a layer of liquid water containing dissolved electrolytes is buried at a depth of the order of 150 km.

The source of the permanent internal field could be magneto-convection, remanent magnetization or dynamo action. Magneto-convection, in which convective motions of a conducting fluid modify an imposed external magnetic field, is expected to generate perturbation fields roughly of the same order of magnitude as the external field. However, the significant difference between Ganymede's internal field strength and the ambient Jovian field strength leads us to reject this proposal (Schubert et al. 1996). Crary and Bagenal (1998) have considered the possibility of remanent magnetization and concluded that the present internal field could be produced by remanent magnetization of a strong ancient dynamo under rather favorable assumptions about the magnetic properties of the rocky materials. Although this possibility cannot be ruled out entirely, an extant self-sustained dynamo action, in which the convective motion of a conducting fluid generates a magnetic field, is more likely to be the source of Ganymede's strong intrinsic field. One possible region that contains electrically conducting fluid that could generate the field is the internal ocean mentioned above. However, the unrealistically large convective speed (>1 m/s) of the ocean required makes the scenario of an internal ocean being the dominant source of the internal field unlikely (Schubert et al. 1996). Therefore, a core dynamo in a liquid metallic core is the most probable source for generating Ganymede's intrinsic magnetic field. Schubert et al. (1996) pointed out that Ganymede's dynamo may be fundamentally different from the geodynamo because the ambient field of Jupiter may establish the orientation of the moon's dipole moment. The existence of a core dynamo has important implications for the moon's interior structure. The Galileo gravity measurements alone could be satisfied if Ganymede were differentiated into a core and a mantle (Anderson et al. 1996). However, the clues provided by the magnetic field measurements make it likely that Ganymede's interior contains a metallic core and a silicate mantle surrounded by an icy shell (Schubert et al. 1996).

A key issue in determining the nature of the internal field is how well the field perturbations resulting from the external currents can be separated from the total observed field. Therefore, an accurate determination of the internal field depends on our knowledge of the external sources, which are dominated by the plasma currents arising from magnetospheric interactions. Such currents are evidenced by the sharp rotations in the observed magnetic field when the spacecraft crossed the magnetospheric boundaries (Kivelson et al. 1996). Because of the field bendback caused by the field-aligned currents, the observed field at high latitudes includes a negative perturbation in the unperturbed flow direction. The inversion analysis (Kivelson et al. 2002) assumed that the field perturbations contributed by the external magnetospheric currents are locally spatially uniform along the spacecraft trajectory near closest approach. Such an assumption is not an unreasonable one. However, it is a simplification that introduces uncertainties in characterizing Ganymede's internal magnetic properties.

A more accurate description of the magnetospheric currents can be provided by numerical simulations of the global magnetosphere, which can self-consistently take into account the interaction of flowing plasma with a magnetized obstacle bounded by a tenuous ionosphere. Ganymede's magnetosphere has been simulated by using single-fluid resistive MHD simulations (Kopp and Ip 2002; Ip and Kopp 2002; Jia et al. 2008) and multi-fluid MHD simulations (Paty and Winglee 2006; Paty et al. 2008). Among these, only the model of Jia et al. (2008) includes the moon's interior in the simulation domain. This approach has the advantage of including the magnetic diffusion effects in the moon's mantle. Furthermore, a high-resolution numerical grid, which is crucial for resolving the fine structure of the magnetospheric boundaries, was used in their model to produce the plasma currents consistent with the observations in both amplitude and location. In a subsequent work, Jia et al. (2009) improved their MHD model by self-consistently taking into account the coupling between the magnetosphere and the moon's ionosphere with a finite Pedersen conductance. The modified conductance is probably more realistic than the high ionospheric conductance used in the earlier model. They found that the simulated magnetosphere in the improved model is in extremely good agreement with the observations on multiple passes. The work therefore suggested that an MHD model can consistently describe the external plasma currents that contribute to the magnetic field measurements in the near-Ganymede environment. The new work also illustrated the sensitivity of global simulations to internal boundary conditions.

Using results (shown in Fig. 10) extracted from an MHD simulation by Jia et al. (2009), we briefly describe the global form of Ganymede's magnetosphere. Consider first the G8 flyby when Ganymede was located very close to Jupiter's central plasma sheet. As shown in Fig. 10(a), the whole magnetosphere forms a roughly cylindrical shape differing from the bullet-like shape of planetary magnetospheres. This is because the sub-Alfvénic ($M_A < 1$) and low plasma β conditions of the typical flow at Ganymede's orbit ($\sim 15 R_J$) imply that the total external pressure is dominated by magnetic pressure which exerts force perpendicular

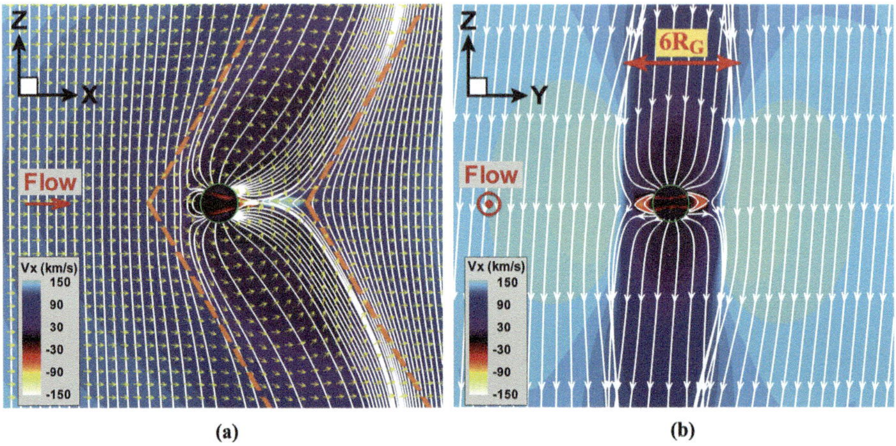

Fig. 10 (**a**) Flows and the projection of field lines (*white solid lines*) in the XZ plane at $Y = 0$. Color represents the V_x contours and unit flow vectors in *yellow* show the flow direction. Results are shown in a Ganymede-centered cartesian coordinates (referred to as "GphiO") in which $\hat{X}$ is along the incident flow direction, $\hat{Y}$ is along the Ganymede-Jupiter vector, positive towards Jupiter, and $\hat{Z}$ is parallel to Jupiter's spin axis. Predicted Alfvén characteristics (*orange dashed lines*) are shown for reference. The projection of the ionospheric flow also is shown as color contours on a circular disk of $r = 1.08\ R_G$ in the center. Flows shown as *red* are return flows from downstream to upstream via closed field lines. (**b**) Same as (**a**) but in the YZ plane at $X = 0$. Flows in *red* are return flows on closed field lines at low latitudes. The figure is adapted from Jia et al. (2009)

to Ganymede's spin axis, whereas planetary magnetospheres are confined by the dynamic pressure of the super-Alfvénic ($M_A > 1$) and supersonic ($M_s > 1$) solar wind flow on the upstream side and by magnetic and thermal pressure on the downstream side. The magnetosphere contains a small closed field line region near the equator and a large polar cap consisting of field lines that link to Jupiter. Those open field lines in the polar cap are tilted with respect to the background field direction in both hemispheres, forming Alfvén wings that mediate the interaction of Ganymede with the plasma and the ionosphere of Jupiter. The front across which the field bends significantly from the unperturbed background field can be described by Alfvén characteristics tilted with respect to the background magnetic field at an angle given by $\theta = \arctan(\frac{v_{flow}}{v_A})$. The magnetosphere has a length of about 8 to 10 R_G in the incident flow direction, whereas the width of the magnetosphere transverse to the flow and the background field is approximately 6 R_G (see Fig. 10(b)). Another feature of Ganymede's magnetosphere differing from planetary magnetospheres is the absence of a bow shock in front of the magnetosphere because the ambient flow normally is submagnetosonic ($M_{ms} < 1$). Nonetheless, as the incident plasma flow approaches Ganymede's magnetosphere, it is slowed by its interaction with compressional magnetosonic waves that propagate upstream (see Fig. 10(a)). Some of the incident flow diverts around the magnetosphere and is accelerated on the flank side (Fig. 10(b)). The ambient plasma has only limited access into the magnetosphere mainly through magnetic reconnection between oppositely oriented magnetic fields on the two sides of Ganymede's low-latitude magnetopause. Plasma that enters the Alfvén wing as a result of reconnection is convected across the polar cap towards the downstream region. Tail reconnection eventually returns part of the flow back towards the moon and ejects the rest down the tail. Within the Alfvén wing, the plasma flow is significantly decelerated because the footprints of the open flux tubes in Ganymede's ionosphere move at a much slower speed than does the background flow. The disturbances

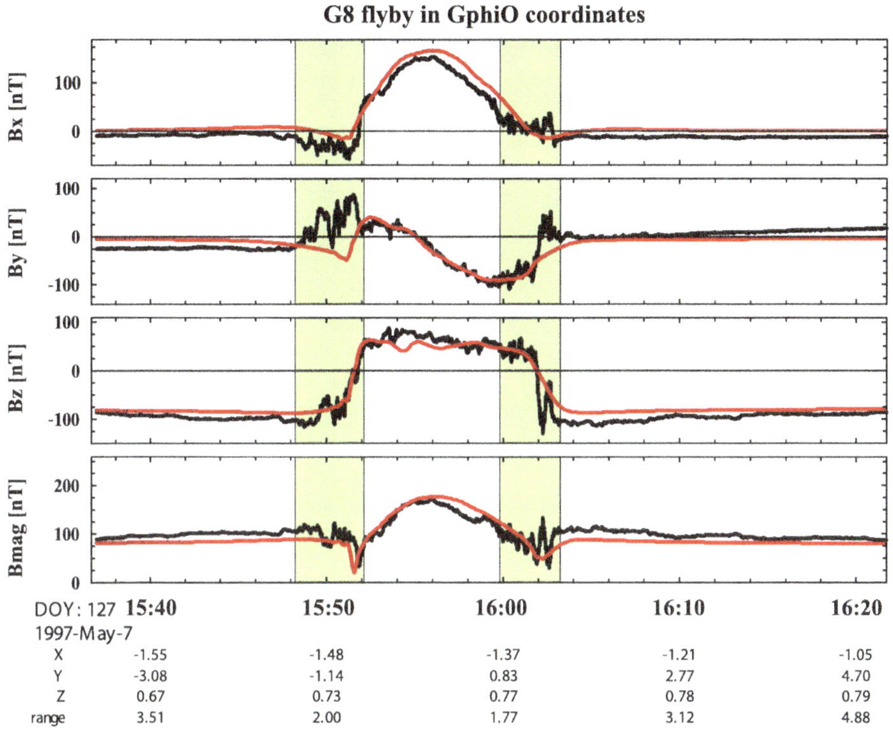

Fig. 11 Magnetic field comparisons between the simulation results and the Galileo observations for the G8 flyby. The magnetic field data are shown in the GphiO coordinate system, which is defined in the figure caption of Fig. 10. *Black solid lines* are the spacecraft measurements while *red traces* are results extracted from a global MHD simulation run using upstream boundary conditions characteristic of the G8 flyby conditions. The marked intervals show large amplitude magnetic fluctuations as observed during both inbound and outbound magnetopause crossings. The figure is reproduced from Jia et al. (2009)

propagate along the magnetic field line carried by Alfvén waves. In the downstream region, the flow is accelerated gradually to the background flow speed by the magnetic tension force exerted on the plasma. Furthermore, a thin current sheet extends several Ganymede radii from the moon's surface in the downstream region, separating magnetic field lines with opposite polarities.

Results of numerical simulations must be taken with caution and must be validated by comparing directly with the measurements before being applied to interpret the observations. A crucial pass for validating the simulation model is the G8 flyby because during this flyby the Galileo spacecraft flew through the upstream region very close to the cusp and the magnetic field and Energetic Particle Detector (EPD) measurements within the magnetosphere provide critical constraints on the magnetospheric configuration (Kivelson et al. 1998; Williams et al. 1997a). In comparison with the observations (shown in Fig. 11), the MHD simulation results accord extremely well with the Galileo magnetometer measurements. The simulation also predicts that near closest approach, Galileo flew into a region of closed field lines inside the magnetosphere, which is consistent with the magnetometer and EPD observations. Also shown in Fig. 11 are large amplitude magnetic field fluctuations present both prior to the entry and after the exit from the magnetosphere. Careful examination of the time evolution of the simulated magnetosphere suggests that boundary fluctuations can be caused by bursty magnetic reconnection at the magnetopause (Jia et al. 2009). Thus it is probable that the fluctuations of the measurements in the vicinity of the boundary arise from bursty reconnection. Ganymede's magnetosphere provides us with an opportunity to investigate the reconnection process in a relatively stable external environment because the plasma at Ganymede's orbit imposes external field and plasma conditions that vary slowly (with the nearly 10.5 hour synodic period of Jupiter's rotation) and the magnetic field remains in a favorable orientation (southward in this case) for reconnection. This is in contrast to the highly fluctuating solar wind that imposes unpredictable variations in plasma and magnetic field on the magnetospheres of the planets it encounters. Combining the observations and the MHD simulation suggests that even under steady external conditions, reconnection appears to be intermittent rather than steady.

The MHD model developed for the G8 pass also reproduces the Galileo observations on other passes for relevant external conditions and the same inner boundary conditions. Because the model can produce a global magnetosphere that matches the observations so faithfully, it can be used to improve the internal field model by providing a realistic description of the external magnetospheric fields. One can assume that the differences between the simulated and the observed magnetic fields can be attributed mainly to the inaccuracy in the internal field model of Kivelson et al. (2002) adopted in the simulation. A correction to the initial field model can be obtained by fitting the residuals between the modeled and the observed fields. The next step would be to iterate by running the simulation with the improved internal field without varying other parameters. The procedure should converge if the assumption of the origin of the model discrepancy is correct and should improve the determination of Ganymede's internal field.

Although refinement of the present internal field model (Kivelson et al. 2002) may, in the future, be achieved with the aid of a good global MHD model, it does not seem possible to establish the presence of either the inductive response or the quadrupole moments unambiguously based on the present observations because of the limitations of available data. For example, a magnetic anomaly in the high-latitude southern hemisphere could have been missed because of the lack of coverage (Fig. 9(a)). Future missions to Ganymede should be designed to obtain the data necessary to distinguish between permanent internal quadrupole moments and an induced dipole moment and to cover the unexplored regions, especially the southern high-latitude region. As in the situation of Europa, long-term (several months or so) continuous magnetic field observations at Ganymede will allow for probing magnetic induction by using multi-frequency signals, including the low frequency signal arising from the moon's orbital eccentricity. Self-consistent global simulations of Ganymede's magnetosphere will have to play a key role in interpreting the observations and fully characterizing the magnetic properties of Ganymede.

6 Moons for which the Situation is Unclear: *Dione and Iapetus*

6.1 Dione

Dione, with a radius of 561.7 km, orbits Saturn in the inner region of Saturn's magnetosphere at a mean radial distance of $\sim$6.3 R_S. This moon is believed to be composed mainly of water ice, however, its relatively high density (1.48 g/cm^3) implies that the moon's interior must contain a significant fraction of dense material like silicate rock. Cassini flew by Dione in 2005 at a distance of $\sim$500 km from its surface. At closest approach, the spacecraft was upstream of Dione and observed only a weak ($\sim$0.5 nT) magnetic field signature suggesting that Dione does not possess any measurable intrinsic field. The observed magnetic field signature is consistent with weak field-line draping around Dione. The origin of field-line draping is not yet clear as either extremely weak plasma loading from a tenuous atmosphere or even the inductive response arising from the low conductivity of ice could explain the draping signature. A downstream flyby is planned for April 2010 which should help discriminate between the two models.

6.2 Iapetus

Similar to Rhea in size, Iapetus is the outermost regular satellite of Saturn, orbiting the planet at a distance of about 60 R_S. The most striking feature of Iapetus is the high contrast in brightness between the leading and trailing hemispheres (Ostro et al. 2006). Cassini made a single relatively close encounter with Iapetus in 2007, flying from the downstream to the flank side at a closest approach altitude of $\sim$2.2 R_I (R_I, radius of Iapetus = 735.6 km). During the flyby, magnetic field perturbations were detected (Leisner et al. 2008). However, because Iapetus was embedded in the solar wind, the observed field perturbations could have been structures in the solar wind only fortuitously observed near the moon. Future close flybys to Iapetus are needed to establish the moon's magnetic properties.

7 Summary

The satellites of Jupiter and Saturn generate magnetic fields in many different ways. Near the moons, magnetic perturbations ubiquitously arise from the interaction between the ambient plasma flow and the moons themselves and/or their atmospheres or ionospheres. Among the moons that have been investigated to this time, Io, Enceladus and Titan serve as substantial sources of neutrals and/or ionized particles that mass-load the background plasma and thus create significant magnetic field perturbations over volumes large compared with their sizes. Moons also generate magnetic fields inductively in response to time-varying external fields provided that they contain extended regions of high electrical conductivity. Periodically varying external conditions are present at the orbits of the Galilean satellites of Jupiter. Indeed, the icy satellites Europa and Callisto have been clearly found to respond inductively to variations at the synodic periods of Jupiter, consistent with the existence of internal oceans. Most of the icy satellites of Saturn, to a good approximation, appear to be magnetically inert although the absence of clear time-varying signals in the ambient field precludes detection of inductive responses on a small number of passes. Magnetic perturbations associated with Saturn's moons are localized mainly in their wake regions where the ambient plasma flow is shadowed by their presence. Ganymede is unique in having a strong intrinsic magnetic field, which most likely originates from a self-sustained dynamo. The internal field forms a mini-magnetosphere in the sub-magnetosonic flow of Jupiter's plasma. Ganymede's magnetosphere shares many features with planetary magnetospheres but differs in ways that reflect its unusual plasma environment. The existence of an internal field has important implications for the structure of Ganymede's interior and yet it seems that little is known about the mechanism responsible for generating the field.

Future missions to Jupiter's system, especially to Europa and Ganymede, are eagerly anticipated. They will acquire more definitive information on properties of the internal oceans and of Ganymede's self-generated magnetic field. In the interim, the ongoing Cassini mission undoubtedly will continue to provide us with surprises and to enrich our knowledge of the satellites of Saturn.

Acknowledgements This research was supported by NASA under grant NNG05GB82G, NNG06GG67G, NNX08AQ46G, NNX06AB91G and NNX08AT48G.

References

M.H. Acuna, J.E.P. Connerney, N.F. Ness, The Z3 zonal harmonic model of Saturn's magnetic field Analyses and implications. J. Geophys. Res. **88**, 8771–8778 (1983). doi:10.1029/JA088iA11p08771

Anderson, B., Magnetic field of Mercury. Space Sci. Rev. (2009)

J.D. Anderson, G. Schubert, Saturn's satellite Rhea is a homogeneous mix of rock and ice. Geophys. Res. Lett. **34**(2). doi:10.1029/2006GL028100

J.D. Anderson, E.L. Lau, W.L. Sjogren, G. Schubert, W.B. Moore, Gravational constraints on the internal structure of Ganymede. Nature **384**, 541–543 (1996)

J.D. Anderson, G. Schubert, R.A. Jacobson, E.L. Lau, W.B. Moore, W.L. Sjo Gren, Distribution of rock, metals, and ices in Callisto. Science **280**, 1573 (1998a). doi:10.1126/science.280.5369.1573

J.D. Anderson, G. Schubert, R.A. Jacobson, E.L. Lau, W.B. Moore, W.L. Sjogren, Europa's differentiated internal structure: inferences from four Galileo encounters. Science **281**, 2019 (1998b). doi:10.1126/science.281.5385.2019

J.D. Anderson, R.A. Jacobson, E.L. Lau, W.B. Moore, G. Schubert, Io's gravity field and interior structure. J. Geophys. Res. **106**, 32,963–32,970 (2001a). doi:10.1029/2000JE001367

J.D. Anderson, R.A. Jacobson, T.P. McElrath, W.B. Moore, G. Schubert, P.C. Thomas, Shape, mean radius, gravity field, and interior structure of Callisto. Icarus **153**, 157–161 (2001b). doi:10.1006/icar.2001.6664

H. Backes, F. Neubauer, M. Dougherty, N. Achilleos, N. Andre, C. Arridge, C. Bertucci, G. Jones, K. Khurana, C. Russell, A. Wennmacher, Titan's magnetic field signature during the first Cassini encounter. Science **308**(5724), 992–995 (2005). doi:10.1126/science.1109763

C. Bertucci, N. Achilleos, M.K. Dougherty, R. Modolo, A.J. Coates, K. Szego, A. Masters, Y. Ma, F.M. Neubauer, P. Garnier, J.-E. Wahlund, D.T. Young, The magnetic memory of Titan's ionized atmosphere. Science **321**, 1475 (2008). doi:10.1126/science.1159780

J.M. Bosqued, N. Lormant, H. Rème, C. d'Uston, R.P. Lin, K.A. Anderson, C.W. Carlson, R.E. Ergun, D. Larson, J. McFadden, M.P. McCarthy, G.K. Parks, T.R. Sanderson, K.-P. Wenzel, Moon-solar wind interaction: First results from the WIND/3DP experiment. Geophys. Res. Lett. **23**, 1259–1262 (1996). doi:10.1029/96GL00303

M.H. Burger, E.C. Sittler, R.E. Johnson, H.T. Smith, O.J. Tucker, V.I. Shematovich, Understanding the escape of water from Enceladus. J. Geophys. Res. **112**, 6219 (2007). doi:10.1029/2006JA012086

M. Carr, M. Belton, C. Chapman, A. Davies, P. Geissler, R. Greenberg, A. McEwen, B. Tufts, R. Greeley, R. Sullivan, J. Head, R. Pappalardo, K. Klaasen, T. Johnson, J. Kaufman, D. Senske, J. Moore, G. Neukum, G. Schubert, J. Burns, P. Thomas, J. Veverka, Evidence for a subsurface ocean on Europa. Nature **391**(6665), 363–365 (1998)

T. Chust, A. Roux, W.S. Kurth, D.A. Gurnett, M.G. Kivelson, K.K. Khurana, Are Io's Alfvén wings filamented? Galileo observations. Planet. Space Sci. **53**, 395–412 (2005). doi:10.1016/j.pss.2004.09.021

D.S. Colburn, R.T. Reynolds, Electrolytic currents in Europa. Icarus **63**, 39–44 (1985). doi:10.1016/0019-1035(85)90019-3

M.R. Combi, K. Kabin, T.I. Gombosi, D.L. DeZeeuw, K.G. Powell, Io's plasma environment during the Galileo flyby: Global three-dimensional MHD modeling with adaptive mesh refinement. J. Geophys. Res. **103**(A5), 9,071–9,081 (1998)

J.E.P. Connerney, Magnetic field of the outer planets. Space Sci. Rev. (2009)

F.J. Crary, F. Bagenal, Remanent ferromagnetism and the interior structure of Ganymede. J. Geophys. Res. **103**, 25,757–25,774 (1998). doi:10.1029/98JE02497

M.K. Dougherty, S. Kellock, D.J. Southwood, A. Balogh, E.J. Smith, B.T. Tsurutani, B. Gerlach, K.-H. Glassmeier, F. Gleim, C.T. Russell, G. Erdos, F.M. Neubauer, S.W.H. Cowley, The Cassini magnetic field investigation. Space Sci. Rev. **114**, 331–383 (2004). doi:10.1007/s11214-004-1432-2

M.K. Dougherty, K.K. Khurana, F.M. Neubauer, C.T. Russell, J. Saur, J.S. Leisner, M.E. Burton, Identification of a dynamic atmosphere at Enceladus with the Cassini magnetometer. Science **311**, 1406–1409 (2006). doi:10.1126/science.1120985

P.D. Feldman, M.A. McGrath, D.F. Strobel, H.W. Moos, K.D. Retherford, B.C. Wolven, HST/STIS ultraviolet imaging of polar aurora on Ganymede. Astrophys. J. **535**, 1,085–1,090 (2000)

L.A. Frank, W.R. Paterson, Return to Io by the Galileo spacecraft: Plasma observations. J. Geophys. Res. **105**, 25,363–25,378 (2000). doi:10.1029/1999JA000460

L.A. Frank, W.R. Paterson, Plasmas observed with the Galileo spacecraft during its flyby over Io's northern polar region. J. Geophys. Res. **107**, 1220 (2002). doi:10.1029/2002JA009240
L.A. Frank, W.R. Paterson, K.L. Ackerson, V.M. Vasyliunas, F.V. Coroniti, S.J. Bolton, Plasma observations at Io with the Galileo spacecraft. Science **274**, 394–395 (1996). doi:10.1126/science.274.5286.394
L.A. Frank, W.R. Paterson, K.L. Ackerson, S.J. Bolton, Outflow of hydrogen ions from Ganymede. Geophys. Res. Lett. **24**(17), 2,151–2,154 (1997a)
L.A. Frank, W.R. Paterson, K.L. Ackerson, S.J. Bolton, Low-energy electrons measurements at Ganymede with the Galileo spacecraft: probes of the magnetic topology. Geophys. Res. Lett. **24**(17), 2,159–2,162 (1997b)
C.K. Goertz, Io's interaction with the plasma torus. J. Geophys. Res. **85**, 2949–2956 (1980). doi:10.1029/JA085iA06p02949
O. Grasset, C. Sotin, F. Deschamps, On the internal structure and dynamics of Titan. Planet. Space Sci. **48**, 617–636 (2000)
D.A. Gurnett, W.S. Kurth, A. Roux, S.J. Bolton, C.F. Kennel, Galileo plasma wave observations in the Io plasma torus and near Io. Science **274**, 391–392 (1996a). doi:10.1126/science.274.5286.391
D.A. Gurnett, W.S. Kurth, A. Roux, S.J. Bolton, C.F. Kennel, Evidence for a magnetosphere at Ganymede from plasma-wave observations by the Galileo spacecraft. Nature **384**, 535–537 (1996b)
J.S. Halekas, S.D. Bale, D.L. Mitchell, R.P. Lin, Electrons and magnetic fields in the lunar plasma wake. J. Geophys. Res. **110**, 7222 (2005). doi:10.1029/2004JA010991
D.T. Hall, D.F. Strobel, P.D. Feldman, M.A. McGrath, H.A. Weaver, Detection of an oxygen atmosphere on Jupiter's moon Europa. Nature **373**, 677 (1995). doi:10.1038/373677a0
D.T. Hall, P.D. Feldman, M.A. McGrath, D.F. Strobel, The far-ultraviolet oxygen airglow of Europa and Ganymede. Astrophys. J. **499**, 475–481 (1998)
C.J. Hansen, L. Esposito, A.I.F. Stewart, J. Colwell, A. Hendrix, W. Pryor, D. Shemansky, R. West, Enceladus' water vapor plume. Science **311**, 1422–1425 (2006). doi:10.1126/science.1121254
R.E. Hartle, E.C. Sittler, K.W. Ogilvie, J.D. Scudder, A.J. Lazarus, S.K. Atreya, Titan's ion exosphere observed from Voyager 1. J. Geophys. Res. **87**, 1383–1394 (1982). doi:10.1029/JA087iA03p01383
T.W. Hill, D.H. Pontius, Plasma injection near Io. J. Geophys. Res. **103**, 19,879–19,886 (1998). doi:10.1029/98JE00945
M. Horanyi, J.A. Burns, D.P. Hamilton, The dynamics of Saturn's E ring particles. Icarus **97**, 248–259 (1992). doi:10.1016/0019-1035(92)90131-P
G. Hulot, N. Olsen, M. Mandea, C. Constable, C. Finley, Magnetic field of the Earth. Space Sci. Rev. (2009)
D.M. Hunten, M.G. Tomasko, F.M. Flasar, R.E. Samuelson, D.F. Strobel, D.J. Stevenson, Titan, in *Saturn*, ed. by T. Gehrels, M.S. Matthews (University of Arizona Press, Tucson 1984), pp. 671–759
W.H. Ip, A. Kopp, Resistive MHD simulations of Ganymede's magnetosphere: 2. Birkeland currents and particle energetics. J. Geophys. Res. **107**(A12), 1491 (2002). doi:10.1029/2001JA005072
X. Jia, R.J. Walker, M.G. Kivelson, K.K. Khurana, J.A. Linker, Three-dimensional MHD simulations of Ganymede's magnetosphere. J. Geophys. Res. **113**, 6212 (2008). doi:10.1029/2007JA012748
X. Jia, R.J. Walker, M.G. Kivelson, K.K. Khurana, J.A. Linker, Properties of Ganymede's magnetosphere inferred from improved three-dimensional MHD simulations. J. Geophys. Res. (2009, submitted)
K.K. Khurana, Euler potential models of Jupiter's magnetospheric field. J. Geophys. Res. **102**, 11,295–11,306 (1997)
K.K. Khurana, M.G. Kivelson, D.J. Stevenson, G. Schubert, C.T. Russell, R.J. Walker, C. Polanskey, Induced magnetic fields as evidence for subsurface oceans in Europa and Callisto. Nature **395**, 777–780 (1998). doi:10.1038/27394
K.K. Khurana, M.G. Kivelson, C.T. Russell, Searching for liquid water in Europa by using surface observatories. Astrobiology **2**(1), 93–103 (2002)
K.K. Khurana, M.K. Dougherty, C.T. Russell, J.S. Leisner, Mass loading of Saturn's magnetosphere near Enceladus. J. Geophys. Res. **112** (2007). doi:10.1029/2006JA012110
K.K. Khurana, C.T. Russell, M.K. Dougherty, Magnetic portraits of Tethys and Rhea. Icarus **193**, 465–474 (2008)
M.G. Kivelson, Does Enceladus govern magnetospheric dynamics at Saturn? Science **311**, 1391–1392 (2006). doi:10.1126/science.1124494
M.G. Kivelson, C.T. Russell, The interaction of flowing plasmas with planetary ionospheres – A Titan–Venus comparison. J. Geophys. Res. **88**, 49–57 (1983). doi:10.1029/JA088iA01p00049
M.G. Kivelson, J.A. Slavin, D.J. Southwood, Magnetospheres of the Galilean satellites. Science **205**, 491–493 (1979). doi:10.1126/science.205.4405.491
M.G. Kivelson, K.K. Khurana, C.T. Russell, R.J. Walker, J. Warnecke, F.V. Coroniti, C. Polanskey, D.J. Southwood, G. Schubert, Discovery of Ganymede's magnetic field by the Galileo spacecraft. Nature **384**, 537–541 (1996)

M.G. Kivelson, K.K. Khurana, R.J. Walker, J.A. Linker, C.R. Russell, D.J. Southwood, C. Polanskey, A magnetic signature at Io: Initial report from the Galileo magnetometer. Science **273**, 337–340 (1996a). doi:10.1126/science.273.5273.337

M.G. Kivelson, K.K. Khurana, R.J. Walker, J. Warnecke, C.T. Russell, J.A. Linker, D.J. Southwood, C. Polanskey, Io's interaction with the plasma torus: Galileo magnetometer report. Science **274**, 396–398 (1996b). doi:10.1126/science.274.5286.396

M. Kivelson, K. Khurana, S. Joy, C. Russell, D. Southwood, R. Walker, C. Polanskey, Europa's magnetic signature: Report from Galileo's pass on 19 December 1996. Science **276**(5316), 1239–1241 (1997)

M.G. Kivelson, J. Warnecke, L. Bennett, S. Joy, K.K. Khurana, J.A. Linker, C.T. Russell, R.J. Walker, C. Polanskey, Ganymede's magnetosphere: Magnetometer overview. J. Geophys. Res. **103**, 19,963–19,972 (1998)

M. Kivelson, K. Khurana, D. Stevenson, L. Bennett, S. Joy, C. Russell, R. Walker, C. Zimmer, C. Polanskey, Europa and Callisto: Induced or intrinsic fields in a periodically varying plasma environment. J. Geophys. Res. **104**(A3), 4609–4625 (1999)

M. Kivelson, K. Khurana, C. Russell, M. Volwerk, R. Walker, C. Zimmer, Galileo magnetometer measurements: A stronger case for a subsurface ocean at Europa. Science **289**(5483), 1340–1343 (2000)

M. Kivelson, K. Khurana, C. Russell, S. Joy, M. Volwerk, R. Walker, C. Zimmer, J. Linker, Magnetized or unmagnetized: Ambiguity persists following Galileo's encounters with Io in 1999 and 2000. J. Geophys. Res. **106**(A11), 26,121–26,135 (2001a)

M.G. Kivelson, K.K. Khurana, C.T. Russell, R.J. Walker, Magnetic signature of a polar pass over Io. AGU Fall Meeting Abstracts (2001b), p. A1

M.G. Kivelson, K.K. Khurana, M. Volwerk, The permanent and inductive magnetic moments of Ganymede. Icarus **157**, 507–522 (2002)

M.G. Kivelson, F. Bagenal, W.S. Kurth, F.M. Neubauger, C. Paranicas, J. Saur, Magnetospheric interactions with satellites, in *Jupiter: The Planet, Satellites and Magnetosphere*, ed. by F. Bagenal, T.E. Dowling, W.B. McKinnon (Cambridge Univ. Press, Cambridge, 2004), pp. 513–536

A.J. Kliore, D.P. Hinson, F.M. Flasar, A.F. Nagy, T.E. Cravens, The ionosphere of Europa from Galileo radio occultations. Science **277**, 355–358 (1997). doi:10.1126/science.277.5324.355

A.J. Kliore, A. Anabtawi, R.G. Herrera, S.W. Asmar, A.F. Nagy, D.P. Hinson, F.M. Flasar, Ionosphere of Callisto from Galileo radio occultation observations. J. Geophys. Res. **107**, 1407 (2002). doi:10.1029/2002JA009365

A.J. Kliore, A.F. Nagy, E.A. Marouf, R.G. French, F.M. Flasar, N.J. Rappaport, A. Anabttawi, S.W. Asmar, D.S. Kahann, E. Barbinis, G.L. Goltz, D.U. Fleischman, D.J. Rochblatt, First results from the Cassini radio occultations of the Titan ionosphere. J. Geophys. Res. **113**, 9317 (2008). doi:10.1029/2007JA012965

A. Kopp, W.H. Ip, Resistive MHD simulations of Ganymede's magnetosphere: 1. Time variabilities of the magnetic field topology. J. Geophys. Res. **107**(A12), 1490 (2002). doi:10.1029/2001JA005071

W.S. Kurth, D.A. Gurnett, A. Roux, S.J. Bolton, Ganymede: A new radio source. Geophys. Res. Lett. **24**, 2167–2170 (1997)

B.N. Lahiri, A.T. Price, Electromagnetic induction in non-uniform conductors, and the determination of the conductivity of the Earth from terrestrial magnetic variations. Phil. Trans. R. Soc. Ser. A **237**(784), 509–540 (1939)

S.A. Ledvina, Y.-J. Ma, E. Kallio, Modeling and simulating flowing plasmas and related phenomena. Space Sci. Rev. **139**, 143–189 (2008). doi:10.1007/s11214-008-9384-6

J.S. Leisner, C.T. Russell, R.J. Strangeway, N. Omidi, M.K. Dougherty, W.S. Kurth, The interior of Iapetus: Constraints provided by the solar wind interaction. AGU Fall Meeting Abstracts (2008), p. C8

J.A. Linker, K.K. Khurana, M.G. Kivelson, R.J. Walker, MHD simulations of Io's interaction with the plasma torus. J. Geophys. Res. **103**(E9), 19,867–19,877 (1998)

Y. Ma, A.F. Nagy, T.E. Cravens, I.V. Sokolov, K.C. Hansen, J.-E. Wahlund, F.J. Crary, A.J. Coates, M.K. Dougherty, Comparisons between MHD model calculations and observations of Cassini flybys of Titan. J. Geophys. Res. **111**, 5207 (2006). doi:10.1029/2005JA011481

Y.-J. Ma, K. Altwegg, T. Breus, M.R. Combi, T.E. Cravens, E. Kallio, S.A. Ledvina, J.G. Luhmann, S. Miller, A.F. Nagy, A.J. Ridley, D.F. Strobel, Plasma flow and related phenomena in planetary aeronomy. Space Sci. Rev. **139**, 311–353 (2008a). doi:10.1007/s11214-008-9389-1

Y.J. Ma, C.T. Russell, A.F. Nagy, G. Toth, C. Bertucci, M.K. Dougherty, F.M. Neubauer, A. Wellbrock, A.J. Coates, P. Garnier, J.-E. Wahlund, T.E. Cravens, F.J. Crary, Time-dependent global MHD simulations of Cassini T32 flyby: from magnetosphere to magnetosheath. J. Geophys. Res. (2008b). doi:10.1029/2008JA013676

M.A. McGrath, E. Lellouch, D.F. Strobel, P.D. Feldman, R.E. Johnson, Satellite atmospheres, in *Jupiter: The Planet, Satellites and Magnetosphere*, ed. by F. Bagenal, T.E.Dowling, W.B. McKinnon (Cambridge Univ. Press, Cambridge, 2004)

L.A. Morabito, S.P. Synnott, P.N. Kupferman, S.A. Collins, Discovery of currently active extraterrestrial volcanism. Science **204**, 972 (1979). doi:10.1126/science.204.4396.972

N.F. Ness, M.H. Acuna, R.P. Lepping, J.E.P. Connerney, K.W. Behannon, L.F. Burlaga, F.M. Neubauer, Magnetic field studies by Voyager 1 – Preliminary results at Saturn. Science **212**, 211–217 (1981). doi:10.1126/science.212.4491.211

N.F. Ness, M.H. Acuna, K.W. Behannon, The induced magnetosphere of Titan. J. Geophys. Res. **87**, 1369–1381 (1982). doi:10.1029/JA087iA03p01369

F.M. Neubauer, Possible strengths of dynamo magnetic fields of the Galilean satellites and of Titan. Geophys. Res. Lett. **5**, 905–908 (1978). doi:10.1029/GL005i011p00905

F.M. Neubauer, Non-linear standing Alfvén wave current system at Io: theory. J. Geophys. Res. **85**, 1,171–1,178 (1980)

F.M. Neubauer, The sub-Alfvénic interaction of the Galilean satellites with the Jovian magnetospheres. J. Geophys. Res. **103**(E9), 19,843–19,866 (1998a)

F.M. Neubauer, Comment on "Interaction of Io with its torus: Does Io have an internal magnetic field?" by Krishan K. Khurana, Margaret G. Kivelson and Christopher T. Russell. Geophys. Res. Lett. **25**, 2349–2350 (1998b). doi:10.1029/98GL01585

F.M. Neubauer, D.A. Gurnett, J.D. Scudder, R.E. Hartle, Titan's magnetospheric interaction, in *Saturn*, ed. by T. Gehrels, M.S. Matthews (University of Arizona Press, Tucson, 1984), pp. 760–787

F.M. Neubauer, H. Backes, M.K. Dougherty, A. Wennmacher, C.T. Russell, A. Coates, D. Young, N. Achilleos, N. André, C.S. Arridge, C. Bertucci, G.H. Jones, K.K. Khurana, T. Knetter, A. Law, G.R. Lewis, J. Saur, Titan's near magnetotail from magnetic field and electron plasma observations and modeling: Cassini flybys TA, TB, and T3. J. Geophys. Res. **111**, 10,220 (2006). doi:10.1029/2006JA011676

K.W. Ogilvie, J.T. Steinberg, R.J. Fitzenreiter, C.J. Owen, A.J. Lazarus, W.M. Farrell, R.B. Torbert, Observations of the lunar plasma wake from the WIND spacecraft on December 27, 1994. Geophys. Res. Lett. **23**, 1255–1258 (1996). doi:10.1029/96GL01069

N. Olsen, K.-H. Glassmeier, X. Jia, Separation of the magnetic field into external and internal parts. Space Sci. Rev. (2009)

S.J. Ostro, R.D. West, M.A. Janssen, R.D. Lorenz, H.A. Zebker, G.J. Black, J.I. Lunine, L.C. Wye, R.M. Lopes, S.D. Wall, C. Elachi, L. Roth, S. Hensley, K. Kelleher, G.A. Hamilton, Y. Gim, Y.Z. Anderson, R.A. Boehmer, W.T.K. Johnson, Cassini RADAR Team, Cassini RADAR observations of Enceladus, Tethys, Dione, Rhea, Iapetus, Hyperion, and Phoebe. Icarus **183**(2), 479–490 (2006). doi:10.1016/j.icarus.2006.02.019

C.J. Owen, R.P. Lepping, K.W. Ogilvie, J.A. Slavin, W.M. Farrell, J.B. Byrnes, The lunar wake at 6.8 R_L: WIND magnetic field observations. Geophys. Res. Lett. **23**, 1263–1266 (1996). doi:10.1029/96GL01354

C. Paty, R. Winglee, The role of ion cyclotron motion at Ganymede: magnetic morphology and magnetospheric dynamics. Geophys. Res. Lett. **33**, L10106 (2006). doi:10.1029/2005GL025273

C. Paty, W. Paterson, R. Winglee, Ion energization in Ganymede's magnetosphere: Using multifluid simulations to interpret ion energy spectrograms. J. Geophys. Res. **113** 1 (2008). doi:10.1029/2007JA012848

D.H. Pontius, T.W. Hill, Enceladus: A significant plasma source for Saturn's magnetosphere. J. Geophys. Res. **111**, 9214 (2006). doi:10.1029/2006JA011674

C.C. Porco, P. Helfenstein, P.C. Thomas, A.P. Ingersoll, J. Wisdom, R. West, G. Neukum, T. Denk, R. Wagner, T. Roatsch, S. Kieffer, E. Turtle, A. McEwen, T.V. Johnson, J. Rathbun, J. Veverka, D. Wilson, J. Perry, J. Spitale, A. Brahic, J.A. Burns, A.D. DelGenio, L. Dones, C.D. Murray, S. Squyres, Cassini observes the active south pole of Enceladus. Science **311**, 1393–1401 (2006). doi:10.1126/science.1123013

T. Rikitake, *Electromagnetism and the Earth's Interior* (Elsevier, Amsterdam, 1966)

E. Roussos, J. Müller, S. Simon, A. Bößwetter, U. Motschmann, N. Krupp, M. Fränz, J. Woch, K.K. Khurana, M.K. Dougherty, Plasma and fields in the wake of Rhea: 3-D hybrid simulation and comparison with Cassini data. Ann. Geophys. **26**, 619–637 (2008)

J. Ruiz, The stability against freezing of an internal liquid-water ocean in Callisto. Nature **412**, 409–411 (2001)

J. Saur, F.M. Neubauer, D.F. Strobel, M.E. Summers, Three-dimensional plasma simulation of Io's interaction with the Io plasma torus: Asymmetric plasma flow. J. Geophys. Res. **104**, 25,105–25,126 (1999)

J. Saur, F.M. Neubauer, J.E.P. Connerney, P. Zarka, M.G. Kivelson, Plasma interaction of Io with its plasma torus, in *Jupiter: The Planet, Satellites and Magnetosphere*, ed. by F. Bagenal, T.E. Dowling, W.B. McKinnon (Cambridge Univ. Press, Cambridge, 2004), pp. 537–560

J. Saur, F.M. Neubauer, K.-H. Glassmeier, Induced magnetic fields in solar system bodies. Space Sci. Rev. (2009)

N. Schilling, K. Khurana, M. Kivelson, Limits on an intrinsic dipole moment in Europa. J. Geophys. Res. **109**, E05006 (2004). doi:10.1029/2003JE002166

N. Schilling, F.M. Neubauer, J. Saur, Time-varying interaction of Europa with the Jovian magnetosphere: Constraints on the conductivity of Europa's subsurface ocean. Icarus **192**(1), 41–55 (2007). doi:10.1016/j.icarus.2007.06.024

G. Schubert, B.R. Lichtenstein, Observations of moon-plasma interactions by orbital and surface experiments. Rev. Geophys. Space Phys. **12**, 592–626 (1974)

G. Schubert, K. Zhang, M.G. Kivelson, J.D. Anderson, The magnetic field and internal structure of Ganymede. Nature **384**, 544–545 (1996)

G. Schubert, J.D. Anderson, B.J. Travis, J. Palguta, Enceladus: Present internal structure and differentiation by early and long-term radiogenic heating. Icarus **188**(2), 345–355 (2007). doi:10.1016/j.icarus.2006.12.012

E.C. Sittler, R.E. Johnson, S. Jurac, J.D. Richardson, M. McGrath, F. Crary, D.T. Young, J.E. Nordholt, Pickup ions at Dione and Enceladus: Cassini plasma spectrometer simulations. J. Geophys. Res. **109**, 1214 (2004). doi:10.1029/2002JA009647

E.C. Sittler, M. Thomsen, R.E. Johnson, R.E. Hartle, M. Burger, D. Chornay, M.D. Shappirio, D. Simpson, H.T. Smith, A.J. Coates, A.M. Rymer, D.J. McComas, D.T. Young, D. Reisenfeld, M. Dougherty, N. Andre, Cassini observations of Saturn's inner plasmasphere: Saturn orbit insertion results. Planet. Space Sci. **54**, 1197–1210 (2006). doi:10.1016/j.pss.2006.05.038

B.A. Smith, L.A. Soderblom, R. Beebe, J. Boyce, G. Briggs, M. Carr, S.A. Collins, T.V. Johnson, A.F. Cook, II, G.E. Danielson, D. Morrison, The Galilean satellites and Jupiter – Voyager 2 imaging science results. Science **206**, 927–950 (1979a). doi:10.1126/science.206.4421.927

B.A. Smith, L.A. Soderblom, T.V. Johnson, A.P. Ingersoll, S.A. Collins, E.M. Shoemaker, G.E. Hunt, H. Masursky, M.H. Carr, M.E. Davies, A.F. Cook, J.M. Boyce, T. Owen, G.E. Danielson, C. Sagan, R.F. Beebe, J. Veverka, J.F. McCauley, R.G. Strom, D. Morrison, G.A. Briggs, V.E. Suomi, The Jupiter system through the eyes of Voyager 1. Science **204**, 951–957 (1979b). doi:10.1126/science.204.4396.951

B.A. Smith, L. Soderblom, R.M. Batson, P.M. Bridges, J.L. Inge, H. Masursky, E. Shoemaker, R.F. Beebe, J. Boyce, G. Briggs, A. Bunker, S.A. Collins, C. Hansen, T.V. Johnson, J.L. Mitchell, R.J. Terrile, A.F. Cook, J.N. Cuzzi, J.B. Pollack, G.E. Danielson, A.P. Ingersoll, M.E. Davies, G.E. Hunt, D. Morrison, T. Owen, C. Sagan, J. Veverka, R. Strom, V.E. Suomi, A new look at the Saturn system – The Voyager 2 images. Science **215**, 504–537 (1982). doi:10.1126/science.215.4532.504

F. Sohl, H. Hussmann, B. Schwentker, T. Spohn, R. Lorenz, Interior structure models and tidal Love numbers of Titan. J. Geophys. Res. **108**(E12), 5130 (2003). doi:10.1029/2003JE002044

D.J. Southwood, M.G. Kivelson, R.J. Walker, J.A. Slavin, Io and its plasma environment. J. Geophys. Res. **85**, 5,959–5,968 (1980)

F. Spahn, J. Schmidt, N. Albers, M. Hörning, M. Makuch, M. Seiß, S. Kempf, R. Srama, V. Dikarev, S. Helfert, G. Moragas-Klostermeyer, A.V. Krivov, M. Sremcević, A.J. Tuzzolino, T. Economou, E. Grün, Cassini dust measurements at Enceladus and implications for the origin of the E ring. Science **311**, 1416–1418 (2006). doi:10.1126/science.1121375

J.R. Spencer, N.M. Schneider, Io on the eve of the Galileo mission. Ann. Rev. Earth Planet. Sci. **24**, 125–190 (1996). doi:10.1146/annurev.earth.24.1.125

J.R. Spencer, J.C. Pearl, M. Segura, F.M. Flasar, A. Mamoutkine, P. Romani, B.J. Buratti, A.R. Hendrix, L.J. Spilker, R.M.C. Lopes, Cassini encounters Enceladus: Background and the discovery of a south polar hot spot. Science **311**, 1401–1405 (2006). doi:10.1126/science.1121661

T. Spohn, G. Schubert, Oceans in the icy Galilean satellites of Jupiter? Icarus **161**, 456–467 (2003). doi:10.1016/S0019-1035(02)00048-9

P.C. Thomas, J.A. Burns, P. Helfenstein, S. Squyres, J. Veverka, C. Porco, E.P. Turtle, A. McEwen, T. Denk, B. Giese, T. Roatsch, T.V. Johnson, R.A. Jacobson, Shapes of the Saturnian icy satellites and their significance. Icarus **190**, 573–584 (2007). doi:10.1016/j.icarus.2007.03.012

R.L. Tokar, R.E. Johnson, T.W. Hill, D.H. Pontius, W.S. Kurth, F.J. Crary, D.T. Young, M.F. Thomsen, D.B. Reisenfeld, A.J. Coates, G.R. Lewis, E.C. Sittler, D.A. Gurnett, The interaction of the atmosphere of Enceladus with Saturn's plasma. Science **311**, 1409–1412 (2006). doi:10.1126/science.1121061

G.L. Tyler, V.R. Eshleman, J.D. Anderson, G.S. Levy, G.F. Lindal, G.E. Wood, T.A. Croft, Radio science investigations of the Saturn system with Voyager 1 – Preliminary results. Science **212**, 201–206 (1981). doi:10.1126/science.212.4491.201

J.H. Waite, M.R. Combi, W.-H. Ip, T.E. Cravens, R.L. McNutt, W. Kasprzak, R. Yelle, J. Luhmann, H. Niemann, D. Gell, B. Magee, G. Fletcher, J. Lunine, W.-L. Tseng, Cassini ion and neutral mass spectrometer: Enceladus plume composition and structure. Science **311**, 1419–1422 (2006). doi:10.1126/science.1121290

D.J. Williams, B. Mauk, R.W. McEntire, Trapped electrons in Ganymede's magnetic field. Geophys. Res. Lett. **24**(23), 2,953–2,956 (1997a)

D.J. Williams, B.H. Mauk, R.W. McEntire, E.C. Roelof, T.P. Armstrong, B. Wilken, S.M.K.J.G. Roederer, T.A. Fritz, L.J. Lanzerotti, N. Murphy, Energetic particle signatures at Ganymede: Implications for Ganymede's magnetic field. Geophys. Res. Lett. **24**(17), 2,163–2,166 (1997b)

R.J. Wilson, R.L. Tokar, M.G. Henderson, T.W. Hill, M.F. Thomsen, D.H. Pontius, Cassini plasma spectrometer thermal ion measurements in Saturn's inner magnetosphere. J. Geophys. Res. (2008) doi:10.1029/2008JA013486

C. Zimmer, K. Khurana, M. Kivelson, Subsurface oceans on Europa and Callisto: Constraints from Galileo magnetometer observations. Icarus **147**(2), 329–347 (2000)

Space Sci Rev (2010) 152: 307–339
DOI 10.1007/s11214-009-9544-3

The Magnetic Field of Mercury

Brian J. Anderson · Mario H. Acuña · Haje Korth · James A. Slavin · Hideharu Uno · Catherine L. Johnson · Michael E. Purucker · Sean C. Solomon · Jim M. Raines · Thomas H. Zurbuchen · George Gloeckler · Ralph L. McNutt Jr.

Received: 13 April 2009 / Accepted: 5 June 2009 / Published online: 9 July 2009

Abstract The magnetic field strength of Mercury at the planet's surface is approximately 1% that of Earth's surface field. This comparatively low field strength presents a number of challenges, both theoretically to understand how it is generated and observationally to distinguish the internal field from that due to the solar wind interaction. Conversely, the small field also means that Mercury offers an important opportunity to advance our understanding both of planetary magnetic field generation and magnetosphere-solar wind interactions. The observations from the Mariner 10 magnetometer in 1974 and 1975, and the MESSENGER Magnetometer and plasma instruments during the probe's first two flybys of Mercury on 14 January and 6 October 2008, provide the basis for our current knowledge of the internal field. The external field arising from the interaction of the magnetosphere with the solar wind is more prominent near Mercury than for any other magnetized planet in the Solar System, and particular attention is therefore paid to indications in the observations of deficiencies in

Mario H. Acuña deceased 5 March 2009.

B.J. Anderson (✉) · H. Korth · R.L. McNutt Jr.
Applied Physics Laboratory, The Johns Hopkins University, Laurel, MD 20723, USA
e-mail: Brian.Anderson@jhuapl.edu

M.H. Acuña · M.E. Purucker
Solar System Exploration Division, NASA Goddard Space Flight Center, Greenbelt, MD 20771, USA

J.A. Slavin
Heliophysics Science Division, NASA Goddard Space Flight Center, Greenbelt, MD 20771, USA

C.L. Johnson · H. Uno
Department of Earth and Ocean Sciences, University of British Columbia, Vancouver, BC V6T 1Z4, Canada

S.C. Solomon
Department of Terrestrial Magnetism, Carnegie Institution of Washington, Washington, DC 20015, USA

G. Gloeckler · J.M. Raines · T.H. Zurbuchen
Department of Atmospheric, Oceanic and Space Sciences, University of Michigan, Ann Arbor, MI 48109, USA

our understanding of the external field. The second MESSENGER flyby occurred over the opposite hemisphere from the other flybys, and these newest data constrain the tilt of the planetary moment from the planet's spin axis to be less than 5°. Considered as a dipole field, the moment is in the range 240 to 270 nT-R_M^3, where R_M is Mercury's radius. Multipole solutions for the planetary field yield a smaller dipole term, 180 to 220 nT-R_M^3, and higher-order terms that together yield an equatorial surface field from 250 to 290 nT. From the spatial distribution of the fit residuals, the equatorial data are seen to reflect a weaker northward field and a strongly radial field, neither of which can be explained by a centered-dipole matched to the field measured near the pole by Mariner 10. This disparity is a major factor controlling the higher-order terms in the multipole solutions. The residuals are not largest close to the planet, and when considered in magnetospheric coordinates the residuals indicate the presence of a cross-tail current extending to within $0.5R_M$ altitude on the nightside. A near-tail current with a density of 0.1 μA/m^2 could account for the low field intensities recorded near the equator. In addition, the MESSENGER flybys include the first plasma observations from Mercury and demonstrate that solar wind plasma is present at low altitudes, below 500 km. Although we can be confident in the dipole-only moment estimates, the data in hand remain subject to ambiguities for distinguishing internal from external contributions. The anticipated observations from orbit at Mercury, first from MESSENGER beginning in March 2011 and later from the dual-spacecraft BepiColombo mission, will be essential to elucidate the higher-order structure in the magnetic field of Mercury that will reveal the telltale signatures of the physics responsible for its generation.

Keywords Mercury · Magnetic field · Magnetosphere · MESSENGER · BepiColombo

1 Introduction

The presence or absence of a magnetic field on a terrestrial planet depends on the interplay of several interior processes. Mercury, the smallest of the inner planets, illustrates the challenges facing an understanding of the origins of planetary magnetism. If Mercury had a pure iron core, thermal history models predict that such a core would now be completely solid, and thus Mercury would have at most a remanent crustal magnetic field (e.g., Solomon 1976). The discovery by Mariner 10 that Mercury has a weak but Earth-like internal magnetic field (Ness et al. 1974, 1975) hinted that the planet's core contains sufficient lighter elements to lower the melting temperature and permit a presently fluid outer core (e.g., Schubert et al. 1988). Accounting for a weak, primarily dipolar field in terms of an Earth-like core dynamo has proved challenging, but a variety of numerical dynamo models have been explored that can predict such a field (Heimpel et al. 2005; Stanley et al. 2005; Christensen 2006; Takahashi and Matsushima 2006; Glassmeier et al. 2007a, 2007b). To test such ideas, considerable effort has gone into extracting as much information as possible about the geometry of the internal field from Mariner 10 observations (Ness 1979; Connerney and Ness 1988; Engle 1997).

The flybys of Mercury by NASA's MErcury Surface, Space ENvironment, GEochemistry, and Ranging (MESSENGER) spacecraft (Solomon et al. 2001, 2007) on 14 January and 6 October 2008 have yielded new magnetic field observations (Anderson et al. 2007, 2008b; Slavin et al. 2008, 2009a, 2009b) and the first plasma ion observations within the planet's magnetosphere (Zurbuchen et al. 2008). As a result of those encounters, we are now in a position to reassess the nature of Mercury's internal magnetic field. Any such consideration necessarily involves a careful treatment of external field sources. The small internal field at Mercury implies that external current systems are particularly important, because they produce magnetic fields at the surface that are comparable in strength to the planetary field (e.g., Slavin and Holzer 1979). This fact, together with the limited trajectories over which in situ observations are available from the Mariner 10 and MESSENGER flybys to date, leads to a situation in which the internal field is difficult to separate from the external and plasma pressure contributions to the magnetic field observations (Connerney and Ness 1988; Korth et al. 2004). In addition to considering what we can deduce about the internal magnetic field, we identify those aspects of the external field description that are most critical for the Mercury system and need to be further understood.

2 Processes Responsible for Magnetic Fields at Mercury

There are three candidate sources of magnetic fields arising within the planet: a core dynamo, crustal magnetization, and induction currents in any electrically conducting regions. The amplitude of Mercury's annual forced libration, recently detected by Earth-based radar observations (Margot et al. 2007), implies that the planet has a fluid outer core. The presence of a light element in an otherwise iron-rich core can permit a presently molten outer core; most specific models of such thermal histories for Mercury have been based on the proposition that sulfur is the principal light element (Schubert et al. 1988; Hauck et al. 2004), although other elements considered as candidate components of the Earth's outer core (e.g., Si, O, H, C) might yield a similar outcome. Early proposals for a remanent origin for Mercury's dipolar internal field (Stephenson 1976; Srnka 1976) were questioned on the grounds that a high specific magnetization would be required and the polarity of the field would have to be stable during the time the crust cooled through the Curie temperature of the relevant magnetic carriers (Schubert et al. 1988). The detection of strong crustal magnetic fields on Mars (Acuña et al. 1999) and the recognition that the thickness of a magnetized crustal layer on Mercury could vary with latitude and longitude (Aharonson et al. 2004) renewed consideration of crustal sources of Mercury's internal field. In addition to fields from these sources, the suite of external currents that must be considered include magnetopause, magnetotail, and perhaps other currents resulting from solar wind plasma interaction that are as yet poorly understood for Mercury. We discuss these processes in turn.

2.1 Internal Field Sources: Crustal, Dynamo, and Induced

The presence of at least a thin shell of molten material in the outer core raises the possibility that the planet supports a dynamo driven by thermal convection or chemical buoyancy (Stevenson 1983; Stanley et al. 2005; Heimpel et al. 2005). A variety of models for such a dynamo have been investigated (Heimpel et al. 2005; Christensen 2006; Stanley et al. 2007; Wicht et al. 2007; Christensen and Wicht 2008), and these models make generally distinct predictions for the long-wavelength structure of the planetary field. A thin-shell dynamo could yield significant non-dipolar structure in the field (e.g., Stanley et al. 2007), which although attenuated at spacecraft altitudes might be diagnostic of a minimum shell thickness. Thick-shell dynamos can also yield a weak field, and a stable but conductive layer at the top of the core could suppress higher-order terms, so such models are generally consistent with an axisymmetric dipolar field at spacecraft altitudes (Christensen 2006; Wicht et al. 2007; Christensen and Wicht 2008). Mineral physics experiments suggest that a stable layer at the top of Mercury's outer core is a plausible hypothesis (Chen et al. 2008).

It has also been proposed that the large-scale field could be due to a dynamo generated by thermoelectric currents along a rough core-mantle boundary (Stevenson 1987; Giampieri and Balogh 2002). Observational constraints on the multipolar structure and dipole axis orientation relative to the rotation axis may be important discriminators among these hypotheses.

Another physical mechanism that may account for the weak global-scale planetary field is crustal remanence (Stephenson 1976; Srnka 1976), known to be the only measurable contributor to the external magnetic field of Mars (Acuña et al. 1999). Mercury's small obliquity and spin-orbit resonance lead to stable, large-scale latitudinal and longitudinal variations in insolation and thus to similar geographical variations in the depth to a specific Curie isotherm at a given time in the planet's thermal history. The crustal fields imparted by a steady internal field during the time that the outer crust cooled through the Curie temperatures of any magnetic minerals present would give rise to an external field with a dominant dipole term and specific relative magnitudes for multipolar components (Aharonson et al. 2004).

Finally, the modest planetary magnetic moment implies both that the conductive core is much larger relative to the magnetosphere than for any other planet of the Solar System and also that the magnetic signatures of induction currents flowing in the core are larger relative to the background core dynamo field. Thus, inductive fields may contribute as much as $\sim$10% to the surface field, depending on variations in the external field (Grosser et al. 2004; Glassmeier et al. 2007a). Since both the external and induced fields are imposed on the core, there may be a feedback such that these external fields act as seed fields for the dynamo over long timescales (Glassmeier et al. 2007b). It is therefore of great interest to identify signatures of induced fields.

2.2 External Current Systems

Because the contribution of the magnetospheric current systems to the total observed magnetic field even near the surface is comparable to the field from internal sources, quantitative understanding of the electric currents associated with solar wind interaction with Mercury's magnetic field is critical to the study of the internal field (Russell et al. 1988). Magnetopause and magnetotail cross-tail currents are known to be present at Mercury. Of these the magnetopause current is the better known, because the magnetopause boundary location can be specified with reasonable accuracy (Slavin and Holzer 1981; Russell et al. 1988) and also because the intensity of the current can be specified from first principles (Tsyganenko 1995).

The cross-tail current is less well understood at Mercury because of its apparent proximity to the planetary surface. Earth-analog models place at least some of the cross-tail current below the surface of the planet (Korth et al. 2004; Tsyganenko and Sitnov 2005). The actual proximity of the tail current to the planet and the down-tail gradient are not well known. Mariner 10 flyby data suggest that the tail current is sufficiently intense close to the planet to reduce the net field to less than one-third of the internal field as close as 700 km from the surface (Ness et al. 1975). For this reason the tail current is one of the key features of magnetosphere magnetic field models customized for Mercury (Giampieri and Balogh 2001; Alexeev et al. 2008).

Prior to the first MESSENGER flybys the presence of local plasmas in Mercury's magnetosphere was inferred from variations in the magnetic field (Christon 1987). We now know that protons are found close to the planet within $0.5 R_M$ altitude (where R_M is Mercury's radius), evidently with sufficient densities to depress the local magnetic field by tens of nT

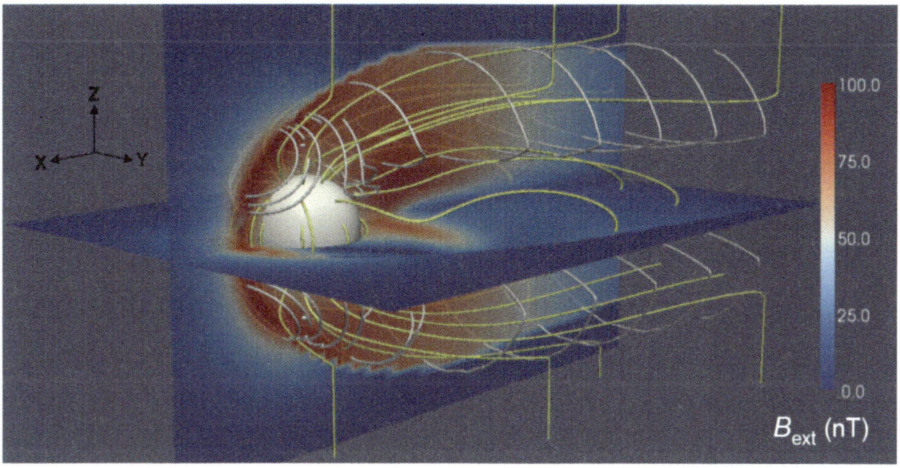

Fig. 1 Magnetic field lines (*yellow*), current density traces (*white*), and external magnetic field magnitude (*color bar*) for the TS04 magnetic field model scaled for Mercury. For the case shown the TS04 model was evaluated for a purely southward IMF of magnitude 10 nT and a solar-wind ram pressure of 20 nPa. An axially aligned dipole with a moment of 250 nT-R_M^3 was used for the planetary field. The model is shown in Mercury-solar-orbital (MSO) coordinates, with X positive toward the Sun, Z positive normal to the orbital plane, and Y completing the right-handed system

(Anderson et al. 2008b; Zurbuchen et al. 2008). The flyby encounters do not allow a comprehensive specification of the plasma distributions, but initial results are at least qualitatively consistent with numerical simulations (Trávníček et al. 2007, 2009) strongly indicating that the plasma distributions at Mercury are very different from those of any other magnetosphere. It is therefore to be expected that the current systems due to plasmas within the magnetosphere will be different at Mercury from those at other magnetospheres.

In addition to these factors, the solar wind and interplanetary magnetic field (IMF) imposed on Mercury's magnetosphere are variable and change both the intensity and configuration of the magnetopause and tail currents and presumably also alter the internal plasma distribution (e.g., Luhmann et al. 1998). The reconfiguration timescale for Mercury's magnetosphere is on the order of tens of seconds to a minute (e.g., Slavin et al. 2007), far shorter than either the transit time of any of the flyby encounters conducted to date or the period of any orbit about the planet. Identification of the appropriate solar wind and IMF conditions to use in attempts to model and correct for the externally generated magnetic fields is an important problem that may ultimately limit knowledge of the internal field structure derived from observations. Nonetheless, a simple extension of conditions of the IMF during the previous solar wind pass should provide approximately a factor of 10 increase in analysis sensitivity over ignoring the external field altogether (Korth et al. 2004).

To illustrate the critical role of the external fields, in Fig. 1 we show the magnetic field and external currents in the Earth-analog TS04 model (Tsyganenko and Sitnov 2005) scaled to Mercury (Korth et al. 2004) and depicted in Mercury-solar-orbital (MSO) coordinates. The fundamental topology of the magnetosphere is evident in this figure. The magnetopause Chapman-Ferraro currents are the closed circular loops on the dayside magnetopause around the magnetic cusps. The dayside is highly compressed, and the magnetic field in the polar regions is topologically linked to the lobes of the magnetotail, which are separated by the cross-tail current. The tail current system flows over the northern and southern lobes and closes across the middle of the tail, flowing from dawn to dusk. In this model the external field contributes over 100 nT to the total field over large regions of space, especially close to the planet. Its accurate specification is obviously crucial to an accurate specification of the planetary magnetic field.

3 Observations from the MESSENGER Flybys

3.1 Magnetic Field Observations Overview

Portions of the trajectories from which magnetic field observations were used for internal field analysis, projected to Mercury body-fixed coordinate latitude and longitude, are shown in Fig. 2. The trajectories include Mariner 10 flybys I and III (M10-I and M10-III), the first two MESSENGER flybys (M1 and M2), and the trajectory planned for the third MESSENGER flyby (M3). The third Mariner 10 encounter provided the only observations to date at high latitudes, and the second MESSENGER flyby yielded the first magnetic field observations from the planet's western hemisphere. The third MESSENGER flyby will cover nearly the same longitudes as M2.

The M1 and M2 flyby trajectories are shown in Fig. 3. Nominal magnetopause and bow-shock boundaries are also shown (Slavin et al. 2009a). For M2 the MESSENGER spacecraft passed inbound farther tailward and outbound somewhat later in the morning than for M1. Magnetic field data for 110 minutes spanning each encounter are depicted in Figs. 4 and 5. From top to bottom the panels show: the field magnitude; the polar angle, θ; the azimuth angle, φ; and the 1–10-Hz passband fluctuation amplitude in nT. The maximum magnetic field was nearly the same for both encounters, 159 nT for M1 and 158 nT for M2, despite the $\sim$180° longitude separation (Fig. 2). As expected from the differences in the trajectories, MESSENGER's inbound bow-shock crossing for M2 occurred earlier relative to closest approach than for M1.

The data for the inbound portions of the passes indicate that for M1 the spacecraft entered into the cross-tail current sheet (CS), whereas for M2 the spacecraft entered directly into the southern magnetic tail lobe (TL). For M1 the field between the magnetopause (MP) and CS remained nominally northward, indicating that the spacecraft was near the center of the cross-tail current. At CS the field began to rotate away from northward to anti-sunward, implying passage from the current sheet into the southern magnetotail lobe, where the field

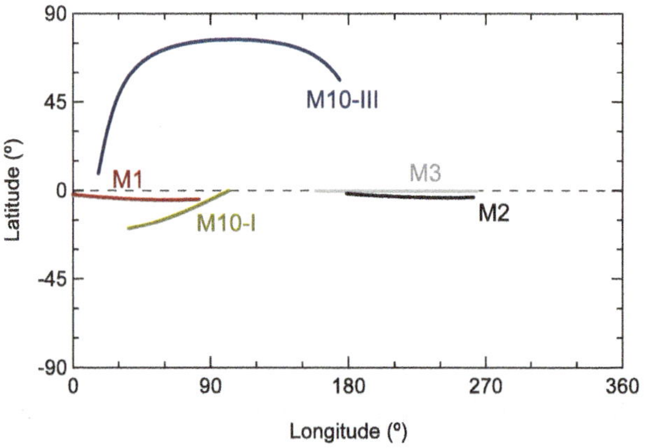

Fig. 2 Trajectories of Mariner 10 flybys M10-I and M10-III and MESSENGER flybys M1, M2, and M3 plotted in Mercury body-fixed (MBF) coordinates. Longitude is positive to the east

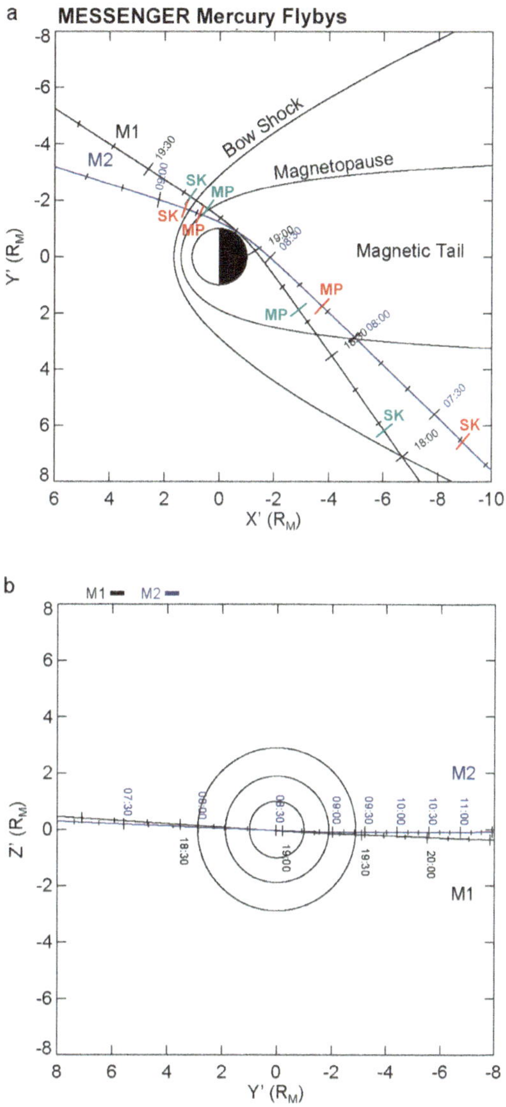

Fig. 3 Trajectories of the first and second MESSENGER flybys of Mercury, denoted M1 and M2, respectively, in MSO coordinates. *Primes* indicate that the system accounts for aberration in the solar wind flow due to Mercury's orbital motion. Panel (**a**) shows the view in the X'–Y' plane looking down from the north, and panel (**b**) shows the view in the Z'–Y' plane looking toward the Sun. Nominal magnetopause and bow-shock boundaries are shown for the equatorial ($Z' = 0$) plane in panel (**a**) and the $X' = 0$ plane in panel (**b**). The observed bow shock (SK) and magnetopause (MP) crossings are marked on panel (**a**) in *green* for M1 and *red* for M2

is anti-sunward. For M2, by contrast, the field was strongly anti-sunward immediately following the inbound MP, indicating direct entry into the magnetotail southern lobe.

The rotation of the field from anti-sunward to northward began at TL for M1 and M2, indicating the transition from the magnetotail lobe to the region dominated by the planet's internal field. For M1 the maximum field magnitude occurred shortly after closest approach (CA), and for M2 the maximum field occurred slightly prior to CA. For both flybys on the outbound leg a relatively sharp drop in field magnitude without a change in direction occurred approximately five minutes prior to MP. We term this transition a dayside boundary layer and denote it as BL. For purposes of estimating the internal planetary magnetic field, the appropriate data ranges are taken from TL to BL because these represent data dominated

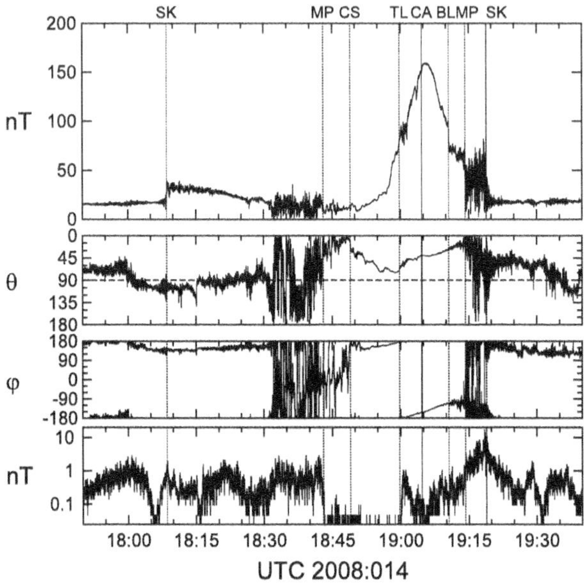

Fig. 4 Magnetic field data from the first MESSENGER flyby presented in MSO coordinates. From top to bottom the panels show: the field magnitude; the polar angle, θ, where $\theta = 0°$ is northward; the azimuth angle, φ, where $\varphi = 0°$ and $90°$ are sunward and duskward, respectively; and the 1–10-Hz band-pass fluctuation amplitude. Magnetic field vectors were sampled every 0.05 s, and the 1–10-Hz band-pass amplitude was evaluated on-board every 1 s from the 0.05-s data. Magnetic boundaries are labeled as follows: SK for bow shock; MP for magnetopause; CS for the tail current sheet; TL for transition out of the tail lobe; CA for closest approach; and BL for entry into a boundary layer

by the planetary field; these ranges do not include observations in the lobes or inside the dayside boundary layer.

For M1 the magnetic field following the outbound MP crossing was somewhat northward both in the magnetosheath and for about 20 minutes after the bow-shock crossing (SK) in the interplanetary medium. For M2, however, the magnetosheath field was strongly southward between MP and SK for both the inbound and outbound portions of the encounter and was southward in the interplanetary medium both before and after the encounter. Thus, the IMF at Mercury was northward as of the outbound passage on M1, whereas for M2 the IMF was likely to have been southward throughout the encounter.

Evidence of magnetospheric dynamics indicates significant differences in Mercury's magnetosphere between M1 and M2. During M1 MESSENGER detected signatures of a modest flux transfer event just outside the inbound tail magnetopause crossing and Kelvin-Helmholtz vortices shortly after the inbound crossing, indicating the occurrence of reconnection and boundary waves analogous to those observed at Earth (Slavin et al. 2008). The first flyby also revealed a boundary layer within and adjacent to the outbound magnetopause crossing, but there were no indications of reconnection on the outbound magnetopause crossing (Anderson et al. 2008b; Slavin et al. 2008). By contrast, data from the second MESSENGER flyby show intense reconnection signatures in the vicinity of both inbound and outbound magnetopause passages (Slavin et al. 2009b). These include flux ropes and a series of traveling compression regions on the inbound leg and two strong reconnection events on the outbound leg, one an intense flux transfer event with a core field stronger than the maximum field observed within the magnetosphere (Slavin et al. 2009b).

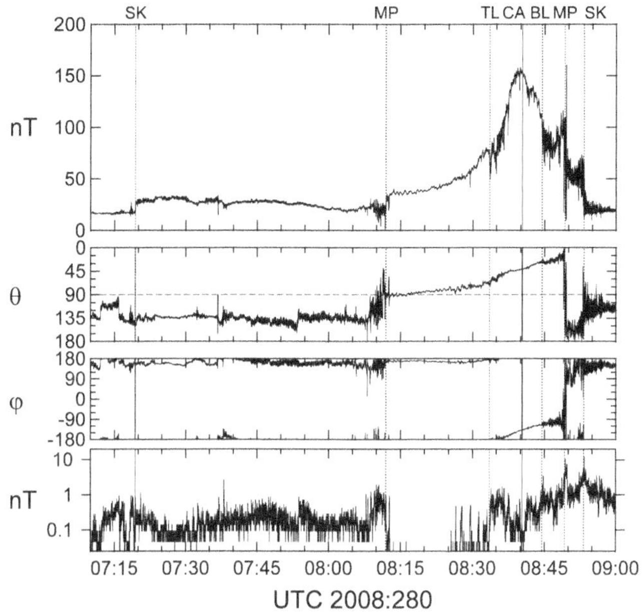

Fig. 5 Magnetic field data from the second MESSENGER flyby in the same format as Fig. 4. The Magnetometer operation and sampling were identical to that for the data shown in Fig. 4. Magnetic boundaries are labeled as in Fig. 4 except that there is no CS boundary in this case

3.2 Plasma Observations

In addition to the Magnetometer, MESSENGER carries a plasma sensor, the Fast Imaging Plasma Spectrometer (FIPS), which measures ions in situ over the energy range from tens of eV to 13.5 keV as part of the Energetic Particle and Plasma Spectrometer (EPPS) instrument (Andrews et al. 2007). The FIPS detector provides coverage over an approximately 1.4π-steradian solid angle, although the useful field of view is somewhat lower because the instrument is obstructed by the spacecraft sunshade and one of the spacecraft's two solar panels. The center of the field of view is somewhat sunward of the perpendicular to the spacecraft-Sun direction. Plasma flow away from the detector look direction will not be measured, and hence in many situations FIPS will record only the high-velocity, super-thermal portion of the plasma distributions. One must therefore always be mindful that differences in flow direction can lead to variations in the FIPS data that are not indicative of corresponding changes in plasma density or temperature.

FIPS proton observations are shown along with expanded views of the magnetic field data near CA in Figs. 6 and 7. The second panels of Figs. 6 and 7 show FIPS proton energy spectra integrated over the FIPS field of view to yield a phase-space density that is normalized to the maximum value in the plot. The third panels of Figs. 6 and 7 show the time series of total proton counts summed over all energies and angles. The phase-space density normalization is different for M1 and M2, but the proton counts are in absolute units, counts per 10 s.

The proton data change at all of the transitions noted from the magnetic field data. At TL for M1, the proton fluxes below 1 keV decreased in approximately two steps, the first at 18:58 UTC prior to TL when the field magnitude began to increase more rapidly, and then

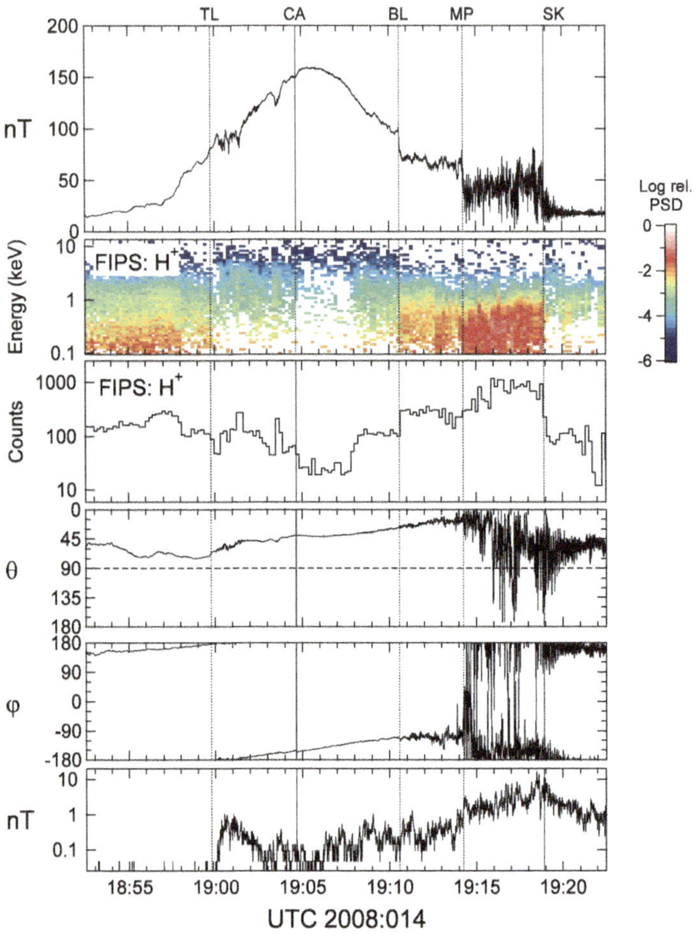

Fig. 6 Magnetic field data together with proton observations from FIPS for 30 minutes spanning the M1 encounter. Transition labels are the same as in Fig. 4. Data between points TL and BL were selected for estimating the internal magnetic field. In addition to the magnetic field data, which are re-plotted in expanded view from Fig. 4, the FIPS proton data show phase-space density (PSD) normalized to the maximum in the interval in *the second panel* and the summed proton FIPS counts in each 10-s integration in *the third panel*

again at TL when the lower-energy fluxes dropped to near the detection threshold. The presence of protons at hundreds of eV prior to TL is consistent with spacecraft residence near the cross-tail current sheet and suggests that the spacecraft remained in the tail plasma sheet until the point labeled TL. For M2 the proton fluxes were very low until TL, consistent with the interpretation that the spacecraft entered directly into the southern tail lobe where densities are expected to be low. Thus, for M2 the TL transition corresponds to the appearance of protons above ∼1 keV in the FIPS data.

Changes in the proton count rates and/or energy spectra occur at the boundary layer, magnetopause, and shock crossings on the outbound legs of both flybys. An increase in the FIPS protons at BL occurred in both cases, though it was greater for M1 than M2. The strong increase in protons at energies below 1 keV at the outbound magnetopause is obvious in both flybys as is the subsequent decrease in signal at the outbound shock crossing. In the

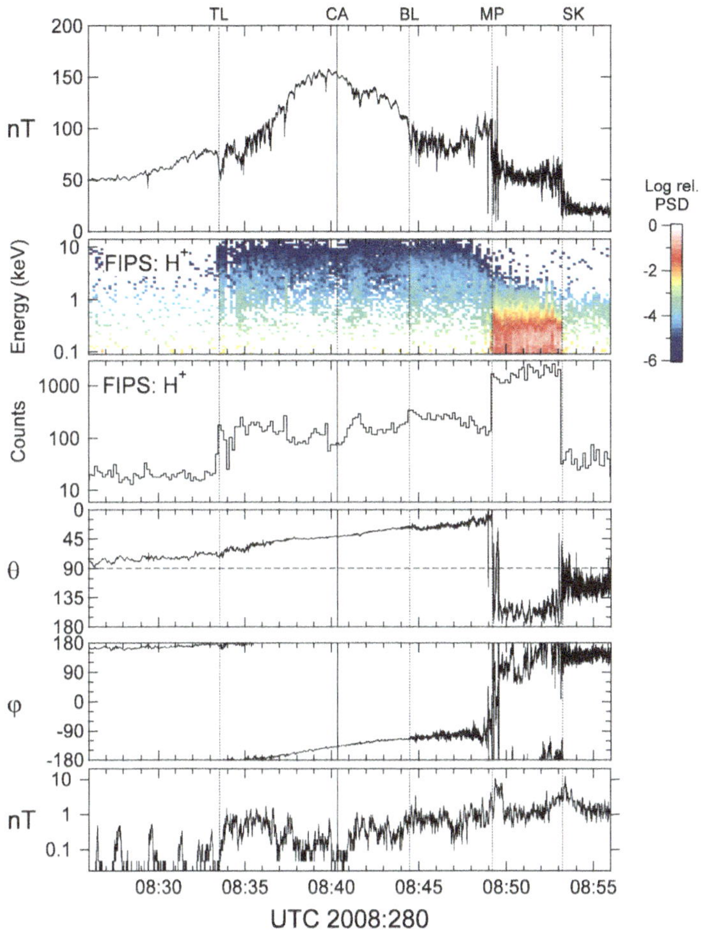

Fig. 7 Magnetic field data together with proton observations from FIPS for 30 minutes spanning the M2 encounter. Transition labels are the same as in Fig. 5. Data between TL and BL were selected for estimating the internal magnetic field. The panels are shown with the same format as in Fig. 6

solar wind just upstream of the bow shock, FIPS did not observe the incident solar wind flow but rather protons reflected upstream from the shock. The changes in the protons therefore correspond with the transitions identified in the Magnetometer data and support the choice of data between TL and BL for analysis of the internal magnetic field. Significantly, they also reveal that ~keV-proton populations permeate Mercury's magnetic field environment near the equator as close to the surface as the M1 and M2 closest approach altitudes of 200 km.

3.3 Implications of Plasma and Solar Wind Environment for Internal Field Estimates

We now focus on the intervals used for estimating the planetary field, between TL and BL in Figs. 6 and 7, paying particular attention to signatures in the data reflecting processes other than internal field sources. We note that both encounters showed sporadic decreases in the field magnitude without significant changes in field direction. The magnitude was generally more variable for M2 than for M1. For M1 there was one particularly prominent decrease in the magnetic field of about 10 nT about 1 minute before CA. This decrease in magnetic field magnitude corresponded to an increase in the FIPS proton count rate by a factor of three, suggesting that the drop in magnetic field was associated with a plasma structure. Prior to this point, from about 19:00 to 19:02 UTC, the magnetic field magnitude was somewhat variable, with multiple downward spikes of ~10 nT.

The field magnitude close to the planet was more variable for M2 than for M1. Nearly coincident with TL on M2 there was a decrease of ~25 nT or nearly one third in the field magnitude coincident with a factor of 5 increase in the proton count rate. The field magnitude varied erratically by up to 20 nT from TL to 18:36:30 UTC. Additional isolated decreases of ~25 nT at 08:37:30 UTC and again of ~15 nT just before 08:40 UTC both coincided with spikes in the proton count rate. A broader local minimum in the field near 08:41:30 UTC was coincident with a correspondingly broad increase in the FIPS proton count rate. The prevalence, intensity, and correlation of these short-timescale magnetic field decreases for both M1 and particularly M2 suggest that the magnetic field even close to the planet is affected by local plasmas and currents corresponding to the pressure gradients in the plasma. Thus, the standard assumption that the volume near the planet is free of local currents (e.g., Backus 1970) does not hold even within a few hundred kilometers of the surface.

There is another difference in character of the magnetic field observed during M2 relative to M1. Other than the relatively sharp field decreases discussed above, the magnetic field between TL and BL on M1 was fairly smooth, whereas for M2 the magnetic field exhibited a train of ~10-nT modulations each lasting ~10 to 20 s. The modulations were most evident after CA, when three or four of these oscillations occurred from ~08:42 to 08:44 UTC, in marked contrast to the smooth field gradient during M1 between CA and BL. As discussed above, the IMF conditions for M2 and M1 were different. The IMF was southward before and after MESSENGER's transit through the magnetosphere for M2, whereas it was northward at the outbound M1 magnetopause crossing. The M2 encounter displayed signatures of intense magnetic reconnection dynamics both in the tail and at the magnetopause with a repetition interval of about 20 to 30 s (Slavin et al. 2009b). The spike in the magnetic field magnitude just after MP, at ~08:49:30 UTC for M2, is the strong flux rope reported by Slavin et al. (2009b). It is possible that the quasi-periodic variations in magnetic field intensity during the M2 pass near the planet are also signatures of this dynamic interaction with the magnetized solar wind plasma.

The inference that local plasmas and the dynamic solar wind interaction make significant contributions to the magnetic field close to the planet has direct implications for estimates of the planetary field from the flyby data. Clearly, the identification of magnetic signatures due to local plasmas and perhaps magnetospheric dynamics implies that crustal signatures will be difficult if not impossible to distinguish from local plasma or dynamic signatures. Moreover, the MESSENGER observations imply that local plasmas and corresponding local magnetization currents are present close to the planet. These local plasmas contribute up to 30-nT signals in the volume of space from which the near-planet magnetic field data were obtained. Thus, one cannot assume that the data were obtained in a current-free volume, and spherical harmonic solutions for the external field must be used with caution because they are strictly applicable only to curl-free fields. We emphasize that the short time-scale plasma signatures indicate the presence of plasma even though we do not attempt to model the short-wavelength signatures of these local plasma phenomena. Large-scale external currents are strongly indicated by the presence of local plasmas, and these long-wavelength signatures are of concern for the spherical harmonic inversion analyses.

In addition, variations of 10 to 20 nT are present that are likely due to dynamic processes and do not reflect spatial structures. That these signatures are found on the M2 pass for which the magnetopause crossings gave evidence for strong, episodic reconnection dynamics suggests that these quasi-periodic oscillations could be due to the dynamic interaction of the magnetosphere with the solar wind. It appears that these signatures permeate the system, and therefore estimates of Mercury's magnetic field will necessarily be subject to residuals of this order. Obviously, fitting these signals as if they were spatial structures in the internal field would be incorrect.

In estimating the internal magnetic field structure of Mercury one must therefore adopt an approach that recognizes the intrinsic limitations that the physical system imposes. First, although there are several obvious common elements between Mercury's magnetosphere and those of other planets, particularly Earth, Mercury's magnetosphere is unlike any system explored to date in many key respects that prevent us from accurately specifying the external field (e.g., Glassmeier 2000). Thus, we use available knowledge to correct for the portions of the external field in which we have the most confidence and then examine the data for signatures of additional current systems that must be considered specifically for the Mercury system. As discussed below, simulation results play a role in guiding this analysis. Second, given the intrinsic limitations of the data in hand and the challenges presented by the Mercury system, the use of higher-order terms in estimating the internal field must be treated with caution. For this discussion, we adopt the view that higher-order terms should be invoked if there is a reason, supported either from observations or simulations, to exclude an external or local current structure as accounting for the observations. Given the nature of the external and local sources of magnetic field, we judge that this posture is necessary to avoid drawing erroneous conclusions about Mercury's internal field.

4 Critical Assessment: External Field Treatment

For Mercury, the importance of external and local sources of magnetic fields for accurate assessment of the internal planetary field is surpassed perhaps only by the difficulty in deriving accurate representations of these external and local sources. Moreover, because of their importance and the evidence for local sources of current even at the lowest altitudes, it is particularly important when including external field corrections for Mercury to be guided by the physics of the magnetosphere and its solar wind interaction. There are three techniques that can be used to estimate the external field. The first is to use a potential formalism that treats the external field much as the internal field, but due to sources outside the sampled volume. The second applies analytical empirical models that use a set of specified current systems that are constrained empirically in location and intensity by observations. Finally, advances in computational capability allow one to contemplate the use of physics-based simulations that obtain the magnetospheric and plasma structures and currents from numerical simulation of fluid and/or particle equations of motion.

4.1 Spherical Harmonic Fitting

One approach commonly used to estimate external field contributions is to use spherical harmonic analysis (SHA) for sources outside the sample volume (Backus 1970; Menke 1989). While this approach is convenient, it makes the implicit assumption that the sample volume is current-free. This cannot be safely assumed for Mercury. Moreover, even with four flybys, the sampling of the magnetospheric volume at Mercury is quite limited. All of the passes are nightside cuts through the system, with M10-I, M1, and M2 passing near the equator and M10-III over the northern pole. The addition of the M2 data require that the external and internal fields are treated in MSO and Mercury body-fixed (MBF) coordinates, respectively, in the same singular value decomposition (SVD) inversion (Uno et al. 2009). Although the M2 trajectory sampled a new range in planetary longitude (Fig. 2), the path in magnetospheric coordinates was similar to that for M1 (Fig. 3). Because the flyby trajectories in MSO coordinates are similar, the constraints on the external currents provided by these data are not particularly robust. Estimates of the external field using SHA yield quite different solutions depending on the choice of maximum spherical harmonic degree for the external field and whether each flyby is considered to have different external fields (Uno 2009). These considerations together with the limitation that the region is not current-free make it difficult to judge the validity of SHA external field solutions. For this discussion, aimed at distinguishing those aspects of Mercury's magnetic field structure that are reasonably robust from those requiring additional study, we choose not to focus on the spherical harmonic approach in dealing with the magnetic field from external sources.

4.2 Magnetospheric Current Models

Fortunately there are fundamental similarities in primary current systems of planetary magnetospheres (cf. Parks 1991, and references therein). The magnetopause and magnetotail currents are necessary consequences of the interaction of the supersonic solar wind with the planetary magnetic field, and all planetary magnetospheres in the solar system have these current systems. The magnetopause current system, or Chapman-Ferraro current layer, separates the magnetic field of the planet from the solar wind environment. For the magnetized planets of our Solar System, the solar wind flow speed is higher than the fast magnetosonic wave speed in the solar wind plasma, so a shock front forms upstream of the magnetopause where the solar wind flow decelerates at a shock bow wave and then diverts around the magnetic obstacle of the planetary field according to magnetohydrodynamics. These processes are generally understood, both observationally and theoretically. Given the shape of the magnetopause boundary and an approximate estimate of the planetary magnetic field (e.g., Sibeck et al. 1991; Shue et al. 1998), one can calculate the magnetopause current and the resulting externally produced magnetic field from first principles (Tsyganenko 1995; Tsyganenko and Sitnov 2005).

The magnetotail current system is the second primary current system directly implied by the structure of the magnetosphere and common to all planetary magnetospheres in the solar system (cf. Lui 1987, and references therein). The north and south polar magnetic flux of the planet is swept in the anti-sunward direction by the solar wind flow to form a pair of magnetic lobes (cf. Fig. 1). In the lobe connected magnetically to the southern magnetic pole, the magnetic field is directed toward the planet while in the other lobe the field is directed away from the planet and is linked to the northern magnetic pole. This configuration requires that there be a current flowing across the magnetotail between the lobes approximately bisecting the magnetotail. For Mercury the south magnetic pole is in the north, so the magnetic field is sunward (anti-sunward) in the lobe magnetically linked to the northern (southern) hemisphere.

The two-lobe magnetotail forms an approximately cylindrical structure that is observed to extend at least 30 times farther in the anti-sunward direction than the sub-solar magnetopause standoff distance measured from the planet center. The magnetotail diameter and the precise location of the cross-tail current sheet both vary between different systems. For systems with a significant tilt between the planetary orbital plane and the magnetic dipole axis, the cross-tail current will be displaced out of the orbital plane toward the pole that points sunward. While all magnetospheres have a magnetotail and a cross-tail current, the north-south location, tilt, intensity, extension toward the planet, and dynamics vary.

Because the size of a magnetosphere scales directly with the planetary dipole moment, one can use empirical models obtained from Earth, for which we have the most in situ observations, to scale quantitatively to other systems (e.g., Luhmann et al. 1998). This scaling applies best for the magnetopause currents, whose location and intensity can be most accurately specified. The magnetotail currents, though certain to exist, are less reliably scaled with this approach. But one can at least use our present estimates of these currents for comparison with the Mercury system.

Other current systems have been identified but are not universally present. Currents associated with internal plasma distributions, e.g., Earth's ring current, or with magnetospheric convection, e.g., Earth's Birkeland and ionospheric currents, vary markedly among systems. For Mercury, we anticipate that the Earth analogy breaks down completely with respect to the ring, Birkeland, and ionospheric currents (Glassmeier 2000). The ring current depends on closed particle drift trajectories around the planet, but Mercury's magnetosphere is so small relative to the planet that no such drift trajectories are expected (Russell et al. 1988). The absence of an ionosphere at Mercury means that Earth analogs for field-aligned currents, which close in the ionosphere, may not apply at Mercury (Slavin et al. 1997; Ip and Kopp 2004).

There are two alternatives for estimating Mercury's external currents with analytical models. The first is to use those portions of Earth models that we have confidence apply to Mercury, namely the magnetopause and tail currents, and scale the models to Mercury (e.g., Luhmann et al. 1998; Korth et al. 2004). The second alternative is to develop analytical and empirical models unique to Mercury (Giampieri and Balogh 2001; Alexeev et al. 2008). Further development of these Mercury models is essential to take full advantage of data to be obtained from orbit around Mercury, first by MESSENGER (Solomon et al. 2007) and then by BepiColombo (Balogh et al. 2007). However, the reliability of these models depends critically on the available data for the Mercury system, which at present are quite limited. We use an Earth analog as a first step in the analysis to guide the identification of additional currents unique to Mercury.

4.3 Physics-Based Simulations

Numerical simulations of Mercury's magnetosphere can inform the derivation of the internal planetary magnetic field in at least two ways. First, they can be used to guide the analysis and interpretation of observations and in that way assist in identification and specification of processes and structures unique to Mercury. Second, they may find application in the inversion of observations to identify higher-order structure in Mercury's magnetic field. This use has already found application in the analysis of the Ganymede system embedded in the Jovian magnetosphere (Jia et al. 2008).

Advances in computational speed have opened new opportunities for first-principles simulations of space plasma systems, and the Mercury magnetosphere system has received particular attention. The fluid approximation using the formalism of magnetohydrodynamics (MHD) has been applied to Mercury with considerable success (e.g., Ip and Kopp 2002). Other investigators have applied a hybrid fluid-kinetic formalism in which kinetic particle simulations are used for ions while the electrons are treated as a fluid (e.g., Trávníček et al. 2007). Fluid MHD simulations offer the advantage of speed and extensive heritage in application to a range of systems, while the hybrid simulations include particle transport processes that may be critical to the Mercury system (e.g., Baumjohann et al. 2006). There is value in applying independent simulation codes using different approaches to the same system. Comparing results from the MHD and hybrid simulations may allow one to identify which processes can be explained from a purely fluid perspective and which depend on ion kinetic processes.

Simulations have already informed our understanding of Mercury's magnetotail and distributions of plasma within the magnetosphere. Both the hybrid and fluid simulations show that Mercury's small magnetic field leads to incidence of solar wind plasmas on the surface in broad regions around magnetic cusps and in equatorial regions on the night side of the planet where solar wind plasma "precipitates" onto the surface (Ip and Kopp 2002; Trávníček et al. 2007, 2009). The hybrid simulations indicate that solar wind plasmas permeate Mercury's magnetosphere even to very low altitudes in the cusp and equatorial regions at least in part following entry in the cusps followed by conventional drift on closed field lines (Trávníček et al. 2007). This process appears to result in structured inclusions of solar wind plasma at very low altitudes, down to the surface, within a planetary radius and near the equator on the nightside (Trávníček et al. 2007). These results are consistent with the MESSENGER magnetic field and plasma observations discussed above. The hybrid simulations also reveal displacement of the cross-tail current toward the north for the negative IMF B_X conditions for M1 and M2 (Trávníček et al. 2007, 2009). Thus, the simulations are already providing a conceptual framework to assist in the interpretation of the available observations.

5 Internal Field Estimation

The MESSENGER observations from the first and second flybys provide the first additional data on the planetary field since the Mariner 10 encounters, and the M2 encounter gave us our first observations of the magnetic field in the western hemisphere of the planet. The MESSENGER encounters are particularly useful because relatively unperturbed data were obtained throughout each encounter, in contrast to the first Mariner 10 flyby, for which a magnetospheric disturbance occurred just after closest approach so that only the inbound portion of that pass can be used in internal field estimations (Ness et al. 1974; Christon 1987). Moreover, closest approach for both MESSENGER flybys was at 200 km altitude, lower than the Mariner 10 encounters. Nonetheless, the third Mariner 10 encounter observations are perhaps the most central to our understanding of Mercury's magnetic field since this flyby offers the only in situ observations to date from the polar regions of the planet.

5.1 Moment Inversions

Following the M1 encounter, Anderson et al. (2008b) assessed the internal field and showed that a pure-dipole representation underestimates the field over the pole while overestimating it near the equator. They compared results for different approaches to correcting the external field and concluded that the planetary moment is most likely in the range 230 to 290 nT-R_M^3. The residuals remained relatively high, 15 to 30 nT, relative to typical planetary moment inversions, but consistent with the signatures of dynamics and local currents. Subsequent analyses assessed the higher-order internal field structure that the data may imply, revealing that the planetary field appears to be dominated by the g_1^0 term (Uno et al. 2009), where g_n^m is the spherical harmonic coefficient for the nth order and mth degree. A search for specific

Table 1 Inversion results for Mercury's magnetic field using observations from Mariner 10 flybys I and III and MESSENGER flybys M1 and M2

Internal model	External field	g_1^0	g_1^1	h_1^1	g_2^0	g_2^1	g_2^2	h_2^1	h_2^2	Residual (nT)	Condition number
1 Dipole	None	−216	−6	14						42	2
2 Dipole	TS04	−240	−1	5						29	2
3 Dipole[a]	SHA	−249	−12	16						30	7
4 Quad.	None	−173	−7	15	−108	−9	−1	16	−17	19	3
5 Quad.	TS04	−213	−4	7	−66	9	4	5	−4	14	3
6 Quad.[a]	SHA	−182	−15	9	−108	10	2	6	−15	15	12
7 Reg.[b]	TS04	−222	12	2	−24	9	9	−6	8	24	n/a
External terms		G_1^0	G_1^1	H_1^1	G_2^0	G_2^1	G_2^2	H_2^1	H_2^2		
3 Dipole[a]	SHA	47	26	8	10	−15	−3	−2	−8		
6 Quad.[a]	SHA	7	−4	−15	−9	−9	−3	2	0.4		

All coefficients are in units of nT for spherical harmonic expansion with distance normalized to a mean Mercury radius. Quad. denotes quadrupole. The coefficients g_n^m and h_n^m are the cosine and sine spherical harmonic coefficients, respectively, of order n and degree m for the terms that decrease with radial distance, hence for the internal sources. The G_n^m and H_n^m are the cosine and sine spherical harmonic coefficients, respectively, for the terms that increase with radial distance, hence for the external sources (cf. Menke 1989)

[a]Results for the spherical harmonic analysis (SHA) treatment for the external field are from Uno (2009)

[b]Results for the regularized solution are from Uno (2009). The g_n^0 terms in the regularized solution are as follows: $g_3^0 = -2$; $g_4^0 = -4$; $g_5^0 = -5$, $g_6^0 = 0$; $g_7^0 = 1$; $g_8^0 = 0$

crustal magnetic field signatures found that the perturbations near CA were not consistent with a crustal magnetization signature (Purucker et al. 2009).

The addition of the M2 data allows an assessment of long-wavelength longitudinal structure in the planetary field. Uno (2009) added the M2 observations to both spherical harmonic analyses in which internal and external fields were co-estimated using SVD (see Sect. 4.1) and to regularized inversions for the internal field after removal of external fields predicted by TS04. The regularized solution yields a dominantly dipolar field, aligned to within 5° of the planetary rotation axis. The results both for the SVD inversions and the regularized solution are summarized in Table 1 together with inversions added here for comparison. Here we add dipole and quadrupole SVD solutions that use either no external field correction or the TS04 external field correction.

The dipole fit results, from internal models 1 through 3 in Table 1, are very similar to previous results after M1 with the difference that the tilt of the dipole from the spin axis is now smaller. The Anderson et al. (2008b) dipole moment fit using the TS04 correction was 229 nT-$R_{\rm M}^3$ with a tilt of 9°, and the SHA external field solution gave a moment of 247 nT-$R_{\rm M}^3$ with a tilt of 12°. The new result using the same inversion, dipole with TS04 correction, gives a dipole moment of 240 nT-$R_{\rm M}^3$ and a tilt of 1°. The SHA external field yields a dipole moment of 250 nT-$R_{\rm M}^3$ and a tilt of 5°. Accounting for the external field even in these approximate ways reduces the residuals.

The results for higher-order terms, models 4 through 7 in Table 1, are also consistent with the previous analyses. The quadrupole solutions and the regularized degree and order

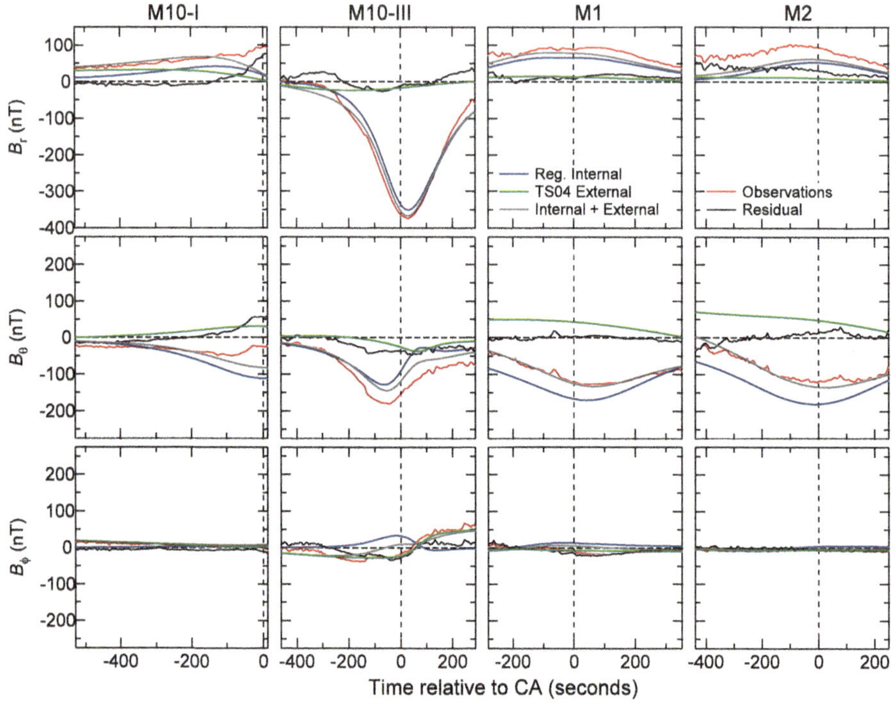

Fig. 8 Overview of Mariner 10 and MESSENGER observations of Mercury's magnetic field. Results are plotted in MBF coordinates: radial (B_r), polar angle $\theta(B_\theta)$, and azimuth angle $\phi(B_\phi)$ versus time relative to closest approach (CA). *Lines* show observations (*red*), internal field model from regularized solution (*blue*), TS04 external field model (*green*), sum of the internal and external models (*grey*), and residuals (*black*). The span of magnetic field values plotted is 550 nT for all three components

8 inversion have lower dipole moments and a quadrupole moment with a magnitude 30% to 60% of the dipole. The quadrupole and regularized inversions that use data corrected for the TS04 external field have smaller higher-order terms than the other two quadrupole inversions. The sum of all g_n^0 from $n = 1$ to 8 for the regularized solution is -255 nT, lower than the sum of g_1^0 and g_2^0 for the quadrupole solutions, 4, 5 and 6, which are -281 nT, -271 nT, and -290 nT, respectively. The residuals for the higher-order fits are 10 to 15 nT lower than the external-field-corrected dipole inversions but are still between 7% and 11% of the dipole term. Note that the magnitude of the residuals for the regularized solution is determined by the weights used in the inversion, and these are conservative (i.e., large) reflecting mainly contributions from uncertainties in the external field correction and short wavelength signals (Sect. 3.3, and see Uno 2009; Uno et al. 2009).

5.2 Residuals: Initial Assessment

We now examine the inversions in detail to understand what features in the data lead to the quadrupole terms and identify the factors contributing to the residuals. The observations, external and internal models, and residuals are shown in Fig. 8 for the regularized solution, model 7, of Table 1. The data are shown in $r-\theta-\phi$ MBF coordinates versus time in seconds relative to CA for each of the flybys. The residuals are shown in Fig. 9 in a similar format with the addition of the bottom row. The net model does a good job of representing the

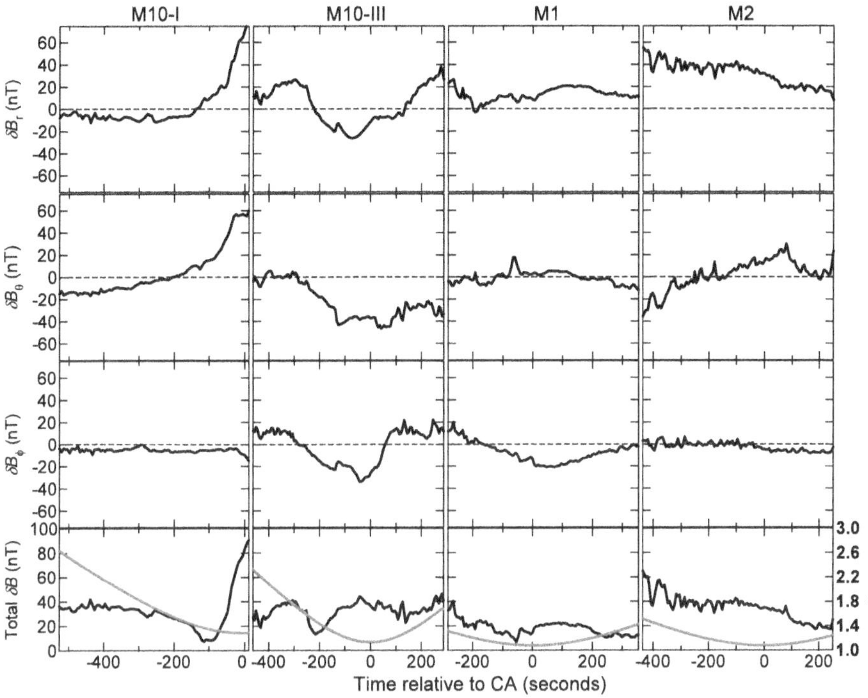

Fig. 9 Magnetic field residuals in MBF spherical coordinates radial (δB_r), polar angle θ (δB_θ), and azimuth angle ϕ (δB_ϕ), for the Mariner 10 and MESSENGER flybys. *In the bottom row* the residual magnitude (δB) is plotted (left-hand axis) together with the planetocentric distance (right-hand axis). Residuals are evaluated relative to the regularized solution with the TS04 external field correction. *Gray traces* show spacecraft radial distance from the planet in units of planetary radii

dominant field components for M10-III, M1, and M2 but cannot fit the M10-I data near closest approach, for which the residuals are nearly 50 nT. In all cases, the TS04 model field is a slowly varying contribution and is everywhere less than ~60 nT. The B_ϕ results for M10-III give the appearance of a TS04 external field that reverses sign shortly after CA, but this is just a coordinate transformation from a uniform field into the azimuthal direction, which reverses as the trajectory crosses near the pole. The observed and modeled B_ϕ for M10-I, M1, and M2 are quite small. Finally, we note that the M10-I B_r and B_θ and the M10-III B_ϕ residuals are largest near CA but that none of the other residuals are largest near CA. The largest residual for M10-I is δB_r near CA, but for M10-III the largest residual is in δB_r about 300 s prior to CA, for M1 it is in δB_θ also at the beginning of the interval about 300 s prior to CA, and for M2 it is in δB_r more than 400 s before CA. The lack of dependence of the residuals on radial distance is also evident in the bottom row of Fig. 9, from which it is difficult to discern a correlation between radial distance and residual magnitude.

5.3 Residuals: Spatial Distributions

To illustrate how the data and trajectories are related we display the data using a different format in Fig. 10. The figure shows the observations in MBF coordinates (top panels) and the residuals from the dipole fit (bottom panels), model 1 of Table 1, to the data without making any external field correction, also in MBF coordinates. The trajectories are shown

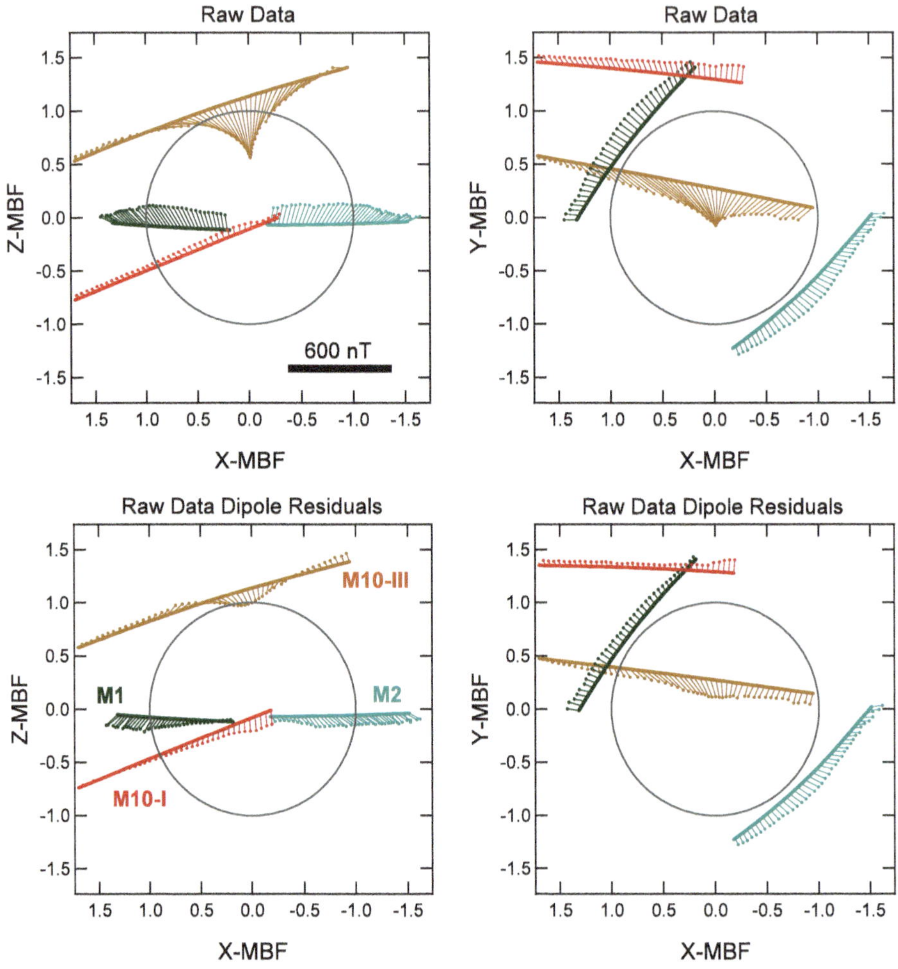

Fig. 10 Overview of Mariner 10 and MESSENGER magnetic field data used for estimating the planetary internal magnetic field. Data and trajectories are shown in MBF coordinates. *Arrows* show the field measured at the point on the trajectory where the arrow originates on the trajectory. The field is projected onto the plane viewed, the X–Z MBF plane *on the left* and the X–Y MBF plane *on the right*. Color coding is as follows: Mariner 10 I (M10-I) is shown in *red*; Mariner 10 III (M10-III) is shown in *tan*; MESSENGER flyby 1 (M1) is shown in *dark green*; MESSENGER flyby 2 (M2) is shown in *light blue*. *The top panels* show the observations prior to any corrections for external or internal field sources. *The bottom panels* show the residuals relative to a centered dipole fit to the observations (model 1 of Table 1). The length of an arrow corresponding to 600 nT is indicated in *the top left panel*

in units of R_M, and the magnetic field data are shown as lines starting from the trajectory in the direction of the field projected onto the plane of the plot. The M10-III flyby data clearly show the magnetic field directed toward the planet over the pole but also have a significant horizontal field over the pole. The M1 and M2 data yield a northward-directed field near the equator but also show a radially outward field, as do the M10-I flyby data. The M10-I observed field is not strongly northward even near closest approach but is primarily radially outward, even nearest the planet.

The departure of the observations from the dipole, model 1 of Table 1, illustrates that a pure dipole cannot fit both the high-latitude and equatorial observations. The M10-III residual from the dipole is still southward over the pole, whereas the M10-I, M1, and M2 residuals are all southward, indicating that the dipole is too weak at the pole and too strong at the equator relative to the observations. The fact that the M10-III observations are too large relative to the M10-I and M1 observations near the equatorial plane to be explained in terms of a dipole was pointed out by Anderson et al. (2008b). The M2 observations confirm that the equatorial field is consistently low even for contrasting IMF conditions. From the Y–X plane view, we see that the radial equatorial field is present in the M10-I, M1, and M2 residuals and that the residual field at M10-III is still significantly horizontal.

To illustrate the role of the TS04 external field correction, in Fig. 11 we show the residuals for fits 1 and 2 of Table 1. The dipole fit residuals without and with the TS04 external field correction are shown in the upper and lower panels, respectively. Considering the Z–X plane first, we see that the external model reduces the southward residuals for M10-I, M1, and M2 but makes the field more northward at M10-III owing primarily to the stronger dipole moment of this fit. The signature of a residual southward field at M10-III near CA persists. In the Y–X plane, the horizontal residual at M10-III is much smaller in the TS04-corrected result, particularly before and after CA. The radial residuals in the equatorial plane are reduced but are still strong for M1 and M2.

Corresponding results of the higher-order internal models are shown in Fig. 12 where we plot the residuals for models 5 and 7 in the same format as Fig. 11. Comparing the TS04-corrected dipole solution (bottom panels of Fig. 11) and TS04-corrected quadrupole solution (top panels of Fig. 12) we see that the quadrupole term resolves the radial residuals in M1 and M2 (Y–X plane) and accounts for the residual B_Y and B_Z at M10-III near CA, but underestimates the northward field at M1 and M2. The regularized TS04-corrected solution (bottom panels of Fig. 12) yields lower north-south residuals at M10-III, M1, and M2 but gives larger residuals in the radial fields at M1, M2, and M10-I near CA and also in B_Y at M10-III. It appears that regularized higher-order solutions cannot simultaneously reduce the residuals in the radial and north-south directions in the equatorial plane, and this was confirmed to be a common characteristic of the inversions via a series of experiments using different choices of weights and misfit levels. The M10-I observations near CA cannot be explained by any of the models. There is also a suggestion of a horizontal component, B_Y in MBF coordinates, in M10-III observations over the pole that cannot be fit with higher-order internal field terms.

That the residuals are not ordered by radial distance (cf. Fig. 9) suggests that their spatial distribution should be considered in a coordinate system appropriate to the external current systems. In Fig. 13 we plot the residuals for models 5 and 7, TS04-corrected quadrupole (top panels) and regularized (bottom panels), in the same format as Fig. 12 but in MSO rather than MBF coordinates. The left-hand panels show the view looking toward the Sun, and the right-hand panels show the view looking southward from above the north pole.

We first focus on the M10-III residuals in the X–Y plane. The M10-III residuals in the X–Y plane are sunward and are somewhat localized to the polar region. These polar-region sunward residuals are indicative of a tilt in the magnetic field such that the lines of force are pulled tailward in a localized region over the polar cap, consistent with magnetospheric convection (e.g., Slavin et al. 2009b) and equivalent to the linked Birkeland field-aligned and horizontal ionospheric current system at Earth (e.g., Cowley 2000; Richmond and Thayer 2000; Anderson et al. 2008a). The possibility of such a system at Mercury has been proposed and is remarkable given that there is no ionosphere to carry the current as readily as at Earth (Glassmeier 1997; Slavin et al. 1997; Ip and Kopp 2004).

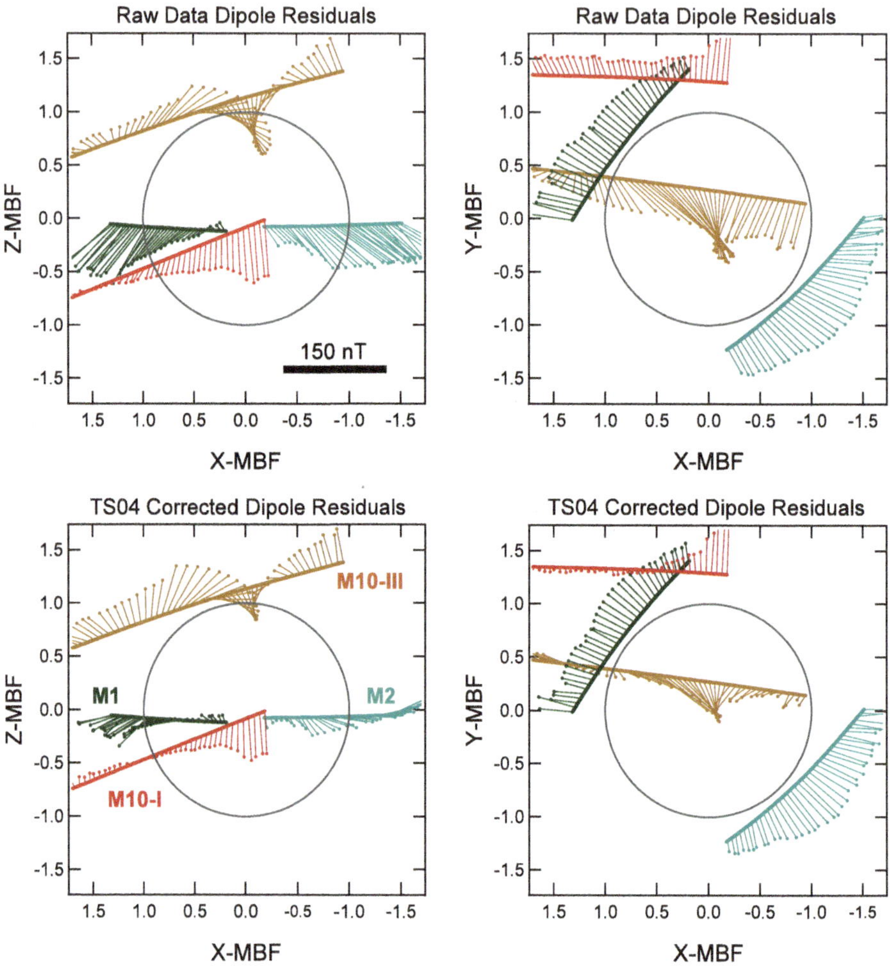

Fig. 11 Overview of Mariner 10 and MESSENGER magnetic field data used for estimating the planetary internal magnetic field in the same format as Fig. 10 except that the length of an arrow corresponding to 150 nT is indicated in *the top left panel*. Data and trajectories are shown in MBF coordinates. *The top panels* show the residuals relative to a centered dipole fit to the observations (model 1 of Table 1), and *the bottom panels* show the residuals relative to a centered dipole fit to the observations that are first corrected for the TS04 external field (model 2 of Table 1)

Considering the M10-I, M1, and M2 residuals, it is evident that all of these data come from the nightside, so that these observations will be strongly influenced by the intensity and location of the cross-tail current system. The pronounced reversal in δB_X and δB_Z near CA for M10-I could reflect a sudden intensification of the cross-tail current. The data used here and selected for internal field modeling are those prior to the strong dynamic variations in the field that have been interpreted as a major substorm (Ogilvie et al. 1977; Christon 1987). The changes in δB_X and δB_Z could reflect the growth phase of the substorm, which at Earth is associated with both motion and intensification of the cross-tail current (cf. Parks 1991). The change observed by M10-I occurred over about 30 s, consistent with the magnetospheric convection timescale at Mercury (Christon 1987; Slavin et al. 2009b).

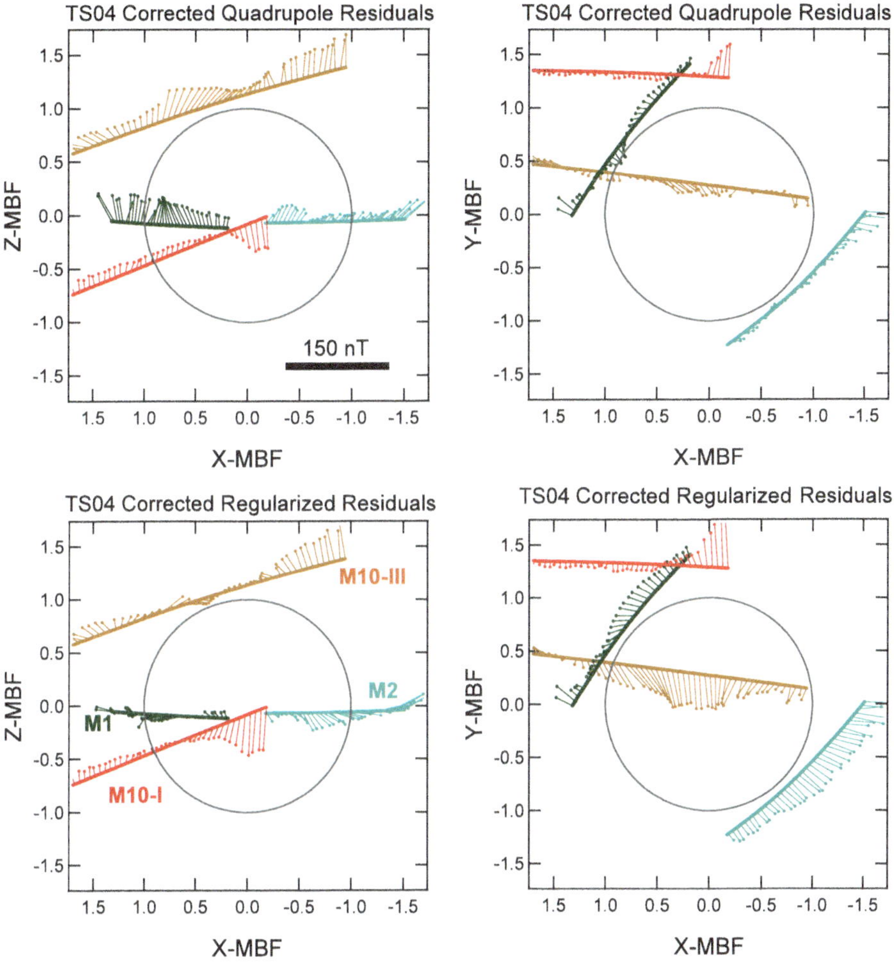

Fig. 12 Overview of Mariner 10 and MESSENGER magnetic field data used for estimating the planetary internal magnetic field in the same format as Fig. 11. The length of an arrow corresponding to 150 nT is indicated in *the top left panel*. Data and trajectories are shown in MBF coordinates. *The top panels* show the residuals relative to a quadrupole fit to the observations, equivalent to an offset dipole, and *the bottom panels* show the residuals relative to the regularized solution of Uno (2009). All solutions are fit to observations that are first corrected for the TS04 external field

The M1 and M2 observations also indicate variability in the tail current system. The δB_X residuals for M1 and M2 are systematically different, with δB_X consistently stronger for M2 than for M1. As noted above, during M1 the MESSENGER spacecraft passed initially into the cross-tail current sheet and then into the plasma sheet of the southern tail lobe, whereas for M2 the spacecraft entered directly into the southern tail lobe. From the Z–Y plane view, however, we see that the trajectories are almost identical relative to the mid-plane of the tail, with both nearly in the $Z = 0$ plane. Thus, the current sheet must have been displaced northward for both M1 and M2, and for M2 the current sheet either was farther northward or the plasma sheet and current sheet were thinner. In addition, the cross-tail current may have been stronger for M2 than for M1.

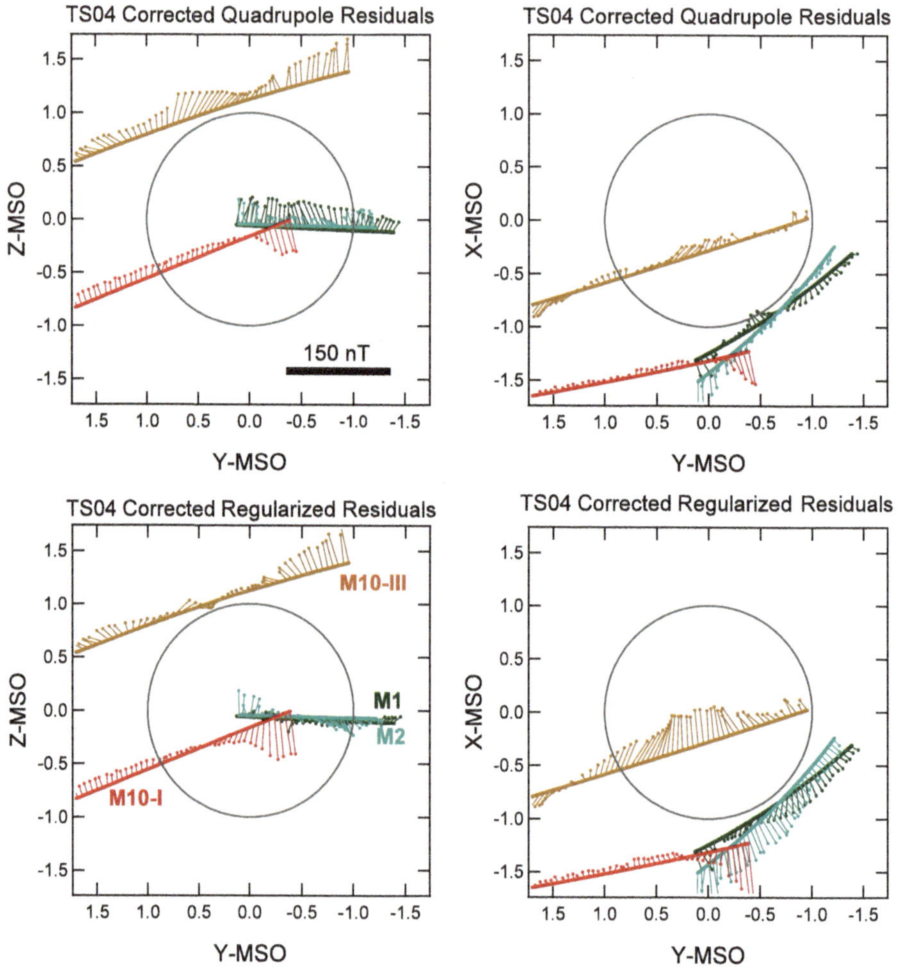

Fig. 13 Overview of Mariner 10 and MESSENGER magnetic field data used for estimating the planetary internal magnetic field in the same format as Fig. 11 but with the data shown in MSO coordinates. *The top panels* show the residuals relative to a quadrupole fit to the observations, model 5 of Table 1, equivalent to an offset dipole, and *the bottom panels* show the residuals relative to the regularized solution of Uno (2009), model 7 of Table 1. All solutions are fit to observations that are first corrected for the TS04 external field

5.4 Implications for Magnetospheric Currents

Since the structure and dynamics of external currents may be responsible for a major portion of the residuals, we now examine the data primarily to assess the magnetospheric current systems. In this we are guided by the hybrid simulations, which indicate an annulus of solar wind plasma within about $0.5R_{\mathrm{M}}$ altitude that extends around the nightside of the planet (Trávníček et al. 2007, 2009). This annulus has a radially inward pressure gradient, which implies an electric current $\mathbf{J} = (\mathbf{B} \times \nabla P)/B^2$, directed from dawn to dusk at midnight, where $\mathbf{B}$ is the vector magnetic field and P is the scalar plasma pressure. The MESSENGER magnetic field and plasma observations presented above, Figs. 6 and 7, provide confirmation that solar wind plasmas are present close to the planet with densities sufficient to significantly perturb the magnetic field. It is possible then, that departures from dipole signatures, particularly in the equatorial plane, may be due primarily to local magnetospheric currents rather than higher-order structure in the internal field. Thus, we now adopt the simplest possible interpretation for the internal field and assess what the remaining signatures imply if considered as due to external and local currents only.

Consider those magnetic sources in the Mercury system that can be identified with greatest confidence. Obviously, the planet possesses at least a dipolar magnetic field and therefore also magnetopause and magnetotail current systems. The higher-order inversions all yield a moment nearly aligned with the planetary rotation axis. Moreover, we are fairly sure that plasmas in the equatorial plane are present and influence the measurements made there. Under this conservative view, it may be that our best measure of the internal dipole is given by the M10-III flyby data. This inference is supported by the hybrid simulations, which do not indicate significant plasma densities over the polar region. We then corrected the M10-III data only for the magnetopause and distant tail currents using the TS04 model as implemented here, and fit a centered, axially aligned dipole to the corrected M10-III data only. This gives a moment of 266 nT-R_{M}^3.

With this simplest possible model for the internal field we consider whether the remaining residuals can be understood in terms of external currents. The residuals of all of the data relative to the M10-III dipole fit are shown in Fig. 14. The top panels show the residuals without an external field correction, and the bottom panels show residuals after subtracting the TS04 external field. The residuals in the bottom panel should reflect uncorrected signatures of external currents relatively close to the planet under the assumption that the signatures of these currents dominate the residuals.

The equatorial passes all indicate radial and southward fields in the magnetotail close to the planet. For M1 and M2 toward the dawn terminator, the external field was directed somewhat more radially than tailward, suggesting that the cross-tail current may wrap around the planet in a manner analogous to the inter-relationship of the Earth's tail and ring-current systems (e.g., Tsyganenko 1995). The southward signatures in M10-I, M1, and M2 near the planet indicate that there is a dawn-to-dusk current just tailward of the spacecraft, perhaps as close as $0.5R_{\mathrm{M}}$ altitude. Hybrid simulations for M1 yielded a neutral sheet hosting such an azimuthal current north of the equator by almost $0.5R_{\mathrm{M}}$ at $1R_{\mathrm{M}}$ altitude, possibly in response to the X-component of the IMF (Trávníček et al. 2009). The IMF X-component was negative, anti-sunward for M2 and M1 (cf. Figs. 4 and 5), but positive for the inbound pass of M10-I, so it is not clear whether this mechanism can account for the tailward field near CA on the M10-I flyby.

We estimate the intensity of this near-tail azimuthal current as follows. If we assume that the M1 and M2 trajectories pass under the near-tail current, then the net change in magnetic field that it generates, δB, is twice the radial field observed, or about 100 nT, since the field switches sign across the current. The linear current sheet density is readily estimated as $\delta B/\mu_0 = 80$ mA/m. If the current sheet is $0.5R_{\mathrm{M}}$ thick, consistent with the hybrid simulations, the current density would be $\sim$0.1 μA/m^2, and if the sheet has a radial extent of $\sim 1R_{\mathrm{M}}$ the total current would be 2×10^5 A. For comparison we show the current density distribution of the TS04 model for nominal solar wind IMF conditions at Mercury in Fig. 15. The plot shows the Y component of the current density in the dawn-dusk meridian. The dayside and tail magnetopause currents are most prominent, but the cross-tail current is clearly evident. In this model, the cross-tail current is appreciable only beyond $X = -2R_{\mathrm{M}}$. The inferred current density for the near-tail current is well within the range of densities required to produce the TS04 tail configuration.

The residuals at M10-III are comparable to those near the nightside equator, so the axial dipole fit, even only to M10-III, is not sufficient to account for those data alone. The primary

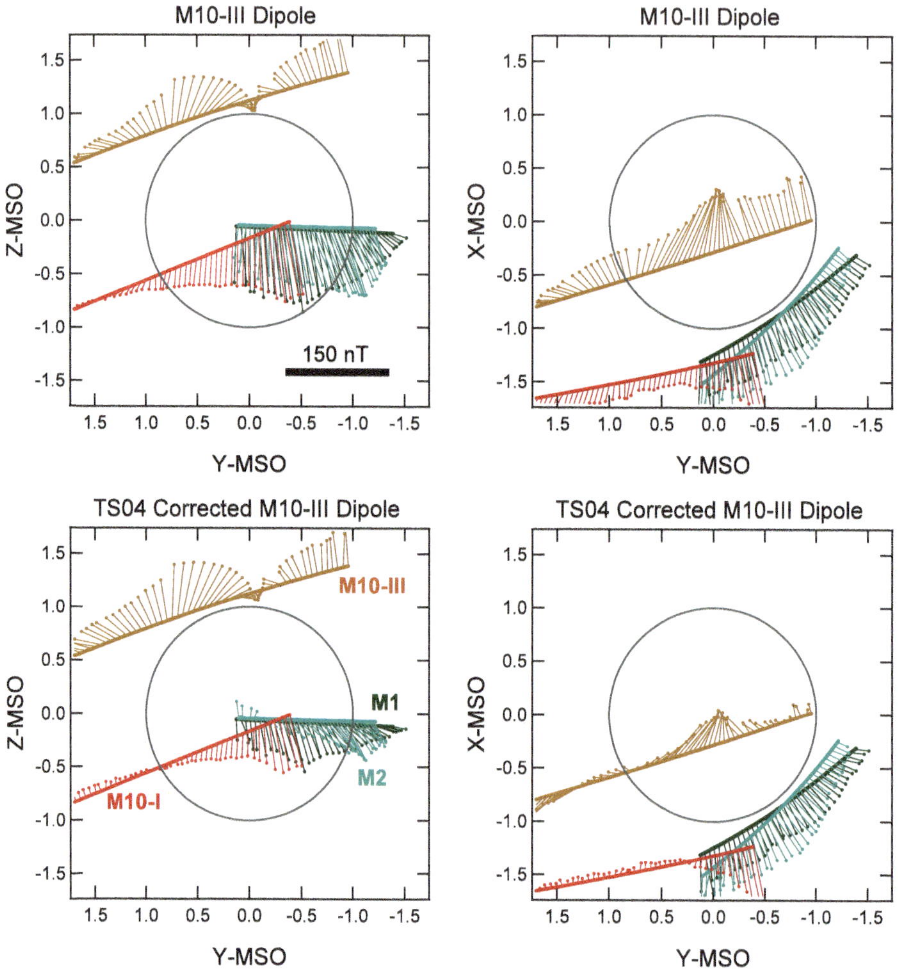

Fig. 14 Overview of Mariner 10 and MESSENGER magnetic field data used for estimating the planetary internal magnetic field in the same format as Fig. 13. *The top panels* show the residuals relative to a M10-III only, axially aligned dipole fit, and *the bottom panels* show the same residuals but also corrected for the TS04 external field

departure is a predominant northward residual. To better understand the departure of the M10-III data from this simple dipole, we show in Fig. 16 the M10-III data corrected for the TS04 external field in Cartesian MSO coordinates together with the centered dipole fit and the residual from the fit. Comparing the observations with the fit shows that the residuals arise because the observed field is more confined to the region near the pole than would be the field from a centered axial dipole field. This is most evident in the Z component, where, although the peaks are the same, near -300 nT for both the fit and the data, the observed field drops to zero more sharply on either side of the peak than does the dipole fit. The dipole field is similarly broader than the observed field in the X and Y components. We note that the distant field of the equatorial azimuthal current identified above cannot explain this discrepancy from the dipole because, although the equatorial current gives a field over the

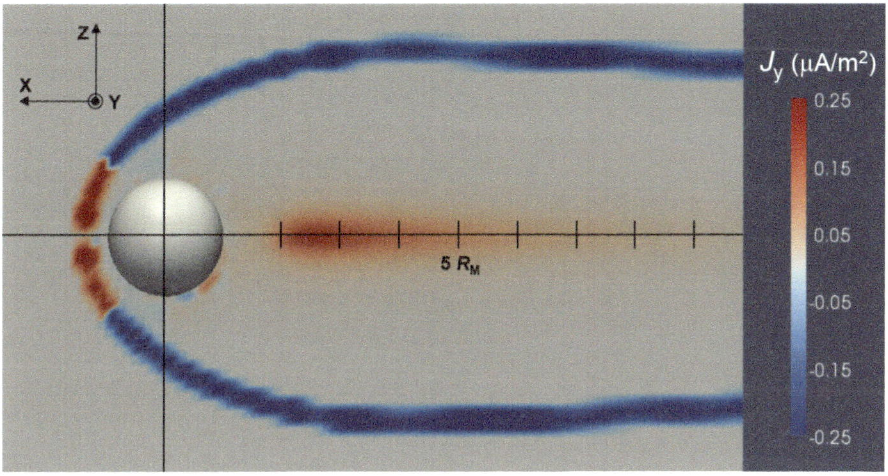

Fig. 15 Current density distribution in the noon-midnight meridian plane calculated from the TS04 model scaled to Mercury. The model is evaluated for the same parameters as that shown in Fig. 1. The resolution of the magnetic field grid from which the currents are computed is $0.15 R_M$. The current densities adjacent to the planet surface are spurious because of numerical errors in calculating J_Y at the inner boundary of the magnetic field model. The high current densities on the magnetopause are thinner layers than resolved in this display, while the cross-tail current is spread over a significant range in Z. Even though the cross-tail current density is comparably low, the total cross-tail current matches the sum of the northern and southern lobe magnetopause currents

pole of the same sense as that from a dipole, it adds a field that is broader than that of the dipole.

The present observations do not provide sufficient information to establish whether this discrepancy is due to structure in the internal field or to the solar wind interaction. The lower-altitude M10-III residuals can be resolved by introducing higher-order terms in the planetary field (e.g., Fig. 8), but the persistence of residuals at higher altitudes (Fig. 9) suggest that this is not a complete explanation. Alternatively, we note that there are irregularities in the observed field, suggesting the possibility that plasma pressure effects are influencing these data as well. Although the polar region itself may be relatively devoid of plasmas, the boundary surface between the polar region and lower latitudes where the plasma is prominent may be a locus of strong pressure gradients and possibly currents. To distinguish between these possibilities we should know where the plasma boundaries are, whether the deviations from a dipolar field are ordered better in local time or by planetary longitude, and whether the field in the south is comparable to that in the north. Although these questions cannot be resolved with the single cut through the polar region by the M10-III flyby, observations from orbit should do so.

Lastly, we note that there is a relatively modest sunward perturbation in the field within about $0.25 R_M$ of the pole. This is indicated not only in Fig. 14 (bottom right) but also by both models 5 and 7 of Table 1, Fig. 13 (right-hand panels). This localized sunward perturbation could be consistent with a set of field-aligned currents, toward the planet in the morning and away from the planet in the evening (Slavin et al. 1997; Ip and Kopp 2004), analogous to the terrestrial Region-1 Birkeland currents (Iijima and Potemra 1976; Anderson et al. 2008a). Judging from Figs. 13 and 14, the perturbation may be as large as ~50 nT, corresponding to a linear current density of ~40 mA/m, which if integrated over $\sim 0.5 R_M$ (e.g., Ip and Kopp 2004) would give a total current of 50,000 A. This is within

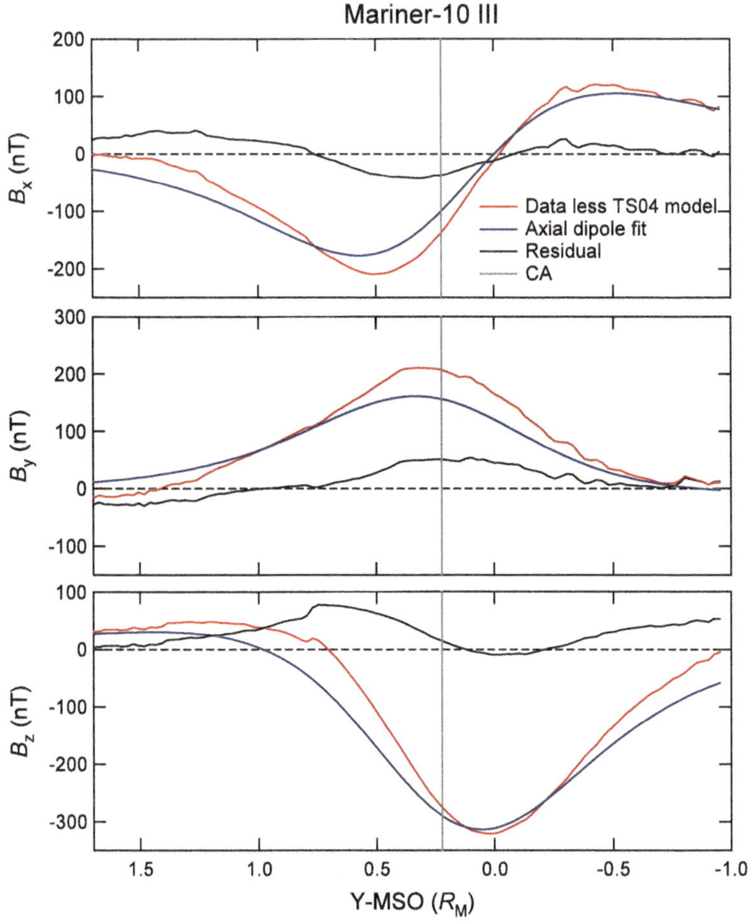

Fig. 16 Mariner 10 III observations corrected for the TS04 external field (*red*) together with an axial dipole fit solely to these data (*blue*) and the residuals (*black*) plotted versus the spacecraft Y-MSO coordinate. The point of closest approach is indicated with the vertical grey line

the range estimated from simulations, though the simulation estimates are highly sensitive to the conductance assumed near the surface where the currents would close (Glassmeier 1997). Data from orbit should definitively establish whether Birkeland currents exist.

6 Assessment Looking Forward

Combining data from all of the flyby encounters with Mercury to date yields somewhat tighter constraints on the planetary dipole moment and clarifies the challenge of separating external contributions from higher-order terms in the internal field. The additional observations from the second MESSENGER flyby constrain the planetary moment to be nearly axially aligned and with a magnitude in the range 240 to 270 nT-R_M^3. The new observations also confirm the presence of a cross-tail current close to the planet, which could account for the radial fields observed near the equator as well as the less strongly northward fields there compared with expectations from a dipole fit to the M10-III data. The stronger dipole moment, near 270 nT-R_M^3, therefore seems more likely. Nonetheless, the M10-III observations over the north pole cannot be understood solely in terms of a dipole field, implying either that the planetary field does possess higher-order structure or that the solar wind interaction influences the polar fields as well. Taken as a whole, then, the observations to date imply that magnetospheric currents close to the planet remain to be understood before definitive conclusions can be made about the structure of the internal field beyond the dipole. Nonetheless, higher-order inversion analysis of these data may prove useful to discriminate some mechanisms for the internal field generation, provided that the higher-order results are recognized as upper limits on the structure of the internal field. This analysis points to specific ways that observations from orbit together with physics-based simulations can be applied to resolve the ambiguities in our present understanding.

6.1 MESSENGER: First Orbital Observations

The MESSENGER spacecraft is on schedule for a final flyby of Mercury in September 2009 and orbit insertion in March 2011 (Solomon et al. 2007). The MESSENGER orbit at Mercury is designed to be highly elliptical, initially with periapsis at 200 km altitude and apoapsis at 15,000 km altitude, $\sim 7.5 R_M$ planetocentric distance, an orbit inclination of $\sim 80°$, and an orbit period of 12 hours (McAdams et al. 2007). The year-long baseline orbital mission will provide over 700 low-altitude passes over the northern polar region providing sampling spanning all local times and planetary longitudes. It will yield our first observations from the dayside magnetosphere of the planet. The orbit cuts through the equatorial region in a near ideal geometry to characterize the inferred equatorial currents and plasma enhancements. The orbit crosses the dayside magnetopause both at high and equatorial latitudes, providing excellent coverage to characterize the persistence of the boundary layer feature. Simulations predict large solar wind densities in the vicinity of the polar cusps (e.g., Trávníček et al. 2007), and the MESSENGER orbit should allow us to definitively establish this feature of the solar wind interaction as well. The MESSENGER observations from orbit will therefore lead to a number of advances key to understanding the internal field.

6.2 The BepiColombo Mission

The more ambitious two-spacecraft BepiColombo mission promises critical advances in definitively establishing the internal field of the planet (Balogh et al. 2007). The Mercury Planetary Orbiter (MPO) will orbit Mercury in a low-altitude polar orbit and provide the first low-altitude magnetic field observations over the southern polar region. The low-eccentricity polar orbit is well suited for the traditional spherical harmonic analyses of internal planetary magnetic fields for which the highly elliptical MESSENGER orbit is not ideal.

The profound influence of the solar wind interaction on the external field and the predominance of the external field contribution even at low altitudes imply that the simultaneous measurement of upstream conditions and the low-altitude magnetic field will be central in fully separating the internal and external field sources. This is an advance that the second BepiColombo spacecraft, the Mercury Magnetosphere Orbiter (MMO), will enable. The MMO spacecraft will provide sampling of the magnetosphere from an orbit different from that of MESSENGER and with a more complete plasma instrumentation package, thereby further advancing our understanding of the magnetosphere. But perhaps equally important for understanding the internal field, MMO will make simultaneous measurements of the solar wind and IMF to complement the low-altitude MPO observations, thus allowing dynamics observed at low altitudes to be related to variations in externally imposed conditions.

These two-point observations should allow not only higher fidelity in separating the external from internal field sources but also an ability to assess the role of fields produced by induced currents in the core (Glassmeier et al. 2007a, 2007b).

6.3 Modeling and Simulations

For the purpose of advancing our understanding of planetary magnetism, these planned observations of Mercury's magnetic field and magnetospheric environment should be matched by corresponding analyses to understand and predict the signatures of different processes generating the internal field (e.g., Zuber et al. 2007). It already appears that the prominent longitudinal structure of a remanent magnetic field (Aharonson et al. 2004) may not be consistent with the observations, though the longitudinal sampling is at present quite limited. Early orbital observations should resolve this question. If the field is dominated by the dipole term, as seems to be most probable, then we need to understand what signatures in the higher-order terms the different dynamo models should produce. Constraining the thickness of the fluid outer core is of great significance for this effort, since until we can constrain this dimension of the dynamo for the internal models, the magnetic field signatures alone may not prove decisive in distinguishing among competing models. The libration, gravity field, and topography of the planet are therefore essential to set limits on the internal structure, thereby helping to constrain the magnetic dynamo (Margot et al. 2007; Solomon et al. 2007; Zuber et al. 2007). In any case, the key to testing different models will be the accuracy of the higher-order terms, and so improved quantitative accuracy in the external field can be expected to impact our ability to discriminate among competing geophysical dynamo models to a degree that is disproportionate to the potentially modest improvement in external field knowledge.

The complexities of the magnetosphere of Mercury are unique in many respects (e.g., Baumjohann et al. 2006; Slavin et al. 2007), and it is to be expected that analogy from other magnetospheres will not apply, especially close to the planet where observations are most relevant to internal field estimation. Moreover, even the orbital sampling that the MESSENGER and BepiColombo missions are anticipated to provide will not yield exhaustive coverage of the magnetospheric volume, so that a quantitative understanding of the magnetosphere and hence the external field will likely require more than observations from orbit to fully resolve the external and internal fields. Physics-based simulations of the interaction will no doubt prove integral to guiding our quantitative specification of the natural system. One obvious approach in coupling the simulations with observations is to integrate the numerical simulations with the internal field inversion so that a physics-based external field is estimated together with the internal field terms (e.g., Jia et al. 2008). Formally we have already done this, because to specify the TS04 external model magnetopause and tail currents one must specify the internal field. We use an iterative process in which we take an initial estimate for the internal field, evaluate the TS04 model, subtract this from the observations, re-estimate the internal field and repeat these steps until the internal solution converges. This process yields an internal solution that typically converges to within one part in 10^4 after as few as four iterations. The same basic approach should be feasible with simulations.

Empirical external field models customized for Mercury will also play important roles for at least two reasons. First, it is likely to remain technically challenging to run a vast number of hybrid or even fluid simulations for the entire range of solar wind and IMF conditions imposed on the system. Thus, the physics-based simulation inversion analysis described above may remain applicable only to selected cases. Second, the physics-based simulations are sensitive to the boundary conditions and numerical diffusion responsible for mimicking reconnection processes. In particular, the conductivity distribution at the surface of the planet is critical to obtaining currents and the distribution of electric fields and flows within the magnetosphere (Glassmeier 1997). To the extent that the assumed boundary conditions are at variance with the natural system, the currents and hence the external field of the simulations will be in error. Empirical models for the magnetospheric magnetic field analogous to those developed for Earth and other magnetized planets (Khurana 1997; Alexeev and Belenkaya 2005; Arridge et al. 2006; Alexeev et al. 2008) customized for Mercury will therefore remain an important tool. They offer the advantages that they are tied primarily to the observations, employ only specified known distributions of current, and can be evaluated rapidly for application to all of the observational data. Numerical simulations could also be used in tandem with the empirical models to inform the current system modules to be included.

The future is therefore very promising for progress on understanding the magnetic field of Mercury. Although many technical challenges remain, the stage appears to be set for a most exciting decade as the MESSENGER and BepiColombo missions return the first observations from orbit around the planet. These observations will in turn spur considerable work in modeling and numerical simulations to quantify our understanding of the processes generating the internal field as well as the dynamic and unique magnetosphere. New observations and powerful new modeling tools can be expected to tease out the secrets of the origin of the innermost planet's enigmatic magnetic field.

Acknowledgements We deeply mourn the loss of our colleague and MESSENGER Co-Investigator Mario Acuña during the preparation of this manuscript; his many contributions to the MESSENGER mission and the space science community live on. Assistance in visualizing magnetic field models from Alexander Ukhorskiy is gratefully acknowledged. The MESSENGER project is supported by the NASA Discovery Program under contracts NASW-00002 to the Carnegie Institution of Washington and NAS5-97271 to the Johns Hopkins University Applied Physics Laboratory. Support is also acknowledged from the NSERC Discovery Grant Program and the MESSENGER Participating Scientist Program (NNX07AR73G).

References

M.H. Acuña, J.E.P. Connerney, N.F. Ness, R.P. Lin, D. Mitchell, C.W. Carlson, J. McFadden, K.A. Anderson, H. Rème, C. Mazelle, D. Vignes, P. Wasilewski, P. Cloutier, Science **284**, 790 (1999)

O. Aharonson, M.T. Zuber, S.C. Solomon, Earth Planet. Sci. Lett. **218**, 261 (2004)

I.I. Alexeev, E. Belenkaya, Ann. Geophys. **23**, 809 (2005)

I.I. Alexeev, E. Belenkaya, S.Yu. Bobrovnikov, J.A. Slavin, M. Sarantos, J. Geophys. Res. **113**, A12210 (2008). doi:10.1029/2008JA013368

B.J. Anderson, M.H. Acuña, D.A. Lohr, J. Scheifele, A. Raval, H. Korth, J.A. Slavin, Space Sci. Rev. **131**, 417 (2007)

B.J. Anderson, H. Korth, C.L. Waters, D.L. Green, P. Stauning, Ann. Geophys. **26**, 671–687 (2008a)

B.J. Anderson, M.H. Acuña, H. Korth, M.E. Purucker, C.L. Johnson, J.A. Slavin, S.C. Solomon, R.L. McNutt Jr., Science **321**, 82 (2008b)

G.B. Andrews, T.H. Zurbuchen, B.H. Mauk, H. Malcom, L.A. Fisk, G. Gloeckler, G.C. Ho, J.S. Kelley, P.L. Koehn, T.W. LeFevere, S.S. Livi, R.A. Lundgren, J.M. Raines, Space Sci. Rev. **131**, 523 (2007)

C.S. Arridge, N. Achilleos, M.K. Dougherty, K.K. Khurana, C.T. Russell, J. Geophys. Res. **111**, A11227 (2006). doi:10.1029/2005JA011574

G. Backus, J. Geophys. Res. **75**, 6339 (1970)

A. Balogh, R. Grard, S.C. Solomon, R. Schulz, Y. Langevin, Y. Kasaba, M. Fujimoto, Space Sci. Rev. **132**, 611 (2007)

W.A. Baumjohann, A. Matsuoka, K.-H. Glassmeier, C.T. Russell, T. Nagai, M. Hoshino, T. Nakagawa, A. Balogh, J.A. Slavin, R. Nakamura, W. Magnes, Adv. Space Res. **38**, 604 (2006)

B. Chen, J. Li, S.A. Hauck II, Geophys. Res. Lett. **35**, L07201 (2008). doi:10.1029/2008GL033311

U.R. Christensen, Nature **444**, 1056 (2006)

U.R. Christensen, J. Wicht, Icarus **196**, 16 (2008)

S.P. Christon, Icarus **71**, 448 (1987)

J.E.P. Connerney, N.F. Ness, in *Mercury*, ed. by F. Vilas, C.R. Chapman, M.S. Matthews (University of Arizona Press, Tucson, 1988), pp. 494–513
S.W.H. Cowley, in *Magnetospheric Current Systems*, ed. by S. Ohtani, R. Fujii, M. Hesse, R.L. Lysak. Geophysical Monograph, vol. 118 (American Geophysical Union, Washington, 2000), pp. 91–106
I.M. Engle, Planet. Space Sci. **45**, 127 (1997)
G. Giampieri, A. Balogh, Planet. Space Sci. **49**, 1637 (2001)
G. Giampieri, A. Balogh, Planet. Space Sci. **50**, 757 (2002)
K.-H. Glassmeier, Planet. Space Sci. **45**, 119 (1997)
K.-H. Glassmeier, in *Magnetospheric Current Systems*, ed. by S. Ohtani, R. Fujii, M. Hesse, R.L. Lysak. Geophysical Monograph, vol. 118 (American Geophysical Union, Washington, 2000), pp. 371–380
K.-H. Glassmeier, J. Grosser, U. Auster, D. Constaninescu, Y. Narita, S. Stellmach, Space Sci. Rev. **132**, 511 (2007a)
K.-H. Glassmeier, H.-U. Auster, U. Motschmann, Geophys. Res. Lett. **34**, L22201 (2007b). doi:10.1029/2007GL031662
J. Grosser, K.-H. Glassmeier, A. Stadelmann, Planet. Space Sci. **52**, 1251 (2004)
S. Hauck, A. Dombard, R. Phillips, S.C. Solomon, Earth Planet. Sci. Lett. **222**, 713 (2004)
M. Heimpel, J. Aurnou, F. Al-Shamali, N. Gomez Perez, Earth Planet. Sci. Lett. **236**, 542 (2005)
T. Iijima, T.A. Potemra, J. Geophys. Res. **81**, 2165 (1976)
W.-H. Ip, A. Kopp, J. Geophys. Res. **107**, 1348 (2002). doi:10.1029/2001JA009171
W.-H. Ip, A. Kopp, Adv. Space Res. **33**, 2172 (2004)
Q. Jia, R.J. Walker, M.G. Kivelson, K.K. Khurana, J.A. Linker, J. Geophys. Res. **113**, A06212 (2008). doi:10.1029/2007JA012748
K.K. Khurana, J. Geophys. Res. **102**, 11295 (1997)
H. Korth, B.J. Anderson, M.H. Acuña, J.A. Slavin, N.A. Tsyganenko, S.C. Solomon, R.L. McNutt Jr., Planet. Space Sci. **54**, 733 (2004)
J.G. Luhmann, C.T. Russell, N.A. Tsyganenko, J. Geophys. Res. **103**, 9113 (1998)
A.T.Y. Lui (ed.), *Magnetotail Physics* (The Johns Hopkins University Press, Baltimore, 1987), 433 pp
J. Margot, S. Peale, R. Jurgens, M. Slade, I. Holin, Science **316**, 710 (2007)
J.V. McAdams, R.W. Farquhar, A.H. Taylor, B.G. Williams, Space Sci. Rev. **131**, 219 (2007)
W. Menke, *Geophysical Data Analysis: Discrete Inverse Theory* (Academic Press, San Diego, 1989), 289 pp
N.F. Ness, K.W. Behannon, R.P. Lepping, Y.C. Whang, K.H. Schatten, Science **185**, 151 (1974)
N.F. Ness, K.W. Behannon, R.P. Lepping, J. Geophys. Res. **80**, 2708 (1975)
N.F. Ness, in *Solar System Plasma Physics*, vol. II, ed. by C.F. Kennel, L.J. Lanzerotti, E.N. Parker (North-Holland, New York, 1979), pp. 185–206
K.W. Ogilvie, J.D. Scudder, V.M. Vasyliunas, R.E. Hartle, G.L. Siscoe, J. Geophys. Res. **82**, 1807 (1977)
G.K. Parks, *Physics of Space Plasmas, An Introduction* (Addison-Wesley, New York, 1991), 538 pp
M.E. Purucker, T.J. Sabaka, S.C. Solomon, B.J. Anderson, H. Korth, M.T. Zuber, G.A. Neumann, Earth. Planet. Sci. Lett. (2009). doi:10.1016/j.epsl.2008.12.017
A.D. Richmond, J.P. Thayer, in *Magnetospheric Current Systems*, ed. by S. Ohtani, R. Fujii, M. Hesse, R.L. Lysak. Geophysical Monograph, vol. 118 (American Geophysical Union, Washington, 2000), pp. 131–146
C.T. Russell, D.N. Baker, J.A. Slavin, in *Mercury*, ed. by F. Vilas, C.R. Chapman, M.S. Matthews (University of Arizona Press, Tucson, 1988), pp. 494–513
G. Schubert, M. Ross, D. Stevenson, T. Spohn, in *Mercury*, ed. by F. Vilas, C.R. Chapman, M.S. Matthews (University of Arizona Press, Tucson, 1988), pp. 429–460
D.G. Sibeck, R.E. Lopez, E.C. Roelof, J. Geophys. Res. **96**, 5489 (1991)
J.-H. Shue, P. Song, C.T. Russell, J.T. Steinberg, J.K. Chao, G. Zastenker, O.L. Vaisberg, S. Kokubun, H.J. Singer, T.R. Detman, H. Kawano, J. Geophys. Res. **103**, 17691 (1998)
J.A. Slavin, R.E. Holzer, Phys. Earth Planet. Inter. **20**, 231 (1979)
J.A. Slavin, R.E. Holzer, J. Geophys. Res. **86**, 11401 (1981)
J.A. Slavin, C.J. Owen, J.E.P. Connerney, S.P. Christon, Planet. Space Sci. **45**, 133 (1997)
J.A. Slavin, S.M. Krimigis, M.H. Acuña, B.J. Anderson, D.N. Baker, P.L. Koehn, H. Korth, S.S. Livi, B.H. Mauk, S.C. Solomon, T.H. Zurbuchen, Space Sci. Rev. **131**, 133 (2007)
J.A. Slavin, M.H. Acuña, B.J. Anderson, D.N. Baker, M. Benna, G. Gloeckler, R.E. Gold, G.C. Ho, R.M. Killen, H. Korth, S.M. Krimigis, R.L. McNutt Jr., L.R. Nittler, J.M. Raines, D. Schriver, S.C. Solomon, R.D. Starr, P. Trávníček, T.H. Zurbuchen, Science **321**, 85 (2008)
J.A. Slavin, B.J. Anderson, T.H. Zurbuchen, D.N. Baker, S.M. Krimigis, M.H. Acuña, M. Benna, S.A. Boardsen, G. Gloeckler, R.E. Gold, G.C. Ho, H. Korth, R.L. McNutt Jr., J.M. Raines, S. Menelaos, D. Schriver, S.C. Solomon, P. Trávníček, Geophys. Res. Lett. **36**, L02101 (2009a). doi:10.1029/2008GL036158
J.A. Slavin, M.H. Acuña, B.J. Anderson, D.N. Baker, M. Benna, S.A. Boardsen, G. Gloeckler, R.E. Gold, G.C. Ho, H. Korth, S.M. Krimigis, R.L. McNutt Jr., J.M. Raines, M. Sarantos, D. Schriver, S.C. Solomon, P. Trávníček, T.H. Zurbuchen, Science **324**, 606 (2009b)

S.C. Solomon, Icarus **28**, 509 (1976)
S.C. Solomon, R.L. McNutt Jr., R.E. Gold, M.H. Acuña, D.N. Baker, W.V. Boynton, C.R. Chapman, A.F. Cheng, G. Gloeckler, J.W. Head III, S.M. Krimigis, W.E. McClintock, S.J. Peale, S.L. Murchie, R.J. Phillips, M.S. Robinson, J.A. Slavin, D.E. Smith, R.G. Strom, J.I. Trombka, M.T. Zuber, Planet. Space Sci. **49**, 1445–1465 (2001)
S.C. Solomon, R.L. McNutt Jr., R.E. Gold, D.L. Domingue, Space Sci. Rev. **131**, 3 (2007)
L.J. Srnka, Phys. Earth Planet. Inter. **11**, 184 (1976)
S. Stanley, J. Bloxham, W. Hutchison, M. Zuber, Earth Planet. Sci. Lett. **234**, 27 (2005)
S. Stanley, M. Zuber, J. Bloxham, Geophys. Res. Lett. **34** (2007). doi:10.1029/2007GL030892.
A. Stephenson, Earth Planet. Sci. Lett. **28**, 454 (1976)
D.J. Stevenson, Rep. Prog. Phys. **46**, 555 (1983)
D.J. Stevenson, Earth Planet. Sci. Lett. **82**, 114–120 (1987)
F. Takahashi, M. Matsushima, Geophys. Res. Lett. **33**, L10202 (2006). doi:10.1029/2006GL02579
P.M. Trávníček, P. Hellinger, D. Schriver, Geophys. Res. Lett. **34**, L05104 (2007). doi:10.1029/2006GL028518.
P.M. Trávníček, P. Hellinger, D. Schriver, D. Herčík, J.A. Slavin, B.J. Anderson, Geophys. Res. Lett. **36**, L07104 (2009). doi:10.1029/2008GL036630.
N.A. Tsyganenko, J. Geophys. Res. **100**, 5599 (1995)
N.A. Tsyganenko, M.I. Sitnov, J. Geophys. Res. **110**, A03208 (2005). doi:10.1029/2004JA010798.
H. Uno, MS Thesis, University of British Columbia, Vancouver, BC, Canada, 2009, 57 pp
H. Uno, C.L. Johnson, B.J. Anderson, H. Korth, S.C. Solomon, Earth. Planet. Sci. Lett. (2009). doi:10.1016/j.epsl.2009.02.032
J. Wicht, M. Mandea, F. Takahashi, U. Christensen, M. Matsushima, B. Langlais, Space Sci. Rev. **132**, 261 (2007)
M.T. Zuber, O. Aharonson, J.M. Aurnou, A.F. Cheng, S.A. Hauck II, M.H. Heimpel, G.A. Neumann, S.J. Peale, R.J. Phillips, D.E. Smith, S.C. Solomon, S. Stanley, Space Sci. Rev. **131**, 105 (2007)
T.H. Zurbuchen, J.M. Raines, G. Gloeckler, S.M. Krimigis, J.A. Slavin, P.L. Koehn, R.M. Killen, A.L. Sprague, R.L. McNutt Jr., S.C. Solomon, Science **321**, 90 (2008)

Space Sci Rev (2010) 152: 341–390
DOI 10.1007/s11214-009-9580-z

Paleomagnetic Records of Meteorites and Early Planetesimal Differentiation

Benjamin P. Weiss · Jérôme Gattacceca · Sabine Stanley · Pierre Rochette · Ulrich R. Christensen

Received: 28 August 2009 / Accepted: 12 October 2009 / Published online: 15 December 2009

Abstract The large-scale compositional structures of planets are primarily established during early global differentiation. Advances in analytical geochemistry, the increasing diversity of extraterrestrial samples, and new paleomagnetic data are driving major changes in our understanding of the nature and timing of these early melting processes. In particular, paleomagnetic studies of chondritic and small-body achondritic meteorites have revealed a diversity of magnetic field records. New, more sensitive and highly automated paleomagnetic instrumentation and an improved understanding of meteorite magnetic properties and the effects of shock, weathering, and other secondary processes are permitting primary and secondary magnetization components to be distinguished with increasing confidence. New constraints on the post-accretional histories of meteorite parent bodies now suggest that, contrary to early expectations, few if any meteorites have been definitively shown to retain records of early solar and protoplanetary nebula magnetic fields. However, recent studies of pristine samples coupled with new theoretical insights into the possibility of dynamo generation on small bodies indicate that some meteorites retain records of internally generated fields. These results indicate that some planetesimals formed metallic cores and early dynamos within just a few million years of solar system formation.

Keywords Paleomagnetism · Planetary magnetic fields · Meteorites · Dynamo models · Differentiation

B.P. Weiss (✉)
Department of Earth, Atmospheric, and Planetary Sciences, Massachusetts Institute of Technology, 54-814, 77 Massachusetts Avenue, Cambridge, MA 02139, USA
e mail: bpweiss@mit.edu

J. Gattacceca · P. Rochette
Centre Européen de Recherche et d'Enseignement des Géosciences de l'Environnement, Aix-Marseille Université-Centre National de la Recherche, PB80, 13545, Aix en Provence cedex 4, France

S. Stanley
Department of Physics, University of Toronto, 60 St. George Street, Toronto, ON M5S 1A7, Canada

U.R. Christensen
Max-Planck-Institut für Sonnensystemforschung, 37191 Katlenburg-Lindau, Germany

1 Introduction

Perhaps the most significant events in planetary history are those responsible for global planetary differentiation. These are the processes of large-scale melting and sequestration of compositionally distinct materials that give rise to a long-lived radially layered structure. Although planetary differentiation is occurring in a localized form even today on volcanically active bodies like Earth, Mars, and Io, the large-scale melting necessary for the formation of metallic cores overlain by silicate mantles and crusts occurred predominantly in the early solar system.

It is now well known that for large (greater than ~2000 km radius) bodies, the gravitational energy of formation exceeds that required to completely melt the bodies, likely leading to the formation of surface magma oceans (Wetherill 1990; Pritchard and Stevenson 2000). Smaller bodies that formed early enough to accrete significant quantities of short-lived nuclides like ^{26}Al should have also experienced radiogenic heating sufficient for large-scale melting (Urey 1955; Hevey and Sanders 2006; Sahijpal et al. 2007). However, these bodies probably melted from the inside-out. Bodies that experienced greater than several tens of weight % melting have the potential to form metallic cores. The formation of cores is evident in the great diversity of iron meteorite groups and in the depletion of siderophile elements in a variety of basaltic achondrites groups (Haack and McCoy 2007; Mittlefehldt 2007). Hf/W chronometry indicates that these cores formed within 0–3 million years (Ma) after the formation of calcium aluminum inclusions (CAIs) (Kleine et al. 2005). These cores would have been initially molten, and if they cooled quickly, they may have convected (Chabot and Haack 2006). Convecting cores may have generated dynamo magnetic fields that could have magnetized the overlying silicate rocks (Weiss et al. 2008a). This magnetization, possibly recorded in meteorite samples today, can be studied by paleomagnetic techniques as a way to infer the history of planetesimal differentiation and field generation.

Paleomagnetic studies could also potentially provide a unique window into understanding early solar magnetic fields generated externally from planetesimal bodies. The T Tauri Sun and protoplanetary nebula are both thought to have been significant field generating sources during the first several Ma of solar system history (Collinson 1994; Balbus 2009). The large-scale steady dynamo field of T Tauri stars is thought to be approximately dipolar with typical surface fields of ~0.1 T that fall off with the inverse cube of distance from the stellar center (Vallée 2003). Early magnetic fields associated with the early Sun and nebula may have slowed the Sun's rotation and permitted continued growth by accretion of disk material. The inner ionized region of the protoplanetary disk is thought to have been unstable to the magnetorotational instability (Balbus 2003, 2009), which likely generated spatially complex fields of up to ~100 μT (Sano et al. 2004; Johansen 2009). The latter process may have been a critical source of turbulent viscosity that in turn is likely required for mass and momentum transfer in the disk and, ultimately, the formation of the Sun and planets. Stellar and MRI-generated fields, as well as residual fields from the parent molecular cloud and transient fields from possible nebular lightning- and impact-generated plasmas, may have also been intimately involved in the formation and/or magnetization of the earliest solar system macroscopic solids, inclusions and chondrules in chondrites (Levy and Araki 1989; Shu et al. 1996, 1997; Crawford and Schultz 1999; Desch and Cuzzi 2000; Desch and Connolly 2002; Joung et al. 2004). Despite their great importance in planet formation, there has been as yet no unambiguous evidence of any of these field sources in meteorites. However, new advances in rock magnetism and magnetic instrumentation suggest that future paleomagnetic studies offer the potential for identifying records of these fields in meteorites and their constituents.

Here we review recent advances in paleomagnetic studies of meteorites thought to be from small planetary bodies and recent theoretical work in small-body differentiation and dynamo generation. Our goal is to provide a detailed overview of the paleomagnetic record that has been studied over the last sixty years beginning with the first investigations of Anyzeski (1949), Levinson-Lessing (1952) and Fonton (1954) and the first detailed analyses of Lovering (1959) and Stacey and Lovering (1959). We examine both the growing modern database as well as revisit older data from a modern paleomagnetic and geomagnetic perspective. We focus on the natural remanent magnetization (NRM) of meteorites and only discuss rock magnetic properties as they relate to the interpretation of NRM. We do not discuss paleomagnetic studies of iron meteorites and pallasites due to their poorly understood magnetic field recording properties (e.g., Guskova 1965b; Brecher and Albright 1977; Nagata et al. 1987). We also do not discuss the extensive work in lunar (reviewed by Fuller 1974, 2007; Hood and Cisowski 1983; Fuller and Cisowski 1987; Dunlop and Ozdemir 1997; Wieczorek et al. 2006) and Martian paleomagnetism (see Rochette et al. 2001, 2005, 2006; Fuller 2007; Acuña et al. 2008) except as they relate to the general context of extraterrestrial paleomagnetism. Although most work on meteorites took place in the 1970s and early 1980s (see previous reviews by Levy and Sonett 1978; Hood and Cisowski 1983; Cisowski 1987; Collinson 1992, 1994; Dunlop and Ozdemir 1997; Fuller 2007; Rochette et al. 2009b), there has recently been a burst of activity brought on by advances in paleomagnetic techniques and instrumentation, an increasingly numerous and diverse sample suite, advances in dynamo theory, and perhaps most importantly, a growing petrologic and geochemical dataset that is providing crucial contextual and geochronological information for understanding the nature and origin of remanent magnetization.

We begin by discussing the technical difficulties specific to paleomagnetic studies of meteorites in Sect. 2. We assume the reader has a working knowledge of paleomagnetism and geomagnetism at the level of Butler (1992). We then review paleomagnetic studies of chondrites and achondrites in Sects. 3 and 4. In Sect. 5, we discuss the theoretical implications of meteorite paleomagnetism for small-body dynamos. We end in Sects. 6 and 7 by discussing outstanding questions and summarizing key conclusions.

2 Challenges in Interpreting the Meteorite Record

2.1 Geologic Context and Paleo-Orientation

The foremost challenge for meteorite paleomagnetic studies is that, unlike in earth science, the parent body, geologic context, and original orientation of nearly all samples are unknown. Other than lunar and Martian meteorites, the only two exceptions are the howardite-eucrite-diogenite (HED) clan, thought be from the asteroid 4 Vesta, and the anomalous ureilite Almahata Sitta, which was recently observed as an F-class asteroid prior to its landing on Earth (see Sect. 4). Even for these samples, the sampling site and orientation are unknown and the parent body is at the moment only barely resolved in telescope images (Thomas et al. 1992). As a result, extraterrestrial paleomagnetic studies have focused on measuring only the magnitude (and not the orientation) of the magnetization vector in order to recover the paleointensity of the field that magnetized the meteorites. This situation is very different from the field of terrestrial paleomagnetism, which has used paleo-orientation measurements to obtain a wealth of information about the Earth's field geometry, field temporal variability (e.g., geomagnetic reversals), and tectonic motions.

2.2 Scarcity of Material and Implications for Demagnetization Methods

Many meteorites are rare and little material is readily available for study. As a result, it is often difficult to obtain the large numbers of mutually oriented samples which are necessary for demonstrating primary magnetization. Those samples that are obtained are typically small in size and can therefore have relatively weak moments and substantial magnetic anisotropy. Another consequence is that often only nondestructive alternating field (AF) methods are permitted for demagnetization and paleointensity studies. AF methods are superior to thermal demagnetization in that they are ideally suited for removing common isothermal remanent magnetization (IRM) overprints from magnets (see Sect. 2.7). They also offer the advantage of not altering the meteorite's mineralogy, which permits the very same subsamples to be further analyzed with rock magnetic techniques and geochronometry. However, unlike thermal methods, AF methods do not unblock thermoremanent NRM in the same way that it was acquired, such that secondary thermal and viscous overprints sometimes cannot be easily isolated and highly accurate ($\sim$10%) paleointensities cannot be measured.

AF methods (particularly those using static field treatments) also can introduce spurious anhysteretic remanent magnetization (ARM) and gyroremanent magnetization (GRM), both of which can mask the underlying NRM and falsely appear to be primary remanence (Collinson 1983; Stephenson 1993; Hu et al. 1998). Spurious acquisition of ARM, in which a directionally apparently random component is increasingly acquired during AF demagnetization due to imperfections in the AF-waveform (Fig. 1), has in fact been known for several decades (see Collinson 1983). This problem is particularly severe for studies of metal-bearing meteorites because the low coercivity of multidomain iron means that the NRM can be quickly masked by ARM-related noise. ARM noise can be reduced by making multiple AF steps at the same or similar peak fields and averaging the resulting directions (e.g., Cisowski 1991; Acton et al. 2007). New advances in measurement automation (Kirschvink et al. 2008) have recently enabled dozens of repeat measurements to be averaged, thereby permitting AF demagnetization of extraterrestrial samples to be carried up to unprecedented levels (Weiss et al. 2008a; Garrick-Bethell et al. 2009).

In contrast to ARM noise, GRM (Hu et al. 1998; Stephenson 1993) is only a recently discovered phenomenon. It is insidious in that it is typically acquired during AF demagnetization as a unidirectional component that could potentially be mistaken as a primary NRM. For example, a study of the angrite NWA 4931 (Weiss et al., unpublished data) found that the majority of subsamples had origin-trending NRM directions after AF demagnetization to $\sim$15 mT. However, some subsamples were susceptible to acquiring weak spurious ARM at AF steps above $\sim$10 mT and, more importantly, strong GRM above $\sim$30 mT (Fig. 1). This was manifested as a dramatic shallowing of the NRM directions and increase in moment during static three-axis AF treatment: GRM is acquired perpendicularly to the final (in this case, vertical) AF axis. Fortunately, GRM-correction methods (Dankers and Zijderveld 1981; Stephenson 1993; Hu et al. 1998), like measuring the moment after each uniaxial AF treatment, mitigate these effects and retrieve origin-trending magnetization (Fig. 1). Furthermore, GRM, because it grows with AF field, can generally be distinguished from primary NRM because it does not produce a characteristic magnetization decaying linearly to the origin. Given that GRM was only discovered in the early 1980s, well after most meteorite and lunar paleomagnetic studies, meteorite paleomagnetic data acquired before this time must be viewed with some caution. In fact, GRM has only been seriously considered in extraterrestrial paleomagnetic studies during just the last two years (Weiss et al. 2008a; Lawrence et al. 2008; Garrick-Bethell et al. 2009).

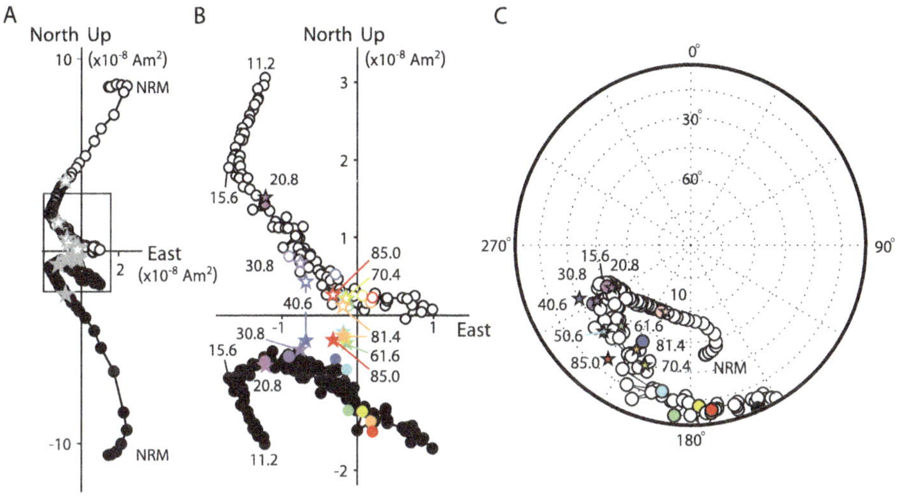

Fig. 1 Spurious ARM and GRM acquisition by angrite NWA 4931 during AF demagnetization. **A** Orthogonal demagnetization plot showing evolution of NRM vector during demagnetization. *Open and closed symbols* represent projections of the endpoint of the magnetization vector on vertical and horizontal planes, respectively. *Circles* = static three-axis (for NRM–AF 85 mT) or single z-axis (for AF 90 to 257.9 mT) demagnetization conducted in the axis order x (north–south), y (east–west), and z (up–down), with moment measurement taken after final (z) step only. *Stars* = NRM directions corrected for GRM using the Zijderveld–Dunlop method (Stephenson 1993) in which moment measurements taken after each of the three uniaxial AF demagnetization steps are averaged. GRM-corrected steps were only acquired every ~10–15 mT. *Boxed area* is magnified in **B**. **B** Close-up of *boxed area* in **A**. Selected demagnetization steps are listed in mT. After each GRM-corrected step, a static three-axis AF step with the same peak field was acquired. For a given AF field, the GRM-corrected and static three-axis AF steps share the same color (*lavender* = 20.8 mT, *purple* = 30.8 mT, *blue* = 40.6 mT, *light blue* = 50.6 mT, *green* = 61.6 mT, *yellow* = 70.4 mT, *red* = 85 mT). **C** Equal area projection showing directions of natural remanent magnetization corresponding to data in **A** and **B**. Color and symbols are the same as in **B**. The directional jitter around the mean NRM direction above ~15 mT is likely a manifestation of weak spurious ARM noise, while the dramatic increase in NRM and directional shallowing above 30 mT for static three-axis AF data are a manifestation of GRM. Weiss et al. (unpublished data)

2.3 Shock

Most meteorites have been shocked to peak pressures in excess of 5–10 GPa. For instance, Martian meteorites have suffered shock pressures around 5–20 GPa (nakhlites) and 30–45 GPa (shergottites) (Fritz et al. 2005) whereas 90% of ordinary chondrites have been shocked to above 5 GPa (Schulze and Stöffler 1997). In contrast, 65% of carbonaceous chondrites have not suffered peak pressures above 5 GPa (Scott et al. 1992).

Shock waves can modify the magnetic record of meteorites in at least three ways (Fig. 2). First, they can partially or completely erase the pre-shock remanent magnetization (e.g., Cisowski et al. 1975; Cisowski and Fuller 1978; Gattacceca et al. 2006; Gilder et al. 2006; Bezaeva et al. 2007). In the presence of an ambient field, they can lead to the acquisition of shock remanent magnetization (e.g., Doell et al. 1970; Wasilewski 1973; Pohl et al. 1975; Cisowski et al. 1975, 1976; Cisowski and Fuller 1978; Pohl and Eckstaller 1981; Gattacceca et al. 2008a; Bezaeva et al. 2009). Finally, above 5–10 GPa, shocks (and even quasistatic pressure) can also permanently modify the intrinsic magnetic properties of rocks, including saturation remanent magnetization, coercivity, susceptibility, and anisotropy of susceptibility and remanence (Pohl and Eckstaller 1981;

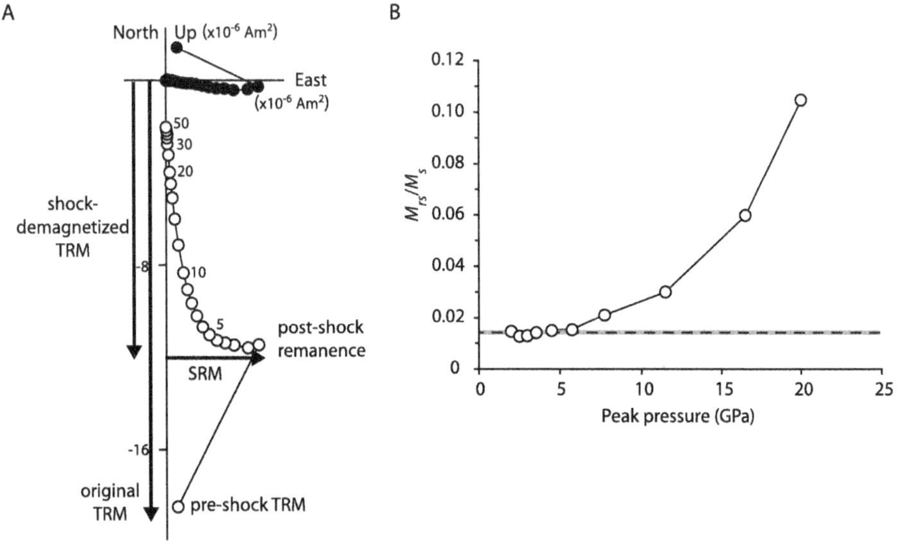

Fig. 2 Effects of shock on remanent magnetization and rock magnetic properties. **A** Orthogonal demagnetization plot showing simultaneous shock demagnetization and remagnetization of a basalt sample. *Open and closed symbols* represent projections of the endpoint of the magnetization vector on vertical and horizontal planes, respectively. The sample, originally carrying a thermoremanent magnetization (TRM) acquired in a 100 μT vertical downward-pointing field, was shocked with a pulsed laser to peak pressures of ∼1 GPa in a 100 μT field horizontal eastward-pointing field. Subsequent AF demagnetization (selected steps shown in mT) isolated two overlapping components of magnetization: a newly acquired SRM superimposed on the partially demagnetized TRM (Gattacceca, unpublished data). **B** Modification of rock magnetic properties of a microdiorite sample shocked with a high-power explosive (M_{rs}: remanent magnetization following exposure to saturating field, M_s: saturation magnetization). *Dashed line* is the mean value for unshocked samples (*grey band* = one s.d.). Adapted from Gattacceca et al. (2007)

Gattacceca et al. 2007; Louzada et al. 2007, 2009; Gilder and Le Goff 2008; Nishioka et al. 2007; Funaki and Syono 2008).

The sensitivity to pressure demagnetization and remagnetization is controlled by the magnetic mineralogy (Kletetschka et al. 2004b; Bezaeva et al. 2007; Louzada et al. 2009). Pyrrhotite has a magnetic phase transition at ∼2.8 GPa (Rochette et al. 2003a), meaning that pyrrhotite-bearing meteorites shocked above this pressures may have lost all or much of the record of their pre-shock remanent magnetization (Rochette et al. 2001; Louzada et al. 2007). This is particularly crucial for basaltic shergottites and Rumuruti chondrites in which pyrrhotite is the main remanence carrier (Rochette et al. 2008, 2009a) and which have generally been shocked to pressures in excess of 5 GPa (Kallemeyn et al. 1996; Fritz et al. 2005). It is also relevant for carbonaceous chondrites that contain pyrrhotite (e.g., Allende) along with FeNi metal and magnetite.

Magnetite is also strongly affected by pressures in the order of a few GPa (Gilder and Le Goff 2008; Gilder et al. 2006; Bezaeva et al. 2009). Presently, the effects of shock on the magnetization and magnetic properties of FeNi metal alloys are still poorly understood (Pohl and Eckstaller 1981). Recent studies have found that the sensitivity to shock decreases from kamacite to taenite to tetrataenite, which correlates with the relative coercivities of these minerals (Bezaeva et al. 2009). FeNi metal may have a phase transition at ∼10–13 GPa (Wasilewski 1973, 1976; Dickinson and Wasilewski 2000).

As a consequence, it appears that the magnetic remanence of shocked meteorites is often unlikely to predate the last major shock event. The major difficulty with this is that nearly all known petrographic shock barometers can only distinguish past shock events with pressures above 4–5 GPa (e.g., Bischoff and Stoffler 1992; Scott et al. 1992). Shock events below 5 GPa leave almost no petrographic imprints but nevertheless can seriously disturb remanent magnetization. In metal-bearing meteorites, anisotropy of magnetic susceptibility is a sensitive shock-induced strain indicator (Gattacceca et al. 2005), but there is a crucial need for developing low-shock quantitative diagnostic techniques in the 0–5 GPa range.

Despite these complications, shocked meteorites should not be immediately rejected for paleomagnetic studies because shock magnetization and demagnetization processes are becoming increasingly well understood. It is now clear that shock remanent magnetization is usually acquired parallel to the ambient field at the time of shock with an intensity proportional to this ambient field (Pohl et al. 1975; Gattacceca et al. 2008a), making it a good recorder of the paleomagnetic field at the time of the last major impact. The main remaining difficulty is that the ambient field at the time of impact may be transiently produced (Crawford and Schultz 1988, 1991, 1993, 1999; Srnka 1977; Doell et al. 1970) or amplified (Hide 1972; Hood 1987; Hood and Artemieva 2008) by the impact itself, therefore making the interpretation of SRM in meteorites a complex topic. On the other hand, for shock pressures above ~40 GPa, the shock-induced temperature increase (Bischoff and Stoffler 1992; Artemieva and Ivanov 2004) is usually high enough to impart a thermoremanent magnetization (TRM) to the rock during slow cooling from post-shock temperatures of ~400–1100°C. The relatively slow thermoremanent magnetization (TRM) acquisition process for such samples is highly unlikely to be magnetized by transient impact-generated fields, which are estimated to last <1 d for the largest basin-forming impacts (Hood and Artemieva 2008) and <100 s for craters <100 km in diameter (Srnka 1977; Crawford and Schultz 1999). For lower shock pressures, localized heterogeneous shock-heating may also be responsible for the acquisition of a TRM, as has been invoked for Martian meteorite ALH 84001 (Weiss et al. 2008b).

In view of the complexity of these shock effects, the main conclusion is that careful selection of unshocked meteorites is the only way to ascertain a possible primary origin of the remanent magnetization (e.g., Weiss et al. 2008a), but that highly shocked and thermally processed samples can play an important subsidiary role in constraining planetary paleomagnetism.

2.4 Inverse Thermoremanent Magnetization

The cubic iron oxide minerals of the magnetite-ulvöspinel series (commonly referred to collectively as titanomagnetite) are important ferromagnetic minerals in carbonaceous chondrites (Herndon et al. 1976; Rochette et al. 2008), angrites (Weiss et al. 2008a), and some Martian meteorites (Rochette et al. 2009a). Nearly stoichiometric magnetite ($Fe_{3-x}Z_xO_4$ with $x < 0.01$–0.02 for a wide range of impurities Z) undergoes two transitions at low temperatures: (1) a magnetic transition at ~130 K when the first magnetocrystalline anisotropy constant passes through zero and (2) a first-order phase transition at ~119 K to a low-temperature monoclinic structure. Nonstoichiometric magnetite will undergo the first transition (usually at a lower temperature that depends on the nature and amount of impurity) but not the second. A number of authors have shown that as a rock warms through each of these transitions in the presence of a magnetic field, it can become magnetized. This "inverse thermoremanent magnetization" (ITRM) could potentially overprint a primary remanence (Dunlop 2006). It has even been suggested that meteorites could be remagnetized by the Earth's magnetic field as they warm up from space temperatures during atmospheric entry.

Meteorites and asteroids will be typically at an equilibrium temperature set by incoming and reradiated solar insolation. For a fast rotating object, the global average equilibrium temperature is given by $T_{eq} = [F_{Sun}(1-A)/4\varepsilon\sigma a_{AU}^2]^{1/4}$ where F_{Sun} is the solar constant (flux at 1 AU), A is the geometric albedo, a_{AU} is the semimajor axis in AU, ε is the emissivity, σ is Stefan–Boltzmann's constant (de Pater and Lissauer 2001). The latitude (λ) dependent equilibrium temperature for a fast rotating object is $T_{eq,l}(\lambda) = T_{eq}[4\cos(\lambda)/\pi]^{1/4}$. For zero albedo and unit emissivity, this means that even at the outer edge of the entire asteroid belt at 3.5 AU, fast rotators have $T_{eq} \sim 150$ K, which is still well above magnetite's isotropic point. However poleward of ~60° latitude in this region of the solar system, temperatures will be below 130 K. Furthermore, samples from the outer solar system, the unilluminated sides of slowly rotating asteroids, and permanently shadowed craters will also be at temperatures below 130 K. On the other hand, because the transfer of meteorites from the asteroid belt and beyond is a gradual process initiated by resonant interactions with the giant planets, the delivery process typically takes millions of years, with even the fastest known scenarios requiring ~100,000 years (Nesvorný et al. 2007). Therefore, by the time any meteoroid reaches the Earth's field, it should long have warmed up through the isotropic point and so will not have acquired an ITRM. The chances of encountering a strong field along the way to Earth are remote given the low number density of objects in the solar system. The most likely effects of warming through these transitions is therefore demagnetization of pre-existing NRM, which would then tend to lead to underestimates of the true paleointensity and hardening of NRM.

2.5 Unusual Ferromagnetic Minerals and Magnetization Acquisition Mechanisms

Two related difficulties with interpreting the meteorite paleomagnetic record are the presence of ferromagnetic minerals with unfamiliar magnetic properties and poorly understood nonthermal NRM acquisition mechanisms (Rochette et al. 2009b). The reduced oxidation state of many classes of meteorites reflects the formation of FeNi alloys (predominantly with the crystallographic structure of α kamacite, but also γ taenite, γ'' tetrataenite, and γ' awaruite) and, in the most reduced meteorites (ureilites, aubrites and enstatite chondrites), suessite [$(Fe, Ni)_3Si$], schreibersite [$(Fe, Ni)_3P$], and cohenite [$(Fe, Ni)_3C$]. The magnetic properties of the latter three phases are not well known.

Another problem is that, depending on the cooling rate and Ni-content, FeNi can form a variety of metastable ferromagnetic phases whose magnetic properties are not only mostly unknown but which acquire NRM via poorly understood phase-transformation thermochemical mechanisms (for a review, see Garrick-Bethell and Weiss 2009). For bulk compositions with >3% Ni and slow cooling, kamacite and taenite continually equilibrate below kamacite's Curie temperature during cooling to produce a phase-transformation CRM in the kamacite. It is not known if such a remanence mechanism retains a memory of TRM acquired at earlier equilibrium states (Dunlop and Ozdemir 1997). For fast cooling, there are six different possible composition-invariant transformations in the Fe-Ni system (Wilson 1994) including the common meteoritic mineral martensite (α_2Fe). How any of these phases magnetize is very poorly understood (e.g., Wasilewski 1974a, 1974b; Wasilewski et al. 2002).

For high-Ni (>~50%) metal, tetrataenite and awaruite form by low-temperature atomic ordering of taenite during cooling of the parent body below 320°C and ~500°C, respectively. Tetrataenite, first described in meteorites (Clarke and Scott 1980), is not known in terrestrial rocks because its formation requires high nickel contents and usually extremely slow cooling (<~1–100°C/Ma). Its extremely high coercivity compared to taenite and kamacite (Nagata and Funaki 1982, 1987; Nagata 1983; Wasilewski 1988; Nagata and Carleton 1989; Gattacceca et al. 2003) makes it a possible stable carrier of early solar system remanent magnetization. Indeed, LL chondrites that are rich in tetrataenite have much more stable NRM than tetrataenite-poor H and L chondrites (Gattacceca and Rochette 2004). However the relationship of tetrataenite's NRM with that of the precursor taenite is largely unknown, which makes the interpretation of tetrataenite-carried remanent magnetization tentative. In fact, tetrataenite formation is a candidate explanation for the small-scale randomness of NRM directions in some meteorites (Wasilewski et al. 2002; Gattacceca et al. 2003).

2.6 Weathering

Meteorites are subjected to low-temperature aqueous alteration after arrival at the Earth's surface. Metal and sulfide phases are rapidly altered to a variety of oxides and oxyhydroxides, while primary oxides like magnetite are more slowly altered to oxyhydroxides and other oxides like maghemite and hematite. The resulting destruction of primary ferromagnetic minerals and the production of new ferromagnetic minerals in the Earth's field can dramatically overprint preterrestrial NRM in meteorites.

Hot desert finds (i.e., from the Sahara and Arabia) and Antarctic meteorites exhibit different styles of weathering that reflects their discovery locations. Hot desert meteorites, which typically have terrestrial ages of 10–20 thousand years (ka) and rarely survive beyond 50 ka, generally weather much more rapidly than those from Antarctica (the latter have mean residence time of finds of $\sim 10^5$ years, with some finds as old as ~2 Ma) (Bland et al. 2000, 2006). As a result, hot desert finds have comparatively smaller abundances of metal and troilite and larger abundances of iron oxides and iron oxyhydroxides. Mössbauer spectroscopy (Bland et al. 1998) and electron microscopy studies (Al-Kathiri et al. 2005) have generally revealed two major classes of magnetic weathering minerals: superparamagnetic goethite and paramagnetic phases including akaganéite and lepidocrocite, and ferromagnetic phases including magnetite, maghemite, and hematite. Magnetite is rare and akaganéite is common in Antarctic meteorites, while the opposite is true for hot desert meteorites. Metal will typically first alter to akaganéite or, where dissolution is very rapid, to magnetite. Pyrrhotite, troilite and silicates may first convert to ferrihydrite. These products may then ultimately transition to goethite, maghemite, hematite and lepidocrocite, with in some cases magnetite remaining as a metastable end product (Bland et al. 2006).

Weathering of metal-bearing meteorites is extremely rapid, particularly in hot desert environments (Fig. 3). Several recent studies of desert weathering of ordinary chondrites meteorites found that a large fraction of the metal (tens of percent) is oxidized within just a few hundred to a few thousand years after landing (Al-Kathiri et al. 2005; Al-Rawas et al. 2007; Bland et al. 1996, 1998; Buchwald and Clarke 1989; Lee and Bland 2004; Lee et al. 2006), with essentially 100% of the metal in meteorites from Oman destroyed within just 20 ka. These studies show that pyrrhotite and troilite are more resistant than metal, taenite is more resistant than kamacite, and magnetite and crystalline silicates (including olivine) are more resistant than both metal and sulfides (Gooding 1986; Al-Kathiri et al. 2005; Lee and Bland 2004). Therefore, most meteorites are far more susceptible to weathering remagnetization than typical magnetite-bearing terrestrial rocks; meteoritic silicate phases can be extremely fresh and show no signs of alteration while the metal can be thoroughly weathered.

Rock magnetic studies reflect the destruction of primary ferromagnetic phases in finds (Rochette et al. 2003b, 2008, 2009a). Progressive weathering often leads to a decrease in

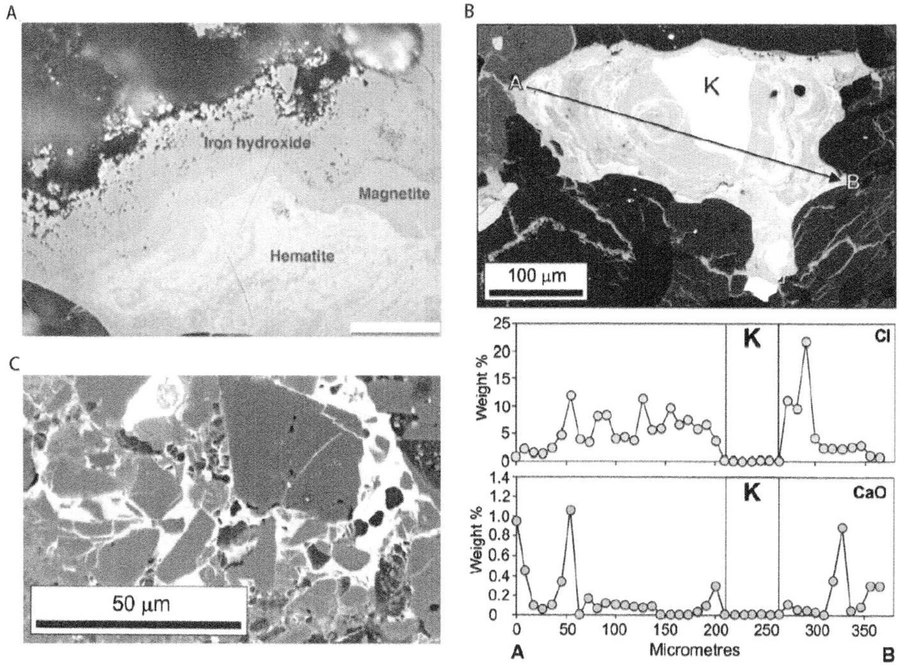

Fig. 3 Weathering of metal and sulfides in chondrites. **A** Backscattered scanning electron microscopy (BSEM) image of hot desert meteorite Al Huqf 010 showing iron oxide and oxyhydroxide weathering products of troilite and kamacite. After (Al-Kathiri et al. 2005). **B** BSEM image of partially weathered kamacite (K) from hot desert meteorite Daraj 014. Concentric layers of weathering products mantle a core of fresh metal. Note light-colored veins of oxyhydroxides within surrounding silicates. Cl and Ca concentrations along the transect **A**–**B**, shown in *lower two panels*, are enhanced in the altered zone and its margins. After (Lee and Bland 2004). **C** BSEM image of weathering-induced brecciation of Antarctic meteorite ALHA 77002. Clasts of olivine, orthopyroxene and feldspar are cemented by Fe-silicate (*lower left hand side*) and a fine-scale intergrowth of iron silicate with iron oxide and oxyhydroxides (*middle right hand side*). After (Lee and Bland 2004)

susceptibility as metal is converted to less magnetic oxides and an increase in NRM as the new phases acquire a crystallization remanent magnetization in the Earth's field (Guskova 1988). This increase in NRM has also been observed in artificial weathering experiments in the laboratory (Kohout et al. 2004).

A robust method for determining whether a meteorite has been remagnetized since arrival on Earth by weathering, viscous magnetization acquired in the Earth's field (e.g., Nagata 1981; Weiss et al. 2008a; Brecher and Arrhenius 1974), or hand magnets (see Sect. 2.7) is the fusion crust baked contact test (Butler 1972; Sugiura and Strangway 1983; Nagata and Funaki 1983; Weiss et al. 2000, 2002, 2008a, 2008b). The basis of this technique is that the fusion crust, the <1 mm thick melted exterior rind typically acquired as a result of heating during passage through Earth's atmosphere, and the immediately adjacent baked interior acquire an approximately unidirectional thermoremanent magnetization (TRM) in Earth's field (Nagata and Sugiura 1977; Nagata 1979d; Weiss et al. 2000, 2002, 2008b). Because the thermal remagnetization zone from atmospheric passage typically only penetrates less than several mm into the interior of stony meteorites, NRM directions of deep interior subsamples should be different from fusion crusted subsamples if there has been no process that has unidirectionally magnetized the en-

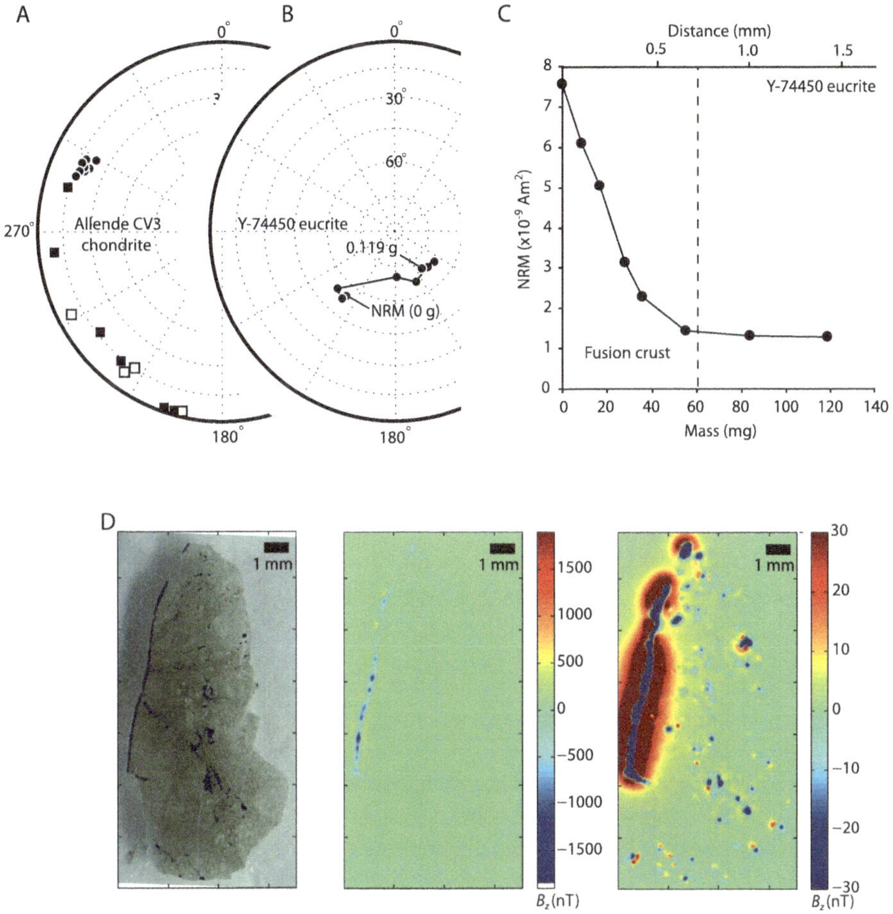

Fig. 4 Fusion crust baked contact for meteorite paleomagnetism. **A** Equal area projection showing directions of NRM of subsamples of Allende meteorite. *Circles* = subsamples from meteorite interior. *Squares* = subsamples from fusion crusted exterior. Adapted from Butler (1972). **B** Equal area projection showing NRM direction of a chip from the Y-74550 eucrite. Fusion crust was progressively scraped off the sample, with the NRM of the remaining chip measured after each scraping step. The NRM direction of the chip changed continuously until the fusion crust was fully removed (after sixth scraping step). Adapted from Nagata (1979d). **C** NRM of Y-74550 chip shown in **B** as a function of cumulative removed mass and depth of removed material. *Dashed line* shows depth of the fusion crust. Adapted from Nagata (1979d). **D** NRM field of martian meteorite ALH 84001. *Left*: Reflected light photograph of 30 μm thin section 227b.2 showing pyroxene (*light brown*), fusion crust *along left side* (*black*) and chromite (*black interior grains*). *Middle*: Vertical component of the NRM field as measured 140 μm above the sample. Positive (out-of-the-page) fields are *red and yellow* and negative (into-the-page) fields are *blue*. The field is dominated by the fusion crust magnetized approximately downward. *Right*: Same as middle except with color scale stretched to show weak, heterogeneously oriented magnetization in the meteorite interior. Adapted from Weiss et al. (2008b).

tire meteorite since its arrival on Earth (Fig. 4). In conducting this test, ratios of NRM/IRM should be measured in concert with measuring NRM directions, because magnets (Sect. 2.7) will often only remagnetize the outer several cm of large meteorites (e.g., Fig. S1B of Weiss et al. 2008a). The proximity of the magnets to the outer portion of a meteorite will produce NRM directions in exterior subsamples that are divergent from deep interior subsamples and give the appearance of a positive baked contact test. High (>∼10%) NRM/IRM ratios in exterior samples can be used to discard such false positives.

2.7 Hand Magnets and Secondary IRM

Even if a meteorite has been demonstrated to not be completely remagnetized by terrestrial weathering, it may still not contain a pre-terrestrial paleomagnetic record as a result of contamination by IRM. Unfortunately, one of the most common techniques for establishing the extraterrestrial nature and meteorite classification of a find is to quickly estimate its metal content by application of a hand magnet. Typical refrigerator magnets produce surface fields of several tens of mT, while rare earth magnets (in wide use among meteorite hunters and dealers) commonly produce surface fields of hundreds of mT. Most meteorites found in hot deserts (e.g., Acfer, Dar al Gani, Dhofar, Northwest Africa, Sahara, meteorites) as well as many meteorite falls have been magnetically contaminated with rare earth magnets (Fig. 5).

Because ferromagnetic minerals in meteorites typically have maximum coercivities ranging from 300 mT (magnetite) to ∼1 T (metal), rare earth magnets (but not common magnets) will destroy most of the NRM at the contact point immediately upon application of the magnet. As the magnet is withdrawn, the contact point will experience a field that changes in direction and weakens in time as the curved fields of the increasingly distant magnet sweep through the point. This blocks different coercivity fractions in different directions, producing a NRM that demagnetizes in curvilinear fashion during AF treatment (Fig. 5). Adjacent locations in the meteorite will experience a different field history and so will be magnetized in a different direction. Therefore the magnet IRM in a meteorite is expected to be spatially nonuniform in direction (as well as in magnitude) throughout the meteorite.

Because of the rapid falloff of fields with distance from hand magnets (proportional to r^{-2} to r^{-3} at the scale of meteorite hand samples), parts of the meteorite that are several cm away from the contact point will retain a high coercivity fraction of their NRM, with the deep interior (>10 cm) nearly unaffected by magnet overprints. Therefore, as discussed in Sect. 2.6, for large (>∼20 cm diameter) meteorite fragments, AF demagnetization of mutually oriented subsamples from a range of depths can be used to isolate NRM in the deep interior of the meteorite (see Fig. S1b of Weiss et al. 2008a; numbers in parentheses next to each subsample in this figure give distance in mm from the fusion crusted exterior). Note that thermal demagnetization, which does not activate the same grain distribution as that affected by an IRM, is *not* very effective at removing IRM overprints (e.g., Lawrence et al. 2008). In the "magnet contact test", subsamples should have more intense NRM due to increasingly strong IRM overprints as the magnet contact point is approached; the unblocking coercivity of this IRM should similarly increase. In the magnet remagnetized zone, the NRM will be spatially nonuniform and exhibit curvilinear components during AF demagnetization (e.g., Fig. 5). The deep interior fraction of a meteorite unaffected by the IRM which contains a primary TRM will have an NRM that is usually far from saturation, spatially uniform in direction, and composed of linear magnetization components that are oriented differently from much of the magnet zone.

3 Paleomagnetic Records in Chondrites

We now turn to the actual paleomagnetic records of meteorites. We begin with the most primitive samples, chondrites. Chondrites are polymict breccias of early solar system solids (refractory inclusions, chondrules, and matrix) that lithified on early formed planetesimals.

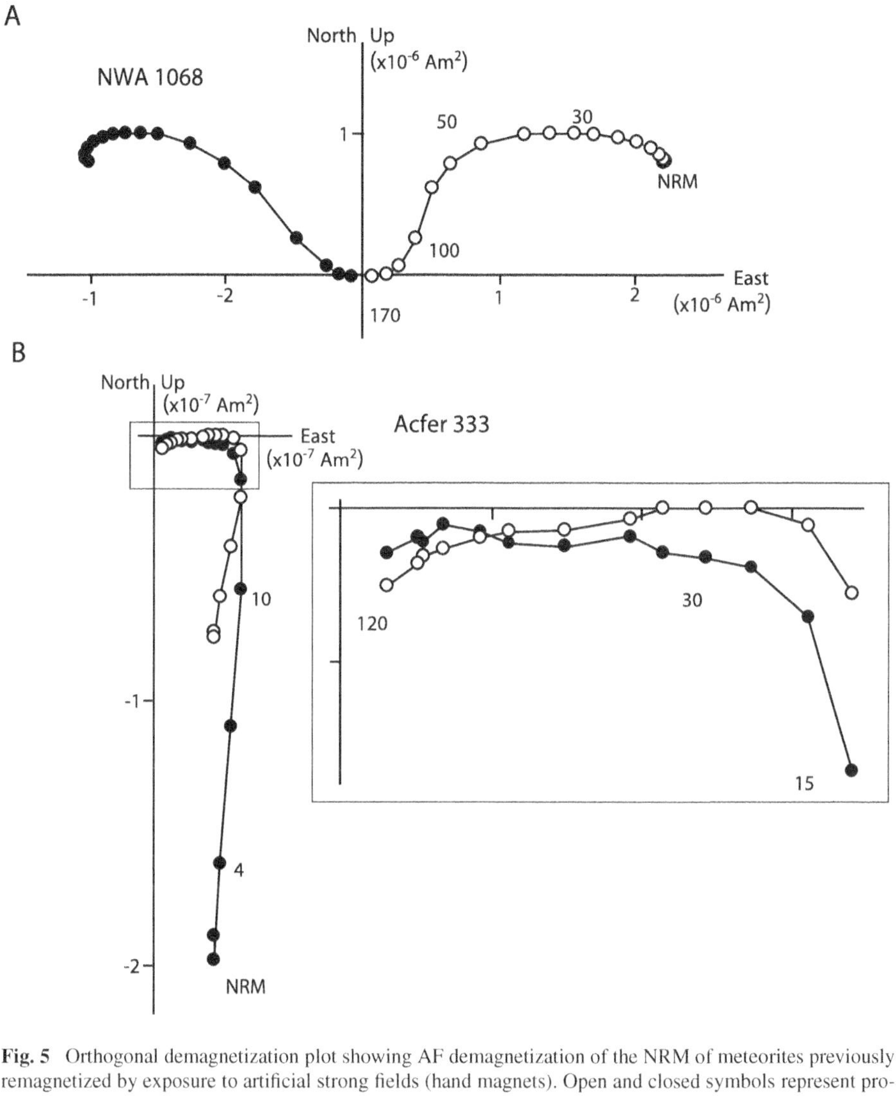

Fig. 5 Orthogonal demagnetization plot showing AF demagnetization of the NRM of meteorites previously remagnetized by exposure to artificial strong fields (hand magnets). Open and closed symbols represent projections of the magnetization vector on vertical and horizontal planes, respectively. Selected AF demagnetization steps are labeled in mT. **A** Shergottite NWA 1068 has been totally remagnetized by strong fields and exhibits the corresponding typical curved demagnetization pattern. It has ratios of NRM to saturation IRM (sIRM) of $\sim$0.11 and derivatives of NRM/sIRM with respect to AF step (known as REM$'$) $\sim$ 0.1 throughout the whole demagnetization process. **B** CO chondrite Acfer 333 has been partially remagnetized by a strong field as indicated by REM$'$ $\sim$ 0.1 in the 0 to 15 mT AF range. However, a high coercivity magnetization component isolated between 30 and 120 mT was not reset by magnet exposure, as indicated by REM$'$ $\sim 2 \times 10^{-3}$ over this AF range. Adapted from Gattacceca and Rochette (2004)

Although they have never experienced subsequent melting, their constituents experienced varying amounts of aqueous alteration and thermal metamorphism in the solar nebular and/or on their parent planetesimals. Because chondrites sample bodies that have never experienced total melting, their magnetic record remains less well understood relative to

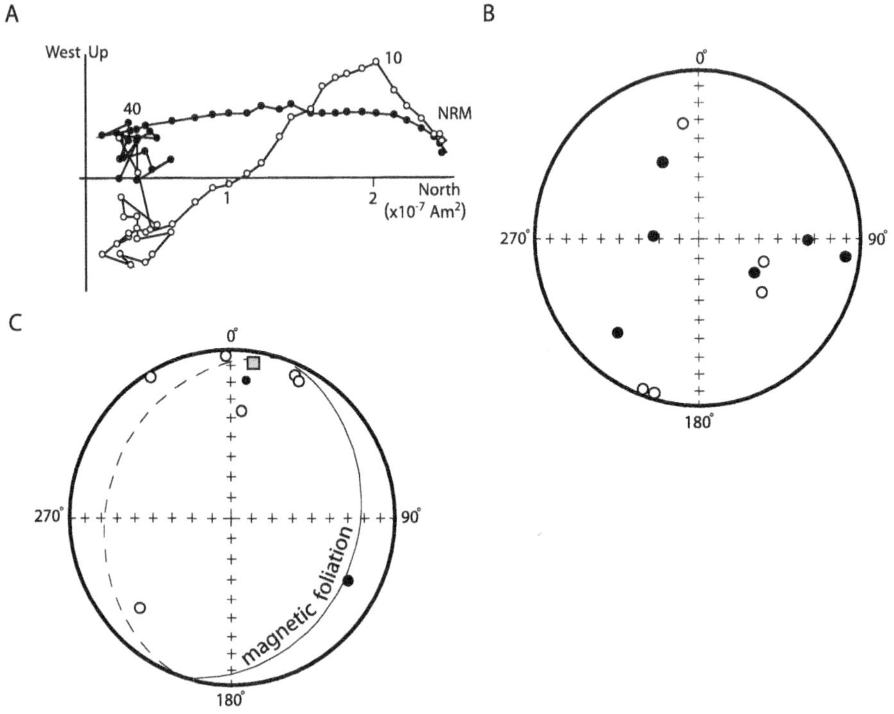

Fig. 6 NRM in the interior of the Bensour LL6 meteorite. **A** Orthogonal demagnetization plot of a ∼ 0.4 g sample of Bensour. Open and solid symbols represent projections of the endpoint of the magnetization vector on vertical and horizontal planes, respectively. Selected demagnetization steps are listed in mT. At least three components of magnetization can be distinguished. **B** Equal area stereographic projection of the medium coercivity (10 to 40 mT) components (corrected for anisotropy of remanence) of 11 mutually oriented samples of Bensour. **C** Equal area stereographic projection of the low coercivity components (NRM to 10 mT) of 8 mutually oriented samples of Bensour. *The grey box* is the magnetic lineation and *the great circle* is the magnetic foliation as determined by anisotropy of magnetic susceptibility and remanence. Adapted from Gattacceca et al. (2003)

igneous meteorites (Sect. 4). A summary of previous paleomagnetic analyses of ordinary and non-ordinary chondrites is given in Tables 1 and 2.

3.1 Ordinary Chondrites

Ordinary chondrites, the most common type of meteorites, have been the target of a large number of paleomagnetic studies starting with Lovering (1959) and Stacey and Lovering (1959). Although most early studies concluded that ordinary chondrites had recorded dynamo-generated or solar extraterrestrial magnetic fields (Stacey and Lovering 1959; Stacey et al. 1961; Lovering 1962), it is now clear that the interpretation of the remanent magnetization of these meteorites is likely to have been substantially altered since formation by subsequent thermal, shock, and low-temperature recrystallization processes.

A major difficulty for studying ordinary chondrites is the low coercivity of their main constituent ferromagnetic phases, multidomain kamacite and taenite. This is a key reason why the NRMs of most H and L chondrites exhibit erratic changes in direction and intensity under AF or thermal demagnetization (Brecher and Ranganayaki 1975; Brecher et al. 1977;

Table 1 Summary of previous paleomagnetic studies of ordinary chondrites. All names and classifications are as listed in the Meteoritical Bulletin Database. Other unofficial names are listed in parentheses. Abstracts and unpublished manuscripts were generally excluded unless they were the sole reference for a given meteorite. Numbers following group identification give petrologic type (Hutchison 2004). The extensive NRM measurements (no demagnetization) of Pochtarev and Guskova (1962), Guskova (1969, 1970) are not listed here

Name	Group	References
ALHA76008 (ALH 768)	H6	Westphal (1986)
Achilles	H5	Larson et al. (1973)
Allegan	H5	Larson et al. (1973)
Avanhandava	H4	Kohout et al. (2006)
Barbotan	H5	Brecher and Leung (1979)
Bath	H4	Westphal and Whitechurch (1983), Kukkonen and Pesonen (1983), Westphal (1986)
Borodino	H5	Guskova (1963)
Brownfield (1937)	H3	Brecher and Leung (1979)
Burdett	H5	Brecher and Leung (1979)
Cavour	H6	Brecher and Leung (1979)
Chamberlin	H5	Kukkonen and Pesonen (1983)
Clovis	H3	Kukkonen and Pesonen (1983)
Farley	H5	Brecher and Leung (1979)
Fleming	H3	Pesonen et al. (1993)
Forest City	H5	Kukkonen and Pesonen (1983)
Gilgoin (Gilgoin Station)	H5	Brecher and Leung (1979)
Gladstone	H4	Weaving (1962a)
Gorlovka	H3	Guskova (1982a)
Hessle	H5	Kukkonen and Pesonen (1983)
Holyoke	H4	Brecher and Leung (1979)
Horace	H5	Weaving (1962a)
Hugoton	H5	Kukkonen and Pesonen (1983)
Indio Rico	H6	Westphal and Whitechurch (1983)
Kargapole	H4	Guskova (1983, 1988)
Kernouve (Clegueric)	H6	Weaving (1962a)
Kesen (Kessen)	H4	Nagata and Sugiura (1977), Sugiura (1977), Kukkonen and Pesonen (1983)
Lancon	H6	Brecher and Ranganayaki (1975)
Leighton	H5	Larson et al. (1973)
Markovka	H4	Guskova (1988)
Metsäkylä	H4	Kukkonen and Pesonen (1983)
Mooresfort	H5	Westphal and Whitechurch (1983)
Morland	H6	Larson et al. (1973)
Mount Browne	H6	Stacey and Lovering (1959), Stacey et al. (1961)
Oakley Stone	H6	Westphal (1986), Westphal and Whitechurch (1983)

Table 1 (*Continued*)

Name	Group	References
Ochansk	H4	Westphal and Whitechurch (1983), Guskova (1963); Brecher and Ranganayaki (1975), Guskova (1976a), Brecher and Leung (1979)
Orlovka	H5	Guskova and Pochtarev (1969)
Petropavlovka	H4	Guskova (1983)
Plainview (1917)	H5	Wasilewski and Dickinson (2000), Wasilewski et al. (2002)
Prairie Dog Creek	H3	Westphal (1986), Westphal and Whitechurch (1983)
Pultusk	H5	Guskova (1963), Guskova (1976a), Brecher and Leung (1979) Westphal and Whitechurch (1983), Westphal (1986), Guskova (1965a)
Quenggouk	H4	Brecher and Ranganayaki (1975)
Raguli	H3	Guskova (1988)
Richardton	H5	Yu et al. (2009)
Rose City	H5	Brecher and Ranganayaki (1975)
Saline	H5	Brecher and Leung (1979)
Seminole	H4	Nagata and Sugiura (1977)
Severny Kolchim	H3	Guskova (1988)
Ställdalen	H5	Guskova (1963), Kukkonen and Pesonen (1983)
Sverdlovsk	H4/5	Guskova (1988)
Tysnes Island	H4	Brecher and Ranganayaki (1975)
Vernon County	H6	Kukkonen and Pesonen (1983)
Y-694	H6	Nagata (1979c)
Y-7301	H5/6	Nagata and Sugiura (1977), Nagata (1979c)
Y-7312	H5	Sugiura (1977)
Y-74014	H6	Westphal (1986)
Y-74371	H4	Sugiura (1977), Westphal (1986)
Y-74640	H6	Westphal (1986)
Y-74647	H5	Sugiura (1977), Nagata (1979c), Westphal (1986)
Yatoor (Nellore)	H5	Weaving (1962a), Guskova (1983)
Yonozu	H4/5	Nagata and Sugiura (1977)
Zhovtnevyi	H6	Guskova (1963, 1976a)
ALHA76009 (ALH 769)	L6	Nagata (1979c), Funaki et al. (1981)
ALHA77260	L3	Nagata and Funaki (1982), Nagata (1983)
Alfianello	L6	Kukkonen and Pesonen (1983), Gattacceca and Rochette (2004)
Andover	L6	Brecher and Ranganayaki (1975)
Aumale	L6	Brecher and Ranganayaki (1975)
Bachmut	L6	Guskova (1983)
Bakhardok	L6	Guskova (1988)
Bald Mountain	L4	Brecher and Ranganayaki (1975)
Barratta	L4	Stacey et al. (1961), Kukkonen and Pesonen (1983)
Bjurböle	L/LL4	Stacey et al. (1961), Brecher and Ranganayaki (1975), Sugiura and Strangway (1982), Kukkonen and Pesonen (1983), Wasilewski and Dickinson (2000), Wasilewski et al. (2002), Acton et al. (2007)

Table 1 (*Continued*)

Name	Group	References
Bluff (a or b?)	L5 or L4?	Guskova (1982a), Kukkonen and Pesonen (1983)
Brewster	L6	Weaving (1962a, 1962b)
Bruderheim	L6	Larson et al. (1973), Kukkonen and Pesonen (1983)
Buschhof	L6	Brecher and Ranganayaki (1975), Kukkonen and Pesonen (1983)
Cabezo de Mayo	L/LL6	Brecher and Ranganayaki (1975)
Calliham	L6	Larson et al. (1973)
Chateau-Renard	L6	Guskova (1963)
Dalgety Downs	L4	Nagata and Sugiura (1977)
Elenovka	L5	Guskova and Pochtarev (1969)
Ergheo	L5	Kukkonen and Pesonen (1983)
Farmington	L5	Stacey et al. (1961), Larson et al. (1973), Guskova (1982a)
FRO 01064	L6	Gattacceca and Rochette (2004)
FRO 01097	L6	Gattacceca and Rochette (2004)
Fukutomi	L5	Nagata and Sugiura (1977)
Gifu (Mino)	L6	Sugiura (1977), Nagata and Sugiura (1977)
Hallingeberg	L3	Gattacceca and Rochette (2004)
Holbrook	L/LL6	Kukkonen and Pesonen (1983)
Homestead	L5	Stacey and Lovering (1959), Stacey et al. (1961)
Julesburg	L3	Gattacceca and Rochette (2004)
Kunashak	L6	Guskova (1976a)
L'Aigle	L6	Guskova (1963, 1976a)
La Lande	L5	Larson et al. (1973)
Ladder Creek	L6	Gattacceca and Rochette (2004)
Leedey	L6	Gattacceca and Rochette (2004)
Long Island	L6	Gattacceca and Rochette (2004)
Marion (Iowa)	L6	Guskova (1983)
McKinney	L4	Guskova (1982a)
Melrose	L5	Larson et al. (1973)
Mezö-Madaras	L3	Guskova and Pochtarev (1969), Sugiura and Strangway (1982)
Mocs	L5-6	Guskova (1963)
Monte Milone	L5	Gattacceca and Rochette (2004)
Ness County	L6?	Larson et al. (1973)
Ozernoe	L6	Guskova (1988)
Pavlograd (Mordvinovka)	L6	Guskova (1963, 1976a)
Pervomaisky	L6	Guskova and Pochtarev (1969)
Potter	L6	Larson et al. (1973)
Rakovka	L6	Guskova (1963, 1976a)
Salla	L6	Kukkonen and Pesonen (1983)
St. Michel	L6	Kukkonen and Pesonen (1983)
Saratov	L4	Guskova (1963, 1976a)
Sevrukovo	L5	Guskova (1963)
Slobodka	L4	Guskova (1963)
Tadjera	L5	Brecher and Ranganayaki (1975)

Table 1 (*Continued*)

Name	Group	References
Tarbagatai	L5	Guskova and Pochtarev (1969), Guskova (1976a)
Tathlith	L6	Gattacceca and Rochette (2004)
Tsarev	L5	Guskova (1982b, 1988)
Utrecht	L6	Brecher and Ranganayaki (1975)
Valkeala	L6	Kukkonen and Pesonen (1983)
Y-7304	L6	Sugiura (1977), Nagata (1979c)
Y-7305	L6	Nagata (1979c)
Y-74191	L3	Sugiura (1977), Nagata (1979c, 1979e), Sugiura and Strangway (1982)
Y-74362	L6	Sugiura (1977), Nagata (1979c, 1979d)
Zavetnoe	L6	Guskova (1963)
Zavid	L6	Brecher and Ranganayaki (1975)
ALHA77304	LL3	Nagata et al. (1986)
Appley Bridge	LL6	Larson et al. (1973)
Arcadia	LL6	Larson et al. (1973)
Beeler	LL6	Gattacceca and Rochette (2004)
Bensour	LL6	Gattacceca et al. (2003), Gattacceca and Rochette (2004)
Chainpur	LL3	Wasilewski and Dickinson (2000), Sugiura and Strangway (1982)
Dhurmsala	LL6	Brecher et al. (1977), Kukkonen and Pesonen (1983), Gattacceca and Rochette (2004)
Ensisheim	LL6	Brecher et al. (1977), Kukkonen and Pesonen (1983)
Guidder	LL5	Pesonen et al. (1993), Gattacceca and Rochette (2004)
Hamlet	LL4	Larson et al. (1973)
Jelica	LL6	Brecher and Ranganayaki (1975)
Kelly	LL4	Larson et al. (1973)
Kilabo	LL6	Gattacceca et al. (2003), Gattacceca and Rochette (2004)
Krymka	LL3	Gattacceca and Rochette (2004)
Lake Labyrinth	LL6	Brecher et al. (1977)
Manbhoom	LL6	Kukkonen and Pesonen (1983), Gattacceca and Rochette (2004)
Melnikovo	LL6	Guskova (1988)
Olivenza	LL5	Larson et al. (1973), Brecher et al. (1977), Gattacceca et al. (2003), Gattacceca and Rochette (2004)
Oued el Hadjar	LL6	Gattacceca and Rochette (2004)
Parnallee	LL3	Brecher et al. (1977)
Saint-Séverin	LL6	Brecher and Ranganayaki (1975), Brecher et al. (1977), Sugiura (1977), Nagata and Funaki (1982), Nagata (1983) Nagata et al. (1986)
Soko-Banja	LL4	Brecher and Ranganayaki (1975), Brecher et al. (1977) Kukkonen and Pesonen (1983)
St. Mesmin	LL6	Brecher et al. (1977), Gattacceca and Rochette (2004)
Vavilovka	LL6	Brecher and Ranganayaki (1975)
Y-7307	LL6[a]	Nagata (1979a, 1979b, 1979e)
Y-74160	LL7	Nagata and Funaki (1982), Nagata (1983, 1993)
Y-74442	LL4	Nagata (1979c)
Y-74646	LL6	Nagata (1979c, 1979d)
Krutikha	OC ung.	Guskova (1988)

[a]Identified as a howardite by Nagata (1979a), Miyamoto et al. (1978)

Table 2 Summary of previous paleomagnetic studies of non-ordinary chondrites. All names and classifications are as listed in the Meteoritical Bulletin Database. Other unofficial names are listed in parentheses. Abstracts and unpublished manuscripts were generally excluded unless they were the sole reference for a given meteorite. Numbers following group identification give petrologic type (Hutchison 2004). CV chondrites are assigned to the reduced (CV_{red}), Allende-like oxidized (CV_A), and Bali-like oxidized (CV_B) subgroups (Weisberg et al. 2006) and CB chondrites are assigned to subtypes "a" (large chondrules) and "b" (small chondrules)

Name	Group	References
Enstatite		
Abee	EH4	Brecher and Ranganayaki (1975), Sugiura and Strangway (1981, 1982, 1983) Kukkonen and Pesonen (1983), Guskova (1985, 1987)
Adhi Kot	EH4	Guskova (1985, 1987)
Atlanta	EL5	Guskova (1985, 1987)
Daniel's Kuil	EL6	Guskova (1985, 1987)
Hvittis	EL6	Brecher and Ranganayaki (1975), Kukkonen and Pesonen (1983), Guskova (1985, 1987)
Indarch	EH4	Sugiura and Strangway (1982), Guskova (1985)
Khairpur	EL6	Brecher and Ranganayaki (1975), Guskova (1985, 1987)
Kota–Kota	EH4	Guskova (1985, 1987)
Neuschwanstein	EL6	Kohout et al. (2010)
Pillistfer	EL6	Guskova (1985, 1987)
St. Mark's	EH5	Guskova (1985, 1987)
Y-691 (Y-a)	EH3	Sugiura and Strangway (1982), Nagata et al. (1975), Nagata (1979c) Sugiura (1977)
Carbonaceous		
Bencubbin	CBa	Guskova (1970)
QUE 94411	CBb	Wasilewski (2000)
Ivuna	CI1	Brecher and Arrhenius (1974), Guskova (1976b)
Orgueil	CI1	Banerjee and Hargraves (1971, 1972), Brecher and Arrhenius (1974), Guskova (1976b, 1978), Nagata (1979b, 1979e), Kukkonen and Pesonen (1983)
Karoonda	CK4	Brecher and Arrhenius (1974), Guskova (1976b), Sugiura (1977), Nagata (1979b, 1979e), Acton et al. (2007)
Kobe	CK4	Funaki and Nakamura (2002)
Y-693 (Y-c)	CK4/5	Nagata et al. (1975), Sugiura (1977), Nagata (1979b, 1979c)
Acfer 331	CM2	Gattacceca and Rochette (2004)
Boriskino	CM2	Guskova (1976b)
Cold Bokkeveld	CM2	Banerjee and Hargraves (1971), Larson et al. (1973), Brecher and Arrhenius (1974), Guskova (1976a, 1976b, 1978)
Haripura	CM2	Brecher and Arrhenius (1974), Guskova (1976b)
Kivesvaara	CM2	Pesonen et al. (1993)
Mighei	CM2	Banerjee and Hargraves (1971, 1972), Larson et al. (1973), Brecher and Arrhenius (1974), Guskova (1976b, 1976a, 1978, 1983), Nagata (1979b, 1979e)

Table 2 (*Continued*)

Name	Group	References
Murchison	CM2	Banerjee and Hargraves (1972), Larson et al. (1973), Brecher and Arrhenius (1974), Guskova (1976b, 1983), Kletetschka et al. (2003)
Murray	CM2	Brecher and Arrhenius (1974), Larson et al. (1973), Guskova (1976b, 1983)
Nawapali	CM2	Larson et al. (1973), Guskova (1976b, 1976a, 1978)
Nogoya	CM2	Banerjee and Hargraves (1971), Guskova (1976b, 1976a)
Y-74662	CM2	Nagata (1979b, 1979c), Brecher (1980), Nagata and Funaki (1989), Nagata et al. (1991), Nagata (1993)
Acfer 333	CO3	Gattacceca and Rochette (2004)
Felix	CO3	Larson et al. (1973)
Kainsaz	CO3	Guskova (1976b, 1978), Larson et al. (1973)
Lancé	CO3	Larson et al. (1973)
Ornans	CO3	Pochtarev and Guskova (1962), Guskova (1976b, 1978), Kukkonen and Pesonen (1983)
Warrenton	CO3	Herndon (1974)
Y-81020	CO3	Nagata and Funaki (1989), Nagata et al. (1991), Nagata (1993)
Renazzo	CR2	Guskova (1976b), Larson et al. (1973), Brecher and Arrhenius (1974)
Allende	CV_A3	Butler (1972), Banerjee and Hargraves (1972), Larson et al. (1973), Brecher and Arrhenius (1974), Sugiura (1977), Guskova (1976b), Lanoix et al. (1977), Guskova (1978), Lanoix et al. (1978), Nagata (1979a, 1979b, 1979e), Sugiura et al. (1979), Wasilewski (1981), Wasilewski and Saralker (1981), Nagata and Funaki (1983), Guskova (1983), Sugiura et al. (1985), Nagata and Funaki (1989), Nagata et al. (1991), Nagata (1993), Funaki (2005), Acton et al. (2007), Acton et al. (2007), Carporzen et al. (2009)
Efremovka	$CV_{red}3$	Guskova (1976b)
Grosnaja	CV_B3	Larson et al. (1973), Guskova (1976b)
Kaba	CV_B3	Larson et al. (1973)
Leoville	$CV_{red}3$	Larson et al. (1973), Sugiura (1977), Nagata and Sugiura (1977), Nagata (1979b), Guskova (1983), Nagata et al. (1991)
Mokoia	CV_B3	Stacey et al. (1961), Brecher and Arrhenius (1974)
Vigarano	$CV_{red}3$	Pochtarev and Guskova (1962), Brecher and Arrhenius (1974), Larson et al. (1973), Guskova (1976b)
Coolidge	C4 ung.	Larson et al. (1973), Brecher and Arrhenius (1974)
EET96026	C4-5 ung.	Gattacceca and Rochette (2004)
Tagish Lake	C2 ung.	Gattacceca and Rochette (2004)
Essebi	C2 ung.	Larson et al. (1973)
Rumuruti-like		
A-881988	R4	Gattacceca and Rochette (2004)
ALH 85151	R3	Gattacceca and Rochette (2004)
PCA 91002	R3-6	Gattacceca and Rochette (2004)
PRE 95411	R3	Gattacceca and Rochette (2004)

Brecher and Leung 1979; Funaki et al. 1981; Sugiura and Strangway 1982; Westphal and Whitechurch 1983; Collinson 1987; Morden 1992a; Morden and Collinson 1992; Gattacceca and Rochette 2004). Despite their low coercivity (1.0 ± 0.8 mT, $n = 15$ for H chondrites and 5.3 ± 4.0 mT, $n = 20$ for L chondrites; Gattacceca's unpublished data), H and L chondrites can have high coercivity of remanence (39.4 ± 40.8 mT, $n = 15$ for H chondrites and 153 ± 124 mT, $n = 20$ for L chondrites; Gattacceca's unpublished data), possibly due to a small amount of tetrataenite (Sect. 2.5). LL ordinary chondrites (e.g., Bensour) possess more stable NRM which is mostly carried by abundant tetrataenite (Collinson 1987; Gattacceca et al. 2003) (Fig. 6). Despite its very high coercivity, this mineral is unfortunately not ideally suited for paleomagnetism because it is a secondary phase whose remanent magnetization cannot be simply related to the magnetization of its precursor taenite grain (see Sect. 2.5).

A second major difficulty with studying ordinary chondrites is the small-scale (down to millimeter) heterogeneity of directions of magnetization in L and LL chondrites (Funaki et al. 1981; Collinson 1987; Morden 1992a; Morden and Collinson 1992; Gattacceca et al. 2003). This heterogeneity is observed almost ubiquitously [for some unusual exceptions, see (Sugiura and Strangway 1982; Westphal and Whitechurch 1983)], even in chondrites that suffered parent body metamorphism to temperatures in excess of the Curie temperatures of kamacite, taenite and tetrataenite. This implies that, at least for these equilibrated chondrites, the magnetization postdates parent body thermal metamorphism. The origin of the small-scale scatter of the NRM directions may be from shock-induced brecciation or related to the formation tetrataenite (see Sect. 2.5). In the Bensour LL6 meteorite, a coherent soft magnetization was also isolated blocked by grains with coercivities below 25 mT and oriented parallel to the magnetic lineation (Gattacceca et al. 2003) (Fig. 6). This magnetization may be shock-related or have a viscous origin. The significance of directional homogeneity observed in some H chondrites (Westphal and Whitechurch 1983) is currently unclear.

A third problem is that ordinary chondrite finds should not be considered for paleomagnetic studies because of rapid weathering of metallic phases and associated remagnetization in the Earth magnetic field (e.g., Sect. 2.6). It is noteworthy that for H and L chondrites, finds have much more stable NRM than falls (Gattacceca and Rochette 2004), which is strong evidence that the NRM of finds was acquired as a crystallization remanent magnetization on Earth through weathering.

In conclusion, it is now clear that the high (>10 μT) paleointensities derived from Thellier–Thellier analyses of H, L and LL chondrites in earlier studies (e.g., Nagata and Sugiura 1977; Westphal and Whitechurch 1983; Pesonen et al. 1993) are overestimates and should almost certainly be discarded, as suggested by Westphal (1986), Pesonen et al. (1993). This is because heating of the samples almost invariably led to destabilization of metallic phases (in particular tetrataenite), and also because the laboratory TRM was often matched to a large and soft component of NRM rather than to any smaller stable components. Paleofields in the several μT range or lower (Brecher and Ranganayaki 1975; Brecher et al. 1977; Gattacceca and Rochette 2004) are more realistic.

Progress in the field of ordinary chondrite paleomagnetism should come from additional detailed studies of selected meteorites. Only falls should be studied. Unequilibrated chondrites are the best candidates for investigating possible pre-accretion magnetization, but the unstable magnetization of H and L chondrites and the unknown magnetization mechanism of tetrataenite make this task difficult. In contrast, the study of equilibrated chondrites heated above their Curie temperatures during parent body metamorphism could lead to a new understanding of the small-scale heterogeneous remanent magnetization and the very low (possibly null) field magnetization processes.

3.2 Enstatite (E) Chondrites

Enstatite chondrites are highly reduced, enstatite-rich, olivine-poor meteorites that form two chemical groups: low-Fe (EL) and high-Fe (EH) (Hutchison 2004; Weisberg et al. 2006). Although enstatite chondrites from both groups have been surveyed paleomagnetically (Guskova 1985, 1987; Nagata et al. 1975; Brecher and Ranganayaki 1975; Sugiura and Strangway 1982), by far the best studied is the Abee EH4 meteorite (Brecher and Ranganayaki 1975; Guskova 1985, 1987; Sugiura and Strangway 1981, 1982, 1983). Its NRM is carried by cohenite and kamacite, with cohenite having a more stable component. NRM directions within a given clast are reasonably grouped whereas they widely differ from clast to clast. The NRM directions of matrix subsamples are also grouped. This positive conglomerate test (e.g., Butler 1992) suggests that the magnetization of the clasts predates brecciation of the meteorite and that the clasts have not been heated above the Curie temperature of cohenite (215°C) after brecciation. In view of the shock history of this meteorite, the clast magnetization is probably a TRM acquired after impact melting. Brecciation caused by subsequent impacts randomized the magnetization directions. However, it is difficult to interpret the very high paleointensity values (in the 100 μT–1 mT range) measured by Sugiura and Strangway. New paleomagnetic studies of both high-grade and low-grade enstatite chondrite are clearly necessary.

3.3 Carbonaceous Chondrites

Carbonaceous chondrites offer the most pristine rock record from the preaccretional phase of the early solar system. CI chondrites, although extensively aqueously altered, provide the closest match to the composition of the sun and are therefore thought to be the least chemically fractionated with respect to the bulk solar system (Anders and Grevesse 1989). CV chondrites are less aqueously altered than CI chondrites and contain the largest and most abundant CAIs, the oldest known solar system solids (Brearley and Jones 1998). Several classes of carbonaceous chondrites also contain abundant chondrules, which are collectively the largest mass of preaccretional samples available and which formed from 0–5 Ma after CAIs (Amelin and Krot 2007; Russell et al. 2006). As a result, it has long been recognized that carbonaceous chondrites and their constituents (refractory inclusions and chondrules) potentially offer records of magnetic fields from the solar nebula and the T Tauri Sun.

Of the eight carbonaceous chondrite groups, the CV and CM chondrites are by far the best studied, followed by CO and CI chondrites (Table 2). CK and CR chondrites are unstudied with the exception of a small number of analyses on Karoonda and Renazzo, respectively. Other than an abstract (Wasilewski 2000) and an NRM measurement by Guskova (1970), we are aware of no published studies of CB chondrites. We are aware of no paleomagnetic data at all for CH and CR chondrites.

Given the great petrologic and geochemical complexity of carbonaceous chondrites, detailed paleomagnetic datasets are necessary in order to come to clear conclusions about the magnetic field record. Extensive paleomagnetic data (e.g., analyses of mutually oriented samples, thermal demagnetization, and fusion crust baked contact tests; see Sect. 2) are only available for Allende (CV3). Therefore we will focus on this meteorite and also briefly discuss the much smaller datasets available for Murchison (CM2), Orgueil (CI1) and Karoonda (CK4).

The CV meteorite Allende, a >2000 kg fall in 1969, is the best studied chondrite of any kind and one of the best paleomagnetically studied rocks in history (nearly two dozen studies by ~9 different groups). Allende and other CV chondrites contain the ferromagnetic phases pyrrhotite, magnetite and metal. Metal is in the form of awaruite ($FeNi_3$)

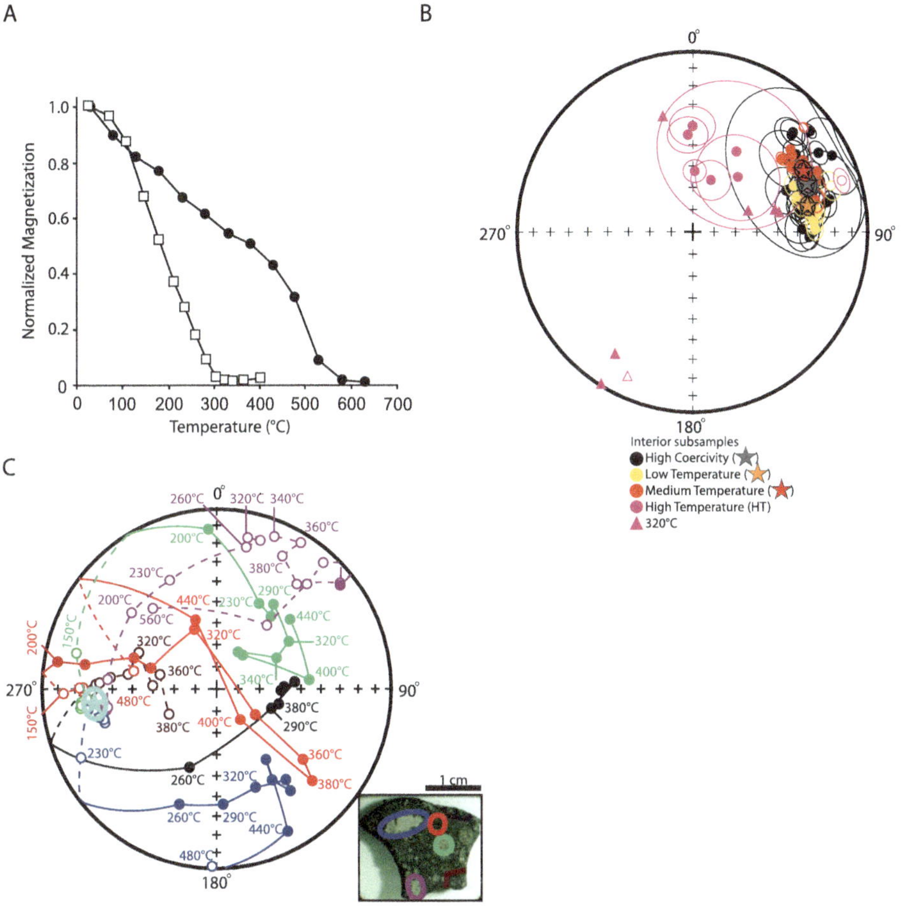

Fig. 7 NRM in the CV carbonaceous chondrite Allende. **A** Remanent magnetization (normalized to the room-temperature value) of bulk samples during thermal demagnetization. *Open symbols* = demagnetization of NRM by Sugiura and Strangway (1985) (their subsample 64). *Closed symbols* = demagnetization of an ARM acquired in a peak ac field of 140 mT and dc bias field of 44 μT by Nagata and Funaki (1983). **B** Equal area stereographic projection showing directions of primary magnetization components of samples from the interior of Allende. *Solid symbols* = lower hemisphere; *open symbols* = upper hemisphere. Ellipsoids are defined as maximum angular deviations associated with the least square fits to sample data. Stars and their ellipsoids represent the average directions and associated 95% confidence intervals. Samples represented by triangles were only thermally demagnetized to 320°C; the directions shown for these are the moments after demagnetization to this temperature (rather than least squares fits). After (Carporzen et al. 2009). **C** Equal area stereographic projection showing directions of NRM during thermal demagnetization of Allende interior subsamples (different parent stone from that shown in **B**). *Inset* shows approximate localizations of subsamples in parent sample prior to extraction. Subsamples are two fine-grained spinel-rich CAIs (*red* and *pink colors*), a porphyritic olivine-pyroxene chondrule (*green color*), an Al-rich chondrule (*blue color*), and two bulk, matrix-rich samples (*brown* and *black colors*). *Light blue* star and ellipsoid are the average NRM direction for the 6 subsamples and associated 95% confidence interval. The bulk samples are demagnetized by <380°C, while CAIs and chondrules retain magnetization directions that move along great circle paths up to ~500°C. After (Carporzen et al. 2009)

in Allende and other oxidized CVs and kamacite in reduced CVs (Krot et al. 1998). Allende shows no evidence of post-accretional shock (above 5 GPa) (Scott et al. 1992), and several fusion crust baked contact tests (e.g., Fig. 4a) have confirmed that it contains highly stable preterrestrial magnetization (Butler 1972; Nagata and Funaki 1983; Carporzen et al. 2009). More than 90% of the stable NRM carried by chondrules, CAIs and matrix is a middle-temperature (MT) component blocked up to only 290°C even though the meteorite contains grains with blocking temperatures up to ~600°C (Fig. 7A). The paleointensity of the magnetizing field which produced the MT magnetization has been estimated to be 10–100 μT (Butler 1972; Banerjee and Hargraves 1972; Sugiura et al. 1979; Nagata 1979b; Wasilewski 1981; Acton et al. 2007; Carporzen et al. 2009) (the anomalously high paleointensities up to 1600 μT of Lanoix et al. 1978 are highly circumspect given the lack of fusion crust tests and mutual subsample orientation). This component is unidirectional throughout the meteorite despite the lithologic heterogeneity and aggregational texture of the meteorite (Figs. 4a and 7b). Therefore, the MT magnetization was certainly acquired following accretion. Although thermal demagnetization above 300°C indicates that chondrules and refractory inclusions may have a much weaker high-temperature (HT) component oriented nonunidirectionally throughout the meteorite (Sugiura et al. 1979; Sugiura and Strangway 1985; Carporzen et al. 2009) (Fig. 7C), the HT component has not yet been demonstrated to decay to the origin and often does not reach a stable direction. Therefore, the possibility of a preaccretional remanence in Allende, which would be expected to be randomly oriented among individual chondrules, is tantalizing but not yet confirmed.

Although Allende's middle-temperature NRM component has been traditionally interpreted to be a record of solar system magnetic fields generated externally to the CV parent body (Levy and Sonett 1978; Sugiura and Strangway 1988; Nagata 1979b; Acton et al. 2007; Cisowski 1987; Stacey 1976), recent petrographic and geochronological data strongly indicate that it was acquired during slow cooling over several Ma ending at least 8 Ma after the formation of CAIs. This requires that the fields were unidirectional with respect to the parent body over long timescales at a time in solar system history after the likely dissipation of the protoplanetary disk and magnetically active Sun. Therefore, it has recently been concluded that the MT magnetization in Allende is most likely the record of an internal core dynamo active at ~4559–4560 Ma (Carporzen et al. 2009). This would indicate that the meteorite is a partially differentiated object with a chondritic crust and a melted interior including metallic core. Although this represents a radically new picture for chondrite parent bodies, it appears to be consistent with a variety of geochemical data and thermal models for the CV body and may provide a solution to several outstanding but seemingly unrelated problems in meteoritics and planetesimal formation (Elkins-Tanton et al. 2009; Carporzen et al. 2009).

The very limited paleomagnetic analyses available for most other carbonaceous chondrites (including other CVs) have also identified NRM, but this magnetization generally appears to be considerably less stable (lower mean destructive field) compared to Allende. For example, while Allende's whole rock NRM only drops by 10–30% following AF demagnetization to 50 mT (Banerjee and Hargraves 1972; Brecher and Arrhenius 1974; Nagata 1979b; Carporzen et al. 2009), the NRMs of whole rock Orgueil (CI1) and Murchison (CM2) drop by about an order of magnitude after AF demagnetization to 50 mT (Banerjee and Hargraves 1972; Larson et al. 1973; Guskova 1976b, 1978, Kukkonen and Pesonen 1983) (Fig. 8). Directional data (Fig. 10 of Brecher and Arrhenius 1974) suggest that Orgueil has been almost completely demagnetized by this AF level (although their analyzed subsamples may be contaminated with fusion crust magnetization). [Note that Orgueil "sample II" of Banerjee and

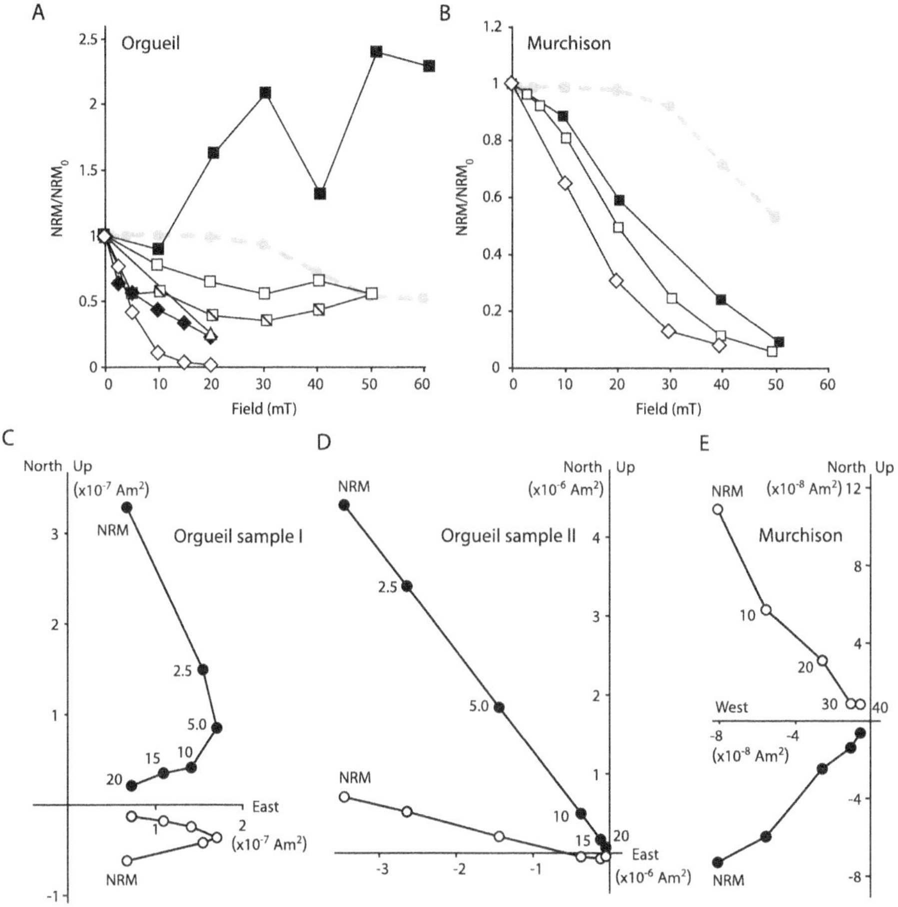

Fig. 8 NRM (normalized to the initial NRM value, NRM$_0$) in the carbonaceous chondrites Orgueil (CI1) and Murchison (CM2). **A** NRM intensity during AF demagnetization of 6 different bulk subsamples from the interior of Orgueil (no fusion crust). *Squares* = data of Guskova (1976b, 1978). *Diamonds* = data of Banerjee and Hargraves (1971) (*closed symbols* = Orgueil subsample I, *open symbols* = Orgueil subsample II), also shown in **C** and **D**. *Triangles* = data of Nagata (1979e). AF demagnetization data for Allende NRM are shown for context (*grey circles and dashed line*). Data of Guskova (1976b, 1978) (particularly above AF 20 mT) may be contaminated with spurious remanence from AF demagnetization, while Orgueil subsample II appears to be contaminated by secondary IRM (which leads to rapid demagnetization). **B** NRM (normalized to the initial NRM value, NRM$_0$) intensity during AF demagnetization of 3 different bulk subsamples from the interior of Murchison (no fusion crust). *Squares* = data of Guskova (1976b). *Circles* = data of Banerjee and Hargraves (1971) (*closed symbols* = Murchison subsample 2, *open symbols* = Murchison subsample 1). *Diamonds* = data of Larson et al. (1973), also shown in **E**. AF demagnetization data for Allende NRM are shown for context (*grey circles and dashed line*). **C, D** Orthogonal demagnetization plot of Orgueil subsamples I and II, respectively, computed from data of Banerjee and Hargraves (1971). Open and solid symbols represent projections of the endpoint of the magnetization vectors on vertical and horizontal planes, respectively. Selected demagnetization steps are listed in mT. These two subsamples are apparently not mutually oriented. **E** Orthogonal AF demagnetization plot of Murchison computed from data of Larson et al. (1973). Open and solid symbols represent projections of the endpoint of the magnetization vector on vertical and horizontal planes, respectively. Selected demagnetization steps are listed in mT

Hargraves 1971, which has ~30 times larger NRM than nearly all other studied Orgueil subsamples—see Fig. 1 of Guskova 1976b and three out of the four Orgueil samples listed in Appendix of Kukkonen and Pesonen 1983—and which demagnetizes straight to the origin (compare Fig. 8D with overprint up to AF 15 mT in Fig. 5B) has likely been remagnetized by a hand magnet or other secondary IRM. Note also that Orgueil subsample 4 and probably also subsample 2 of Guskova 1976b, 1978 may have acquired substantial spurious remanence during AF demagnetization (e.g., Sect. 2.2) as indicated by increasing moment magnitude in Fig. 8A coupled with directional changes toward one of the orthogonal measurement axes in Fig. 4 of Guskova 1976b.]

The NRMs of chondrules from Karoonda (CK4) are demagnetized by 96% by just AF 20 mT. Whole rock Karoonda samples are similarly unstable (Brecher and Arrhenius 1974). Murchison was also found to be much less resistant to thermal demagnetization analyses to 150°C than Allende or Orgueil (Banerjee and Hargraves 1972). Zero-field cycling of these magnetite-rich meteorites to 77 K leads to large changes in NRM intensity (Brecher and Arrhenius 1974; Kletetschka et al. 2003) (with whole rock samples from Karoonda demagnetized by 95%), indicating that Karoonda, Orgueil, and Murchison are highly susceptible to remagnetization and demagnetization effects from thermal cycling in space through magnetite's Verwey transition (see Sect. 2.4). Magnetic viscosity experiments (Kletetschka et al. 2003) suggest nearly all of Murchison's NRM could be accounted for as a VRM acquired in the Earth's field since its arrival on Earth in 1969.

Furthermore, given that Murchison and Orgueil were likely never heated even to ~150°C (Busemann et al. 2007; Cody et al. 2008) since formation, blocking-temperature relations for magnetite (Pullaiah et al. 1975) and kamacite (Garrick-Bethell and Weiss 2009) indicate that these minerals are unlikely to have retained a primary TRM from the early solar system (4.5 Ga). Only a crystallization remanent magnetization acquired during the formation of magnetite, recently dated to be 4560.3 ± 0.4 Ma (Hohenberg et al. 2000) (essentially contemporaneous with the likely age of Allende's NRM Carporzen et al. 2009), could have persisted from this early time. On the other hand, Karoonda, which has been heavily altered and metamorphosed to estimated temperatures of 500–600°C (Matza and Lipschutz 1977), could in principle retain an ancient TRM, although the age of metamorphism is currently poorly constrained at ~4500 Ma (Podosek 1970, 1971; Hohenberg et al. 2000). In any case, as described above, the limited data for these meteorites cannot rule out the hypothesis that they formed in near zero-field conditions (see Sect. 7.2). Similarly, Tagish Lake (C2 ungrouped) has a ratio of NRM to saturation IRM (sIRM) of ~3 to 9×10^{-4} (for 8 samples with masses between 2 to 6000 mg), among the lowest ever measured for any meteorite (Gattacceca's unpublished data), indicating that the high paleointensity published by Gattacceca and Rochette (2004) for this meteorite is attributable to contamination by a hand magnet. Such a scenario would be consistent with the standard paradigm of primitive, undifferentiated parent bodies and a lack of external field sources at the late time when they were thermally and chemically processed. A key way to further test this zero-field hypothesis is to demonstrate that these meteorites can acquire artificial laboratory magnetization (e.g., TRM or ARM) that is much more stable and intense than their NRM (similar to approach of Garrick-Bethell and Weiss 2009 and Lawrence et al. 2008 for lunar rocks).

3.4 Rumuruti-like (R) Chondrites

Rumuruti-like chondrites are a recently recognized highly oxidized, matrix-rich, almost metal-free chondrite group. The paleomagnetism of five bulk Rumuruti chondrites (all Antarctic finds) was studied by Gattacceca and Rochette (2004). All samples had a stable remanent magnetization mostly carried by pyrrhotite, with the exception of the magnetite-bearing A-881988 meteorite (Rochette et al. 2008). NRM normalization techniques indicated similar paleointensities around ~5 μT. In view of the pressure-induced magnetic phase transition of pyrrhotite at 2.8 GPa (e.g., Rochette et al. 2003a) and the typical peak shock pressures of Rumuruti chondrites (shock stage S2, with peak pressures >5 GPa), the NRM of Rumuruti chondrites is unlikely to predate the last major impact and must be regarded either as a shock remanent magnetization or, for highly-shocked samples, thermoremanence acquired during post-impact cooling. In either case, the paleointensity estimates suggest the existence of a (possibly transient in the SRM hypothesis) magnetic field at the surface of the Rumuruti parent body at the time of the impact event. Additional detailed paleomagnetic studies of Rumuruti chondrites are needed to infer the origin of this magnetic field.

3.5 Kakangari-like (K) Chondrites

Kakangari-like chondrites are another recently identified chondrite group distinguished by their high matrix and metal contents, oxidation state between H and enstatite chondrites and distinctive oxygen isotopic composition (Weisberg et al. 1996; Krot et al. 2007). We are aware of no paleomagnetic studies of these meteorites.

4 Paleomagnetic Records in Small-Body Stony Achondrites and Mesosiderites

4.1 The Howardite-Eucrite-Diogenite (HED) Clan

The members of the howardite-eucrite-diogenite clan form the most numerous and diverse of the small-body basaltic achondrite groups. Other than lunar and Martian meteorites and an anomalous ureilite (see Sect. 4.4), they are the only achondrites whose parent body (the asteroid 4 Vesta) has been confidently identified. This makes them key samples for studying the early global differentiation, petrogenesis, and the early thermal evolution of planetesimals. However, essentially all known HED meteorites were metamorphosed ≥800°C, brecciated and/or shocked (Metzler et al. 1995; Yamaguchi et al. 1996) at least several tens of Ma (and up to several hundred Ma) after their formation (Bogard and Garrison 2003; Kleine et al. 2005b; Kunz et al. 1995). Therefore, they certainly cannot contain records of early nebular or T Tauri fields. Given that metallic cores are only expected to be convective on Vesta-sized planetesimals for up to <~100 Ma (Weiss et al. 2008a; Elkins-Tanton et al. 2009), only HEDs with the oldest ^{40}Ar/^{39}Ar ages could potentially have retained magnetization from any internal core dynamo. Table 3 lists previous paleomagnetic studies of HEDs and other small-body achondrites.

The dominant ferromagnetic mineral in most HED meteorites is kamacite (Collinson 1994; Collinson and Morden 1994; Rochette et al. 2009a). Analysis of four unbrecciated eucrites found possibly coherent but generally extremely weak NRM, consistent with paleointensities of only order 1–5 μT (Cisowski 1991). The magnetization directions of two mutually oriented subsamples from each of these meteorites appear to be approximately unidirectional (divergent by < ~30°). Two of these meteorites, PCA 82502 and Moore County, have been dated with ^{40}Ar/^{39}Ar chronometry at 4.506 ± 0.033 and 4.48 ± 0.03 Ga, respectively. The former has in fact the oldest known precise ^{40}Ar/^{39}Ar age of any eucrite. At face value, this would seem to indicate that there were weak (several μT) magnetic fields on the HED parent body in the vicinity of these meteorites ~60–80 Ma after the formation of the parent body and solar system. However, because fusion crust baked contact tests

Table 3 Summary of previous paleomagnetic studies of small-body achondrites (excluding irons, pallasites, lunar and Martian meteorites). All names and classifications are as listed in the Meteoritical Bulletin Database with one exception (see below). Other unofficial names are listed in parentheses. Abstracts and unpublished manuscripts were generally excluded unless they were the sole reference for a given meteorite. E = eucrite, D = diogenite, H = howardite, mmict = monomict breccia, pmict = polymict breccia, cm = cumulate unbrecciated, unbr = unbrecciated. Letter and number for mesosiderites give petrologic class and metamorphic grade, respectively (see Hutchison 2004)

Name	Group/Petrologic class	References
Angrites		
A-881371		Weiss et al. (2008a)
Angra dos Reis		Weiss et al. (2008a)
D'Orbigny		Weiss et al. (2008a)
Aubrites		
Bishopville		Brecher et al. (1979), Guskova (1976a, 1984)
Cumberland Falls		Stacey et al. (1961), Larson et al. (1973)
Norton County		Pochtarev and Guskova (1962), Guskova (1976a, 1984), Kukkonen and Pesonen (1983), Gattacceca and Rochette (2004)
Peña Blanca Spring		Gattacceca and Rochette (2004)
Pesyanoe		Pochtarev and Guskova (1962), Guskova (1976a, 1984)
Howardites, eucrites, and diogenites		
DaG 684	E	Gattacceca and Rochette (2004)
ALHA76005	E	Nagata (1979a)
ALHA77302	E-pmict	Nagata (1980), Nagata and Dunn (1981)
ALHA78040	E-pmict	Nagata (1980)
ALHA81001	E-mmict[a]	Cisowski (1991)
FRO 97045	E-pmict	Gattacceca and Rochette (2004)
Juvinas	E-mmict	Pochtarev and Guskova (1962), Guskova (1976a), Brecher et al. (1979), Kukkonen and Pesonen (1983), Gattacceca and Rochette (2004)
Millbillillie	E-mmict	Morden (1992b)
Moore County	E-cm	Lovering (1959), Cisowski (1991)
Nobleboro	E-pmict	Brecher et al. (1979)
Pasamonte[b]	E-pmict	Brecher et al. (1979)
Petersburg	E-pmict[c]	Collinson and Morden (1994)
PCA 82502	E-unbr	Nagata (1979c), Collinson and Morden (1994)
Sioux County	E-mmict	Brecher et al. (1979), Kukkonen and Pesonen (1983), Collinson and Morden 1994
Stannern	E-mmict	Pochtarev and Guskova (1962), Guskova (1976a), Brecher et al. (1979), Kukkonen and Pesonen (1983)
Y-74159	E-mmict	Nagata (1979a, 1979b, 1979c, 1979d, 1979e)
Y-74450	E-pmict	Nagata (1979a, 1979b, 1979c, 1979d, 1979e)
Y-791195	E-cm	Nagata (1979c), Cisowski (1991)
ALHA77256	D	Brecher (1980)
EETA79002	D	Collinson and Morden (1994)
Johnstown	D	Brecher et al. (1979), Kukkonen and Pesonen (1983), Collinson and Morden (1994)

Table 3 (*Continued*)

Name	Group/Petrologic class	References
Roda	D	Brecher et al. (1979), Collinson and Morden (1994)
Shalka	D	Brecher et al. (1979), Collinson and Morden (1994)
Tatahouine	D	Brecher et al. (1979), Gattacceca and Rochette (2004)
Y-74013	D	Sugiura (1977), Nagata (1979a, 1979b, 1979c, 1979e)
Y-74037	D	Nagata (1979d, 1979e)
Y-74097	D	Nagata (1979a, 1979b, 1979e)
Y-74648	D	Nagata (1979a, 1979b, 1979e)
Y-75032	D	Nagata (1979a, 1979b, 1979d, 1979e)
Y-692 (b)	D	Nagata (1979a, 1979b, 1979c, 1979e)
EET 87503	H	Collinson and Morden (1994)
Kapoeta	H	Brecher et al. (1979), Collinson and Morden (1994)
Le Teilleul	H	Brecher et al. (1979), Collinson and Morden (1994)
Luotolax	H	Guskova (1976a)
Pavlovka	H	Pochtarev and Guskova (1962), Guskova (1976a), Brecher et al. (1979)
Yurtuk	H	Pochtarev and Guskova (1962), Guskova (1976a)
Y-7308	H	Nagata (1979c, 1979d)
Mesosiderites		
Bondoc	B4	Larson et al. (1973)
Clover Springs	A2	Larson et al. (1973)
Crab Orchard	A1	Guskova (1965b), Kukkonen and Pesonen (1983)
Estherville	A3/4	Guskova (1965b), Kukkonen and Pesonen (1983), Collinson (1991)
Hainholz	A4	Guskova (1965a, 1965b), Larson et al. (1973)
Mincy	B4	Larson et al. (1973)
Morristown	A3	Guskova (1969), Kukkonen and Pesonen (1983)
Ureilites		
ALHA77257		Nagata (1979e), Brecher (1980), Nagata (1980)
Dyalpur		Brecher (1980), Guskova (1982a)
Goalpara		Larson et al. (1973), Brecher and Fuhrman (1979)
Haverö		Neuvonen et al. (1972), Brecher and Fuhrman (1979), Kukkonen and Pesonen (1983)
Kenna		Brecher and Fuhrman (1979), Guskova (1982a)
Novo-Urei		Pochtarev and Guskova (1962), Larson et al. (1973), Guskova (1976a, 1982a), Kukkonen and Pesonen (1983)
Ungrouped		
GRA 06129		Shearer et al. (2008)
Ibitira[d]		Brecher (1980), Cisowski (1991)

[a]Unbrecciated according to Delaney et al. (1984), Cisowski (1991)

[b]New oxygen isotopic data (Scott et al. 2009) suggest Pasamonte may sample a parent distinct from the other HED meteorites in this table

[c]Classified as a howardite by Collinson and Morden (1994)

[d]Although Ibitira is classified as a eucrite in the Meteoritical Bulletin Database, we follow Mittlefehldt (2007) in classifying it as ungrouped

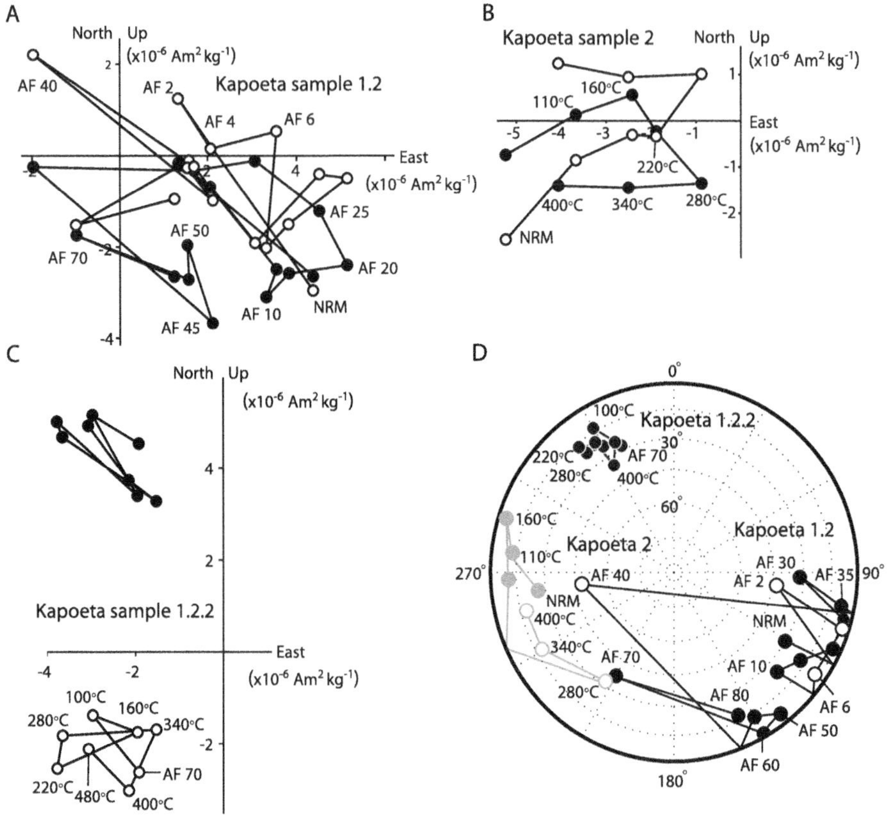

Fig. 9 NRM in three mutually oriented subsamples from the interior of the howardite Kapoeta. **A** Orthogonal projection of the NRM vector of sample 1.2 during AF demagnetization as computed from data of Collinson and Morden (1994). *Open and solid symbols* represent projections of the magnetization vector on vertical and horizontal planes, respectively. Selected demagnetization steps are listed in mT. **B** Orthogonal projection of the NRM vector of sample 2 during thermal demagnetization as computed from data of Collinson and Morden (1994). *Open and solid symbols* represent projections of the magnetization vector on vertical and horizontal planes, respectively. Selected demagnetization steps are listed in °C. **C** Orthogonal projection of the NRM vector of sample 1.2.2 during thermal demagnetization as computed from data of Collinson and Morden (1994). *Open and solid symbols* represent projections of the magnetization vector on vertical and horizontal planes, respectively. Selected demagnetization steps are listed in °C. Sample had been AF demagnetized to 70 mT prior to beginning thermal demagnetization. **D** Equal area stereographic projection of the NRM of each sample shown in **A**–**C** during AF or thermal demagnetization (*black symbols* = sample 1.2, *dark grey symbols* = samples 1.2.2, and *light grey symbols* = sample 2). *Open and solid symbols* represent projections of the magnetization vector on *upper and lower hemispheres*, respectively. Selected demagnetization steps are listed in mT or °C. Samples are mutually oriented

(Sect. 2.6), tests for spurious ARM and GRM remanence acquisition during AF demagnetization (Sect. 2.2), tests for magnetic viscosity, and detailed analysis of shock effects (Sect. 2.3) have not yet been conducted for these samples, this conclusion must be currently regarded as provisional.

The majority of HEDs are breccias and their magnetic carriers have been further strained after brecciation (Gattacceca et al. 2008b). Paleomagnetic analyses of mutually oriented subsamples of most brecciated and unbrecciated HEDs have observed NRMs with widely

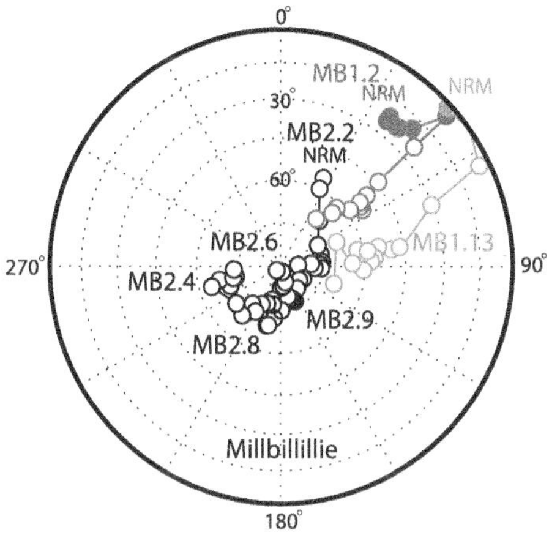

Fig. 10 NRM in seven mutually oriented subsamples from the interior of the eucrite Millbillillie. Shown is an equal area stereographic projection of the magnetization of each sample during AF demagnetization. *Open and solid symbols* represent projections of the magnetization vector on upper and lower hemispheres, respectively. Samples were demagnetized up to 100 mT and reached a stable, nearly vertical final direction. Sample names are listed next to each curve. Samples MB1.13 (*light grey symbols*) and MB1.2 (*dark grey symbols*) have overprints mostly removed by 100 mT, above which their directions are close to that of MB2.2, MB2.4, MB2.6, MB2.8, and MB2.9 (*black symbols*). Adapted from Morden (1992b)

scattered directions and that AF demagnetize erratically (Fig. 9) (Brecher et al. 1979; Collinson and Morden 1994). For the unbrecciated samples, such magnetization is clearly not a primary thermoremanent record. For the brecciated samples, the origin of the NRM is currently ambiguous: the scattered magnetization directions provide a positive conglomerate test demonstrating that the magnetization predates brecciation, indicate poor magnetic recording properties, and/or near-zero fields during the last remagnetization event. The hypothesis of pre-brecciation NRM can be distinguished from the other two mechanisms via identification of origin-trending, unidirectional magnetization within single clasts (e.g., Sugiura and Strangway 1983), a test which the howardite Kapoeta clearly fails (Fig. 9).

In stark comparison, seven mutually oriented subsamples of the Millbillillie eucrite impact melt breccia (Yamaguchi et al. 1994) exhibit a unidirectional magnetization component blocked to at least 100 mT and apparently acquired in a 6–37 μT paleofield (Morden 1992b) (Fig. 10). Given the uncertainty in the paleointensity methodology, this is within error of the paleointensities from unbrecciated eucrites discussed above. However, again, no baked contact tests, tests for spurious ARM and GRM remanence acquisition, and magnetic viscosity tests have been conducted. Assuming future such analyses confirm that the NRM is a primary thermoremanence, then ^{40}Ar/^{39}Ar chronometry likely dates this magnetization to the time of impact-induced heating at 3.55 ± 0.02 Ga (Yamaguchi et al. 1994). Although Morden (1992b) interpreted Millbillillie's paleomagnetism as evidence for a core dynamo, this age is so young that it is hard to account for the NRM by anything other than either a remanent crustal field on Vesta or, much less likely, an impact-generated field or close approach to another magnetized body.

4.2 Ibitira

Ibitira is an unbrecciated, vesicular metabasalt that fell in 1957 (Mittlefehldt 2005). Although it was long thought to be an unbrecciated eucrite, oxygen isotopic and major element data indicate it may be from a basaltic asteroid other than Vesta (Wiechert et al. 2004; Mittlefehldt 2005). Following metamorphism to peak temperatures of 1100°C, it cooled slowly until it reached the ~300°C closure temperature for the $^{40}Ar/^{39}Ar$ system at 4.495 ± 0.015 Ga (Bogard and Garrison 1995). Following metamorphism, it was shocked to ~10–30 GPa (Steele and Smith 1976). These shock effects combined with the 4.495 Ga age of thermoremanence acquisition (which is towards the end of the expected timescale of early dynamo activity; see Sect. 6) may explain why both paleomagnetic studies of mutually oriented subsamples Ibitira found highly nonunidirectional remanence (Cisowski 1991; Brecher 1980). Therefore, Ibitira's paleomagnetic record does not convincingly indicate the presence of a paleomagnetic field.

4.3 Mesosiderites

Mesosiderites are breccias containing clasts of basalt and metal in a fine-grained matrix. They are thought to be the products of the collision of a metal-rich planet onto the basaltic surface of one or more differentiated bodies (Mittlefehldt et al. 1998) or else the impact-induced mixing of the metallic core and silicate mantle on single bodies (Scott et al. 2001). The mineralogy, chemical composition, and oxygen isotopic composition of the silicates suggest that mesosiderites were derived from a parent body similar to (Rubin and Mittlefehldt 1993) or the same as (Greenwood et al. 2006) that of the HED meteorites. Because mesosiderites are shocked, polymict impactites that have experienced multiple episodes of crystallization, melting and metamorphism (Rubin and Mittlefehldt 1993), they make challenging targets for paleomagnetism. Several mesosiderites have been analyzed with AF demagnetization (Larson et al. 1973), but detailed paleomagnetic data have only been reported for Estherville. Analysis of 17 mutually oriented fragments of matrix and metal from Estherville demonstrate found magnetization directions are essentially randomly distributed (Collinson 1991). The major ferromagnetic minerals are kamacite and tetrataenite. Given the complex history of this meteorite, the fact that the matrix samples are themselves composites of smaller clasts, and the unknown method by which tetrataenite acquires NRM, it is not clear how to interpret these data. In any case, the fact that Estherville and other mesosiderites underwent extremely slow cooling (0.1–0.5°C Ma^{-1}) below 500–700°C (Mittlefehldt and Garrison 1998) means that if their NRM is a TRM, it must have been acquired long after any early putative core dynamo or early solar/disk had decayed.

4.4 Ureilites

The ureilites, the second largest group amongst achondrites, are carbon-rich, ultramafic, granular mostly unbrecciated rocks (Mittlefehldt 2007). They contain a variety of ferromagnetic phases, including predominantly kamacite (Rowe et al. 1975; Rochette et al. 2009a) and minor taenite, martensite (in shocked samples), sulfide, suessite (in the shocked North Haig breccia) (Keil et al. 1982), schreibersite, and cohenite (Berkley et al. 1980; Rubin 1997). Recent spectral observations of the meteoroid 2008 TC3 and recovered meteorite samples following its impact on Earth have provisionally tied at least some ureilites to F-type asteroids (Jenniskens et al. 2009). Despite their great number, we are aware of NRM studies of only six ureilites (Larson et al. 1973; Brecher 1980; Brecher and Fuhrman 1979; Nagata 1979e, 1980; Guskova 1982a). We will focus here on Haverö (the only studied fall) and Goalpara, both of which show little evidence of weathering. The dominant ferromagnetic mineral in both meteorites appears to be kamacite (Rowe et al. 1975). Goalpara has been shocked to at least 60 GPa (estimated post-shock temperatures of at least 1100°C)

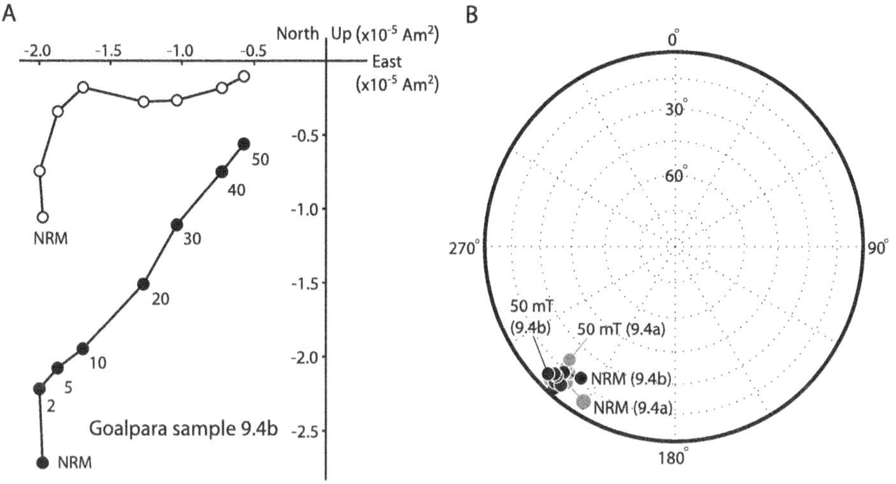

Fig. 11 NRM in two mutually oriented subsamples from the interior of the ureilite Goalpara. **A** Orthogonal projection of the NRM vector of subsample 9.4b during AF demagnetization as computed from data of Brecher and Fuhrman (1979). *Open and solid symbols* represent projections of the magnetization vector on vertical and horizontal planes, respectively. Selected demagnetization steps are listed in mT. **B** Equal area projection showing directions of NRM vector for subsamples 9.4a (*grey circles*) and 9.4b (*black circles*; data also shown in **A**). Selected demagnetization steps are listed in mT. Data from Brecher and Fuhrman (1979)

(Carter et al. 1968; Bischoff and Stoffler 1992) and then apparently thermally annealed. Such an event should have completely thermally remagnetized the rock. Brecher and Fuhrman's analyses of two mutually oriented samples of Goalpara identified a strong NRM with two components: a weak, soft component up to 5 mT and a second dominant component stable up to at least 50 mT that was unidirectional across two mutually oriented subsamples (Brecher and Fuhrman 1979) (Fig. 11). Very similar results were obtained by Larson et al. (1973) for a single chip of Goalpara. Laboratory TRM given to these two subsamples demagnetized at a rate very similar to the high coercivity NRM component. A paleointensity experiment using the AF demagnetization of TRM method (Schwarz 1969; Schwarz and Symons 1970) yielded a paleofield value of 140 μT. Using their measured ratio of NRM to sIRM, we calculate a total REM paleointensity of 34 μT (assuming that the paleofield in microteslas = NRM/sIRM ×3000 Gattacceca and Rochette 2004), which is within error of the previous AF of TRM value given the factor of ~3 uncertainties of the REM method (see Kletetschka et al. 2003, 2004a, 2006; Gattacceca and Rochette 2004; Yu 2006; Yu et al. 2007).

Brecher and Fuhrman's (1979) Haverö sample 9.1 has an NRM that is 58% of sIRM and is 20–50 times stronger per unit mass than the two samples studied by Kukkonen and Pesonen (1983); AF demagnetization to 40 mT reduces its intensity by two orders of magnitude. Since no fusion crust baked contact test was conducted, we are suspicious that this subsample may have been remagnetized by a hand magnet. This suspicion also extends to their samples of Kenna, which also have NRMs that are between 13–34% of sIRM and which decay much more rapidly than a laboratory TRM during AF demagnetization. Therefore, at the moment Goalpara is the only ureilite whose paleomagnetism is a good candidate for a thermoremanent record of substantial past magnetic fields. The slow cooling of this meteorite suggests these fields were steady over long timescales and therefore not likely to be of impact origin. The source of this field is at present a mystery given the lack of good geochronological constraints on ureilites. The primitive nature of ureilites, which show little evidence of siderophile depletion and have highly heterogeneous oxygen isotopic compositions, at present provides little evidence of a possible metallic core on the parent body.

4.5 Aubrites

Aubrites are highly reduced, coarse-grained brecciated achondrites with affinities to enstatite chondrites. The main magnetic carrier in these achondrites is kamacite, although minor amounts of taenite, tetrataenite, schreibersite, and cohenite have been reported (Easton 1986; Rubin 1997; Rochette et al. 2009a). Paleomagnetic studies have only been reported for five aubrites (Stacey et al. 1961; Larson et al. 1973; Brecher et al. 1979; Guskova 1984; Gattacceca and Rochette 2004). Norton County and Pesyanoe possess stable remanent magnetization isolated between 10 and 50 mT. If interpreted as primary TRM acquired during cooling on the aubrite parent body, they indicate a stable paleofield of ~10 μT during cooling below 700°C. However, in view of the brecciated nature of aubrites and their complex shock history (e.g., Sect. 2.3), interpretation of their NRM requires additional studies.

4.6 Angrites

The angrites are a group of twelve basaltic achondrites from an unknown parent body. They are distinguished from other achondrites by three features which make them superb targets for paleomagnetic analysis. Firstly, one of the most important ferromagnetic carriers is magnetite (Weiss et al. 2008a; Rochette et al. 2009a), a mineral whose rock magnetic properties and mode of magnetization acquisition are both simpler and far better understood than those of iron-nickel minerals (for a review, see Dunlop and Ozdemir 1997). Secondly, due to their unusually high U/Pb ratios, they have extremely precise and ancient Pb/Pb ages ranging from ages of 4564–4558 Ma (Amelin 2008; Markowski et al. 2007; Zartman et al. 2006). Thirdly, their state of preservation since final cooling is excellent: they show no evidence for subsequent shock, brecciation, and parent body weathering (Kurat et al. 2004; Prinz et al. 1977; Mittlefehldt et al. 1998; McKay et al. 1988; Kuehner et al. 2006; Yanai 1994) and their (U-Th)/He ages are within error of their Pb/Pb ages for all but two angrites (Busemann et al. 2006). Angrites also have the advantage that their cooling rates (which range from $0.3^{\circ}\mathrm{C\,y^{-1}}$ to $10^{\circ}\mathrm{C\,h^{-1}}$) are sufficiently slow to rule out impact-generated or other transient field-generation processes but far faster than that required for tetrataenite formation (in any case, tetrataenite is not favored by the low-nickel content of iron metal in angrites).

Recent paleomagnetic analyses of mutually oriented subsamples of the angrites D'Orbigny (Pb/Pb age 4563.3 ± 0.1 Ma) and Angra dos Reis (Pb/Pb age 4557.1 ± 0.1 Ma) (ages from Amelin 2008) identified stable, unidirectional remanent magnetization (Weiss et al. 2008a) (Fig. 12). A fusion crust baked contact test and magnetic viscosity analyses demonstrate that the magnetization in Angra dos Reis is preterrestrial. The lack of shock effects and terrestrial weathering, the high AF stability of the NRM, tests for spurious ARM and GRM acquisition, and low-temperature data indicating the lack of magnetite's Verwey transition collectively indicate that the magnetization is likely a primary TRM dating back to the early solar system. The relatively late age of the meteorite and its slow cooling rate indicate the NRM was acquired too late in solar system history and over too long a time period to have been the product of an external magnetic field from the young sun or protoplanetary nebula. AF paleointensity analyses suggest the field was of order 10–20 μT. Collectively, these data leave magnetization by a core dynamo field as the only compelling hypothesis for producing

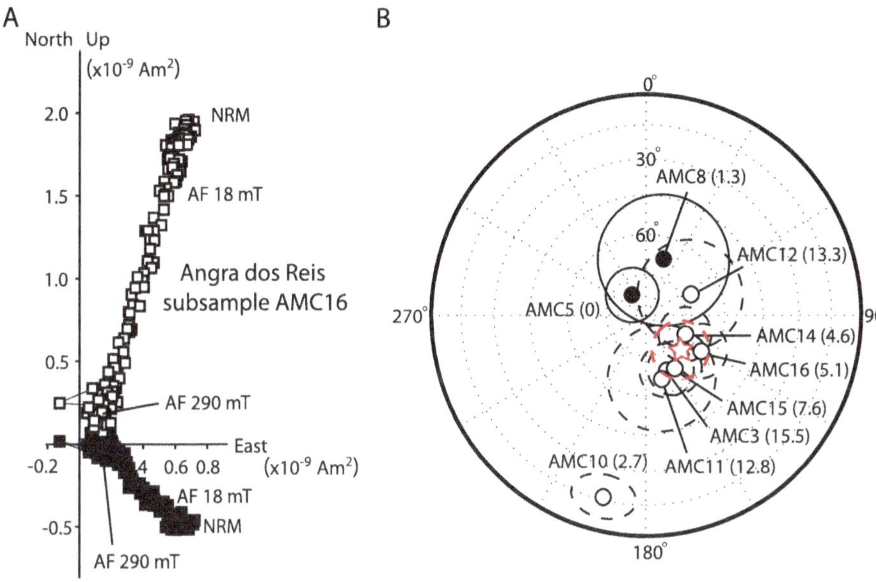

Fig. 12 NRM of the angrite Angra dos Reis. **A** Orthogonal projection showing evolution of NRM vector during AF demagnetization. *Open and closed symbols* represent projections of the endpoint of the magnetization vector on vertical and horizontal planes, respectively. Peak fields for selected AF demagnetization steps are labeled. One main high coercivity component is visible for this subsample. Other interior subsamples have the same HC component sometimes weakly overprinted by a low coercivity secondary component. **B** Fusion crust baked contact test on Angra dos Reis parent sample AMNH. Shown are HC magnetization directions of mutually oriented subsamples ranging from the fusion crusted exterior to the interior (including subsample in **A**), plotted on an equal area stereographic projection. *Closed* (*open*) *symbols* represent projections of vector directions onto *lower* (*upper*) *hemisphere*. Ellipses give estimated orientation uncertainty (either maximum angular deviation of least squares fit or estimated sample positioning uncertainty, whichever is larger). Distance from fusion crust in millimeters is listed next to each sample. Only sample AMC5 contains fusion crust. The seven remaining samples are from the interior, with AMC8 and AMC10 apparently baked by atmospheric passage. Fisher mean direction (*red star*) and associated 95% uncertainty confidence estimate ($\alpha_{95} = 10.7°$) are shown for interior subsamples. The shallow depth of divergent magnetization directions (<3 mm) and the fact that measured samples have NRM/sIRM < 1% throughout their full coercivity range indicate that the exterior has been thermally remagnetized by atmospheric passage rather than isothermally remagnetized by a magnet (see Sects. 2.6 and 2.7). Adapted from Weiss et al. (2008a)

the meteorite's NRM (Weiss et al. 2008a). This is consistent with Hf/W chronometry, which indicates a metallic core formed on the angrite parent body within 3 Ma of solar system formation (Markowski et al. 2007; Kleine et al. 2009).

4.7 Brachinites

Brachinites are olivine-rich cumulates intermediate between differentiated meteorites and primitive achondrites (see below). No paleomagnetic studies have yet been conducted on confirmed brachinites, although recent preliminary paleomagnetic analyses of the ungrouped achondrite GRA 06129, which has oxygen isotopic affinities to brachinites, observed a stable NRM of currently unknown origin (Shearer et al. 2008).

Table 4 Small-body parameters used in the calculations of core heat flux, velocities and magnetic fields. For the electrical conductivity, we use a minimum value from Secco and Schloessin (1989) and we infer the thermal conductivity using the Wiedemann-Franz law. The remaining parameters are from Weiss et al. (2008a). The variables are defined in the text

k	$20\ \mathrm{W\,m^{-1}\,K^{-1}}$
α	$10^{-4}\ \mathrm{K^{-1}}$
ρ_c	$5000–8000\ \mathrm{kg\,m^{-3}}$
r_c	10–350 km
$g = 4\pi G r_c \rho_c$	$0.04–2.3\ \mathrm{m\,s^{-2}}$
T	1273 K
C_P	$800\ \mathrm{J\,kg^{-1}\,K^{-1}}$
σ	$6 \times 10^5\ \mathrm{S\,m^{-1}}$
Ω	$2.4 \times 10^{-5}–3 \times 10^{-4}\ \mathrm{s^{-1}}$

4.8 Primitive Achondrites

Primitive achondrites are equilibrated, metamorphosed samples that generally have only moderately fractionated chondritic elemental compositions and are not highly differentiated (Mittlefehldt 2007). We are aware of no paleomagnetic analyses of either of the two primitive achondrites clans (acapulcoite-lodranites or winonaite-IAB-iron silicate inclusions) nor of any ungrouped primitive achondrites (e.g., Zag).

5 Dynamo Generation on Small Bodies

The strong primary remanent magnetization present in several angrites (and provisionally identified in other achondrites) may have important implications for the nature of their parent bodies. Most workers consider the angrite parent body to have been an asteroid-sized planetesimal (Mittlefehldt 2007; Mittlefehldt et al. 1998, 2002) although an origin on ancient Mercury has also been discussed (Irving and Kuehner 2007; Ruzicka and Huston 2006). Assuming angrites are remanents of an early solar-system planetesimal, their strong magnetic fields may be the result of cooling in a dynamo-generated field. Here we discuss the feasibility of dynamo generation on small bodies early in solar system history. Using a range of characteristics plausible for these small bodies, we determine whether certain basic criteria for dynamos are met, as well as estimate the resulting core magnetic field strengths.

We consider a body that has differentiated into a core, mantle and crust (Hevey and Sanders 2006) and determine whether its core is susceptible to dynamo action. We will assume that advection is thermally driven (as opposed to compositionally or mechanically driven) such that the core must have a superadiabatic temperature profile in order to generate a dynamo. The heat flux conducted down the core adiabat is given by:

$$F_{\mathrm{cond}} = \frac{k\alpha g T}{C_p} \tag{1}$$

where k is thermal conductivity, α is the thermal expansion coefficient, g is the gravitational acceleration at the core-mantle boundary, C_p is the specific heat and T is the temperature at the core-mantle boundary. The conductive heat flux depends on the core radius r_c and core density ρ_c through the gravitational acceleration g. Using the range of parameter values for small bodies given in Table 4, we find the conductive heat flux lies in the range 1.3×10^{-4} to $7.3 \times 10^{-3}\ \mathrm{W/m^2}$.

Thermal evolution modeling has shown that superadiabatic heat fluxes in the range 0.05–0.5 W/m^2 lasting for millions of years were possible for bodies ranging from 70 to 500 km in radius with parameter values similar to those shown in Table 4 (Weiss et al. 2008a). Additionally, cooling rates of parent bodies of iron meteorites have been estimated to be of the order $dT/dt = 2$–6600 K/Ma from metallographic analyses (Chabot and Haack 2006; Yang and Goldstein 2006; Yang et al. 2008, 2009). The corresponding heat flux at the surface of the metallic core before the onset of crystallization:

$$F = \frac{\rho_c r_c C_P}{3} \frac{dT}{dt}$$

is ~0.0008 to 90 W/m^2 for parameters in Table 4, which encompasses the range from thermal evolution models. These are the extreme bounds combining the above range of dT/dt estimates with the full range of assumed core radii. For a Vesta-sized body with $r_c = 100$ km and $dT/dt = 100$ K/Ma, a representative heat flux value of 0.5 W/m^2 is obtained. It therefore seems likely that small-body cores can maintain thermal convective motions to drive a dynamo early in their histories. It is possible for dynamo action to occur in a body with subadiabatic heat flux if the fluid motions are dominated by mechanical stirring or compositional convection. In a small body, the thermodynamic efficiency of thermal convection is low and compositionally driven convection can be much more efficient in magnetic field generation (Nimmo 2009). However, because these processes are harder to quantify due to uncertainties in the composition of the core and the nature of possible stirring mechanisms, we will ignore these contributions to the fluid motions and only consider a thermally driven dynamo. As such, these calculations strictly apply to dynamos on bodies prior to crystallization of the core. After crystallization has begun, such an approach is conservative if asteroid cores crystallize from the inside outwards, but would overly favor dynamos if cores instead crystallize from the outside inwards (Williams 2009). Because of uncertainties in the phase relations and thermodynamic properties of iron-sulfur alloys neither crystallization regime can be excluded. The way crystallization proceeds in an asteroidal core is a key unresolved problem for understanding dynamos in small bodies.

Producing motions in the core does not guarantee a dynamo. The motions must be three-dimensional and have significantly complex morphologies (for a review of planetary dynamo theory, see Kono and Roberts 2002). We will assume that the convective motions present in these small bodies meet this morphology criterion (as they do in Earth's core). The fluid flows must also be sufficiently rapid to ensure the field is regenerated faster than its ohmic decay. This requires that the magnetic Reynolds number

$$Re_m = \sigma \mu U L$$

exceed a critical value, where σ is electrical conductivity, μ is magnetic permeability, U is a characteristic velocity, and L is a characteristic length scale (e.g., core radius).

Various estimates of lower bounds for the magnetic Reynolds number, Re_m^{crit}, are found in the literature (Roberts and Gubbins 1987). Arguments based on energetics obtain the bound $Re_m^{\text{crit}} e_{\max} > \pi^2$, where $e_{\max}$ is the maximum eigenvalue of the rate of strain tensor. Another bound gives $Re_m^{\text{crit}} > \pi$ where the velocity is based on the maximum velocity in the core. These bounds are lower limits, so the actual critical magnetic Reynolds number may be larger than this. In numerical dynamo models, typically a critical value around 50 is found. However, because these models cannot work in the parameter regime appropriate for planetary and small-body cores, the numerical estimates may not be representative either. In the following discussion, we consider a critical value of 50 as fairly conservative and allow for the possibility that self-sustained dynamos may operate at $Re_m > 10$.

To estimate the magnetic Reynolds number in our small-body cores, we use the core radius as a characteristic length scale and the electrical conductivity for small-body cores given in Table 4. To determine a characteristic velocity, we use various scaling laws applicable for different force balances in planetary cores. For comparison, the Earth's core ($L \sim 3500$ km) has $Re_m \sim 10^3$ if the secular variation of the field is used to determine the characteristic flow velocities ($U \sim 0.5$ mm/s), suggesting that the Earth has a fairly supercritical magnetic Reynolds number.

We consider three different scaling laws to determine the characteristic velocities for our small bodies. For more information on these and other potential scaling laws, see the paper in this volume by Christensen (2009).

(i) U_{MAC}: This velocity estimate is based on a magnetostrophic balance argument. If the dominant force balance in the core is between the magnetic, Archimedean and Coriolis (MAC) forces, then the velocity scales as

$$U_{\mathrm{MAC}} \sim \left(\frac{2\pi G \alpha r_c F_{\mathrm{conv}}}{C_P \Omega} \right)^{0.5}$$

where Ω is the core angular rotation rate and F_{conv} is the convective heat flux (Stevenson 2003).

(ii) U_{CIA}: This velocity estimate comes from a balance of the non-geostrophic components of the Coriolis, non-linear inertial, and Archimedean (CIA) forces (Aubert et al. 2001; Christensen and Aubert 2006). Geostrophy (Coriolis forces are balanced by pressure gradients) is removed by taking the curl of the momentum equation and therefore eliminating the pressure term resulting in the following velocity estimate:

$$U_{\mathrm{CIA}} \sim \left(\frac{2\sqrt{2\pi} G \alpha}{C_P} \right) \left(\frac{r_c^3 F_{\mathrm{conv}}^2}{\Omega} \right)^{1/5}.$$

(iii) U_{ML}: This velocity estimate is from mixing length theory and assumes a balance between the non-linear inertial and Archimedean forces. It may be appropriate if the convection is highly turbulent and rotation is not a dominant force in the core (Stevenson 2003):

$$U_{\mathrm{ML}} \sim \left(\frac{4\pi G \alpha r_c^2 F_{\mathrm{conv}}}{C_P} \right)^{1/3}$$

where we have assumed that the characteristic length scale is the core radius.

Figure 13 shows regions of parameter space (our varying parameters are the core radius r_c, rotation rate Ω, and convective heat flux F_{conv}) where the magnetic Reynolds number is supercritical for the different velocity scalings. We find a range of supercritical Re_m from all three scalings suggesting that supercritical magnetic Reynolds numbers, and hence dynamos, are quite feasible in small bodies. In general, U_{ML} and U_{CIA} are 1–2 orders of magnitude larger than U_{MAC} and so the magnetic Reynolds numbers based on U_{MAC} are smaller. This is expected since fast rotation (which is assumed in the magnetostrophic balance) inhibits convective motions.

We can also use scaling laws to estimate magnetic field strengths generated by small-body dynamos. There are many scaling laws to choose from in the literature (for a summary

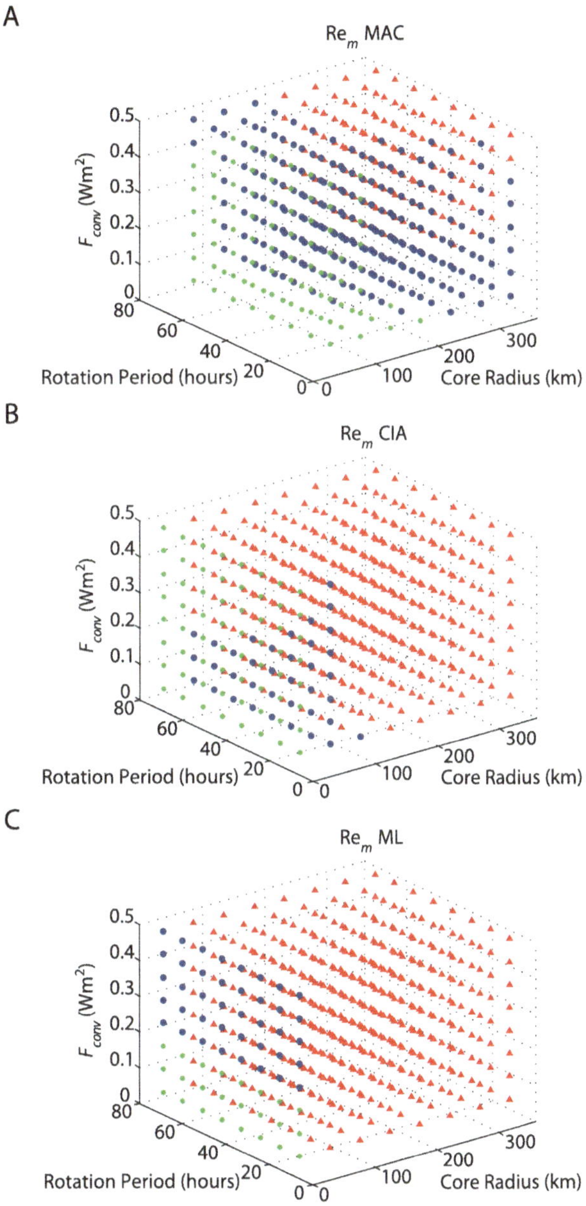

Fig. 13 Parameter values resulting in supercritical Re_m for various velocity scalings. Parameter values resulting in $1 \leq Re_m \leq 10$ are shown in *green stars* (probably no dynamo), $10 < Re_m \leq 50$ are shown in *blue circles* (perhaps a dynamo), and $Re_m > 50$ are shown in *red triangles* (dynamo likely). **A** Uses velocity estimates from the MAC balance, **B** Uses velocity estimates from the CIA balance, and **C** uses mixing length velocity estimates. Core radius is in km, rotation period is in hours, and F_{conv} is in W m^{-2}

see the paper by Christensen in this issue Christensen 2009). We chose two different magnetic field scaling laws related to the force balances from which we derived our velocity estimates. In both of these estimates, we used a core density value of 8000 kg/m^3:

(i) B_{MAC}: This estimate assumes the same force balance as the U_{MAC} estimate (magnetostrophic balance). In this case, if Lorentz and Coriolis forces are comparable, the magnetic field strength is given by:

$$B_{\mathrm{MAC}} \sim (2\Omega\rho_c\mu r_c U_{\mathrm{MAC}})^{0.5}$$

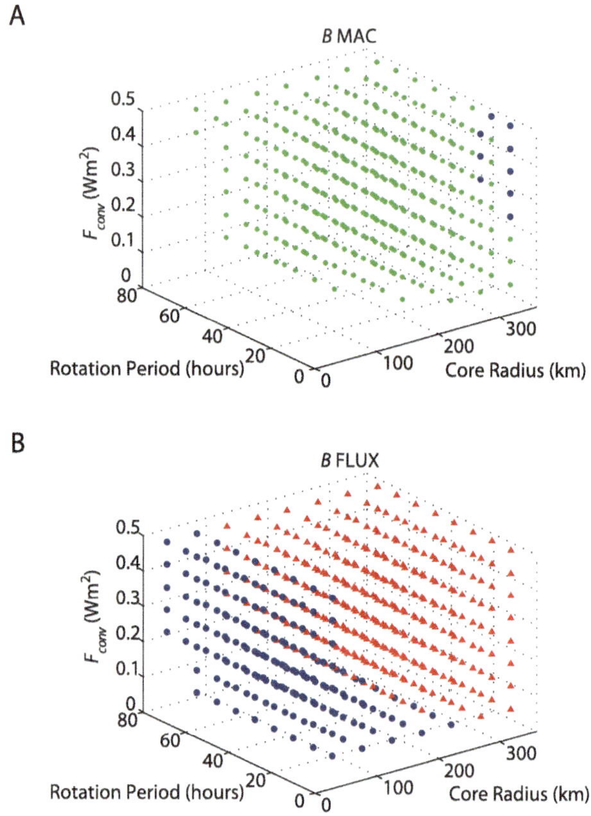

Fig. 14 Core surface magnetic field strengths for parameter values that result in $Re_m > 10$ for the MAC balance (**A**) and the FLUX balance (**B**). Parameter values resulting in $1 \leq B \leq 100$ μT are shown in *green stars*, $100 < B \leq 1000$ μT are shown in *blue circles*, and $B > 1000$ μT are shown in *red triangles*. Core radius is in km, rotation period is in hours, and F_{conv} is in W m^{-2}

(ii) B_{FLUX}: This estimate assumes the magnetic field strength is determined by the available power (from the convective heat flux F_{conv}) to drive the velocity field:

$$B_{\text{FLUX}} \sim \left(\frac{8\pi \mu G \alpha f_{\text{ohm}} \rho_c}{C_P} \frac{r_c^2 F_{\text{conv}}}{U} \right)^{1/2}$$

where f_{ohm} is the ratio of available power that is lost through ohmic dissipation and is a large fraction of 1 (so we approximate $f_{\text{ohm}} = 1$). This magnetic field scaling requires a velocity estimate. Since the mixing length velocities were of the same order of magnitude as the CIA based velocities but with a larger variation, we chose the mixing length velocity in the above scaling (i.e., the range of values found using the U_{CIA} estimates is covered in the range of values found using the U_{ML} values).

Figure 14 demonstrates the range of core magnetic field strengths for the small-body parameters that produced supercritical magnetic Reynolds numbers (where we chose our critical magnetic Reynolds number to be 10). The MAC balance scaling predicts core magnetic fields in the range 0.1–150 μT. The power balance scaling predicts core magnetic field strengths in the range 100–2600 μT.

In the case of the angrites, paleomagnetic studies have determined that the magnetizing field had an intensity of ~20 μT (Weiss et al. 2008a). In order to compare observations of magnetic field strengths in meteorites to our small-body estimates, the planetary surface magnetic field strength must be determined from the core magnetic field strength. This requires that we take into account two factors: (i) the ratio of poloidal field strength B_p at the CMB to total field strength B_{tot} in the core, and (ii) the distance from the surface to the core. Assuming $B_p \sim 0.1 B_{tot}$ (based on Earth observations and dynamo models), and a range of core to surface radius ratios (see Weiss et al. 2008a), we find small-body parameters that have both supercritical Re_m and $B_{sur} > 20$ μT, and hence are capable of explaining angrite paleomagnetism. It is therefore quite plausible that angrite paleomagnetism was the product of a dynamo operating in a small parent body early in solar system history.

In addition to the angrites, the estimated range of small-body parameters capable of sustaining dynamo action early in solar system history suggests that many remnants from this time period (such as asteroids and other meteorites) might contain remanent crustal magnetization. With the number of bodies in the asteroid belt, measurements of asteroid crustal magnetic fields could provide a wealth of information to constrain planetary dynamo processes, the critical magnetic Reynolds number, and scaling laws. Our testbed for planetary dynamo processes could increase from the current value of approximately 10 planets and large moons, to hundreds of small bodies.

For example, we consider three of the largest asteroids in the asteroid belt, two of which are soon to be visited by the Dawn mission: Vesta ($r \sim 250$ km), Pallas ($r \sim 270$ km) and Ceres ($r \sim 487$ km). Spectroscopic observations indicate that Vesta has a basaltic surface (McCord et al. 1970) and Hf/W chronometry of HED meteorites indicates it formed a metallic core within 3–4 Ma of solar system formation (Kleine et al. 2004). Although Ceres and Pallas both appear to have chondritic surfaces (McCord and Sotin 2005; Schmidt et al. 2009), a recent geophysical analysis of Ceres could not rule out the possibility that it is a partially differentiated object with small metallic core. It is reasonable that at least Vesta, and possibly Ceres or Pallas, could have core radii in the range of favorable dynamo conditions we considered above and hence that they have remanent crustal fields from dynamo action in their early history. Unfortunately, the Dawn mission is not carrying a magnetometer and therefore will not be able to measure magnetic fields directly. However, the possibility of using surface space weathering characteristics to infer the presence of crustal magnetic fields has been suggested (Vernazza et al. 2006).

Assuming that the conditions for dynamo operation in Vesta or Ceres are satisfied, scaling laws can be used to estimate the topology of the magnetic field. The ratio of inertial to Coriolis forces has been found to have a controlling influence on field geometry: when rotational forces dominate, the field tends to be dipolar, but when inertial forces are relatively important, the magnetic field is dominated by higher multipoles with a fairly white spectrum at the surface of the dynamo (Sreenivasan and Jones 2006; Christensen and Aubert 2006). The ratio of the two forces is expressed by a Rossby number: $Ro = U/(\Omega L)$. Using Vesta's rotation rate of $\Omega = 3 \times 10^{-4}$ s^{-1}, assuming a core heat flux of 300 mW m^{-2}, resulting in a characteristic CIA velocity scaling of 1 mm/s, and assuming a core radius of 100 km (consistent with astronomical observations and meteorite data for Vesta Ruzicka et al. 1997; Ruzicka et al. 2001; Hilton 2002) as a length scale results in $Ro \sim 10^{-5}$, much less than one and suggesting the magnetic field should be dipolar. However, Christensen and Aubert (2006) found that a local Rossby number defined with a characteristic length scale l of the flow, $Ro_l = U/(\Omega l)$, provides better discrimination between dynamo regimes, which is dipolar for $Ro_l < 0.12$ and multipolar for $Ro_l > 0.12$. Olson and Christensen (2006) give a dependence of the local Rossby number on various physical parameters derived from numerical dynamo models, $Ro_l \approx F_{\text{conv}}^{1/2} \eta^{1/5} \kappa^{-1/5} \nu^{-1/3} r_c^{-1/3} \Omega^{-7/6}$ where $\eta = (\sigma\mu)^{-1}$ is the magnetic diffusivity, $\kappa = k/(\rho C_P)$ is the thermal diffusivity and ν is the kinematic viscosity. Using parameters from Table 4 along with $\nu = 10^{-6}$ m^{2}s^{-1}, the local Rossby number is estimated to be approximately 0.5, and hence should operate in the multipolar regime. There is a large degree of uncertainty associated with this statement, because the estimated value does not far exceed the transitional value of 0.12 and some of the parameter values used are highly uncertain. In addition, the scaling law of Olson and Christensen (2006) for Ro_l is purely empirical and depends on viscosity and thermal diffusivity, whose values in the numerical models on which the scaling has been based are far removed from those in planetary cores.

Our illustration using characteristic values for Vesta or Ceres demonstrates that measurements of asteroidal magnetic fields by future missions could provide a wealth of information on asteroid interiors as well as provide useful constraints for planetary dynamo theory.

6 Open Questions

6.1 Origin of Heterogeneous Magnetization

A puzzling aspect of the paleomagnetism of meteorites is the small-scale (down to millimeters in some meteorites) heterogeneity of NRM directions. As discussed above, this heterogeneity has been observed in ordinary chondrites, enstatite chondrites, HED achondrites, as well as in a Martian meteorite (Kirschvink et al. 1997; Collinson 1997; Weiss et al. 2008b) and lunar rocks (Collinson 1985). Several mechanisms could potentially explain this unusual phenomenon: (1) low-temperature accretion of previously magnetized magnetic grains, (2) brecciation by impacts, or (3) tetrataenite formation. However, none of these mechanisms is applicable to all the above-mentioned meteorites. Mechanism 1 does not hold for meteorites that were metamorphosed to temperatures in excess of the Curie temperatures of these grains (for instance, the Bensour meteorite in Gattacceca et al. 2003). Mechanism 2 does not hold for unbrecciated meteorites (e.g., Ibitira in Cisowski 1991 and some of the ordinary chondrites in Morden 1992a). Moreover in some brecciated meteorites, the randomness of NRM is observed at scales smaller than that of the brecciation. Mechanism 3 does not hold for tetrataenite-free meteorites (e.g., HED meteorites in Collinson and Morden 1994). As a consequence, it seems unavoidable that small scale NRM scatter has a different origin in the different meteorites in which it was observed. This scatter not only has important implications for the mode of origin and age of the NRM, but also for the accuracy of paleointensity methods: measurements of bulk samples with primary but nonunidirectional TRM (e.g., due to cold brecciation), would only place lower limits on the true fine-scale NRM and paleointensity (Wasilewski et al. 2002; Weiss et al. 2008b).

6.2 Zero-Field Magnetization

A fourth related explanation for nonunidirectional NRM (Sect. 6.1) is that the magnetization was acquired in an extremely weak magnetic field with interplanetary intensity (~nT) and therefore is not a robust paleomagnetic record. Being able to determine when a rock has such "zero-field magnetization" is critical since it serves as the null hypothesis for extraterrestrial paleomagnetism. As described above, samples with candidate zero-field primary magnetization include Murchison (CM2), Orgueil (CI1), and Tagish Lake (C2 ungrouped) the Ibitira achondrite, and possibly some HED and ordinary chondrites.

Two key difficulties with identifying zero-field magnetization are that all ferromagnetic grains will have spontaneous moments that cannot be demagnetized and also that the demagnetization process (particularly via AF methods; Sect. 2.2) always introduce some spurious remanence. There have been some previous theoretical attempts to predict expected NRM intensities from zero-field effects (e.g., Irving et al. 1961; Dickson 1962; Kristjansson 1973; Brecher 1976), but these have thus far been unable to quantitatively account for the relatively strong remanence levels in most meteorites. Such calculations need to be more thoroughly pursued in the future.

6.3 Origin of Young Magnetization

Magnetization has been identified in a number of meteorites with relatively young $^{40}Ar/^{39}Ar$ ages (e.g., the Millbillillie eucrite; Sect. 4.1). If this magnetization is confirmed to be a primary record dating back to time of $^{40}Ar/^{39}Ar$ closure, this does not leave many possible explanations for the magnetic field source. One possibility is that of magnetization by remanent crustal magnetic anomalies. This hypothesis could be tested for young eucrites via observations of space weathering effects on Vesta (Vernazza et al. 2006). A second possibility is SRM acquired in an impact-generated or -amplified magnetic field.

7 Conclusions

- Demonstration of preterrestrial TRM in meteorites can be accomplished via a combination of fusion crust baked contact tests, observation of unidirectional magnetization directions from analysis of mutually oriented subsamples, tests for viscous magnetic contamination in the Earth's field and spurious remanence acquisition during demagnetization, demonstration of the lack of shock effects, analysis of thermal stability of NRM carriers over solar system history, and precise geochronometry and thermochronometry.
- Such analyses have been performed in paleomagnetic studies of only a few meteorites. As a result, most meteorites have not yet been definitively shown to contain TRM records of early solar system fields. Two important exceptions are the angrites, a pristine basaltic achondrite group, and Allende, an extremely well-studied CV carbonaceous chondrite. Both meteorites appear to record parent body magnetic fields instead of early external magnetic fields from the T Tauri sun and protoplanetary nebula. The magnetization of several other meteorites including CI and ordinary chondrites is provisionally consistent with a lack of magnetic fields present during their formation.
- Theoretical analyses indicate that many early planetesimals were likely capable of generating short-lived core dynamos. The intensity and timing of these fields are consistent with paleomagnetic observations of angrites and Allende. Therefore paleomagnetism provides paleogeophysical evidence for planetesimal differentiation within just several Ma of solar system formation.

Acknowledgements B.P. Weiss thanks the NASA Mars Fundamental Research Program, the NASA Lunar Advanced Science and Exploration Research Program, the NASA Lunar Science Institute, the NASA Planetary Major Equipment Program, the NSF Instrumentation and Facilities Program, the Charles E. Reed Faculty Initiatives Award fund, the MIT-France Seed Funds program and the Victor P. Starr Professorship fund for support. J. Gattacceca was supported by grant ANR-05-JCJC-0133 and the MIT-France Seed Funds program.

References

G. Acton, Q.Z. Yin, K.L. Verosub, L. Jovane, A. Roth, B. Jacobsen, D.S. Ebel, J. Geophys. Res. **112** (2007). doi:10.1029/2006JB004655

M.H. Acuña, G. Kletetschka, J.E.P. Connerney, in *The Martian Surface: Composition, Mineralogy, and Physical Properties*, ed. by J.F. Bell (Cambridge University Press, Cambridge, 2008), pp. 242–262

A. Al-Kathiri, B.A. Hofmann, A.J.T. Jull, E. Gnos, Meteorit. Planet. Sci. **40**, 1215–1239 (2005)

A.D. Al-Rawas, A.M. Gismelseed, A.A. Yousif, M.E. Elzain, M.A. Worthing, A. Al-Kathiri, E. Gnos, B.A. Hofmann, D.A. Steele, Planet. Space Sci. **55**, 859–863 (2007)

Y. Amelin, Geochim. Cosmochim. Acta **72**, 221–232 (2008)

Y. Amelin, A.N. Krot, Meteorit. Planet. Sci. **42**, 1321–1335 (2007)

E. Anders, N. Grevesse, Geochim. Cosmochim. Acta **53**, 197–214 (1989)

V. Anyzeski, Pop. Astron. **57**, 249–250 (1949)

N. Artemieva, B. Ivanov, Icarus **171**, 84–101 (2004)

J. Aubert, D. Brito, H.-C. Nataf, P. Cardin, J.-P. Masson, Phys. Earth Planet. Inter. **128**, 51–74 (2001)

S.A. Balbus, Annu. Rev. Astron. Astrophys. **41**, 555–597 (2003)

S.A. Balbus, in *Physical Processes in Circumstellar Disks Around Young Stars*, ed. by P. Garcia (University of Chicago Press, Chicago, 2009, in press)

S.K. Banerjee, R.B. Hargraves, Earth Planet. Sci. Lett. **10**, 392–396 (1971)

S.K. Banerjee, R.B. Hargraves, Earth Planet. Sci. Lett. **17**, 110–119 (1972)

J.L. Berkley, G.J. Taylor, K. Keil, G.E. Harlow, M. Prinz, Geochim. Cosmochim. Acta **44**, 1579–1597 (1980)

N.S. Bezaeva, P. Rochette, J. Gattacceca, R.A. Sadykov, V.I. Trukhin, Geophys. Res. Lett. **34**, L23202 (2007). doi:23210.21029/22007GL031501

N.S. Bezaeva, J. Gattacceca, P. Rochette, R.A. Sadukyov, V.I. Trukhin, Phys. Earth Planet. Inter. (2009, in press)

A. Bischoff, D. Stoffler, Eur. J. Mineral. **4**, 707–755 (1992)

P.A. Bland, F.J. Berry, T.B. Smith, S.J. Skinner, C.T. Pillinger, Geochim. Cosmochim. Acta **60**, 2053–2059 (1996)

P.A. Bland, F.J. Berry, C.T. Pillinger, Meteorit. Planet. Sci. **33**, 127–129 (1998)

P.A. Bland, A.W.R. Bevan, A.J.T. Jull, Quat. Res. **53**, 131–142 (2000)

P.A. Bland, M.E. Zolensky, G.K. Benedix, M.A. Sephton, in *Meteorites and the Early Solar System II*, ed. by D.S. Lauretta, H.Y. McSween (The University of Arizona Press, Tucson, 2006), pp. 853–867

D.D. Bogard, D.H. Garrison, Geochim. Cosmochim. Acta **59**, 4317–4322 (1995)

D.D. Bogard, D.H. Garrison, Meteorit. Planet. Sci. **38**, 669–710 (2003)

A.J. Brearley, R.H. Jones, in *Planetary Materials*, ed. by J.J. Papike (Mineralogical Society of America, Washington, 1998), pp. 3-1–3-398

A. Brecher, Earth Planet. Sci. Lett. **29**, 131–145 (1976)

A. Brecher, Lunar Planet. Sci. **11**, 106–108 (1980)

A. Brecher, L. Albright, J. Geomag. Geoelectr. **29**, 379–400 (1977)

A. Brecher, G. Arrhenius, J. Geophys. Res. **79**, 2081–2106 (1974)

A. Brecher, M. Fuhrman, Moon Planets **20**, 251–263 (1979)

A. Brecher, L. Leung, Phys. Earth Planet. Inter. **20**, 361–378 (1979)

A. Brecher, R.P. Ranganayaki, Earth Planet. Sci. Lett. **25**, 57–67 (1975)

A. Brecher, J. Stein, M. Fuhrman, Moon **17**, 205–216 (1977)

A. Brecher, M. Fuhrman, J. Stein, Moon Planets **20**, 265–279 (1979)

V.F. Buchwald, R.S. Clarke, Am. Mineral. **74**, 656–667 (1989)

H. Busemann, S. Lorenzetti, O. Eugster, Geochim. Cosmochim. Acta **70**, 5403–5425 (2006)

H. Busemann, C.M.O.D. Alexander, L.R. Nittler, Meteorit. Planet. Sci. **42**, 1387–1416 (2007)

R.F. Butler, Earth Planet. Sci. Lett. **17**, 120–128 (1972)

R.F. Butler, *Paleomagnetism: Magnetic Domains to Geologic Terranes* (Blackwell Scientific Publications, Boston, 1992), p. 319

L. Carporzen, B.P. Weiss, D.S. Ebel, J. Gattacceca, D.L. Shuster (2009, submitted)

N.L. Carter, C.B. Raleigh, P.S. DeCarli, J. Geophys. Res. **73**, 5439–5461 (1968)

N.L. Chabot, H. Haack, in *Meteorites and the Early Solar System II*, ed. by D.S. Lauretta, H.Y. McSween (University of Arizona, Tucson, 2006), pp. 747–771

U.R. Christensen, Space Sci. Rev. (2009, this issue)

U.R. Christensen, J. Aubert, Geophys. J. Int. **166**, 97–114 (2006)

C.S. Cisowski, in *Geomagnetism*, vol. 2, ed. by J.A. Jacobs (Academic Press, London, 1987), pp. 525–560

C.S. Cisowski, M.D. Fuller, Y.M. Wu, M.F. Rose, P. Wasilewski, Proc. Lunar Planet. Sci. Conf. **6**, 3123–3141 (1975)

S.M. Cisowski, Earth Planet. Sci. Lett. **107**, 173–181 (1991)

S.M. Cisowski, M. Fuller, J. Geophys. Res. **83**, 3441–3458 (1978)
S.M. Cisowski, J.R. Dunn, M. Fuller, Y.M. Wu, M.F. Rose, P.J. Wasilewski, Proc. Lunar Sci. Conf. **7**, 3299–3320 (1976)
R.S. Clarke, E.R.D. Scott, Am. Mineral. **65**, 624–630 (1980)
G.D. Cody, C.M.O.D. Alexander, H. Yabuta, A.L.D. Kilcoyne, T. Araki, H. Ade, P. Dera, M. Fogel, B. Militzer, B.O. Mysen, Earth Planet. Sci. Lett. **272**, 446–455 (2008)
D.W. Collinson, *Methods in Rock Magnetism and Paleomagnetism* (Chapman and Hall, New York, 1983), p. 503
D.W. Collinson, Earth Moon Planet. **33**, 31–58 (1985)
D.W. Collinson, Earth Planet. Sci. Lett. **84**, 369–380 (1987)
D.W. Collinson, Meteoritics **26**, 1–10 (1991)
D.W. Collinson, Adv. Space Res. **8**, 227–237 (1992)
D.W. Collinson, Philos. Trans. R. Soc. Lond. A **349**, 197–207 (1994)
D.W. Collinson, Meteorit. Planet. Sci. **32**, 803–811 (1997)
D.W. Collinson, S.J. Morden, Earth Planet. Sci. Lett. **126**, 421–434 (1994)
D.A. Crawford, P.H. Schultz, Nature **336**, 50–52 (1988)
D.A. Crawford, P.H. Schultz, J. Geophys. Res. **96**, 18807–18817 (1991)
D.A. Crawford, P.H. Schultz, Int. J. Impact. Eng. **14**, 205–216 (1993)
D.A. Crawford, P.H. Schultz, Int. J. Impact. Eng. **23**, 169–180 (1999)
P.H.M. Dankers, J.D.A. Zijderveld, Earth Planet. Sci. Lett. **53**, 89–92 (1981)
I. de Pater, J.J. Lissauer, *Planetary Sciences* (Cambridge University Press, Cambridge, 2001), p. 528
J.S. Delaney, M. Prinz, C.E. Nehru, C.P. Stokes, Lunar Planet. Sci. **15**, 212–213 (1984)
S.J. Desch, H.C. Connolly, Meteorit. Planet. Sci. **37**(2), 183–207 (2002)
S.J. Desch, J.N. Cuzzi, Icarus **143**, 87–105 (2000)
T.L. Dickinson, P. Wasilewski, Meteorit. Planet. Sci. **35**(1), 65–74 (2000)
G.O. Dickson, J. Geophys. Res. **67**, 4943–4945 (1962)
R.R. Doell, C.S. Gromme, A.N. Thorpe, F.E. Senftle, Science **167**, 695–697 (1970)
D.J. Dunlop, J. Geophys. Res. **111**, B12S02 (2006). doi:10.1029/2006JB004572
D.J. Dunlop, O. Ozdemir, *Rock Magnetism: Fundamentals and Frontiers* (Cambridge University Press, New York, 1997), p. 573
A.J. Easton, Meteoritics **21**, 79–93 (1986)
L.T. Elkins-Tanton, B.P. Weiss, M.T. Zuber (2009, submitted)
S.S. Fonton, Meteoritika **11**, 121–122 (1954)
J. Fritz, N. Artemieva, A. Greshake, Meteorit. Planet. Sci. **40**, 1393–1411 (2005)
M. Fuller, Rev. Geophys. Space Phys. **12**, 23–69 (1974)
M. Fuller, S.M. Cisowski, in *Geomagnetism*, vol. 2, ed. by J.A. Jacobs (Academic Press, Orlando, 1987), pp. 307–455
M.D. Fuller, in *Encyclopedia of Geomagnetism and Paleomagnetism*, ed. by D. Gubbins, E. Herrero-Bervera (Springer, Dordrecht, 2007), pp. 788–801
M. Funaki, J. Magn. Soc. Jpn. **29**, 918–925 (2005)
M. Funaki, N. Nakamura, Antartic Meteorites **XXVII**, 21–22 (2002)
M. Funaki, Y. Syono, Meteorit. Planet. Sci. **43**, 1–13 (2008)
M. Funaki, T. Nagata, K. Momose, Mem. Natl. Inst. Polar Res. **20**(Spec. Issue), 300–315 (1981)
I. Garrick-Bethell, B.P. Weiss, Earth Planet. Sci. Lett. (2009, submitted)
I. Garrick-Bethell, B.P. Weiss, D.L. Shuster, J. Buz, Science **323**, 356–359 (2009)
J. Gattacceca, P. Rochette, Earth Planet. Sci. Lett. **227**, 377–393 (2004)
J. Gattacceca, P. Rochette, M. Bourot-Denise, Phys. Earth Planet. Inter. **140**, 343–358 (2003)
J. Gattacceca, P. Rochette, M. Denise, G. Consolmagno, L. Folco, Earth Planet. Sci. Lett. **234**, 351–368 (2005)
J. Gattacceca, M. Boustie, B.P. Weiss, P. Rochette, E.A. Lima, L.E. Fong, F.J. Baudenbacher, Geology **34**, 333–336 (2006)
J. Gattacceca, A. Lamali, P. Rochette, M. Boustie, L. Berthe, Phys. Earth Planet. Inter. **162**, 85–98 (2007)
J. Gattacceca, L. Berthe, M. Boustie, F. Vadeboin, P. Rochette, T. De Resseguier, Phys. Earth Planet. Inter. **166**, 1–10 (2008a)
J. Gattacceca, P. Rochette, M. Gounelle, M. van Ginneken, Earth Planet. Sci. Lett. **270**, 280–289 (2008b)
S.A. Gilder, M. Le Goff, Geophys. Res. Lett. **35**, L10302 (2008). doi:10310.11029/12008GL033325
S.A. Gilder, M. Le Goff, J.-C. Chervin, High Pressure Res. **26**, 539–547 (2006)
J.L. Gooding, in *International Workshop on Antarctic Meteorites*. LPI Technical Report, vol. 86-01 (Lunar and Planetary Science Institute, Houston, 1986), pp. 48–54
R.C. Greenwood, I.A. Franchi, A. Jambon, J.A. Barrat, T.H. Burbine, Science **313**, 1763–1765 (2006)
Y.G. Guskova, Geomagn. Aeron. **3**, 308–312 (1963)

Y.G. Guskova, Meteoritika **26**, 60–65 (1965a)
Y.G. Guskova, Geomagn. Aeron. **5**, 91–96 (1965b)
Y.G. Guskova, Meteoritika **29**, 116–127 (1969)
Y.G. Guskova, Meteoritika **30**, 74–87 (1970)
Y.G. Guskova, *The Magnetic Properties of Meteorites*. NASA Technical Translation TT F-792 (National Aeronautics and Space Administration, Washington, 1976a)
Y.G. Guskova, Meteoritika **35**, 78–86 (1976b)
Y.G. Guskova, Meteoritika **37**, 144–151 (1978)
Y.G. Guskova, Meteoritika **40**, 61–66 (1982a)
Y.G. Guskova, Meteoritika **41**, 44–49 (1982b)
Y.G. Guskova, Meteoritika **42**, 119–137 (1983)
Y.G. Guskova, Meteoritika **43**, 155–160 (1984)
Y.G. Guskova, Meteoritika **44**, 111–118 (1985)
Y.G. Guskova, Meteoritika **46**, 94–96 (1987)
Y.G. Guskova, Meteoritika **47**, 137–142 (1988)
Y.G. Guskova, V.I. Pochtarev, in *Meteorite Research*, ed. by P.M. Millman (Reidel, Dordrecht, 1969), pp. 633–637
H. Haack, T.J. McCoy, in *Treatise on Geochemistry*, vol. 1, ed. by H.D. Holland, K.K. Turekian (Elsevier, Amsterdam, 2007), pp. 1–22
J.M. Herndon, The Magnetization of Carbonaceous Chondrites, in Geochemistry, PhD Texas, A&M University (1974), p. 114
J.M. Herndon, M.W. Rowe, E.E. Larson, D.E. Watson, Earth Planet. Sci. Lett. **29**, 283–290 (1976)
P.J. Hevey, I.S. Sanders, Meteorit. Planet. Sci. **41**, 95–106 (2006)
R. Hide, Moon **4**, 39 (1972)
J.L. Hilton, in *Asteroids III*, ed. by W.F. Bottke, A. Cellino, P. Paolicchi, R.P. Binzel (University of Arizona Press, Tucson, 2002), pp. 103–112
C.M. Hohenberg, O.V. Pravdivtseva, A. Meshik, Geochim. Cosmochim. Acta **64**, 4257–4262 (2000)
L.L. Hood, Geophys. Res. Lett. **14**, 844–847 (1987)
L.L. Hood, N.A. Artemieva, Icarus **193**, 485–502 (2008)
L.L. Hood, C.S. Cisowski, Rev. Geophys. Space Phys. **21**, 676–684 (1983)
S. Hu, E. Appel, V. Hoffmann, W.W. Schmahl, S. Wang, Geophys. J. Int. **134**, 831–842 (1998)
R. Hutchison, *Meteorites: A Petrologic, Chemical, and Isotopic Synthesis* (Cambridge University Press, Cambridge, 2004), p. 506
A.J. Irving, S.M. Kuehner, in *Workshop on Chronology of Meteorites and the Early Solar System*, abstract #4050 (2007)
E. Irving, P.M. Stott, M.A. Ward, Philos. Mag. **6**, 225–241 (1961)
P. Jenniskens, M.H. Shaddad, D. Numan, S. Elsir, A.M. Kudoda, M.E. Zolensky, L. Le, G.A. Robinson, J.M. Friedrich, D. Rumble, A. Steele, S.R. Chesley, Nature **458**, 485–488 (2009)
A. Johansen, in *Cosmic Magnetic Fields: From Planets, to Stars and Galaxies*, ed. by K.G. Strassmeier, A.G. Kosovichev, J.E. Beckman. Proc. IAU Symp., vol. 259 (2009), pp. 119–128
M.K.R. Joung, M.-M. Mac Low, D.S. Ebel, Astrophys. J. **606**, 532–541 (2004)
G.W. Kallemeyn, A.E. Rubin, J.T. Wasson, Geochim. Cosmochim. Acta **60**, 2243–2256 (1996)
K. Keil, J.L. Berkley, L.H. Fuchs, Am. Mineral. **67**, 126–131 (1982)
J.L. Kirschvink, A.T. Maine, H. Vali, Science **275**, 1629–1633 (1997)
J.L. Kirschvink, R.E. Kopp, T.D. Raub, Geochem. Geophys. Geosyst. **9**, Q05Y01 (2008). doi:10.1029/2007GC001856
T. Kleine, K. Mezger, C. Münker, H. Palme, A. Bischoff, Geochim. Cosmochim. Acta **68**, 2935–2946 (2004)
T. Kleine, K. Mezger, H. Palme, E. Scherer, C. Munker, Geochim. Cosmochim. Acta **69**, 5805–5818 (2005)
T. Kleine, B. Bourdon, A.J. Irving, Lunar Planet. Sci. **40**, abstract #2403 (2009)
G. Kletetschka, T. Kohout, P.J. Wasilewski, Meteorit. Planet. Sci. **38**, 399–405 (2003)
G. Kletetschka, M.H. Acuna, T. Kohout, P.J. Wasilewski, J.E.P. Connerney, Earth Planet. Sci. Lett. **226**, 521–528 (2004a)
G. Kletetschka, J.E.P. Connerney, N.F. Ness, M.H. Acuna, Meteorit. Planet. Sci. **39**, 1839–1848 (2004b)
G. Kletetschka, M.D. Fuller, T. Kohout, P.J. Wasilewski, E. Herrero-Bervera, N.F. Ness, M.H. Acuna, Phys. Earth Planet. Inter. **154**, 290–298 (2006)
T. Kohout, F. Donadini, L.J. Pesonen, M. Uehara, Geophysica **46**, 3–19 (2010)
T. Kohout, G. Kletetschka, M. Kobr, P. Pruner, P.J. Wasilewski, Phys. Chem. Earth **29**, 885–897 (2004)
T. Kohout, G. Kletetschka, L.J. Pesonen, P.J. Wasilewski, Lunar Planet. Sci. **37**, abstract #1601 (2006)
M. Kono, P.H. Roberts, Rev. Geophys. **40**, 1013 (2002). doi:1010.1029/2000RG000102
L. Kristjansson, Tellus **25**, 300–304 (1973)

A.N. Krot, M.I. Petaev, E.R.D. Scott, B.-G. Choi, M.E. Zolensky, K. Keil, Meteorit. Planet. Sci. **33**, 1065–1085 (1998)
A.N. Krot, K. Keil, E.R.D. Scott, C.A. Goodrich, M.K. Weisberg, in *Treatise on Geochemistry*, vol. 1, ed. by H.D. Holland, K.K. Turekian (Elsevier, Amsterdam, 2007), pp. 1–52
S.M. Kuehner, A.J. Irving, T.E. Bunch, J.H. Wittke, G.M. Hupe, A.C. Hupe, Lunar Planet. Sci. **37**, abstract #1344 (2006)
I.T. Kukkonen, L.J. Pesonen, Bull. Geol. Soc. Finland **55**, 157–177 (1983)
J. Kunz, M. Trieloff, K.D. Bobe, D. Stöffler, E.K. Jessberger, Planet. Space Sci. **43**, 527–543 (1995)
G. Kurat, M.E. Varela, F. Brandstatter, G. Weckwerth, R.N. Clayton, H.W. Weber, L. Schultz, E. Wasch, M.A. Nazarov, Geochim. Cosmochim. Acta **68**, 1901–1921 (2004)
M. Lanoix, D.W. Strangway, G.W. Pearce, Proc. Lunar Sci. Conf. **8**, 689–701 (1977)
M. Lanoix, D.W. Strangway, G.W. Pearce, Geophys. Res. Lett. **5**, 73–76 (1978)
E.E. Larson, D.E. Watson, J.M.R. Herndon, J. Geomag. Geoelectr. **25**, 331–338 (1973)
K. Lawrence, C. Johnson, L. Tauxe, J. Gee, Phys. Earth Planet. Inter. (2008). doi:10.1016/j.pepi.2008.1005.1007
M.R. Lee, P.A. Bland, Geochim. Cosmochim. Acta **68**, 893–916 (2004)
M.R. Lee, C.L. Smith, S.H. Gordon, M.E. Hodson, Meteorit. Planet. Sci. **41**, 1123–1268 (2006)
F.Y. Levinson-Lessing, in *Selected Works*, vol. 3 (USSR Academy of Sciences Press, 1952), pp. 363–366
E.H. Levy, S. Araki, Icarus **81**, 74–91 (1989)
E.H. Levy, C.P. Sonett, in *Protostars and Planets: Studies of Star Formation and of the Origin of the Solar System*, ed. by T. Gehrels (University of Arizona Press, Tucson, 1978), pp. 516–532
K.L. Louzada, S.T. Stewart, B.P. Weiss, Geophys. Res. Lett. **34**, L05204 (2007). doi:05210.01029/02006GL027685
K.L. Louzada, S.T. Stewart, B.P. Weiss, J. Gattacceca, N.S. Bezaeva, Earth Planet. Sci. Lett. (2009, submitted)
J.F. Lovering, Am. J. Sci. **257**, 271–275 (1959)
J.F. Lovering, in *Researches on Meteorites*, ed. by C.B. Moore (Wiley, New York, 1962), pp. 179–197
A. Markowski, G. Quitte, T. Kleine, A.N. Halliday, M. Bizzarro, A.J. Irving, Earth Planet. Sci. Lett. **262**, 214–229 (2007)
S.D. Matza, M.E. Lipschutz, Geochim. Cosmochim. Acta **41**, 1398–1401 (1977)
T.B. McCord, J.B. Adams, T.V. Johnson, Science **168**, 1445–1447 (1970)
T.B. McCord, C. Sotin, J. Geophys. Res. **110**, E05009 (2005). doi:05010.01029/02004JE002244
G. McKay, D. Lindstrom, S.-R. Yang, J. Wagstaff, Lunar Planet. Sci. **19**, 762–763 (1988)
K. Metzler, K.D. Bobet, H. Palme, B. Spettel, D. Stöffler, Planet. Space Sci. **43**, 499–525 (1995)
D.W. Mittlefehldt, Meteorit. Planet. Sci. **40**, 665–677 (2005)
D.W. Mittlefehldt, in *Treatise on Geochemistry*, vol. 1, ed. by H.D. Holland, K.K. Turekian (Elsevier, Amsterdam, 2007), pp. 1–40
D.W. Mittlefehldt, D.H. Garrison, Geochim. Cosmochim. Acta **62**, 1459–1468 (1998)
D.W. Mittlefehldt, T.J. McCoy, C.A. Goodrich, A. Kracher, in *Planetary Materials*, ed. by J.J. Papike (Mineralogical Society of America, Chantilly, 1998), pp. 4-1–4-195
D.W. Mittlefehldt, M. Killgore, M.T. Lee, Meteorit. Planet. Sci. **37**, 345–369 (2002)
M. Miyamoto, H. Takeda, K. Yanai, Mem. Natl. Inst. Polar Res. **8**(Spec. Issue), 170–184 (1978)
S.J. Morden, Phys. Earth Planet. Inter. **71**, 189–204 (1992a)
S.J. Morden, Meteoritics **27**, 560–567 (1992b)
S.J. Morden, D.W. Collinson, Earth Planet. Sci. Lett. **109**, 185–204 (1992)
T. Nagata, Proc. Lunar Planet. Sci. Conf. **10**, 2199–2210 (1979a)
T. Nagata, Phys. Earth Planet. Inter. **20**, 324–341 (1979b)
T. Nagata, Mem. Natl. Inst. Polar Res. **12**(Spec. Issue), 238–249 (1979c)
T. Nagata, Mem. Natl. Inst. Polar Res. **15**(Spec. Issue), 253–272 (1979d)
T. Nagata, Mem. Natl. Inst. Polar Res. **15**(Spec. Issue), 280–293 (1979e)
T. Nagata, Mem. Natl. Inst. Polar Res. **17**(Spec. Issue), 233–242 (1980)
T. Nagata, Phys. Earth Planet. Inter. **26**, 125–133 (1981)
T. Nagata, Proc. Lunar Planet. Sci. Conf. **13**, A779–A784 (1983)
T. Nagata, in *Primitive Solar Nebula and Origin of Planets*, ed. by H. Oya (Terra Scientific Publishing Company, Tokyo, 1993), pp. 89–103
T. Nagata, B.J. Carleton, Proc. Jpn. Acad. B **65**, 121–124 (1989)
T. Nagata, J.R. Dunn, Mem. Natl. Inst. Polar Res. **20**(Spec. Issue), 333–344 (1981)
T. Nagata, M. Funaki, Mem. Natl. Inst. Polar Res. **25**(Spec. Issue), 222–250 (1982)
T. Nagata, M. Funaki, Mem. Natl. Inst. Polar Res. **30**(Spec. Issue), 403–434 (1983)
T. Nagata, M. Funaki, Mem. Natl. Inst. Polar Res. **46**(Spec. Issue), 245–262 (1987)
T. Nagata, M. Funaki, Antartic Meteorites **XIV**, 160–163 (1989)

T. Nagata, M. Funaki, J.A. Danon, Mem. Natl. Inst. Polar Res. **41**(Spec. Issue), 364–381 (1986)
T. Nagata, N. Sugiura, Phys. Earth Planet. Inter. **13**(4), 373–379 (1977)
T. Nagata, N. Sugiura, F.C. Schwerer, Mem. Natl. Inst. Polar Res. **5**(Spec. Issue), 91–110 (1975)
T. Nagata, J.A. Danon, M. Funaki, Mem. Natl. Inst. Polar Res. **46**(Spec. Issue), 263–282 (1987)
T. Nagata, M. Funaki, H. Kojima, Proc. NIPR Symp. Antarct. Meteorites **4**, 390–403 (1991)
D. Nesvorný, D. Vokrouhlický, W.F. Bottke, B. Gladman, T. Häggström, Icarus **188**, 400–413 (2007)
K.J. Neuvonen, B. Ohlson, H. Papunen, T.A. Häkli, P. Ramdohr, Meteoritics **7**, 515–531 (1972)
F. Nimmo, Geophys. Res. Lett. **36**, L10201 (2009). doi:10.1029/2009GL037997
I. Nishioka, M. Funaki, T. Sekine, Earth Planets Space **59**, e45–e48 (2007)
P. Olson, U.R. Christensen, Earth Planet. Sci. Lett. **250**, 561–571 (2006)
L.J. Pesonen, M. Terho, I.T. Kukkonen, Proc. NIPR Symp. Antarct. Meteor. **6**, 401–416 (1993)
V.I. Pochtarev, Y.G. Guskova, Geomagn. Aeron. **2**, 626–634 (1962)
F.A. Podosek, Geochim. Cosmochim. Acta **84**, 341–365 (1970)
F.A. Podosek, Geochim. Cosmochim. Acta **35**, 157–178 (1971)
J. Pohl, A. Eckstaller, Lunar Planet. Sci. **12**, 851–853 (1981)
J. Pohl, U. Bleil, U. Hornemann, J. Geophys. **41**, 23–41 (1975)
M. Prinz, K. Keil, P.F. Hlava, J.L. Berkley, C.B. Gomes, W.S. Curvello, Earth Planet. Sci. Lett. **35**, 317–330 (1977)
M.E. Pritchard, D.J. Stevenson, in *Origin of the Earth and Moon*, ed. by R.M. Canup, K. Righter (The University of Arizona Press, Tucson, 2000), pp. 179–196
G. Pullaiah, G. Irving, K.L. Buchan, D.J. Dunlop, Earth Planet. Sci. Lett. **28**, 133–143 (1975)
P.H. Roberts, D. Gubbins, in *Geomagnetism*, vol. 2, ed. by J.A. Jacobs (Academic Press, London, 1987), pp. 185–249
P. Rochette, J.P. Lorand, G. Fillion, V. Sautter, Earth Planet. Sci. Lett. **190**(1–2), 1–12 (2001)
P. Rochette, G. Fillion, R. Ballou, F. Brunet, B. Ouladdiaf, L. Hood, Geophys. Res. Lett. **30**, 1683 (2003a)
P. Rochette, L. Sagnotti, M. Bourot-Denise, G. Consolmagno, L. Folco, J. Gattacceca, M.L. Osete, L. Pesonen, Meteorit. Planet. Sci. **38**, 251–268 (2003b)
P. Rochette, J. Gattacceca, V. Chevrier, V. Hoffmann, J.P. Lorand, M. Funaki, R. Hochleitner, Meteorit. Planet. Sci. **40**, 529–540 (2005)
P. Rochette, J. Gattacceca, V. Chevrier, P.E. Mathé, M. Menvielle, M.S. Team, Astrobiology **6**, 423–436 (2006)
P. Rochette, J. Gattacceca, L. Bonal, M. Bourot-Denise, V. Chevrier, J.-P. Clerc, G. Consolmagno, L. Folco, M. Gounelle, T. Kohout, L. Pesonen, E. Quirico, L. Sagnotti, A. Skripnik, Meteorit. Planet. Sci. **43**, 959–980 (2008)
P. Rochette, J. Gattacceca, M. Bourot-Denise, G. Consolmagno, L. Folco, T. Kohout, L. Pesonen, L. Sagnotti, Meteorit. Planet. Sci. **44**, 405–427 (2009a)
P. Rochette, B.P. Weiss, J. Gattacceca, Elements **5**, 223–228 (2009b)
M.W. Rowe, J.M. Herndon, E.E. Larson, D.E. Watson, NASA Report, NASA-CR-141143 (1975)
A.E. Rubin, Meteorit. Planet. Sci. **32**, 231–247 (1997)
A.M. Rubin, D.W. Mittlefehldt, Icarus **101**, 201–212 (1993)
S.S. Russell, L. Hartmann, J. Cuzzi, A.N. Krot, M. Gounelle, S. Weidenschilling, in *Meteorites and the Early Solar System II*, ed. by D.S. Lauretta, H.Y. McSween (University of Arizona, Tucson, 2006), pp. 233–251
A. Ruzicka, M. Huston, in *69th Annual Meteoritical Society Meeting*, abstract #5080 (2006)
A. Ruzicka, G. Snyder, L.A. Taylor, Meteorit. Planet. Sci. **32**, 825–840 (1997)
A. Ruzicka, G. Snyder, L.A. Taylor, Geochim. Cosmochim. Acta **65**, 979–997 (2001)
S. Sahijpal, P. Soni, G. Gupta, Meteorit. Planet. Sci. **42**, 1529–1548 (2007)
T. Sano, S.-I. Inutsuka, N.J. Turner, J.M. Stone, Astrophys. J. **605**, 321–339 (2004)
B.E. Schmidt, P.C. Thomas, J.M. Bauer, J.-Y. Li, L.A. McFadden, M.J. Mutchler, S.C. Radcliffe, A.S. Rivkin, C.T. Russell, J.W. Parker, S.A. Stern, Science **326**, 275–278 (2009)
H. Schulze, D. Stöffler, Meteoritics **32**, A116 (1997)
E.J. Schwarz, J. Geomag. Geoelectr. **21**, 669–667 (1969)
E.J. Schwarz, D.T.A. Symons, J. Geophys. Res. **75**, 6631–6640 (1970)
E.R.D. Scott, K. Keil, D. Stoffler, Geochim. Cosmochim. Acta **56**, 4281–4293 (1992)
E.R.D. Scott, H. Haack, S.G. Love, Meteorit. Planet. Sci. **36**, 869–881 (2001)
E.R.D. Scott, R.C. Greenwood, I. Franchi, I.S. Sanders, Geochim. Cosmochim. Acta **73**, 5835–5853 (2009)
R.A. Secco, H.H. Schloessin, J. Geophys. Res. **94**, 5887–5894 (1989)
C.K. Shearer, P.V. Burger, C.R. Neal, Z. Sharp, L.E. Borg, L. Spivak-Birndorf, M. Wadhwa, J.J. Papike, J.M. Karner, A.M. Gaffney, J. Shafer, B.P. Weiss, J.W. Geissman, V.A. Fernandes, Am. Mineral. **93**, 1937–1940 (2008)

F.H. Shu, H. Shang, T. Lee, Science **271**(5255), 1545–1552 (1996)
F.H. Shu, H. Shang, A.E. Glassgold, T. Lee, Science **277**(5331), 1475–1479 (1997)
B. Sreenivasan, C.A. Jones, Geophys. J. Int. **164**, 467–476 (2006)
L.J. Srnka, Proc. Lunar Sci. Conf. **8**, 785–792 (1977)
F.D. Stacey, Annu. Rev. Earth Planet. Sci. **4**, 147–157 (1976)
F.D. Stacey, J.F. Lovering, Nature **183**, 529–530 (1959)
F.D. Stacey, J.F. Lovering, L.G. Parry, J. Geophys. Res. **66**, 1523–1534 (1961)
I.M. Steele, J.V. Smith, Earth Planet. Sci. Lett. **33**, 67–78 (1976)
A. Stephenson, J. Geophys. Res. **98**, 373–381 (1993)
D.J. Stevenson, Earth Planet. Sci. Lett. **208**, 1–11 (2003)
N. Sugiura, J. Geomag. Geoelectr. **29**, 519–539 (1977)
N. Sugiura, D.W. Strangway, Proc. Lunar Planet. Sci. Conf. **12**, 1243–1256 (1981)
N. Sugiura, D.W. Strangway, Mem. Natl. Inst. Polar Res. **25**(Spec. Issue), 260–280 (1982)
N. Sugiura, D.W. Strangway, Earth Planet. Sci. Lett. **62**, 169–179 (1983)
N. Sugiura, D.W. Strangway, Proc. Lunar Planet. Sci. Conf. **15**, C729–C738 (1985)
N. Sugiura, D.W. Strangway, in *Meteorites and the Early Solar System*, ed. by J.F. Kerridge, M.S. Mathews (University of Arizona Press, Tucson, 1988), pp. 595–615
N. Sugiura, M. Lanoix, D.W. Strangway, Phys. Earth Planet. Inter. **20**, 342–349 (1979)
N. Sugiura, T. Matsui, D.W. Strangway, Proc. Lunar Planet. Sci. Conf. **16**, 831–832 (1985)
P.C. Thomas, R.P. Binzel, M.J. Gaffey, A.D. Storrs, E.N. Wells, B.H. Zellner, Science **277**, 1492–1495 (1992)
H.C. Urey, Proc. Natl. Acad. Sci. USA **41**, 127–144 (1955)
J.-P. Vallée, New Astron. Rev. **47**, 85–168 (2003)
P. Vernazza, R. Brunetto, G. Strazzulla, M. Fulchignoni, P. Rochette, N. Meyer-Vernet, I. Zouganelis, Astron. Astrophys. **451**, L43–L46 (2006)
P.J. Wasilewski, Moon **6**, 264–291 (1973)
P. Wasilewski, Moon **9**, 335–354 (1974a)
P. Wasilewski, Moon **11**, 313–316 (1974b)
P. Wasilewski, Phys. Earth Planet. Inter. **11**, P5–P11 (1976)
P. Wasilewski, Phys. Earth Planet. Inter. **26**(1–2), 134–148 (1981)
P. Wasilewski, Phys. Earth Planet. Inter. **52**, 150–158 (1988)
P. Wasilewski, Lunar Planet. Sci. **31**, abstract #1454 (2000)
P. Wasilewski, T. Dickinson, Meteorit. Planet. Sci. **35**, 537–544 (2000)
P.J. Wasilewski, C. Saralker, Proc. Lunar Planet. Sci. Conf. **12**, 1217–1227 (1981)
P. Wasilewski, M.H. Acuna, G. Kletetschka, Meteorit. Planet. Sci. **37**, 937–950 (2002)
B. Weaving, Geochim. Cosmochim. Acta **26**, 451–455 (1962a)
B. Weaving, Geophys. J. R. Astron. Soc. **7**, 203–211 (1962b)
M.K. Weisberg, M. Prinz, R.N. Clayton, T. Mayeda, M.M. Grady, I. Franchi, C.T. Pillinger, G.W. Kallemeyn, Geochim. Cosmochim. Acta **60**, 4253–4263 (1996)
M.K. Weisberg, T.J. McCoy, A.N. Krot, in *Meteorites and the Early Solar System II*, ed. by D.S. Lauretta, H.Y. McSween (University of Arizona, Tucson, 2006), pp. 19–52
B.P. Weiss, J.L. Kirschvink, F.J. Baudenbacher, H. Vali, N.T. Peters, F.A. MacDonald, J.P. Wikswo, Science **290**, 791–795 (2000)
B.P. Weiss, H. Vali, F.J. Baudenbacher, J.L. Kirschvink, S.T. Stewart, D.L. Shuster, Earth Planet. Sci. Lett. **201**, 449–463 (2002)
B.P. Weiss, S. Berdahl, L.T. Elkins-Tanton, S. Stanley, E.A. Lima, L. Carporzen, Science **322**, 713–716 (2008a)
B.P. Weiss, L.E. Fong, H. Vali, E.A. Lima, F. Baudenbacher, Geophys. Res. Lett. **35**(23), L23207 (2008b)
M. Westphal, Phys. Earth Planet. Inter. **43**, 300–306 (1986)
M. Westphal, H. Whitechurch, Phys. Earth Planet. Inter. **31**, 1–9 (1983)
G.W. Wetherill, Annu. Rev. Earth Planet. Sci. **18**, 205–256 (1990)
U.H. Wiechert, A.N. Halliday, H. Palme, D. Rumble, Earth Planet. Sci. Lett. **221**, 373–382 (2004)
M.A. Wieczorek, B.L. Jolliff, A. Khan, M.E. Pritchard, B.P. Weiss, J.G. Williams, L.L. Hood, K. Righter, C.R. Neal, C.K. Shearer, I.S. McCallum, S. Tompkins, B.R. Hawke, C. Peterson, J.J. Gillis, B. Bussey, Rev. Mineral. Geochem. **60**, 221–364 (2006)
Q. Williams, Earth Planet. Sci. Lett. **284**, 564–569 (2009)
E.A. Wilson, ISIJ Int. **34**, 615–630 (1994)
A. Yamaguchi, H. Takeda, D.D. Bogard, D.H. Garrison, Meteoritics **29**, 237–245 (1994)
A. Yamaguchi, G.J. Taylor, K. Keil, Icarus **124**, 97–112 (1996)
K. Yanai, Proc. NIPR Symp. Antarct. Geosci. **7**, 30–41 (1994)
J. Yang, J.I. Goldstein, Geochim. Cosmochim. Acta **70**, 3197–3215 (2006)

J. Yang, J.I. Goldstein, E.R.D. Scott, Geochim. Cosmochim. Acta **72**, 3043–3061 (2008)
J. Yang, J.I. Goldstein, J.R. Michael, P.G. Kotula, Lunar Planet. Sci. **40**, abstract #1186 (2009)
Y. Yu, Earth Planet. Sci. Lett. **250**, 27–37 (2006)
Y. Yu, L. Tauxe, J.S. Gee, Earth Planet. Sci. Lett. **162**, 244–248 (2007)
Y. Yu, S.-J. Doh, W. Kim, K. Min, Phys. Earth Planet. Inter. **177**, 12–18 (2009)
R.E. Zartman, E. Jagoutz, S.A. Bowring, Lunar Planet. Sci. **38**, abstract #1580 (2006)

Space Sci Rev (2010) 152: 391–421
DOI 10.1007/s11214-009-9581-y

Induced Magnetic Fields in Solar System Bodies

Joachim Saur · Fritz M. Neubauer · Karl-Heinz Glassmeier

Received: 5 April 2009 / Accepted: 12 October 2009 / Published online: 12 December 2009

Abstract Electromagnetic induction is a powerful technique to study the electrical conductivity of the interior of the Earth and other solar system bodies. Information about the electrical conductivity structure can provide strong constraints on the associated internal composition of planetary bodies. Here we give a review of the basic principles of the electromagnetic induction technique and discuss its application to various bodies of our solar system. We also show that the plasma environment, in which the bodies are embedded, generates in addition to the induced magnetic fields competing plasma magnetic fields. These fields need to be treated appropriately to reliably interpret magnetic field measurements in the vicinity of solar system bodies. Induction measurements are particularly important in the search for liquid water outside of Earth. Magnetic field measurements by the Galileo spacecraft provide strong evidence for a subsurface ocean on Europa and Callisto. The induction technique will provide additional important constraints on the possible subsurface water, when used on future Europa and Ganymede orbiters. It can also be applied to probe Enceladus and Titan with Cassini and future spacecraft.

Keywords Electromagnetic induction · Magnetic fields · Solar system bodies

1 Motivation and Background

The value of the electrical conductivity in naturally occurring matter varies enormously by more than 20 orders of magnitude from highly conductive metals to nearly perfect isolators.

J. Saur (✉) · F.M. Neubauer
Institut für Geophysik und Meteorologie, Universität zu Köln, Albertus-Magnus-Platz, 50923 Cologne, Germany
e-mail: saur@geo.uni-koeln.de

K.-H. Glassmeier
Institut für Geophysik und extraterrestrische Physik, TU Braunschweig, 38106 Braunschweig, Germany

K.-H. Glassmeier
Max Planck Institut für Sonnensystemforschung, Max-Planck-Strasse 2, 37191 Katlenburg-Lindau, Germany

Measurements of the electrical conductivity can thus be discriminative and well suited to characterize properties and spatial structures of matter in general or planetary bodies in our case. Knowledge of the electrical conductivity combined with cosmo-chemical, geological, gravitational and other geophysical information provides strong constraints on the structure of planetary bodies.

The change of magnetic flux through a conductor or the motion of a conductor through a magnetic field generates electric fields and thus also electric currents in the conductor. The changing electric fields and the electric currents are the source of secondary magnetic fields, also called *induced magnetic fields*. The primary time variable magnetic fields are called *inducing magnetic fields*. Studies of the inducing and induced magnetic fields are the core principle of the induction technique, through which the electrical conductivity can be determined or constrained. This technique finds also ample application in engineering and day-to-day experience. Each time we pass airport security, temporally changing electromagnetic fields are exposed upon us. Possession of significantly conducting objects, such as metal devices, generate secondary, i.e., induced magnetic fields. Airport security measures the secondary magnetic fields as their diagnostic means for metal objects.

Note, if we refer to induced magnetic fields, we refer to the aforementioned electromagnetic induction process, but not to a change of the magnetic induction **B** due to magnetization **M** of matter. In the latter case the magnetic field strength **H** is modified by changes in the magnetic susceptibility χ_m such that $\mathbf{B} = \mu_0(\mathbf{H} + \mathbf{M}) = \mu_0(1 + \chi_m)\mathbf{H} = \mu_r\mu_0\mathbf{H}$. For diamagnetic and paramagnetic materials the relative magnetic permeability μ_r is close to unity and $\chi_m \approx 0$. This even holds in crustal layers of the Earth with ferromagnetic constituents. It is also a good assumption in metallic cores of planetary bodies with temperatures above the Curie temperature. In this paper we consequently assume $\mu \approx \mu_0$. Also note, in accordance with standard practice in the literature, in this chapter we refer to **B** as magnetic field, even though its is technically labeled magnetic induction.

The electromagnetic induction technique is a well developed and established method at Earth. It is extensively applied to study the electrical conductivity of the Earth at different scales. The induction technique uses magnetic field measurements (and sometimes also electric field measurements) obtained on the Earth's surface or from satellites such as Magsat, Ørsted, and Champ. Time variable inducing magnetic fields are either naturally provided by time dependent effects in the Earth's ionosphere and magnetosphere or man made, for example, by transmitters particularly developed for the induction method or by low-frequency radio waves operated for communications, but used by geophysicists for measuring inductive responses. The scales which are studied in geophysics by the induction technique vary from global scales to study the Earth's crust and parts of the Earth's mantle to small scale objects down to dimensions of meters and less, e.g., water pipes, aquifers, waste sites, brines. Information on the spatial conductivity distribution provides in general constraints on chemical composition and temperature. The induction technique at Earth is well developed and mature at all fronts, i.e., in its physical and theoretical understanding, its technical applications, its numerical analysis of the measured data, as well as its associated geological interpretation of the derived electrical conductivities. Not surprisingly, also a gigantic set of scientific references exists in this field, which can be tapped into by the planetary community (e.g., Parkinson 1983; Schmucker 1985; Olsen 1999; Constable and Constable 2004). Applying the induction technique to other solar system bodies is in its infancy compared to the level of sophistication reached at Earth. However, the induction technique applied to solar system bodies is often faced with new challenges not common at Earth and to be discussed in this chapter.

The general aim in the application of the induction technique to solar system bodies is to determine the conductivity structure of their interior. One of the key drivers for this application is the search for liquid water in our solar system. Liquid water is considered to be one of the few essential building blocks for life as we know it. Surface layers of several planetary bodies in the outer solar system consist of frozen water with the possibility of liquid water under the ice crust due to internal heating. Such subsurface liquid water layers under an ice crust also exist at Earth under the Antarctic ice, e.g. Lake Vostok and at least 145 smaller lakes in the same region (e.g., Siegert et al. 2001, 2005). Frozen and liquid water have nearly the same density and thus cannot be distinguished directly by gravity measurements. However the electrical conductivity between frozen water and liquid saline water differs by several orders of magnitude making the induction technique suitable for distinction. The induction method finds also application in the study of other planetary properties, e.g. the existence and properties of a metallic core, which is e.g. uncertain in the case of the Earth moon.

In Sect. 2 we summarize some basic physical principles of the induction technique and also discuss particular challenges this technique meets at solar system bodies. In the subsequent sections we then give a guided tour of solar system objects where induction plays a role. Particular emphasise is given to Jupiter's satellite Europa, where the induction technique is, arguably, further developed than for any other solar system body, with the exception of Earth.

2 Basic Theory: Earth and New Problems at Planetary Bodies

Before we discuss the state of the art of induction at planetary bodies, we repeat some of the basic theory of induction, which has been developed historically for the Earth. We begin with the underlying equations for induction in a solid state or a conductive fluid at rest (Sect. 2.1) and then focus on induction for spherical geometry (Sect. 2.2). Afterwards we generalize the description of induction to a moving conductive fluid (Sect. 2.3).

2.1 Basic Equations

The fundamental equation to describe the induction process in a solid state body or in a conductive fluid at rest is the induction equation in a material with electrical conductivity σ

$$\frac{\partial \mathbf{B}}{\partial t} = -\nabla \times \left(\frac{1}{\sigma \mu_0} \nabla \times \mathbf{B} \right) \tag{1}$$

which reads for constant conductivity

$$\frac{\partial \mathbf{B}}{\partial t} = \frac{1}{\sigma \mu_0} \Delta \mathbf{B}. \tag{2}$$

These equations can be derived from Maxwell's equations and the simplest form of the isotropic Ohm's law

$$\mathbf{j} = \sigma \mathbf{E} \tag{3}$$

for the electric current density $\mathbf{j}$ and the electric field $\mathbf{E}$. In the derivation of (1) and (2), we retain the conductive currents (3), but neglect the displacement current in Ampere's law. With this assumption, the magnetic field diffuses into the conductive material rather than propagates into it. This assumption holds if the time scales of the temporal variation expressed by a frequency ω obey

$$\omega \ll \sigma \mu_0 c^2 \tag{4}$$

with c the speed of light. Assuming a characteristic time scale of one minute, this conditions is fulfilled if the conductivity is larger than $\sim 10^{-12}$ S m^{-1}. This is well fulfilled for any known natural object in the solar system. Most time scales under consideration at solar system bodies are even longer than one minute. In the derivation of (2) it is also assumed that the magnetic permeability μ can be approximated by μ_0 (see Sect. 1).

Useful time and length scales for characterization of the induction process for objects of size L are the diffusion time

$$T_{\text{diff}} = \sigma \mu_0 L^2 \tag{5}$$

and the frequency dependent skin depth

$$\delta(\omega) = \sqrt{\frac{2}{\sigma \mu_0 \omega}} \approx \frac{1}{2}\sqrt{\rho T}\ \text{km} \tag{6}$$

with the resistivity $\rho = 1/\sigma$ and the period $T = 2\pi/\omega$. The last term on the right-hand side in (6) holds for SI-units and approximates the skin depth in km.

Since the induction equation is linear, we can decompose the time dependent primary inducing magnetic field $\mathbf{B}(\mathbf{x}, t)$ as a superposition of the real parts of fields with individual frequencies ω in the form

$$\mathbf{B}(\mathbf{x}, t) = \tilde{\mathbf{B}}(\mathbf{x}) e^{i\omega t}. \tag{7}$$

Substituting (7) into (2) and introducing the complex wave vector $k = \pm(1+i)\sqrt{\mu_0 \sigma \omega/2}$ yields the Helmholtz equation

$$k^2 \tilde{\mathbf{B}}(\mathbf{x}) = \Delta \tilde{\mathbf{B}}(\mathbf{x}). \tag{8}$$

2.2 Induction in Spherical Geometry

The induction equation (2) is a linear parabolic partial differential equation. Well developed tools for its solution exist in mathematical physics. In spherical coordinates it has been solved on various levels of complexity for spherical distributions of the electrical conductivity (e.g., Lahiri and Price 1939; Rikitake 1966; Parkinson 1983). In this subsection, we will discuss several solutions for spherical geometry.

2.2.1 Limit of Nearly Infinite Conductivity Case

One of the most simple solutions and in several cases already a very useful first order description for some planetary satellites is given in the case when the conductivity and the thickness of the conducting shell D are large enough and the period T of the inducing field is small enough to obey the inequality (Neubauer 1999; Kivelson et al. 1999)

$$(\delta/D)^2 = \frac{1}{\pi}(T/T_{\text{diff}})^2 \ll 1. \tag{9}$$

In this case, the planetary body acts like a superconductor and time variable components obeying (9) are excluded from the conducting sphere by surface currents just inside the surface of the sphere. For real superconductors this effect is called the Meissner–Ochsenfeld effect and holds for any frequency ω. For non-superconductors in contrast, steady-state or very low-frequency fields with $\omega = 2\pi/T \to 0$ can fully penetrate the conductor.

In case the external time-varying inducing field $\mathbf{B_0(t)}$ is spatially homogeneous and in the limit of nearly infinite conductivity $\sigma \to \infty$, the induced, i.e. the secondary, field is a dipole magnetic field

$$\mathbf{B}_{\text{ind}}^{\infty}(t) = \frac{\mu_0}{4\pi}\left(3\left(\mathbf{r}\cdot\mathbf{M}(t)\right)\mathbf{r} - r^2\mathbf{M}(\mathbf{t})\right)/r^5 \tag{10}$$

with

$$\mathbf{M}(t) = -\frac{2\pi}{\mu_0}\mathbf{B_0}(t)r_0^3 \tag{11}$$

and the outer radius r_0 of the conducting sphere or shell.

2.2.2 Single Layer with Finite Conductance

A very useful and still fairly easy to handle model is the single shell model with a constant conductivity (Lahiri and Price 1939; Parkinson 1983; Zimmer et al. 2000). We assume one shell with conductivity σ, outer radius r_0 and inner radius r_1. The conductivity is zero everywhere else inside the planetary body of radius R.

Now we further assume that the exterior inducing field can be written as a potential field $\mathbf{B}_0 = -\nabla U_e$. This holds in regions with negligible electric current, i.e. when $\nabla \times \mathbf{B} = \mu_0 \mathbf{j} = \mathbf{0}$. The planetary bodies under considerations in this paper generally consist of poorly conductive surface layers. In these cases the potential representation is valid on and within the surface layers of the planetary bodies. The potential field can then be expanded into spherical harmonics $S_n^m(\Theta, \phi)$, with colatitude Θ and longitude ϕ (Rikitake 1966; Parkinson 1983; Olsen et al. 2009). If the conductivity distribution is radially symmetric, the resulting induced fields and its associated potentials $U_i^{n,m}$ are of the same degree n and order m as the inducing field and its associated potentials $U_e^{n,m}$ (Parkinson 1983; Schmucker 1985). This largely simplifies the analysis. For most planetary bodies radial symmetry of the electrical conductivity is a reasonable first order assumption, and we can therefore consider the inducing potential of a particular n and m without any loss of generality as the real part of

$$U_e^{n,m} = RB_e\left(\frac{r}{R}\right)^n S_n^m(\Theta, \phi)e^{i\omega t} \tag{12}$$

with the complex coefficient B_e. The full magnetic potential $U = U_i + U_e$ is a linear superposition of all contributions of degree n and order m and of all frequencies ω. The inducing fields can be of varying nature for planetary bodies, e.g. spatially constant or inhomogeneous. It can be also both, periodic and aperiodic. An aperiodic inducing signal can still be formally approximated in a temporally limited interval by a Fourier superposition in frequency space. We will discuss these effects in the following chapter where we cover the planetary bodies individually.

The induction equation (2) must be solved inside the shell and also outside the shell subject to the following boundary conditions for the total time-varying field: (i) $\mathbf{B}$ must be continuous across the boundaries of each shell. The normal component is continuous, and the tangential component is continuous for the assumption $\mu = \mu_0$ (see Sect. 1). (ii) **B** must not be infinite at the center of the body, $\mathbf{r} = 0$, and (iii) **B** must be asymptotically equal to the external field far away from the body ($r \gg R$).

The total potential $U^{n,m}$ outside of the shell is the potential of the inducing $U_e^{n,m}$ and the induced fields $U_i^{n,m}$ for frequency ω and spherical harmonics S_n^m, characterized by the complex coefficients B_e and B_i

$$U^{n,m} = r_0\left(B_e\left(\frac{r}{r_0}\right)^n + B_i\left(\frac{r_0}{r}\right)^{n+1}\right)S_n^m(\Theta,\phi)e^{i\omega t}. \tag{13}$$

The solution of (2) or (8) inside the conducting shell only contains poloidal magnetic fields which can be described as a sum of the product of Bessel functions for the radial dependence with spherical harmonics for the longitudinal and latitudinal dependence (Rikitake 1966; Parkinson 1983). Application of the boundary conditions (i) to (iii) at the inner and outer parts of the shell constrains the ratio of the complex amplitudes of the inducing to the induced field to

$$\frac{B_i}{B_e} = -\left(\frac{n}{n+1}\right)\frac{\xi J_{n+3/2}(r_0 k) - J_{-n-3/2}(r_0 k)}{\xi J_{n-1/2}(r_0 k) - J_{-n+1/2}(r_0 k)} \tag{14}$$

with

$$\xi = \frac{(r_1 k)J_{-n-3/2}(r_1 k)}{(2n+1)J_{n+1/2}(r_1 k) - (r_1 k)J_{n-1/2}(r_1 k)}. \tag{15}$$

J_m are the Bessel function of first kind and order m. Zimmer et al. (2000) derived a similar solution under the simplifying assumption that the inducing field is spatially homogeneous, i.e., $n = 1$.

The real and imaginary part of $B_i/B_e = A\exp(i\Phi)$ provide the amplitude A and the phase Φ so that the induced field outside of the planetary body is related to the induced field of a perfectly conducting sphere (see (10))

$$\mathbf{B}_{\text{ind}}(t) = Ae^{i\Phi}\mathbf{B}_{\text{ind}}^{\infty}(t) = A\mathbf{B}_{\text{ind}}^{\infty}\left(t + \frac{\Phi}{\omega}\right). \tag{16}$$

For finite conductivity the induced field is weaker than the primary field and its phase lags the primary inducing field, i.e. $A < 1$ and $-\pi/2 \le \Phi < 0$. Note, A is calculated here at the surface of the conducting shell r_0 and not at the surface of the satellite R. For spherical symmetric conductivities, which is a premise for (12), an external inducing field $n = 1$, i.e. a spatially constant field, induces a dipole field.

In geophysics, several methods have been developed to describe the geomagnetic variations at Earth, i.e. the inductive response of the Earth's conductive layers to temporally changing magnetic fields in the Earth's ionosphere and magnetosphere (Schmucker 1985). One of these methods compares similar to the mathematical procedure of (12) to (15) the ratio of the internal, i.e. induced, and external, i.e. inducing, parts of the magnetic potentials. This ratio is commonly called in geophysics the Q-response function and is a frequency dependent transfer function that connects the internal and the external coefficients of the field expansion in spherical harmonics of order n (Schmucker 1985; Olsen 1999; Constable and Constable 2004):

$$Q_n(\omega) \equiv \frac{\iota_n^m}{\epsilon_n^m} = \frac{g_n^m - ih_n^m}{q_n^m - is_n^m} = \frac{B_i}{B_e} \tag{17}$$

where $\iota_n^m = g_n^m - ih_n^m$ are the internal, and $\epsilon_n^m = q_n^m - is_n^m$ are the external potential coefficients. For more details on the complex notation of the magnetic potential and its coefficients see chapter by Olsen et al. (2009) in this issue.

Attempts to estimate the conductivity profile as a function of depth at Earth based on satellite observations, so far, rely on the assumption that the external (inducing) source is of magnetospheric origin with a spatial structure dominated by the spherical harmonics S_1^0, i.e. a spatially constant but time-varying field (Olsen 1999; Constable and Constable 2004). In the absence of currents at satellite altitude and after removal of other contributions, such as the dynamo field from the core, the crustal and ionospheric magnetic fields, the remaining field can simply be written as the potential

$$U^{1,0}(t) = r_0\left[\iota_1^0(t)\left(\frac{r_0}{r}\right)^2 + \epsilon_1^0(t)\left(\frac{r}{r_0}\right)\right]S_1^0(\Theta). \tag{18}$$

According to Gauss, the observed fields can then be separated in external (ϵ) and internal (ι) parts. Fourier transformation of the coefficients $\epsilon(t)$ and $\iota(t)$ into frequency space ω renders the frequency dependent transfer function $Q_1(\omega)$. Constable and Constable (2004), for example, use corresponding periods from less than one day up to roughly 0.4 year. With inversion techniques subsequently, models of the conductivity-depth profile can be constructed based on $Q_1(\omega)$ (Olsen 1999; Constable and Constable 2004).

2.2.3 *Induction in Sphere with Multi-layered Conductivity*

In the previous subsection a single shell with finite conductivity was assumed. The solution can be extended to multiple layers with constant conductivities in each shell. The solution of (2) in each shell together with the boundary conditions (i) to (iii) lead to recursive formula found by Srivastava (1966). Schilling et al. (2007) derive a solution for a three layer model representing Europa's ocean, mantle and core. Models with non constant conductivity date back to as early as Lahiri and Price (1939) who assumed a conductivity increasing with an inverse power law of the radial distance from the center of the Earth.

2.3 Role of a Dynamic Plasma Environment

Solar system bodies are generally embedded in a dynamically changing magneto-plasma. All magnetic field measurements (except for the Earth and the Moon) have been performed in the past (and will still be performed in the foreseeable future) during satellite flybys or from satellites in orbit. In cases where the objects do not possess a substantial atmosphere (such as Europa, Ganymede, Callisto, or Enceladus), the objects are usually fully surrounded by plasma starting at the surface of the bodies. In other cases where objects possess a substantial atmosphere (such as Titan), no plasma reaches the objects surfaces. In all these cases the measurements are still gathered (above the surface or above the lower atmosphere) in a plasma environment which often contains significant electric current. Therefore the magnetic field cannot be written as a potential field and the separation in internal and external contributions as developed by Gauss (see chapter by Olsen et al. 2009) is not possible any more and new approaches need to be developed. The following approach is based on Schilling et al. (2007, 2008).

Plasma effects and their associated electric currents, e.g. in the satellites' ionospheres or Alfvén wings, cause strong magnetic field perturbations which can be equal or larger than the internally induced fields. In the analysis of magnetic field measurements to determine internally induced fields, the external fields therefore cannot be neglected. The external currents close to the satellites can additionally be time variable and can contribute to induction as well, while induction, vice versa, can modify the plasma flow and magnetic fields outside of the planetary object.

The magnetic field in a plasma or within an electrically conductive fluid, cannot be described by an induction equation for solid state matter of the form (2) based on Ohm's law (3). In particular, the plasma velocity $\mathbf{v}$ with respect to the planetary objects introduces a motional electric field which changes Ohm's law to

$$\mathbf{j} = \sigma(\mathbf{E} + \mathbf{v} \times \mathbf{B} + \mathbf{T}_{\text{add}}). \tag{19}$$

In a plasma Ohm's law is in general quite complex (e.g., Baumjohann and Treumann 1996) and outside the realm of this chapter. Any additional terms in Ohm's law are represented in (19) in the term $\mathbf{T}_{\text{add}}$. Their nature and relative importance depend on the particular plasma physical environment around the planetary object. The conductivity σ in (19) is in the simplest case isotropic and can, e.g., be due to the conductivity of the object's ionosphere. Together with Faraday's induction law, (19) leads to an induction equation of the form

$$\frac{\partial \mathbf{B}}{\partial t} = \nabla \times (\mathbf{v} \times \mathbf{B}) - \nabla \times (\eta \nabla \times \mathbf{B} - \mathbf{T}_{\text{add}}) \tag{20}$$

with the magnetic diffusivity $\eta = 1/(\mu_0 \sigma)$. Note, (20), reduces to (1) inside a non-moving part of a planetary body where $\mathbf{v} = 0$. No measurements of the fluid velocities in possible, conductive subsurface oceans exist. But the velocities are assumed to be small and to only weakly contribute to the exterior magnetic field. For example, Tyler (2008) discusses velocities of less than 1 m/s within Europa's possible ocean. Ocean tides at Earth generate small poloidal fields of the order of 1–10 nT, which can be observed from space and allow monitoring of the ocean flows (Tyler et al. 2003). Flows of extraterrestrial conductive oceans might similarly be sounded through their induced magnetic fields.

To comprehensibly describe the magnetic field in the vicinity of a planetary body, the induction equation (2) in the interior and the induction equation (20) for a plasma in the exterior need to be solved simultaneously. When the interior of a planetary body is liquid and moves with appreciable velocity $\mathbf{v}$, then an induction equation of the form in (20) needs to be solved in the interior as well. In addition, the plasma momentum equation together with the continuity and the energy equations in the exterior need to be solved self-consistently as well. The plasma environment can in principal also be described by non-fluid models, such as hybrid or full kinetic models. Due to the highly non-linear nature of these sets of plasma equations, they can be solved in their full extent only numerically. A straightforward, but in most cases unfeasible procedure would be to self-consistently solve the plasma equations and the induction equation inside and outside of the planetary body. The difficulty stems from the different time scales of the processes, which will make simulations unrealistically long.

The magnetic field

$$\mathbf{B}(\mathbf{x}) = \langle \mathbf{B}(\mathbf{x}) \rangle + \delta \mathbf{B}(\mathbf{x}, \mathbf{t}) \tag{21}$$

consists in general of a time independent $\langle \mathbf{B}(\mathbf{x}) \rangle$ and a time dependent component $\delta \mathbf{B}(\mathbf{x}, \mathbf{t})$. For a better understanding of the interaction in between the different processes that generate magnetic field in this complex environment, as well for laying out a framework how to numerically solve the set of equations, it is useful to decompose the magnetic field $\mathbf{B}$ into different components:

$$\mathbf{B} = \mathbf{B}_0 + \mathbf{B}_P + \delta\mathbf{B}_{\text{ind}}(\delta\mathbf{B}_0) + \delta\mathbf{B}_{\text{ind}}(\delta\mathbf{B}_P). \tag{22}$$

$\mathbf{B}_0$ is the component of an external background field, e.g. the magnetic field of a planetary magnetosphere that surrounds a planetary moon at some distance from it. The second contribution is the magnetic field $\mathbf{B_P}$ produced by plasma processes close to the body, e.g. through the currents in the bodies ionospheres or Alfvén wings. The third and fourth contributions are the internally induced magnetic field $\delta\mathbf{B}_{\text{ind}}(\delta\mathbf{B}_0)$, $\delta\mathbf{B}_{\text{ind}}(\delta\mathbf{B}_P)$ due to the time-varying external background field $\delta\mathbf{B}_0$ and due to the time-varying plasma fields $\delta\mathbf{B}_P$, respectively. The induced fields $\delta\mathbf{B}_{\text{ind}}(\delta\mathbf{B}_0)$ and $\delta\mathbf{B}_{\text{ind}}(\delta\mathbf{B}_P)$ are potential fields outside of the planetary body, and the magnetospheric field $\mathbf{B_0}$ can usually be approximated by a potential field near the satellites as well.

With the aim to use induction to study properties of the planetary interior, the fields due to the external plasma interaction $\mathbf{B_P} + \delta\mathbf{B}_{\text{ind}}(\delta\mathbf{B}_P)$ can be considered a systematic measurement error that needs to be considered. In the absence of a significant time-varying externally inducing magnetic fields $\delta\mathbf{B}_0$ (such as in the Saturn system), induction signals $\delta\mathbf{B}_{\text{ind}}(\delta\mathbf{B}_P)$ might be generated due to time variable plasma magnetic fields $\delta\mathbf{B}_P$. These time variable plasma magnetic fields can be, for example, at Enceladus due to temporal changes in its plume activity, or at Titan a result of an eclipse or due to temporal variations in the conductivity distribution of the ionosphere. In these latter cases, $\mathbf{B_P(t)} + \delta\mathbf{B}_{\text{ind}}(\delta\mathbf{B}_P)$ changes from measurement error to measurement target, and $\mathbf{B_P(t)} + \delta\mathbf{B}_{\text{ind}}(\delta\mathbf{B}_P)$ establishes a possibility for the usage of the induction technique.

In cases of the Jovian and many other satellites, fortunately, the time scales that control the plasma in the exterior and the induction in the interior separate (Neubauer 1998). The time scale that controls the plasma interaction are on the order of minutes, while the time scales for induction are on the order of the planetary rotation period. Thus the planetary induction field can be considered quasi-stationary on time scales of the plasma interaction. But the plasma interaction reaches a quasi steady state as well on these time scales. The plasma interaction during one planetary rotation can therefore be described by a consecutive set of quasi-stationary plasma states during which the induction field is quasi-stationary as well.

If the primary, inducing field is due to planetary rotation, the following iterative procedure to self-consistently solve the induction equation in the interior and exterior together with the other plasma equations in the exterior based on the separation of time scales might be applied:

(a) First, the time interval spanning one rotation period is divided into small subsets, during which the plasma interaction and the induction fields can be considered quasi steady state, respectively. In each small subset, we then solve the plasma interaction in the exterior including the induction magnetic fields due to the changing planetary magnetic field, i.e. we compute $\mathbf{B}_0$, $\mathbf{B_P}$ and $\delta\mathbf{B}_{\text{ind}}(\delta\mathbf{B}_0)$. The latter expression is calculated with (14) and (15). These calculations result in a set of three-dimensional magnetic fields at different times, i.e. subsets, during one rotation period.

(b) In this set of plasma interactions models, we now calculate the temporally constant and the temporally varying magnetic field contributions for each point in space. The temporally constant field $\langle\mathbf{B(x)}\rangle$ is the temporally averaged field over one rotation period. The temporally changing field $\delta\mathbf{B(x, t)}$ is the deviation from the averaged field (see (21)).

(c) The total time variable fields $\delta\mathbf{B}(\mathbf{x}, \mathbf{t})$ now also contain time variable plasma fields, which generate time variable plasma induced magnetic fields $\delta\mathbf{B}_{\text{ind}}(\delta\mathbf{B}_P)$. These plasma induced fields are computed in this step of the iteration process. Therefore we represent the time variable plasma field $\delta\mathbf{B}_P$ as a potential field U_e (see (12)) at or just below the surface of the satellite (where a potential representation of the magnetic field is possible). Equations (14) and (15) subsequently provide the plasma induced magnetic fields for given U_e. Note, the time variable plasma fields are highly inhomogeneous in contrast to the time variable planetary magnetic field near the satellites and thus higher order contributions n in the spherical expansions in (14) and (15) are important.

(d) In the next iteration step, we calculate similar to (a) the plasma interaction and its associated plasma magnetic field $\mathbf{B_P}$ within each small quasi-steady-state subset, but we now include, additionally to $\mathbf{B_0}$ and $\delta\mathbf{B}_{\text{ind}}(\delta\mathbf{B}_0)$, the plasma induced field $\delta\mathbf{B}_{\text{ind}}(\delta\mathbf{B}_P)$ computed in the previous step (c).

This cycle of iterations (b) to (d) are repeated until convergence is reached, i.e. induction and plasma interaction in each individual subset do not change from one iteration to the next one any more.

The procedure described above can be called forward-modeling similar to the nomenclature in applied geophysics. In forward-modeling, induction and the plasma interaction are self-consistently calculated for a given set of model parameters, which are here the conductivity structure in the satellite's interior, and background plasma and neutral properties outside. The results are compared with the observations. Then a series of such forward-models with different model parameters are carried out to find a parameter set that explains the observations best.

A new level of modeling will be to construct an inverse-model. In inverse-modeling the whole forward-model is inverted to directly solve for the internal parameters, which provide the best, e.g. least mean square, fit to the measurements.

3 Moon

Next to the Earth, the Moon and probably Europa are the solar system bodies best understood in terms of electromagnetic induction. As the Moon does not possess a global magnetic field generated by an active dynamo (Stevenson 2003), the interaction of the Moon with the interplanetary medium is much different from the terrestrial case. The Moon is always immersed in the ever changing magnetized solar wind plasma or is passing through the magnetotail of the Earth. No major magnetosphere with internal electric currents is developing, which implies that the Moon is directly exposed to the time-varying, external magnetic field of the solar wind except perhaps for regions with strong magnetic anomalies (Dyal et al. 1975). As expected of even a poorly conducting object, the electromagnetic response of the Moon to these temporal fluctuations of the interplanetary magnetic field and its motion through the magnetotail is non-vanishing. Numerous studies on this response were conducted (e.g., Blank and Sill 1969; Sonett et al. 1971; Dyal and Parkin 1973; Vanyan and Egorov 1975; Sonett 1982). Thus only a short summary is given here to demonstrate the potential of induction studies once sufficient observations are available.

Observational conditions for lunar induction studies are still much better than for any other solar system body, except Earth. Surface magnetic field measurements from the Apollo lunar surface magnetometer (Dyal and Gordon 1973) allow a detailed comparison with magnetic field measurements made onboard a spacecraft such as Explorer 35. A transfer function

$H(f)$ can be defined by

$$H(f) \cdot B_0(f) = B_0(f) + B_{\text{ind}}(f) \tag{23}$$

where $B_0(f)$ is the Fourier component at frequency f of the external, inducing field of solar wind or magnetospheric origin, and B_{ind} the induced magnetic field. The ratio of induced to inducing field can be derived from actual observations at the Moon's surface and onboard an orbiting spacecraft. It can be interpreted using the Lahiri–Price theory referred to above. For a single layer reaching from the lunar surface to a depth of half the lunar radius and with a conductance of $\sigma = 10^{-2}$ S/m the transfer function at 10^{-4} Hz is about 1.4.

Hobbs et al. (1983) presented a detailed analysis of the available day-side magnetometer data and derived the transfer function shown in Fig. 1. An apparent conductivity can be determined from the equation (e.g., Khan et al. 2006)

$$\sigma_a(f) = \frac{4H^2(f)}{2\pi f \mu_0 R^2_{\text{Moon}}}, \tag{24}$$

where R_{Moon} is the lunar radius. The frequency dependent skin depth (6) can be used to develop a depth sounding method. Low-frequency external field contributions penetrate deep into the Moon and sound its deep structures, while higher frequency signals can only penetrate a short distance and thus sound only shallow structures. Therefore, this frequency

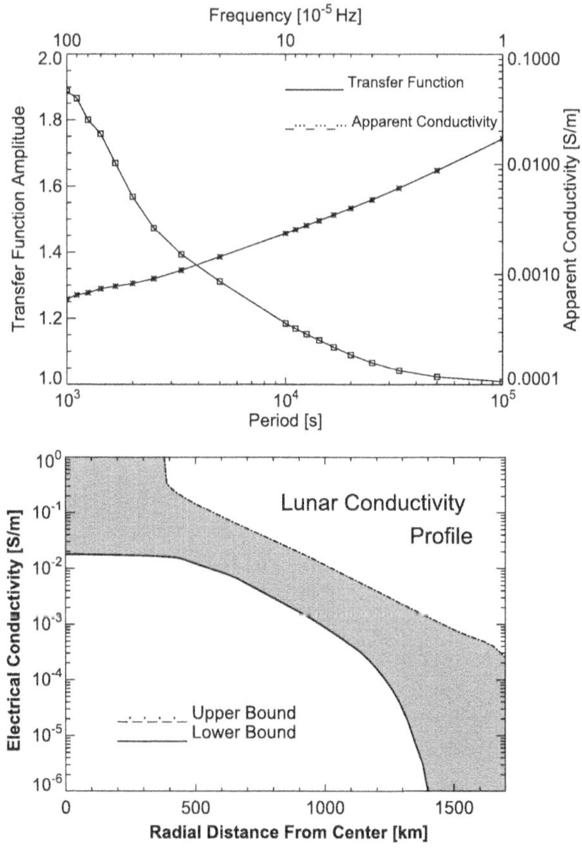

Fig. 1 A Lunar electromagnetic transfer function estimate and its associated apparent conductivity curve, based on results of Hobbs et al. (1983) (*upper panel*). The *bottom panel* displays a best-effort electric conductivity model of the Moon as derived by Hood et al. (1982)

sounding allows a first view into the object under discussion. In case of the Moon the transfer function indicates an increasing electrical conductivity with depth.

However, further modeling and inversion efforts are necessary to infer the real conductivity structure. The bottom panel of Fig. 1 shows the electrical conductivity structure of the Moon as a function of radial distance from its center derived by Hood et al. (1982). Upper and lower bounds for the conductivity are given. The conductivity was found to rise from 10^{-4}–10^{-3} S/m at a few hundred km depth to roughly 10^{-2} –1S/m at about 1100 km depth. The study by Hood et al. (1982) also allows for a metallic lunar core with a radius of about 360 km and a conductivity $\sigma > 10^2$ S/m. Knowing the conductivity profile furthermore enables one to constrain the compositional and thermal structure of the Moon (e.g., Khan et al. 2006). Electromagnetic induction studies prove to be a powerful tool to study planetary interiors.

4 Mercury

Mercury is an enigmatic planet, also with respect to induction effects. The planetary magnetic field is relatively weak with a current best estimate of its magnetic dipole moment of $M = -(230\text{–}290)$ nT $(4\pi/\mu_0)R_M^3$, where $R_M = 2439$ km is the Hermean planetary radius. The negative sign denotes a polarity as for the current terrestrial dipole moment (Anderson et al. 2008). This weak magnetic field is nevertheless able to withstand the fast flowing solar wind plasma and generates a planetary magnetosphere first discovered by Ness et al. (1974). However, this magnetosphere is small with a mean planetocentric magnetopause stand-off distance of only $1.7R_M$ (Siscoe and Christopher 1975). Due to the small size of the magnetosphere, which is filled by the planet to a large extent, surface magnetic fields of the order of 100 nT generated by magnetopause currents are comparable to the dynamo generated planetary magnetic field of Mercury (e.g., Glassmeier et al. 2007b). As the magnetosphere is highly dynamic these large external fields cause major induction effects in the electrically conducting planetary interior.

Already Hood and Schubert (1979) and Suess and Goldstein (1979) discussed such induction effects and pointed out that induction generated magnetic fields support the planet in withstanding the solar wind. A more detailed study on induction effects at Mercury was presented by Grosser et al. (2004). Based on a potential field model of the magnetospheric magnetic field following (Voigt 1981), assuming periodic variations of the magnetospheric size and shape, and using the above outlined theoretical approach, especially (14), they showed that the induced magnetic field may well reach surface values of up to 20 nT, which is about 10% of the dynamo generated magnetic field. As an example, Fig. 2 displays simulated magnetic field measurements made along a polar orbit around Mercury. The electrical conductivity distribution within the planet is spherically symmetric with a core conductivity of $\sigma = 10^6$ S/m. A core radius of 1860 km is assumed; the mantle is modeled as an electrically non-conductive shell. The magnetic field reaches a maximum value of 390 nT, out of which about 290 nT are attributed to the dynamo generated field, 170 nT to magnetospheric currents, and about 20 nT to the induction effects.

The induced contribution, of course, depends on the assumed electrical conductivity distribution. Not much is known about the actual conductivity distribution, but future magnetic field measurements at planet Mercury (Anderson et al. 2007; Glassmeier et al. 2009) will allow to constrain future models of the conductivity distribution.

Induction caused by temporal variations of the external magnetic field is not the only induction effect of importance at Mercury. Not only a change of magnetic flux through a

Fig. 2 Modelled magnetic field measurements along an elliptic polar orbit around Mercury (day-side periherm $1.15 R_M$, nightside apoherm $4 R_M$) (after Grosser et al. 2004)

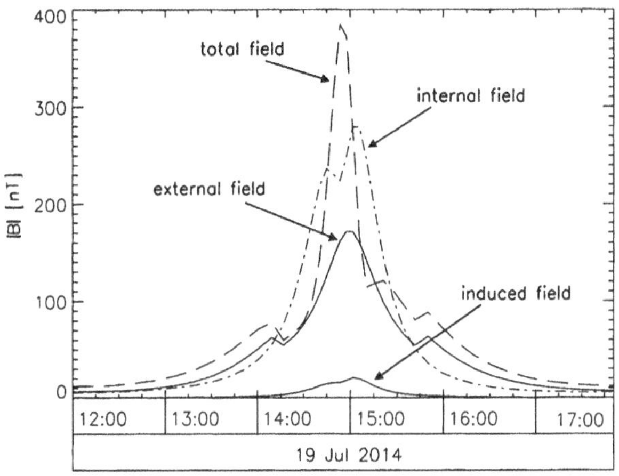

conductor causes induction, but also the motion of a conductor through a magnetic field. This situation is probably also realized at Mercury. Current Hermean dynamo models (e.g., Wicht et al. 2007) assume a fluid planetary core. As this core is immersed in a large magnetic field of magnetospheric origin induction effects due to the relative motion of the conductive fluid to the field are to be expected. As a possible effect of the external field on the fluid Glassmeier et al. (2007a) and Heyner et al. (2009) study the inductive influence of the magnetospheric field on the dynamo process in the planetary interior (described by an equation of the form (20) with a particular choice of T_{add} and externally inducing magnetic field contributions, see (22)). As the external field, generated by the solar wind plasma-planetary magnetic field interaction, is always directed anti-parallel to the dynamo generated dipole moment a negative feedback is possible. A feedback dynamo is thus an interesting possibility to explain the observed weak magnetic field at Mercury.

5 Jupiter's Satellites

5.1 Basic Principle and the Nature of Inducing Fields

Jupiter's large icy satellites Europa and Callisto are two of the few solar system objects where induction has been observed to take place (Khurana et al. 1998; Neubauer 1998; Kivelson et al. 2000). The third large icy satellite of Jupiter, Ganymede, shows hints of induction, which are however inconclusive (Kivelson et al. 2002). At Io no induction signatures have been observed. Io's magnetic field environment is dominated by very strong plasma interaction fields compared to the other Galilean satellites (Kivelson et al. 2004; Saur et al. 2002).

The induction is generated by a time periodic magnetic field, whose primary origin is the inclination of Jupiter's magnetic dipole moment with respect to Jupiter's rotation axis by 9.6°. The inclined magnetic moment rotates with Jupiter similar to the beacon of a light house. The inclined moment can be decomposed in a component parallel and a component perpendicular to the spin axis (see Fig. 3). The component parallel to the spin axis generates a dipole magnetic field symmetric with respect to the rotation axis and thus does not produce a time variable magnetic field. The component of the magnetic moment perpendicular to the spin axis (red arrow at Jupiter in Fig. 3) generates a magnetic field at the locations of

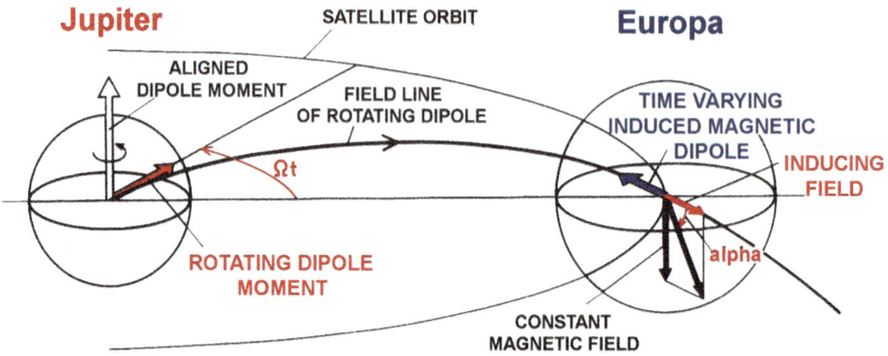

Fig. 3 Principle of induction at Jupiter's satellite Europa after Neubauer (1999)

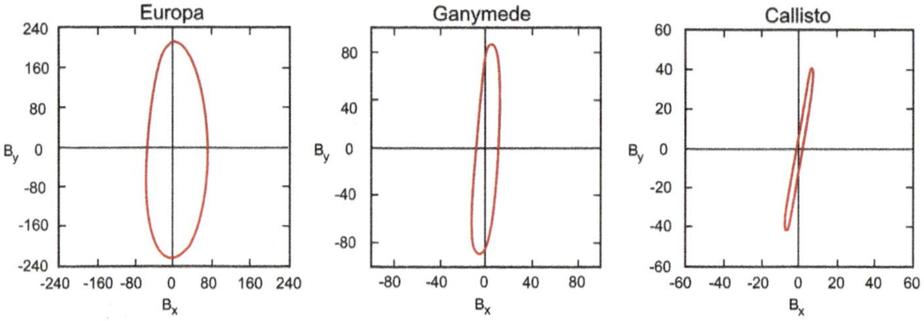

Fig. 4 Hodogram of time-varying magnetic fields at Europa, Ganymede, and Callisto combined from Khurana et al. (1998), Kivelson et al. (2002). The magnetic field components are given in a local **I**nteraction coordinate **S**ystem for each satellite (IS-System). Its origin is in the center of an individual satellite, the x-axis is along the orbital direction of the satellite (i.e. in the azimuthal direction of a Jupiter centered coordinate system), the y-axis is towards the planet (i.e. in the opposite direction of the radial direction in a Jupiter centered system) and the z-axis complements a right-handed coordinate system (i.e. in the opposite direction of the latitudinal component in a Jupiter centered system)

Jupiter's satellites, which changes direction by 360° within one synodic rotation period of Jupiter (red arrow at Europa in Fig. 3). Due to the existence of a conductive layer of most likely saline liquid water under the ice crust of Jupiter's large icy satellites the time periodic inducing field generates a secondary induced field (the associated magnetic moment of this induced field is shown as blue arrow in Fig. 3).

Jupiter's large icy satellites are always within Jupiter's magnetosphere and are thus constantly subject to the time-varying fields. The magnetospheric field $\mathbf{B_0}$ (in the sense of the nomenclature of (22)) has in addition to the inclined dipole several additional components. Jupiter's internal magnetic field possesses higher order moments (Connerney et al. 1998; Khurana et al. 2004). More importantly, magnetospheric effects become increasingly relevant with increasing distance from Jupiter. Due to Jupiter's fast rotation and strong mass sources inside its magnetosphere a dense equatorial plasma sheet forms which strongly stretches the magnetospheric fields and enhances the relative contribution of the radial magnetic field component. The time-varying magnetic fields $\delta\mathbf{B_0}$ over a full synodic rotation period at the location of Europa, Ganymede and Callisto are shown in Fig. 4. If only Jupiter's inclined dipole field were to contribute to the inducing field, the curves in Fig. 4 would be ellipses with ratio 1:2. The plasma or current sheet effects flatten the ellipses, which is most pronounced for Callisto.

Next to the primary inducing period, the synodic rotation period of Jupiter (11.23, 10.53, 10.18 h, for Europa, Ganymede, and Callisto, respectively), there are several other periods of relevance (Neubauer 1998). There are first of all periods associated with multiples of the synodic rotation frequency. An other, lower and distinct period is introduced by the rotation period of the satellites around Jupiter (3.55, 7.16, 16.69 days for Europa, Ganymede, and Callisto, respectively). The slightly eccentric orbits of Europa, Ganymede and Callisto introduce even in a perfectly symmetric magnetic field a time variable field in the rest frames of the orbiting satellites. The component with the largest inducing amplitude for this frequency is the B_z-component. In a dipole field the z-component varies by $\delta B \approx 6\, e\, B_z$, with B_z the magnetic field at the location of the satellite and e the eccentricity of the satellite. Note the z-direction does not have a (or only a very small) temporally changing component due to the synodic rotation period of Jupiter. With increasing distance from Jupiter, local time effects of the magnetosphere due to solar wind effects are the other main contributor to time dependent magnetic fields seen at the rotation period of the satellites. Other potential periods which generate very low-frequency magnetic fields are the rotation period of the sun, which establishes quasi-periodic sector structures in the solar wind, and the 11/22 years solar cycle which controls the overall dynamics of the solar wind.

5.2 Europa

Jupiter's satellite Europa is one of the planetary objects after Earth with the highest developed understanding of induction (Neubauer 1998, 1999; Khurana et al. 1998; Kivelson et al. 1999; Zimmer et al. 2000; Schilling et al. 2007, 2008). Therefore we will cover Europa in more detail compared to other satellites in this chapter.

The Galileo spacecraft was the first space probe to provide measurements of the magnetic environment near Europa during several close flybys. During flybys where Europa was outside of the Jovian current sheet, Khurana et al. (1998) and Kivelson et al. (1999, 2000) find that the observed magnetic field perturbations contain a clear dipole contribution consistent with induction due to a homogeneous time-varying field caused by Jupiter's synodic rotation period. For the E04 flyby, the observed magnetic fields together with modeled induced fields in a perfectly conducting sphere are displayed in Fig. 5, left (Kivelson et al. 1999). Subsequent Europa flybys at different system III longitudes, i.e. different orientations of the inducing field, unambiguously demonstrate that the dipole perturbations cannot be due to a permanent internal dipole moment, but are due to induction in an electrically conducting layer close to Europa's surface (Kivelson et al. 2000). The measurements provide an upper limit for a fixed permanent internal field of 25 nT in comparison with the magnitude of the induced fields of the order of 100 nT (Schilling et al. 2004).

A more refined model of the inductive response by Zimmer et al. (2000) includes a single conductive shell with a finite conductivity. Fitting their model to the observed magnetic fields from several flybys outside of Jupiter's current sheet, Zimmer et al. (2000) derive a lower limit for the amplitude $A = 0.7$ of the inductive response. Zimmer et al. (2000) also find that the conductive layer is located at a depth of less than 200 km and requires a conductivity of at least 0.06 S/m.

Europa also possesses a thin oxygen atmosphere (Hall et al. 1995) and an ionosphere (Kliore et al. 1997). Even though (Zimmer et al. 2000) rule out that the conductivity in Europa's ionosphere accounts for the observed inductive response, the interaction of Jupiter's magnetosphere with Europa's ionosphere (conductance of ~50 S) generates a current system of ~500 000 Ampere through Europa's atmosphere (Saur et al. 1998; Neubauer 1980,

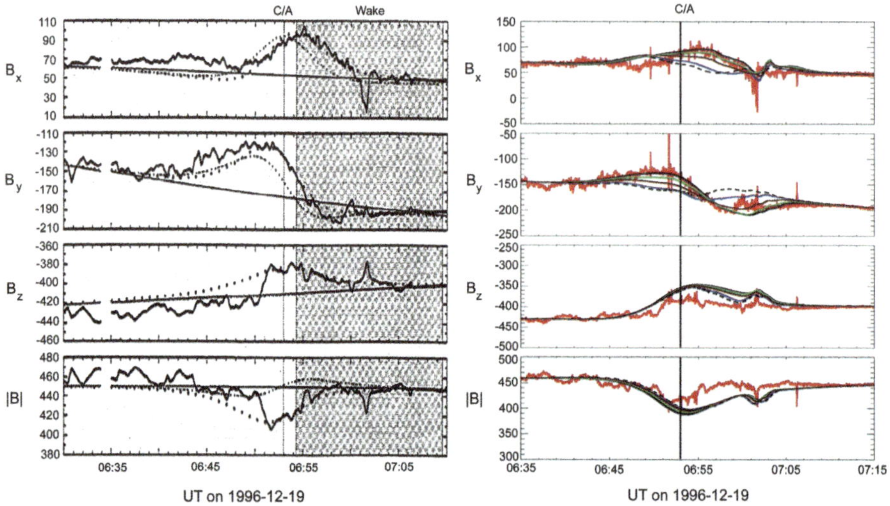

Fig. 5 Measured and model magnetic fields for the Europa E04 flyby. *Left* (after Kivelson et al. 1999): Observed magnetic field (*thick solid curves*); magnetospheric background magnetic field (*thin solid lines*); sum background field and induced dipole due a perfectly conducting sphere (*dashed curve/thick dots*); background field plus best dipole fit (*dotted curve*). *Right* (after Schilling et al. 2007): Observed magnetic field (*red curve*); model fields including plasma currents and induction for varying conductivities (*other curves*)

1998; Saur 2004). These plasma currents also produce significant magnetic field perturbations, which possess time variable components and contribute to the observed magnetic field perturbations near Europa. Schilling et al. (2007, 2008) created a model which includes self-consistently induction in a three-layered conductive shell model in the interior of Europa and the plasma currents in the atmosphere/ionosphere of Europa and its surrounding plasma. Results of this model are shown in Fig. 5 (right) for different values of the conductivity in the outer shell, which represents Europa's liquid ocean. When the plasma effects are included, a conductivity of at least 0.25 S/m in a near subsurface layer of thickness 100 km is required to explain the observed magnetic field perturbations (Schilling et al. 2008). When Europa is in the center of the magnetospheric current sheet the plasma magnetic field perturbations become even larger and start to dominate the overall magnetic field signature and thus need to be modeled in detail for a reliable interpretation of the internal magnetic field contributions.

As discussed in Sect. 1, measurements of the electrical conductivity can provide strong constraints on the internal structure of solar system bodies in conjunction with other geophysical or geological information. Galileo gravity data verify that Europa is differentiated into an outer H_2O layer about 100 km thick (Anderson et al. 1998), but the data cannot determine the current physical state of this layer (liquid or solid). Europa's surface is relatively young with an average age of $\sim$50 Myr (Pappalardo et al. 1999). Its surface is much less heavily cratered than its two neighboring satellites, Ganymede and Callisto, and displays a wealth of geological phenomena such as ridges, chaotic regions, and cycloids possibly due to flexing of the ice crust under tidal forces (Pappalardo et al. 1999). Even though these features are naturally explained by a subsurface ocean of liquid water, they cannot rule out the possibility that all of the surface morphologies could be due to processes in warm, soft ice with only localized or partial melting. However, the derived electrical conductivity of 0.25 S/m cannot be achieved by frozen or soft ice, but only by liquid saline water in the context of

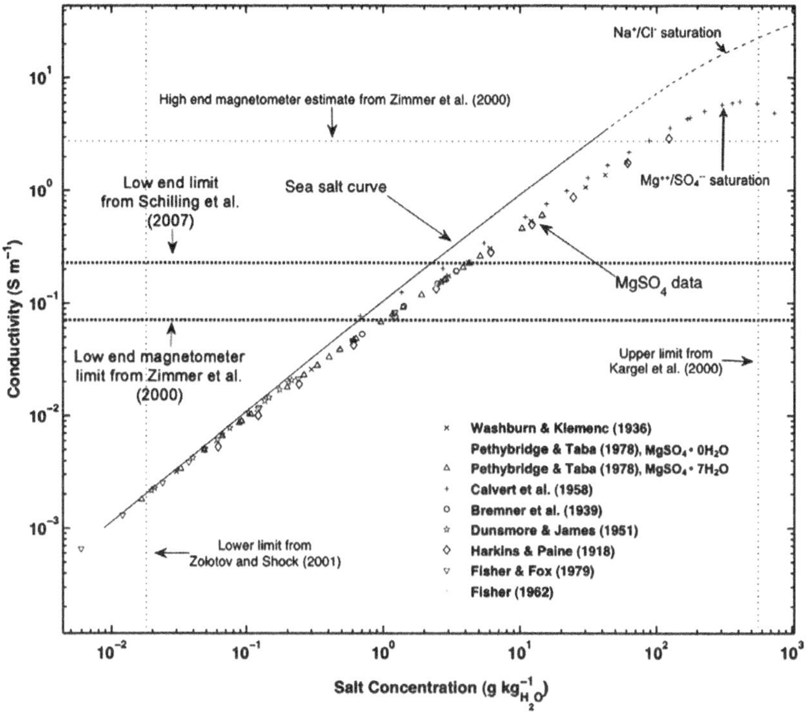

Fig. 6 Salt content in Europa's ocean after Hand and Chyba (2007) including lower limits of the electrical conductivity from Zimmer et al. (2000) and Schilling et al. (2007)

the discussed constraints of this paragraph. Most theoretical studies (Kargel et al. 2000; Zolotov and Shock 2001; McKinnon and Zolensky 2003) as well as aquateous leaching experiments (Fanale et al. 2001) suggest $MgSO_4$ and not NaCl, as the dominant salt in Europa's ocean. In Fig. 6, we show the conductivity as a function of the expected salt concentration after Hand and Chyba (2007) together with the conductivity constraints by Zimmer et al. (2000) and Schilling et al. (2007).

The currently available magnetic field observations obtained during several Europa flybys, however, only allow to constrain the amplitude A of the induction signal caused the inducing field due to Jupiter's synodic rotation period (Zimmer et al. 2000). The magnitude of the phase $|\Phi|$ is close to zero and no lower limit larger than zero could be derived. With only these measurements the value of the conductivity σ and the thickness h of the conducting shell cannot be determined simultaneously. In Fig. 7, we show isolines of the amplitude A of the induced field as a function of thickness and conductivity calculated with the three layer model of Schilling et al. (2007). The black lines are for induction due to the synodic rotation of Jupiter with frequency $\Omega_{J,\mathrm{syn},E}$ and the red lines due to the orbital movements of Europa with frequency Ω_E. A similar calculation with the single layer model of Zimmer et al. (2000) is shown in Khurana et al. (2002). For sufficiently small thickness h and conductivity σ, the iso-contours of A are straight lines and a function of the product σ times h. In this case measurements of the amplitude A as a function of frequency, i.e. $A(\Omega_{J,\mathrm{syn},E})$ and $A(\Omega_E)$, will not help to separate σ and h. However, the case is different when the red and the black lines are not parallel any more. This effect is maximum in between the regions $h/\delta(\Omega_{J,\mathrm{syn},E}) = 1$ and $h/\delta(\Omega_E) = 1$ shown as dashed lines in Fig. 7 with the skin depth δ

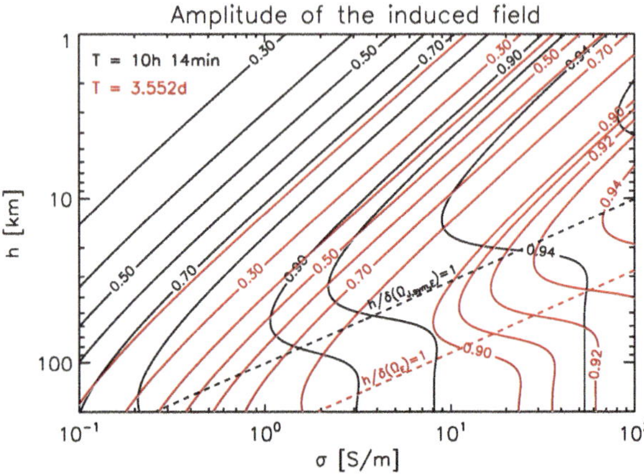

Fig. 7 Multi-frequency sounding for Europa. Values for the amplitude A of the induced signal for varying shell thickness h and conductivity σ are calculated with the three layer model of Schilling et al. (2008). *Black solid lines* are for induction due to the synodic rotation period of Jupiter, and *red solid lines* are due to the orbital period of Europa

(see (6)). In the context of the fact that the thickness and conductivity cannot be resolved independently with one single frequency, it is noteworthy to point out that the same product of conductivity σ times thickness h does in general not lead to the same induction signal A (as sometimes mistakenly assumed).

With magnetic field measurements from a future Europa (or Ganymede) orbiter low-frequency inducing fields such as the rotation period of Europa (or Ganymede) can possibly be used as well. Lower frequencies generally probe deeper into a conducting shell and thus can be used within some range in parameter space of σ and h to resolve both values independently. An orbiter with an approximate global, i.e. 4 π coverage of magnetic field measurements as close as possible to the surface and with measurements spanning several rotation periods of Europa, the depth of the ocean as well as asymmetries in its spatial extent might be determined.

With an orbiter and sufficient mission lifetime, lower frequencies such as those discussed in Sect. 5.1 might be used to more deeply probe the satellites, possibly down to the satellites' cores. In case the external inducing and the internal induced fields obtained by an orbiter can be disentangled, the $Q(\omega)$-transfer function (see Sect. 2.2.2) can be used, similar to observations at Earth, to establish more sophisticated models of the internal conductivity structure as a function of depth. The separation of the internal and external induction fields might be undertaken with a method pioneered by Schilling et al. (2007) as discussed in Sect. 2.3.

5.3 Ganymede

Ganymede, the largest satellite in our solar system, possesses a permanent internal magnetic field with an equatorial surface strength of ~750 nT (Kivelson et al. 1996, 1997; Gurnett et al. 1996). Its magnetic moment is tilted by 176° with respect to its spin axis, i.e. roughly anti-parallel to the moment of Jupiter. Jupiter's magnetospheric field at the location of Ganymede is on the order of 100 nT. This field strongly varies in time, but is always smaller than Ganymede's surface field. Due to the relative amplitudes and orientations of Ganymede's internal field and Jupiter's magnetospheric field, Ganymede possesses a mini-magnetosphere within Jupiter's magnetosphere (Kivelson et al. 1996, 1997; Neubauer 1998; Kopp and Ip 2002; Paty and Winglee 2004; Jia et al. 2008).

Kivelson et al. (2002) investigate the magnetic field measurements from Galileo's close Ganymede flybys for signatures of induction due to a possible subsurface ocean. The temporally variable magnetic field components due to Jupiter's rotation period are shown in Fig. 4.

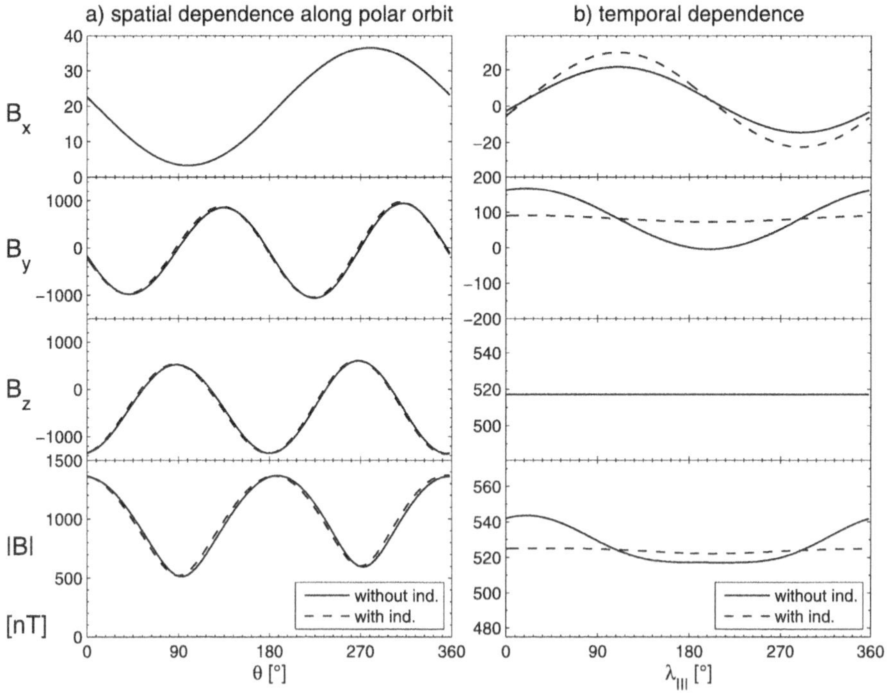

Fig. 8 Magnetic field at 100 km altitude above Ganymede's surface with and without induction (*dashed* and *solid lines*, respectively). (**a**) Magnetic field on a circular orbit along the $0°/180°$ meridian at $\lambda_{\rm III} = 200°$. (**b**) Magnetic field at fixed point $y = R_G + 100$ km during one synodic rotation period of Ganymede. R_G is the radius of Ganymede. The y-axis points towards Jupiter, the x-axis is in orbital direction of Ganymede and the z-axis completes a right-handed coordinate system

Ganymede's strong magnetic field dominates possibly induced fields close to Ganymede. Kivelson et al. (2002) find that, both, contributions of quadrupole moments from a permanent internal field or an induced field fit the observations equally well. Thus the existence of an induced field at Ganymede remains inconclusive. The core difficulty stemming from single flybys is to distinguish between spatial and temporal variations. The spatial variations of the magnetic field along Ganymede's $0°/180°$ meridian at an altitude of 100 km and at a fixed time, i.e. a fixed position of $\lambda_{\rm III} = 200°$ in Jupiter's magnetosphere, are shown in Fig. 8a. The magnetic fields displayed as solid lines are calculated from a superposition of a permanent internal field including quadrupole contributions and the magnetospheric field at the orbit of Ganymede. Values to calculate both fields are from Kivelson et al. (2002). The dashed lines include additionally the induced field for the case of a maximum response, i.e. a very good conductor very close to the surface. In Fig. 8a the induced contribution is barely visible due to the large permanent components. The problem in analyzing such a field is that the induced field could be in principle also attributed to a permanent component. Figure 8b shows the temporal variation along a synodic rotation period of Ganymede at a fixed point (100 km above the surface on the sub-Jovian side). For the B_y-component, the contribution from the induced field suppresses the temporal variations of the magnetospheric field by ~80 nT. Induction enhances the B_x-component at this particular position. Thus an induction signal can be identified with sufficient temporal coverage provided adequate knowledge of the temporal changing magnetospheric field. Magnetic field measurements on a future Ganymede orbiter will thus establish/rule out the existence of an induced field and thus help to settle the question of the existence of a subsurface ocean.

Similarly to Mercury, the very long period inducing fields at Ganymede may also be involved in the dynamo action in Ganymede.

5.4 Callisto

Callisto, the second largest of the Galilean satellites is only partially differentiated and posses the oldest surface of all Galilean satellites. It is heavily cratered and geologically inactive. Therefore it was quite surprising when the Galileo spacecraft magnetic field measurements near Callisto were observed to be consistent with induction possibly due to a liquid subsurface ocean in Callisto's interior (Neubauer 1998; Khurana et al. 1998; Kivelson et al. 1999; Zimmer et al. 2000). Near the orbit of Callisto, the radial component of Jupiter's magnetic field is the dominating time variable component (i.e. the B_y-component in Fig. 4). It is directed outward or inward depending on whether Callisto is below or above the current sheet. For the cases when Callisto is outside of the current sheet, Zimmer et al. (2000) compared modeled induced magnetic fields from a single shell model to measurements by the Galileo spacecraft. The authors find that the conductivity in the model shell needs to exceed 0.02 S/m at a depth of less than 300 km. The larger depth of the potentially associated subsurface ocean compared to the smaller depth at Europa could be the reason why Callisto's surface is old and geological unaltered. An ocean much closer to the surface, as at Europa, will more strongly drive geological and tectonic near-surface processes.

6 Saturn's Satellites

6.1 General

One of the major discoveries of the early Pioneer 11 and Voyager flyby missions to Saturn in 1979–1981 was the detection of a highly axisymmetric planetary magnetic field derived within the limitations of the flyby trajectories (Acuña and Ness 1980; Smith et al. 1980). These results were confirmed by the analysis of the first Cassini magnetic field data (Dougherty et al. 2005).

At Saturn a small periodic planetary magnetic field portion with a period apparently near the probable rotational period of the planet's interior has been observed mainly in the magnetospheric B_φ-component but also in the other components (Southwood and Kivelson 2007). The apparent periodicity is slowly varying in time but with a rate of change currently not very well understood. In addition, the weak variation of the periodic field with distance from Saturn seems not to be compatible with an internal origin (Espinosa and Dougherty 2000; Espinosa et al. 2003).The periodic field component is weak compared with the axisymmetric portion in the inner magnetosphere. With the definitions in (22) out to Rhea the contribution of the magnetospheric field $\mathbf{B_0}(t)$ to the inducing field $\mathbf{B_i}$ is then composed of periodic fields with $\sim$10 nT peak-to-peak amplitude (Giampieri et al. 2006; Southwood and Kivelson 2007) with a slowly varying period near $\sim$10.5 h and aperiodic fields due to internally and externally driven magnetospheric dynamics. In contrast to the situation at Jupiter's satellite Europa (Schilling et al. 2007) and also at Ganymede and Callisto, where the inducing fields are dominated by the time-varying magnetospheric component, the dominant sources of the inducing fields $\mathbf{B_i}$ at Saturn's satellites Enceladus and Titan are the substantial temporal variations in their total plasma magnetic fields $\mathbf{B_P}(t)$ defined as $\delta\mathbf{B_P}(t)$ instead of the small magnetospheric contributions. This can be deduced from model computations reported in the next sections on Enceladus and on Titan. Although the periodic field component of magnetospheric origin is weak compared with the axisymmetric portion in the inner magnetosphere, it becomes comparable further out, i.e. at Titan's orbit. However, at Titan it is not of much help because of Titan's almost complete shielding by its thick atmosphere- ionosphere system.

While the induction cases of Enceladus and Titan will be discussed in the next two Subsections in some detail, a few words on induction in the remaining satellites of Saturn seem to be warranted. No substantial atmosphere has been detected at these satellites. Thus inducing fields will not be modified noticeably by an atmosphere-ionosphere system, i.e. $\delta\mathbf{B_P}(t) \approx 0$. The small inducing fields of magnetospheric origin out to Rhea have already been mentioned. Outside of Titan's orbit only aperiodic fields are available for induction studies. A study of induction effects would require at least a carefully chosen sequence of flybys. Looking for possible targets for induction studies Rhea seems to be the best candidate (Hussmann et al. 2006) for the existence of an ocean. Finally, we mention for completeness that permanent internal magnetic field components have not been seen at any of the Saturnian satellites.

6.2 Enceladus

After early hints for a geologically active interior of Enceladus by the Voyager imaging experiment (Smith et al. 1982) an active gas plume above the southern polar region interacting strongly with the magnetospheric plasma was initially observed by the Cassini magnetometer experiment (Dougherty et al. 2006). This was followed by important subsequent observations and interpretations by remote sensing and in-situ experiments onboard Cassini (Porco et al. 2006; Hansen et al. 2006; Waite et al. 2006; Tokar et al. 2006). Although a full discussion of this interesting subject is outside the scope of this book, the following picture relevant for this section has emerged. The jets of mostly water vapor issue from locations on the so-called "tiger stripes", which are linear features in the southern polar region of Enceladus. Several sources, which vary in time, have been identified. The origin of the water vapor is still controversial between sublimation from internal clathrate surfaces and evaporation from the liquid–vapor interface in nozzle-like vents (Schmidt et al. 2008). The energy source is also not clear yet. In addition to water vapor the plumes contain gases like CO_2, CO or N_2, CH_4 etc. and ice particles which maintain the E-ring (Waite et al. 2006; Spahn et al. 2006). The composition of the partly contaminated ice and dust grains originating at Enceladus suggest the water bodies below the surface of Enceladus are "contaminated" by cations and anions due to leaching of the "rocky" part of the interior (Postberg et al. 2009). This would then make them "visible" to electromagnetic induction studies. The same is true if an innermost rock-metal core of Enceladus were large and conductive enough.

Electromagnetic induction studies can thus be used to observationally investigate electrolytically conducting bodies of water ranging from a global ocean to a more localized pond below the surface and possible a core of Enceladus, which are problems at the heart of the contemporary discussion of Enceladus' interior and plume mechanisms. For these studies sufficiently quickly varying inducing fields $\mathbf{B}_i(t) \sim \delta\mathbf{B}_P(t)$ are necessary. The inducing fields $\mathbf{B}_i(t)$ drive induced fields $\mathbf{B}_{\text{ind}}$, which have to be separated from the inducing field in a proper analysis of the data. For a proper assessment, some information on the electrical conductivity would be helpful. With the only quantitative model of the chemistry of an ocean below Enceladus' surface, which is in contact with a rocky core, Zolotov (2007) has obtained salinities up to 20 g/kg H_2O, which is just below 35 g/kg H_2O for the terrestrial ocean. Thus except for differences in chemical composition we take the conductivity of the

Table 1 EM saturation periods for two models of water bodies and $A = 0.7$

Type of water body	Parameters			
	displacement from Enceladus center towards south pole in km	outer radius of water body in km	inner radius of water body in km	saturation period T in s
Global ocean, centered spherical shell	0	247.1	147.1	4.7×10^4
Local pond displaced sphere	197.1	50.0	0	1.9×10^3

terrestrial ocean of 5 S/m (Telford 1993) as an upper limit for the expected conductivity of the ocean of Enceladus. Table 1 then gives the periods T of harmonic variations of $\delta\mathbf{B}_P(t)$, for which the induction response starts to saturate, i.e. the amplitude of the induced field reaches $A = 0.7$ (see (16)). The periods T where saturation is reached are given in Table 1 for a model of a global ocean (Schubert et al. 2007) with an upper limit ocean thickness of 100 km and spherical shell geometry and a localized spherical pond of radius 50 km below the south pole. The crustal thickness is 5 km and $A = 0.7$ at the south pole in the latter case and constant everywhere in the former case. Note that at lower values of A the results will also depend on the core and mantle conductivity.

Only limited information is available on temporal variations of the plume-plasma interaction from observations. Variations of the oxygen cloud in Saturn's E-ring on time scales of months have been derived from observations by Esposito et al. (2005),whereas local neutral gas observations (Waite et al. 2006) point to spatial and temporal variations near Enceladus, which are difficult to separate unfortunately. The combination of neutral gas observations and magnetic field observations for Enceladus flybys E0, E1 and E2 lead to an order of magnitude decrease of the rate of mass loss of Enceladus from E0 to E1 (Saur et al. 2008), which were twenty days apart, and insignificant change over the longer time interval between E1 and E2. (These three flybys are also referred to as 03En, 04En, 11En and occurred on 17 February 2005, 9 March 2005 and 14 July 2005, respectively.) However, the temporal variations may be severely under-sampled, i.e. the changes between E0, E1, and E2 may be the expression of much faster variations. Interesting predictions for the control of plume activity by tides at the orbital period of Enceladus of $T_E = 1.184 \times 10^5$ s due to its driven orbital eccentricity have been made (Hurford et al. 2007; and subsequent oral presentations; Smith-Konter and Pappalardo 2008) without observational confirmation at the time of the writing of this manuscript, but consistent with the variations in Saur et al. (2008). This is except for the possibility that at these short periods the results of the model by Saur et al. (2008) may be modified by induction effects, which have not been incorporated in the models yet. The tidal variations may also involve harmonics of the orbital period of Enceladus. But even if the plume activity were stable in time the plasma interaction would vary appreciably because of the variation of the magnetospheric plasma (Gurnett et al. 2007) with the synodic rotation period $T_{S,\mathrm{syn},E} = 5.55 \times 10^4$ s of Saturn in the rest frame of Enceladus. The investigation of local ponds would require even shorter periods, i.e. shorter than approximately the saturation period in Table 1 unless the conductivity is higher than expected. It is interesting to note that the saturation period for a strongly conducting global ocean is not much less than T_E. Thus temporal variations of $\delta\mathbf{B}_P(t)$ at the tidal period T_E could drive induction in a global ocean at the assumed upper limit of conductivity into saturation, particularly, if higher tidal harmonics come into play. The synodic rotation period $T_{S,\mathrm{syn},E}$ would be sufficient by itself. After establishing the temporal variations of the plume-plasma interaction and further refinement of the model by Saur et al. (2007, 2008) it will be possible to compute $\delta\mathbf{B}_P(t)$ as a function of time and location in three dimensions. The problem of a global ocean (see Table 1) could then be solved in the same way as by Schilling et al. (2007) for Europa. A few Enceladus flybys as provided by the Cassini mission may be sufficient for this case. A different approach would be to observe the magnetic field by an array of magnetometers on Enceladus' surface or from a number of low orbits. This seems to be beyond missions in the near future, however.

6.3 Titan

The objectives of electromagnetic induction studies at Titan are to investigate the properties of an ocean below the surface and a rock and iron core (Grasset et al. 2000; Sohl et al. 2003) of this largest satellite of Saturn. After prediction of a water ocean in models of the interior with the ocean sandwiched between layers of ice first indications of the existence of an ocean came from variations in Titan's spin rate from Cassini RADAR observations (Lorenz et al. 2008). In contrast to Enceladus no chemical model of the ocean has been published so-far. The "anti-freeze" ingredient ammonia is generally taken as an ocean constituent or ammonium sulfate more recently (Grindrod et al. 2008). From the leaching of rocky material in the far geological past the ocean should still contain cations and anions for sufficient electrical conductivity. If the fourth harmonic of the orbital period of four days is taken as the exciting period and an amplitude factor of $A = 0.8$ is required for detection, a subsurface ocean would need to have an electrical conductivity of ~9 S/m. For comparison, typical values for terrestrial sea water are 5 S/m.

The physical mechanism leading to inducing fields $\mathbf{B}_P(t)$ at Titan is quite different from Enceladus because of a very different neutral atmosphere–ionosphere configuration. The first detailed study of Titan's neutral atmosphere was done during the Voyager mission, which showed it to have a surface pressure of 1.5 bars and to consist mainly of nitrogen (Hunten et al. 1984). An ionosphere was marginally discovered. Detailed studies of the neutral atmosphere were made since the beginning of the Cassini mission establishing the abundance of methane and minor constituents. Earlier detailed modeling work of the plasma interaction by Keller et al. (1994) showed that the ionosphere of Titan is dominated by photoionization and collisional ionization by photoelectrons. Collisional ionization by hot magnetospheric electrons is secondary in importance. These theoretical results were later confirmed by results from the Cassini mission (Backes et al. 2005; Cravens et al. 2005; Ma et al. 2006; Simon et al. 2006). The day-side or photoionization hemisphere of Titan is centered around the sunward direction. On the other hand the magnetospheric electrons are carried inward towards Titan by the magnetospheric flow which is generally considered to be approximately corotational in the equatorial plane of Saturn. During an orbital revolution of Titan around Saturn with Saturnian local time (SLT) varying through 24 hours the angle α between the flow direction and the solar direction will then vary from $\sim 0°$ to $\sim 180°$ and back, where the details depend on the season. At the equinoxes α varies from 0° at SLT = 18:00 h to 180° at SLT = 6:00 h. At the solstices the corresponding variation will be reduced due to the orbital obliquity of 26.7° of Titan and Saturn with respect to Saturn's orbital plane. Figure 9 illustrates the situation in the rotating frame of Titan with x in the corotational flow direction, y towards Saturn and z along the rotational axis under equinox conditions. The day-side hemisphere is also indicated in the Figure as it rotates around the

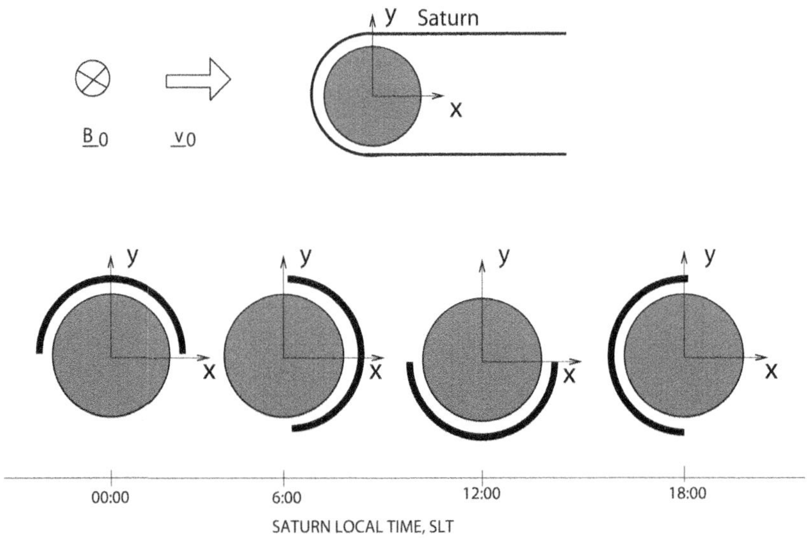

Fig. 9 Schematic view of the varying geometry of the sunlit hemisphere of Titan as a function of Saturn Local Time for solar equinox conditions; $\mathbf{B_0}$ and $\mathbf{v_0}$ are the incident magnetic field vector and velocity vector, respectively

z-axis in the clockwise sense. The sunlit hemisphere's symmetry axis, i.e. the solar direction, is tilted by up to $\pm 26.7°$ between the equinoxes with polar night and polar day condition prevailing at the winter and summer poles, respectively.

Ionization in the umbra region, approximately a cylinder of $R_T + 800$ km radius is then due to the inflow of magnetospheric electrons, heating of electrons by heat flux along the field lines and a minor contribution from photoelectrons (with Titan radius $R_T = 2575$ km). A deep minimum in electron density then exists in a broad region around the anti-subsolar point. The resulting distribution of electron density N has an important influence on the magnetic field configuration around Titan described by the induction (20) in Sect. 2. The relative role of transport and diffusion is determined by the magnetic diffusivity η, which is proportional to $1/N$ and the electron-neutral collision frequency ν_e which is controlled by n_n, the neutral gas density. Far from Titan frozen-in field conditions prevail in the incoming flow because of the small collision frequency ν_e. The magnetospheric field $\mathbf{B_0}$ away from Titan is typically characterised by a small component $\mathbf{B_{0,x}}$, negative $\mathbf{B_{0,z}} \sim -3$ nT and $\mathbf{B_{0,y}}$ of a similar maximum magnitude but a negative sign for northern summer and a positive one for southern summer (see Arridge et al. 2008).

$\mathbf{B_0}$ varies with time over a wide range of periods. Neglecting secular variations in Saturn's magnetospheric field we can expect strong seasonal variations at the orbital period of Saturn of 30 years and smaller variations at Titan's orbital period and aperiodic variations due to external magnetospheric dynamics.

For the discussion of the interaction between $\mathbf{B_0}$ and the time-varying ionosphere in Titan's frame of reference we assume $\mathbf{B_0}$ to be constant over time scales up to the quasi-stationarity scales of one hour. We can then discuss the magnetic field under these assumptions. On the sunlit side the magnetic diffusivity remains low down to the ionospheric maximum at ~1200 km altitude at the subsolar point and somewhat higher elsewhere. Inside the magnetic diffusivity increases rapidly. The magnetic field forms a clear magnetic ionopause according to modeling and Cassini observations with at most a small minimum magnetic field magnitude below the ionopause. Because of the much lower electron densities magnetic diffusion is much more important on the night side, i.e. in the umbra. The combined action of magnetic diffusion and inflow of plasma leads to the strongest appearance of magnetic fields below the ionosphere in the lower neutral atmosphere at less than 800 km altitude, say. When Titan is at SLT ~6.00 h, we expect the maximum inducing field perpendicular to the x-axis to occur close to or on the $(-x)$-axis, i.e. at the center of the ramside hemisphere. As Titan moves along its orbit during a Titan day for any given point fixed to Titan the inducing field is at most very small on the sunlit side, grows in the umbra, where it reaches a maximum near its center, to fall off towards the sunlit side again.

It is clear from the discussion including Fig. 9 that a constant incident $\mathbf{B_0}$ leads to an inducing field below the ionosphere, which at any position (x, y, z) can be described by a Fourier series in time with the fundamental frequency Ω_T derived from a Titan day. In reality $\mathbf{B_0}$ will not be constant in time but vary at many frequencies Ω. For a given frequency Ω in $\mathbf{B_0}$ and Ω_T describing the rotation of the solar energy source in Fig. 9 the interaction between the plasma flow and the magnetic field described by (20) and the full set of fluid (or kinetic) equations (not given in this manuscript) then leads to the inducing fields $\mathbf{B_i}(t)$ below the ionosphere. At any given location $\mathbf{B_i}(t)$ must then be composed of a component constant in time, harmonic components at frequencies Ω and Ω_T and multiples thereof and frequencies abs$(\Omega_T \pm \Omega)$ etc. due to the non-linearities in the interaction. There will be contributions to Ω due to the SLT-variation of the magnetosphere and more importantly due to the synodic rotation period of Saturn $T_{S,\text{syn},T} = 10.8$ h as seen from Titan. Interestingly, there will not only be small variations in $\mathbf{B_0}(t)$ but also in the rotating plasma properties at $T_{S,\text{syn},T}$ enhancing the inducing signal at the relatively high corresponding frequency (Arridge et al. 2008). This leads to lower conductivity requirements of about 2.2 S/m for the ocean with some remaining weak dependence on mantle properties. All frequency peaks will be smeared by aperiodic dynamic variations. In the frame of Titan there will also be a small component of the magnetic field that is constant in time and therefore penetrates Titan completely. The induction investigation could be done using an array of magnetometers on the surface of Titan over several Titan days. A less demanding approach could use an orbiter with repeated measurements at the pericenter $x \approx -900$ km and $y \approx 0$, $z \approx 0$ and other locations distributed around Titan with measurements over at least several Titan days. The data analysis will also require straightforward 3D-modeling along the lines of Schilling et al. (2007).

Finally, there is always the possibility of induction studies using spacecraft encounters occurring under serendipitous conditions like strong Saturnian magnetospheric dynamics in response to the solar wind dynamics. In addition, observations near Titan at times when the sun is eclipsed by Saturn would be very useful around equinox conditions.

7 Uranus, Neptune and Beyond

Compared to the Saturn system, the Uranus and Neptune systems are paradises for induction studies. Uranus' axial tilt, i.e. the angle between its spin axis and the normal to its orbital plane, is 97.8°. Uranus' magnetic field can be represented to first order by a dipole moment of 2.3×10^4 nT $(4\pi/\mu_0)R_U^3$, where $R_U = 25{,}600$ km is Uranus' planetary radius (Ness et al. 1986). The dipole is offset from the center of the planet by $0.3R_U$. It is also inclined with respect to Uranus' spin axis by 60°. The rotation period of the magnetic field of 17.3 h is used to define Uranus' rotation period (Ness et al. 1986). Uranus' satellites Ariel, Umbriel, Titania and Oberon have diameters larger than 1000 km and small inclination and eccentricity. Up to now no magnetic field data are available near these satellites, but future spacecraft measurements will be able to use the electromagnetic induction technique to probe their interior electrical conductivity structure and thus provide constraints in the search for liquid water. Possible oceans within these satellites are discussed in Hussmann et al. (2006). Due to the off-center dipole moment and the inclined moment particularly strong inducing fields at the synodic rotation period of the satellites and harmonics will be available to sound their interiors.

Neptune's axial tilt is 28.3°. Its magnetic field can be represented to first order by a dipole moment of 1.3×10^4 nT $(4\pi/\mu_0)R_N^3$, where $R_N = 24{,}800$ km is Neptune's planetary radius. The dipole moment is offset from the center of the planet by 0.55 R_N. It is also inclined with respect to Neptune's spin axis by 47° (Ness et al. 1989).

Neptune's largest satellite Triton, the only moon larger than 1000 km in the Neptunian system, is a particularly interesting body. It has an inclination of 156.8°, but nearly zero eccentricity, while Neptune's other satellites have nearly zero inclination and low eccentricity. Triton's larger than 90° degree inclination, which corresponds to a retrograde motion, implies that it is not a natural satellite of Neptune, but captured, possible originating from the Kuiper-Belt. The Voyager 2 spacecraft observed geyser activity on Triton which together with models of Triton's interior suggest that Triton might possess a subsurface ocean of liquid water (Smith et al. 1989; Stevenson 2002; Hussmann et al. 2006). Triton's large inclination, and Neptune's offset and inclined dipole, provide inducing frequencies at both Triton's orbital period (5.88 d) and Neptune's synodic rotation period (14.46 h) and harmonics. In Fig. 10, we display the magnetic field in the vicinity of Triton during an arbitrary time window of 15 days starting January 1, 2020. The x-component measured in a Triton based coordinate system is along Triton's orbital motion, the y-component points towards Neptune, and the z-component complements a right-handed coordinate system. Both inducing

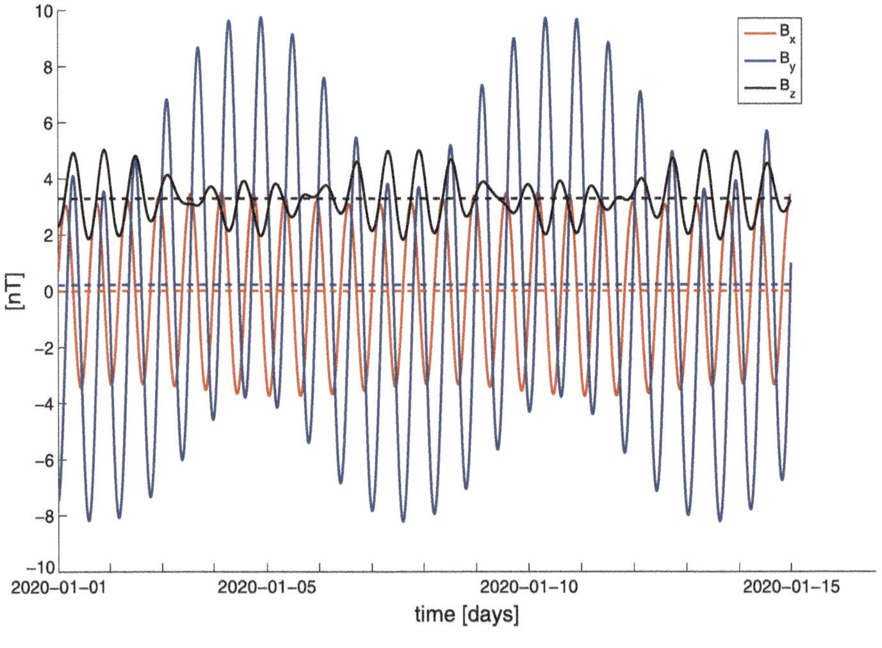

Fig. 10 Neptune's magnetic field at the location of Triton during 15 days starting arbitrarily at January 1, 2020. The magnetic field is calculated with the multipole magnetic field model of Ness and Connerney (1995)

frequencies are clearly visible in Fig. 10. The time averaged fields are shown as dashed lines. The average B_z is ~3 nT and often smaller in magnitude than the time variable components. The strong time variable components at both frequencies might resolve the thickness and the conductivity of a possible ocean independently (see discussion in Sect. 5.2).

Possible subsurface oceans are also being discussed for large Kuiper-Belt objects such as Pluto, Charon, Eris, Sedna (Stevenson 2002; Hussmann et al. 2006). These bodies are immersed in the solar wind (based on the current state of knowledge). Therefore no immediately obvious time periodic field is available that can be used in future spacecraft flybys. Two spacecraft near these objects would allow for an observational setup where one determines the time-varying field while the other one, closer to the body, could measure the inductive response. Asteroids might be similarly sounded with the induction method.

Induction possibly also plays an important role for extra-solar planets. Planets very close to the central star might be exposed to large time-varying stellar magnetic fields. Laine et al. (2008) discuss the possibility that Ohmic dissipation due to induction caused by short period planets could be large enough to inflate the planet and significantly contribute to mass loss. The Ohmic heating however does not provide a significant heat source in the possible subsurface ocean of the solar system satellites discussed in this chapter when compared to tidal heating in the icy crust and the inner parts of the satellites (Hussmann et al. 2002; Tovie et al. 2005). For the solar system bodies, tidal dissipation of Rossby waves within the possible ocean might also significantly contribute to the heating of the oceans (Tyler 2008).

Finally, the induction technique can also be applied to all six non-discussed planets of our solar system, similar to how it is applied at Earth to probe the interior of the Earth (e.g., Schmucker 1985; Olsen 1999; Constable and Constable 2004). This could be achieved by orbiting spacecraft at small radial distances from the planets.

Acknowledgements We thank Nico Schilling, Nils Olsen, Stefan Duling and Mario Seufert for valuable contributions to this chapter. We appreciate helpful comments by the referee on the presentation of the manuscript. JS acknowledges financial support by Deutsche Forschungsgesellschaft. The work by KHG was financially supported by the German Ministerium für Wirtschaft und Technologie and the German Zentrum für Luft- und Raumfahrt under contract 50 QW 0602.

References

M.H. Acuña, N.F. Ness, The magnetic field of Saturn–Pioneer 11 observations. Science **207**, 444–446 (1980)

B.J. Anderson, M.H. Acuña, D.A. Lohr, J. Scheifele, A. Raval, H. Korth, J.A. Slavin, The magnetometer instrument on MESSENGER. Space Sci. Rev. **131**, 417–450 (2007). doi:10.1007/s11214-007-9246-7

B.J. Anderson, M.H. Acuña, H. Korth, M.E. Purucker, C.L. Johnson, J.A. Slavin, S.C. Solomon, R.L. McNutt, The structure of Mercury's magnetic field from MESSENGER's first flyby. Science **321**, 82 (2008). doi:10.1126/science.1159081

J.D. Anderson, G. Schubert, R.A. Jacobson, E.L. Lau, W.B. Moore, W.L. Sjogren, Europa's differentiated internal structure: Inferences from four Galileo encounters. Science **281**, 2019–2022 (1998)

C. Arridge, N. Andre, N. Achilleos, et al., Thermal electron periodicities at 20rs in Saturn's magnetosphere. Geophys. Res. Lett. **35**, 15 (2008)

H. Backes, et al., Titan's magnetic field signature during the first Cassini encounter. Science **308**, 992–995 (2005)

W. Baumjohann, R.A. Treumann, *Basic Space Plasma Physics* (Imperial College Press, London, 1996)

J.L. Blank, W.R. Sill, Response of the Moon to the time-varying interplanetary magnetic field. J. Geophys. Res. **74**, 736–743 (1969). doi:10.1029/JA074i003p00736

J.E.P. Connerney, M.H. Acuna, N.F. Ness, T. Satoh, New models of Jupiter's magnetic field constrained by the Io Flux Tube footprint. J. Geophys. Res. **103**, 11929–11939 (1998)

S. Constable, C. Constable, Observing geomagnetic induction in magnetic satellite measurements and associated implications for mantle conductivity. Geochem. Geophys. Geosyst. **5** (2004). doi:10.1029/2003GC000634

T.E. Cravens, et al., Titan's ionosphere: Model comparisons with Cassini Ta data. Geophys. Res. Lett. **32**, 12108 (2005)
M.K. Dougherty, N. Achilleos, N. Andre, C.S.A.A. Balogh, C. Bertucci, et al., Cassini magnetometer observations during Saturn orbit insertion. Science **307**, 1266–1270 (2005)
M.K. Dougherty, K.K. Khurana, F.M. Neubauer, C.T. Russell, J. Saur, J.S. Leisner, M. Burton, Identification of a dynamic atmosphere at Enceladus with the Cassini Magnetometer. Science **311**, 1406 (2006)
P. Dyal, D.I. Gordon, Lunar surface magnetometers. IEEE Trans. Magn. **9**, 226–231 (1973). doi:10.1109/TMAG.1973.1067650
P. Dyal, C.W. Parkin, Global electromagnetic induction in the Moon and planets. Phys. Earth Planet. Inter. **7**, 251–265 (1973). doi:10.1016/0031-9201(73)90052-6
P. Dyal, C.W. Parkin, D. W.D., Magnetism and the interior of the moon. Rev. Geophys. Space Phys. **12**, 568–591 (1975)
S. Espinosa, M. Dougherty, Periodic perturbations in Saturn's magnetic field. Geophys. Res. Lett. **27**, 2785–2788 (2000)
S. Espinosa, D. Southwood, M. Dougherty, How can Saturn impose its rotation period in a noncorotating magnetosphere? J. Geophys. Res. **108**, 11–1 (2003)
L.W. Esposito, J.E. Colwell, K. Larsen, et al., Ultraviolet imaging spectroscopy shows an active Saturnian system. Science **307**, 1251 (2005)
F.P. Fanale, Y.P. Li, E. Decarlo, C. Farley, S. Sharma, K. Horton, An experimental estimate of Europa's "ocean" composition independent of Galileo orbital remote sensing. J. Geophys. Res. **106**, 14595–14600 (2001)
G. Giampieri, M. Dougherty, C.T. Smith, E.J. Russell, A regular period for Saturn's magnetic field that may track its internal rotation. Nature **441**, 62–64 (2006)
K.H. Glassmeier, H.U. Auster, U. Motschmann, A feedback dynamo generating Mercury's magnetic field. Geophys. Res. Lett. **34**, 22201–22205 (2007a). doi:10.1029/2007GL031662
K.H. Glassmeier, J. Grosser, U. Auster, D. Constantinescu, Y. Narita, S. Stellmach, Electromagnetic induction effects and dynamo action in the Hermean system. Space Sci. Rev. **132**, 511–527 (2007b). doi:10.1007/s11214-007-9244-9
K.H. Glassmeier, H. Auster, D. Heyner, K. Okrafka, C. Carr, G. Berghofer, B.J. Anderson, A. Balogh, W. Baumjohann, P.J. Cargill, U. Christensen, M. Delva, M. Dougherty, K. Fornaçon, T.S. Horbury, E.A. Lucek, W. Magnes, M. Mandea, A. Matsuoka, M. Matsushima, U. Motschmann, R. Nakamura, Y. Narita, I. Richter, K. Schwingenschuh, H. Shibuya, J.A. Slavin, C. Sotin, B. Stoll, H. Tsunakawa, S. Vennerstrom, J. Vogt, T. Zhang, The fluxgate magnetometer of the BepiColombo planetary orbiter. Planet. Space Sci. (2009)
O. Grasset, C. Sotin, F. Deschamps, On the internal structure and dynamics of titan. Planet. Space Sci. **48**, 617–636 (2000)
P.M. Grindrod, A.D. Fortes, F. Nimmo, et al., The long-term stability of a possible aqueous ammonium sulfate ocean inside titan. Icarus **197**, 137–151 (2008)
J. Grosser, K.H. Glassmeier, A. Stadelmann, Induced magnetic field effects at planet Mercury. Planet. Space Sci. **52**, 1251–1260 (2004). doi:10.1016/j.pss.2004.08.005
D.A. Gurnett, W.S. Kurth, A. Roux, S.J. Bolton, C.F. Kennel, Galileo plasma wave observations in the Io plasma torus and near Io. Science **274**, 391–392 (1996)
D.A. Gurnett, et al., The variable rotation period of the inner region of Saturn's plasma disk. Science **316**, 442–445 (2007)
D.T. Hall, D.F. Strobel, P.D. Feldman, M.A. McGrath, H.A. Weaver, Detection of an oxygen atmosphere on Jupiter's moon Europa. Nature **373**(6516), 677–679 (1995)
K.P. Hand, C.F. Chyba, Empirical constraints on the salinity of the European ocean and implications for a thin ice shell. Icarus **189**, 424–438 (2007)
C.J. Hansen, et al., Enceladus' water vapor plume. Science **311**, 1422–1425 (2006)
D. Heyner, D. Schmitt, J. Wicht, K.H. Glassmeier, H. Korth, U. Motschmann, Concerning the initial temporal evolution of a Hermean feedback dynamo. Earth Planet. Sci. Lett. (2009)
B.A. Hobbs, L.L. Hood, F. Herbert, C.P. Sonett, An upper bound on the radius of a highly electrically conducting lunar core. J. Geophys. Res. **88**, 97 (1983). doi:10.1029/JB088iS01p00B97
L. Hood, G. Schubert, Inhibition of solar wind impingement on Mercury by planetary induction currents. J. Geophys. Res. **84**, 2641–2647 (1979)
L.L. Hood, F. Herbert, C.P. Sonett, The deep lunar electrical conductivity profile—Structural and thermal inferences. J. Geophys. Res. **87**, 5311–5326 (1982). doi:10.1029/JB087iB07p05311
R. Hunten, et al., *Saturn* (Univ. of Arizona Press, Tucson, 1984)
T. Hurford, P. Helfenstein, G. Hoppa, R. Greenberg, B. Bills, Eruptions arising from tidally controlled periodic openings of rifts on Enceladus. Nature **447**, 292–294 (2007)

H. Hussmann, T. Spohn, K. Wieczerkowski, Thermal equilibrium states of Europa's ice shell: Implications for internal ocean thickness and surface heat flow. Icarus **156**, 143–151 (2002)
H. Hussmann, F. Sohl, T. Spohn, Subsurface oceans and deep interiors of medium-sized outer planet satellites and large trans-Neptunian objects. Icarus **185**, 258–273 (2006)
X. Jia, R. Walker, M. Kivelson, K. Khurana, J. Linker, Three-dimensional MHD simulations of Ganymede's magnetosphere. J. Geophys. Res. **113**, 06212 (2008)
J. Kargel, J. Kaye, J. Head, G. Marion, R. Sassen, O.P. Ballesteros, S. Grant, D. Hogenboom, Europa's crust and ocean: Origin, composition and the prospects for life. Icarus **148**, 368–390 (2000)
C. Keller, T.E. Cravens, L. Gan, One-dimensional multispecies hydrodynamic models of the ramside ionosphere of titan. J. Geophys. Res. **99**, 6511–6525 (1994)
A. Khan, J.A.D. Connolly, N. Olsen, K. Mosegaard, Constraining the composition and thermal state of the moon from an inversion of electromagnetic lunar day-side transfer functions. Earth Planet. Sci. Lett. **248**, 579–598 (2006). doi:10.1016/j.epsl.2006.04.008
K.K. Khurana, M.G. Kivelson, D.J. Stevenson, G. Schubert, C.T. Russell, R.J. Walker, C. Polanskey, Induced magnetic fields as evidence for subsurface oceans in Europa and Callisto. Nature **395**, 777–780 (1998)
K.K. Khurana, M.G. Kivelson, C.T. Russell, Searching for liquid water in Europa by using surface observations. Astrobiol. J. **2**, 93–103 (2002)
K.K. Khurana, et al., The configuration of Jupiter's magnetosphere, in *Jupiter*, ed. by F. Bagenal (Cambridge Univ. Press, Cambridge, 2004), pp. 593–616, Chap. 24
M.G. Kivelson, K.K. Khurana, R.J. Walker, C.T. Russell, J.A. Linker, D.J. Southwood, C. Polanskey, A magnetic signature at Io: Initial report from the Galileo magnetometer. Science **273**, 337–340 (1996)
M.G. Kivelson, K.K. Khurana, S. Joy, C.T. Russell, D.J. Southwood, R.J. Walker, C. Polanskey, Europa's magnetic signature: Report from Galileo's first pass on 19 December 1996. Science **276**, 1239–1241 (1997)
M.G. Kivelson, K.K. Khurana, D.J. Stevenson, L. Bennett, S. Joy, C.T. Russell, R.J. Walker, C. Zimmer, C. Polanskey, Europa and Callisto: Induced or intrinsic fields in a periodically varying plasma environment. J. Geophys. Res. **104**(A3), 4609–4625 (1999)
M.G. Kivelson, K.K. Khurana, C.T. Russell, M. Volwerk, J. Walker, C. Zimmer, Galileo magnetometer measurements: A stronger case for a subsurface ocean at Europa. Science **289**(5483), 1340–1343 (2000)
M.G. Kivelson, K.K. Khurana, M. Volwerk, The permanent and inductive magnetic moments of Ganymede. Icarus **157**, 507–522 (2002)
M.G. Kivelson, F. Bagenal, F.M. Neubauer, W. Kurth, C. Paranicas, J. Saur, Magnetospheric interactions with satellites, in *Jupiter*, ed. by F. Bagenal (Cambridge Univ. Press, Cambridge, 2004), pp. 513–536, Chap. 21
A.J. Kliore, D.P. Hinson, F.M. Flasar, A.F. Nagy, T.E. Cravens, The ionosphere of Europa from Galileo radio occultations. Science **277**(5324), 355–358 (1997)
A. Kopp, W. Ip, Resistive MHD simulations of Ganymede's magnetosphere 1. Time variabilities of the magnetic field topology. J. Geophys. Res. **107**, 41–11490 (2002)
B.N. Lahiri, A.T. Price, Electromagnetic induction in non-uniform conductors. Philos. Trans. R. **784**(A 237), 509–540 (1939)
R. Laine, D. Lin, S. Dong, Interaction of close-in planets with the magnetosphere of their host stars. 1. Diffusion, Ohmic dissipation of time-dependent field, planetary inflation, and mass loss. Astrophys. J. **685**, 521–542 (2008)
R.D. Lorenz, B.W. Stiles, R.L. Kirk, et al., Titan's rotation reveals an internal ocean and changing zonal winds. Science **319**, 1649 (2008)
Y. Ma, et al., Comparisons between MHD model calculations and observations of Cassini flybys of Titan. J. Geophys. Res. **111**, 05207 (2006)
W. McKinnon, M. Zolensky, Sulfate content of Europa's ocean and shell: Evolutionary considerations and some geological and strobiological implications. Astrobiology **3**, 879–897 (2003)
N.F. Ness, J.E.P. Connerney, Neptune's magnetic field and field-geometric properties, in *Neptune and Triton*, ed. by D.P. Cruikshank (Univ. of Arizona Press, Tucson, 1995), pp. 141–168
N.F. Ness, K.W. Behannon, R.P. Lepping, Y.C. Whang, K.H. Schatten, Magnetic field observations near Mercury: preliminary results from Mariner 10. Science **185**, 131–135 (1974)
N.F. Ness, M.H. Acuña, K.W. Behannon, L.F. Burlaga, J. Connerney, R.P. Lepping, F.M. Neubauer, Magnetic fields at Uranus. Science **233**, 85–89 (1986)
N.F. Ness, M.H. Acuña, L.F. Burlaga, J. Connerney, R.P. Lepping, F.M. Neubauer, Magnetic fields at Neptune. Science **246**, 1473–1478 (1989)
F.M. Neubauer, Nonlinear standing Alfvén wave current system at Io: Theory. J. Geophys. Res. **85**(A3), 1171–1178 (1980)
F.M. Neubauer, The sub-Alfvénic interaction of the Galilean satellites with the Jovian magnetosphere. J. Geophys. Res. **103**(E9), 19843–19866 (1998)

F.M. Neubauer, Alfvén wings and electromagnetic induction in the interiors: Europa and Callisto. J. Geophys. Res. **104**(A12), 28671 (1999)
N. Olsen, Induction studies with satellite data. Surv. Geophys. **20**, 309–340 (1999)
N. Olsen, et al., Separation of magnetic field into external and internal parts, in *Planetary Magnetism* (Springer, Berlin, 2009)
R.T. Pappalardo, et al., Does Europa have a subsurface ocean? Evaluation of the geological evidence. J. Geophys. Res. **104**, 24015–24056 (1999)
W. Parkinson, *Introduction of Geomagnetism* (Scottish Academic Press, Edinburgh, 1983)
C. Paty, R. Winglee, Multi-fluid simulations of Ganymede's magnetosphere. Geophys. Res. Lett. **31**, 24806 (2004)
C. Porco, et al., Cassini observes the active south pole of Enceladus. Science **311**, 1393–1401 (2006)
F. Postberg, S. Kempf, J. Schmidt, N. Brillantov, A. Beinsen, B. Abel, U. Buck, R. Srama, Sodium salts in E-ring ice grains from an ocean below the surface of Enceladus. Nature **459**, 1098–1101 (2009)
T. Rikitake, *Electromagnetism and the Earth's Interior* (Elsevier, Amsterdam, 1966)
J. Saur, A model for Io's local electric field for a combined Alfvénic and unipolar inductor far-field coupling. J. Geophys. Res. **109**, 01210 (2004)
J. Saur, D.F. Strobel, F.M. Neubauer, Interaction of the Jovian magnetosphere with Europa: Constraints on the neutral atmosphere. J. Geophys. Res. **103**(E9), 19947–19962 (1998)
J. Saur, F.M. Neubauer, D.F. Strobel, M.E. Summers, Interpretation of Galileo's Io plasma and field observations: The J0, I24, I27 flybys, and close polar passes. J. Geophys. Res. **107**(A12), 1422 (2002). doi:10.1029/2001JA005067
J. Saur, F.M. Neubauer, N. Schilling, Hemisphere coupling in Enceladus' asymmetric plasma interaction. J. Geophys. Res. **112**, 11209 (2007). doi:10.1029/2007JA012479
J. Saur, N. Schilling, F.M. Neubauer, et al., Evidence for temporal variability of Enceladus' gas jets: Modeling of Cassini observations. Geophys. Res. Lett. **35**, 20105 (2008)
N. Schilling, K.K. Khurana, M.G. Kivelson, Limits on an intrinsic dipole moment in Europa. J. Geophys. Res. **109**, 05006 (2004)
N. Schilling, F.M. Neubauer, J. Saur, Time-varying interaction of Europa with the Jovian magnetosphere: Constraints on the conductivity of Europa's subsurface ocean. Icarus **192**, 41–55 (2007)
N. Schilling, F.M. Neubauer, J. Saur, Influence of the internally induced magnetic field on the plasma interaction of Europa. J. Geophys. Res. **113**, 03203 (2008)
J. Schmidt, N. Brilliantov, F. Spahn, S. Kempf, Slow dust in Enceladus' plume from condensation and wall collisions in tiger stripe fractures. Science **451**, 685–688 (2008)
U. Schmucker, Magnetic and electric fields due to electromagnetic induction by external sources, in *Landolt-Börnstein New-Series, 5/2b* (Springer, Berlin–Heidelberg, 1985), pp. 100–125
G. Schubert, J. Anderson, B. Travis, J. Palguta, Enceladus: Present internal structure and differentiation by early and long-term radiogenic heating. Icarus **188**, 345–355 (2007)
M. Siegert, J. Ellis-Evans, M. Tranter, C. Mayer, J.R. Petit, A. Salamatin, J. Priscu, Physical, chemical and biological processes in Lake Vostok and other Antarctic subglacial lakes. Nature **416**, 603–609 (2001)
M. Siegert, S. Carter, I. Tabacco, S. Popov, D. Blankenship, A revised inventory of Antarctic subglacial lakes. Antarct. Sci. **17**, 453–460 (2005)
S. Simon, A. Bößwetter, T. Bagdonat, U. Motschmann, K.H. Glassmeier, Plasma environment of Titan: a 3-D hybrid simulation study. Ann. Geophys. **24**, 1113–1135 (2006)
G. Siscoe, L. Christopher, Variations in the solar wind stand-off distance at Mercury. Geophys. Res. Lett. **2**, 158–160 (1975)
B.A. Smith, L. Soderblom, R. Batson, P. Bridges, J. Inge, H. Masursky, A new look at the Saturn system: The voyager 2 images. Science **215**, 505–537 (1982)
B.A. Smith, et al., Voyager 2 at Neptune: imaging science results. Science **246**, 1422–1449 (1989)
E. Smith, L. Davis, D. Jones, P. Coleman, D. Colburn, P. Dyal, C. Sonett, Saturn's magnetic field and magnetosphere. Science **207**, 407–410 (1980)
B. Smith-Konter, R.T. Pappalardo, Tidally driven stress accumulation and shear failure of Enceladus' tiger stripes. Icarus **198**, 435 (2008)
F. Sohl, H. Hussmann, B. Schwentker, T. Spohn, R.D. Lorenz, Interior structure models and tidal love numbers of titan. J. Geophys. Res. **108**, 4–1 (2003)
C.P. Sonett, Electromagnetic induction in the moon. Rev. Geophys. Space Phys. **20**, 411–455 (1982)
C.P. Sonett, P. Dyal, C.W. Parkin, D.S. Colburn, J.D. Mihalov, B.F. Smith, Whole body response of the moon to electromagnetic induction by the solar wind. Science **172**, 256–258 (1971)
D.J. Southwood, M.G. Kivelson, Saturnian magnetospheric dynamics: Elucidation of a camshaft model. J. Geophys. Res. **112**, 12 (2007)
F. Spahn, et al., Cassini dust measurements at Enceladus and implications for the origin of the E-ring. Science **311**, 1416–1418 (2006)

B.J. Srivastava, Theory of the magnetotelluric method of a spherical conductor. Geophys. J. R. Astron. Soc. **11**, 373–387 (1966)

D. Stevenson, Planetary oceans. Sky Telesc. **104**, 38–44 (2002)

D.J. Stevenson, Planetary magnetic fields. Earth Planet. Sci. Lett. **208**, 1–2 (2003). doi:10.1016/S0012-821X(02)01126-3

S.T. Suess, B.E. Goldstein, Compression of the Hermean magnetosphere by the solar wind. J. Geophys. Res. **84**, 3306–3312 (1979)

W.M. Telford, *Applied Geophysics* (Cambridge University Press, Cambridge, 1993)

R. Tokar, et al., The interaction of the atmosphere of Enceladus with Saturn's plasma. Science **311**, 1409–1412 (2006)

G. Tovie, A. Mocquet, C. Sotin, Tidal dissipation within large icy satellites: Applications to Europa and Titan. Icarus **177**, 534–549 (2005)

R.H. Tyler, S. Maus, H. Lühr, Satellite observations of magnetic fields due to ocean tidal flow. Science **239**, 239–241 (2003)

R.H. Tyler, Strong ocean tidal flow and heating on moons of the outer planets. Nature **456**, 770–772 (2008)

L.L. Vanyan, I.V. Egorov, Electromagnetic induction in the moon. Moon **12**, 253–275 (1975). doi:10.1007/BF02629697

G.H. Voigt, A mathematical magnetospheric field model with independent physical parameters. Planet. Space Sci. **29**, 1–20 (1981). doi:10.1016/0032-0633(81)90134-3

J. Waite, et al., Cassini ion and neutral mass spectrometer: Enceladus plume composition and structure. Science **311**, 1419–1409 (2006)

J. Wicht, M. Mandea, F. Takahashi, U.R. Christensen, M. Matsushima, B. Langlais, The origin of Mercury's internal magnetic field. Space Sci. Rev. **132**, 261–290 (2007). doi:10.1007/s11214-007-9280-5

C. Zimmer, K. Khurana, M. Kivelson, Subsurface oceans on Europa and Callisto: Constraints from Galileo magnetometer observations. Icarus **147**, 329–347 (2000)

M.Y. Zolotov, An oceanic composition on early and today's Enceladus. Geophys. Res. Lett. **34**, 23 (2007)

M.Y. Zolotov, E. Shock, Composition and stability of salts on the surface of Europa and their oceanic origin. J. Geophys. Res. **106**, 32815–32827 (2001)

Space Sci Rev (2010) 152: 423–447
DOI 10.1007/s11214-009-9582-x

The Interior Structure, Composition, and Evolution of Giant Planets

Jonathan J. Fortney · Nadine Nettelmann

Received: 12 May 2009 / Accepted: 13 October 2009 / Published online: 19 December 2009

Abstract We discuss our current understanding of the interior structure and thermal evolution of giant planets. This includes the gas giants, such as Jupiter and Saturn, that are primarily composed of hydrogen and helium, as well as the "ice giants," such as Uranus and Neptune, which are primarily composed of elements heavier than H/He. The effect of different hydrogen equations of state (including new first-principles computations) on Jupiter's core mass and heavy element distribution is detailed. This variety of the hydrogen equations of state translate into an uncertainty in Jupiter's core mass of $18M_{\oplus}$. For Uranus and Neptune we find deep envelope metallicities up to 0.95, perhaps indicating the existence of an eroded core, as also supported by their low luminosity. We discuss the results of simple cooling models of our solar system's planets, and show that more complex thermal evolution models may be necessary to understand their cooling history. We review how measurements of the masses and radii of the nearly 50 transiting extrasolar giant planets are changing our understanding of giant planets. In particular a fraction of these planets appear to be larger than can be accommodated by standard models of planetary contraction. We review the proposed explanations for the radii of these planets. We also discuss very young giant planets, which are being directly imaged with ground- and space-based telescopes.

Keywords Giant planet interiors · Exoplanets

1 Introduction

In order to understand the formation of giant planets, and hence, the formation of planetary systems, we must be able to determine the interior structure and composition of giant

Both authors contributed equally to this work.

J.J. Fortney (✉)
Department of Astronomy and Astrophysics, University of California, Santa Cruz, CA, USA
e-mail: jfortney@ucolick.org

N. Nettelmann
Institut fur Physik, Universitat Rostock, Rostock, Germany
e-mail: nadine.nettelmann@uni-rostock.de

planets. Jupiter and Saturn, our solar system's gas giants, combine to make up 92% of the planetary mass of our solar system. Giant planets are also vast natural laboratories for simple materials under high pressure in regimes that are not yet accessible to experiment. With the recent rise in number and stunning diversity of giant planets, it is important to understand these planets as a class of astronomical objects.

We would like to understand the answers to basic questions about the structure and composition of these planets. Are gas giants similar in composition to stars, predominantly hydrogen and helium with a small mass fraction of atoms more massive than helium of only $\sim$1%? If these planets are enhanced in "heavy elements" (the Z component) relative to stars, are these heavy elements predominantly mixed into the hydrogen-helium (H-He) envelope, or are they mainly found in a central core? If a dense central core exists, how massive is it, what is its state (solid or liquid), and is it distinct or diluted into the above H-He envelope? Can we understand if a planet's heavy element mass fraction depends on that of its parent star? What methods of energy transport are at work in the interiors of these planets? Does this differ between the gas giants and the ice giants? Can we explain a planet's observable properties such as the luminosity and radius at a given age?

New data on the atmospheric composition or gravitational fields of our solar system's giant planets comes quite rarely, with long intervals between space missions that gather these precious data sets. We are therefore at the mercy of both our own creativity, as we search for new ideas to explain the data we have, and at the mercy of technology, which allows us to push first decade of our new century is seeing a number of important advances in both the experiment and theory of materials at high pressure, so that we are in a better position to answer some of our questions outlined above.

Giant planets have long been of interest to physicists because they are natural laboratories of hydrogen and helium in the megabar to gigabar pressure range, at temperatures on the order of 10^4 K, which at the high pressure end is outside the realm of experiment. The data that we use to shape our understanding of giant planets comes from a variety of sources. Laboratory data on the equation of state (EOS, the pressure-density-temperature relation) of hydrogen, helium, "ices" such at water, ammonia, and methane, silicate rocks, and iron serve as the initial inputs into models. Importantly, data are only available over a small range of phase space, so that detailed theoretical EOS calculations are critical to understanding the behavior of planetary materials at high pressure and temperature. Within the solar system, spacecraft data on planetary gravitational fields allows us to place constraints on the interior density distribution for Jupiter, Saturn, Uranus, and Neptune.

The year 1995 was Earth-shattering to the field of giant planets, as the first extrasolar giant planet 51 Peg b (Mayor and Queloz 1995) and also the first bona fide brown dwarf, Gliese 229B (Nakajima et al. 1995), were discovered. In particular the close-in orbit of 51 Peg b led to questions regarding its history, structure, and fate (Guillot et al. 1996; Lin et al. 1996). Four years later, the first transiting planet, HD 209458b (Charbonneau et al. 2000; Henry et al. 2000), was found to have an inflated radius of $\sim$1.3 Jupiter radii ($R_{\rm J}$), confirming that proximity to a parent star can have dramatic effects on planetary evolution (Guillot et al. 1996). However, the detections of nearly 50 additional transiting planets (as of May 2009) has raised more questions than it has answered. For exoplanets, we often must make due with little information on planetary structure, namely a planet's mass and radius only. For these planets, what we lack in detailed knowledge about particular planets, we can make up for in number.

Much further from their parent stars, young luminous gas giant planets are being directly imaged from the ground and from space (Kalas et al. 2008; Marois et al. 2008). For these planets, planetary thermal emission is detected in a few bands, and a planet's mass determination rests entirely on comparisons with thermal evolution models, which aim to predict a planet's luminosity and spectrum with time. As the initial conditions for planetary cooling are uncertain, the luminosity of young planets is not yet confidently understood (Marley et al. 2007; Chabrier et al. 2007).

In this paper we first discuss in some detail results of structural models of Jupiter, our "standard" example for gas giant planets. We then look at similar models for Uranus and Neptune. Our discussion then moves to calculations of the thermal evolution of our solar system's giant planets. We then discuss current important issues in modeling exoplanets, and how these models compare to observations of transiting planets, as well as directly imaged planets. We close with a look at the future science of extrasolar giant planets (EGPs).

2 Core Mass and Metallicity of Jupiter, Uranus, & Neptune

2.1 Introduction

In this section we address the core mass and metallicity of Jupiter, Uranus and Neptune. In Sect. 2.4 we compare results for Jupiter obtained with different equations of state which are described in Sect. 2.2. In Sect. 2.5 a large range of Uranus and Neptune structure models is presented that are consistent with the observed gravity data. Based on these models we discuss in Sect. 3.1 the traditional concept of a rocky or icy core which is often used to derive implications for the formation process.

2.2 EOS of H, He, and Metals

2.2.1 Matter Inside the Giant Planets in the Solar System

Gas giant planets such as Jupiter and Saturn do not consist of *gas* and icy giant planets such as Uranus and Neptune not of *ice*. The gaseous phase of hydrogen, which is the predominant element of gas giant planets, becomes a non-ideal fluid at densities $\rho > 0.01$ g/cm^3 (Saumon et al. 1995). In Jupiter, this hydrogen density is reached in the outer 0.01% of the total mass, and in Saturn in the outer 0.1%. Similarly, the ice I phase of water in Uranus and Neptune is left after only 0.02% and the liquid phase after 0.2% of the outer mass shell due to adiabatically rising temperature. The assumption of an adiabatic temperature gradient is important to the construction of state-of-the-art interior models (Saumon and Guillot 2004; Militzer et al. 2008) and is supported by diverse observations. (See Sect. 2.3.) This moderate rise of temperature accompanied with fast rising pressure towards deeper layers causes matter in giant planet interiors to transform to a warm, dense fluid, characterized by ionization, strong ion coupling and electron degeneracy. In Jupiter-size and Saturn-size planets, hydrogen, maybe helium too—depending on the EOS, metallizes giving rise to a strong magnetic field; in Neptune-size planets, water prefers (depending on the entropy) the ionically conducting superionic phase or the plasma phase (Nettelmann et al. 2008b).

Laboratory experiments for the EOS of warm dense matter are very challenging. To date, the EOS of H is well constrained below $\sim$0.3 g/cm^3 and below $\sim$25 GPa (0.25 Mbar) by precise gas gun shock compression experiments. See Saumon and Guillot (2004) for an overview of data from compression experiments of deuterium and French et al. (2009a) for water. At larger densities and pressures however, as relevant for planetary interiors, experimental data have large error bars and single-shock data (Hugoniots) bend towards higher temperature regions in the phase diagram than are relevant for solar system giant planets. High-precision multi-shock experiments are urgently required to constrains the EOS of hydrogen. Until then, there is much space for theoretical EOS. Next we will describe seven EOS that are consistent with experimental EOS data, and have been applied to Jupiter models as well. Five of them are based on the chemical picture of distinct species interacting via specific effective pair potentials, and the other two are based on the physical picture of electrons and nuclei interacting via Coulomb forces (see, e.g., Saumon et al. 1995).

2.2.2 Hydrogen EOS

Sesame: The H-EOS *Sesame 5251* is the deuterium EOS 5263 scaled in density as developed by Kerley (1972). It is built on the assumption of three phases: a molecular solid phase, an atomic solid phase, and a fluid phase that takes into account chemical equilibrium between molecules and atoms and ionization equilibrium of the fluid phase of atoms. A completely revised version by Kerley (2003) includes, among many other improvements, fits to more recent shock compression data resulting into larger (smaller) compressibility at $\sim$0.5 (10) Mbar. In this article we call this improved version *H-Sesame-K03*. Saumon and Guillot (2004) patched the original version at pressures between 100 bar and 0.4 Mbar with another EOS in order to reproduce the gas gun data and call this version *H-Sesame-p*.

H-SCVH-i: This widely used EOS omits the astrophysically irrelevant region of cold dense solid hydrogen and relies on the free-energy-minimization technique throughout the ρ–T region that is relevant for giant planets and low-mass stars. As in the fluid phase of Sesame 5251, it takes into account the species H_2, H, H^+, and e. But at the transition to metallic hydrogen, thermodynamic instabilities are found and considered as a first-order phase transition, the *Plasma Phase Transition*. In an alternative version, *H-SCVH-i*, the instabilities are smoothed out by careful interpolation between the molecular and the metallic phase. Details are given in Saumon et al. (1995).

LM-SOCP and LM-H4: These EOS are modifications of the simple linear mixing model of Ross (see, e.g., Holmes et al. 1995). It assumes the total Helmholtz free energy F of a system of H_2 molecules and metallic H as linear superposition of the single components' free energies F_{mol} and F_{met}, respectively. The original EOS was constructed to fit the gas gun data by adjusting the effective molecular pair potential, and to fit their low reshock temperatures by addition of a fitting term F_{fit} in the total free energy. This term causes a region where $\nabla_{\mathrm{ad}} < 0$ along the Jupiter isentrope. Saumon and Guillot (2004) avoided this behavior by taking into account electron screening in the metallic component (LM-SOCP) or by admixing of D_4 chains as an additional species (LM-H4).

DFT-MD: Applying density-functional molecular dynamics (FVT-MD) to simultaneous simulation of H and He nuclei (100 H and 9 He nuclei in periodic boundary conditions), Militzer et al. (2008) were the first to provide an EOS including H/He mixing effects for a broad range of densities ρ and temperatures T relevant for Jupiter's interior. They used the CPMD code with Troullier-Martins norm-conserving pseudopotentials and the VASP code with projector augmented wave pseudopotentials to generate EOS data at $\rho \geq 0.2$ g/cm^3 and $T \geq 500$ K and used classical Monte Carlo simulations at smaller densities. Other H/He mass mixing ratios other than 0.2466 were realized by diminishing the density along the J-isentrope in accordance with He EOS data.

H-REOS: For hydrogen densities $0.2 \leq \rho \leq 9$ g/cm^3 and temperatures $1000 \leq T \leq 30000$ K, Nettelmann et al. (2008a) also use the VASP code developed by Kresse and Hafner, Kresse and Hafner (1993a, 1993b) and Kresse and Furthmüller (1996). The main differences in the calculation of EOS data for H/He mixtures compared to DFT-MD EOS by Militzer et al. (2008) are (i) inclusion of finite temperature effects on the electronic subsystem by Fermi weighting of the occupation of bands before minimizing the electronic energy density functional (FT-DFT), (ii) separate FT-DFT molecular dynamics calculations for H and He with subsequent linear mixing, (iii) application of FVT^+ (French et al. 2009b) to generate H EOS data at lower densities and higher temperatures. FVT^+ combines fluid variational theory, a minimization method for the free energy of neutral hydrogen, with Pade formulas for fully ionized hydrogen taking into account ionization equilibrium. While FVT^+ predicts a plasma phase transition between 0.27 and 0.5 g/cm^3, H-REOS does not, since it transitions smoothly from FVT^+ to FT-DFT-MD data below 0.2 g/cm^3.

Other EOS: It is interesting to note that there are H-EOS that do not give acceptable Jupiter models, indicating an invalid ρ–P relation at those pressures where the Jovian gravity field is most sensitive to the internal mass distribution. Among such EOS are LM-B (Saumon and Guillot 2004) and FVT^+. In this sense, Jupiter interior models serve as a check of EOS data in the warm dense matter regime.

2.2.3 He EOS

Helium equations of state used together with the hydrogen equations of state described above are listed in Table 1. The He EOS *He-SCVH* is described in Saumon et al. (1995), *He-Sesame-K04* in Kerley (2004a), *He-REOS* in Nettelmann et al. (2008a), and *DFT-MD* in Vorberger et al. (2007), respectively. Relative differences in pressure and internal energy along relevant isotherms are within ~30%, comparable to those of the H EOSs. With an average H/He particle number ratio below 1/10, the effect of the He EOS on giant planet interior models lies less in its ρ–P relation but more in its mixing ability with hydrogen and the possibility of He sedimentation. This topic is addressed in Sect. 2.6.

2.2.4 EOS of Metals

Diverse EOS of heavy elements are used to represent *metals* within Jupiter's envelope and core. Saumon and Guillot (2004) take the Sesame EOS 7154 of water to represent ices[1] (I), and the Sesame EOS 7100 of dry sand to represent rocks (R) with an upper limit of the R-component of 4% in the envelope. Nettelmann et al. (2008a) either scale He-REOS in density by a factor of four (He4) or use *H2O-REOS*. This new EOS of water is a combination of accurate ice I and liquid water data, FT-DFT-MD data at densities and temperatures relevant for giant planet interiors, and Sesame 7150 at small densities and high temperatures with interpolated regions in between to smoothly join these different data sets. They assume a rocky core using the fit-formula to experimental rock data below 2 Mbar by Hubbard and Marley (1989). Rocks lead to roughly 50% less massive cores than ices. Kerley (2004b) represents the core material by SiO_2. For metals in the envelope, he assumes an initial composition of O, C, N, and S of relative solar abundance in the outer region with the addition of Si and Fe in the inner region of Jupiter. For a given enrichment factor, the chemical equilibrium abundances of molecules formed by these species in a H/He mixture is calculated and the corresponding EOS tables of the occurring components are added linearly to the H/He EOS.

[1]The label *ice* refers to a mixture of H_2O, CH_4, and NH_3 that are supposed to have been in an ice phase during protoplanetary core formation.

2.3 Construction of Interior Models: Constraints and Methods

Constraints For interior models of the solar system giant planets, in general the following observational constraints are taken into account: the total mass M, the equatorial radius R_{eq}, the 1-bar temperature T_1, the angular velocity ω, the gravitational moments J_{2n}, in particular J_2 and J_4, the atmospheric He mass fraction Y_1, and occasionally the atmospheric abundances of volatile species, except oxygen. Due to low atmospheric temperatures, O, if present, is believed to condense out as H_2O clouds at higher pressures deeper inside the planets. These pressures have not yet been reached by observation, such that an observed O abundance is believed to not be indicative for the overall abundance in the envelope. On the other hand, the measured supersolar abundances of other volatiles are generally explained by the dissolution of volatile-rich icy planetesimals that were captured by the young forming planet, implying a supersolar overall water abundance. In the absence of representative data, the O abundance is usually assumed of the order of other volatiles abundances (Kerley 2004b). The mean He content, $\bar{Y}$, cannot be observed, but from solar evolution theory in accordance with observational data for the sun, a value of $Y = 0.275 \pm 0.01$ is generally accepted as a constraint for planet interior models (Fortney and Hubbard 2003). Beside the uncertainties in the equation of state, the error bars of the observables give rise to broad sets of models for a single planet.

Methods The luminosity is an important observable for evolution models, as described below in Sect. 2.6. For structure models, it is important in the sense that it gives a hint of the temperature profile. The high intrinsic luminosities of Jupiter, Saturn, and Neptune for instance strongly point towards an adiabatic, convective interior on large scales, since energy transport by radiation or conduction are too inefficient to account for the observed heat flux (Hubbard 1968). This is because of frequent collisions in the dense interior and strong molecular absorption in the less dense outer region. Convection, which will tend to homogenize the planet, leads to an adiabatic temperature gradient. In the absence of a convection barrier, the envelope of a giant planet can be assumed adiabatic (isentropic) and homogeneous, where the entropy is fixed by T_1 (Hubbard 1973).

Given M_p, to reproduce R_p one has to either make the additional assumption of a core of heavy elements, or to choose a particular envelope metallicity Z, since Jupiter and Saturn are smaller in radius than pure H-He planets (e.g., Demarcus 1958; Podolak and Cameron 1974). Thus the radius fixes the core mass M_{core} or, alternatively, Z. This property is used to derive a core mass or metallicity of transiting extrasolar planets, since only the mass and radius can be measured. Furthermore, the *Voyager* and *Galileo probe* measurements give $Y_1 < \bar{Y}$ for Jupiter and Saturn, implying either an inhomogeneous interior, or $\bar{Y}$ below the cosmological value, or a mixing barrier dividing the interior into a He-depleted outer envelope with $Y = Y_1$ and a He-enriched inner envelope. Most modelers prefer the last scenario. There are several possibilities for where to locate the layer boundary, characterized by the transition pressure P_{12} between the outer (layer 1) and inner (layer 2) envelope, depending on the mechanism causing the He discontinuity. Candidates are a first-order phase transition, e.g. a plasma phase transition of H whose existence is still a matter of debate, and H/He phase separation with He sedimentation. For practical purposes, P_{12} can be varied within a reasonable range around 3 Mbar. For Uranus and Neptune, Y_1 is consistent with $\bar{Y}$ within the observational error bars.

While ω enters the equations to be solved explicitly, the gravitational moments J_2 and J_4 have to be adjusted within an iterative procedure and thus require two further free parameters. These can be the metallicities Z_1 and Z_2 in the two envelope layers. More generally,

Table 1 Overview of Jupiter model series

Name (EOS)	H-EOS	He-EOS	Z-EOS	type	Ref. (J)
SCVH-I-99	H-SCVH-I	He-SCVH	He-SCvH	$Y_1 < Y_2$ $Z_1 \neq Z_2$	(1)
SCVH-I-04	H-SCVH-I	He-SCVH	Sesame 7154, Sesame 7100	$Y_1 < Y_2$ $Z_1 = Z_2$	(2)
LM-SOCP	LM-SOCP	"	"	"	"
LM-H4	LM-H4	"	"	"	"
Sesame-p	H-Sesame-p	"	"	"	"
Sesame-K04	Sesame-K03	Sesame-K04	linear mixture of H_2O, CH_4, NH_3, C, N, O, H_2S, S, SiO_2, Fe	$Y_1 < Y_2$ $Z_1 < Z_2$	(3)
LM-REOS	H-REOS	He-REOS	H_2O-REOS, He4-REOS	$Y_1 < Y_2$ $Z_1 < Z_2$	(4)
DFT-MD	DFT-MD	DFT-MD	CH_4, H_2O	$Y_1 = Y_2$ $Z_1 = Z_2$	(5)

References for Jupiter models: (1) Guillot (1999), (2) Saumon and Guillot (2004), (3) Kerley (2004b), (4) Nettelmann et al. (2008a), (5) Militzer et al. (2008). In all cases: $Y_1 = Y_{\rm atm} = 0.238$

the parameters Z_1, Z_2, and $M_{\rm core}$ are used to adjust J_2, J_4, and $R_{\rm eq}$ (Chabrier et al. 1992; Guillot 1999; Nettelmann et al. 2008a). Other authors do not allow for a discontinuity of metals (Saumon and Guillot 2004; Militzer et al. 2008). An argument in favor of $Z_1 = Z_2$ is large-scale convection of the hot, young planet; an argument in favor of $Z_1 \neq Z_2$ is core-accretion formation with inhomogeneous planetesimal delivery in the envelope leading to early formation of a convection barrier, due to mean molecular weight gradients. However, if remnant planetesimal gradients are present, it is unlikely that they could be characterized simply by one number, Z_2. Furthermore composition gradients inhibiting convection would void the assumption of an adiabatic interior.

Table 1 gives an overview about the Jupiter model series and the EOS used therein, the underlying different structure type assumptions (discontinuities in Y and Z). We present and discuss results for Jupiter's core mass and heavy element abundance in the following subsection.

2.4 Results: Core Mass and Metallicity of Jupiter

Figure 1 shows the resulting mass of the core and the mass M_Z of metals in the envelope(s) found by different authors using the diverse EOS as listed in Table 1. Note that all these solutions have $\bar{Y} = 0.275 \pm 0.01$ except DFT-MD models, which have $\bar{Y} = 0.238$. To better compare these solutions, enhancing $\bar{Y}$ by 0.03 to 0.27 in the latter solutions requires replacing $\sim 9 M_\oplus$ of metals by He. In this case, DFT-MD models have metal-free envelopes. To avoid this problem, Militzer et al. (2008) suggest a He layer above the core due to He sedimentation yielding rocky core masses of 5–$9 M_\oplus$, instead of 14–$18 M_\oplus$, in better agreement with all other solutions. The other extreme of high envelope metallicity, up to $37 M_\oplus$, is found using LM-SOCP or SCvH-I-99. To show the effect of the EOS of metals, models using the He EOS scaled in density by a factor of 4 (He4) and using water for metals are presented. Heavier elements, i.e. magnesium-silicates, would give even lower M_Z values. We conclude from this figure that the choice of composition and EOS for the metals has a large effect on the envelope metallicity and a small effect on the core. If these EOS reflect

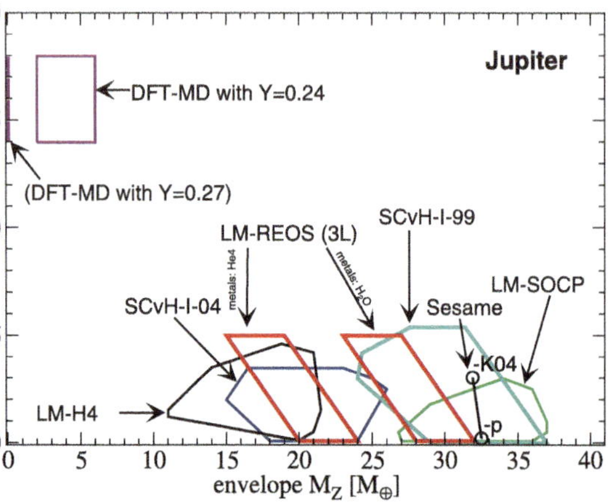

Fig. 1 Mass of the core and of heavy elements within the envelope. Each box represents the solutions found using a particular equation of state as listed in Table 1. In the case of DFT-MD EOS models which originally have $\bar{Y} = 0.238$, we also indicate the position if 3% ($\sim 9 M_\oplus$) of metals are replaced by He in order to have $Y = 0.27$, in accordance with all other models in this figure

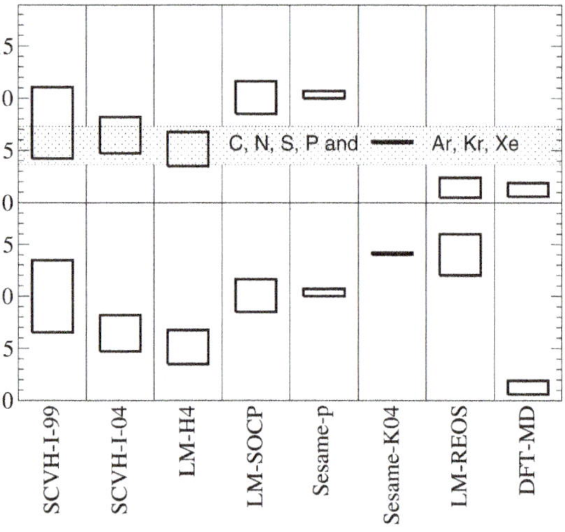

Fig. 2 Mass fraction of heavy elements in the outer envelope (Z_1) and the inner envelope (Z_2) of Jupiter interior models using the different equations of state described in Sect. 2.2. *The dotted region* shows the atmospheric metallicity *if* the O abundance is similar to the values measured for C, N, S and some noble gases, i.e. 2–4 times solar

our current knowledge, we conclude that the interior of Jupiter is badly constrained with a possible core mass ranging from 0 to $18 M_\oplus$ and an envelope heavy element (Z) mass from 0 to $37 M_\oplus$. If these large uncertainties are taken at face value, a prediction about Jupiter's formation process is highly unreliable.

Figure 2 shows the mass fraction of metals in the two envelopes for the same EOSs as in Fig. 1. Models without a discontinuity of metals have $Z_1 = Z_2$ per definition. For tentative evaluation of these results, Z_1 is compared with the range of atmospheric abundances of some volatile species, where we used two assumptions. The first is that O atoms are as abundant as the species C, N, S and Ar, Kr, Xe, i.e. 2–4 × solar (Mahaffy et al. 2000), and the second is a mass fraction equivalent of 1× solar $\simeq$1.9%. As stated in Sect. 2.3, the real O abundance x_O in Jupiter might be much higher than the measured value of 30% of the solar value due to condensation of water above 20 bar, where the *Galileo* probe stopped working (Wong et al. 2004). If however $x_O \ll x_{C,N,S,P}$, then the lower boundary of the dotted region in Fig. 2 would sink, otherwise if $x_O \gg x_{C,N,S,P}$, then the upper boundary would rise.

For interior modeling, there are several assumptions that affect the resulting envelope metallicity. All EOS except DFT-MD use the simplifying linear mixing approximation to combine H- and He-EOS. In case of the DFT-MD EOS however, which takes into account the mixing effect by simultaneous simulation of H and He atoms, an up to 5% volume enhancement (density decrement) is found compared to linear mixing (Vorberger et al. 2007) at pressures and temperatures typical for Jupiter's deep outer envelope, where J_4 is most sensitive to the metallicity. Compensating for this reduction in density of the H/He subsystem requires a corresponding enhancement in metals. Thus the calculated Z_1 values might increase by up to 5 percentage points, except for DFT-MD EOS. Furthermore, Jupiter's cloud patterns are known to rotate on cylinders with different velocities as a function of latitude. If differential rotation extends into the interior, the gravitational moments calculated by assuming rigid body rotation have to be corrected. Zharkov and Trubitsyn (1978) suggest a small correction of 0.5% for J_2 and 1% for J_4 based on observations of atmospheric winds (see also Hubbard 1982); Liu et al. (2008) predict a penetration depth of deep-zonal winds down only $0.04R_J$, supporting an only slight effect on the low-degree gravitational harmonics. Militzer et al. (2008) on the other hand invoke interior winds penetrating 10% into Jupiter's envelope in order to match J_4, which otherwise would differ from the observed value by more than two standard deviations. Applying the same correction necessary for DFT-MD models on interior models using LM-REOS, which exhibits the smallest Z_1/Z_2 ratio (see Fig. 2), gives $Z_1/Z_2 > 1$.

We conclude that future spacecraft-based measurements are desirable in order to constrain the envelope metallicity and, consequently, narrow the set of H/He equations of state currently offered. Among the most helpful observations we suggest a measurement of Jupiter's O abundance at pressures between 10 and 60 bar (above and below the liquid water to vapor transition along the isentrope), and a determination of deep-zonal winds by measuring high-order harmonics. NASA's forthcoming *Juno Mission* will indeed measure these harmonics, as well as constrain the deep water and ammonia abundances from microwave spectra (Matousek 2007).

2.5 Results: Core Mass and Metallicity of Uranus and Neptune

We apply the same method used for Jupiter interior calculations with LM-REOS for three-layer models of Uranus and Neptune. Planet models consist of a two-layer envelope and core. Envelope metals are represented by water and the core consists of rocks. Uranus and Neptune have large observational error bars of J_4 of 10% and 100%, respectively. Results are shown in Fig. 3. For a given transition pressure P_{12} between the two envelopes composed of mixtures of H/He and water (layer 1 and 2), the solutions move along almost straight lines, and changing P_{12} causes a parallel shift of the line. Decreasing Z_1 requires a higher inner envelope metallicity (Z_2) in order to match J_2. Simultaneously, the mass of the core (layer 3) shrinks with $M_{\rm core} = 0$ defining the maximal possible Z_2 value for a given layer boundary. Here, pure water envelopes are not allowed. Replacing a H/He mass fraction of 5% (10%, 12%) by the molecular weight of CH_4 results in a H_2O/CH_4 mass ratio of 0.6 (0.2, 0), but those models with an inner envelope of pure 'icy' composition have not been calculated here. On the other hand, replacing some H_2O by rocks will result into a higher H/He fraction, and in the more realistic case of a solar ice/rock ratio of ~2.7, H/He free deep envelopes are not possible. These results are in good agreement with those by Hubbard and Marley (1989).

Most Uranus and all Neptune models presented here have also a significant heavy element (water) enrichment in the outer envelope ($P < P_{12}$). An upper limit of Uranus' Z_1 is given by the requirement to meet J_4; for Neptune, the large error bar of J_4 allows for

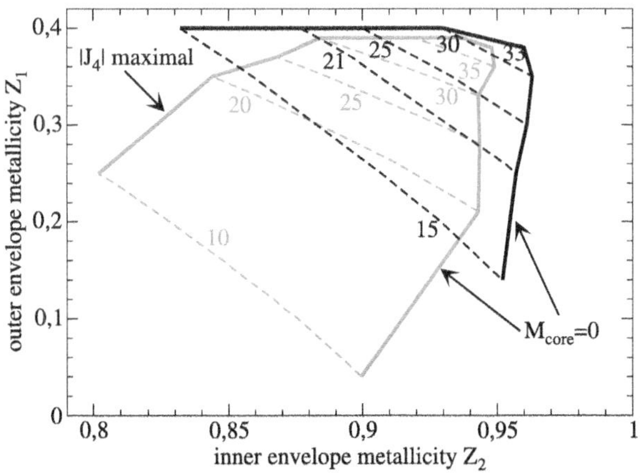

Fig. 3 Mass fraction of metals in the outer envelope (Z_1) and in the inner envelope (Z_2) of three-layer models of Uranus (*grey*) and Neptune (*black*). *The thick solid lines* indicate the range where solutions have been found. Numbers at *dashed lines* give the transition pressure P_{12} in GPa, and *dashed lines* show the behavior of solutions if P_{12} is kept constant and J_4 is varied within the 1σ-error. Increasing Z_1 increases $|J_4|$. No Uranus solutions are found above *the upper thick line*. Neptune's J_4-error bar is large, so we stopped arbitrarily at $Z_1 = 40\%$. Decreasing Z_1 results into higher Z_2 values and smaller core masses. Below the lower *thick lines*, no solutions exists. These models are based on LM-REOS using water for metals

even higher outer envelope metallicities than 0.4. In any case, all models have a pronounced heavy element discontinuity. No Uranus (Neptune) models are found with $P_{12} > 38$ (33) GPa because of $M_{\rm core} \to 0$. We did not calculate models with $P_{12} < 10$ GPa, since this discontinuity is perhaps caused by the transition from molecular water to ionic dissociated water, which occurs around 20 GPa (French et al. 2009a).

Uranus and Neptune are very similar planets with respect to their core mass and total heavy element enrichments (Hubbard et al. 1995), and are very different planets with respect to their internal heat fluxes, as well as to observed molecular species. While C in both planets is about 30–60 times solar, CO and HCN have been detected in Neptune, but not in Uranus (Gautier et al. 1995), likely indicating the absence of efficient convective transport in Uranus. Convection can be inhibited by a steep compositional gradient or by a region with sufficiently high conductivity. Calculations by Guillot et al. (1994) suggest the presence of such a radiative region at 1000 K in Uranus.[2] Since the temperature in a radiative layer rises less than in the adiabatic case, this explanation for Uranus' small heat flux tends to smaller present-day central temperatures. At layer boundaries induced by a steep compositional gradient on the other hand, the temperature rises faster than in the adiabatic case leading to higher present-day central temperatures. One step forward to decide as to the more appropriate scenario could be a calculation of cooling curves using non-adiabatic temperature gradients. The good agreement of Neptune cooling curves based on two adiabatic, homogeneous layers of pure H/He and water (see below, and M. Ikoma, *personal communication* 2008) with the present luminosity possibly shows us that Neptune's structure may not necessarily be extremely complex.

[2]These calculations should be revisited in light of since-discovered strong opacity sources in the deep atmosphere of Jupiter, which close its previously postulated radiative window (Guillot et al. 2004).

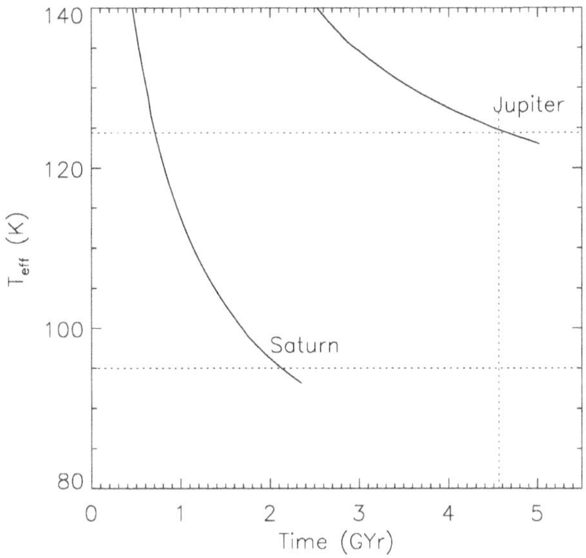

Fig. 4 Homogeneous evolutionary models of Jupiter and Saturn, adapted from Fortney and Hubbard (2003). The solar system's age as well as the T_{eff} of Jupiter and Saturn are shown with *dotted lines*

2.6 Results: Evolution of Jupiter and Saturn

Our understanding of the evolution of Jupiter and Saturn is currently imperfect. The most striking discrepancy between theory and reality is Saturn's luminosity. Saturn's current luminosity is over 50% greater than one predicts using a homogeneous evolution model, with the internally isentropic planet radiating over time both its internal energy and thermalized solar radiation. This discrepancy has long been noted (Pollack et al. 1977; Grossman et al. 1980; Guillot et al. 1995; Hubbard et al. 1999). Homogeneous evolutionary models of Saturn tend to reach an effective temperature of 95.0 K (Saturn's current known T_{eff}) in only 2.0–2.7 Gyr, depending on the hydrogen-helium equation of state (EOS) and atmosphere models used. However, purely homogeneous models appear to work well for Jupiter. Figure 4 shows homogeneous evolutionary models for both planets from Fortney and Hubbard (2003). It has also long been believed that the most promising route to resolving this discrepancy is the possible phase separation of neutral helium from liquid metallic hydrogen in the planet's interior, beginning when Saturn's effective temperature reached 100–120 K (Stevenson and Salpeter 1977b, 1977a). This sinking of "helium rain" can be an appreciable energy source.

Fortney and Hubbard (2003) tested a variety of high-pressure H/He phase diagrams that had been published since the mid 1970's. Of particular note, they found that the phase diagram of Hubbard and Dewitt (1985), which is essentially the same as that of Stevenson (1975), is inapplicable to the interiors of Jupiter and Saturn, if helium phase separation is Saturn's only additional energy source. As Fig. 5 shows, this phase diagram prolongs Saturn's cooling only 0.8 Gyr, even in the most favorable circumstance that all energy liberated is available to be radiated, and does not instead go into heating the planet's deep interior.

Fortney and Hubbard (2003) next inverted the problem to derive an ad-hoc phase diagram that could simultaneously explain Saturn's current luminosity as well as its current atmospheric helium abundance (Conrath and Gautier 2000). The helium abundance is depleted relative to the Sun, and is consistent with helium being lost to deeper regions of liquid metallic hydrogen at Mbar pressures. The ad-hoc phase diagram forced helium that rained out to fall all the way down to Saturn's core, thereby liberating a significant amount of

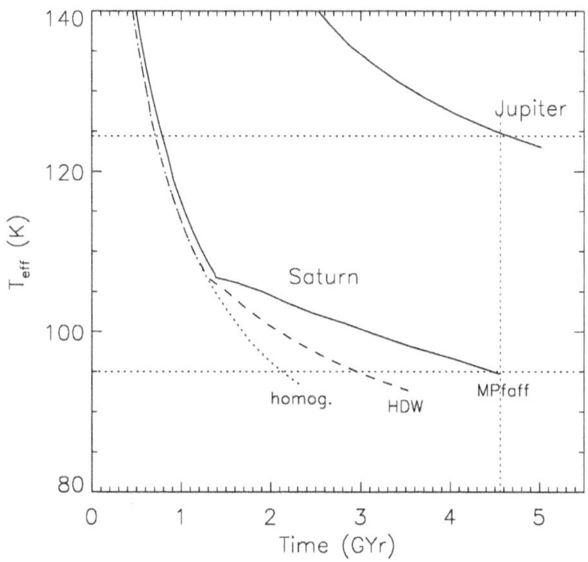

Fig. 5 Evolutionary models of Saturn including helium phase separation, adapted from Fortney and Hubbard (2004). "HDW" uses the H/He phase diagram of Hubbard and Dewitt (1985), which allows immiscible helium to redissolve at higher pressures and hotter temperatures in the liquid metallic hydrogen. "MPfaff" is an ad-hoc phase diagram that forces immiscible helium to rain down to Saturn's core

gravitational potential energy. In light of the new first principles calculations of H/He phase diagrams (Lorenzen et al. 2009; Morales et al. 2009), thermal evolution models of Jupiter and Saturn should now be revisited.

2.7 Results: Evolution of Uranus and Neptune

In the previous section we have seen that homogeneous evolution models work well for Jupiter, but not for Saturn , yielding cooling times that are too short. In this section we will see that homogeneous evolution models work fairly well for Neptune, but certainly not for Uranus, yielding cooling times too long to be consistent with the age of the solar system.

The general results of Sects. 2.6 and 2.7 is based on solving the common energy balance equation

$$L - L_{\odot} = L_{\rm int} \tag{1}$$

where $L(t) = 4\pi R^2(t)\sigma T_{\rm eff}(t)^4$ is the luminosity (mostly measured as flux in the mid infrared) of the planet attributed to an effective temperature $T_{\rm eff}$. Here, $L_{\odot}(t) = 4\pi R^2(t) \times \sigma T_{\rm eq}(t)^4$ is the luminosity due to only to thermalized and reradiated absorbed solar flux, as parameterized by the equilibrium temperature $T_{\rm eq}$, the $T_{\rm eff}$ that planet would have in case of no intrinsic luminosity, $L_{\rm int}(t)$. Taking into account cooling and ongoing gravitational contraction as energy sources to supply the radiative losses, we can write

$$L_{\rm int}(t) = -\int_0^M dm\, T(m,t)\frac{\partial s(m,t)}{\partial t}, \tag{2}$$

where $T(m,t)$ is the internal temperature profile at time t and $s(m,t)$ is the specific entropy. With a relation between the $T_{\rm eff}$ and the atmospheric temperature at say, 1 bar (see, e.g., Burrows et al. 1997 for detailed atmosphere models for warmer planets), (1) and (2) can be converted into a single differential equation for $T_{\rm eff}(t)$. Often an arbitrary initial condition is used (see Sect. 4.3) and the early $T_{\rm eff}$ drops very quickly, such that planets "forget" their

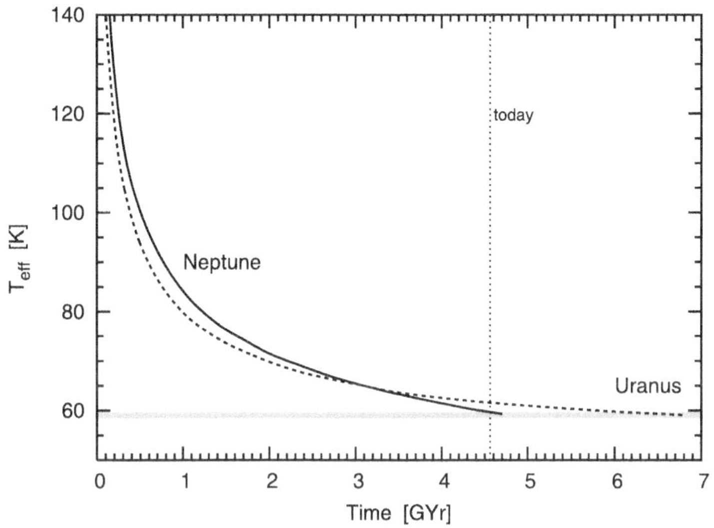

Fig. 6 Homogeneous evolutionary models of Uranus (*dashed*) and Neptune (*solid*). The underlying interior models are among those presented in Fig. 3. Notably, the real Uranus is underluminous as compared to the model. The solar system's age is shown (*dotted line*) and the *grey bar* indicates the present T_{eff}s

initial conditions. Other cooling curves can be obtained only by introducing an additional free parameter or by assuming a birth that is colder than the arbitrarily hot start.

In an earlier investigation of Uranus and Neptune cooling models, Hubbard et al. (1995) for instance assume mean values for the internal temperature and the specific heat $c_v = Tds/dT$ of the planetary material, neglect the relatively small contributions to the intrinsic luminosity from current gravitational contraction, and allowed a cold start. Alternatively, they introduce a variable fraction of the thermal heat content that contributes to the intrinsic luminosity, i.e. that lies within the convectively unstable, homogeneous region of the planet. Based on these assumptions they find that both Uranus and Neptune's cooling time would exceed the age of the Solar system with a larger deviation of some gigayears for Uranus, necessitating either a cold start or a large fraction of the interior not contributing to L_{int}.

Qualitatively the same result of $\Delta t_{cool}^{(U)} \gg \Delta t_{cool}^{(N)} > 4.56$ GYr was reported by M. Ikoma (*personal communication* 2008) for fully differentiated models with three homogeneous layers (rock core, ice layer, H/He envelope) using diverse ice equations of state. However, such a centrally condensed interior structure is not consistent with the gravity field data of Uranus and Neptune, as discussed above. In this work we present in Fig. 6 evolution tracks based on interior models within the sets of acceptable present-day solutions from Fig. 3. The Uranus model in Fig. 6 has $P_{12} = 25$ GPa, $Z_1 = 0.35$, $Z_2 = 0.887$, $M_{core} = 1.48 M_\oplus$, and the Neptune model has $P_{12} = 21$ GPa, $Z_1 = 0.37$, $Z_2 = 0.896$, $M_{core} = 1.81 M_\oplus$. The thick grey line indicates the uncertainty of their present day T_{eff} of 59.1 and 59.3 K, respectively. Note that while $T_{eff}^{(U)} \simeq T_{eff}^{(N)}$ and $R^{(U)} \simeq R^{(N)}$, we have $L^{(U)} \simeq L^{(N)}$, but $L_{int}^{(U)} < L_{int}^{(N)}$ because of (1) and $L_\odot^{(U)} > L_\odot^{(N)}$ (as Neptune is less irradiated). With the same underlying relation between effective and atmospheric temperature and the same equation of state (LM-REOS), homogeneous cooling of Neptune gives roughly an age of 4.6 Gyr, but for Uranus of $\approx$2.5 Gyr more.

Obviously, the results $\Delta t_{\rm cool}^{(U)} \gg 4.56$ Gyr and $\Delta t_{\rm cool}^{(N)} > 4.56$ Gyr appear insensitive to the details of the structure model and of the equations of state used. Hence we must call into question the assumption of convective envelope(s) beneath all of these models.

Convection can be inhibited by a steep compositional gradient or by a region with sufficiently high conductivity. Calculations by Guillot et al. (1994) suggest the presence of such a radiative region at 1000 K in Uranus. Since the temperature in a radiative layer rises less than in the adiabatic case, this explanation for Uranus' small heat flux tends to *smaller* present-day central temperatures, and hence, to relatively low initial temperatures (cold start). At layer boundaries induced by a compositional gradient on the other hand, the temperature rises faster than in the adiabatic case leading to *higher* present-day central temperatures. In that case, heat from the initial hot start remains restored in deep shells and is prevented from escaping to the surface reducing the total $L_{\rm int}$. In this picture, the smaller intrinsic luminosity of Uranus arises from a more extended convectively stable region or from a colder start compared with Neptune. Both possibilities can potentially be explained by different characteristics of giant impacts during formation. (See Hubbard et al. (1995) for a detailed discussion.) Furthermore, both possibilities are not in contradiction to the apparent similarity of the interior models presented in Sect. 2.5, since J_2 and J_4 are not unique with respect to the density distribution on small scales.

In Sect. 2.6 we have seen that gravitational settling of immiscible material tends to lengthen the cooling time of Saturn by some gigayears; equivalently, redistribution of water from the inner envelope to the outer H/He envelope due to immiscibility offers an explanation for Uranus' low $T_{\rm eff}$. One step forward could be a calculation of cooling curves using non-adiabatic temperature gradients and heat transport through diffusive layers, and the calculation of material properties of gas-ice-rock mixtures.

3 Discussion

3.1 The Concept of the Core Mass

In Sect. 2.4 we presented results for the core mass and metallicity of Jupiter (Uranus and Neptune: Sect. 2.5) assuming a core composed of rocks or ices (U and N: 100% rocks) and metals in the H/He envelopes being ice or ice-rock mixtures (U and N: H_2O). These approximations for Uranus and Neptune have been applied also by Fortney and Hubbard (2003) on Saturn evolution models. Other Jupiter and Saturn models not presented here, e.g. by Chabrier et al. (1992), assumed for the core a central agglomeration of rocks overlayed by an ice shell. Such assumptions can be considered *state-of-the-art*.

In Fig. 7 we show a collection of model derivations of Jupiter's core mass derived by a variety of authors over the past 35 years. The spread is large. Generally, as our understanding of H/He under high pressure has (presumably) improved, core masses have fallen. Notably, in the 1970s and 1980s, a variety of groups used a variety of different H/He EOSs to compute structure models. From the mid 1990s to mid 2000s, essentially only the Saumon et al. (1995) EOS was used, predominantly by T. Guillot. We have now finally entered the era of first-principles calculations of H and He EOSs, and the behavior of this diagram over the coming years will be quite interesting. Since the very nature of a well-behaved layered planet is only an assumption, in the following we also look at more complex diluted cores. With gravity field data alone, it is not possible to differentiate between these simple and more complex models.

A common feature of Uranus and Neptune models is a large inner envelope metallicity, in our case up to ~0.95 in mass, bringing it close to an ice shell. The small rocky core

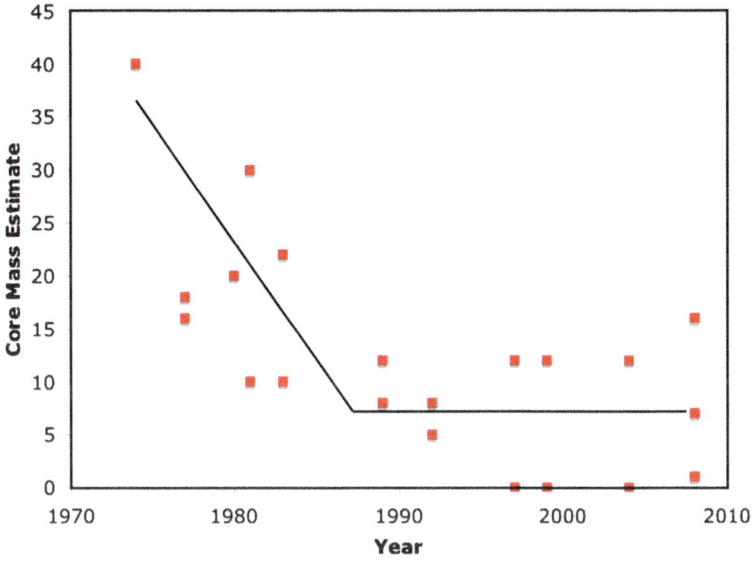

Fig. 7 Jupiter's core mass, as derived by many different authors, at various times since the early 1970s

of Uranus and Neptune models, together with this almost-ice shell, resembles a large core. With 0–2$M_\oplus$ central rocks and 9–12$M_\oplus$ of envelope H_2O in Uranus (12–14.5$M_\oplus$ in Neptune), this gives a central mass of heavy elements of ~11.5$M_\oplus$ for Uranus and ~14.5$M_\oplus$ for Neptune, since larger rocky cores are accompanied by smaller Z_2 values. For brevity, we call this mass $M_{23,Z}$, the mass of the Z-component in layers 2 and 3. It is in good agreement with the core mass predicted by the core accretion formation models (CAF) models by Pollack et al. (1996). More recent CAF models however by Alibert et al. (2005) predict significantly smaller core masses of ~6$M_\oplus$ for Jupiter and Saturn. Uranus' and Neptune's $M_{23,Z}$ is larger than Jupiter's $M_{\rm core}$ (except if using DFT-MD, which gives 14–18$M_\oplus$). An obvious consequence is the following hypothesis: All solar system giant planets formed by CAF with an initial core mass of ~5–15$M_\oplus$. A deviation of their present core mass from this value indicates dissolving of initial core material within the deep interior, and does not indicate an inconsistency with CAF.

This dissolving of core material may have happened in the early hot stages of the planet's evolution or within a continuous, slowly progressing process. To explain Jupiter's relatively small derived core, Saumon and Guillot (2004) suggest a larger mixing of core material in Jupiter than in Saturn due to a larger gas accretion rate during formation; in this sense, the high metallicity of Uranus' and Neptune's inner envelope implies weak core erosion and thus a small gas accretion rate in agreement with their small derived total gas fraction. A small Jupiter core today can also be explained by continuous, slow erosion. If the protocore contained ice, this ice at present Jupiter core conditions of ~20000 K and >40 Mbar would be in the plasma phase (French et al. 2009a) which is soluble with hydrogen. However, we do not know how fast such an ice-enriched H/He/ice mixture can be redistributed by convection. Instead, a deep layer of H/He/ice can form which is stable against convection due to a compositional gradient. Note that an extended compositional gradient is not a preferred solution because of Jupiter's large heat flux, which strongly points to large-scale convection.

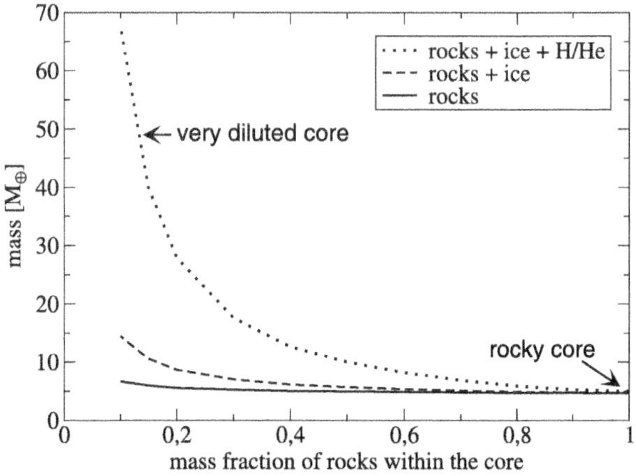

Fig. 8 Core mass of Jupiter assuming an isothermal core of H/He, water, and rocks with varying rock mass fraction. For all underlying models, the water to H/He mass ratio in the core is the same as in the inner envelope. The rock mass fraction in the core is varied between 1.0 (usual rocky core) and 0.1 (very diluted core). *Solid line*: mass of rocks in the core, *dashed line*: mass of rocks and water in the core, *dotted line*: total core mass

Within a simplified Jupiter model we can examine if a central region containing rocks, ice, and H/He can have $\sim 10 M_\oplus$ of heavy elements. For this examination we use LM-REOS. We assume a central region containing H/He and H_2O in the same relative fraction as in the usual deep envelope and vary the fraction of rocks in the central region. The result is shown in Fig. 8.

It turns out that for rock mass fractions $X_{R,\text{core}}$ between 100 and 60% in the central region, the mass $M_{Z,\text{core}}$ of heavy elements is essentially unaffected. In order to obtain $M_{Z,\text{core}} > 10 M_\oplus$ decreasing $X_{R,\text{core}}$ down to $< 20\%$ is required. These models have $> 30 M_\oplus$ H/He in the central region, the pressure at the core-mantle boundary decreases from 39 to 23 Mbar, and the core region growth from $\sim 1 R_\oplus$ to $> 3 R_\oplus$. The larger core region tends to enhance J_2 which in turn forces the fitting procedure to smaller inner envelope metallicities Z_2. In order to keep $|J_4|$ at a constant value, which decreases with smaller Z_2, Z_1 must become larger by some ΔZ_1. For $X_{R,\text{core}} < 0.2$ we find $\Delta Z_1 > 50\%$. This should be kept in mind when evaluating models obtained with different EOS as presented in Sect. 2.4.

3.2 Summary and Conclusions

Since the pioneering work of Demarcus (1958) over 50 years ago, it has been clear that Jupiter is composed predominantly of H and He. But its content and distribution of heavy elements is still a matter of debate, despite great efforts to precisely measure its gravity field and huge advances in high-pressure experiments for H.

On the observational side, the unknown extent of differential rotation into the interior has given room to a variety of re-interpretations of the measured J_4 value. J_4 is an important quantity that strongly influences the distribution of metals in the envelope. Whether or not homogeneous envelope models are consistent with J_4 depends on the EOS. Accurate higher order moments from the *Juno* mission might greatly advance our understanding of Jupiter's differential rotation, thereby constraining interior models.

Neglecting differential rotation, J_2 and J_4 and the EOS allow one to restrict Jupiter's core mass to 0–$7 M_\oplus$ and the envelope metallicity to 11–$37 M_\oplus$; including differential rotation, this uncertainty rises to $M_\text{core} = 0\text{–}18 M_\oplus$ and $M_Z = 2\text{–}37 M_\oplus$ with $M_\text{core} + M_Z = 12 - 37 M_\oplus$.

For *Uranus and Neptune* we obtain a deep envelope metallicity of 0.80–0.95. Larger fractions of rock (or ices lighter than H_2O) would shift this range towards smaller (higher) values. These models resemble a slightly eroded ice-rock core of $\sim 11 M_\oplus$ (U) and $\sim 15 M_\oplus$ (N) below a thin, ice-enriched H/He layer.

Eroded core models of Jupiter give ice-rock core masses below $10 M_\oplus$ unless the core is assumed to be very diluted. This would indicate partial redistribution of core material into Jupiter's envelope. Any prediction of Jupiter's formation process from its present core mass is highly unreliable.

4 Exoplanets

4.1 Current Explanations for Large Radii of Gas Giants

As discussed in early sections, the standard cooling theory for giant planets (e.g., Hubbard et al. 2002) envisions an adiabatic H/He envelope, likely enhanced in heavy elements, on top of a distinct heavy element core, likely composed of ices and rocks. It is the radiative atmosphere that serves as the bottleneck for interior cooling and contraction. The effects of modest Jovian-like stellar irradiation on cooling models of Jupiter was investigated by Hubbard (1977). The ways in which strong stellar irradiation retards the contraction and interior cooling of giant planets was first worked out by Guillot et al. (1996). The high external radiation keeps the atmosphere quite hot (1000–2000 K) and drives a shallow radiative temperature gradient deep into the atmosphere, to pressures of ~ 1 kbar. A shallower dT/dP gradient in the atmosphere, compared to an isolated planet, means that the flux carried through the atmosphere must be necessarily reduced. Atmospheric pressure-temperature profiles at a variety of incident flux levels are shown in Fig. 9. Note that this incident flux itself does not directly effect the interior of the planet—the stellar flux is calculated to be wholly absorbed at pressures less than ~ 5 bar (Iro et al. 2005). This means that these planets (which have inflated radii up to $1.8 R_J$) must reside in close-in orbits for their entire lives. If they had

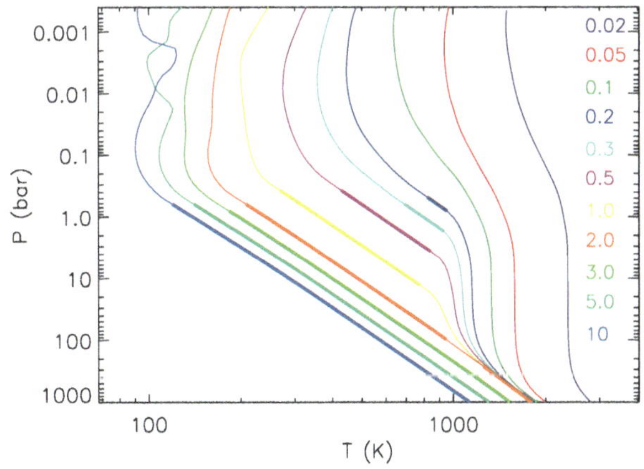

Fig. 9 Pressure-temperature profiles for 4.5 Gyr Jupiter-like planets ($g = 25$ m/s^2, $T_{\rm int} = 100$ K) from 0.02 to 10 AU from the Sun. Distance from the Sun in AU is color coded along *the right side of the plot*. *Thick lines* are convective regions, while *thin lines* are radiative regions. Planets closer to the Sun have deeper atmospheric radiative zones. The profiles at 5 and 10 AU show deviations that arise from numerical noise in the chemical equilibrium table near condensation points, but this has a negligible effect on planetary evolution

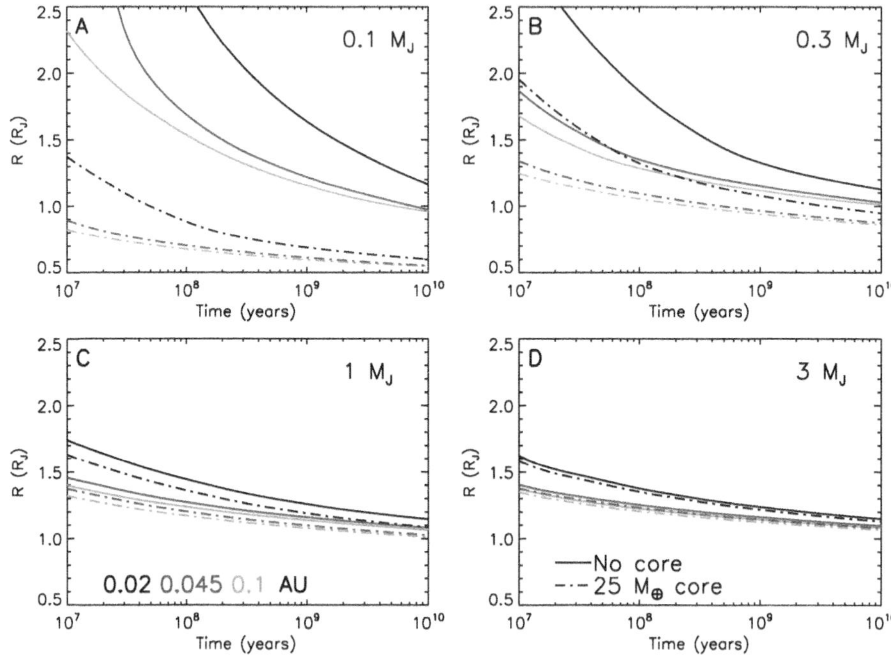

Fig. 10 Planetary radii as a function of time for masses of $0.1M_J$ ($32M_\oplus$, **A**), $0.3M_J$ (**B**), $1.0M_J$ (**C**), and $3.0M_J$ (**D**). The three shades code for the three different orbital separations from the Sun, shown in (**C**). *Solid lines* indicate models without cores and *dash-dot lines* indicate models with a core of $25M_\oplus$

previously cooled at 5 AU, and were brought in very recently, their radii would be $\sim 1R_J$, similar to Jupiter, with a very small increase in radius just due to a puffed up atmosphere (Burrows et al. 2000).

The upshot of this shallow atmospheric temperature gradient is that a smaller flux from the deep interior can be carried through the atmosphere—the cooling of the interior (and hence, contraction) is slowed, compared to the isolated case (Guillot and Showman 2002; Baraffe et al. 2003; Burrows et al. 2003). The effects of irradiation of 0.02, 0.045, and 0.1 AU from a constant luminosity Sun are shown in Fig. 10, using the models of Fortney et al. (2007). For Jupiter-mass planets, radii of $1.2R_J$ are expected at gigayear ages. Nevertheless, as seen in Fig. 11, many planets have radii in excess of $1.2R_J$, and most receive irradiation far below that expected at 0.02 AU. Explaining the large radii has been a major focus of exoplanet research for several years. Below we briefly review the previous work.

- Tidal dissipation in a giant planet's interior can produce heating that would slow or stop planetary contraction. Bodenheimer et al. (2001) proposed that the radius of HD 209458b could be explained by non-zero orbital eccentricity, forced by an unseen additional planetary companion. This eccentricity would then be tidally damped, perhaps for gigayears. For HD 209458b and other planets, this is potentially ruled out by the timing of the secondary eclipse (e.g., Deming et al. 2005), which indicates an eccentricity of zero. Interest in tides continues, however. Jackson et al. (2008a) have shown the orbits of hot Jupiters are still decaying due to the tide raised on the star by the planet, and that tidal heating in the not-to-distant past could have been appreciable (Jackson et al. 2008b). Levrard et al. (2009) have followed up on this work and shown that nearly all detected transiting planets

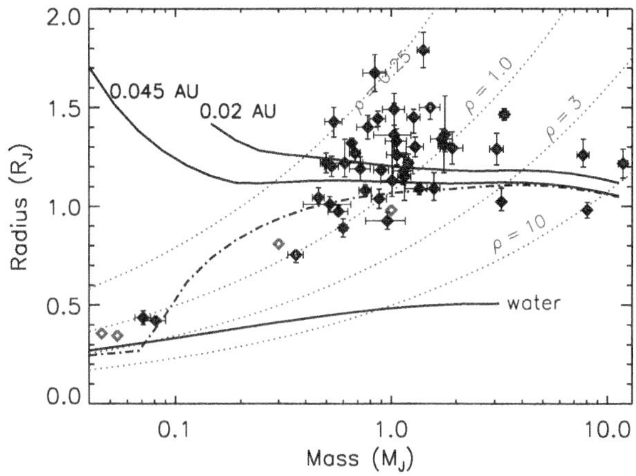

Fig. 11 The masses and radii of known transiting exoplanets, with error bars. Planetary radii from the models of Fortney et al. (2007) at 4.5 Gyr. *The solid-curve* models are for planets at 0.02 AU and 0.045 AU, with a composition of pure H/He (which is likely unrealistic given the structure of Jupiter & Saturn), and includes the transit radius effect. Models with a *dash-dot curve* at 0.045 AU include $25 M_\oplus$ of heavy elements (50/50 ice and rock) in a distinct core. *The lower solid curve* is for pure water planets. *Diamonds* without error bars are solar system planets

will eventually fall into their parent stars. Recently Ibgui and Burrows (2009) and Miller et al. (2009) have extended the Jackson et al. (2008b) work by computing the first hot Jupiter contraction models that explicitly couple tidal heating to the thermal evolution of giant planets. While tidal heating should be very important for some systems, it likely cannot explain all of the inflated planets.

- Guillot and Showman (2002) proposed that a small fraction ($\sim$0.5–1%) of absorbed stellar flux is converted to kinetic energy (winds) and dissipated at a depth of tens of bars by, e.g., the breaking of atmospheric waves. This mechanism would presumably effect all hot Jupiters to some degree. While this mechanism is attractive, much additional work is needed to develop it in detail, as Burkert et al. (2005) did not find this dissipation in their simulations. Direct simulation of these atmospheres in 3D over long time scales is computationally expensive.
- Baraffe et al. (2004) found that HD 209458b could be in the midst of extreme envelope evaporation, leading to a large radius, and we are catching the planet at a special time in its evolution. The authors themselves judged this to be very unlikely. Current models of atmospheric escape from hot Jupiters (e.g., Murray-Clay et al. 2009) find evaporation rates much lower than those previously assumed by Baraffe and collaborators, which were based on earlier work.
- Winn and Holman (2005) found that HD 209458b may be stuck in a Cassini state, with its obliquity turned over at 90 degrees, which leads to a tidal damping of obliquity over gigayear ages. Additional work by Levrard et al. (2007) and Fabrycky et al. (2007) have cast serious doubt on this mechanism for HD 209458b and all close-in planets. This work was recently reviewed in some detail by Peale (2008).
- Burrows et al. (2007) propose that atmospheres with significantly enhanced opacities (10× that of a solar mixture) would stall the cooling and contraction of the planetary interior, leading to larger radii at gigayear ages. This would be due to, for example, a large underestimation of the true opacities in these atmospheres (see also Ikoma et al. 2006). Spectra of hot Jupiter atmospheres will either support or refute this (currently ad-hoc) possibility. We note that if the H/He envelope were wholly 10× solar in *metallicity*, the increased molecular weight of the H/He-dominated envelope would entirely negate this high-opacity effect (Hansen and Barman 2007; Guillot 2008).
- Chabrier and Baraffe (2007), independently following along the lines of a hypothesis from Stevenson (1985), suggest that gradients in heavy elements (such as from core dredge-up or dissolution of planetesimals) could suppress convection and cooling in the H/He

envelope, leading to large radii at gigayear ages. This double diffusive convection (where there are gradients in both temperature and composition) occurs in the Earth's oceans. These diffusive layers could, however, be quite fragile, and 3D simulations of this process are required, under conditions relevant to giant planet interiors. Note that this effect could be present for planets at any orbital distance.

- Hansen and Barman (2007) suggest that if mass loss due to evaporation leads to a preferential loss of He vs. H (perhaps due to magnetic fields confining H^+), that planets could be larger than expected due to a smaller mean molecular weight. This mechanism would also presumably effect all hot Jupiters to some degree. However, Guillot (2008) has shown that some planets are still larger than can be accommodated by pure hydrogen composition.
- Arras and Socrates (2009) very recently postulated that a thermal tide in the atmosphere of hot Jupiters could lead to energy dissipation in their atmospheres, thereby potentially leading to inflated radii. However, there appear to be problems with the implementation in this work, and Gu and Ogilvie (2009) perform a somewhat similar analysis and find very weak energy dissipation.

We note that a planetary radius-inflation mechanism that would affect all hot Jupiters is quite reasonable. Since giant planets are expected to be metal rich (Jupiter and Saturn are 5–20% heavy elements) a mechanism that would otherwise lead to large radii could easily be canceled out by a large planetary core or a supersolar abundance of heavy elements in the H/He envelope in most planets (Fortney et al. 2006). Planets that appear "too small" are certainly expected and are relatively easy to account for due to a diversity in internal heavy element abundances (Guillot et al. 2006; Burrows et al. 2007).

Some of these inflation mechanisms should scale with stellar irradiation level or with orbital separation, while others do not. Therefore, a premium should be placed on finding transiting planets farther from their parent stars. All but two of the known transiting planets have orbits of only 1–6 days. However, the French *CoRoT* and American *Kepler* missions have the potential to find transiting giant planets out to 0.2 AU and 1 AU, respectively. *CoRoT* has already announced 5 planets in close-in orbits, and *Kepler* just launched in March 2009. The orbital separation limits on these missions are due entirely to the length of time these telescopes will stare at a given patch of sky—the longer the time duration, the longer the planetary orbital period that can be seen to have multiple transits. This science will continue to expand, and the future is bright.

4.2 The Expanding Field of Exo-Neptunes

The transits of planets GJ 436b (Gillon et al. 2007) and HAT-P-11b (Bakos et al. 2009) have opened the field of direct characterization of Neptune-class planets in addition to Jupiter- and Saturn-class. This is extremely exciting. Two things that we have immediately learned from merely a measured mass and radius are that: (1) These planets *must* have H/He envelopes (they cannot be purely heavy elements), but that these envelopes are probably only 10–20% of the planet's mass, similar to Uranus and Neptune. (2) That these two planets are not likely to be remnants of evaporated gas giants. Baraffe et al. (2006) had calculated that Neptune-mass planets that are evaporation remnants should have large radii around $\sim 1 R_J$, due to a tenuous remaining gaseous envelope, while these two planets have radii less than $0.5 R_J$. Hubbard et al. (2007) have also shown that the mass function of observed radial velocity exoplanets is nearly independent of orbital distance. If evaporation were important, one might expect a deficit of close-in Saturn-mass planets, which would be easier to evaporate than more massive giants. However, it will take a statistically interesting number of transiting planet detections before we can claim to see trends in these lower mass planets.

4.3 Young Gas Giant Planets

As discussed in the previous section, there is in wide use a model for the cooling and contraction of gas giant planets that is now being tested in a variety of cases at Gyr ages. It is clear from giant planet formation theories that these planets are hot, luminous, and have larger radii at young ages, and they contract and cool inexorably as they age. However, since the planet formation process is not well understood *in detail*, we understand very little about the initial conditions for the planets' subsequent cooling. Since the Kelvin-Helmholtz time is very short at young ages (when the luminosity is high and radius is large) it is expected that giant planets forget their initial conditions quickly. This idea was established with the initial Jupiter cooling models in the 1970s (Graboske et al. 1975; Bodenheimer 1976).

Since our solar system's giant planets are thought be 4.5 Gyr old, there is little worry about how thermal evolution models of these planets are effected by the unknown initial conditions. The same may not be true for very young planets, however. Since giant planets are considerably brighter at young ages, searches to directly image planets now focus on young stars. At long last, these searches are now bearing fruit (Chauvin et al. 2005; Marois et al. 2008; Kalas et al. 2008). It is at ages of a few million years where understanding the initial conditions and early evolution history is particularly important, if we are to understand these planets. Traditional evolution models, which are applied to both giant planets and brown dwarfs, employ an arbitrary starting point. The initial model is large in radius, luminosity, and usually fully adiabatic. The exact choice of the starting model is usually thought to be unimportant, if one is interested in following the evolution for ages greater than 1 Myr (Burrows et al. 1997; Chabrier and Baraffe 2000).

We will now briefly discuss how these models are used. Thermal evolution models, when coupled to a grid of model atmospheres, aim to predict the luminosity, radius, $T_{\rm eff}$, thermal emission spectrum, and reflected spectrum, as a function of time. When a planetary candidate is imaged, often only the apparent magnitude in a few infrared bands are known, at least initially. If the age of the parent star can be estimated (itself a tricky task) then the observed infrared magnitudes can be compared with calculations of model planets for various masses, to estimate the planet's mass, which is not an observable quantity unless some dynamical information is also known. It is not known if these thermal evolution models are accurate at young ages—they are relatively untested, which has been stressed by Baraffe et al. (2002) for brown dwarfs and Marley et al. (2007) for planets. Indeed, Stevenson (1982) had stressed that these cooling models "... cannot be expected to provide accurate information on the first 10^5–10^8 years of evolution because of the artificiality of an initially adiabatic, homologously contracting state".

Marley et al. (2007) examined the issue of the accuracy of the arbitrary initial conditions (termed a "hot start" by the authors) by using initial conditions for cooling that were not arbitrary, but rather were given by a leading core accretion planet formation model (Hubickyj et al. 2005). The core accretion calculation predicts the planetary structure at the end of formation, when the planet has reached its final mass. The Marley et al. (2007) cooling models use this initial model for time zero, and subsequent cooling was followed as in previously published models. Figure 12 shows the resulting evolution. The cooling curves are dramatically different, yielding cooler (and smaller) planets. The initial conditions are not quickly "forgotten," meaning that the cooling curves do not overlap with the arbitrary start models for 10^7 to 10^9 years. What this would mean, in principle, is that a mass derived from "hot start" evolutionary tracks would significantly underestimate the true mass of a planet formed by core accretion.

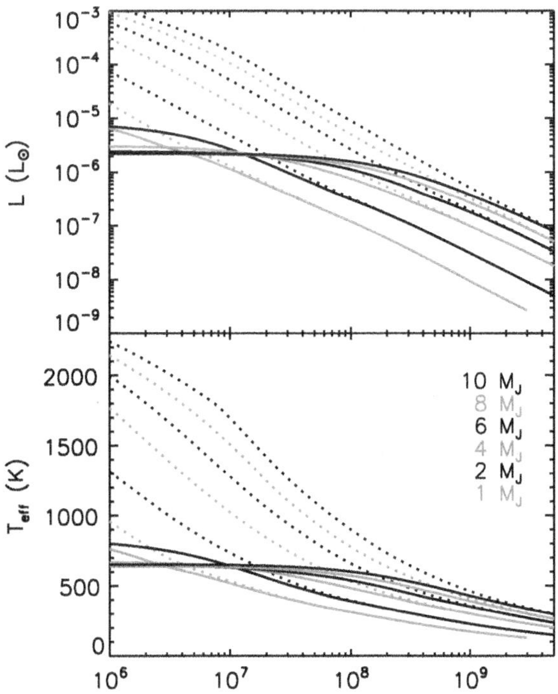

Fig. 12 Models from Marley et al. (2007) of the thermal evolution of giant planets from 1 to $10M_J$. *The dotted curves* are standard "hot start" models with an arbitrary initial condition, and *the solid curves* use as an initial condition the core accretion formation models of Hubickyj et al. (2005)

Certainly one must remember that a host of assumptions go into the formation model which yields the starting point for evolution, so it is unlikely that these new models are quantitatively correct. However, they highlight that much additional work is needed to understand the energetics of the planet formation process. The Hubickyj et al. (2005) models yield relatively cold initial planets because of an assumption that accreting gas is shocked and readily radiates away this energy. The end result is that the accreted gas is of relatively low specific entropy, leading to a low luminosity starting point for subsequent evolution. Significant additional work on multi-dimensional accretion must be done, as well as on radiative transfer during the accretion phase, before we can confidently model the early evolution.

Another issue, which is more model independent, is that since the planet formation by core accretion may take ~1–5 Myr to complete, it is likely incorrect to assume that a parent star and its planets are coeval. This will be particularly important for young systems. If a planetary candidate with given magnitudes is detected, overestimating its age (since it would be younger than its parent star) would lead to an overestimation of its mass. Thankfully, it appears that detections of young planets are now beginning to progress quickly, which will help to constrain these models.

4.4 Conclusions: Exoplanets

Since the information that we can gather about interiors of the solar system's giant planets is inherently limited, advances in understanding giant planets as classes of astronomical objects will likely rest on the characterization of a large number of exoplanets. While for any particular planet, the amount of knowledge to be gleaned is relatively small, this can be overcome by the shear numbers of these planets. Therefore, in the future, some of this work will necessarily have to be statistical in nature. This has already begun to some degree, as Fressin et al. (2007, 2009) have analyzed current transit surveys to derive the yields and giant planet properties from these observations.

Understanding the mass-radius relation of giant planets as a function of orbital distance is a critically important question. What is causing the large planetary radii and how does it scale with distance? The French *CoRoT* mission should be able to detect planets out to 0.2 AU, and the American *Kepler* mission out to 1 AU, due to its longer time baseline. Any planets found in these wider orbits will be critical data points. After *Kepler*, it is not at all clear when, *if ever*, we may have access to precise radii and masses and giant planets for planets in orbits of months to years.

The direct imaging of giant planets is now ramping up and allows us to sample additional parameter space—mostly young, massive planets far from their parent stars. Determining the physical properties of these planets in eras not long after their formation will allow us to better understand planet formation and thermal evolution.

References

Y. Alibert, O. Mousis, C. Mordasini, W. Benz, Astrophys. J. **626**, L57 (2005)
P. Arras, A. Socrates, Astrophys. J., astro-ph/0901.0735 (2009, submitted)
G.Á. Bakos, G. Torres, A. Pál, J. Hartman, G. Kovács, R.W. Noyes, D.W. Latham, D.D. Sasselov, B. Sipőcz, G.A. Esquerdo, D.A. Fischer, J.A. Johnson, G.W. Marcy, R.P. Butler, H. Isaacson, A. Howard, S. Vogt, G. Kovács, J. Fernandez, A. Moór, R.P. Stefanik, J. Lázár, I. Papp, P. Sári, Astrophys. J. (2009, submitted). 0901.0282 [astro-ph]
I. Baraffe, G. Chabrier, F. Allard, P.H. Hauschildt, Astron. Astrophys. **382**, 563 (2002)
I. Baraffe, G. Chabrier, T.S. Barman, F. Allard, P.H. Hauschildt, Astron. Astrophys. **402**, 701 (2003)
I. Baraffe, F. Selsis, G. Chabrier, T.S. Barman, F. Allard, P.H. Hauschildt, H. Lammer, Astron. Astrophys. **419**, L13 (2004)
I. Baraffe, Y. Alibert, G. Chabrier, W. Benz, Astron. Astrophys. **450**, 1221 (2006)
P. Bodenheimer, Icarus **29**, 165 (1976)
P. Bodenheimer, D.N.C. Lin, R.A. Mardling, Astrophys. J. **548**, 466 (2001)
A. Burkert, D.N.C. Lin, P.H. Bodenheimer, C.A. Jones, H.W. Yorke, Astrophys. J. **618**, 512 (2005)
A. Burrows, M. Marley, W.B. Hubbard, J.I. Lunine, T. Guillot, D. Saumon, R. Freedman, D. Sudarsky, C. Sharp, Astrophys. J. **491**, 856 (1997)
A. Burrows, T. Guillot, W.B. Hubbard, M.S. Marley, D. Saumon, J.I. Lunine, D. Sudarsky, Astrophys. J. **534**, L97 (2000)
A. Burrows, D. Sudarsky, W.B. Hubbard, Astrophys. J. **594**, 545 (2003)
A. Burrows, I. Hubeny, J. Budaj, W.B. Hubbard, Astrophys. J. **661**, 502 (2007)
G. Chabrier, I. Baraffe, Annu. Rev. Astron. Astrophys. **38**, 337 (2000)
G. Chabrier, I. Baraffe, Astrophys. J. **661**, L81 (2007)
G. Chabrier, D. Saumon, W.B. Hubbard, J.I. Lunine, Astrophys. J. **391**, 817 (1992)
G. Chabrier, I. Baraffe, F. Selsis, T.S. Barman, P. Hennebelle, Y. Alibert, in *Protostars and Planets V*, ed. by B. Reipurth, D. Jewitt, K. Keil (2007) p. 623
D. Charbonneau, T.M. Brown, D.W. Latham, M. Mayor, Astrophys. J. **529**, L45 (2000)
G. Chauvin, A.-M. Lagrange, C. Dumas, B. Zuckerman, D. Mouillet, I. Song, J.-L. Beuzit, P. Lowrance, Astron. Astrophys. **438**, L25 (2005)
B.J. Conrath, D. Gautier, Icarus **144**, 124 (2000)
W.C. Demarcus, Astron. J. **63**, 2 (1958)
D. Deming, S. Seager, L.J. Richardson, J. Harrington, Nature **434**, 740 (2005)
D.C. Fabrycky, E.T. Johnson, J. Goodman, Astrophys. J. **665**, 754 (2007)
J.J. Fortney, W.B. Hubbard, Icarus **164**, 228 (2003)
J.J. Fortney, W.B. Hubbard, Astrophys. J. **608**, 1039 (2004)
J.J. Fortney, M.S. Marley, J.W. Barnes, Astrophys. J. **659**, 1661 (2007)
J.J. Fortney, D. Saumon, M.S. Marley, K. Lodders, R.S. Freedman, Astrophys. J. **642**, 495 (2006)
M. French, T.R. Mattsson, N. Nettelmann, R. Redmer, Phys. Rev. B **79**, 054107 (2009a)
M. French, T.R. Mattsson, N. Nettelmann, R. Redmer, Phys. Rev. B **79**, 054107 (2009b)
F. Fressin, T. Guillot, V. Morello, F. Pont, Astron. Astrophys. **475**, 729 (2007)
F. Fressin, T. Guillot, N. Lionel, Astron. Astrophys. **504**, 605 (2009). 0901.3083 [astro-ph]

D. Gautier, B.J. Conrath, T. Owen, I. de Pater, S.K. Atreya, in *Neptune and Triton*, ed. by D.P. Cruikshank, M.S. Matthews, A.M. Schumann (1995) p. 547
M. Gillon, F. Pont, B.-O. Demory, F. Mallmann, M. Mayor, T. Mazeh, D. Queloz, A. Shporer, S. Udry, C. Vuissoz, Astron. Astrophys. **472**, L13 (2007)
H.C. Graboske, R.J. Olness, J.B. Pollack, A.S. Grossman, Astrophys. J. **199**, 265 (1975)
A.S. Grossman, J.B. Pollack, R.T. Reynolds, A.L. Summers, H.C. Graboske, Icarus **42**, 358 (1980)
P.-G. Gu, G.I. Ogilvie, Mon. Not. R. Astron. Soc. **395**, 422 (2009). arXiv:0901.3401
T. Guillot, Planet. Space Sci. **47**, 1183 (1999)
T. Guillot, Phys. Scripta T **130**, 014023 (2008)
T. Guillot, A.P. Showman, Astron. Astrophys. **385**, 156 (2002)
T. Guillot, D. Gautier, G. Chabrier, B. Mosser, Icarus **112**, 337 (1994)
T. Guillot, G. Chabrier, D. Gautier, P. Morel, Astrophys. J. **450**, 463 (1995)
T. Guillot, A. Burrows, W.B. Hubbard, J.I. Lunine, D. Saumon, Astrophys. J. **459**, L35 (1996)
T. Guillot, D.J. Stevenson, W.B. Hubbard, D. Saumon, *The interior of Jupiter* (Jupiter. The Planet, Satellites and Magnetosphere) (2004) p. 35
T. Guillot, N.C. Santos, F. Pont, N. Iro, C. Melo, I. Ribas, Astron. Astrophys. **453**, L21 (2006)
B.M.S. Hansen, T. Barman, Astrophys. J. **671**, 861 (2007)
G.W. Henry, G.W. Marcy, R.P. Butler, S.S. Vogt, Astrophys. J. **529**, L41 (2000)
N.C. Holmes, M. Ross, W.J. Nellis, Phys. Rev. B **52**, 15835 (1995)
W.B. Hubbard, Astrophys. J. **152**, 745 (1968)
W.B. Hubbard, Space Sci. Rev. **14**, 424 (1973)
W.B. Hubbard, Icarus **30**, 305 (1977)
W.B. Hubbard, Icarus **52**, 509 (1982)
W.B. Hubbard, H.E. Dewitt, Astrophys. J. **290**, 388 (1985)
W.B. Hubbard, M.S. Marley, Icarus **78**, 102 (1989)
W.B. Hubbard, M. Podolak, D.J. Stevenson, in *Neptune and Triton*, ed. by D.P. Kruikshank (Univ. of Arizona Press, Tucson, 1995), p. 109
W.B. Hubbard, T. Guillot, M.S. Marley, A. Burrows, J.I. Lunine, D.S. Saumon, Planet. Space Sci. **47**, 1175 (1999)
W.B. Hubbard, A. Burrows, J.I. Lunine, Annu. Rev. Astron. Astrophys. **40**, 103 (2002)
W.B. Hubbard, M.F. Hattori, A. Burrows, I. Hubeny, Astrophys. J. Lett. **658**, L59 (2007)
O. Hubickyj, P. Bodenheimer, J.J. Lissauer, Icarus **179**, 415 (2005)
L. Ibgui, A. Burrows, Astrophys. J. **700**, 1921 (2009). arXiv:0902.3998
M. Ikoma, T. Guillot, H. Genda, T. Tanigawa, S. Ida, Astrophys. J. **650**, 1150 (2006)
N. Iro, B. Bezard, T. Guillot, Astron. Astrophys. **436**, 719 (2005)
B. Jackson, R. Greenberg, R. Barnes, Astrophys. J. **678**, 1396 (2008a)
B. Jackson, R. Greenberg, R. Barnes, Astrophys. J. **681**, 1631 (2008b)
P. Kalas, J.R. Graham, E. Chiang, M.P. Fitzgerald, M. Clampin, E.S. Kite, K. Stapelfeldt, C. Marois, J. Krist, Science **322**, 1345 (2008)
G.I. Kerley, Los Alamos Scientific Laboratory Report LA-4776 (1972)
G.I. Kerley, Sandia Report, SAND2003-3613, Sandia National Laboratory, Albuquerque, NM 87185 (2003)
G.I. Kerley, Research Rep., KTS-04-2 (2004a)
G.I. Kerley, Research Rep., KTS-04-1 (2004b)
G. Kresse, J. Furthmüller, Phys. Rev. B **54**, 11169 (1996)
G. Kresse, J. Hafner, Phys. Rev. B **47**, 558 (1993a)
G. Kresse, J. Hafner, Phys. Rev. B **48**, 13115 (1993b)
B. Levrard, A.C.M. Correia, G. Chabrier, I. Baraffe, F. Selsis, J. Laskar, Astron. Astrophys. **462**, L5 (2007)
B. Levrard, C. Winisdoerffer, G. Chabrier, Astrophys. J. **692**, L9 (2009)
D.N.C. Lin, P. Bodenheimer, D.C. Richardson, Nature **380**, 606 (1996)
J. Liu, P.M. Goldreich, D.J. Stevenson, Icarus **196**, 653 (2008)
W. Lorenzen, B. Holst, R. Redmer, Phys. Rev. Lett. **102**, 115701 (2009)
P.R. Mahaffy, H.B. Niemann, A. Alpert, S.K. Atreya, J. Demick, T.M. Donahue, D.N. Harpold, T.C. Owen, J. Geophys. Res. **105**, 15061 (2000)
M.S. Marley, J.J. Fortney, O. Hubickyj, P. Bodenheimer, J.J. Lissauer, Astrophys. J. **655**, 541 (2007)
C. Marois, B. Macintosh, T. Barman, B. Zuckerman, I. Song, J. Patience, D. Lafreniere, R. Doyon, Science **322**, 1348 (2008)
S. Matousek, Acta Astronaut. **61**, 932 (2007)
M. Mayor, D. Queloz, Nature **378**, 355 (1995)
B. Militzer, W.B. Hubbard, J. Vorberger, I. Tamblyn, S.A. Bonev, Astrophys. J. **688**, L45 (2008)
N. Miller, J.J. Fortney, B. Jackson, Astrophys. J. **702**, 1413 (2009)

M.A. Morales, E. Schwegler, D.M. Ceperley, C. Pierleoni, S. Hamel, K. Caspersen, PNAS **106**, 1324 (2009). arXiv:0903.0980
R. Murray-Clay, E. Chiang, N. Murray, Astrophys. J. **693**, 23 (2009). arXiv:0811.0006
T. Nakajima, B.R. Oppenheimer, S.R. Kulkarni, D.A. Golimowski, K. Matthews, S.T. Durrance, Nature **378**, 463 (1995)
N. Nettelmann, B. Holst, A. Kietzmann, M. French, R. Redmer, D. Blaschke, Astrophys. J. **683**, 1217 (2008a)
N. Nettelmann, U. Kramm, R. Redmer, R. Neuhäuser, Astron. Astrophys. **683**, 1217 (2008b)
S.J. Peale, in *Astronomical Society of the Pacific Conference Series*, vol. 398, ed. by D. Fischer, F.A. Rasio, S.E. Thorsett, A. Wolszczan (2008) p. 281
M. Podolak, A.G.W. Cameron, Icarus **22**, 123 (1974)
J.B. Pollack, A.S. Grossman, R. Moore, H.C. Graboske, Icarus **30**, 111 (1977)
J.B. Pollack, O. Hubickyj, P. Bodenheimer, J.J. Lissauer, M. Podolak, Y. Greenzweig, Icarus **124**, 62 (1996)
D. Saumon, G. Chabrier, H.M.V. Horn, Astrophys. J. Suppl. **99**, 713 (1995)
D. Saumon, T. Guillot, Astrophys. J. **609**, 1170 (2004)
D.J. Stevenson, Phys. Rev. B **12**, 3999 (1975)
D.J. Stevenson, Planet. Space Sci. **30**, 755 (1982)
D.J. Stevenson, Icarus **62**, 4 (1985)
D.J. Stevenson, E.E. Salpeter, Astrophys. J. Suppl. **35**, 239 (1977a)
D.J. Stevenson, E.E. Salpeter, Astrophys. J. Suppl. **35**, 221 (1977b)
J. Vorberger, I. Tamblyn, B. Militzer, S. Bonev, Phys. Rev. B **75**, 024206 (2007)
J.N. Winn, M.J. Holman, Astrophys. J. Lett. **628**, L159 (2005)
M.H. Wong, P.R. Mahaffy, S.K. Atreya, H.B. Niemann, T.C. Owen, Icarus **171**, 153 (2004)
V.N. Zharkov, V.P. Trubitsyn, *Physics of Planetary Interiors*. Astron. Astrophys. Series (Pachart, Tucson, 1978)

Space Sci Rev (2010) 152: 449–500
DOI 10.1007/s11214-009-9587-5

Thermal Evolution and Magnetic Field Generation in Terrestrial Planets and Satellites

Doris Breuer · Stephane Labrosse · Tilman Spohn

Received: 20 May 2009 / Accepted: 2 November 2009 / Published online: 12 January 2010

Abstract Of the terrestrial planets, Earth and Mercury have self-sustained fields while Mars and Venus do not. Magnetic field data recorded at Ganymede have been interpreted as evidence of a self-generated magnetic field. The other icy Galilean satellites have magnetic fields induced in their subsurface oceans while Io and the Saturnian satellite Titan apparently are lacking magnetic fields of internal origin altogether. Parts of the lunar crust are remanently magnetized as are parts of the crust of Mars. While it is widely accepted that the magnetization of the Martian crust has been caused by an early magnetic field, for the Moon alternative explanations link the magnetization to plasma generated by large impacts. The necessary conditions for a dynamo in the terrestrial planets and satellites are the existence of an iron-rich core that is undergoing intense fluid motion. It is widely accepted that the fluid motion is caused by convection driven either by thermal buoyancy or by chemical buoyancy or by both. The chemical buoyancy is released upon the growth of an inner core. The latter requires a light alloying element in the core that is enriched in the outer core as the solid inner core grows. In most models, the light alloying element is assumed to be sulfur, but other elements such as, e.g., oxygen, silicon, and hydrogen are possible. The existence of cores in the terrestrial planets is either proven beyond reasonable doubt (Earth, Mars, and Mercury) or the case for a core is compelling as for Venus and the Moon. The Galilean satellites Io and Ganymede are likely to have cores judging from Galileo radio tracking data of the gravity fields of these satellites. The case is less clear cut for Europa. Callisto is widely taken as undifferentiated or only partially differentiated, thereby lacking an iron-rich core. Whether or not Titan has a core is not known at the present time. The terrestrial planets that do have magnetic fields either have a well-established inner core with known radius and density such as Earth or are widely agreed to have an inner core such as Mercury. The absence of an inner core in Venus, Mars, and the Moon (terrestrial bodies that lack fields) is not as well established although considered likely. The composition of the Martian core may be close

D. Breuer (✉) · T. Spohn
DLR, Institute of Planetary Research, Rutherfordstr. 2, 12489 Berlin, Germany
e-mail: doris.breuer@dlr.de

S. Labrosse
Laboratoire des sciences de la Terre, ENS Lyon, 46 Allée d'Italie, 69364 Lyon Cedex 07, France

to the Fe–FeS eutectic which would prevent an inner core to grow as long as the core has not cooled to temperatures around 1500 Kelvin. Venus may be on the verge of growing an inner core in which case a chemical dynamo may begin to operate in the geologically near future. The remanent magnetization of the Martian and the lunar crust is evidence for a dynamo in Mars' and possibly the Moon's early evolution and suggests that powerful thermally driven dynamos are possible. Both the thermally and the chemically driven dynamo require that the core is cooled at a sufficient rate by the mantle. For the thermally driven dynamo, the heat flow from the core into the mantle must by larger than the heat conducted along the core adiabat to allow a convecting core. This threshold is a few mW m^{-2} for small planets such as Mercury, Ganymede, and the Moon but can be as large as a few tens mW m^{-2} for Earth and Venus. The buoyancy for both dynamos must be sufficiently strong to overcome Ohmic dissipation. On Earth, plate tectonics and mantle convection cool the core efficiently. Stagnant lid convection on Mars and Venus are less efficient to cool the core but it is possible and has been suggested that Mars had plate tectonics in its early evolution and that Venus has experienced episodic resurfacing and mantle turnover. Both may have had profound implications for the evolution of the cores of these planets. It is even possible that inner cores started to grow in Mars and Venus but that the growth was frustrated as the mantles heated following the cessation of plate tectonics and resurfacing. The generation of Ganymede's magnetic field is widely debated. Models range from magneto-hydrodynamic convection in which case the field will not be self-sustained to chemical and thermally-driven dynamos. The wide range of possible compositions for Ganymede's core allows models with a completely liquid near eutectic Fe–FeS composition as well as models with Fe inner cores or cores in with iron snowfall.

Keywords Magnetic field generation · Thermal evolution · Terrestrial planets · Satellites

1 Introduction

Magnetic fields of internal origin are widely held to be characteristic of planets and at least the large satellites (for a recent review, see Connerney 2007). Magnetized meteorites suggest that even planetesimals may have had their own self-generated fields (Weiss et al. 2008). Of the terrestrial planets and satellites, Mercury, Earth, and Ganymede are known to have largely dipolar internally generated magnetic fields although there has been some debate of whether or not Ganymede's field is really self-sustained (e.g., Crary and Bagenal 1998). Mars has a remanently magnetized crust in mostly the southern hemisphere (Acuña et al. 1998). This remanent magnetization is evidence for an earlier self-generated field (Connerney et al. 2004). The Moon also has crust units with remanent magnetization, some of them located suspiciously close to the antipodes of major impact basins (Mitchell et al. 2008). While the magnetization may be taken as evidence of an ancient magnetic field for the Moon, alternative explanations have been given. One of the alternative explanations invokes plasma generated upon impact where the great basins are formed (e.g., Hood and Vickery 1984). It has been argued that Venus should have had an early magnetic field, also (Stevenson et al. 1983). The high surface temperature would have annealed any remanent magnetization since the temperature is above the Curie temperature (the magnetic blocking temperature) of most remanently magnetizable minerals.

Generation of a magnetic field requires an electrically conducting shell within a planet and motion within that shell. In the terrestrial planets and by the satellites, this region is agreed to be the fluid iron-rich core at the center. For recent reviews of Earth and planetary dynamos and convection in the core, see Roberts (2007), Christensen and Wicht (2007), Jones (2007), and Busse and Simitev (2007). There may be a solid inner core; the growth of which may provide a buoyancy flux that may drive the dynamo. The buoyancy in this case derives from a difference in composition between the solid inner core and the fluid outer core. Light alloying elements such as sulfur and oxygen (most models consider sulfur) tend to be expelled from the solidifying core and concentrate in the fluid outer core. In addition to a chemical buoyancy, flux from the inner core or alternatively to that buoyancy flux, thermal buoyancy may drive the flow. The thermal buoyancy results from a sufficiently large temperature difference between the core and the rocky mantle surrounding the core.

Regardless of whether chemical or thermal buoyancy drive core convection, the existence of a dynamo and the magnetic field strength are strongly dependent on the heat transfer rate through the mantle of the planet. The thermal buoyancy flux is directly related to the heat flow extracted by the mantle from the core. For a chemically driven dynamo, the rate of inner core growth and, therefore, the rate of buoyancy release depends on the core cooling rate and, therefore, on the heat flow from the core. The heat flow from the core is directly dependent on the vigor of convection in the mantle and the mantle heat transfer rate. The latter can be calculated as a function of time from planet thermal history models.

Thermal history models have been pioneered by Schubert and others (e.g., Schubert et al. 1979) in the late 70s and early 80s. These models use a parameterization of the heat transfer rate by convection through the planetary mantle and are usually based on a relation between the Nusselt number—a dimensionless measure of heat flow—and the Rayleigh number—a dimensionless measure of convective vigor. The boundary layer theory has been used to establish that relation (e.g., Schubert et al. 2001) but empirical data have also been used (e.g., Booker and Stengel 1978). More recently, it has been recognized that the classical boundary layer theory must be modified to include the effects of stagnant lids in fluids with strong temperature dependent viscosities (Davaille and Jaupart 1993; Moresi and Solomatov 1995; Grasset and Parmentier 1998).

Stevenson et al. (1983) were the first to discuss the thermal evolution of the core and the dynamo for the terrestrial planets. These authors included a discussion of the growth of an inner core and a simple estimate of the magnetic field strength based on the power dissipated in the core. Stevenson et al. concluded that the Earth should have had a thermally driven dynamo in its early evolution when the core was completely liquid. Freeze out of an inner core started after a few billion years of cooling and the dynamo switched from thermally driven to chemically driven. Venus, as these authors concluded, would have had an early magnetic field driven by a thermal dynamo that ceased to operate as the heat flow from the core dropped below the value that would also be supported by heat conduction along the core adiabat. A similar explanation was offered to explain the lack of a magnetic field of Mars that was later refined by Schubert and Spohn (1990). Mercury has a magnetic field because the thin mantle cools the core rapidly enough to allow the growth of an inner core.

These early models have not included the effects of stagnant lids on the cooling of the planet. In the late 90s of the previous century, the magnetism of the Galilean satellites was explored by the Galileo spacecraft. At the time of this writing, the Cassini mission is actively exploring the Saturnian satellites. In addition, new laboratory data on the melting relations of iron and iron alloys have become available and first super-Earth Exoplanets have been detected. It is thus timely to revisit the subject of the magnetism of the terrestrial planets and to review the progress of research in the field since Stevenson et al. (1983), revise their findings where necessary, and include a discussion of the Moon, the Galilean satellites, and Titan. In an outlook, we briefly speculate on the magnetic properties of Earth-like Exoplanets.

2 Structure Sets the Stage

It is widely accepted that the source region of self-generated magnetic fields in terrestrial planets and satellites is the iron-rich core. It is also widely accepted that the terrestrial planets are differentiated into iron-rich cores, silicate largely peridotitic mantles, and basaltic crusts. (The Earth's crust is divided between a granitic continental crust and a basaltic oceanic crust.) For the satellites, the situation is less well established. For instance, it is still debated whether or not the Moon and Titan have iron-rich cores (see Sohl and Schubert 2007, and Hussmann et al. 2007 for reviews) and the Jovian satellite Callisto is widely agreed to be undifferentiated or only partly differentiated.

The interior structure of a planet can be deduced from geodetic and geophysical data (gravity, topography, rotation, seismic, and magnetic field) but cosmochemical data are often providing conclusive evidence when the geophysical data are not allowing unique solutions. The recently published Volume 10 of the *Treatise on Geophysics* collects a number of excellent reviews of relevance on planetary geophysics and geodesy (e.g., Sohl and Schubert 2007; Wieczorek 2007; Van Holst 2007; Lognonne and Johnson 2007; Connerney 2007; Hussmann et al. 2007; Breuer and Moore 2007). Unfortunately, seismological data are only available for the Earth and the Moon. Although plans for establishing seismic networks on other terrestrial planets, in particular on Mars, have been discussed for decades, network missions have yet to be selected and implemented by the space agencies.

The most useful and widely available data for determining the core radius of a planet or satellite are the mass M, the radius r_p, and the quadrupole moment J_2 of the gravity field. For a planet in hydrostatic equilibrium, the Radau–Darwin relation holds and the latter is directly proportional to the moment of inertia C of the planet about its rotational axis

$$J_2 \sim \frac{C}{Mr_\mathrm{p}^2} \tag{1}$$

(see, e.g., Zharkov and Trubitsyn 1978; Hubbard 1984). Since the moment of inertia C is the integral over mass times distance from the rotation axis squared, C/Mr_p^2 is a dimensionless measure of the mass distribution in the planet. It can be compared to the average moment of inertia of a spherical model planet I

$$I \equiv \int_0^M r^2\,\mathrm{d}m, \tag{2}$$

where r is the radial distance of the mass element $\mathrm{d}m$ from the rotation axis. The dimensionless average moment of inertia I/Mr_p^2 is related to C/Mr_p^2 via

$$\frac{I}{Mr_\mathrm{p}^2} = \frac{C}{Mr_\mathrm{p}^2} - \frac{2}{3}J_2 \tag{3}$$

(Sohl et al. 2005). I/Mr_p^2 is the constraint to be used for interior structure modeling.

If the planet is not in hydrostatic equilibrium, J_2 is proportional to a linear combination of the principal moments of inertia and additional data such as the precession rate of the planet must be used to calculate C and I. In any case, a reasonable constraint can only be derived for a fast enough rotator for which J_2 is a clear measure of the dynamic flattening of the planet. This can be estimated by comparing the value of J_2 with the ratio between

Table 1 Planetary radii, moment of inertia factors, core radii, and densities of terrestrial planets and satellites. The planetary radii are from Landolt-Börnstein (2009), the moment of inertia factor for the Earth is from Lodders and Fegley (1998), for Mars and the Moon from Sohl and Schubert (2007), for the Galilean satellites from Schubert et al. (2004). The core radii and densities are discussed in the text. Note that the value for Venus is a representative value estimated by rescaling the Earth to Venus

	r_p [km]	C/Mr_p^2	I/Mr_p^2	r_c [km]	ρ_c [10^3 kg m^{-3}]
Earth	6371.01	0.3307	0.3293	3480 (oc)	10.9
				1220 (ic)	12.9
Venus	6052.878	?	?	*3089*	11
Mars	3390.500	0.3650 ± 0.0012	0.3635	1400–1900	5.5–8.5
Mercury	2439.700	?	?	1700–2300	5.1–8.1
Moon	1737.064	0.3931 ± 0.0002	0.3930	150–400	5.1–8.1
Io	1821.	0.37824 ± 0.00022	0.3770	650–950	5.1–8.1
Europa	1560.	0.346 ± 0.005	0.3460	200–700	5.1–8.1
Ganymede	2634.	0.3115 ± 0.0028	0.3114	650–900	5.1–8.1
Callisto	2400.	0.3549 ± 0.0042	0.3549	No	
Titan	2575.	?	?	0–800	5.1–8.1

the centrifugal acceleration measured at the equator of the planet to the acceleration due to gravity at the equator

$$q \equiv \frac{\omega^2 r_p^3}{GM}. \tag{4}$$

For a planet in hydrostatic equilibrium, J_2 is about $0.5q$. For Venus, J_2 is two orders of magnitude larger than q. Reliable data are available for the Earth and Mars. For the Galilean satellites, the Galileo mission has provided values of J_2 and J_{22}. Values of C/MR^2 haven been derived from these under the assumption of hydrostatic equilibrium and interior structure models have been calculated. Unfortunately, no data are yet available for Titan.

In the following, we will briefly summarize what is known about the core and inner core radii and densities for the terrestrial planets and satellites. Parameter values have been collected in Table 1.

For the Earth, the combined evidence from seismology and gravity (Dziewonski and Anderson 1981; Kennett et al. 1995) gives a core of 3480 km radius and a mass of 1.932×10^{24} kg. The core is layered with a liquid outer core and a solid inner core. The inner core has a radius of 1220 km. The average densities of the outer and inner cores as given by Olson (2007) are 10.9×10^3 kg m^{-3} and 12.9×10^3 kg m^{-3}, respectively. The density increase across the boundary is 0.5×10^3 kg m^{-3} to 1×10^3 kg m^{-3} (Souriau 2007) and is attributed to a difference in the specific volumes of solid and liquid iron and to a difference in chemistry between the outer and inner cores. The outer core is widely held to contain about 10 weight-% of light alloying elements with oxygen, sulfur, and silicon as the most likely candidates. The boundary between the cores is the liquidus of the alloy and the light alloying elements are expelled from the inner core as the latter grows through core freezing (see, e.g., Vocadlo 2007, and references therein).

The interior structure of Venus is much less well constrained. There are no seismic data available for the planet. The technological challenge of placing seismometers on the surface of the planet is overwhelming given the average surface temperature of about 740 K, the pressure of 95 bar, and the presence of highly corrosive SO_2 in the atmosphere. The gravity field and the topography have been measured by Doppler tracking of the Magellan spacecraft (Sjogren et al. 1997) and by radar altimetry (Rappaport et al. 1999) but the moment of inertia factor cannot be reliably estimated because of the slow rotation rate of the planet and the dominance of nonhydrostatic effects on the J_2 coefficient. It is widely assumed (see Sohl and Schubert 2007, for a review) that the interior structure of Venus is Earth-like with a core, a mantle and a crust. Since the intrinsic density of Venus is smaller than that of Earth (BVSP 1981), the core is relatively small; 0.51 planetary radii as compared with 0.55 planetary radii for Earth. Doppler tracking data of the Magellan spacecraft suggest that the core is liquid; at least that there is a liquid outer core (Yoder 1997). Stevenson et al. (1983) have suggested that the Venusian core is lacking a growing inner core to explain the absence of a present-day magnetic field. Their rescaling of the Earth to the smaller size and mass of Venus while accounting for its high surface temperature resulted in a central pressure and temperature similar to the temperature and pressure at the surface of Earth's present day inner core. The core of Venus—according to that model—may start freezing within the next few million years.

The interior structure of Mars is better constrained although useful seismic data have not been collected from the surface of that planet as well. The Viking landers in the 70s did carry seismometers but these did not deliver useful data. Thus, the interior structure of Mars is mostly constrained by gravity, topography, and rotation data and by the chemistry of the SNC meteorites. Highly accurate gravity and topography data have been obtained using laser altimetry and two-way Doppler tracking of the Mars Global Surveyor (MGS) spacecraft (Smith and Zuber 1996; Zuber et al. 2000; Lemoine et al. 2001). The parameters describing the orientation of the rotation axis and its variation with time have been obtained by combining Viking and Mars Pathfinder tracking data (Folkner et al. 1997) and later by including MGS and Mars Odyssey data (Konopliv et al. 2006). The best available value for C/Mr^2 is 0.3650 ± 0.0012 (Yoder et al. 2003) from which a value of $I/Mr^2 = 0.3635 \pm 0.0012$ has been derived (Sohl et al. 2005). The latter value also includes minor corrections for the Tharsis bulge. The improved values of the moment of inertia factors are smaller than previous values and suggest stronger concentrations of the mass toward the center than in, e.g., the models of Sohl and Spohn (1997). However, due to the non-uniqueness of any interpretation of gravity data and due to uncertainties in crust thickness and crust, mantle, and core densities the value of the Martian core radius is still uncertain. Most published models (e.g., Sohl and Spohn 1997; Sohl et al. 2005; Bertka and Fei 1997) have core radii between 1400 and 1900 km and core densities between 5500 and 8400 kg m^{-3}. Models that satisfy the chemistry of the SNC meteorites in addition to the gravity data tend to have core radii around 1600 km and densities of about 6000 kg m^{-3}. These models suggest that the core of Mars has about 14 wt.-% sulfur, which is close to the eutectic Fe–FeS composition at Martian core pressure.

The state of the Martian core is not known with certainty, as well as the answer to the question of whether or not there is a solid inner core. However, analysis of 3 years of MGS tracking data suggests that at least an outer layer of the core is liquid (Yoder et al. 2003). Schubert and Spohn (1990) have argued for the simplest explanation for the absence of a present-day magnetic field to be the absence of an inner core driving a chemical dynamo. The results of the laboratory measurements of Fei et al. (1997) of the core liquidus temperature together with estimates of the actual core temperature by, e.g., Williams and Nimmo (2004) and Breuer and Spohn (2006) also suggest that the Martian core is entirely liquid.

The deep interior structure of Mercury is about as well constrained from geophysical data as that of Venus. The high density of the planet suggests that the core is large, about 1800 km radius or 80% of the radius of the planet (e.g., Siegfried and Solomon 1974; BVSP

1981; Spohn et al. 2001a). Harder and Schubert (2001) have argued for an even larger core that would require a substantial amount of volatiles in the core. Most cosmochemists (e.g., Woods et al. 1981), however, would argue for a refractory composition of the planet and a mostly iron core. The existence of a magnetic field measured by the Mariner 10 spacecraft (Ness et al. 1975) suggests that there is at least an outer liquid core and the thin mantle of the planet suggests that cooling has resulted in a substantial solid inner core (e.g., Stevenson et al. 1983).

Peale (1976, 1988) has outlined a method that would allow calculating the moment of inertia factor of the planet without having to assume hydrostatic equilibrium. The method would also allow proving the existence of a liquid outer core and an estimate of the moment of inertia factor of the core itself. The latter can be used to constrain the size of the inner core. The model is based on the observation that Mercury occupies a resonant Cassini state with the rotation axis and the normal to the ecliptic plane being coplanar. Measurement of the amplitude of physical libration together with the obliquity and the gravity field coefficients J_2 and J_{22} will allow the determination of C/Mr^2 as well as the ratio between C_{m} and C where C_{m} is the moment of inertia attributed to the solid part of the planet. The ratio is 0.5 and smaller in case there is a liquid core or outer core shell while it will be about 1 in case the planet is completely solid. Margot et al. (2007) have recently succeeded in measuring the libration rate from earthbound radar speckle observations. The data already suggest a liquid core layer. The gravity coefficients are expected to be measured by the Messenger (Solomon et al. 2001) and BepiColombo (Anselmi and Scoon 2001) missions.

The lunar interior has been probed with seismometers but an iron core has not been detected. The geometry of the seismic stations on the Moon and the foci of deep lunar quakes did not allow for rays that probed the core to be recorded (for a recent review of planetary and lunar seismology, see Lognonne and Johnson 2007). The seismic data are consistent with a core of 170 to 360 km (Nakamura et al. 1974). The averaged moment of inertia factor I/Mr^2 has been determined with improved accuracy from Doppler tracking of the Lunar Prospector spacecraft and from lunar laser ranging data by Konopliv et al. (2001) to 0.3931 ± 0.0002. This value together with the mass of the planet suggests a core radius of 220 km to 450 km. This range is consistent with independent lunar electromagnetic induction data (Hood et al. 1999) and joint inversions of seismic and gravity data (e.g., Khan et al. 2004) and with the chemical models of Kuskov and Kronrod (2001). The densities consistent with the radii given above vary between roughly $5100\ \mathrm{kg\,m^{-3}}$ and $8100\ \mathrm{kg\,m^{-3}}$. The former value is typical for a Fe–FeS eutectic composition while the latter suggest a pure iron core. The lunar laser ranging data suggest that the core is liquid (Yoder 1981; Dickey et al. 1994; Williams et al. 2001).

The Galileo mission has provided useful gravity data on the interior structure of the Galilean satellites. For discussions and overviews, see Sohl et al. (2002), Schubert et al. (2004), and Hussmann et al. (2007). Two-way Doppler tracking of the spacecraft has provided values for the J_{22} quadrupole gravity field coefficient from near-equatorial fly-bys. Assuming hydrostatic equilibrium, J_2 can be calculated from

$$J_2 = \frac{10}{3} J_{22}. \tag{5}$$

For Io, polar fly-bys have allowed an independent determination of J_2. The measured values of J_2 and J_{22} are consistent with hydrostatic equilibrium. Note, however, that (5) is required for equilibrium but not sufficient to prove it. Again, assuming hydrostatic equilibrium, values of C/Mr^2 have been calculated for Io (Anderson et al. 2001a), Europa (Anderson et al. 1998), Ganymede (Anderson et al. 1996), and Callisto (Anderson et al. 2001b). The core radii for Io vary between 650 km for a purely iron core and 950 km for a Fe–FeS eutectic core (Schubert et al. 2004). The densities would be 8090 kg m^{-3} and 5150 kg m^{-3}, respectively. The large availability of sulfur in the outer parts of the solar system and the large amounts of sulfur on Io's surface (e.g., Lewis 1982) favors sulfur-rich models of the satellite's core although it must be said that the composition of the cores of the Galilean satellites is not known.

The core radii of the icy satellites of Jupiter (Europa, Ganymede, and Callisto) are even less certain because the significant share of ice as a third component in addition to rock and iron adds to the nonuniqueness of the models. Estimates of the core radius of Europa vary between roughly 200 and 700 km depending on the composition of the core and the thickness of the ice layer (Sohl et al. 2002). The former value is consistent with an iron rich core while the latter is consistent with a eutectic Fe-FeS core and densities as given for Io above. For Ganymede, core radii vary between 650 km and 900 km, similar to the values for Io and with similar values for the range of core densities (Sohl et al. 2002). The ice shell in these models is 900 km thick which supports a rule-of-thumb structural model for Ganymede as an Io plus an ice shell (see also Kuskov and Kronrod 2001). The presence of a magnetic field of deep internal origin for Ganymede suggests that at least an outer shell of Ganymede's core is liquid.

Callisto is unusual among the larger satellites of Jupiter. The moment of inertia data suggests that the satellite is incompletely differentiated and, therefore, lacks a core (Anderson et al. 2001a; Sohl et al. 2002; Schubert et al. 2004). Nagel et al. (2004) have provided a model for the incomplete differentiation of the satellite. It has been suggested that Titan may also be incompletely differentiated (e.g., Lunine and Stevenson 1987; Grasset et al. 2000) although possibly to a different degree. Interpretation of the tracking data of the Cassini spacecraft is made difficult by the large fly-by distance and the effects of the dense atmosphere of the satellite and a value for the moment of inertia factor is still not available. Models must be based on the mass and radius alone and allow wide ranges of interior structure between models similar to Callisto and models that are similar to Ganymede with a substantial core.

3 The Core Is the Stage

The magnetic field of a terrestrial planet is most readily maintained against losses by diffusion by flow in its iron-rich core. This statement results from the induction equation for a conducting fluid that reads

$$\frac{\partial \mathbf{B}}{\partial t} = \nabla \times (\mathbf{u} \times \mathbf{B}) - \nabla \times (\eta_{\mathrm{m}} \nabla \times \mathbf{B}), \tag{6}$$

with η_{m}, the magnetic diffusivity, **B** the magnetic induction, and **u** the fluid velocity. (Vector quantities are bold face.) In the absence of flow, this equation reduces to a diffusion equation which implies that any initial magnetic field would decay on a timescale of d^2/η_{m}, with d the characteristic length-scale on which the field varies. With η_{m} approximately 5×10^{-6} m^2 s^{-1} a timescale of the order of 15 kyr is derived for the Earth taking the core radius as the length scale. This time is much shorter than the lifetime of the Earth's magnetic field. As a consequence, a sufficiently vigorous fluid flow is required to maintain dynamo action. The magnetic Reynolds number, $R_{\mathrm{m}} = Ud/\eta_{\mathrm{m}}$ measures the ratio between the two terms on the right-hand side of the induction equation. Some flows are known to not generate magnetic fields irrespective of the value of R_{m} (these are subject to so called antidynamo theorems), but for many flows dynamo action occurs whenever R_m exceeds the critical value of order 100. The magnetic diffusivity is linked to the electrical conductivity σ by $\eta_m = 1/\mu_0\sigma$ where μ_0 is the magnetic permeability. Since rocky planet mantles are insulators, their magnetic Reynolds numbers are too small to allow dynamo action which leaves the liquid cores as the source regions for the magnetic fields.

Several driving forces are possible to maintain the flow in the core. The most likely ones, and the only ones considered in this paper, are convection flows. The other main possibility is mechanical stirring caused either by a change in the rotation vector of the mantle or by tidal deformation of the mantle which can lead to elliptic instabilities (Kerswell 2002). Precession driven dynamos fall in the first category and the flow also comes from an elliptic instability. Tilgner (2005) showed that dynamo action can be produced by such a flow, although this requires the basic state to be neutrally stratified. In a planet core, this implies that the temperature follows an isentrope and that the composition is uniform, which can only be obtained through vigorous convection. Whether or not precession plays a role, it must be considered as an effect additional to convection.

3.1 Core Heat and Solute Transfer

Convection is the major player in the dynamics of planetary interiors and our understanding of this problem benefits from the important physics literature on the subject. A major difference between well-controlled convection experiments and planetary interiors is that in the latter case, the energy available to drive the flow comes from the long term evolution of the planet and the decay of radioactive isotopes. Therefore, the driving power is constantly evolving and an assumption of statistical steady state is not warranted. Fortunately, the low viscosity of liquid iron ensures a dynamics that is much faster than the evolution timescale of the planet, mostly controlled by sluggish convection in the solid mantle and separation of timescales is usually assumed to hold. The short term dynamics are considered to maintain the core in an average state close to isentropic and well mixed and the long term evolution of this average state provides the energy to drive convection.

3.1.1 Convection

Broadly defined, convection is a transport mechanism that occurs by macroscopic motion of fluid, independently of the origin of the flow. In planetary interiors, the flow originates from unstable density stratifications maintained by buoyancy sources at the boundaries—the planet's surface or the surface of its core, and in the bulk interior. When both compositional and temperature variations contribute to buoyancy, thermo-compositional or thermo-chemical convection arises. The motion resulting from the initially unstable stratification redistributes this buoyancy and tends to erase the initial stratification. In the absence of buoyancy sources that are somehow maintained, the system simply comes to rest after sufficient redistribution has been attained. The planetary cores can convect beyond an initial overturn because buoyancy sources are maintained by the slow thermal evolution of the planet.

In the case of planetary cores, two types of buoyancy contribute to convective motion, thermal and compositional buoyancy. Thermal buoyancy results from thermal expansion in a hot core in contact with a colder mantle. In addition, the core may be heated by the decay of radioactive elements. Compositional buoyancy results from the core being an alloy of metal (iron and nickel) and some lighter elements; the most discussed ones being S, Si, and O (Poirier 1994) and, in the most classical scenario, the crystallization of an inner core that is denser than the liquid outer core. Conservation of chemical species implies a flux of light elements from the bottom of the outer core which provides chemical buoyancy able to drive flow in the liquid outer core. The origin of this chemical buoyancy and the crystallization of the inner core is therefore linked to the cooling of the core.

As a mode of heat or solute transfer, convection competes with diffusion. In the case of a solute, diffusion is very inefficient and can usually be neglected. On the other hand, metals are good thermal conductors and thermal diffusion may play an important role in planetary cores. Heat transfer in a solid inner core is thought to occur by diffusion (Yukutake 1998). But even in liquid outer cores, diffusion is relevant in boundary layers and in the bulk of the interior as heat conduction along the isentropic temperature gradient.

In planetary cores, convection is assumed to maintain an isentropic (or adiabatic) temperature profile that can be calculated from

$$\frac{\mathrm{d}T}{\mathrm{d}z} = \frac{\alpha T g}{C_\mathrm{p}} \tag{7}$$

with T denoting temperature, z depth, α the thermal expansion coefficient, g the acceleration of gravity, and C_p the specific heat at constant pressure. In fact, the isentropic temperature gradient has to be exceeded for convection to occur. Generally speaking, thermal convection occurs when the imposed temperature gradient exceeds a minimum value, or equivalently when a suitably defined dimensionless number, the Rayleigh number, is larger than a critical value. The exact definition of this number depends on the mode of heating which will be discussed further below in Sect. 4.1.1. In the case of compressible fluids, the relevant temperature gradient is that in excess of the isentropic value (e.g., Jeffreys 1930; Spiegel and Veronis 1960; Hewitt et al. 1975). Because of the low viscosity of liquid iron, a very small excess temperature suffices to start vigorous convection. With the thermal conductivity of iron typically being ten times larger than that of rock (Stacey and Anderson 2001), the heat flow along the isentropic temperature profile (hereafter termed the isentropic heat flow) can be a large fraction of the total heat flow out of the core. For example, the isentropic heat flow at the top of the Earth's core is thought to be about 5–7 TW or about 8 to 12 $\mathrm{mW\,m^{-2}}$ (e.g., Labrosse 2003), a nonnegligible fraction of the total heat loss of the Earth of 46 TW (Jaupart et al. 2007).

3.1.2 Diffusion

Even when convection dominates in the bulk of the core, the vertical velocity has to go to zero at the horizontal boundaries. This means that radial diffusion near the boundary must balance the convective heat flow and leads to the formation of boundary layers where the radial temperature gradient is larger than in the interior. However, the thicknesses of these boundary layers are inversely proportional to the Rayleigh number and in the core are usually thought to be quite small. An estimate can be obtained for the case of the Earth's core by using the secular variation of the magnetic field. The flow velocity at the top of the core can be estimated from the secular variation of the Earth's magnetic field (see Hulot et al. 2002, for a recent calculation) and the typical velocity, about $10^{-4}\ \mathrm{m\,s^{-1}}$, allows an estimate of the total buoyancy that drives it, that is, the relative density anomalies $\delta\rho/\rho \sim 10^{-9}$ (e.g., Braginsky and Roberts 1995; Labrosse et al. 1997). Assuming these density anomalies to result from temperature anomalies and using $\alpha \sim 10^{-5}\ \mathrm{K^{-1}}$, we get $\delta T \sim 10^{-4}$ K. These small lateral temperature variations result from the destabilization of boundary layers and the total temperature differences across them are of the same order of magnitude. Assuming a super-isentropic heat flow at the core-mantle boundary (CMB) of, say, $\delta Q = 4\pi r_c^2 k\delta T/\delta z \approx$ 1 TW (r_c being the radius of the core), one can obtain a typical thickness for the boundary layer of $\delta z \sim 0.8$ m. This shows that the boundary layers that must exist in the core are very thin with very small total super-isentropic temperature differences. The reason is the very low viscosity of liquid iron (Poirier 1988; Terasaki et al. 2001) which likely applies to other planetary cores as well. Note that the boundary layers in the mantle are many orders of magnitude thicker as will be addressed further below.

Thermal diffusion can play an important role in wider regions of the core if the heat flow at the CMB is lower than the isentropic heat flow. For the Earth, the heat flow from the core may be approximately equal to the isentropic heat flow and thermal convection at the top of the core is not necessarily guaranteed. This problem has been recognized by earlier workers in the field (Loper 1978a, 1978b; Stevenson 1983) who proposed that compositional convection would act against thermal stratification in this case and maintain the average state close to isentropic. This requires compositional convection to transport downward the excess heat conducted along the isentrope and would lead to a much lower efficiency of the dynamo. Labrosse et al. (1997) and Lister and Buffett (1998) proposed instead that a stably stratified layer of about 100 km thickness through which heat would be transferred by thermal diffusion could form at the top of the core. Even if the total heat flow at the CMB were above the isentropic value, there could be regions in the core where a local thermal stratification would develop and their effect on the global core dynamics could be important. Also, because of fluctuations of the CMB heat flow imposed by the mantle on a time-scale of about 400 Myr (e.g., Grignè et al. 2005; Nakagawa and Tackley 2005; Labrosse and Jaupart 2007), one may speculate about periods in time during which thermal stratification would expand from the top of the core into the deeper interior and periods where it would shrink (Buffett 2007). The effect of such a scenario on the generation of the magnetic field has not been fully considered yet in the case of the Earth and it could be important for other planets as well.

Christensen (2006) proposed a similar scenario for Mercury, where inefficient heat transport by the mantle makes the CMB heat flow small. The main effect of such a stable layer at the top of the core is to filter the magnetic field that is produced underneath. The magnetic field observed outside of the planet is then smaller and of larger scale than that in the convectively active region.

Compositional stratification is also possible at the top of the core through the accumulation of light elements (Braginsky 1993, 2006). Braginsky (2006) calls this layer the stratified ocean of the core because the positive compositional buoyancy of such a layer compared to its thermal buoyancy would be similar to the negative buoyancy of the ocean in the atmosphere. It is difficult to test the existence of this layer but it would help to explain an incomplete chemical equilibration between the core and the mantle (Labrosse et al. 2007) which otherwise would result in a core containing too many light elements to be reconciled with seismological observations (Asahara et al. 2007). If this layer is formed by extraction (possibly exsolution, see below) of light elements from the bulk of the core, an additional buoyancy source would be provided that would help drive the dynamo.

3.2 Energy and Entropy Balances

In this section, the global balances for energy and entropy are written for planetary cores, in the case of an Earth-like structure, that is with an inner core growing from the center outward. As discussed below, more exotic situations can be encountered in other planets but their thermodynamics have not been completely worked out yet.

The thermal evolution of planetary cores is controlled by an equation for energy conservation in which the total heat loss Q_{CMB}, set by mantle convection (Sect. 4), is balanced by the sum of several sources:

$$Q_{\mathrm{CMB}} = Q_{\mathrm{ICB}} + Q_{\mathrm{C}} + Q_{\mathrm{L}} + E_{\xi} + Q_{\mathrm{H}} \tag{8}$$

with Q_{ICB} the heat flow from the inner core, Q_{C} the secular cooling term associated with the heat capacity, Q_{H} the radiogenic heating term in heat producing elements are contained in the core, Q_{L} the latent heat produced by core freezing, and E_{ξ} the compositional energy arising from redistribution of solute in a spatially varying chemical potential. The full expressions for these terms, which have been derived from a number of papers (e.g., Buffett et al. 1992, 1996; Braginsky and Roberts 1995; Lister and Buffett 1995; Labrosse et al. 1997, 2001; Labrosse 2003; Gubbins et al. 2004; Nimmo et al. 2004; Nimmo 2007), will only be briefly explained.

The secular cooling term always exists and represents the change of internal energy of the core with time. Using temperature as thermodynamic variable (instead of entropy which appears as a simpler choice when dealing with the average state of the core), one writes the secular cooling term as

$$Q_{\mathrm{C}} = -\int_{v} \rho C_{\mathrm{p}} \frac{\partial T}{\partial t} \mathrm{d}V. \tag{9}$$

The time derivative introduced here must be understood as a derivative with respect to the long term evolution. Indeed, the evolution of planetary cores involves at least two very different timescales. The first one is short and is related to the dynamics of the dynamo and the second one is related to the long term evolution. These two scales are vastly different and their separation was explained in detail by Braginsky and Roberts (1995). When dealing with the long term evolution, as is done here, the dynamical processes are averaged out and only considered as ways of maintaining the core close to its reference state. It is usually assumed that vigorous convection maintains the core close to isentropic on average. In this case, the isentropic temperature profile (see (7)) can be used to compute Q_{C}.

The isentrope is the solution of a differential equation and depends on the boundary condition, which is the only part varying with time in the isentrope. If the core is entirely liquid, the temperature at either the center or the CMB can serve as a boundary condition. In standard evolution scenarios, however, the inner core starts to crystallize at the center (Jacobs 1953) and it appears more natural to use the value at the center, T_{C} as the time-varying parameter. In this case, the secular cooling term takes the form

$$Q_{\mathrm{eC}} = -P_{\mathrm{eC}} \frac{\mathrm{d}T_{\mathrm{c}}}{\mathrm{d}t} \tag{10}$$

with P_{eC} a parameter having the units of $J\ \mathrm{K}^{-1}$ which can be computed from the isentropic profile and the density profile. When an inner core has started to crystallize, a natural boundary condition appears: the temperature at the phase boundary is equal to the liquidus of the outer core. This temperature also varies with time for two reasons. First, as the inner core grows, the pressure of its upper boundary (commonly called the inner core boundary, ICB) decreases, leading to a decrease of the liquidus temperature in the Earth's core. (In smaller planets, the liquidus temperature may increase with pressure; see below.) The other reason is linked to the time evolution of the core composition. The other reason is linked to the time evolution of the core composition. Due to the partitioning of light elements in the outer core during fractional crystallization at the ICB, the liquidus temperature decreases with time (Stevenson et al. 1983; Lister 2003; Gubbins et al. 2004; Nimmo 2007). Both variations are linked to the growth of the inner core and the ICB boundary temperature may be replaced with the radius of the inner core r_{ic}. For this reason, when an inner core is present and growing, the secular cooling term of the energy equation can be written as

$$Q_{IC} = P_{IC}(c)\frac{dc}{dt}, \tag{11}$$

where $P_{IC}(r_{ic})$ is a function of the inner core radius r_{ic} and has the unit of $J\ m^{-1}$. The volume V of integration is the total volume of the core, including the inner core, which explains the absence of an explicit heat flow from the inner core. Strictly speaking, this heat flow should be computed by solving a diffusion equation for the inner core (Labrosse et al. 1997) but owing to the small size of the inner core, this term can be approximated by extending the isentrope into the inner core.

The case of the latent heat of freezing is straightforward and the rate of energy release by this process is directly proportional to the growth rate of the inner core. It simply reads as

$$Q_L = T_L(c)\rho\Delta S 4\pi c^2\frac{dc}{dt} \equiv P_L(c)\frac{dc}{dt}, \tag{12}$$

with ΔS the entropy of freezing.

The compositional energy is related to the secular change of composition of the outer core in which a gradient of chemical potential is maintained by gravity and reads (Lister and Buffett 1995; Braginsky and Roberts 1995)

$$E_\xi = \int_{OC} (\mu - \mu_{ICB})\frac{d\xi}{dt}\, dV \tag{13}$$

with μ the chemical potential, and ξ the mass fraction of light elements. The volume of integration is the outer core volume only. Since the core is assumed to be well mixed at all times, $d\xi/dt$ is also uniform and E_ξ is found to be proportional to it. The profile of chemical potential is prescribed in the reference state in much the same way as that of the temperature. The change of composition is due to fractionation upon inner core freezing and is therefore proportional to the rate of inner core growth and this term, as well as the previous ones, can be written as

$$E_\xi = P_\xi(c)\frac{dc}{dt}. \tag{14}$$

Finally, the radiogenic heating is independent of the inner core size, and simply reads

$$Q_H(t) = M_C\sum_{i=1}^{N} h_i e^{-\lambda_i t} \tag{15}$$

summing the N contributions of all radioactive isotopes assumed to be present in the core, each one of which releases heat at a rate h_i per unit mass at time $t = 0$ and having a decay constant λ_i. The mass considered here, M_C, is the total mass of the core, which implies that any chemical exchange with the mantle is neglected. The partitioning of radioactive elements between the outer and the inner core is of no consequence in this calculation.

The energy balance does not include any contribution from the magnetic field, thus it cannot be used to discuss dynamo generation. This comes from the smallness of the magnetic energy (Gubbins 1977) and a well-known result of the thermodynamics of convective flows (Hewitt et al. 1975) that dissipation is internally balanced by the work of buoyancy forces. For this reason, Aubert et al. (2009) define a Rayleigh number for the dynamo based on the total dissipation, which can be obtained using the entropy balance of the core:

$$\frac{Q_{\mathrm{CMB}}}{T_{\mathrm{CMB}}} = \frac{Q_{\mathrm{ICB}} + Q_{\mathrm{L}}}{T_{\mathrm{ICB}}} + \frac{Q_{\mathrm{H}}}{T_{\mathrm{H}}} + \frac{Q_{\mathrm{C}}}{T_{\mathrm{C}}} + \frac{Q_{\mathrm{D}}}{T_{\mathrm{D}}}, \tag{16}$$

where each of the heat sources from the energy balance equation (8) enters divided by the effective temperature at which it is provided and where $Q_{\mathrm{D}}/T_{\mathrm{D}}$ is the total internal dissipation:

$$\frac{Q_{\mathrm{D}}}{T_{\mathrm{D}}} = \int_{\mathrm{OC}} \left[k \left(\frac{\nabla T}{T} \right)^2 + \frac{\boldsymbol{\tau} : \nabla \mathbf{u}}{T} \right] \mathrm{d}V + \int_{V\infty} \frac{\mathbf{J}^2}{\sigma T} \mathrm{d}V \tag{17}$$

with J the electrical current density and σ the electrical conductivity. The ohmic dissipation term in (17) is integrated over all space because the magnetic field produced inside the core spreads out to the universe and its time variation induces currents in all conductors, which produces entropy. The energy required to maintain this dissipation is transported from the core along the Poynting vector. In the case of the Earth, the mantle is usually assumed to be insulating, which is equivalent to neglecting its contribution to the ohmic heating. The inner core is likely as conducting as the outer core but, in the case of the Earth again, its volume is small and its contributions to the balance equations can be approximated reasonably well by assuming its temperature profile to be isentropic (Labrosse et al. 2001). In this case, the energy and entropy balances are simplified by extending over the entire core (Labrosse 2003; Lister 2003). For other planets such as Mercury, the situation can be more complex, with a larger inner core (Stanley et al. 2005) or with a thick stagnant layer at the top of the core (Christensen 2006).

The equations for the energy and entropy balances can be combined to eliminate the total heat loss (Q_{CMB}) and arrive at an efficiency equation that relates the total dissipation to all energy sources:

$$\begin{aligned} Q_{\mathrm{D}} = {} & \frac{T_{\mathrm{D}}}{T_{\mathrm{CMB}}} \left(1 - \frac{T_{\mathrm{CMB}}}{T_{\mathrm{C}}} \right) Q_{\mathrm{C}} + \frac{T_{\mathrm{D}}}{T_{\mathrm{CMB}}} \left(1 - \frac{T_{\mathrm{CMB}}}{T_{\mathrm{ICB}}} \right) (Q_{\mathrm{L}} + Q_{\mathrm{ICB}}) \\ & + \frac{T_{\mathrm{D}}}{T_{\mathrm{CMB}}} \left(1 - \frac{T_{\mathrm{CMB}}}{T_{\mathrm{H}}} \right) Q_{\mathrm{H}} + \frac{T_{\mathrm{D}}}{T_{\mathrm{CMB}}} E_{\xi}. \end{aligned} \tag{18}$$

The value of T_{D} is difficult to estimate but since the CMB is likely the coldest place of the core, the ratio $T_{\mathrm{D}}/T_{\mathrm{CMB}} > 1$. Therefore, the efficiency factor for converting heat into work is larger than that of a Carnot engine (Hewitt et al. 1975). This comes from the fact that—contrary to Carnot engines—work is performed inside the convective region and the dissipative heating is not lost but can contribute to fuel motion. Equation (18) also shows that the conversion of compositional energy is not affected by the Carnot factor and has a larger efficiency than heat sources.

All the terms on the right-hand side of (18) can be computed from the reference state of the core in much the same way as the energy terms, except for the factor T_{D}. Each term, except for the radiogenic heat source term, can be linked to the inner core and is proportional to its growth rate. The total dissipation $Q_{\mathrm{D}}/T_{\mathrm{D}}$ contains three contributions. The contribution from thermal diffusion (chemical diffusion should also be included but is generally neglected with good reason), $\int k(\nabla T/T)^2 \mathrm{d}V$ is easily computed from the basic isentropic state. The remaining two contributions are associated with the work of buoyancy forces and contain a contribution from viscous friction, which is usually neglected (see Braginsky and Roberts 1995, for a justification), and a contribution from ohmic dissipation which is relevant to the dynamo problem:

$$\Phi = T_D \int \frac{J^2}{\sigma T} \mathrm{d}V \tag{19}$$

with $\mathbf{J} = \nabla \times \mathbf{B}/\mu_0$ the electrical current density and σ the electrical conductivity. The relationship of J to B involves a spatial derivative so that any estimate of the former from the latter is plagued by large uncertainties at small length-scales. Also, B contains both a poloidal and a torroidal component. The torroidal component is confined to the planetary cores and cannot be measured directly. Therefore, the ohmic dissipation rate is badly constrained for the Earth and even more so for all other planets.

3.3 Snow and Exsolution

The energy and entropy balances discussed in the preceding sections are based on the assumption of an isentropic reference state in the core. The main argument supporting this assumption is that, if the dynamics is sustained by both thermal and compositional destabilizing gradients, convection will tend to homogenize the extensive state variables responsible for the motion, entropy, and composition. Intensive variables like temperature and chemical potential can obviously not be homogenized and their gradients establish a disequilibrium that maintains motion. The assumption of widely uniform composition and entropy is consistent with a very large Rayleigh number (values of 10^{29} and more are considered) and the smallness of the lateral variations of temperature at the top of the Earth's core supports this assumption.

Recent experimental work (Fei et al. 1997, 2000; Chudinovskikh and Boehler 2007) have identified two important aspects of the Fe–FeS phase diagram that apply to the smaller Earth-like planets and satellites (e.g., Mars, Moon, Mercury, the Galilean satellites, and Titan): (1) at pressures lower than 14 GPa the eutectic melting temperature decreases with increasing pressure, and (2) at pressures lower than 40 GPa the eutectic sulfur concentration decreases with increasing pressure.

A melting temperature with a negative slope will have profound implications. Other than at the elevated pressure of the Earth's core where the eutectic melting temperature increases with pressure, Fe may precipitate at the CMB rather than in the center and may fall as iron snow (Fig. 9 below shows a schematic representation of the Fe–FeS melting diagram and compares a snow model for Ganymede with a convectional inner core growth model). Note, however, that experimental data are only available to date for pure iron and eutectic Fe–FeS. Models require interpolations between these two curves that must remain speculative to some extent. For instance, it is not certain at which composition the positive slope of the Fe-rich liquidus will turn into a negative one. Bland et al. (2008) argue that the transition may occur at 3 weight-% of sulfur.

Two scenarios for a Fe-snowing core can be envisioned depending on whether the core is in thermal equilibrium or not: In equilibrium (or close to equilibrium), the sinking iron will initially remelt at greater depths. As a consequence, the concentration of Fe will increase with depth in the precipitation zone, a stable density gradient will form, and the liquidus and core temperature profile will become co-linear. The layer below the precipitation zone will be well mixed and will become increasingly iron rich with time. With further cooling, the precipitation zone will grow in thickness. When it extends through the entire core, a solid inner core will form upon further cooling. In the second scenario, rapid crystal sinking—possibly supported by convective overturn induced by lateral differences in crystal concentration—and slow melting kinetics may prevent equilibrium (Stewart et al. 2007). A metastable agglomerate of iron will then form a solid iron inner core.

The iron snow model has been applied to Mars (Stewart et al. 2007), Mercury (Chen et al. 2008), and Ganymede (Hauck et al. 2006; Bland et al. 2008). Whether or not iron snow can drive a dynamo in these planets is still unclear, however, and controversially discussed. No magneto-hydrodynamic dynamo models driven by iron snow have been published yet.

A further variant of the snowing core model has been discussed in particular for Ganymede. If the composition of the core were sulfur-rich, iron sulfide (FeS) would precipitate instead of Fe. Solid FeS has a density lower than that of liquid Fe–FeS in the pressure range of Ganymede's core (Fei et al. 1995; Balog et al. 2003) and would float upward from the deep core to form a solid FeS layer on top of the core (Hauck et al. 2006). Hauck et al. (2006) speculate that this process may also drive a dynamo but the model has not been studied in much detail.

Another possibility for compositional convection introduced by Stevenson (1983) and presented at several conferences thereafter is related to exsolution of light elements such as Si, O, S, Mg upon cooling of the core. The solubility of light elements in iron generally decreases with temperature. Considering the large amount of cooling of the core since the formation of the Earth (as discussed in Sect. 5.1), it is conceivable that light elements were exsolved if the core started not too far from saturation. On the other hand, the pressure of equilibration between core and mantle materials is between 20 and 40 GPa (e.g., Thibault and Walter 1995; Li and Agee 1996), much lower than the pressure in the core. It is thus possible that during differentiation liquid iron blobs incorporated light elements at low pressures that would later not be in equilibrium at the high pressures of the core. Experimental studies of core-mantle chemical interaction (Knittle and Jeanloz 1991; Goarant et al. 1992; Ozawa et al. 2008), however, tend to show that the core could readily dissolve a larger amount of light elements than is consistent with its observed density. The exsolution scenario, therefore, seems unlikely to work for the Earth, at least if the most commonly discussed ones (Si, O, S) are considered. However, should light elements reach saturation (Stevenson mentions Mg), the density difference involved would be large and could drive strong flows and possibly a dynamo.

4 The Mantle Reigns the Core

4.1 Mantle Convection

In the foregoing section, we have discussed how dynamo action depends on the cooling of the core. Since the core heat must flow through the mantle, it is heat transfer through the mantle that regulates the cooling of the core. A discussion of magnetic field generation in planetary cores must, therefore, include a discussion of mantle heat transfer and the thermal evolution of the planet. It is widely agreed that mantle heat transfer is mostly by convection and heat conduction. To model heat transfer by convection two methods are mostly used: the first is based on numerical solutions of the hydrodynamic field equations. The second uses semiempirical relations between the heat transfer rate and the vigor of mantle convection.

4.1.1 Finite Amplitude Convection Models

With finite amplitude convection models in 2D or 3D geometry, sets of nonlinear differential equations are solved that include conservation of mass, energy, and momentum balance equations and an equation of state subject to suitable boundary conditions. These models yield velocity and temperature fields as functions of time, and thus provide details of the mantle flow and the associated heat transport. In most convection models, the mantle is considered as an incompressible and highly viscous fluid for which inertia forces in the momentum equation can be neglected. Moreover, the Boussinesq approximation is most often taken, for which the density is assumed constant except for in the buoyancy term. An extended version of the Boussinesq approximation includes the effects of viscous heating as well as adiabatic cooling and heating (see Schubert et al. 2001). (Note that the term adiabatic is conventionally used in the convection literature instead of isentropic. An adiabatic state is equivalent to the isentropic state in the absence of contributions to entropy other than from heat flow across the boundaries.) The mantle may be compositionally layered or mantle layers may be separated by phase transition boundaries. Some models (e.g., Stegman et al. 2003a, 2003b) consider gradients in composition and include mass transfer equations to model thermo-chemical convection. In the following, we present the non-dimensional equations for Boussinesq thermal convection models (compositional variations are not considered).

The conservation of mass, momentum, and energy equations read

$$\nabla \cdot \mathbf{u}' = 0, \tag{20}$$

$$\nabla p' = \nabla \cdot \left(\eta'\left(\nabla \mathbf{u}' + \{\nabla \mathbf{u}'\}^{\mathrm{T}}\right)\right) + Ra T' \mathbf{e}_{\mathrm{r}}, \tag{21}$$

$$\frac{\partial T'}{\partial t'} + \mathbf{u}' \nabla T' = \nabla^2 T' + \frac{Ra_{\mathrm{H}}}{Ra}, \tag{22}$$

where all primed quantities are nondimensionalized and scaled, ∇ denotes the nabla operator, η' the viscosity, $\mathbf{u}'$ the velocity vector, p' the dynamic pressure, $\{\ \}^{\mathrm{T}}$ the tensor transpose, T' the temperature, t' the time, and $\mathbf{e}_{\mathrm{r}}$ the unit vector in radial direction. The Rayleigh number measures the buoyancy term relative to the retarding effects of viscosity and buoyancy loss by heat conduction, and thus the strength of the convection. For a fluid heated from below, it is defined as

$$Ra = \frac{\alpha \rho_{\mathrm{m}} g \Delta T d^3}{\kappa \eta_{\mathrm{ref}}} \tag{23}$$

and for a fluid heated from within as

$$Ra_{\mathrm{H}} = \frac{\alpha \rho_{\mathrm{m}} g Q_{\mathrm{m}} d^5}{k \kappa \eta_{\mathrm{ref}}} \tag{24}$$

with ρ_{m} the mantle density, κ the thermal diffusivity, $d = r_{\mathrm{p}} - r_{\mathrm{c}}$ the mantle or layer thickness, r_{p} the planetary radius, r_{c} the core radius, ΔT the super-adiabatic temperature difference across the convecting mantle, η_{ref} the reference viscosity, Q_{m} the volumetric heating rate in the mantle (heat produced by the decay of radioactive elements), and k the thermal conductivity. The ratio of Ra_{H}/Ra is a dimensionless measure of the amount of internal heating.

Typical boundary conditions are free-slip mechanical boundary conditions at the core-mantle interface (CMB) and no slip or free-slip conditions at the surface. For the temperature boundary conditions, a constant temperature is usually assumed at the surface and a constant temperature or a constant heat flow at the CMB.

The viscosity is the most important parameter for understanding the role of mantle convection in transporting heat. The temperature dependence of the viscosity acts as a thermostat to regulate the mantle temperature. In addition to the temperature dependence, the rheology in a planetary mantle can be described by two main creep mechanisms: diffusion creep and dislocation creep. For the case of diffusion creep, the solid behaves as a Newtonian fluid where the viscosity is independent of the applied shear stresses. In contrast, for dislocation creep, the solid behaves as a non-Newtonian fluid where viscosity varies with the applied shear stress. Indeed, viscosity tends to decrease with increasing shear stress, often nonlinearly. It is not certain which creep mechanism is valid in terrestrial mantles. Most laboratory studies of mantle deformation have concluded that dislocation creep is the applicable deformation mechanism in the upper Earth mantle and diffusion creep in the lower mantle (Schubert et al. 2001). However, this observation is not consistent with post-glacial rebound studies that favor diffusion creep also for the upper mantle. Moreover, laboratory experiments have suggested that the pressure-dependence of viscosity cannot be neglected for a terrestrial mantle (Karato and Rubie 1997). Thus, the viscosity of a terrestrial mantle can be described with the following Arrhenius relationship:

$$\eta = \frac{\mu^n}{2A}\left(\frac{1}{\tau}\right)^{n-1}\left(\frac{h}{B^*}\right)^m \exp\left(\frac{E+pV}{RT}\right) \tag{25}$$

with μ the shear moduls (~80 GPa). A is a preexponential factor, τ the shear stress, h the average grain size, m the grain size exponent, B^* the length of the Burgers vector (~0.5 nm), E the activation energy, p the pressure, V the activation pressure, and R the gas constant. For a Newtonian rheology, n is equal to 1 and for a non-Newtonian rheology a typical value of n is 3.5. For most numerical studies, however, an exponential viscosity law, termed the Frank–Kamenetskii approximation, is used.

$$\eta = \frac{f}{\tau^{n-1}}\big(\exp(-\gamma T') + \ln(\Delta\eta_\mathrm{p})(r'_\mathrm{p} - r')\big), \tag{26}$$

where f is a constant, γ is related to (25) through $\gamma = E/RT_i'^2$ for purely temperature-dependent viscosity, and $\Delta\eta_\mathrm{p}$ is the viscosity increase due to pressure.

Convection Regimes The convection pattern and the heat transfer rate through a planetary mantle vary significantly with the vigor of convection measured by the Rayleigh number and the viscosity variation $\Delta\eta = \exp(\gamma)$ across the mantle. Depending on the viscosity contrast, convection may reach to the top (and cold) surface or a stagnant lid may form underneath that surface. Four different regimes have been identified (e.g., Hansen and Yuen 1993; Solomatov 1995; Trompert and Hansen 1998; Huettig 2009) (Fig. 1):

The *mobile regime* is typical for convection in an isoviscous fluid or for convection in a fluid with sufficiently low viscosity contrast. It is characterized by the flow reaching all the way up to the surface. The surface is said to be fully mobilized in this regime.

The *sluggish regime,* sometimes called transitional regime, is typical for convection in fluids with moderate viscosity variation and for dominantly bottom heated convection in plane layers. The surface is almost stagnant here due to the high viscosity of the fluid near the cold surface.

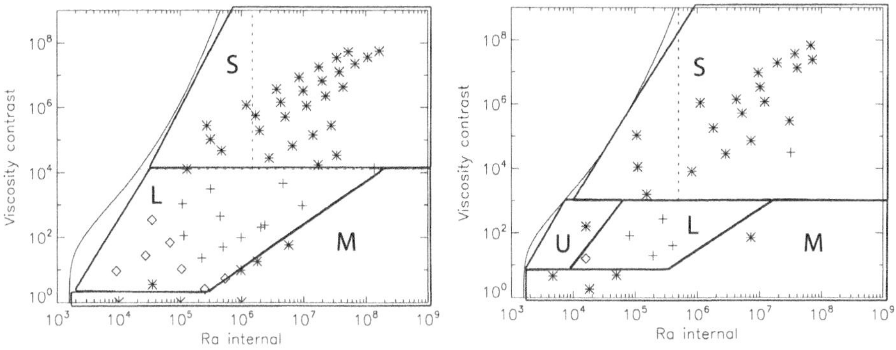

Fig. 1 Convection platform regimes are shown in the Ra_{H} (see (24)) and $\Delta\eta$ parameter plane for internally heated spherical shells with temperature dependent viscosity (*left*) and temperature and pressure dependent viscosity (*right*). (The pressure increases linearly thorough the layer and the variation of the viscosity due to pressure variation is a factor of 100.) M denotes the mobile regime, U the sluggish regime, L the low-degree regime, and the S the stagnant-lid regime. *Crosses* mark degree-one, *diamonds* degree-two, and *stars* higher degrees of convection. The *vertical dashed line* divides the stagnant lid regime further into a steady regime (to the left of the line) and a time dependent regime (to the right of the line) (From Huettig 2009)

Table 2 Flow law parameters for olivine (Karato and Wu 1993). Dry refers to water-free and wet to water-saturated conditions, respectively

	Dislocation creep		Diffusion creep	
	Dry	Wet	Dry	Wet
A [s^{-1}]	3.5×10^{22}	2.0×10^{18}	8.7×10^{15}	5.3×10^{15}
N	3.5	3.0	1.0	1.0
M	0	0	2.5	2.5
E [kJ mol^{-1}]	540	430	300	240
V [cm^3 mol^{-1}]	10–25	10–20	6	5
h [m]	–	–	10^{-2}–1	10^{-2}–1

The *low-degree regime* is typical of convection in a fluid mostly heated from within and in spherical shells. This regime may entirely replace the sluggish regime in fluids with no or pressure dependence of viscosity or may replace it at high Rayleigh numbers for pressure dependent viscosity. The regime is characterized by long wavelengths flow patterns.

The *stagnant-lid regime* occurs for strongly temperature dependent viscosity fluids. The near surface fluid here is completely stagnant and does not participate in the convection. It forms a lid and most of the viscosity variation occurs in the lid. The fluid underneath the lid is almost isoviscous. The thickness of the lid increases with increasing viscosity contrast but is also a function of the heat flow through the entire layer. Since the lid is stagnant, heat transfer through the lid is by thermal conduction.

Applying the parameter values assumed characteristic of the rheology of mantle rock (Table 2), the viscosity contrast across the mantles of the terrestrial planets is several orders of magnitude. Thus, all terrestrial planetary bodies even the Earth should fall into the stagnant lid regime. The existence of plate tectonics on Earth cannot be understood from the temperature dependence of mantle viscosity alone and requires the consideration of additional factors such as strain localization and weakening to form plate boundaries. For the Earth, this is achieved by faulting in the brittle and elastic upper part of the lithosphere and by ductile shear localization below this layer. The latter is thought to be the most important mechanism for generating weak plate boundaries and occurs through grain size reduction, void-volatile weakening or viscous dissipation (for a review, see Tackley 2000a). Two simplified rheology models—strain rate weakening (e.g., Bercovici 1998) and plastic yielding (e.g., Tackley 2000b)—have been used to model plate tectonics. In strain rate weakening, stress initially increases with increasing strain rate and reaches a maximum at some critical rate. In plastic yielding, the viscosity is rapidly decreased beyond a transition stress. Although substantial improvements in the modeling of plate tectonics have been made in recent years, convection models with self-consistent plate generation are still in their infancy. The models are, for instance, not capable of simulating one-sided subduction, i.e., the subduction of one plate underneath another, a common feature on Earth and do not use realistic values of the yield strength.

Thermal Evolution Models with 2D or 3D Convection Models Numerical solutions of (20) to (22) are often used to study the steady state or quasi-steady state of mantle flow. The typical model runs then determine, e.g., the flow structure, the heat flow (Nusselt number), the boundary layer thicknesses, and the temperature distribution. Steady state models, however, cannot be directly used to infer the thermal evolution of terrestrial planets. Instead they are useful to explore the various convection regimes and derive appropriate scaling laws. These can be used to parameterize the convective heat transfer rate as discussed in Sect. 4.1.2 below.

Other numerical solutions include time dependence to model the thermal evolution of terrestrial planets (e.g., Steinbach and Yuen 1994; Konrad and Spohn 1997; Conzelmann 1999; Spohn et al. 2001c; Buske 2006; Nakagawa and Tackley 2004; Xie and Tackley 2004; Butler et al. 2005; Costin and Butler 2006; Ziethe et al. 2009; Keller and Tackley 2009). These models require some modifications with respect to steady-state models in the parameters and the boundary conditions, in particular, if the core is to be included. The internal heating rate Q_m cannot be taken constant but should decrease exponentially with time ($Q_\mathrm{m} = Q_0 \exp(-\sigma_\mathrm{r} t)$) with the average decay rate of the radioactive elements σ_r and the initial value Q_0. Furthermore, the temperature at the CMB is not constant but decreases according to how much heat is transported out of the core by the mantle. To obtain the core-mantle boundary temperature, an energy equation for the core needs to be solved (compare also (29) below). The time rate of change of internal energy of the core is equated to the heat flow out of the core. If inner growth is neglected, the following dimensionless equation results

$$\frac{\mathrm{d}T'_\mathrm{cmb}}{\mathrm{d}t'} = \frac{3}{\varepsilon_\mathrm{c} r'_\mathrm{c}} \frac{\rho_\mathrm{m} C_\mathrm{m}}{\rho_\mathrm{c} C_\mathrm{c}} \frac{\mathrm{d}T'_\mathrm{cmb}}{\mathrm{d}r'}, \tag{27}$$

where T'_cmb is the nondimensional core-mantle boundary temperature, C_m and C_c the mantle and core heat capacities, respectively, r'_c the nondimensional core radius, and ε_c is the ratio between the core temperature that is representative of the internal energy of the core and T_cmb.

4.1.2 Parameterized Convection Models

Most thermal evolution models of terrestrial planets that have been calculated to date use the approach of parameterized convection, however. These models use parameterizations of the convective heat transfer rate derived from finite amplitude convection models or from laboratory experiments (see Schubert et al. 2001 for a discussion). The method is restricted to the calculation of global properties such as the mean mantle and core temperature, the

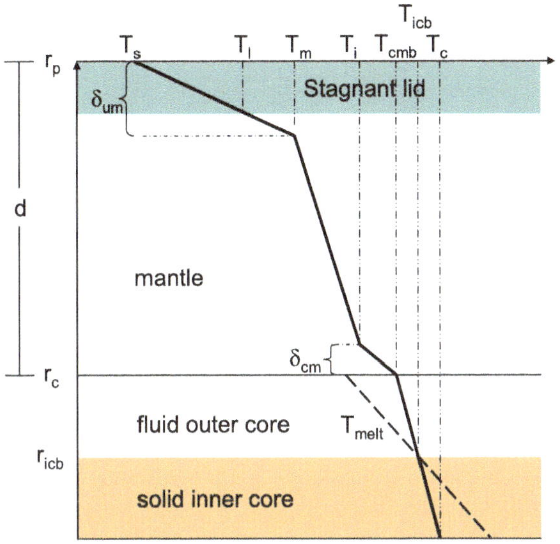

Fig. 2 Thermal structure of the interior of a planet or satellite with notations used in the text. The *bluish region* is the stagnant lid, the *bold solid line* marks the temperature profile, the *long dashed line* marks the melting temperature in the core, and the *orange region* is the solid inner core. *Dash-dotted lines* indicate specific temperature and layer thicknesses. T_s is the surface temperature. T_l is the temperature at the base of the stagnant lid and T_m is the temperature at the base of the upper thermal boundary layer. T_i is the temperature at the top of the lower thermal boundary layer and T_{cmb} is the temperature at the core-mantle boundary which coincides with the base of the lower thermal boundary layer of mantle convection. T_{icb} is the temperature at the top of the inner core and T_c is the central temperature. δ_{um} is the thickness of the upper and δ_{cm} that of the lower thermal boundary layer of mantle convection. r_p, r_c, and r_{icb} are the radii of the surface, the core, and the inner core, respectively, and d denotes the thickness of the mantle

average surface and core-mantle heat flow, the lithosphere thickness and the mean mantle velocity as functions of time only. These models have the advantage of not being limited by available computer power and extensive parameter searches can be performed.

In the following, we briefly present the basic equations and methods used for parameterized convection models. In the next section, we will discuss applicable scaling laws. Detailed discussions of the equations and the methods used to derive them can be found in the literature cited below. Some models include the growth of a crust by mantle partial melting and differentiation and the associated redistribution of radioactive elements (e.g., Hauck et al. 2002, 2004; Breuer and Spohn 2003; Schumacher and Breuer 2006; Grott and Breuer 2008). These, however, are not considered in the present work.

The basic equations setting up a thermal evolution model are energy balance equations for the mantle and the core (see also Fig. 2 for a definition of relevant variables and parameters). The energy equation for the mantle is

$$\rho_m C_m V_m \varepsilon_m \frac{dT_m}{dt} = -q_m A_m + Q_m V_m \tag{28}$$

with V_m and A_m the volume and the surface of the mantle, respectively. T_m is the mantle temperature at the bottom of the thermal boundary layer, ε_m is the ratio between the mantle temperature that is representative of the internal energy of the mantle and T_m and q_m is the heat flow out of the mantle.

The energy equation of the core is

$$\rho_c C_c V_c \varepsilon_c \frac{dT_{cm}}{dt} = -q_c A_c + (L + E_g)\frac{dm_{ic}}{dt} \tag{29}$$

with V_c and A_c the volume and the surface of the core, respectively. T_{cm} is the temperature at the core-mantle boundary, ε_c is the ratio between the core temperature that is representative of the internal energy of the core and T_{cm} and q_c is the heat flow out of the core. The second term on the right side describes the energy released upon growth of an inner core of mass m_{ic}, with the latent heat L and the gravitational energy E_G that is released if the inner core composition is denser than the outer core. To solve for dm_{ic}/dt, the melting temperature and the adiabat of the core as a function of pressure are required. The radius of the inner core is at a melting temperature (compare Fig. 2 and Breuer et al. 2007).

The heat flow out of the convecting mantle can be calculated from

$$q_m = Nu\frac{k\Delta T}{d} = k\frac{\Delta T_{sm}}{\delta_{um}} \tag{30}$$

with Nu the Nusselt number, i.e., the dimensionless heat flux out of the convecting layer, δ_{um} the thicknesses of the upper thermal boundary layer (including the stagnant lid), $\Delta T = \Delta T_{sm} + \Delta T_{cm}$ with $\Delta T_{sm} = T_m - T_s$ the temperature difference across the upper thermal boundary, $\Delta T_{cm} = T_{cm} - T_i$ the temperature difference through the lower thermal boundary layer, T_s the surface temperature, and T_i the temperature at the top of the lower thermal boundary layer (Fig. 2).

The temperature T_i is given by the adiabatic temperature increase through the mantle with $T_i = T_m + \Delta T_{ad} = T_m + \alpha g T_m(d - \delta_{um} - \delta_{cm})/C_m$ where δ_{cm} is the thickness of the lower thermal boundary layer. The adiabatic temperature gradient can be calculated from (7).

The core-mantle heat flow q_c can be calculated from a local stability criterion (e.g., Stevenson et al. 1983; Deschamps and Sotin 2000)

$$q_c = \frac{k\Delta T_{cm}}{\delta_{cm}}, \tag{31}$$

$$\delta_{cm} = \left(\frac{\eta_{cm}\kappa Ra_\delta}{\alpha\rho g\Delta T_{cm}}\right)^{1/3}, \tag{32}$$

where η_{cm} is the average viscosity of the core-mantle boundary and Ra_δ is the lower thermal boundary layer Rayleigh number.

To solve the equations above, we need to apply scaling laws from finite amplitude models which will be discussed in the next section.

Scaling Laws for Isoviscous Convection (Plate Tectonic Planet) Thermal history models for the plate tectonics planet Earth usually apply the scaling laws derived for fluids with constant viscosity which seem to work quite well (Schubert et al. 2001). Although the viscosity in the Earth varies similarly with temperature as in Mars or Venus, this is in line with the mobile lid regime requiring small viscosity variations. Apparently, from the (limited) point of view of the theory of convection with temperature dependent viscosity, Earth behaves as if it were a constant viscosity planet; what causes the Earth to behave that way is still not completely understood.

From classical boundary layer theory, we have (e.g., Turcotte and Oxburgh 1967; Roberts 1979)

$$Nu = a \cdot Ra^{\beta}, \tag{33}$$

where as applied to the mantle $Nu \equiv q_{\mathrm{m}} d / k \Delta T$. Equivalently, the thickness of the upper boundary layer is

$$\delta_{\mathrm{um}} = d \left(\frac{Ra_{\mathrm{cr}}}{Ra} \right)^{-\beta}. \tag{34}$$

The constants a and β need to be determined from either numerical finite amplitude convection calculations or experiments and depend on the boundary conditions and the heating mode. For instance, $a = 0.258$ and $\beta = 0.321$ for free slip boundary conditions at both the upper and lower boundary, $a = 0.336$ and $\beta = 0.252$ for no-slip and free slip conditions at the upper and lower boundary, respectively, and $a = 0.339$ and $\beta = 0.223$ for no-slip condition at both the upper and lower boundaries (Deschamps and Sotin 2000).

Ra_{cr} in classical boundary layer theory is the critical Rayleigh number for the onset of convection. Empirical scaling laws find a factor of proportionality that is numerically close to the critical Rayleigh number of about 500 for the onset of convection in plane layers and spherical shells (Schubert et al. 2001).

For constant viscosity and free slip boundary conditions, the critical Rayleigh number for the lower thermal boundary layer, Ra_{δ}, can be taken as constant with a value of 2.46 although a slight dependence on the Rayleigh number has been observed (Deschamps and Sotin 2000). Note that this value is considerably smaller than the one suggested by the experiments of Booker and Stengel (1978) of 2000 and used for instance by Stevenson et al. (1983). This is likely caused by the no-slip boundary condition in the Booker and Stengel experiments.

Scaling Laws for Strongly Temperature Dependent Viscosity (Stagnant Lid Planet) The heat transfer scaling laws used for models of the thermal evolution of one-plate (or stagnant lid) planets have evolved considerably over the past few decades. Early models applied the scaling laws (33) and (34) for fluids with constant viscosity (e.g., Sharpe and Peltier 1978, 1979; Schubert 1979; Schubert et al. 1979; Stevenson et al. 1983) but used the temperature at the base of the lithosphere (about 1100 K) as surface temperature T_{s}. Later, the models were modified to include the effects of a growing lithosphere by Schubert and Spohn (1990). Their model as well as the later models of Spohn (1991) and Schubert et al. (1992) solved the one-dimensional heat conduction equation with a Stefan-like boundary condition for the rate of growth of the lithosphere. The heat flow from the mantle was calculated using the conventional constant viscosity parameterization. The base of the lithosphere in these models is an isotherm assumed to be characteristic for the transition from viscous to rigid response to loads applied over geologic timescales. This lithosphere can be identified with a rheological lithosphere with a representative value of the isotherm of around 1100 K.

The scaling laws for stagnant lid convection that are used in present models of the thermal evolution of one-plate planets were pioneered by Christensen (1985), Solomatov (1995), Davaille and Jaupart (1993), and Grasset and Parmentier (1998). The scaling laws differ slightly between the parameterizations but give similar results and in all cases the Nusselt number depends on the viscosity contrast in addition to its dependence on the Rayleigh number. We use the formulation of Solomatov (1995) in the following. The *Nu*–*Ra* scaling law for stagnant lid convection is then

$$Nu = a \Theta^{-c} Ra_{\mathrm{i}}^{\beta} \tag{35}$$

with Ra_i the Rayleigh number based on the viscosity at the base of the upper thermal boundary layer and the Frank–Kamenetskii parameter

$$\Theta \equiv \frac{E}{RT_\mathrm{m}^2}\Delta T = \gamma \Delta T. \tag{36}$$

The constants a, c, and β depend on the rheology (Newtonian or non-Newtonian), the heating mode (bottom, from within, or mixed) and the geometry. For example, $a = 0.67$, $b = 1.33$, and $\beta = 0.333$ for internal heating, Newtonian rheology and spherical geometry (Reese et al. 2005). Parameter values for a variety of models have been compiled from numerical calculations by Solomatov and Moresi (2000).

The temperature at the base of the stagnant lid T_l is not constant as in the growing lithosphere models described above but is a function of the temperature of the underlying mantle T_m, and the rate of change of viscosity with temperature:

$$T_\mathrm{l} = T_\mathrm{m} - a_\mathrm{rh}\left(\frac{\mathrm{d}\ln\eta}{\mathrm{d}T}\right)^{-1} = T_\mathrm{m} - a_\mathrm{rh}\frac{RT_\mathrm{m}^2}{E} \tag{37}$$

with the constant $a_\mathrm{rh} = \ln 10 \approx 2.21$ for a Newtonian fluid (e.g., Solomatov 1995). Equation (37) has been derived by observing that in stagnant lid convection most of the viscosity variation occurs in the lid while the convecting layer underneath supports a viscosity variation by only about a factor of 10.

The lower thermal boundary layer critical Rayleigh number Ra_δ has been found to depend on Ra_i (Deschamps and Sotin 2000)

$$Ra_\delta = 0.28 Ra_\mathrm{i}^{021}. \tag{38}$$

Thermal Evolution: Characteristics for Plate Tectonic and Stagnant Lid Planets If we apply the parameterized convection models described above to generic planets in either the stagnant lid or plate tectonic regime, we find characteristically different cooling behaviors (compare Fig. 3). We use parameters applicable to Mars in this comparison (see Breuer and Spohn 2006 for the chosen parameter values).

The stagnant-lid planet cools mainly from above by thickening its lithosphere (stagnant lid plus upper thermal boundary layer), to about 300 km in the present model. The interior remains hot—the mantle temperature increases first and decreases thereafter but after 4.5 Ga has about the same value as the initial temperature (The isotherm moves deeper into the interior of the planet). The core-mantle boundary temperature also increases early on but after 4.5 Ga is about 50 K below its starting value. The interior of a plate tectonics planet, on the other hand, cools quite efficiently. The present day mantle temperature is about 450 K lower than the initial temperature and the core-mantle boundary cools by 600 K. The thermal boundary layer at the surface reaches a thickness that is only a few tens of kilometers smaller than the stagnant lid thickness. These differences in the efficiency of interior cooling have substantial consequences for the thermal evolution of the core. For the one-plate planet, the core temperature is about 550 K above the core temperature of the plate tectonics planet. As a consequence, core freezing and a compositionally driven dynamo will start much earlier for the plate tectonics planet. In addition, an early thermal dynamo would be active for a longer time in a planet with plate tectonics than in a planet with stagnant lid convection. In general, a planet undergoing plate tectonics is more likely to generate a magnetic field than a planet with stagnant lid convection.

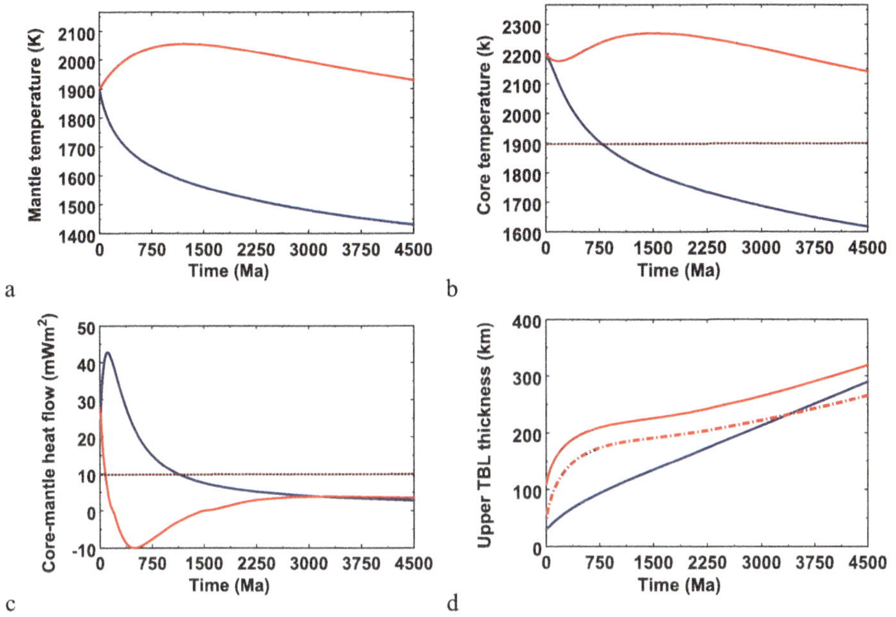

Fig. 3 Mantle temperature (**a**), core-mantle temperature (**b**), core-mantle heat flow (**c**), and thickness of the upper thermal boundary layer (**d**) are shown as functions of time for two tectonic mechanisms: stagnant lid (*red line*) and plate tectonics (*blue line*). The *dashed horizontal line in panel* (**b**) indicates the temperature at the CMB below which an inner core starts to grow at the center. The *dashed horizontal line in panel* (**c**) indicates the heat flow conducted along the isentrope (or adiabat) above which the core cools through thermal convection. The *solid red line in panel* (**d**) is the thickness of the upper thermal boundary layer including the stagnant lid and the *dashed red line* is the stagnant lid thickness

5 Specific Evolution Models

5.1 Earth

The Earth is obviously the most studied of the terrestrial planets and serves as a test case for theories. Yet, many parts are still not understood and there is room for improvements. Many constraints are available for the Earth, concerning the structure, the present dynamics and the long term evolution of the core.

In terms of structure, the most important aspect is the existence of a solid inner core and a liquid outer core with the size of the inner core being very well known (Sect. 2). Moreover, the average stratification of seismic velocities, density, gravity, isentropic bulk modulus in the core is rather well constrained from seismology which helps to fix most parameters in the average state. Recent advances on the structure of the base of the mantle, in particular, the discovery of a new post-perovskite mineralogical phase may explain the seismic discontinuities in the "D" layer and may allow an estimate of the heat flow out of the core (Hernlund et al. 2005; Lay et al. 2008).

In terms of present dynamics, the Earth magnetic field has been systematically measured for a few centuries and high precision global maps have been obtained using satellites since 1980. Moreover, the determination of the time derivative of the magnetic field, the so-called secular variation, allows computing maps of the flow velocity at the top of the core (e.g., Hulot et al. 2002). This constrains the order of magnitude of fluctuations around the long term evolution of the averaged state and justifies the use of a separation of time scales as discussed in Sect. 3 above. Moreover, the time dependent magnetic field can be used to constrain dynamo models and to help derive scaling laws (Christensen and Tilgner 2004; Christensen and Aubert 2006; Aubert et al. 2009).

Finally, the long term evolution is constrained by paleomagnetic data that provide some insight into the characteristics of the main magnetic field in the past. The first important result from these studies is a proof of the existence of an early magnetic field. Tarduno et al. (2007) showed convincingly the existence of a magnetic field 3.2 Ga ago, which to date is the oldest well-documented record of the Earth's magnetic field. Paleomagnetic data are difficult to come by; first because the number of available samples decreases with increasing age and second because old samples have often been magnetically reworked. Nevertheless, the available data show that the magnetic field has been in existence without significant (more than 100 Myr) interruption since that time. This observation provides a strong constraint on scenarios for Earth core evolution. For earlier times, beyond 3.2 Ga. b.p., no records are available which does not prove the absence of a magnetic field, of course. A proof of absence would be difficult to come by. Constraints could actually come from the chemistry of the lunar soil: Ozima et al. (2005) found that the amount of nitrogen in the lunar soil can best be explained by the implantation of gases from the Earth atmosphere. This would be easiest if it occurred before the geomagnetic field was strong enough to protect the atmosphere from erosion by interaction with the solar wind. They suggest that before 3.9 Gyr ago, the geomagnetic field was very weak or even nonexistent.

Many models for the evolution of the Earth's core and the geodynamo have been proposed over the years. Two approaches are possible, both with pros and cons. The heat flow at the CMB is controlled by mantle convection and one would ideally develop a coupled self-consistent model to compute the evolution of both the core and the mantle. In addition to the obvious advantage of self-consistency, such a model would allow to use constraints other than those directly related to the core, such as the heat flow at the Earth's surface. A problem with this approach is that Earth's mantle convection by itself is not fully understood and the thermal evolution of the Earth is a subject of debate (e.g., Korenaga 2006; Jaupart et al. 2007; Labrosse and Jaupart 2007, for recent discussions). The heat flow at the CMB is highly uncertain (Lay et al. 2008) and the physics that controls it is controversially discussed (e.g., Labrosse 2002; Moore 2008). Therefore, these models may combine two ill-understood issues and discrepancies between model results and observations could result from both individually and from their interactions. The other option is to impose an evolution of the heat flow at the CMB and study the resulting evolution of the core.

The most comprehensive early study of the coupled mantle-core evolution is undoubtedly that by Stevenson et al. (1983) who calculated thermal evolution models for the mantles and cores of the terrestrial planets. They found that a thermal dynamo most likely was active in the early evolution of the Earth and that the inner core should have started to crystallize some 1.9–2.1 Gyr ago in order to reach the proper size at the present time. The onset of the inner core crystallization is accompanied by the start of compositional convection which leads to a higher ohmic dissipation (Sect. 3.2). Assuming that the large scale magnetic field scales as a square root of the ohmic dissipation, they cautiously speculate about a sharp increase of the core magnetic field upon inner core nucleation. Stevenson et al. caution that the change in the core field—if it occurred—would not necessarily be observable at the surface because of the unknown distribution of magnetic energy between the torroidal and poloidal components of the field. Mollett (1984) used a similar approach and included the effect of radioactivity in the core. The model approach has recently been revisited (Nimmo et al. 2004; Buffett 2002) in view of the progress in dynamo theory that has since been achieved.

As has already been discussed in the preceding section, these models use a constant viscosity parameterization of mantle convection. The increase of computing power allowed the development of fully self-consistent calculations in which a cooling core model is coupled to a dynamical mantle convection model (Nakagawa and Tackley 2004, 2005). These authors find that heat sources in the core equivalent to 100 ppm potassium would be required to sustain a dynamo for 3.5 Ga and to arrive at the presently observed inner core radius. However, their model still has periods where the dynamo is inactive because of fluctuations of the CMB heat flow that occasionally drops below the adiabatic heat flow value. Including too much radiogenic heating would prevent growth of the inner core, or even make it remelt thereby frustrating compositional convection. Butler et al. (2005) and Costin and Butler (2006) also studied a coupled model with a full dynamical mantle and specifically included the effect of a stable layer at the base of the mantle enriched in radioactive elements as well as the effect of potassium in the core. They had the same difficulties in maintaining a dynamo throughout Earth's history while getting the inner core to reach the proper size at the present time. These authors also argue for some potassium in the core to help to solve the issue.

The difficulties faced by self-consistent approaches using either parameterized or full dynamical mantle convection models come mostly from the complexity of mantle convection. Therefore, the other approach of concentrating on the core is still valuable to test simple ideas on its evolution. In this approach, the heat flow at the CMB is imposed a priori, either constant or varying with time, and the evolution of the core and the dynamo is computed. Buffett et al. (1992) showed how to compute the growth of the inner core for a given heat flow at the CMB and found a rapid growth rate. Labrosse et al. (2001) further showed that the onset time for inner core crystallization is the solution of an equation that can be solved for any given heat flow history and concentration of radioactive elements. Because the inner core is a small fraction of the core, about 6% of its mass, the energy needed to be extracted from the core to grow to its present size is small and the required growth can be accomplished in less than 2 Ga. This time can, in principle, be lengthened if radiogenic heating is allowed in the core (e.g., Labrosse et al. 2001; Gubbins et al. 2004). However, a problem arises when the Ohmic dissipation rate is to be maintained. Since radiogenic heat is less efficiently converted into Ohmic dissipation (e.g. Roberts et al. 2003; Lister and Buffett 1995; Gubbins et al. 2003) than gravitational energy released upon inner core growth, the gain in Ohmic dissipation associated with the inclusion of radiogenic heat cannot compensate the loss in Ohmic dissipation associated with a smaller inner core growth rate. Thus, a certain level of Ohmic dissipation would require a certain growth rate, and thus determine the age of the inner core (Labrosse 2003; Nimmo et al. 2004). However, one should note that the rate of Ohmic dissipation required to explain the Earth's magnetic field is still uncertain.

Figure 4 shows the evolution of the radius of the inner core as function of time for an imposed constant CMB heat flow of 9 TW and without radiogenic heating in the core. The energy equation (8), in which all terms can be written as a function of the radius of the inner core r_{ic} multiplied by its growth rate (provided the heat flow at the ICB is included in the secular cooling of the outer core, a very good approximation for a small inner core like that of the Earth, Labrosse et al. 2001), is used to compute the evolution of r_{ic} and of each term in the equation, starting from the present situation backward in time. Then the efficiency equation (18) is used to obtain the evolution of the Ohmic dissipation and each of its contributions. The time evolution of the energy and entropy balances for the same calculation are shown in Fig. 5. All the physical parameters and the details of the calculation can be obtained from Labrosse (2003).

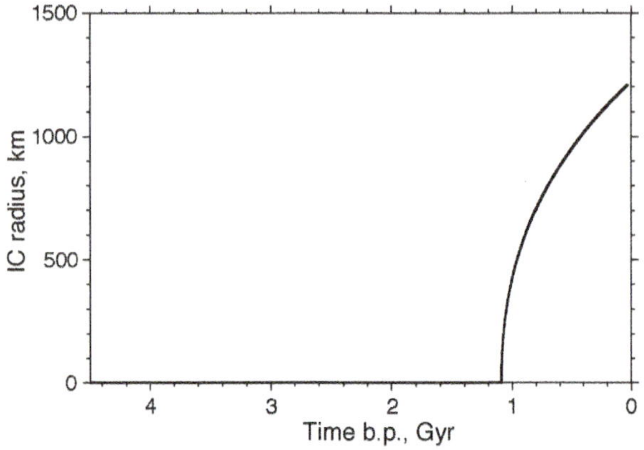

Fig. 4 Radius of the inner core as function of time for a constant CMB heat flow of 9 TW

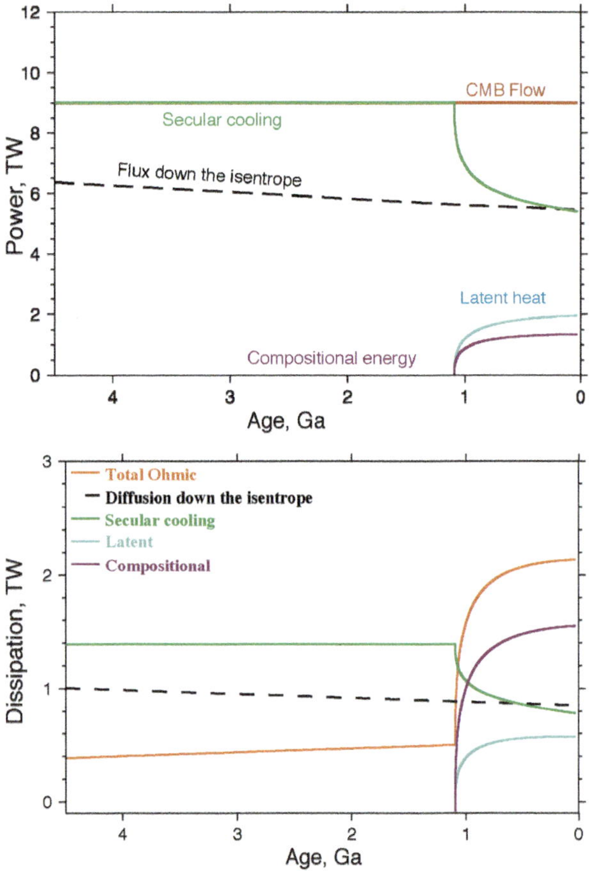

Fig. 5 Evolution of the power (*upper panel*) and entropy (*lower panel*) balances for a constant heat flow at the CMB of 9 TW. Based on the model of Labrosse (2003)

The growth of the inner core radius is found to follow a power law, $c(t) = c_0(t/t_0)^\beta$, with $1/3 < \beta < 1/2$ in this model and the age of the inner core is found to be about 1 Ga. The energy balance shows that when the inner core starts to crystallize, the secular cooling rate decreases as a consequence of the release of gravitational energy and latent heat that are

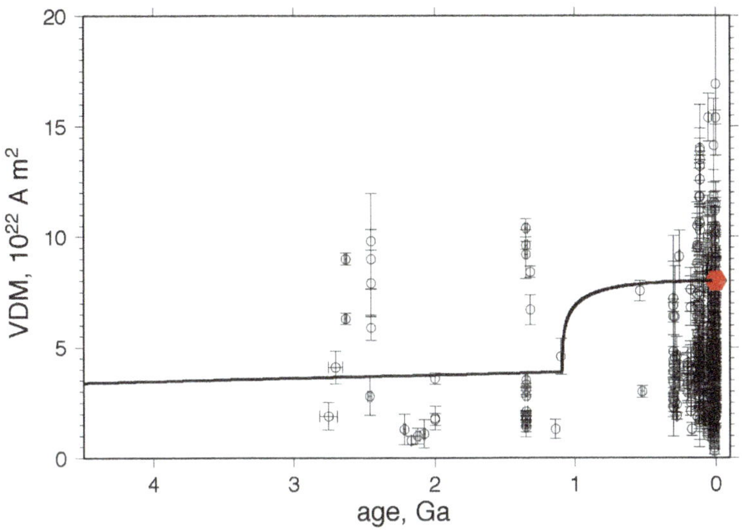

Fig. 6 Dipole moment of the Earth's magnetic field calculated for the models of Figs. 4 and 5 and compared to paleomagnetic data. The *red hexagon* is the present value and the dipole is assumed to scale as the square root of the ohmic dissipation

each presently of order 1 TW. Although gravitational energy contributes the smallest part of the power budget, it provides the largest contribution to the Ohmic dissipation as has been discussed in Sect. 3.2.

The onset of inner core growth results in a sharp increase in Ohmic dissipation, by a factor 4. It is then tempting to speculate on a sharp increase of the magnetic field that would allow the detection of the onset of inner core growth in the paleomagnetic record. However, the link between the Ohmic dissipation rate and the strength of the dipole, which dominates the paleomagnetic signal, is not obvious. The Ohmic dissipation rate is $\phi =\propto J^2 \propto (\nabla \times \mathbf{B})^2$, which means that for a magnetic field spectrum $B(l)$—with length scale l—$\phi(l) \sim B^2(l)/l^2$. If the increase of Ohmic dissipation occurred without affecting the shape of the spectrum, the total dissipation rate would be proportional to the square of the dipole field. On the other hand, dissipation usually occurs at small length-scales and the total dissipation rate could occur by increasing the energy at the smallest scales without changing the large scale dipole field. Stevenson et al. (1983) discussed the former option whereas Stevenson (1983) argued in favor of a large scale magnetic field that was independent of the Ohmic dissipation rate with an Elsasser number close to 1. More recently, the systematic investigation of dynamo models has allowed to develop scaling relationships in a rather wide parameter space (Christensen and Tilgner 2004; Christensen 2006; Christensen et al. 2009; Aubert et al. 2009) suggesting a RMS core magnetic field that scales with the power of $1/3$ of the Ohmic dissipation. This is a slightly smaller power than the scaling exponent of $1/2$ previously used. Figure 6 shows the evolution of the dipole field of Figs. 4 and 5 assuming that it scales as a square root of the dissipation rate and choosing the parameters such that the present dipole moment is obtained. The assumed power law thus maximizes the amplitude of the variations that are still smaller than the short term fluctuations found in the paleomagnetic record. This indicates that a detection of the onset of inner core growth is unlikely to result from studies of paleomagnetic intensities.

Aubert et al. (2009) extended the study of the scaling of the average magnetic field to that of the dipole moment measured at the surface of the planet. They found that the onset of inner core crystallization, although making the Ohmic dissipation larger does not influence greatly the dipole moment because the magnetic field is generated deeper in the core for a compositional dynamo. The two effects balance out to make the dipole almost constant with time. This finding further complicates attempts to interpret Earth's magnetic record in terms of its core dynamics.

5.2 Venus

Since the first passage of Mariner 2 by Venus in 1962 at a distance of 6.6 planetary radii, it was clear that Venus does not have an Earth-like magnetosphere. The upper limit to the dipole moment obtained from the Pioneer Venus Orbiter placed the Venus intrinsic magnetic field at less than about 10^{-5} times that of Earth (Russell et al. 1979a, 1979b).

The lack of a present dynamo does not imply that Venus never had an intrinsic magnetic field although we have no information that relates directly to the past history of the field. The surface temperature of about 740 K is close to the Curie temperature of the main magnetic carrier minerals (magnetite, hematite, and pyrrhotite) of 850, 940, and 600 K, respectively (e.g., Dunlop and Ozdemir 1997). Thus, temperature in the crust—possibly except for a thin near surface layer—is expected to be above the Curie point and any remanent crustal magnetic field remaining from an early period of dynamo activity should be weak if existent at all. It is important to note that the slow rotation of Venus (a Venus day of $\sim$243 Earth days is almost equal to the length of its year of $\sim$224 Earth days and its sense of rotation is retrograde) does not exclude dynamo action. This appears to have been first noted by Hide (1956) and was recently reconfirmed by Christensen and Aubert (2006).

Thermal evolution models predict that there was an early magnetic field for Venus with a dipole moment of the same order as Earth's early field for about the first three billion years of Venus' history (Stevenson et al. 1983). During that time, thermal convection had driven the dynamo as in the other terrestrial planets. After the CMB heat flow had fallen below the adiabatic heat flow, the magnetic field ceased and did not rejuvenate; the core was too hot and the pressure was too low for inner core growth. The evolution models of Stevenson et al. (1983) were calculated using the constant viscosity parameterization and by simulating the stagnant lid through an increased, constant surface temperature of 1075 K. This is the temperature at which silicate rock may be sufficiently weak to flow. The models cool more efficiently than models based on the scaling laws for stagnant lid convection. Assuming that Venus was in a stagnant lid regime throughout the entire evolution, the phase of early dynamo action would have most likely been shorter than predicted by Stevenson et al. because they should have overestimated core cooling.

One alternative scenario for the present lack of a Venus magnetic field is connected to the global crustal resurfacing event about 300–800 Ma ago (Schaber et al. 1992; McKinnon et al. 1997). Stevenson (2002) suggested that Venus's field ceased at that time due to a transition from a plate tectonics to a stagnant lid regime. The transition would have been followed by a period of warming of the mantle and the core and thermal convection and dynamo action in the core would have stopped. Even if Venus had an inner core at the time, inner core growth would have ceased with a warming mantle. The inner core may have started even to decrease in radius. The dynamo cannot be driven by chemical buoyancy in these circumstances.

If, on the other hand, the resurfacing was the consequence of a catastrophic overturn of the mantle or of lithosphere delamination (e.g., Turcotte 1993), Venus may have had a short-lived magnetic field during these times. During the catastrophic overturn, the interior

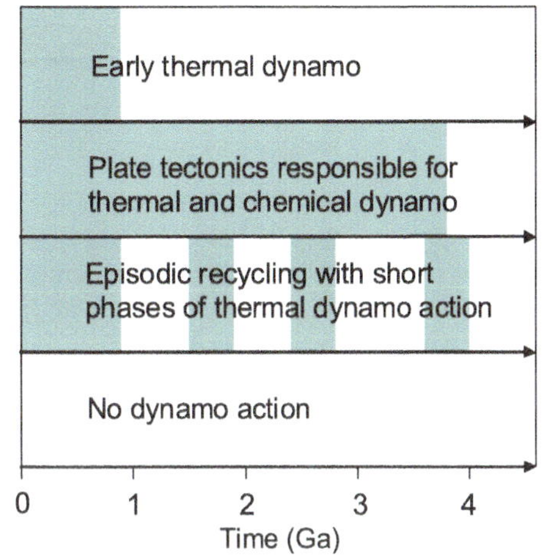

Fig. 7 Four possible scenarios for the timing of internal magnetic field generation in Venus. The *bluish regions* indicate the time of dynamo generation. For further explanations, see the text

may have been cooled strongly and it is possible that the heat flow out of the core exceeded the critical isentropic core heat flow. This could have resulted in a short period of thermal convection and dynamo action in the core. Figure 7 summarizes the various scenarios of magnetic evolution postulated for Venus.

5.3 Mars

Mars today has no internally generated magnetic field, but the presence of a strong magnetization of the oldest parts of the crust (Acuña et al. 1998, 1999, 2001; Connerney et al. 1999) suggests that the planet generated a magnetic field early in its history. The inferred crustal magnetization of up to $\sim 10^{-30}$ A m^{-1}, is an order of magnitude stronger than that of continental rock on Earth (Toft and Arkani-Hamed 1992; Arkani-Hamed and Dyment 1996). The magnetization is comparable in magnitude to the remanent magnetization of fresh extrusive basalt at oceanic ridges (Bleil and Petersen 1983).

One of the most effective ways to magnetize crustal rock is by thermo-remanent magnetization (TRM, Langlais et al. 2009, this issue). With TRM, the magnetization is produced when rock cools below the Curie temperature in the presence of a magnetizing field. There is a trade-off between the concentration of magnetic carriers and the strength of the magnetic field: The lower the magnetic field the more magnetic carriers are required to explain an observed magnetization. Assuming an early Martian magnetic field similar in strength to the present day Earth field, the required concentration of magnetic carriers in the Martian crust would be comparable to that in extrusive basalt. However, there is ample evidence that the FeO content of the Martian mantle is about twice that of the Earth's mantle (Sohl and Spohn 1997; Sanloup et al. 1999). Whether this high concentration of FeO translates to a high concentration of magnetic minerals depends on the oxidation state of the Martian mantle and lower parts of the crust. For a high concentration of magnetic minerals, the early magnetic field could have been smaller than that of the present Earth. In any case, the strength of the Martian magnetic anomalies suggests the existence of an intrinsic Martian magnetic field 0.1 to 10 times as strong as that of the present Earth (Ness et al. 1999, Mitchell et al. 1999, 2001).

The timing of the dynamo places constraints on the thermal evolution of the planet. The surface distribution of the magnetic anomalies suggests that the magnetization event predated the formation of the Hellas and Argyre basins (Connerney et al. 1999, 2004), roughly

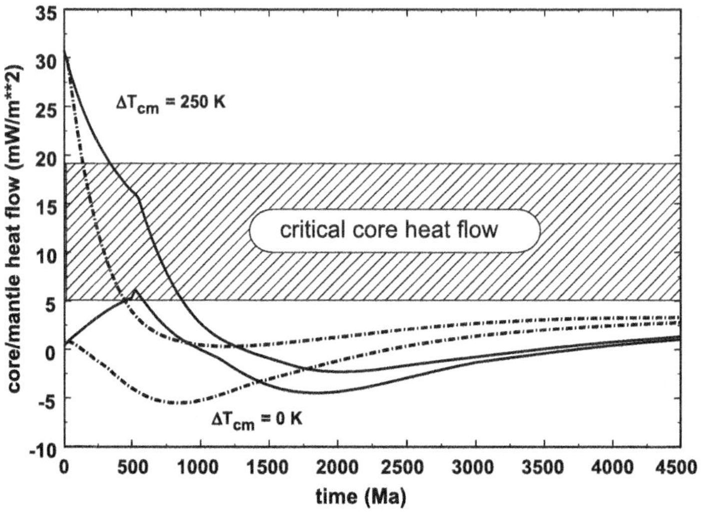

Fig. 8 Core-mantle heat flow as a function of time for models of Mars with early plate tectonics (*solid line*) and with stagnant lid convection throughout the entire evolution (*dash-dotted line*). The initial temperature differences assumed across the core-mantle boundary are $\Delta T_{\mathrm{cm}} = 0$ and $\Delta T_{\mathrm{cm}} = 250$ K (from Breuer and Spohn 2003). An early magnetic field can be generated if the heat flow out of the core is larger than the critical core heat flow. Estimates of the latter range between 5 and 19 mW m^{-2}

at the end of the Early Noachian (Head et al. 2001). It has been alternatively suggested that the dynamo turned on after the giant impact craters formed implying that the remanent magnetization of the crust was acquired later in Martian history (Schubert et al. 2000). This model, however, lacks an explanation for the strong magnetization of parts of the oldest crust.

An early dynamo is consistent with the results of most thermal evolution models that consider the magnetic field history (e.g., Schubert and Spohn 1990; Spohn et al. 1998, 2001b; Hauck and Phillips 2002; Breuer and Spohn 2003, 2006; Williams and Nimmo 2004). The models suggest an early thermal dynamo driven by rapid cooling of a core initially superheated with respect to the mantle (Fig. 8). In the subsequent evolution, the core remains liquid with a heat flow below the critical adiabatic value, and thus no dynamo action is observed. The core may remain completely molten if it contains at least ~5 wt.-% sulfur and assuming a dry mantle rheology (Williams and Nimmo 2004). A core sulfur concentration of 14.2 wt.-% has been suggested based on chemical analyses of the SNC meteorites (Dreibus and Wänke 1985; McSween 1985).

It has been noted by Nimmo and Stevenson (2000) that a superheated core may not necessarily result from early differentiation and core formation. An alternative scenario thus assumes early rapid cooling of the core by plate tectonics and a transition to single-plate tectonics after about 500 Ma (Nimmo and Stevenson 2000; Stevenson 2001). Connerney et al. (2005) have also argued for early plate tectonics. They noted two parallel great faults in the Terra Meridiani region with an offset of magnetic field contours similar to transform faults that occur in oceanic crust on Earth (Connerney et al. 2005). During the phase of plate tectonics, the interior cools efficiently and a thermal dynamo possibly followed by a chemical dynamo is conceivable. Soon after the transition to stagnant lid convection, however, the mantle warms and dynamo action stops (Fig. 8). Another model links the cessation of the Martian dynamo to the large impacts that formed the Hellas and Argyre basins. Roberts et al. (2009) suggested that impact heating associated with the largest basins (diameters >2500 km) could have caused the heat flow at the core-mantle boundary to decrease significantly by 10–40% which would have resulted in the cessation of the Martian dynamo.

One of the main characteristics of the Martian surface is its crustal dichotomy with the old heavily cratered crust in the southern hemisphere and a superficially younger crust in the northern hemisphere. This crustal dichotomy is accompanied by a difference in the magnetic field intensities in the two hemispheres: The northern hemisphere shows weak magnetic signals, whereas the southern hemisphere contains both strong and weak fields (Langlais et al. 2004). Explaining this difference by suggesting that the surface of the northern hemisphere is younger than the southern hemisphere and that it formed after the dynamo turned off is not entirely satisfying. The crust underneath the surface of the northern hemisphere is most likely as old as the crust of the southern hemisphere (Frey 2006). Thus, assuming a typical axial-dipole-dominated magnetic field and the northern and southern crust of similar composition and age, one would expect crustal fields of similar strength in both hemispheres. The magnetic dichotomy, therefore, requires post-dynamo mechanisms in the northern hemisphere that would have allowed removing its early remanent magnetization. Possible explanations are hydrothermal alteration (Solomon et al. 2005) or demagnetization resulting from early large impacts (e.g., Rochette et al. 2003; Mohit and Arkani-Hamed 2004). The strong correlation between the crustal and magnetic dichotomy also in their time of formation (both are believed to have been formed during the first 700 Ma) suggests a connection between the two processes. The crustal dichotomy if formed by internal dynamic processes requires a hemispheric-scale pattern of mantle flow. Several models exist to explain a degree-1 pattern that would also allow sufficiently vigorous core convection to sustain a short-lived dynamo. These include mantle convection in the presence of a deep perovskite phase transition layer close to the core-mantle boundary (Weinstein 1995; Harder and Christensen 1996; Buske 2006), mantle convection with strong radial viscosity variations (Zhong and Zuber 2001; Roberts and Zhong 2006), early magma ocean crystallization resulting in mantle overturn (Elkins-Tanton et al. 2003, 2005), or superplumes resulting from destabilization of the mantle lower thermal boundary layer (Ke and Solomatov 2006).

Degree-1 convection in the early evolution may even shape the magnetic field morphology at the surface. Stanley et al. (2008) show that variations of the core-mantle heat flow as a consequence of a low degree convection pattern could result in a single-hemisphere dynamo. This dynamo would produce strong magnetic fields in the southern hemisphere in case the heat flow out of the core below the southern hemisphere were stronger than that below the northern hemisphere. Such a dynamo would not need any post-dynamo processes to explain the observed magnetic field dichotomy and could also have had its impact on early atmospheric loss processes (Kulikov et al. 2007).

Stewart et al. (2007) have recently proposed that Mars—upon further cooling—may enter the snowing core regime described in more detail in Sect. 3.3. Whether or not convection driven by the falling iron crystals can drive a dynamo is not settled at this point in time, however.

5.4 Mercury

Mercury is exceptional among the terrestrial planets due to its large iron core that comprises about 80% of the planet's radius (e.g., Spohn et al. 2001a) and the presence of a weak magnetic field (Ness et al. 1975). The finding of a weak dynamo field by Mariner 10 has now been confirmed by MESSENGER (Anderson et al. 2008)

It is generally agreed that the Hermean magnetic field is likely generated in the core (Schubert et al. 1988), although a crustal source of the field can presently not be excluded

(Aharonson et al. 2004). Thus, Mercury's magnetic field suggests that the planet's core should at least contain a fluid outer core layer. This conjecture has recently been supported by measurements of variations in Mercury's spin axis (Margot et al. 2007). A totally fluid core—without an inner core—is difficult to reconcile with the interior structure, thermal history models, and the magnetic field. The interior structure models suggest the mantle to be comparatively thin, about 600 km, and thus heat may escape comparatively rapidly from the core. The heat flow from the core that is obtained from thermal history models (e.g., Stevenson et al. 1983; Schubert et al. 1988; Hauck et al. 2004; Buske 2006) is below the critical isentropic heat flow of about 11 $\mathrm{mW\,m^{-1}\,K^{-1}}$ (Schubert et al. 1988) and although the field is rather weak thermal convection in the core seems to be insufficient to drive the dynamo. It is even possible that the present mantle is not convecting and heat is only transported by conduction (Hauck et al. 2004). Thus, compositional buoyancy generated during inner core growth is likely necessary to produce the present day field.

As with the other terrestrial planets, the most likely candidate for the light alloying element in the core is sulfur but its concentration is basically unknown. Early planet formation models concluded that the Hermean core should contain none or only a small amount (a few percent) of sulfur because light elements would not condense at the orbital distance of Mercury close to the sun (e.g., Lewis 1972). Other formation models suggest efficient radial mixing of the accreting material (e.g., Wetherill 1988). Volatiles could thus have been transported from outer regions of the nebula toward the sun with the consequence of a possibly stronger concentration of sulfur in the Hermean core. Thermal evolution models together with the observed lobate scarps that suggest a small planetary contraction of about 2 km since the heavy bombardment of the planet (e.g., Strom et al. 1975) place some additional constraints on the concentration of sulfur in the core. Parameterized convection models by Hauck et al. (2004) assuming a non-Newtonian mantle rheology suggest a value between 5 and 8 wt.-% sulfur whereas the 2D and 3D thermal convection models with Newtonian rheology by Conzelmann (1999) and Buske (2006) can explain the observations with only 2 wt.-% of sulfur. Because of uncertainties in the radiogenic heat source concentration, the mantle rheology and the liquidus, it is difficult to derive firm bounds on the sulfur concentration from these models, however (Breuer et al. 2007).

Thermal evolution models (Stevenson et al. 1983; Schubert et al. 1988; Conzelmann 1999; Hauck et al. 2004; Buske 2006) suggest the following possible scenario for a dynamo-generated magnetic field. In the early evolution—possibly until 1–2 Ga b.p.—the dynamo would have been driven by thermal convection. The thermal dynamo would require the core to have been superheated with respect to the mantle (e.g., Breuer and Spohn 2003). It is possible that the dynamo ceased because the heat flow from the core became subadiabatic before an inner core started to grow. The onset time of inner core growth depends on various mantle convection parameters but increases with increasing sulfur content in the core (Hauck et al. 2004; Breuer et al. 2007). Since the time of the onset of inner core growth the Hermean dynamo should have been active until present if Mercury continued to cool steadily.

Early models of a hydromagnetic dynamo had problems, though, to explain the observed field strength which was found to be too weak in comparison with predictions (Stevenson et al. 1983; Schubert et al. 1988). Recent models have demonstrated that dynamos both in thin outer core shells (relative inner core size larger than about 0.8, Stanley et al. 2005) or in thick core shells (Heimpel et al. 2005) can produce weak magnetic fields, possibly consistent with Mercury's field. An alternative model of a thick shell dynamo has recently been suggested by Christensen (2006). In this model, the dynamo operates deep within the shell and is overlain by a stagnant fluid layer. A nonconvecting outer core layer may arise as a consequence of a subadiabatic CMB heat flow as has been obtained in the thermal evolution models of Hauck et al. (2004) and Buske (2006). The time-dependent magnetic field produced by the dynamo—as was shown by Christensen (2006) would then be strongly attenuated by the skin effect in the nonconvecting upper layer of the outer core.

Other alternative proposals for a weak magnetic field in Mercury include the thermoelectric dynamo (Stevenson 1987; Giampieri and Balogh 2002), the feedback dynamo (Glassmeier et al. 2007), and the snowing core dynamo (Chen et al. 2008). The thermoelectric dynamo makes use of a thermally-derived electromotive force set up at a distorted core-mantle boundary. Such a dynamo requires topography variations of the core-mantle boundary of the order of one kilometer due to mantle convection. The feedback dynamo model considers the feedback between magnetopause currents and a core dynamo. As Glassmeier et al. (2007) have shown for a simple alpha-omega dynamo model, the feedback between a magnetopause close enough to the planet (possibly closer than 2 planetary radii) and the dynamo may result in a weak external field. A variant of the snowing core dynamo model (see Sect. 3.3) has also been proposed for Mercury. Mercury is small enough such that its core is in the relevant pressure range (smaller than 40 GPa). As the inner core grows in a Fe–FeS core, sulfur becomes enriched in the outer core. The outer core composition may come close to the eutectic even if the overall sulfur content of the core is small (a few percent) and iron snow crystals may form near the CMB and sink toward the interior. Whether or not a dynamo could work driven by the falling iron snow is not well established as we have noted before, neither is it clear what the magnetic moment would be. Depending on details of the model, it is conceivable, however, that the power dissipated in the core under these circumstances will be relatively small.

The presently available magnetic field and other geophysical data do not allow a conclusion as to which dynamo may work in the planet. It is not even certain that an inner core exists although most workers in the field would consider that likely. New insights into the magnetic field evolution and the dynamo mechanism of Mercury are expected from future Mercury missions (e.g., Messenger and BepiColombo). In particular, the simultaneous measurements planned for the two spacecraft on the BepiColombo mission promise a better understanding of the sources of the Hermean magnetic field.

5.5 Moon

The Moon presently does not generate an internal magnetic field but paleomagnetic data, combined with radiometric ages of Apollo samples, suggests that a field of possibly 10^4 nT existed at 4.0 Ga decreasing to a few thousand nT at 3.2 Ga (Stephenson et al. 1975; Cisowski and Fuller 1986). Runcorn (1975) has argued that the lunar rocks were magnetized at the time of their formation by a field of internal origin. The easiest explanation for such a field is the operation of a dynamo in an iron-rich lunar core (e.g., Konrad and Spohn 1997; Spohn et al. 2001c; Stegman et al. 2003a, 2003b). The core of the Moon is, however, small (see Table 1 and the discussion in Sect. 2) which has been used as an argument against a dynamo (Runcorn 1975).

Some workers doubt that an internal dynamo is required to explain the magnetization of the Moon, and favor an alternative idea. The observed magnetic signature is suggested to be generated in association with large impacts during the early history (Hide 1972; Hood and Vickery 1984; Hood and Huang 1991; Hood et al. 2001). In this model, magnetization should be concentrated diametrically opposite to major impact basins in the presence of an ambient field. Satellite observations from the Lunar Prospector mission show concentrations of crustal magnetization at the antipodes of some but not all of the large impact structures (Purucker 2008).

Assuming an internal origin of the lunar magnetic field, one must be aware that the paleomagnetic data show a gap of about 500 Ma in the remanent magnetization of the lunar rock. There exist two alternative explanations for this gap: (1) the remanent magnetization of older basaltic rock was destroyed by impact gardening and/or the volcanic activity during the first 500 Ma was very low or (2) the internal field started about 500 Ma after core formation.

An early magnetic field has been suggested from thermal evolution models incorporating 2D and 3 D convection with strongly temperature dependent rheology (Konrad and Spohn 1997; Spohn et al. 2001c). These models show that a thermally driven dynamo might have been active from the beginning of core formation up to about 3 Ga if the core was superheated by about 200 K with respect to the mantle. The models, however, neglect chemical stratification of the mantle due to the freezing of an early magma ocean.

A late onset of a dynamo may have been indirectly caused by the magma ocean (Alley and Parmentier 1998; Stegman et al. 2003a, 2003b). After crystallization of the magma ocean, a dense ilmenite and pyroxene cumulate located just underneath the KREEP layer may have sunk to the deep lunar interior and part of the KREEP layer may have been carried to depth together with the ilmenite layer. This mixed layer strongly enriched in radiogenic elements would have encircled the lunar core thermally insulating it from the rest of the mantle. Heat would have been trapped in the core and prevented the core from cooling and from powering a dynamo. The radiogenic heat sources would have heated the thermal blanket until it became thermally unstable and started to rise to the surface. With the removal of the thermal blanket, the core became able to convect vigorously and could have produced a short-lived magnetic field. An essential element of the model is the timing. The length of time that it takes the thermal blanket to heat up and rise back toward the surface is broadly consistent with two important events in early lunar history—the eruption of the mare basalts onto the lunar surface and the magnetization of lunar rocks. Even so, the model has some shortcomings. A Rayleigh–Taylor instability requires a substantial density difference between the ilmenite layer and the mantle for the ilmenite layer to sink rapidly enough into the deep interior (Parmentier et al. 2002). It is then difficult to heat this layer subsequently such that thermal buoyancy can overcome the compositional density difference barrier.

5.6 Galilean Satellites and Titan

One of the surprising discoveries of the Galileo mission was the detection of a dipolar, self-generated magnetic field at Ganymede (Kivelson et al. 1996), with an equatorial surface strength of about 720 nT (Kivelson et al. 2002). No magnetic signals caused by permanent internal dipoles were found at the other major satellites of Jupiter, Io, Europa, and Callisto. Rather, Europa and Callisto have time varying fields that were interpreted as being caused by induction in subsurface oceans. An induced field component was also suggested for Ganymede since the dipolar field cannot fully explain the data. A small remaining component can either be explained as a quadrupole component of the self-generated magnetic field but best by an induced field (Kivelson et al. 2002). Cassini to date did not find evidence of magnetic fields—induced or self-sustained—for Titan (Backes et al. 2005) although it has been speculated to have an ocean (e.g., Hussmann et al. 2007, for a review).

The Galileo Ganymede magnetic field data were interpreted by Schubert et al. (1996) to conclude that the satellite should have a core and a dynamo. Other magnetic source processes, such as magneto-convection (e.g., Sarson et al. 1997), and remanent magnetism (Crary and Bagenal 1998) were ruled out mostly on the basis of the measured field strength that was too large to be explained by the alternatives.

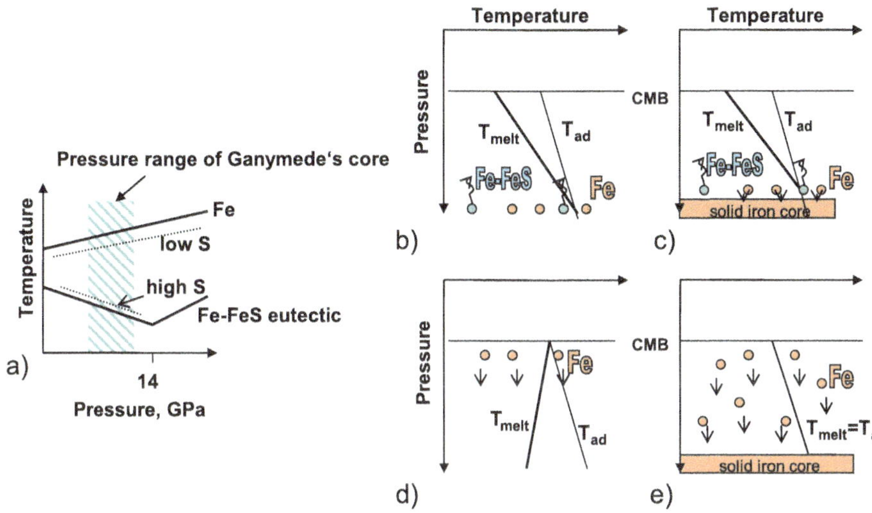

Fig. 9 Schematic diagram showing the liquidus temperatures of Fe and eutectic Fe–FeS with pressure (**a**) and sketches possible snowing core scenarios (**b**)–(**e**). In the pressure range of Ganymede's core, the eutectic temperature of Fe–FeS decreases with pressure. As a consequence, $\mathrm{d}T_{\mathrm{melt}}/\mathrm{d}P$ is positive for small sulfur concentrations and negative for sulfur concentrations close to the eutectic composition. Panels (**b**)–(**e**) show the isentrope (T_{ad}) and the melting temperature (T_{melt}) with pressure for two different cases: for small sulfur concentrations (**b**) and (**c**) and for close to eutectic concentrations (**d**) and (**e**). From (**b**) to (**c**) and (**d**) to (**e**), the core is cooling. For small sulfur concentrations (**b**), (**c**), precipitation starts in the center of the core and core convection is driven by the buoyancy of the light Fe–FeS liquid that forms close to the ICB (similar to the classical compositional driven convection in the Earth's core). For near eutectic sulfur concentrations (**d**), (**e**), precipitation of Fe starts close to CMB (**d**). Sinking iron snow may remelt and produce a gradient in the concentration of Fe (Fe content increases with pressure) until the melting temperature becomes parallel to the isentropic temperature gradient—a solid inner iron core can then form (**d**). Precipitation of Fe at that stage may occur in the entire fluid core and may enforce compositional convection (adapted from Hauck et al. 2006)

A compositionally driven dynamo has been suggested for Ganymede by McKinnon (1996), Spohn and Breuer (1998), Hauck et al. (2006) and Bland et al. (2008), with the latter two discussing various forms of the snowing core model (compare Sect. 3.3). The snow model is particularly relevant for Ganymede because its core falls into the most relevant pressure range below 14 GPa. A summary of possible models and a schematic representation of the core melting diagram is given in Fig. 9. Depending on the sulfur content, the core composition may be on the Fe-rich side of the eutectic or on the FeS-rich side and fluid motion may be induced by either Fe-snow sinking from regions close to the core-mantle boundary or by FeS-snow floating upward from the deep core. The sulfur content of the core is basically unknown. Kuskov and Kronrod (2001) demonstrate that a eutectic composition of about 20 weight-% as well as a low sulfur core is consistent with the geophysical data. However, models of the Jovian nebula and the dominance of sulfur on the surface of Io argue against a sulfur poor composition (e.g., Schubert et al. 2004, for a review). The unknown oxidation state of the satellite interiors during core formation precludes an exact knowledge of the composition of the cores of the satellites (e.g., Scott et al. 2002), however, even if the bulk composition were known. A further uncertainty in the modeling comes from the unknown temperature and thermal history of the core and details of the core power balance.

Both Hauck et al. (2006) and Bland et al. (2008) consider power balances for the dynamo similar to (8) above and assume that the core is initially molten. Both require the silicate mantle to cool the core sufficiently rapidly to arrive at a sufficiently large cooling power Q_c. Sufficient cooling is possible if the rheology of the mantle is close to that of a wet olivine mantle. In addition, both require a freezing core with latent heat and gravitational energy contributing to the power balance. For iron-rich compositions, Hauck et al. (2006) allow a maximum of 13 weight-% S to arrive at a freezing core. Bland et al. (2008) argue that latent heat in the iron snowing core model would not be available to drive the dynamo since it would be released immediately below the CMB and should be effectively removed. Therefore, Bland et al. accept models only where the latent heat is released deep in the core. These authors conclude that this would require the core to have less than 3 weight-% sulfur. At least the latter bound on the sulfur content is difficult to reconcile with cosmochemical models.

A purely thermal dynamo has recently been suggested by Kimura et al. (2009). These authors assume an initially cold interior (initial core temperature close to the eutectic temperature of about 1250 K) that would have heated with time as a consequence of radiogenic decay and only recently would have started to cool by thermal convection. The latter would drive the dynamo. The model would work best for a core with a composition close to the eutectic. In that case, the radius of the core would be maximized and the thickness of the mantle minimized, allowing most efficient cooling. One should note, however, that Kimura et al. (2009) assume a critical isentropic heat flow of only 1 mW m^{-2} which is even smaller than the minimum possible value considered by Hauck et al. (2006) and Bland et al. (2008).

The difficulty of generating a dynamo in Ganymede either through thermal or chemical convection led Stevenson (1996) and Showman et al. (1997) to suggest that a period of tidal heating in Ganymede's past may have blocked the cooling of the core for some time, and thus may have enabled the generation of the present-day field. The necessary tidal heating would have been the consequence of a resonance passage that resulted in an excitation of the satellite's eccentricity (Showman and Malhotra 1997). If sufficient tidal dissipation occurred in Ganymede's silicate mantle, an increasing temperature would have prevented the core from cooling until the resonance passage ended. After escape from the resonance, rapid mantle and core cooling would have triggered dynamo action via thermal and/or compositional convection. Bland et al. (2008), however, argue that tidal heating in Ganymede's mantle may have never sufficed to cause the thermal runaway necessary to prevent core cooling. Another variant of the delayed core cooling model is that of Spohn and Breuer (1998) who suggest that Ganymede may have accreted cold (with temperature decreasing with depth as in classical accretion scenarios) and differentiation and core formation may have been caused by warming of the interior over a few Ga as a consequence of radiogenic heating and heat trapping. The core would thus have formed late and core cooling and dynamo action may be a recent feature for this satellite.

The lack of a present internal dynamo in Io, Callisto, and perhaps Titan can be more easily understood. Callisto's moment-of-inertia factor suggests a largely undifferentiated (or partly differentiated) satellite—and there would be no iron-rich core in which a dynamo could be generated. A similar explanation may apply to Titan if it could be confirmed that it does not have a self-sustained magnetic field. Models of the early evolution and differentiation of the satellite (e.g., Kirk and Stevenson 1987; Lunine and Stevenson 1987; Tobie et al. 2005) suggest that it did not differentiate to form an iron-rich core.

The case for Europa is less clear, in particular when comparing Europa with Ganymede. It is widely believed that Europa has an interior structure similar to that of Ganymede with a core, a silicate mantle, and an ice layer (Sohl et al. 2002), although the absence of a self-sustained magnetic field makes the case for a core in Europa less compelling. As a consequence, one would expect a similar thermal and magnetic evolution for both satellites. Thermal evolution models suggest that it is even more likely for Europa to generate an internal field if the same set of parameter values for mantle rheology and radioactive heat source density is used as for Ganymede (Breuer et al. 2008). Accepting core convection in Ganymede, the question than poses itself of how to explain the absence of core convection in Europa. Three possible explanations are: (1) Europa has no iron core. This is (marginally) consistent with the gravity observation but leaves the question open why Ganymede should have fully differentiated while Europa did not. (2) A higher concentration of light elements in Europa's core. Taking sulfur as a point in case, Europa may have more sulfur (on the iron-rich side of the eutectic), in which case more cooling would be required to freeze the core. (3) Tidal heating in the silicate mantle of Europa. This is certainly possible if the mantle is partially molten which is suggested at least for the early evolution in thermal evolution models (Breuer et al. 2008). In fact, only a few times the present-day radiogenic heating rate would be required to possibly frustrate dynamo action. This much tidal heat is consistent with the models of Hussman and Spohn (2004).

Finally, we note that the absence of a magnetic field of Io is best explained by tidal heating in the mantle blocking the heat flow from the core (Wienbruch and Spohn 1995; see also Schubert et al. 2004, for a discussion).

6 Summary and Discussion

A magnetic field of internal origin—as we have discussed in the present paper—is a distinguishing feature between terrestrial planets and satellites that may reveal properties of their deep interiors. Necessary conditions for the existence of these fields are electrically conducting regions in the interior with material flow, a sufficiently strong power source to drive a dynamo, and sufficiently strong rotation to promote helicity of the flow and large scale magnetic field generation (compare Busse and Simitev 2007; Christensen and Wicht 2007; Stanley and Glatzmeier 2009; Christensen 2009). The candidate regions in terrestrial planets and satellites are the iron-rich cores. Salty water oceans in icy satellites could in principle also serve as dynamo regions but at least for the icy satellites of Jupiter this possibility has been ruled out on the basis of considerations of the power necessary to drive a dynamo (e.g., Schubert et al. 1996).

The existence of cores in Earth, Mars, and Mercury are proven beyond any reasonable doubt. Although there is no conclusive evidence for an iron-rich core in Venus, it is reasonable to assume a core for that planet as well because of its similarity in size and mass to the Earth. For the satellites, the situation is not as clear cut. The available data suggest that the lunar core is small and it cannot be completely ruled out that there is no iron core at all. Strong cases can be made for cores in Io and Ganymede; but models without cores are possible for Europa, albeit not required. Callisto is likely to lack a core. The available evidence suggests that this satellite is incompletely differentiated. However, it must be kept in mind that the models of the interior structures of these satellites assume hydrostatic equilibrium, an assumption that may be problematic given their small sizes and masses. No conclusive evidence for or against cores in Titan, Triton, Pluto, and Vesta is available. There is hope that the Cassini mission will eventually provide data for Titan and the Dawn and New Horizons missions may do so for Vesta and Pluto, respectively.

Rotation is a necessary condition for dynamo action to promote helicity and large scale magnetic field. Balancing the Lorentz and Coriolis forces has resulted in a semi-empirical prediction of the dipole moments of the planets known as Busse's law (Busse 1976). It is also sometimes postulated that the lack of a magnetic field of Venus may be explained by its small rotation rate. However, Hide (1956) already noted that even the Coriolis force in Venus' core is strong enough to effect large scale flow. Stevenson et al. (1983) have used a power balance for the core to estimate the dipole moment. The recent work by Christensen and Aubert (2006) has confirmed the notion that the strengths of the planetary magnetic fields should be dependent on the power balance of the core and only in extreme cases on the planetary rotation rate.

The power balance of the cores is largely dependent on the heat transfer rate in the mantle. Regardless of whether the dynamo is driven by thermal or compositional buoyancy, the core must be cooling to induce convection. This requires the heat flow from the core to exceed the heat flow conducted along the core adiabat (or isentrope) for thermal convection. The critical heat flow depends as we have discussed on the temperature of the core, gravity, and the core thermodynamic and transport properties, in particular, the thermal expansion coefficient, the density, the specific heat, and the thermal conductivity. The adiabatic temperature gradient can also be cast in terms of the Grüneisen parameter and the bulk modulus. In any case, the threshold heat flow increases with planetary mass from a few mW m^{-2} for Ganymede to reach a few tens of mW m^{-2} for Earth and Venus. No threshold for the core-mantle heat flow is expected for compositional convection. The latent heat must be removed but it will simply reduce the rate of inner core growth. Inner core growth requires, however, that the temperature in the core falls below the liquidus which in turn depends on pressure and core composition. Both, the thermal and the compositionally driven dynamo will have to overcome Ohmic dissipation (e.g., Gubbins et al. 2003; Christensen and Tilgner 2004).

The rate of heat transfer in the mantle depends largely on mantle rheology, its physical, thermodynamic and transport properties, and its composition. We will not discuss conduction here since conduction is unlikely to provide sufficient cooling on planetary scales. Rather, we will restrict ourselves to discuss convection. Among the physical properties, the most important ones are the thickness of the mantle and gravity. While the thickness of the mantle cancels out in constant viscosity mantle convection heat transfers scaling laws, it is retained in variable viscosity scaling laws. Gravity immediately affects the vigor of mantle convection. More important than these parameters are the rheology parameters. Since the viscosity of rock is strongly temperature dependent, the viscosity can vary by many orders of magnitude and the heat transfer rate can vary by substantial factors. Moreover, the temperature dependence of viscosity provides for a thermostat effect and determines the style of mantle convection. In addition to the temperature dependence, the rheology is dependent on the chemistry of the planetary mantle, in particular on the concentration of volatiles. Although it is not well established why the Earth has plate tectonics and the other planets have not, it is widely held that at least part of the answer lies with the availability of water on the planet. In any case, numerical simulations of mantle convection have shown that plate tectonics can cool the deep interior of a planet much more effectively than stagnant lid convection. A simple reasoning shows why: For plate tectonics, the surface plates participate in the convection and the convective heat transfer cycle operates between the surface temperature and the core-mantle boundary temperature. In stagnant lid convection, the convective heat transfer cycle operates between the temperature below the lid (about 1100 K) and the core-mantle boundary temperature.

Although there are a large number of uncertain parameter values and, therefore, a large number of possible models, a simple model can account for the differences in the magnetism of the terrestrial planets. The two planets, Earth and Mercury, that do have self-sustained magnetic fields at the present time have inner cores. This has been proven for the Earth and is likely for Mercury given its presumed refractory composition (e.g., Woods et al. 1981; Lewis 1988). Both planets have comparatively large heat transfer rates through their mantles; the Earth because of its plate tectonics and Mercury because of its thin mantle. In addition, the liquidus temperature in the Hermean core should be comparatively large, given its presumed refractory composition. The other two, Mars and Venus, may lack inner cores. Their core to mantle radii ratios are around 0.5 (as compared to 0.8 for Mercury) and both are likely to lack plate tectonics. In addition, Mars' core has a low liquidus temperature if it is indeed close to eutectic in composition. Both planets may have had thermally driven dynamos in their early evolutions but the heat flow from the mantle may have become subcritical during their later evolutions (e.g., Stevenson et al. 1983; Schubert and Spohn 1990; Breuer and Spohn 2003; Hauck et al. 2004). However, more exotic models are also possible (e.g., Stevenson 2001). For instance, it is possible that these planets do have inner cores that are not growing at the present time. Thermal history calculations show that the core of Mars may have been warming after cooling through an early phase of plate tectonics (Nimmo and Stevenson 2000; Breuer and Spohn 2003) and Venus may undergo some episodic form of mantle convection with surface foundering events throughout its history (e.g., Turcotte 1993).

The Moon may be explained in a simple scenario just like Mars and Venus. If it has a core, than the relative core radius at 0.2 is smaller than that of Mars and Venus and heat transfer through the mantle should be even smaller. Thus, the Moon may have had an early dynamo that died a few hundred million years to a billion years after formation (Konrad and Spohn 1997; Spohn et al. 2001c).

The interesting case is Ganymede. Although the multitude of parameters certainly allows the construction of models that predict a dynamo for the satellite, the task becomes more difficult if the model has to explain the lack of fields on Mars and Europa at the same time. If Ganymede is to work like Mars and Venus, then the heat transfer from the core is small. The ratio between the core radius and the radius of the silicate shell is about 0.5 just as for the two planets. In addition, Ganymede has a thick ice shell that would further insulate the deep interior. Ganymede's core is also likely to be sulfur rich, around 21 wt.-% if it is chondritic in composition. This will reduce its likelihood to freeze an inner core.

In Fig. 10, we graphically display the possible evolution of the simple model. The plot qualitatively displays the planets in a plane defined by the concentration of light elements in the core as an indicator of the time needed to cool to the liquidus temperature (the higher the concentration, the longer the time needed to cool to the liquidus) and the heat transfer rate through the mantle. Figure 10a shows the situation for the planets in the early solar system. Most of the cores are liquid and convecting but Mercury has begun freezing an inner core mostly because of its lack of light constituents. The planets have magnetic fields, thermally driven for Venus, Earth, and Mars, and compositionally driven for Mercury. Figure 10b portrays the present-day situation. Mars and Venus still have liquid cores but the thermally driven dynamo has died off. Earth is growing an inner core and has a compositionally driven dynamo. Mercury has almost frozen its core and the inner core growth rate has substantially decreased. Accordingly, consistent with Stanley et al. (2005) or Christensen (2006), its magnetic field has weakened. In Fig. 10c, some time into the future of the solar system, Mars and Venus have begun freezing inner cores. Both planets may thus have restarted their dynamos this time compositionally driven. Earth has frozen most of the core; the growth rate has decreased as has the strength of the dynamo. Mercury has frozen the core completely.

It is interesting to speculate on Earth-like Exoplanets. There is no simple way to predict whether or not such a planet will have a magnetic field. In view of its habitability a long lasting field would be advantageous to allow the evolution of life on the surface. The evidence from the solar system certainly suggests that plate tectonics will be helpful. The scientific

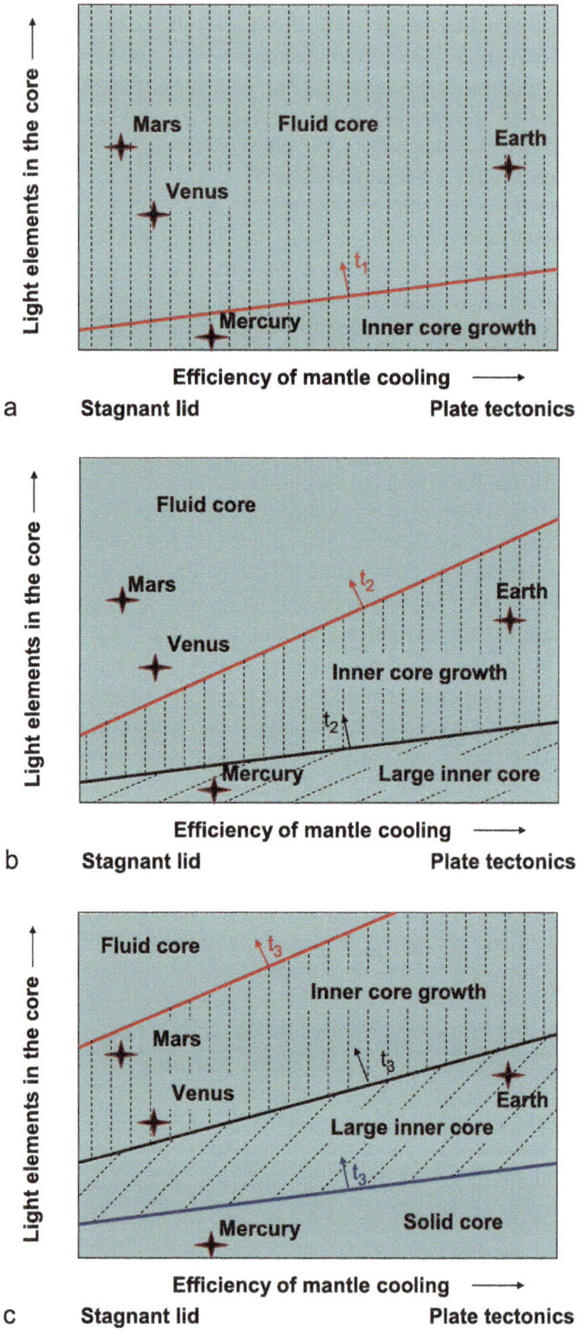

Fig. 10 Simple model of the evolution of the magnetism of the terrestrial planets. *Vertical dashed areas* are regions where the dynamos are driven by compositional convection resulting from a growing inner core. In the *hashed area*, the models have large inner cores with small growth rates and weak magnetic fields. The *red line* marks the onset of inner core growth. The *black line* marks the transition to large inner cores. The *blue line* indicates the cessation of the dynamo for a complete frozen core. Time increases from (**a**) to (**c**) where (**a**) indicates a time in the early evolution, (**b**) the present day, and (**c**) a time in the future. The positions of the planets are indicated with stars and depend on the assumed amount of light elements in the core and the efficiency of cooling. Note that the borders between regimes are, of course, schematic. For further explanation, see the text

community presently debates whether or not super Earths would be expected to have plate tectonics or not. The debate has not settled with, e.g., O'Neill and Lenardic (2007) arguing against and Valencia et al. (2007) and Valencia and O'Connell (2009) arguing for plate tectonics. It is to be expected that the threshold for a thermally driven dynamo will increase with pressure because the adiabatic gradient is expected to increase but the variation of the melting point depression due to, e.g., sulfur beyond 25 GPa is unknown.

A better understanding of the magnetism of the terrestrial planets and satellites certainly requires a better knowledge of material properties and their variations with pressure and temperature. This can be achieved—at least to some extent—by progress in laboratory techniques. However, even more important than that—it seems to us—would be *in-situ* data from planets and satellites. These include geodetic and geophysical data from hitherto unexplored bodies such as Vesta and Pluto but also and perhaps even more importantly conclusive data on the cores of the closer to home planets Mars, Venus, and Mercury; how large are their cores and their inner cores, if there are any. And, last but not least, a better exploration of the magnetic fields of Mars and Mercury.

References

M.H. Acuña, J.E.P. Connerney, N.F. Ness, R.P. Lin, D. Mitchell, C.W. Carlson, J. McFadden, K.A. Anderson, H. Rème, C. Mazelle, D. Vignes, P. Wasilewski, P. Cloutier, Magnetic field and plasma observations at Mars: Initial results of the Mars global surveyor mission. Science **279**, 1676–1680 (1998)

M.H. Acuña, J.E.P. Connerney, N.F. Ness, R.P. Lin, D. Mitchell, C.W. Carlson, J. McFadden, K.A. Anderson, H. Rème, C. Mazelle, D. Vignes, P. Wasilewski, P. Cloutier, Global distribution of crustal magnetism discovered by the Mars global surveyor MAG/ER experiment. Science **284**, 790–793 (1999)

M.H. Acuña, J.E.P. Connerney, P. Wasilewski, R.P. Lin, D. Mitchell, K.A. Anderson, C.W. Carlson, J. McFadden, H. Rème, C. Mazelle, D. Vignes, S.J. Bauer, P. Cloutier, N.F. Ness, Magnetic field of Mars: Summary of results from the aerobraking and mapping orbits. J. Geophys. Res. **106**(E10), 23,403–23,418 (2001)

O. Aharonson, M.T. Zuber, S.C. Solomon, Crustal remanence in an internally magnetized non-uniform shell: A possible source for Mercury's magnetic field? Earth Planet. Sci. Lett. **218**, 261–268 (2004)

K.M. Alley, E.M. Parmentier, Numerical experiments on thermal convection in a chemically stratified viscous fluid heated from below: implications for a model of lunar evolution. Phys. Earth Planet. Int. **108**, 15–32 (1998)

B.J. Anderson, M.H. Acuña, H. Korth, M.E. Purucker, C.L. Johnson, J.A. Slavin, S.C. Solomon, R.L. McNutt, The structure of Mercury's magnetic field from MESSENGER's first flyby. Science **321**, 82–85 (2008). doi:10.1126/science.1159081

J.D. Anderson, E.L. Lau, W.L. Sjogren, G. Schubert, W.B. Moore, Gravitational constraints on the internal structure of Ganymede. Nature **384**, 541–543 (1996)

J.D. Anderson, E.L. Lau, W.L. Sjogren, G. Schubert, W.B. Moore, Europa's differentiated internal structure: Inferences from two Galileo encounters. Science **276**, 1236–123 (1998)

J.D. Anderson, R.A. Jacobson, T.P. McElrath et al., Shape, mean radius, gravity field and interior structure of Callisto. Icarus **153**, 157–161 (2001a). doi:10.1006/icar.2001.6664

J.D. Anderson, R.A. Jacobson, E.L. Lau, W.B. Moore, G. Schubert, Io's gravity field and interior structure. J. Geophys. Res. **106**(E12), 32,963–32,969 (2001b)

A. Anselmi, G. Scoon, BepiColombo, ESA's Mercury Cornerstone mission. Planet. Space Sci. **49**, 1409–1420 (2001)

J. Arkani-Hamed, J. Dyment, Magnetic potential and magnetization contrasts of Earth's lithosphere. J. Geophys. Res. **101**(B5), 11,401–11,426 (1996)

Y. Asahara, D.J. Frost, D.C. Rubie, Partitioning of FeO between magnesiowustite and liquid iron at high pressures and temperatures: Implications for the composition of the Earth's outer core. Earth Planet. Sci. Lett. **257**, 435–449 (2007)

J. Aubert, S. Labrosse, C. Poitou, Modelling the paleo-evolution of the geodynamo. Geophys. J. Int. **179**, 1414–1428 (2009). doi:10.1111/j.1365-246X.2009.04361.x

H. Backes, F.M. Neubauer, M.K. Dougherty, N. Achilleos, N. André, C.S. Arridge, C. Bertucci, G.H. Jones, K.K. Khurana, C.T. Russell, A. Wennmacher, Titan's magnetic field signature during the first Cassini encounter. Science **308**, 992–995 (2005)

P.S. Balog, R.A. Secco, D.C. Rubie, D.J. Frost, Equation of state of liquid Fe-10 wt.% S: Implications for the metallic cores of planetary bodies. J. Geophys. Res. **108**(B2), 2124 (2003). doi:10.1029/2001JB001646

Basaltic Volcanism Study Project (BVSP), *Basaltic Volcanism on the Terrestrial Planets* (Pergamon, New York, 1981).

D. Bercovici, Generation of plate tectonics from lithosphere-mantle flow and void-volatile self-lubrication. Earth Planet. Sci. Lett. **154**, 139–151 (1998)
C.M. Bertka, Y. Fei, Mineralogy of the Martian interior up to core–mantle boundary pressures. J. Geophys. Res. **102**, 5251–5264 (1997)
M.T. Bland, A.P. Showman, G. Tobie, The production of Ganymede's magnetic field. Icarus **198**, 384–399 (2008)
U. Bleil, N. Petersen, Variations in magnetization intensity and low-temperature titanomagnetite oxidation of ocean floor basalts. Nature **301**, 384–388 (1983)
J.R. Booker, K.C. Stengel, Further thoughts on convective heat transport in a variable viscosity fluid. J. Fluid Mech. **86**, 289–291 (1978)
S.I. Braginsky, Mac-oscillations of the hidden ocean of the core. J. Geomag. Geoelectr. **45**, 1517–1538 (1993)
S.I. Braginsky, Formation of the stratified ocean of the core. Earth Planet. Sci. Lett. **243**, 650–656 (2006)
S.I. Braginsky, P.H. Roberts, Equations governing convection in Earth's core and the geodynamo. Geophys. Astrophys. Fluid Dyn. **79**, 1–97 (1995)
D. Breuer, W.B. Moore, Dynamics and thermal history of the terrestrial planets, the Moon, and Io, in *Planets and Moons*, ed. by T. Spohn. Treatise on Geophysics, vol. 10 (Elsevier, Amsterdam, 2007), pp. 299–348
D. Breuer, T. Spohn, Early plate tectonics versus single-plate tectonics: Evidence from the magnetic field history and crust evolution. J. Geophys. Res.-Planets **108**(E7), 5072 (2003). doi:10.1029/20002JE001999
D. Breuer, T. Spohn, Viscosity of the Martian mantle and its initial temperature: Constraints from crust formation history and the evolution of the magnetic field. Planet. Space Sci. **54**, 153–169 (2006)
D. Breuer, S.A. Hauck, M. Buske, M. Pauer, T. Spohn, Interior evolution of Mercury. Space Sci. Rev. **132**(2–4), 229–260 (2007). doi:10.1007/s11214-007-9228-9
D. Breuer, H. Hussmann, T. Spohn, The magnetic dichotomy of the Galilean satellites Europa and Ganymede, in *Geophysical Research Abstracts*, European Geosciences Union—General Assembly, Vienna (Austria), 2008-04-13–2008-04-18, abstract 07511, 2008
B.A. Buffett, Estimates of heat flow in the deep mantle based on the power requirements for the geodynamo. Geophys. Res. Lett. **29**, 7 (2002)
B. Buffett, Core-mantle interactions, in *Treatise on Geophysics*, ed. by G. Schubert (Elsevier, Amsterdam, 2007), pp. 345–358
B.A. Buffett, H.E. Huppert, J.R. Lister, A.W. Woods, Analytical model for solidification of the Earth's core. Nature **356**, 329–331 (1992)
B.A. Buffett, H.E. Huppert, J.R. Lister, A.W. Woods, On the thermal evolution of the Earth's core. J. Geophys. Res. **101**, 7989–8006 (1996)
M. Buske, Dreidimensionale thermische Evolutionsmodelle für das Innere von Mars und Merkur, Doktorarbeit, IMPRS, Katlenburg-Lindau, 2006
F.H. Busse, Phys. Earth Planet. Inter. **12**, 350–358 (1976)
F.H. Busse, R. Simitev, Planetary dynamos, in *Planets and Moons*, ed. by T. Spohn. Treatise on Geophysics, vol. 10 (Elsevier, Amsterdam, 2007), pp. 281–290
S. Butler, W. Peltier, S. Costin, Numerical models of the Earth's thermal history: Effects of inner-core solidification and core potassium. Phys. Earth Planet. Inter. **152**, 22–42 (2005)
B. Chen, J. Li, S.A. Hauck, Non-ideal liquidus curve in the Fe–S system and Mercury's snowing core. Geophys. Res. Lett. **35** (2008). doi:10.1029/2008GL033311
U.R. Christensen, Thermal evolution models for the Earth. J. Geophys. Res. **90**, 2995–3007 (1985)
U.R. Christensen, A deep dynamo generating Mercury's magnetic field. Nature **444**, 1056–1058 (2006). doi: 10.1038/nature05342
U. Christensen, Dynamo scaling laws and applications to the planets. Space Sci. Rev. (2009). doi:10.1007/s11214-009-9553-2, this issue
U.R. Christensen, J. Aubert, Geophys. J. Int. **166**(1), 97–114 (2006)
U.R. Christensen, A. Tilgner, Power requirement of the geodynamo from ohmic losses in numerical and laboratory dynamos. Nature **429**, 169–171 (2004)
U.R. Christensen, J. Wicht, Numerical dynamo simulations, in *Core Dynamics*, ed. by P. Olson. Treatise on Geophysics, vol. 8 (Elsevier, Amsterdam, 2007), pp. 245–282
U.R. Christensen, V. Holzwarth, A. Reiners, Energy flux determines magnetic field strength of planets and stars. Nature **457**, 167–169 (2009)
L. Chudinovskikh, R. Boehler, Eutectic melting in the system Fe–S to 44 GPa. Earth Planet Sci. Lett. **257**, 97–103 (2007). doi:10.1016/j.epsl.2007.02.024
S.M. Cisowski, M. Fuller, Lunar paleointensities via the IRMs normalization method and the early magnetic history of the Moon, in *Origin of the Moon*, ed. by W.K. Hartmann, R.J. Phillips, G.J. Taylor (Lunar and Planetary Science Institute, Houston, 1986), pp. 411–424
J.E.P. Connerney, Planetary magnetism, in *Planets and Moons*, ed. by T. Spohn. Treatise on Geophysics, vol. 10 (Elsevier, Amsterdam, 2007), pp. 243–280

J.E.P. Connerney, M.H. Acuña, P. Wasilewski, N.F. Ness, H. Rème, C. Mazelle, D. Vignes, R.P. Lin, D. Mitchell, P. Cloutier, Magnetic lineations in the ancient crust of Mars. Science **284**, 794–798 (1999)
J.E.P. Connerney, M.H. Acuña, N.F. Ness, T. Spohn, G. Schubert, Mars crustal magnetism. Space Sci. Rev. **111**(1–2), 1–32 (2004)
J.E.P. Connerney, M.H. Acuña, N.F. Ness, G. Kletetschka, D.L. Mitchell, R.P. Lin, H. Reme, From the cover: Tectonic implications of Mars crustal magnetism. Proc. Natl. Acad. Sci. **102**, 14970–14975 (2005)
V. Conzelmann, Thermische Evolution des Planeten Merkur berechnet unter Anwendung verschiedener Viskositätsgesetze. Ph.D. Thesis, Westfälische Wilhelms-Universität Münster, 1999
S.O. Costin, S.L. Butler, Modelling the effects of internal heating in the core and lowermost mantle on the earth's magnetic history. Phys. Earth Planet. Inter. **157**, 55–71 (2006)
F.J. Crary, F. Bagenal, Remanant ferromagnetism and the interior structure of Ganymede. J. Geophys. Res. **103**, 25757–25773 (1998)
A. Davaille, C. Jaupart, Transient high-Rayleigh number thermal convection with large viscosity variations. J. Fluid Mech. **253**, 141–166 (1993)
F. Deschamps, C. Sotin, Inversion of two-dimensional numerical convection experiments for a fluid with a strongly temperature-dependent viscosity. Geophys. J. Int. **143**, 204–218 (2000)
J.O. Dickey, P.L. Bender, J.E. Faller et al., Lunar laser ranging: A continuing legacy of the Apollo program. Science **265**, 482–490 (1994)
G. Dreibus, H. Wänke, Mars: A volatile-rich planet. Meteoritics **20**, 367–382 (1985)
D.J. Dunlop, O. Ozdemir, *Rock Magnetism: Fundamentals and Frontiers* (Cambridge University Press, Cambridge, 1997), p. 573
A.M. Dziewonski, D.L. Anderson, Preliminary reference Earth model. Phys. Earth Planet. Inter. **25**, 297–356 (1981)
L.T. Elkins-Tanton, E.M. Parmentier, P.C. Hess, Magma ocean fractional crystallization and cumulate overturn in terrestrial planets: Implications for Mars. Meteorit. Planet. Sci. **38**(12), 1753–1771 (2003)
L.T. Elkins-Tanton, S. Zaranek, E.M. Parmentier, Early magnetic field and magmatic activity on Mars from magma ocean overturn. Earth Planet. Sci. Lett. **236**, 1–12 (2005)
Y. Fei, C.T. Prewitt, H. Mao, C.M. Bertka, Structure and density of FeS at high pressure and high temperature and the internal structure of Mars. Science **268**, 1892–1894 (1995)
Y. Fei, C.M. Bertka, L.W. Finger, High-pressure iron-sulfur compound, Fe_3S_2, and melting relations in the Fe–FeS system. Science **275**, 1621–1623 (1997)
Y. Fei, J. Li, C.M. Bertka, C.T. Prewitt, Structure type and bulk modulus of Fe_3S, a new iron–sulfur compound. Am. Mineral. **85**, 1830–1833 (2000)
W.M. Folkner, C.F. Yoder, D.N. Yuan, E.M. Standish, R.A. Preston, Interior structure and seasonal mass redistribution of Mars from radio tracking of Mars Pathfinder. Science **278**, 1749–1752 (1997)
H.V. Frey, Impact constraints on the age and origin of the lowlands of Mars. Geophys. Res. Lett. **33**, L08S02 (2006). doi:10.1029/2005GL024484
G. Giampieri, A. Balogh, Mercury's thermoelectric dynamo model revisited. Planet. Space Sci. **50**, 757–762 (2002)
K.H. Glassmeier, H.-U. Auster, U. Motschmann, A feedback dynamo generating Mercury's magnetic field. Geophys. Res. Lett. **34**, L22201 (2007). doi:10.1029/2007GL031662
F. Goarant, F. Guyot, J. Peyronneau, J.-P. Poirier, High-pressure and high-temperature reactions between silicates and liquid iron alloys, in the diamond anvil cell, studied by analytical electron microscopy. J. Geophys. Res. **97**, 4477–4487 (1992)
O. Grasset, E.M. Parmentier, Thermal convection in a volumetrically heated, infinite Prandtl number fluid with strongly temperature-dependent viscosity: Implications for planetary thermal evolution. J. Geophys. Res. **103**, 18171–18181 (1998)
O. Grasset, C. Sotin, F. Deschamps, On the internal structure and dynamics of Titan. Planet. Space Sci. **48**, 617–636 (2000)
C. Grignè, S. Labrosse, P.J. Tackley, Convective heat transfer as a function of wavelength: Implications for the cooling of the Earth. J. Geophys. Res. **110**, B03409 (2005). doi:10.1029/2004JB003376
M. Grott, D. Breuer, The evolution of the martian elastic lithosphere and implications for crustal and mantle rheology. Icarus **193**, 503–515 (2008). doi:10.1016/j.icarus.2007.08.015
D. Gubbins, Energetics of the Earth's core. J. Geophys. **43**, 453–464 (1977)
D. Gubbins, D. Alfè, G. Masters, D. Price, M.J. Gillan, Can the Earth's dynamo run on heat alone? Geophys. J. Int. **155**, 609–622 (2003)
D. Gubbins, D. Alfè, G. Masters, D. Price, M.J. Gillan, Gross thermodynamics of 2-component core convection. Geophys. J. Int. **157**, 1407–1414 (2004)
U. Hansen, D.A. Yuen, High Rayleigh number regime of temperature-dependent viscosity convection and the Earth's early thermal history. Geophys. Res. Lett. **20**, 2191–2194 (1993)
H. Harder, U. Christensen, A one-plume model of Martian mantle convection. Nature **380**, 507–509 (1996)

H. Harder, G. Schubert, Sulfur in Mercury's core? Icarus **151**, 118–122 (2001)
S.A. Hauck, R.J. Phillips, Thermal and crustal evolution of Mars. J. Geophys. Res. **107**(E7) (2002). doi:10.1029/2001JE001801
S.A. Hauck, A.J. Dombard, R.J. Phillips, S.C. Solomon, Internal and tectonic evolution of Mercury. Earth Planet. Sci. Lett. **222**(3–4), 713–728 (2004)
S.A. Hauck, J.M. Aurnou, A.J. Dombard, Sulfur's impact on core evolution and magnetic field generation on Ganymede. J. Geophys. Res. **111** (2006)
J.W. Head, R. Greeley, M.P. Golombek et al., Geological processes and evolution. Space Sci. Rev. **96**(1/4), 263–292 (2001)
M.H. Heimpel, J.M. Aurnou, F.M. Al-Shamali, N. Gomez Perez, A numerical study of dynamo action as a function of spherical shell geometry. Earth Planet. Sci. Lett. **236** (2005). doi:10.1016/j.epsl.2005.04.032
J.W. Hernlund, C. Thomas, P.J. Tackley, Phase boundary double crossing and the structure of Earth's deep mantle. Nature **434**, 882–886 (2005). doi:10.1038/nature03472
J.M. Hewitt, D.P. McKenzie, N.O. Weiss, Dissipative heating in convective flows. J. Fluid Mech. **68**, 721–738 (1975)
R. Hide, The hydrodynamics of the Earth's core. Phys. Chem. Earth **1**, 94–137 (1956)
R. Hide, Comments on the Moon's magnetism. Moon **4**, 39 (1972)
L. Hood, Z. Huang, Formation of magnetic anomalies antipodal to lunar impact basins: Two dimensional model calculations. J. Geophys. Res. **96**, 9837–9846 (1991)
L.L. Hood, D.L. Mitchell, R.P. Lin, M.H. Acuna, A.B. Binder, Initial measurements of the lunar induced magnetic dipole moment using Lunar Prospector magnetometer data. Geophys. Res. Lett. **26**, 2327–2330 (1999)
L. Hood, A. Vickery, Magnetic field amplification and generation in hypervelocity meteorid impacts with application to lunar paleomagnetism, in *Proceedings of the Lunar and Planetary Science Conference 15th*, Part 1. J. Geophys. Res. **89**(supplement), C211–C223 (1984)
L.L. Hood, A. Zakharian, J. Halekas et al., Initial mapping and interpretation of lunar crustal magnetic anomalies using Lunar Prospector magnetometer data. J. Geophys. Res. **106**, 27825–27839 (2001)
W.B. Hubbard, in *Planetary Interiors* (Van Nostrand–Reinhold, New York, 1984)
C. Huettig, Scaling laws for internally heated mantle convection, Ph.D. Thesis, Westfälische Wilhelms-Universität Münster, 2009
G. Hulot, C. Eymin, B. Langlais, M. Mandea, N. Olsen, Small-scale structure of the geodynamo inferred from Oersted and Magsat satellite data. Nature **416**, 620–623 (2002)
H. Hussman, T. Spohn, Thermal-orbital evolution of Io and Europa. Icarus **171**, 391–410 (2004)
H. Hussmann, C. Sotin, J.I. Lunine, Interiors and evolution of Icy satellites, in *Planets and Moons*, ed. by T. Spohn. Treatise on Geophysics, vol. 10 (Elsevier, Amsterdam, 2007), pp. 509–540
J.A. Jacobs, The Earth's inner core. Nature **172**, 297–298 (1953)
C. Jaupart, S. Labrosse, J.-C. Mareschal, Temperatures, heat and energy in the mantle of the Earth, in *Mantle Dynamics*. Treatise of Geophysics, vol. 7 (Elsevier, Amsterdam, 2007), pp. 253–303
H. Jeffreys, The instability of a compressible fluid heated below. Proc. Camb. Philos. Soc. **26**, 170–172 (1930)
C.A. Jones, Thermal and compositional convection in the outer core, in *Core Dynamics*, ed. by P. Olson. Treatise on Geophysics, vol. 8 (Elsevier, Amsterdam, 2007), pp. 131–185
S. Karato, D.C. Rubie, Towards an experimental study of deep mantle rheology: A new multi-anvil specimen assembly for deformation studies under high pressure and temperatures. J. Geophys. Res. **102**, 20,111–20,122 (1997)
S. Karato, P. Wu, Rheology of the upper mantle. Science **260**, 771–778 (1993)
Y. Ke, V.S. Solomatov, Early transient superplumes and the origin of the Martian crustal dichotomy. J. Geophys. Res. **111** (2006). doi:10.1029/2005JE002631
T. Keller, P.J. Tackley, Towards self-consistent modelling of the Martian dichotomy: The influence of low-degree convection on crustal thickness distribution. Icarus **202** (2009). doi:10.1016/j.icarus.2009.03.029
B.L.N. Kennett, E.R. Engdahl, R.P. Buland, Constraints on seismic velocities in the Earth from travel times. Geophys. J. Int. **122**, 108–124 (1995)
R. Kerswell, Elliptical instability. Ann. Rev. Fluid Mech. **34**, 83–113 (2002)
A. Khan, K. Mosegaard, J.G. Williams, P. Lognonné, Does the Moon possess a molten core? Probing the deep lunar interior using results from LLR and Lunar Prospector. J. Geophys. Res. **109**(E09), E09007 (2004). doi:10.1029/2004JE002294
J. Kimura, T. Nakagawa, K. Kurita, Size and compositional constraints of Ganymede's metallic core for driving an active dynamo. Icarus (2009). doi:10.1016/-j.icarus.2009.02.026
R.L. Kirk, D.J. Stevenson, Thermal evolution of a differentiated Ganymede and implications for surface features. Icarus **69**, 91–134 (1987)
M.G. Kivelson, K.K. Khurana, C.T. Russell, R.J. Walker, J. Warnecke, F.V. Coroniti, C. Polanskey, D.J. Southwood, G. Schubert, Discovery of Ganymede's magnetic field by the Galileo Spacecraft. Nature **384**, 537–541 (1996)

M.G. Kivelson, K.K. Khurana, M. Volwerk, The permanent and inductive magnetic moments of Ganymede. Icarus **157**, 507–522 (2002)

E. Knittle, R. Jeanloz, Earth's core-mantle boundary: Results of experiments at high pressures and temperatures. Science **251**, 1438–1443 (1991)

A.S. Konopliv, S.W. Asmar, E. Carranza, W.L. Sjogren, D.N. Yuan, Recent gravity models as a result of the Lunar Prospector mission. Icarus **150**, 1–18 (2001)

A.S. Konopliv, C.F. Yoder, E.M. Standish, D.N. Yuan, W.L. Sjogren, A global solution for the Mars static and seasonal gravity, Mars orientation, Phobos and Deimos masses, and Mars ephemeris. Icarus **182**, 23–50 (2006)

W. Konrad, T. Spohn, Thermal history of the Moon: Implications for an early core dynamo and post-accretional magmatism. Adv. Space Res. **19**(10), 1511–1521 (1997)

J. Korenaga, Archean geodynamics and the thermal evolution of the Earth, in *Archean Geodynamics and Environments*, ed. by K. Benn, J. Mareschal, K.C. Condie. Geophysical Monograph Series, vol. 164 (American Geophysical Union, Washington, 2006), pp. 7–32

Y.N. Kulikov, H. Lammer, H.I.M. Lichtenegger, T. Penz, D. Breuer, T. Spohn, R. Lundin, H.K. Biernat, A comparative study of the influence of the active young sun on the early atmospheres of Earth, Venus, and Mars. Space Sci. Rev. **129**(1–3), 207–243 (2007). doi:10.1007/s11214-007-9192-4

O.L. Kuskov, V.A. Kronrod, Core sizes and internal structure of Earth's and Jupiter's satellites. Icarus **151**, 204–227 (2001)

S. Labrosse, Hotspots, mantle plumes and core heat loss. Earth Planet. Sci. Lett. **199**, 147–156 (2002)

S. Labrosse, Thermal and magnetic evolution of the Earth's core. Phys. Earth Planet. Inter. **140**, 127–143 (2003)

S. Labrosse, C. Jaupart, Thermal evolution of the earth: Secular changes and fluctuations of plate characteristics. Earth Planet. Sci. Lett. **260**, 465–481 (2007)

S. Labrosse, J.-P. Poirier, J.-L. Le Mouîl, On cooling of the Earth's core. Phys. Earth Planet. Inter. **99**, 1–17 (1997)

S. Labrosse, J.-P. Poirier, J.-L. Le Mouîl, The age of the inner core. Earth Planet. Sci. Lett. **190**, 111–123 (2001)

S. Labrosse, J.W. Hernlund, N. Coltice, A crystallizing dense magma ocean at the base of Earth's mantle. Nature **450**, 866–869 (2007)

A. Landolt-Börnstein, Terrestrial planets and satellites: Planetary interiors, in *Numerical Data and Functional Relationships in Science and Technology*. New Series Group VI: Astronomy and Astrophysics, 4: Astronomy, Astrophysics, and Cosmology (Subvolume B Solar System) (Springer, Berlin, 2009), pp. 200–224. ISBN 978-3-540-88054-7. ISSN 0942-8011

B. Langlais, M.E. Purucker, M. Mandea, Crustal magnetic field of Mars. J. Geophys. Res. **109**, 37 (2004). doi:10.1029/2003JE002048

B. Langlais, V. Lesur, M.E. Purucker, J.P.E. Connerney, M. Mandea, Crustal magnetic fields of terrestrial planets. Space Sci. Rev. (2009). doi:10.1007/s11214-009-9557-y, this issue

T. Lay, J. Hernlund, B.A. Buffett, Core-mantle boundary heat flow. Nat. Geosci. **1**, 25–32 (2008)

F.G. Lemoine, D.E. Smith, D.D. Rowlands et al., An improved solution of the gravity field of Mars (GMM-2B) from Mars Global Surveyor. J. Geophys. Res. **106**, 23359–23376 (2001)

J.S. Lewis, The temperature gradient in the solar nebula. Science **186**, 440–443 (1972)

J.S. Lewis, Io: Geochemistry of sulfur. Icarus **50**, 103–114 (1982)

J.S. Lewis, Origin and composition of Mercury, in *Mercury*, ed. by F. Vilas, C.R. Chapman, M.S. Matthews (Univ. of Arizona Press, Tucson, 1988), pp. 651–666

J. Li, C.B. Agee, Geochemistry of mantle-core differentiation at high pressure. Nature **381**, 686–689 (1996)

J.R. Lister, Expressions for the dissipation driven by convection in the Earth's core. Phys. Earth Planet. Inter. **140**, 145–158 (2003)

J.R. Lister, B.A. Buffett, The strength and efficiency of the thermal and compositional convection in the geodynamo. Phys. Earth Planet. Inter. **91**, 17–30 (1995)

J.R. Lister, B.A. Buffett, Stratification of the outer core at the core-mantle boundary. Phys. Earth Planet. Inter. **105**, 5–19 (1998)

K. Lodders, B. Fegley, *The Planetary Scientist's Companion* (Oxford University Press, Oxford, 1998)

P. Lognonne, C. Johnson, Planetary seismology, in *Planets and Moons*, ed. by T. Spohn. Treatise on Geophysics, vol. 10 (Elsevier, Amsterdam, 2007), pp. 69–122

D.E. Loper, The gravitationally powered dynamo. Geophys. J. R. Astron. Soc. **54**, 389–404 (1978a)

D.E. Loper, Some thermal consequences of a gravitationally powered dynamo. J. Geophys. Res. **83**, 5961–5970 (1978b)

J.I. Lunine, D.J. Stevenson, Clathrate and ammonia hydrates at high pressure: Application to the origin of methane on Titan. Icarus **70**, 61–77 (1987)

J.L. Margot, S.J. Peale, R.F. Jurgens, M.A. Slade, I.V. Holin, Large longitude libration of mercury reveals a molten core. Science **316**(5825), 710–714 (2007)
W.B. McKinnon, Core evolution in the icy satellites, and the prospects for dynamo-generated magnetic fields. Bull. Am. Astron. Soc. **28**, 1076 (1996)
W.B. McKinnon, K. Zahnle, B.A. Ivanov, H.J. Melosh, Cratering on Venus: Modeling and observations, in *Venus II*, ed. by S.W. Bougher, D.M. Hunten, R.J. Phillips (University of Arizona Press, Tucson, 1997), pp. 969–1014
H.Y. McSween Jr., SNC meteorites: Clues to Martian petrologic evolution? Rev. Geophys. **23**, 391–416 (1985)
D.L. Mitchell, R.P. Lin, H. Rème, M.H. Acuña, P.A. Cloutier, N.F. Ness, Crystal magnetospheres observed in the Martian night hemisphere, in *American Astronomical Society, DPS Meeting #31*, abstracts #59.04, Bull. Am. Astron. Soc. **31**, 1584 (1999)
D.L. Mitchell, R.P. Lin, C. Mazelle, H. Rème, P.A. Cloutier, J.E.P. Connerney, M.H. Acuña, N.F. Ness, Probing Mars' crustal magnetic field and ionosphere with the MGS electron reflectometer. J. Geophys. Res. **106**(E10), 23,419–23,428 (2001)
D.L. Mitchell, J.S. Halekas, R.P. Lin, S. Frey, L.L. Hood, M.H. Acuña, Global mapping of lunar crustal fields by Lunar Prospector. Icarus **194**, 401–409 (2008)
P.S. Mohit, J. Arkani-Hamed, Impact demagnetization of the Martian crust. Icarus **168** (2004). doi:10.1016/j.icarus.2003.12.005
L.-N. Moresi, V.S. Solomatov, Numerical investigation of 2D convection with extremely large viscosity variations. Phys. Fluids **7**, 2154–2162 (1995)
S. Mollett, Thermal and magnetic constraints on the cooling of the Earth. Geophys. J. R. Astron. Soc. **76**, 653–666 (1984)
W.B. Moore, Heat transport in a convecting layer heated from within and below. J. Geophys. Res. **113** (2008)
K. Nagel, D. Breuer, T. Spohn, A model for the interior structure, evolution, and differentiation of Callisto. Icarus **169**, 402–412 (2004)
T. Nakagawa, P.J. Tackley, Effects of thermo-chemical mantle convection on the thermal evolution of the Earth's core. Earth Planet. Sci. Lett. **220**, 107–119 (2004)
T. Nakagawa, P.J. Tackley, Deep mantle heat flow and thermal evolution of the Earth's core in thermochemical multiphase models of mantle convection. Geochem. Geophys. Geosyst. **6**, Q08003 (2005). doi:10.1029/2005GC000967
Y. Nakamura, G. Latham, D. Lammlein, M. Ewing, F. Duennebier, J. Dorman, Deep lunar interior inferred from recent seismic data. Geophys. Res. Lett. **1**, 137–140 (1974)
N.F. Ness, K.W. Behannon, R.P. Lepping, Y.C. Whang, Magnetic field of Mercury confirmed. Nature **255**, 204 (1975)
N.F. Ness, M.H. Acuña, J. Connerney, P. Wasilewski, C. Mazelle, J. Sauvaud, D. Vignes, C. D'Uston, H. Reme, R. Lin, D.L. Mitchell, J. McFadden, D. Curtis, P. Cloutier, S.J. Bauer, MGS magnetic fields and electron reflectometer investigation: Discovery of paleomagnetic fields due to crustal remanence. Adv. Space Res. **23**(11), 1879–1886 (1999)
F. Nimmo, Energetics of the core, in *Treatise on Geophysics*, vol. 8, ed. by G. Schubert (Elsevier, Amsterdam, 2007), pp. 31–66
F. Nimmo, D. Stevenson, Influence of early plate tectonics on the thermal evolution and magnetic field of Mars. J. Geophys. Res. **105**, 11,969–11,979 (2000)
F. Nimmo, G.D. Price, J. Brodholt, D. Gubbins, The influence of potassium on core and geodynamo evolution. Geophys. J. Int. **156**, 263–376 (2004)
P. Olson, Overview on core dynamics, in *Treatise on Geophysics*, vol. 8, ed. by G. Schubert (Elsevier, Amsterdam, 2007), pp. 1–31. Chap. 8.01
C. O'Neill, A. Lenardic, Geological consequences of super-sized Earths. Geophys. Res. Lett. **34**, L19204 (2007). doi:10.1029/2007GL030598
H. Ozawa, K. Hirose, M. Mitome, Y. Bando, N. Sata, Y. Ohishi, Chemical equilibrium between ferropericlase and molten iron to 134 GPa and implications for iron content at the bottom of the mantle. Geophys. Res. Lett. **35** (2008)
M. Ozima, K. Seki, N. Terada, Y.N. Miura, F.A. Podosek, H. Shinagawa, Terrestrial nitrogen and noble gases in lunar soils. Nature **436**, 655–659 (2005)
E.M. Parmentier, S. Zhong, M. Zuber, Gravitational differentiation due to initial chemical stratification: Origin of lunar asymmetry by the creep of denseKREEP? Earth Planet Sci. Lett. **201**, 473–480 (2002)
S.J. Peale, Inferences from the dynamical history of Mercury's rotation. Icarus **28**, 459–467 (1976)
S.J. Peale, The rotational dynamics of Mercury and the state of its core, in *Mercury*, ed. by F. Vilas, C.R. Chapman, M.S. Matthews (University of Arizona Press, Tucson, 1988), pp. 461–493
J.-P. Poirier, Transport properties of liquid metals and viscosity of the Earth's core. Phys. Earth Planet. Inter. **92**, 99–105 (1988)

J.-P. Poirier, Light elements in the Earth's core: A critical review. Phys. Earth Planet. Inter. **85**, 319–337 (1994)
M.E. Purucker, A global model of the internal magnetic field of the Moon based on Lunar Prospector magnetometer observations. Icarus **197** (2008). doi:10.1016/j.icarus.2008.03.016
N.J. Rappaport, A.S. Konopliv, A.B. Kucinskas, P.G. Ford, An improved 360 degree and order model of Venus topography. Icarus **139**, 19–31 (1999)
C.C. Reese, V.S. Solomatov, J.R. Baumgardner, Scaling laws for time dependent stagnant lid convection in a spherical shell. Phys. Earth Planet. Inter. **149**, 361–370 (2005)
G.O. Roberts, Fast viscous Bernard convection. Geophys. Astrophys. Fluid Dyn. **12**, 235–272 (1979)
P.H. Roberts, Theory of the geodynamo, in *Core Dynamics*, ed. by P. Olson. Treatise on Geophysics, vol. 8 (Elsevier, Amsterdam, 2007), pp. 67–105
J.H. Roberts, S. Zhong, Degree-1 convection in the Martian mantle and the origin of the hemisphere dichotomy. J. Geophys. Res. **111**, E06013 (2006)
J.H. Roberts, R.J. Lillis, M. Manga, Giant impacts on early Mars and the cessation of the Martian dynamo. J. Geophys. Res. **114**, E04009 (2009). doi:10.1029/2008JE003287
P.H. Roberts, C.A. Jones, A.R. Calderwood, Energy fluxes and ohmic dissipation in the Earth's core, in *Earth's Core and Lower Mantle*, ed. by C.A. Jones, A.M. Soward, K. Zhang (Taylor & Francis, London, 2003), pp. 100–129
P. Rochette, G. Fillion, R. Ballou, F. Brunet, B. Ouladdiaf, L.L. Hood, High pressure magnetic transition in pyrrhotite and impact demagnetization on Mars. Geophys. Res. Lett. **30** (2003). doi:10.1029/2003GL017359
S.K. Runcorn, An ancient lunar magnetic dipole field. Nature **253**, 701–703 (1975)
C.T. Russell, R.C. Elphic, J.A. Slavin, Initial Pioneer Venus magnetic field results: Dayside observations. Science **203**, 745 (1979a)
C.T. Russell, R.C. Elphic, J.A. Slavin, Initial Pioneer Venus magnetic field results: Nightside observations. Science **205**, 114 (1979b)
C. Sanloup, A. Jambon, P. Gillet, A simple chondritic model of Mars. Phys. Earth Planet. Inter. **112**, 43–54 (1999)
G.R. Sarson, C.A. Jones, K. Zhang, G. Schubert, Magneto-convection dynamos and the magnetic fields of Io and Ganymede. Science **276**, 1106–1108 (1997)
G.G. Schaber, R.G. Strom, H.J. Moore et al., Geology and distribution of impact craters on Venus: What are they telling us? J. Geophys. Res. **97**, 13,257–13,301 (1992)
G. Schubert, Ann. Rev. Earth Planet. Sci. **7**, 289 (1979)
G. Schubert, T. Spohn, Thermal history of Mars and the sulfur content of its core. J. Geophys. Res. **95**, 14095–14104 (1990)
G. Schubert, P. Cassen, R.E. Young, Subsolidus convective cooling histories of terrestrial planets. Icarus **38**, 192–211 (1979)
G. Schubert, M.N. Ross, D.J. Stevenson, T. Spohn, Mercury's thermal history and the generation of its magnetic field, in *Mercury*, ed. by F. Viulas, C.R. Chapman, M.S. Matthews (Univ. Press of Arizona, Tuscon, 1988), pp. 514–561
G. Schubert, S.C. Solomon, D.L. Turcotte, M.J. Drake, N.H. Sleep, Origin and thermal evolution of Mars, in *Mars*, ed. by H.H. Kieffer, B.M. Jakosky, C.W. Snyder, M.S. Matthews (Univ. Press of Arizona, Tucson, 1992), pp. 147–183
G. Schubert, K. Zhang, M.G. Kivelson, J.D. Anderson, The magnetic field and internal structure of Ganymede. Nature **384**, 544–545 (1996)
G. Schubert, C.T. Russell, W.B. Moore, Timing of the Martian dynamo. Nature **408**, 666–667 (2000)
G. Schubert, D.L. Turcotte, P. Olson, *Mantle Convection in the Earth and Planets* (Cambridge Univ. Press, Cambridge, 2001), 940 pp.
G. Schubert, J.D. Anderson, T. Spohn, W.B. McKinnon, Interior composition, structure and dynamics of the Galilean satellites, in *Jupiter*, ed. by F. Bagnell, T. Dowling, W.B. McKinnon. The Planet, Satellites and Magnetosphere (Cambridge University Press, Cambridge, 2004), pp. 281–306
S. Schumacher, D. Breuer, Influence of a variable thermal conductivity on the thermochemical evolution of Mars. J. Geophys. Res. **111**, E02006 (2006). doi:10.1029/2005JE002429
H.P. Scott, Q. Williams, F.J. Ryerson, Experimental constraints on the chemical evolution of large icy satellites. Earth Planet. Sci. Lett. **203**, 399–412 (2002)
H.N. Sharpe, W.R. Peltier, Parameterized mantle convection and the Earth's thermal history. Geophys. Res. Lett. **5**, 737–740 (1978)
H.N. Sharpe, W.R. Peltier, A thermal history model for the Earth with parameterized convection. Geophys. J. R. Astron. Soc. **59**, 171–203 (1979)
A.P. Showman, R. Malhotra, Tidal evolution into the Laplace resonance and the resurfacing of Ganymede. Icarus **127**, 93–111 (1997)

A.P. Showman, D.J. Stevenson, R. Malhotra, Coupled orbital and thermal evolution of Ganymede. Icarus **129**, 367–383 (1997)

R.W. Siegfried, S.C. Solomon, Icarus **23**, 192–205 (1974)

W.L. Sjogren, W.B. Banerdt, P.W. Chodas et al., The Venus gravity field and other geodetic parameters, in *Venus II: Geology, Geophysics, Atmosphere, and Solar Wind Environment*, ed. by S.W. Bougher, D.M. Hunten, R.J. Philips (Univ. Press of Arizona, Tucson, 1997), pp. 1125–1162

D.E. Smith, M.T. Zuber, The shape of Mars and the topographic signature of the hemispheric dichotomy. Science **271**, 184–188 (1996)

F. Sohl, G. Schubert, Interior structure, composition, and mineralogy of the terrestrial planets, in *Planets and Moons*, ed. by T. Spohn. Treatise on Geophysics, vol. 10 (Elsevier, Amsterdam, 2007), pp. 27–68

F. Sohl, T. Spohn, The structure of Mars: Implications from SNC-Meteorites. J. Geophys. Res. **102**, 1613–1635 (1997)

F. Sohl, T. Spohn, D. Breuer, K. Nagel, Implications from Galileo observations on the interior structure and chemistry of the Galilean satellites. Icarus **157**, 104–119 (2002)

F. Sohl, G. Schubert, T. Spohn, Geophysical constraints on the composition and structure of the Martian interior. J. Geophys. Res. **110**, E12008 (2005). doi:10.1029/2005JE002520

V.S. Solomatov, Scaling of temperature- and stress-dependent viscosity. Phys. Fluids **7**, 266–274 (1995)

V.S. Solomatov, L.N. Moresi, Scaling of time-dependent stagnant lid convection: Application to small-scale convection on Earth and other terrestrial planets. J. Geophys. Res. **105**, 21795–21817 (2000)

S.C. Solomon, R.L. McNutt Jr., R.E. Gold, M.H. Acuña, D.N. Baker, W.V. Boynton, C.R. Chapman, A.F. Cheng, G. Gloeckler, J.W. Head III, S.M. Krimigis, W.E. McClintock, S.L. Murchie, S.J. Peale, R.J. Phillips, M.S. Robinson, J.A. Slavin, D.E. Smith, R.G. Strom, J.I. Trombka, M.T. Zuber, Planet. Space Sci. **49**, 1445–1465 (2001)

C.S. Solomon, O. Aharonson, J.M. Aurnou et al., New perspectives on ancient Mars. Science **307**, 1214–1220 (2005)

A. Souriau, Deep Earth structure—The Earth's cores, in *Seismology and the Structure of the Earth*, ed. by A.M. Dziewonski, B.A. Romanowicz. Treatise on Geophysics, vol. 1 (Elsevier, Amsterdam, 2007), pp. 655–693

E.A. Spiegel, G. Veronis, On the Boussinesq approximation for a compressible fluid. Astrophys. J. **131**, 442–447 (1960)

T. Spohn, Mantle differentiation and thermal evolution of Mars, Mercury, and Venus. Icarus **90**, 222–236 (1991)

T. Spohn, D. Breuer, Implications from Galileo observations on the interior structure and evolution of the Galilean satellites, in *Planetary Systems: The Long View*, ed. by L.M. Celnikier, J. Tran Thanh Van (Editions Frontiers, 1998), pp. 135–145

T. Spohn, F. Sohl, D. Breuer, Mars. Astron. Astrophys. Rev. **8**, 181–235 (1998)

T. Spohn, F. Sohl, K. Wieczerkowski, V. Conzelmann, The interior structure of Mercury: What we know, what we expect from BepiColombo. Planet. Space Sci. **49**, 1561–1570 (2001a)

T. Spohn, M.A. Acunã, D. Breuer et al., Geophysical constraints on the evolution of Mars. Space Sci. Rev. **96**, 231–262 (2001b)

T. Spohn, W. Konrad, D. Breuer, R. Ziethe, Lunar volcanism induced by heat advected from the lower mantle: Results of 2D and 3D mantle convection calculations. Icarus **149**, 54–65 (2001c)

F.D. Stacey, O.L. Anderson, Electrical and thermal conductivities of Fe–Ni–Si alloy under core conditions. Phys. Earth Planet. Inter. **124**, 153–162 (2001)

S. Stanley, Glatzmeier, Dynamo models for planets other than Earth. Space Sci. Rev. (2009). doi:10.1007/s11214-009-9573-y, this issue

S. Stanley, J. Bloxham, W.E. Hutchinson, M.T. Zuber, Thin shell dynamo models consistent with Mercury's weak observed magnetic field. Earth Planet. Sci. Lett. **234**, 27–38 (2005). doi:10.1016/j.epsl.2005.02.040

S. Stanley, L. Elkins-Tanton, M.T. Zuber, E.M. Parmentier, Mars' paleomagnetic field as the result of a single-hemisphere dynamo. Science **321**(5897), 1822–1825 (2008)

D.R. Stegman, A.M. Jellinek, S.A. Zatman, J.R. Baumgardner, M.A. Richards, An early lunar core dynamo driven by thermochemical mantle convection. Nature **421**, 143–146 (2003a)

V. Steinbach, D. Yuen, Effects of depth-dependent properties on the thermal anomalies produced in flush instabilities from phase transitions. Phys. Earth Planet. Inter. **86**, 165–183 (1994)

D.R. Stegman, A.M. Jellinek, S.A. Zatman, J.R. Baumgardner, M.A. Richards, An early lunar core dynamo driven by thermochemical mantle convection. Nature **421**(6919), 143–146 (2003b)

A. Stephenson, S.K. Runcorn, D.W. Collinson, On changes in the intensity of the ancient lunar magnetic field, in *Proceedings of the 6th Lunar Science Conference*. Geochim. Cosmochim. Acta (1975)

D.J. Stevenson, Planetary magnetic fields. Rep. Prog. Phys. **46**, 555–620 (1983). doi:10.1088/0034-4885/46/5/001

D.J. Stevenson, Mercury's magnetic field: A thermoelectric dynamo? Earth Planet. Sci. Lett. **82**, 114 (1987)
D.J. Stevenson, When Galileo met Ganymede. Nature **384**, 511–512 (1996)
D.J. Stevenson, Mars core and magnetism. Nature **412**, 214–219 (2001)
D.J. Stevenson, Possible connections between the history of the Venus magnetic field and observable features, American Geophysical Union, Spring Meeting 2002, abstract #P21A-08 (2002)
D. Stevenson, T. Spohn, G. Schubert, Magnetism and thermal evolution of the terrestrial planets. Icarus **54**, 466–489 (1983)
A.J. Stewart, M.W. Schmidt, W. van Westrenen, C. Liebske, Mars: A new core-crystallization regime. Science **316**, 1323–1325 (2007)
R.G. Strom, N.J. Trask, J.E. Guest, J. Geophys. Res. **80**, 2478–2507 (1975)
P.J. Tackley, The quest for self-consistent incorporation of plate tectonics in mantle convection, in *History and Dynamics of Global Plate Motions*, ed. by M.A. Richards, R. Gordon, R. van der Hilst. Geophysical Monograph Series, vol. 121 (American Geophysical Union, Washington, 2000a)
P.J. Tackley, Self-consistent generation of tectonic plates in time-dependent, three-dimensional mantle convection simulations, Part 1: Pseudoplastic yielding. Geochem. Geophys. Geosyst. **1** (2000b)
J.A. Tarduno, R.D. Cottrell, M.K. Watkeys, D. Bauch, Geomagnetic field strength 3.2 billion years ago recorded by single silicate crystals. Nature **446**, 657–660 (2007)
H. Terasaki, T. Kato, S. Urakawa, K. Funakoshi, A. Suzuki, T. Okada, M. Maeda, J. Sato, T. Kubo, S. Kasai, The effect of temperature, pressure, and sulfur content on viscosity of the Fe–FeS melt. Earth Planet. Sci. Lett. **190**, 93–101 (2001)
Y. Thibault, M.J. Walter, The influence of pressure and temperature on the metal-silicate partition coefficients of nickel and cobalt in a model C1 chondrite and implications for metal segregation in a deep matma ocean. Geochim. Cosmochim. Acta **59**, 991–1002 (1995)
A. Tilgner, Precession driven dynamos. Phys. Fluids **17**, 034104 (2005)
G. Tobie, A. Mocquet, C. Sotin, Tidal dissipation within large icy satellites: Europa and Titan. Icarus **177**, 534–549 (2005)
P.B. Toft, J. Arkani-Hamed, Magnetization of the Pacific Ocean lithosphere deduced from Magsat data. J. Geophys. Res. **97**, 4387–4406 (1992)
R.A. Trompert, U. Hansen, On the Rayleigh number dependence of convection with strongly temperature dependent viscosity. Phys. Fluids **10**(2), 351–360 (1998)
D.L. Turcotte, An episodic hypothesis for Venusian tectonics. J. Geophys. Res. **98**, 17,061–17,068 (1993)
D.L. Turcotte, E.R. Oxburgh, Finite amplitude convective cells and continental drift. J. Fluid. Mech. **28**, 29–42 (1967)
D. Valencia, R.J. O'Connell, Convection scaling and subduction on Earth and super-Earths. Earth Planet. Sci. Lett. **286**, 492–502 (2009)
D. Valencia, R.J. O'Connell, D. Sasselov, Inevitability of plate tectonic on super-Earths. Astrophys. J. **670**, L45–L48 (2007)
T. Van Holst, The rotation of the terrestrial planets, in *Planets and Moons*, ed. by T. Spohn. Treatise on Geophysics, vol. 10 (Elsevier, Amsterdam, 2007), pp. 123–164
L. Vocadlo, Mineralogy of the Earth—The Earth's core: Iron and iron alloys, in *Mineral Physics*, ed. by D. Price. Treatise on Geophysics, vol. 2 (Elsevier, Amsterdam, 2007), pp. 91–120
S.A. Weinstein, The effects of a deep mantle endothermic phase change on the structure of thermal convection in silicate planets. J. Geophys. Res. **100**(E6), 11,719–11,728 (1995)
B. Weiss, J. Berdahl, L. Elkins-Tanton, S. Stanley, E.A. Lima, L. Carpozen, Magnetism on the angrite parent body and the early differentiation of planetesimals. Science **332**, 713–716 (2008)
G.W. Wetherill, Acculmulation of mercury from planetesimals, in *Mercury*, ed. by F. Viulas, C.R. Chapman, M.S. Matthews (Univ. Press of Arizona, Tucson, 1988), pp. 514–561
M.A. Wieczorek, Gravity and topography of the terrestrial planets, in *Planets and Moons*, ed. by T. Spohn. Treatise on Geophysics, vol. 10 (Elsevier, Amsterdam, 2007), pp. 165–206
U. Wienbruch, T. Spohn, A self sustained magnetic field on Io? Planet. Space Sci. **9**, 1045–1057 (1995)
J.G. Williams, D.H. Boggs, C.F. Yoder, J.T. Ratcliff, J.O. Dickey, Lunar rotational dissipation in solid body and molten core. J. Geophys. Res. **106**, 27,933–27,968 (2001)
J.-P. Williams, F. Nimmo, Thermal evolution of the Martian core: Implications for an early dynamo. Geology **32**, 97–100 (2004)
D.T. Woods, J.M. Shull, C.L. Sarazin, Astrophys. J. **249**, 399 (1981)
S. Xie, P.J. Tackley, Evolution of helium and argon isotopes in a convecting mantle. Phys. Earth Planet. Inter. **146**, 417–439 (2004)
C.F. Yoder, The free librations of a dissipative Moon. Philos. Trans. R. Soc. Lond. Ser. A **303**, 327–338 (1981)
C.F. Yoder, Venusian spin dynamics, in *Venus II: Geology, Geophysics, Atmosphere, and Solar Wind Environment*, ed. by S.W. Bougher, D.M. Hunten, R.J. Philips (Univ. Press of Arizona, Tucson, 1997), pp. 1087–1124

C.F. Yoder, A.S. Konopliv, D.N. Yuan, E.M. Standish, W.M. Folkner, Fluid core size of Mars from detection of the solar tide (1993). doi:10.1126/science.1079645
C.F. Yoder, A.S. Konopliv, D.N. Yuan, E.M. Standish, W.M. Folkner, Fluid core size of Mars from detection of the solar tide. Science **300**, 299–303 (2003). doi:10.1126/science.1079645
T. Yukutake, Implausibility of thermal convection in the Earth's solid inner core. Phys. Earth Planet. Inter. **108**, 1–13 (1998)
R. Ziethe, K. Seiferlin, H. Hiesinger, Duration and extent of lunar volcanism: Comparison of 3D convection models to mare basalt ages. Planet. Space Sci. **57**(7), 784–796 (2009). doi:10.1016/j.pss.2009.02.002
V.N. Zharkov, V.P. Trubitsyn, *Physics of Planetary Interiors* (Pachart, Tucson, 1978)
S. Zhong, M.T. Zuber, Degree-1 mantle convection and the crustal dichotomy on Mars. Earth Planet. Sci. Lett. **189**, 75–84 (2001)
M.T. Zuber, S.C. Solomon, R.J. Phillips et al., Internal structure and early thermal evolution of Mars from Mars Global Surveyor topography and gravity. Science **287**, 1788–1793 (2000)

Space Sci Rev (2010) 152: 501–542
DOI 10.1007/s11214-010-9638-y

Theory and Modeling of Planetary Dynamos

J. Wicht · A. Tilgner

Received: 8 June 2009 / Accepted: 15 February 2010 / Published online: 18 March 2010

Abstract Numerical dynamo models are increasingly successful in modeling many features of the geomagnetic field. Moreover, they have proven to be a useful tool for understanding how the observations connect to the dynamo mechanism. More recently, dynamo simulations have also ventured to explain the surprising diversity of planetary fields found in our solar system. Here, we describe the underlying model equations, concentrating on the Boussinesq approximations, briefly discuss the numerical methods, and give an overview of existing model variations. We explain how the solutions depend on the model parameters and introduce the primary dynamo regimes. Of particular interest is the dependence on the Ekman number which is many orders of magnitude too large in the models for numerical reasons. We show that a minor change in the solution seems to happen at $E = 3\times 10^{-6}$ whose significance, however, needs to be explored in the future. We also review three topics that have been a focus of recent research: field reversal mechanisms, torsional oscillations, and the influence of Earth's thermal mantle structure on the dynamo. Finally we discuss the possibility of tidally or precession driven planetary dynamos.

Keywords Dynamo · Planets · Numerical model · Reversals

1 Introduction

Numerical dynamo simulations have flourished during the last 15 years. The work by Glatzmaier and Roberts (1995a) marked a first highlight after previous authors had proven the general validity of the concept (Zhang and Busse 1988). Since then, several new numerical codes have been developed that mostly aim at modeling the geodynamo. In a more

J. Wicht (✉)
Max-Planck-Institut für Sonnensystemforschung, 37191 Kaltenburg-Lindau, Germany
e-mail: wicht@mps.mpg.de

A. Tilgner
Institut für Geophysik, Universität Göttingen, 37077 Göttingen, Germany

recent development, dynamo simulations also attempt to explain the surprisingly different magnetic fields of the other planets in our solar system (Stanley and Bloxham 2004; Stanley et al. 2005; Takahashi and Matsushima 2006; Christensen 2006).

Numerical dynamos adequately model the strength and large scale geometry of several of the planetary fields in our solar system. Many details of the geomagnetic field as well as several aspects of its dynamics are also replicated successfully. This seems surprising since the numerical limitations force dynamo modelers to run their simulations at parameters that are far away from realistic values. For example, the fluid viscosity is generally many orders of magnitude too large in order to damp the small scale turbulent structures that can not be resolved numerically. There is therefore some doubt that dynamo simulations operate in the magnetostrophic regime thought to be characteristic for planetary dynamos. Recent scaling analysis, however, suggest that these doubts may no be warranted and that the success of the simulations is not coincidental (Christensen [doi:10.1007/s11214-009-9553-2], this issue).

Dynamo simulations have also proven to be a useful tool for understanding the magnetic fields and interior dynamics of planets. They are constantly refined to become more realistic and allow to explore an increasing number of phenomena on various length and time scales.

Several publications provide extensive overviews of the different aspects in numerical dynamo simulations (Braginsky and Roberts 1995; Jones 2000, 2007; Glatzmaier 2002) and discuss their success in modeling the geomagnetic field (Kono and Roberts 2002; Christensen and Wicht 2007). Here, we concentrate on more recent developments that mainly concern the geodynamo.

After a brief introduction into the fundamentals of the dynamo process in Sect. 2 we describe the mathematical formulation and the numerical methods employed for the simulations in Sect. 3. Section 4 and Sect. 5 provide an overview of solutions in non-magnetic convection and in the full dynamo problem, respectively. The following three chapters cover issues that were a focus of recent geodynamo simulations and highlight the power of today's numerical models. Section 6 discusses the dynamics of magnetic field reversals; a subject where numerical simulations are indispensable since paleomagnetic data provide very little insight. The implementation of the thermal boundary conditions imposed by Earth's lower mantle is an interesting attempt to make geodynamo simulations more realistic. We discuss the possible consequences in Sect. 8. Torsional oscillations are thought to form an important part of the decal geomagnetic field variations. They can also serve as a proxy for the magnetostrophic regime planetary dynamos are thought to operate in. Section 7 briefly reviews the topic and examines whether torsional oscillation can be found in dynamo simulations. In Sect. 9 we explore the possibility of tidally and precession driven dynamos. Section 10 closes the paper with a conclusion.

2 Fundamentals

Many review articles and textbooks summarize the aspects of magnetohydrodynamics and dynamo theory relevant for planetology, so that only a brief survey is appropriate here. A very readable introduction is provided by Davidson (2001). More mathematical treatments can be found in Moffat (1978) and Roberts (1987).

According to present day convictions, the induction equation of magnetohydrodynamics adequately represents magnetic field evolution in planetary interiors. This equation determines the behavior of magnetic field inside a liquid conductor in motion and derives from Maxwell's equations under the assumption that all motions are slow compared with the velocity of light and assuming Ohm's law is valid. The Hall effect for instance is excluded.

One furthermore needs to assume that the material is characterized by a constant magnetic permeability. Thus no ferromagnetism is allowed. This chapter will deal with the simple case in which all magnetic material properties are negligible and the conductor has spatially uniform electrical conductivity σ. The induction equation for the magnetic field $\mathbf{B}(\mathbf{r}, t)$ then reads

$$\frac{\partial}{\partial t}\mathbf{B} + \nabla \times (\mathbf{B} \times \mathbf{U}) = \frac{1}{\mu\sigma}\nabla^2\mathbf{B} \tag{1}$$

where μ is the magnetic permeability of vacuum and $\mathbf{U}(\mathbf{r}, t)$ is the velocity of the conductor. To pose a mathematically meaningful problem, boundary conditions must also be specified. Vacuum boundary conditions are frequently used, which is appropriate if one models the mantle surrounding the core as a perfect insulator.

The induction equation is more useful in dimensionless form. If one expresses all lengths in multiples of a length L characterizing the size of the volume filled with liquid conductor, and all times in multiples of L/u where u is the characteristic velocity of the conductor, one arrives at

$$\frac{\partial}{\partial t}\mathbf{B} + \nabla \times (\mathbf{B} \times \mathbf{U}) = \frac{1}{\mathrm{Rm}}\nabla^2\mathbf{B} \tag{2}$$

in which all quantities (length, time, velocity, magnetic field) are dimensionless. The magnetic Reynolds number

$$\mathrm{Rm} = \mu\sigma L\mathrm{u} \tag{3}$$

is a rough measure for the ratio of magnetic induction to Ohmic dissipation.

The induction equation can be solved as an independent kinematic problem or as part of a full self consistent approach. In the kinematic problem, $\mathbf{U}$ is prescribed, so that the dynamo equation (2) becomes an eigenproblem. One can then explore at which critical magnetic Reynolds number the magnetic field starts to grow.

In the full problem that we will discuss in the following section, additional equations such as the Navier-Stokes equation have to be used in order to compute the velocity field $\mathbf{U}$ from a prescribed driving mechanism, as for example thermal convection. The velocity field is affected via the Lorentz force by the magnetic field it generates. This back reaction will limit the exponential field growth found in the kinematic problem. The self-consistent approach is a very intricate problem which can only be solved numerically in most cases of interest. The kinematic dynamo problem is much simpler in that it does not prescribe the driving mechanism but the velocity field itself, which is considered as a given. This approach is useful if one wants to find out whether a certain motion is capable of dynamo action at all.

Thanks to its simplicity, kinematic theory gave rise to several useful notions. There are a few mathematically rigorous statements, known as antidynamo theorems, about situations in which no dynamo action can occur. For example, it is known that a purely axisymmetric field cannot result from dynamo action. In terms of positive results, there are two well established scenarios for field amplification. The first scenario relies on the Alfvén theorem which states that in a perfect conductor, magnetic field lines are frozen into the fluid. Two fluid particles lying on a magnetic field line will stay connected by a magnetic field line forever. Stretching of field lines locally increases the magnetic energy density. Strongest field amplification thus occurs in flows with Lagrangian chaos, in which two initially nearby fluid parcels separate exponentially in the course of time. This mechanism obviously only operates at high magnetic Reynolds numbers.

The second scenario is in principle not restricted to any particular range of Rm, but its theory is best developed for small Rm. The so called α-effect relates the magnetic field and the induction term (the second term in the induction equation above) at large scales. This formalism circumvents the need to compute the magnetic field at small scales. The effects of small scales are subsumed in the α-effect, which in turn is expressed mathematically in the form of the α-tensor. The entries in this tensor can in most circumstances only be chosen heuristically. But there is one well studied example of a 2D periodic flow first investigated by G.O. Roberts (1972) for which the α-tensor can be computed asymptotically for small Rm. This flow is helical, and one can give a handwaving description of the amplification mechanism in terms of magnetic flux tubes twisted by helical vortices. Since rotating flows always contain helicity, and because the magnetic Reynolds numbers in planetary dynamos tend to be moderate, most interpretations of these dynamos (or their numerical models) use the concept of an α-effect in a helical flow. On the other hand, it should be pointed out that helicity is not a prerequisite for dynamo action, even in the restricted class of 2D periodic dynamos. However, non-helical 2D periodic dynamos need a higher Rm than comparable helical dynamos (Tilgner 2004).

3 Mathematical Model

3.1 Basic Equations

Self-consistent dynamo models simultaneously solve for the convective fluid flow and magnetic field generation in a rotating spherical shell (see Fig. 1). The shell is confined by an outer boundary at radius r_o which represent the core-mantle boundary in terrestrial planets or the transition to an electrically insulating outer envelope in the gas planets. An inner boundary at $r = r_i$ models the surface of the solid iron cores of terrestrial planets or the rocky cores in the gas giants. The problem is solved in a reference frame corotating with the planetary rotation about the z-axis. The convective motions are driven by density differences due to variations in chemical composition and in temperature. Compositional convection arises in terrestrial planets that have a growing inner core. The lighter constituents mixed into the iron/nickel core alloy have a smaller solubility in the solid than in the liquid phase. They are therefore rejected at the inner-core freezing front and give rise to compositional density differences. The latent heat associated with the phase change provides an important source

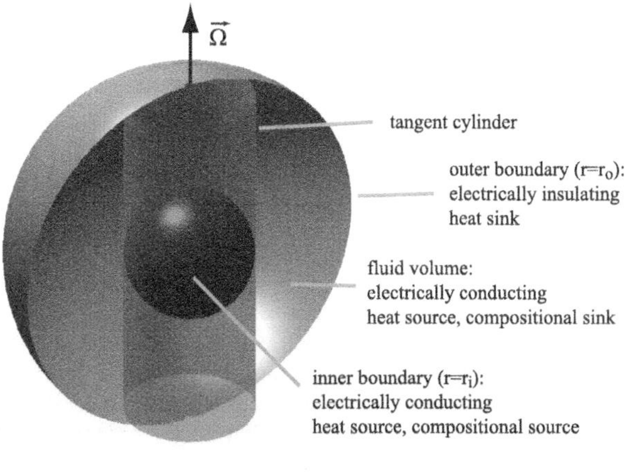

Fig. 1 Geometry of a typical dynamo model for a terrestrial planet with a solid inner core

for thermal convection. Other potential sources are secular cooling and the heat production from radiogenic elements. Only the temperature variations that exceed the adiabatic gradient have to be considered in the models since only these give rise of convection.

Dynamo codes solve for the disturbances around a reference state which is assumed to be hydrostatic, well mixed, and has no magnetic field. The statement 'well mixed' implies that the convection is strong enough to practically homogenize the chemical composition and entropy. The reference state is thus adiabatic. This approach is justified since the disturbances are indeed very small. The convective temperature disturbances, for example, amount to only 10^{-6} times the adiabatic temperature drop across Earth's outer core (Braginsky and Roberts 1995; Jones 2007).

The planets secular cooling and differentiation not only provide the necessary heat and compositional fluxes to drive the dynamo but will also cause the references state to evolve. This evolution is so slow that it can be neglected on the time scales typical dynamo models are concerned with (Labrosse 2003; Lister and Buffett 1995).

Developing the general set of equation for the dynamo problem is beyond the scope of this paper; we refer to Braginsky and Roberts (1995). Here, we restrict ourselves to the simplified system provided by the Boussinesq approximation where the dissipation number,

$$\mathrm{Di} = -\frac{L}{\bar{T}} \frac{\partial \bar{T}}{\partial r}, \tag{4}$$

and the compressibility parameter,

$$\mathrm{Co} = -\frac{L}{\bar{\rho}} \frac{\partial \bar{\rho}}{\partial r}, \tag{5}$$

are assumed to vanish. The Boussinesq approximation thus neglects the temperature and density variations in the reference state. Here, $L = r_o - r_i$ is the depth of the shell and bars over the variables refer to the reference state. Di and Co are identical to the inverse reference temperature and density scale heights, two alternative parameters sometimes used in dynamo theory.

In Earth's outer core, the density increases by about 25% from top to bottom which translates to a compressibility parameter of $\mathrm{Co} \approx 0.2$. The relative temperature increase is of comparable magnitude and yields $\mathrm{Di} \approx 0.3$ (Anufriev et al. 2005). Similar values can be expected for the dynamo regions of the other terrestrial planets. While both these non-dimensional numbers are not very small, they nevertheless indicate that the Boussinesq approximation may yield a fair approach. Since it considerably simplifies the problem, the Boussinesq approximation has been employed by the great majority of numerical dynamo simulations, even for gas planets where the applicability is questionable.

Neglecting the temperature and density variation in the reference state not only implies that $\bar{\rho}$ is homogeneous and constant, the same also holds for the other physical properties: the electrical conductivity σ, the thermal expansivity α, the dynamic viscosity η, the thermal conductivity k, the magnetic permeability μ, and the heat capacity c_p. The Boussinesq approximation moreover has the consequence that viscous heating and Joule heating are neglected and that only the density variations due to temperature and compositional differences in the buoyancy force are retained.

In the following, we describe only one possible non-dimensional form for the set of equations which largely follows Christensen and Aubert (2006). The readers should be aware that there are several different ways of choosing the scaling factors and the non-dimensional parameters, which often leads to confusion when comparing different models. We use L as the length scale, Ω^{-1} as the time scale, the codensity jump Δc across the shell as the codensity scale, and $(\bar{\rho}\mu)^{1/2}\Omega L$ as the magnetic scale.

The mathematical dynamo model in the Boussinesq approximation then comprises the Navier-Stokes equation

$$\frac{d\mathbf{U}}{dt} = -\nabla P - 2\hat{\mathbf{z}} \times \mathbf{U} + \mathrm{Ra}^{\star}\frac{r}{r_o}C\hat{\mathbf{r}} + (\nabla \times \mathbf{B}) \times \mathbf{B} + \mathrm{E}\nabla^2\mathbf{U}, \tag{6}$$

the induction equation

$$\frac{\partial \mathbf{B}}{\partial t} = \nabla \times (\mathbf{U} \times \mathbf{B}) + \frac{\mathrm{E}}{\mathrm{Pm}}\nabla^2\mathbf{B}, \tag{7}$$

the codensity equation

$$\frac{dC}{dt} = \frac{\mathrm{E}}{\mathrm{Pr}}\nabla^2 C + q, \tag{8}$$

the continuity equation

$$\nabla \cdot \mathbf{U} = 0, \tag{9}$$

and the magnetic continuity equation

$$\nabla \cdot \mathbf{B} = 0. \tag{10}$$

Here d/dt stands for the substantial time derivative $\partial/\partial t + \mathbf{U} \cdot \nabla$.

The model is controlled by five dimensionless numbers: Ekman number

$$\mathrm{E} = \frac{\nu}{\Omega L^2}, \tag{11}$$

modified Rayleigh number

$$\mathrm{Ra}^{\star} = \frac{\bar{g}_o \Delta c}{\Omega^2 L}, \tag{12}$$

Prandtl number

$$\mathrm{Pr} = \frac{\nu}{\kappa}, \tag{13}$$

magnetic Prandtl number

$$\mathrm{Pm} = \frac{\nu}{\lambda}, \tag{14}$$

and aspect ratio

$$a = r_i/r_o. \tag{15}$$

We have introduced three diffusivities here: the magnetic diffusivity $\lambda = 1/\sigma\mu$, the viscous diffusivity $\nu = \eta/\bar{\rho}$, and the thermal diffusivity $\kappa = k/(c_p\bar{\rho})$. $\bar{g}_o$ is the references gravity at the outer boundary. $\mathrm{Ra}^{\star}$ is connected with the more classical Rayleigh number

$$\mathrm{Ra} = \frac{\bar{g}_o \Delta c L^3}{\kappa\nu} \tag{16}$$

via $\mathrm{Ra}^{\star} = \mathrm{Ra} E^2 \mathrm{Pr}^{-1}$.

Above, we have used an additional simplification that has been adopted in several dynamo models of terrestrial planets: The density variations due to temperature and composition are combined into one variable, the codensity

$$c = \alpha T + \gamma \chi. \tag{17}$$

Here, χ describes the compositional disturbances around the reference state compositions $\bar{\chi} = m_L/(m_L + m_H)$. We assume a simplified binary model composition with heavy constituents (iron, nickel) of total mass m_H and light constituents (sulfur, oxygen, carbon) of total mass m_L. The compositional expansivity is given by $\gamma = -\bar{\rho}(\rho_H - \rho_L)/\rho_H \rho_L$, where ρ_H and ρ_L are the densities of heavy and light elements in the liquid core, respectively. Note that T in (17) refers to the dimensional temperature and that the dimensionless codensity in (8) is given by $C = c/\Delta c$.

Describing the evolution of temperature and composition by the combined equation (8) assumes that both quantities have similar diffusivities. This seems like a daunting simplification since the chemical diffusivity may be three orders of magnitude smaller than the thermal one (Braginsky and Roberts 1995). The approach is often justified with the argument that the small scale turbulent mixing, which can not be resolved by the numerical codes, may result in larger effective turbulent diffusivities of comparable magnitude (Braginsky and Roberts 1995). Consequently, the 'turbulent' Prandtl numbers would then be of order one. Some studies suggest that the differences in diffusivities may have interesting implications (Busse 2002) and the potential effects on planetary dynamo simulations remains to be explored.

In Jupiter, the pressure increases by over a factor of 200 over the convective gas envelopes which we consider to start at 1000 bar (Evonuk 2008). This demonstrates that the Boussinesq approximation does not apply in gas planets. The implementation of the density variations in a dynamo model and the possible effects that go along with such a strong density stratification are, for example, discussed in Braginsky and Roberts (1995), in Anufriev et al. (2005), and in (Stanley and Glatzmaier [doi:10.1007/s11214-009-9573-y], this issue).

3.2 Boundary Conditions and Driving Modes

The differential equations comprising the mathematical dynamo problem can only be solved when supplemented with an adequate set of boundary conditions. For the flow **U**, either rigid or free slip conditions are used. In both cases the radial flow component is forced to vanish and the differences lie in the horizontal components. They must match the motion of the boundary for rigid conditions, for example the solid body rotation of an inner iron core around the planetary rotation axis. For free slip conditions the horizontal components of the viscous stress are force to vanish which allows for a discontinuity in the horizontal flow itself. Rigid flow conditions are most appropriate for the dynamo regions of terrestrial planets and the boundary to a rocky core in gas planets.

In Jupiter and Saturn, the outer boundary of the dynamo region is marked by the hydrogen transition from the electrically insulating to the conducting state. In Uranus and Neptune, ionically conducting liquid ice layers seem the most likely candidates for the dynamo regions (Stanley and Bloxham 2006). Lorentz forces and higher densities are thought to slow down the dynamics in the conducting layers. Since the insulating layer can thus more easily adapt to the conducting layer, stress free outer boundary conditions seem appropriate. However, the dynamical separation of the conducting and insulating envelopes in gas planets is little understood. For example in Saturn and Jupiter, the conductivity decreases only gradually with pressure (Liu et al. 2008) which questions the strict separation of the two layers.

Some authors (Kuang and Bloxham 1997; Busse and Simitev 2005) prefer to employ free slip conditions when modeling the geodynamo in order to exclude the too thick boundary layers in the numerical models. We come back to this point in Sect. 5.3.

In terrestrial planets, the conductivity of the mantle is orders of magnitudes lower than that of the core. Like in the gas planets, the magnetic field must therefore match a potential field at the outer boundary. The same applies at the boundary to an electrically insulating inner core. For modeling the magnetic field in conducting solid inner cores a simplified induction equation (7) has to be solved that includes diffusive effects and the advection due to the inner core rotation (Christensen and Wicht 2007). Inner core and outer core magnetic fields have to match at the interface.

Most dynamo models employ the simplifying codensity description and use either fixed flux or fixed codensity conditions. This is done for simplicity and has no physical basis. Kutzner and Christensen (2000) suggest that the solutions are not very sensitive to the boundary conditions as long as the inner core boundary is the main codensity source. Their examination was restricted to laterally homogeneous conditions in a few cases and more work is needed to clarify the influence of thermal and compositional boundary conditions in the future. We outline the proper conditions for gas and terrestrial planets in the following, assuming that two separate equations of the form (8) may be used to describe the evolution of temperature T and composition χ, respectively.

Fixed temperature conditions seem appropriate for the outer boundary of gas planets, provided the dynamics of the insulating layer is vigorous enough to homogenize the temperature. At the inner boundary, a fixed heat flux $F_i^T = f_i\, k\, \partial T/\partial r$ should model secular cooling and potential radioactive heating of the rocky core. Here, f_i is the inner core surface, and $\partial T/\partial r$ refers to the spherical symmetric contribution, the other contributions being zero.

In terrestrial planets, the appropriate boundary conditions for temperature and composition are more complex. The inner core is a source for heat and compositional codensity represented by the respective fluxes (see Fig. 1). Since the chemical core elements cannot penetrate the mantle the composition flux has to vanish here: $F_o^\chi = 0$. The outer thermal boundary condition is imposed by the slowly evolving mantle which changes on time scales of tens to hundreds of million years. The much faster changing core is virtually isothermal on these time scales so that horizontal differences in the lower mantle temperature translate into a cmb heat flux amplitude and pattern. The resulting laterally inhomogeneous temperature boundary layer at the base of the mantle is sometimes identified with the D″ layer. We discuss the possible consequences for the dynamo process in Sect. 8.

The thermal and compositional boundary conditions at the interface to a growing inner iron core are more involved. The compositional flux F_i^χ and heat flux F_i^T released from the inner core boundary are proportional to the inner core growth rate. The growth rate in turn depends on the local cooling rate $\bar{\rho} c_p dT/dt$ and compositional rate of change $d\chi/dt$, i.e. it depends on how fast the outer core convection removes heat and light elements (Braginsky and Roberts 1995). These relations yield two equations that connect temperature flux and compositional flux to dT/dt and $d\chi/dt$ and form the appropriate boundary conditions (Braginsky and Roberts 1995; Glatzmaier and Roberts 1996a; Jones and Roberts 2000). They allow for lateral variations in the flux from the inner core that translate to differences in the inner core growth rate. Such variations are also permitted in codensity formulations that impose a fixed codensity at the inner boundary. Aubert et al. (2008a) use such a dynamo model to show that the variations may explain hemispherical differences in the inner-core seismic signal, as we will discuss in Sect. 8.

The homogeneous volumetric source/sink term q in (8) determines the ratio of volumetric driving to bottom driving and, strictly speaking, is an additional system parameter. It serves to model volumetric sources like secular cooling and radioactive heating and has the additional task to prevent an evolution of the reference state by balancing the compositional flux F_i^X with a sink term (see Fig. 1). The most commonly used setup employs $q = 0$ and thus models a purely thermal heat flux from the inner core boundary. Consequently, the codensity (17) could be replaced by αT in the mathematical model which yields a form of equation more typically found in the dynamo literature. Choosing $q \neq 0$ will add volumentric sources which increase the outer boundary heat flux. When inner and outer boundary flux conditions are used they have to fulfill $F_o = -F_i + Vq$, where V is the volume of the shell. For fixed codensity conditions, the inner boundary flux F_i and/or the outer boundary flux F_o are controlled by the vigor of convection, i.e. by the Rayleigh number and the other system parameters. Kutzner and Christensen (2002) model pure chemical convection by imposing a vanishing flux $F_o = 0$ at r_o, by choosing $q < 0$, and by using a fixed codensity condition at r_i. For gas planets and terrestrial planets without a growing inner core $q > 0$ models the homogeneous secular cooling or a homogeneous radiogenic heating as the only driving forces. Whether radiogenic heating is sizable or can be neglected is still a matter of debate (Labrosse [doi:10.1007/s11214-010-9630-6], this issue). Secular cooling can be modeled by a uniform decrease of an adiabatic reference temperature throughout the core (Labrosse 2003; Nimmo 2007). The errors made by approximating the inner core temperatures by an adiabat are very small. The associated inner core boundary condition is a fixed heat flux $F_i = V_i q$, where V_i is the inner core volume. Note that we have neglected a potential difference of the radiogenic heat source densities in inner and outer core.

A deviation from fixed codensity conditions requires a different codensity scale and thus a different Rayleigh number than (12) (Kutzner and Christensen 2002; Aubert et al. 2008a). For example, when imposing a dimensional flux density f at either boundary a possible codensity scale is Lf/κ which yields the Rayleigh number:

$$\mathrm{Ra}' = \frac{\bar{g}_o f}{\Omega^2 \kappa}. \tag{18}$$

Further discussions of the different driving sources and the respective boundary conditions in terrestrial planets can be found elsewhere (Lister and Buffett 1995; Wicht et al. 2007) (Labrosse [doi:10.1007/s11214-010-9630-6], this issue).

3.3 Parameters, Force Balance, and Scaling

Table 1 compares the non-dimensional parameters for Earth with those for some typical dynamo simulations that we will explore in more detail in the following sections. Obviously, dynamo simulations operate at Ekman numbers and magnetic Prandtl numbers which are orders of magnitude too large. The reason is, that the turbulent small scale flow thought to be present in planetary dynamo regions can not be resolve with today's computers. The larger Ekman numbers used in the simulations reflect an increase in viscosity which helps to damp away the smaller scale flows.

The obvious differences in parameters complicate a direct comparison of the simulation results with the planetary magnetic fields. The numerical models are therefore typically discusses in term of dimensionless numbers that quantify important dynamo properties. Like the dimensionless parameters introduced in Sect. 3, these have the advantage of combining different physical into quantities that better characterize the dynamics. The Ekman number

Table 1 List of parameters and properties for some of the dynamo simulations presented here. Reynolds number Rm Elsasser number Λ, and local Rossby number Ro_ℓ are based on time average rms energies in the dynamo region. The relative dipole strength D at the outer boundary is also a time averaged value. Earth values are takes from (Jones 2007), the Ro_ℓ value is suggested by Christensen and Aubert (2006). Earth's Rayleigh number is hard to constrain but thought to be very much supercritical (Gubbins 2001)

Name	E	Ra	Pm	Pr	Ro	Rm	Λ	Ro_ℓ	D	reversing?
E3a	10^{-3}	1.10×10^{-1}	5	1	9×10^{-3}	46	9	3.4×10^{-2}	0.8	no
E3b	10^{-3}	4.50×10^{-1}	10	1	4×10^{-2}	392	25	1.1×10^{-1}	0.3	yes
E5b	3×10^{-5}	1.08×10^{-1}	1	1	10^{-2}	405	8	9.7×10^{-2}	0.6	no
E5c	3×10^{-5}	1.35×10^{-1}	1	1	2×10^{-2}	607	4	1.4×10^{-1}	0.09	yes
E5F	3×10^{-5}	5.04×10^{-2}	1	1	8×10^{-3}	260	4	6.5×10^{-2}	0.8	no
E6	3×10^{-6}	9.00×10^{-3}	0.5	1	2×10^{-3}	261	3	2.3×10^{-2}	0.7	no
Earth	10^{-15}		3×10^{-7}	1	2×10^{-6}	500	1	0.09	0.3–0.6	yes

can be interpreted as a measure for the relative importance of viscous forces compared to the Coriolis force. Two additional non-dimensional numbers quantify the relative importance of inertial effects and magnetic forces in the Navies-Stokes equation (6), once again in comparison to the Coriolis force. These are the Rossby number

$$\mathrm{Ro} = \frac{\mathrm{u}}{L\Omega} \tag{19}$$

and the Elsasser number

$$\Lambda = \frac{\mathrm{b}^2}{\bar{\rho}\Omega\mu\lambda}. \tag{20}$$

u and b refer to typical dimensional flow and magnetic field amplitudes. Note that the Rossby number Ro is identical to the non-dimensional flow amplitude in the scaling chosen here. The magnetic Reynolds number $\mathrm{Rm} = \mathrm{u}L/\lambda$ has already been introduced in Sect. 2 as an important measure for the ratio of magnetic induction to diffusion.

Planetary dynamos are thought to operate in the magnetostrophic regime. This refers to a specific first order force balance in the Navier-Stokes equation between Coriolis force, pressure gradient, Lorentz force, and buoyancy. Viscous forces and inertial forces are deemed negligible in comparison because the Ekman numbers and Rossby numbers of planetary dynamo regions are small. An Elsasser number of order one testifies that the Lorentz force contributes to the first order balance. The magnetostrophic balance can be understood as an extension of the geostrophic balance in non-magnetic systems where Coriolis force and pressure gradient constitute the main force balance with buoyancy contributing to the radial accelerations.

Elsasser number, Ekman number, and Rossby number can only provide rough estimates of the true force balances and may actually fail in many instances. The Rossby number seems to do a particularly bad job in estimating the importance of the non-linear advection $\mathbf{U} \cdot \nabla\mathbf{U}$ in the Navier-Stokes equation (6). This non-linearity can give rise to so-called Reynolds stresses that transport energy from smaller to larger scales or vice versa. Reynolds stresses are the result of a statistically persistent correlation between non-axisymmetric flow components. Very prominent examples for their potential power are the fierce zonal winds observed on Jupiter and Saturn (Christensen 2002; Heimpel et al. 2005). The length scale dependence, introduced by the ∇ operator, can thus not be neglected when estimating the importance of non-linear advection. Christensen and Aubert (2006) therefore suggest to use the local Rossby number

$$\mathrm{Ro}_\ell = \frac{\mathrm{u}}{\Omega \ell}, \tag{21}$$

instead of Ro. The length scale ℓ represents a weighted average base on the spherical harmonic decomposition of the flow:

$$\ell = r_o \pi / \langle n \rangle \tag{22}$$

with

$$\langle n \rangle = \frac{\sum_n n E_k(n)}{\sum_n E_k(n)}, \tag{23}$$

where $E_k(n)$ is the kinetic energy carried by all modes with spherical harmonic degree n.

Local Rossby number Ro_ℓ, Elsasser number Λ, and magnetic Reynolds number Rm are not input parameters but measure properties of the solution. Rather than stating absolute values of magnetic field strength or flow speed dynamo model results are typically discussed in terms of Elsasser number Λ and magnetic Reynolds number Rm. Realistic values of Λ and Rm are thought synonymous with realistic values of b and u. Concerning the magnetic field strength this implies that b depends on Ω, $\bar{\rho}$, μ, and λ since $\mathrm{b} = (\Lambda \Omega \bar{\rho} \mu \lambda)^{1/2}$. Some authors rescale the dimensionless field strength in their solutions by assuming realistic values for all these four properties (Wicht 2005; Christensen 2006; Wicht et al. 2009). The length scale is provided via a realistic shell thickness L, and the magnetic diffusion time $t_\lambda = L^2/\lambda$ typically serves to rescale time. However, Ekman number and magnetic Prandtl number of a numerical model already fix the ratio between the rotation period and t_λ to $\mathrm{E}/\mathrm{Pm} = \lambda / \Omega L^2$ which is typically many orders of magnitude too large. The described field strength scaling therefore amounts to an extrapolation of E/Pm, sometimes called the magnetic Ekman number E_λ, to the much smaller planetary values (Glatzmaier 2002). Realistic flow amplitudes are assumed for realistic magnetic Reynolds numbers when t_λ serves to rescale time.

Christensen and Aubert (2006) realized that the Elsasser number in their numerical simulations is not necessarily of order one but varies over three orders of magnitude. The extrapolation for many of their model would therefore yield unrealistic field strengths despite that fact that they have very similar characteristics otherwise. Wicht and Christensen (2010) argue that the Elsasser number may not always be a good proxy for the force balance in the dynamo region and demonstrate that the numerical models may still obey the magnetostrophic force balance for Elsasser numbers significantly larger than one. Motivated by their findings, Christensen and Aubert (2006) proceed to develop a scaling where the magnetic field and flow amplitudes depend on the heat flux out of the dynamo region (Christensen and Aubert 2006). This scaling seems to do a better job in subsuming the simulation results and correctly predicts the field strength for several planetary dynamos and even for the dynamos of fast rotating stars (Christensen et al. 2009). We refer to (Christensen [doi:10.1007/s11214-009-9553-2], this issue) for a review of the various scaling laws that have been suggested over the years.

3.4 Model Variations

The most commonly adopted dynamo modeling approach, which we will refer to as the 'standard model' in the following, solves the full set of equations in the Boussinesq approach, has an Earth like inner core size, employs a codensity description, and uses rigid flow and fixed codensity boundary conditions. The inner core has either been treated as an electrical insulator or a conductor and Christensen et al. (2001) defines a benchmark for both cases. Minor variations concern the use of different flow and codensity boundary conditions. Some more involved variations characterize several other types of convection driven dynamo models.

The Kuang and Bloxham model (Kuang and Bloxham 1997, 1999) adopts the Boussinesq approximation, uses stress free boundary conditions, and neglects the non-axisymmetric inertial terms in the Navier-Stokes equation in order to possibly better approach the magnetostrophic force balance. The axisymmetric inertial terms are retained to allow for the torsional oscillations discussed in Sect. 7. Another feature of this model is the use of hyperdiffusivity, a numerical trick that increases the diffusivity with spherical harmonic degree l to stronger damp small scale contributions. The Kuang & Bloxham model has been adopted for several publications since (Kuang 1999; Stanley and Bloxham 2004, 2006; Stanley et al. 2005, 2007).

An early Glatzmaier and Roberts model neglects all inertial terms in a Boussinesq approach (Glatzmaier and Roberts 1995a, 1995b) and also employs hyperdiffusion. The later Glatzmaier and Roberts model retains compressible effects in an anelastic approximation, allows for axisymmetric inertia and also uses hyperdiffusivity. It furthermore comprises several advanced features like separate equations for describing the evolution of temperature and composition and a model of inner-core growth along with the appropriate temperature and composition boundary conditions (3.2). The influence of these advanced features has not been properly explored so far. This later Glatzmaier and Roberts model has been employed in several studies during the last years (Glatzmaier and Roberts 1996a, 1996b; Glatzmaier et al. 1999; Coe et al. 2000; Roberts and Glatzmaier 2001; Ogden et al. 2006).

The Busse group models employ the Boussinesq approach, typically use stress free boundary conditions, and are driven a mixture of volumetric and bottom buoyancy sources (Ardes et al. 1996; Wicht and Busse 1997; Grote et al. 1999, 2000; Grote and Busse 2000; Simitev and Busse 2005; Busse and Simitev 2006, 2008). The Kageyama type model describes the dynamo medium as an ideal gas and retains weak compressibility effects (Kageyama and Sato 1995, 1997; Kageyama et al. 1999, 2008; Ochi et al. 1999; Li et al. 2002; Kageyama and Yoshida 2005; Nishikawa and Kusano 2008).

3.5 Numerical Methods

The most common method for solving the numerical dynamo problem is a pseudo-spectral approach. More recently, several authors have adopted so-called local methods which promise to be better adapted to massive parallel computing.

In the pseudo-spectral implementations, all variables are defined on a numerical grid and also have a global spectral representation. Spherical harmonic functions $Y_{lm}(\theta, \phi)$ are the obvious choice for the spectral representation in colatitude θ and longitude ϕ where l and m denote degree and order. Chebychev polynomials $C_n(r)$ are typically chosen for the spectral representation in radius. When applied appropriately, they provide a denser radial grid near the inner and outer boundaries where boundary layers may have to be sampled (Glatzmaier 1984; Christensen and Wicht 2007).

The equation system is time-stepped in spherical harmonic and radial space (l, m, r) using a mixed implicit/explicit scheme where non-linear terms as well as the Coriolis force are treated explicitely in an Adams-Bashforth time step. This guarantees that all spherical harmonic modes (l, m) decouple. A Crank-Nicolson implicit scheme completes the time integration for the remaining contributions. The nonlinear terms are evaluated on the local grid space, which requires a transform from the (l, m, r) to the (θ, ϕ, r) representation and back. Fast Fourier transforms can be applied for the longitudinal dependence. The Gauss-Legendre transformation employed in the latitudinal direction are significantly more time consuming and can considerably slow down the computation for highly resolved cases.

All partial derivatives are evaluated in spectral space which guarantees a high degree of exactness. Local approaches use finite difference or finite element methods whose lower precision can require considerably denser grids (Christensen et al. 2001). Another advantage of the pseudo-spectral codes is the straightforward implementation of the magnetic boundary conditions at the outer boundary. Local methods either use the artificial condition that the horizontal field has to vanish at the interface or have to rely on additional numerical procedures (Matsui and Okuda 2005; Wicht et al. 2009).

Pseudo spectral methods use a poloidal/toroidal decomposition of flow and magnetic field. For example, the flow decomposition reads

$$\mathbf{U} = \nabla \times \hat{\mathbf{r}} w + \nabla \times \nabla \hat{\mathbf{r}} v, \tag{24}$$

where w and v are the toroidal and poloidal flow potentials, respectively. The continuity equations (9) and (10) are then fulfilled automatically. Local approaches, however, typically use primitive variables, i.e. the individual flow and magnetic field components, in order to avoid the additional derivatives introduced by (24). Special numerical measures are then required to assure that the continuity equations are fulfilled (Harder and Hansen 2005). Local methods also face the problem that the grid points in the natural longitude/latitude grid come very close near the poles which may cause numerical instabilities. These instabilities can either be damped (Kageyama and Sato 1995, 1997) or other grids have to be used that provide a more even coverage (Kageyama and Yoshida 2005; Harder and Hansen 2005).

The disadvantage of the pseudo-spectral approaches is the large communication overhead between the individual processors in a parallel computer. Each processor needs the full solution information in every time step. In the local methods a processor that represents a numerical grid point or volume element only requires the information from processors representing neighboring grid points. The local methods have not yet proven to be faster than pseudo-spectral methods in the dynamo context, but this may change on massively parallel computing systems with several thousand processors. A recent overview of local methods can be found in Wicht et al. (2009). More information on the pseudo-spectral method can be found in Glatzmaier (1984) and Christensen and Wicht (2007). The dynamo benchmarks defined by Christensen et al. (2001) serve as test cases for the various methods; an update can be found in Wicht et al. (2009).

4 Convective Flow Dynamics

Am important consequence of the strong Coriolis force in planetary dynamo regions is formulated by the Taylor-Proudman theorem: The system seeks to minimize flow variations in the direction of the planetary rotation axis. A geostrophic flow is therefore generally two-dimensional and takes the form of convection columns that are aligned with the planetary rotation axis. In a spherical container, however, convective motions can not be strictly geostrophic since they will be diverted when meeting the boundaries. This gives rise to a secondary north-south meridional flow component.

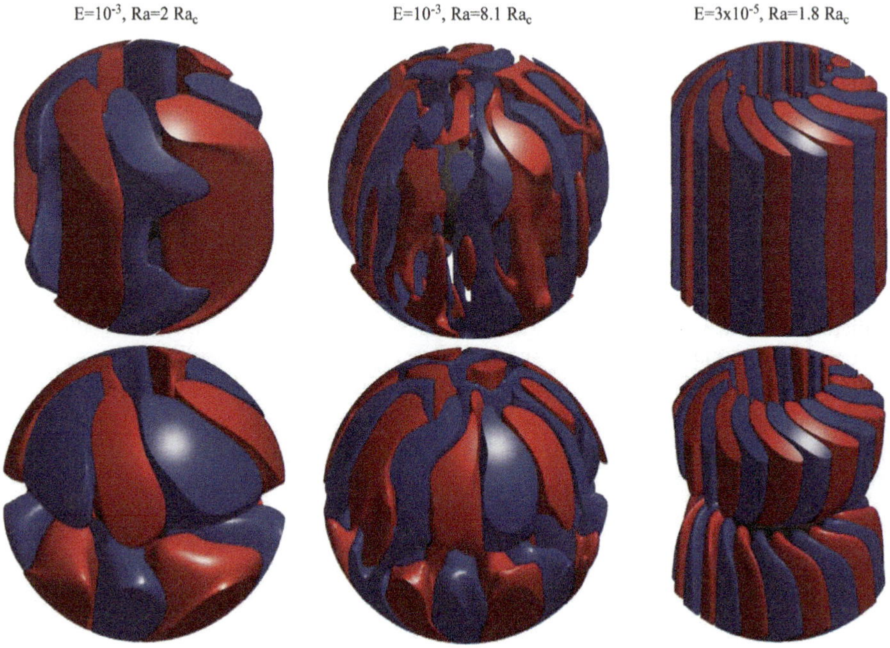

Fig. 2 Flow in non-magnetic convection simulations at different parameter combinations. *The top row* shows positive (*red*) and negative (*blue*) iso-surfaces of the axial vorticity, *the bottom row* shows positive (*red*) and negative (*blue*) iso-surfaces of the z-component of the flow. The Prandtl number is unity in all cases

Figure 2 illustrates the flow structure in non-magnetic convection simulations at the Ekman numbers $\mathrm{E} = 10^{-3}$ (left and middle) and $\mathrm{E} = 3 \times 10^{-5}$ (right). Iso-surfaces of the axial vorticity $\hat{\mathbf{z}} \cdot (\nabla \times \mathbf{U})$ visualize the convective columns that dominate the dynamics in all cases. The stronger Coriolis force influence clearly promotes a more two-dimensional geostrophic configuration in the lower Ekman number model. Increased buoyancy forces and nonlinear interactions, on the other hand, lead to a less geostrophic flow at higher Rayleigh numbers. The lower panels in Fig. 2 show the axial flow component (z-component) to visualize the meridional circulation which is directed away from the equatorial plane in convective columns that rotate in prograde direction (cylones, red) but converges at the equatorial plane in retrograde rotating columns (anti-cyclones, blue). Thus, the z component of the helicity $h = \mathbf{U} \cdot (\nabla \times \mathbf{U})$ has predominantly one sign in the northern hemisphere and the opposite sign in the southern hemisphere. Helicity is known to play an important role in the dynamo process and we will outline in Sect. 5.1 how its large scale coherence is responsible for creating a dipole dominated magnetic field.

Figure 2 demonstrates that the decrease of the Ekman number goes along with a smaller length scale. The reason becomes apparent when considering the linear onset of convection (Roberts 1968; Busse 1970): While the main part of the Coriolis force is balanced by pressure gradients, viscous forces must balance a remaining smaller part in order to facilitate convection (Zhang and Schubert 2000). As the Ekman number is lowered, the balance can only be maintained by further decreasing the length scale in order to keep viscous effects large enough. The convection sets in with a critical wave number m_c that scales like $m_c \sim E^{-1/3}$ in the asymptotic limit for infinitely small Ekman number. The classical critical Rayleigh number Ra_c, following definition (16), grows like $\mathrm{Ra}_c \sim E^{-4/3}$ because of the increasingly impeding action of the Coriolis force. For the two Ekman numbers illustrated in Fig. 2 the critical wave number grows from $m_c = 4$ at $\mathrm{E} = 10^{-3}$ to $m_c = 10$ at $\mathrm{E} = 3 \times 10^{-5}$ while the classical critical Rayleigh number increases from $\mathrm{Ra}_c = 5.5 \times 10^4$ to $\mathrm{Ra}_c = 2.8 \times 10^6$.

Close to onset, the convection is periodic in azimuth (according to m_c) and also symmetric with respect to the equator. When the Rayleigh number is increased beyond its critical value, the system undergoes a succession of symmetry breaking that is accompanied by a decrease in length scale and a growing complexity in time behavior. The azimuthal symmetry is lost first, then the equatorial symmetry. Solutions that are symmetric in azimuth simply drift around the rotation axis, but this simple time dependence is superseded by oscillatory and finally chaotic behavior once the azimuthal symmetry is broken. The second column in Fig. 2 shows a solution at $\mathrm{Ra} = 4.5\,\mathrm{Ra}_c$ with chaotic time behavior and broken azimuthal and equatorial symmetry. The flow remains, nevertheless, predominantly equatorially symmetric at the parameters explored in typical planetary dynamo simulations owed to the prevailing importance of the Coriolis force (Taylor-Proudman theorem). The succession of time dependencies is illustrated in Fig. 4.

Decreasing Ekman number and increasing Rayleigh number both promote primarily smaller azimuthal length scales. The length scale perpendicular to the rotation axis is much less effected and the convective columns therefore change from a rounder to a more sheet like shape. The length scale in the direction of the rotation axis also seems to vary only little over the Rayleigh numbers explored in typical dynamo simulations where the Coriolis force continues to impose a dominantly geostrophic structure.

The tangent cylinder is an imaginary boundary that is aligned with the rotation axis and touches the inner-core equator. It separates the shell into three regions with distinct dynamical behavior: the two regions inside the tangent cylinder above and below the inner core and the region outside the tangent cylinder. The columnar convection discussed above is restricted to the latter region. Inside the tangent cylinder, the convective motions are more effectively prohibited by the Taylor-Proudman effect since gravity roughly acts in the direction of the rotation axis here. Consequently, the convection starts somewhat later at Rayleigh numbers which are a few time Ra_c and it takes the form of local plumes or upwellings rather than columns (Tilgner and Busse 1997). The associated lateral temperature differences can drive strong thermal winds inside the tangent cylinder (Aubert 2005; Sreenivasan and Jones 2006). Thermal winds obey a balance between the curl of the Coriolis force and the buoyancy term in the Navier-Stokes equation (6):

$$2\frac{\partial \mathbf{U}}{\partial z} = -\mathrm{Ra}^\star \frac{r}{r_o} \nabla \times (T\hat{\mathbf{r}}). \tag{25}$$

Thermal zonal winds are thus driven by longitudinal temperature differences and show a characteristic z-variation which is distinctly non-geostrophic.

Figure 9a illustrates the zonal flow structure typical for convection simulations at higher Rayleigh numbers. Outside the tangent cylinder, Reynolds stresses drive a retrograde inner and a basically prograde outer flow (Christensen 2002; Heimpel et al. 2005; Aubert 2005). Thermal winds clearly dominate the region inside the tangent cylinder and are responsible for the weaker north south variation in the outer region. We will discuss the zonal flows further in comparison with the dynamo cases in Sect. 5.4.

The impact of inertial effects scales with the inverse of the Prandtl number. The above described scenarios apply for moderate Prandtl numbers of about one and mildly larger. For large Prandtl numbers, the bending of the columns and the associated Reynolds stresses become smaller. At Prandtl numbers below one, however, the Reynolds stress driven zonal winds can become so strong that the associated shear virtually stops the convection every now and then. This results in a particular time dependence called relaxation oscillation by Grote and Busse (2001). Also, the convection inside the tangent cylinder sets in at lower Rayleigh numbers for smaller Prandtl numbers. At very low Prandtl numbers, however, the dynamics is ruled by thermal inertial waves or pure inertial waves which have too short time scales, in the order of days, to affect the dynamo process (Zhang and Schubert 2000).

5 Numerical Dynamo Solutions, an Overview

5.1 Fundamental Dynamo Mechanism

The relatively simple structure of the benchmark dynamo (Christensen et al. 2001) allows to unravel the details of the dynamo process and to understand how it shapes the field at the top of the dynamo region. The magnetic field is produced in an α^2 mechanism, a terminology that goes back to mean field dynamos (Sect. 2) and here simply means that the large scale poloidal and toroidal magnetic fields are created by the action of small scale flow on small scale magnetic field. This is likely the dominant process for all dynamo solutions presented here. An alternative is the $\alpha\omega$ mechanism where the toroidal field is created by shear in the zonal rotation ω. It may, for example, be responsible for the field production in dynamo models with stress free boundary conditions that allow for larger zonal flows in the Kuang and Bloham or the Busse group models (Sect. 3.4).

The particular α^2 mechanism presented here has been envisioned by Olson et al. (1999) and was later confirmed in a detailed analysis by Wicht and Aubert (2005) and Aubert et al. (2008b). Figure 3 shows radial magnetic fields at the outer boundary r_o in the top panels and iso-surfaces of the axial vorticity along with magnetic fieldlines in the lower panels. The thickness of the fieldlines is proportional to the local magnetic field strength and their color indicates the field direction: red and blue stand for radially outward and inward pointing fields, respectively. The left column of Fig. 3 shows the benchmark II dynamo (model E3a) where the magnetic field production cycle can more clearly be discerned. The radial outflow between cyclonic (red) and anti-cyclonic columns (blue) grabs a north-south oriented fieldline around the equatorial plane and stretches it towards the outer boundary. This produces strong inverse radial field on either side of the equatorial plane that shows as prominent thick fieldlines in the foreground of Fig. 3. The pairwise inverse equatorial field patches often found at the outer boundary of many dynamo simulations are a result of this process.

The fieldlines are then wrapped around the anti-cyclones and are subsequently advected and stretched in northward and southward direction on both sides of the equator, respectively. Responsible for the latter action are the secondary meridional or axial flows discussed in Sect. 4 which are directed away from the equatorial plane in anti-cyclones but converge at the equatorial plane in cyclones. This is where the z-component of the helicity comes into play. The coherent separation of the opposing fields that largely cancels in the equatorial region is vital for producing the global dipole field. The imaginary cycle ends with another north-south oriented fieldline that amplifies the starting line and thereby compensates Ohmic decay. Since the anti-cyclonic columns are the more active part in the above described process the resulting magnetic structure has been called a magnetic anti-cyclones by Aubert et al. (2008b).

Flows converging where cyclones come close to the outer boundary further shape the magnetic field by advectively concentrating the background dipole field. The location of

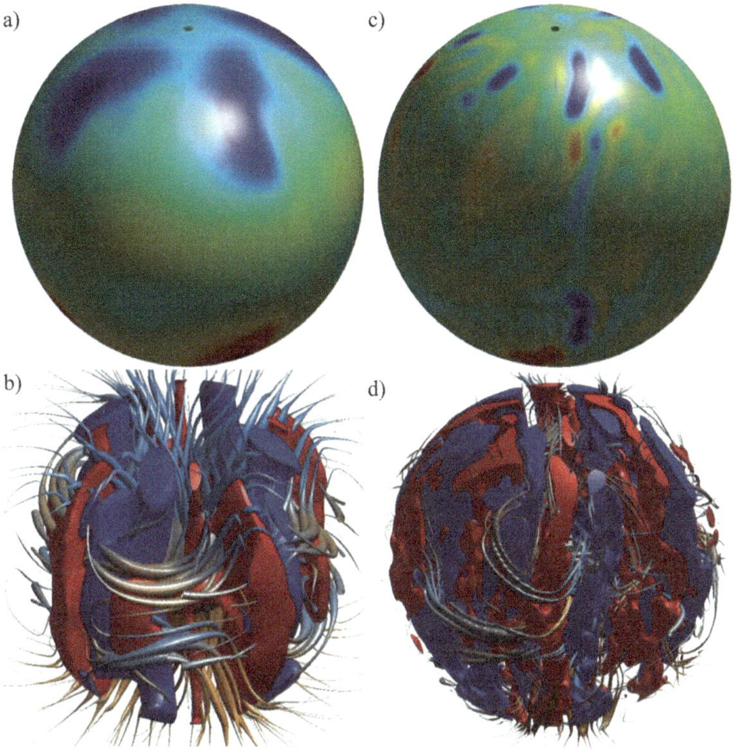

Fig. 3 The top panels (**a**) and (**c**) show snapshots of the radial magnetic field at the outer boundary r_o for the model E3a (dynamo benchmark II) and model E3b which undergoes magnetic field reversals. *Red* and *blue* indicates radially outward and inward field, respectively. *The lower panels* show positive (*red*) and negative (*blue*) iso-surfaces of the axial vorticity along with magnetic fieldlines to visualize the dynamo process. The thickness of the fieldlines is proportional to the local magnetic energy, *the color* indicates the radial direction

the resulting mid to high latitude normal flux spots on the outer boundary roughly indicates where the convective columns touch the tangent cylinder and may be used as a proxy for the inner core size (Stanley et al. 2007). Inside the tangent cylinder, the magnetic field is weakened by advective processes due to the meridional circulation that is directed away from the pole and converges into the cyclonic columns.

The right panels in Fig. 3 show the configuration during a stable polarity epoch in the reversing higher Rayleigh number model E3b. While the structure is significantly more complex and more difficult to interpret than in the benchmark, similar features can nevertheless be discerned: (1) inverse field associated with radial outflows in the equatorial region, (2) the stretching of fieldlines towards the equator in cyclones but towards the poles in anticyclones, (3) the concentration of normal polarity field where flows converge into cyclones, and (4) the weakening of magnetic field inside the tangent cylinder. A feature typical for dynamos at larger Rayleigh numbers is the production of stronger azimuthal toroidal magnetic field in the outer parts of the equatorial plane. A prominent example can seen in the left equatorial region in panel (d) of Fig. 3. Strong toroidal field is produced by azimuthal flows mostly associated with outward stretching flow cyclones. This toroidal field together with the concentrated normal polarity field patches and the equator ward stretching of fieldlines associated with cyclonic columns form the magnetic cyclones introduced Aubert et al. (2008b).

5.2 Dynamo Regimes

Figure 4a illustrates how the properties of a numerical dynamo model depend on Rayleigh and magnetic Prandtl number in a standard model. Ekman number and Prandtl number are kept fixed at $E = 10^{-3}$ and $P = 1$, respectively. The diagram includes the benchmark case II (Christensen et al. 2001) which has been marked by a large square in panel (a) and is listed as model E3a in Table 1. Increasing the Rayleigh number is equivalent to increasing the thermal driving while larger magnetic Prandtl numbers represent larger electrical conductivities. Increasing either parameter can contribute to pushing the magnetic Reynolds number beyond the critical value for onset of dynamo action; there is therefore a certain tradeoff between the two.

The different symbols in Fig. 4 indicate the time behavior. Solutions that simply drift in azimuth (squares) are superseded by oscillating (triangles) and finally chaotic solutions (circles) when the Rayleigh number is increased. The time dependence basically follows the behavior of the purely convective solutions (grey symbols) including the associated symmetry breaking. Exceptions can be found at low Rayleigh numbers where the drifting and oscillatory solutions are replaced by chaotic solutions for larger magnetic Prandtl numbers. The highly symmetric solutions with simple time dependencies occupy only a very small fraction of the parameter space at low Rayleigh numbers and thus seem of little importance for planetary dynamos.

The colored contours in Fig. 4a show the dipolarity measure D, the relative rms dipole field strength at the top of the dynamo region. When the Rayleigh is increased, the dynamo

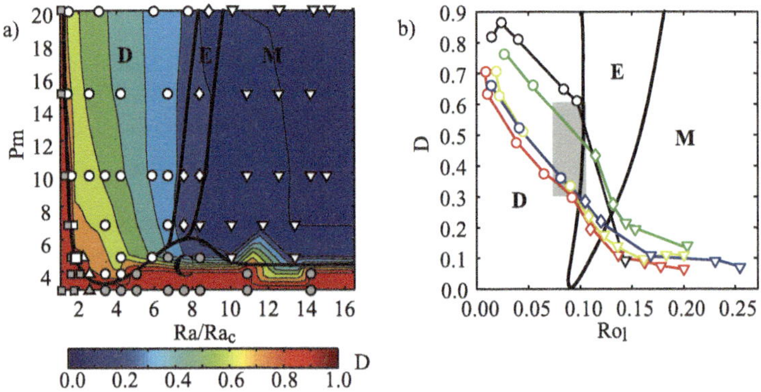

Fig. 4 Regime diagrams that illustrates the transition from stable dipole dominated (regime **D**) to constantly reversing multipolar dynamos (regime **M**). Earth-like rarely reversing models (regime **E**) can be found at the transition. Panel (**a**) is based on standard models with $E = 10^{-3}$. *Grey symbols* mark non-magnetic convective solutions (regime **C**). Different symbols code the time dependence: *squares* = drifting, *upward pointed triangles* = oscillatory, *circles* = chaotic, *diamonds* = Earth-like rarely reversing, downward pointed *triangles* = constantly reversing. Panel (**b**) shows the dependence of the dipolarity measure D on the local Rossby number Ro_ℓ for five different dynamo models. *Blue* and *red*: models from panel (**a**) with Pm = 10 and Pm = 20, respectively; *yellow*: identical parameters to the blue model but with chemical boundary conditions; *green*: $E = 3\times10^{-4}$, Pm = 3, chemical boundary conditions; *black*: $E = 3\times10^{-5}$, Pm = 1, fixed temperature conditions. The Prandtl number is unity in all cases. Chemical convection refers to codensity models that force the codensity flux to vanish at the outer boundary while a fixes codensity is used at the inner boundary (Kutzner and Christensen 2002). *The grey rectangle* shows the range of estimated values for the geodynamo (Wicht et al. 2009)

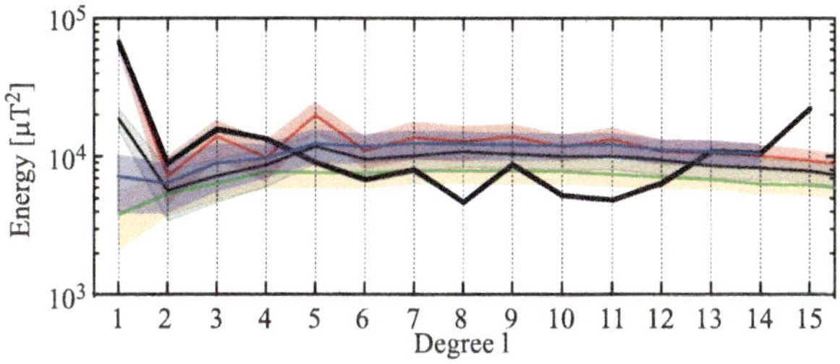

Fig. 5 Time averaged magnetic energy spectra for three dynamo simulations at $E = 3 \times 10^{-4}$, $Pr = 1$, and $Pm = 3$ that use chemical boundary conditions. These models contribute to *the green curve* in Fig. 4 and show solutions in regime **D** ($Ro_\ell = 0.07$, *red line*), in regime **M** ($Ro_\ell = 0.13$, *green line*), and in regime **E** ($Ro_\ell = 0.21$, *black and green lines*). *The black line* depicts a stable polarity epoch while *the green line* depicts a transitional epoch of this reversing model. The time averaging has been performed over periods equivalent to several 10 000 yr in each case. Transparent colored bands in the with of the standard deviation illustrate the time variability. *The thick black line* shows the spectrum for the geomagnetic field model POMME for the year 2000 (Maus et al. 2006)

changes from a dipole dominated regime **D** to a multipolar regime **M** where the dipole component has lost its special character (Kutzner and Christensen 2002; Christensen and Aubert 2006; Olson and Christensen 2006; Wicht et al. 2009). While the dipole never reverses in regime **D** it continuously switches polarity in regime **M**. Earth like reversing behavior (regime **E**), where the dipole still dominates in a time mean sense and polarity transitions, are relatively rare events can be found at the transition. We assume that a model qualifies for this regime when the time averaged dipolarity D is larger than 0.2 and the magnetic pole spends less than a tenth of the time further than 45° degree away from the closest geographic pole. This definition is somewhat arbitrary (Wicht et al. 2009) but nevertheless demonstrates that the Earth-like regime occupies only a small part of the parameter space.

Figure 5 compares time averaged magnetic energy spectra of cases in regimes **D**, **E**, and **M** for $E = 3 \times 10^{-4}$ with the spectrum of a geomagnetic field model for the year 2000 (Maus et al. 2006). The spectra show the magnetic energy carried by the spherical harmonic contributions of degree l at the outer boundary (cmb). The simulations have been scaled by assuming realistic values for Ω, $\bar{\rho}$, μ, and λ (Sect. 3.3) to facility the comparison with the geomagnetic model. Spectra during the transitional periods in regime **E** show the same low dipole contribution as those in regime **M**. Kutzner and Christensen (2002) thus conclude that the models in regime **E** briefly venture into the configuration that is assumed more or less constantly for the larger Rayleigh numbers cases in regime **M**.

Kutzner and Christensen (2002) and Christensen and Aubert (2006) demonstrate that this basic regime diagram also holds for smaller Ekman numbers with two interesting differences. First, as the Ekman number is decreased the boundary between the dipolar and the multipolar regimes moves further out to higher values of Ra/Ra_c. Second, the critical magnetic Prandtl number Pm_c at which dynamo action is still possible decreases. For the simulations at $E = 10^{-3}$ shown in Fig. 4 no dynamos are found below $Pm_c = 4$ (all magnetic field decayed at $Pm = 3$); the values goes down to $Pm_c = 0.1$ at $E = 10^{-5}$ (Christensen and Aubert 2006). This phenomenon has been examined by Takahashi and Matsushima (2006) who suggest a connection with the magnetic Reynolds number. They argue that an effective magnetic Reynolds number could be considered which accounts for the different length scales in the induction equation:

$$Rm_E = \frac{UL^2}{\lambda \ell} = Rm \frac{L}{\ell}. \tag{26}$$

This definition assumes a dipole dominated magnetic field with length scale L and an α-effect that operates on the flow length scale ℓ (22). Since ℓ decreases with the Ekman number one can get away with smaller values of Rm and still maintain the necessary effective magnetic Reynolds number for dynamo action. Smaller values of Rm translate into larger magnetic diffusivities or smaller magnetic Prandtl numbers for fixed values of the flow amplitude U.

Christensen and Aubert (2006) argue that the transition from the dipolar to the multipolar regime should more appropriately by discussed in terms of the local Rossby number Ro_ℓ introduced in (21). They examined a suite of dynamo simulations at various parameter combinations and found that the transition always takes place around $\mathrm{Ro}_\ell \approx 0.12$ (also see Christensen [doi:10.1007/s11214-009-9553-2], this issue). This is also true for all the models presented here and has been confirmed for many others (Olson and Christensen 2006; Takahashi et al. 2008a; Wicht et al. 2009). The examination of Christensen and Aubert (2006) was restricted to dynamos with fixed temperature boundary conditions but also seems to apply to other driving modes and thermal boundary conditions (Olson and Christensen 2006; Wicht et al. 2009; Aubert et al. 2009) with some exceptions that we discuss at the end of this chapter. Figure 4b compares the respective regime diagram for five dynamo models that vary in thermal boundary condition, magnetic Prandtl number, and Ekman number. Only the Ekman number seems to have a significant impact here: The transition steepens when the Ekman number is decreased. At lower Ekman numbers the solutions remain very dipolar right up to the transitions to the multipolar regime. The D values only become Earth-like in regime **E** (Wicht et al. 2009). The heat-flux based scaling suggested by Christensen and Aubert (2006) places Earth at about $\mathrm{Ro}_\ell \approx 0.09$ (Christensen [doi:10.1007/s11214-009-9553-2], this issue) which is nicely compatible with the fact that numerical simulations show Earth-like reversals at comparable Ro_ℓ values.

The dependence on the local Rossby number indicates that inertia, or more precisely the nonlinear advection term in the Navier-Stokes equation (6), plays an important role for the transition from dipolar to multipolar dynamos. The exact reason, however, is not understood nor do we know why the transition should happen at this particular value of Ro_ℓ. Wicht et al. (2009) suggest that the mixing due to nonlinear advection counteracts the ordering influence of the Coriolis force. As a consequence, the dynamo looses its ability to maintain a globally correlated large scale field. We come back to this point in Sect. 6. The value of about $\mathrm{Ro}_\ell \approx 0.09$ suggested for Earth questions the common conclusion that inertia in negligible in planetary dynamo regions. Using the relation $\mathrm{Ro}_\ell = \mathrm{Ro}L/\ell$ to estimated the involved length scale renders $\ell = 4 \times 10^{-5}$ which is equivalent to 100 m in Earth's core. It is hard to conceive that the flow on such a small length scale should influence the dipole behavior. Clearly, more research is need to elucidate the meaning of the local Rossby number.

Sreenivasan and Jones (2005) explore what happens when Prandtl number and magnetic Prandtl number are varied while Rayleigh number and Ekman number are kept fixed. Decreasing Pr increases the importance of inertia in the system and therefore offers an alternative route into the regime where inertia play a more significant role. The study by Sreenivasan and Jones (2005) confirms that the growing influence of inertial forces gradually decreases the symmetry of the solutions and leads from dipole dominated magnetic fields to configurations where the dipole component is significantly less important. Their simulations were not run for long enough to establish whether these latter solutions would actually reverse.

Some authors report deviations from the above described regime scenario. Li et al. (2002) and Nishikawa and Kusano (2008) present Kageyama type models where magnetic field reversals set in at relatively low Rayleigh numbers when the magnetic Prandtl number is increased (see also McMillan and Sarson 2003). This is at odds with the minor influence illustrated in Fig. 4 where a rise in Pm can provoke a more complex time dependence and a less dipolar field but never causes reversals. Both group of authors employ parameters similar to the ones explored in Fig. 4 but retain weak compressibility effects instead of using the Boussinesq approximation. Whether this really explains the differences remains to be explored.

The work of Simitev and Busse (2009) offers another interesting view point. Like Sreenivasan and Jones (2005) these authors explore the Prandtl number dependence of their Busse group dynamo model. They find dipole dominated stable solutions for larger Prandtl numbers, where inertia is less important, and reversing multipolar solutions for small Prandtl numbers. For Prandtl numbers around $\mathrm{Pr} \approx 1$, however, both types of solutions can be found depending on the starting condition of the simulation. A similar case is reported by Christensen and Aubert (2006). The possible coexistence of different attractors is a common property of non-linear systems and needs to be explored further in the dynamo context.

Under certain conditions, non-dipolar dynamos may already be preferred at low Rayleigh numbers right at the onset of dynamo action. For example, Kutzner and Christensen (2002) find multipolar solutions for a standard model which, however, is driven exclusively by internal heat sources. Grote et al. (1999) and Grote and Busse (2000) report Busse group models that are dominated by the magnetic quadrupole term or mainly operate in one hemisphere. These cases are to some degree driven by volumetric buoyancy sources which suggest that this driving type promotes non-dipolar dynamos. Aubert et al. (2009), on the other hand, find clearly dipole dominated purely volumetric driven dynamos. The reason for the differences remains unclear but the existence of two parallel attractors once more offers a possible explanation (Simitev and Busse 2009). A more extensive discussion on non-dipolar dynamos can be found in Stanley and Glatzmaier [doi:10.1007/s11214-009-9573-y], this issue).

5.3 Magnetic Field Structure

Figure 6 demonstrates how the magnetic field structure changes when the Ekman number is decreased with representative snapshots. Model E3b is the larger Ekman number model ($\mathrm{E} = 10^{-3}$) we depicted to explain the dynamo mechanism (Fig. 3). Models E5b, E5F, and E5c share the low Ekman number $\mathrm{E} = 3\times10^{-5}$ as well as identical Prandtl and magnetic Prandtl numbers ($\mathrm{Pr} = 1$, $\mathrm{Pm} = 1$). Model E5b has a local Rossby number of $\mathrm{Ro}_\ell = 0.97$ where reversals can be expected, but we did not observe any such event in the limited time we where able to compute. The larger Rayleigh number in Model E5c increases the local Rossby number to $\mathrm{Ro}_\ell = 0.14$; consequently the dipole is weak and reverses frequently. Model E5F has a lower Rayleigh number that places it safely into the dipole dominated regime at $\mathrm{Ro}_\ell = 0.065$ and is the only case where stress free boundary conditions are employed here. Finally, model E6 has the lowest Ekman number of $\mathrm{E} = 3\times10^{-6}$ and a moderate Rayleigh number. The dipole component is rather strong and the local Rossby number small at $\mathrm{Ro}_\ell = 0.023$. We imposed a four fold azimuthal symmetry in model E6 in order to reduce the numerical costs. Parameters and properties of all cases are listed in Table 1.

The left column of Fig. 6 shows the radial field at the outer boundary. The comparison between models E3b and E5b once more reveals the smaller scales found at lower Ekman numbers but also shows many similarities: (1) The strong normal polarity patches close to where the tangent cylinder touches the outer boundary, (2) the weakened field inside the tangent cylinder, (3) inverse field patches inside the tangent cylinder, and (4) pairwise inverse field patches around the equator. When the Ekman number is reduced from $\mathrm{E} = 3\times10^{-5}$ in model E5b to $\mathrm{E} = 3\times10^{-6}$ in model E6, however, interesting differences appear. The normal

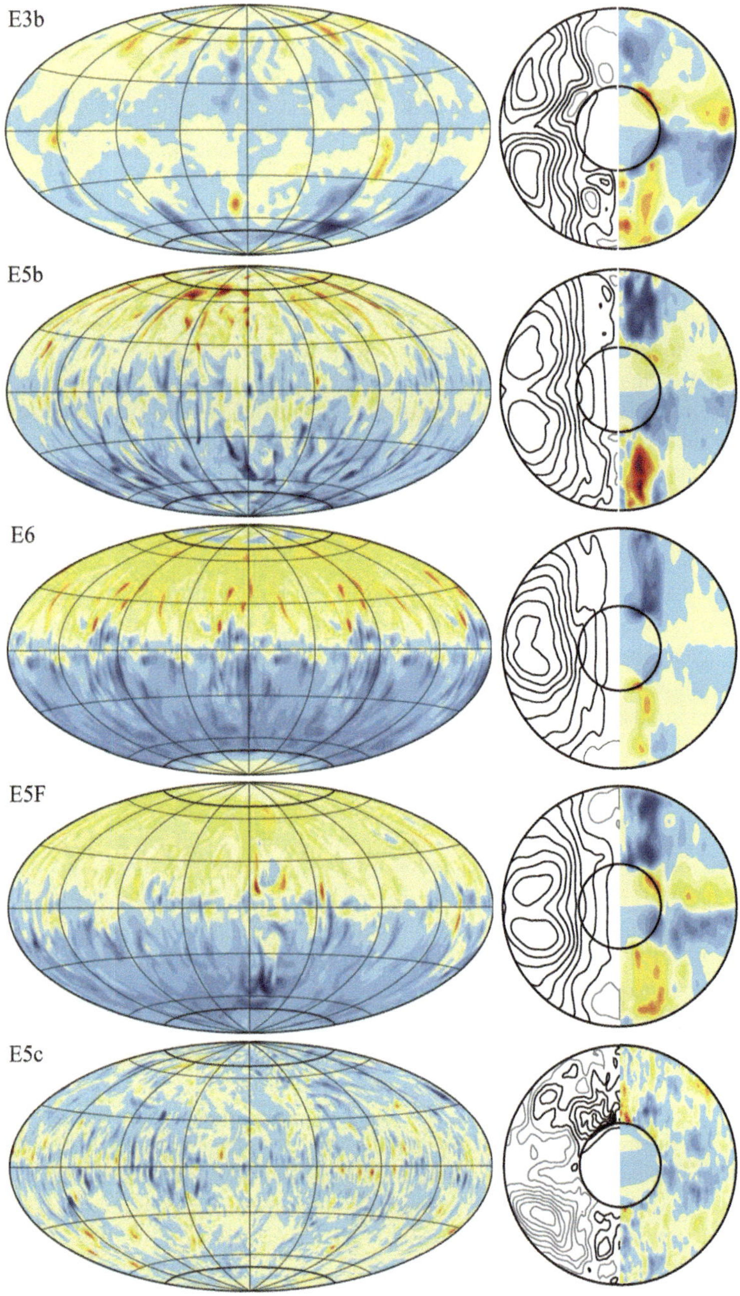

Fig. 6 Radial magnetic field at the outer boundary (*left panels*), axisymmetric magnetic field lines (*right panels, left half*) and axisymmetric toroidal field (*right panels, right half*) for five dynamo models. See Table 1 for parameters. *Blue* (*yellow/red*) *colors* refer to radially inward (outward) or retrograde (prograde) field directions, respectively. The field direction is clockwise (counter-clockwise) along *black* (*grey*) fieldlines

polarity field is more evenly distributed over both hemispheres, respectively, and normal polarity patches also dominate at mid to lower latitudes. The pairwise inverse field patches around the equator are now limited to a thin latitude band. The stress free model E5F shows a very similar configuration. The Coriolis force still seems to be too little influential at larger Ekman numbers. At model E5b either the stronger impact of inertial forces due to the larger Rayleigh number and/or the somewhat larger viscosity may still prevent the lower latitude field to be more Earth-like. The former reason seems more likely, since the viscous boundary friction is already quire low. Dynamo simulations by Takahashi et al. (2008a) show a similar transition at comparable Ekman numbers.

Figure 7 compares the filtered outer boundary field in models E5b and E5F with the geomagnetic field model GUFM for the epoch 1990 (Jackson et al. 2000). The filter simulates the effect of Earth's crustal magnetization that screens the internal field beyond spherical harmonic degree $l = 12$. Both numerical models agree with the geomagnetic field in several properties: the dipole dominance, the typical high latitude normal polarity patches, the weakened or inverse field inside the tangent cylinder, and the presence of inverse patches in the equatorial region. The strong normal polarity patches close to the equator, that are evident in Fig. 7, seem to be typical for the historic geomagnetic field (Jackson 2003; Jackson and Finlay 2007). Whether they also form an important characteristic of the fundamental geodynamo process and would prevail in a long time scale average is unknown. The low latitude normal patches are much weaker in our models E6 and E5F and inverse patches are less abundant than in the historic geomagnetic field. Whether this changes at larger Rayleigh numbers where reversals can be expected remains to be explored. The strong nor-

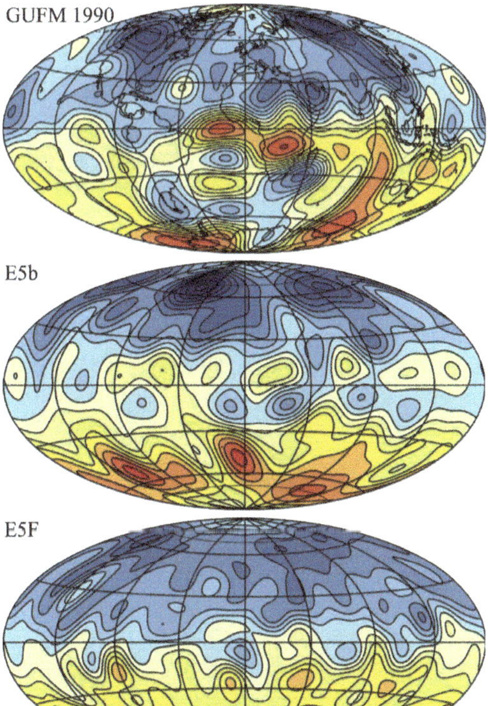

Fig. 7 Comparison of the radial magnetic field at the core-mantle boundary for the 1990 GUFM model (Jackson et al. 2000) and filtered snapshots at the outer boundary off models E5b and E5F. Figure 6 shows the full numerical resolution for both numerical models. We have inverted the field direction here, to ease the comparison with the geomagnetic model

mal polarity patches close the tangent cylinder still remain prominent in the filtered outer boundary field in models E6 and E5F (see Fig. 7 for the latter case). Close-by inverse field patches evident in the full resolution fields shown in Fig. 6 reduce the impact of the low to mid latitude normal patches in the filtered field version.

The change in the dynamo mechanism responsible for the difference in field geometry in unknown. Except for the multipolar model E5c the axisymmetric poloidal fieldlines and the axisymmetric azimuthal toroidal fields shown in Fig. 6 are remarkable similar. Inverse fieldlines around the equatorial plane seem a common feature but become weaker for lower Ekman numbers which is likely a condition for the prevailing normal polarity cmb field at lower latitudes. The strong axisymmetric azimuthal toroidal field found inside the tangent cylinder in many dynamo models at larger Rayleigh numbers is produce by the thermal winds via an ω effect.

5.4 Flow Structure

Figure 8 illustrates the flow structures that correspond to the magnetic field snapshots shown in Fig. 6. The comparison of flow and magnetic field structures for model E3c reveals that the latter is of smaller scale. This is owed to the larger magnetic Prandtl number of $\mathrm{Pm} = 10$ in this model which means that the viscous diffusivity is an order of magnitude larger than the magnetic diffusivity. Figure 8 once more demonstrates that the convective features become progressively thinner and more sheet-like as the Ekman number is decreased. However, the sheets seem to break into smaller entities in the high Rayleigh number multipolar model E5c.

It is difficult to assess how the magnetic field changes the flow in order to saturate its growth. A simple primary effect is that the magnetic field decrease the rms flow amplitude U. A comparison between the convective columns in Fig. 2 and Fig. 3 also reveals that the cyclones are much smaller than the anti-cyclones in the presence of a magnetic field. This is due to the fact that the fieldlines tend to pass directly through the cyclones but stay more clear of the anti-cyclones (Aubert and Wicht 2004). However, the difference seems to vanish at larger Rayleigh numbers and smaller Ekman numbers. More research is required to elucidate the true saturation mechanism which may actually be quite subtle (Jones 2000).

Scenarios of magnetic field saturation are sometimes discussed in connection with a run-away growth of the magnetic field and the jump from a weak field to a strong field branch. On the weak field branch, the Lorentz force is too small to balance the Coriolis force significantly. Suppose however, that a field strength fluctuation leads to a slightly better cancelation between Lorentz and Coriolis force. This would somewhat alleviate the viscous force from balancing the Coriolis force to facilitate convection. The flow could assume a larger scale and would become more vigourous. This in turn produces a yet stronger magnetic field resulting in a runaway that ends on the strong field branch where the magnetostrophic balance is established at an Elsasser number of order one.

The promoting effects of magnetic fields on convection have extensively been explored in magnetostrophic models where the magnetic field is imposed rather than produced by a dynamo effect (see Wicht et al. 2009, for a recent overview). The change in flow length scale and the runaway growth have also been observed in simplified self-consistent dynamo simulations that employ Cartesian geometries (Rotvig and Jones 2002; Stellmach and Hansen 2004). These simulations suggest that the Ekman number has to be small enough for these effects to be significant, probably as low as $\mathrm{E} \leq 10^{-5}$ (Stellmach and Hansen 2004). Though today's self-consistent spherical dynamo models have reached Ekman numbers as low as $\mathrm{E} = 10^{-6}$ the predicted large scale change and the

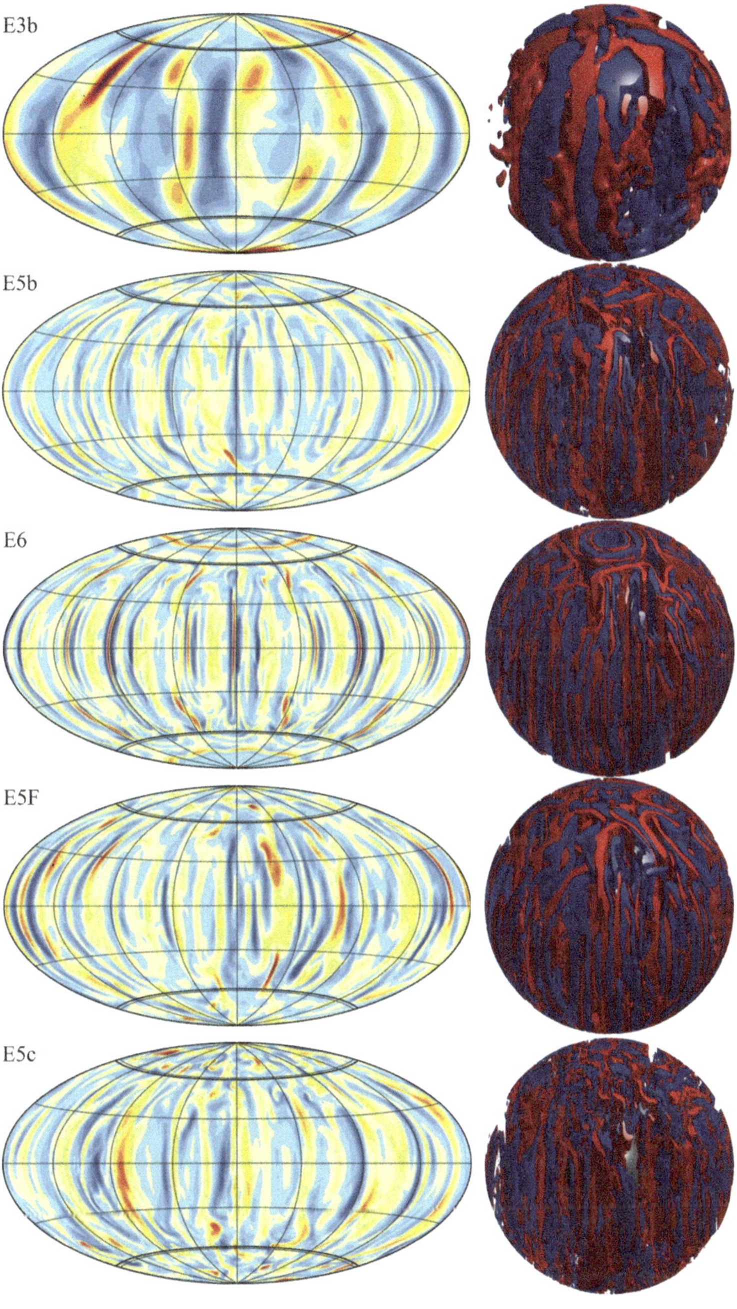

Fig. 8 Radial flow at mid shell $r = r_i + (r_o - r_i)/2$ (*left*) and iso-surfaces of axial vorticity (*right*) for the models depicted in Fig. 6. *Yellow/red* (*blue*) contours stand for radially outward (inward) radial flows. *Red* (*blue*) iso-surfaces depict prograde (retrograde) rotating features

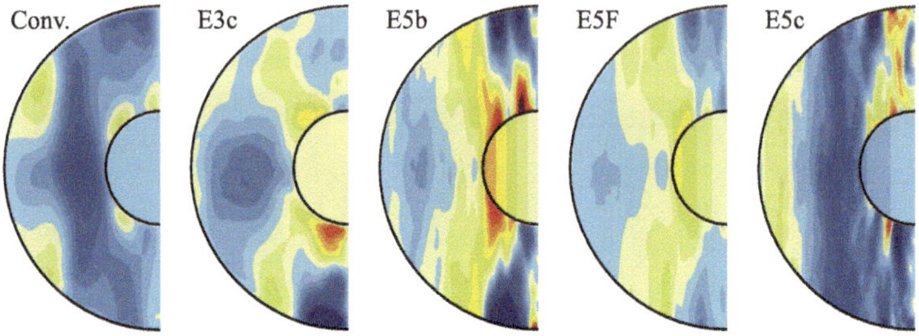

Fig. 9 Snapshots of he zonal flow in four different dynamo models and a non-magnetic convection simulation (Conv.) that uses the same parameters as model E3c. The contour steps are identical for the left tow panels and for the right three panels, respectively. The snapshots show typical large scale flow configurations. The smaller scale details may vary with time, in particular inside the tangent cylinder where the convective plumes become rather time dependent at larger Rayleigh numbers (cases E3c and E5c)

jump between two distinct branches has not been observed. Most if not all of these models would probably qualify as strong field cases. (Takahashi et al. 2008a) report that the mean flow scale ℓ (22) increases by only 20% when comparing a dynamo and a non-magnetic simulation at $E = 10^{-6}$. The rather weak influence on the flow scale may be attributed to the fact that dynamos tend to adopt a configuration where the interaction between shear and magnetic field is minimized (Aubert and Wicht 2004; Aubert 2005; Jones 2007): the fieldlines wrap around the convective features rather than passing through them and mostly arrange themselves perpendicular to the shear which minimizes their interaction with the flow.

Figure 9 shows snapshots of the zonal flow that illustrate typical configuration in several numerical models. The leftmost panel depicts a non-magnetic case with the same parameters as the large Ekman model E3c. The comparison between the zonal flow in these two cases reveals a significant difference outside the tangent cylinder. Here, Reynolds stresses rule in the non-magnetic case while thermal winds dominate in the magnetic case. This effect has first been reported by Aubert (2005) who finds that the zonal components of Coriolis and Lorentz force balance to a good degree in dynamo models. The associated relaxation of the Taylor-Proudman theorem allows an increased heat transport in the equatorial region and inside the tangent cylinder. The resulting temperature gradients are responsible for the stronger thermal winds that dominate the zonal flow in dynamo simulations. The comparison of the zonal flows for models E5b and E5F in Fig. 9 shows that the main differences can be found inside the tangent cylinder where the higher Rayleigh number in the former model drives stronger thermal winds. Outside the tangent cylinder, the solutions are remarkably similar. We have analyzed the viscous forces acting on the zonal flows in model E5b and find that the viscous drag at the boundaries is comparable to the bulk friction. This may explain the minor differences between the stress free and the rigid boundary model.

The last panel in Fig. 9 shows the zonal flow in the larger Rayleigh number multipolar case E5c. The zonal flow is back to a Reynolds stress influenced configuration outside the tangent cylinder which suggests that the Lorentz force less efficiently balances the Coriolis force when the dipole component is weak. Inside the tangent cylinder, the structure is more complex. Possibly, Reynolds stress effects (Heimpel et al. 2005) and thermal wind effects are now at work here. The high latitude retrograde flow found close to the outer boundary in the dipole dominated cases is also a typical feature in core flow models derived from geomagnetic secular variation data (Eymin and Hulot 2005). Aurnou et al. (2003) and Aubert

(2005) show that the amplitude of these flows can be scaled to Earth-like values based on the heat flux through the outer boundary as discussed in Sect. 3.3.

6 Reversals Mechanisms

Magnetic field reversals are the most drastic changes in the geomagnetic field. Paleomagnetic data quantify several of their properties but provide essentially no information about their causes and internal dynamics. Numerical dynamo models try to fill this void with extensive parameter studies and detailed analysis of simulated polarity transitions. We have outlined in Sect. 5.2 that many dynamo simulations show reversals when the convection is driven strongly enough to make the inertial forces sufficiently influential which seems to happen around a local Rossby numbers of $\mathrm{Ro}_\ell \approx 0.1$.

At which parameters reversals set in exactly is hard to determine numerically. Since the likelihood for reversals seems to increase with the local Rossby number (Wicht et al. 2009) and the duration of stable polarity epochs varies stochastically, models at lower Ro_ℓ values may simply be categorized as non-reversing because the simulated time span is too short. Wicht et al. (2009) explored several dynamo models at $\mathrm{E} = 10^{-3}$ and $\mathrm{Pm} = 10$ where the computation of extremely long time spans could be afforded. No reversal was found during more than 140 magnetic diffusion times (equivalent to more than 17 Myr for Earth) in a model with $\mathrm{Ro}_\ell = 0.08$ while a model with $\mathrm{Ro}_\ell = 0.10$ underwent 4 reversals during 87 magnetic diffusion times (equivalent to more than 10 Myr for Earth). (Both cases entered Fig. 4.) This seems to support the rule of thumb that reversals start to appear around $\mathrm{Ro}_\ell = 0.1$ but it can not be excluded that reversals just become extremely unlikely for smaller local Rossby numbers.

Several authors have observed that numerical reversal processes correlate with an increased equatorially symmetric magnetic field component (Li et al. 2002; Takahashi et al. 2005, 2007, Nishikawa and Kusano 2008). This is typically accompanied by an increase of the equatorially asymmetric kinetic field which is required to convert equatorially asymmetric magnetic field into its symmetric counterpart. However, significant violations of either symmetry can already be observed for dynamos that never seem to reverse (Wicht et al. 2009). The breaking of equatorial and azimuthal symmetries are therefore necessary but not sufficient conditions. The same is true for the plume like convection inside the tangent cylinder that may also play a role in polarity transitions (Rotvig 2009).

Aubert et al. (2008b) have identified distinct magnetic features that are responsible for producing significant amounts of inverse magnetic field. These features are tied to flow upwellings or plumes and have therefore been named magnetic upwellings by these authors. They fall into two distinct categories: those that rise inside the tangent cylinder and those that prefer low to mid latitudes. The geomagnetic field shows inverse field patches in both regions.

Inside the tangent cylinder, the magnetic upwellings are connected to the convective plumes we briefly touched on in Sect. 4 (Aurnou et al. 2003; Aubert et al. 2008b). Sreenivasan and Jones (2006) show that the size and number of these plumes depends on the background magnetic field strength. Fewer but thicker plumes can be observed for stronger fields. This provides an interesting feedback mechanism whose role for the inverse field production has not yet been fully explored. The plumes inside the tangent cylinder can significantly reduce the background dipole field and tilt it to a limited extent.

More instrumental for field reversals, however, are the low to mid latitude magnetic upwellings. We have outlined in Sect. 5.1 that the outflows in the equatorial region produce a

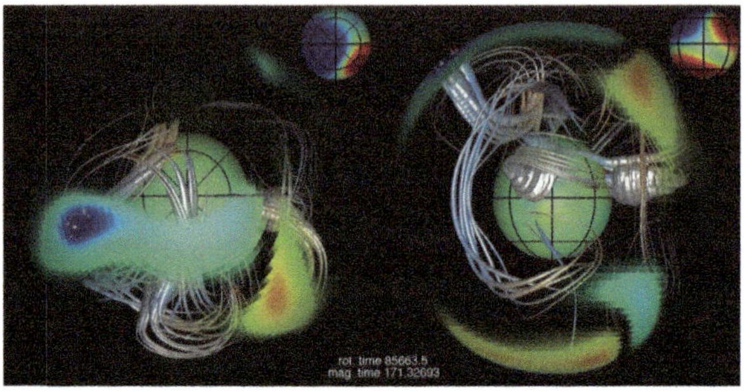

Fig. 10 Magnetic fieldlines during a reversal visualized with the DMFI tool (Aubert et al. 2008b). The thickness of the fieldlines has been scaled with the local magnetic energy. North-polar and side view are shown at left and right, respectively. The inner core is represented by the central sphere. Color coded radial magnetic field is shown on the inner core, the outer boundary (only dominant field patches), and at a level representing Earth's surface in the upper right corner of *both panels*. The snapshot depicts two magnetic upwellings in the northern hemisphere that have already created enough inverse field to significantly tilt the dipole. A magnetic-anticyclone can been seen in the left hemisphere

large amount of inverse field even in very simple dynamo models that never reverse. Here, inverse field of both polarities is produce at either side of the equatorial plane so that the effects largely cancel. In magnetic upwellings, the north/south coherence is obviously violated and inverse field is produced predominantly on one side of the equator (Aubert et al. 2008b; Wicht and Olson 2004). This corresponds to the loss of equatorial asymmetry in the magnetic field and of equatorial symmetry in the flow field described by some authors (Li et al. 2002; Takahashi et al. 2007; Coe and Glatzmaier 2006; Nishikawa and Kusano 2008).

The typical time scale of the magnetic upwellings is about one kyr, but their duration, amplitude, and frequency seem to vary stochastically. Weaker magnetic field excursions can be caused by one magnetic upwelling of average amplitude. Full polarity reversals or strong global excursions with predominantly inverse field directions require a particularly fierce and long lasting upwelling to sufficiently cancel the normal polarity field. Alternatively, several upwellings can team up to perform the job. Both scenarios are unlikely which explains the rareness of these events. The simulations suggest that the cancelation of the normal polarity field is the essential process that enables reversals. It leaves a largely undecided field polarity and once the magnetic upwellings have ceased a subsequent restart of the 'normal' dynamo mechanism may produce field of either polarity, rendering the event a reversal or an excursion (Kutzner and Christensen 2002).

A detailed analysis of reversal mechanisms that considers the 3d structure of flow and magnetic field is complex and has therefore been limited to a few cases, mostly at large Ekman numbers. The importance of magnetic upwellings for field reversals has most clearly been demonstrated at $E = 10^{-2}$ and $E = 2\times10^{-2}$ in the models by Wicht and Olson (2004) and Aubert et al. (2008b), respectively. Magnetic upwellings remain important at $E = 3\times10^{-4}$, but so far no detailed reversal analysis has been performed at this Ekman number (Aubert et al. 2008b). Takahashi et al. (2007) report that the inverse field in their model at $E = 2\times10^{-5}$ is part of a strong magnetic loop that correlates with a deformed convective column. Whether and how these features relate to magnetic upwellings remains unclear. Common to most reversal simulations is the importance of strong inverse magnetic field patches at low to mid latitudes in the beginning phase of a reversal (Driscoll and Olson 2009). This agrees with the findings in the paleomagnetic model for the Matuyama-Bruñes reversal by Leonhardt and Fabian (2007).

Busse and Simitev (2008) describe a completely different reversal scenario which is based on a Parker dynamo wave: inverse field starts to grow on either side of the equatorial plane and propagates towards the poles (Parker 1955; Simitev and Busse 2005). A Parker wave is a purely magnetic linear oscillatory process that does not rely on any flow changes. Busse and Simitev (2008) find quasi oscillatory reversals in the toroidal magnetic field while the poloidal field varies to a much lesser extend. Only occasionally are the oscillations strong enough to cause a global excursion of field reversal. This explains their rareness in comparison to the proposed Parker wave period of 40 kyr for Earth.

We have already discussed in Sect. 5.2 that the scaling based on the heat flux out of Earth core places the geodynamo at $Ro_\ell = 0.09$ and thus into the regime **E** where Earth like reversals are found in many simulations (Christensen and Aubert 2006; Olson and Christensen 2006). Should the weakly compressible models that seem to model reversals at smaller local Rossby numbers provide a vital alternative here (Li et al. 2002; Nishikawa and Kusano 2008; McMillan and Sarson 2003)?

We refer to Glatzmaier and Coe (2007) and Wicht et al. (2009) for a more detailed analysis of simulated reversals and a comparison with paleomagnetic findings.

7 Torsional Oscillations

Torsional oscillations are a specific form of flow variations that concern the rotation of geostrophic cylinders. Like the tangent cylinder, these cylinders are aligned with the planetary rotation axis. Their rotation is the only truly geostrophic flow component and also determines the total angular momentum of the fluid. Decadal variations in Earth's length-of-day are commonly attributed to an exchange of angular momentum between core and mantle. This is strongly supported by the reasonable agreement between these variations and changes in the estimated core angular momentum (Jault et al. 1988; Jault and LeMouël 1989; Jackson et al. 1993; Jackson 1997; Jault 2003; Amit and Olson 2006). Torsional oscillations have a decadal time scale which is significantly shorter than the flow turnover time $t_U \approx 120\,\mathrm{y}$ thought to be characteristic for the dynamo process. They may thus not only form an important part of the faster magnetic field variations but are also a likely cause for the decadal length-of-day variations (Braginsky 1970). Bloxham et al. (2002) suggest that torsional oscillations may also cause geomagnetic jerks, a rapid change in the geomagnetic secular variation signal on a yearly time scale. Moreover, they can provide valuable information in the internal dynamics of the dynamo region.

When integrating the azimuthal component of the Navier-Stokes equation over a geostrophic cylinder Coriolis force and pressure force drop out. Only Lorentz force, the nonlinear inertial term, and viscous effects therefore contribute to the acceleration of these cylinders. The latter two contributions are generally thought to be negligible in the magnetostrophic regime which leaves the Lorentz force as the only contribution in the first order force balance. Since this could lead to an awkward unbalanced acceleration Taylor (1963) concluded that the azimuthal Lorentz forces must largely cancel when integrated over the cylinders. In recognition of Taylor's work, the respective Lorentz force integral has been named the Taylor integral, the cancelation is referred to as Taylorization, and the dynamo configuration that is established by the cancelation is called a Taylor state.

Variations in the degree of Taylorization at an individual cylinder lead to an acceleration of this cylinder with respect to its neighbors. This shears B_s, the magnetic component perpendicular to the rotation axis (Taylor 1963), and thereby produces an azimuthal magnetic field. The Lorentz forces associated with this newly created field oppose the shear according to Lenz's law and therefore provide a restoring mechanism. The resulting oscillations are called torsional oscillations since the magnetic fieldlines effectively act as torsional springs in this process (Braginsky 1970). These oscillations travel as one-dimensional Alfvén waves in s direction with a characteristic Alfvén speed that is proportional to the rms B_s value over the geostrophic cylinders (Braginsky 1970).

Lorentz forces, gravitational forces, and viscous effects potentially couple the geostrophic cylinders to the inner core and the mantle. The coupled system can be approximated by an eigensystem that has been analyzed by several authors (Braginsky 1970; Buffett and Mound 2005; Mound and Buffett 2005; Dumberry and Mound 2008). The eigenperiods of this system are also proportional to the rms B_s value and lie in the range of some years to decades for Earth (Braginsky 1970; Buffett and Mound 2005). Viscous effects, Ohmic effects, the coupling to the inner core, but in particular the coupling to the mantle can significantly damp torsional oscillations on time scales comparable to the oscillation period (Zatman and Bloxham 1997, 1999; Buffett and Mound 2005; Dumberry and Mound 2008). This could prevent the excitation of the individual standing eigenmodes that were envisioned by (Braginsky 1970) but would still allow for the one-dimensional traveling Alfvén waves to survive for a limited time (Buffett and Mound 2005). We refer to Zatman and Bloxham (1999) for an extensive discussion of excitation and damping mechanisms.

The magnetostrophic balance, Taylorization, Taylor state, and torsional oscillations are intimately linked. Though the importance of torsional oscillations for the dynamo process itself may be minor, their presence can serve as an important indicator for the fact that viscous forces and inertial forces are indeed small in a planetary dynamo region. Moreover, torsional oscillations can provide estimates of the rms B_s value and thus valuable information on the magnetic field inside the dynamo region which is not directly accessible by other observations. This highlights the potential importance of the work by Zatman and Bloxham (1997) and Bloxham (2002), who claim to have identified torsional oscillations of the predicted time scales in the geomagnetic field. In a recent study, however, Dumberry and Mound (2008) show that the damping due to the coupling to the mantle may actually be too strong for oscillations to be excited sufficiently. The decadal time scale, identified in core flow inversions and length-of-day variations, will then more likely reflect the dynamics of the excitation mechanism rather than the predicted eigenperiod (Dumberry and Mound 2008). Very likely, this excitation mechanism is the dynamo process which would thus manifest itself on much shorter time scales than the overturn time t_U.

Several authors have tried to identify torsional oscillations in dynamo simulations. Kuang (1999) and Dumberry and Bloxham (2003) show that viscous and inertial forces are still too influential in their Kuang and Bloxham model at $E = 10^{-4}$. Busse and Simitev (2005) report to have found torsional oscillations at an Ekman number of $E = 2\times10^{-5}$. Wicht and Christensen (2010) examine several dynamo models at different parameters and confirm that Taylorization and torsional oscillations go along with each other and can be observed when the viscous forces are small enough: No torsional oscillations can be identified at $E = 3\times10^{-4}$, first traces appear at $E = 3\times10^{-5}$, and the clearest signatures are present in their lowest Ekman number model at $E = 3\times10^{-6}$ referred to as E6 here. Figure 11 shows the zonal flow amplitude of geostrophic cylinders for a selected time span in model E6 where three waves are clearly discernable. The white lines mark the 'path' a wave would take when traveling with the Alfvén speed predicted for the propagation of torsional oscillations. The good agreement with the true propagation is an important test for the origin of these waves. The fact that the waves are rather short lived events demonstrates that the damping is large. While the strong damping of geomagnetic torsional oscillations is likely provided by the coupling to the mantle viscous effects may play this role in the numerical simulation.

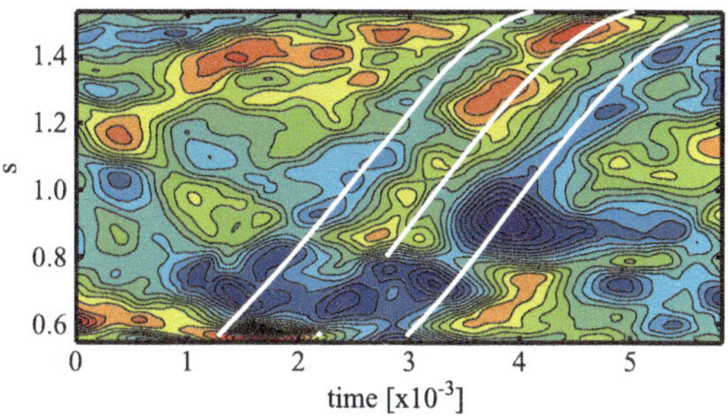

Fig. 11 Velocity of geostrophic cylinders in a selected time span of model E6. *The white lines* mark torsional oscillations that propagate with the predicted Alphvén velocity. The time is given in multiples of the magnetic diffusion time here

Wicht and Christensen (2010) also find that Reynolds stresses are too influential and prevent significant torsional oscillations as well as an efficient Taylorization at local Rossby number values around $\mathrm{Ro}_\ell = 0.1$ where the simulations start to show reversals. Translated to Earth this would mean the identification of torsional oscillations in the outer core by Zatman and Bloxham (1997) and the value of $\mathrm{Ro}_\ell = 0.09$ predicted by Christensen and Aubert (2006) are not compatible.

8 Mantle Control

Neither the longitudinal nor the equatorial symmetry are broken in the dynamo formalism developed in Sect. 3.1. The cmb field should therefore be perfectly axisymmetric and equatorially asymmetric when averaged over periods that are much longer than the typical time scale of magnetic field production, the convective overturn time of $t_U \approx 120\,\mathrm{yr}$. There are, however, some indications that both these symmetries are broken in the geomagnetic field; we mainly discuss deviations from the axial symmetry in the following.

A particularly prominent persistent longitudinal feature are normal polarity flux patches in the radial magnetic field under Canada and Siberia with respective counterparts at comparable longitudes and latitudes south of the equator. They show up in paleomagnetic field models covering up to 5 Myr (Gubbins and Kelly 1993; Kelly and Gubbins 1997; Johnson and Constable 1995; Carlut and Courtillot 1998; Johnson et al. 2003) and can also be identified in archeomagnetic (Korte and Constable 2006) and historic (Jackson et al. 2000; Bloxham 2002) data (see Fig. 7).

These patches are reminiscent of the normal polarity patches close but outside the tangent cylinder in many dynamo simulations. However, as long is the longitudinal symmetry is not broken, these patches move around and also cease and emerge in an irregular fashion. Imposing a laterally varying cmb heat flux is a straight forward way to break both the axial and the equatorial symmetry. Seismic tomography data can be used to derive a geophysically motivated pattern by translating faster than average seismic velocities in the lowermost mantle (D''-layer) into colder temperatures. Since the cmb is basically isothermal, colder temperatures yield a higher heat flux from the core to the mantle.

The procedure results in the typical 'tomographic' pattern shown in Fig. 12a of increased heat flux beneath the circum-Pacific rim which is often associated with the remains of colder subducted plates. Particularly high flux can be found beneath central America, eastern Siberia, and west of Australia (Masters et al. 2000). Aubert et al. (2008a) suggest that the relative amplitude of the lateral variation, $f^\star = (f_{\max} - f_{\min})/2f_0$, may amount to about $f^\star = 0.3$ for Earth. However, neither the mean heat flux f_o nor the tomographic variation pattern are particularly well constrained. Earth's mean cmb heat flux may actually

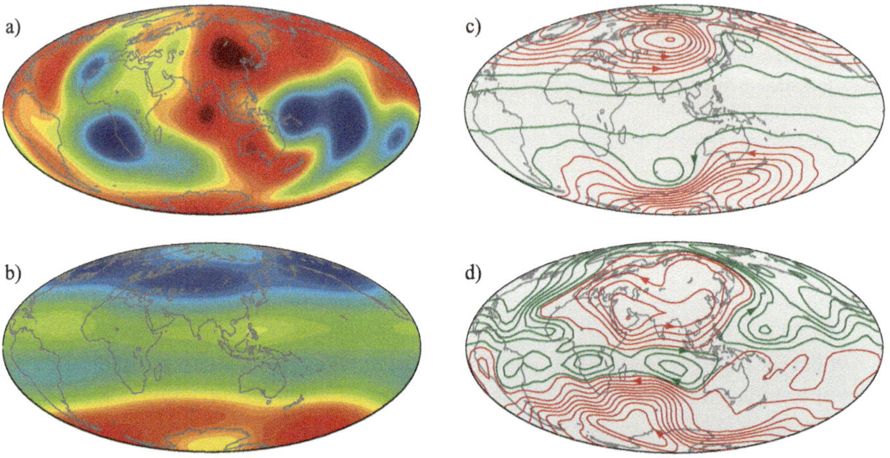

Fig. 12 (**a**) cmb heat flux pattern derived from a seismic tomography model for Earth's mantle (Masters et al. 2000), (**b**) time averaged radial magnetic field at the outer boundary in the dynamo model by Aubert et al. (2008a) that employs the pattern shown in panel (**a**) as the outer boundary condition, (**c**) associated time averaged horizontal flow underneath the viscous Ekman layer, (**d**) time averaged flow field (Amit and Olson 2006) based on a geomagnetic field model that covers the years 1840–1990 (Jackson et al. 2000)

be close enough to being adiabatic so that $f^\star$ can assume values larger than one and the heat flux would actually be subadiabatic in some regions. Several authors report that such a situation kills off the dynamo action in their numerical model (Glatzmaier et al. 1999; Olson and Christensen 2002; Willis et al. 2007) but this seems not to be the case in the simulations by Takahashi et al. (2008b).

The tomographic heat flux condition indeed promotes persistent magnetic flux patches at comparable locations as those found in the geomagnetic field (Olson and Christensen 2002; Christensen and Olson 2003; Gubbins et al. 2007; Aubert et al. 2008a). Figure 12b shows the results from Aubert et al. (2008a) which can be compared with the geomagnetic field model in Fig. 7. Figure 12 demonstrates that at least the northern hemispherical patches suggestively correlate with the two persistent flow cyclones that emerge in dynamo simulations employing tomographic thermal boundary condition. An associated down flow advectively collects the normal polarity field in a process similar to the one operating in the convective cyclones discussed in Sect. 5.1. The situation is less clear for the southern hemisphere.

Aubert et al. (2008a) averaged over a period equivalent to 0.7 Myr to yield this result. Olson and Christensen (2002) report that the magnetic field can be drastically different in snap shots and that the patches start to show up more clearly in their model when averaging over time spans equivalent to 2000 yr or 17 convective turnover times. Is it a coincidence that these patches are so prominent in the historic geomagnetic field and seem to have moved little over the last 400 years (Jackson et al. 2000)? Willis et al. (2007) explore under which condition their dynamo model shows more persistent field morphologies that are quasi locked to the mantle. A not too small value of $f^\star \approx 0.3$ seems to be essential, but also a rather low Rayleigh number: $\mathrm{Ra} = 1.5\,\mathrm{Ra}_c$ at $\mathrm{E} = 2.2\times10^{-4}$ in their case. This yields a dynamo with a relatively simple large scale solution that would never undergo field reversals. Also, the magnetic Reynolds number and thus the flow amplitude seem to be on the low side. Other authors use significantly more supercritical Rayleigh numbers (Olson and Christensen 2002; Aubert et al. 2008a).

There is some hope that the averaging time required to recover the mantle influence in the fluid flow underneath Earth's cmb is shorter than for the magnetic field. The smaller length scale of the flow yields a shorter mixing time so that convective features are generally less persistent than magnetic structures (Huy et al. 2000). The underlying mantle control therefore has a better chance to reveal itself for shorter averaging periods. Aubert et al. (2007) explore this question with several dynamo simulations, calculating the correlation coefficients between averages of the flow at the top of the free stream over historic time spans with those over long time spans. Long refers to several hundred convective turnover times, while three overturn times, roughly 360 yr for Earth, qualify as historic. This is about the time span for which geomagnetic models can provide an idea of the flow at the top of Earth's core. The correlation coefficient between the historic and the long averages reach encouraging values of 0.5 or higher, in particular for larger values of $f^{\star}$. This supports the conclusion by Amit and Olson (2006) that the geomagnetic flows models may already reveal a mantle signature.

The lateral cmb heat flux variations alone would drive a pure thermal winds system. Aubert et al. (2007) demonstrate that the time averaged horizontal flows in their simulations are still dominated by thermal winds even though the models involve a strong component of core convection. However, the effects of deeper convection mix with the boundary induced features in a complex way (Olson and Christensen 2002; Amit and Olson 2006; Aubert et al. 2007). The latter are modified and possibly shifted westward with respect to the thermal mantle structure, and in some regions the deeper core convection seem to dominate altogether. This is particularly true for the zonal flows inside and close to the tangent cylinder (Amit and Olson 2006) that we discussed in Sect. 5.4. Amit and Olson (2006) remark that the time averaged flow fields show a significant equatorial asymmetry, those based on the historic geomagnetic field (Fig. 12) as well as the long-time averages from dynamo simulations that use the topographic boundary conditions. At least in the latter case, this must be due to the cmb heat flux pattern.

How deep do the persistent flow structures reach into the core? The equatorial asymmetry implies that the flow cyclones that have been identified underneath the cmb can not be the straight forward manifestation of convective cyclones that go all the way through the outer core. Aubert et al. (2008a) suggest that at least the cyclones beneath Asia may reach down to the inner core and increases the removal of heat and light elements from it's eastern hemisphere. This could help to explain why seismic waves that pass through the top hundred 100 km of this hemisphere are faster, more isotropic, and more attenuated (Tanaka and Hamaguchi 1997; Niu and Wen 2001; Yu and Wen 2006). Laboratory experiments have shown that the inner-core growth rate influences the structure of the hexagonal-closed-pack (hcp) inner-core iron: slower iron solidification favors wider spaced and more oriented dendritic growth (Bergman 2003). The lower density yields a slower seismic velocities and the larger degree of orientation yields a higher seismic anisotropy since hcp iron crystals have a seismically faster axis. A faster solidification of the inner core's eastern hemisphere, promoted by the more efficient removal of heat and light elements, could therefore explain the variation in the seismic signal by causing a denser and more random crystal structure in this region.

Freezing the outer 100 km of Earth's inner core would take 100–300 Myr when assuming typical growth rates of 0.3–0.9 mm/yr (Labrosse et al. 2001). It seems not implausible that the larger scale cmb heat flux pattern remained more or less unaltered during such a time span since mantle structures typically change on time scales of some 10 Myr to 100 Myr. Gravitational forces between inner core and mantle could provide the necessary locking between the inner core and the persistent cyclone (Aubert et al. 2008a).

The potential mantle control of the dynamo mechanism is also interesting in the context of magnetic field reversals. Glatzmaier et al. (1999) and Kutzner and Christensen (2004) show that the cmb heat flux pattern can influence the reversal rate in a numerical dynamo model. That this effect may indeed play a role in the geomagnetic field is supported by the fact that Earth's reversal rate seems to change over the time scales of mantle convection (Loper 1992). Besides comparing cases with homogeneous and tomographic cmb heat flux Glatzmaier et al. (1999) as well as Kutzner and Christensen (2004) also impose other patterns to explore their influence. Kutzner and Christensen (2004) find that reversals are promoted when the heat flux in the equatorial region in increased while the heat flux in the polar regions seems to have no influence on the reversal rate. The higher equatorial heat flux increases the convective flow vigor outside the tangent cylinder. It may thus simply yield larger Ro_ℓ values and thereby increase the reversals likelihood. Glatzmaier et al. (1999), however, come to an opposing conclusion. In their model the heat flux in the polar regions is more influential: An increase promotes stronger convective plumes inside the tangent cylinder. The associated stronger thermal winds increase the toroidal fields that are created at the boundary to the tangent cylinder. These stronger toroidal fields in turn yield stronger poloidal fields in general and a stronger dipole field in particular which is less prone to reversals (Glatzmaier and Roberts 1997). The toroidal and poloidal field production inside and close to the tangent cylinder plays little role in the standard model employed by Kutzner and Christensen (2004).

Some paleomagnetic studies also claim that during field reversals the magnetic pole preferentially follows two longitude bands which are about 180° apart and roughly coincide with the areas of increased cmb heat flow in the tomographic models (Laj et al. 1991; Merrill and McFadden 1999). We have discussed in Sect. 6 that the dipole field decreases during a reversal and subsequently recovers with opposite polarity. The transitional pole positions therefore reflect virtual geomagnetic poles (VGP) rather than true dipole poles and depend on the site the paleomagnetic data originate from. Coe et al. (2000) and Kutzner and Christensen (2004) show that the preferred bands can largely be recovered in a numerical dynamo model when imposing a tomographic cmb heat flux. The increased flux underneath the Pacific rim drives somewhat stronger convective up and down streams in these regions which in turn promote the appearance of more intense magnetic flux patches of both polarities. These patches may dominate the magnetic field during transitional times sufficiently enough to largely determine the VGPs for paleomagnetic sites in their respective sphere of influence.

9 Tidally Driven and Precession Driven Dynamos

Bullard (1949) apparently was the first to consider tidal deformation of the core mantle boundary and precession of the rotation axis as possible energy sources for the geodynamo. Tides never looked like a plausible driving agent in the Earth, but it cannot be ruled out that precession is driving the geodynamo. In this section, we look at how precession or tides can in principle drive a dynamo and whether the parameters of the planets in the solar system are such that any of these effects is likely to occur.

Let us first assume that a planetary core is adiabatically stratified so that we do not need to worry about buoyancy forces. It helps to next take the point of view of a fluid dynamicist and to consider tidal deformation or precession rate as a control parameter which can be increased at will starting from zero. As long as the control parameter is small, the excited flow is laminar. Tides and precession can conveniently be discussed together because the

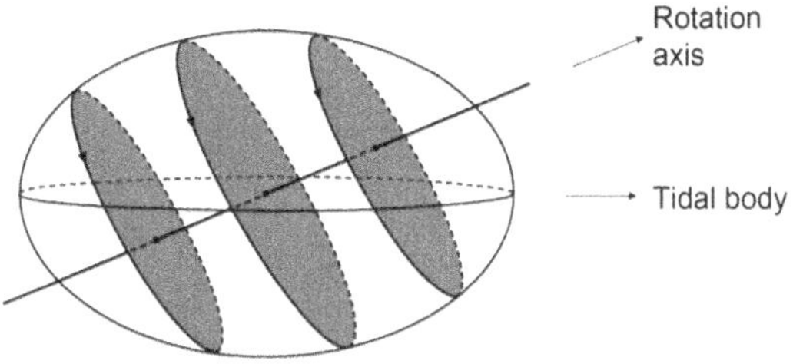

Fig. 13 Streamlines for laminar flow driven by tides or precession. The axis labels are for tidal flow. For the case of precession, there is no tidal body, the oblique axis should be labeled "rotation axis of the fluid", and the rotation axis of the mantle is vertical

laminar flows are similar in both cases. Let us first look at tides: A planet rotates about an axis which is also an axis of symmetry. The liquid core of the planet, viewed from an inertial frame of reference, rotates about the same axis at the same rate. Fluid particles move on circular paths. If one now adds tidal deformation to the picture, the rotation axis is no longer an axis of symmetry and fluid particles circulate on elliptical streamlines (see Fig. 13). It is also noteworthy that the rotation axis is not joining the centers of the ellipses.

Let us now turn to precession. At zero precession rate, core and mantle rotate again about the same axis. If the mantle starts to precess, the core will have to follow. In a steady state, the rotation axes of core and mantle may differ, but they have to precess at the same rate about the same precession axis. One then arrives at the same streamline pattern as in Fig. 13, except that the rotation axis of the planet is now vertical in the figure and still is a symmetry axis of the planet. In a gas planet, the same figure applies, with the "core mantle boundary" replaced by a "surface of constant pressure".

Known antidynamo theorems nearly exclude that the flow in Fig. 13 is a dynamo. Streamlines in Fig. 13 lie in planes. There is an antidynamo theorem stating that dynamo action in that case is impossible provided that conductor-vacuum interfaces are also parallel to those planes, which is not the case in Fig. 13. Even though dynamo action is not excluded by antidynamo theorems, it seems unlikely that the flow of Fig. 13 is a dynamo at magnetic Reynolds numbers of planetary relevance, if at all.

Something needs to happen to the laminar precessional or tidal flow before it becomes interesting for the dynamo effect. If one increases the control parameter, instabilities will occur. Instabilities lead to a more complicated, three dimensional flow potentially capable of dynamo action. Our knowledge about instabilities is more complete for precession than for tides. Instabilities are best studied close to their onset, i.e. in a narrow range of control parameters, which are easier to fine tune in numerical simulations than in experiments. A sphere is a much more convenient geometry for numerics than an ellipsoid. Tidal effects cannot be dissociated from ellipsoidal boundaries, whereas precessing spheres and ellipsoids produce similar instabilities.

Three types of instabilities are known from precessing spheres (Lorenzani and Tilgner 2001). The first occurs in the boundary layers and stays confined to them, so that it has little effect on the flow globally. The second type occurs in internal shear layers. It is not understood why these internal shear layers appear on top of the flow in Fig. 13 except for the strongest of them (Busse 1968), but they are a robust feature of both experiments and simulations. The third type is a consequence of the streamline geometry in Fig. 13: Streamlines are elliptical, giving rise to the so called "elliptical instability" (Kerswell 2002), and there is differential motion between adjacent planes in which the ellipses lie, giving rise to shear instabilities. The third type is by far the most accessible to theoretical analysis and is understood best. However, early experiments conveyed the intuition that the second type is the most important (Malkus 1968; Vanyo et al. 1995). Clearly, different instabilities set in

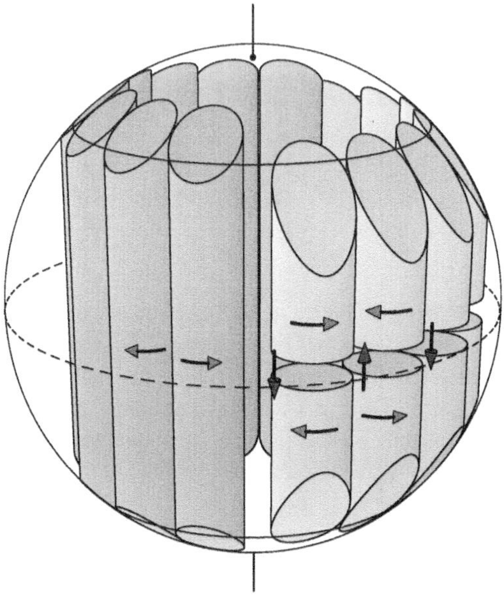

Fig. 14 Sketch of the structure of unstable precession driven flow (from Lorenzani and Tilgner 2001)

first depending on the control parameters and we do not know at present what to expect for planetary values.

Figure 14 is a sketch of the elliptical instability as it was observed in precession driven flow. Similar to convection, columnar vortices drifting around the rotation axis are excited. In convection, instability starts from an axisymmetric ground state, but it starts from a non-axisymmetric flow in the precessional or tidal case. For that reason, convection excites one set of columns or one wave, whereas precession or tides excite two of them. In the schematic Fig. 14, the azimuthal wave number of the two sets of vortices differs by 1 (it would have to differ by 2 in the instability of an equatorial tide). One of the two sets has qualitatively the same structure (and in particular the same helicity distribution) as thermal convection.

The elliptical instability also occurs in the simplest tidal flow inside a planet, caused by a moon in the equatorial plane. We know for this geometry a rigorous upper bound for the growth rate of elliptical instabilities in inviscid fluids (Kerswell 2002). This upper bound is proportional to the ellipticity of the streamlines if the tidal deformation is small. We can thus proceed as follows to determine whether an elliptical instability is likely in any given planet: From the masses of planet and moon, M_{Planet} and M_{Moon}, the radius of the planet R_{Planet}, and the distance from the planet to the moon, R_{Moon}, one computes the amplitude of the equilibrium tide h_T at the outer edge of the conducting region of the planet which has radius R_c. In a planet of uniform density, which is assumed in this simple estimate, the ellipticity of the tidal deformation is independent of depth and one finds:

$$\frac{h_T}{R_c} = \frac{1}{2}\frac{M_{\text{Moon}}}{M_{\text{Planet}}}\left(\frac{R_{\text{Planet}}}{R_{\text{Moon}}}\right)^3 \tag{27}$$

h_T/R_c is listed in Table 2 and gives (apart form a prefactor of order 1) the upper bound of the growth rate in multiples of $1/\Omega$, where Ω is the angular frequency of rotation of the planet. Exact growth rates of elliptical instabilities tend to be of the same order as this upper bound, so that we go on with h_T/R_c as a representative value for the growth rate. This growth needs to overcome viscous dissipation. In a flow with Ekman number E, defined as $\mathrm{E} = \nu/(\Omega R_c^2)$,

Table 2 Some planetary parameters as defined and used in the text

Planet	Moon	R_c [m]	M_{Planet} [kg]	R_{Moon} [m]	M_{Moon} [kg]	h_T/R_c
Earth	Moon	3.47×10^6	5.97×10^{24}	3.84×10^8	7.35×10^{22}	2.81×10^{-8}
Jupiter	Io	5.71×10^7	1.90×10^{27}	4.22×10^8	8.91×10^{22}	1.59×10^{-7}
Saturn	Titan	2.53×10^7	5.68×10^{26}	1.22×10^9	1.35×10^{23}	1.43×10^{-8}
Uranus	Ariel	2.05×10^7	8.70×10^{25}	1.91×10^8	1.44×10^{21}	1.94×10^{-8}
Neptune	Triton	1.98×10^7	1.03×10^{26}	3.55×10^8	3.40×10^{23}	5.31×10^{-7}

ν being the kinematic viscosity, the dissipation rate is on the order of $E^{1/2}$ in the presence of no slip boundaries and on the order of E for free slip boundaries. Commonly accepted values for the Earth's core are $E = 10^{-14} - 10^{-15}$. It is more shaky to assign an Ekman number to a gas planet because ν is depth dependent, but E around 10^{-20} is a reasonable value. At any rate, an elliptical instability is likely in all gas planets, but unlikely inside the Earth. Coexistence with convective flows does not change the argument because of a separation of time scales (Tilgner 2007). It is more difficult to take a turbulent viscosity into account with any confidence (Fabijonas and Holm 2003).

A more elaborate calculation is required for precession. Precession is most important in planets with a big moon well outside the equatorial plane. This singles out the Earth (in which precessional instability is plausible but whether it actually happens depends on the exact value of E) and the Neptune/Triton pair (in which precessional instability is likely).

All these considerations were about the onset of instability. Once the new flow has grown, it will interact with convection if it is present. The mutual influence of both flows has not been studied yet. On the other hand, it is known that precession driven unstable flow can maintain a magnetic field (Tilgner 2005).

10 Conclusion

Three dimensional self-consistent numerical dynamo simulations are capable of reproducing many features of the geomagnetic field. Some of its basic properties are fairly robust and can be found over a large range of parameters. A dominant axial dipole component, strong flux patches at latitudes close to where the tangent cylinder touches the outer boundary, and weaker or inverse field inside the tangent cylinder can already be found in rather simple large scale models with Ekman numbers of order $E = 10^{-3}$. If the Rayleigh number is chosen high enough to yield a sufficiently strong inertial force at a local Rossby number around $Ro_\ell = 0.1$, these dynamos also show Earth-like reversals.

Though Ekman number and magnetic Prandtl number are far away from realistic values even simple models seem able to capture the fundamental dynamics correctly. In order to translate the dimensionless simulations to absolute physical values the results have to be extrapolated to realistic parameters. The simplicity of the scaling laws that have been suggested for the extrapolation (Christensen and Aubert 2006; Amit and Olson 2008; Takahashi et al. 2008a; Christensen et al. 2009; Christensen [doi:10.1007/s11214-009-9553-2], this issue) support the view that the numerical simulations, planetary dynamos, and even dynamos of fast rotating stars share the same fundamental dynamo processes. The fact that the small scale flow dynamics is certainly not captured correctly seems to play a secondary role. Even the simple larger Ekman number dynamos can thus prove useful, for example for unraveling the fundamental dynamo action and for understanding and predicting dipole behavior including magnetic field reversals.

However, our analysis of the Ekman number dependence reveals that some detail features may indeed be more Earth like at lower Ekman numbers. The low latitude magnetic field at the outer boundary may become more similar to the historic magnetic field for $E \leq 3 \times 10^{-5}$. More studies are needed to understand this issue and to explore the parameter dependence. Also, torsional oscillations, which may form an important part of the decadal geomagnetic field variations, only start to appear in simulations with $E \leq 3 \times 10^{-5}$. And even for $E = 3 \times 10^{-6}$ they travel much slower than in Earth (Wicht et al. 2009). While these aspects seem secondary as far as the fundamental dynamo process is concerned, they may prove essential for modeling and explaining Earth's smaller scale field structure and its evolution on time scales of decades to centuries.

A recent simulations by Kageyama et al. (2008) at $E = 10^{-6}$ indicates a more drastic regime change: Sheetlike convective columns are restricted to the region closer to the tangent cylinder and the magnetic field is rather small scale and non-dipolar. However, the results are inconclusive since the magnetic field is still developing and may only be transient.

The influence of the magnetic Prandtl number has not been explored very extensively so far owed to the fact that lower magnetic Prandtl numbers can only be assumed at low Ekman numbers. Unfortunately, low E and low Pm values both increase the numerical costs of the computations. Pm provides a measure for the ratio between the small length scales in flow and magnetic field, respectively. The low Pm value around 10^{-6} in planetary dynamo regions mean that the magnetic field is insensitive to the small scale turbulent flow structures. Numerical dynamo simulations have not ventured below $Pm = 0.1$ and have thus not really explored the effects of this scale separation (Schaeffer and Cardin 2006).

For a less generic prediction of the evolution of individual features in Earth's magnetic field the additional problem remains that the outer boundary field of a numerical simulations generally never exactly matches the geomagnetic field. Data assimilations techniques can help here. Kuang et al. (2009) demonstrate that imposing the history magnetic field from 1900 to 2000 in 20 yr intervals to a dynamo simulation can reduce the forecast error by 30%. They show that the data assimilation technique gradually changes the dynamic state of the outer core and that its influence reaches progressively deeper into the core over time. The use of more realistic thermal boundary conditions imposed by Earth's mantle is another attempt to make the numerical simulations more realistic. Whether this can already have an effect on the historic time scale of decades to centuries remains dubious but the long time scale influence certainly yields intriguing results.

The coming years will probably see dynamo simulations at Ekman numbers of 10^{-7} and magnetic Prandtl numbers of order 10^{-2}. Naturally, this will largely depend on the computing power available for dynamo simulations. The better scalability of the newer local numerical methods (Wicht et al. 2009) may help to reach this goal on massively parallel computing systems. The disadvantage of these ambitious simulations is that the small scale solutions are difficult to interpret and that the runs cover comparatively short periods. This will make a detailed analysis of the internal dynamics impossible, likely preclude the simulation of reversals, and rule out an exploration of the long-term mantel control.

Not too long ago, convection driven geodynamo simulations where deemed realistic when they produced a dipole dominated magnetic field. The fact that they can also undergo reversals was seen as a final prove for their validity (Glatzmaier and Roberts 1995b). Since then, the models have been refined to replicate more and more features of the geomagnetic field and to provide an explanation for the observations. Several spherical self-consistent dynamo codes have been developed over the last years and some have been made freely available to other researchers. At least the models with larger Ekman numbers around $E = 10^{-3}$ already run sufficiently fast on a multi-core PC so that the numerical costs also no longer pose a barrier. Does this leave room for the highly idealized and parameterized models that were so common in the past: kinematic dynamos where simple flow fields are prescribed or mean field approaches where the small scale dynamo action is parameterized all together? These models may still remain useful where really fundamental questions are addressed or the numerical limitations still prevent the use of the self-consistent full approaches. One example is the statistical exploration of the reversal likelihood. The maximum number of reversals found in a self-consistent full model is about 160 (Wicht et al. 2009). Parameterized models, however, can afford to simulate thousands of reversals and therefore offer a much better statistics (Ryan and Sarson 2007; Schmitt et al. 2001). In addition, they allow to vary the parameters to explore which combination provides the most Earth-like statistics. Another example is the possible coupling of the magnetospheric and the internal dynamo at Mercury which has so far not been explored in a self-consistent full model. The simple parameterized approach by Glassmeier et al. (2007) suggests that this may indeed yield interesting results and explain the planet's low magnetic field strength. However, the results of the parameterized models should always be interpreted with care and should be reconciled with full models as far as possible.

Self-consistent full numerical dynamo models are increasingly successful in explaining the magnetic field of the other planets in our solar system. Variations of the inner-core size, the driving mode, and the outer thermal boundary condition seem to play key roles here. We refer the interested reader to Stanley and Glatzmaier [doi:10.1007/s11214-009-9573-y], this issue). Tidally and precession driven dynamos offer an interesting alternative, but the respective numerical models have a long way to go to catch up with convection driven dynamo simulations.

References

H. Amit, P. Olson, Phys. Earth Planet. Inter. **155**, 120 (2006)
H. Amit, P. Olson, Phys. Earth Planet. Inter. **166**, 226 (2008)
A.P. Anufriev, C.A. Jones, A.M. Soward, Phys. Earth Planet. Inter. **152**, 163 (2005)
M. Ardes, F. Busse, J. Wicht, Phys. Earth Planet. Inter. **99**, 55 (1996)
J. Aubert, J. Fluid Mech. **542**, 53 (2005)
J. Aubert, H. Amit, G. Hulot, Phys. Earth Planet. Inter. **160**, 143 (2007)
J. Aubert, H. Amit, G. Hulot, P. Olson, Nature **454**, 758 (2008a)
J. Aubert, J. Aurnou, J. Wicht, Geophys. J. Int. **172**, 945 (2008b)
J. Aubert, S. Labrosse, C. Poitou (2009). doi:10.1111/j.1365-246X.2009.04361.x
J. Aubert, J. Wicht, Earth. Planet. Sci. Lett. **221**, 409 (2004)
J. Aurnou, S. Andreadis, L. Zhu, P. Olson, Earth Planet. Sci. Lett. **212**, 119 (2003)
M. Bergman, J. Cryst. Growth **255**, 204 (2003)
J. Bloxham, Geophys. Res. Lett. **39** (2002). doi:10.1029/2001GL014543
J. Bloxham, S. Zatman, M. Dumberry, Nature **420**, 65 (2002)
S. Braginsky, Geomag. Aeron. **10**, 1 (1970)
S. Braginsky, P. Roberts, Geophys. Astrophys. Fluid Dyn. **79**, 1 (1995)
B.A. Buffett, J.E. Mound, J. Geophys. Res. **110**, B08104 (2005)
E. Bullard, Proc. R. Soc. A **197**, 433 (1949)
F. Busse, J. Fluid Mech. **33**, 739 (1968)
F. Busse, J. Fluid Mech. **44**, 441 (1970)
F.H. Busse, Geophys. Res. Lett. **29**, 070000 (2002)
F.H. Busse, R. Simitev, in *Fluid Dynamics and Dynamos in Astrophysics and Geophysics*, ed. by A.M. Soward, C.A. Jones, D.W. Hughes, N.O. Weiss, (2005), p. 359
F.H. Busse, R.D. Simitev, Geophys. Astrophys. Fluid Dyn. **100**, 341 (2006)
F.H. Busse, R.D. Simitev, Phys. Earth Planet. Inter. **168**, 237 (2008)
J. Carlut, V. Courtillot, Geophys. J. Int. **134**, 527 (1998)
U. Christensen, J. Aubert, Geophys. J. Int. **116**, 97 (2006)
U. Christensen, P. Olson, Phys. Earth Planet. Inter. **138**, 39 (2003)

U. Christensen, J. Wicht, in *Treatise on Geophysics*, ed. by P. Olson. Core Dynamics, vol. 8 (Elsevier, Amsterdam, 2007), p. 245
U.R. Christensen, J. Fluid Mech. **470**, 115 (2002)
U.R. Christensen, Nature **444**, 1056 (2006)
U.R. Christensen, J. Aubert, F.H. Busse et al., Phys. Earth Planet. Inter. **128**, 25 (2001)
U.R. Christensen, V. Holzwarth, A. Reiners, Nature **457**, 167 (2009)
R. Coe, L. Hongre, A. Glatzmaier, Philos. Trans. R. Soc. Lond. A **358**, 1141 (2000)
R.S. Coe, G.A. Glatzmaier, Geophys. Res. Lett. **33**, 21311 (2006)
P. Davidson, *An Introduction to Magnetohydrodynamics* (Cambridge University Press, Cambridge, 2001)
P. Driscoll, P. Olson, Earth Planet. Sci. Lett. **282**, 24 (2009)
M. Dumberry, J. Bloxham, Phys. Earth Planet. Inter. **140**, 29 (2003)
M. Dumberry, J.E. Mound, J. Geophys. Res. **113**, B03102 (2008)
M. Evonuk, Astrophys. J. **673**, 1154 (2008)
C. Eymin, G. Hulot, Phys. Earth Planet. Inter. **152**, 200 (2005)
B. Fabijonas, D. Holm, Phys. Rev. Lett. **90**, 124501 (2003)
K. Glassmeier, H. Auster, U. Motschmann, Geophys. Res. Lett. **34**, 22201 (2007)
G. Glatzmaier, J. Comput. Phys. **55**, 461 (1984)
G. Glatzmaier, Annu. Rev. Earth Planet. Sci. **30**, 237 (2002)
G. Glatzmaier, R. Coe, in *Treatise on Geophysics*, ed. by P. Olsen. Core Dynamics, vol. 8 (Elsevier, Amsterdam, 2007), p. 283
G. Glatzmaier, R. Coe, L. Hongre, P. Roberts, Nature **401**, 885 (1999)
G. Glatzmaier, P. Roberts, Phys. Earth Planet. Inter. **91**, 63 (1995a)
G. Glatzmaier, P. Roberts, Nature **337**, 203 (1995b)
G. Glatzmaier, P. Roberts, Physica D **97**, 81 (1996a)
G. Glatzmaier, P. Roberts, Science **274**, 1887 (1996b)
G. Glatzmaier, P. Roberts, Contemp. Phys. **38**, 269 (1997)
E. Grote, F. Busse, Phys. Rev. **62**, 4457 (2000)
E. Grote, F. Busse, Fluid Dyn. Res. **28**, 349 (2001)
E. Grote, F. Busse, A. Tilgner, Phys. Rev. E **60**, 5025 (1999)
E. Grote, F. Busse, A. Tilgner, Phys. Earth Planet. Inter. **117**, 259 (2000)
D. Gubbins, Phys. Earth Planet. Inter. **128**, 3 (2001)
D. Gubbins, P. Kelly, Nature **365**, 829 (1993)
D. Gubbins, A.P. Willis, B. Sreenivasan, Phys. Earth Planet. Inter. **162**, 256 (2007)
H. Harder, U. Hansen, Geophys. J. Int. **161**, 522 (2005)
M. Heimpel, J. Aurnou, J. Wicht, Nature **438**, 193 (2005)
M.L. Huy, M. Mandea, L.L. Jean-Mouël, A. Pais, Earth Planets Space **52**, 163 (2000)
A. Jackson, Phys. Earth Planet. Inter. **103**, 293 (1997)
A. Jackson, Nature **424**, 760 (2003)
A. Jackson, J. Bloxham, D. Gubbins, in *Dynamics of Earth's Deep Interior and Earth Rotation* (1993), p. 97
A. Jackson, C. Finlay, in *Treatise on Geophysics*, ed. by M. Kono. Geomagnetism, vol. 5 (Elsevier, Amsterdam, 2007), p. 147
A. Jackson, A. Jonkers, M. Walker, Philos. Trans. R. Soc. Lond. A **358**, 957 (2000)
D. Jault, in *Earth's Core and Lower Mantle*, ed. by C.A. Jones, A.M. Soward, K. Zhang (2003), p. 56
D. Jault, C. Gire, J.-L. LeMouël, Nature **333**, 353 (1988)
D. Jault, J.-L. LeMouël, Geophys. Astrophys. Fluid Dyn. **48**, 273 (1989)
C. Johnson, C. Constable, Geophys. J. Int. **122**, 489 (1995)
C. Johnson, C. Constable, L. Tauxe, Science **300**, 2044 (2003)
C. Jones, Philos. Trans. R. Soc. Lond. A **358**, 873 (2000)
C. Jones, in *Treatise on Geophysics*, ed. by P. Olson. Core Dynamics, vol. 8 (Elsevier, Amsterdam, 2007), p. 131
C. Jones, P. Roberts, J. Fluid Mech. **404**, 311 (2000)
A. Kageyama, T. Miyagoshi, T. Sato, Nature **454**, 1106 (2008)
A. Kageyama, M. Ochi, T. Sato, Phys. Rev. Lett. **82**, 5409 (1999)
A. Kageyama, T. Sato, Phys. Plasmas **2**, 1421 (1995)
A. Kageyama, T. Sato, Plasma Phys. Contr. Fusion **39**, 83 (1997)
A. Kageyama, M. Yoshida, J. Phys. Conf. Ser. **16**, 325 (2005)
P. Kelly, D. Gubbins, Geophys. J. Int. **128**, 315 (1997)
R. Kerswell, Annu. Rev. Fluid Mech. **34**, 83 (2002)
M. Kono, P. Roberts, Rev. Geophys. **40**, 1013 (2002). doi:10.1029/2000RG000102
M. Korte, C. Constable, Geophys. J. Int. (2006). doi:10.1111/j.1365-246X.2006.03088.x
W. Kuang, Phys. Earth Planet. Inter. **116**, 65 (1999)

W. Kuang, J. Bloxham, Nature **389**, 371 (1997)
W. Kuang, J. Bloxham, J. Comput. Phys. **153**, 51 (1999)
W. Kuang, A. Tangborn, Z. Wei, T. Sabaka, Geophys. J. Int. **179**, 1458 (2009)
C. Kutzner, U. Christensen, Geophys. Res. Lett. **27**, 29 (2000)
C. Kutzner, U. Christensen, Phys. Earth Planet. Inter. **131**, 29 (2002)
C. Kutzner, U. Christensen, Geophys. J. Int. **157**, 1105 (2004)
S. Labrosse, Phys. Earth Planet. Inter. **140**, 127 (2003)
S. Labrosse, J.-P. Poirier, J.-L. Le Mouël, Earth Planet. Sci. Lett. **190**, 111 (2001)
C. Laj, A. Mazaud, R. Weeks, Nature **351**, 447 (1991)
R. Leonhardt, K. Fabian, Earth Planet. Sci. Lett. **253**, 172 (2007)
J. Li, T. Sato, A. Kageyama, Science **295**, 1887 (2002)
J. Lister, B. Buffett, Phys. Earth Planet. Inter. **91**, 17 (1995)
J. Liu, P.M. Goldreich, D.J. Stevenson, Icarus **196**, 653 (2008)
D.E. Loper, Geophys. Res. Lett. **19**, 25 (1992)
S. Lorenzani, A. Tilgner, J. Fluid Mech. **447**, 111 (2001)
V. Malkus, Science **160**, 259 (1968)
G. Masters, G. Laske, H. Bolton, A. Dziewonski, in *AGU Monograph*, ed. by S. Karato, F.A.R. Lieberman, G. Masters, L. Stixrude. Earth's Deep Interior: Mineral Physics and Tomography from the Atomic to the Global Scale, vol. 117 (American Geophysical Union, Washington, 2000), p. 63
H. Matsui, H. Okuda, Int. J. Comput. Fluid Dyn. **19**, 15 (2005)
S. Maus, M. Rother, C. Stolle et al., Geochem. Geophys. Geosyst. **7**, 7008 (2006)
D.G. McMillan, G.R. Sarson, 2003, *AGU Fall Meeting Abstracts*, C280+
R. Merrill, P. McFadden, Rev. Geophys. **37**, 201 (1999)
H. Moffat, *Magnetic Field Generation in Electrically Conducting Fluids* (Cambridge University Press, Cambridge, 1978)
J.E. Mound, B.A. Buffett, J. Geophys. Res. **110**, B08103 (2005)
F. Nimmo, in *Treatise on Geophysics*, ed. by P. Olson. Core Dynamics, vol. 8 (Elsevier, Amsterdam, 2007), p. 31
N. Nishikawa, K. Kusano, Phys. Plas. **15**, 082903 (2008)
F. Niu, L. Wen, Nature **410**, 1081 (2001)
M.M. Ochi, A. Kageyama, T. Sato, Phys. Plas. **6**, 777 (1999)
D.E. Ogden, G.A. Glatzmaier, R.S. Coe, Geophys. Astrophys. Fluid Dyn. **100**, 107 (2006)
P. Olson, U. Christensen, Geophys. J. Int. **151**, 809 (2002)
P. Olson, U. Christensen, Earth Planet. Sci. Lett. **250**, 561 (2006)
P. Olson, U. Christensen, G. Glatzmaier, J. Geophys. Res. **104**, 10,383 (1999)
E. Parker, Astrophys. J. **121**, 293 (1955)
G. Roberts, Philos. Trans. R. Soc. A **271**, 411 (1972)
P. Roberts, Philos. Trans. R. Soc. Lond., A **263**, 93 (1968)
P. Roberts, in *Irreversible Phenomena and Dynamical Systems Analysis in Geosciences*, ed. by C. Nicolis, G. Nicolis (Reidel, Dordrecht, 1987), p. 73
P. Roberts, G. Glatzmaier, Geophys. Astrophys. Fluid Dyn. **94**, 47 (2001)
J. Rotvig, Phys. Earth Planet. Inter. **176**, 69 (2009)
J. Rotvig, C. Jones, Phys. Rev. E **66**, 056308 (2002)
D.A. Ryan, G.R. Sarson, Geophys. Res. Lett. **34**, 2307 (2007)
N. Schaeffer, P. Cardin, Earth Planet. Sci. Lett. **245**, 595 (2006)
D. Schmitt, M.A.J.H. Ossendrijver, P. Hoyng, Phys. Earth Planet. Inter. **125**, 119 (2001)
R. Simitev, F. Busse, J. Fluid Mech. **532**, 365 (2005)
R.D. Simitev, F.H. Busse, Europhys. Lett. **85**, 19001 (2009)
B. Sreenivasan, C.A. Jones, Geophys. Res. Lett. **32**, L20301 (2005)
B. Sreenivasan, C.A. Jones, Geophys. Astrophys. Fluid Dyn. **100**, 319 (2006)
S. Stanley, J. Bloxham, Nature **428**, 151 (2004)
S. Stanley, J. Bloxham, Icarus **184**, 556 (2006)
S. Stanley, J. Bloxham, W. Hutchison, M. Zuber, Earth Planet. Sci. Lett. **234**, 341 (2005)
S. Stanley, M.T. Zuber, J. Bloxham, Geophys. Res. Lett. **34**, 19205 (2007)
S. Stellmach, U. Hansen, Phys. Rev. E **70**, 056312 (2004)
F. Takahashi, M. Matsushima, Geophys. Res. Lett. **33**, L10202 (2006)
F. Takahashi, M. Matsushima, Y. Honkura, Science **309**, 459 (2005)
F. Takahashi, M. Matsushima, Y. Honkura, Earth Planets Space **59**, 665 (2007)
F. Takahashi, M. Matsushima, Y. Honkura, Phys. Earth Planet. Inter. **167**, 168 (2008a)
F. Takahashi, H. Tsunakawa, M. Matsushima, N. Mochizuki, Y. Honkura, Earth Planet. Sci. Lett. **272**, 738 (2008b)

S. Tanaka, H. Hamaguchi, J. Geophys. Res. **102**, 2925 (1997)
J. Taylor, Proc. R. Soc. Lond. A **274**, 274 (1963)
A. Tilgner, Geophys. Astrophys. Fluid Dyn. **98**, 225 (2004)
A. Tilgner, Phys. Fluids **17**, 034104 (2005)
A. Tilgner, in *Treatise on Geophysics*, ed. by G. Schubert (Elsevier, Amsterdam, 2007), p. 207
A. Tilgner, F.H. Busse, J. Fluid Mech. **332**, 359 (1997)
J. Vanyo, P. Wilde, P. Cardin, P. Olson, Geophys. J. Int. **121**, 136 (1995)
J. Wicht, Geophys. J. Int. **162**, 371 (2005)
J. Wicht, J. Aubert, GWDG-Bericht **68**, 49 (2005)
J. Wicht, F. Busse, Geophys. Astrophys. Fluid Dyn. **86**, 103 (1997)
J. Wicht, U. Christensen, Geophys. J. Int. (2010, accepted for publication)
J. Wicht, M. Mandea, F. Takahashi et al., Space Sci. Rev. **132**, 261 (2007)
J. Wicht, P. Olson, Geochem. Geophys. Geosyst. **5** (2004). doi:10.1029/2003GC000602
J. Wicht, S. Stellmach, H. Harder, in *Geomagnetic Field Variations—Space-Time Structure, Processes, and Effects on System Earth*. Springer Monograph, ed. by K. Glassmeier, H. Soffel, J. Negendank (Springer, Berlin, 2009), p. 107
A.P. Willis, B. Sreenivasan, D. Gubbins, Phys. Earth Planet. Inter. **165**, 83 (2007)
W.-C. Yu, L. Wen, Earth Planet. Sci. Lett. **245**, 581 (2006)
S. Zatman, J. Bloxham, Nature **388**, 760 (1997)
S. Zatman, J. Bloxham, Geophys. J. Int. **138**, 679 (1999)
K. Zhang, G. Schubert, Annu. Rev. Fluid Mech. **32**, 409 (2000)
K.-K. Zhang, F. Busse, Geophys. Astrophys. Fluid Dyn. **44**, 33 (1988)

Space Sci Rev (2010) 152: 543–564
DOI 10.1007/s11214-009-9546-1

Laboratory Dynamo Experiments

Gautier Verhille · Nicolas Plihon · Mickael Bourgoin · Philippe Odier · Jean-François Pinton

Received: 22 March 2009 / Accepted: 29 May 2009 / Published online: 7 July 2009

Abstract Since the turn of the century, experiments have produced laboratory fluid dynamos that enable a study of the effect in controlled conditions. We review here magnetic induction processes that are believed to underlie dynamo action, and we present results of these dynamo experiments. In particular, we detail progress that have been made through the study of von Kármán flows, using gallium or sodium as working fluids.

Keywords Magnetic fields · Magnetohydrodynamics · Dynamo · Experiments · Instabilities · Turbulence

1 Introduction

Although dynamo research is essentially motivated by observations from planetary and stellar dynamos, the conditions that prevail in such natural objects cannot be reproduced in the laboratory. Experiments and numerics can only be run in quite different parameter regimes, but they both provide useful insights into the features of natural dynamos. Experiments cannot be rotated as fast as real systems do, convective motion cannot be as strong, etc., but they run with real fluids and probe quantities that cannot be accessed from remote observations of natural dynamos. Numerical experiments record the complete dynamical fields in space and time (but at quite removed parameter values) while laboratory experiments probe a limited part of the velocity and magnetic fields $(\mathbf{u}, \mathbf{B})$. The two approaches are complementary and have been associated in most recent works. Numerical development are reported in several contributions to this volume. We focus here on the specific issues involved in the actual implementation of an experimental dynamo and recall the findings of recent studies. The reader

G. Verhille · N. Plihon · M. Bourgoin · P. Odier · J.-F. Pinton (✉)
Laboratoire de Physique de l'École Normale Supérieure de Lyon, CNRS & Université de Lyon, 69364 Lyon, France
e-mail: pinton@ens-lyon.fr

M. Bourgoin
Laboratoire des Ecoulements Geophysiques et Industriels, CNRS/UJF/INPG UMR5519, BP53, 38041 Grenoble, France

is also referred to reviews that have recently been published (Magnetohydrodynamics 2002; C.R. Acad. Sci. 2008; Stefani et al. 2008; Pétrélis et al. 2007).

In order for a dynamo to be self-sustained, the production of induced currents by fluid motions must overcome the resistive Joule dissipation. This condition sets an instability threshold, requiring that the magnetic Reynolds number of the flow ($R_M = UL/\lambda$) exceeds a critical value R_M^c (U and L are characteristic velocity and length and λ is the magnetic diffusivity). An energy-based criterion for generation inside a sphere gives a lower bound $R_M^c > \pi^2$. For more complex geometries, and taking into account real flow structure there is no general expression for R_M^c. On the contrary, theorems prevent dynamo action for too simple geometries (Moffatt 1978). Assuming R_M^c values of the order of 10 to 100 (a value often quoted for the Earth), and given the fact that all liquid metals have magnetic Prandtl numbers (ratio of kinetic to magnetic diffusivities) of the order of $P_M \sim 10^{-5}$, one realizes that dynamo flows are associated with huge kinetic Reynolds number values $R_V = R_M/P_M$, typically exceeding 10^6. Such high R_V values are associated with fully developed turbulence, an observation that raises several central issues for dynamo experiments.

Turbulence is often synonymous of (a) disordered motions, and (b) strongly diffusive features. In this context, (a) means that the specific motions that favor dynamo action may be disrupted by the randomness of the flow. In addition, the small magnetic Prandtl number values impose that the Joule resistive scale is very much larger than the hydrodynamic viscous length. It is therefore tempting to perform some kind of sub-scale average of the action of the turbulent velocity field. Then (b) leads to an effective magnetic diffusivity that could be much larger than the molecular value. These considerations have raised doubts on the very existence of fully turbulent dynamos (Schekochihin et al. 2004).

One alternative is to engineer flow configurations that will preserve flow patterns which are essential for the dynamo generation. The design of the Riga and Karlsruhe experiments have been made to ensure that the time-averaged flow field resembles laminar flow that are kinematic dynamo solutions (Magnetohydrodynamics 2002; C.R. Acad. Sci. 2008). These pioneering studies have validated the principle of a *fluid* dynamo, and have shown many fundamental dynamo properties. The threshold for dynamo action has been found to be in good agreement with predictions, showing the predominance of the large scales in their dynamo processes.

Another possibility, often explored in the geophysics community, is provided by strongly rotating flows. In this case, the Proudman-Taylor constraint may be able to prevent the development of strong three-dimensional turbulent fluctuations. This effect may even be strengthened by the generation of a dynamo dipole with its axis parallel to the rotation vector. Experiments in rotating Couette flows are studied in Grenoble (Schmitt et al. 2008) and Maryland (Sisan et al. 2004). Preliminary studies have not shown self-generation, but have pointed to the existence of *waves* in these strong rotating and magnetized flows: inertial, magneto-rotational, Alfvèn, etc. The role of these waves regarding dynamo self-generation has yet to be elucidated.

The VKS experiments have shown that it is possible to generate a dynamo from fully turbulent motions. Its characteristics have not been predicted by studies based on the time-averaged flow pattern, although it is believed that helicity and differential rotation do play a leading role. The existence of fully turbulent motions has a major impact on the power requirement of the experiment. In the limit of very high R_V values, the hydrodynamic power consumption (below dynamo threshold) scales as $P = \rho L^2 U^3$ (ρ is the fluid density), leading to magnetic Reynolds number $R_M = \mu\sigma(PL/\rho)^{1/3}$ (μ and σ are the magnetic permeability and electrical conductivity of the fluid, so that $\lambda = 1/\mu\sigma$ is its magnetic diffusivity). Engineering difficulties typically scale with the size L of the experiment, while operational costs are best associated with the power input P. Then, the above scaling shows that in order to reach high R_M values, one should use a fluid with the best electrical conductivity and lowest density (hence the use of liquid sodium), with size and power consumption only contributing to the one-third power. $R_M \sim 10$ requires power inputs of the order 100 kW. As a result, it is easily realized that highly overcritical regimes will not be achieved in laboratory studies using conventional fluids (Spence et al. 2009): experiments aim at reaching for the neighborhood of the threshold for dynamo onset. In addition, it implies that once a geometry has been chosen, small variations in R_M^c can have a huge impact on the power or the size of the flow needed to reach self-generation. As a result, all experiments have been thoroughly optimized to have the lowest possible critical magnetic Reynolds number. This is a noteworthy peculiarity of the study of this hydromagnetic instability: energy and engineering constrains are such that one is not able to increase at will the control parameter of the instability; one must also choose conditions such that the threshold is within the capacity of the selected setup. In this respect, the details of flow entrainment and the adjustment of boundary conditions have been essential to all dynamo experiments so far.

We will first review the main induction processes that have been evidenced in fully turbulent MHD flows, before discussing the findings of dynamo experiments. We illustrate most features using measurements in the swirling flow generated in the gap between coaxial impellers inside a cylinder—the von Kármán flow geometry, which we have studied using gallium or sodium as working fluid.

2 Magnetic Induction and Dynamo Issues

Magnetohydrodynamics solves the coupled set of induction and momentum equations:

$$\partial_t \mathbf{B} = \nabla(\mathbf{u} \times \mathbf{B}) + \lambda \Delta \mathbf{B}, \tag{1}$$

$$\partial_t \mathbf{u} + (\mathbf{u} \cdot \nabla)\mathbf{u} = -\frac{\nabla p}{\rho} + \nu \Delta \mathbf{u} + \frac{\mathbf{j} \times \mathbf{B}}{\rho} + \mathbf{f}, \tag{2}$$

where $\lambda = 1/\mu_0 \sigma$ is the magnetic diffusivity of the fluid with density ρ, $\mathbf{j}$ the electrical current, and $\mathbf{f}$ is the forcing term (which may include Coriolis or buoyancy forces if rotation/convective effects must be included). Flows of liquid metals as considered in the experiments discussed here are incompressible so that the velocity and magnetic fields are divergence free. Boundary conditions correspond to no-slip for the hydrodynamic field (the velocity at the boundary is equal to the velocity of the boundary), with the magnetic boundary conditions set by the characteristics of the boundary (electrical conductivity σ_b, magnetic permittivity μ_b). Magnetic boundary conditions can be implemented by setting continuity conditions at the surface, e.g. $[\mathbf{n}.\mathbf{B}] = 0$, $[\mathbf{n} \times \mathbf{H}] = \mathbf{j}_s$ where $\mathbf{n}$ is the normal to the surface, $\mathbf{j}_s$ are eventual surface currents and [.] stands for 'jump across the surface'. Alternatively, the induction equation can be extended to the entire domain (encompassing regions outside the flow—with the condition that $\mathbf{B} \to 0$ at infinity), and inhomogeneities are incorporated into Ohm's law:

$$\nabla \times \frac{\mathbf{B}}{\mu(\mathbf{r})} = \sigma(\mathbf{r})(\mathbf{E} + \mathbf{u} \times \mathbf{B}), \tag{3}$$

where the magnetic permeability and electrical conductivity depend on position. Taking the curl of the above equation yields the induction equation in an inhomogeneous medium:

$$\partial_t \mathbf{B} = \nabla \times (\mathbf{u} \times \mathbf{B}) + \lambda \Delta \mathbf{B} + \nabla \times (\lambda \nabla \ln \mu \times \mathbf{B}) - \nabla \lambda \times (\nabla \times \mathbf{B}) \tag{4}$$

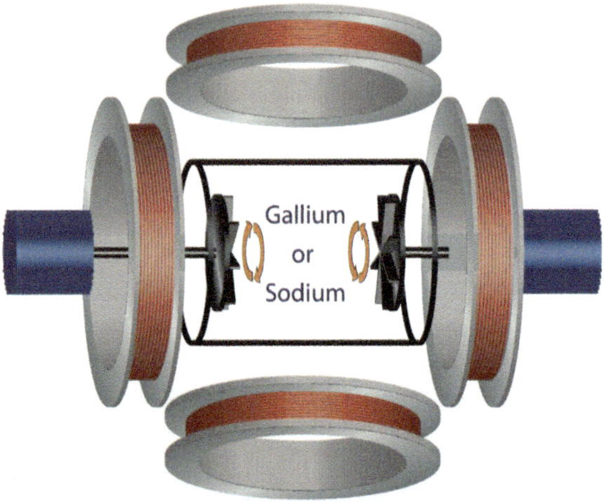

Fig. 1 Schematics of induction measurements in the von Kármán geometry (Zandbergen and Dijkstra 1987). Flows are generated by the independent rotation of 2 coaxial impellers (co- or counter-rotating) in a cylinder—the aspect ratio is one, with a cylinder diameter equal to the distance between the impellers. Depending on the configurations, the resulting mean flows have azimuthal and meridional profiles similar to the $s_{1,2}t_{1,2}$ types considered by Dudley and James (1989) for dynamo generation. Note that these flows are strongly turbulent (Mordant et al. 1998; Pinton et al. 1998; Volk et al. 2006b); discussions in this section are restricted to the mean (time-averaged) features. External coils are set to apply either axial or transverse fields, and Hall probes inserted inside the flow vessel allow the measurement of the magnetic fields induced by the fluid motions

with $\lambda = 1/\mu\sigma$, and it will be pointed out below that boundary conditions are essential for the interpretation of measurements of magnetic induction (Bullard and Gubbins 1977; Moffatt 1978; Bourgoin 2003).

2.1 Induction Processes

Despite efforts that are now almost a century old, research has so far failed to establish sufficient conditions for dynamo action—some necessary conditions have been provided by anti-dynamo theorems. In this context, work has been aimed at uncovering efficient induction processes that could co-operate towards dynamo generation. In these studies an external field $\mathbf{B}^A$ is applied, and one analyses magnetic response, *i.e.* the induced field $\mathbf{B}^I$. For simplicity, the applied field is often homogeneous. We review in this section essential mechanisms that have been uncovered, with examples drawn from our own studies of the von Kármán flows generated inside a cylinder by the rotation of coaxial impellers, cf. Fig. 1. In these studies the applied field is low, so that the effect of the Lorentz force can essentially be neglected in the fluid momentum equation. The magnetic response probes the structure of the velocity gradients and boundary conditions.

2.1.1 Shear and ω-effect

One mechanism is the shearing of magnetic field lines by velocity gradients, *i.e.* with $\mathbf{B}^I$ having its source in the $B^A \partial_A \mathbf{v}$ term of the induction equation, where $\partial_A \cdot$ stands for 'gradient along the A-direction'. For steady state conditions at low magnetic Reynolds number, the induction equation then leads to a linear dependence of B^I with R_M, a feature that can be used to estimate an intrinsic magnetic Reynolds number from experimental data as

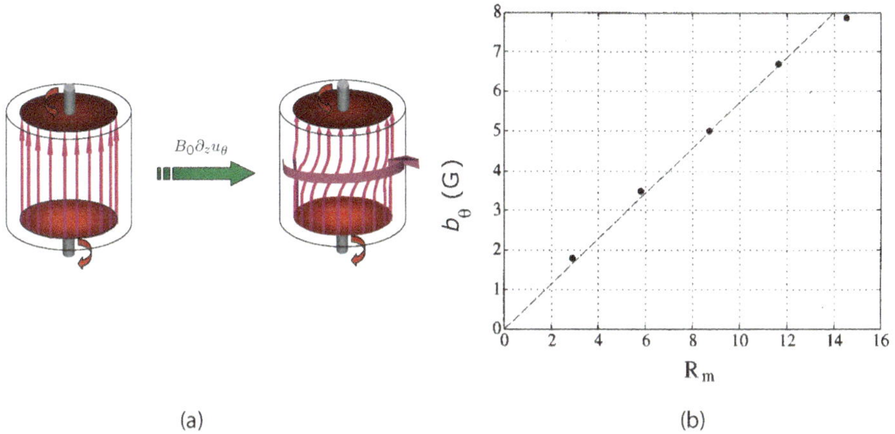

Fig. 2 Experimental verification of ω-effect in the von Kármán flow. Gallium is the working fluid. An axial magnetic field is applied with external Helmholtz coils, and the azimuthal induced field is measured by a Hall probe in the mid-plane and inside the flow. Applied field 33 G. (**a**) Sketch of the effect; (**b**) measurement of the induced field with $R_M = 2\pi R^2\Omega/\lambda$—where R is the radius of the cylindrical vessel and Ω the rotation rate of the impellers. Plot from Odier et al. (1998)

$R_M^I \equiv B^I/B^A$ (Martin et al. 2000). This induction process is called the ω-effect when the velocity gradient comes from differential rotation, *e.g.* from a variation of a rotation rate along its axis. An example is provided in the von Kármán setup when the flow is driven by counter rotation of the impellers. A layer with differential rotation forms in the mid-plane and the twisting of axially applied magnetic field lines generates an azimuthal component—cf. Fig. 2.

Another simple instance of this situation is when an electrically conducting plate rotates above a similar one at rest: if an external field is applied parallel to the axis of rotation, the differential rotation generates an induced field in the azimuthal direction. The solid rotor dynamo experiment devised in the 1960s by Lowes and Wilkinson is based on this principle (Lowes and Wilkinson 1963, 1968): it couples two such induction effects in an 'ω^2-dynamo'.

2.1.2 Helicity and the 'Parker-Effect'

Another crucial mechanism is the 'stretch and twist' effect (Parker 1955): helical motions can deform initially straight magnetic field lines into loops which are associated with induced currents parallel to the applied field, $\mathbf{j}^I \propto \mathbf{B}^A$. This non-linear effect was first tested experimentally by Steenbeck et al. (1968) in an arrangement with interlaced channels—note that in this original setup the local helicity $h = \mathbf{u} \cdot (\nabla \times \mathbf{u})$ is strictly zero everywhere, so that the magnetic field diffusion is essential in this process.

This Parker-effect has also been observed directly in von Kármán experiments (Pétrélis et al. 2003; Bourgoin et al. 2004a): the flow is generated inside a cylinder by the rotation of one disk at one end of the cylinder; it acts as a centrifugal pump so that fluid is drawn in a swirling motion along the cylinder axis. As expected the induced magnetic field varies quadratically with the flow velocity, $B^I \sim B^A R_M^2$ (dashed line in the measurements of Fig. 3). However, at high R_M another effect sets in: the expulsion of the applied field from coherent eddies (see below) which causes a saturation in the Parker-induced magnetic field.

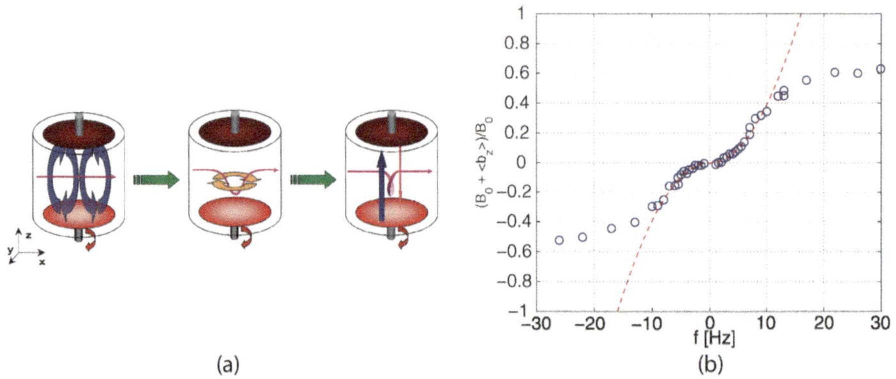

Fig. 3 Experimental verification of the Parker-effect in the von Kármán sodium flow. A magnetic field perpendicular to the cylinder (along the x-axis) is applied with external coils and the axial component of induced field is measured by a Hall probe inserted in the mid-plane, inside the flow. (**a**) Sketch of the effect; (**b**) measurement of the induced field variation with the rotation rate of the impellers, from Pétrélis et al. (2003). The *dashed (red) line* is a quadratic fit; departure at large rotation rates (> 10 Hz) is due to the expulsion of the applied field by the flow rotation

2.1.3 Coherent Vortex Motion and Expulsion

Rotational flow motion was involved in the two above examples of ω and Parker effects. However, another well-known effect is the expulsion of magnetic field lines from regions with closed streamlines (Moffatt 1978) (Chap. 3). If a magnetic field is initially applied transverse to a coherent vortex, then closed loops form and decay in such a way as to gradually expel all magnetic flux from any region in which the streamlines are closed. This effect is related to the traditional skin penetration of conventional electromagnetism (in the reference frame rotating with the vortex). It is has been clearly evidenced in the von Kármán swirling flow with the impellers *co-rotating* so as to generate a coherent axial vortex (Simand et al. 2000).

A sketch of the effect and corresponding experimental measurements are shown in Fig. 4, taken from Odier et al. (2000). The field decays near the rotation axis, and correspondingly field lines are compressed outside of the vortex core. This effect has also been clearly observed during the growing stage of the Riga dynamo: the Earth magnetic field is first expelled from the swirling motion (Gailitis et al. 2003) and then the dynamo field grows.

2.1.4 Electrical Boundary Conditions

Boundary conditions play an important role, particularly in the way induced currents flow in the system. As a result, induced magnetic fields measured by a local probe come from an overall distribution of induced currents rather than from a local deformation of applied magnetic field lines. This can be clearly illustrated again using the von Kármán geometry with counter-rotating impellers and a transverse applied field—Fig. 5(a1). The deformation associated with differential rotation corresponds to an axial current sheet in the mid-plane, as in Fig. 5(a2, 3). In turn, these currents generate induced field near the insulating boundary (shown in Fig. 5(a4)) at a right angle from the applied field (Bourgoin 2003; Bourgoin et al. 2004a).

This effect, which we term the 'BC-effect', has its origin in the $\nabla\lambda \times (\nabla \times \mathbf{B})$ term in (4). It varies linearly with the magnetic Reynolds number as does its source—the dif-

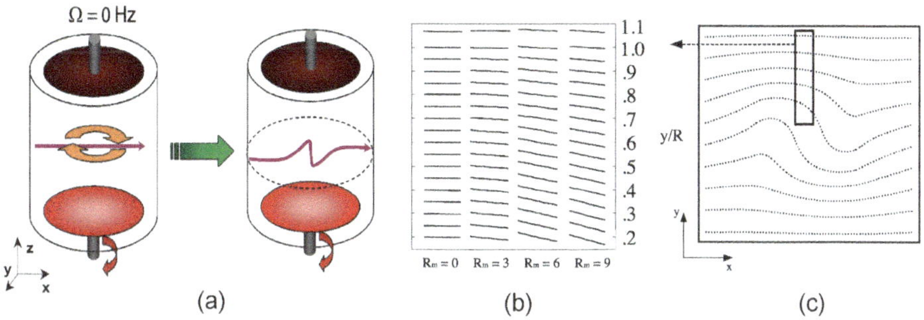

Fig. 4 Experimental verification of magnetic expulsion by coherent vortex motions. Gallium is the working fluid. A uniform magnetic field perpendicular to the cylinder (along the x-axis) is applied with external coils and the evolution of its orientation along a radius is studied as the magnetic Reynolds number is increased. (**a**) Sketch of the torsion of the applied field lines; (**b**) evolution of magnetic field orientation in the measurement (the lines give the local direction of the magnetic field inside the vessel), from Odier et al. (2000); (**c**) equivalent numerical simulation at $R_M = 20$ from Weiss (1966). The boxed region is the one probed experimentally in (**b**), at increasing R_M values

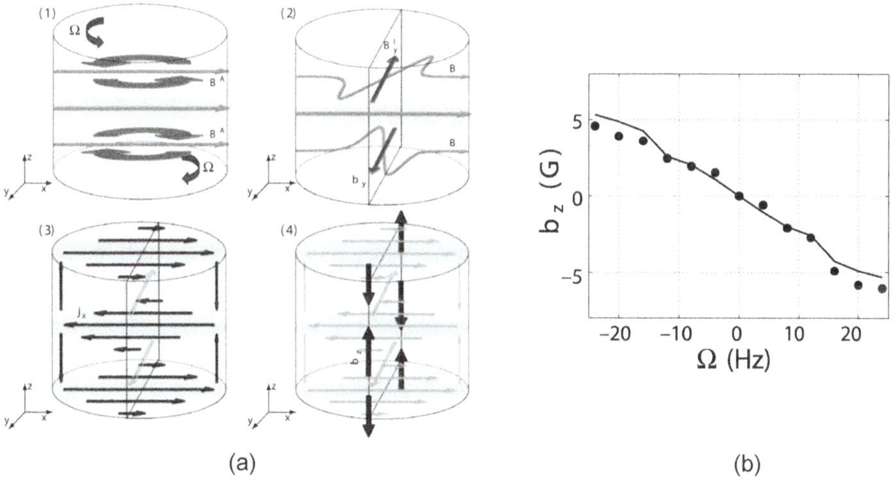

Fig. 5 Effect of boundary conditions. (**a**) Sketch of the BC-effect for a transverse applied field (33 gauss) B^A along the x-axis in a counter rotating von Kármán flow: (**1**) initial field and discs rotations; (**2**) differential rotation creates an induced component B_y; (**3**) induced current sheets responsible for the generation of B_y; (**4**) axial field B_z generated at the wall $y = R$, due to the discontinuity in electrical conductivity. (**b**) Measurements of the induced axial magnetic field by BC-effect for a transverse applied field along the x-axis with both counter-rotating discs in von Kármán experiment, from Bourgoin et al. (2004a)

ferential rotation, and the effect has been accurately treated in several numerical simulations (Bourgoin et al. 2004b; Stefani et al. 2006; Xu et al. 2008). Inhomogeneities in the magnetic boundary conditions, which appear to favour dynamo action (Monchaux et al. 2007) have been considered in numerical studies (Avalos-Zuniga and Plunian 2005; Gissinger et al. 2008) but need to be probed in more details experimentally.

2.1.5 Discussion

The above induction mechanisms have been the building blocks of dynamo experiments so far—and this has motivated their presentation. A detailed discussion of induction in MHD is outside the scope of this review, but the following remarks may be helpful in regards to experimental investigations:

- At low R_M, induction is linear in the velocity gradients and can thus be used to probe them. This is a very useful feature for the study of hydrodynamics in liquid metals, for which velocimetry techniques are not as well developed as for conventional liquids. For instance, it has recently been used in the Grenoble and Maryland experiments for the study of waves in spherical Couette flows (Kelley et al. 2007; Schmitt et al. 2008).
- Induction from uniform applied field may mask non-local features of the induction process, for which magnetic field advection is as important as line deformation. In addition, the presence of the Laplacian term in the induction equation performs some kind of averaging since at low P_M the resistive scale is much larger than the turbulence viscous cut-off. Advection effects from localized magnetic field sources have been studied in the VKS experiment (Volk et al. 2006a); field transport can reach distances larger than the flow integral scale and cause significant magnetic intermittency when turbulence is present.
- It has been observed in several instances (cf. Gailitis et al. 2000; Ravelet et al. 2007) that the induced magnetic field within the flow can reach values higher than the applied field. However, these situations were not necessarily associated with dynamo self-generation. In other words, flows can have very efficient field amplification factors without dynamo property (an analytic example is the case of a plane flow with diverging streamlines from a point source which would strongly amplify any toroidal magnetic field). In fact, an experimental criterion for the proximity of a dynamo threshold is still lacking. The decay time of applied magnetic field pulses has been studied by the Maryland group (Peffley et al. 2000), but in conditions where dynamo generation was not reached so it is not yet known how these decay times would diverge near onset.

2.2 Mean-Field MHD

We now discuss separately the case of the mean-field MHD approach and its consequences. The reason is that it proposes an efficient treatment of the effects of turbulence and has been very successfully used in the modelling of stellar dynamos—e.g. Dikpati and Gilman (2001). There has thus been some strong motivation to test it experimentally.

We first recall the basics of mean-field MHD, readers being referred to Moffatt (1978), Krause and Rädler (1980), Rädler and Stepanov (2006) for a more complete presentation. One splits the velocity and magnetic fields into an ensemble average and a fluctuating component: $\mathbf{u} = \overline{\mathbf{u}} + \mathbf{u}'$ and $\mathbf{B} = \overline{\mathbf{B}} + \mathbf{B}'$,

$$\partial_t \overline{\mathbf{B}} = \nabla \times (\overline{\mathbf{u}' \times \mathbf{B}'}) + \nabla \times (\overline{\mathbf{u}} \times \overline{\mathbf{B}}) + \lambda \Delta \overline{\mathbf{B}}, \tag{5}$$

where the mean electromotive force $\mathcal{E} = \overline{\mathbf{u}' \times \mathbf{B}'}$ can be computed after solving the equation for the fluctuating part of the magnetic field:

$$\partial_t \mathbf{B}' = \nabla \times (\mathbf{u}' \times \mathbf{B}') - \nabla \times \mathcal{E} + \nabla \times (\overline{\mathbf{u}} \times \mathbf{B}') + \nabla \times (\mathbf{u}' \times \overline{\mathbf{B}}) + \lambda \Delta \mathbf{B}'. \tag{6}$$

In the above equation, one then usually invokes scale separation between flow sizes at which velocity gradients are effective and the global scale at which magnetic effects are considered.

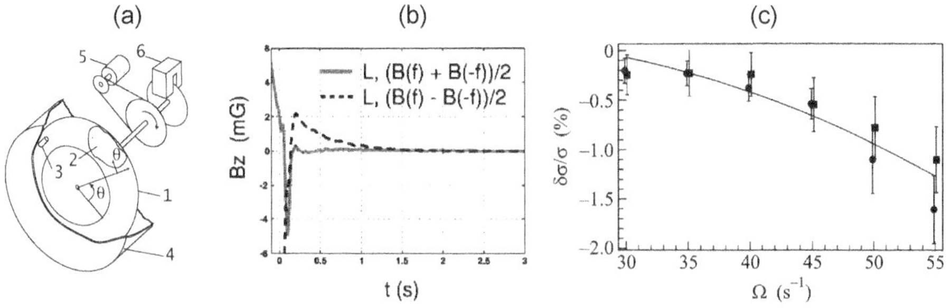

Fig. 6 Investigation of mean-field contributions. (**a**) Sketch of the Perm torus experiment (1: vessel, 2: test section, 3: Hall probe, 4: external shield, 5: motors, 6: braking device). (**b**) Study of possible α-mechanisms: response to a DC applied field with its direction parallel to the large scale helicity in the spin-down regime: the *red curve* has the symmetry of the alpha effect and is concentrated during the initial phase of motion, while the *black dashed curve* is the linear induction and follows the flow decay—figure from Stepanov et al. (2006). (**c**) Study of possible β effect, using the induction from an AC applied field. The plot shows the change in effective electrical conductivity with the initial rotation rate of the torus, from Denisov et al. (2008)

One then makes the assumption that the mean e.m.f. can be expanded in terms of the mean magnetic field $\overline{\mathbf{B}}$ and its spatial derivatives. For very general situations in the presence of a mean flow this is quite a complex task, often guided by symmetry considerations (Rädler and Stepanov 2006). Usually, the first two terms of the expansion are retained, corresponding to

- the so-called α-effect, *i.e.* the possibility to generate induced currents which are parallel to the large scale field, via an emf $\mathcal{E} = \alpha(\mathbf{u})\overline{\mathbf{B}}$. For instance, such currents have the ability to generate a poloidal magnetic field from an azimuthal one (as cooperative Parker effect from small scales), a very interesting feature for the modelization of natural dynamos.
- the 'β-effect', which corresponds to a contribution to the induced mean emf related to the gradients of $\overline{\mathbf{B}}$, $\mathcal{E} \propto \beta(\mathbf{u})(\nabla \times \overline{\mathbf{B}})$. This term represents a potential additional diffusion of the magnetic field, over the molecular Joule effect.

In cases where the flow geometry is prescribed in a simple enough form, the α term can be computed analytically. Roberts (1972) computed this α term for a flow, consisting of an array of columns in which the motions are helical with alternate axial and angular velocities, but unchanged helicity. This effect is the basis of the Karlsruhe dynamo experiment (Busse 1992; Tilgner 1997), for which both the scale separation and the helical nature of the flow are enforced by the design of the tubes that guide the flow. One can thus conclude that a cooperative effect of small-scale helical motions is experimentally observed in situations where the helical motions are strictly enforced.

The picture is somewhat different in flows where helicity evolves freely. For homogeneous (but not mirror-symmetric) turbulence, α can be computed in terms of the helicity spectrum of the flow (Moffatt 1978). For more complex geometries (in particular when mean flow motions are also present) a formal derivation of the α tensor has been performed using symmetry arguments (Rädler and Stepanov 2006). Some contributions have been tested experimentally in the Perm spin-down experiment (Frick et al. 2002; Denisov et al. 2001), displayed in Fig. 6(a): liquid Gallium is spun inside a torus which is suddenly halted; as the liquid moves with respect to the vessel, it flows past mechanical diverters which impart a global screw motion—then helicity is often assumed to cascade down to smaller scales (Chen et al. 2003). Analysis of the magnetic field induced by an applied toroidal field is shown in Fig. 6(b) where the induced field is monitored in the spin-down regime (Stepanov et al. 2006). The α value measured experimentally is much lower than predictions derived by simple arguments of the mean-field theory based on small-scale helicity; however, it has the correct order of magnitude if one includes effects associated with the inhomogeneity of velocity gradients during spin-down (Stepanov et al. 2006). In addition, the main contribution traces back to the large scale inhomogeneity of the velocity gradients, rather than to the small scale helicity distribution.

Direct measurements of the β effect have been performed in two experiments. In a first one at very moderate turbulent Reynolds number, an increase of the molecular diffusion by a few percent has been claimed (Reighard and Brown 2001). A more recent, and far more detailed experimental study, has been performed by the Perm group. It shows a correction to the molecular magnetic diffusivity of 1% at most (Denisov et al. 2008)—Fig. 6(c). The scaling with magnetic Reynolds number was found to be in agreement with mean-field theory expressions, but the amplitude again much smaller than predicted. Finally, an indirect estimation of the effect of turbulence on magnetic diffusivity is provided by the dynamo onset in the Karlsruhe and Riga experiments: in both cases the threshold was found to be in excellent agreement with predictions based on a laminar mean flow, *i.e.* neglecting small-scale turbulent fluctuations. The field at saturation however, was observed to be in good agreement with a balance of the Lorentz force with the pressure term and turbulent fluctuations (inertial term) (Pétrélis and Fauve 2001).

There is thus experimental indications that for flows of liquid metals at moderate magnetic Reynolds numbers, the effects that can be attributed to small-scale turbulence are actually smaller than predicted by the mean-field MHD theory. It does not mean that turbulence does not play a role in experimental dynamos. On the contrary, we will show that the VKS dynamo cannot be attributed to the mean (time-averaged) flow motions. In addition, induction measurements made in the Madison experiment (Spence et al. 2006) have directly evidenced the existence of a turbulent emf: an induced dipole moment has been measured in response to an axisymmetric magnetic field—with Cowling's theorem (Cowling 1933) showing that it cannot be attributed to axisymmetric flow motions. These are indications that, at least for low P_M fluids, the turbulent and non-stationary fluctuations in the vicinity of the large scales may play a dominant role in the induction processes.

2.3 A Synthetic Dynamo

We conclude this section with an example of a dynamo based on an arrangement initially proposed by Bullard (1955), and in which the dynamo cycle is viewed as a series of magnetic induction steps. An initial magnetic seed field $\mathbf{B}_0$, transported and stretched by the velocity gradients gives rise to an induced magnetic field component $\mathbf{B}_1$, which in turn generates an induced field $\mathbf{B}_2$, etc. until eventually the contribution after n steps $\mathbf{B}_n$ reinforces $\mathbf{B}_0$ (Bourgoin et al. 2004b). If this feedback process is efficient enough, $\mathbf{B}_0$ is self-sustained (it is the neutral mode of the dynamo instability).

In the spirit of Bullard's design, we use the differential rotation in the von Kármán flow with two counter rotating disks. It advects and stretches an externally applied axial field B_z, generating a toroidal component B_θ. This induced field is used to drive a power source which generates the current in the Helmholtz coils creating B_z—cf. Fig. 7. Hence, part of the dynamo cycle is generated by an external feed-back: one prescribes the mechanism by which a toroidal magnetic field generates an induced poloidal one. The feedback loop from the induced toroidal field to the applied axial one has an adjustable gain which selects the magnetic Reynolds number for dynamo onset. The flow turbulence is included in the poloidal to toroidal conversion (ω-effect) and has a leading role. Another feature of this arrangement is that saturation of the dynamo, in its present form, is not due to the modification of the flow field, but rather to the limit value $I_{\max}$ of the currents in the external coils.

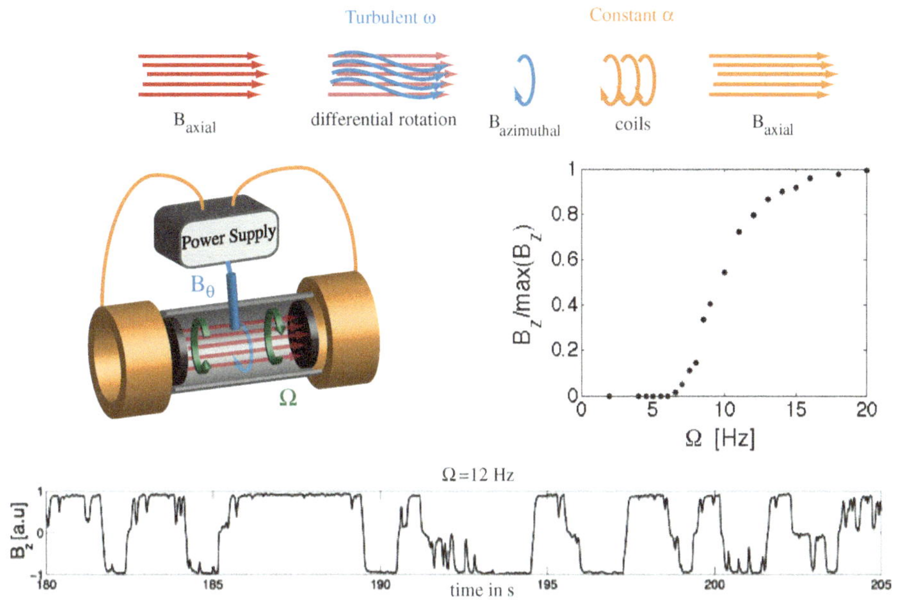

Fig. 7 The Bullard von-Kármán dynamo. (*top*): Principle of dynamo feed-back loop with actual implementation (*middle left*). The *mid-right curve* is the bifurcation curve of the self-sustained magnetic field as the rotation rate of the impellers increase. The *bottom figure* shows an example of time signal, with reversals of the dipole field, figure from Bourgoin et al. (2006)

The effect of turbulence are thus best isolated in the vicinity of onset (Bourgoin et al. 2006; Verhille et al. 2009):

(i) the bifurcation to dynamo proceeds via an on-off scenario (Sweet et al. 2001),
(ii) the bifurcation is very much dependent on the geometry and dynamics of the von Kármán flow fluctuations. Both homopolar or reversing dynamos have been observed depending on the presence of additive noise in the induction process—see Fig. 7. This is in agreement with recent models of on-off bifurcations (Aumaitre et al. 2005, 2006; Aumaitre and Pétrélis, 2006).

3 Dynamo Experiments

3.1 The Riga Experiments

The arrangement in Riga (Fig. 8(a, b)) is inspired by the Ponomarenko kinematic dynamo (Ponomarenko 1973; Gailitis and Freibergs 1976), generated by the helical motions in an infinite stationary conductor. With the first experiments made in the 80s, this swirling flow configuration has been thoroughly optimized (Gailitis 1996): back flow characteristics, addition of an external layer of sodium at rest, length of the main channel, poloidal to toroidal velocity ratio, etc. Generation and saturation of a time-dependent dynamo was first observed during summer 2000.

For a fluid in axial translation at velocity U_z while rotating at speed U_θ in a cylinder of radius R, the threshold for (Ponomarenko) self-generation in magnetic Reynolds number $R_M = R(U_\theta^2 + U_z^2)^{1/2}/\lambda$ is $R_M^c \sim 17.7$, and the bifurcation is a Hopf one: the magnetic field

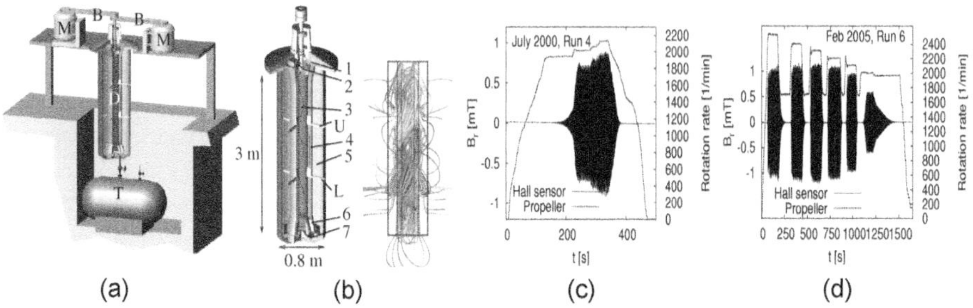

Fig. 8 The Riga dynamo experiment and its neutral mode. (**a**) Sketch of the facility. M—Motors. B—Belts. D—Central dynamo module. T—Sodium tank. (**b**) Sketch of the central module. 1—Guiding blades. 2—Propeller. 3—Helical flow region without any flow-guides, flow rotation is maintained by inertia only. 4—Back-flow region. 5—Sodium at rest. 6—Guiding blades. 7—Flow bending region. Associated simulated magnetic eigenfield. The gray scale indicates the vertical components of the field. (**c**) and (**d**): Two experimental runs carried out in July 2000 and in February 2005. Rotation rate of the motors, and magnetic field measured at one Hall external sensor plotted vs. time. Figures from Stefani et al. (2008) (Figs. 5 & 6)

at onset is oscillatory. This is very close to the threshold (17.6) observed in the experiment (Fig. 8(c, d)), with the additional following observations (Gailitis et al. 2001, 2002a, 2002b):

- optimization studies have shown that lowest threshold values are obtained when $U_\theta/U_z \sim 1$, *i.e.* for a unit ratio of 'poloidal' to 'toroidal' velocities—a feature common to all dynamo experiments so far,
- the layer of sodium at rest around the flow leads to a decrease in R_M^c,
- the main mechanism of magnetic field saturation lies in the downward breaking of the differential rotation between the innermost helical flow and the back-flow region (Gailitis et al. 2004).

3.2 The Karlsruhe Experiment

Like the previous one, the Karlsruhe experiment is designed to replicate a velocity field with a topology that is known to generate a dynamo. Here, the flow field traces back to a calculation by Roberts (1972) for a periodic array of vortices with the same helicity. This arrangement was later adapted by F. Busse for possible scenario of the Earth dynamo, a possibility that motivated the Karlsruhe experiment. In contrast to the Riga design, the Karlsruhe dynamo is a two-scale experiment. Scale separation is achieved using a quasi-periodic columnar vortex structure which forms an array of motions with like-sign helicity—this is the 'small scale' motion (Fig. 9(a)). The magnetic field, on the other hand, can develop on the larger scale of the whole experiment. In practice each of the 52 helical vortex generators is made of 2 concentric channels: one in which the flow is purely axial, and a surrounding one in which the fluid is guided in helical motion.

A very comprehensive review of the experiment and its findings can be found in Müller et al. (2006). Dynamo action was obtained for critical values of the magnetic Reynolds number $Rm^c = \alpha_\perp L/\lambda \in [8.4, 9.3]$ (here the α parameter is used as a velocity characteristic scale, and L is the overall cross-section of the experiment). This is again very close to the value (8.2) predicted using several approaches based on a laminar flow structure. One may note here again the two-scale structure of this dynamo: almost identical prediction have been obtained using the complete flow or a mean-field approach (Tilgner and Busse 2002; Rädler et al. 2002; Tilgner 2002).

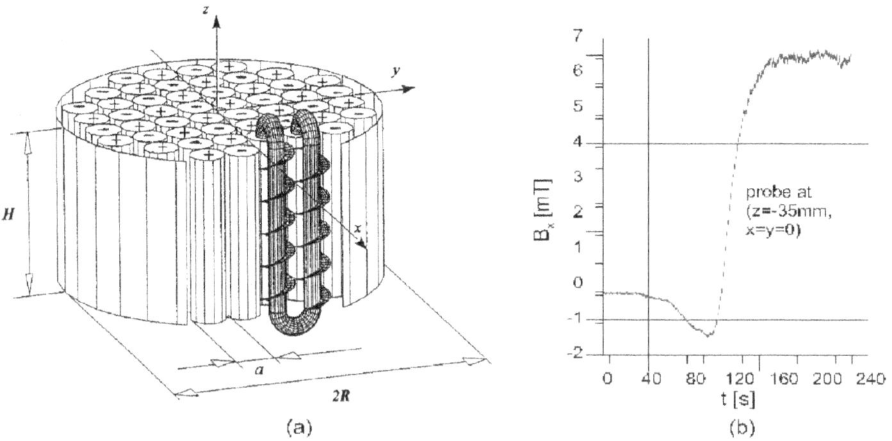

Fig. 9 The Karlsruhe dynamo experiment. (**a**) Central part of the facility. The module consists of 52 spin-generators, each containing a central tube with non-rotating flow and an outer tube where the flow is forced on a helical path. (**b**) Self-excitation and saturation in the Karlsruhe dynamo experiment. Hall sensor signals of B_x in the inner bore of the module. Figure taken from Stefani et al. (2008)

The bifurcation is supercritical, with a (statistically) stationary magnetic field generated at onset (Fig. 9(b)). The magnetic field generated is of dipolar character, oriented perpendicular to the axis of the spin generators ('an equatorial dipole'). Measurement of the pressure and flow rates in the channels indicates an additional dissipation above onset which varies linearly with the flow rate, with an order of magnitude of about 10 kW at 10% over threshold (Müller et al. 2006, 2008). As the dynamo field grows, a reduction of the flow velocity in the spin generators is also observed, however, the detailed mechanisms of the saturation of the instability are not known.

3.3 VKS Experiment

The Karlsruhe and Riga experiments have validated the principle of fluid dynamos. In each flow the kinetic Reynolds number is huge, but the overall behavior is in good agreement with the predictions of quasi-laminar approximations. They have shown that the dynamo generation is controlled by the topology and dynamics of the large scale flow. Secondary bifurcations or more complex magnetic regimes have not been observed. The VKS experiment has been designed to keep essential ingredients (shear and helicity) while at the same time allowing for more freedom to the hydrodynamic flow—and hence to the magnetic field dynamics.

The flow is generated by rotating two impellers inside a cylindrical copper vessel—details are given in Fig. 10. The impellers can be independently driven up to typically 27 Hz by motors with a total of 300 kW available power. In all configurations reported here, the impellers are manufactured from soft iron. When both impellers rotate at the same frequency $F_1 = F_2$, the forcing is symmetric with respect to any rotation R_π of π around any radial axis in its equatorial plane ($x = 0$). Otherwise, when $F_1 \neq F_2$, the system is no longer R_π-symmetric. For simplicity, we will also refer these situations as 'symmetric/asymmetric' cases.

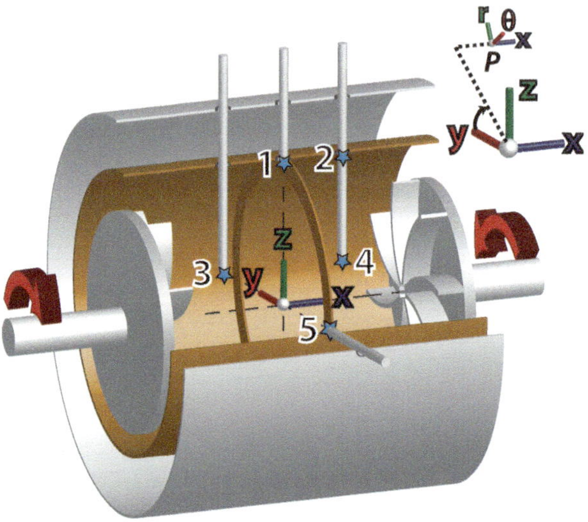

Fig. 10 Sketch of the VKS flow configuration. The flow is generated by rotating two impellers of radius 154.5 mm, 371 mm apart in a thin cylindrical copper vessel, $2R = 412$ mm in inner diameter and 524 mm in length. The impellers are fitted with 8 curved blades of height $h = 41.2$ mm; in most experimental runs, the impellers are rotated so that the blades move in a 'non-scooping' direction, defined as the positive direction by arrows. The flow is surrounded by sodium at rest contained in another concentric cylindrical copper vessel, 578 mm in inner diameter and 604 mm long. An oil circulation in this thick copper vessel maintains a regulated temperature in the range 110–160°C. In the mid plane between the impellers one can attach a thin annulus inner diameter 350 mm and thickness 5 mm. The impellers that generate the flow have been machined from pure soft iron ($\mu_r \sim 100$). Positions 1 to 5 correspond to points where time recordings of the magnetic field have been made, using 3D Hall probes. Figure from Monchaux et al. (2009)

3.3.1 A Statistically Steady Turbulent Dynamo

For symmetric forcing, one observes that, as the rotation rate of the impellers $F = F_1 = F_2$ is increased above 17 Hz, the magnetic field inside the flow develops strong fluctuations and its main component (in the azimuthal direction at the probe location) grows and saturates to a mean values up to 100 G—Fig. 11. This value is about 100 times larger than the ambient magnetic field in the experimental hall, from which the flow volume is not shielded. The hundred-fold increase is also one order of magnitude larger than the induction effects and field amplification previously recorded in the VKS experiment with externally applied magnetic field, either homogeneously over the flow volume (Bourgoin et al. 2002) or localized at the flow boundary (Volk et al. 2006a). The most salient features of the dynamo observed with a symmetric forcing are the following (Monchaux et al. 2007, 2009; Aumaître et al. 2008):

- it appears via a supercritical bifurcation at $R_M^c \sim 32$, generating a statistically steady magnetic field,
- the geometry of the dynamo field is mainly that of an axial dipole,
- opposite polarities of the dipole have been observed as R_M is increased above threshold (in agreement with the expected $\mathbf{B} \to -\mathbf{B}$ symmetry of the equations) but once a direction of the dipole has been chosen at onset, no secondary bifurcation is observed as R_M is increased to its maximum accessible value ($R_M^{\max} \sim 50$),
- the amplitude of the field at saturation is in good agreement with a balance of the Lorentz force with the non-linear term in the Navier-Stokes equation (Pétrélis and Fauve 2001).

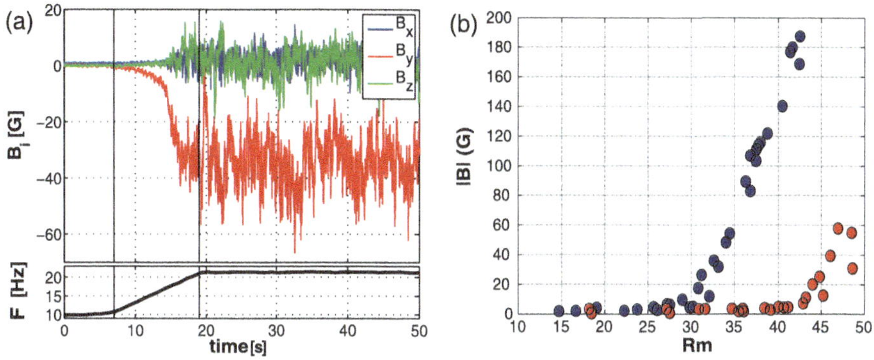

Fig. 11 Self-generation in the VKS experiment. (**a**) Growth of the magnetic field as the rotation rate of the impellers $F = F_1 = F_2$ is increased from 10 to 22 Hz. Three components of the magnetic field recorded at location 1. (**b**) Magnetic field amplitude $\langle B^2 \rangle^{1/2}$ recorded at location 3; (*blue circles*): counter-rotating impellers at equal rotation rates, in the positive direction shown in Fig. 9. (*red circles*): Impellers counter-rotating in the opposite direction, i.e. with the blades on the impellers moving in a 'scooping' or negative direction. Changes in the efficiency of the stirring are taken into account in the definition of R_M: $R_M = K\mu_0\sigma R^2 F$ with $K = K+ = 0.6$ in the normal, positive direction of rotation and $K = K- = 0.7$ in the opposite direction. Figure from Monchaux et al. (2009)

The observation that the magnetic field geometry for the symmetric VKS forcing is mainly dipolar has important implications. It strongly differs from the prediction of kinematic calculations based on the topology of the mean von Kármán flow which tends to favor a transverse dipole (Marié et al. 2003; Ravelet et al. 2005; Bourgoin et al. 2004b). In addition, Cowling theorem (Moffatt 1978) implies that it has not been generated by the mean flow motions alone. The axial symmetry, however, is what would be expected from an $\alpha - \omega$ dynamo. The differential rotation induced by the counter-rotation of the impellers has the ability to generate an azimuthal field component from an axial magnetic field (Bourgoin et al. 2002, 2004a). The conversion by the α-effect of a toroidal field into a poloidal one is often thought to rely on helical flow motions, but the source of the α term in VKS is yet unclear. Several mechanisms have been proposed (Laguerre et al. 2008; Monchaux et al. 2009) and further measurements are underway. A point which deserves clarification concerns the role of the ferromagnetic impellers which are used to drive the flow. Dynamo action has not been reached in identical conditions for impellers manufactured from stainless steel instead of soft iron. Further studies will have to establish (i) whether the use of iron impellers only lowers the threshold of the same dynamo that would exist with non-ferromagnetic drives, (ii) if, instead, the modification of the boundary conditions favor an axisymmetric mode compared to the equatorial dipole predicted from kinematic dynamo simulations based on the time-averaged flow; (iii) more generally, the role of high permeability of the impellers on the generation of a poloidal magnetic field from a toroidal component induced by differential rotation.

3.3.2 Dynamical Regimes for an Asymmetric Forcing

When the flow is forced with the impellers rotating at different rates, studies in water-prototypes have shown that global rotation is imparted to the flow (Marié 2003; Monchaux 2007): there are strong similarities between the von Kármán flow forced by impellers rotating respectively at F_1 and F_2 in the laboratory frame or by impellers rotating at $(F_1 + F_2)/2$

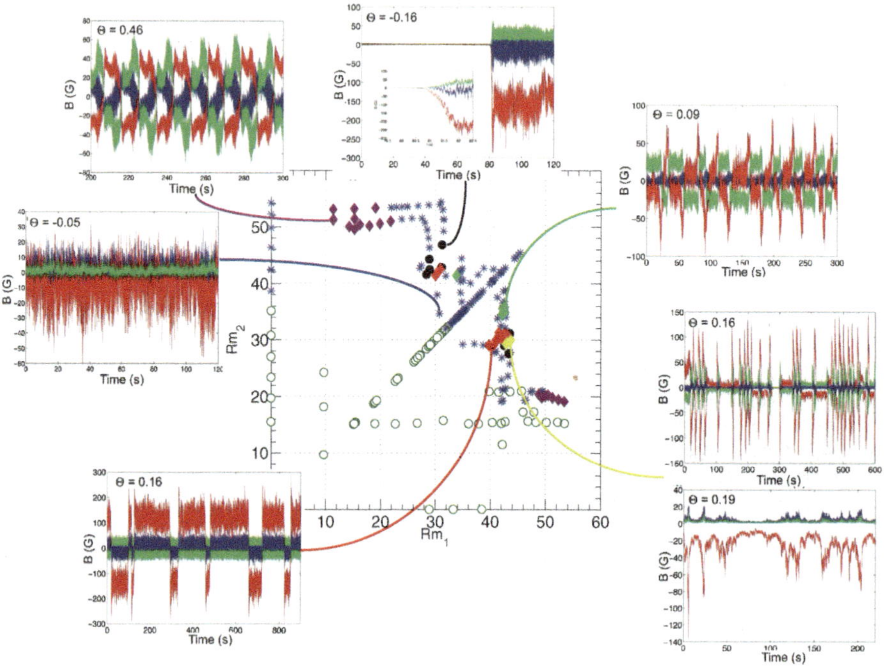

Fig. 12 Dynamical regimes observed in the von Kármán sodium experiment when the impellers driving the flow rotate at varying rates: the coordinates in the main plot are the magnetic Reynolds numbers built from the velocity of each impeller. The *insets* give examples of time signals of the 3 components of the magnetic field recorded in the mid-plane of the cylinder. Figure extracted from data in Monchaux et al. (2009)

in a frame rotating at $(F_1 - F_2)/2$. In addition, this asymmetric differential forcing brakes the R_π symmetry.

The added degree of freedom gives access to a variety of dynamos with complex dynamical regimes—they are shown in Fig. 12 using two independent Reynolds numbers built on the rotation rates of each impeller (Monchaux et al. 2009). Regions where (statistically) stationary dynamos are generated alternate with regions for which the magnetic field is time dependent. Of particular interest is the region near $|\theta| \equiv 2|F_1 - F_2|/(F_1 + F_2) = 0.16$ where reversals of the dipole field are observed at irregular time intervals. This regime, which also includes excursions, bears some similarity with the behavior of the geodynamo (Berhanu et al. 2007). Another very intriguing regime is reached in the same region where the dynamo has sudden 'bursts' between high and low fields states—Fig. 12 (right part). Both regimes are observed in a region where the flow has also two possible states (Monchaux et al. 2006; Ravelet et al. 2008).

Further studies (Monchaux et al. 2008) have shown that these regimes can be interpreted as the development and interaction of few dynamo modes (essentially a dipole and a quadrupole) when the flow is forced asymmetrically. The proximity of dipole and quadrupole modes in $\alpha - \omega$ dynamos has been pointed out by many numerical studies (Kutzner and Christensen 2004; Morin 2004), and few-modes interactions have long been used to ascribe dynamical features to the dynamo instability (Nozières 1978). For the case of the VKS dynamos, the onset of oscillatory behavior and the occurrence of random reversals, have recently been described using a low dimensional model (Pétrélis and Fauve 2008).

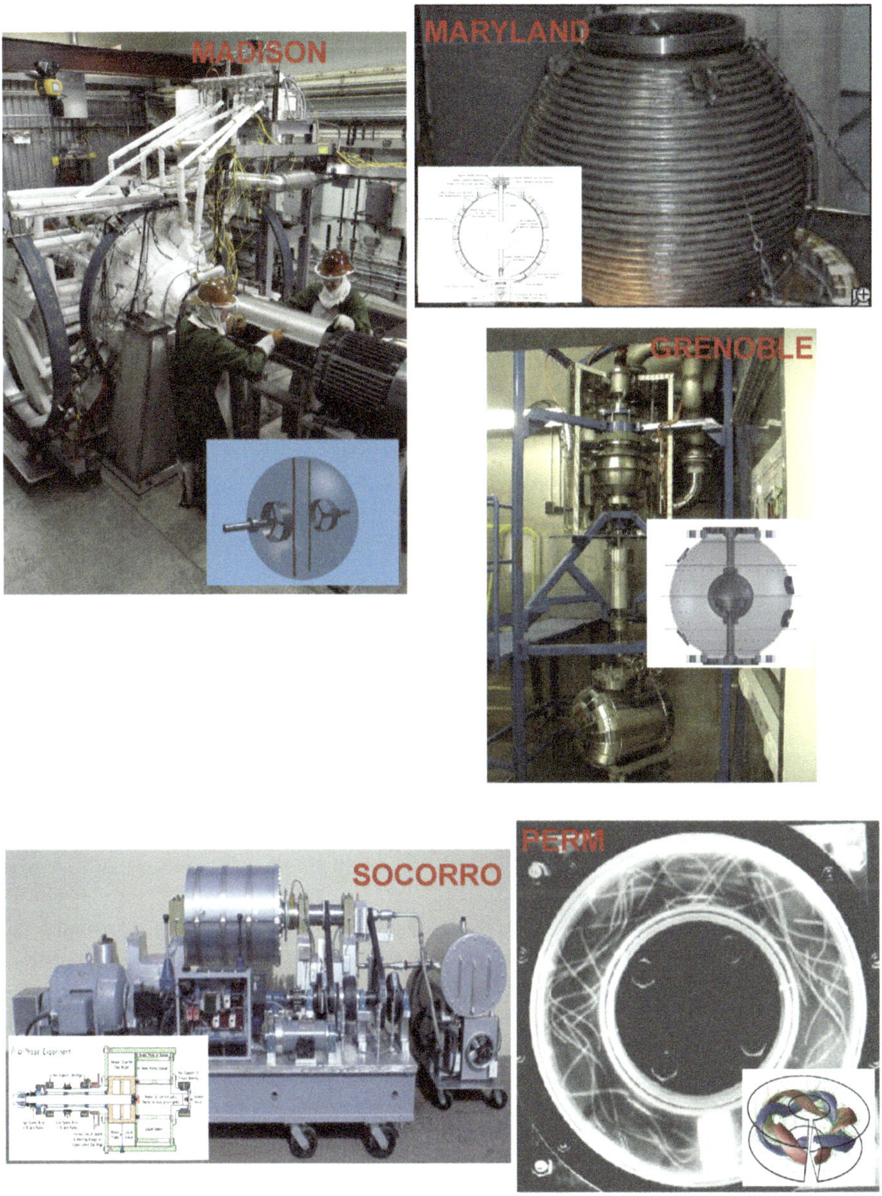

Fig. 13 Other sodium experiments operated or in preparation, aimed at studying the dynamo instability. Starting from top/left and clockwise: the Madison dynamo experiment at Univ. Wisconsin, USA; the 3-Meter system being installed at University of Maryland, USA; the DTS spherical Couette flow of the Grenoble team, France; the spin-down helical flow in the Perm torus, Russia; the cylindrical Couette flow in Socorro, New Mexico, USA

3.4 Related Dynamo Experiments

Several other experiments, shown in Fig. 13, are operated or in preparation world-wide.

- A setup similar to the VKS arrangement, but in a spherical volume, is studied by the group of Cary Forest at University of Wisconsin (USA). It has shown the possibility that turbulent motions induce an axial dipole from an applied external field (Spence et al. 2006).
- The Complex Dynamics group headed by Daniel Lathrop at University of Maryland (USA) has operated a large variety of sodium experiments, showing in particular the influence of the driving on the induction efficiency (Peffley et al. 2000). The most recent developments have been made in spherical Couette flows with applied magnetic fields, showing coupled flow and magnetic field patterns consistent with a magneto-rotational instability (Sisan et al. 2004). A large spherical Couette experiment (with an outer sphere 3 meters in diameter) is planned; it will reach the highest accessible magnetic Reynolds numbers of all current experiments.
- A spherical Couette experiment is run in Grenoble (France) by the group of H.-C. Nataf, D. Jault and Ph. Cardin. The inner sphere contains a strong permanent magnet with the purpose of studying flows and magnetic regimes in conditions closer to the ones that prevail in the Earth. The first measurements have evidenced a large variety of magneto-inertial waves (Schmitt et al. 2008).
- At the Institute for Continuous Media Mechanics in Perm (Russia), the team of Peter Frick has designed a spin-down helical flow inside a torus (Frick et al. 2002; Denisov et al. 2001). Compared to other experiments the flow is strongly non-stationary: a strongly anisotropic turbulence develops and decays, as the fast-rotating torus is suddenly put to halt (Noskov et al. 2009). This flow has the potential to sustain a dynamo in the transient relaxation, and repeated realizations are expected to help understand the magnetic-velocity field interactions during the phases of growth, saturation and decay.
- In Socorro (New Mexico, USA), the team led by Stirling Colgate has designed a Couette flow operated with sodium, with the aim of generating an $\alpha\omega$ dynamo. Characterization of hydrodynamic flow transition have been carried out (Beckley 2002).

There are also several experiments intended to study the Magneto-Rotational Instability (MRI) in Couette flows. One is developed in Obninsk (Russia), in collaboration with the Kurchatov Institute in Moscow (Velikhov 2006). Another is in Rossendorf, operated by the group of Gunter Gerbeth and Frank Stefani. This experiment has recently shown an MRI in the form of a travelling wave (Stefani et al. 2007).

4 Concluding Remarks

Several features are shared by experimental dynamos operated so far with progressively less constrained motions—since the pioneering works of Lowes and Wilkinson in Cambridge in the 60s. All have observed that the bifurcations is supercritical, and subcritical bifurcation are yet to be evidenced experimentally. Boundary conditions have been essential, particularly in order to shift the critical magnetic Reynolds number R_M^c within the range accessible in the chosen experimental devices. Much further studies are necessary in order to understand the role of turbulence in the generation of the magnetic field and its saturation. With a growing number of dynamo experiments worldwide, and with the narrowing gap with numerical simulations, the next decade appears very promising.

Acknowledgements The authors gratefully acknowledge the contribution of an anonymous referee. Bullard-von Kármán and standard von Kármán experiments are operated at ENS-Lyon. The von Kármán Sodium experiments are designed and run by the VKS collaboration on the premises of the CEA—

Direction de l'Energie Nucléaire in Cadarache. The VKS team includes François Daviaud, Arnaud Chiffaudel, Sébastien Aumaître, Bérengère Dubrulle, Florent Ravelet, Romain Monchaux, Louis Marié at CEA—Département de Sciences de la Matière; Stephan Fauve, François Pétrélis, Nicolas Mordant, Michael Berhanu at Ecole Normale Supérieure de Paris and Jean-François Pinton, Philippe Odier, Mickael Bourgoin, Romain Volk, Nicolas Plihon at Ecole Normale Supérieure de Lyon. Fruitful discussions with members of the VKS team are gratefully acknowledged.

This work was supported by CEA, CNRS and ANR NT05-1 42110, ANR-08-BLAN-0039-02.

References

MHD dynamo experiments. Magnetohydrodynamics **38**(1/2) (2002), special issue

J.-F. Pinton, R. Moreau (eds.), The dynamo effect, experimental progress, geo and astrophysical challenges. C.R. Acad. Sci. **9**(7) (2008)

S. Aumaitre, F. Pétrélis, Modification of instability processes by multiplicative noises. Eur. J. Phys. B **51**, 357 (2006)

S. Aumaitre, F. Pétrélis, K. Mallick, Low frequency noise controls on-off intermittency of bifurcating systems. Phys. Rev. Lett. **95**, 064101 (2005)

S. Aumaitre, K. Mallick, F. Pétrélis, Effects of the low frequencies of noise on on-off-bifurcations. J. Stat. Phys. **123**, 909 (2006)

S. Aumaître, M. Berhanu, M. Bourgoin, A. Chiffaudel, F. Daviaud, B. Dubrulle, S. Fauve, L. Marié, R. Monchaux, N. Mordant, Ph. Odier, F. Pétrélis, J.-F. Pinton, N. Plihon, F. Ravelet, R. Volk, The VKS experiment: turbulent dynamical dynamos. C.R. Acad. Sci. **9**(7), 689 (2008)

R. Avalos-Zuniga, F. Plunian, Influence of inner and outer walls electromagnetic properties on the onset of a stationary dynamo. Eur. Phys. J. B **47**(1), 127 (2005)

H.F. Beckley, Measurement of annular Couette flow stability at the fluid Reynolds number $Re = 4.4\ 10^6$: the fluid dynamic precursor to a liquid sodium $\alpha\omega$ dynamo. PhD Dissertation, New Mexico Institute of Mining and Technology, 2002

M. Berhanu, R. Monchaux, M. Bourgoin, M. Moulin, Ph. Odier, J.-F. Pinton, R. Volk, S. Fauve, N. Mordant, F. Pétrélis, A. Chiffaudel, F. Daviaud, B. Dubrulle, C. Gasquet, L. Marié, F. Ravelet, Magnetic field reversals in an experimental turbulent dynamo. Europhys. Lett. **77**, 59007 (2007)

M. Bourgoin, Etudes en magnétohydrodynamique, application à l'effet dynamo. PhD Thesis, Ecole Normale Supérieure de Lyon, 2003. http://tel.archives-ouvertes.fr/tel-00008302/en/

M. Bourgoin, L. Marié, F. Pétrélis, C. Gasquet, A. Guigon, J.-B. Luciani, M. Moulin, F. Namer, J. Burguete, A. Chiffaudel, F. Daviaud, S. Fauve, Ph. Odier, J.-F. Pinton, Magnetohydrodynamics measurements in the von Kármán sodium experiment. Phys. Fluids **14**, 3046–3058 (2002)

M. Bourgoin, R. Volk, P. Frick, S. Kripechenko, P. Odier, J.-F. Pinton, Induction mechanisms in von Kármán swirling flows of liquid Gallium. Magnetohydrodynamics **40**(1), 13–31 (2004a)

M. Bourgoin, P. Odier, J.-F. Pinton, Y. Ricard, An iterative study of time independent induction effects in magnetohydrodynamics. Phys. Fluids **16**(7), 2529–2547 (2004b)

M. Bourgoin, R. Volk, N. Plihon, P. Augier, Ph. Odier, J.-F. Pinton, A Bullard von Kármán dynamo. New J. Phys. **8**, 329 (2006)

E.C. Bullard, The stability of a homopolar dynamo. Proc. Camb. Philos. Soc. **51**, 744 (1955)

E.C. Bullard, D. Gubbins, Generation of magnetic fields by fluid motions of global scale. Geophys. Astrophys. Fluid Dyn. **8**, 43 (1977)

F.H. Busse, Dynamo theory of planetary magnetism and laboratory experiments, in *Evolution of Dynamical Structures in Complex Systems*, ed. by R. Friedrich, A. Wunderlin (Springer, Berlin, 1992), pp. 359–384

Q.N. Chen, S.Y. Chen, G.L. Eyink, The joint cascade of energy and helicity in three-dimensional turbulence. Phys. Fluids **15**(2), 361–374 (2003)

T.G. Cowling, The magnetic field of sunspots. Mon. Not. R. Astron. Soc. **94**, 39 (1933)

S. Denisov, V. Noskov, A. Sukhanovskiy, P. Frick, Unsteady turbulent spiral flows in a circular channel. Fluid Dyn. **36**(5), 734–742 (2001)

S.A. Denisov, V.I. Noskov, R.A. Stepanov, P.G. Frick, Measurements of turbulent magnetic diffusivity in a liquid-Gallium flow. JETP Lett. **88**(3), 167–171 (2008)

M. Dikpati, P.A. Gilman, Flux-transport dynamos with α-effect from global instability of tachocline differential rotation: a solution for magnetic parity selection in the sun. Astrophys. J. **559**(1), 428–442 (2001)

N.L. Dudley, R.W. James, Time-dependent kinematic dynamos with stationary flows. Proc. R. Soc. Lond. Ser. A **425**, 407 (1989)

P. Frick, V. Noskov, S. Denisov, S. Khripchenko, D. Sokoloff, R. Stepanov, A. Sukhanovsky, Non-stationary screw flow in a toroidal channel: way to a laboratory dynamo experiment. Magnetohydrodynamics **38**, 136–155 (2002)

A. Gailitis, Project of a liquid sodium MHD dynamo experiment. Magnetohydrodynamics **32**, 58–62 (1996)
A. Gailitis, Ya. Freibergs, Theory of a helical MHD dynamo. Magnetohydrodynamics **12**, 127–129 (1976)
A. Gailitis, O. Lielausis, S. Dement'ev, E. Platacis, A. Cifersons, G. Gerbeth, T. Gundrum, F. Stefani, M. Christen, H. Hänel, G. Will, Detection of a flow induced magnetic field eigenmode in the Riga dynamo facility. Phys. Rev. Lett. **84**(19), 4365 (2000)
A. Gailitis et al., Magnetic field saturation in the Riga dynamo experiment. Phys. Rev. Lett. **86**, 3024–3027 (2001)
A. Gailitis, O. Lielausis, E. Platacis, G. Gerbeth, F. Stefani, On back-reaction effects in the Riga dynamo experiment. Magnetohydrodynamics **38**, 15–26 (2002a)
A. Gailitis, O. Lielausis, E. Platacis, G. Gerbeth, F. Stefani, *Colloquium*: Laboratory experiments on hydromagnetic dynamos. Rev. Mod. Phys. **74**, 973–990 (2002b)
A. Gailitis, O. Lielausis, E. Platacis, G. Gerbeth, F. Stefani, The Riga dynamo experiment. Surv. Geophys. **24**(3), 247–267 (2003)
A. Gailitis, O. Lielausis, E. Platacis, G. Gerbeth, F. Stefan, Riga dynamo experiment and its theoretical background. Phys. Plasmas **11**(5), 2838 (2004)
C. Gissinger, A. Iskakov, S. Fauve, E. Dormy, Effect of magnetic boundary conditions on the dynamo threshold of von Karman swirling flows. Europhys. Lett. **82**(2), 29001 (2008)
D.H. Kelley, S.A. Triana, D.S. Zimmerman, A. Tilgner, D.P. Lathrop, Inertial waves driven by differential rotation in a planetary geometry. GAFD **101**(5–6), 469–487 (2007)
F. Krause, K.-H. Rädler, *Mean Field Magnetohydrodynamics and Dynamo Theory* (Pergamon Press, New York, 1980)
C. Kutzner, U.R. Christensen, Simulated geomagnetic reversals and preferred virtual geomagnetic pole paths. Geophys. J. Int. **157**(3), 1105–118 (2004)
R. Laguerre, C. Nore, A. Ribeiro, J. Leorat, J.L. Guermond, F. Plunian, Impact of impellers on the axisymmetric magnetic mode in the VKS2 dynamo experiment. Phys. Rev. Lett. **101**(10), 104501 (2008), correction in Phys. Rev. Lett. **101**(21), 219902 (2008)
F.J. Lowes, I. Wilkinson, Geomagnetic dynamo: a laboratory model. Nature **198**, 1158–1160 (1963)
F.J. Lowes, I. Wilkinson, Geomagnetic dynamo: an improved laboratory model. Nature **219**, 717–718 (1968)
L. Marié, J. Burguete, F. Daviaud, J. Léorat, Numerical study of homogeneous dynamo based on experimental von Kármán type flows. Eur. Phys. J. B **33**, 469–485 (2003)
L. Marié, Transport de moment cinétique et de champ magnétique par un écoulement tourbillonnaire turbulent: influence de la rotation. PhD thesis, Université de Paris 7, 2003. http://tel.archives-ouvertes.fr/tel-00007755/en/
A. Martin, P. Odier, J.-F. Pinton, S. Fauve, Effective permeability in a binary flow of liquid gallium and iron beads. Eur. Phys. J. B **18**, 337–341 (2000)
H.K. Moffatt, *Magnetic Field Generation in Electrically Conducting Fluids* (Cambridge University Press, Cambridge, 1978)
R. Monchaux, Mécanique statistique et effet dynamo dans un écoulement de von Kármán turbulent. PhD thesis Université Diderot, Paris 7, 2007. http://tel.archives-ouvertes.fr/tel-00199751/en/
R. Monchaux, F. Ravelet, B. Dubrulle, F. Daviaud, Properties of steady states in turbulent axisymmetric flows. Phys. Rev. Lett. **96**(12), 124502 (2006)
R. Monchaux, M. Berhanu, M. Bourgoin, M. Moulin, Ph. Odier, J.-F. Pinton, R. Volk, S. Fauve, N. Mordant, F. Pétrélis, A. Chiffaudel, F. Daviaud, B. Dubrulle, C. Gasquet, L. Marié, F. Ravelet, Generation of magnetic field by dynamo action in a turbulent flow of liquid sodium. Phys. Rev. Lett. **98**, 044502 (2007)
R. Monchaux, M. Berhanu, S. Aumaître, A. Chiffaudel, F. Daviaud, B. Dubrulle, F. Ravelet, M. Bourgoin, Ph. Odier, J.-F. Pinton, N. Plihon, R. Volk, S. Fauve, N. Mordant, F. Pétrélis, Chaotic dynamos generated by a turbulent flow of liquid sodium. Phys. Rev. Lett. **101**, 074502 (2008)
R. Monchaux, M. Berhanu, S. Aumaître, A. Chiffaudel, F. Daviaud, B. Dubrulle, L. Marié, F. Ravelet, S. Fauve, N. Mordant, F. Pétrélis, M. Bourgoin, P. Odier, J.-F. Pinton, N. Plihon, R. Volk, The VKS experiment: turbulent dynamical dynamos. Phys. Fluids **21**, 035108 (2009)
N. Mordant, J.-F. Pinton, F. Chilla, Characterization of turbulence in a closed flow. J. Phys. II (France) **7**(11), 1729 (1998)
V. Morin, Instabilités et bifurcations associées à la modélisation de la géodynamo. PhD thesis, Université Diderot, Paris 7, 2004. http://tel.archives-ouvertes.fr/tel-00011484/fr/
U. Müller, R. Stieglitz, S. Horanyi, Experiments at a two-scale dynamo test facility. J. Fluid Mech. **552**, 419 (2006)
U. Müller, R. Steiglitz, F.H. Busse, A. Tilgner, The Karlsruhe two-scale dynamo experiment. C.R. Phys. **9**, 729–740 (2008)
V. Noskov, R. Stepanov, S. Denisov, P. Frick, G. Verhille, N. Plihon, J.-F. Pinton, Dynamics of a turbulent spin-down flow inside a torus. Phys. Fluids **21**, 045108 (2009)

Ph. Nozières, Reversals of the Earth's magnetic field: an attempt at a relaxation model. Phys. Earth Planet. Inter. **17**, 55–74 (1978)
P. Odier et al., Advection of a magnetic field by a turbulent swirling flow. Phys. Rev. E **58**(6), 7397–7401 (1998)
P. Odier, J.-F. Pinton, S. Fauve, Magnetic induction by coherent vortex motion. Eur. Phys. J. B **16**, 373 (2000)
E.N. Parker, Hydromagnetic dynamo models. Astrophys. J. **122**, 293 (1955)
N.L. Peffley, A.B. Cawthorne, D.P. Lathrop, Toward a self-generating magnetic dynamo: The role of turbulence. Phys. Rev. E **61**(5), 5287 (2000)
F. Pétrélis, S. Fauve, Saturation of the magnetic field above the dynamo threshold. Eur. Phys. J. B **22**, 273–276 (2001)
F. Pétrélis, S. Fauve, Chaotic dynamics of the magnetic field generated by dynamo action in a turbulent flow. J. Phys., Condens. Matter **20**, 494203 (2008)
F. Pétrélis, M. Bourgoin, L. Marié, A. Chiffaudel, S. Fauve, F. Daviaud, P. Odier, J.-F. Pinton, Non linear induction in a swirling flow of liquid sodium. Phys. Rev. Lett. **90**(17), 174501 (2003)
F. Pétrélis, N. Mordant, S. Fauve, On the magnetic fields generated by experimental dynamos. Geophys. Astrophys. Fluid Dyn. **101**(3–4), 289 (2007)
J.-F. Pinton, F. Plaza, L. Danaila, P. Le Gal, F. Anselmet, On velocity and passive scalar scaling laws in a turbulent swirling flow. Physica D **122**(1–4), 187 (1998)
Yu.B. Ponomarenko, Theory of the hydromagnetic generator. J. Appl. Mech. Tech. Phys. **14**, 775–779 (1973)
K.-H. Rädler, M. Rheinhardt, E. Apstein, On the mean-field theory of the Karlsruhe dynamo experiment I. Kinematic theory. Magnetohydrodynamics **38**, 41–71 (2002)
K.H. Rädler, R. Stepanov, Mean electromotive force due to turbulence of a conducting fluid in the presence of mean flow. Phys. Rev. E **73**(5), 056311 (2006)
F. Ravelet, A. Chiffaudel, F. Daviaud, J. Léorat, Toward an experimental von Kármán dynamo: Numerical studies for an optimized design. Phys. Fluids **17**, 117104 (2005)
F. Ravelet, R. Volk, A. Chiffaudel, F. Daviaud, B. Dubrulle, R. Monchaux, M. Bourgoin, P. Odier, J.-F. Pinton, M. Berhanu, S. Fauve, N. Mordant, F. Petrelis, Magnetic induction in a turbulent flow of liquid sodium: mean behaviour and slow fluctuations. arXiv:0704.2565 (2007)
F. Ravelet, A. Chiffaudel, F. Daviaud, Supercritical transition to turbulence in an inertially driven von Karman closed flow. J. Fluid Mech. **601**, 339 (2008)
A.B. Reighard, M.R. Brown, Turbulent conductivity measurements in a spherical liquid sodium flow. Phys. Rev. Lett. **86**(13), 2794 (2001)
G.O. Roberts, Dynamo action of fluid motions with two-dimensional periodicity. Philos. Trans. R. Soc. Lond. A **271**, 411–454 (1972)
A.A. Schekochihin, E.L. Haugen, A. Brandenburg, C. Cowley, J.L. Maron, J.C. McWilliiam, The onset of small scale turbulent dynamo at low magnetic Prandtl numbers. Astrophys. J. **625**, 115 (2004)
D. Schmitt, T. Alboussiere, D. Brito, P. Cardin, N. Gagniere, D. Jault, H.-C. Nataf, Rotating spherical Couette flow in a dipolar magnetic field: experimental study of magneto-inertial waves. J. Fluid Mech. **604**, 175–197 (2008)
C. Simand, F. Chillà, J.-F. Pinton, Study of inhomogeneous turbulence in the closed flow between corotating disks. Europhys. Lett. **49**, 336 (2000)
D. Sisan et al., Experimental observation and characterization of the magnetorotational instability. Phys. Rev. Lett. **93**(11), 114502 (2004)
E.J. Spence, M.D. Nornberg, C.M. Jacobson, R.D. Kendrick, C.B. Forest, Observation of a turbulence-induced large scale magnetic field. Phys. Rev. Lett. **96**, 055002 (2006)
E.J. Spence, K. Reuter, C.B. Forest, A spherical plasma dynamo experiment. arXiv:0901.3406 (2009)
M. Steenbeck et al., Experimental discovery of the electromotive force along the external magnetic field induced by a flow of liquid metal (α-effect). Sov. Phys. Dokl. **13**, 443 (1968)
F. Stefani, M. Xu, G. Gerbeth, F. Ravelet, A. Chiffaudel, F. Daviaud, J. Léorat, Ambivalent effects of added layers on steady kinematic dynamos in cylindrical geometry: application to the VKS experiment. Eur. J. Mech. B, Fluids **25**(6), 894 (2006)
F. Stefani, T. Gundrum, G. Gerbeth, G. Rüdiger, J. Szklarski, R. Hollerbach, Experiments on the magnetorotational instability in helical magnetic fields. New J. Phys. **9**, 295 (2007)
F. Stefani, A. Gailitis, G. Gerbeth, Magnetohydrodynamic experiments on cosmic magnetic fields. Z. Angew. Math. Mech. **88**, 930 (2008)
R. Stepanov, R. Volk, S. Denisov, P. Frick, V. Noskov, J.-F. Pinton, Induction, helicity and alpha effect in a toroidal screw flow of liquid gallium. Phys. Rev. E **73**, 046310 (2006)
D. Sweet, E. Ott, J.M. Finn, T.M. Antonsen Jr., D.P. Lathrop, Blowout bifurcations and the onset of magnetic activity in turbulent dynamos. Phys. Rev. E **63**, 066211 (2001)
A. Tilgner, A kinematic dynamo with a small scale velocity field. Phys. Lett. A **226**, 75–79 (1997)

A. Tilgner, Numerical simulation of the onset of dynamo action in an experimental two-scale dynamo. Phys. Fluids **14**, 4092–4094 (2002)
A. Tilgner, F.H. Busse, Simulation of the bifurcation diagram of the Karlsruhe dynamo. Magnetohydrodynamics **38**, 35–40 (2002)
E.P. Velikhov, Magneto-rotational instability in differentially rotating liquid metals. Phys. Lett. A. **356**, 216–221 (2006)
G. Verhille, N. Plihon, G. Fanjat, R. Volk, M. Bourgoin, J.-F. Pinton, Large scale fluctuations and dynamics of the Bullard-von Kármán dynamo. Geophys. Astrophys. Fluid Dyn. (2009, submitted)
R. Volk, R. Monchaux, M. Berhanu, F. Ravelet, A. Chiffaudel, F. Daviaud, B. Dubrulle, S. Fauve, N. Mordant, Ph. Odier, F. Pétrélis, J.-F. Pinton, Transport of magnetic field by a turbulent flow of liquid sodium. Phys. Rev. Lett. **97**, 074501 (2006a)
R. Volk, Ph. Odier, J.-F. Pinton, Fluctuation of magnetic induction in von Karman swirling flows. Phys. Fluids **18**(8), 085105 (2006b)
N.O. Weiss, The expulsion of magnetic flux by eddies. Proc. R. Soc. Lond., Ser. A **293**, 310 (1966)
M. Xu, F. Stefani, G. Gerbeth, The integral equation approach to kinematic dynamo theory and its application to dynamo experiments in cylindrical geometry. J. Comput. Phys. **227**(17), 8130 (2008)
P.J. Zandbergen, D. Dijkstra, Von Kármán swirling flows. Annu. Rev. Fluid Mech. **19**, 465 (1987), and references therein

Space Sci Rev (2010) 152: 565–590
DOI 10.1007/s11214-009-9553-2

Dynamo Scaling Laws and Applications to the Planets

U.R. Christensen

Received: 6 February 2009 / Accepted: 8 June 2009 / Published online: 7 July 2009

Abstract Scaling laws for planetary dynamos relate the characteristic magnetic field strength, characteristic flow velocity and other properties to primary quantities such as core size, rotation rate, electrical conductivity and heat flux. Many different scaling laws have been proposed, often relying on the assumption of a balance of Coriolis force and Lorentz force in the dynamo. Their theoretical foundation is reviewed. The advent of direct numerical simulations of planetary dynamos and the ability to perform them for a sufficiently wide range of control parameters allows to test the scaling laws. The results support a magnetic field scaling that is not based on a force balance, but on the energy flux available to balance ohmic dissipation. In its simplest form, it predicts a field strength that is independent of rotation rate and electrical conductivity and proportional to the cubic root of the available energy flux. However, rotation rate controls whether the magnetic field is dipolar or multipolar. Scaling laws for velocity, heat transfer and ohmic dissipation are also discussed. The predictions of the energy-based scaling law agree well with the observed field strength of Earth and Jupiter, but for other planets they are more difficult to test or special pleading is required to explain their field strength. The scaling law also explains the very high field strength of rapidly rotating low-mass stars, which supports its rather general validity.

Keywords Planetary magnetic fields · Geodynamo · Dynamo models

1 Introduction

The observed magnetic field strength B_p at the surface of those planets in the solar system that have presently an active dynamo varies over a wide range, from approximately 400 nT at Mercury to 500,000 nT at Jupiter. For the dipole moment, which is given by $M \approx B_p R_p^3$ for a dominantly dipolar field with R_p the planetary radius, the range is even larger, between 4×10^{12} Tm^3 (Mercury) and 1.5×10^{20} Tm^{-3} (Jupiter). The primary aim of a dynamo scaling theory is to explain the field strength in terms of fundamental properties of the planet

U.R. Christensen (✉)
Max-Planck-Institut für Sonnensystemforschung, 37191 Katlenburg-Lindau, Germany
e-mail: christensen@mps.mpg.de

Table 1 Proposed scaling laws

#	Rule	Author	Remark
1	$B_p R_p^3 \propto (\rho \Omega R_p^5)^a$	e.g. Russell (1978)	magnetic Bode law
2	$B^2 \propto \rho \Omega^2 R_c^2$	Busse (1976)	
3	$B^2 \propto \rho \Omega \sigma^{-1}$	Stevenson (1979)	Elsasser number rule
4	$B^2 \propto \rho R_c^3 q_c \sigma$	Stevenson (1984)	at low energy flux
5	$B^2 \propto \rho \Omega R_c^{5/3} q_c^{1/3}$	Curtis and Ness (1986, modified)	mixing length theory
6	$B^2 \propto \rho \Omega^{3/2} R_c \sigma^{-1/2}$	Mizutani et al. (1992)	
7	$B^2 \propto \rho \Omega^2 R_c$	Sano (1993)	
8	$B^2 \propto \rho \Omega^{1/2} R_c^{3/2} q_c^{1/2}$	Starchenko and Jones (2002)	MAC balance
9	$B^2 \propto \rho R_c^{4/3} q_c^{2/3}$	Christensen and Aubert (2006)	energy flux scaling

or its dynamo region. It is not immediately obvious which quantities play the key role, but candidates are the radius R_c of the electrically conducting fluid core of the planet, the conductivity σ and density ρ of the core, the rotation rate Ω and the convected energy flux q_c in the core. Also, it is not clear if a single scaling law is applicable to all magnetic planets. This may require that their dynamos are qualitatively similar and differ only in specific parameter values. As we will see, there are probably sufficiently severe differences between planetary dynamos so that the application of the same scaling law to all of them is not straightforward.

The interest in a scaling theory for planetary magnetic fields is twofold. In terms of the theoretical understanding of planetary magnetism, a well-established scaling law is an essential part of a comprehensive dynamo theory. On the more practical side, such a law would allow to make predictions for the paleo-fields of the solar system planets, at times when control parameters (e.g. q_c, Ω) had been different and when now extinct dynamos were active in Mars or other planetary bodies. This has possible applications to the question of atmospheric evolution and planetary habitability (Dehant et al. 2007).

Over the past decades space missions provided data on the first-order magnetic field properties of the planets. At the same time, scaling laws for their field strength have been proposed with a rather confusing diversity (Table 1). The 'magnetic Bode law' (#1 in the Table 1), suggesting a relation between the magnetic dipole moment and the angular momentum of the planet, is purely empirical. Most other scaling laws are based on an assumed force balance between Coriolis force and Lorentz force, but make different assumptions on the characteristic velocity or length scales in the dynamo. Rules #4 and #9 in Table 1 consider the energy flux that is available to balance ohmic dissipation.

In Table 1, B usually refers to the characteristic magnetic field strength inside the dynamo. It is assumed that the strength of the exterior dipole field is proportional to B. The predictions of the proposed scaling laws have been compared with the observed planetary fields and fair or good agreement has been claimed in every case. Given the diversity of the scaling laws, it seems quite surprising that they all fit the observations more or less well. One reason is that in many cases the comparison has been made in terms of the dipole moment, which means that the predicted field strength is multiplied by the cube of the radius. The pitfalls of such a procedure has been discussed by Cain et al. (1995) for the case of the magnetic Bode law. Suppose that for a set of hypothetical planets the surface field strength B_p is uncorrelated with ρ, Ω and R_p, and that all these quantities vary within some range. When B_p is multiplied by R_p^3 and $\rho\Omega$ is multiplied by R_p^5 to obtain the magnetic moment and the angular momentum, respectively, a correlation is found between the logarithms of the two moments (with a most probable exponent $a = 3/5$ in Table 1). Obviously this has no physical basis other than that big planets are more likely than small planets to have large values in properties that depend on high powers of the radius.

A new perspective on planetary scaling laws has been opened by the advent of numerical dynamo models (Glatzmaier and Roberts 1995; Christensen and Wicht 2007). The ability to run simulations for a large number of model cases that cover a substantial range of the values of control parameters allows to test scaling laws against the numerical results. One point of concern is that in these models the viscosity and the thermal diffusivity are far larger than they are in planetary cores and, therefore, the models may be in a different dynamical regime.

This paper has three main sections. In the next one various theoretical approaches for scaling laws are discussed. In Sect. 3 theoretical predictions are compared with the results from numerical dynamo models. The emphasis is on scaling laws of magnetic field strength, but we also address the scaling of velocity and the rate of ohmic dissipation. Also, the question of what controls the magnetic field morphology is of interest. In Sect. 4 the scaling laws are compared with the known magnetic properties of solar system planets and also with observational data for certain classes of stars.

2 Theory

Before considering principles for scaling laws, some fundamental requirements and assumptions for planetary dynamos are reiterated. A fluid region of sufficient thickness D with a sufficiently high electrical conductivity σ must exist inside the planet. The fluid must move with a sufficiently large velocity U, so that the magnetic Reynolds number

$$Rm = \frac{UD}{\lambda} \tag{1}$$

exceeds a critical value Rm_{crit} ($\lambda = 1/\mu_o\sigma$ is the magnetic diffusivity). The flow pattern must also be suitable for dynamo action, which requires a certain complexity. Although dynamos may also exist for flows that are excited by precession or by tides (Wicht and Tilgner 2009, this issue), we assume here an origin by thermal or compositional buoyancy forces. Also, we take a spherical shell geometry for the dynamo region and assume that Coriolis forces play a significant part in the force balance of the fluid and influence the pattern of convection. With these assumptions the requirement for 'flow complexity' seems to be satisfied and self-sustained dynamo action is possible above $Rm_{crit} \approx 50$ (Christensen and Aubert 2006). We will also assume that the fluid can be treated as incompressible (Boussinesq approximation), although modifications that apply to the case of strong density stratification are also discussed.

2.1 Magnetostrophic Force Balance

The equation of motion for an incompressible, rotating and electrically conducting fluid driven by thermal buoyancy in the presence of a magnetic field is (Gubbins and Roberts 1987):

$$\rho\left(\frac{\partial \mathbf{u}}{\partial t} + (\mathbf{u}\cdot\nabla)\mathbf{u}\right) + 2\rho\Omega\mathbf{e}_z \times \mathbf{u} + \nabla P = \rho\nu\nabla^2\mathbf{u} + \rho\alpha g T\,\mathbf{e_r} + \mathbf{j}\times\mathbf{B}, \tag{2}$$

where $\mathbf{u}$ is velocity, Ω rotation rate, ρ density, P non-hydrostatic pressure, ν kinematic viscosity, α thermal expansivity, g gravity, T temperature, $\mathbf{B}$ magnetic field, $\mathbf{j} = \mu_o^{-1}\nabla \times \mathbf{B}$ current density, r radius and z the direction parallel to the rotation axis. The terms in Eq. 2 describe in order the linear and non-linear parts of inertial forces, Coriolis force, pressure gradient force, viscous force, buoyancy force and Lorentz force.

A very simple estimate for the relative magnitude of the various forces in the Earth's core is obtained by assuming the following characteristic values (in SI-units): $U \approx 10^{-4}$, $\Omega \approx 7 \times 10^{-5}$, $\rho \approx 10^4$, $B \approx 5 \times 10^{-3}$, and for the characteristic length scale associated with the various spatial derivatives $\ell \approx 3 \times 10^5$. The viscous force and the inertial force turn out to be much smaller than the Lorentz force and the Coriolis force, whose magnitudes are comparable (magnetostrophic balance). The magnitude of temperature anomalies driving the flow is hard to estimate *a priori*, but it is assumed that buoyancy also contributes at leading order. The resulting force balance is called the MAC-balance (from **M**agnetic, **A**rchimedean and **C**oriolis forces).

One of the simplest scaling rules for the magnetic field based on the magnetostrophic force balance is the Elsasser number rule. In order to derive it, we relate the characteristic current density J in the Lorentz force term in Eq. 2 to the characteristic velocity U by the generalized Ohm's law, $\mathbf{j} = \sigma(\mathbf{E} + \mathbf{u} \times \mathbf{B})$. Ignoring the electric field $\mathbf{E}$, this results in $J \propto \sigma U B$ and the order of the Lorentz force is $JB \propto \sigma U B^2$, while that of the Coriolis force is $2\rho\Omega U$. Their ratio is the Elsasser number, in which the velocity drops out:

$$\Lambda = \frac{\sigma B^2}{2\rho\Omega}. \tag{3}$$

Since both forces are assumed to be of the same order, the Elsasser number should be of order one, which leads to the prediction for the magnetic field strength given by rule #3 in Table 1. The Elsasser number rule is supported by the theory of rotating magnetoconvection, i.e. convection in an imposed magnetic field. Convection is inhibited both by rotational effects and by magnetic effects when they are considered in isolation. However, for an imposed field of simple geometry it has been shown that the combination of both effects will actually reduce the impeding influence and that this constructive interplay is most efficient when the Elsasser number is around one (Chandrasekhar 1961; Stevenson 1979). Applied to dynamos, it is argued that as long as the magnetic field is weak ($\Lambda \ll 1$), any field growth will intensify convection, meaning more efficient dynamo action and further increase of the field. Field growth at $\Lambda \gg 1$ would weaken convection, hence it is assumed that the field equilibrates at an Elsasser number of one.

The Elsasser rule has found widespread acceptance because of its plausible theoretical basis and because reasonable estimates of the magnetic field strength in the cores of the Earth and some other planets put the Elsasser number in the range of 0.1–10 (Stevenson 2003). Possible caveats are: (1) The effect on convection of the complex magnetic field that might be expected in a dynamo could be different from that of the simple field assumed in magnetoconvection. (2) The assumption $J \propto UB$ may not hold because velocity and magnetic field could be largely aligned with each other. (3) The Coriolis forces could be balanced to a large degree by a pressure gradient force (nearly geostrophic flow) and only the residual must be balanced by electromagnetic or other forces. Furthermore, Stevenson (1983, 1984) pointed out that the Elsasser number rule (as well as several other proposed scaling laws) ignore the energy requirement for maintaining a magnetic field against ohmic dissipation. He suggested that the Elsasser number rule is only applicable when sufficient energy is available.

Other scaling rules based on a magnetostrophic force balance (#2,5,6,7 in Table 1) have been derived by expressing the current density in the Lorentz force term by the curl of **B** and by assuming a characteristic length scale for the field ℓ_B. Often $\ell_B \approx R_c$ is set. This gives the order of the Lorentz force as $\mu_o^{-1} B^2 R_c^{-1}$. Balancing it with the Coriolis force one obtains

$$B^2 \propto \mu_o \rho U \Omega R_c. \tag{4}$$

The choice of R_c for ℓ_B is certainly too large, and Starchenko and Jones (2002) assume it to be 0.02 R_c. But as long as ℓ_B is a fixed fraction of R_c, this will only affect the prefactor in the scaling law. A weak point is that we have no particular reason to assume ℓ_B to be constant. Typically it should vary with the magnetic Reynolds number as $\ell_B/R_c \propto Rm^{-1/2}$ (e.g. Galloway et al. 1978), although Starchenko and Jones (2002) argue that for large Rm the magnetic length scale may approach a constant value. A scaling rule for the velocity U is needed in order to obtain from Eq. 4 a law that relates the magnetic field to the fundamental physical quantities. In some cases, e.g. rule #6 in Table 1, an educated guess had been made, but in other cases the estimate for U has been based on some scaling theory or other. This will be addressed in the next section.

2.2 Velocity Scaling

We will consider three different scaling rules for the characteristic velocity U, based on force balance arguments: Mixing length theory, a MAC-balance estimate, and an estimate based on the triple force balance Coriolis force, inertia, and buoyancy.

The mixing length theory is frequently applied for highly turbulent convection in stars (Kippenhahn and Weigert 1990). Assuming a characteristic length scale ℓ on which convective mixing of momentum and entropy (or temperature in the Boussinesq limit) occur, and balancing the nonlinear inertia term and the buoyancy term in Eq. 2, results in $\rho U^2/\ell \propto \rho\alpha g T'$, where T' stands for the characteristic temperature fluctuations away from the radially averaged temperature distribution. In convection upwelling and downwelling flow are positively correlated with positive and negative values of T', respectively, hence the convected heat flux can be written as $q_c \propto \rho c_p U T'$, where c_p is the specific heat capacity. Introducing the temperature scale height $H_T = c_p/(\alpha g)$, the velocity scales as

$$U \propto \left(\frac{\ell q_c}{\rho H_T} \right)^{1/3}. \tag{5}$$

Note that the temperature scale height does not enter because the adiabatic temperature lapse would be of interest *per se*, but because the ratio ℓ/H_T describes the thermodynamic efficiency of the conversion of heat into kinetic energy through the work done by buoyancy forces. In stellar theory it is usually assumed that the mixing length ℓ is of the order of the pressure (or density) scale height. In planets the core radius R_c or the depth of the convection layer D are taken for ℓ, because here they are smaller than the density scale height (Stevenson 2003). Combining Eqs. 4 and 5 results in the magnetic field scaling rule suggested by Curtis and Ness (1986) when H_T is fixed. When we take into account that for fixed α and c_p gravity and H_T^{-1} scale as $\rho_c R_c$, we obtain rule #5 in Table 1. A caveat against this rule is that planetary dynamos are probably not in a regime described by a mixing length balance, where the energy generated by the work of buoyancy forces is transported through an inertial cascade to small length scales where viscous dissipation kicks in.

In the MAC balance regime the buoyancy forces do work against Lorentz forces and generate magnetic energy. Because the Lorentz force is assumed to be of the same order as the Coriolis force, we can balance the latter against buoyancy forces (Starchenko and Jones 2002), i.e. $\rho\Omega U \propto \rho g\alpha T'$, and obtain

$$U \propto \left(\frac{q_c}{\rho\Omega H_T}\right)^{1/2}. \tag{6}$$

Inserting this into Eq. 4 and using $H_T^{-1} \propto \rho R_c$ leads to rule #7 in Table 1.

In rapidly rotating convection the flow is organized in quasi-geostrophic columns that run parallel to the rotation axis through the entire fluid shell. The Coriolis force is balanced to a large degree by the pressure gradient and the fraction that must be balanced by other forces could be much smaller than $\rho\Omega U$. Using a vorticity equation, which is obtained by applying the curl to Eq. 2, eliminates pressure forces. In the Boussinesq case the non-linear inertia term, Coriolis force term and the buoyancy term are, in this order, $\rho\nabla \times (\mathbf{u} \times (\nabla \times \mathbf{u}))$, $2\rho\Omega\partial\mathbf{u}/\partial z$, and $\rho\alpha g\nabla \times T\mathbf{e}_r$, where $\nabla \times \mathbf{u}$ is vorticity and $\mathbf{e}_r$ the radial unit vector. Aubert et al. (2001) considered a balance between these three terms in the vorticity equation (CIA-balance from **C**oriolis, **I**nertia, **A**rchimedean). The characteristic length scale ℓ associated with the differential operators in the inertial term and in the buoyancy term is assumed to be the same. Typically $\ell \ll R_c$, whereas a much larger scale L is associated with the z-derivative in the Coriolis term because of the quasi-geostrophic flow structure. ℓ must not be assumed, unlike in mixing length theory. The triple balance provides two conditions, which allows to derive scaling rules both for U and for ℓ:

$$U \propto \left(\frac{q_c}{\rho H_T}\right)^{2/5}\left(\frac{L}{\Omega}\right)^{1/5}, \tag{7}$$

$$\ell \propto \left(\frac{UL}{\Omega}\right)^{1/2}. \tag{8}$$

In the Boussinesq case we can set $L \approx R_c$, but we note here that in the case of density stratification the Coriolis term in the vorticity equation is modified, which implies that the length scale L is then of the order of the density scale height H_ρ. Equation 7 fits the convection velocity found in laboratory experiments (Aubert et al. 2001) and numerical simulations (Christensen 2002) of rapidly rotating convection. But it is less clear if it applies to magnetohydrodynamic flow, where according to the usual assumption Lorentz forces replace the inertial forces as a leading-order contributor to the force balance and might lead to a break-up of the quasi-geostrophic flow structure.

2.3 Power-Based Magnetic Field Scaling

Aside from satisfying the relevant force balance, any dynamo must be thermodynamically consistent. In particular, the ohmic dissipation associated with a magnetic field of certain strength and geometry cannot exceed that part of the energy flux that is available for conversion to other forms of energy. A discussion of basic principles of the energetics of Earth's core can be found in Breuer et al. (2009, this issue). For estimating the variation of magnetic field strength during the evolutionary history of a planet, Stevenson et al. (1983) assumed a linear dependence of B^2 on the energy flux. Stevenson (1984) suggested that in cases of a comparatively low energy flux, the available energy limits the magnetic field strength to values that correspond to an Elsasser number $\Lambda < 1$, whereas at high flux the field saturates at $\Lambda \approx 1$. His prediction for the low-flux regime results in rule #4 in Table 1. Based on concepts laid down in Stacey (1977),

Schubert et al. (1988) arrived at basically the same rule, which was applied to model the evolution of Mercury's magnetic field and in later papers (Schubert and Spohn 1990; Leweling and Spohn 1997) that of Mars. From numerical dynamo simulations Christensen and Aubert (2006) conclude that the energy flux controls the field strength in a wide range of fluxes, however, they arrive at a different dependence of field strength on heat flux and conductivity (rule # 9).

The thermodynamically available power per unit volume is q_c/H_T and the ohmic dissipation is $D = (\nabla \times \mathbf{B})^2/(\sigma\mu_o^2) \propto \lambda B^2/(2\mu_o \ell_B^2)$. When f_{ohm} is the fraction of the available power that is converted to magnetic energy and eventually lost by ohmic dissipation, the magnetic energy density $E_m = B^2/(2\mu_o)$ in the dynamo scales as

$$\frac{B^2}{2\mu_o} \propto f_{ohm} \frac{\ell_B^2}{\lambda} \frac{q_c}{H_T}. \tag{9}$$

ℓ_B^2/λ is equivalent to the ohmic dissipation time $\tau_\lambda = E_m/D$. Christensen and Aubert (2006) assume that τ_λ scales with the magnetic Reynolds number as

$$\tau_\lambda \propto L^2/\lambda Rm^{-1} = L/U. \tag{10}$$

Here, L is again the relevant 'large' length scale (R_c for planetary applications). We can now use some scaling law for the velocity to obtain the magnetic field scaling. The mixing length rule (Eq. 5) results in a particularly simple magnetic field scaling

$$\frac{B^2}{2\mu_o} \propto f_{ohm} \rho^{1/3} \left(\frac{q_c L}{H_T}\right)^{2/3}, \tag{11}$$

which leads to rule #9 in Table 1 when setting $L = R_c$ and $H_T^{-1} \propto \rho R_c$. What makes this scaling law unique is that it depends neither on the rotation rate, nor on the electrical conductivity. Using instead of the mixing length velocity the prediction from the MAC-balance or the CIA-theory, we obtain, respectively,

$$\frac{B^2}{2\mu_o} \propto f_{ohm} L \left(\frac{\rho \Omega q_c}{H_T}\right)^{1/2} \tag{12}$$

i.e. rule #8, or

$$\frac{B^2}{2\mu_o} \propto f_{ohm} \rho^{2/5} \Omega^{1/5} L^{4/5} \left(\frac{q_c}{H_T}\right)^{3/5}. \tag{13}$$

Aside from a somewhat weaker dependence on the energy flux than in Eq. 11, they predict a weak dependence (Eq. 13) or moderate dependence (Eq. 12) on the rotation rate.

The convected energy flux q_c and the temperature scale height H_T are not constant with radius in a planetary dynamo. Also, for Jupiter the radial variation of ρ can perhaps not be neglected. When the density stratification is strong (as its is in stars) the density scale height is a more appropriate measure for L than the radius of the dynamo region. Christensen et al. (2009) assume that a representative magnetic field strength is obtained by averaging Eq. 11 over the shell volume V. By introducing a reference value q_o for the heat flux (the flux at the outer boundary of the dynamo) and by normalizing density with its mean value $\bar{\rho}$, they condense the averaging of radially variable properties into an efficiency factor

$$F^{2/3} = \frac{1}{V} \int_{r_i}^{R} \left(\frac{q_c(r)}{q_o} \frac{L(r)}{H_T(r)}\right)^{2/3} \left(\frac{\rho(r)}{\bar{\rho}}\right)^{1/3} 4\pi r^2 dr, \tag{14}$$

where r_i is the inner radius of the spherical shell and $L = \min(D, H_\rho(r))$ is assumed. F is found to be of order one for a wide class of objects from planets to stars (Christensen et al. 2009). The scaling law (Eq. 11) now becomes

$$\frac{B^2}{2\mu_o} = c f_{ohm} \bar{\rho}^{1/3} (F q_o)^{2/3}, \tag{15}$$

where c is a constant prefactor.

When convection is partly driven by compositional rather than by thermal buoyancy, q_c/H_T must be augmented by the work done by the chemical flux in an appropriate way. In order to keep the formalism simple, Christensen et al. (2009, online supplementary information) replace the compositional flux by an equivalent effective convected heat flux.

3 Scaling Laws Versus Model Results

In this section the proposed scaling laws will be compared with the results of direct numerical simulations of convection-driven dynamos in rotating spherical shells. In these models the equation of motion together with the magnetic induction equation and an equation for the advection and diffusion of heat (or light element concentration) are solved, subject to appropriate boundary conditions (Christensen and Wicht 2007; Wicht and Tilgner 2009).

A rather large data basis of dynamo model results has been built over time by the author and coworkers (Christensen et al. 1999, 2009; Christensen and Tilgner 2004; Christensen and Aubert 2006). The data set is homogeneous in the sense that the basic model is simple and the same in all simulations, with driving of convection by a fixed temperature contrast ΔT between the inner and the outer boundaries of a rotating spherical shell, no-slip boundary conditions and a fixed ratio of inner boundary radius to outer boundary radius of 0.35. In Christensen et al. (2009) models with different thermal boundary conditions, meant to represent a proxy for compositional convection, are also included. The parameters describing rotation rate, viscosity, magnetic and thermal diffusivity and convective driving are varied within a fairly wide range. All the simulations have been run for a sufficient time to reach an equilibrated state. The time-averaged *rms*-values of velocity and magnetic field strength have been calculated for the equilibrium state and are taken for the characteristic values U and B. Averages of other values of interest, for example f_{ohm}, have also been recorded.

The model results have been calculated in terms of non-dimensional quantities. Non-dimensionalizing the governing equations results in a set of four control parameters. The Ekman number describes the ratio of viscous forces to Coriolis forces

$$E = \frac{\nu}{\Omega D^2}. \tag{16}$$

The modified Rayleigh number is the ratio of driving buoyancy to retarding rotational forces

$$Ra^* = \frac{\alpha g_o \Delta T}{\Omega^2 D}. \tag{17}$$

The Prandtl number is the ratio of viscosity to thermal diffusivity κ,

$$Pr = \frac{\nu}{\kappa}, \tag{18}$$

and the magnetic Prandtl number is

$$Pm = \frac{\nu}{\lambda}. \tag{19}$$

We note that none of the scaling laws discussed so far predicts a dependence on viscosity ν or thermal diffusivity κ. In order to test them, the definition of the Rayleigh number in Eq. 17, which does not contain any diffusivity (viscosity being also subsumed under this term), is therefore more advantageous than the conventional definition $Ra = \alpha g_o \Delta T D^3/(\kappa\nu)$. The two Rayleigh numbers are related by $Ra^* = RaE^2Pr^{-1}$. Because there are three different diffusion constants in the dynamo equations, ν, κ and λ, the other three control parameters must necessarily contain diffusivities.

The scaling laws discussed above have been formulated in terms of the convected heat flux q_c, for which decent estimates exist for several planets, whereas the driving (superadiabatic) temperature contrast ΔT is unknown. Christensen and Aubert (2006) therefore introduce a flux Rayleigh number, given by $Ra_Q^* = Ra(Nu-1)E^3Pr^{-2}$, which is likewise independent of viscosity and diffusivities. The Nusselt number Nu is the ratio of heat flux in the convecting state to the conductive heat flux. Unless when flux boundary conditions are used, Ra_Q^* is therefore not a primary parameter, but is derived from the solution. The flux Rayleigh number is proportional to a non-dimensional form of the available energy flux (or convective power). For different driving modes the constant of proportionality is different. Therefore, Christensen et al. (2009) use instead of a Rayleigh number a related parameter that is directly equivalent to the available energy flux (see also Aubert et al. 2009):

$$Fq^* = \frac{Fq_o}{\rho\Omega^3 R_c^3}. \tag{20}$$

Here, F is the efficiency factor defined by Eq. 14 and q_o is the reference value of the heat flux. For Boussinesq models with $r_i/R = 0.35$ and fixed ΔT the integration of Eq. 14 leads to $F = 0.88R_c/H_T^o$ when taking for q_o the heat flux on the outer boundary (H_T^o is calculated with gravity g_o at R_c). In cases of a fixed flux q_i on the inner boundary and zero flux on the outer boundary, which is a proxy for composition convection, $F = 0.45R_c/H_T^o$ when setting the reference flux to $q_o = q_i r_i^2/R_c^2$. In both cases the heat flux must be corrected for the non-convected part, by multiplying it with $(Nu-1)/Nu$ in the Boussinesq case.

Below we will take the non-dimensional form of the available heat flux Fq^* as control parameter, and an equivalent non-dimensional form of the velocity and the magnetic field strength, to compare the various proposed scaling laws with the numerical results.

3.1 Velocity Scaling

The appropriate nondimensional form for expressing the velocity is the Rossby number

$$Ro = \frac{U}{\Omega R_c}. \tag{21}$$

In terms of the Rossby number and the non-dimensional available heat flux, the three scaling laws for the velocity discussed in Sect. 2.2 can all be written in the form

$$Ro \propto (Fq^*)^\alpha, \tag{22}$$

where the exponent α is $1/3$ in case of the mixing length scaling, $2/5$ for the CIA-balance scaling and $1/2$ for the MAC-balance scaling.

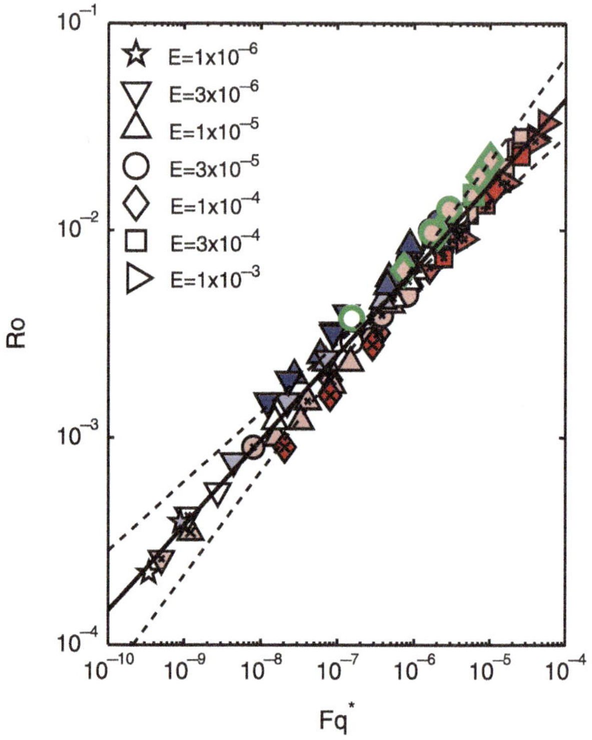

Fig. 1 Rossby number versus available power for dynamos models with dipole dominated magnetic field. *Pm* is color-coded (*dark blue* $Pm \leq 0.1$, *white* $Pm = 1$, *dark red* $Pm \geq 10$, paler shading for intermediate values). $Pr = 1$, except where indicated by a *cross* ($Pr > 1$) or circle ($Pr < 1$) inside main symbol. Black-edged symbols are for fixed ΔT, green-edged symbols for fixed heat flux at r_i and zero flux at R_c. Slopes of broken lines are $1/3$ and $1/2$, respectively, for the best-fitting line it is 0.411. The preexponential constant is 1.23 (1.05 for a fit with a forced exponent of 0.400)

In Fig. 1 the Rossby number is plotted versus Fq^* for the numerical model results. Here we restrict the data set to dynamos that generate a dipole dominated magnetic field, meaning that the dipole field strength on the outer boundary is at least 35% of the combined field strength in harmonic degrees 1–12. Also, only dynamos with a Nusselt number $Nu > 2$ are plotted, to exclude cases in which convection is not fully developed.

The data are well fitted by Eq. 22. Result for different values of the Ekman number and the two Prandtl numbers are collapsed on a single line. These are the parameters that involve the viscosity and diffusivities, therefore they do not play a first-order role in the dynamo models. Only with respect to the magnetic Prandtl number a slight bias seems to exist in the fit (see Christensen and Aubert 2006). The fitting exponent of 0.41 agrees best with the prediction from the CIA-balance. The lines with slopes of $1/3$ and $1/2$ show clear systematic deviations.

3.2 Magnetic Energy Scaling

The appropriate form of the non-dimensional magnetic energy is given by

$$E_m^* = \frac{E_m}{\rho \Omega^2 R_c^2}. \tag{23}$$

E_m^* is proportional to the squared Lorentz number defined by Christensen and Aubert (2006) for measuring magnetic field strength. The non-dimensional form of Eqs. 11–13 is

$$E_m^* = c f_{ohm} (Fq^*)^{\alpha}, \tag{24}$$

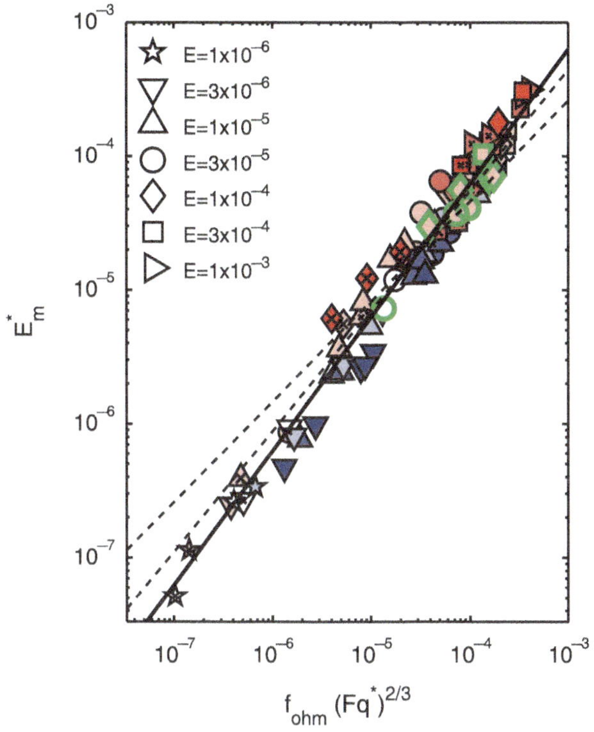

Fig. 2 Non-dimensional magnetic energy density versus available power for dynamos models with dipole-dominated magnetic field. See Fig. 1 for explanation of symbols. The slope of the fitting line has been set to 1.0, the best-fitting pre-factor is $c = 0.63$. The two broken lines refer to exponents α in Eq. 24 of 0.6 and 0.5, respectively

where the exponent α is $2/3$ for the mixing-length prediction, 0.5 for the MAC-balance prediction and 0.6 for the prediction assuming a CIA-balance. In Fig. 2 we plot the magnetic energy against the right-hand side of Eq. 24, setting $\alpha = 2/3$. The best-fitting exponent for the energy flux is 0.677 (the slope in the figure has been set to one). The good fit strongly supports the scaling law Eq. 11, in which the magnetic field strength is independent of rotation rate. An exponent $\alpha = 0.6$ appears marginally compatible with the numerical results, but $\alpha = 0.5$, which corresponds to rule #8 in Table 1, is ruled out. Again, the fact that data for different values of the Ekman number and the Prandtl number are collapsed on a single line shows that diffusivities do not play a first-order role in the numerical dynamos. A weak dependence on the magnetic Prandtl number is found again—cases with high Pm (red symbols in Fig. 2) tend to plot slightly above the fitting line and those with low Pm (blue symbols) tend to plot below the line.

Although a $2/3$-power law for the magnetic energy seems to be well established by the model results, we note a discrepancy—Eq. 11 has been derived using the $1/3$-power law dependence of velocity on the available energy flux, whereas the numerical results clearly support a $2/5$-power law (Sect. 3.1). Takahashi et al. (2008) confirmed for independent dynamo simulations in a similar parameter range the $2/5$-power law for the velocity and the $2/3$-power law for the magnetic energy. In Sect. 3.5 the apparent discrepancy between velocity scaling and field scaling will be discussed further.

3.2.1 Different Non-Dimensionalization Schemes and Test of Other Rules

The non-dimensionalization of magnetic energy and of convected heat flux used in the previous section (Eqs. 20 and 23) is not unique. The scales involve the square and the cube of the rotation frequency Ω, which does not affect the magnetic field strength according to Eq. 15. Therefore, one may suspect that the good agreement in Fig. 2 is at least partly feigned, similar to the case of the 'magnetic Bode law' discussed in the introduction. To test this, different schemes for making Eq. 15 non-dimensional are explored. This also offers the opportunity to directly test other proposed magnetic field scaling rules listed in Table 1.

Dimensional analysis shows that the most general form of an energy density scale is given by

$$E_m^{scale} = \rho \Omega^p \nu^q \lambda^r \kappa^s R^{2p-2}, \tag{25}$$

with the corresponding scale for the available heat flux being

$$q^{scale} = \rho \Omega^{3p/2} \nu^{3q/2} \lambda^{3r/2} \kappa^{3s/2} R^{3p-3}. \tag{26}$$

The exponents must satisfy $p + q + r + s = 2$, but are otherwise arbitrary. In Sect. 3.2 the scaling $(p, q, r, s) = (2, 0, 0, 0)$ had been used ('rotational' scaling). In Fig. 3 four other options are shown. The different schemes of non-dimensionalizing re-shuffle the data points, but they are always decently fit by a line with slope one (fixed to this value in the plots in accord with Eq. 15). Slightly different slopes marginally improve the fits. In Fig. 3a 'magnetic scaling' is used, i.e. magnetic energy density is non-dimensionalized with $\rho\lambda^2 R^2$. This may suffer from the same caveat as the rotational scaling, because the variables are normalized with the second and third power of λ, which does not appear naturally in the scaling law. For the scaling in Fig. 3b the non-dimensional magnetic energy density is just equal to the Elsasser number Λ. It varies over three orders of magnitude, showing that the assumption $\Lambda \approx 1$ (rule #3) does not hold for the set of dynamo models.

In Fig. 3c the available energy flux is scaled in such a way that it becomes proportional to the ratio of the flux Rayleigh number divided by the critical Rayleigh number Ra_c (assuming $Ra_c \propto \mathrm{E}^{-4/3}$ and ignoring the influence of Pr on Ra_c). This way of non-dimensionalization is the least susceptible for creating a spurious correlation, because it leads to the smallest spread in the data. Finally, the scheme used in Fig. 3d makes the non-dimensional available flux equivalent to the 'energy flux number' Φ, proposed by Stevenson (1984). Stevenson suggested that a change of regime occurs around $\Phi = 3000$ (or $\Phi^{2/3} \approx 200$; note that $\Phi^{2/3}$ is plotted in Fig. 3d), with $\Lambda \approx 0.003\Phi$ at small Φ and $\Lambda \approx 1$ at large Φ. No indication for a regime change is found. Rule #4 (Table 1), which is supposed to be applicable at low flux, is equivalent to a simple linear dependence of magnetic energy on the available flux when they are normalized using the magnetic scaling (Fig. 3a). Although thc best-fitting exponent of 0.70 is slightly larger than 2/3, an exponent of one is not compatible with the results.

So far, we rejected rules #3, 4 and 8 as incompatible with the numerical data. Rule #2 corresponds to a constant value of magnetic energy in the magnetic scaling, but this actually varies by four orders of magnitude (Fig. 3a). Rule #5 predicts a slope of $1/3$ in the rotational scaling, which is clearly far off (Fig. 2). Rule #6 is equivalent to the constancy of dimensionless magnetic energy when scaled with $(p, q, r, s) = (3/2, 0, 1/2, 0)$, but this is found to vary within a range of 250. The problem with rule #7 is that is contains a hidden length scale ℓ_α, being the constant of proportionality between the α-parameter in the turbulent mean-field theory used by Sano (1993) and Ω, and which should be added on the right-hand side of rule #7. Sano assumes a fixed value of ℓ_α for the various planets, but the fundamental MHD equations do not support the existence of another length scale that is independent from the other properties. Assuming $\ell_\alpha \propto R_c$ leads of course to rule #2. In summary, we find that all scaling rules listed in Table 1 are incompatible with the numerical dynamo results, with the exception of #9. That is no surprise, since this rule as been suggested on the basis of numerical data in the first place.

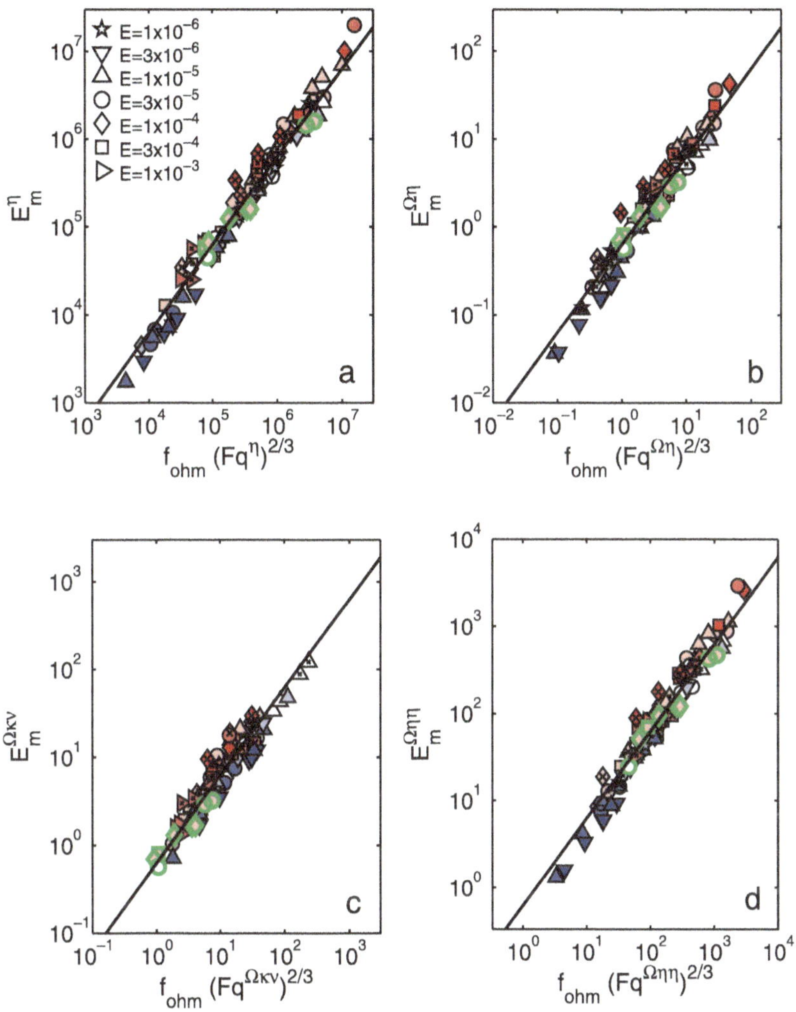

Fig. 3 Magnetic energy versus available power for dynamo models with dipole-dominated fields using different non-dimensionalization schemes. The exponents (p, q, r, s) in the scale values of magnetic energy and the convected flux are in (**a**) (0, 0, 2, 0), in (**b**) (1, 0, 1, 0), in (**c**) (8/9, −2/9, 0, 4/3) and in (**d**) (2/3, 0, 4/3, 0). See Fig. 1 for explanation of symbols

3.3 Field Topology and Reversals

Simply speaking, the numerical dynamos can be classified into those that generate a dipole dominated magnetic field and those with a multipolar field. We use the ratio of the mean dipole field strength at the surface of the dynamo to the combined field strength in harmonic degrees from 1 to 12 as measure for the degree of dipolarity. For the present geomagnetic field, this ratio is 0.65 based on the POMME model (Maus et al. 2006). Christensen and Aubert (2006) proposed that the dipolarity is controlled by a local Rossby number

$$Ro_\ell = \frac{U}{\Omega \ell}, \tag{27}$$

Fig. 4 Degree of dipolarity versus the local Rossby number. See Fig. 1 for explanation of symbols. The likely range for the dipolarity of the geodynamo outside times of reversals and excursions of 0.45–0.75 is marked in grey

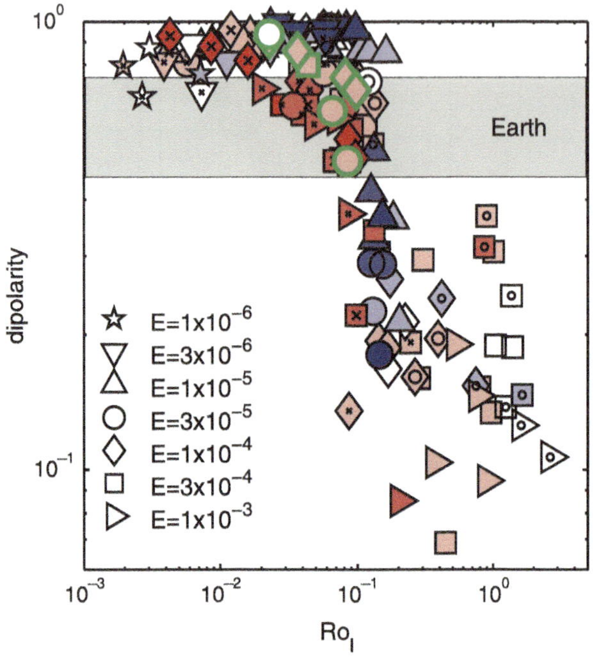

where the characteristic length scale ℓ of the flow is derived from the kinetic energy spectrum. Ro_ℓ seems a more appropriate measure for the ratio of inertial forces to the Coriolis force than the Rossby number defined with the global length scale. Sreenivasan and Jones (2006) showed that dynamo models become less dipolar when inertial forces become stronger relative to the Coriolis force. Christensen and Aubert (2006) found in their models that the transition from dipole-dominated dynamos to multipolar dynamos occurs at a critical local Rossby number of 0.12, irrespective of the values of the Ekman number and the Prandtl numbers. An updated version of their diagram is shown in Fig. 4.

Olson and Christensen (2006) confirmed the local Rossby number criterion for the selection of the magnetic field morphology, including also dynamo simulations from the literature. They observed that numerical dynamos that are dipole dominated but show occasional field reversals are only found in the range 0.06–0.12 for Ro_ℓ, i.e. close to the transition point to multipolar dynamos. Wicht et al. (2009) put the range for Earth-like reversal behaviour at slightly higher values of Ro_ℓ. In order to express the local Rossby number in terms of the fundamental control parameters, Olson and Christensen (2006) found it necessary to invoke a dependence on all four parameters. Their suggested scaling law is written here in terms of the non-dimensional available energy flux

$$Ro_\ell = 0.11(Fq^*)^{1/2} E^{-1/3} Pr^{1/5} Pm^{-1/5}. \tag{28}$$

It provides a good fit to the numerical data, but is (so far) purely empirical without an obvious theoretical explanation. Dynamos with different thermal boundary conditions or source-sink distributions may have different values for the critical local Rossby number (Olson and Christensen 2006; Aubert et al. 2009). In dynamo models with stress-free boundaries Simitev and Busse (2005) found a variety of magnetic field pattern; whether their selection is also governed by the value of Ro_ℓ is not clear.

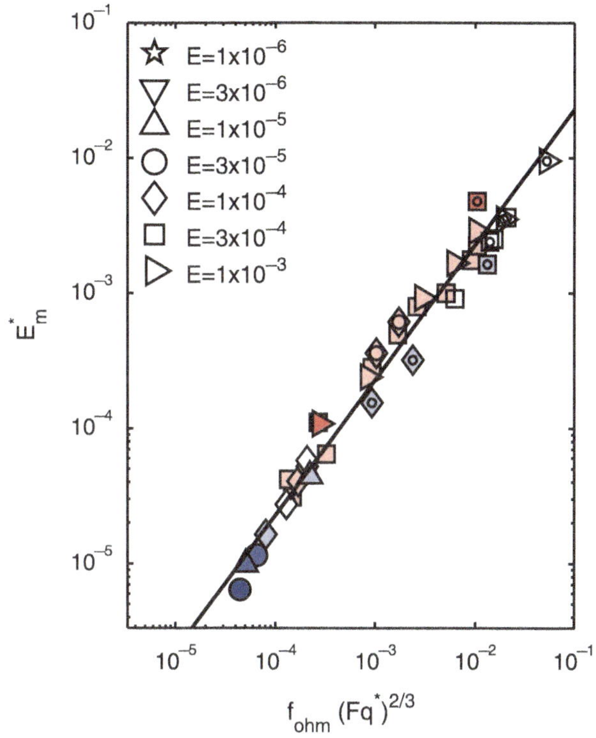

Fig. 5 Non-dimensional magnetic energy density versus 2/3 power of available energy for dynamos generating a multipolar magnetic field. See Fig. 1 for explanation of symbols

3.4 Field Scaling for Multipolar Dynamos

So far, the magnetic field scaling law in Eq. 11 has been tested only for numerical dynamos in the dipole dominated regime. The scaling arguments made in Sect. 2.3 should in principle apply independent of the magnetic field topology. In fact, plotting magnetic energy density against the available flux for the subgroup of multipolar numerical dynamos (dipolarity <0.35) shows again good agreement with the 2/3-power law (Fig. 5). The prefactor c is significantly lower than in case of the dipolar dynamos, 0.23 instead of 0.63. Such difference can be expected because of the different magnetic power spectra. In the non-dipolar case a relatively smaller fraction of the total magnetic energy resides in large spatial scales that hardly contribute to dissipation, therefore at the same value of the magnetic Reynolds number the ohmic dissipation is larger in the multipolar case.

3.5 Scaling of Ohmic Dissipation

Christensen and Tilgner (2004) used a set of numerical dynamo models to derive a scaling law for the ohmic dissipation time τ_λ. An inverse dependence on the magnetic Reynolds number, $\tau_\lambda = 0.27D^2/\lambda Rm^{-1}$, provides a decent fit to their numerical results. They also found that the fit could be improved by invoking an additional dependence on the magnetic Prandtl number Pm (or, equivalently, on the hydrodynamic Reynolds number). However, they rejected the latter because including results for the ohmic dissipation in the Karlsruhe dynamo experiment supported a simple dependence on Rm alone.

Here we revisit the scaling law for the ohmic dissipation time using the now much larger data basis of numerical dynamo results. A fit of the normalized dissipation time

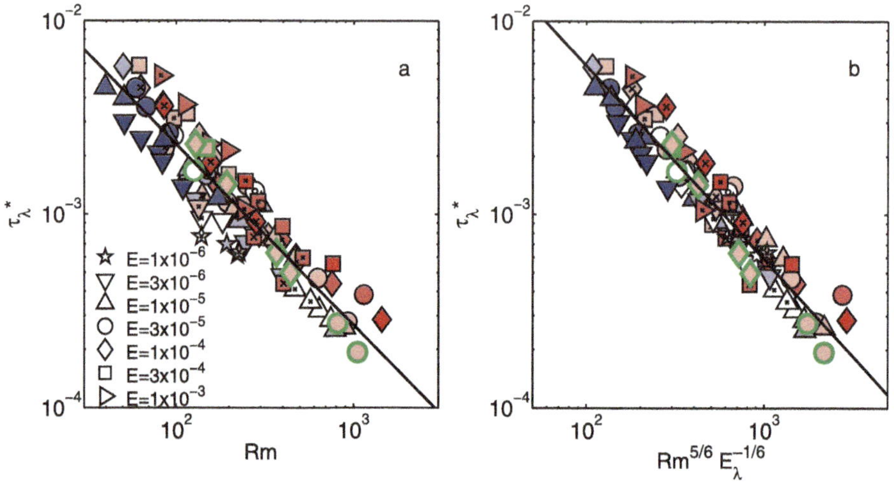

Fig. 6 Fit for ohmic dissipation time (scaled by D^2/λ), (**a**) against magnetic Reynolds number, (**b**) against a combination of magnetic Reynolds number and magnetic Ekman number. Slope −0.93 in (**a**) and −1.0 in (**b**). Symbols as in Fig. 1

$\tau_\lambda^* = \lambda/D^2\tau_\lambda$ versus Rm results in a slightly different exponent of −0.93. Also the standard error of the fit (0.29 with respect to fitting $\ln \tau_\lambda^*$ vs. $\ln Rm$) has increased compared to the original analysis (Fig. 6a). Three different options for a two-parameter fit are tested. Including Pm improves the fit marginally (standard error 0.25). Using the Ekman number as second parameter, the misfit is reduced to 0.17. However, here we concentrate on the magnetic Ekman number $E_\lambda = E/Pm$ as additional parameter, which does not lead to quite the same variance reduction, but still improves the fit (Fig. 6b). It is given by

$$\tau_\lambda^* = 0.59 Rm^{-5/6} E_\lambda^{1/6}, \tag{29}$$

with a standard error of 0.21.

Replacing the dependence of dissipation time on Rm^{-1} with that given by Eq. 29, whose dimensional form is

$$\tau_\lambda = \ell_B^2/\lambda \propto (D/U)^{5/6}\Omega^{-1/6}, \tag{30}$$

reconciles the discrepancy between the exponents found from the numerical data fit for the velocity and the magnetic field, respectively. Using Eqs. 9 and 30 together with the velocity from the CIA-balance (Eq. 7) exactly recovers the 2/3-power law for the magnetic field with its independence from Ω (Eq. 11). The weak dependence on rotating rate in Eq. 30 is plausible, at least for in the numerical models with $Pm \approx 1$. More rapid rotation leads to smaller flow scales in the dynamo (Takahashi et al. 2008). A more small-scaled flow pattern can be expected to be more efficient in generating a small-scaled magnetic field than a large-scaled flow at the same value of Rm. Whether the same argument applies to real planetary dynamos with $Pm \ll 1$ is less clear, because here the smallest scales in the flow do not affect the magnetic field, which is homogenized at these scales by the large magnetic diffusivity.

3.6 Strength of The External Field

Equation 24 predicts the rms magnetic field strength inside the dynamo. But only the long-wavelength part of the magnetic field at the top of the dynamo can be inferred directly from observations. For planets other than Earth only the dipole and the lowest-order multipoles are known. Olson and Christensen (2006) tested an equivalent rule to Eq. 24 for the rms-strength of the dipole component of the field on the outer boundary and found a decent agreement for dynamo models in the dipolar regime ($Ro_\ell < 0.12$), although the scatter is larger it is than it is for the internal field. In the multipolar regime the scatter becomes very large.

Some earlier theories for the geodynamo assumed that the toroidal component of the magnetic field in the dynamo, which is invisible from the outside, might me much stronger than the poloidal field, similar to what is assumed for the solar dynamo. If this were correct, it would be very difficult to compare scaling law predictions for the internal dynamo field with the observable part of the field. However, all numerical geodynamo models show roughly equal strength of the poloidal and toroidal field components. Therefore, we assume here that the ratios between the internal (poloidal plus toroidal) field and the poloidal outside field found in geodynamo models are representative for planetary dynamos.

The ratio of the rms internal field strength to the dipole strength on the outer boundary has been found in the range between 3 and 10 for dipolar dynamos with a fixed ΔT (Christensen and Aubert 2006). The ratio between internal field strength to the total dynamo surface field strength is around 3.5. However, dynamo models with different thermal boundary conditions and those with a stably stratified layer at the top can have larger ratios between internal field and external dipole field (Aubert et al. 2009; Christensen et al. 2009). The uncertainty in this factor may preclude in some cases a meaningful test of the scaling law based on the observed planetary magnetic field.

3.7 Secular Variation Time Scales

Christensen and Tilgner (2004) used geodynamo models to determine the parameter dependence of the characteristic time scales of secular variation. For the field at a given harmonic degree n, it is defined as the ratio between the degree power of the field and of its time derivative: $\tau_n^{sec} = (B_n^2/\dot{B}_n^2)^{1/2}$. In the models, $\tau_n^{sec} = \tau^{sec}/n$ fits the spectral dependence well for $n > 1$. In the recent geomagnetic field a somewhat steeper decrease on τ_n^{sec} with n has been found (Holme and Olson 2006). For the master coefficient τ^{sec} Christensen and Tilgner (2004) determine an inverse dependence on the magnetic Reynolds number from fitting the numerical data, which is rewritten here as

$$\tau^{sec} = 3.4\frac{D^2}{\lambda}Rm^{-1} = 3.4\frac{D}{U}. \tag{31}$$

3.8 Heat Transport

With respect to planetary dynamos, knowing the dependence of the Nusselt number Nu on the Rayleigh number Ra (or on Fq^*) is of limited practical interest, because it is hardly possible to quantify from observation the tiny superadiabatic temperature contrast that drives convection. However, the scaling law for the Nusselt number, $Nu \propto Ra^\beta$, is of great theoretical interest in convection theory. A possible break in slope β is indicative of a change in the dynamical regime, with likely consequences for magnetic field generation in case of a dynamo.

Christensen (2002) and Christensen and Aubert (2006) found for rotating convection and for dynamos a value β of the order 6/5, surprisingly large compared to the well-established value $\beta = 2/7$ in non-rotating turbulent convection. King et al. (2009) showed for rotating convection that the large value of β applies only when the rotational constraints on the flow are strong and the Ekman boundary layer is thinner than the thermal boundary layer. Above a transition value for the Rayleigh number, $Ra_t = 1.4E^{-7/4}$, the slope of Nu vs. Ra approaches the classical 2/7 value. They estimated that in the Earth's core Ra is close to Ra_t. The same break in slope occurs in convective dynamo models (E. King, J. Aurnou, personal communications) and is more or less coincident with the change from dipolar to multipolar dynamos. Whether the criterion of the local Rossby number (Sect. 3.3) or that of the transitional Rayleigh number better captures the change in magnetic field geometry remains to be determined.

4 Application to the Planets

4.1 Magnetic Field Strength of Solar System Planets

The only rule that passes the test against numerical model results is Eq. 15, which predicts a dependence of the magnetic field strength on the cubic root of the available power and its independence from the rotation rate. But we face several problems when we want to test the scaling rule against observational data.

We need to know the convected heat flux in the dynamo. The core heat flow can be estimated for Earth within a factor of three (Lay et al. 2008) and for the gas planets it can be deduced from the planet's excess infrared radiation (e.g. Ingersoll et al. 1975). For other terrestrial planets and for Ganymede only crude estimates based on thermal evolution models are available. In these planets the core heat flow is controlled by the sluggish solid state convection in the overlying silicate mantle. The removal of internal heat is more efficient in the presence of plate tectonics compared to cases where mantle convection occurs below a rigid lid. Furthermore, only the convected part of the heat flux is relevant. The fraction of heat that is conducted along an adiabatic temperature gradient must be deducted from the total heat flow. For the gas planets this is not an important issue, but in the terrestrial planets the conducted heat flux is a significant part of the total flux. In some planets it may actually be the only means of heat transport. When a growing solid inner core exists, compositional driving of convection must be accounted for. This can be done by using buoyancy flux instead of the convected heat flux, where the former may have a thermal and a compositional contribution.

Our scaling rule predicts the mean field strength inside the dynamo, but only the field at or above the planetary surface is accessible to observation. The ratios between the internal field and the exterior magnetic field found in numerical dynamo models serve as a guide, but this is fraught with some uncertainty. Also, the scaling law has been tested for a geometry and distribution of buoyancy forces appropriate for the present geodynamo, i.e. a deep fully convecting spherical shell, with a strong flux of buoyancy emanating from the inner core. Other planetary dynamos may operate in a different way, i.e. in a thin shell, with a different distribution of buoyancy sources, and possibly below or above a stably stratified fluid layer. As long as the same basic principles of a power-controlled field strength apply to these dynamos, the efficiency factor as calculated by Eq. 14 should take care of such modifications. However, the spatial spectrum of the magnetic field and the factor relating internal field strength to exterior field strength can be different for the different types of dynamos (Aubert et al. 2009).

Jupiter's magnetic field and the geomagnetic field are similar in terms of the dipole tilt and the ratio between dipole power and the power in low-degree multipoles, suggesting that the dynamos in the two planets are generically similar. Christensen and Aubert (2006)

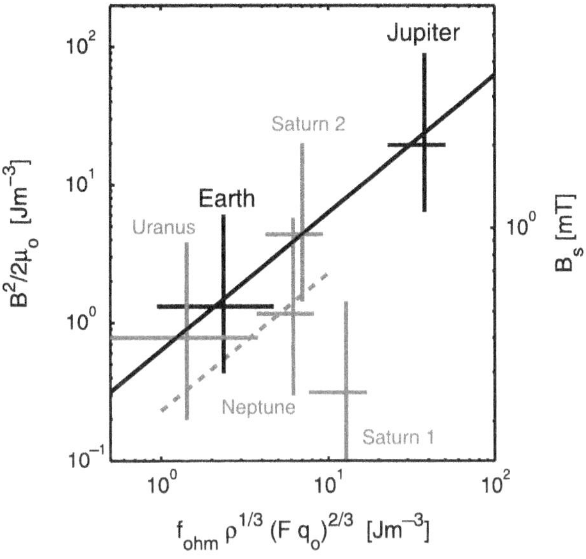

Fig. 7 Magnetic energy density in planetary dynamos vs. the 2/3-power of the available heat flux. The solid line is the prediction of Eq. 15, with $c = 0.63$ taken from the fit to dipolar numerical dynamos, the broken line has $c = 0.23$ found for multipolar dynamos. The equivalent total field strength at the top of the dynamo is plotted on the axis to the right. See text for explanations of the two Saturn cases. Mercury and Ganymede are not included because of the available energy flux is not observationally constrained

and, in a more refined analysis including a detailed calculation of the efficiency factor F, Christensen et al. (2009) compared the predictions of Eq. 15 with the observed fields of Earth and Jupiter. For Earth $F = 0.27 - 0.52$ and for Jupiter $F = 1.1$ is found. In Fig. 7 their result is reproduced, along with estimates for other solar system planets. The internal field strengths deduced from the surface fields of Earth and Jupiter match the predictions well (see Christensen et al. 2009, including online supplementary information for assumptions and error estimates).

For other solar system planets the compliance with our preferred scaling rule is much less clear. But this is also true for the Elsasser number rule, which works well for Earth and Jupiter, but fails for Mercury, Uranus, Neptune (Stevenson 2003), and perhaps marginally for Saturn. We will now discuss these planets one by one.

Mercury's field is enigmatically weak. The heat flow from the core is unknown, but explaining the weak external field on the basis of Eq. 15 by an extremely low energy flux would not be a viable option. The corresponding velocity, obtained via Eq. 7, would imply a subcritical magnetic Reynolds number. Mercury's dynamo may be very different from the geodynamo (Stanley and Glatzmaier 2009, this issue). For example, it may be restricted to a thin fluid shell and generate a strong (hidden) toroidal field, but a weak (observable) poloidal field (Stanley et al. 2005). In the model of Christensen (2006) the dynamo operates in a deep convecting sublayer covered by a thick stagnant fluid region, which arises because the heat flux at Mercury's core-mantle boundary is probably significantly lower than the flux that can be conducted along an adiabatic temperature gradient. Both models could explain the weak field observed outside the planet. At the same time the strength of the internal magnetic fields may comply with the scaling rule of the available energy flux, although this would be very difficult to verify since most of the dynamo field is hidden from observation.

Venus and *Mars* have no active dynamo at present. The heat flow in these planets is probably significantly less than it is in the Earth because of a lack of plate tectonics. In particular, the heat flow in the core may be less than the flux that can be conducted down the adiabatic gradient, so that the core would be thermally stable. The convected heat flow q_c is then zero. In contrasts to Mercury, these planets may have failed to nucleate an inner core. Compositional convection is then unavailable and no dynamo exists. In Mars the early dynamo, whose existence is suggested by the strong magnetization of the martian crust in the southern hemisphere, would have been powered by a superadiabatic core heat flux during the first couple of hundred million years of the planet's history.

Saturn is similar to Jupiter in its internal structure, which comprises a large core region of metallic hydrogen (Fortney and Nettelmann 2009, this issue). It also has a high internal heat flow. But Saturn's magnetic field is extremely axisymmetric and 20 times weaker than Jupiter's field at the surface. Assuming the top of the dynamo region at a pressure of 1.3 Mbar, which corresponds to $R_c = 0.62R_p$, the scaling rule significantly overpredicts the observed field strength (Fig. 7, Saturn 1). To explain the unexpectedly high heat flux compared to the prediction of evolution models, Stevenson (1980) proposed that at the top of the metallic hydrogen region helium becomes immiscible and precipitates, creating a compositionally stratified layer. Differential rotation in such a convectively stable electrically conducting layer above the dynamo would filter out non-axisymmetric magnetic field components (Stevenson 1982; Christensen and Wicht 2007) and can explain the geometry of Saturn's external field. The thickness of the stably stratified layer, if it exists, is highly uncertain. A deeper dynamo means a lower effective energy flux and, for geometrical reasons, a stronger field at the top of the dynamo for a given surface field. By tuning the upper boundary of the active dynamo region to $R_c = 0.40R_p$, Saturn can be brought to agreement with the theoretical prediction (Saturn 2 in Fig. 7).

In *Uranus* and *Neptune* the dynamo resides in a fluid region of 'ices' that extends to $R_c \approx 0.75R_p$ and has an ionic electrical conductivity (Fortney and Nettelmann 2009, this issue). The magnetic fields of the two planets are similar, with nearly Earth's field strength at the planetary surface, strong dipole tilt, and quadrupole and octupole components that are comparable in strength to the dipole (Russell and Dougherty 2009, this issue). But differences exist for the internal heat flow; for Neptune it is $q_o = 430 \pm 46$ mW m^{-2}, whereas for Uranus it is only 42 ± 47 mW m^{-2} (Peale and Conrath 1991). In the case of Uranus, zero heat flux is within the error bar, but is not compatible with the existence of a convective dynamo. Estimating the internal field strength is difficult, because at the top of the dynamo the quadrupole and octupole components are stronger than the dipole. For the estimates used in Fig. 7 the field models by Holme and Bloxham (1996) are downward continued to R_c. As a rough guide, the total field strength at the dynamo surface is taken to be five times the typical strength in harmonics such as $n = 2$, 3 or 4. This ratio is found in multipolar dynamo models at values of the magnetic Reynolds appropriate for Uranus and Neptune (see below). A mean field strength of 0.4–0.5 mT at the top of the dynamo is inferred and is scaled up by a factor of 3.5 to arrive at the interior field strength. Setting the F-factor to one, the estimated internal field strength is in fair agreement with the theoretical prediction within the uncertainties (Fig. 7). However, the interior of Uranus and Neptune may in part be stably stratified (Hubbard et al. 1995), and host thin-shell dynamos quite unlike the geodynamo (Stanley and Bloxham 2004), which adds to the other uncertainties.

Ganymede's dynamo requires that the small iron core is convecting at present. Estimates of the available energy flux are highly uncertain. Nonetheless, applying the power-based scaling rule, Hauck et al. (2006) showed that plausible estimates for the buoyancy flux lead to a magnetic field strength in agreement with observation.

4.2 Comparison with Magnetic Fields of Rapidly Rotating Stars

The test of the energy flux scaling vis-à-vis the observed planetary field strength is only partly successful. Earth and Jupiter fit well, but for the other planets the uncertainties are large, or particular assumptions must be made that are plausible, but not based on hard

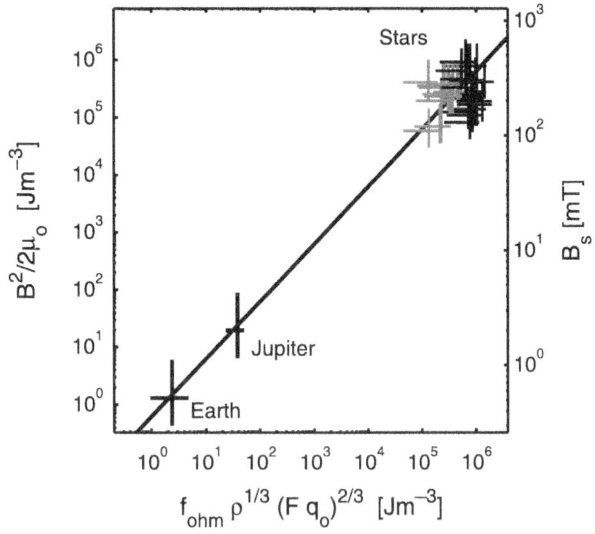

Fig. 8 Similar to Fig. 7 but including rapidly rotating stars main-sequence stars in black and T Tauri stars in grey

constraints. From this comparison alone, it could hardly be claimed the energy flux scaling is superior to the Elsasser number rule. The unsatisfactory situation can be alleviated by including into the comparison other cosmic objects with spherical shell convection-driven dynamos. The solar dynamo is probably too different from planetary dynamos to lend itself to a direct comparison, because much of the field generation is thought to occur at the intense shear layer (tachocline) at the bottom of the solar convection zone. Also, in the slowly rotating Sun Coriolis forces play less of a role than in planetary dynamos (Jones et al. 2009, this issue). Christensen et al. (2009) considered two groups of stars, whose surface magnetic field has been determined spectroscopically through the Zeeman effect. One are low-mass main-sequence stars and the other are very young contracting (T Tauri) stars, with rotation periods of less than 4 days. These objects are fully convective and lack a tachocline (or are nearly fully convective). The energy flux and the magnetic field strength is much larger than it is for planets, yet, the stars seem to follow exactly the same scaling rule as the Earth and Jupiter (Fig. 8). We note that the bold line in Fig. 8 is *not* a fit to the observations, but the slope and the prefactor are derived from theory and numerical models.

4.3 Flow Velocity, Secular Variation and Ohmic Dissipation

The dimensional version of Eq. 22, with an exponent of 2/5 and the prefactor derived from the models inserted, is

$$U = 1.05(Fq_o/\rho)^{2/5}(\Omega R_c)^{-1/5}. \tag{32}$$

Assuming for *Earth's* core an effective superadiabatic heat flux of at the core-mantle boundary of 100 mW m^{-2}, the predicted velocity is ≈ 2 mm s^{-1} (the effective heat flux comprises also compositional driving of the dynamo Christensen et al. 2009). For $\lambda = 1.3$ m^2 s^{-1} this corresponds to a magnetic Reynolds number of 3500, somewhat high compared to other estimates. Using the observed secular variation time of the geomagnetic field $\tau^{sec} \approx 500$ yr in Eq. 31 provides a handle to derive the characteristic velocity in the Earth's core independent of assumptions that enter into inversion models for the flow at the top of Earth's core

(Holme 2007). The result, 0.5 mm s^{-1} or $Rm \approx 800$, is similar to estimates obtained from core flow models. The discrepancy between the two velocity estimates could be resolved by a slightly larger exponent than 2/5 in Eq. 21; a value of 0.45 would be marginally consistent with the data in Fig. 1 and lead to a predicted $U \approx 1$ mm s^{-1}. Also, the energy flux in Earth's core could be significantly smaller than the value assumed above. Christensen and Aubert (2006) had argued for a rather small buoyancy flux (energy flux), using the 'observed' velocity at the top of the core and solving Eq. 32 for Fq_o. The predicted magnetic field strength for a low energy flux dynamo may still be compatible within the uncertainties (but see also Aubert et al. 2009, for a discussions of high-energy and low-energy scenarios).

The amount of ohmic dissipation in the Earth's core has been a contentious issue, with estimates ranging between 0.1 and 3.5 TW (Buffett 2002; Roberts et al. 2003; Labrosse 2003; Gubbins et al. 2003; Christensen and Tilgner 2004). We note that according to our favoured scaling theory for the field strength, the ohmic dissipation is simply related to the available energy flux by $D = f_{ohm} Fq_o$. With the range for Fq_o considered in Christensen et al. (2009) and setting $f_{ohm} = 0.75 - 1$, the ohmic dissipation is between 1.1 and 15 TW. This is rather large compared to previous estimates, and in contrast to the range of 0.2–0.5 TW obtained by Christensen and Tilgner (2004) on the basis of the scaling law for the ohmic dissipation time with the inverse magnetic Reynolds number. The revised scaling for τ_λ (Eq. 29) leads, at the same $Rm \approx 800$ as considered in Christensen and Tilgner (2004) and for an Earth-like value of the magnetic Ekman number, to a dissipation time that is five times shorter (five times higher dissipation). This brings the dissipation back in line with the above estimate at a low available energy flux. High energy flux goes along with a higher value of Rm, as discussed before, which shortens the dissipation time further.

In *Jupiter* and *Saturn* the predicted velocities are in the range 10–20 mm s^{-1}. Assuming $\lambda \approx 4$ m^2 s^{-1}, the magnetic Reynolds number is of order 10^5. Secular variations have not been observed at other planets, except that Russell et al. (2001) reported a change of the tilt of Jupiter's dipole by 0.5° over 25 yrs, i.e., at a rate comparable to that of Earth's dipole. Because both the size of the dynamo and the predicted characteristic velocity are an order of magnitude larger in Jupiter compared to the geodynamo, the secular variation time scales should in fact be the same according to Eq. 31. NASA's Juno mission may constrain the secular variation of Jupiter's field in the future.

For *Uranus* and *Neptune* the predicted velocities are 3 mms^{-1} and 8 mms^{-1}, respectively. For $\lambda \approx 400$ m^2s^{-1} (Nellis et al. 1988) and a deep convection shell with $D \approx 13{,}000$ km, the respective magnetic Reynolds numbers are 110 and 300. This is not very far above the critical value for the onset of dynamo action of about 50. Assuming that convection is restricted to a thin shell overlying a stably stratified interior might bring Rm dangerously close to the critical value, in particular for Uranus.

4.4 Field Topology and Reversals

Using Eq. 28, Olson and Christensen (2006) made a crude estimate for the local Rossby number in the different planets. For Earth the predicted value is 0.1, i.e. close to the transition between the dipolar and multipolar regime and in a range where occasional dipole reversals have been found in numerical models. But while the predictions seems to agree well with the known behaviour of the geomagnetic field, there is a problem with such high value of Ro_ℓ. The associated length scale ℓ is only 100 m—we note that it is less than the length scale of a few kilometers predicted by the CIA-balance (Eq. 8). Even if the flow in the Earth's core carried significant energy in 100-m-sized eddies, it is difficult to understand how they could affect the magnetic field, which is homogenized by diffusion at such small scales. A possible mechanism is a strong back-reaction of the small eddies on the circulation at larger scales through Reynolds stress (inertial) effects. This is typical in rotating convection and leads, for example, to the excitation of vigorous zonal jet flow.

For the gas planets, and in particular for Uranus and Neptune, Olson and Christensen (2006) estimated values of Ro_ℓ an order of magnitude smaller than for the Earth's core. Hence one would expect an axially dipolar magnetic field at Uranus and Neptune, which is in obvious contrast to the strong dipole tilt and large quadrupole and octupole contributions. A thin shell dynamo above a stably stratified conducting core might explain the multipolar nature of the magnetic field (Stanley and Bloxham 2004, 2006), but it could run into the problem of a subcritical magnetic Reynolds number, as mentioned above. Only for Mercury a local Rossby number larger than one is expected because of its slow rotation, which puts Mercury into the multipolar dynamo regime. The existence of a multipolar field inside Mercury's dynamo, despite the dipole dominance of the field at Mercury's surface, plays in fact an essential role in the model by Christensen (2006) to explain the low external dipole moment.

So far, the support for the rule of the local Rossby number as criterion for the selection of the magnetic field topology is rather scanty from the comparison with the planets. It must also be kept in mind that the rule for calculating Ro_ℓ (Eq. 28) is purely empirical, obtained by fitting model data, and that it requires huge extrapolations to apply it to the planets. However, some support for the validity of the rule comes again from studying the magnetic fields of low-mass stars. Mapping the field of several such stars by a technique called Zeeman-Doppler tomography, Donati et al. (2008) found a trend from multipolar to axially dipolar geometry when the Rossby number (defined in a somewhat analogous to the local Rossby number used here) becomes smaller.

5 Discussion and Outlook

In the past, scaling laws for planetary dynamos have been suggested on the basis of fairly general theoretical principles. In recent years, it has become possible to compare scaling laws with the results of direct numerical simulations of the dynamo. One apprehension has been that these dynamo models, running at parameter values vastly different from planetary dynamos, might be in a different dynamical regime and in particular being strongly influenced by viscosity. But this fear seems to be unfounded—if it were true the scaling for the Rossby number and the non-dimensional magnetic field strength in the models should depend on the Ekman number, but no such dependence is found. A concern that must be taken more seriously is that the numerical simulations assume a magnetic Prandtl number of order one, whereas the planetary value is of order 10^{-6}. The disparity of the small length scales in the flow and small scales of the magnetic field, which exists in planetary dynamos, is not accounted for in the models. There are hints in the numerical data for an influence of the magnetic Prandtl number on the velocity and magnetic field scaling laws (Christensen and Aubert 2006). The dependence on Pm is weak, but because of the large range over which the magnetic Prandtl number must be extrapolated, it would lead to significant differences if it persisted to very low values of Pm. From this point of view, it seems more urgent to develop dynamo models at low magnetic Prandtl number than models with a very low Ekman number. But the tasks are not independent—to obtain a dipolar dynamo model at low Pm, the Ekman number must also be made very small. Otherwise the strong driving of the flow, which is required in a low-Pm dynamo, would lead to a dominance of inertial forces and to a multipolar magnetic field (Christensen and Aubert 2006). Truely low magnetic Prandtl numbers will be hard to reach in numerical simulations, and insights may come from future laboratory dynamo experiments with significant net rotation.

The agreement of the energy-flux based magnetic field scaling law with the numerical results (Figs. 2 and 3), and its agreement with observation for objects with presumably simple dynamos (Fig. 8), provide rather strong support for its validity (simple dynamos are those operating in deep convective layers without shielding by stagnant regions). The magnetic field strength does not seem to be controlled by a force balance, but rather by the energetics of the planet (or star). This does not mean that the flow in the dynamo would not obey a balance in which Lorentz forces, Coriolis forces and buoyancy forces play an important role. But the force balance may be more subtle than what is assumed in simple scaling arguments. For example, pressure gradients could cancel most of the Coriolis force, or magnetic field and velocity could be strongly aligned, so that induction effects are small and $\sigma U B^2$ is an overestimate for the order of the Lorentz force. Also, inertial forces that kick in at small flow scales may be more important than previously assumed.

The rule of the local Rossby number for the selection of the field topology seems to be on less solid ground than the scaling rule for the field strength. On the one hand, the local Rossby number rule is supported by a large number of dynamo simulations and it seems plausible that the ratio of inertial forces to Coriolis forces can have an effect on the magnetic field. On the other hand, it is not understood in detail why strengthening of the inertial force leads to a severe change in the magnetic field configuration. While inertial forces are clearly important in some of the dynamo models, it is not clear how they can become influential in planetary dynamos. Furthermore, the rule for estimating the local Rossby number at planetary parameter values, which are vastly different from model values, is also not beyond doubt. The local Rossby number rule has rather limited support from a comparison of its prediction with natural dynamos. For the two planets with a multipolar field geometry, Uranus and Neptune, the local Rossby number is not larger and probably smaller than for Earth or Jupiter. Nonetheless, so far only the local Rossby number rule has been suggested as a general criterion for the selection of the magnetic field geometry and it can be considered as a viable hypothesis. At any rate, the qualitative statement that more rapid rotation favours dipolar dynamos seems fairly robust.

Further studies are needed into types of planetary dynamos that differ from those in Earth and Jupiter. For dynamos in thin shells, dynamos with a different distribution of buoyancy sources than considered so far, or dynamos with stably stratified layers, the prefactors in the scaling laws or the critical numbers (for Ro_ℓ, Rm) must be determined. While this is amenable to numerical simulations (e.g. Aubert et al. 2009), unfortunately the details of the dynamo structure in Mercury, Saturn, Uranus or Neptune are poorly known. The factors relating the internal field strength to that of the observed external field must also be reevaluated. They could be more variable for multipolar dynamos than for dipolar ones (compare Fig. 4). For these reasons the application of scaling laws to planets with 'special' dynamos may remain tentative.

The scaling law based on energy flux provides a handle for estimates of the strength of Earth's paleofield (Aubert et al. 2009) and for the magnetic fields of now extinct dynamos, for example in early Mars or in large differentiated asteroids (Weiss et al. 2008, 2009, this issue). It also allows to estimate the magnetic field strength of extrasolar planets. For a relatively young supermassive planet of several Jupiter masses, Christensen et al. (2009) obtained a surface field an order of magnitude stronger than Jupiter's field. The strong field would imply a much higher intensity of non-thermal radio emissions than in the case of Jupiter (Zarka 2007). This could lead to the new detection of such planets. From observations of the radiowave frequency spectrum, which has a cutoff at the electron cyclotron frequency near the planet's surface $\omega_c \propto B_P$, the field strength could be determined.

References

J. Aubert, D. Brito, H.C. Nataf, P. Cardin, J.P. Masson, Phys. Earth Planet. Inter. **128**, 51 (2001)
J. Aubert, S. Labrosse, C. Poitou, Geophys. J. Int. (2009, submitted)
D. Breuer, S. Labrosse, T. Spohn, Space Sci. Rev. (2009, this issue)
B.A. Buffett, Geophys. Res. Lett. **29** (2002). doi:1.12029/2001GL014649
F.H. Busse, Phys. Earth Planet. Inter. **12**, 350 (1976)
J.C. Cain, P. Beaumont, W. Holter, Z. Wang, H. Nevanlinna, J. Geophys. Res. **100**, 9439 (1995)
S. Chandrasekhar, *Hydrodynamic and Hydromagnetic Stability* (Oxford University Press, Oxford, 1961)
U.R. Christensen, J. Fluid Mech. **470**, 115 (2002)
U.R. Christensen, Nature **444**, 1056 (2006)
U.R. Christensen, J. Aubert, Geophys. J. Int. **166**, 97 (2006)
U.R. Christensen, V. Holzwarth, A. Reiners, Nature **457**, 167 (2009)
U.R. Christensen, P. Olson, G.A. Glatzmaier, Geophys. J. Int. **138**, 393 (1999)
U.R. Christensen, A. Tilgner, Nature **429**, 169 (2004)
U.R. Christensen, J. Wicht, in *Treatise of Geophysics, vol. 8: Core Dynamics*, ed. by G. Schubert (Elsevier, Amsterdam, 2007), pp. 245–282
S.A. Curtis, N.F. Ness, J. Geophys. Res. **91**, 11003 (1986)
V. Dehant, H. Lammer, Y.N. Kulikov et al., Space Sci. Rev. **129**, 279 (2007)
J.-F. Donati, J. Morin, P. Petit et al., Mon. Not. R. Astron. Soc. **390**, 545 (2008)
J.J. Fortney, N. Nettelmann, Space Sci. Rev. (2009, this issue)
D.J. Galloway, M.R.E. Proctor, N.O. Weiss, J. Fluid Mech. **87**, 243 (1978)
G.A. Glatzmaier, P.H. Roberts, Nature **337**, 203 (1995)
D. Gubbins, D. Alfè, G. Masters, G.D. Price, M.J. Gillan, Geophys. J. Int. **155**, 609 (2003)
D. Gubbins, P. Roberts, in *Geomagnetism, vol. 2*, ed. by J.A. Jacobs (Academic Press, London, 1987), pp. 1–183
S.A. Hauck, J.M. Aurnou, A.J. Dombard, J. Geophys. Res. **111**, E09008 (2006)
R. Holme, in *Treatise of Geophysics, vol. 8: Core Dynamics*, ed. by G. Schubert (Elsevier, Amsterdam, 2007), pp. 107–130
R. Holme, J. Bloxham, J. Geophys. Res. **101**, 2177 (1996)
R. Holme, N. Olson, Geophys. J. Int. **166**, 518 (2006)
W.B. Hubbard, M. Podolak, D.J. Stevenson, in *Neptune and Triton*, ed. by D.P. Cruikshank (University of Arizona Press, Tucson, 1995), pp. 109–138
A.P. Ingersoll, G. Münch, G. Neugebauer et al., Science **188**, 472 (1975)
C.A. Jones, M.J. Thompson, S. Tobias, Space Sci. Rev. (2009, this issue)
E. King, S. Stellmach, J. Noir, U. Hansen, J. Aurnou, Nature **457**, 301 (2009)
R. Kippenhahn, A. Weigert, *Stellar Structure and Evolution* (Springer, Berlin, 1990)
S. Labrosse, Phys. Earth Planet. Inter. **140**, 127 (2003)
T. Lay, J. Hernlund, B.A. Buffett, Nature Geosci. **1**, 25 (2008)
M. Leweling, T. Spohn, Planet. Space Sci. **45**, 1389 (1997)
S. Maus, M. Rother, C. Stolle et al., Geochem. Geophys. Geosys. **7**, Q07008 (2006)
H. Mizutani, T. Yamamoto, A. Fujimura, Adv. Space Res. **12**, 265 (1992)
W.J. Nellis, D.C. Hamilton, N.C. Holmes et al., Science **240**, 779 (1988)
P. Olson, U.R. Christensen, Earth Planet. Sci. Lett. **250**, 561 (2006)
J.C. Peale, B.J. Conrath, J. Geophys. Res. **96**, 18921 (1991) supplement
P.H. Roberts, C.A. Jones, A. Calderwood, in *Earth's Core and Lower Mantle*, ed. by C.A. Jones, A.M. Soward, K. Zhang (Taylor & Francis, London, 2003), pp. 100–129
C.T. Russell, Nature **272**, 147 (1978)
C.T. Russell, M.K. Dougherty, Space Sci. Rev. (2009, this issue)
C.T. Russell, Z.J. Yu, K.K. Khurana, M.G. Kivelson, Adv. Space Res. **28**, 897 (2001)
Y. Sano, J. Geomag. Geoelectr. **45**, 65 (1993)
G. Schubert, M.N. Ross, D.J. Stevenson, T. Spohn, in *Mercury*, ed. by F. Vilas, C.R. Chapman, M.S. Matthews (University of Arizona Press, Tucson, 1988), pp. 651–666
G. Schubert, T. Spohn, J. Geophys. Res. **95**, 14095 (1990)
R. Simitev, F.H. Busse, J. Fluid Mech. **532**, 365 (2005)
B. Sreenivasan, C.A. Jones, Geophys. J. Int. **164**, 467 (2006)
F.D. Stacey, *Physics of the Earth* (Wiley, New York, 1977)
S. Stanley, J. Bloxham, Nature **428**, 151 (2004)
S. Stanley, J. Bloxham, Icarus **184**, 556 (2006)
S. Stanley, J. Bloxham, W.E. Hutchison, Earth Planet. Sci. Lett. **234**, 341 (2005)
S. Stanley, G. Glatzmaier, Space Sci. Rev. (2009, this issue)

S.V. Starchenko, C.A. Jones, Icarus **157**, 426 (2002)
D.J. Stevenson, Geophys. Astrophys. Fluid Dyn. **12**, 139 (1979)
D.J. Stevenson, Science **208**, 746 (1980)
D.J. Stevenson, Geophys. Astrophys. Fluid Dyn. **21**, 113 (1982)
D.J. Stevenson, Rep. Prog. Phys. **46**, 555 (1983)
D.J. Stevenson, Astronom. Nachr. **305**, 257 (1984)
D.J. Stevenson, Earth Planet. Sci. Lett. **208**, 1 (2003)
D.J. Stevenson, T. Spohn, G. Schubert, Icarus **54**, 466 (1983)
F. Takahashi, M. Matsushima, Y. Honkura, Phys. Earth Planet. Inter. **167**, 168 (2008)
B.P. Weiss, J.S. Berdahl, L. Elkins-Tanton et al., Science **322**, 713 (2008)
B.P. Weiss, J. Gattacceca, S. Stanley, P. Rochette, U.R. Christensen, Space Sci. Rev. (2009, this issue)
J. Wicht, A. Tilgner, Space Sci. Rev. (2009, this issue)
J. Wicht, S. Stellmach, H. Harder, in *Geomagnetic Field Variations*, ed. by K.H. Glassmeier, H. Soffel, J.F.W. Negendank (Springer, Berlin, 2009), pp. 107–158
P. Zarka, Planet. Space Sci. **55**, 598 (2007)

Space Sci Rev (2010) 152: 591–616
DOI 10.1007/s11214-009-9579-5

The Solar Dynamo

**Chris A. Jones · Michael J. Thompson ·
Steven M. Tobias**

Received: 24 June 2009 / Accepted: 12 October 2009 / Published online: 19 December 2009

Abstract Observations relevant to current models of the solar dynamo are presented, with emphasis on the history of solar magnetic activity and on the location and nature of the solar tachocline. The problems encountered when direct numerical simulation is used to analyse the solar cycle are discussed, and recent progress is reviewed. Mean field dynamo theory is still the basis of most theories of the solar dynamo, so a discussion of its fundamental principles and its underlying assumptions is given. The role of magnetic helicity is discussed. Some of the most popular models based on mean field theory are reviewed briefly. Dynamo models based on severe truncations of the full MHD equations are discussed.

Keywords Solar magnetism · Solar dynamo · Sunspots · Solar cycles · Solar interior · Helioseismology

1 Observations

1.1 Sunspots and Solar Magnetic Activity

Sunspots are perhaps the most obvious manifestation of solar magnetism (Thomas and Weiss 2008). Sunspots were observed already by the ancients, and were shown by Galileo to be features on the Sun itself; but the observational evidence that sunspots possess magnetic fields was established only a hundred years ago by Hale (1908). The number of sunspots, or the area occupied by them, follows an irregular 11-year activity cycle (lower panel of

C.A. Jones (✉) · S.M. Tobias
Department of Applied Mathematics, University of Leeds, Leeds LS2 9JT, UK
e-mail: cajones@maths.leeds.ac.uk

S.M. Tobias
e-mail: smt@maths.leeds.ac.uk

M.J. Thompson
School of Mathematics and Statistics, University of Sheffield, Sheffield S3 7RH, UK
e-mail: michael.thompson@sheffield.ac.uk

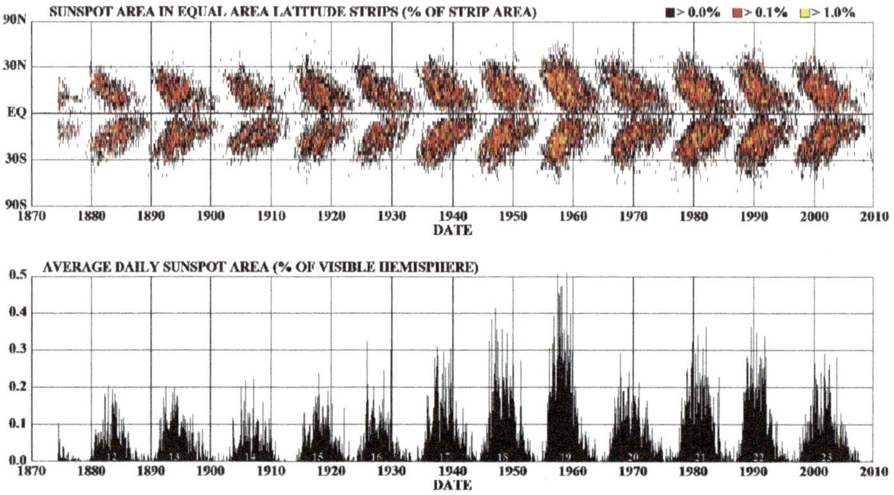

Fig. 1 Solar-cycle activity on the Sun since 1874. The lower panel shows the daily sunspot area, as a percentage of the visible hemisphere that is covered by sunspots and averaged over individual solar rotations. The butterfly diagram in the upper panel shows the corresponding incidence of sunspots as a function of latitude and time. At the beginning of a new cycle, spots appear around latitudes of $\pm 30°$. The activity zones spread until they extend to the equator, and then gradually die away, disappearing at the equator as the first spots of the next cycle appear at higher latitudes (courtesy of D.H. Hathaway)

Fig. 1). Sunspots exist for anything from a few days to more than two solar rotation periods (i.e. more than two months). Early in the solar cycle, new sunspots occur preferentially at mid-latitudes on the Sun, and occur successively closer to the solar equator as the cycle progresses. Plotting the locations of sunspots as a function of latitude and time thus gives rise to the so-called butterfly diagram (upper panel of Fig. 1).

Sunspots are cooler and darker than the rest of the solar photosphere (the visible surface of the Sun), with typical sunspot temperatures being 4000 K compared with the photosphere's 6000 K. The outer 30 per cent by radius of the Sun constitutes the convection zone, where convective motions of the plasma transport heat from the deeper interior to the solar surface. But the strong magnetic fields in sunspots inhibits the convective motions (a typical strength at the centre of a spot is 2–3000 G and the plasma beta is of order unity), causing the spots to be cooler and hence darker. Actually the structure of a sunspot is rather complex and beautiful, with a dark inner region known as the umbra and a surrounding region called the penumbra which exhibits a filamentary structure and is quite dynamic, as revealed for example by the exquisite observations from the Hinode satellite (e.g. Jurčak and Bellot Rubio 2008).

Further work by Hale and his collaborators (Hale et al. 1919) concerning the polarity of sunspot magnetic fields and their distribution and orientation in pairs established the basis for what are now known as Hale's Laws and Joy's Law. Sunspots occur in bipolar pairs consisting of a leading spot and a trailing spot, aligned approximately parallel to the solar equator. Hale's Laws are that the magnetic polarities of the leading and trailing spot in a pair are opposite, with the polarities being reversed in the two hemispheres; and that the polarities reverse in successive solar cycles. Joy's Law is that the line joining leading and trailing spots in a pair tends to be inclined at an angle of about 4° to the equator. Taking the change in polarity into account, it is seen therefore that the quasi-period of the Sun's magnetic cycle as revealed by sunspots is about 22 years.

As a theoretical aside, we may note the generally accepted theoretical picture is that bipolar sunspot pairs are caused by toroidal field in the form of magnetic flux tubes rising through the convection zone and forming so-called Omega loops. These loops emerge through the surface, the two regions of intersection of the loops with the photosphere being identified with the bipolar sunspot pair (e.g. Fan 2004).

Sunspots typically occur within larger magnetic complexes called active regions. There is a reported tendency for active regions to recur at the same longitudinal position on the Sun during a solar cycle and even from one cycle to another: these locations are called active longitudes. Above active regions one observes magnetic loops or arcades of loops in the overlying chromosphere and in the corona which is the outer atmosphere of the Sun. These loops are made manifest by emission from the plasma confined within them. Occasional explosive reconfigurations of these magnetic structures, caused by magnetic reconnection, give rise to flares and coronal mass ejections (CMEs).

As well as the sunspot fields, there is magnetic field on a wide range of smaller scales. Smaller bipolar structures called ephemeral regions appear at all latitudes and show little cycle variation. A discovery of recent years is that the Sun has a lot of small scale magnetic flux that has been dubbed the solar 'magnetic carpet'. The flux that comprises the magnetic carpet is constantly emerging and disappearing (by subduction or reconnection), such that the magnetic carpet is renewed over a timescale of order one day. Whether this magnetic flux is the product of local small-scale dynamo action or whether it is recycled old magnetic flux is a matter of debate.

The Sun also possesses a weak (of order 1–2 G) global magnetic field that is predominantly dipolar. In the polar regions, the global field rises to around 10 G. Like the sunspot field, the polar field reverses its polarity roughly every 11 years. The polar field reversal occurs at about the time that the azimuthal sunspot field reaches its peak strength, at about 15–20° latitude (Parker 2007).

It may be noted that other observables that are related to the solar magnetic field also vary over the solar cycle: these include the 10.7 cm radio flux and the solar irradiance (particularly at short UV wavelengths).

Solar-cycle studies can be extended further back in time using proxy data. The magnetic field from the Sun deflects galactic cosmic rays. In the Earth's atmosphere these rays produce radioactive isotopes such as ^{14}C, which is preserved in trees, and ^{10}Be, which is preserved in polar ice-caps. The abundances of these isotopes therefore vary in antiphase with the solar magnetic field. Studies of both the ice-core record and tree-ring data allow the 11-yr cycle to be traced back over hundreds of years, while its envelope modulation can be traced back over millennia (see for example the review by Weiss and Thompson 2009).

From about 1645 to 1715, soon after Galileo had made the first telescopic studies of sunspots, there were very few sunspots seen on the solar disc and those that were seen were almost all in the Sun's southern hemisphere. This period is known as the Maunder Minimum (Eddy 1976; Ribes and Nesme-Ribes 1993). Thus in this period the sunspot cycle seems to be broken. Yet ice-core studies show that a cyclical modulation continued through the Maunder Minimum, indicating that the Sun's large-scale magnetic field continued even though there were few sunspots (Beer et al. 1998). The north-south symmetry of the sunspot distribution was soon re-established after the end of the Maunder Minimum.

Studies of the proxy datasets indicate that the Maunder Minimum is only the latest of the grand minima that have occurred in solar activity over time. Conversely, the present era of rather large-amplitude solar cycles (cf. Fig. 1) is a grand maximum, though this is perhaps now coming to an end (Abreu et al. 2008).

1.2 Solar Rotation and Helioseismology

It is generally accepted that the strong toroidal field that gives rise to sunspots is produced by a large-scale dynamo. Differential rotation of the solar plasma is likely an important ingredient of that dynamo. The Sun's rotation at the surface can be observed directly by spectroscopic Doppler measurements and by observing the motion of sunspots and other tracers. These observations show that the surface rotation period at the solar equator is about 25 days, increasing with latitude to more than a month at high latitudes. There is some variation in the periods obtained from different methods of measurement (see Beck 1999 for a comprehensive review).

In recent years helioseismology has provided a unique view of conditions in the solar interior, including how the rotation varies with depth and latitude. Global helioseismology uses observations of global resonant oscillations of the Sun. These modes are set up by essentially acoustic waves, which are generated by convective turbulence in the upper part of the convection zone and which bounce around in the solar interior. Because the Sun is nearly spherical, the horizontal structure of the modes thus formed is described by spherical harmonics, and the resonant frequencies are determined by the properties of the solar interior. Rotation of the solar interior causes the modes set up by eastward- and westward-propagating waves to have slightly different frequencies. After measuring this 'frequency splitting' for modes that are sensitive to different ranges of latitude and depth within the solar interior, inverse techniques can be used to infer how the rotation varies with latitude and depth. Moreover, measurements at different times reveal small temporal variations in the rotation rate over periods of months and years.

Energy is generated by fusion reactions in the inner core of the Sun. This energy is transported outwards through the inner 70 percent of the solar interior by radiative transport, and thereafter by convective heat transport until one approaches very close to the surface. Helioseismology has demonstrated that the transition to convective heat transport which we refer to as the base of the convection zone, is located at a fractional radius $r/R_\odot = 0.713 \pm 0.003$ (where R is the radius of the Sun) (Christensen-Dalsgaard et al. 1991; Basu 1998). The location of the base of the convection zone appears to be independent of latitude (Basu and Antia 2001).

The above measurement refers to the location of the essentially adiabatically stratified envelope, which may include a region of convective overshooting where the motions are sufficiently rapid to establish an adiabatic thermal stratification. The extent of overshooting is relevant to dynamo models, since an overshoot region may provide a place where magnetic flux is stored and amplified before rising to the surface. Modelling of the convection in stars suggests that beneath the adiabatically stratified overshoot region there will be a rather sharp transition to subadiabatic stratification: if that is the case, helioseismology indicates that the extent of such overshooting is small, probably no more than one tenth of the pressure scale height (Basu and Antia 1994; Monteiro et al. 1994; Christensen-Dalsgaard et al. 1995). A larger extent of overshooting would be consistent with the seismology if the transition were smoother: such a model has been proposed by Rempel (2004).

Helioseismology has determined the rotation profile of a substantial part of the solar interior: this is illustrated in Fig. 2. Beneath the base of the convection zone the solution is consistent with solid-body rotation, whereas in the convection zone itself there is a differential rotation with faster rotation at low latitudes and slow rotation in the polar regions. The rotation rate of the radiative interior matches the envelope rotation rate at mid-latitudes. There is thus a region of shear flow near the base of the convection zone, which is known as

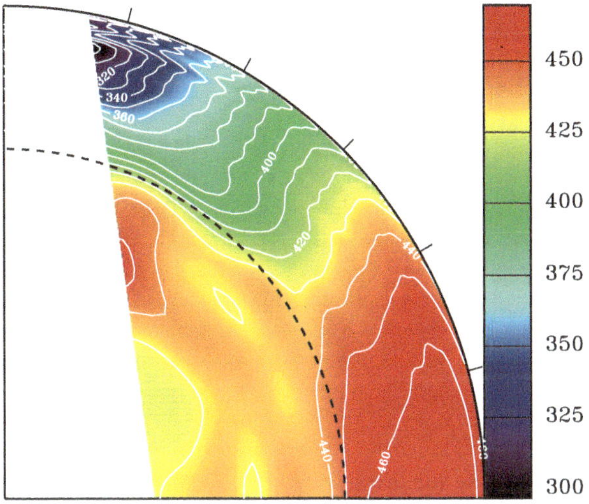

Fig. 2 The internal rotation of the Sun, as determined from observations by the MDI instrument on board the SOHO satellite. The solar equator is along the horizontal axis, the pole along the vertical axis. Values of $\Omega/2\pi$ are shown, in nHz. *The dashed line* indicates the base of the convection zone, and *tick marks* are at 15° intervals in latitude (from Thompson et al. 2003)

the tachocline, at both low and high latitudes. This is widely believed to be where the large-scale solar dynamo that generates the toroidal sunspot fields is located. There is another shear layer near the surface.

The helioseismic results concerning the tachocline region have been summarised in the review by Christensen-Dalsgaard and Thompson (2007). Kosovichev (1996) obtained the first quantitative results about the location and thickness of the tachocline. He adopted a functional dependence $\Phi(r)$ for the transition in depth of the latitudinal differential rotation of the form

$$\Phi(r) = \frac{1}{2}[1 + \mathrm{erf}(2(r - r_0)/w)], \tag{1}$$

where erf is the error function. This provides a continuous step function, with characteristic width w and centre location r_0. Kosovichev found that the tachocline thus defined is centred at $r/R_\odot = 0.692 \pm 0.05$ and has a width $w/R_\odot = 0.09 \pm 0.04$. The modern result of Basu and Antia (2003), converted to the same form as that used by Kosovichev, is that the centre location $r/R_\odot$ of the tachocline varies from 0.692 ± 0.002 at low latitudes to 0.710 ± 0.002 at 60° latitude. The corresponding tachocline widths $w/R_\odot$ at those latitudes are 0.033 ± 0.007 and 0.076 ± 0.010. Thus the tachocline is prolate, and its width is greater at higher latitudes. At low latitudes the tachocline is essentially wholly beneath the base of the convection zone, whereas at high latitudes it is roughly centred upon it.

As already noted above, helioseismology is able to reveal also the temporal variation of the rotation profile, on timescales of months and years. Superimposed on the mean rotation profile there are weak but apparently coherent bands of zonal flow that migrate from mid-latitudes to the equator and which extend beneath the surface at least one-third of the convection-zone depth (Howe et al. 2000a). These bands essentially track the migration of the active latitudes where sunspots are located, though the precise causal link between the magnetic activity and the zonal flows is unclear. At high latitudes too there is small but significant variation of the rotation rate, which appears to extend to all depths in the convection zone (Basu and Antia 2001; Vorontsov et al. 2002).

If the seat of the solar dynamo is in the tachocline region then one might hope to detect solar cycle-length variations in the angular velocity there, but none have been seen: there have though been reports of quasi-periodic variations of the angular velocity near the base of the convection zone with a period of 1.3–1.4 yrs (Howe et al. 2000b). It is interesting though that there are apparent small changes in the wave speed near the base of the convection zone which are well correlated with solar activity (Baldner and Basu 2008; see also Serebryanskiy and Chou 2005).

One further dynamical ingredient that may be important to the operation of the solar dynamo, as for example in the flux-transport dynamo models reviewed by Dikpati and Gilman (2009), is the north-south meridional circulation. Such flows have been measured in the near-surface layers using local helioseismic techniques: the flows are poleward and largely steady over a timescale of years. In the outer two per cent or so of the solar radius, the poleward flow speed is about 20–30 m s^{-1} and independent of depth (Haber et al. 2002). There is also evidence that in the approach to the last solar maximum the flow in the northern hemisphere developed a counter-cell at mid-latitudes. Unfortunately, the helioseismic evidence for what the meridional circulation is at greater depths is as yet inconclusive.

2 Modelling the Solar Dynamo

The observations discussed in the previous section demonstrate that stellar magnetic fields display variability on an extremely wide range of space and time scales. Thus the modelling of the dynamics of this magnetic field must be able to identify the processes that lead to dynamo action on these scales and formulate the mechanisms by which these scales can interact, or at least attempt to parameterise the effects of the scales that have not been included in the theory. This is a difficult task. Essentially, as we shall argue below, the level of turbulence in solar and stellar interiors leads to extreme parameter regimes in which the theory must be applied and makes progress extremely slow.

2.1 Difficulties for Dynamo Modelling: The Solar Parameter Regime

The equations that govern magnetic field generation in the solar and stellar interiors are well-known. Essentially they take the form

$$\begin{aligned}&\rho(\partial_t \mathbf{u} + \mathbf{u}\cdot\nabla\mathbf{u} + 2\boldsymbol{\Omega}\times\mathbf{u})\\ &\quad = -\nabla p + \frac{1}{\mu_0}(\nabla\times\mathbf{B})\times\mathbf{B} + \mu\left(\nabla^2\mathbf{u} + \frac{1}{3}\nabla(\nabla\cdot\mathbf{u})\right) + \mathbf{F}, \qquad (2)\end{aligned}$$

$$\partial_t \mathbf{B} = \nabla\times(\mathbf{u}\times\mathbf{B}) + \eta\nabla^2\mathbf{B}, \qquad (3)$$

$$\partial_t \rho + \nabla\cdot(\rho\mathbf{u}) = 0, \qquad (4)$$

$$\nabla\cdot\mathbf{B} = 0, \qquad (5)$$

where ρ is the density of the plasma, $\mathbf{u}$ is the velocity, $\mathbf{B}$ is the magnetic field, μ_0 is the permeability of free space, $\boldsymbol{\Omega}$ is the rotation vector and $\mathbf{F}$ is the forcing in the momentum equation that drives the fluid flow. Here μ is the dynamic viscosity and η is the magnetic diffusivity, both assumed constant. In general in stellar interiors the driving is from buoyancy forces that lead to thermally driven convection. Hence for a self-consistent solution, the momentum equation and induction equation above must be coupled to an energy equation and an equation of state (where usually the plasma is taken to be an ideal gas in which $p = \mathcal{R}\rho T$, with T the temperature and $\mathcal{R}$ the gas constant).

Given that the equations for magnetic field generation via fluid flow are well-known, the difficulties in describing the field generation processes occur owing to the fact that it is

Table 1 Typical values of dimensionless parameters in the Sun

		Base of convection zone	Photosphere
Rayleigh number	$Ra = g\alpha\beta L^4\rho/\kappa\mu$	10^{20}	10^{16}
Reynolds number	$Re = UL/\nu$	10^{13}	10^{12}
Magnetic Reynolds number	$Rm = UL/\eta$	10^{10}	10^{6}
Prandtl number	$Pr = \nu/\kappa$	10^{-7}	10^{-7}
Magnetic Prandtl number	$Pm = \nu/\eta$	10^{-3}	10^{-6}
Rossby number	$Ro = U/2\Omega L$	0.1–1	10^{-3}–0.4
Ekman number	$Ek = \nu/2\Omega L^2$	10^{-14}	10^{-15}–10^{-13}
Mach number	$M = U/c_s$	10^{-4}	1

Here g is the local gravity, α the thermal expansion coefficient, β is the superadiabatic temperature gradient, and κ the thermal diffusivity

impossible directly to solve these equations in the parameter regimes that occur in stellar interiors with the current computational resources. As for planetary and geodynamo theory, the difficulties arise because one or more of the non-dimensional parameters is extremely large. However for dynamo theory relating to field generation in moderately rotating stars, the difficulties arise in different places in the equations than for planetary models. The review of the solar dynamo by Ossendrijver (2003) gives estimated numerical values for the relevant non-dimensional parameters both at the base of the solar convection zone and at the photosphere (the visible surface). These are included in Table 1 for completeness.

The set of equations described above together with the parameter set detailed in Table 1 in theory give all the information necessary to determine the dynamics. As noted above the difficulties in determining the dynamics of magnetic fields in stellar interiors are manifested in different ways to those for the interiors of planets. By far the most troublesome of these parameter values for the solution of the dynamo system is the extreme values of the fluid and magnetic Reynolds numbers, Re and Rm. The large Reynolds number ensures that the flows that drive the dynamo are extremely turbulent with a large range of spatial and temporal scales. These are largely unconstrained and interact in highly nonlinear manner. Perhaps more troubling from a dynamo perspective, is that the magnetic Reynolds number in this turbulent environment remains extremely large, leading to the efficient generation of magnetic fields on extremely small scales, as we shall see. This is in contrast to the environment in planetary interiors where although the fluid Reynolds number is usually large and the flow turbulent, the flow is constrained by rotation (which itself provides a problem for modellers). Moreover in planetary interiors Rm is moderate (probably only a few times above critical) and so the process of field *generation* can be explored in the correct parameter regime. In short, whilst the modelling of planetary dynamos is plagued by the inability to get the correct magnetostrophic balance in the momentum equation, solar and stellar dynamo theorists are still arguing about the correct solutions to the turbulent induction equation!

2.2 Large and Small Scale Dynamo Action

The observations described in the introduction were focused on the behaviour of the systematic global solar magnetic field. High resolution observations of the magnetic fields in the Sun also reveal the existence of small-scale unsystematic magnetic fields at the solar surface (e.g. Régnier et al. 2008), the dynamics of which appears to be largely decoupled from that of the solar cycle. This "magnetic carpet" is also believed to be generated by dynamo action—convective motions at the solar surface lead to the generation of fields there. The scale of these fields is comparable with or smaller than the largest scale convective motions. This co-existence of magnetic fields on a wide-range of scales with very different properties has led to the separation of solar dynamo theory into two strands, small-scale dynamo theory (sometimes termed fluctuation dynamo theory) in which the field is generated on scales smaller (or of the same size) than those of the turbulent eddies, and large-scale dynamo theory, which is concerned with the systematic generation of fields at a scale larger than that of the turbulence. It is not clear that this distinction between large and small-scale dynamos is at all useful. However, historically a different range of techniques has been applied to the two problems and for this reason they have remained largely separated.

The bulk of this article is concerned with the theory for the generation of the systematic solar activity cycle. However we begin by discussing recent advances in small-scale turbulent dynamo theory (see Tobias 2009 for a more thorough discussion). It is fair to say that much more is understood about the dynamics of small-scale dynamos than large-scale ones. The dramatic recent increase in computational power coupled with the development of efficient numerical algorithms have led to major breakthroughs in our understanding. Numerical simulations of turbulent dynamos driven by Boussinesq convection (see e.g. Cattaneo 1999; Cattaneo and Hughes 2006; Tobias et al. 2008) have demonstrated that this turbulent flow is capable of generating small-scale magnetic field if the magnetic Reynolds number Rm—the non-dimensional measure of the rate of stretching to diffusion—is large enough. A number of local models have followed these initial calculations. The role of compressibility has been included in a number of models (see e.g. Stein and Nordlund 2002; Vögler and Schüssler 2007). It has been suggested that strong stratification can inhibit dynamo action (Stein and Nordlund 2002) although it appears that this is not the case, providing the magnetic Reynolds number is large enough. The role of shear, penetration and rotation in modifying small-scale dynamos has also been systematically investigated (Tobias et al. 2008). There is also still some doubt as to whether a small-scale dynamo can operate when the fluid Reynolds number Re is much larger than Rm, the so-called small magnetic Prandtl number limit—the appropriate limit for the solar interior. Although this issue is not settled (and is exceedingly difficult to settle via numerical computation, see e.g. Iskakov et al. 2007) the indications are that these small-scale dynamos can survive efficiently in this limit (Boldyrev and Cattaneo 2004; Tobias and Cattaneo 2008).

One consistent result that emerges from these simulations is that for high magnetic Reynolds number dynamos (i.e. ones for which $Rm \gg Rm_c$ with Rm_c the critical value) the field that is generated is dominated by small scales—even in the presence of rotation, although the simulations are capable of generating systematic fields when Rm is close to its critical value (see e.g. Brandenburg et al. 2008, Käpylä et al. 2008). This is an important result and has been discussed in detail (Cattaneo and Hughes 2006; Hughes and Cattaneo 2008). The apparent predominance of small-scale magnetic fields at high Rm in turbulent flows is a central problem of turbulent dynamo theory and is an issue that we shall return to in our discussion of large-scale field generation below.

2.3 Physical Effects and Approaches to Modelling the Solar Dynamo

For the rest of this article we discuss the current status of the theory for the generation of the systematic global solar field. The theory has developed from along a number of parallel strands, utilising a combination of basic theory, turbulence modelling and numerical computation. The range of approaches will be outlined below, before a detailed discussion of each in subsequent sections. In this section we discuss the basic physical effects that can lead to systematic dynamo action.

Conceptually it is convenient to discuss the dynamics of a global (large-scale) axisymmetric magnetic field. It is then possible to decompose the field (in spherical polar coordinates) into toroidal and poloidal parts so that $\boldsymbol{B} = B_\phi \boldsymbol{e}_\phi + \nabla \times (A\boldsymbol{e}_\phi)$; here $B_\phi(r,\theta)$ is the toroidal (zonal) field and $A(r,\theta)$ is the vector potential for the poloidal (meridional) field. It is well known that the presence of differential rotation $\boldsymbol{u}_\phi(r,\theta)$ naturally leads to the generation of B_ϕ from A—toroidal field is generated efficiently from poloidal field via gradients in angular velocity. This action of differential rotation is often termed the Ω-effect and is uncontroversial. If this were the only ingredient of the dynamo then the axisymmetric field would ultimately decay as demonstrated by Cowling (1933) in his famous anti-dynamo theorem. However in a landmark paper Parker (1955) argued that turbulent (small-scale) convective motions in the solar interior could produce small-scale (non-axisymmetric) magnetic fields. Furthermore, he argued that the net effect of the small-scale flows interacting with the small-scale magnetic fields could be so as to produce a net electromotive force (e.m.f.) that is capable of regenerating the axisymmetric poloidal field from the axisymmetric toroidal field. This non-trivial effect was formalised in the landmark paper of Steenbeck et al. (1966) which introduced mean-field electrodynamics. This formalism has been the basis of solar and stellar dynamo theory for more than forty years and has been investigated in detail (as described in the next section).

Although much attention has focused on the mean-field formalism (both in terms of examining the applicability and the consequences of the theory) other approaches to the dynamo problem have been investigated. The breathtaking advances in computational resources and algorithms have enabled the construction of global dynamo models which, although in parameter regimes many orders of magnitude away from the conditions inside a star, are beginning to give insights into the operation of stellar dynamos. These global models are reviewed in Sect. 5. A final approach is to use the techniques of nonlinear dynamics and bifurcation theory to investigate the mathematical structure of the underlying equations and to construct simple (low-order) models that lead to an understanding of the possible complicated dynamics of stellar dynamos. These models will also be discussed in Sect. 5.

3 Mean Field Dynamos: Theory

The subject of mean-field dynamo theory can be divided into two main areas. First, there is the theory itself, the conditions under which it is valid, its relationship to turbulence theory, and its extension to include nonlinear effects. Second, there are the applications of mean-field dynamo theory, the solar and stellar models that have been developed using the α-effect term in the induction equation, which is the essential new ingredient provided by the theory. There is surprisingly little interaction between these two areas of research. A vast body of work on α-effect dynamo models exists, using many different geometries, both with and without nonlinear terms in the equations, and with many different spatial locations of the α and Ω effects. In contrast, much less has been done on the fundamental assumptions underlying mean-field dynamo theory, so that it is still controversial to what extent the models using mean-field concepts can be relied upon. As we see below, there are some clearly defined limits where mean-field dynamo theory is certainly correct, but unfortunately these limits are very far from the conditions in the Sun, stars and galaxies where they are applied. This does not necessarily mean that mean-field dynamo theory gives wrong answers for astrophysical dynamos, but it does mean that such theories cannot be relied on very strongly. They may suggest possible scenarios for the behaviour of stellar and galactic dynamos, and provide a framework in which observations can be interpreted, but unlike Newton's laws of motion, they cannot provide definite predictions of future behaviour.

The theory developed in the 1960's by Krause, Rädler and Steenbeck (see e.g. Krause and Rädler 1980) was essentially a linear theory. Independently, Braginsky (1975) developed his theory of the nearly axisymmetric dynamo, which contains similar ideas, though based on a different averaging procedure. Vainshtein and Cattaneo (1992) questioned whether the classical mean-field theory could survive intact when nonlinear effects were included. They argued that because of the very high magnetic Reynolds number in most astrophysical objects, the α-effect would be quenched by the nonlinear effect of even rather small magnetic fields, thus throwing into doubt the relevance of the α-effect model in astrophysics. Subsequently, this negative result has been questioned, but the whole issue of the reliability of the α-effect model is still controversial. To date there has not been much support for mean-field dynamo theory from direct numerical simulations of the MHD equations. Proponents of mean-field theory correctly point out that these numerical simulations are not in the astrophysical parameter regime either, but nevertheless it remains a matter of concern that there is so little agreement between these two main lines of research into dynamo theory.

3.1 Averaging the Dynamo Equations

The basic idea of mean-field theory is to split the magnetic field and the flow into mean and fluctuating parts,

$$\mathbf{B} = \overline{\mathbf{B}} + \mathbf{B}', \qquad \mathbf{u} = \overline{\mathbf{u}} + \mathbf{u}'. \tag{6}$$

The mean part is thought of as the large scale average field and velocity, the fluctuating part is thought of as a small scale turbulent contribution. We also need an averaging procedure, discussed in more detail in Sect. 3.2 below. The fundamental property of the averaging is that the fluctuating quantities on their own average to zero, but products of fluctuating quantities can have a non-zero average. So

$$\overline{\mathbf{B}'} = \overline{\mathbf{u}'} = 0. \tag{7}$$

However the averaging process is defined, it must obey the averaging rules, that is the sum of the average of a number of terms must be the average of their sum, averaging commutes with differentiating both with respect to space and time, and averaging an averaged quantity leaves it unchanged. There are several different ways of defining the average. If the turbulence is small-scale and the mean field we are interested in is large-scale, then there is *scale separation*. We can average over an intermediate length-scale

$$\overline{F}(\mathbf{x}, t) = \int F(\mathbf{x} + \xi, t), g(\xi)\, d^3\xi, \qquad \int g(\xi)\, d^3\xi = 1. \tag{8}$$

We choose the weight function g to go to zero at large distances on the intermediate length scale, but g is constant over small distances, so fluctuations average out but mean field doesn't,

$$\int F'(\mathbf{x} + \xi, t)\, g(\xi)\, d^3\xi = 0, \qquad \int \overline{F}(\mathbf{x} + \xi, t) g(\xi)\, d^3\xi = \overline{F}. \tag{9}$$

Braginsky (1975) suggested another method of averaging, over a coordinate, usually the azimuthal angle ϕ in spherical systems. So then $\overline{\mathbf{B}}$ is just the axisymmetric part of the field. In Cartesian models, with gravity and possibly rotation in the z-direction, it is common to define the average over both the horizontal coordinates so that $\overline{\mathbf{B}}$ is a function of z only. This method of defining the average has become more popular recently, as it is suited to comparison with simulations. It is quite different from the scale separation method of defining the average; in general the nonaxisymmetric components of the flow will have a similar length-scale to the axisymmetric components.

We now insert (6) into the induction (3) and average to obtain the mean-field induction equation

$$\frac{\partial \overline{\mathbf{B}}}{\partial t} = \nabla \times (\overline{\mathbf{u}} \times \overline{\mathbf{B}}) + \nabla \times (\overline{\mathbf{u}' \times \mathbf{B}'}) + +\eta \nabla^2 \overline{\mathbf{B}}. \tag{10}$$

Note that the terms involving products of a mean and a fluctuating quantity average to zero, but crucially the term involving products of fluctuating quantities can have a non-zero mean part. The fluctuating field $\mathbf{B}'$ and the fluctuating turbulent part of the velocity $\mathbf{u}'$ can be systematically aligned to give a non-zero mean e.m.f.

$$\mathcal{E} = \overline{\mathbf{u}' \times \mathbf{B}'}. \tag{11}$$

The great advantage of this new mean e.m.f. term in the mean induction equation (10) is that Cowling's theorem no longer applies, so we can find simple axisymmetric dynamos as solutions. This releases us from the constraint of having to solve fully three-dimensional solutions of the dynamo equations, which invariably involve the construction of complex numerical computer programs, which is both difficult and time-consuming. In view of this, it is not surprising that most work on dynamo theory has used the mean field equations with a non-zero mean e.m.f. The mean e.m.f. term creates poloidal field from toroidal field, and hence closes the dynamo loop. We already have the differential rotation stretching out poloidal field into toroidal field, and the mean e.m.f. gives a mechanism for converting that toroidal field into new poloidal field. However, to exploit this we must have a rational way of determining what the mean e.m.f. actually is.

3.2 Evaluation of the Mean e.m.f.

If we subtract the mean-field equation (10) from the full equation (3), we obtain the equation for the fluctuating field $\mathbf{B}'$,

$$\begin{aligned} \frac{\partial \mathbf{B}'}{\partial t} &= \nabla \times (\overline{\mathbf{u}} \times \mathbf{B}') + \nabla \times (\mathbf{u}' \times \overline{\mathbf{B}}) + \nabla \times \mathcal{G} + \eta \nabla^2 \mathbf{B}', \\ \mathcal{G} &= \mathbf{u}' \times \mathbf{B}' - \overline{\mathbf{u}' \times \mathbf{B}'}. \end{aligned} \tag{12}$$

There are a number of different approaches to dealing with this equation. In the classical mean-field approach, Krause and Rädler (1980), we view $\mathbf{u}'$ as a given turbulent velocity independent of $\mathbf{B}'$ and $\overline{\mathbf{B}}$. In the Braginsky (1975) approach, the equation of motion is linearised, and $\mathbf{u}'$ and $\mathbf{B}'$ are solved simultaneously. It is also possible to make assumptions about the relation between $\mathbf{u}'$ and $\mathbf{B}'$ through the equation of motion, one such being the EDQNM model (Pouquet et al. 1976), referred to again in Sect. 3.6 below.

In the classical mean-field approach, $\mathbf{u}'$ is given so (12) is a linear equation in $\mathbf{B}'$ with a forcing term $\nabla \times (\mathbf{u}' \times \overline{\mathbf{B}})$. There are then two possibilities. Equation (12) may have non-trivial solutions when $\overline{\mathbf{B}} = 0$, in which we say there is a *small-scale dynamo*. The alternative is that there is no small-scale dynamo, in which case $\mathbf{B}' \to 0$ as $t \to \infty$ if $\overline{\mathbf{B}} = 0$. If this is the case, $\mathbf{B}'$ will be entirely created by the forcing from the action of the turbulence on the mean field. Then $\mathbf{B}'(\mathbf{x})$ will be proportional to $\overline{\mathbf{B}}$, but only if the turbulence has a short correlation length will $\mathbf{B}'(\mathbf{x})$ depend on only the local value $\overline{\mathbf{B}}(\mathbf{x})$, more generally it will be a non-local dependence, i.e. involving an integral containing $\overline{\mathbf{B}}(\mathbf{x}')$. However, if there is a short correlation length, we can evaluate $\mathbf{B}'$ and hence $\mathcal{E}$ as a Taylor series

$$\mathcal{E}_i = \mathcal{E}_i^0 + a_{ij}\overline{\mathbf{B}}_j + b_{ijk}\frac{\partial \overline{\mathbf{B}}_k}{\partial x_j} + \cdots, \tag{13}$$

where the tensors a_{ij} and b_{ijk} depend on $\mathbf{u}'$ and $\overline{\mathbf{u}}$ and the term $\mathcal{E}_i^0$ is driven by the small-scale dynamo, if any. This equation is really the essence of mean-field dynamo theory. As we see below, it usually further assumed that the tensors appearing in (13) are isotropic, and it is this equation which forms the basis of the vast majority of mean-field dynamo models.

Many assumptions have gone into deriving this key result (13). When are they likely to be valid? The first assumption was that $\mathbf{u}'$ is given. It is now widely recognised that actually it will be affected by $\mathbf{B}'$ and $\overline{\mathbf{B}}$. This leads to α-quenching, discussed in more detail in Sect. 3.6 below. If there is a small-scale dynamo, then $\mathbf{B}'$ can grow independently of any mean field, and will reach a level such that the small-scale Lorentz force affects the small-scale flow sufficiently to prevent further growth. This leads to the zeroth order term $\mathcal{E}_i^0$ in (13), whose possible form was discussed by Rädler (1976) and subsequently by Yoshizawa et al. (2000), see also references therein. In most applications, however, this zeroth order term is omitted.

In the original concept of mean-field dynamo theory, it was envisaged there would be scale separation, in which case the appropriate length scale for the fluctuating field is not the integral scale d of the whole dynamo region, but the much smaller turbulent correlation length, say ℓ. Now $Rm' = u'\ell/\eta$, and there is some hope that Rm' will be too small to allow a small-scale dynamo, but $Rm = ud/\eta$ is large enough to allow large-scale field to grow. Under these circumstances, $\mathbf{B}'$ is simply linearly forced by $\overline{\mathbf{B}}$, and mean-field theory with $\mathcal{E}_i^0 = 0$ will be valid. So when may we expect scale separation to occur? Unfortunately not in the Sun, because the observational evidence suggests that turbulence occurs in the form of 'granules', flows that last some minutes of time and with a spatial scale of 10^3 km, which would give a large Rm'; not as large as the Rm for the whole Sun, but still large enough to generate a small scale magnetic field by dynamo action. There is not much evidence of scale separation in simulations either, with the possible exception of very rapidly rotating convection (Stellmach and Hansen 2004). Here tall thin convecting columns of fluid can occur, in which the horizontal scale is significantly less than the vertical scale, but this is unlikely to be the case in the Sun, which rotates comparatively slowly. It is therefore not usually possible to justify applying mean-field dynamo theory to the solar dynamo in any rigorous sense. It may be that mean-field models still have relevance to the Sun even though scale separation does not occur, but we must treat any mean field results with caution. To be believable, there must be strong, independent evidence supporting such models.

3.3 Tensor Representation of $\mathcal{E}$

We now consider the possible forms of the tensors occurring in (13).

a_{ij} *tensor*: we split this into a symmetric part $(a_{ij} + a_{ji})/2 = \alpha_{ij}$ and the antisymmetric part, $(a_{ij} - a_{ji})/2 = \epsilon_{ijk}A_k$. Then from this tensor we obtain

$$\mathcal{E} = \alpha_{ij}\overline{\mathbf{B}}_j - \mathbf{A} \times \overline{\mathbf{B}}. \tag{14}$$

We already have a term $\overline{\mathbf{u}} \times \overline{\mathbf{B}}$ in the induction equation (10), so the $\mathbf{A}$ term just modifies the mean velocity. Along the principal axes, the symmetric part in general has 3 different components, but for isotropic turbulence

$$\alpha_{ij} = \alpha\delta_{ij}, \tag{15}$$

leading to the usual mean-field α-effect dynamo equation,

$$\frac{\partial \overline{\mathbf{B}}}{\partial t} = \nabla \times (\overline{\mathbf{u}} \times \overline{\mathbf{B}}) + \nabla \times \alpha \overline{\mathbf{B}} + + \eta \nabla^2 \overline{\mathbf{B}}. \tag{16}$$

b_{ijk} *tensor*: first we split the tensor $\partial_j \overline{\mathbf{B}}_k$ into symmetric and antisymmetric parts, so

$$\partial_j \overline{\mathbf{B}}_k = (\nabla \overline{\mathbf{B}})^s - \frac{1}{2} \epsilon_{jkm} (\nabla \times \overline{\mathbf{B}})_m. \tag{17}$$

The symmetric part is believed to be less important than the antisymmetric part. The antisymmetric part combines with b_{ijk} to give a second rank tensor $b_{ijk}\epsilon_{jkm}$ (the summation convention applies), and this second rank tensor has a symmetric and an antisymmetric part, giving

$$\mathcal{E}_i = -\beta_{ij} (\nabla \times \overline{\mathbf{B}})_j - \epsilon_{ijk} \delta_j (\nabla \times \overline{\mathbf{B}})_k. \tag{18}$$

The δ-effect term has been discussed (e.g. Krause and Rädler 1980), but most interest has focused on the β-effect, especially the case where the β_{ij} tensor is isotropic, $\beta_{ij} = \beta \delta_{ij}$, as it then gives a term identical in form to the diffusion term. We can now write the mean e.m.f. in the form most commonly used in mean-field theory

$$\mathcal{E} = \alpha \overline{\mathbf{B}} - \beta \nabla \times \overline{\mathbf{B}}. \tag{19}$$

If we further assume β is constant in space, we obtain

$$\frac{\partial \overline{\mathbf{B}}}{\partial t} = \nabla \times (\overline{\mathbf{u}} \times \overline{\mathbf{B}}) + \nabla \times \alpha \overline{\mathbf{B}} + (\eta + \beta) \nabla^2 \overline{\mathbf{B}}. \tag{20}$$

The β-term just enhances the diffusivity from a laminar value η to a turbulent value $\eta + \beta$. It is normally argued that $\beta \gg \eta$, so the diffusivity is greatly enhanced by the turbulence. Of course it would be nice to use this enhanced diffusivity to claim that the small scale turbulence does indeed have a low Rm', but this is scarcely a logical argument, since scale separation was invoked to explain the appearance of the β term in the first place.

3.4 Helicity and the α-Effect

The tensor approach to mean-field dynamo theory leads to a large number of unknown quantities for the tensor components, which is rather unsatisfactory as one might suspect that models with almost any desired property could be constructed by choosing these coefficients sufficiently carefully. A more physical approach (Moffatt 1978) is to assume that the turbulence is a random superposition of waves

$$\mathbf{u}' = Re\{\mathbf{u} \exp i(\mathbf{k} \cdot \mathbf{x} - \omega t)\}. \tag{21}$$

With short correlation lengths, the mean velocity term in the induction equation can be removed by working in moving frame to give

$$\begin{array}{cccc} \dfrac{\partial \mathbf{B}'}{\partial t} = (\overline{\mathbf{B}} \cdot \nabla) \mathbf{u}' + & \nabla \times (\mathbf{u}' \times \mathbf{B}' - \overline{\mathbf{u}' \times \mathbf{B}'}) & + \eta \nabla^2 \mathbf{B}', & \\ O(B'/\tau) \qquad O(\overline{B}u'/\ell) & O(B'u'/\ell) & O(\eta B'/\ell^2). & \end{array} \tag{22}$$

If the small-scale magnetic Reynolds number $u'\ell/\eta$ is small, the awkward curl term is negligible. This is the first order smoothing assumption,

$$\frac{\partial \mathbf{B}'}{\partial t} = (\overline{\mathbf{B}} \cdot \nabla)\mathbf{u}' + \eta \nabla^2 \mathbf{B}'. \tag{23}$$

This implies $\mathbf{B}' \ll \overline{\mathbf{B}}$, almost certainly not true in the Sun. Then inserting (21) into (23), we obtain

$$\mathbf{B}' = Re\left\{ \frac{i(\mathbf{k} \cdot \overline{\mathbf{B}})\mathbf{u}}{\eta k^2 - i\omega} \exp i(\mathbf{k} \cdot \mathbf{x} - \omega t) \right\}. \tag{24}$$

Now we can evaluate $\mathcal{E}$,

$$\mathcal{E} = \overline{\mathbf{u}' \times \mathbf{B}'} = \frac{1}{2} \frac{i\eta k^2 (\mathbf{k} \cdot \overline{\mathbf{B}})}{\eta^2 k^4 + \omega^2} (\mathbf{u}^* \times \mathbf{u}), \tag{25}$$

where $*$ denotes complex conjugate, equivalent to

$$\alpha_{ij} = \frac{1}{2} \frac{i\eta k^2}{\eta^2 k^4 + \omega^2} k_j \epsilon_{imn} u_m^* u_n. \tag{26}$$

For any particular wave, the α-tensor is not isotropic, but if we have a random collection of waves with **u** and **k** in random directions, the sum of all the contributions to the α-effect will be isotropic, so $\alpha_{ij} = \alpha \delta_{ij}$.

We now consider the helicity

$$H = \overline{\mathbf{u}' \cdot \nabla \times \mathbf{u}'} = \frac{1}{2} i\mathbf{k} \cdot (\mathbf{u}^* \times \mathbf{u}). \tag{27}$$

Taking the trace of (26) gives

$$\alpha = -\frac{1}{3} \frac{\eta k^2 H}{\eta^2 k^4 + \omega^2}. \tag{28}$$

So, under the first order smoothing approximation, the mean e.m.f. is proportional to the helicity of the turbulence. Since helicity is the scalar product of velocity and vorticity, a flow with non-zero helicity must consist of helical motion. If an element of fluid rises and rotates about the vertical as it does so, the flow has helicity. To obtain a non-zero α-effect, there must be a systematic bias to the rotation, for example if rising fluid in the northern hemisphere predominantly rotates clockwise, the net helicity is negative, so from (28), positive α is generated in the northern hemisphere. It is not required that every rising element of fluid in the northern hemisphere rotates clockwise, only that on average more rotate clockwise than anticlockwise. This type of asymmetry is most easily generated by the effect of rotation on the convection, and indeed rotating convection invariably produces non-zero helicity. It also usually produces helicity of opposite sign in the two hemispheres, so most models of the α-distribution make it antisymmetric across the equator.

Another way of relating the helicity to the mean e.m.f. is to assume that the correlation time of the turbulent fluctuating part is short, the so-called short sudden approximation. Now in (22) the time-derivative term just balances the first term on the right provided $u' < \ell/\tau$, and now we assume Rm' large, so that the diffusive term is also negligible. Again assuming the turbulence has no preferred direction, this leads to

$$\alpha = -\frac{\tau}{3} \overline{\mathbf{u}' \cdot \nabla \times \mathbf{u}'} \tag{29}$$

so again α is proportional to the helicity.

These results suggest that the value of α will be anticorrelated with helicity. However, they all assume either first-order smoothing or short sudden turbulence, neither of which is easy to justify in solar conditions. Courvoisier et al. (2006) looked numerically at the α produced by a variety of flows where neither of these assumptions applied, and found that there was no robust correlation of α with helicity. Instead, they found that α could even change sign as Rm' varied. It seems that the relationship of helicity and α is quite model dependent, and only in limits unlikely to apply in the Sun is the relation clear-cut. Schrinner et al. (2007) have also performed numerical simulations in the rapidly rotating case, and find that the α-tensor (14) is highly anisotropic.

3.5 Helicity and the Parker Loop Mechanism

Despite the difficulties in justifying mean field theory and the α-effect in the Sun, observations of sunspots and their associated magnetic active regions suggest that magnetic field is being brought to the surface by rising fluid coming up from below. A sunspot pair is created when an azimuthal loop of magnetic field rises through the solar photosphere. The vertical field impedes convection, reducing the heat transport and hence producing a relatively dark spot. Joy's law, based on sunspot observations over a long time, says that sunspot pairs are systematically tilted, with the leading spot being nearer the equator. Assuming flux was created as azimuthal flux deep down, this suggests that the loop has indeed twisted through a few degrees as it rose, and in a sense that implies negative helicity in the northern hemisphere. A rising loop of flux being twisted by helical motion is illustrated in Fig. 3a. Note that the induced current is antiparallel to the original magnetic field. Parker (1955) realised the possible significance of this result, and it has subsequently become incorporated into many dynamo models. Note that it is important that the amount of twist is small. If it is too great, as in Fig. 3b, the loop can be twisted through a large angle and the induced current no longer simply points in the 'correct' direction. The currents induced are no longer coherent, and will average out to give only a negligible net current. In the first order smoothing approximation, the $\mathbf{B}'$ field generated by the turbulence is always small compared to the original $\overline{\mathbf{B}}$ field so the twist is necessarily small. Possibly the same effect is achieved in the Sun by the field rising quite rapidly, so the field does not have time to rotate by more than a small angle before it emerges at the surface. This suggests that it might be possible for mean field theory to work in the Sun, even though the assumptions needed to guarantee its validity do not hold. It is though less clear that this mechanism can generate α throughout the convection zone, and indeed in many models, e.g. the Babcock-Leighton model (Leighton 1969) and the Parker (1993) interface dynamo, the regions where the Ω and α effects occur are spatially separated.

3.6 Nonlinear Effects and α-Quenching

The mean-field induction equation (20) is linear in $\overline{\mathbf{B}}$, so it predicts fields either grow or decay exponentially. To determine how the field saturates, and the field strength at which this occurs, we must include nonlinear effects. The only possible nonlinearity is the Lorentz force in the equation of motion (2), though there are a number of different ways in which this term could achieve magnetic saturation. Perhaps the simplest is that the Lorentz force could act to stop the differential rotation, the Ω-quenching mechanism. Indeed, simulations of convection in rotating fluids often show a large reduction in differential rotation once a dynamo generated magnetic field permeates the fluid. The magnetic interior of the gas giant Jupiter also appears to be rotating rather uniformly, in strong contrast to the non-magnetic outer regions where there are fast flowing jet flows, i.e. significant differential

Fig. 3 The Parker loop mechanism, after Moffatt 1978. Rising fluid with positive helicity lifts up a line of magnetic field and twists it. (a) With small twist, an induced current with a component antiparallel to the field is produced. (b) With large twist, the induced currents can be in any direction, cancellation will occur and the net induced current may be small

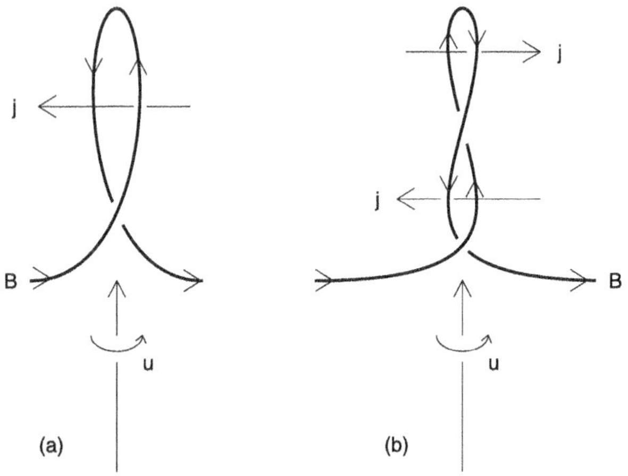

rotation. However, there is some doubt as to whether this mechanism operates in the Sun, as helioseismology has not found any eleven year cycle signal in the angular velocity near the tachocline (but there is evidence for a cycle-dependent variation in wave speed in that region: see Baldner and Basu 2008), where the Ω-effect is believed to be most effective. Another possible mechanism is buoyancy-quenching, nonlinear limiting through magnetic buoyancy. When the field gets to a certain strength, it rises up through the buoyancy effect (see e.g. Parker 1979) and emerges at the photosphere, where it gets destroyed. Yet another possibility is that the magnetic field has a rather subtle effect on the stretching properties of the fluid. Fast dynamo simulations (e.g. Galloway and Proctor 1992) suggest that field is created where neighbouring points in the fluid are separated rapidly by local shears in the fluid. Magnetic field linking such points is then rapidly stretched leading to strong field growth. If this field is reinforced by twisting or folding, this field growth can be sustained. However, the Lorentz force might reduce the stretching properties, or disrupt the folding and twisting necessary for the dynamo process. Unfortunately, these fluid properties can be hard to quantify. Two flows may appear quite similar when viewed in a numerical simulation, but one may be an efficient dynamo and the other give no dynamo action. So it is quite plausible that magnetic field may affect the flow field in such a way as to reduce any further field growth to zero. In the context of mean-field dynamos, this idea may be expressed as the magnetic field affecting the helicity of the flow, and hence reducing the effective value of α, so this mechanism is referred to as α-quenching.

The earliest models of α-quenching (Jepps 1975; Weiss et al. 1984) assumed that α could be written

$$\alpha = \frac{\alpha_0}{1 + \lambda \overline{\mathbf{B}}^2 / B_0^2}, \qquad B_0^2 = \mu \rho u^2. \tag{30}$$

B_0 is called the equipartition field strength, because it is the field strength at which the magnetic energy per unit volume equals the kinetic energy per unit volume. If the dimensionless parameter λ is taken as order unity, (30) gives a reasonable value for the magnetic field strength, though as there is still considerable uncertainty about the actual field strength in the deep interior, there is still substantial latitude in the estimates.

A more sophisticated estimate of how α might be affected by the Lorentz force was given by Pouquet et al. (1976), using a turbulence closure model known as the eddy-damped quasi-normal Markovian approximation (EDQNM). As with all turbulence closure models, there is no rigorous way of justifying such approximations, but they do use plausible assumptions about the statistical nature of the turbulence. Interestingly, they found (as did Gruzinov and Diamond 1994) that (29) is replaced by

$$\alpha = -\frac{\tau}{3}\left(\overline{\mathbf{u}' \cdot \nabla \times \mathbf{u}'} - \frac{1}{\rho}\overline{\mathbf{j}' \cdot \mathbf{B}'}\right). \tag{31}$$

This formula includes the Lorentz force into the expression for α so that nonlinearity directly reduces the effective α. Its range of validity has been discussed by Proctor (2003). Naturally, this expression for α again depends on some first order smoothing approximation or short sudden approximation, as it gives the usual anticorrelation of α and helicity in the weak field limit, questioned by Courvoisier et al. (2006).

In the early 1990's, Vainshtein and Cattaneo (1992) suggested that at large magnetic Reynolds numbers the parameter λ in (30) should be $O(Rm)$, not $O(1)$. As Rm is very large in the Sun, there is a huge difference between these estimates, the Vainshtein and Cattaneo estimate only allowing a very small mean field. Since there does seem to be a large reasonably large coherent mean magnetic field, as evidenced by the solar cycle and direct flux measurements, this very active α-quenching came to be known as *catastrophic* α-quenching. There has been much discussion of possible ways of avoiding the catastrophic α-quenching dilemma without abandoning the whole concept of mean-field dynamo theory. One way is to postulate that the regions where the α-mechanism operates are physically different from those where the Ω-effect occurs. Then it is possible that α operates where the fields are relatively small, so avoiding the strong quenching, but Ω operates in the strong field region near the tachocline where α-quenching is irrelevant. This is the basic idea of Parker's (1993) interface dynamo (see also Charbonneau and MacGregor 1997), and is also part of the philosophy of flux-transport dynamo models (Dikpati and Charbonneau 1995), though to avoid α-quenching, if it exists, very low values for the mean-field are required in the convection zone.

3.7 Magnetic Helicity and α-Quenching

How does the catastrophic α-quenching come about? Vainshtein and Cattaneo (1992) performed numerical simulations which suggested $\lambda \sim Rm$, but they also had a simple explanation of why this comes about. At large Rm fluid flow expels magnetic flux out of eddies, into the relatively stagnant regions between them. This is known as flux expulsion (Weiss 1966). Galloway et al. (1977) showed that the resulting flux ropes have thickness $Rm^{-1/2}$ times the integral length scale, and the field is much stronger in these ropes than in the mean field. The Lorentz force in the dynamo process acts when there is *local* equipartition between kinetic and magnetic energy, but this is when the locally very large fluctuating field is in equipartition with the kinetic energy. The mean field is then much smaller than the equipartition energy. The root of the problem is that at large Rm the fluctuating small scale field is actually much larger than the mean field, not smaller as in the low Rm first order smoothing scenario. It is when this small scale field generates significant Lorentz force that the dynamo saturates.

Further insight into the nature of catastrophic α-quenching through the concept of magnetic helicity was obtained by Gruzinov and Diamond (1994). Because $\nabla \cdot \boldsymbol{B} = 0$, we can write

$$\boldsymbol{B} = \nabla \times \mathbf{A}, \qquad \nabla \cdot \mathbf{A} = 0,$$

where $\mathbf{A}$ is called the vector potential. The condition $\nabla\cdot\mathbf{A}=0$ is called the Coulomb gauge, and is necessary to specify $\mathbf{A}$ because otherwise we could add on the grad of any scalar. We can compute $\mathbf{A}$ explicitly from $\boldsymbol{B}$ using the Biot-Savart integral,

$$\mathbf{A}(\mathbf{x}) = \frac{1}{4\pi}\int \frac{\mathbf{y}-\mathbf{x}}{|\mathbf{y}-\mathbf{x}|^3} \times \boldsymbol{B}(\mathbf{y})\, d^3y,$$

the integral being over all space. The induction equation in terms of $\mathbf{A}$ is then

$$\frac{\partial \mathbf{A}}{\partial t} = (\mathbf{u}\times\nabla\times\mathbf{A}) + \eta\nabla^2\mathbf{A} + \nabla\phi, \quad \nabla^2\phi = -\nabla\cdot(\mathbf{u}\times\boldsymbol{B}), \tag{32}$$

and the magnetic helicity is defined as

$$H_m = \int \mathbf{A}\cdot\boldsymbol{B}\, d^3x = \langle \mathbf{A}\cdot\boldsymbol{B}\rangle, \tag{33}$$

the integral being over the dynamo region, so the angle brackets denote the volume integral. The significance of magnetic helicity is that it is a conserved quantity in the non-dissipative limit. In objects with large Rm, magnetic helicity will not be conserved exactly, but one may expect that it can only change relatively slowly, on a diffusive timescale possibly enhanced by turbulence. Using some vector identities, we obtain

$$\left\langle \frac{D}{Dt}(\mathbf{A}\cdot\mathbf{B})\right\rangle = \int \nabla\cdot[\mathbf{B}(\phi+\mathbf{u}\cdot\mathbf{A})] + \nabla\cdot[\mathbf{A}\times(\eta\nabla\times\mathbf{B})]\, d^3x - 2\eta\mu\langle\mathbf{B}\cdot\mathbf{J}\rangle. \tag{34}$$

The first divergence term can usually be eliminated by defining ϕ suitably. The second divergence term gives a resistive surface term and can usually be ignored. So the total magnetic helicity is conserved if η is small (large Rm). The quantity $\mathbf{B}\cdot\mathbf{J}$ is called the current helicity, and it too is of interest. There have even been some recent attempts to measure its value on the solar surface (Sokoloff et al. 2008). In a steady state, (34) shows that the current helicity must average to zero.

We now split the magnetic helicity into its mean and fluctuating parts and perform the same manipulations as before (for details see e.g. Jones 2008, Sect. 5.3)

$$\frac{\partial}{\partial t}\overline{\mathbf{A}'\cdot\mathbf{B}'} = -2\overline{\mathbf{B}}\cdot\overline{\mathbf{u}'\times\mathbf{B}'} - 2\eta\mu\overline{\mathbf{j}'\cdot\mathbf{B}'} + \text{divergence terms} \tag{35}$$

so ignoring the divergence terms, which only involve surface effects, the small scale helicity also decays because of the current helicity, but there is an additional term involving the mean e.m.f. $\overline{\mathcal{E}} = \overline{\mathbf{u}'\times\mathbf{B}'}$. This allows transfer between the small scale and large scale magnetic helicities. Now using (19) in a steady state

$$\alpha\overline{\mathbf{B}}^2 - \mu\beta\overline{\mathbf{B}}\cdot\overline{\mathbf{J}} = -\mu\eta\overline{\mathbf{j}'\cdot\mathbf{B}'}. \tag{36}$$

There are two different ways of interpreting (36). If the length scale ℓ of the fluctuations is short compared to the mean field length scale d, the β term is $O(\ell/d)$ compared to the α term, so the β term can be ignored, and the balance is between the α term and the molecular diffusion term on the right-hand side. In this case α is limited by the molecular diffusivity unless $|\mathbf{B}'| \gg |\overline{\mathbf{B}}|$, the opposite of the normal assumption of mean field theory. Alternatively, one can argue that although the β term is nominally smaller than the α in mean-field theory, in practice the terms may be comparable, so the primary balance is between the α and β

terms in (36), with the diffusive term on the right-hand side being negligible. Then α is limited not by the very small inverse magnetic Reynolds number, but instead by ℓ/d (e.g. Brandenburg and Subramanian 2005).

If the Lorentz force limits α by reducing the helicity, we might also expect the turbulent diffusion to be reduced. This is called β-quenching, and it is currently an active research topic (e.g. Gruzinov and Diamond 1994; Petrovay and Zsargo 1998; Keating et al. 2008). These authors point out that if β quenching is significant, it is likely that β will be strongly anisotropic. In view of the discussion above, if β is strongly quenched, then we may expect α to be strongly quenched also.

4 Mean Field Dynamo Models

4.1 Dynamo Waves

The simplest and perhaps the most instructive solutions of the mean-field equations are the dynamo waves (Parker 1955). Since we are no longer constrained by Cowling's theorem and the related antidynamo theorems, growing waves in two dimensions are possible. We can therefore look for axisymmetric dynamos. We can even simplify further by adopting a local Cartesian model with the y-coordinate representing the azimuthal direction ϕ. Then

$$\mathbf{B} = (-\partial A/\partial z, B, \partial A/\partial x), \qquad \mathbf{u} = (-\partial\psi/\partial z, u_y, \partial\psi/\partial x) \tag{37}$$

and the induction equation can be split into a 'meridional' part

$$\frac{\partial A}{\partial t} + \frac{\partial(\psi, A)}{\partial(x, z)} = \alpha B + \eta\nabla^2 A, \tag{38}$$

and an 'azimuthal' part

$$\frac{\partial B}{\partial t} + \frac{\partial(\psi, B)}{\partial(x, z)} = \frac{\partial(A, u_y)}{\partial(x, z)} - \nabla\cdot(\alpha\nabla A) + \eta\nabla^2 B. \tag{39}$$

The simplest case is when $\psi = 0$, and α is constant, and the flow $u_y = \mathbf{U}'\cdot\mathbf{x}$, a constant shear. Then waves $A = \exp(\sigma t + i\mathbf{k}\cdot\mathbf{x})$ are possible, and the dispersion relation is

$$(\sigma + \eta k^2)^2 = i\alpha\hat{\mathbf{y}}\cdot\mathbf{U}'\times\mathbf{k} + \alpha^2 k^2. \tag{40}$$

If the shear is zero, and provided α is large enough to overcome diffusion, this gives growing and decaying stationary solutions known as the α^2 dynamo, while if the shear dominates and the α^2 term is comparatively small, there are also growing and decaying solutions known as the $\alpha\Omega$ dynamo. Naturally, the growing solutions dominate and eventually the field saturates due to some form of α-quenching as discussed above. The main solar magnetic field propagates towards the equator during the cycle, so almost all mean-field solar dynamo models are of $\alpha\Omega$ type as these give propagating waves rather than a steady dynamo. The dimensionless combination $D = \alpha|\mathbf{U}'|d^3/\eta^2$ where d is a typical size of the dynamo region (often taken as the depth of the convection zone) is called the dynamo number, and in confined geometry there is a critical D which must be exceeded for onset.

Note that it is only the component of shear perpendicular to the wavevector $\mathbf{k}$ that can drive an $\alpha\Omega$ propagating dynamo wave. So if we want dynamo waves propagating towards the equator, as observed in the solar cycle, we need a shear in the radial direction. Also, if the product of α and the shear has the wrong sign, the growing dynamo wave may propagate towards the poles rather than the equator. Actually, at high latitudes, there is some evidence of poleward waves, which might be explained in mean-field terms by a change in the sign of α near the equator, or alternatively by a change in the sign of the shear. The need for radial shear to give latitudinal wave propagation has focused attention on regions of the Sun which have a strong radial shear. There are two such regions, the tachocline and the shallow region immediately below the photosphere. Most models place the shear in the tachocline, but not all, see Brandenburg (2005). The theory is not changed greatly when we move into axisymmetric spherical shell geometry, and it is possible to get models with the main features of the solar dynamo provided the shear and the distribution of α are chosen appropriately.

4.2 Interface Dynamo

Parker (1993) suggested that the shear occurred in the tachocline and the α-effect in the convection zone. This separation is desirable because then the strong fields in the tachocline might not suppress the α-effect in the convection zone so much, and the tachocline is known to be a region of strong shear. It also seems broadly consistent with the twisting of rising flux tubes in the convection zone being responsible for the α-effect. Numerical models with plausible α-distributions, with e.g. a latitudinal dependence $\cos\theta \sin^2\theta$ (see e.g. Chan et al. 2008 for a recent interface model dynamo) can lead to models which give a solar-like butterfly diagram. This is encouraging, but generally the results to date do seem to be rather dependent on the form of α selected; for example Markiel and Thomas (1999) found that with a latitudinal dependence $\cos\theta$ and a differential rotation profile consistent with helioseismology, it was difficult to obtain a realistic butterfly diagram.

4.3 Flux Transport Dynamos

These models go back to the ideas of Babcock (1961) and Leighton (1969), and are similar to the interface dynamo concept, in that they are $\alpha\Omega$ dynamos in which the shear is in the tachocline while the α-effect is in the convection zone, and indeed may be due to the observed twisting of field in line with Joy's law. So far, this is quite similar to the interface dynamo, but the new feature of flux transport dynamos is that the poloidal flux is transported back to the tachocline by a meridional circulation rather than by diffusion alone (Dikpati and Charbonneau 1995). Such circulations do exist on the Sun. At least in the northern hemisphere, there is an observed poleward flow of up to 20 m s^{-1} and presumably there must be a return flow at depth from mass conservation. If the return flow occurs near the tachocline, it is likely to be considerably slower than this because of the large variation of density with depth in the Sun. The meridional flow also appears to be rather time-dependent. The most attractive feature of the flux-transport model is that the essential ingredients all have some observational basis; the shear in the tachocline is observed by helioseismology, the generation of poloidal field from toroidal field is seen in Joy's law, and some meridional circulation certainly exists. There are, however, some drawbacks to these models. The poloidal field generated by the observed process is small scale, and yet somehow it must be rather coherent and large-scale when it reaches the tachocline, to start the next cycle. It is not entirely clear how this happens, and indeed there are difficulties seeing how the rather slow downwelling meridional circulation can counteract magnetic buoyancy, which tends to make field concentrations rise rather than fall. Mathematical models of flux transport dynamos usually focus on the induction equation, and it is not yet clear whether they are fully consistent with the equation of motion. Nevertheless, they have been used to make predictions of the strength of the next solar cycle (Dikpati et al. 2006), and it will be of interest to see over the next few years how well these predictions compare with observations.

4.4 Tachocline Based Dynamos

The convection zone of the Sun is a highly turbulent region, and the nearer the photosphere, the faster the convection driven velocities get. Despite the hopes of mean field dynamo theory, it seems probable that large scale field gets shredded into small scale field by this process, and even though this shredded field may have a non-zero average, giving a mean poloidal field component, it is not so clear how this small scale field is eliminated to get the coherent field necessary for the shear in the tachocline to produce the rather coherent toroidal field that seems to emerge at the surface. If it were the case that the field emerging at the surface had both polarities, with one slightly greater than the other, this problem might not arise, but actually the emerging active regions are overwhelmingly of one sign at a particular time during the cycle. In the models, the coherent fields are produced through the action of a large turbulent diffusivity, but there is little evidence of this process at work on the solar surface.

In view of this difficulty, a number of authors (e.g. Spiegel and Weiss 1980) have proposed that dynamo generation occurs in the tachocline and lower convection zone only (see Tobias 2005). On this view, what we see on the surface is merely a rather scrambled view of this process as the generated field rises through the convection zone to the surface. Effects such as Joy's law and meridional circulation are then essentially irrelevant to the dynamo process but are merely secondary effects affecting the observed field only. The dynamo is then confined to the comparatively stable regions either just below or at the base of the convection zone, making it easier to understand how such coherent fields are generated.

On this picture we can make use of the tachocline shear to generate large toroidal fields as in the other models, but the poloidal field must be produced by some three-dimensional instability in the tachocline region. This could be stimulated either by magnetic buoyancy (Brandenburg and Schmitt 1998; Thelen 2000; Thelen and Cattaneo 2000) or a shear-driven mechanism (e.g. Cline et al. 2003). Such dynamos require simultaneous solution of the equation of motion and the induction equation, and are therefore mathematically more challenging than traditional mean field models. Perhaps for this reason they are comparatively unexplored. A potential problem for such models is that some diffusion is necessary for dynamos to work, although the amount of diffusion necessary can be quite small. In the stably stratified parts of the tachocline, turbulence may be strongly damped and so magnetic diffusion limited to a very low near molecular value. It is not known whether this is sufficient to produce coherent fields.

5 Alternative Approaches: Low-order models and Global Computational Models

As noted above, both the underlying theory and models of mean field theory are poorly constrained. This leads to the construction of a number of plausible scenarios for dynamo action in the Sun all of which can be made to fit the observational data. For this reason a number of other approaches to understanding the solar dynamo problem have been suggested. These range from attempting to understand the underlying mathematical properties of the nonlinear partial differential equations via the construction of low-order models through to the construction of global computational models of dynamo action in turbulently convecting, moderately rotating spherical shells. Both of these approaches, like the mean-field models above, have their strengths and weaknesses.

There is a strong tradition, dating back to the pioneering work of Lorenz (1963), of trying to understand the dynamics of complicated nonlinear partial differential equations (PDEs) using sets of ordinary differential equations (ODEs). For the case of the solar dynamo, these "low-order models" can not give any useful information about the mechanism for the generation of large-scale modes for the solar magnetic field—rather they do yield an understanding of the possible interactions of this magnetic field once it has been generated. In this manner they complement the other approaches that seek to understand how large-scale magnetic fields emerge from a turbulent background.

In particular low-order models have focused on the interaction of large-scale modes of different symmetries (say dipole and quadrupole symmetries) and also the interaction of the magnetic fields with large-scale flows (primarily driven by the Lorentz force in the momentum equation). Both of these interactions lead to modulation of the basic activity cycle in both amplitude and symmetry about the equator. For a comprehensive review of the observational evidence and theoretical understanding of modulational processes in solar and stellar dynamos see Tobias (2002) and for more details on the importance of intermittency see Weiss and Thompson (2009) or Spiegel (2009).

Deterministic low-order models demonstrate how nonlinear interactions can lead to modulation of cyclic activity. These models may be constructed in two ways. The first involves the truncation of the relevant partial differential equations (in this case either the mean-field equations or the full MHD equations). This approach demonstrates that the nature of the modulation can depend on the precise form of the nonlinearity of the original PDEs. For nonlinear interactions of the magnetic field with the differential rotation, the dynamics reduces to a complex generalisation of the Lorenz equations (Jones et al. 1985). If the magnetic field acts back on a dynamic alpha-effect then the structure of the low-order model is different (Ashwin et al. 1999; Covas et al. 2001). The method of truncation is open to the criticism that the dynamics may be dependent on the level of truncation used—this is certainly true of the Lorenz equations, which are derived as a truncation of the PDEs describing convection. An alternative is to derive low-order models using normal-form theory and the underlying symmetries of the equations. This has the advantage that the dynamics found is structurally stable and robust (at least close to the initial bifurcation). Both types of modulational interactions have been investigated within this framework. The interaction of magnetic modes with differential rotation is well-described by the normal form for the saddle-node/Hopf bifurcation (Tobias et al. 1995). Here the transition to chaos as the dynamo number is increased occurs via a sequence of Hopf and secondary Hopf bifurcations and the breakdown of a torus. The interaction of modes of different symmetries (dipole and quadrupole) can also lead to interesting nonlinear dynamics (Landsberg and Knobloch 1996). The interaction between these two mechanisms for modulation can be studied by combining these models in a sixth order system of ODEs (Knobloch et al. 1998). The behaviour of this model may be extremely complicated, but can be compared in detail with the model of the corresponding PDEs (Tobias 1996, 1997; Beer et al. 1998). In particular this model is able to explain the phenomenon of "flipping" where the symmetry of the magnetic field changes from dipolar to quadrupolar (or vice-versa) in a Grand Minimum.

Of course the information available from low-order modelling is limited. For information about the generational processes themselves computational models must be used. These fall largely into two main categories—local and global models. Local models are designed to reach extreme parameter regimes and to shed light on turbulent dynamo action and the utility (or otherwise) of mean-field electrodynamics. Some of these models have been described in Sect. 3, with reference to particular issues of mean-field theory. Global models are constructed to examine the generation of large-scale modes in a solar context. This approach is similar to that used so successfully for the geodynamo (and planetary dynamos), however the level of success for stellar dynamos is debatable. The reason, as noted right at the start of this paper, is of course the level of turbulence in stellar interiors. Computational models therefore have profound difficulty in reaching the correct parameter regime not only in the momentum equation (as is the case for the geodynamo) but also for the induction equation itself. A comprehensive review of the formulation and results of global models is beyond the scope of this paper; the interested reader should see Miesch and Toomre (2009) or Brun and Rempel (2009). Here we summarise the main findings of the models, which have grown in sophistication from the pioneering models of Gilman (1983). The models all examine the dynamo properties of convection in a rotating spherical shell. The initial Boussinesq models of Gilman were rapidly adapted to include the anelastic approximation (Glatzmaier 1985) which enabled some of the density contrast of stellar interiors to be included in the hydrodynamics. The advent of massively parallel machines has enabled such models to be pushed further into the regime of turbulence and this is the approach taken by Boulder group. The interesting results from these simulations may be summarised as follows. As predicted by theory and local computations, turbulent convection in a rotating shell is capable of producing strong small-scale magnetic fields. The higher Rm the greater the predominance of small-scales (Brun et al. 2004). If rotating turbulent convection is the only ingredient in the dynamo then it seems to be difficult to generate any systematic large-scale fields. However a large-scale field may be generated via the inclusion in the model of a relatively laminar tachocline with a strong shear (Browning et al. 2006). This large-scale field is dipolar and largely steady and is stronger than the small-scale fields that are seen in the convecting layer of the simulation. Further investigations with more realistic parameters are necessary, though it is clear that this approach used judiciously with advances in the underlying theory promises to enhance our understanding of the solar dynamo a great deal in the future.

References

J.A. Abreu, J. Beer, F. Steinhilber, S.M. Tobias, N.O. Weiss, For how long will the current grand maximum of solar activity persist? Geophys. Res. Lett. **35**, L20109 (2008). doi:10.1029/2008GL035442

P. Ashwin, E. Covas, R. Tavakol, Transverse instability for non-normal parameters. Nonlinearity **12**, 563–577 (1999)

H.W. Babcock, The topology of the Sun's magnetic field and the 22-year cycle. Astrophys. J. **133**, 572–587 (1961)

C.S. Baldner, S. Basu, Solar cycle related changes at the base of the convection zone. Astrophys. J. **686**, 1349–1361 (2008)

S. Basu, Effects of errors in the solar radius on helioseismic inferences. Mon. Not. R. Astron. Soc. **298**, 719–728 (1998)

S. Basu, H.M. Antia, Effects of diffusion on the extent of overshoot below the solar convection zone. Mon. Not. R. Astron. Soc. **269**, 1137–1144 (1994)

S. Basu, H.M. Antia, A study of possible temporal and latitudinal variations in the properties of the solar tachocline. Mon. Not. R. Astron. Soc. **324**, 498–508 (2001)

S. Basu, H.M. Antia, Changes in solar dynamics from 1995 to 2002. Astrophys. J. **585**, 553–565 (2003)

J.G. Beck, A comparison of differential rotation measurements. Solar Phys. **191**, 47–70 (1999)

J. Beer, S.M. Tobias, N.O. Weiss, An active Sun throughout the Maunder minimum. Solar Phys. **181**, 237–249 (1998)

S. Boldyrev, F. Cattaneo, Magnetic-field generation in Kolmogorov turbulence. Phys. Rev. Lett. **92**, 144501 (2004)

S.I. Braginsky, Nearly axisymmetric model of the hydromagnetic dynamo of the Earth. Geomagn. Aeron. **15**, 122–128 (1975)

A. Brandenburg, The case for a distributed solar dynamo shaped by near-surface shear. Astrophys. J. **625**, 539–547 (2005)

A. Brandenburg, D. Schmitt, Simulations of an alpha-effect due to magnetic buoyancy. Astron. Astrophys. **338**, L55–L58 (1998)

A. Brandenburg, D. Subramanian, Astrophysical magnetic fields and nonlinear dynamo theory. Phys. Rep. **417**, 1–209 (2005)

A. Brandenburg, K.-H. Rädler, M. Rheinhardt, P.J. Kapyla, Magnetic diffusivity tensor and dynamo. Astrophys. J. **676**, 740–751 (2008)

M.K. Browning, M.S. Miesch, A.S. Brun, J. Toomre, Dynamo action in the solar convection zone and tachocline: pumping and organization of toroidal fields. Astrophys. J. **648**, L157–L160 (2006)

A.S. Brun, M.S. Miesch, J. Toomre, Global-scale turbulent convection and magnetic dynamo action in the solar envelope. Astrophys. J. **614**, 1073–1098 (2004)

A.S. Brun, M. Rempel, Large scale flows in the solar convection zone. Space Sci. Rev. **144**, 151–173 (2009)

F. Cattaneo, On the origin of magnetic fields in the quiet photosphere. Astrophys. J. **515**, L39–L42 (1999)

F. Cattaneo, D.W. Hughes, Dynamo action in a rotating convective layer. J. Fluid Mech. **553**, 401–418 (2006)

K.H. Chan, X. Liao, K. Zhang, A three-dimensional multilayered spherical dynamic interface dynamo using the Malkus-Proctor formulation. Astrophys. J. **682**, 1392–1403 (2008)

P. Charbonneau, K.B. MacGregor, Solar interface dynamos II. Linear, kinematic models in spherical geometry. Astrophys. J. **486**, 502 (1997)

J. Christensen-Dalsgaard, M.J. Thompson, Observational results and issues concerning the tachocline, in *The Solar Tachocline*, ed. by D.W. Hughes, R. Rosner, N.O. Weiss (Cambridge University Press, Cambridge, 2007), pp. 53–85

J. Christensen-Dalsgaard, D.O. Gough, M.J. Thompson, The depth of the solar convection zone. Astrophys. J. **378**, 413–437 (1991)

J. Christensen-Dalsgaard, M.J.P.F.G. Monteiro, M.J. Thompson, Helioseismic estimation of the convective overshoot in the Sun. Mon. Not. R. Astron. Soc. **276**, 283–292 (1995)

K.S. Cline, N.H. Brummell, F. Cattaneo, Dynamo action driven by shear and magnetic buoyancy. Astrophys. J. **599**, 1449–1468 (2003)

A. Courvoisier, D.W. Hughes, S.M. Tobias, The alpha effect in a family of chaotic flows. Phys. Rev. Lett. **96**, 034503 (2006)

T.G. Cowling, The Magnetic Field of Sunspots. Mon. Not. R. Astron. Soc. **94**, 39–48 (1933)

E. Covas, R. Tavakol, P. Ashwin, A. Tworkowski, J.M. Brooke, In-out intermittency in PDE and ODE models. Chaos **11**, 404–409 (2001)

M. Dikpati, P. Charbonneau, A Babcock-Leighton flux transport dynamo with solar-like differential rotation. Astrophys. J. **518**, 508–520 (1995)

M. Dikpati, G. deToma, P.A. Gilman, Predicting the strength of solar cycle 24 using a flux transport dynamo-based tool. Geophys. Res. Lett. **33**, L05102 (2006)

M. Dikpati, P.A. Gilman, Flux-transport solar dynamos. Space Sci. Rev. **144**, 67–75 (2009)

J.A. Eddy, The Maunder minimum. Science **192**, 1189–1202 (1976)

Y. Fan, Magnetic fields in the solar convection zone. Living Rev. Solar Phys. **1** (2004). http://www.livingreviews.org/lrsp-2004-1

D.J. Galloway, M.R.E. Proctor, N.O. Weiss, Formation of intense magnetic fields near the surface of the Sun. Nature **266**, 686–689 (1977)

D.J. Galloway, M.R.E. Proctor, Numerical calculations of fast dynamos for smooth velocity fields with realistic diffusion. Nature **356**, 691–693 (1992)

P.A. Gilman, Dynamically consistent nonlinear dynamos driven by convection in a rotating spherical shell. II—Dynamos with cycles and strong feedbacks. Astrophys. J. Suppl. **53**, 243–268 (1983)

G.A. Glatzmaier, Numerical simulations of stellar convective dynamos. II—Field propagation in the convection zone. Astrophys. J. **291**, 300–307 (1985)

A.V. Gruzinov, P.H. Diamond, Self-consistent theory of mean-field electrodynamics. Phys. Rev. Lett. **72**, 1651–1653 (1994)

D.A. Haber, B.W. Hindman, J. Toomre, R.S. Bogart, R.M. Larsen, F. Hill, Evolving submerged meridional circulation cells within the upper convection zone revealed by ring-diagram analysis. Astrophys. J. **570**, 855–864 (2002)

G.E. Hale, On the probable existence of a magnetic field in sunspots. Astrophys. J. **28**, 315–343 (1908)

G.E. Hale, F. Ellerman, S.B. Nicholson, A.H. Joy, The magnetic polarity of sun-spots. Astrophys. J. **49**, 153–185 (1919)

R. Howe, J. Christensen-Dalsgaard, F. Hill, R.W. Komm, R.M. Larsen, J. Schou, M.J. Thompson, J. Toomre, Deeply penetrating banded zonal flows in the solar convection zone. Astrophys. J. **533**, L163–L166 (2000a)

R. Howe, J. Christensen-Dalsgaard, F. Hill, R.W. Komm, R.M. Larsen, J. Schou, M.J. Thompson, J. Toomre, Dynamic variations at the base of the solar convection zone. Science **287**, 2456–2460 (2000b)
D.W. Hughes, F. Cattaneo, The alpha-effect in rotating convection: size matters. J. Fluid Mech. **594**, 445–461 (2008)
A.B. Iskakov, A.A. Schekochihin, S.C. Cowley, J.C. McWilliams, M.R.E. Proctor, Numerical demonstration of fluctuation dynamo at low magnetic Prandtl numbers. Phys. Rev. Lett. **98**, 208501 (2007)
S.A. Jepps, Numerical models of hydromagnetic dynamos. J. Fluid Mech. **67**, 625–646 (1975)
C.A. Jones, Dynamo theory, in *Dynamos*, ed. by P. Cardin, L.F. Cugliandiolo (Ecole de Physique de les Houches, Elsevier, Amsterdam, 2008)
C.A. Jones, N.O. Weiss, F. Cattaneo, Nonlinear dynamos: a complex generalization of the Lorenz equations. Physica D **14D**, 161–176 (1985)
J. Jurčak, L.R. Bellot Rubio, Penumbral models in the light of Hinode spectropolarimetric observations. Astron. Astrophys. **481**, L17–L20 (2008)
P.J. Käpylä, M.J. Korpi, A. Brandenburg, Large-scale dynamos in turbulent convection with shear. Astron. Astrophys. **491**, 353–362 (2008)
S.R. Keating, L.J. Silvers, P.H. Diamond, On crossphase and the quenching of the turbulent diffusion of magnetic fields in two dimensions. Astrophys. J. Lett. **678**, L137 (2008)
E. Knobloch, S.M. Tobias, N.O. Weiss, Modulation and symmetry changes in stellar dynamos. Mon. Not. R. Astron. Soc. **297**, 1123–1138 (1998)
A.G. Kosovichev, Helioseismic constraints on the gradient of angular velocity at the base of the solar convection zone. Astrophys. J. **469**, L61–L64 (1996)
F. Krause, K.-H. Rädler, *Mean Field Magnetohydrodynamics and Dynamo Theory* (Pergamon Press, New York, 1980)
A.S. Landsberg, E. Knobloch, Oscillatory bifurcation with broken translation symmetry. Phys. Rev. E **53**, 3579–3600 (1996)
R.B. Leighton, A magneto-kinematic model of the solar cycle. Astrophys. J. **156**, 1–26 (1969)
E.N. Lorenz, Deterministic nonperiodic flow. J. Atmos. Sci. **20**, 130–141 (1963)
J.A. Markiel, J.H. Thomas, Solar interface dynamo profiles with a realistic rotation profile. Astrophys. J. **523**, 827–837 (1999)
M.S. Miesch, J. Toomre, Turbulence, magnetism, and shear in stellar interiors. Annu. Rev. Fluid Mech. **41**, 317–345 (2009)
H.K. Moffatt, *Magnetic Field Generation in Electrically Conducting Fluids* (Cambridge University Press, Cambridge, 1978)
M.J.P.F.G. Monteiro, J. Christensen-Dalsgaard, M.J. Thompson, Seismic study of overshoot at the base of the solar convective envelope. Astron. Astrophys. **283**, 247–262 (1994)
M. Ossendrijver, The solar dynamo. Astron. Astrophys. Rev. **11**, 287 (2003)
E.N. Parker, Hydromagnetic dynamo models. Astrophys. J. **122**, 293 (1955)
E.N. Parker, *Cosmical Magnetic Fields, their Origin and Activity* (Clarendon Press, Oxford, 1979)
E.N. Parker, A solar dynamo surface wave at the interface between convection and nonuniform rotation. Astrophys. J. **408**, 707 (1993)
E.N. Parker, Solar dynamo, in *Encyclopedia of Geomagnetism and Paleomagnetism*, ed. by D. Gubbins, E. Herrero-Bervera (Springer, Dordrecht, 2007), p. 178
K. Petrovay, J. Zsargo, On the validity of quasi-linear kinematic mean-field electrodynamics in astrophysical flows. Mon. Not. R. Astron. Soc. **296**, 245 (1998)
A. Pouquet, U. Frisch, J. Léorat, Strong MHD helical turbulence and the nonlinear dynamo effect. J. Fluid Mech. **77**, 321–354 (1976)
M.R.E. Proctor, Dynamo processes: the interaction of turbulence and magnetic fields, in *Stellar Astrophysical Fluid Dynamics*, ed. by M.J. Thompson, J. Christensen-Dalsgaard (Cambridge university Press, Cambridge, 2003)
K.-H. Rädler, Mean-field magnetohydrodynamics as a basis of solar dynamo theory, in *IAU Symposium 1971*, ed. by V. Bumba, J. Kleczek (Dordrecht-Holland, Dordrecht, 1976), p. 323
S. Régnier, C.E. Parnell, A.L. Haynes, A new view of quiet-Sun topology from Hinode/SOT. Astron. Astrophys. **484**, L47–L50 (2008)
M. Rempel, Overshoot at the base of the solar convection zone: a semianalytical approach. Astrophys. J. **607**, 1046–1064 (2004)
J.C. Ribes, E. Nesme-Ribes, The solar sunspot cycle in the Maunder minimum AD-1645 to AD-1715. Astron. Astrophys. **276**, 549–563 (1993)
A. Serebryanskiy, D.-Y. Chou, Comparison of solar cycle variations of solar p-mode frequencies from GONG and MDI. Astrophys. J. **633**, 1187–1190 (2005)
M. Schrinner, K.H. Rädler, D. Schmitt, M. Rheinhardt, U.R. Christensen, Mean-field concept and direct numerical simulations of rotating magnetoconvection and the geodynamo. Geophys. Astrophys. Fluid Dyn. **101**, 81–116 (2007)

D. Sokoloff, H. Zhang, K.M. Kuzanyan, V.N. Obridko, D.N. Tomin, V.N. Tutubalin, Current helicity and twist as two indicators of the mirror asymmetry of solar magnetic fields. Solar Phys. **248**, 17–28 (2008)
E.A. Spiegel, Chaos and intermittency in the solar cycle. Space Sci. Rev. **144**, 25–51 (2009)
E.A. Spiegel, N.O. Weiss, Magnetic activity and variations in solar luminosity. Nature **287**, 616 (1980)
M. Steenbeck, F. Krause, K.-H. Rädler, A calculation of the mean electromotive force in an electrically conducting fluid in turbulent motion under the influence of Coriolis forces. Z. Naturforsch. **21a**, 369–376 (1966)
R.F. Stein, A. Nordlund, Solar surface magneto-convection and dynamo action, in *SOLMAG 2002. Proceedings of the Magnetic Coupling of the Solar Atmosphere Euroconference and IAU Colloquium 188*, 11–15 June 2002, Santorini, Greece, ed. by H. Sawaya-Lacoste. ESA SP, vol. 505 (ESA Publications Division, Noordwijk, 2002), pp. 83–89
S. Stellmach, U. Hansen, Cartesian convection driven dynamos at Low Ekman number. Phys. Rev. E. **70**, 056312:1–16 (2004)
J.-C. Thelen, Non-linear α–ω-dynamos driven by magnetic buoyancy. Mon. Not. R. Astron. Soc. **315**, 165–183 (2000)
J.-C. Thelen, F. Cattaneo, Dynamo action driven by convection; the influence of magnetic boundary conditions. Mon. Not. R. Astron. Soc. **315**, L13–L17 (2000)
J.H. Thomas, N.O. Weiss, *Sunspots and Starspots* (Cambridge University Press, Cambridge, 2008)
M.J. Thompson, J. Christensen-Dalsgaard, M.S. Miesch, J. Toomre, The internal rotation of the Sun. Annu. Rev. Astron. Astrophys. **41**, 599–643 (2003)
S.M. Tobias, Grand minima in nonlinear dynamos. Astron. Astrophys. **307**, L21 (1996)
S.M. Tobias, The solar cycle: parity interactions and amplitude modulation. Astron. Astrophys. **322**, 1007–1017 (1997)
S.M. Tobias, Modulation of solar and stellar dynamos. Astron. Nachr. **323**, 417–423 (2002)
S.M. Tobias, The solar tachocline: Formation, stability and its role in the solar dynamo, in *Fluid Dynamics and Dynamos in Astrophysics and Geophysics*, ed. by A.M. Soward, C.A. Jones, D.W. Hughes, N.O. Weiss (CRC Press, Boca Raton, 2005), p. 193
S.M. Tobias, The Solar Dynamo: The role of penetration, rotation and shear on convective dynamos. Space Sci. Rev. **144**, 77–86 (2009)
S.M. Tobias, N.O. Weiss, V. Kirk, Chaotically modulated stellar dynamos. Mon. Not. R. Astron. Soc. **273**, 1150–1166 (1995)
S.M. Tobias, F. Cattaneo, Dynamo action in complex flows: the quick and the fast. J. Fluid Mech. **601**, 101–122 (2008)
S.M. Tobias, F. Cattaneo, N.H. Brummell, Dynamo action with penetration, rotation and shear. Astrophys. J. **685**, 596 (2008)
S.I. Vainshtein, F. Cattaneo, Nonlinear restrictions on dynamo action. Astrophys. J. **393**, 165–171 (1992)
S.V. Vorontsov, J. Christensen-Dalsgaard, J. Schou, V.N. Strakhov, M.J. Thompson, Helioseismic measurement of solar torsional oscillations. Science **296**, 101–103 (2002)
A. Vögler, M. Schüssler, A solar surface dynamo. Astron. Astrophys. **465**, L43–L46 (2007)
N.O. Weiss, The expulsion of magnetic flux by eddies. Proc. R. Soc. Lond. A **293**, 310 (1966)
N.O. Weiss, M.J. Thompson, The solar dynamo. Space Sci. Rev. **144**, 53–66 (2009)
N.O. Weiss, F. Cattaneo, C.A. Jones, Periodic and aperiodic dynamo waves. Geophys. Astrophys. Fluid Dyn. **30**, 305–341 (1984)
A. Yoshizawa, H. Kato, N. Yokoi, Mean field theory interpretation of solar polarity reversal. Astrophys. J. **537**, 1039–1053 (2000)

Space Sci Rev (2010) 152: 617–649
DOI 10.1007/s11214-009-9573-y

Dynamo Models for Planets Other Than Earth

Sabine Stanley · Gary A. Glatzmaier

Received: 6 May 2009 / Accepted: 4 August 2009 / Published online: 28 October 2009

Abstract Observations from planetary spacecraft missions have demonstrated a spectrum of dynamo behaviour in planets. From currently active dynamos, to remanent crustal fields from past dynamo action, to no observed magnetization, the planets and moons in our solar system offer magnetic clues to their interior structure and evolution. Here we review numerical dynamo simulations for planets other than Earth. For the terrestrial planets and satellites, we discuss specific magnetic field oddities that dynamo models attempt to explain. For the giant planets, we discuss both non-magnetic and magnetic convection models and their ability to reproduce observations of surface zonal flows and magnetic field morphology. Future improvements to numerical models and new missions to collect planetary magnetic data will continue to improve our understanding of the magnetic field generation process inside planets.

Keywords Planetary magnetic fields · Planetary dynamos · Numerical dynamo models

PACS 91.25.Cw · 91.25.Za · 96.12.Hg · 96.12.Pc · 96.15.Gh · 96.15.Nd

1 Introduction

In the past four decades, magnetometers have flown on missions to every planet in the solar system. Magnetic field observations have demonstrated the diversity of planetary magnetic fields and provided vital information on planetary interiors. Earth, Jupiter, Saturn, Uranus, Neptune, Ganymede and most likely Mercury have active dynamos generating magnetic

S. Stanley (✉)
Department of Physics, University of Toronto, 60 St. George St., Toronto, ON, M5S1A7, Canada
e-mail: stanley@physics.utoronto.ca

G.A. Glatzmaier
Earth and Planetary Sciences Department, Earth and Marine Sciences Building,
University of California, Santa Cruz, CA 95064, USA
e-mail: glatz@es.ucsc.edu

fields. Earth, Moon, Mars, and possibly Mercury, have remanent crustal fields due to dynamo action in their ancient pasts. For more details on planetary magnetic fields, we refer the reader to papers by Hulot et al., Langlais et al., Dougherty and Russell, Jia et al., and Anderson in this issue.

Over the past two decades, numerical dynamo simulations have been used to investigate planetary magnetic fields. Although the majority of these simulations were investigations of Earth's dynamo, studies of other planetary dynamos have flourished in recent years. In this paper, we review dynamo models for planets other than Earth. We discuss the differences in planetary magnetic fields which modelers attempt to explain as well as the differences in planetary interiors which need to be included in the models. We will concentrate on the aspects of dynamo modeling that are pertinent to our objectives here. For further details on the theory and modeling of dynamos, we refer the reader to the papers by Wicht and Tilgner and by Christensen in this issue.

Numerical dynamo models discretize and evolve the equations governing magnetic field generation in an electrically conducting fluid spherical shell. The system is governed by the following equations:

- Conservation of mass:

$$\frac{D\rho}{Dt} = -\rho(\nabla \cdot \mathbf{u}) \tag{1}$$

  where ρ is density, D/Dt is the Lagrangian derivative, and $\mathbf{u}$ is velocity.
- The magnetic induction equation:

$$\frac{\partial \mathbf{B}}{\partial t} = \nabla \times (\mathbf{u} \times \mathbf{B}) - \nabla \times (\lambda \nabla \times \mathbf{B}) \tag{2}$$

  where $\mathbf{B}$ is magnetic field, and $\lambda = (\sigma\mu_0)^{-1}$ is the magnetic diffusivity where σ is electrical conductivity and μ_0 is the magnetic permeability.
- The momentum equation:

$$\rho\frac{D\mathbf{u}}{Dt} + 2\rho\boldsymbol{\Omega} \times \mathbf{u} = -\nabla P + \rho\mathbf{g} + \mathbf{J} \times \mathbf{B} + \nabla \cdot \tau \tag{3}$$

  where $\boldsymbol{\Omega}$ is the angular velocity of the rotating frame of reference, P is pressure, $\mathbf{g}$ is the gravitational acceleration, $\mathbf{J}$ is current density, and τ is the deviatoric stress tensor.
- A prescribed equation of state (EOS):

$$\rho = \rho(T, P) \tag{4}$$

  where T is temperature. The EOS is typically chosen to be the liquid EOS for the fluid cores of terrestrial planets and the perfect gas EOS for the atmospheres and shallow interiors of giant planets.
- The energy equation:

$$\rho C_p \frac{DT}{Dt} - \alpha_T T \frac{DP}{Dt} = \frac{J^2}{\sigma} + \tau : \nabla\mathbf{u} + H - \nabla \cdot \mathbf{q} \tag{5}$$

  where C_P is the specific heat at constant pressure, α_T is the coefficient of thermal expansion, H is the volumetric heat sources, and $\mathbf{q}$ is heat flux.

Aside from assuming a single component fluid, we have kept the equations quite general. All planetary dynamo models solve these equations, but different approximations, boundary conditions, stability and parameter regimes may be used in simulating the different planets.

The timescales associated with fast acoustic and seismic waves in the core are not important for dynamo processes. Therefore, models usually filter out these waves by making either the Boussinesq or anelastic approximation, both of which take $\partial\rho/\partial t = 0$ in the conservation of mass equation and instead, solve for the evolution of density via (3)–(5). The Boussinesq approximation goes further by assuming a constant density except for in the buoyancy term and also assuming that density is only a weak function of temperature. The Boussinesq approximation is technically only valid for planets where the density scale height is much larger than the depth of the convective shell being considered. It is relatively acceptable for the terrestrial planets. For example, the largest terrestrial planet, Earth, has $r_o \approx 0.4H_T$ where $H_T = C_P/\alpha_T g$ is the average density scale height in the outer fluid core and r_o is the core radius.

For the giant planets, the anelastic approximation is the better approximation since the convective shell thicknesses can be much larger than the scale height. However, some giant planet dynamo models use the Boussinesq approximation as a simplification if they are interested in understanding certain mechanisms or features of the magnetic field that may not be a direct consequence of the density stratification of the fluid. The local density scale height is also a strong function of radius, being orders of magnitude larger in the deep interior than near the surface of a giant planet. Therefore, it can be argued that the Boussinesq approximation is justified in the very deep interior of a giant planet.

In the following sections, we discuss various planetary dynamo models. Section 2 deals with the two terrestrial planets other than Earth for which we have evidence of current or past dynamo action. The dynamo models for these planets are usually concerned with explaining anomalous (i.e. non-Earth-like) magnetic observations and the solutions usually appeal to some combination of core geometry, convective stability or mantle influence. Because we have no observations of time variation or core fluid flows for these terrestrial planets, the only core properties we can compare are magnetic field strength and morphology. Luckily, these two characteristics present enough puzzles to keep dynamo modelers busy. In Sect. 3 we review both non-magnetic and magnetic convection models for the giant planets. These planets provide us with additional model constraints in the form of observations of fluid flows in the outer layers of the planet. In Sect. 4 we discuss the known satellite dynamos and in Sect. 5 we conclude with a discussion of the current status and future of planetary dynamo modeling.

2 Terrestrial Planet Dynamos

The four terrestrial planets present a complete spectrum of dynamo states. Earth has a presently active dynamo generating an axial-dipole dominated field and paleomagnetic studies demonstrate that the dynamo has been around for at least the last three billion years. In contrast, Earth's sister planet Venus does not presently have an active dynamo, nor any evidence for remanent crustal fields implying dynamo action in its past, although better data will be required to completely rule out crustal magnetism on Venus. It is surprising that these two planets, so similar in internal structure and composition, present such different dynamo states. This difference is most likely due to the planets' abilities to cool their cores, clearly demonstrating the importance of the mantle in driving the dynamos.

The two smaller terrestrial planets, Mercury and Mars, have evidence for dynamo action sometime in their histories (Mercury currently and Mars in its early history). Both planets present magnetic mysteries which dynamo modelers must address and we discuss each planet individually below.

2.1 Mercury

Mercury's magnetic field was observed during the first and third flybys of the Mariner 10 mission in the mid-1970's (Ness et al. 1975, 1976). Recent flybys of Mercury by the MESSENGER spacecraft have confirmed these measurements and fit the observations with a dipole moment of $\sim$ 230–290 nT-R_M^3, where R_M is Mercury's radius (Anderson et al. 2008). Little information on Mercury's magnetic field morphology is presently available aside from the dipole moment and possibly, the quadrupole moment, but more constraints are likely once MESSENGER begins taking measurements from orbit in 2011. Libration observations demonstrate that Mercury's core is at least partially liquid, suggesting that the magnetic field may be the result of an active dynamo (Margot et al. 2007). Another possibility, that the field is a remanent crustal field, requires lateral inhomogeneities in the crust (Stephenson 1975; Srnka 1976; Aharonson et al. 2004). Due to the purpose of this article, we will only consider the active dynamo possibility here.

If we assume Mercury's dynamo works much like Earth's dynamo, we can estimate the expected strength of the resulting magnetic field in the core. Independent scalings based on energetics and magnetostrophic balance suggest that Mercury's dynamo would produce a core magnetic field intensity of the order 10^5–10^7 nT (Stevenson 1987; Schubert et al. 1988). Assuming an Earth-like field partitioning between toroidal and poloidal magnetic fields, this implies Mercury's surface field should be of order 4×10^3–4×10^6 nT, much stronger than the observed value of $\sim$ 260 nT.

Dynamo models for Mercury have focused on explaining the weakness of the observed field. Some reasoning or mechanism is proposed for why Mercury's dynamo should be non-Earth-like, and models are constructed to demonstrate that the proposed mechanism can produce a weaker Mercury-like surface field.

Several models evoke core geometry in their reasoning. Stanley et al. (2005) demonstrate that dynamos operating in thin shells with sufficiently low Rayleigh numbers such that convection only onsets outside the tangent cylinder can produce weak surface fields. In these models, toroidal fields are maintained strong throughout the core by differential rotation between the inner and outer cores, but poloidal fields are inefficiently regenerated by the weak convection in the limited area outside the tangent cylinder (Fig. 1(a)). This results in a weaker observed surface field. This model allows the generation of a strong field dynamo, but a non-Earth like field partitioning between poloidal and toroidal fields, thereby explaining why the observed (poloidal) fields are weaker than anticipated.

Takahashi and Matsushima (2006) have also published thin-shell dynamo models that produce weak dipole surface fields, but the reason for the weak fields is different. In their models, when a sufficiently large Rayleigh number is used such that convection is vigorous inside the tangent cylinder, smaller-scale non-dipolar fields dominate in the core. Because the power in smaller-scale modes decreases with distance faster than larger-scale modes, the small scale modes are much weaker at the surface and the observed surface field is dominated by its weak dipole component (Fig. 1(b)). Although this appears to be a promising mechanism, one issue with the models is that the inner core is made electrically insulating. The authors state that when they use conducting inner cores in their models, stronger dipolar fields result, suggesting that Mercury may not be explainable with this mechanism.

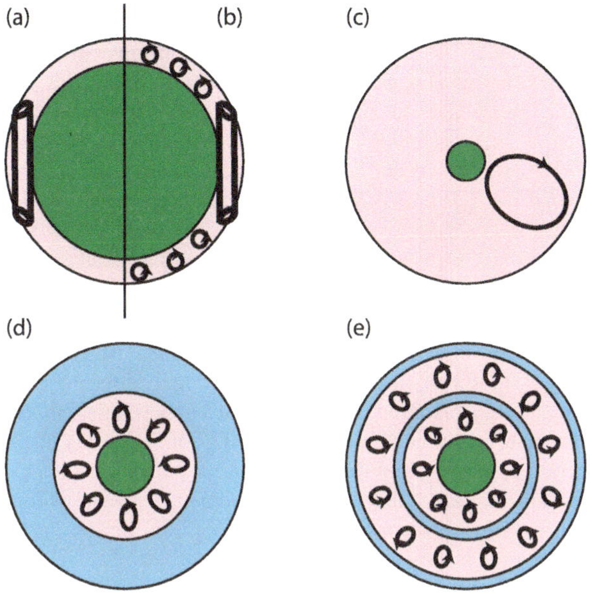

Fig. 1 Mercury dynamo model geometries. (**a**): Stanley et al. (2005) thin shell model with convection only outside the tangent cylinder, (**b**) Takahashi and Matsushima (2006) thin shell model with convection throughout the core, (**c**) Heimpel et al. (2005a) very thick shell model with isolated convection plume, (**d**) Christensen (2006) surrounding stable layer model, (**e**) Vilim et al. (2008) double dynamo model. A meridional slice is shown in (**a**) and (**b**) where the vertical line is the rotation axis, (**c**–**e**) are equatorial slices. The solid inner core is shown in *green*, stably-stratified layers are shown in *blue* and convectively unstable layers are shown in *pink*. In (**a**) and (**b**), long vertical cylinders represent the convection rolls outside the tangent cylinder whereas smaller circles represent the more 3-D convection pattern inside the tangent cylinder. In (**c**)–(**e**) slices of the convection rolls are represented as ellipses

At the other extreme in geometry, Heimpel et al. (2005a) rely on a very thick shell geometry to explain Mercury's weak observed field. Their numerical dynamo simulations which incorporate a very small solid inner core result in a single-plume mode of convection. This weaker convection generates weaker poloidal magnetic field explaining Mercury's weak observed surface field (Fig. 1(c)).

Other models for Mercury's weak surface field employ stable stratification (Christensen 2006; Christensen and Wicht 2008). Observations of Mercury's surface heat flow suggest it is possible that the outer portion of Mercury's core is thermally stratified and hence stable to convection. Numerical models which employ a thick outer stable layer in the core produce a weak surface field even though the field in the deep dynamo source region is strong (Fig. 1(d)). This is because the surrounding stable layer acts to attenuate the field through the skin effect, preferentially removing smaller-scale fields. Due to the slower rotation rate in Mercury, scaling laws suggest that Mercury's dynamo should be in the non-dipolar regime. Therefore, these models are affected both by the faster decay of the smaller scale features (since the dynamo operates deeper in the core) and the skin effect due to the stable layer.

Vilim et al. (2008) presents another Mercury model which relies on a non-Earth-like core stratification. Recent experiments have demonstrated that the Fe-S system produces non-ideal behaviour at Mercury-like pressures and temperatures for sulfur concentrations in the range 7–12 wt% (Chen et al. 2008). This non-ideal behaviour results in the dissolution of Fe from S at various locations in the core depending on the sulfur concentration. For example, it is possible that Mercury's core is being driven by the freezing out of an Fe snow at either the outer boundary of the core or mid-way through the core, or both. If the layer is at mid-depth, then the freezing out of Fe also results in the release of a more buoyant S rich fluid above the layer which can also drive convection. Dynamo models operating in this geometry demonstrate that it is possible to reproduce Mercury's weak magnetic field through this mechanism (Fig. 1(e)). The two dynamo regions (one above and one below the mid-freezing layer) can drive dynamos that produce fields of opposite sign thereby diminishing the observed surface field strength.

Another model proposes a magnetospheric feedback mechanism to explain Mercury's weak surface field (Grosser et al. 2004; Glassmeier et al. 2007a, 2007b; Heyner et al. 2009). Because Mercury has a small magnetosphere (owing to its weak magnetic field) and is deeply embedded in the solar wind (owing to its proximity to the Sun), magnetospheric currents can generate significant magnetic fields affecting dynamo action in Mercury's core. Chapman-Ferraro currents, resulting from interactions between the internal dynamo and the solar wind, can generate an ambient field of order 50 nT at Mercury's surface. Kinematic dynamo models show that this ambient field is opposite in polarity to the field generated in the core, resulting in cancellation of field in the core. This mechanism is termed a "feedback" because the small magnetosphere results in these opposite polarity ambient fields which results in perpetuating the weak magnetosphere.

A thermoelectric dynamo was an early proposed alternative to the normal dynamo explanation for Mercury (Stevenson 1987; Giampieri and Balogh 2002). In this model, topography on the core-mantle boundary results in a poloidal thermoelectric current and associated toroidal field. Helical motions due to weak convection then act on the produced toroidal field to generate the weak observed poloidal field. Since this model does not rely on convection alone to drive the dynamo, it doesn't suffer from the same large scaling estimates for the intensity of the produced field. However, it does require good electrical conductivity of the lower mantle for the thermocurrents to close. It would be interesting to see numerical simulations of a thermoelectric dynamo to determine the feasibility of explaining Mercury's magnetic field.

When Mercury's magnetic field was first observed, it was difficult to explain the field through a dynamo mechanism. It now appears that several mechanisms are possible, so the current problem is distinguishing which mechanism is operating in Mercury. Determining Mercury's inner core size from upcoming observations will be challenging and so it will be difficult to eliminate potential dynamo mechanisms by gathering independent (not related to magnetic field) data on Mercury's core geometry. However, it may be possible to distinguish between the different mechanisms by examining characteristic features of the resulting magnetic field such as the observed surface magnetic power spectrum, locations of strong flux spots, and secular variation. Future data from the MESSENGER (Zuber et al. 2007) and Bepi-Colombo (Wicht et al. 2007) missions may therefore be able to distinguish between the models.

The dynamo models make various predictions for the surface magnetic power spectrum. The Stanley et al. (2005) thin shell model and Heimpel et al. (2005a) thick shell model produce significant power in the dipole and quadrupole components whereas the Vilim et al. (2008) double-dynamo model produces larger octupole power compared to the quadrupole mode. The Takahashi and Matsushima (2006) model produces strong power in several non-dipolar modes, whereas the Christensen (2006) model has little power in any mode but the dipole and quadrupole. The thermoelectric dynamo power spectrum depends on the scale of topography on the CMB.

Information may also be gathered by the location and secular variation of small-scale flux spots. Small scale intense flux spots are limited to equatorial regions in the Stanley et al. (2005) and Vilim et al. (2008) models. In contrast, the Heimpel et al. (2005a) thick shell model produces strong normal flux spots at higher latitudes. These models also experience secular variation. The Christensen (2006) model suggests the fields should have little energy in higher harmonics and very slow secular variation due to the skin effect of the surrounding stable layer.

2.2 Mars

Mars' remanent crustal magnetic field, globally mapped by Mars Global Surveyor (Acuna et al. 1999), is most likely the result of an active dynamo in Mars' early history. Several mysteries surround the field including its hemispheric difference in field strengths, the magnetic carriers in the rocks and the driving force and timing of the dynamo. It is most likely that the dynamo was active sometime between 4.5–3.9 Ga since the large impact basins from the late-heavy bombardment (~ 3.9 Ga) are demagnetized, the majority of Tharsis (< 3.9 Ga) is demagnetized, and the ancient Martian meteorite ALH84001 has a magnetization age of 3.9–4.1 Ga (Weiss et al. 2002). Thermal evolution models have demonstrated that Mars could have maintained a super-adiabatic temperature gradient for millions to hundreds of millions of years sometime before 3.9 Ga if, for example, the core was initially superheated (Breuer and Spohn 2003; Williams and Nimmo 2004), magma ocean overturn resulted in the placement of cold cumulates at the core-mantle boundary (Elkins-Tanton et al. 2005), or plate tectonics was active in Mars' early history (Nimmo and Stevenson 2000).

If Mars possessed a super-adiabatic temperature gradient, then a past thermal dynamo is feasible. Other motions such as compositional convection as the inner core freezes and releases a light element (Stevenson 2001) or tidal motions due to interactions of Mars with an orbiting body (Arkani-Hamed et al. 2008) do not require a super-adiabatic core, further broadening the range of parameters for which an active dynamo could exist on early Mars. Martian interior models show that the core radius is approximately half the planetary radius, but aside from tidal Love numbers suggesting that some portion of the core is still liquid (Yoder et al. 2003), no information on the size of the solid inner core is available. Since Mars' dynamo has not been active since its early history, its likely that it either has no inner core, or that the inner core is no longer growing.

Mars dynamo models are geared towards understanding the mysteries surrounding Mars' remanent magnetic field. A recent model addresses the difference in field intensity between the hemispheres (Stanley et al. 2008). Maps of the crustal magnetism demonstrate that the intensity of the fields correlate with the crustal hemispheric dichotomy. This suggests that whatever process produced the dichotomy in crustal thickness may be related to the dichotomy in magnetic fields. Both endogenic and exogenic formation scenarios for the crustal dichotomy have been presented in the literature. In the endogenic models, a degree-1 mantle circulation is responsible for the different hemispheric crustal thicknesses. This degree-1 pattern can result from a variety of scenarios such as mantle phase transitions (Weinstein 1995; Harder 1998; Breuer et al. 1998), radial viscosity variations in the mantle (Zhong and Zuber 2001; Roberts and Zhong 2006), magma ocean overturn (Elkins-Tanton et al. 2003, 2005) or boundary layer instabilities resulting from a superheated core (Ke and Solomatov 2006). In the exogenic models, a large glancing impact in the northern hemisphere is responsible for the excavation of northern hemisphere crust (Wilhelms and Squyres 1984; Frey and Shultz 1988; Andrews-Hanna et al. 2008; Marinova et al. 2008; Nimmo et al. 2008).

Both the endogenic and exogenic dichotomy formation scenarios have implications for the temperatures at Mars' core-mantle boundary (CMB). In the endogenic models, if the

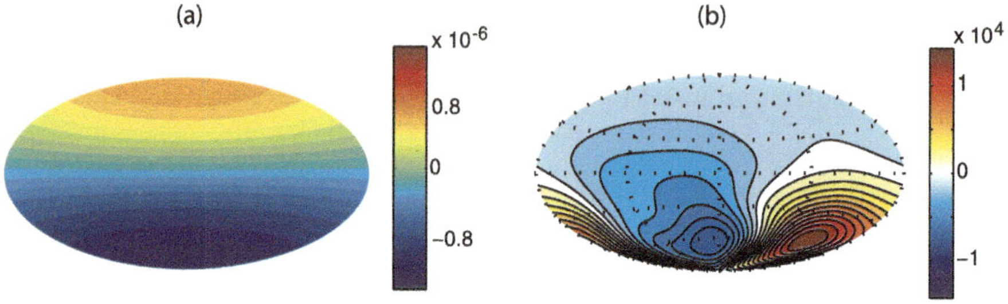

Fig. 2 (**a**): The degree-1 fixed heat flux boundary condition for the Stanley et al. (2008) Mars dynamo model. The superadiabatic heat flux is largest out of the southern hemisphere. Units are W/m^2 and a positive direction is into the core (hence outward heat flux at the CMB is negative). A thermal conductivity $k = 40$ W/mK was used to dimensionalize the superadiabatic heat flux in the models, however, dimensionalization is somewhat misleading due to the numerically unavoidable parameter regime used in the simulation. (**b**) The surface radial magnetic field for the Stanley et al. (2008) Mars dynamo model. Units are in nT

upwelling manifests in the northern hemisphere (resulting in basal crustal erosion and hence a thinner northern hemisphere crust) then the temperature on Mars' northern hemisphere CMB is hotter than in the southern hemisphere (where cold mantle material is falling onto the CMB). In the exogenic models, thermal perturbations due to the large impacts in the northern hemisphere can penetrate to the CMB under the impact (Watters et al. 2009). Both dichotomy formation scenarios can therefore result in a larger heat flux from the southern hemisphere CMB compared to the northern hemisphere CMB.

Stanley et al. (2008) implement the degree-1 CMB heat flux boundary variations implied by the dichotomy formation scenarios in a numerical dynamo model to determine the effects on the magnetic field (Fig. 2(a)). The result is a single-hemisphere dynamo where magnetic field generation is concentrated and strongest in the southern hemisphere. Due to the reduced heat flux from the northern CMB, the northern hemisphere of the core is sub-adiabatic and convection is most strongly driven in the southern hemisphere. The strong thermal winds in the model break the Proudman-Taylor constraint (Proudman 1916; Taylor 1917), although in the southern hemisphere, signs of rotational influences are still present. The resulting surface magnetic field is non-dipolar, non-axisymmetric and stronger in the southern hemisphere (Fig. 2(b)). This suggests that the reason Mars' crustal magnetic field is concentrated in the southern hemisphere may be because the dynamo produced strongest fields in the southern hemisphere.

The cessation of the Martian dynamo is another mystery that dynamo modelers have addressed. It is generally assumed that dynamo action ceased when the driving forces weakened enough such that the Rayleigh number became sub-critical. However, dynamo analytics (Childress and Soward 1972) show that a sub-critical dynamo is possible in the "strong field" regime where the Coriolis, Lorentz and buoyancy forces balance in the momentum equation. Essentially, the inhibiting effects of rotation and magnetic fields can offset each other resulting in convection at Rayleigh numbers below the critical value if only rotation were present. Recently, Kuang et al. (2008) produced sub-critical dynamo action in a spherical shell model suggesting that Mars' dynamo may have lasted longer than age estimates based on requiring a super-critical Rayleigh number. Their models also demonstrate that once the driving force decreases below the threshold for dynamo action in the sub-critical regime, it is unlikely to start up again unless the Rayleigh number increases by about 25 percent, suggesting it is hard to restart the dynamo. Their Mars dynamo, which maintains a relatively strong axial-dipolar dominated field in the super-critical regime, becomes more non-dipolar with frequent reversals in the sub-critical regime.

Another possibility for the termination of the Martian dynamo was suggested by Glatzmaier et al. (1999). The Mars dynamo may have ended by going through a series of dipole reversals and not completely recovering after each one. This occurs in one of their dynamo simulations that has a heat flux imposed at the CMB that is greatest at mid-latitude and least at the equator and poles (note that this is a different CMB heat flux variability than that of the Stanley et al. (2008) work discussed above). After several hundred thousand years with a strong magnetic field (starting with a homogeneous CMB heat flux condition) this case experiences a spontaneous dipole reversal with the usual drop in magnetic dipole intensity but fails to fully recover to the original field intensity after the reversal. Roughly a hundred thousand years later it reverses again and again recovers only to about 10% of its pre-reversal integrated magnetic energy. The unfavorable CMB heat flux condition, which forces more heat to be convected to mid-latitude instead of the preferred equatorial and polar regions, apparently destroys the more efficient fluid flow patterns for the dynamo mechanism.

3 Giant Planet Dynamos

All four of our solar system giant planets possess active dynamos, but the fields produced by the gas giants are markedly different from the ice giant fields. Whereas Jupiter and Saturn possess axial-dipole-dominated fields, Uranus' and Neptune's fields are dominated by multipolar components without a preference for axisymmetry (Connerney 1993). Aside from this difference in field morphologies, other mysteries surrounding the giant planet magnetic fields include:

- What mechanism is responsible for the zonal bands in the giant planet atmospheres, do they represent deeper zonal flows that play a role in dynamo generation, and why are they different for the different planets?
- Why is Saturn's dynamo producing such an axisymmetric field?
- How can Uranus and Neptune be generating magnetic fields with such low surface heat flows?

3.1 Overview of the Different Types of Models for Giant Planets

Most studies of the internal dynamics of giant planets have focused on explaining the maintenance of the observed surface flows and predicting the structure and extent of these flows below the surface. The flows in the deep interior, where the fluid is electrically conducting, maintain magnetic fields by shearing and twisting the existing field at a rate sufficient to offset magnetic diffusion. Since density increases with depth, convective velocity decreases with depth and electrical conductivity increases with depth (for our solar system giants). Therefore, magnetic induction by the flows and the resulting Lorentz forces on the fluid are mainly important within a range of depths in a giant planet where fluid velocity and electrical conductivity are both sufficiently large.

To be credible, a dynamo simulation of convection and magnetic field generation in the outer part of a giant planet should at least produce fluid flows near the non-conducting surface similar to those observed on the particular giant planet being simulated. This provides a valuable constraint that, for example, those who simulate convective dynamos in terrestrial planets do not have. To test how realistic are the flows and fields well *below* the surface in a dynamo simulation one can check if the total rate of entropy production by ohmic heating within the convection zone

$$\int (J^2/\sigma T)dV \tag{6}$$

is, on average, no larger than the rate entropy flows out at the top minus the rate it flows in at the bottom of the dynamo region, i.e., no larger than the *observed luminosity* times $(T_{top}^{-1} - T_{bot}^{-1})$ (Backus 1975; Hewitt et al. 1975). Here, J is the simulated electric current density and σ and T are the prescribed electrical conductivity and temperature, respectively.

We will therefore first review some non-magnetic modeling studies of fluid flows in giant planets before discussing simulations of dynamically-consistent convective dynamos. There have been several two dimensional (2D) modeling studies of vortices in the shallow atmosphere of giant planets. However, since we are interested in planetary dynamos, here we consider only three-dimensional (3D) global simulations. We will also focus on zonal winds (i.e., the axisymmetric part of the longitudinal velocity) because they dominate the flow at the surface and because their extension into the semi-conducting region below the surface plays a critical role in maintaining the toroidal magnetic field.

The surface zonal winds on our giant *gas* planets, Jupiter and Saturn, advect the latitudinally-banded clouds observed on the surfaces of these planets (Sanchez-Lavega et al. 2002; Porco et al. 2003). These longitudinally-averaged winds alternate in latitude between eastward directed (i.e., prograde) and westward directed (i.e., retrograde) relative to a rotating frame of reference (Fig. 3), which is chosen to be that of the global magnetic field based on observed radio emissions. Note, that this reference frame does not represent the mean angular velocity at the cloud tops nor the rotating frame in which the total angular momentum of the planet vanishes. It also does not necessarily represent the average angular velocity of the deep interior. It approximates the mean angular velocity of the region in which the dynamo operates and could have a phase velocity relative to this region. In addi-

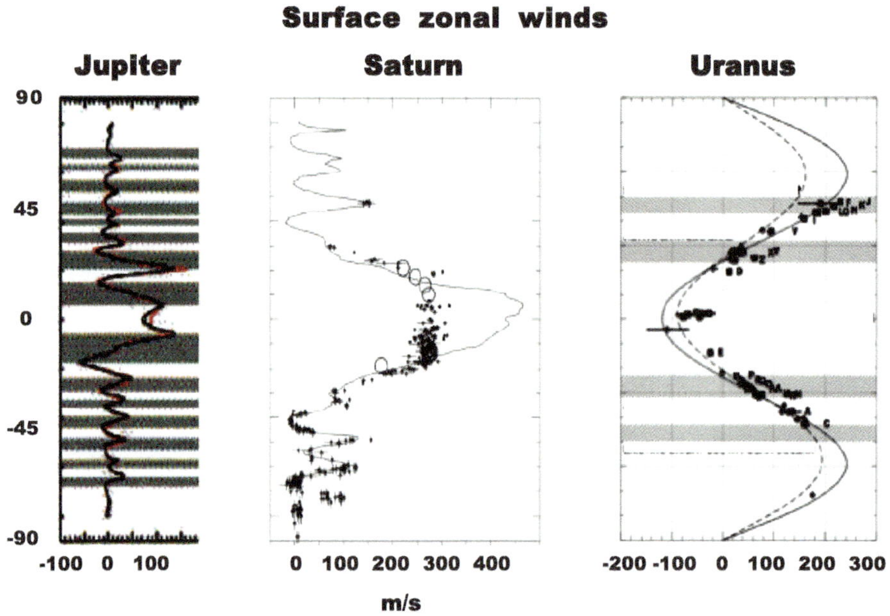

Fig. 3 Plots of the observed zonal winds vs. latitude on Jupiter, Saturn and Uranus. Positive velocities are prograde; negative are retrograde. *Hubble Space Telescope* measurements (*dots*) of Saturn's zonal winds from 1996–2002 show that the peak equatorial jet has decreased by roughly a factor of two since the measurements by the *Voyager Mission* (*solid line*) from 1980–1981. (Sanchez-Lavega et al. 2002; Porco et al. 2003; Hammel et al. 2005)

tion, like the magnetic fields in the sun and the Earth, it is not necessarily constant in time (Giampieri et al. 2006) because the field is maintained by time dependent magnetohydrodynamics. The equatorial jets on Jupiter and Saturn are prograde relative to this chosen frame of reference. Although the zonal wind pattern on Jupiter has been fairly constant for the past three decades, the intensity of the equatorial jet on Saturn has decreased by roughly a factor of two since the time of the *Voyager Mission* (Fig. 3). This may have been an actual change in wind speed or a change in the level of the clouds, which are measured to calculate the speed.

The *ice* giants, Uranus and Neptune, also display a surface differential rotation rate in latitude. However, unlike the gas giants and more like the zonal winds in the Earth's atmosphere, they have a broad retrograde equatorial jet and only one prograde zonal flow at high latitude in each hemisphere (e.g., Hammel et al. 2005).

Many attempts have been made to explain the maintenance of these surface zonal wind patterns and predict their amplitudes and patterns below the surface, i.e., the differential rotation. They can be maintained by Coriolis forces resulting from a thermally-driven meridional circulation (i.e., a "thermal wind") or by the transport of longitudinal momentum by radial and latitudinal flows (i.e., the convergence of Reynolds stress). The axisymmetric parts of viscous and magnetic forces also play a role; however, these forces usually inhibit differential rotation in planets.

Most studies of zonal winds in giant planets have come from two scientific modeling groups: the atmospheric climate modeling community and the geodynamo and solar dynamo modeling communities. Both groups have attempted to use modified versions of their traditional computer models. Modelers from the atmospheric community usually prescribe an atmosphere stratified in density and temperature based on estimates of what these are for the relevant planet. However, their global models are based on the assumption that the observed surface winds are maintained exclusively by the dynamics within the very shallow weather layer. The "shallow-water" model simplifications are based on large-scale horizontal flows and ignore the small-scale buoyantly-driven flows in radius. For reviews of (non-magnetic) climate models see, for example, Dowling (1995) and Showman et al. (2007).

By placing an impermeable lower surface just below the clouds, as is the case for the Earth, these atmospheric models completely ignore the convection in the deep interiors of these fluid bodies. The *Galileo* probe, however, measured strong zonal wind down to a pressure of 20 bars, well below the weather layer driven by solar insolation, with no indication of the wind becoming weaker with depth (Atkinson et al. 1998). The existence of a global magnetic field provides indirect evidence of significant flows at much greater depths. That is, since the magnetic dipole decay time for a body with the size, electrical conductivity and temperature of Jupiter's deep interior is roughly only a million years, a dynamo likely exists in its electrically-conducting interior, driven by deep convection and zonal winds.

Modelers from the geodynamo community use 3D deep convection models to investigate the structure and maintenance of zonal winds (i.e., differential rotation) without making shallow-water type approximations. However, most have assumed no stratification of density, a fairly good approximation for the Earth's fluid core but not for the outer regions of giant planets. A few studies, however, do account for both deep convection zone and large density stratification, similar to modeling studies that have been conducted in the solar dynamo community for the past quarter century. Although these models capture more of the correct physics of the problem, they too need improvements. See, for example, Dormy et al. (2000), Kono and Roberts (2002) and Glatzmaier (2002) for reviews of the geodynamo models.

The shallow-atmosphere community and the deep-convection community have mostly ignored each other and have seldom referenced each others papers. However, since deep-convection *dynamo* models are being judged on their ability to maintain banded zonal winds at their outer boundaries like those observed in the shallow cloud layers of our gas giants, it is important to discover and understand the mechanism that maintains these winds. In particular, it is important to know if this mechanism is seated mainly within the shallow atmosphere, as the atmospheric community claims, or mainly within the deep convection zone as claimed by that community. A critical issue for studies of our solar system giant planet dynamos is to know the depth to which the strong winds penetrate. Also, although the outer atmospheres of our gas giants are relatively cold and have very low electrical conductivity, the outer atmospheres of the extrasolar gas giants in close orbit around their parent stars are probably partially ionized by the stellar radiation. Therefore, the structures of the flows in these electrically conducting atmospheres is very important for studies of the surface dynamos of these planets.

The very different approaches and approximations to modeling the internal dynamics of giant planets have several subtle, but seldom discussed, effects on the respective results. We discuss the resulting deficiencies below as we describe some of the non-magnetic modeling results. Then we discuss some simulations of convective magnetohydrodynamic (MHD) dynamos for gas giants and ice giants. These dynamically self-consistent solutions of thermal convection and magnetic field generation have only been produced using deep convection models.

3.2 Atmospheric Models

Some atmospheric models of giant planets rely on a thermal wind scenario (e.g., Allison 2000) for which a meridional circulation (the axisymmetric north-south flow) drives the zonal winds (the axisymmetric east-west flow) via Coriolis forces. That is, axisymmetric flow toward the equator moves fluid further away from the planet's rotation axis, increasing its moment of inertia, and, to the extent that angular momentum is conserved, causes its angular velocity to decrease. Likewise, flow toward a pole increases the fluid's angular velocity. This naturally results in a retrograde zonal wind in the equatorial region relative to the global mean angular velocity (e.g., Williams 1978; Cho and Polvani 1996), as is the case for the Earth's lower atmosphere and for our ice giants. Note, that a latitudinally-banded pattern of zonal winds with a prograde equatorial jet can be obtained by tuning a prescribed heating function distributed in latitude and radius that continually nudges the temperature toward a profile that drives the desired zonal wind pattern (e.g., Williams 2003). However, such mathematical fits to surface observations do not provide a dynamically-consistent prediction or explanation for the maintenance of an observed zonal wind structure.

Some of the most sophisticated gas giant modeling attempts within the atmospheric community have been made using general circulation models (GCMs). These 3D global models, which were originally designed and used for studies of the Earth's climate and weather, represent a shallow spherical shell of density-stratified gas with a depth less than a percent of the planetary radius. They typically parameterize the radiation transfer using a Newtonian heating/cooling method, which simply nudges the local temperature toward a prescribed temperature profile based on an assumed timescale. GCMs also have "shallow-water" simplifications based on the assumption that the length and velocity scales in radius are small compared to those in the horizontal direction. For example, they assume local hydrostatic equilibrium instead of solving the radial component of the full momentum equation. That is, the radial pressure gradient is forced to always exactly balance the weight of the fluid above. Consequently, the Coriolis, buoyancy and viscous forces and the divergence of Reynolds stress are all neglected in the radial direction. Therefore, instead of evolving the radial component of the fluid velocity via the momentum equation, as is done for the horizontal components, it is calculated by determining what it needs to be to satisfy mass conservation, hydrostatic balance and the equation of state once the horizontal divergence of the horizontal flows is updated (at each grid point and time step).

This has worked sufficiently well for GCMs used to study the large scale flows in the Earth's atmosphere. However, the Earth's atmosphere *does* have an impermeable lower boundary; the atmosphere of a giant planet, on the other hand, extends smoothly into a dense, convecting interior (e.g., Bodenheimer et al. 2003; Guillot 1999). The radial velocity associated with such a grid cell in a GCM represents the spatial average of many, much stronger, small-scale up and down drafts within the entire cell; it is not meant to be a representative value of the actual dynamical velocity in the Earth's atmosphere. When averaged this way, these radial velocities are small relative to the horizontal winds. Therefore, most GCMs also neglect the contribution to the Coriolis and Reynolds stress terms in the *longitudinal* component of the momentum equation that are due to radial velocity.

Solving for the radial flow via the divergence/convergence of horizontal flows, instead of driving it with buoyancy and Coriolis forces, misses critical aspects of the dynamics of deep rotating convection. This is apparent in profiles of the flow and temperature produced by giant planet GCMs for which vortices parallel to the planetary rotation axis are not prominent; such vortices are a dominant feature of flows seen in laboratory experiments and 3D simulations of deep rotating convection. In addition, inertial oscillations, which are driven by Coriolis restoring forces, are also poorly represented in GCMs because of the neglect of Coriolis forces in radius due to longitudinal flows and in longitude due to radial flows. The neglect of buoyancy to drive radial velocity and the neglect of Coriolis and Reynolds stress terms involving the radial velocity have prevented GCM studies from self-consistently producing prograde equatorial zonal winds.

Lian and Showman (2009) have recently produced much more promising GCM simulations that model the effects of moist convection. Although moist convection is highly parameterized in their model, it increases the vigor of the radial flows by releasing latent heat where water vapor condenses. That is, when fluid containing water vapor is advected upward it condenses at some lower-temperature level and releases latent heat. This causes thermal expansion and horizontal divergence, which pulls up more fluid. Unlike previous GCM simulations, their model can produce, without prescribing ad hoc heating patterns, banded zonal surface flows with a prograde equatorial jet (Fig. 4) qualitatively similar to those of our gas giants. These profiles are maintained due to the convergence of Reynolds stress, including that due to the product of radial and longitudinal velocities, which previous GCM studies have neglected. They argue that the latitudinal widths of the zonal wind bands are determined by a "Rhines effect" (Rhines 1975), which predicts a local length scale proportional to the square root of the zonal wind speed. This arises when the advection in latitude of radial vorticity is balanced by the generation of radial vorticity due to the latitudinal gradient of the radial component of the planetary rotation rate.

Lian and Showman (2009) can also produce a retrograde equatorial jet (Fig. 4), like those of our ice giants, when they prescribe a larger source of water vapor, which effectively increases the vigor of the convective flows. The change from prograde to retrograde zonal flow in the equatorial region when convective driving increases relative to Coriolis effects has been demonstrated in deep convection models and shown to also depend on the ratio of viscous to thermal diffusion and on the degree of density stratification (Glatzmaier and Gilman 1982). Although the Lian-Showman model still suffers from the hydrostatic shallow-water

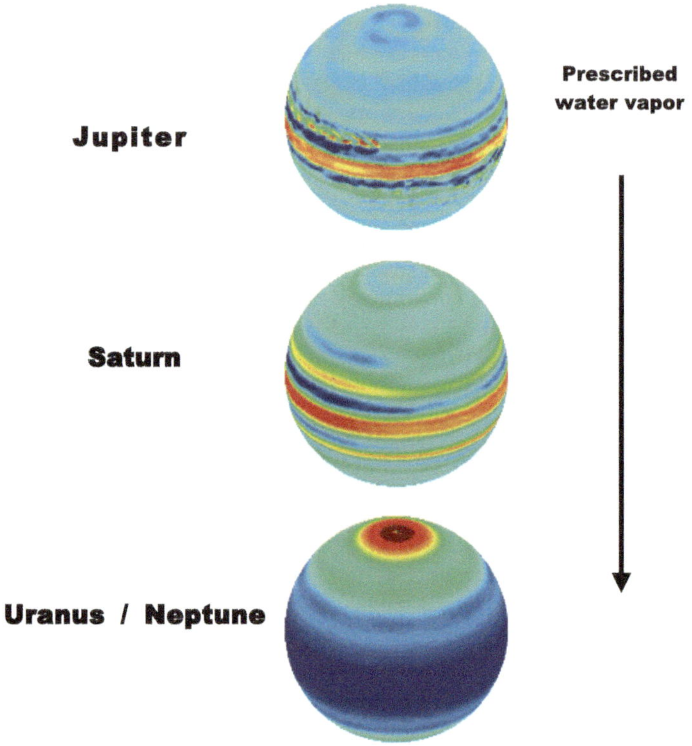

Fig. 4 Snapshots of the longitudinal velocity in GCM simulations with moist convection. *Reds* represent prograde winds; *blues* retrograde. These simulations vary, from "Jupiter" to "Uranus/Neptune" by increasing the amount of prescribed water vapor in the model. (Lian and Showman 2009)

simplifications mentioned above, it is beginning to capture some of the important dynamics seen in 3D deep convection models.

3.3 Non-magnetic Deep Convection Models

It may be the case that the zonal winds on giant planets are confined to the shallow surface layers and driven by heat sources and moist convection there without significant influence from the deep convection below. However, hydrostatic shallow-water type models, including GCMs, assume this from the start instead of allowing the full dynamics to demonstrate it. Deep convection models can also produce banded zonal winds with either prograde or retrograde equatorial jets, without neglecting the dynamics of the deep interior. They are designed to capture the 3D dynamics within a deep rotating fluid shell but not necessarily the 2D dynamics in a shallow atmosphere. Therefore, they neglect moist convection and approximate radiative transfer as a thermal diffusive process.

The deep convection models of the geodynamo and solar dynamo communities are what the atmospheric community would call "non-hydrostatic" models. That distinction is not made in the dynamo communities because all 3D convection and dynamo models are non-hydrostatic. There was never a reason to make shallow-water type approximations for these models because the convection zone depths for the geodynamo and solar dynamo are comparable to the radii of the respective bodies, as they are for giant planets. Therefore, all three components of the momentum equation are solved, putting the radial component of velocity on the same footing as the horizontal components. This is really needed to capture the full dynamics of deep rotating convection; i.e., the entire Coriolis force and Reynolds stress are used and the buoyancy force drives convection.

The full representation of these forces in deep convection models captures the generation of vorticity parallel to the planetary rotation axis as fluid moves relative to this axis. When Coriolis and pressure gradient forces nearly balance everywhere and all other forces are relatively small, the classic Proudman-Taylor theorem (Proudman 1916; Taylor 1917) predicts that rotating, incompressible, laminar fluid tends to flow within planes parallel to the equatorial plane. Within such a convective column or vortex, rising fluid generates negative vorticity relative to the rotating frame of reference and sinking fluid generates positive vorticity. This results in a prograde-propagating Rossby-like wave because positive vorticity is generated on the prograde sides of positive vortices and likewise negative vorticity is generated on the prograde sides of negative vortices. Since the effect is greatest near the surface, rising fluid parcels within the vortices curve to the east and sinking curve to the west. This causes rising fluid to be correlated with eastward flow and sinking fluid to be correlated with westward flow. The resulting convergence of this nonlinear Reynolds stress maintains a zonal wind (i.e., differential rotation) with a prograde equatorial jet near the surface. In such a case, a meridional circulation, much weaker than the zonal wind, is maintained by the Coriolis forces resulting from the zonal wind. If, on the other hand, buoyancy forces dominate over Coriolis forces, the curving of rising and sinking fluid is much less correlated and the thermal wind, with a retrograde equatorial jet, is maintained by Coriolis forces resulting from the meridional circulation (e.g., Glatzmaier and Gilman 1982; Aurnou et al. 2007). Of course this assumes convective velocities are large enough to make a relatively significant Reynolds stress. If they are not, as is the case for the geodynamo, the zonal flow in the equatorial region tends to be retrograde even if Coriolis forces dominate over buoyancy forces (e.g., Kono and Roberts 2002).

Two basic types of deep convection models have been used to study giant planets: those that ignore the background density stratification and assume a liquid equation of state (Boussinesq models) and those that account for density stratification and use a gas equation of state (anelastic models). There is a fundamental difference between the way these two types of models maintain differential rotation. Models that assume a constant background density rely on the classic vortex-stretching mechanism (Busse 1983, 2002) to generate vorticity and maintain differential rotation. This relies on the existence of convective columns aligned parallel to the planetary rotation axis and spanning from the outer (impermeable) surface in the northern hemisphere to that in the southern hemisphere. Negative vorticity (i.e., an anti-cyclone) is generated in rising fluid (within such a column when outside the tangent cylinder to the inner boundary) because conservation of mass and incompressibility require rising fluid to spread out normal to the axis as it approaches the sloping impermeable outer boundary. The resulting Coriolis torque generates the vorticity, which drives a prograde propagating Rossby-like wave (as described above) because of the spherical shape of the outer boundary. Likewise, sinking fluid is stretched parallel to the axis because of the sloping boundaries, generating positive vorticity (i.e., a cyclone). Under the special geostrophic conditions mentioned above, with enough viscosity to maintain relatively laminar flow and large diameter columns, this mechanism is able to maintain differential rotation via the convergence of Reynolds stress as explained above (e.g. in simulations: Gilman and Miller 1981; Christensen 2002; Heimpel et al. 2005b; and in laboratory experiments: Busse and Carrigan 1976; Hart et al. 1986; Manneville and Olson 1996; Aubert et al. 2001; Aurnou 2007).

Fig. 5 A snapshot of the longitudinal winds maintained on the outer and inner boundaries in a simulation of Jupiter using a Boussinesq, deep rotating convection model. *Reds* represent prograde winds; *blues* retrograde. (Heimpel et al. 2005b)

Most published studies of convection in giant planets have been based on the Boussinesq approximation and most have simulated only the outer 10 to 20% in radius of the planet (see papers by Wicht and Tilgner and by Christensen in this issue). The resulting differential rotation manifests itself on the surface as a zonal wind, beautifully-banded in latitude, with a strong prograde equatorial jet qualitatively similar to those of Jupiter and Saturn. An example by Heimpel et al. (2005b) is shown in Fig. 5.

Heimpel et al. (2005b) argue that the widths of the bands in latitude are determined by a modified "Rhines effect" due to the approximate balance of the advection of vorticity by zonal wind and the stretching of fluid along the axes of convective columns due to the spherical geometry of the outer impermeable boundary. This Coriolis stretching effect decreases with latitude, unlike the Coriolis effect of shallow-water GCMs. Therefore, this mechanism maintains a *prograde* equatorial jet. According to the authors, the location of the inner impermeable boundary in their model determines the width and amplitude of the equatorial jet.

In addition to the prograde equatorial jets of Jupiter and Saturn, the retrograde equatorial jets of Uranus and Neptune have been investigated using Boussinesq deep convection models. Aurnou et al. (2007) show that the direction of the equatorial jet in their models depends on the relative importance of Coriolis and buoyancy forces as Glatzmaier and Gilman (1982) demonstrated with anelastic models. As discussed above, when Coriolis forces dominate over buoyancy forces a prograde equatorial jet, similar to those seen on the gas giants, is maintained. If instead buoyancy forces become comparable to or exceed Coriolis forces, a retrograde equatorial jet is maintained.

Although beautiful results have been obtained by these deep-convection constant-density simulations, an obvious question is how realistic is the simulated structure of the convection and the maintenance of differential rotation when density stratification is ignored. The density in giant planets varies in radius by many orders of magnitude, especially in the outer part of the planet (e.g., Bodenheimer et al. 2003; Guillot 2005). For the Boussinesq approximation to be valid, the depth of the convection zone needs to be small relative to the local density scale height. However, since the top of the convection zone is at a pressure of roughly 1 bar, Jupiter's outer 10% in radius, for example, spans about eight density scale heights. That is, the expansion of rising fluid and contraction of sinking fluid have first-order effects on the fluid flows in the outer regions of giant planets. In addition, because the interiors of giant planets have very small viscosity, convective columns (vortices) in these planets likely have extremely small diameters and the flows are probably strongly turbulent, not laminar. Under these conditions, it is unlikely that such long thin convective columns would develop and stretch uniformly while spanning many density scale heights from the northern to southern boundaries (Glatzmaier et al. 2009). Instead, many short, disconnected vortices or vortex sheets would likely exist, on which the sloping outer boundary would have little effect.

To capture the effects of a significant density stratification and a gas equation of state, for which density perturbations depend on both temperature and pressure perturbations, some deep convection models employ the anelastic approximation. As for GCMs and Boussinesq models, sound waves are naturally filtered out in anelastic models in order to use much larger numerical time steps. This approximation to the fully compressible system of equations is valid when the fluid velocity is small compared to the local sound speed and thermodynamic perturbations are small relative to the mean (spherically symmetric) profiles, as is likely the case within giant planets. Background (reference state) profiles in radius of density, temperature and pressure are prescribed that are hydrostatic and usually adiabatic; this reference state may also evolve in time. The 3D time-dependent thermodynamic perturbations, which are neither hydrostatic nor adiabatic, are solved relative to this state. Various initial reference states have been employed: for example, polytropic solutions to the Lane-Emden equation (e.g., Hansen and Kawaler 1994) and polynominal fits to one-dimensional evolutionary models of giant planets (e.g., Guillot 2005). The equation of state for the outer part of a gas giant can be approximated as a perfect gas, i.e., reference state pressure is proportional to density times temperature. However, the deep interior is a non-relativistic degenerate electron gas (e.g., Guillot et al. 2004), with reference state pressure proportional to density to the 5/3 power (e.g., Hansen and Kawaler 1994).

Recently Jones and Kuzanyan (2009) produced anelastic simulations to study the internal convection and differential rotation of density-stratified gas giants. In one of their studies they compare solutions in the outer 30% in radius; one case spans five density scale heights (i.e., a factor of 148 in density) and another spans only a tenth of a scale height (essentially Boussinesq). Both of these cases have a broad equatorial prograde jet, with relatively little zonal wind at higher latitudes (Fig. 6). They clearly demonstrate the effects of density stratification. In the strongly stratified case the amplitude of the convection and differential rotation decrease significantly with depth; whereas the amplitudes are much more uniform in depth for the nearly constant density case.

One of us (Glatzmaier) has simulated the internal dynamics of giant planets using a modified version of an anelastic solar dynamo model (Glatzmaier 1984). Figure 7 shows snapshots of two recent anelastic simulations that use the same density background profile, which spans five density scale heights. These two cases also have the same thermal diffusivities and the same velocity and thermal boundary conditions. However, viscous diffusivity for case 2 is ten times smaller than that for case 1. This decrease in viscosity increases both buoyancy and Coriolis forces by roughly the same factor. The more viscous case (1) maintains a shallow prograde jet in the equatorial region with a peak velocity of 140 m/s (relative to the rotating frame) and essentially solid-body rotation inside a cylinder tangent to this jet (i.e., not tangent to the inner impermeable boundary); the amplitude of the convection drops two orders of magnitude from the outer to the inner regions. The more turbulent case (2) has

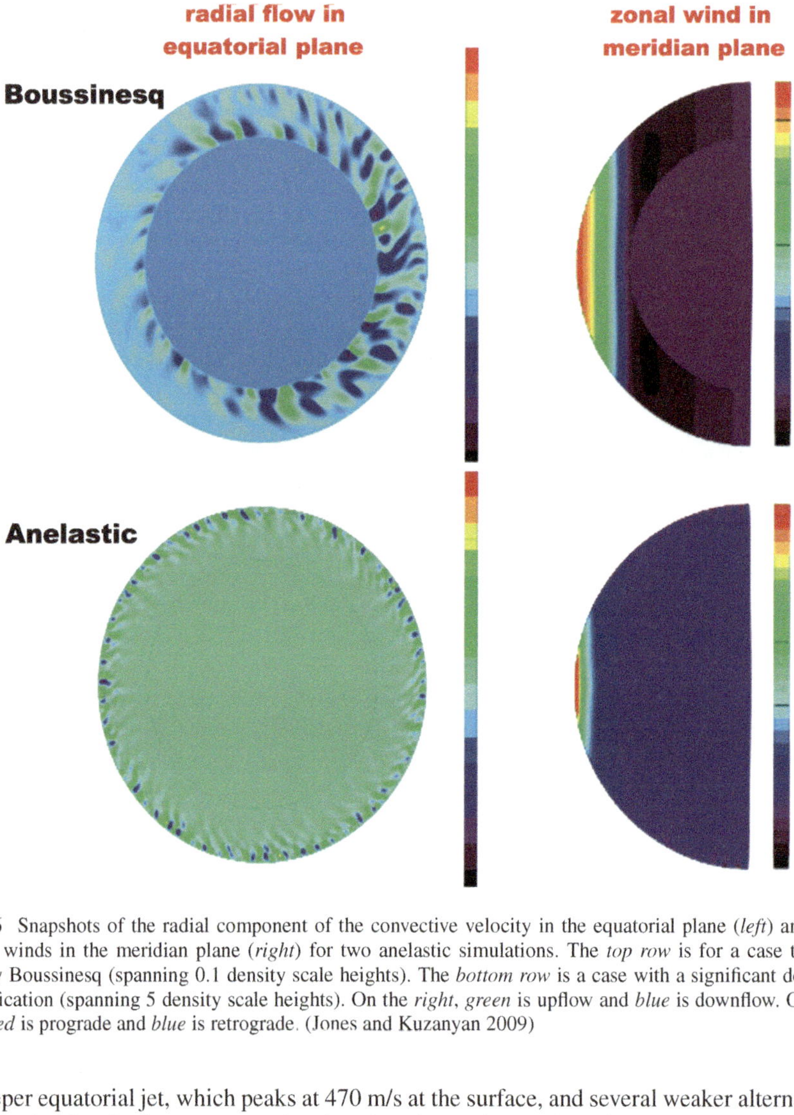

Fig. 6 Snapshots of the radial component of the convective velocity in the equatorial plane (*left*) and the zonal winds in the meridian plane (*right*) for two anelastic simulations. The *top row* is for a case that is nearly Boussinesq (spanning 0.1 density scale heights). The *bottom row* is a case with a significant density stratification (spanning 5 density scale heights). On the *right*, *green* is upflow and *blue* is downflow. On the *left*, *red* is prograde and *blue* is retrograde. (Jones and Kuzanyan 2009)

a deeper equatorial jet, which peaks at 470 m/s at the surface, and several weaker alternating zonal wind bands that penetrate the deep interior and extend to the poles; the convective amplitude drops only about one order of magnitude with depth. Jones and Kuzanyan (2009) see qualitatively similar results. Starchenko and Jones (2002) argue that the amplitude of the convective flows deep within gas giants should be about a cm/s or less in order to maintain roughly the same total convective heat flow as observed at the surface. One of their assumptions is that rising fluid is always hot and sinking is cold, which is reasonable for laminar convection. However, for strongly turbulent convection, especially when dominated by vortices, this need not be the case. Therefore, one could expect the average amplitude of the flows to increase, without increasing the net heat flow, when viscosity is decreased. This effect, however, may be greater near the surface where the flow is more turbulent.

Although long, thin and straight convective columns spanning the entire interior are not precluded in these anelastic simulations, they seldom develop when the flow is even weakly turbulent. Unlike the Boussinesq models, however, such structures are not needed

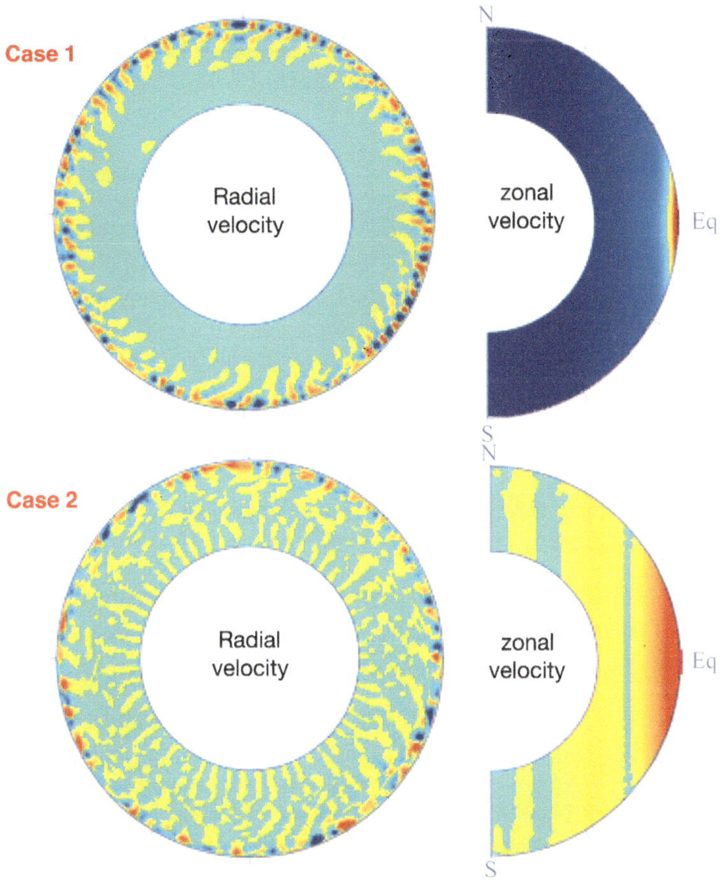

Fig. 7 Snapshots of the radial component of the convective velocity in the equatorial plane (*left*) and the zonal winds in the meridian plane (*right*) for two anelastic simulations, both spanning five density scale heights. Case 2 has viscosity ten times smaller than that of Case 1. *Reds* and *yellows* are upflow (*left*) and prograde (*right*); *blues* are downflows (*left*) and retrograde (*right*). (Simulations by G.A. Glatzmaier)

to maintain a differential rotation pattern with a prograde equatorial jet at the surface. That is, instead of being generated by a stretching torque, local vorticity (parallel to the planetary rotation axis) is generated mainly by compressional torque due to the expansion of rising fluid and contraction of sinking fluid (Glatzmaier and Gilman 1981; Glatzmaier et al. 2009). A prograde propagating Rossby-like wave is now driven because density decreases outward, not because the boundaries are spherical; and the persistent tilt in longitude of rising and sinking fluid, required for the convergence of prograde angular momentum in the surface equatorial region occurs because the density scale height decreases with increasing radius (at least in the gas giants). This mechanism for maintaining differential rotation exists even for short isolated vortices rising and sinking within strongly turbulent environments. The resulting pattern of differential rotation, in radius and latitude, depends on the details of the thermal, compositional and density stratifications in radius.

One important distinction to recognize is the difference between these *non-axisymmetric* convective vortices and the *axisymmetric* zonal flows centered on the planetary rotation axis (i.e., the "constant-on-cylinders" differential rotation). The former tend to be relatively small scale; whereas the latter are broad and, at least in the equatorial region, do extend through the interior from the northern surface to the southern surface. This difference exists because the zonal flow has no component in radius and therefore does not experience the density stratification.

3.4 Magnetic Deep Convection Models

3.4.1 Gas Giants

Having discussed the various 3D simulations of convective structures and zonal flows that may occur in giant planets, we now consider the types of magnetic fields these flows of electrically-conducting fluid could maintain by shearing and twisting existing fields. Models of the geodynamo and solar dynamo typically prescribe a constant electrical conductivity. Several computational studies of dynamos in gas giants have also assumed a constant background density and electrical conductivity (see papers by Wicht and Tilgner and by Christensen in this issue). However, more realistic models of gas giants need to account for the large stratifications in density and electrical conductivity in the outer semi-conducting region, which have first order effects on the structure of the generated magnetic field. The electrical conductivity, for example, likely increases exponentially with depth before becoming constant below the molecular-metallic hydrogen phase transition (e.g., Guillot et al. 2004). That is, in the outer semi-conducting region magnetic diffusivity is not constant, but increases rapidly with radius (Liu et al. 2008).

The effects of significant stratifications in electric conductivity and density have been demonstrated in an anelastic dynamo simulation of a Jupiter-like planet (Glatzmaier 2005). This deep convection zone model spans the outer 80% in radius; the inner 20% is assumed to be a solid core. Constant viscous and thermal diffusivities are prescribed, with the ratio of viscous to thermal being 0.01. The prescribed electrical conductivity increases exponentially with depth by three orders of magnitude in the outer semi-conducting region, up to a constant value in the lower "metallic hydrogen" region. However, the amplitude of convection decreases with depth because of the density stratification. Therefore, magnetic energy is very small in the deep interior and instead peaks in the lower semi-conducting region, where electrical conductivity and convective velocity are both high enough to efficiently generate magnetic field. The results are sensitive to the number of density scale heights represented, the radial profile of electrical conductivity, the vigor of the turbulent convection and the relative effects of Coriolis forces.

Guided by this deep-shell anelastic simulation, Glatzmaier (2005) also simulated a shallower convective dynamo to study the dynamics of the semi-conducting region. Only the outer 20% of the planetary radius is simulated. This generic gas-giant model has a radius, a rotation rate and radial profiles of density, temperature and pressure that are roughly averages of what are obtained in one-dimensional models of Jupiter and Saturn (Guillot 2005). In terms of traditional non-dimensional parameters, the (classic) Rayleigh number is 10^9, Ekman number is 5×10^{-7} and Prandtl number is 3×10^{-2}. The simulation spans about 6 years, more than 3,000,000 numerical time steps, the last half of these at a spatial resolution of 768 grid points in latitude, 768 in longitude and 241 in radius.

Due to the density stratification, the convective velocity near the transition depth is typically an order of magnitude smaller than it is near the surface and two orders of magnitude smaller than the surface zonal winds. Small-scale isolated vortex structures exist in the convection zone. Differential rotation persists throughout the convection zone as latitudinally-alternating angular velocity constant on cylinders coaxial with the rotation axis (Fig. 8).

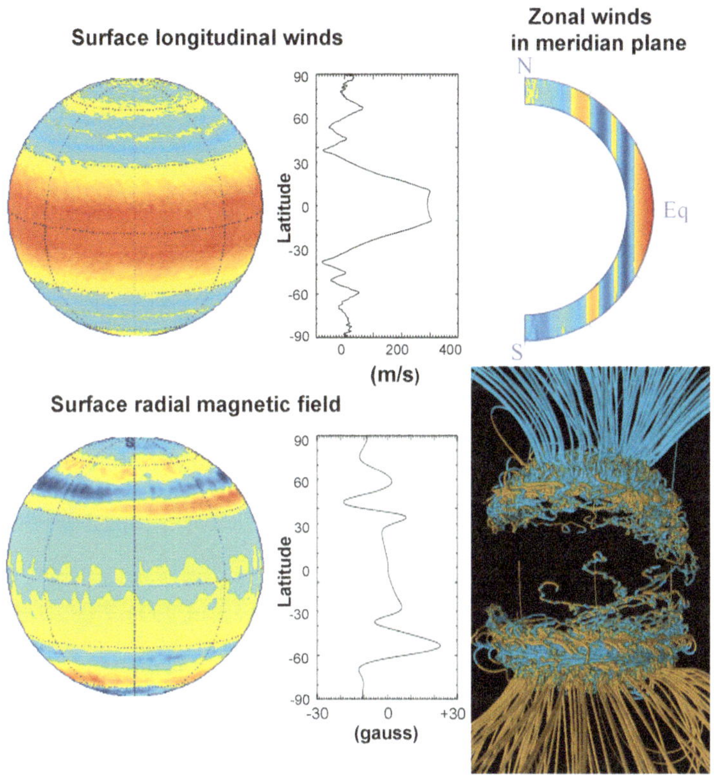

Fig. 8 Snapshots of an anelastic simulation of a gas giant dynamo. *Top row*: The longitudinal component of velocity at the surface, its zonal average vs. latitude and the zonal velocity displayed in the meridian plane. *Reds* and *yellows* are eastward up to 300 m/s; *blues* are westward relative to the rotating frame up to 100 m/s. *Bottom row*: The radial component of the magnetic field at the surface, its zonal average vs. latitude and the 3D field illustrated with lines of force. *Reds* and *yellows* are outward directed; *blues* are inward. *Orange lines* of force are directed outward; *blue* are directed inward. Typical intensity within the convection zone is a few hundred gauss. (Glatzmaier 2005)

This appears on the surface as a banded axisymmetric zonal wind profile. The differential rotation in radius is maintained by the density-stratification mechanism (Glatzmaier et al. 2009) because the density scale height decreases with radius, as discussed above. The peak zonal wind velocities at the surface decrease with latitude because the density scale height, measured normal to the rotation axis, naturally decreases with latitude. The strength and latitudinal extent of the equatorial jet at the surface (Fig. 8), which here is more like Saturn's than Jupiter's (Fig. 3), is sensitive to the prescribed density stratification, viscosity and the ratio of buoyancy to Coriolis forces. Maintaining a differential rotation profile more like Jupiter's will likely require a more turbulent simulation, which would be achieved by decreasing viscosity while increasing spatial and temporal resolution.

Although convection kinetic energy peaks near the surface in this simulation, magnetic induction is most efficient in the lower quarter of the simulated convection zone where the electrical conductivity is greatest. There the differential rotation shears the poloidal field into toroidal (east-west) fields. To limit this shear (and the resulting ohmic dissipation), poloidal field tends to align parallel to surfaces of constant angular velocity, as suggested by Ferraro's iso-rotation law (Ferraro 1937). This makes the radial field at the surface significantly larger

Table 1 Surface magnetic field structures. A selection of the low-degree (ℓ) axisymmetric ($m = 0$) Gauss coefficients (in units of gauss) for the magnetic fields of Jupiter (Connerney et al. 1998), Saturn (Giampieri and Dougherty 2004) and the simulation by Glatzmaier (2005) (averaged in time). The sign of the axial dipole (g_1^0) defines the dipole polarity

Gauss coefficients			
ℓ	Jupiter g_ℓ^0	Saturn g_ℓ^0	Simulation g_ℓ^0
1	4.205	0.212	−2.770
2	−0.051	0.016	−1.041
3	−0.016	0.028	0.715
4	−0.168		−1.058

at high latitudes, as seen in Fig. 8. Also seen is the banded pattern of the radial component of the field at the surface, related to the banded pattern of the angular velocity.

The simulation's dipole moment is 1.1×10^{20} Tm^3, somewhat less than Jupiter's (1.5×10^{20} Tm^3) and larger than Saturn's (4.2×10^{18} Tm^3). More detailed structure of the magnetic field at and beyond the surface is usually described in terms of Gauss coefficients in a spherical harmonic expansion (e.g., Connerney et al. 1998). Table 1 lists a few of the lowest degree, ℓ, coefficients (i.e., largest scales) from fits to observations of the fields of Jupiter and Saturn and from the snapshot of the simulated field illustrated in Fig. 8. Better observations are needed to describe the planetary fields beyond degree 4 (see the paper by Dougherty and Russell in this issue); the computer simulation, in comparison, updates over 130,000 Gauss coefficients every numerical time step. Only the axisymmetric coefficients ($m = 0$) are listed here since the non-axisymmetric coefficients depend on the choice of the longitude zero and are more time dependent. However, it is seen that even the first four axisymmetric coefficients for the simulation do not agree with those for either planet, other than being of the same order of magnitude and having an axial dipole (g_1^0) that dominates. Note that its sign merely reflects the dipole polarity, which for the simulation is currently opposite of the present polarities for Jupiter and Saturn. The real problem is that many more than four degrees are needed to well describe even the axisymmetric part of the radial surface field vs latitude; the plot of this requires Gauss coefficients up to at least degree 15 before the high-latitude peaks become prominent as they are in Fig. 8. For example, the degree 9 coefficient is −1.087 gauss.

Future missions to Jupiter and Saturn that orbit much closer to their surfaces will provide more accurate estimates of higher degree magnetic structure and might detect a banded magnetic field structure as this simulation predicts. Gravity measurements in such missions could also test the banded cylindrical pattern of differential rotation deep below the surface (Hubbard 1990).

Since the amplitudes of the simulated zonal wind and magnetic field at the surface are somewhat similar to those of Jupiter and Saturn, it is instructive to check if the amplitudes and 3D geometrical configurations of the flows and fields well below the surface satisfy the constraint in (6) regarding the total rate of entropy production due to ohmic heating. Although time dependent, in general this constraint is satisfied in this simulation because the amplitude of the non-axisymmetric flow decreases with depth and because of the tendency for the field to satisfy Ferraro's iso-rotation law (Ferraro 1937; Glatzmaier 2008).

However, the spatial resolution of this simulation is not sufficient to capture small scale eddies and vortices near the surface. Actually, no deep convection simulation has yet even

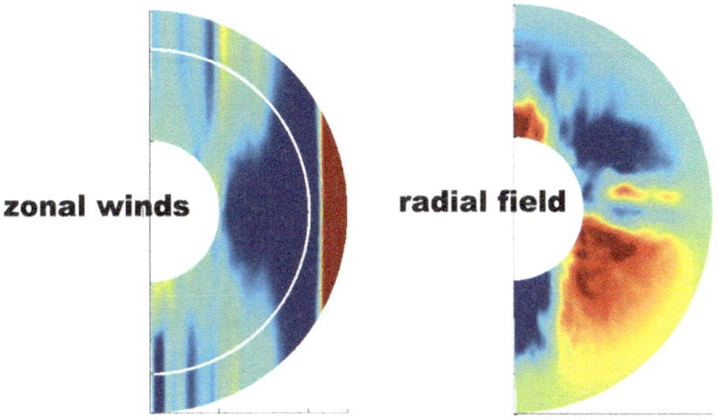

Fig. 9 Snapshots of a Boussinesq simulation of a gas giant dynamo. *Left*: Zonally-averaged longitudinal velocity in the meridian plane. *Reds* are prograde; *blues* are retrograde. *Right*: Zonally-averaged radial magnetic field in the meridian plane. *Reds* are outward; *blues* are inward. (Heimpel and Gomez Perez 2008)

produced a large vortex at the surface like the "Great Red Spot". This may require a representation of radiative transfer and moist convection near the surface, as is done in GCMs.

Another gas giant dynamo model has recently been developed by Heimpel and Gomez Perez (2008) based on their Boussinesq deep-convection model. In their model the convection zone represents the outer two-thirds in radius of the planet. Although density stratification is still neglected, this model has a prescribed electrical conductivity that increases exponentially by three orders of magnitude from the surface, R, down to radius $0.8\ R$. Their electrical conductivity is constant in the highly-conducting metallic-hydrogen region below this radius. An intense magnetic field is maintained in this inner metallic region where the resulting Lorentz forces suppress the zonal wind amplitude (Fig. 9). The dominant feature in the outer low-conductivity region is a strong prograde equatorial jet. The magnetic field at the surface of their simulation is dominantly dipolar and much weaker than the field in the deep interior below $0.8\ R$.

This simulation nicely demonstrates how an intense magnetic field can inhibit differential rotation. However, the field in the deep interior may be too intense because the convective flows there have amplitudes similar to those near the surface due to the model's lack of density stratification.

Dynamo models have also been used to investigate the axisymmetry of Saturn's magnetic field. Although Jupiter's (and Earth's) magnetic fields are dipole dominated, they still contain some observable amount of non-axisymmetry. For example the dipole tilts (ratio of equatorial dipole to axial dipole) for Earth and Jupiter are approximately 10 degrees (Stevenson 1983). In contrast, Saturn observations by Pioneer 11, Voyager I, II and Cassini obtain dipole tilts of < 0.1–1 degrees and a similar lack of non-axisymmetry in the higher moments (Acuna and Ness 1980; Ness et al. 1981; Acuna et al. 1981, 1983; Smith et al. 1980; Davis and Smith 1990; Giampieri and Dougherty 2004). This almost perfect axisymmetry seems to contradict Cowling's theorem (Cowling 1933) which states that a dynamo cannot maintain a purely axisymmetric field. However, Cowling's theorem applies to the dynamo source region, so there is no contradiction if some mechanism can be found that axisymmetrizes the observed field outside the planet.

Stevenson (1980, 1982, 1983) suggested that a helium rain-out layer in Saturn could produce such a mechanism. At conditions in Saturn's interior where hydrogen attains a significant conductivity, helium becomes immiscible in hydrogen and would therefore rain-out of the hydrogen. This would result in a thin stably-stratified layer surrounding the deep dynamo region. Stevenson proposed that differential rotation in this layer due to thermal winds could shear out the non-axisymmetry in the field. In Stevenson's model, the thermal winds result because of pole to equator temperature differences at Saturn's surface due to solar insolation. Kinematic dynamo models have investigated the axisymmetrizing effects of stably-stratified layers surrounding the dynamo (Love 2000; Schubert et al. 2004). They found that flows in surrounding stable layers do not necessarily act to axisymmetrize the field. Depending on the interior field morphology and on the geometry of the stable layer zonal flows, a variety of field symmetries were produced.

Dynamic dynamo models have also investigated the effect of surrounding stable layers in an effort to explain Saturn. Christensen and Wicht (2008) found that with a thick stable layer, a highly axisymmetric field can result for certain parameter values. The average dipole tilt in their model is 1.5 degrees. However, their models work with a much thicker stable layer (~ 0.4 core radii) than the thin helium-rain out layer initially proposed. Thin stably-stratified layers have also been investigated by Stanley and Mohammadi (2008). These models demonstrated that coupling between the stable and unstable layers can act to destabilize the dynamo and result in more non-axisymmetry. However, these models do not drive flows in the stable layer via thermal winds the way Stevenson envisioned.

A recent model by Stanley (2008) employs an outer thermal boundary condition on the dynamo model to mimic the effect of laterally varying solar insolation, or alternatively, the thermal perturbations naturally produced by convection in Saturn's non-metallic outer layers (Aurnou et al. 2008). They find that the surface magnetic fields can be axisymmetrized through these zonal flows, but only when the boundary thermal perturbations result in thermal winds matching the sign and morphology of those already occurring in the deeper interior. The necessary profile to produce axisymmetrization is the expected profile for solar insolation or atmospheric convection scenarios. This suggests that Stevenson's mechanism may explain the axisymmetrization of Saturn's magnetic field. Of course, since these models employ the Boussinesq approximation and do not model the outer layers of Saturn, they are intended more to study the specific mechanism for Saturn axisymmetrization rather than produce a complete model of Saturn's interior flows. In addition, vigorous convection may easily overshoot through a thin stably-stratified layer, both from below and from above, completely dominating any thermal wind driving by solar insolation at the surface.

An important question regarding the axisymmetrization of Saturn's magnetic field and regarding the dynamo mechanism in general for giant planets is how deep do the observed zonal winds extend below the surface, i.e., what is the differential rotation in radius and latitude. Liu et al. (2008) attempt to answer this question by doing a steady-state, axisymmetric, kinematic scale-analysis. By estimating the amount of electric current that would be generated within the interior with their prescribed differential rotation shearing their prescribed internal magnetic field and using their estimate of the depth-dependent electrical conductivity they get an estimate of the total amount of ohmic heating as a function of the depth of the differential rotation. Assuming that this ohmic heating cannot exceed the observed luninosity, they predict that the zonal winds extend no deeper than 0.96 of Jupiter's radius and 0.85 of Saturn's radius.

This is a bold attempt to answer a difficult question, especially without the aid of a 3D self-consistent convective dynamo simulation. However, their analysis has several uncertainties (Glatzmaier 2008; Jones and Kuzanyan 2009). As shown by Backus (1975), Hewitt et al. (1975), the total ohmic heating can exceed the luminosity because the bulk of the dissipation occurs at a much higher temperature than the surface temperature. Also, as shown by Ferraro (1937) for the conditions assumed by Liu et al. (2008), if the internal poloidal field is everywhere parallel to surfaces of constant angular velocity there would be no electric current generated and therefore no ohmic heating produced. Of course, some current needs to be generated to maintain a global magnetic field and so the dynamo mechanism is intimately connected with the differential rotation below the surface. How well the field obeys the Ferraro iso-rotation law and allows deep-seated zonal winds in gas giants is still an open question.

3.4.2 Ice Giants

The magnetic fields of Uranus and Neptune are non-axially, non-dipolar dominated (see papers by Dougherty and Russell and by Fortney in this issue). It turns out that generating axial-dipolar dominated dynamo models is relatively easy when working in an Earth-like thick shell geometry with standard boundary conditions and driving forces. Even observations suggest that axial-dipolar dominated fields are the norm since both Earth and Jupiter have similar magnetic field morphologies although vastly different interior structure and composition. Explaining the ice giants' anomalous fields has therefore resided in understanding what could be different inside these planets to result in such strange magnetic fields.

Kinematic dynamo models investigating field symmetries show that axial dipolar, axial quadrupolar, and equatorial dipolar magnetic fields can result from very similar flows, suggesting that it may not be difficult to generate non-axial-dipole dominated fields (Gubbins et al. 2000). Indeed, although the majority of Earth-like geometry dynamo models produce axially dipolar dominated fields, non-dipolar, non-axisymmetric dynamo models have been found (Grote et al. 1999, 2000; Grote and Busse 2000; Ishihara and Kida 2000; Aubert and Wicht 2004). These solutions are found in isolated regions of parameter space and it remains unknown whether these solutions would exist in the planetary parameter regime (although one has to admit that this is also the case for the axial-dipole dominated dynamo models). However, there are two fundamental issues in using these models to explain Uranus' and Neptune's fields: (i) These non-axial-dipolar dominated dynamos produce much simpler fields than observed in Uranus and Neptune. They are dominated by a specific component of the large scale field whereas Uranus and Neptune have significant power in several modes (e.g. dipole, quadrupole, octupole, axisymmetric and non-axisymmetric). (ii) These models do not match the structure of the ice giant interiors determined from their low heat flows (Podolak et al. 1991; Hubbard et al. 1995). It is therefore unclear as to whether these simpler models can explain Uranus' and Neptune's fields. However, they are certainly important in understanding mechanisms of non-dipolar non-axial field generation.

Kutzner and Christensen (2002) found that non-dipolarity increases with Rayleigh number, suggesting that one could explain Uranus and Neptune by saying they convect more vigorously than the other planets. However, based on observations of Uranus' and Neptune's low heat flows, it seems unlikely that the ice giants are more supercritical than the gas giants.

Uranus' and Neptune's low heat flows resulted in an early proposition to explain their magnetic fields. Hubbard et al. (1995) demonstrated that the observed heat flux is inconsistent with whole-planet convection. They showed that the heat flows imply that only the outer 0.4 Uranus radii and 0.6 Neptune radii of the planets are convectively unstable. They suggest that the stably-stratified interior regions can explain the low heat flows and possibly also the anomalous magnetic fields since this stratification results in a thin-shell dynamo geometry. The stratification is due to compositional gradients in the interiors of the planets which are not unreasonable considering the composition and thermal evolution of the ice giants.

Stanley and Bloxham (2004, 2006) investigated whether dynamos operating in thin layers surrounding stably-stratified interiors could produce Uranus and Neptune like fields. They found that non-dipolar, non-axisymmetric fields resulted with similar surface power spectra to the observations from these planets. This is due to the influence of the stably stratified interior on the stability of the dipole component in these models. Thin shell models surrounding insulating solid interiors, conducting solid interiors and stably-stratified fluid interiors were investigated. It was found that, although the fluid flows in these models are relatively similar in morphology and intensity, insulating solid inner core models and stably-stratified inner region models produced magnetic fields with a very different morphology than the conducting solid inner core models. In the conducting solid inner core models, the anchoring effect due to the inner core on magnetic field lines stabilized the dynamo producing a strong dipolar field. When the electrical conductivity of the inner core was removed, this anchoring effect was reduced, and in combination with the thinner shell geometry, resulted in smaller scale (non-dipolar, non-axisymmetric) fields that reversed continuously. Models with thin shells surrounding stably-stratified interiors produced similar fields to the insulating solid inner core models because the fluid inner core can mimic an insulating solid core in its response to electromagnetic stress. The stably-stratified interior loses its anchoring influence because the magnetic fields can differentially move the fluid, whereas they cannot do so in a solid conducting interior.

These dynamo models only considered the deep regions of the planet where the pressures are high enough such that the electrical conductivity is substantial enough to drive a dynamo. The outer layers of the planet are not modeled. Models by Gomez-Perez and Heimpel (2007) instead consider the outer regions of the planets as well by incorporating a radially variable electrical conductivity. They find that the deeper regions of the planets which have the larger conductivity have much weaker zonal flows than the outer, relatively more insulating layers. The higher magnetic diffusivities in these models, combined with the surrounding zonal flows, can result in non-dipolar, non-axisymmetric fields. Similar to the Stanley and Bloxham (2004) models, these models produce complex fields with similar power spectra as those of Uranus and Neptune observations. However, they do not adhere to the interior geometry constrained by the heat flow observations.

3.5 Extrasolar Planet Models

More than 300 extrasolar giant planets have been detected during the past decade using Doppler measurements of the periodic orbital velocities of their sun-like parent stars. A small number have also been detected using photometry during their transits in front of and behind their parent stars. Extrasolar terrestrial planets, because of their much smaller mass, have been much more difficult to detect.

The extrasolar giants that have been discovered have small orbits because of an observational selection effect due to the correspondingly short orbital periods. Having orbital radii as small as 0.05 AU, compared to Jupiter's 5 AU orbital radius, makes the incident stellar radiation four orders of magnitude greater than that received by Jupiter from our sun. The small orbital radii also mean large tidal effects; so it is assumed that the rotation and orbital periods of these planets are nearly synchronous (e.g., Goldreich and Soter 1966). That is, one hemisphere is always facing the parent star and the other is always hidden from it. This external heating, especially on the dayside, likely produces a very stable thermal stratification down to much greater pressures in these "Hot Jupiters" than the roughly 1 bar pressure level that marks the top of the convection zone in our solar system giants. This extreme dayside heating and nightside cooling must also drive strong atmospheric circulations. What structure these winds have and how deep they extend have been key questions on which several modeling groups have attempted to shed light.

Many modeling studies of the non-magnetic surface dynamics on Hot Jupiters have come from the atmospheric climate modeling community. Like the GCMs for solar system giants, these models completely neglect convection in the deep interior and the zonal winds it maintains. For a nice review of these see Showman et al. (2007).

These models predict wind speeds up to 10^5 m/s and atmospheric temperatures up to 10^4 K, which means that in some regions these winds can be supersonic. GCMs do not simulate supersonic flows; nor do anelastic models. This problem requires a model that solves the fully-compressible radiative-hydrodynamic equations. The price one pays is that the numerical time step needs to be much smaller in order to resolve sound waves. Dobbs-Dixon and Lin (2008) have used such a model to study 3D winds in a Hot Jupiter. This model simulates roughly the outer 8% in radius of the planet and does not make the shallow-water simplifications employed in GCMs. Their model treats the radiation transfer using a flux-limited method that smoothly varies from the diffusion limit (for an optically thick region) to the streaming limit (for an optically thin region). The simulated winds reach a Mach number of 2.7. The zonal wind pattern has a broad prograde equatorial jet and mid-latitude retrograde jets. They argue how important it is for this problem to solve the full 3D momentum equation and employ a self-consistent treatment of the radiation transfer. However, they do not address the dynamics of the deep interior.

Evonuk and Glatzmaier (2009) have developed an anelastic model to study convection and zonal winds of the deep interior of an extrasolar gas giant, assuming a negligible magnetic field. The model is a fully-convective density-stratified rotating fluid *sphere*; but the model's top boundary is placed below the shallow weather layer. Vigorous convection is driven throughout the model planet by a heat source in the central region meant to represent the heat generated by the slow contraction of the planet. When the planetary rotation is relatively weak compared to buoyancy, the preferred mode of convection is a dipolar flow through the center of the planet; a greater rotational influence results in strong zonal flows that extend deep into the interior. The winds at the model's surface have a broad prograde equatorial jet and mid-latitude retrograde jets.

What have not yet been studied are the types of magnetic fields that could be generated in extrasolar gas giants. The proximity of Hot Jupiters to their parent stars suggests that the outer atmospheres of these planets are partially ionized and therefore electrically conducting. The type of dynamos that operate in these supersonic surface flows is a fascinating unstudied problem. Ideally, for Hot Jupiters, the dynamos of both the parent star and the gas giant should be simulated simultaneously with robust treatments of the gravitational, radiation and magnetic interactions.

3.6 Future Model Improvements

Future convective dynamo models of giant planets should include more realistic radial stratifications of density and electrical conductivity. They should solve the full momentum equation, as done in deep convection models, and should include radiative transfer and moist convection using schemes near the surface that are at least as self-consistent and sophisticated as those employed in GCMs of the Earth's atmosphere. Higher spatial resolution will also be needed so viscosity and thermal conductivity can be reduced and more strongly turbulent convection can be simulated.

The velocity boundary conditions prescribed in these models also need to be improved. The top boundary should not continue to be a fixed impermeable boundary, which artificially forces radial flow into horizontal flow. The vertical coordinate could instead be the column mass, as in one dimensional planetary evolutionary models, or some variation of pressure, as in GCMs. Alternatively, if one wishes to continue using radius as the vertical coordinate, permeable "free surface" boundaries should be employed within a stratosphere.

The treatment of the very deep interior should also be improved. Some extrasolar giant planets (and possibly also Jupiter) likely have no solid core or even a stably-stratified fluid core at some time in their evolution. Models of these planets require a full sphere instead of a spherical shell. As discussed above, preliminary full-sphere rotating, stratified, convection calculations have been produced (Evonuk and Glatzmaier 2009); however, these need to be run as a dynamo to see how the field and flow interact in the deep interior constrained by the rate of entropy production by ohmic heating.

Other complications that need to be addressed in simulation studies of giant planets are the hydrogen phase transition and helium settling (Stevenson and Salpeter 1977; Guillot et al. 2004), double diffusive convection (e.g., Stellmach et al. 2009) and tidal effects and inertial modes (e.g., Guillot et al. 2004; Tilgner 2007).

4 Satellite Dynamos

At least two moons in our solar system have evidence for dynamo action sometime during their history. Observations from the Galileo mission demonstrated that Ganymede has an intrinsic field most likely due to an active dynamo (Kivelson et al. 1996). Data from various missions showed that Earth's Moon, on the other hand, possesses localized crustal magnetic fields which may be evidence of a past dynamo (Fuller and Cisowski 1987; Halekas et al. 2001). Further information on lunar magnetism comes from the magnetization of Apollo samples allowing constraints to be placed on the timing of the lunar dynamo (Cisowski et al. 1983; Garrick-Bethel et al. 2009). These two bodies maintain some of the most intriguing questions in planetary magnetic fields.

For Ganymede, the Galileo mission measured a dipole moment, but no information on the higher multipoles is available. The data can be modeled relatively well by an appropriately scaled Earth-dynamo model. However, the question remains as to how a small body like Ganymede can maintain a present-day fluid outer core with strong convection (although perhaps this is not as surprising in light of the fact that Mercury does as well). The presence of sulfur in Ganymede's core is employed to explain this, and the possibility that the driving force for Ganymede is freezing of iron snow below the CMB (Hauck et al. 2006) suggest that there may be work for dynamo modelers yet. In addition, the orbital resonances of the four Galilean satellites may provide alternative driving mechanisms for Ganymede in its past, as well as an additional heat source maintaining Ganymede's fluid outer core. A mean-field model for Ganymede investigated the effects of Jupiter's magnetic field in aiding the generation of Ganymede's field by providing a source field for the convective motions to interact with Sarson et al. (1997). They found that although Jupiter's seed magnetic field is not required to maintain Ganymede's dynamo, it may influence the polarity of the resulting field.

If the Moon's remanent magnetic field is the result of rocks cooling in the presence of an ancient dynamo (rather than due to impact magnetization, Hood and Artemieva 2008), then it must contain a metallic core; a piece of information current interior models cannot verify. Thermal evolution models for the Moon suggest that it can maintain a short-lived dynamo early in its history, or slightly later in its history if the core is insulated early on by a radioactive thermal blanket (Stegman et al. 2003). Impact magnetization and demagnetization have likely played a role is shaping the morphology of the crustal field, but not enough information is available to constrain the dynamo generated field from the crustal remanent fields, and hence no dynamo models. However, one aspect of the Moon which dynamo modelers should investigate is the importance of mechanical stirring due to the stronger tidal forces in the past.

5 Conclusions

Since magnetic fields are generated in a planet's deep interior and extend beyond the surface where they are observable, magnetic fields can act as important probes of planetary interior structure and dynamics. Planetary dynamo models therefore, not only investigate the magnetic field generation process, but also tell us about regions of the planet difficult to study with other means. Current planetary dynamo models have opened a window into planetary cores, providing insights into core fluid flows, convective stability and thermal evolution. However, much work is still needed before the planetary dynamo process is fully understood and we can take full advantage of the implications of planetary magnetic field observations.

The main improvements needed are in numerical modeling methods and computational power and better observational and experimental data on magnetic fields, interior structure and dynamics. Current and near-future planetary missions such as MESSENGER, Bepi-Colombo, Juno and Grail will hopefully provide us with new constraints on the planetary dynamo process.

Acknowledgement G.G. thanks NASA for support under grants NNG05GG69G and NNX09AD89G. S.S. thanks NSERC for support under the Discovery Grants program.

References

M.H. Acuna, N. Ness, The magnetic field of Saturn: Pioneer 11 observations. Science **207**, 444–446 (1980)
M.H. Acuna, J. Connerney, N. Ness, Topology of Saturn's main magnetic field. Nature **292**, 721–726 (1981)
M.H. Acuna, J. Connerney, N. Ness, The Z_3 zonal harmonic model of Saturn's magnetic field: Analyses and implications. J. Geophys. Res. **88**, 8771–8778 (1983)
M.H. Acuna et al., Global distribution of crustal magnetization discovered by the Mars Global Surveyor MAG/ER experiment. Science **284**, 790–793 (1999)
O. Aharonson, M. Zuber, S. Solomon, Crustal remanence in an internally magnetized non-uniform shell: a possible source for Mercury's magnetic field? Earth Planet. Sci. Lett. **218**, 261–268 (2004)
M. Allison, A similarity model for the windy jovian thermocline. Planet. Space Sci. **48**, 753–774 (2000)
B. Anderson et al., The structure of Mercury's magnetic field from MESSENGER's first flyby. Science **321**, 82–85 (2008)
J. Andrews-Hanna, M. Zuber, W. Banerdt, The Borealis basin and the origin of the martian crustal dichotomy. Nature **453**, 1212–1215 (2008)
J. Arkani-Hamed, B. Seyed-Mahmoud, K.D. Aldridge, R.E. Baker, Tidal excitation of elliptical instability in the Martian core: Possible mechanism for generating the core dynamo. J. Geophys. Res. **113** (2008). doi:10.1029/2007JE002982
D.H. Atkinson, J.B. Pollack, A. Seiff, The Galileo Probe Doppler Wind Experiment: measurement of the deep zonal winds on Jupiter. J. Geophys. Res. **103**, 22911–22928 (1998)
J. Aubert, J. Wicht, Axial vs. equatorial dipolar dynamo models with implications for planetary magnetic fields. Earth Planet. Sci. Lett. **221**, 409–419 (2004)
J. Aubert, D. Brito, H.-C. Nataf, P. Cardin, J.-P. Masson, A systematic experimental study of rapidly rotating spherical convection in water and liquid gallium. Phys. Earth Planet. Inter. **128**, 51–74 (2001)
J.M. Aurnou, Planetary core dynamics and convective heat transfer scaling. Geophys. Astrophys. Fluid Dyn. **101**, 327–345 (2007)
J.M. Aurnou, M. Heimpel, J. Wicht, The effects of vigorous mixing in a convective model of zonal flow on the ice giants. Icarus **190**, 110–126 (2007)
J. Aurnou, M. Heimpel, L. Allen, E. King, J. Wicht, Convective heat transfer and the pattern of thermal emission on the gas giants. Geophys. J. Int. **173**, 793–801 (2008)
G.E. Backus, Gross thermodynamics of heat engines in the deep interior of the Earth. Proc. Natl. Acad. Sci. Wash. **72**, 1555–1558 (1975)
P. Bodenheimer, G. Laughlin, D.N.C. Lin, On the radii of extrasolar giant planets. Astrophys. J. **592**, 555–563 (2003)

D. Breuer, T. Spohn, Early plate tectonics versus single-plate tectonics on Mars: Evidence from magnetic field history and crust evolution. J. Geophys. Res. **108** (2003). doi:10.1029/2002JE001999
D. Breuer, D.A. Yuen, T. Spohn, S. Zhang, Three dimensional models of Martian mantle convection with phase transitions. Geophys. Res. Lett. **25**, 229–232 (1998)
F.H. Busse, A model of mean zonal flows in the major planets. Geophys. Astrophys. Fluid Dyn. **23**, 153–174 (1983)
F.H. Busse, Convective flows in rapidly rotating spheres and their dynamo action. Phys. Fluids **14**, 1301–1314 (2002)
F.H. Busse, C.R. Carrigan, Laboratory simulation of thermal convection in rotating planets and stars. Science **191**, 81–83 (1976)
B. Chen, J. Li, S. Hauck, Non-ideal liquidus curve in the Fe-S system and Mercury's snowing core. Geophys. Res. Lett. **35** (2008). doi:10.1029/2008GL033311
S. Childress, A. Soward, Convection-driven hydromagnetic dynamo. Phys. Rev. Lett. **29**, 837–839 (1972)
J.Y.-K. Cho, L.M. Polvani, The morphogenesis of bands and zonal winds in the atmospheres on the giant outer planets. Science **273**, 335–337 (1996)
U.R. Christensen, Zonal flow driven by strongly supercritical convection in rotating spherical shells. J. Fluid Mech. **470**, 115–133 (2002)
U. Christensen, A deep dynamo generating Mercury's magnetic field. Nature **444**, 1056–1058 (2006)
U. Christensen, J. Wicht, Models of magnetic field generation in partly stable planetary cores: Applications to Mercury and Saturn. Icarus **196**, 16–34 (2008)
S. Cisowski, D. Collinson, S. Runcorn, A. Stephenson, M. Fuller, A review of lunar paleointensity data and implications for the origin of lunar magnetism. J. Geophys. Res. **88**, 691–704 (1983)
J. Connerney, Magnetic fields of the outer planets. J. Geophys. Res. **98**, 18659–18679 (1993)
J. Connerney, M.H. Acuna, N.F. Ness, T. Satoh, New models of Jupiter's magnetic field constrained by the Io flux tube footprint. J. Geophys. Res. **103**, 11929–11939 (1998)
T. Cowling, The magnetic field of sunspots. Mon. Not. R. Astron. Soc. **94**, 39–48 (1933)
L. Davis, E. Smith, A model of Saturn's magnetic field based on all available data. J. Geophys. Res. **95**, 15257–15261 (1990)
I. Dobbs-Dixon, D.N.C. Lin, Atmospheric dynamics of short-period extrasolar gas giant planets I. Dependence of nightside temperature on opacity. Astrophys. J. **673**, 513–548 (2008)
E. Dormy, J.-P. Valet, V. Courtillot, Numerical models of the geodynamo and observational constraints. Geochem. Geophys. Geosyst. **1** (2000), paper 2000GC000062
T.E. Dowling, Dynamics of Jovian atmospheres. Ann. Rev. Fluid Mech. **27**, 293–334 (1995)
L. Elkins-Tanton, E. Parmentier, P. Hess, Magma ocean fractional crystallization and cumulate overturn in terrestrial planets: Implications for Mars. Meteorit. Planet. Sci. **38**, 1753–1771 (2003)
L. Elkins-Tanton, S. Zaranek, E. Parmentier, P. Hess, Early magnetic field and magmatic activity on Mars from magma ocean cumulate overturn. Earth Planet. Sci. Lett. **236**, 1–12 (2005)
M. Evonuk, G.A. Glatzmaier, Thermal convection in a 3D rotating density-stratified giant planet without a core. Geophys. J. Int. (2009, under review)
V.C.A. Ferraro, The nonuniform rotation of the Sun and its magnetic field. Mon. Not. R. Astron. Soc. **97**, 458–472 (1937)
H. Frey, R. Shultz, Large impact basins and the mega-impact origin for the crustal dichotomy on Mars. Geophys. Res. Lett. **15**, 229–232 (1988)
M. Fuller, S. Cisowski, Lunar paleomagnetism, in *Geomagnetism*, vol. 2, ed. by J. Jacobs (Academic Press, San Diego, 1987), pp. 307–456
I. Garrick-Bethel, B. Weiss, D. Shuster, J. Buz, Early lunar magnetism. Science **323**, 356–359 (2009)
G. Giampieri, A. Balogh, Mercury's thermoelectric dynamo model revisited. Planet. Space Sci. **50**, 757–762 (2002)
G. Giampieri, M. Dougherty, Rotation rate of Saturn's interior from magnetic field observations. Geophys. Res. Lett. **31** (2004). doi:10.1029/2004GL020194
G. Giampieri, M.K. Dougherty, E.J. Smith, C.T. Russell, A regular period for Saturn's magnetic field that may track its internal rotation. Nature **441**, 62–64 (2006)
P.A. Gilman, J. Miller, Dynamically consistent nonlinear dynamos driven by convection in a rotating spherical shell. Astrophys. J. Suppl. **46**, 211–238 (1981)
K.H. Glassmeier, H.U. Auster, U. Motschmann, A feedback dynamo generating Mercury's magnetic field. Geophys. Res. Lett. **34** (2007a). doi:10.1029/2007GL031662
K.H. Glassmeier et al., Electromagnetic induction effects and dynamo action in the Hermean system. Space Science Rev. **132**, 511–527 (2007b)
G.A. Glatzmaier, Numerical simulations of stellar convective dynamos I. The model and method. J. Comput. Phys. **55**, 461–484 (1984)

G.A. Glatzmaier, Geodynamo simulations—How realistic are they? Ann. Rev. Earth Planet. Sci. **30**, 237–257 (2002)

G.A. Glatzmaier, A saturnian dynamo simulation. American Geophysical Union Fall Meeting 2005, San Francisco, CA (2005)

G.A. Glatzmaier, A note on Constraints on deep-seated zonal winds inside Jupiter and Saturn. Icarus **196**, 665–666 (2008)

G.A. Glatzmaier, P.A. Gilman, Compressible convection in a rotating spherical shell. III. Analytic model for compressible vorticity waves. Astrophys. J. Suppl. **45**, 381–388 (1981)

G.A. Glatzmaier, P.A. Gilman, Compressible convection in a rotating spherical shell. V. Induced differential rotation and meridional circulation. Astrophys. J. **256**, 316–330 (1982)

G.A. Glatzmaier, R.S. Coe, L. Hongre, P.H. Roberts, The role of the Earth's mantle in controlling the frequency of geomagnetic reversals. Nature **401**, 885–890 (1999)

G.A. Glatzmaier, M. Evonuk, T.M. Rogers, Differential rotation in giant planets maintained by density-stratified turbulent convection. Geophys. Astrophys. Fluid Dyn. **103**, 31–51 (2009)

P. Goldreich, S. Soter, Q in the solar system. Icarus **5**, 375–389 (1966)

N. Gomez-Perez, M. Heimpel, Numerical models of zonal flow dynamos: an application to the ice giants. Geophys. Astrophys. Fluid Dyn. **101**, 371–388 (2007)

J. Grosser, K.H. Glassmeier, A. Stadelmann, Induced magnetic field effects at planet Mercury. Planet. Space Sci. **52**, 1251–1260 (2004)

E. Grote, F. Busse, Hemispherical dynamos generated by convection in rotating spherical shells. Phys. Rev. E **62**, 4457–4460 (2000)

E. Grote, F. Busse, A. Tilgner, Convection-driven quadrupolar dynamos in rotating spherical shells. Phys. Rev. E **60**, 5025–5028 (1999)

E. Grote, F. Busse, A. Tilgner, Regular and chaotic spherical dynamos. Phys. Earth Planet. Int. **117**, 259–272 (2000)

D. Gubbins, C. Barber, S. Gibbons, J. Love, Kinematic dynamo action in a sphere II. Symmetry selection. Proc. R. Soc. Lond. A **456**, 1669–1683 (2000)

T. Guillot, A comparison of the interiors of Jupiter and Saturn. Planet. Space Sci. **47**, 1183–1200 (1999)

T. Guillot, The interiors of giant planets: models and outstanding questions. Ann. Rev. Earth Planet. Sci. **33**, 493–530 (2005)

T. Guillot, D.J. Stevenson, W.B. Hubbard, D. Saumon, The interior of Jupiter, in *Jupiter: The Planet, Satellites and Magnetosphere*, ed. by F. Bagenal, T. Dowling, W. McKinnon (Cambridge Univ. Press, Cambridge, 2004)

J. Halekas et al., Mapping of crustal magnetic anomalies on the lunar near side by the Lunar Prospector electron reflectometer. J. Geophys. Res. **106**, 27841–27852 (2001)

H.B. Hammel, I. de Pater, S. Gibbard, G.W. Lockwood, K. Rages, Uranus in 2003: zonal winds, banded structure, and discrete features. Icarus **175**, 534–545 (2005)

C.J. Hansen, S.D. Kawaler, *Stellar Interiors: Physical Principles, Structure, and Evolution. Astronomy and Astrophysics Library* (Springer, Berlin, 1994)

J.E. Hart, G.A. Glatzmaier, J. Toomre, Spacelaboratory and numerical simulations of thermal convection in a rotating hemispherical shell with radial gravity. J. Fluid Mech. **173**, 519–544 (1986)

H. Harder, Phase transitions and the three-dimensional planform of thermal convection in the Martian mantle. J. Geophys. Res. **103**, 16775–16797 (1998)

S. Hauck, J. Aurnou, A. Dombard, Sulfur's impact on core evolution and magnetic field generation on Ganymede. J. Geophys. Res. **111** (2006). doi:10.1029/2005JE002557

M. Heimpel, N. Gomez Perez, Numerical models of the transition from zonal flow to dynamo action in Jupiter and Saturn. American Geophysical Union Fall Meeting 2008, P11B-1276, San Francisco, CA (2008)

M. Heimpel, J. Aurnou, F. Al-Shamali, N. Gomez-Perez, A numerical study of dynamo action as a function of spherical shell geometry. Earth Planet. Sci. Lett. **236**, 542–557 (2005a)

M. Heimpel, J. Aurnou, J. Wicht, Simulation of equatorial and high-latitude jets on Jupiter in a deep convection model. Nature **438**, 193–196 (2005b)

J.M. Hewitt, D.P. McKenzie, N.O. Weiss, Dissipative heating in convective flows. J. Fluid Mech. **68**, 721–738 (1975)

D. Heyner et al., Concerning the initial temporal evolution of a Hermean feedback dynamo. Earth Planet. Sci. Lett. (2009, submitted)

L. Hood, N. Artemieva, Antipodal effects of lunar basin-forming impacts: Initial 3D simulations and comparisons with observations. Icarus **193**, 485–502 (2008)

W.B. Hubbard, Gravitational signature of Jupiter's deep zonal flows. Icarus **137**, 357–359 (1990)

W. Hubbard, M. Podolak, D. Stevenson, Interior of Neptune, in *Neptune and Triton*, ed. by D. Cruickshank (University of Arizona Press, Tucson, 1995), pp. 109–138

N. Ishihara, S. Kida, Axial and equatorial magnetic dipoles generated in a rotating spherical shell. J. Phys. Soc. Jpn. **69**, 1582–1585 (2000)
C.A. Jones, K.M. Kuzanyan, Compressible convection in the deep atmospheres of giant planets. Icarus (2009, in press)
Y. Ke, V. Solomatov, Early transient superplumes and the origin of the Martian crustal dichotomy. J. Geophys. Res. **111** (2006). doi:10.1029/2005JE002631
M. Kivelson et al., Discovery of Ganymede's magnetic field by the Galileo spacecraft. Nature **384**, 537–541 (1996)
M. Kono, P.H. Roberts, Recent geodynamo simulations and observations of the geomagnetic field. Rev. Geophys. **40**, 4-1–53 (2002)
W. Kuang, W. Jiang, T. Wang, Sudden termination of Martian dynamo? Implications from subcritical dynamo simulations. Geophys. Res. Lett. **35** (2008). doi:10.1029/2008GL034183
C. Kutzner, U. Christensen, From stable dipolar towards reversing numerical dynamos. Phys. Earth Planet. Int. **131**, 29–45 (2002)
Y. Lian, A.P. Showman, Generation of zonal jets by moist convection on the giant planets. Icarus (2009, under review)
J. Liu, P.M. Goldreich, D.J. Stevenson, Constraints on deep-seated zonal winds inside Jupiter and Saturn. Icarus **196**, 506–517 (2008)
J. Love, Dynamo action and the nearly axisymmetric magnetic field of Saturn. Geophys. Res. Lett. **27**, 2889–2892 (2000)
J.B. Manneville, P. Olson, Banded convection in rotating fluid spheres and the circulation of the jovian atmosphere. Icarus **122**, 242–250 (1996)
J. Margot, S. Peale, R. Jurgens, M. Slade, I. Holin, Large longitude libration of Mercury reveals a molten core. Science **316**, 710–714 (2007)
M. Marinova, O. Aharonson, E. Asphaug, Mega-impact formation of the Mars hemispheric dichotomy. Nature **453**, 1216–1219 (2008)
N. Ness, K. Behannon, R. Lepping, Y. Whang, Magnetic field of Mercury confirmed. Science **255**, 204–206 (1975)
N. Ness, K. Behannon, R. Lepping, Y. Whang, Observations of Mercury's magnetic field. Icarus **28**, 479–488 (1976)
N. Ness et al., Magnetic field studies by Voyager 1: Preliminary results at Saturn. Science **212**, 211–217 (1981)
F. Nimmo, D. Stevenson, Influence of early plate tectonics on the thermal evolution and magnetic field of Mars. J. Geophys. Res. **105**, 11969–11979 (2000)
F. Nimmo, S. Hart, D. Korycansky, C. Agnor, Implications of an impact origin for the martian hemispheric dichotomy. Nature **453**, 1220–1223 (2008)
M. Podolak, W. Hubbard, D. Stevenson, Models of Uranus' interior and magnetic field, in *Uranus*, ed. by J. Bergstralh, E. Minor, M. Matthews (University of Arizona Press, Tucson, 1991), pp. 29–61
C.C. Porco, R.A. West, A. McEwen et al., Cassini imaging of Jupiter's atmosphere. satellites, and rings. Science **299**, 1541–1547 (2003)
J. Proudman, On the motion of solids in a liquid possessing vorticity. Proc. R. Soc. Lond. A **92**, 408–424 (1916)
P.B. Rhines, Waves and turbulence on a beta-plane. J. Fluid Mech. **69**, 417–443 (1975)
J. Roberts, S. Zhong, Degree-1 convection in the Martian mantle and the origin of the hemispheric dichotomy. J. Geophys. Res. **111** (2006). doi:10.1029/2005JE002668
A. Sanchez-Lavega, S. Perez-Hoyos, J.F. Rojas, R. Hueso, R.G. French, A strong decrease in Saturn's equatorial jet at cloud level. Nature **423**, 623–625 (2002)
G. Sarson, C. Jones, K. Zhang, G. Schubert, Magnetoconvection dynamos and the magnetic fields of Io and Ganymede. Science **276**, 1106–1108 (1997)
G. Schubert, M. Ross, D. Stevenson, T. Spohn, Mercury's thermal history and the generation of its magnetic field, in *Mercury*, ed. by F. Vilas, C. Chapman, M. Matthews (University of Arizona Press, Tucson, 1988), pp. 429–460
G. Schubert, K. Chan, X. Liao, K. Zhang, Planetary dynamos: Effects of electrically coonducting flows overlying turbulent regions of magnetic field generation. Icarus **172**, 305–315 (2004)
A.P. Showman, K. Menou, J.Y.-K. Cho, Atmospheric circulation of hot Jupiters: a review of current understanding, in *APS Conf. Ser.*, ed. by F. Rasio et al. (APS, San Francisco, 2007)
E. Smith et al., Saturn's magnetic field and magnetosphere. Science **207**, 407–410 (1980)
L. Srnka, Magnetic dipole moment of a spherical shell with TRM acquired in a field of internal origin. Phys. Earth Planet. Int. **11**, 184–190 (1976)
S. Stanley, A dynamo model of Saturn's extremely axisymmetric internal magnetic field. Eos Trans. AGU **89**(23), Jt. Assem. Suppl., Abstract GP31A-02 (2008)

S. Stanley, J. Bloxham, Convective-region geometry as the cause of Uranus' and Neptune's unusual magnetic fields. Nature **428**, 151–153 (2004)
S. Stanley, J. Bloxham, Numerical dynamo models of Uranus' and Neptune's magnetic fields. Icarus **184**, 556–572 (2006)
S. Stanley, A. Mohammadi, Effects of an outer thin stably stratified layer on planetary dynamos. Phys. Earth Planet. Int. **168**, 179–190 (2008)
S. Stanley, J. Bloxham, W. Hutchison, M. Zuber, Thin shell dynamo models consistent with Mercury's weak observed magnetic field. Earth Planet. Sci. Lett. **234**, 27–38 (2005)
S. Stanley, L. Elkins-Tanton, M. Zuber, E.M. Parmentier, Mars' paleomagnetic field as the result of a single-hemisphere dynamo. Science **321**, 1822–1825 (2008)
S.V. Starchenko, C.A. Jones, Typical velocities and magnetic field strengths in planetary interiors. Icarus **157**, 426–435 (2002)
D. Stegman, A. Hellinek, S. Zatman, J. Baumgardner, M. Richards, An early lunar core dynamo driven by thermochemical mantle convection. Nature **421**, 143–146 (2003)
S. Stellmach, P. Garaud, A. Traxler, N. Brummell, T. Radko, 3D simulations of layer formation by compositionally driven double diffusive convection. Science (2009, under review)
A. Stephenson, Crustal remanence and the magnetic moment of Mercury. Earth Planet. Sci. Lett. **28**, 454–458 (1975)
D. Stevenson, Saturn's luminosity and magnetism. Science **208**, 746–748 (1980)
D. Stevenson, Reducing the non-axisymmetry of a planetary dynamo and an application to Saturn. Geophys. Astrophys. Fluid Dyn. **21**, 113–127 (1982)
D. Stevenson, Planetary magnetic fields. Rep. Prog. Phys. **46**, 555–620 (1983)
D. Stevenson, Mercury's magnetic field: a thermoelectric dynamo? Earth Planet. Sci. Lett. **82**, 114–120 (1987)
D. Stevenson, Mars' core and magnetism. Nature **412**, 214–219 (2001)
D.J. Stevenson, E.E. Salpeter, The phase diagram and transport properties of hydrogen-helium fluid planets. Astrophys. J. Suppl. **35**, 221–237 (1977)
F. Takahashi, M. Matsushima, Dipolar and non-dipolar dynamos in a thin shell geometry with implications for the magnetic field of Mercury. Geophys. Res. Lett. **33** (2006). doi:10.1029/2006GL025792
G. Taylor, Motion of solids in fluids when the flow is not irrotational. Proc. R. Soc. Lond. A **93**, 99–113 (1917)
A. Tilgner, Zonal wind Driven by inertial modes. Phys. Rev. Lett. **99** (2007). doi:10.1103/PhysRevLet.99.194501
R. Vilim, S. Stanley, S. Hauck, Dynamo generation in the presence of iron snow zones: Application to mercury's weak surface field. Eos Trans. AGU **89**(53), Fall Meet. Suppl., Abstract U21A-0004 (2008)
W. Watters, M. Zuber, B. Hager, Thermal perturbations caused by large impacts and consequences for mantle convection. J. Geophys. Res. **114** (2009). doi:10.1029/2007JE002964
S.A. Weinstein, The effects of a deep mantle endothermic phase change on the structure of thermal convection in silicate planets. J. Geophys. Res. **100**, 11719–11728 (1995)
B. Weiss et al., Records of an ancient Martian magnetic field in ALH84001. Earth Planet. Sci. Lett. **201**, 449–463 (2002)
J. Wicht et al., The origin of Mercury's internal magnetic field. Space Sci. Rev. **132**, 261–290 (2007)
D. Wilhelms, S. Squyres, The Martian hemispheric dichotomy may be due to a giant impact basin. Nature **309**, 138–140 (1984)
G.P. Williams, Planetary circulations: 1. Barotropic representation of Jovian and terrestrial turbulence. J. Atmos. Sci. **35**, 1399–1424 (1978)
G.P. Williams, Jovian dynamics. Part III: multiple, migrating, and equatorial jets. J. Atmos. Sci. **60**, 1270–1296 (2003)
J. Williams, F. Nimmo, Thermal evolution of the Martian core: Implications for an early dynamo. Geology **32**, 97–100 (2004)
C. Yoder, A. Konopliv, D. Yuan, E. Standish, W. Folkner, Fluid core size of Mars from detection of the solar tide. Science **300**, 299–303 (2003)
S. Zhong, M. Zuber, Degree-1 mantle convection and the crustal dichotomy on Mars. Earth Planet. Sci. Lett. **189**, 75–84 (2001)
M. Zuber et al., The geophysics of Mercury: Current status and anticipated insights from the MESSENGER mission. Space Sci. Rev. **131**, 105–132 (2007)

Space Sci Rev (2010) 152: 651–664
DOI 10.1007/s11214-009-9572-z

Planetary Magnetic Fields: Achievements and Prospects

David J. Stevenson

Received: 6 July 2009 / Accepted: 26 July 2009 / Published online: 28 October 2009

Abstract The past decade has seen a wealth of new data, mainly from the Galilean satellites and Mars, but also new information on Mercury, the Moon and asteroids (meteorites). In parallel, there have been advances in our understanding of dynamo theory, new ideas on the scaling laws for field amplitudes, and a deeper appreciation on the diversity and complexity of planetary interior properties and evolutions. Most planetary magnetic fields arise from dynamos, past or present, and planetary dynamos generally arise from thermal or compositional convection in fluid regions of large radial extent. The relevant electrical conductivities range from metallic values to values that may be only about one percent or less that of a typical metal, appropriate to ionic fluids and semiconductors. In all planetary liquid cores, the Coriolis force is dynamically important. The maintenance and persistence of convection appears to be easy in gas giants and ice-rich giants, but is not assured in terrestrial planets because the quite high electrical conductivity of an iron-rich core guarantees a high thermal conductivity (through the Wiedemann-Franz law), which allows for a large core heat flow by conduction alone. This has led to an emphasis on the possible role of ongoing differentiation (growth of an inner core or "snow"). Although planetary dynamos mostly appear to operate with an internal field that is not very different from $(2\rho\Omega/\sigma)^{1/2}$ in SI units where ρ is the fluid density, Ω is the planetary rotation rate and σ is the conductivity, theoretical arguments and stellar observations suggest that there may be better justification for a scaling law that emphasizes the buoyancy flux. Earth, Ganymede, Jupiter, Saturn, Uranus, Neptune, and probably Mercury have dynamos, Mars has large remanent magnetism from an ancient dynamo, and the Moon might also require an ancient dynamo. Venus is devoid of a detectable global field but may have had a dynamo in the past. Even small, differentiated planetesimals (asteroids) may have been capable of dynamo action early in the solar system history. Induced fields observed in Europa and Callisto indicate the strong likelihood of water oceans in these bodies. The presence or absence of a dynamo in a terrestrial body (including Ganymede) appears to depend mainly on the thermal histories and energy sources of these bodies, especially the convective state of the silicate mantle and the existence and history of a growing inner solid core. As a consequence, the understanding of planetary

D.J. Stevenson (✉)
California Institute of Technology, 150-21, Pasadena, CA 91125, USA
e-mail: djs@gps.caltech.edu

magnetic fields depends as much on our understanding of the history and material properties of planets as it does on our understanding of the dynamo process. Future developments can be expected in our understanding of the criterion for a dynamo and on planetary properties, through a combination of theoretical work, numerical simulations, planetary missions (MESSENGER, Juno, etc.) and laboratory experiments.

Keywords Planets · Magnetism · Dynamos

1 Overview

Magnetic fields are everywhere in the universe. In particular, they are often a characteristic of planets and most of the planets in our solar system have substantial fields. In many planets, the cause of this field is electrical currents deep within the body and its presence and behavior tells us something about the physical state and dynamics of the material deep within the planet. Indeed, the magnetic field is one the few ways of probing the interior structure. Moreover, the field can usually be determined remotely (i.e., by an orbiting or flyby spacecraft). Many conventional geophysical techniques for determining interior structure (e.g., seismology) are not readily accessible from orbit or flyby. In some cases, the field is only present as remanent magnetism (the "permanent" magnetism of minerals in the outer part of a solid body) but even then it may be telling us about past dynamics of the deep interior. In a few cases (notably the Galilean satellites Europa and Callisto) there are induced fields arising from the time variation of an external field. These are also telling us something important about the body. My focus here is on the information magnetic fields provide us about history and structure of bodies in our solar system, not just the planets as conventionally defined but also satellites and even small bodies (e.g., asteroids). The absence of a present-day large field (e.g Venus, Mars) is just as interesting as its presence and the nature of the field when present (i.e., its harmonic spectrum, possible time variability) is also of great interest.

In many respects, this chapter is an update of a recent review (Stevenson 2003). However, a remarkable amount of relevant new work and some new observations have occurred since then and this necessitates not merely an update but also some new perspectives. Additionally, this chapter offers some views on the future of the field, part of which is linked to future spacecraft missions or extrasolar planet observations and part of which is linked to future directions in theory and in understanding the material properties and dynamics of planets. The chapter ends with a commentary on each planetary body (including hypothetical planets in other planetary systems.)

2 Observed Fields

Details of planetary observations are well covered elsewhere. Here, the intent is to make a summary of the current observational situation, with comments on the particular distinctive features of these observations. See Table 1.

3 The Nature of Dynamos

The central idea for understanding large, planetary scale magnetic fields is the hydromagnetic dynamo. The essence of a dynamo lies in electromagnetic induction: The creation of

Table 1 Observed magnetic fields (*based on* Stevenson 2003)

Planet or satellite	Observed surface field (in Tesla, approximate)	Comments and interpretation
Mercury	2×10^{-7}	Not well characterized or understood yet, but MESSENGER data suggest a dynamo
Venus	$<10^{-8}$ (global); no useful constraint on local fields	No dynamo at present. Small remanence is possible
Earth	5×10^{-5}	Core dynamo
Moon	Patchy; (10^{-9}–10^{-7} T) No global field	Ancient dynamo? Remanent magnetism is related to impact. More data needed
Mars	Patchy but locally strong (10^{-9}–10^{-4} T); may field	Ancient dynamo, Strong remanent magnetism
Jupiter	4.2×10^{-4}	Dynamo (extends to near surface). Earthlike dipole tilt.
Io	$<10^{-6}$?	Complex (deeply imbedded in Jovian field. Data do not require a dynamo
Europa	10^{-7}	Induction response (Salty Water Ocean)
Ganymede	2×10^{-6}	Dynamo likely. May also exhibit an induction response (like Europa and Callisto)
Callisto	4×10^{-9}	Induction response (Salty Water Ocean)
Saturn	2×10^{-5}	Dynamo (deep down?). Field appears to be spin-axisymmetric
Titan	$<10^{-7}$	No evidence for a dynamo or internal induction response.
Uranus	2×10^{-5}	Dynamo with large dipole tilt and quadrupole
Neptune	2×10^{-5}	Dynamo with large dipole tilt and quadrupole

emf and associated currents and field through the motion of conducting fluid across magnetic field lines (Moffatt 1978; Parker 1979). This can expressed mathematically through the combination of Ohm's law, Ampere's law and Faraday's law of induction (often called the *induction equation*):

$$\partial \mathbf{B}/\partial t = \lambda \nabla^2 \mathbf{B} + \nabla \mathbf{x}(\mathbf{v} \times \mathbf{B}) \qquad (1)$$

where $\mathbf{B}$ is the magnetic field, $\mathbf{v}$ is the fluid motion (relative to a rigidly rotating frame of reference, the normal choice for planetary fluid dynamical problems) and $\lambda \equiv 1/\mu_0 \sigma$ is known as the *magnetic diffusivity* (μ_0 is the permeability of free space, $4\pi \times 10^{-7}$ in SI units and σ is the electrical conductivity in S/m, and assumed constant). If there is no fluid motion ($\mathbf{v} = \mathbf{0}$) then the field will undergo free ("diffusive") decay on a timescale $\tau \sim L^2/\pi^2\lambda \sim$ (3000 yr) $\cdot (L/1000\ \text{km})^2 \cdot (1\ \text{m}^2\ \text{sec}^{-1}/\lambda)$ where L is some characteristic length scale of the field, no more than the radius of the electrically conducting region (the core). In terrestrial planets, the electrical conductivity corresponds to liquid metallic iron, modified by alloying with other elements (e.g., sulfur). This corresponds to $\sigma \sim 5 \times 10^5$ S/m and $\lambda \sim 2$ m^2/sec (cf. Merrill et al. 1996) but the uncertainties on this value remain large (roughly a factor of two, sometimes more) depending on pressure and assumed composition. In gas giants, shock wave experiments suggest that hydrogen attains the lowest conductivities appropriate to metals ($\sigma \sim 2 \times 10^4$ to 2×10^5 S/m, $\lambda \sim 5$ to 50 m^2/sec) at pressure $P \sim 1.5$ Megabar and $T \sim a$ few thousand degrees (Nellis 2000). This corresponds to the conditions at 0.8 of Jupiter's radius or 0.5 of Saturn's radius. The dynamo in Jupiter may operate at radii beyond the peak conductivity reached in these experiments (this is discussed further in the sections on Jupiter and Saturn). Shock wave experiments (Nellis et al. 1997) suggest that an "ice" mixture (dominated by water, but containing many ionic species) will reach conductivities of $\sigma \sim 1 \times 10^4$ S/m ($\lambda \sim 100$ m^2/sec), conditions met in Uranus and Neptune at around 0.7 of their radii.

In each case, the free decay time is much less than the age of the solar system. For example, in Earth's core, this timescale is ten thousand years or so. The fact that free decay times are geologically short means that if a planet has a large field *now* then it must have a means of generating the field *now*; it cannot rely on some primordial field or pre-existing field.

Dimensional analysis of the induction equation above immediately suggests that the importance of a flow is characterized by the *magnetic Reynolds number* $R_m \equiv vL/\lambda$ where v is a characteristic fluid velocity and L is a characteristic length scale of the motions or field (e.g., the core radius). The existence of solutions in which the field does not decay to zero after a long time depends on R_m and a value exceeding $\sim$10 or 100 is thought sufficient, but this is a vague criterion: Which velocity and how is it determined? We return to this below.

4 The Nature of Planets and Their Evolution

Planets are conveniently categorized according to their primary constituents (De Pater and Lissauer 2001; Stevenson 2002; Guillot 1999). Planets and their satellites are not distinguished because satellites are subject to the same planetary processes if they are sufficiently large (>1000 km radius, roughly). *Terrestrial planets* (Mercury, Venus, Earth, Moon, Mars, and Io) consist primarily of materials that condense at high temperatures: oxides and silicates of iron and magnesium, together with metallic iron. The high density and lower melting point of iron alloys relative to silicates generally lead us to expect that these bodies form metallic iron-rich cores. These cores are generally at least partially liquid, even after 4.5 billion years of cooling, because at least one of the core-forming constituents (sulfur) lowers the freezing point of the iron alloy below the operating (convecting) temperature of the overlying mantle. If the sulfur content is small then the fluid region of a core may be thin. This persistence of a liquid layer arises from the eutectic nature of the phase diagram and, more generally, the fact that one must go to a temperature lower than that needed to get heat out through the mantle in order to produce *complete* freezing. It is for this reason that the presence or absence of a dynamo should not be thought of as related to the presence or absence of an outer liquid core but rather to the vigor of motions in that layer. This argument only fails for very small bodies (even smaller than Earth's moon). *Gas giants* (Jupiter and Saturn) have hydrogen as their major constituent. They may possess "Earthlike" central cores but this may have little bearing on understanding their magnetic fields. Freezing is a non-issue. *Ice giants* (Uranus and Neptune) contain a hydrogen-rich envelope but their composition is rich in H_2O, CH_4 and NH_3 throughout much of the volume, extending out to perhaps $\sim$80% of their radii. Freezing in these bodies seems unlikely given that their interiors are mixtures, but would perhaps be marginally possible if one thought that the freezing curve of *pure* water were relevant (cf. French et al. 2009). *Large icy satellites and solid icy planets* (Ganymede, Callisto, Titan, Triton, Pluto; also Europa as a special case) contain both ice (predominantly H_2O) and rock. They may be differentiated into an Earthlike structure (silicate rock and possibly an iron-rich core), overlain with varying amounts of primarily water ice, or (as in the case of Callisto) the ice and rock may be partly mixed. Europa is a special case because the water rich layer is relatively small and may be mostly liquid.

Planets differ from small masses of the same material because of the action of gravity and the difficulty of eliminating heat on billion year time scales. Gravity causes pressure, which can modify the thermodynamic and phase equilibrium behavior of the constituents. This is why bodies rich in materials that are poor conductors at low pressures (e.g., hydrogen, water) may nonetheless have high conductivity at depth. In giant gas and ice planets, the heat of formation is sufficient to guarantee fluidity and convection. In terrestrial planets, the difficulty of eliminating the heat of formation and subsequent radioactive heat generation leads to unavoidably large internal temperatures, usually sufficient to guarantee fluidity of a metallic core, and sustained mantle convection. Terrestrial core convection may not be easily sustained, however, because the heat carried by conduction alone is typically within a factor of two of the expected core heat flow.

Some of the issues can be appreciated by considering the simple case of a generic planet in which the heat flow in the proposed dynamo region arises primarily from cooling, and no phase changes (e.g., freezing or gravitational differentiation) take place. (In terrestrial planets, a major source of *surface* heat flow is radioactive decay, but the radioactive elements are thought not to reside in the core. In giant planets, cooling from a primordial hot state probably dominates at all levels, though gravitational differentiation may also contribute significantly; Guillot 1999). In this approximation, and assuming that the core cools everywhere at the about the same rate, we have

$$F_{\text{total}}(r) = -\rho_c C_p r (dT_c/dt)/3 \tag{2}$$

where $F_{\text{total}}(r)$ is the total heat flow at radius r, ρ_c is the mean core density, C_p is the specific heat, T_c is the mean core temperature and t is time. In fluid cores, the viscosity is so small that it plays a negligible role in the criterion for convection (totally unlike the case for convection in solid silicate mantles). To an excellent approximation, the condition for convection is that the heat flow must exceed that which can be carried by conduction along an adiabat:

$$F_{\text{total}} > F_{\text{cond,ad}} \equiv k\alpha \text{Tg}(r)/C_p \quad \Leftrightarrow \quad \text{thermal convection} \tag{3}$$

where k is the thermal conductivity, α is the coefficient of thermal expansion, and $g(r)$ is the gravitational acceleration at radius r. If the heat flow were less than this value then the core would be stably stratified (vertically displaced fluid elements would tend to oscillate). We can approximate $g(r)$ by $4\pi G\rho_c r$ where G is the gravitational constant. Notice that both F_{total} and $F_{\text{cond,ad}}$ are linear in r in this approximation, so the comparison of their magnitudes will be the same independent of planet size and location in the core. From this, we obtain a critical cooling rate that must be exceeded for convection. It is typically about 100 K/Ga for parameters appropriate to Earth's core and may be as large as 300 or 400 K/Ga for smaller (but Earthlike) cores. e.g., Ganymede. It is substantially lower for giant gas or ice planets, where the conductivity is lower. For Earth's core, a cooling rate like 100 K/Ga corresponds to a heat flow at the top of the core of around 20 mW/m^2.

From condensed matter physics, we also have the Wiedemann-Franz "law" (e.g., Poirier 1991);

$$k/\sigma T \equiv L \approx 2 \times 10^{-8} \text{ W Ohm/K}^2 \tag{4}$$

where L is called the Lorenz number. This applies to a metal in which the electrons dominate both the heat and charge transport and is accurate to better than a few tens of percent. Combined with (3) this implies an *upper* bound to the electrical conductivity in order that thermal convection take place. For nominal parameter choices, this upper bound is roughly the actual value of the electrical conductivity in earth's core. This makes the important point that high electrical conductivity may indirectly prevent a dynamo! (The use of Wiedemann-Franz is specifically for terrestrial planets: In gas and ice giants, we have independent estimates for thermal conductivity that show that the heat flow along the adiabat is much smaller than the actual heat flow.)

If the core is cooling and the central temperature drops below the liquidus for the core alloy, then an inner core will nucleate. In Earth, we know from seismic evidence that the core is $\sim$10% less dense than pure iron and many suggestions have been offered for the identity of the light elements that are mixed with the iron (Poirier 1994; Gessmann et al. 2001). As the inner core freezes, it is likely that some or all of these light elements are partially excluded from the crystal structure. The introduction of light elements into the lowermost core fluid will tend to promote convection and cause mixing throughout all or most of the outer core, provided the cooling is sufficiently fast (Gubbins 1977; Loper 1978; Labrosse et al. 2001; Buffett and Bloxham 2002). Latent heat release at the inner core-outer core boundary will also contribute to the likelihood of convection. However, inner core growth permits outer core convection even when the heat flow through the core-mantle boundary is less (perhaps much less) than the heat carried by conduction along an adiabat. In this regime, the temperature gradient is very slightly less steep than adiabatic and the compositional convection carries heat *downwards*. The total heat flux is still outwards, of course, since the heat carried by conduction is large. This state is possible because the buoyancy release associated with the compositional change exceeds the work done against the unfavorable thermal stratification.

It is possible but not certain that terrestrial planets require inner core growth in order to sustain a dynamo at the present epoch. It does *not* follow that there is a one-to-one correspondence between presence of an inner core and presence of a dynamo. One can have an inner core without a dynamo (conceivably present Mars if the cooling of the core is insufficiently rapid or absent). One can also imagine a dynamo without a growing inner core (conceivably early Earth or other bodies early in their history) if the core were then cooling much more rapidly than now. Partial freeze-out of light material from the core is also a possible dynamo driving mechanism (Buffett et al. 2000) and has been advocated by this author at recent conferences. There may also be more complicated phase diagrams that allow freeze-out away from either the top or bottom of the core; this is mentioned in the summaries offered below for specific planets; especially Mercury and Mars.

5 Dynamo Theory and Dynamo Scaling

Numerical and analytical work suggest that a dynamo will exist if the fluid motions have certain desired features and the magnetic Reynolds number R_m exceeds about 10 or 100 (Roberts and Glatzmaier 2000; Jones 2000; Busse 2000; Gubbins 2001; Christensen et al. 2001, 2009). It seems likely that fluid motions of the desired character arise naturally in a convecting fluid (irrespective of the source of fluid buoyancy), provided the Coriolis force has a large effect on the flow, i.e., Rossby number $Ro \equiv v/2\Omega L < 1$ where Ω is the planetary rotation rate. This is easily satisfied for any plausible fluid motion of interest, even for slowly rotating planets such as Venus.

Although the dynamo mechanism is much studied, it is still imperfectly understood, despite the recent dramatic advances in numerical simulation referenced above. In particular, we do not know the quantitatively precise sufficient conditions for the existence of a planetary dynamo. How can we assess the value for the velocity v that enters into the "typical" estimate of magnetic Reynolds number? One possible estimate comes from mixing length theory (Clayton 1968; Stevenson 1979, 1987a):

$$v_{ml} \sim 0.3(l F_{\text{conv}}/\rho H_T)^{1/3} \tag{5}$$

where v_{ml} is the predicted velocity, l is the "mixing length" (plausibly the size of the core), $F_{\text{conv}} = F_{\text{total}} - F_{\text{cond.ad}}$, and $H_T \equiv C_p/\alpha g$ is the temperature scale height, not enormously larger than the core radius except in the limit of small bodies. An alternative estimate, plausibly more relevant if a dynamo is operating, assumes that buoyancy, Coriolis and Lorentz forces are comparable (Jones 2000). In this magnetostrophic regime,

$$v_{\text{mac}} \sim (F_{\text{conv}}/\rho\Omega H_T)^{1/2} \tag{6}$$

and this is typically an order of magnitude or so smaller than v_{ml}. Note that slow rotation is favorable (i.e. increases convective velocity). We can also envisage estimates intermediate between (5) and (6) in which the dependence on rotation is intermediate between inverse square-root and no dependence. As discussed in Stevenson (2003) both parameterizations (but especially the simple mixing length theory choice) have the property that the convective velocity rises rapidly once F is positive, because of the cube root and square-root behaviors, respectively. As a consequence, it follows that except for small bodies or bodies of low electrical conductivity (the ice giants, perhaps), the issue of sufficiently vigorous convection for a dynamo is almost identical to the issue of whether convection is possible at all. There is only a narrow range of conditions for which the convection is present but insufficiently vigorous for dynamo action. Of course, these arguments remain plausible rather than rigorous and one awaits a more quantitative assessment of this important question.

The expected value of the field has long fascinated people. There are two kinds of arguments that can be made on this question (with intermediates of these two extremes also possible). One extreme is to view the field amplitude as being a strictly dynamical issue, involving force balances but not (directly) involving the vigor of the convection (the buoyancy flux). In this picture, it has been argued that the expected field magnitude *inside the region of field generation* is given by Elsasser number $\Lambda \equiv \sigma B^2/2\rho\Omega$ of order unity, which implies $B \sim (2\rho\Omega/\sigma)^{1/2}$ where ρ is the fluid density. This is approximately satisfied by the values listed in Table 1 of observed fields (and see Stevenson 2003, for more details), especially if one allows that the field inside the dynamo region may be larger than at the top of the dynamo region by a factor of a few. The exception may be Uranus and Neptune, although downward extrapolation of their fields is difficult because they are not predominantly dipolar. The testing of this expectation is made difficult by quite large uncertainties in some parameters.

Recently, Christensen et al. (2009) made a persuasive case for a very different kind of scaling; in effect one in which there is a proportionality between the energy in the field and the "nominal" kinetic energy estimate provide by mixing length theory (5). A remarkable feature of this scaling (field proportional to the cube root of buoyancy flux, or $B^2/2\mu_0 \sim f\rho v_{ml}^2$ where f is a nearly universal dimensionless number) is that it does not give any dependence of field strength on either the rotation rate or the magnetic diffusivity. The evidence in favor of this scaling is partly theoretical but also from some stars, where there is a much larger buoyancy flux and a much larger observed field. This assumes that these stars operate in the same dynamo regime as planets. The new proposal is actually the same as the "scaling" suggested based on energy considerations by Stevenson et al. (1983) in their discussion of how Earth's field might change through geologic time depending on the presence or absence of an inner core. It must be stressed, however, that this old suggestion was not based on deeply considered arguments of how dynamos actually operate. The Elsasser number criterion can also be made consistent with the energy budget (at least in principle) by assigning most of the dissipation to higher harmonics of the field. The new scaling proposal should be regarded as a major development, given the arguments advanced in its support. Depending on one's point of view, this very different scaling law does as well, or as poorly, as the one based on Elsasser number. But certainly the theoretical evidence (numerical dynamo models) argues against a strict constancy of Elsasser number, so it seems likely that the Elsasser number "rule" is not well-justified.

6 Field Geometries

External to the planet and the large currents responsible for most of the field, the magnetic field **B** can be written as the gradient of a scalar potential that satisfies Laplace's equation. In the standard way, we can identify general solutions to Laplace's equation in terms $\propto Y_{lm} r^{-(l+1)}$ for internal sources, where Y_{lm} is a spherical harmonic, r is the distance from the center of the planet $l = 1$ is the dipole, $l = 2$ is the quadrupole and so on. Terms with $m = 0$ represent spin-axisymmetric components (if we choose the pole of coordinates to be the geographically defined pole of planet rotation), so (for example) $l = 1$ and $m = \pm 1$ represents the tilt of the dipole and the longitude of that tilted dipole. Planetary fields are sometimes described as "tilted, offset dipoles" but this is misleading at best. There is no fundamental significance to a dipole: A current distribution of finite extent will typically produce many additional harmonics. It might be imagined that all harmonics are comparably important at the core radius. However, many bodies have fields that are predominantly dipolar, in the sense that the power in the higher harmonics is significantly smaller than that in the dipole component, when evaluated at the core radius. For *Earth, Jupiter and Saturn* (and probably *Ganymede*, maybe also *Mercury*), the field is predominantly dipolar. The tilt of the dipole relative to the rotation axis is of order 10 degrees for Jupiter and Earth and near zero for Saturn. For *Uranus and Neptune*, the field is about equally dipole and quadrupole and the tilt of the dipole is 40–60 degrees. Evidently, Uranus and Neptune represent a different class of dynamos. The peculiarities of these planets are described below and attributed to distinctive features of their internal structure.

7 Induction Fields

The requirement for a significant induction field is much less restrictive than for a dynamo (Zimmer et al. 2000). The conductivity can be much smaller and the fluid does not have to be in motion (it can be even be a solid). For an external field that varies as $\exp[i\omega t]$, and a thin, conducting shell of thickness d and radius R, there will be a large induction response if the electromagnetic skin depth $(\lambda/\omega)^{1/2} < (Rd)^{1/2}$. For example, a layer of low-pressure salty water (such as Earth's oceans, with $\lambda \sim 10^6\ \mathrm{m^2/sec}$) will satisfy this for a thickness of order 10 km and $\omega \sim 2 \times 10^{-4}$ (corresponding to the frequency of Jupiter's tilted dipole field as it sweeps by Europa). A plausible estimate for R_m in such an ocean is 10^{-3} so there is no significant *internal* induction effect. The observed fields of Europa, Callisto are consistent with an externally induced induction field, and the most likely conductor is salty water (Stevenson 2000).

8 A Survey of the Planets

Mercury has been determined to have a liquid outer core (Margot et al. 2007) and some models predict that this core could continue to convect and perhaps sustain a dynamo (Stevenson et al. 1983; Schubert et al. 1988). Some models suggest that an additional energy source may be needed (e.g., Williams et al. 2007) but there are additional complications that may arise from new evidence of a more complicated Fe–S phase diagram than previously assumed; one that allows for the formation of "snow" (Chen et al. 2008). Mercury is nonetheless an enigma because the observed field is over an order of magnitude smaller than the field strength predicted for either scaling law described above. There are at least four possibilities: permanent magnetism (e.g., Aharonson et al. 2004), an exotic non-dynamo explanation such as thermoelectric currents (Stevenson 1987b; Giampieri and Balogh 2002), a dynamo that produces much larger internal (e.g., toroidal) fields than the observed external fields (Stanley et al. 2005), or a dynamo that for some reason fails to reach the expected field amplitude, for example by operating at greater depths in a core that is layered or partially stratified (e.g. Christensen 2006; Christensen and Wicht 2008). The conventional dynamo explanation is most likely, and is compatible with the earliest results from MESSENGER (Anderson et al. 2008), suggesting a relatively simple field geometry.

Venus is likely to have a liquid outer core (with or without an inner core) but has no dynamo at present. The predicted dynamo field is over two orders of magnitude larger than the observational upper bound. Equation (6) suggests that slow rotation may be good for dynamos (provided the Coriolis force remains dynamically important, as it is for all planets), so if Venus were like Earth in all respects except for its rotation then it would have no difficulty exceeding this upper bound. The most probable interpretation is that the liquid core of Venus does not convect. This could arise because there is no inner core (Stevenson et al. 1983) or because the core is currently not cooling. The absence of an inner core is plausible if the inside of Venus is hotter than the corresponding pressure level of Earth. This can arise because Earth has plate tectonics, which eliminates heat more efficiently than a stagnant lid form of mantle convection. Alternatively (or in addition), Venus' core may not be cooling at present because it is undergoing a transition in convective style following a resurfacing event ~700 Ma ago (Schubert et al. 1998). In this scenario, Venus had a dynamo in the past. Since the surface rocks are at a temperature below the blocking temperature of likely carriers of remanent magnetism, a small paleofield is marginally possible.

Earth remains imperfectly understood, a humbling reminder of the dangers of claiming an understanding of other planets. Growth of the inner core is thought essential for sustaining convection and sufficient energy to run the dynamo field (see references cited earlier for the need for compositional convection). Doubts have been expressed about whether Earth's field can be sustained for its known history (at least 3.5 Ga) if the inner core has existed for only of order a couple of billion years. An additional energy source may be needed; potassium-40 has been suggested. See Labrosse et al. (2001) for a discussion of this. Another possibility is that cooling rates of the lower mantle have been underestimated for earlier epochs.

Moon probably has a core that is at least partially liquid (Stevenson 1983; Williams et al. 2001). It has patches of strong crustal magnetization that may have been acquired following impacts and compression of conducting plasma at the antipode (Hood et al. 2001). It is not known whether the pre-impact field was necessarily a global field of the kind that only a dynamo produces. Even if it is a dynamo, it may (uniquely among planets in our solar system) have arisen through mechanical stirring of the inner core (Williams et al. 2001). Rapid cooling of a boundary layer immediately above the core mantle boundary might also conceivably maintain a dynamo for some time. The latest paleomagnetic results (Garrick-Bethell et al. 2009) make the demands on a dynamo less stringent than previously thought, but clearly more data are needed: This new work also casts doubt on much of the previous paleomagnetic work.

Mars had an ancient dynamo, probably in the period prior to 4.0 Ga (Acuna et al. 2001; Stevenson 2001). There are three possibilities for why this dynamo existed and then died:

(1) Core cooling decreased to the point where conductive heat loss dominated (but no inner core formed); Stevenson et al. (1983).
(2) Mars underwent a change in convective style, from an efficient mode (e.g., plate tectonics) to the currently observed stagnant lid mode. This would cause the mantle and core to stop cooling and turn off core convection and the dynamo (Nimmo and Stevenson 2000). This model would work irrespective of whether Mars has an inner core.
(3) The core of Mars froze sufficiently so that the remaining fluid region was too thin to sustain a dynamo. However, Mars may also be more complicated than these simple models; e.g. Stewart et al. (2007); and more needs to be known about paleointensities (cf. Weiss et al. 2008b). Unorthodox alternatives exist to conventional dynamos, e.g. Arkani-Hamed (2009).

Some *asteroids* underwent differentiation early in solar system history and had liquid cores that may have convected vigorously for a short period of time (of order a million years). These bodies may have had dynamos that were then responsible for the observed meteoritic paleomagnetism (Weiss et al. 2008a; Nimmo 2009). This exciting possibility deserves more study since it may give us some information on the lower limit of size needed for a dynamo.

Jupiter may have dynamo generation out to levels where hydrogen is only a semiconductor, perhaps 80 to 85% of the planet radius. However, coupling of the flows with the field may persist out to larger radii (Liu et al. 2008). There is a hint of complex field structure in the observations of the aurora (Grodent et al. 2008). Despite the central importance of Jupiter in our solar system there is not yet a successful numerical simulation that looks like Jupiter's observed field and contains the essential physics (e.g. very large variation in electrical conductivity and density).

Io exhibits no convincing evidence of a dynamo and no simple inductive response (Kivelson et al. 2001). Although Io has a metallic core, it might not be undergoing much long-term cooling if the mantle is heated steadily by tides. Recent astrometric data (Lainey et al. 2009) suggest that Io is spiraling in towards Jupiter but is in thermal equilibrium (i.e., as much heat is escaping as is being tidally generated). This suggests a non-steady–state regime for the thermal history but does not offer an immediate explanation for the absence of a dynamo.

Europa has a clear signature of an induction field (Zimmer et al. 2000) and no evidence of a permanent dipole. The induction field can be explained by a water ocean of similar conductivity to Earth's oceans, provided this ocean has a thickness exceeding $\sim$10 km. No other plausible source of the required conductivity has been suggested.

Ganymede has a clear signature of a permanent dipole (Kivelson et al. 1998). A permanent magnetism explanation is conceivable but unlikely, and the most reasonable interpretation is a dynamo in the metallic core. A liquid Fe–S core is expected in Ganymede. Nonetheless, this dynamo is surprising, partly because of Ganymede's size but mainly because of the difficulty in sustaining convection in such a small body. The presence of large amounts of sulfur and large ^{40} K mantle heating may help. There may also be a much smaller induction signal from a water ocean.

Callisto has a clear induction signal (Zimmer et al. 2000); explained by a salty water ocean that underlies the low pressure (phase I) of water ice layer, around 150 to 200 km in depth. This ocean is expected because of radioactive heating alone.

Saturn may have a dynamo very similar to that of Jupiter, but more deep-seated (the reason for the smaller surface field). It may be overlain by a region that greatly reduces the non-spin axisymmetric components, perhaps explaining the small observed dipole tilt (Stevenson 1982; Christensen and Wicht 2008).

Titan has an observed upper bound to the field that is less than the field expected for a Ganymede-like dynamo. It may have a water ocean and thus produce an induction signal, potentially detectable by Cassini. However, this will be more difficult to detect because Saturn lacks a significant dipole tilt, so the time variable part of Saturn's field is much smaller than that for Jupiter.

Uranus and *Neptune* are very similar in structure and in field strength and geometries. Their very different obliquities are evidently irrelevant to understanding their fields. Although it seems likely that high-pressure ionic (not metallic) water can provide the desired conductivity for a dynamo, it is marginal and the observed field strength seems smaller than expected . This raises the question of whether these planets are actually generating their fields deeper down. Quadrupolar dynamos are permitted by dynamo theory, and the dynamo activity might be limited to a thin shell (Hubbard et al. 1995). Models of this kind have been developed (Stanley and Bloxham 2004, 2006).

Triton and Pluto might possibly have water-ammonia oceans and might therefore be capable of induction signals, to the extent that they are subjected to small, time varying external magnetic fields.

Extrasolar Giant Planets can be expected to be convective at depth, and to have the conductivities sufficient for dynamo action. It is possible that "hot Jupiters" have field lines that connect the planet to the neighboring star; this could have observational consequences.

9 The Future

Future developments in this field depend on four things: More observations, more dynamo simulations, more lab data and more synthesis.

Observational priorities include: Mercury, a determination of whether Venus has any (small spatial scale) field, lunar paleomagnetism, a detailed correlation of Mars magnetism and geology and ages of surface units, and better spatial and time resolution of the Jovian field. We can expect the MESSENGER mission and the follow-on European effort (Bepi Colombo) to do an excellent characterization of Mercury's field environment, though the relatively large magnetospheric effects may limit precise determination of the purely internal part. We also expect Juno (launch in 2011, arrival at Jupiter in 2016) to do a spectacular job on Jupiter; indeed the field will be better characterized than Earth's dynamo (because there is no confusion from a crustal field). As described above, we may be making progress on lunar and meteoritic paleomagnetism, and this can also help our understanding of dynamos.

Dynamo simulations continue to benefit from Moore's law and creep ever upwards in spatial and temporal resolution (e.g., Kageyama et al. 2008). However, dynamo theory requires clever ideas as well as merely brute force improvement of the parameter regime. In particular, we need a useful criterion for planetary dynamos. The definition of useful is this: Given perfect knowledge of all the planetary physical parameters, what is the minimal convective heat flow or buoyancy flux needed for sustaining a dynamo? What field amplitude is then expected?

Lab data are essential to understand the transport properties of various cosmically important mixtures as well as the alloying properties relevant to Earth's core. Significant new developments (described above) suggest that we need a better understanding of the phase diagram in particular. We still do not know for sure whether Earth's core contains significant radioactive heating. Laboratory simulations of dynamos also play a role by testing our understanding of the relevant fluid dynamics.

Synthesis requires knowing the regimes and scaling behaviors of mantle convection, with and without plate tectonics and mantle layering (both for silicates and for ice). It also requires understanding the extent of mixing deep within giant gas and ice planets. Ultimately, these issues cannot be separated from the big questions of how planets form, differentiate and evolve. The spectacular explosion in work on extrasolar planets can be expected to affect our thinking on these issues in the coming decade.

Acknowledgements I thank a reviewer for useful comments and corrections. My primary research support is the NASA Planetary Geology and Geophysics program.

References

M.H. Acuna, J.E.P. Connerney, P. Wasilewski et al., Magnetic field of Mars: Summary of results from the aerobraking and mapping orbits. J. Geophys. Res. Planet. **106**, 23403–23417 (2001)

O. Aharonson, M.T. Zuber, S.C. Solomon, Crustal remanence in an internally magnetized non-uniform shell: a possible source for Mercury's magnetic field? Earth Planet. Sci. Lett. **218**, 261–268 (2004)

B.J. Anderson, M.H. Acuna, H. Korth et al., The structure of Mercury's magnetic field from MESSENGER's first flyby. Science **321**, 82–85 (2008)

J. Arkani-Hamed, Did tidal deformation power the core dynamo of Mars? Icarus **201**, 31–43 (2009)

B.A. Buffett, J. Bloxham, Energetics of numerical geodynamo models. Geophys. J. Int. **149**, 211–224 (2002)

B.A. Buffett, E.J. Garnero, R. Jeanloz, Sediments at the top of Earth's core. Science **290**, 138–1342 (2000)

F.H. Busse, Homogeneous dynamos in planetary cores and in the laboratory. Annu. Rev. Fluid Mech. **32**, 383–408 (2000)

B. Chen, J. Li, S.A. Hauck, Non-ideal liquidus curve in the Fe–S system and Mercury's snowing core. Geophys. Res. Lett. **35**, L07201 (2008)

U.R. Christensen, A deep dynamo generating Mercury's magnetic field. Nature **444**, 1056–1058 (2006)

U.R. Christensen, J. Wicht, Models of magnetic field generation in partly stable planetary cores: applications to Mercury and Saturn. Icarus **196**, 16–34 (2008)

U.R. Christensen, J. Aubert, P. Cardin et al., A numerical dynamo benchmark. Phys. Earth Planet. Int. **128**, 25–34 (2001)

U.R. Christensen, V. Holzwarth, A. Reiners, Energy flux determines magnetic field strength of planets and stars. Nature **457**, 167–169 (2009)

D.D. Clayton, *Principles of Stellar Evolution and Nucleosynthesis* (McGraw-Hill, New York, 1968), 256 pp.

I. De Pater, J.J. Lissauer, *Planetary Sciences* (Cambridge University Press, New York, 2001), 528 pp.

M. French, T.R. Mattsson, N. Nettelmann et al., Equation of state and phase diagram of water at ultrahigh pressures as in planetary interiors. Phys. Rev. B **79** (2009). Article Number: 054107

I. Garrick-Bethell, B.P. Weiss, D.L. Shuster et al., Early Lunar magnetism. Science **323**, 356–359 (2009)

C.K. Gessmann, B.J. Wood, D.C. Rubie, M.R. Kilburn, Solubility of silicon in liquid metal at high pressure: implications for the composition of the Earth's core. Earth Planet. Sci. Lett. **184**, 367–376 (2001)

G. Giampieri, A. Balogh, Mercury's thermoelectric dynamo model revisited. Planet. Space Sci. **50**, 757–762 (2002)

D. Grodent, B. Bonfond, J.C. Gerard et al., Auroral evidence of a localized magnetic anomaly in Jupiter's northern hemisphere. J. Geophys. Res. **113**, A09201 (2008)

D. Gubbins, Energetics of the Earth's core. J. Geophys. **47**, 453–464 (1977)

D. Gubbins, The Rayleigh number for convection in the Earth's core. Phys. Earth Planet. Int. **128**, 3–12 (2001)

T. Guillot, Interiors of giant planets inside and outside the solar system. Science **286**, 72–77 (1999)

L.L. Hood, A. Zakharian, J. Halekas et al., Initial mapping and interpretation of lunar crustal magnetic anomalies using Lunar Prospector magnetometer data. J. Geophys. Res. Planet. **106**, 27825–27839 (2001)

W.B. Hubbard, M. Podolak, D.J. Stevenson, The interior of Neptune, in *Neptune*. Un. Arizona Space Science Series (1995), pp. 109–138

C.A. Jones, Convection-driven geodynamo models. Philos. Trans. Roy. Soc. A **358**, 873–897 (2000)

A. Kageyama, T. Miyagoshi, T. Sato, Formation of current coils in geodynamo simulations. Nature **454**, 1106–1109 (2008)

M.G. Kivelson, J. Warnecke, L. Bennett et al., Ganymede's magnetosphere: magnetometer overview. J. Geophys. Res. Planet. **103**, 19963–19972 (1998)
M.G. Kivelson, K.K. Khurana, C.T. Russell et al., Magnetized or unmagnetized: ambiguity persists following Galileo's encounters with Io in 1999 and 2000. J. Geophys. Res. Space **106**, 26121–26135 (2001)
S. Labrosse, J.-P. Poirier, J.-L. Le Mouel, The age of the inner core. Earth Planet. Sci. Lett. **190**, 111–123 (2001)
V. Lainey, J.E. Arlot, O. Karatekin et al., Strong tidal dissipation in Io and Jupiter from astrometric observations. Nature **459**, 957–959 (2009)
J.-J. Liu, P. Goldreich, D.J. Stevenson, Ohmic dissipation constraint on deep-seated zonal winds in Jupiter and Saturn. Icarus **196**, 653–664 (2008)
D.E. Loper, Some thermal consequences of a gravitationally powered geodynamo. J. Geophys. Res. **83**, 5961–5970 (1978)
J.L. Margot, S.J. Peale, R.F. Jurgens et al., Large longitude libration of Mercury reveals a molten core. Science **316**, 710–714 (2007)
R.T. Merrill, M.W. McElhinney, P.L. McFadden, *The Magnetic Field of the Earth* (Academic Press, New York, 1996), 531 pp.
H.K. Moffatt, *Magnetic Field Generation in Electrically Conducting Fluids* (Cambridge University Press, New York, 1978), 336 pp.
W.J. Nellis, Metallization of fluid hydrogen at 140 GPa (1.4 Mbar): implications for Jupiter. Planet. Space Sci. **48**, 671–677 (2000)
W.J. Nellis, N.C. Holmes, A.C. Mitchell et al., Equation of state and electricalconductivity of "synthetic Uranus", a mixture of water, ammonia, and isopropanol, at shock pressure up to 200 GPa (2 Mbar). J. Chem. Phys. **107**, 9096–9100 (1997)
F. Nimmo, Energetics of asteroid dynamos and the role of compositional convection. Geophys. Res. Lett. **36**, L10201 (2009)
F. Nimmo, D.J. Stevenson, Influence of early plate tectonics on the thermal evolution and magnetic field of Mars. J. Geophys. Res. Planet. **105**, 11969–11979 (2000)
E.N. Parker, *Cosmical Magnetic Fields: Their Origin and Their Activity* (Clarendon Press and Oxford University Press, New York, 1979), 841 pp.
J.-P. Poirier, *Introduction to the Physics of the Earth's Interior* (Cambridge University Press, New York, 1991), p. 191
J.-P. Poirier, Light-elements in the Earth's outer core—a critical-review. Phys. Earth Planet. Int. **85**, 319–337 (1994)
P. Roberts, G.A. Glatzmaier, Geodynamo theory and simulations. Rev. Modern Phys. **72**, 1081–1123 (2000)
G. Schubert, M.N. Ross, D.J. Stevenson, T. Spohn, T. Mercury's thermal history and the generation of its magnetic field, in *Mercury*, ed. by C. Chapman et al. (University of Arizona Press, Tucson, 1988), pp. 429–460
G. Schubert, V.S. Solomatov, P.J. Tackley, D.L. Turcotte, Mantle Convection and the thermal evolution of Venus, in *Venus II*, ed. by S.W. Bougher, D.M. Hunten, R.J. Phillips (University of Arizona Press, Tucson, 1998), pp. 429–460
S. Stanley, J. Bloxham, Convective-region geometry as the cause of Uranus' and Neptune's unusual magnetic fields. Nature **428**, 151–153 (2004)
S. Stanley, J. Bloxham, Numerical dynamo models of Uranus' and Neptune's magnetic fields. Icarus **184**, 556–572 (2006)
S. Stanley, J. Bloxham, W.E. Hutchison, M.T. Zuber, Thin shell dynamo models consistent with Mercury's weak observed magnetic field. Earth Planet. Sci. Lett. **234**, 27–38 (2005)
D.J. Stevenson, Turbulent thermal convection in the presence of rotation and a magnetic field: a heuristic theory. Geophys. Astrophys. Fluid Dyn. **12**, 139–169 (1979)
D.J. Stevenson, Reducing the non-axisymmetry of a planetary dynamo and an application to Saturn. Geophys. Astrophys. Fluid Dyn. **21**, 113–127 (1982)
D.J. Stevenson, Planetary magnetic fields. Rep. Progr. Phys. **46**, 555–620 (1983)
D.J. Stevenson, Limits on lateral density and velocity variations in the Earth's outer core. Geophys. J. Roy. Astron. Soc. **88**, 311–319 (1987a)
D.J. Stevenson, Mercury's magnetic field—a thermoelectric dynamo. Earth Planet. Sci. Lett. **82**, 114–120 (1987b)
D.J. Stevenson, Europa's ocean—the case strengthens. Science **289**, 1305–1307 (2000)
D.J. Stevenson, Mars' core and magnetism. Nature **412**, 214–219 (2001)
D.J. Stevenson, Introduction to planetary interiors, in *Proceedings of the International School of Physics "Enrico Fermi"* (Italian Physical Society, Bologna, 2002)
D.J. Stevenson, Planetary magnetic fields. Earth Planet. Sci. Lett. **208**(2003), 1–11 (2003)

D.J. Stevenson, T. Spohn, G. Schubert, Magnetism thermal evolution of the terrestrial planets. Icarus **54**, 466–489 (1983)

A.J. Stewart, M.W. Schmidt, W. van Westrenen, C. Liebske, Mars: a new core-crystallization regime. Science **316**, 1323–1325 (2007)

B.P. Weiss, J.S. Berdahl, L. Elkins-Tanton et al., Magnetism on the angrite parent body and the early differentiation of planetesimals. Science **322**, 713–716 (2008a)

B.P. Weiss, L.E. Fong, H. Vali et al., Paleointensity of the ancient Martian magnetic field. Geophys. Res. Lett. **35**, L23207 (2008b)

J.G. Williams, D.H. Boggs, C.F. Yoder et al., Lunar rotational dissipation in solid body and molten core. J. Geophys. Res. Planet. **106**, 27933–27968 (2001)

J.P. Williams, O. Aharonson, F. Nimmo, Powering Mercury's dynamo. Geophys. Res. Lett. **34**, L21201 (2007)

C. Zimmer, K.K. Khurana, M.G. Kivelson, Subsurface oceans on Europa and Callisto: constraints from Galileo magnetometer observations. Icarus **147**, 329–347 (2000)

Space Science Series of ISSI

1. R. von Steiger, R. Lallement and M.A. Lee (eds.): *The Heliosphere in the Local Interstellar Medium*. 1996 ISBN 0-7923-4320-4
2. B. Hultqvist and M. Øieroset (eds.): *Transport Across the Boundaries of the Magnetosphere*. 1997 ISBN 0-7923-4788-9
3. L.A. Fisk, J.R. Jokipii, G.M. Simnett, R. von Steiger and K.-P. Wenzel (eds.): *Cosmic Rays in the Heliosphere*. 1998 ISBN 0-7923-5069-3
4. N. Prantzos, M. Tosi and R. von Steiger (eds.): *Primordial Nuclei and Their Galactic Evolution*. 1998 ISBN 0-7923-5114-2
5. C. Fröhlich, M.C.E. Huber, S.K. Solanki and R. von Steiger (eds.): *Solar Composition and its Evolution – From Core to Corona*. 1998 ISBN 0-7923-5496-6
6. B. Hultqvist, M. Øieroset, Goetz Paschmann and R. Treumann (eds.): *Magnetospheric Plasma Sources and Losses*. 1999 ISBN 0-7923-5846-5
7. A. Balogh, J.T. Gosling, J.R. Jokipii, R. Kallenbach and H. Kunow (eds.): *Co-rotating Interaction Regions*. 1999 ISBN 0-7923-6080-X
8. K. Altwegg, P. Ehrenfreund, J. Geiss and W. Huebner (eds.): *Composition and Origin of Cometary Materials*. 1999 ISBN 0-7923-6154-7
9. W. Benz, R. Kallenbach and G.W. Lugmair (eds.): *From Dust to Terrestrial Planets*. 2000 ISBN 0-7923-6467-8
10. J.W. Bieber, E. Eroshenko, P. Evenson, E.O. Flückiger and R. Kallenbach (eds.): *Cosmic Rays and Earth*. 2000 ISBN 0-7923-6712-X
11. E. Friis-Christensen, C. Fröhlich, J.D. Haigh, M. Schüssler and R. von Steiger (eds.): *Solar Variability and Climate*. 2000 ISBN 0-7923-6741-3
12. R. Kallenbach, J. Geiss and W.K. Hartmann (eds.): *Chronology and Evolution of Mars*. 2001 ISBN 0-7923-7051-1
13. R. Diehl, E. Parizot, R. Kallenbach and R. von Steiger (eds.): *The Astrophysics of Galactic Cosmic Rays*. 2001 ISBN 0-7923-7051-1
14. Ph. Jetzer, K. Pretzl and R. von Steiger (eds.): *Matter in the Universe*. 2001 ISBN 1-4020-0666-7
15. G. Paschmann, S. Haaland and R. Treumann (eds.): *Auroral Plasma Physics*. 2002 ISBN 1-4020-0963-1
16. R. Kallenbach, T. Encrenaz, J. Geiss, K. Mauersberger, T.C. Owen and F. Robert (eds.): *Solar System History from Isotopic Signatures of Volatile Elements*. 2003 ISBN 1-4020-1177-6
17. G. Beutler, M.R. Drinkwater, R. Rummel and R. von Steiger (eds.): *Earth Gravity Field from Space – from Sensors to Earth Sciences*. 2003 ISBN 1-4020-1408-2
18. D. Winterhalter, M. Acuña and A. Zakharov (eds.): *"Mars" Magnetism and its Interaction with the Solar Wind*. 2004 ISBN 1-4020-2048-1
19. T. Encrenaz, R. Kallenbach, T.C. Owen and C. Sotin: *The Outer Planets and their Moons* ISBN 1-4020-3362-1
20. G. Paschmann, S.J. Schwartz, C.P. Escoubet and S. Haaland (eds.): *Outer Magnetospheric Boundaries: Cluster Results* ISBN 1-4020-3488-1
21. H. Kunow, N.U. Crooker, J.A. Linker, R. Schwenn and R. von Steiger (eds.): *Coronal Mass Ejections* ISBN 978-0-387-45086-5

22. D.N. Baker, B. Klecker, S.J. Schwartz, R. Schwenn and R. von Steiger (eds.): *Solar Dynamics and its Effects on the Heliosphere and Earth* ISBN 978-0-387-69531-0
23. Y. Calisesi, R.-M. Bonnet, L. Gray, J. Langen and M. Lockwood (eds.): *Solar Variability and Planetary Climates* ISBN 978-0-387-48339-9
24. K.E. Fishbaugh, P. Lognonné, F. Raulin, D.J. Des Marais and O. Korablev (eds.): *Geology and Habitability of Terrestrial Planets* ISBN 978-0-387-74287-8
25. O. Botta, J.L. Bada, J. Gomez-Elvira, E. Javaux, F. Selsis and R. Summons (eds.): *Strategies of Life Detection* ISBN 978-0-387-77515-9
26. A. Balogh, L. Ksanfomality and R. von Steiger (eds.): *Mercury* ISBN 978-0-387-77538-8
27. R. von Steiger, G. Gloeckler and G.M. Mason (eds.): *The Composition of Matter* ISBN 978-0-387-74183-3
28. H. Balsiger, K. Altwegg, W. Huebner, T.C. Owen and R. Schulz (eds.): *Origin and Early Evolution of Comet Nuclei, Workshop honouring Johannes Geiss on the occasion of his 80th birthday* ISBN 978-0-387-85454-0
29. A.F. Nagy, A. Balogh, T.E. Cravens, M. Mendillo and I. Mueller-Wodarg (eds.): *Comparative Aeronomy* ISBN 978-0-387-87824-9
30. F. Leblanc, K.L. Aplin, Y. Yair, R.G. Harrison, J.P. Lebreton and M. Blanc (eds.): *Planetary Atmospheric Electricity* ISBN 987-0-387-87663-4
31. J.L. Linsky, V. Izmodenov, E. Möbius and R. von Steiger (eds.): *From the Outer Heliosphere to the Local Bubble: Comparison of New Observations with Theory* ISBN 978-1-4419-0246-7
32. M.J. Thompson, A. Balogh, J.L. Culhane, Å. Nordlund, S.K. Solanki and J.-P. Zahn (eds.): *The Origin and Dynamics of Solar Magnetism* ISBN 978-1-4419-0238-2
33. U.R. Christensen, A. Balogh, D. Breuer and K.-H. Glaßmeier (eds.): *Planetary Magnetism* ISBN 978-1-4419-5900-3
34. C.W.F. Everitt, M.C.E. Huber, R. Kallenbach, G. Schäfer, B.F. Schutz and R.A. Treumann (eds.): *Probing The Nature of Gravity: Confronting Theory and Experiment in Space* ISBN 978-1-4419-1361-6

GPSR Compliance

The European Union's (EU) General Product Safety Regulation (GPSR) is a set of rules that requires consumer products to be safe and our obligations to ensure this.

If you have any concerns about our products, you can contact us on ProductSafety@springernature.com

In case Publisher is established outside the EU, the EU authorized representative is:

Springer Nature Customer Service Center GmbH
Europaplatz 3
69115 Heidelberg, Germany

Batch number: 10755458

Printed by Printforce, the Netherlands